16th International Symposium on Particles, Strings and Cosmology 2010

(PASCOS 2010)

Journal of Physics: Conference Series Volume 259

Valencia, Spain
19 – 23 July 2010

Editors:

Susana Cabrera
Martin Hirsch
Vasiliki Mitsou
Carlos Munoz
Sergio Pastor
Maria Amparo Tortola
Jose W. F. Valle
O. Vives

ISBN: 978-1-61782-248-3
ISSN: 1742-6588

Printed from e-media with permission by:

Curran Associates, Inc.
57 Morehouse Lane
Red Hook, NY 12571

Printed by Curran Associates, Inc. (2011)

For permission requests, please contact the Institute of Physics
at the address below.

Institute of Physics
Dirac House, Temple Back
Bristol BS1 6BE UK

Phone: 44 1 17 929 7481
Fax: 44 1 17 920 0979

techtracking@iop.org

Additional copies of this publication are available from:

Curran Associates, Inc.
57 Morehouse Lane
Red Hook, NY 12571 USA
Phone: 845-758-0400
Fax: 845-758-2634
Email: curran@proceedings.com
Web: www.proceedings.com

TABLE OF CONTENTS

Journal of Physics: Conference Series **259** (2010) 011001 doi:10.1088/1742-6596/259/1/011001

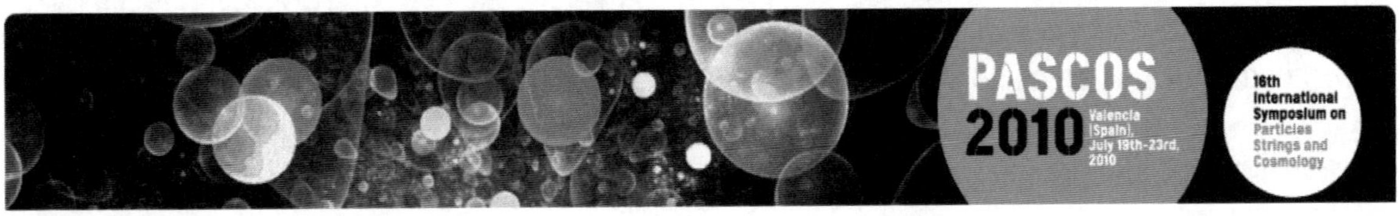

Preface

The XVI Symposium in the *Particles, Strings and Cosmology* (PASCOS) series took place on 19–23 July 2010, in the historic city of Valencia, and was hosted by the Instituto de Física Corpuscular (IFIC), the largest particle physics laboratory of the Spanish National Research Council (CSIC), jointly operated with the University of Valencia. The PASCOS series of annual symposia is dedicated to the latest advances on the study of the forces that govern the elementary constituents of matter – the microcosm – and their effects on the understanding of the Universe at large – the macrocosm. Indeed the basic principles of uncertainty and mass-energy equivalence imply that when one probes deep inside the subatomic scale, one inevitably excites states of very high energy and mass which were copiously produced at the Big Bang. Recreating these particles in the laboratory is tantamount to tracing back the very early history of the universe. The interface of particle physics, string theory and cosmology has indeed become a highly active field of research at the frontier of human knowledge and the PASCOS meetings aim to bring together researchers from the three areas so as to facilitate the exchange of ideas and to identify possible synergies. The series started in the mid-nineties in the United States, where the first events took place. However it has by now become truly global, having circulated through India, South Korea, the United Kingdom, Canada and Germany. The aim of the conference was to review the recent progress in particle physics, string theory and cosmology, promoting the exchange of ideas and discussing future prospects. With the startup of LHC and the launch of the Planck satellite as well as many other experiments under way or planned, PASCOS2010 looked at an exciting future, giving theorists an opportunity to prepare for this wealth of new data and the stringent tests to which they will subject the existing theories. While the conferences in this series have traditionally been aimed mainly at theorists, PASCOS2010 had a stronger emphasis on new results and future experiments.

The venue for PASCOS 2010 was the Fundación Universidad Empresa (ADEIT), a modern building, located in the historical city centre of Valencia and very close to the conference hotels and to most historic buildings and monuments. The scientific programme consisted of 26 invited Plenary Talks and 99 contributions, all available online at http://pascos2010.astroparticles.es/. Young researchers were encouraged to submit abstracts, many of which were subsequently selected by the organizing committee for oral presentation in parallel sessions. This year's symposium, attended by over 160 participants from all over the globe, was of particular significance, as it came in the wake of first data from the LHC as well as the Planck satellite. In general, plenary sessions were held in the mornings, with afternoons devoted to parallel sessions focussing on the three areas of particles, strings and cosmology.

Official Inauguration and Public Talk

The Opening of the Symposium took place in the main Auditorium of ADEIT, starting with a brief welcome address by Professor José W F Valle, Chair of the Organizing Committee, followed by addresses of Professor Pedro Carrasco, Research Vice-Rector of the University of València, Professor José Pío Beltran, representing the President of the CSIC, and Professor Francisco Botella, Director of IFIC.

There was also a very successful popular science talk by Professor Carlos Frenk, Director of the

Journal of Physics: Conference Series **259** (2010) 011001 doi:10.1088/1742-6596/259/1/011001

Institute for Computational Cosmology at Durham University in the United Kingdom on "The Structure of the Universe". This as well as the opening of the PASCOS2010 conference was covered by the local media, and a short summary is also published in the CERN Courier.

Valencia, October 2010
The Editors
Susana Cabrera, Martin Hirsch,Vasiliki Mitsou, Carlos Muñoz, Sergio Pastor, María Amparo Tórtola, José W. F. Valle and Óscar Vives

Conference Committees

International Advisory Committee

Jonathan Bagger	Johns Hopkings U.
Wilfried Buchmülller	DESY
Paul Frampton	North Carolina U.
John F. Gunion	UC Davis
Jihn E. Kim	Seoul National U.
Robert C. Myers	Perimeter Institute
Pran Nath	Northeastern U.
Stuart Raby	Ohio State U.
Durga P. Roy	Tata Institute
Joseph Silk	Oxford U.
Michael T. Vaughn	Northeastern U.
Kameshwar C. Wali	Syracuse U.

Local Organizing Committee

Susana Cabrera	IFIC, CSIC / U. Valencia
Martin Hirsch	IFIC, CSIC / U. Valencia
Vasiliki Mitsou	IFIC, CSIC / U. Valencia
Carlos Muñoz	IFT, CSIC / U. Autónoma de Madrid
Sergio Pastor	IFIC, CSIC / U. Valencia
María Amparo Tórtola	IFIC, CSIC / U. Valencia
José W. F. Valle	IFIC, CSIC / U. Valencia - Chairman
Óscar Vives	Dpto. Física Teórica and IFIC, CSIC / U. Valencia

Scientific Secretaries

M. Sofiane Boucenna	IFIC, CSIC / U. Valencia
Joel Jones	Dpto. Física Teórica and IFIC, CSIC / U. Valencia
Laslo Reichert	IFIC, CSIC / U. Valencia
Avelino Vicente	IFIC, CSIC / U. Valencia

e-mail: `pascos2010@ific.uv.es`
web page: `http://pascos2010.astroparticles.es`

Journal of Physics: Conference Series **259** (2010) 011001 doi:10.1088/1742-6596/259/1/011001

Sponsors

Consejo Superior de Investigaciones Científicas (CSIC), Instituto de Física Corpuscular (IFIC), Universitat de València (UVEG), Ministerio de Ciencia e Innovación (MICINN), Proyecto Multimessenger Approach for Dark Matter Detection (MULTIDARK), Proyecto Centro Nacional de Física de Partículas, Astropartículas y Nuclear (CPAN)

List of Participants

PARTICIPANT	INSTITUTION
Amine Ahriche	LPTh, Univ of Jijel
Alexey Anisimov	Univ Bielefeld
Matti Antola	Univ Helsinki, HIP
Ignatios Antoniadis	CERN
Stefan Antusch	MPI Physics, Munich
Alberto Aparici Benages	IFIC (CSIC-Univ València)
Carolina Arbeláez	IFIC (CSIC-Univ València)
Paul Archer	Univ Sussex
Cristian Armendariz	Syracuse Univ
Takehiko Asaka	Niigata Univ
Amjad Ashoorioon	Uppsala Univ
Spasimir Balev	CERN
Igor Bandos	Ikerbasque, & EHU Bilbao
Neil Barnaby	CITA Toronto
Florian Bauer	Univ Barcelona
Jochen Baumann	MPI Physics, Munich
Silvio Bonometto	Milano-Bicocca Univ
Lotfi Boubekeur	IFIC (CSIC-Univ València)
Sofiane Boucenna	IFIC (CSIC-Univ València)
Carolin B. Bräuninger	TU Munich
Clare Burrage	DESY Hamburg
Susana Cabrera	IFIC (CSIC-Univ València)
Fabrizio Cei	Univ Pisa
Lucio Cerrito	Queen Mary, Univ London
Diego V. Chialva	NORDITA
Kiwoon Choi	KAIST
Jodi Cooley	Southern Methodist Univ
Siba P. Das	IFIC (CSIC-Univ València)
Reinier de Adelhart T.	Nikhef Amsterdam
Jeffrey de Jong	Oxford Univ
Massimiliano Lattanzi	Univ Roma "La Sapienza"
Roberto A. Lineros	Univ Torino
Maurício B. Magro	CUFSA, Santo André
Michal Malinsky	KTH Stockholm
Jelena Maricic	Drexel Univ
Justo Martín-Albo	IFIC (CSIC-Univ València)
Carlos Mayoral	IFIC (CSIC-Univ València)
Anupam Mazumdar	Lancaster Univ
Daan Meerburg	Univ Amsterdam
Alessandro Melchiorri	Univ Roma "La Sapienza"
Elena Melkumova	Moscow State Univ
Olga Mena	IFIC (CSIC-Univ València)
Gennaro Miele	Univ Naples "Federico II"
Masato Minamitsuji	Kwansei-Gakuin Univ
Shiraz Minwalla	TIFR Mumbai
Vassiliki Mitsou	IFIC (CSIC-Univ València)
Myriam Mondragón	UNAM
María Moreno Llácer	IFIC (CSIC-Univ València)
Stefano Morisi	IFIC (CSIC-Univ València)
Aldo Morselli	INFN Roma Tor Vergata
Carlos Muñoz	IFT (CSIC-UAM)
Marco Nardecchia	SISSA Trieste
José Navarro-Salas	IFIC (CSIC-Univ València)
Marek Olechowski	Univ Warsaw
Muzeyyen P. Ozdemir	Student
Ezgi Paçal	Student
Antonio Padilla	Univ Nottingham
Heinrich Päs	TU Dortmund
Juan P. Palacios	Univ Zürich
Antonio Palazzo	IFIC (CSIC-Univ València)

João de Mello Neto	Fed. Univ Rio de Janeiro
Valentina De Romeri	Univ Torino
Frank Deppisch	Univ Manchester
Antonio Dobado	Univ Complutense Madrid
Manuel Drees	Bonn Univ
Bhaskar Dutta	Texas A&M Univ
John Ellison	Univ California, Riverside
Francisco Escrihuela	IFIC (CSIC-Univ València)
João Esteves	CFTP-IST Lisbon
António Figueiredo	CFTP-IST Lisbon
Renato Fonseca	CFTP-IST Lisbon
Nicolao Fornengo	Univ Torino & INFN
Paul Frampton	UNC-Chapel Hill
Urbano França	IFIC (CSIC-Univ València)
Roberto Franceschini	EPFL Lausanne
Carlos S. Frenk	Durham Univ
Andrei V. Frolov	Simon Fraser Univ
Damien P. George	Nikhef Amsterdam
Pradipta Ghosh	Indian Assoc. Cult. Science
Andrea Giammanco	Univ Cath Louvain
Inés Gil Botella	CIEMAT
Mario E. Gómez	Univ Huelva
Razvan Gornea	LHEP Bern
Srubabati Goswami	Physical Research Lab
Luca Grisa	UMass, Amherst
Bohdan Grzadkowski	Univ Warsaw
John F. Gunion	Univ California, Davis
Rhiannon Gwyn	King's College London
Howard E. Haber	Univ California, Santa Cruz
Richard Hawkings	CERN
Jonathan Hays	Imperial College London
Juan Herrero García	IFIC (CSIC-Univ València)
Martin Hirsch	IFIC (CSIC-Univ València)
Robert Hodgkinson	IFIC (CSIC-Univ València)
Andreas Hohenegger	MPI Nuc. Phys., Heidelberg
Stephan J. Huber	Univ Sussex
Christopher Jackson	Univ Manchester
Matti Jarvinen	CP3-Origins, SDU
Filipe R. Joaquim	CERN-TH
Joel Jones-Pérez	IFIC (CSIC-Univ València)
Rolf Kappl	MPI Physics, Munich
Muge Karagoz	Oxford Univ
Clemens Kiessig	MPI Physics, Munich
Jihn E. Kim	Seoul National University
Steve F. King	Univ Southampton
Olaf Kittel	Univ Granada
Bhavik Kodrani	Physical Research Lab
Laura Koponen	Tähdet ja avaruus magazine
Alexey Koshelev	Vrije Universiteit Brussel
Philipp M. Kostka	MPI Physics, Munich
Valentin Kozlov	Karlsruhe Inst Technology
Bumseok Kyae	Pusan Nat'l Univ
Grigoris Panotopoulos	IFIC (CSIC-Univ València)
Sergio Pastor	IFIC (CSIC-Univ València)
Eduardo Peinado	IFIC (CSIC-Univ València)
Manuel Peña Jiménez	IFIC (CSIC-Univ València)
Armando Pérez	IFIC (CSIC-Univ València)
Teresa Pérez	IFIC (CSIC-Univ València)
Ofelia Pisanti	Univ Naples "Federico II"
Werner Porod	Univ Würzburg & IFIC
Marieke Postma	Nikhef Amsterdam
Giovanni M. Pruna	Univ Southampton, NExT
Laslo Reichert	IFIC (CSIC-Univ València)
Francesco Riggi	INFN Catania
Eduardo Rodríguez	Univ Católica Concepción
Jorge C. Romão	CFTP-IST Lisbon
D.P. Roy	TIFR Mumbai
Vadim Rusu	Fermilab
Bjorn H. Samset	Univ Oslo
Eusebio Sánchez	CIEMAT
Mayly Sanchez	Iowa State Univ / ANL
Miguel A. Sanchis	IFIC (CSIC-Univ València)
Francesco Sannino	CP3-Origins, SDU
Arcadi Santamaria	IFIC (CSIC-Univ València)
Dario Schalla	TU Dortmund
Kai Schmidt-Hoberg	TU Munich
Qaisar Shafi	Univ Delaware
Cristoforo Simonetto	TU Munich
Anupam Singh	L.N. Mittal I.I.T.
Navin Sivanandam	Univ Texas, Austin
Elif Siyel	Student
Edwin J. Son	CQUeST, Sogang Univ
Miguel Sopena	Univ Sussex
Michel Sorel	IFIC (CSIC-Univ València)
Yoichiro Suzuki	Kamioka Obs, Univ Tokyo
Dylan Tanner	King's College London
Marco Taoso	IFIC (CSIC-Univ València)
Ricard Tomàs	Univ Hamburg
Máriam Tórtola	IFIC (CSIC-Univ València)
Minas Tsoukalas	N.T.U.A.
Kimmo Tuominen	CP3-Origins, SDU
Paula Tuzón	IFIC (CSIC-Univ València)
Ángel Uranga	IFT (CSIC-UAM) & CERN
Kunihito Uzawa	Kindai Univ
José W.F. Valle	IFIC (CSIC-Univ València)
John D. Vergados	Univ Ioannina & Tübingen
Avelino Vicente	IFIC (CSIC-Univ València)
Oscar Vives	IFIC (CSIC-Univ València)
Kameshwar C. Wali	Syracuse Univ
Christoph Weniger	DESY Hamburg
Jose Wudka	Univ California, Riverside
Masahide Yamaguchi	Tokyo Inst Technology
Yared Yamaoka	Duke Univ

Journal of Physics: Conference Series **259** (2010) 011001 doi:10.1088/1742-6596/259/1/011001

Conference Photographs

Status and prospects from the LHC

Richard Hawkings
CERN, CH 1211 Geneva 23, Switzerland
E-mail: richard.hawkings@cern.ch

Abstract. This article reviews the status of the CERN Large Hadron Collider and associated experiments as of July 2010. After a brief discussion of the progress in accelerator and experiment commissioning, the LHC physics landscape is presented, together with a selection of the experimental results achieved so far. Finally the prospects for the 2010-11 LHC physics run are reviewed, with an emphasis on possible discoveries in the Higgs and supersymmetry sectors.

1. The Large Hadron Collider and experiments

The Large Hadron Collider (LHC) is a high-luminosity proton-proton and heavy ion collider installed in the 27 km tunnel that formerly housed the CERN LEP electron-positron collider. The LHC and its associated experiments are designed to shed light on many of the unsolved questions in fundamental physics, some of whose solutions predict new phenomena at the TeV energy scale. After around twenty years of design, construction, installation and commissioning, the LHC and experiments were finally ready to begin operations in September 2008, when part of the accelerator was damaged in a commissioning accident caused by a faulty magnet interconnect. A year of repair and consolidation followed, and the LHC began operations in earnest in November 2009, with first collisions at a centre-of-mass energy $\sqrt{s} = 900\,\mathrm{GeV}$ happening on 23rd November, and first 7 TeV collisions on 31st March 2010.

The LHC interaction points host four major experiments—ATLAS, CMS, LHCb and ALICE, and several smaller ones. ATLAS and CMS are large general-purpose detectors designed to explore the full range of LHC proton-proton physics, including both searches for new particles and precision measurements. The LHCb experiment is dedicated to exploring the physics of heavy flavours (b and c quarks), in particular the violation of charge-parity (CP) symmetry. ALICE is a dedicated heavy ion experiment, focused on studying the very high multiplicity of particles produced in the collisions of lead nuclei during special LHC running periods. CMS and ATLAS will also participate in these runs, the first of which is scheduled for late 2010.

After having achieved the first pilot 7 TeV pp collisions, the main goal of the accelerator team for 2010 is to fully commission the machine and increase the intensity step-by-step, ideally reaching an instantaneous luminosity of $L = 10^{32}\,\mathrm{cm^{-2}s^{-1}}$ by the end of the 2010 running period. This should open the way to accumulating $1\,\mathrm{fb^{-1}}$ in 2011, before the accelerator is shut down in 2012 for work on improving the magnet interconnects ready for $\sqrt{s} = 14\,\mathrm{TeV}$ running in 2013. The step-by-step approach has resulted in alternating periods of machine commissioning and physics data accumulation, as clearly visible in the integrated luminosity history plots shown in Figure 1. The machine team started with a small number of colliding bunches with 2×10^{10} p/bunch, then commissioned the 'squeeze' (tightening the focus of the beams at the interaction

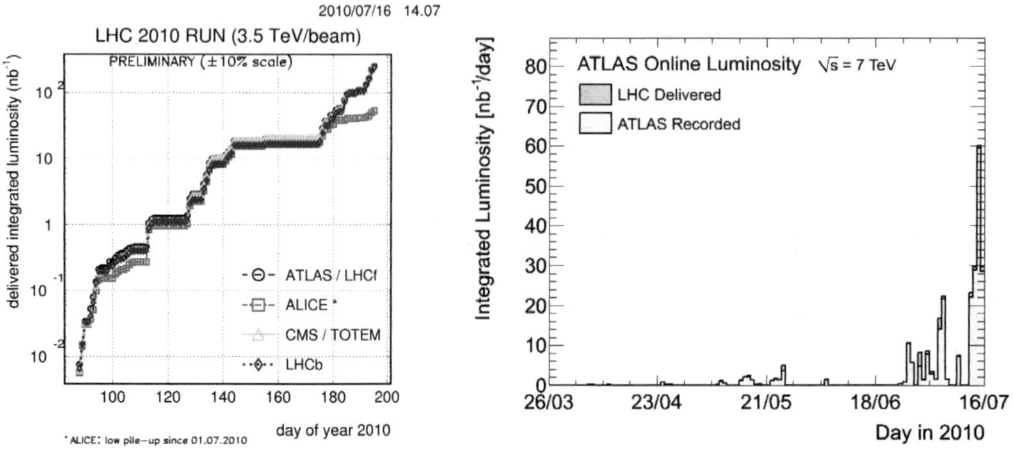

Figure 1. Accumulated integrated luminosity delivered to the LHC experiments in 2010 (left), and daily integrated luminosity delivered to and recorded by ATLAS (right).

points) to increase the collision rate. An intense period of commissioning with nominal intensity bunches (1×10^{11} p/bunch) then followed, with the number of bunches being increased to 13 per beam in July, giving instantaneous luminosities of about $1 \times 10^{30}\,\mathrm{cm^{-2}s^{-1}}$. At the time of the PASCOS 2010 conference, over $200\,\mathrm{nb^{-1}}$ had been delivered to each experiment.

A key parameter for safe machine operation is the stored energy per beam; in July this had reached about 1 MJ per beam, around the maximum energy seen in other proton accelerators such as HERA and the Tevatron, and enough to cause significant damage to the accelerator if the beam is lost in an uncontrolled way. The next step therefore involves a period of sustained, stable running to gain operational experience and build confidence in the machine protection systems, before moving onto larger numbers of bunches. These will be injected in closely spaced bunch trains, with the aim of having routine operation with several hundred bunches by the end of 2010. This is still far below the LHC design goal of 2808 bunches/beam for an instantaneous luminosity of $L = 10^{34}\,\mathrm{cm^{-2}s^{-1}}$, but should allow a data sample of $O(100\,\mathrm{pb^{-1}})$ to be accumulated by the end of October, before the LHC switches to lead ions for a short heavy-ion commissioning run in November.

2. The LHC physics landscape

Most of the new physics signals which ATLAS and CMS are searching for have very small production cross-sections, implying that they will give rise to rare events which have to be detected amongst an overwhelming background of Standard Model physics processes. The rates for some typical processes at the LHC design energy and luminosity are given in Table 1, from which it can be seen for example that a 100 GeV Higgs boson is expected to be produced about once in every 10^{10} collisions. Much of the early data analysis work at LHC therefore focuses on building up a thorough understanding of the more common Standard Model physics processes, and how they are reconstructed and measured in the detectors. Many interesting studies can be performed with these data, and some of the processes also serve as 'standard candles' allowing the calibration of the detectors to be verified and improved.

Although the full LHC discovery potential will only be reached after several years of high-luminosity running, even the $O(100\,\mathrm{pb^{-1}})$ expected in 2010 will allow a discovery reach significantly beyond that of the Fermilab Tevatron in many areas. Current Tevatron search limits are typically based on $O(5\,\mathrm{pb^{-1}})$ of $p\bar{p}$ data at $\sqrt{s} = 1.96\,\mathrm{TeV}$, so the LHC enjoys a factor

Table 1. Approximate expected rates for some physics processes in LHC pp collisions at $\sqrt{s} = 14\,\text{TeV}$ and $L = 10^{34}\,\text{cm}^{-2}\text{s}^{-1}$.

Process	Rate ($\sqrt{s} = 14\,\text{TeV}$)
Inelastic pp collision	10^9 Hz
b-quark pair production	10^6 Hz
Jet production, $E_T > 250\,\text{GeV}$	10^3 Hz
$W \to \ell\nu$	10^2 Hz
Top-quark pair production	10 Hz
Standard Model Higgs, $m_H = 100\,\text{GeV}$	0.2 Hz

3.6 advantage in energy, balanced by a expected factor of approximately 50–100 less integrated luminosity in 2010, improving to 5–10 at the end of 2011. The potential gain over the Tevatron depends on the ratio of parton luminosities at the energy scale X appropriate for the production of a state of mass M_X, and whether that state is produced primarily in gluon-gluon ($gg \to X$) or quark-antiquark ($q\bar{q} \to X$) processes [1]. For example, the production of a 160 GeV Higgs boson is dominated by gg and gains a factor 15 in cross-section from the increased centre-of mass energy. Top quark pair production ($m_X = 2 \times 175\,\text{GeV}$) becomes gg dominated at LHC, and gains a factor of 20 compared to the $q\bar{q}$-dominated production at the Tevatron. A possible 1 TeV Z' state gains a factor 100, opening up new territory even with $100\,\text{pb}^{-1}$ of LHC data. There are therefore significant chances for a discovery even with early LHC data, if the new physics lies just beyond the Tevatron search limits.

3. Commissioning the experiments

Once first collisions were delivered, initially at $\sqrt{s} = 900\,\text{GeV}$, one run at 2.36 TeV and then at 7 TeV, the experiments focused on completing the commissioning of their detectors, trigger and offline computing systems. These had already been intensively exercised over the preceding few years with cosmic rays and other dedicated tests, but benefited greatly from the gradual increase in LHC luminosity over the first weeks and months of accelerator commissioning.

The tracking and particle identification detectors began to see the 'zoo' of Standard Model hadrons from the first LHC runs; $K^0_s \to \pi^+\pi^-$, $\phi \to K^+K^-$, $\Lambda^0 \to p^+\pi^-$ and even $\Omega^- \to \Lambda^0 K^-$ were quickly reconstructed. Similarly, the calorimeters began to see signatures such as $\pi^0 \to \gamma\gamma$, $\eta, \omega \to \pi^+\pi^-\pi^0$ and later $J/\psi \to e^+e^-$, together with $J/\psi \to \mu\mu$ in the muon detectors. As well as acting as commissioning signals, precise measurements of the production rates of all these particles are useful in tuning Monte Carlo models. The first physics runs also allowed the experiments to start commissioning their missing transverse energy (E_T^{miss}) reconstruction—this is based on the vector sum of calorimeter energy deposits resolved into the transverse plane. Since the initial state of colliding protons has no significant momentum component transverse to the beam direction, larges values of E_T^{miss} indicate 'unbalanced' events, with transverse momentum being carried away by undetected particles (such as neutrinos). The E_T^{miss} measurement is of vital importance in many analyses, but requires that 'fake' E_T^{miss} from inefficient or noisy calorimeter cells, miscalibration or cracks first be thoroughly understood.

All the LHC experiments employ complex multi-stage trigger systems to select the most interesting pp collisions online and reduce the rate to around 200 Hz (in the case of ATLAS and CMS) for recording and offline analyses. Since the actual pp collision rate already exceeded 200 Hz after the first few runs, it has been crucial to commission the trigger systems as soon as possible to keep up with the data-taking rates. These triggers are typically based on a first level

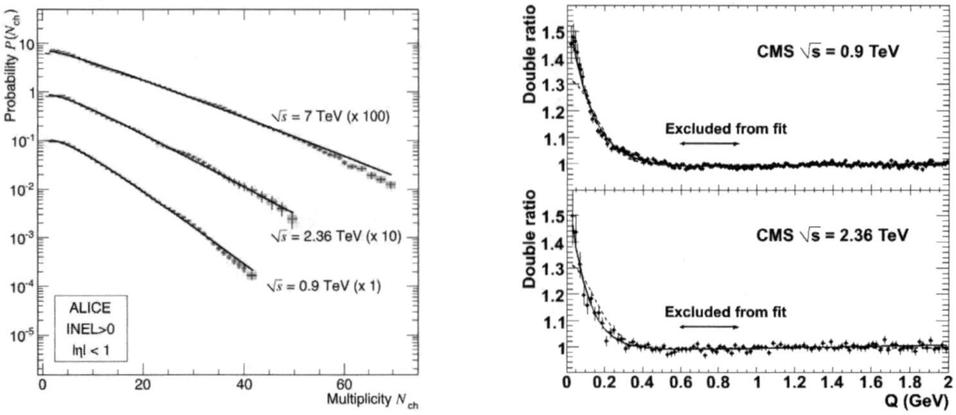

Figure 2. Charged-particle multiplicity distributions for different centre-of-mass energies $\sqrt{s}$ measured by ALICE (left), and Bose-Einstein correlations measured at $\sqrt{s} = 0.9$ and 2.36 TeV by CMS (right).

of hardware triggers, making use of only a subset of event information, followed by software-based high-level triggers (HLTs) running on farms of standard computers. The experiments started with low-threshold level-1 triggers and the HLTs running in pass-through mode (*i.e.* not rejecting any events), allowing the efficiencies of higher-threshold trigger chains to be studied on unbiased data samples before they need to be enabled in rejection mode. More selective triggers have been implemented step-by-step, with the aim of always filling the available data recording bandwidth irrespective of the LHC luminosity.

The worldwide LHC Grid computing effort has also worked very well in the initial data-taking phase, with $O(1\,\mathrm{GB/s})$ of data being recorded and distributed to the > 100 Tier-1 and Tier-2 sites. Data has been analysed by 1000s of individual users throughout the geographical spread of the collaborations, in parallel with data reprocessing with improved calibrations and the large-scale production of Monte Carlo simulation samples.

4. Minimum bias and QCD measurements

The most basic process in proton-proton collisions is $pp \to X$, composed of a mixture of elastic, diffractive and inelastic processes, collectively referred to as 'minimum bias'. In practice, the exact mixture of processes sampled depends on the experiments' triggers, which are typically sensitive to activity on either one or both sides of the detector. All of the LHC experiments have measured the charged particle multiplicity in such collisions, as a function of pseudorapidity η and particle p_T; measurements from ALICE of the multiplicity distribution in pp collisions at 0.9, 2.36 and 7 TeV are shown in Figure 2(left) [2, 3]. As expected, the overall event charged particle multiplicity increases with $\sqrt{s}$; however existing Monte Carlo models tuned at the Tevatron tend to underestimate the overall multiplicity and predict a harder p_T spectrum than in the data. Work has already started on incorporating first LHC data in such model tunes, which should lead to a better description of this data.

CMS and ALICE have also shown measurements of Bose-Einstein correlations (BEC) in minimum bias events. These correlations are manifested as an enhancement of the density of pairs of identical bosons (e.g. $\pi^+\pi^+$) close in phase space, *i.e.* with four-momentum $p_1 \approx p_2$. They can be characterised by studying the particle pair density dN/dQ where $Q = \sqrt{-(p_1 - p_2)^2}$ with respect to a reference distribution $dN/dQ_{\rm ref}$ constructed to be free of

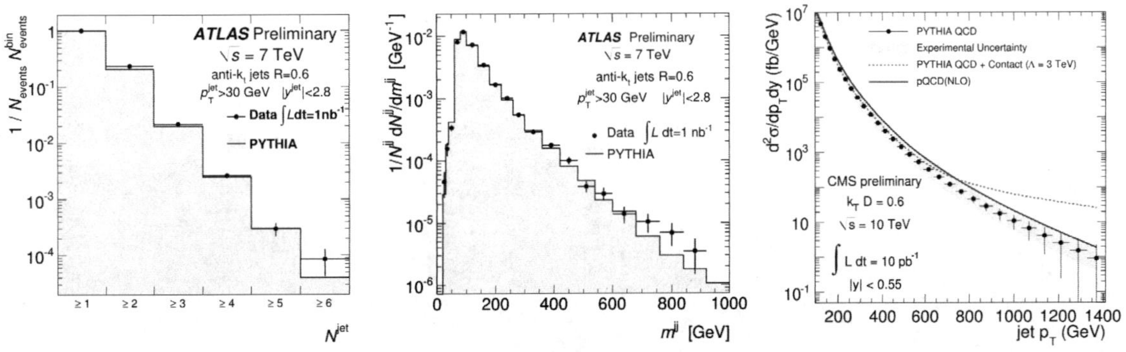

Figure 3. Distributions of jet multiplicity (left) and dijet invariant mass (centre) in $1\,\mathrm{nb}^{-1}$ of ATLAS 7 TeV data; expected jet p_T spectrum in $10\,\mathrm{pb}^{-1}$ of simulated 10 TeV CMS data, showing the prediction of NLO QCD and the effect of a contact interaction of scale $\Lambda = 3\,\mathrm{TeV}$ (right).

two-particle correlations. The reference distribution can be constructed from the data by taking opposite charged pairs (*e.g.* $\pi^+\pi^-$) after correcting for resonance decays, by reflecting particles from the opposite hemisphere, by mixing events, or by using a Monte Carlo simulation sample without BEC. Figure 2(right) shows an example from CMS of BEC measured at $\sqrt{s} = 0.9$ and 2.36 TeV with respect to combination of reference samples [4]. The correlations can be parameterised in terms of the size and strength of a source emitter, and both CMS and ALICE [2] data show that the emitter size increases as a function of the number of charged particles in the event.

Hard interactions between the quarks and gluons in the colliding protons produce events with characteristic dijet and multijet structure. These interactions dominate the event sample with large total transverse energy, and were seen right from the first LHC runs at $\sqrt{s} = 7\,\mathrm{TeV}$. The jets are reconstructed primarily in the experiments' calorimeters, and provide an essential tool in understanding and calibrating the calorimeter response, as well as a sensitive probe of QCD models. Figure 3 shows measurements of the multiplicity of jets reconstructed in $1\,\mathrm{nb}^{-1}$ of 7 TeV ATLAS data [5], together with the dijet invariant mass spectrum, compared to the predictions of the Pythia [6] leading-order Monte Carlo generator. Once the jet energy scale systematic is fully understood, by using *e.g.* energy balance in dijet and later γ-jet events, these distributions will be used to set limits on possible deviations from the behaviour expected from QCD. For example, Figure 3(right) shows a CMS simulation study of the effect of a contact interaction term with scale $\Lambda = 3\,\mathrm{TeV}$ [7]; this produces a clear enhancement at high dijet invariant mass which would be visible already with a few pb^{-1} of data.

As well as the hard interaction giving rise to quark and gluon jets, the experiments are beginning to study the 'underlying event' which comes from other components of the *pp* interaction, *e.g.* initial and final state gluon radiation, the remnants of the proton which do not participate in the hard scattering, and multiple parton interactions. The properties of the underlying event are expected to be approximately independent of the activity in the hard scattering. This hypothesis can be studied by using the leading (highest-p_T) track to define an event direction in the transverse plane correlated to the hard scattering, and looking at the distributions of charged particles in towards, transverse and away regions with respect to this direction, as shown in Figure 4(left). Densities of charged particles in these regions have been measured by ATLAS [8], and are shown in Figure 4(right). It can be seen that the particle density as a function of p_T reaches a plateau in the transverse region, whilst it continues to increase in the

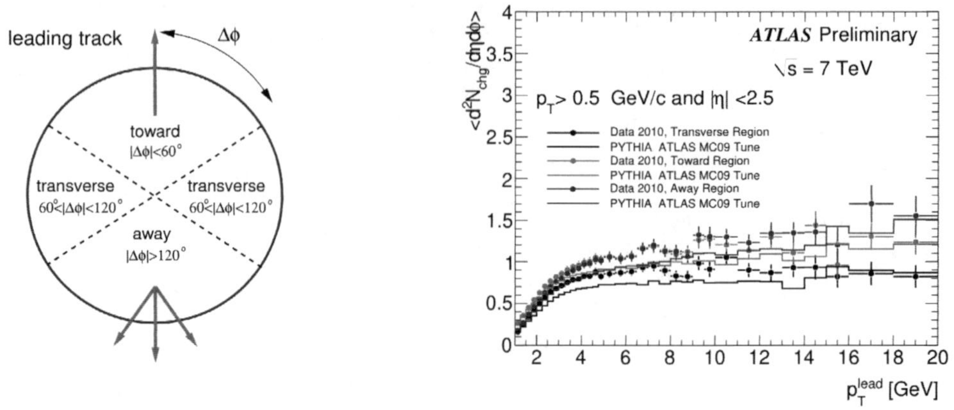

Figure 4. Definitions of the towards, away and transverse regions in underlying event analysis (left); measurements of charged particle density as a function of p_T of the leading track in these three regions, from ATLAS $\sqrt{s} = 7\,\text{TeV}$ data.

towards and away regions, which are populated more strongly by the hard scattering. However, other variables in the transverse region are correlated with the leading track p_T, indicating possible contributions from components other than the beam remnant. These studies can be used to tune Monte Carlo models and are important in understanding the background to jets in high p_T events.

5. Early flavour physics

The large $b\bar{b}$ production cross-section at LHC ($O(100\,\mu\text{b})$ at $\sqrt{s} = 7\,\text{TeV}$) makes the LHC a true b-factory, with b-hadron production rates in excess of any previous collider. The dedicated LHCb experiment has been optimised to fully exploit the potential of LHC for precision b (and c) physics, in particular in the area of CP-violation studies. Many of these studies require the study of b-hadron decays involving J/ψ mesons, and LHCb has already shown reconstructed $J/\psi \to \mu\mu$ decays in the first few nb^{-1} of data, as shown in Figure 5(left). These mesons are produced both directly ('prompt' J/ψ production) and in the decay of b-hadrons, and the two components can be separated by reconstructing the proper decay time of the J/ψ from measurements of its lifetime and distance from the event primary vertex. The latter relies on precise tracking measurements using the LHCb VELO silicon vertex detector. Figure 5(right) shows that the separate contributions can already be separated, with the exponential decay of the b-hadron component allowing the average b-lifetime to be measured [9]. With more statistics, these samples will allow the study of quarkonium production, spectroscopy and polarisation, and b-physics in both inclusive and exclusive modes. The $J/\psi \to \mu\mu$ decays are also very useful for studying the detector calibration, alignment and magnetic field maps.

LHCb is also starting to reconstruct exclusive charm decays, for example D^* mesons via the decay $D^{*+} \to D^0\pi^+$, with $D^0 \to K^-\pi^+$ or the rarer decay mode $D^0 \to K^-K^+$. Due to the small D^*-D mass difference, the π^+ produced directly in the D^* decay produces a very narrow peak in the $m_{\pi^+K^-\pi^+} - m_{K^-\pi^+}$ mass spectrum and also tags the flavour (D^{*+} or D^{*-}) of the parent meson. Studying the time-dependent decay rates of D^* into the various final states provides access to D^0-$\bar{D}^0$ mixing and CP-violation observables. LHCb has already reconstructed several 100s of the rare $D^0 \to K^+K^-$ mode, and should have a sample ten times larger than that of BaBar with $100\,\text{pb}^{-1}$ [10]. The experiment is also starting to reconstruct significant numbers of semi-exclusive semileptonic b-hadron decays such as $B \to D^0\mu\nu X$ and $B_s \to D_s\mu\nu X$. These

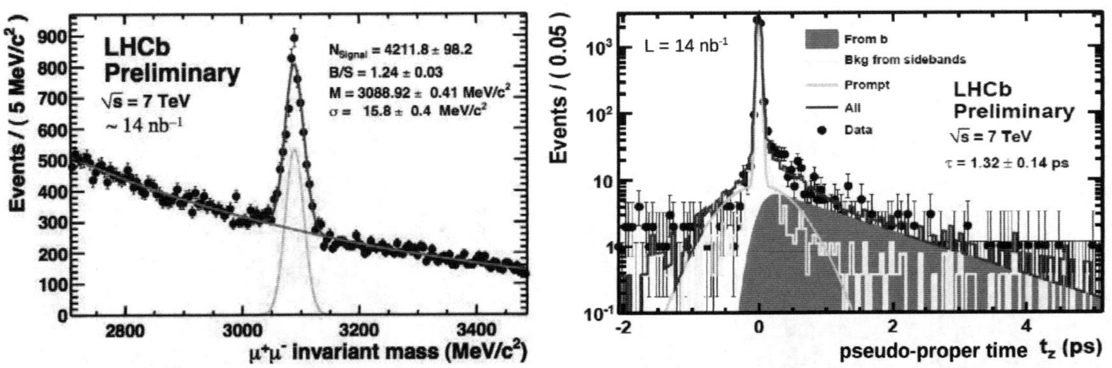

Figure 5. Invariant mass (left) and proper time (right) distributions of $J/\psi \to \mu\mu$ decays reconstructed by LHCb in $14\,\text{nb}^{-1}$ of $\sqrt{s} = 7\,\text{TeV}$ data.

will provide a complementary measurement of the $b\bar{b}$ production cross-section to that using $b \to J/\psi X$ decays, access to B^0-$\bar{B}^0$ oscillations measurements and calibration of flavour tagging algorithms [10].

6. Vector boson and top quark production

Beyond the production of high-p_T jets, the next early physics milestone for ATLAS and CMS is the observation of the electroweak vector bosons W and Z, through their leptonic decays $W \to \ell\nu$ and $Z \to \ell\ell$ (ℓ represents an electron or muon). W-boson production is the dominant source of high-p_T leptons, which tend to be more isolated than the low-p_T leptons produced from the decays of b- and c-flavoured hadrons in jets. The production of a neutrino ν in W boson decay also gives rise to significant missing transverse energy E_T^{miss}, which can be used to enhance the W signal.

ATLAS and CMS both observed W production in the first few nb^{-1} of data, using selections based on requiring isolated high-p_T leptons and significant E_T^{miss}. Since the longitudinal (parallel to the beam axis) component of the neutrino momentum cannot be measured, the W mass cannot be reconstructed. Instead, the experiments make use of the transverse mass m_T, defined as:

$$m_T = \sqrt{2p_T^\ell p_T^\nu (1 - \cos(\phi^\ell - \phi^\nu))}$$

where the E_T^{miss} vector is used as an estimate of the neutrino transverse momentum p_T^ν. Figure 6 shows distributions of m_T measured by ATLAS in the electron channel [11], and CMS in the muon channel [12]; clear Jacobian peaks around the W mass are visible, smeared due to the use of m_T rather than m. The numbers of observed events are consistent with the expectations from the Standard Model.

The Z boson production rate is about a factor 10 smaller than that for W, and the first few events have been observed in both $Z \to ee$ and $Z \to \mu\mu$. As can be seen from Figure 7 showing the mass distribution of Z boson candidates in CMS [12], the expected background is much smaller than for W, and the selected event sample is dominated by the continuum $q\bar{q} \to Z/\gamma^* \to \ell\ell$ Drell-Yan process, which has a resonant peak at the Z mass. Due to the clean signature with low background and the well-determined Z mass, with more statistics such event samples will be extremely useful for calibrating the detectors' energy response and uniformity. They can also be used for measuring the efficiency of both trigger and offline reconstruction/identification algorithms, by selecting events using only one lepton from the Z, and studying the properties of the other lepton in an unbiased way (the so-called 'tag and probe' technique).

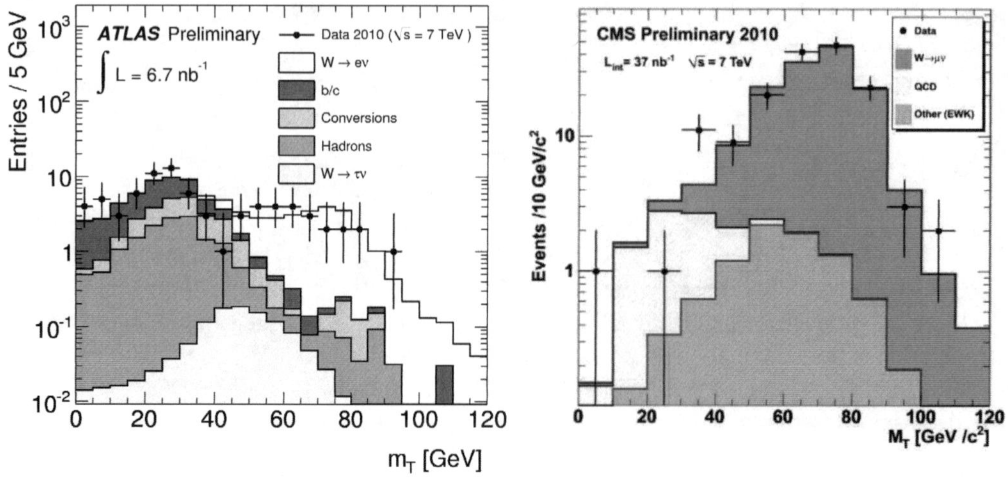

Figure 6. Distributions of the W boson transverse mass m_T, measured in early data by ATLAS in the electron channel (left) and CMS in the muon channel (right).

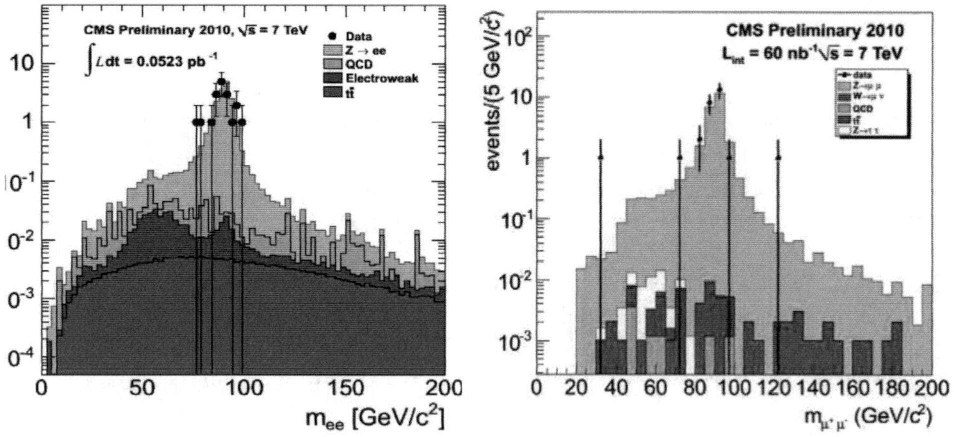

Figure 7. Distributions of the Z candidate mass reconstructed by CMS in ee (left) and $\mu\mu$ (right) channels.

With $O(3000)$ W and $O(300)$ Z bosons per lepton flavour expected in each pb^{-1} of data, many precision measurements should become possible already in 2010. For example, the production rates of W^+ and W^- bosons are not equal because of the dominance of u over d valence quarks in the proton. Predictions for the asymmetry $A(\eta)$ defined as:

$$A(\eta) = \frac{N(W^+ \to \ell^+\nu) - N(W^- \to \ell^-\nu)}{N(W^+ \to \ell^+\nu) + N(W^- \to \ell^-\nu)}$$

depend in detail on the parton density functions (PDFs) assumed for the proton, and the level of precision achievable with $100\,\mathrm{pb}^{-1}$ is already enough to improve on current PDF uncertainties, see Figure 8(left) [13]. W and Z bosons can also be produced in association with one or more hard jets—each extra jet 'costing' approximately a factor of α_s. With $1\,\mathrm{fb}^{-1}$, 1000s of events will be produced up to Z+4 jets, see Figure 8(right), and the resulting distributions of jet

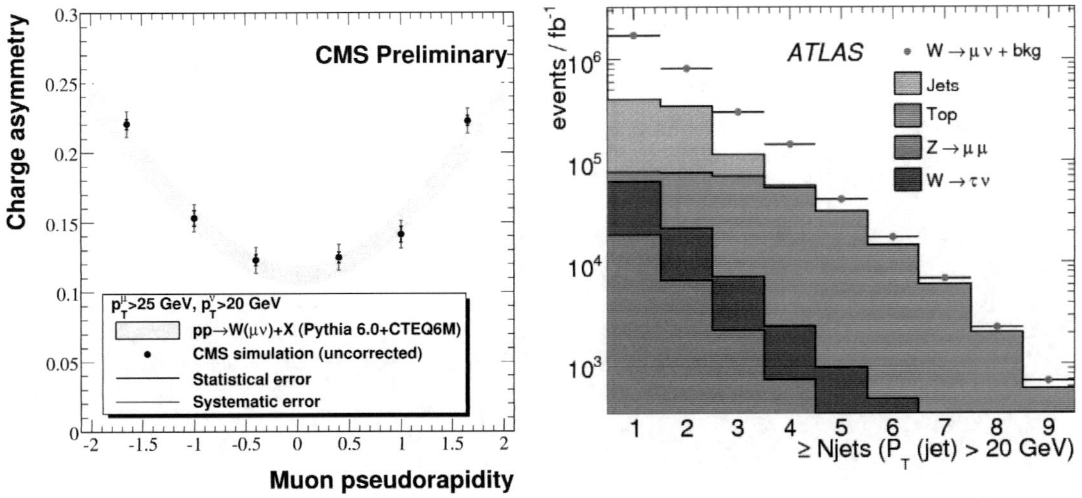

Figure 8. Expected $W \to \mu\nu$ charge asymmetry measurement as a function of η, assuming $100\,\mathrm{pb}^{-1}$ of CMS data at $\sqrt{s} = 10\,\mathrm{TeV}$ (left); expected jet multiplicity distribution in association with W boson production in $1\,\mathrm{fb}^{-1}$ of simulated ATLAS data at $\sqrt{s} = 10\,\mathrm{TeV}$ (right).

multiplicities and momenta can be used to provide sensitive tests of QCD and Monte Carlo models [14]. These events are also a significant background to many analyses of final states involving high p_T leptons, jets and E_T^{miss}, for example top quark physics and many new physics searches.

Beyond the W and Z bosons, the last and heaviest known outpost of the Standard Model is the top quark, so far seen only at the Tevatron. The top quark is expected to be produced most copiously in pairs ($t\bar{t}$), and decays almost exclusively as $t \to Wb$. The $t\bar{t}$ decay topologies are therefore determined by the decay modes of the W, and include the dilepton state $t\bar{t} \to \ell\nu b\,\ell\nu b$ ($\ell = e$ or μ) with a branching ratio of 5 %, and the semileptonic final state $t\bar{t} \to \ell\nu b\,jjb$ with a branching ratio of 30 %. Both decay modes give rise to events with isolated leptons, E_T^{miss} and jets, some with b-flavour. The e-μ dilepton mode is the cleanest, as can be seen in Figure 9(left), which shows the expected composition of e-μ dilepton events as a function of jet multiplicity in a CMS simulation study at $\sqrt{s} = 10\,\mathrm{TeV}$. However, at $\sqrt{s} = 7\,\mathrm{TeV}$, only a few dilepton events are expected per pb^{-1} of data. The semileptonic final states have higher rate, and also allow the reconstruction of the hadronic top mass m_{jjb} from three of the four jets—this will provide confirmation that the observed object is the 173 GeV top quark also seen at Fermilab. This state suffers from higher background, principally from W+jet events, as can be seen from the ATLAS simulation study shown in Figure 9(right). The background can be significantly reduced if b-tagging can be applied, only a small fraction of the W+jets background having b-flavoured jets. As of July 2010, both ATLAS and CMS were eagerly awaiting the first $t\bar{t}$ candidate events.

Once the initial $t\bar{t}$ signal is established and the production cross-section measured (to an expected precision of 10–20 % with $100\,\mathrm{pb}^{-1}$), the LHC precision top physics program can begin. With the $1\,\mathrm{fb}^{-1}$ expected in 2011, the top quark sample will significantly exceed that available at the Tevatron, allowing *e.g.* the top quark mass to be measured with a statistical error well below 1 GeV using the m_{jjj} distribution or a full kinematic fit. The key will be to control systematic errors to the same level, especially those related the jet energy scale, for both light and b-flavoured jets. With enough data statistics, alternative techniques less dependent on jet energies will also be used, *e.g.* use of the lepton p_T and b-decay length distributions, or the very

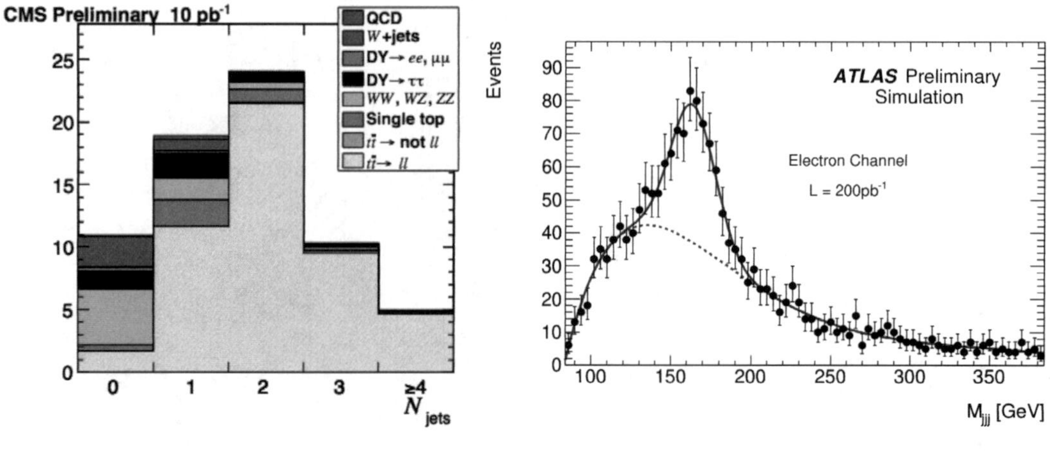

Figure 9. Expected jet multiplicity distribution in e-μ dilepton events in $10\,\mathrm{pb}^{-1}$ of $\sqrt{s} = 10\,\mathrm{TeV}$ simulated CMS data (left); expected 3-jet invariant mass distribution in the electron plus 4-jet selection in $200\,\mathrm{pb}^{-1}$ of $\sqrt{s} = 10\,\mathrm{TeV}$ simulated ATLAS data.

clean measurements possible in $t \to b \to J\psi$ decays.

Electroweak single top production will also be intensively studied in the first years at LHC; as at the Tevatron, complex multivariate techniques will be required to extract the signal from the dominant W+jets (with heavy flavour) and $t\bar{t}$ backgrounds. Simulations studies [15] indicate that $200 - 1000\,\mathrm{pb}^{-1}$ will be required at $\sqrt{s} = 7\,\mathrm{TeV}$ to establish the dominant t-channel production mechanism, but background characterisation studies can start already with the 2010 data.

7. Higgs, supersymmetry and beyond

The prospects for an early discovery of the Standard Model Higgs boson at LHC depend crucially on the Higgs mass m_H. The first chance will come using the $H \to WW$ decay mode in the mass range $150 < m_H < 180\,\mathrm{GeV}$, where the Higgs boson is produced either in the gluon fusion mode, or via the fusion of W or Z bosons produced in the initial state (vector boson fusion or VBF). The analysis requires two high p_T leptons and large E_T^{miss} from the W decays, and a veto on additional jets in the central region to suppress $t\bar{t}$ background. The sensitivity in the VBF mode can be enhanced by requiring additional forward jets from the quarks radiating the bosons. The remaining background is dominated by WW diboson and W+jets. Figure 10(left) shows the significance achievable in CMS with $1\,\mathrm{fb}^{-1}$ at $\sqrt{s} = 7\,\mathrm{TeV}$ [16]; a 3–5σ signal is expected for $150 < m_H < 180\,\mathrm{GeV}$.

For larger m_H, the $H \to ZZ \to 4\ell$ channel is most promising—with sufficient data statistics, the Higgs signal stands out clearly over the continuum background from ZZ diboson production. But for $1\,\mathrm{fb}^{-1}$, the sensitivity is limited to cross-sections a few times larger than that of the Standard Model, see Figure 10(right). An intriguing possibility is that of a 4th generation of Standard Model-like quarks: in this case, the $gg \to H$ production cross-section is enhanced by a factor 9, irrespective of the t' quark mass, rendering the Higgs boson observable up to about $m_H = 400\,\mathrm{GeV}$ [16]. At low m_H below the $H \to WW$ threshold, a variety of channels contribute, including $H \to \gamma\gamma$. In this channel, the Higgs is visible as a small peak over a large continuum background from direct $\gamma\gamma$ production and γ-jet production with a misidentified jet. With $1\,\mathrm{fb}^{-1}$, a sensitivity of 4–6 times the Standard Model cross-section can be achieved depending on m_H [16, 17].

Early LHC data also offers some sensitivity to the neutral Higgs bosons of the Minimal

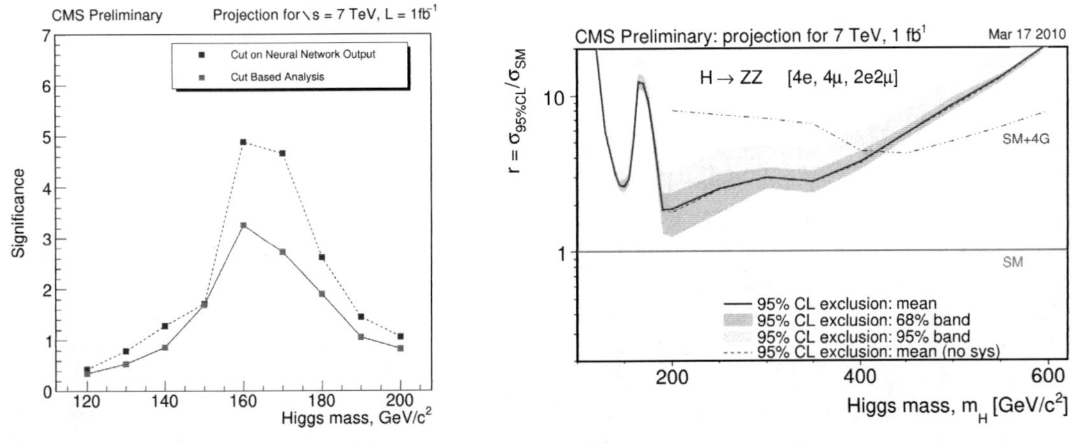

Figure 10. Expected significance of a Standard Model $H \to WW$ signal (left) and ratio of expected 95 % confidence level exclusion cross-section to the Standard Model Higgs production cross-section in the $H \to ZZ \to 4\ell$ channel (right), as a function of m_H, assuming $1\,\mathrm{fb}^{-1}$ of CMS data at $\sqrt{s} = 7\,\mathrm{TeV}$.

Supersymmetic Standard Model (MSSM) if the Higgs vacuum expectation value ratio $\tan\beta$ is large. In this case, the A, H and h are produced preferentially in association with b-quarks, and can be seen in the $A, H, h \to \tau\tau$ decay mode. With $1\,\mathrm{fb}^{-1}$, low mass Higgs bosons can be seen for $\tan\beta > 20$ [16].

In the most widely-studied R-parity conserving supersymmetry (SUSY) models, squark and gluino pairs will be copiously produced at LHC, decaying into a variety of final states. These are characterised by multiple jets, E_T^{miss}due to the escaping lightest supersymmetric particle which is neutral, non-interacting and stable, and sometimes one more leptons depending on the decay mode. The most important issues in early SUSY searches are a good performance of E_T^{miss} measurements, to minimise 'fake' E_T^{miss} caused by detector problems or resolution tails, and the understanding of Standard Model backgrounds, particularly those from $t\bar{t}$ and W+jets. These background processes often contribute in 'pathological' ways which are difficult to simulate, *e.g.* when one lepton is missed or badly reconstructed, so need to be evaluated using data control samples wherever possible. Studies show that sensitivities to 500 GeV squarks and gluinos, just beyond the Tevatron limits, can be achieved with $100\text{-}500\,\mathrm{pb}^{-1}$ using final states with and without leptons [16, 18].

An alternative strategy involves looking for events with two like-sign leptons, also in association with jets and E_T^{miss}. Such selections tend to have lower signal efficiency, but also much lower Standard Model backgrounds, mainly from $t\bar{t}$ events where the second lepton comes from a b-hadron decay, or is a misreconstructed fake. This technique also offers the possibility for significant improvements over the current Tevatron limits with $1\,\mathrm{fb}^{-1}$ [16], but in all cases, LHC data at $\sqrt{s} = 14\,\mathrm{TeV}$ will be needed to probe for SUSY particles at the 1 TeV mass scale.

Many other exotic physics scenarios can be probed with the early LHC data. For example, several models predict high-mass W or Z-like objects which decay in similar ways to the Standard Model bosons: $W' \to \ell\nu$ and $Z' \to \ell\ell$. Limits at the Tevatron are now constrained by the available centre-of-mass energy to around 1 TeV, and these should be overtaken with 7 TeV LHC data samples of around $20\,\mathrm{pb}^{-1}$ (W') and $100\,\mathrm{pb}^{-1}$ (Z'). Models with large extra dimensions predict an excess of high-mass diphoton events, where the sensitivity beyond the Tevatron can be achieved with as little as $50\,\mathrm{pb}^{-1}$ [16]. Finally, searches for resonances or other deviations in the dijet mass spectrum can provide sensitivity to quark compositeness or excited quarks, again in the first year of LHC datataking.

8. Conclusions

After many years of development, construction, installation and commissioning, the LHC physics era has finally begun. The commissioning of the LHC accelerator at $\sqrt{s} = 7\,\text{TeV}$ is going very well, and no major performance limitations have been encountered so far. The LHC team has a clear plan towards operation at an instantaneous luminosity of $L = 10^{32}\,\text{cm}^{-2}\text{s}^{-1}$ near the end of 2010, and an integrated luminosity of $1\,\text{fb}^{-1}$ per experiment in 2011, before a major shutdown in 2012 to prepare for 14 TeV operation. The experiments are working very well, triggering and recording data with high efficiency, and analysing it effectively using the worldwide Grid computing environment.

The first LHC physics results have already been derived, with measurements of basic event properties in the new energy regime. The heavy flavour and jet physics programs have started, W and Z boson production observed and ATLAS and CMS are on the verge of observing top quark production. The integrated luminosity profile is such that first results on new physics searches can be expected already at the end of 2010, whilst extending the Higgs boson search beyond the current Tevatron limits will also require the 2011 data sample.

References

[1] A.D. Martin, W.J. Stirling, R.S. Thorne, G. Watt, Parton distributions for the LHC, Eur. Phys. J. C. (2009) 63: 189; W.J. Stirling, private communication.
[2] P. Kuijer on behalf of the ALICE collaboration, Status report of the ALICE experiment, talk given at the 102nd LHCC meeting, 7th July 2010, `http://indico.cern.ch/conferenceDisplay.py?confId=98442`.
[3] Charged-particle multiplicity measurement in proton-proton collisions at $\sqrt{s} = 7\,\text{TeV}$ with ALICE at LHC, Eur. Phys. J. C. (2010) 68: 345.
[4] CMS collaboration, First measurement of Bose-Einstein correlations in proton-proton collisions at $\sqrt{s} = 0.9$ and 2.36 TeV at the LHC, Phys. Rev. Lett. 105 (2010) 032001.
[5] ATLAS collaboration, Observation of energetic jets in pp collisions at $\sqrt{s} = 7\,\text{TeV}$ using the ATLAS experiment at the LHC, ATLAS-CONF-2010-043.
[6] T. Sjöstrand et al, JHEP 05, 026 (2006).
[7] CMS collaboration, Initial measurement of the inclusive jet cross-section at 10 TeV with CMS, CMS-PAS-QCD-08-001.
[8] ATLAS collaboration, Track-based underlying event measurements in pp collisions at $\sqrt{s} = 900\,\text{GeV}$ and 7 TeV with the ATLAS detector at LHC.
[9] S. Hansmann-Menzemer on behalf of the LHCb collaboration, Status report of LHCb, talk given at the 102nd LHCC meeting, 7th July 2010, `http://indico.cern.ch/conferenceDisplay.py?confId=98442`.
[10] O. Schnieder on behalf of the LHCb collaboration, LHCb first physics results, talk given at Physics at the LHC, DESY, Hamburg, June 7–12 2010.
[11] ATLAS collaboration, Observation of $W \to \ell\nu$ and $Z \to \ell\ell$ production in proton-proton collisions at $\sqrt{s} = 7\,\text{TeV}$ with the ATLAS detector.
[12] C. Leonidopoulos on behalf of the CMS collaboration, Status report of the CMS experiment, talk given at the 102nd LHCC meeting, 7th July 2010, `http://indico.cern.ch/conferenceDisplay.py?confId=98442`.
[13] CMS collaboration, Muon differential cross-section and charge asymmetry in inclusive $pp \to W \to \mu\nu X$ production at $\sqrt{s} = 10\,\text{TeV}$, CMS-PAS-EWK-09-003.
[14] ATLAS collaboration, Expected performance of the ATLAS experiment: Detector, trigger and physics, CERN-OPEN-2008-020.
[15] ATLAS collaboration, Search for single top events using early data in ATLAS at the LHC, ATLAS-PHYS-PUB-2010-003; CMS collaboration, Prospects for the measurement of the single-top t-channel cross section in the muon channel with $200\,\text{pb}^{-1}$ of CMS data at 10 TeV.
[16] CMS collaboration, The CMS physics reach for searches at 7 TeV, CMS NOTE-2010/008.
[17] ATLAS collaboration, ATLAS sensitivity prospects for Higgs boson production at the LHC running at 7 TeV, ATLAS-PHYS-PUB-2010-009.
[18] ATLAS collaboration, Prospects for Supersymmetry discovery based on inclusive searches at 7 TeV centre-of-mass energy with the ATLAS detector, ATLAS-PHYS-PUB-2010-010.

Journal of Physics: Conference Series **259** (2010) 012002 doi:10.1088/1742-6596/259/1/012002

LHC phenomenology of supersymmetric models beyond the MSSM

Werner Porod[a,b]

[a] Institut für Theoretische Physik und Astrophysik, Universität Würzburg, D-97074 Würzburg, Germany

[b] AHEP Group, Instituto de Física Corpuscular - C.S.I.C., Universitat de València, E-46071 València, Spain

E-mail: porod@physik.uni-wuerzburg.de

Abstract. We discuss various phenomenological aspects of superssymmetric models beyond the MSSM. A particular focus is on models which can correctly explain neutrino data and the possiblities of LHC to identify the underlying scenario.

1. Introduction

Supersymmetric extensions of the standard model (SM) are promising candidates for new physics at the TeV scale [1, 2] as the solve several short-comings of the Standard Model (SM). The Minimal Supersymmetric Standard Model (MSSM) solves the hierarchy problem of the SM [3], leads to a unification of the gauge couplings [4, 5] and introduces several candidates for dark matter depending on how SUSY is broken [6, 7]. Its phenomenology with respect to present and future colliders has been widely explored, see e.g. [8, 9].

However, similarly as the SM it needs additional ingredients to explain neutrino data, e.g. by either incooperating heavy new particles giving rise to tiny neutrino masses via the seesaw mechanism [10] or via the braking of R-parity [11]. Moreover, a new problem arises in the MSSM not present in the SM: the superpotential contains a parameter with dimension mass, namely the so called μ parameter which gives masses to the Higgs bosons and higgsinos. From a purely theoretical point of view, the value of this parameter is expected to be either of the order of the GUT/Planck scale or exactly zero, if it is protected by a symmetry. For phenomenological aspects, however, it has to be of the order of the scale of electroweak symmetry breaking (EWSB) and it has to be non-zero to be consistent with experimental data. This discrepancy is the so called μ-problem of the MSSM [12].

In this paper we discuss various supersymmetric models addressing at least one of these two topics focusing on features of their phenomenology which can be tested at the LHC and which differ from the usual MSSM phenomenology. We will take as main guideline the requirement that neutrino data are correctly explained. We will first discuss briefly the case of Dirac neutrinos and afterwards discuss models leading to Majorana neutrinos, both in the context of conserved R-parity. In the third part we cover models with broken R-parity and their phenomenology at the LHC.

Journal of Physics: Conference Series **259** (2010) 012002 doi:10.1088/1742-6596/259/1/012002

2. Dirac neutrinos

Technically the easiest way to obtain masses for neutrinos is by introducing a new Yukawa coupling similar to the ones for the SM fermions. In this case the MSSM has to be extended by additional right handed neutrinos, which are gauge singlets with respect to the SM gauge group, and the superpotential of the MSSM

$$W_{MSSM} = \hat{H}_d \hat{L} Y_e \hat{E}^c + \hat{H}_d \hat{Q} Y_d \hat{D}^c + \hat{H}_u \hat{Q} Y_u \hat{U}^c - \mu \hat{H}_d \hat{H}_u \tag{1}$$

has to be extended by the term

$$W_{\nu^c} = \hat{H}_u \hat{L} Y_\nu \hat{\nu}^c . \tag{2}$$

Here $\hat{H}_d$, $\hat{H}_u$, $\hat{Q}$, $\hat{D}^c$, $\hat{U}^c$, $\hat{L}$ and $\hat{E}^c$ are the MSSM superfields containing the Higgs bosons, quarks, leptons and their supersymmetric partners. In addition the superfield $\hat{\nu}^c$ for the right-handed neutrino and the right-sneutrino has been introduced. The corresponding Yukawa coupling Y_ν^{ij} has to be tiny, of the order 10^{-12} and smaller, to explain correctly neutrino masses in the sub-eV range as required by data [13]. This in turn implies that one is essentially left with the usual MSSM phenomenology, except for the case where the right-sneutrino is the lightest supersymmetric particle (LSP): in this case all decays are as in the MSSM down to the next to lightest supersymmetric particle (NLSP) which eventually decays into the LSP. The corresponding width of the last step is proportional to $|Y_\nu|^2$ and thus rather small implying at the LHC decay lengths ranging from $O(cm)$ up to $O(km)$ [14]. Consequently the signal of such a scenario is very two long-lived particles in each event which can even appear as stable particles in a typical detector of high energy collider experiments. Detailed studies have been performed for the cases of a stau [15] NLSP and a stop [16] NLSP demonstrating that LHC should be able to identity such scenarios.

3. Majorana neutrinos via the seesaw mechanism

A possibility to obtain tiny neutrino masses while having at the same time sizable Yukawa couplings are seesaw scenarios where very heavy new particles are postulated inducing the so-called Weinberg operator [17, 18]

$$\frac{f_{\alpha\beta}}{\Lambda}(H_u L_\alpha)(H_u L_\beta) \Rightarrow (m^\nu)_{\alpha\beta} = \frac{f_{\alpha\beta} v_u^2}{2\Lambda} \tag{3}$$

when integrating out the heavy degrees of freedom. Here Λ is a measure of the scale for these new particles and $f_{\alpha\beta}$ is usually a combination of different Yukawa couplings. One can show, that at tree-level only three possibilities exist to realize such scenarios [19]. Type-I is the well-known case of the exchange of a heavy fermionic singlet usually denoted ν_R [10, 20, 21, 22]. Type-II corresponds to the exchange of a scalar $SU(2)$ triplet [23, 24]. In seesaw type-III one adds (at least two) fermionic $SU(2)$ triplets to the field content of the SM [25]. In the last two models particles charged under the SM gauge group are added and which correspond to incomplete $SU(5)$ representations. As they would destroy the nice feature of gauge coupling unification one often adds at the seesaw scale(s) additional particles to obtain complete $SU(5)$ representations to maintain the feature of gauge coupling unification. A detailed discussion including the embedding in $SU(5)$ models can be found in e.g. in ref. [26].

Below we present only the various superpotential for briefness. In addition there will also be the corresponding soft SUSY terms which, however, reduce at the electroweak scale to the MSSM ones once the states with masses at the seesaw scale(s) are integrated out and, thus, are not discussed further. These additional terms only contribute to threshold corrections at the seesaw scale(s) and their effect is negligible if one requires universal boundary conditions for all soft terms [27]. Here we will assume common soft SUSY breaking at the GUT-scale

M_G to specify the spectrum at the electroweak scale: a common gaugino mass $M_{1/2}$, a common scalar mass m_0 and the trilinear coupling A_0 which gets multiplied by the corresponding Yukawa couplings to obtain the trilinear couplings in the soft SUSY breaking Lagrangian. In addition the sign of the μ parameter is fixed as well as $\tan\beta = v_u/v_d$ at the electroweak scale where v_d and v_u are the the vacuum expectation values (vevs) of the neutral component of H_d and H_u, respectively.

3.1. Supersymmetric seesaw type-I

In this class of models one postulates very heavy right-handed neutrinos yielding the following superpotential below the GUT-scale:

$$W_I = W_{MSSM} + W_{\nu^c} + \frac{1}{2}\hat{\nu}^c M_R \hat{\nu}^c \tag{4}$$

For the neutrino mass matrix one obtains the well-known formula

$$m_\nu = -\frac{v_u^2}{2} Y_\nu^T M_R^{-1} Y_\nu . \tag{5}$$

Being complex symmetric, the light Majorana neutrino mass matrix in eq. (5) is diagonalized by a unitary 3×3 matrix U [23]

$$\hat{m}_\nu = U^T \cdot m_\nu \cdot U \,. \tag{6}$$

Inverting the seesaw equation eq. (5) allows to express Y_ν as [28]

$$Y_\nu = \sqrt{2}\frac{i}{v_u}\sqrt{\hat{M}_R} \cdot R \cdot \sqrt{\hat{m}_\nu} \cdot U^\dagger , \tag{7}$$

where the $\hat{m}_\nu$ and $\hat{M}_R$ are diagonal matrices containing the corresponding eigenvalues. R is in general a complex orthogonal matrix. We assume $R = 1$ implying that Y_ν contains only "diagonal" products $\sqrt{M_i m_i}$.

3.2. Supersymmetric seesaw type-II

In seesaw models of type II one adds a scalar $SU(2)$ triplet T to generate neutrino masses. As this triplet carries also hypercharge one has to embed it in a 15-plet of $SU(5)$ which has under $SU(3) \times SU_L(2) \times U(1)_Y$ the following decomposition [29]

$$\begin{aligned} \mathbf{15} &= S + T + Z \\ S &\sim (6,1,-\tfrac{2}{3}), \qquad T \sim (1,3,1), \qquad Z \sim (3,2,\tfrac{1}{6}). \end{aligned} \tag{8}$$

In supersymmetric models one adds a pair 15 and $\overline{15}$ to avoid anomalies below the GUT-scale. Below the GUT scale in the $SU(5)$-broken phase the superpotential reads

$$\begin{aligned} W_{II} &= W_{MSSM} + \frac{1}{\sqrt{2}}(Y_T \widehat{L}\widehat{T}_1\widehat{L} + Y_S \widehat{D}^c \widehat{S}_1 \widehat{D}^c) + Y_Z \widehat{D}^c \widehat{Z}_1 \widehat{L} \\ &+ \frac{1}{\sqrt{2}}(\lambda_1 \widehat{H}_d \widehat{T}_1 \widehat{H}_d + \lambda_2 \widehat{H}_u \widehat{T}_2 \widehat{H}_u) + M_T \widehat{T}_1 \widehat{T}_2 + M_Z \widehat{Z}_1 \widehat{Z}_2 + M_S \widehat{S}_1 \widehat{S}_2 \end{aligned} \tag{9}$$

where the states with index 1 (2) belong the 15-plet ($\overline{15}$-plet). The second term in eq. (9) is responsible for the generation of the neutrino masses yielding

$$m_\nu = -\frac{v_2^2}{2}\frac{\lambda_2}{M_T} Y_T . \tag{10}$$

Note that

$$\hat{Y}_T = U^T \cdot Y_T \cdot U \tag{11}$$

i.e. Y_T is diagonalized by the same matrix as m_ν.

In addition there are the couplings Y_S and Y_Z, which in principle are not determined by any low-energy data. In the calculation of lepton flavour violating observables both Yukawa couplings, Y_T and Y_Z, contribute. Having a GUT model in mind we require for the numerical discussion later the $SU(5)$ boundary conditions $Y_T = Y_S = Y_Z$ at the GUT scale. As long as $M_Z \sim M_S \sim M_T \sim M_{15}$ gauge coupling unification will be maintained. The equality need not be exact for a successful unification. In our numerical studies we have taken into account the different running of these mass parameters but we decouple them all at the scale $M_T(M_T)$ because the differences are small.

3.3. Supersymmetric seesaw type-III

In the case of a seesaw model type III one needs new fermions Σ at the high scale being in the adjoint representation of $SU(2)$. They have to be embedded in a 24-plet to obtain a complete $SU(5)$ representation which decomposes under $SU(3) \times SU_L(2) \times U_Y(1)$ as

$$\begin{aligned} 24_M &= (1,1,0) + (8,1,0) + (1,3,0) + (3,2,-5/6) + (3^*,2,5/6) \\ &= \widehat{B}_M + \widehat{G}_M + \widehat{W}_M + \widehat{X}_M + \widehat{\bar{X}}_M \end{aligned} \tag{12}$$

The fermionic components of $(1,1,0)$ and $(1,3,0)$ have exactly the same quantum numbers as ν^c and Σ. Thus, the 24_M always produces a combination of the type-I and type-III seesaw. In the $SU(5)$ broken phase the superpotential reads

$$\begin{aligned} W_{III} &= W_{MSSM} + \widehat{H}_u(\sqrt{2}\widehat{W}_M Y_N - \sqrt{\frac{3}{10}}\widehat{B}_M Y_B)\widehat{L} + \widehat{H}_u \widehat{\bar{X}}_M Y_X \widehat{D}^c \\ &+ \frac{1}{2}\widehat{B}_M M_B \widehat{B}_M + \frac{1}{2}\widehat{G}_M M_G \widehat{G}_M + \frac{1}{2}\widehat{W}_M M_W \widehat{W}_M + \widehat{X}_M M_X \widehat{\bar{X}}_M \end{aligned} \tag{13}$$

As above we use at the GUT scale the boundary condition $Y_N = Y_B = Y_X$ and $M_B = M_G = M_W = M_X$. Integrating out the heavy fields yields the following formula for the neutrino masses at the low scale:

$$m_\nu = -v_u^2 \left(\frac{3}{10} Y_B^T M_B^{-1} Y_B + \frac{1}{2} Y_W^T M_W^{-1} Y_W \right). \tag{14}$$

The boundary conditions at M_G imply that at the seesaw scale(s) one still has $M_B \simeq M_W$ and $Y_B \simeq Y_W$ so that one can write in a good approximation

$$m_\nu = -v_u^2 \frac{4}{5} Y_W^T M_W^{-1} Y_W \tag{15}$$

and one can use the corresponding decomposition for Y_W as discussed in section 3.1.

3.4. Effect of the heavy particles on the MSSM spectrum

The appearance of charged particles at scales between the electroweak scale and the GUT scale leads to changes in the beta functions of the gauge couplings [29, 30]. In the MSSM the corresponding values at 1-loop level are $(b_1, b_2, b_3) = (33/5, 1, -3)$. In case of one 15-plet the additional contribution is $\Delta b_i = 7/2$ whereas in case of 24-plet it is $\Delta b_i = 5$. This results in case of type II in a total shift of $\Delta b_i = 7$ for the minimal model and in case of type III in $\Delta b_i = 15$ assuming 3 generations of 24-plets. This does not only change the evolution of the gauge couplings but also the evolution of the gaugino and scalar mass parameters with profound

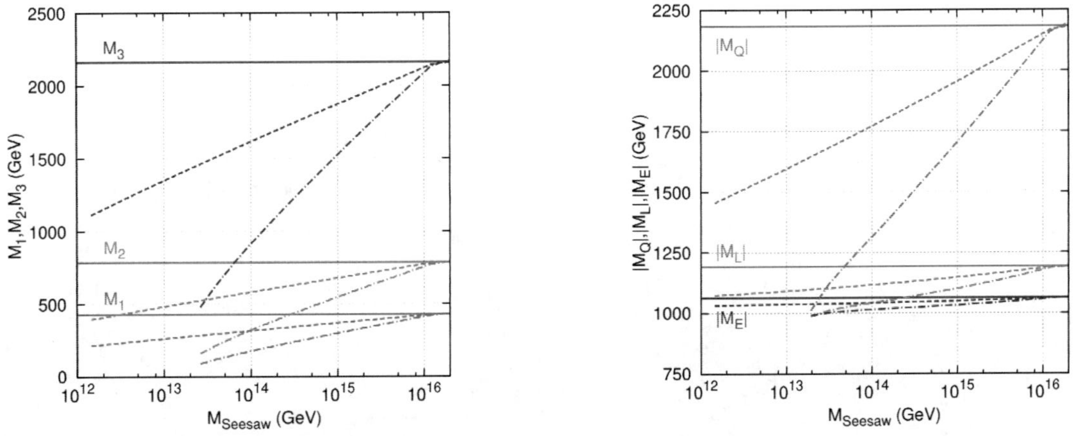

Figure 1. Mass parameters at $Q = 1$ TeV versus the seesaw scale for fixed high scale parameters $m_0 = M_{1/2} = 1$ TeV, $A_0 = 0$, $\tan\beta = 10$ and $\mu > 0$. The full lines correspond to seesaw type I, the dashed ones to type II and the dash-dotted ones to type III. In all cases a degenerate spectrum of the seesaw particles has been assumed.

implications on the spectrum [30, 31]. Additional effects on the spectrum of the scalars can be present if some of the Yukawa couplings get large [31, 32, 33]. In figure 1 we exemplify this by showing the values of selected mass parameters at $Q = 1$ TeV versus the seesaw scale for fixed high scale parameters $m_0 = M_{1/2} = 1$ TeV, $\tan\beta = 10$ and $A_0 = 0$ where we have assumed that the additional Yukawa couplings are small. As expected, the effects in case of models of type II and III are larger the smaller the corresponding seesaw-scale is. The scalar mass parameters shown are of the first generation and, thus, the results are nearly independent of $\tan\beta$ and A_0.

Note that in all three model types the ratio of the gaugino mass parameters is nearly the same as in the usual mSUGRA scenarios but the ratios and differences of the sfermion mass parameters change [30, 31]. One can form four 'invariants' where at least at the 1-loop level the dependence on $M_{1/2}$ and m_0 is rather weak, e.g. $(m^2_{\tilde{L}} - m^2_{\tilde{E}})/M^2_1$, $(m^2_{\tilde{Q}} - m^2_{\tilde{E}})/M^2_1$, $(m^2_{\tilde{D}} - m^2_{\tilde{L}})/M^2_1$ and $(m^2_{\tilde{Q}} - m^2_{\tilde{U}})/M^2_1$. Here one could replace M_1 by any of the other two gaugino masses which simply would amount in an overall rescaling. In figure 2 we show these 'invariants' in the leading-log approximation at 1-loop order to demonstrate the principal behaviour for seesaw type II with a pair of **15**-plets and seesaw type III with three **24**-plets. Due to the larger change in the beta-coefficient the effect is more pronounced in case of type III models. From this one concludes that in principle one has a handle to get information on the seesaw scale for given assumptions on the underlying neutrino mass model and assuming universal boundary conditions at the GUT scale. For the type-I, i.e. singlets only, of course $\Delta b_i = 0$ and no change with respect to mSUGRA are expected. A detailed discussion can be found e.g. in [31].

3.5. Lepton flavour violation in the slepton sector and signals at the LHC

Additional effects beside the ones on the spectrum discussed above can occur due to lepton flavour mixing entries in the sleptons mass matrix which are induced due to RGE effects by the additional Yukawa couplings. From a one-step integration of the RGEs one gets, assuming mSUGRA boundary conditions, a first rough estimate for the lepton flavour violating entries in the slepton mass parameters:

$$m^2_{\tilde{L},ij} \simeq -\frac{a_k}{8\pi^2}\left(3m_0^2 + A_0^2\right)\left(Y^{k,\dagger} L Y^k\right)_{ij} \quad (16)$$

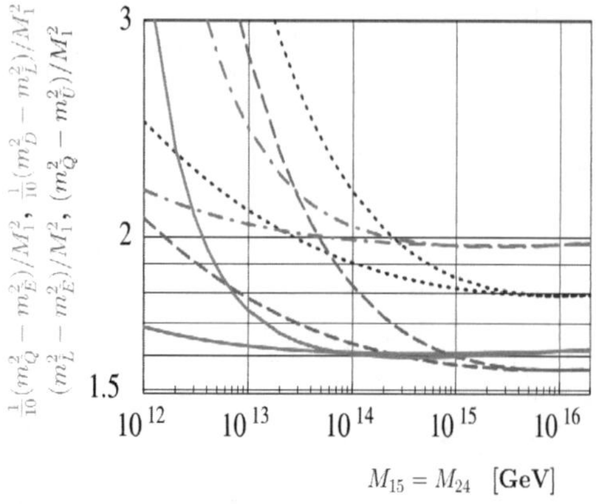

Figure 2. Four different "invariant" combinations of soft masses versus the mass of the **15**-plet *or* **24**-plet, $M_{15} = M_{24}$. The calculation is at 1-loop order in the leading-log approximation. The lines running faster up towards smaller M are for type-III seesaw, the lower ones are for type-II seesaw.

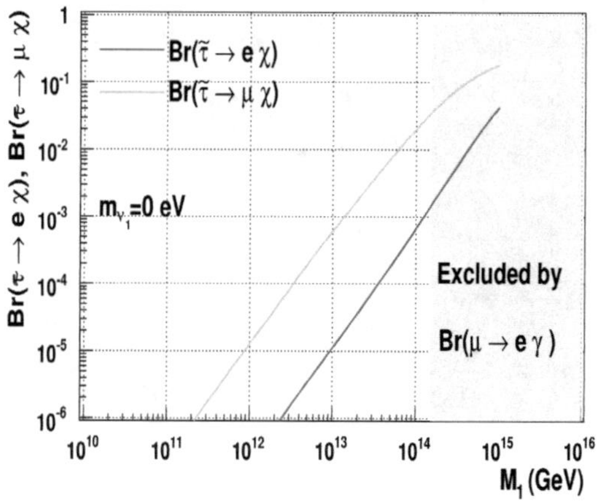

Figure 3. Flavour violating decays of the heavier stau for degenerate right-handed neutrinos, $M_0 = 90$ GeV, $M_{1/2} = 400$ GeV, $A_0 = 0$ GeV, $\tan\beta = 10$, $\mu > 0$), from ref. [34]

$$A_{l,ij} \simeq -\frac{3a_k}{16\pi^2} A_0 \left(Y_E Y^{k,\dagger} L Y_N^k \right)_{ij} \tag{17}$$

for $i \neq j$ in the basis where Y_E is diagonal and $L_{ij} = \log(M_G/M_i)\delta_{ij}$. The corresponding coefficient a_k for seesaw type k are: $a_I = 1$, $a_{II} = 6$ and $a_{III} = 9/5$. All models have in common that the predict negligible flavour violation for the R-sleptons $m^2_{\tilde{E},ij} \simeq 0$. Sizable entries for the these parameters would be a clear hint for a left-right symmetric extension of the MSSM [35].

These flavour mixing entries induce on the one hand rare lepton decays, e.g. $\mu \to e\gamma$, and on the other hand lepton flavour violating decays of sleptons and neutralinos. As a typical example we show in fig. 3 the flavour violating decays of the heavier stau in a seesaw I scenario. One sees that branching ratios of at most a few per-cent can be reached close to the region excluded by $\mu \to e\gamma$ where Y_ν gets sizable. Also in case of seesaw type II one obtains similar results [31]. At the LHC however one will not be able to identify the individual branching ratios but one has to consider the complete cascade starting from the production and taking into account all steps of the various decays, e.g. $\tilde{q}_L \to q\tilde{\chi}^0_2 \to q l_i^\pm \tilde{l}_j^\mp \to q l_i^\pm l_k^\mp \tilde{\chi}^0_1$. As typical examples we display in fig. 4 the obtainable cross section containing a $\mu^\pm\tau^\mp$ pair in the final state in case of seesaw models of type I and II for various values of m_0 as a function of $M_{1/2}$. Here the signal has been maximized by choosing the Yukawa couplings such that $BR(\mu \to e\gamma) = 10^{-12}$ [36]. One sees that one gets at most signals up to $O(20 - 30)$ fb which clearly requires a large luminosity to be detected.

4. R-parity violation

We now turn to the possibility that low-energy supersymmetry itself may provide the origin of neutrino mass [37, 11], for a review see Ref. [38]. Usually one assumes that R-parity, defined as $(-1)^{3B+L+2S}$, is an exact symmetry under which all superpartners are odd and SM particles even. However the terms that break R-parity are allowed by supersymmetry as well as the SM gauge invariance. Expressed as superfields, they have the form $\hat{L}\hat{H}_u$, $\hat{L}\hat{L}\hat{E}^c$, $\hat{Q}\hat{L}\hat{D}^c$ and $\hat{U}^c\hat{D}^c\hat{d}^c$.

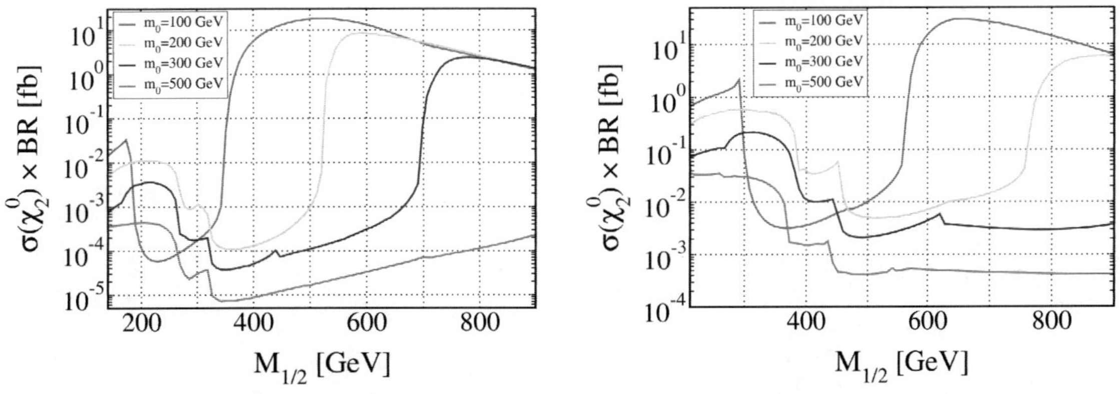

Figure 4. Cross section $\sigma(pp \to \tilde{\chi}_2^0) \times BR(\chi_2^0 \to \sum_{i,j} \tilde{l}_i l_j \to \mu^\pm \tau^\mp \tilde{\chi}_1^0)$ as a function of $M_{1/2}$ for various values of m_0, $A_0 = 0$, $\tan\beta = 10$, $\mu > 0$. Left side, seesaw type I with degenerate ν_R, right side seesaw type II with $\lambda_1 = 0.02$, $\lambda_2 = 0.5$; from ref. [36].

If all four terms are present proton decay becomes very rapid. This problem is circumvented by simply forbidding the last term, e.g. by using baryon triality or a similar symmetry [39, 40]. The remaining three terms break lepton number explicitly implying that a combination of tree and loop diagrams in these models can lead to realistic neutrino masses and mixings. From the point of view of collider physics, there is an important implication of LSP decay as at least some of its decay properties are correlated to neutrino physics since the same couplings governing neutrino physics also lead to visible decays of the LSP.

We will exemplify this by focusing on bilinear R_P breaking, for discussion of tri-linear $\not{R}_p$ see for example [41, 42] and for the so-called $\mu\nu$SSM see for example [43]. The absence of tri-linear terms could be explained, for example, if bilinear R-parity breaking is the effective low-energy limit of some spontaneous $\not{R}_p$ model, see below.

4.1. Explicit bilinear R-parity violation

The superpotential of the bilinear $\not{R}_p$ model can be written as

$$\mathcal{W} = W_{MSSM} + \epsilon_i \widehat{L}_i \widehat{H}_u. \tag{18}$$

In addition, one must include bilinear $\not{R}_p$ soft supersymmetry breaking terms

$$V_{\text{soft}} = \epsilon_i B_i \tilde{L}_i H_u + V_{\text{soft}}^{MSSM}. \tag{19}$$

These terms induce mixings between the MSSM Higgs bosons and the left scalar neutrinos which in consequence also obtain vevs v_i once electro-weak symmetry is broken. Usually one trades the B_i by the v_i using the corresponding tad-pole equations which are a consequence of eq. (19) as the connections to neutrino physics become more apparent.

The effective neutrino mass matrix at tree-level can then be cast into a very simple form

$$m_{\nu,ij} = -\frac{m_\gamma}{4\det(M_{\chi^0})}\Lambda_i \Lambda_j \tag{20}$$

The "photino" mass parameter is defined as $m_\gamma = g^2 M_1 + g'^2 M_2$, $det(M_{\chi^0})$ is the determinant of the $(4,4)$ neutralino mass matrix and $\Lambda_i \equiv \epsilon_i v_d + v_i \mu$ are the "alignment parameters".

Due to the projective nature of eq. (20) the other two neutrino masses are generated only at 1-loop order. Generally the most important contributions come from loops with sbottoms and

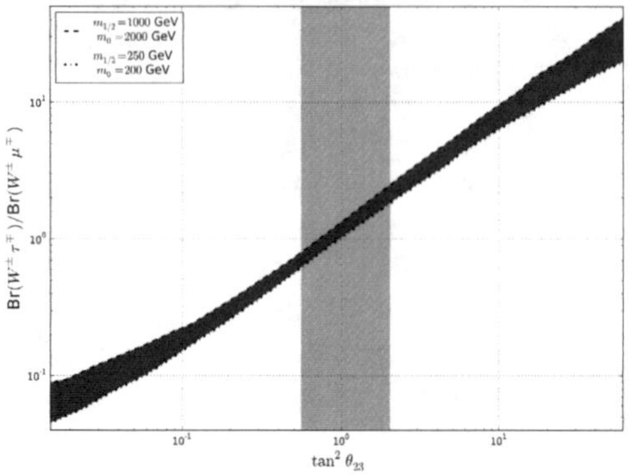

Figure 5. Ratio of semi-leptonic branching ratios, $\mathrm{BR}(\chi_1^0 \to \mu q'\bar{q})$ over $\mathrm{Br}(\chi_1^0 \to \tau q'\bar{q})$ as a function of the atmospheric neutrino angle calculated within bilinear $\not R_p$ SUSY, see ref. [49].

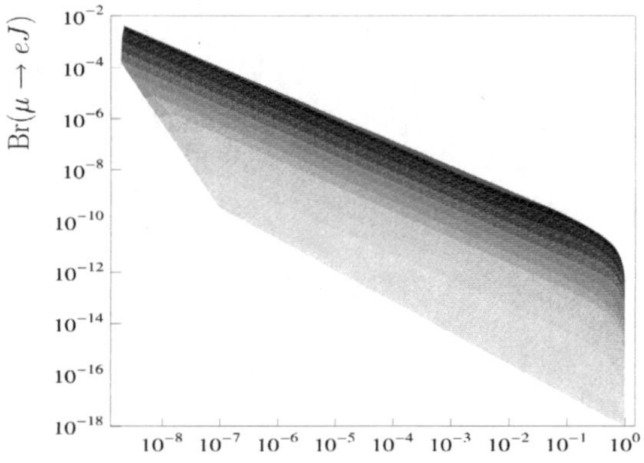

Figure 6. Branching ratio $\mathrm{BR}(\mu \to eJ)$ versus visible lightest neutralino decay. $\mu \to eJ$ and $\chi_1^0 \to J\nu$ are correlated, see ref. [50].

staus [44, 45]. However, there exist also parameter regions in the general $\not R_p$ MSSM where the sneutrino-anti-neutrino loop gives a sizeable contribution [46, 47, 48]. One finds that in order to explain the observed neutrino mixing angles one requires certain relations among the $\not R_p$ parameters to be satisfied [44], e.g. the maximal atmospheric angles requires $\Lambda_\mu \simeq \Lambda_\tau$.

Once R-parity is broken the LSP decays. The decays of a neutralino LSP have been studied in [51, 49]. Decay lengths for the neutralino are approximately fixed once the neutrino masses are fitted to experimental data. Typical lengths range from tens of cm for very light neutralinos to sub-millimeter for neutralinos of several hundred GeV [49]. One of the most exciting aspects of bilinear $\not R_p$, however, is the fact that once neutrino angles are fitted to the values required [13] by the neutrino oscillation data, the ratios of LSP decay branching ratios are fixed and correlate with the observed neutrino mixing angles, as illustrated for example in fig. 5. Measurements at the LHC should allow to test this prediction, if signals of SUSY are found [52].

Within $\not R_p$ SUSY any supersymmetric particle can be the LSP. It has been shown that within bilinear $\not R_p$ correlations between the measured neutrino angles and ratios of LSP decays can be found for all LSP candidates [53, 54, 55]. Thus, it is possible to exclude the minimal bilinear $\not R_p$ model experimentally at the LHC.

4.2. Spontaneous R-parity violation

In spontaneous R-parity violation models [56, 37, 57] R-parity violation results from the minimization of the Higgs potential through nonzero sneutrino vacuum expectation values. If lepton number is ungauged this implies the existence of a Nambu-Goldstone boson - the majoron. However, a doublet majoron is ruled out by LEP measurements of the Z width [58]. Hence, viable spontaneous R-parity breaking models must be characterized by two types of sneutrino vevs, those of right and left sneutrinos, singlets and doublets under $\mathrm{SU(3)} \otimes \mathrm{SU(2)_L} \otimes \mathrm{U(1)_Y}$ respectively [59, 60]. These obey the "vev-seesaw" relation $v_L v_R \sim Y_\nu m_W^2$ where Y_ν is the small Yukawa coupling that governs the strength of R–parity violation [59, 60].

In this case the majoron is so weakly coupled that bounds from LEP and astrophysics [61] are easily satisfied. For example, the superpotential of [59] can be written as

$$W = \widehat{H}_u \widehat{Q} Y_u \widehat{U} + \widehat{H}_d \widehat{Q} Y_d \widehat{D} + \widehat{H}_d \widehat{L} Y_e \widehat{E} + \widehat{H}_u \widehat{L} Y_\nu \widehat{\nu}^c - h_0 \widehat{H}_d \widehat{H}_u \widehat{\Phi} + h \widehat{\Phi} \widehat{\nu}^c \widehat{S} + \frac{\lambda}{3!} \widehat{\Phi}^3. \quad (21)$$

The first three terms are the usual MSSM Yukawa terms. The terms coupling the lepton doublets to $\widehat{\nu}^c$ fix lepton number. The coupling of the field $\widehat{\Phi}$ with the Higgs doublets generates an effective μ-term a lá the Next to Minimal Supersymmetric Standard Model (NMSSM) [62, 63, 64]. Note, that $v_R = \langle \tilde{\nu}^c \rangle \neq 0$ generates effective bilinear terms $\epsilon_i = Y_\nu^i v_R$ and that v_R, v_S and v_{L_i} violate lepton number and R-parity spontaneously. Neutrino oscillation data enforce $v_{L_i}^2 \ll v_R^2$ and $v_{L_i}^2 \ll v_d^2 + v_u^2$ implying the majoron is mainly a singlet in this model [65, 66].

The existence of the majoron affects the phenomenology at colliders mainly in two ways: (i) the lightest Higgs can decay invisibly into two majorons [65, 66]. (ii) Also the decays of the lightest neutralino are affected, since the new decay channel $\chi_1^0 \to J\nu$ is invisible at colliders. In ref. [67, 68] it has been shown that this mode can be close to 100% for certain parameter combinations implying that a large luminosity at the LHC might be required to detect R-parity violation. However, as can be seen in fig. 6 $chi_1^0 \to J\nu$ is correlated with the decay $\mu \to Je$, thus allowing to probe for a complementary part of parameter space [50] with low energy experiments.

Spontaneous R-parity violation can also be obtained by enlarging the gauge group [69, 70, 71, 72, 73, 74, 75]. In this case the majoron would be the longitudinal component of an additional neutral heavy vector boson which can be produced in s-channel processes at the LHC.

4.3. LHC studies

In the following we mainly focus on mSUGRA models which are augmented by bilinear R-parity breaking parameters at the electroweak scale implying that one has eleven free parameters, namely m_0, $M_{1/2}$, $\tan\beta$, sign(μ), A_0, ϵ_i, and Λ_i. In order to fit current neutrino oscillation data, the effective strength of R-parity violation must be small. This implies that supersymmetric particle spectra are expected to be the same as in the conventional R-parity conserving model, and that processes involving single production of SUSY states [76] are negligible at the LHC and that the only differences occur due to the decays of the LSP.

In these scenarios the lightest neutralino is the LSP and its main decay channels are $\tilde{\chi}_1^0 \to \nu \ell^+ \ell^-$ with $\ell = e$, μ denoted by $\ell\ell$; $\tilde{\chi}_1^0 \to \nu \tau^+ \tau^-$, called $\tau\tau$; $\tilde{\chi}_1^0 \to \tau \nu \ell$, called $\tau\ell$. $\tilde{\chi}_1^0 \to \nu q \bar{q}$ denoted jj; $\tilde{\chi}_1^0 \to \tau q' \bar{q}$, called τjj; $\tilde{\chi}_1^0 \to \ell q' \bar{q}$, called ℓjj; $\tilde{\chi}_1^0 \to \nu b\bar{b}$, which we denote by bb; $\tilde{\chi}_1^0 \to \nu b\bar{b}$, which we denote by bb; $\tilde{\chi}_1^0 \to \nu\nu\nu$.

In ref. [77] a comparison has been performed between the reach of LHC for R-parity violating SUSY using the same cuts as for R-parity conserving models [78]. The main topologies are: Inclusive jets and missing transverse momentum; zero lepton, jets and missing transverse momentum; one lepton, jets and missing transverse momentum; opposite sign lepton pair, jets and missing transverse momentum; same sign lepton pair, jets and missing transverse momentum; tri-leptons, jets and missing transverse momentum; multi-leptons, jets and missing transverse momentum. Due to the reduced missing energy the all-inclusive channel will have a reduced reach in the parameter space. However, the decays of the neutralino increase the multiplicities of the multi-lepton channel. As an example we display in Fig. 7 the LHC reach in the three– and multi–lepton channels with/without R–parity conservation for an integrated luminosity of 100 fb^{-1}. These constitute the best standard channels for the discovery of bilinear R-parity violation.

The sizable decay length of the neutralino is quite useful as this topology has little, if any, background expected at the LHC. This feature has been exploited in ref. [77, 79], where a comparison has been performed between the reach of LHC for R-parity violating SUSY including explicitly the displaced vertex topologies. In figure 8 we present the displaced vertex reach in the m_0-$M_{1/2}$ plane for $\tan\beta = 10$, $\mu > 0$, $A_0 = -100$ GeV. As one can see form this figure, the LHC will be able to look for the displaced vertex signal up to $m_{1/2}$ ~800 (1000) GeV for a large range of m_0 values and an integrated luminosity of 10 (100) fb^{-1}. Notice that the reach in this channel is rather independent of m_0. However, this signal disappears in the region where the stau is the LSP due to its rather short lifetime.

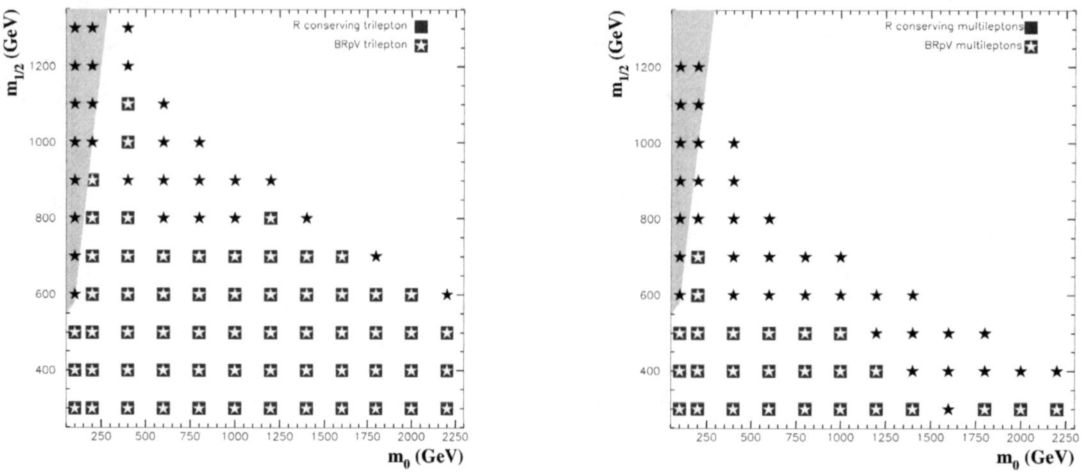

Figure 7. LHC discovery potential in the three lepton channel (top panel) and the multi-lepton one (bottom panel) for $A_0 = -100$ GeV, $\tan\beta = 10$, $\mu > 0$ and an integrated luminosity of 100 fb^{-1}, from ref. [77].

Next we discuss a tantalizing possibility, namely a double discovery at the LHC: (i) find evidence for supersymmetry, and (ii) uncover the Higgs boson. There are regions in parameter space where $\tilde{\chi}_1^0$ may have a sizeable branching ratio up to 22% into the channel νh^0 where h^0 is the lightest Higgs boson [80]. This would lead to displaced vertices containing two b-jets as a characteristic signature for Higgs production at the LHC [80]. The displaced vertex signal implies that also LHCb will have good sensitivity for such scenarios in particular in case of final states containing muons such as $\tilde{\chi}_1^0 \to \nu\mu^+\mu^-$.Figure 9 demonstrates that the ATLAS and CMS experiments will be able to look for the signal up to $M_{1/2} \sim 700$ (900) GeV for a LHC integrated luminosity of 10 (100) fb^{-1}. The hatched region in Fig. 9 indicates the LHCb reach for 10 fb^{-1}. Due to the strong cut on the pseudo–rapidity required by this experiment the reach for 2 fb^{-1} is severely depleted and only a small region of the parameter space is covered.

4.4. Trilinear R-parity breaking

Bilinear R-parity violation is essentially equivalent to tri-linear R-parity breaking with the superpotential

$$W_{tri} = \tfrac{1}{2}\lambda_{ijk}\widehat{L}_i\widehat{L}_j\widehat{E}_k + \lambda'_{ijk}\widehat{L}_i\widehat{Q}_j\widehat{D}_k \tag{22}$$

where the tri-linear couplings have the following structures

$$\lambda_{ijk} \simeq \frac{\epsilon_i}{\mu}Y_e^{jk}\ ,\ \lambda'_{ijk} \simeq \frac{\epsilon_i}{\mu}Y_d^{jk} \tag{23}$$

Obviously the phenomenology will be very similar if tri-linear R-parity violation is close to this structure. In the case of significant deviations from this structure is realized, one gets new interesting signatures. For example there exists light stau LSP scenarios where $\tilde{\tau}_1$ decays dominantly via 4-body decays such as $\tilde{\tau}_1 \to \tau^-\mu^-u\bar{d}$ with long lifetimes leading to displaced vertices [81]. Another interesting signals are the resonant production of sleptons as discussed in [82, 83] or associated production of single sleptons with t-quarks [84].

Journal of Physics: Conference Series **259** (2010) 012002 doi:10.1088/1742-6596/259/1/012002

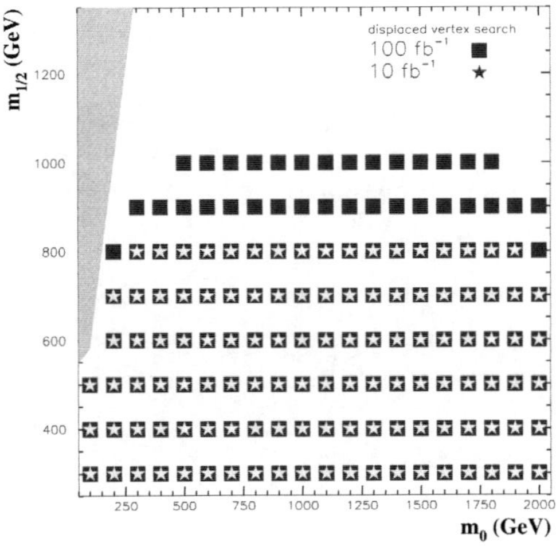

Figure 8. Discovery reach for displaced vertices channel in the m_0-$M_{1/2}$ plane for $\tan\beta = 10$, $\mu > 0$, $A_0 = -100$ GeV. The stars (squares) stand for points where there are more than 5 displaced vertex signal events for an integrated luminosity of 10 (100) fb^{-1}. The marked grey (green) area on the left upper corner is the region where the stau is the LSP, from ref. [77].

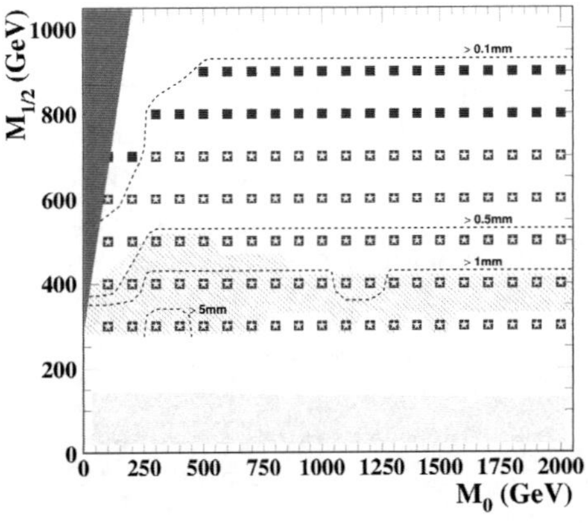

Figure 9. LHC reach for Higgs search in them_0-$M_{1/2}$ plane for $\tan\beta = 10$, $A_0 = -100$ GeV, and $\mu > 0$. The yellow stars (blue squares) show the reach for an integrated luminosity of 10 (100) fb^{-1} and the hatched region the reach of LHCb for an integrated luminosity of 10 fb^{-1}. The (yellow) shaded region in the bottom is excluded by direct LEP searches, while the (red) upper–left area represents the region with a stau as LSP; the black lines delimit different regimes of LSP decay length, from ref. [80].

An interesting question is to which extent one can measure deviations from the hierarchical structure above, e.g. the coupling λ'_{211} can still be of order 0.1. It has been shown in [85] that in such a case one LHC will be able to measure such couplings of such a strength with an accuracy of about 10%.

5. Summary

We have discussed aspects of various extensions of the MSSM taking the explanation of neutrino data as guide-line. As a first model the case has been considered where one adds a Dirac mass term for the neutrinos with tiny Yukawa couplings. In this case one obtains the usual MSSM phenomenology except for the case where the right-sneutrino is the LSP because in this case the NLSP will be very long-lived and one gets as signature quasi-stable particles at the LHC.

In case of seesaw models one finds that in case of type II and type III models the spectrum can be quite different when compared to the usual mSUGRA scenarios as a consequence of additional charged particles between the seesaw scale(s) and the GUT scale. These differences are the larger the lower the seesaw scale and affect in particular the masses of the sfermions. One can study four ratios at the LHC which can give information on the GUT scale if one assumes universal boundary conditions for the soft SUSY breaking parameters at the GUT scale. Moreover, there are parameter regions in all seesaw models where lepton flavour violating signals can be found in SUSY cascade decays. However, for this in general a rather high luminosity in the order of 100 fb^{-1} or higher is required.

Supersymmetric models also offer the possibility to explain neutrino data via R-parity violation. Here we have mainly focused on models where R-parity is broken by bilinear terms as this is sufficient to explain neutrino data and to explore the relevant LHC phenomenology. Moreover, this class of models can be obtained as effective model in case of spontaneous R-parity breaking. Bilinear R-parity breaking implies correlations between neutrino data and decays properties of the LSP, e.g. in case of an neutralino LSP the ratio $BR(\tilde{\chi}_1^0 \to W^-\tau^+)/BR(\tilde{\chi}_1^0 \to W^-\mu^+) \simeq \tan^2\theta_{atm}$ where θ_{atm} is the mixing angle related to the the atmospheric neutrino sector. In addition one finds very often that the LSP life time is sufficiently small to produce a displaced vertex in a typical collider experiments.

Acknowledgments

This work has been supported by the DFG, project number PO-1337/1-1, the Alexander von Humboldt Foundation and the Spanish grant FPA2008-00319/FPA.

References

[1] Nilles H P 1984 *Phys. Rept.* **110** 1–162
[2] Haber H E and Kane G L 1985 *Phys. Rept.* **117** 75–263
[3] Witten E 1981 *Nucl. Phys.* **B188** 513
[4] Dimopoulos S, Raby S and Wilczek F 1981 *Phys. Rev.* **D24** 1681–1683
[5] Ibanez L E and Ross G G 1981 *Phys. Lett.* **B105** 439
[6] Ellis J R, Hagelin J S, Nanopoulos D V, Olive K A and Srednicki M 1984 *Nucl. Phys.* **B238** 453–476
[7] Steffen F D 2009 *Eur. Phys. J.* **C59** 557–588 (*Preprint* arXiv:0811.3347)
[8] Weiglein G *et al.* (LHC/LC Study Group) 2006 *Phys. Rept.* **426** 47–358 (*Preprint* hep-ph/0410364)
[9] Nath P *et al.* 2010 *Nucl. Phys. Proc. Suppl.* **200-202** 185–417 (*Preprint* arXiv:1001.2693) "
[10] Minkowski P 1977 *Phys. Lett.* **B67** 421
[11] Hall L J and Suzuki M 1984 *Nucl. Phys.* **B231** 419
[12] Kim J E and Nilles H P 1984 *Phys. Lett.* **B138** 150
[13] Schwetz T, Tortola M A and Valle J W F 2008 *New J. Phys.* **10** 113011 (*Preprint* arXiv:0808.2016)
[14] de Gouvea A, Gopalakrishna S and Porod W 2006 *JHEP* **11** 050 (*Preprint* hep-ph/0606296)
[15] Gupta S K, Mukhopadhyaya B and Rai S K 2007 *Phys. Rev.* **D75** 075007 (*Preprint* hep-ph/0701063)
[16] Choudhury D, Gupta S K and Mukhopadhyaya B 2008 *Phys. Rev.* **D78** 015023 (*Preprint* arXiv:0804.3560)
[17] Weinberg S 1979 *Phys. Rev. Lett.* **43** 1566–1570
[18] Weinberg S 1980 *Phys. Rev.* **D22** 1694
[19] Ma E 1998 *Phys. Rev. Lett.* **81** 1171–1174 (*Preprint* hep-ph/9805219)
[20] Yanagida T KEK lectures, 1979 Ed. O. Sawada and A. Sugamoto (KEK, 1979)
[21] Gell-Mann M, Ramond P and Slansky R 1979 Print-80-0576 (CERN)
[22] Mohapatra R N and Senjanovic G 1980 *Phys. Rev. Lett.* **44** 912
[23] Schechter J and Valle J W F 1980 *Phys. Rev.* **D22** 2227
[24] Cheng T P and Li L F 1980 *Phys. Rev.* **D22** 2860
[25] Foot R, Lew H, He X G and Joshi G C 1989 *Z. Phys.* **C44** 441
[26] Borzumati F and Yamashita T 2009 (*Preprint* arXiv:0903.2793)
[27] Kang S K, Morozumi T and Yokozaki N 2010 (*Preprint* arXiv:1005.1354)
[28] Casas J A and Ibarra A 2001 *Nucl. Phys.* **B618** 171–204 (*Preprint* hep-ph/0103065)
[29] Rossi A 2002 *Phys. Rev.* **D66** 075003 (*Preprint* hep-ph/0207006)
[30] Buckley M R and Murayama H 2006 *Phys. Rev. Lett.* **97** 231801 (*Preprint* hep-ph/0606088)
[31] Hirsch M, Kaneko S and Porod W 2008 *Phys. Rev.* **D78** 093004 (*Preprint* arXiv:0806.3361)
[32] Calibbi L, Frigerio M, Lavignac S and Romanino A 2009 *JHEP* **12** 057 (*Preprint* arXiv:0910.0377)
[33] Biggio C and Calibbi L 2010 (*Preprint* arXiv:1007.3750)
[34] Hirsch M, Valle J W F, Porod W, Romao J C and Villanova del Moral A 2008 *Phys. Rev.* **D78** 013006 (*Preprint* arXiv:0804.4072)
[35] Vicente A 2010 (*Preprint* arXiv:1010.0799)
[36] Esteves J N *et al.* 2009 *JHEP* **05** 003 (*Preprint* arXiv:0903.1408)
[37] Ross G G and Valle J W F 1985 *Phys. Lett.* **B151** 375
[38] Hirsch M and Valle J W F 2004 *New J. Phys.* **6** 76 (*Preprint* hep-ph/0405015)
[39] Dreiner H K 1997 (*Preprint* hep-ph/9707435)
[40] Dreiner H K, Luhn C and Thormeier M 2006 *Phys. Rev.* **D73** 075007 (*Preprint* hep-ph/0512163)

[41] Allanach B *et al.* (R parity Working Group) 1999 (*Preprint* hep-ph/9906224)
[42] Barbier R *et al.* 2005 *Phys. Rept.* **420** 1–202 (*Preprint* hep-ph/0406039)
[43] Bartl A, Hirsch M, Vicente A, Liebler S and Porod W 2009 *JHEP* **05** 120
[44] Hirsch M, Diaz M A, Porod W, Romao J C and Valle J W F 2000 *Phys. Rev.* **D62** 113008 (*Preprint* hep-ph/0004115)
[45] Diaz M A, Hirsch M, Porod W, Romao J C and Valle J W F 2003 *Phys. Rev.* **D68** 013009 (*Preprint* hep-ph/0302021)
[46] Grossman Y and Haber H E 2001 *Phys. Rev.* **D63** 075011 (*Preprint* hep-ph/0005276)
[47] Dedes A, Rimmer S and Rosiek J 2006 *JHEP* **08** 005 (*Preprint* hep-ph/0603225)
[48] Dedes A, Haber H E and Rosiek J 2007 *JHEP* **11** 059 (*Preprint* arXiv:0707.3718)
[49] Porod W *et al.* 2001 *Phys. Rev.* **D63** 115004
[50] Hirsch M, Vicente A, Meyer J and Porod W 2009 *Phys. Rev.* **D79** 055023 (*Preprint* arXiv:0902.0525)
[51] Mukhopadhyaya B, Roy S and Vissani F 1998 *Phys. Lett.* **B443** 191–195
[52] De Campos F *et al.* 2010 (*Preprint* arXiv:1006.5075)
[53] Hirsch M, Porod W, Romao J C and Valle J W F 2002 *Phys. Rev.* **D66** 095006 (*Preprint* hep-ph/0207334)
[54] Restrepo D, Porod W and Valle J W F 2001 *Phys. Rev.* **D64** 055011
[55] Hirsch M and Porod W 2003 *Phys. Rev.* **D68** 115007 (*Preprint* hep-ph/0307364)
[56] Aulakh C S and Mohapatra R N 1982 *Phys. Lett.* **B119** 136
[57] Santamaria A and Valle J W F 1989 *Phys. Rev.* **D39** 1780–1783
[58] Amsler C *et al.* (Particle Data Group) 2008 *Phys. Lett.* **B667** 1
[59] Masiero A and Valle J W F 1990 *Phys. Lett.* **B251** 273–278
[60] Romao J C, Santos C A and Valle J W F 1992 *Phys. Lett.* **B288** 311–320
[61] Raffelt G G Chicago, USA: Univ. Pr. (1996) 664 p
[62] Barbieri R, Ferrara S and Savoy C A 1982 *Phys. Lett.* **B119** 343
[63] Nilles H P, Srednicki M and Wyler D 1983 *Phys. Lett.* **B120** 346
[64] Chamseddine A H, Arnowitt R L and Nath P 1982 *Phys. Rev. Lett.* **49** 970
[65] Hirsch M, Romao J C, Valle J W F and Villanova del Moral A 2004 *Phys. Rev.* **D70** 073012 (*Preprint* hep-ph/0407269)
[66] Hirsch M *et al.* 2006 *Phys. Rev.* **D73** 055007 (*Preprint* hep-ph/0512257)
[67] Hirsch M and Porod W 2006 *Phys. Rev.* **D74** 055003 (*Preprint* hep-ph/0606061)
[68] Hirsch M, Vicente A and Porod W 2008 *Phys. Rev.* **D77** 075005 (*Preprint* arXiv:0802.2896)
[69] Kuchimanchi R and Mohapatra R N 1993 *Phys. Rev.* **D48** 4352–4360 (*Preprint* hep-ph/9306290)
[70] Huitu K and Maalampi J 1995 *Phys. Lett.* **B344** 217–224 (*Preprint* hep-ph/9410342)
[71] Huitu K, Maalampi J and Puolamaki K 1999 *Eur. Phys. J.* **C6** 159–166
[72] Fileviez Perez P and Spinner S 2009 *Phys. Lett.* **B673** 251–254 (*Preprint* arXiv:0811.3424)
[73] Everett L L, Fileviez Perez P and Spinner S 2009 *Phys. Rev.* **D80** 055007 (*Preprint* arXiv:0906.4095)
[74] Barger V, Fileviez Perez P and Spinner S 2009 *Phys. Rev. Lett.* **102** 181802 (*Preprint* arXiv:0812.3661)
[75] Ji X, Mohapatra R N, Nussinov S and Zhang Y 2008 *Phys. Rev.* **D78** 075032
[76] Nogueira P, Romao J C and Valle J W F 1990 *Phys. Lett.* **B251** 142–149
[77] de Campos F *et al.* 2008 *JHEP* **05** 048 (*Preprint* arXiv:0712.2156)
[78] Baer H, Mizukoshi J K and Tata X 2000 *Phys. Lett.* **B488** 367–372 (*Preprint* hep-ph/0007073)
[79] de Campos F, Eboli O J P, Magro M B and Restrepo D 2009 *Phys. Rev.* **D79** 055008 (*Preprint* arXiv:0809.0007)
[80] de Campos F, Eboli O J P, Magro M B, Restrepo D and Valle J W F 2009 *Phys. Rev.* **D80** 015002 (*Preprint* arXiv:0809.1637)
[81] Dreiner H K, Grab S and Trenkel M K 2009 *Phys. Rev.* **D79** 016002 (*Preprint* arXiv:0808.3079)
[82] Dreiner H K, Richardson P and Seymour M H 2001 *Phys. Rev.* **D63** 055008 (*Preprint* hep-ph/0007228)
[83] Moreau G, Perez E and Polesello G 2001 *Nucl. Phys.* **B604** 3–31 (*Preprint* hep-ph/0003012)
[84] Bernhardt M A, Dreiner H K, Grab S and Richardson P 2008 *Phys. Rev.* **D78** 015016 (*Preprint* arXiv:0802.1482)
[85] Choudhury D, Godbole R M and Polesello G 2002 *JHEP* **08** 004 (*Preprint* hep-ph/0207248)

Technicolor and Beyond: Unification in Theory Space

Francesco Sannino

CP^3-Origins, University of Southern Denmark, Odense M. Denmark.

E-mail: sannino@cp3.sdu.dk, sannino@cp3-origins.net

Abstract. I will briefly review the salient features of models of dynamical electroweak symmetry breaking together with the traditional extensions needed to provide masses to the standard model fermions in absence of fundamental scalars. The idea walking idea is introduced according to which one should carefully take into account the effects of the extended technicolor dynamics on the technicolor dynamics itself. The interplay between the four fermion interactions stemming from the extended technicolor interactions and the technicolor model can strongly enhance the anomalous dimension of the mass of the techniquarks allowing to decouple the Flavor Changing Neutral Currents problem from the one of the generation of the large top mass. I will also review the Minimal Walking Technicolor (MWT) models.

In the second part of this review I consider the interesting possibility to marry supersymmetry and technicolor. The reason is to provide a unification of different extensions of the standard model. For example, this means that one can recover, according to the parameters and spectrum of the theory distinct extensions of the standard model, from supersymmetry to technicolor and unparticle physiscs. A surprising result is that a minimal (in terms of the smallest number of fields) supersymmetrization of the MWT model leads to the maximal supersymmetry in four dimensions, i.e. $\mathcal{N}=4$ SYM.

1. The Need to Go Beyond

The standard model (SM) of particle interactions passes a large number of experimental tests. Yet we know that it cannot be the ultimate model of nature since it fails to explain the origin of the matter-antimatter asymmetry and does not provide phenomenologically viable dark matter candidates. Several extensions of the SM have been proposed, and two stand out in the quest of a better theory: Supersymmetry and Technicolor (TC). We will review here the basics of TC, the problems intrinsically related to the early models of ETC interactions and propose some concrete solutions. We, in fact, suggest that the best models of TC having a chance to solve the top mass problem together with the suppression of the Flavor Changing Neutral Currents problem, must take into account the effects of the four-fermion interactions stemming from the unspecified ETC dynamics on the TC dynamics. We will argue that when this effect is taken properly into account it will lead to *ideal walking* models which are phenomenologically viable.

In the second part of this review I will adopt a more *liberal* attitude towards possible extensions of the SM while still embracing the view that the new extension of the SM should be natural. We will introduce the new concept of *Unification in Theory Space*, according to which our new model will feature supersymmetry and a new gauge theory coupled to the SM a lá TC, but also supersymmetric.

Journal of Physics: Conference Series **259** (2010) 012003 doi:10.1088/1742-6596/259/1/012003

2. Dynamical Electroweak Symmetry Breaking

Even in complete absence of the Higgs sector in the standard model (SM) the electroweak symmetry breaks [1] due to the condensation of the following quark bilinear in QCD:

$$\langle \bar{u}_L u_R + \bar{d}_L d_R \rangle \neq 0 \ . \tag{1}$$

and its contribution to the W-gauge boson mass is:

$$M_W = \frac{gF_\pi}{2} \sim 29\text{MeV} \ , \tag{2}$$

with $F_\pi \simeq 93$ MeV the pion decay constant. This contribution is, of course, too small when compared to the actual value of the M_W mass that one typically neglects it.

According to the original idea of technicolor [2, 3] one augments the SM with another gauge interaction similar to QCD but with a new dynamical scale of the order of the electroweak one. It is sufficient that the new gauge theory is asymptotically free and has global symmetry able to contain the SM $SU_L(2) \times U_Y(1)$ symmetries. It is also required that the new global symmetries break dynamically in such a way that the embedded $SU_L(2) \times U_Y(1)$ breaks to the electromagnetic abelian charge $U_Q(1)$. The dynamically generated scale will then be fit to the electroweak one. Note that, except in certain cases, dynamical behaviors are typically nonuniversal which means that different gauge groups and/or matter representations will, in general, posses different dynamics.

The simplest example of technicolor theory is the scaled up version of QCD, i.e. an $SU(N_{TC})$ nonabelian gauge theory with two Dirac Fermions transforming according to the fundamental representation or the gauge group. We need at least two Dirac flavors to realize the $SU_L(2) \times SU_R(2)$ symmetry of the SM discussed in the SM Higgs section. One simply chooses the scale of the theory to be such that the new pion decaying constant is:

$$F_\pi^{TC} = v_{\text{weak}} \simeq 246 \text{ GeV} \ . \tag{3}$$

This model correctly accounts for the electroweak symmetry breaking but unfortunately these first attempts are disfavored by precision data.

We have recently proposed several minimal working models [4, 5, 6, 7, 8, 9, 10, 11, 12] passing precision data and possessing interesting dynamics relevant for collider phenomenology [13, 14, 15, 16] and cosmology [17, 18, 19, 20, 21, 22, 23, 24, 25, 26, 27, 28, 29, 30, 31, 32, 11, 33, 34, 35, 36, 37] [1].

These models are also being investigated via first principle lattice simulations [39, 40, 41, 42, 43, 44, 45, 46, 47, 48, 49, 50, 51, 52, 53, 54, 55, 56]. An up-to-date review is Ref. [60].

2.1. Standard Model Fermion Masses

Since in a purely technicolor model the Higgs is a composite particle the Yukawa terms, when written in terms of the underlying technicolor fields, amount to four-fermion operators. The latter can be naturally interpreted as low energy operators induced by a new strongly coupled gauge interaction emerging at energies higher than the electroweak theory. These type of theories have been termed extended technicolor interactions (ETC) [61, 62].

Here we describe the simplest ETC model in which the ETC interactions connect the chiral symmetries of the techniquarks to those of the SM fermions.

[1] Note that if dark matter arises as a pseudo Goldstone boson in technicolor models (the TIMP) [20, 21, 30, 31, 32, 11, 33] they should not be confused with the scaled up version of the ordinary baryon envisioned earlier [17, 18, 19]. For example one striking feature is that the pseudo Goldstone Bosons can be sufficiently light to be produced at the Large Hadron Collider experiment at CERN.

Let's start with the case in which the ETC dynamics is represented by a $SU(N_{ETC})$ gauge group with:

$$N_{ETC} = N_{TC} + N_g \ , \tag{4}$$

and N_g is the number of SM generations. In order to give masses to all of the SM fermions, in this scheme, one needs a condensate for each SM fermion. This can be achieved by using as technifermion matter a complete generation of quarks and leptons (including a neutrino right) but now gauged with respect to the technicolor interactions.

The ETC gauge group is assumed to spontaneously break N_g times down to $SU(N_{TC})$ permitting three different mass scales, one for each SM family. This type of technicolor with associated ETC is termed the *one family model* [63]. The heavy masses are provided by the breaking at low energy and the light masses are provided by breaking at higher energy scales. This model does not, per se, explain how the gauge group is broken several times, neither is the breaking of weak isospin symmetry accounted for. Schematically one has $SU(N_{TC}+3)$ which breaks to $SU(N_{TC}+2)$ at the scale Λ_1 providing the first generation of fermions with a typical mass $m_1 \sim 4\pi(F_\pi^{TC})^3/\Lambda_1^2$ at this point the gauge group breaks to $SU(N_{TC}+1)$ with dynamical scale Λ_2 leading to a second generation mass of the order of $m_2 \sim 4\pi(F_\pi^{TC})^3/\Lambda_2^2$ finally the last breaking $SU(N_{TC})$ at scale Λ_3 leading to the last generation mass $m_3 \sim 4\pi(F_\pi^{TC})^3/\Lambda_3^2$.

Without specifying an ETC one can write down the most general type of four-fermion operators involving technicolor particles Q and ordinary fermionic fields ψ. Following the notation of Hill and Simmons [64] we write:

$$\alpha_{ab}\frac{\bar{Q}\gamma_\mu T^a Q\bar{\psi}\gamma^\mu T^b\psi}{\Lambda_{ETC}^2} + \beta_{ab}\frac{\bar{Q}\gamma_\mu T^a Q\bar{Q}\gamma^\mu T^b Q}{\Lambda_{ETC}^2} + \gamma_{ab}\frac{\bar{\psi}\gamma_\mu T^a\psi\bar{\psi}\gamma^\mu T^b\psi}{\Lambda_{ETC}^2} \ , \tag{5}$$

where the Ts are unspecified ETC generators. After performing a Fierz rearrangement one has:

$$\alpha_{ab}\frac{\bar{Q}T^a Q\bar{\psi}T^b\psi}{\Lambda_{ETC}^2} + \beta_{ab}\frac{\bar{Q}T^a Q\bar{Q}T^b Q}{\Lambda_{ETC}^2} + \gamma_{ab}\frac{\bar{\psi}T^a\psi\bar{\psi}T^b\psi}{\Lambda_{ETC}^2} + \dots \ , \tag{6}$$

The coefficients parametrize the ignorance on the specific ETC physics. To be more specific, the α-terms, after the technicolor particles have condensed, lead to mass terms for the SM fermions

$$m_q \approx \frac{g_{ETC}^2}{M_{ETC}^2}\langle\bar{Q}Q\rangle_{ETC} \ , \tag{7}$$

where m_q is the mass of e.g. a SM quark, g_{ETC} is the ETC gauge coupling constant evaluated at the ETC scale, M_{ETC} is the mass of an ETC gauge boson and $\langle\bar{Q}Q\rangle_{ETC}$ is the technicolor condensate where the operator is evaluated at the ETC scale. Note that we have not explicitly considered the different scales for the different generations of ordinary fermions but this should be taken into account for any realistic model.

The β-terms of Eq. (6) provide masses for pseudo Goldstone bosons and also provide masses for techniaxions [64]. The last class of terms, namely the γ-terms of Eq. (6) induce flavor changing neutral currents. For example it may generate the following terms:

$$\frac{1}{\Lambda_{ETC}^2}(\bar{s}\gamma^5 d)(\bar{s}\gamma^5 d) + \frac{1}{\Lambda_{ETC}^2}(\bar{\mu}\gamma^5 e)(\bar{e}\gamma^5 e) + \dots \ , \tag{8}$$

where s, d, μ, e denote the strange and down quark, the muon and the electron, respectively. The first term is a $\Delta S = 2$ flavor-changing neutral current interaction affecting the $K_L - K_S$ mass difference which is measured accurately. The experimental bounds on these type of operators together with the very *naive* assumption that ETC will generate these operators with γ of order one leads to a constraint on the ETC scale to be of the order of or larger than 10^3 TeV [61]. This should be the lightest ETC scale which in turn puts an upper limit on how large the ordinary fermionic masses can be. The naive estimate is that one can account up to around 100 MeV mass for a QCD-like technicolor theory, implying that the top quark mass value cannot be achieved.

The second term of Eq. (8) induces flavor changing processes in the leptonic sector such as $\mu \to e\bar{e}e, e\gamma$ which are not observed. It is clear that, both for the precision measurements and the fermion masses, that a better theory of the flavor is needed. For the ETC dynamics interesting developments recently appeared in the literature [65, 66, 67, 68]. We note that nonperturbative chiral gauge theories dynamics is expected to play a relevant role in models of ETC since it allows, at least in principle, the self breaking of the gauge symmetry. Recent progress on the phase diagrams of these theories has appeared in [60].

2.2. *Ideal Walking*

To better understand in which direction one should go to modify the QCD dynamics we analyze the TC condensate. The value of the technicolor condensate used when giving mass to the ordinary fermions should be evaluated not at the technicolor scale but at the extended technicolor one. Via the renormalization group one can relate the condensate at the two scales via:

$$\langle \bar{Q}Q \rangle_{ETC} = \exp\left(\int_{\Lambda_{TC}}^{\Lambda_{ETC}} d(\ln\mu)\gamma(\alpha(\mu))\right) \langle \bar{Q}Q \rangle_{TC} \ , \tag{9}$$

where γ is the anomalous dimension of the techniquark mass-operator. The boundaries of the integral are at the ETC scale and the TC one. For TC theories with a running of the coupling constant similar to the one in QCD, i.e.

$$\alpha(\mu) \propto \frac{1}{\ln\mu} \ , \quad \text{for } \mu > \Lambda_{TC} \ , \tag{10}$$

this implies that the anomalous dimension of the techniquark masses $\gamma \propto \alpha(\mu)$. When computing the integral one gets

$$\langle \bar{Q}Q \rangle_{ETC} \sim \ln\left(\frac{\Lambda_{ETC}}{\Lambda_{TC}}\right)^{\gamma} \langle \bar{Q}Q \rangle_{TC} \ , \tag{11}$$

which is a logarithmic enhancement of the operator. We can hence neglect this correction and use directly the value of the condensate at the TC scale when estimating the generated fermionic mass:

$$m_q \approx \frac{g_{ETC}^2}{M_{ETC}^2}\Lambda_{TC}^3 \ , \qquad \langle \bar{Q}Q \rangle_{TC} \sim \Lambda_{TC}^3 \ . \tag{12}$$

The tension between having to reduce the FCNCs and at the same time provide a sufficiently large mass for the heavy fermions in the SM as well as the pseudo-Goldstones can be reduced if the dynamics of the underlying TC theory is different from the one of QCD. The computation of the TC condensate, at different scales, shows that if the dynamics is such that the TC coupling does not *run* to the UV fixed point but rather slowly reduces to zero one achieves a net enhancement of the condensate itself with respect to the value estimated earlier. This can be achieved if the theory has a near conformal fixed point. This kind of dynamics has been denoted of *walking* type. In this case

$$\langle \bar{Q}Q \rangle_{ETC} \sim \left(\frac{\Lambda_{ETC}}{\Lambda_{TC}}\right)^{\gamma(\alpha^*)} \langle \bar{Q}Q \rangle_{TC} \ , \tag{13}$$

which is a much larger contribution than in QCD dynamics [69, 70, 71, 72]. Here γ is evaluated at the would be fixed point value α^*. Walking can help resolving the problem of FCNCs in technicolor models since with a large enhancement of the $\langle \bar{Q}Q \rangle$ condensate the four-fermi operators involving SM fermions and technifermions and the ones involving technifermions are enhanced by a factor of $\Lambda_{ETC}/\Lambda_{TC}$ to the γ power while the one involving only SM fermions is not enhanced.

There are several issues associated with the original idea of walking:

- Since the number of flavors cannot be changed continuously it is not possible to get arbitrarily close to the lower end of the conformal window. This applies to the technicolor theory *in isolation* i.e. before coupling it to the SM and without taking into account the ETC interactions.
- It is hard to achieve large anomalous dimensions of the fermion mass operator even near the lower end of the conformal window for ordinary gauge theories.
- It is not always possible to neglect the interplay of the four fermion interactions on the TC dynamics.

We have argued in [73] that it is possible to *solve* simultaneously all the problems above by consistently taking into account the effects of the four-fermion interactions on the phase diagram of strongly interacting theories for any matter representation as function of the number of colors and flavors. A positive effect is that the anomalous dimension of the mass increases beyond the unity value at the lower boundary of the new conformal window and can get sufficiently large to yield the correct mass for the top quark. We have also shown that the conformal window, for any representation, shrinks with respect to the case in which the four-fermion interactions are neglected.

We made the further *unexpected* discovery that when the extended TC sector, responsible for giving masses to the standard model fermions, is sufficiently strongly coupled the TC theory, in isolation, must feature an infrared fixed point in order for the full model to be phenomenologically viable and correctly break the electroweak symmetry.

Using the new phase diagram we showed that the minimal walking TC models are the archetypes of models of dynamical electroweak symmetry breaking.

The presence of the four-fermion interactions allows to get sufficiently near the lower end of the conformal window using as continuous variable the strength of the four-fermion interactions.

3. Minimal Technicolor

The simplest models of TC which are shown to pass the precision tests were put forward recently [4, 7]. To be concrete we describe here the MWT [74] extension of the SM.

The gauge group is $SU(2)_{TC} \times SU(3)_C \times SU(2)_L \times U(1)_Y$ and the field content of the technicolor sector is constituted by four techni-fermions and one techni-gluon all in the adjoint representation of $SU(2)_{TC}$. The model features also a pair of Dirac leptons, whose left-handed components are assembled in a weak doublet, necessary to cancel the Witten anomaly [75] arising when gauging the new technifermions with respect to the weak interactions. Summarizing, the fermionic particle content of the MWT is given explicitly by

$$Q_L^a = \begin{pmatrix} U^a \\ D^a \end{pmatrix}_L , \qquad U_R^a \, , \quad D_R^a, \quad a = 1,2,3 \; ; \quad L_L = \begin{pmatrix} N \\ E \end{pmatrix}_L , \qquad N_R \, , \; E_R \, . \tag{14}$$

The following generic hypercharge assignment is free from gauge anomalies:

$$\begin{aligned} Y(Q_L) &= \frac{y}{2} \, , & Y(U_R, D_R) &= \left(\frac{y+1}{2}, \frac{y-1}{2}\right) \, , \\ Y(L_L) &= -3\frac{y}{2} \, , & Y(N_R, E_R) &= \left(\frac{-3y+1}{2}, \frac{-3y-1}{2}\right) \, . \end{aligned} \tag{15}$$

Journal of Physics: Conference Series **259** (2010) 012003 doi:10.1088/1742-6596/259/1/012003

The global symmetry of this TC theory is $SU(4)$, which breaks explicitly to $SU(2)_L \times U(1)_Y$ by the natural choice of the EW embedding [4, 7]. EWSB is triggered by a fermion bilinear condensate and the vacuum choice is stable against the SM quantum corrections [76]. This model can feature very light composite Higgs scalars as advocated in [5, 7, 77] and supported by the recent analysis performed by Natale's group [78, 79, 80, 81, 82, 83].

An explicit construction of an extended TC type model addressing the problem of giving mass to the third generation of quarks and the new generation of leptons appeared in [84].

A less natural model introducing a novel scalar (*bosonic TC*) mimicking the effects of the extended TC interactions has also been introduced in [15, 85] following the pioneering work of Simmons [86], Kagan and Samuel [87], and Carone [88, 89]. More recently this type of models have been investigated also in [90, 91, 85]. Interesting related work can be also found in [92, 93]. Eventhough these models are phenomenologically viable, they suffer from a SM-like fine tuning and are therefore unnatural. Supersymmetric TC has been considered [94, 95] as a way to naturalize bosonic TC. Another possibility would be to imagine the new scalars also to be composite of some new strong dynamics.

4. Minimal Supersymmetric Conformal Technicolor: Unification in Theory Space

We investigate now the supersymmetrized version of the MWT. The resulting model provides a natural ultraviolet completion of the model introduced in [15]. The appeal of supersymmetry (SUSY) resides in its higher level of space-time symmetries as well as in its often praised natural link to string theory. The most investigated route to introduce SUSY has been to supersymmetrize the SM and then invoke some mechanism to break SUSY again, given that no sign of the superpartners has yet been observed in experiments.

We fuse the basic features of TC and SUSY to construct a natural supersymmetric TC theory possessing several interesting theoretical and phenomenological features. Besides symmetry arguments another equally relevant reason to look for a supersymmetric TC extension of the SM is linked to the fact that the generation of the SM fermion masses is less involved than in models with total absence of scalars, although still natural.

The supersymmetric TC idea was put forward in [94], though the phenomenological viability of these early models seemed difficult to achieve. An important difference with our model is that the underlying supersymmetric and TC theories, which can be resumed by decoupling, either the TC fields or the superpartners by sending their masses to infinity, are both phenomenologically viable.

The basic properties of the model are:

- The model possesses the highest degree of four-dimensional space-time symmetry compatible with experiments.
- The model can interpolate between already studied extensions of the SM at the TeV scale, such as unparticle physics [96, 97], MWT [4, 7], and MSSM (see [98] for a review). Hence the models can be used as a well defined laboratory to investigate different theoretical ideas.
- The model is a UV complete theory in which fermions naturally acquire mass.
- The models possess a clear and direct link to string theory in such a way that AdS/CFT techniques [99] are readily applicable to realistic extensions of the SM.

We start with the observation that the fermionic and gluonic spectrum of MWT fits perfectly in an $\mathcal{N}=4$ supermultiplet, provided that we also include three scalar superpartners. In fact the $SU(4)$ global symmetry of MWT is nothing but the well known $SU(4)_R$ R symmetry of the $\mathcal{N}=4$ Super Yang-Mills (4SYM) theory. Having at hand already a great deal of the spectrum of 4SYM we explore the possibility of using this theory as a natural candidate for supersymmetric TC.

We gauge part of the $SU(4)_R$ global symmetry of the supersymmetric TC theory in order to couple the new supersymmetric sector to the weak and hypercharge interactions of the SM. We choose to do this in such a way that the model can still preserve $\mathcal{N} = 1$ SUSY. To this end one of the four Weyl technifermions is identified with the techni-gaugino and should be a singlet under the SM gauge group. The only possible candidates for this role are $\bar{U}_R$ and $\bar{D}_R$, for $y = \mp 1$ respectively: we arbitrarily choose $y = 1$ and identify $\bar{D}_R$ with the techni-gaugino. With this choice the charge assignments of the particles in Eq.(14) under $SU(2)_{TC} \times SU(3)_C \times SU(2)_L \times U(1)_Y$ are

$$Q_L \sim \left(\mathbf{3}, \mathbf{1}, \mathbf{2}, \frac{1}{2}\right), \; \bar{U}_R \sim (\mathbf{3}, \mathbf{1}, \mathbf{1}, -1), \; \bar{D}_R \sim (\mathbf{3}, \mathbf{1}, \mathbf{1}, 0),$$
$$L_L \sim \left(\mathbf{1}, \mathbf{1}, \mathbf{2}, -\frac{3}{2}\right), \; \bar{N}_R \sim (\mathbf{1}, \mathbf{1}, \mathbf{1}, 1), \; \bar{E}_R \sim (\mathbf{1}, \mathbf{1}, \mathbf{1}, 2). \tag{16}$$

Based on these assignments we then define the scalar and fermion components of the $\mathcal{N} = 4$ superfields via

$$\left(\tilde{U}_L,\, U_L\right) \in \Phi_1, \quad \left(\tilde{D}_L,\, D_L\right) \in \Phi_2, \quad \left(\tilde{\bar{U}}_R,\, \bar{U}_R\right) \in \Phi_3, \quad (G,\, \bar{D}_R) \in V, \tag{17}$$

where we used a tilde to label the scalar superpartner of each fermion. We indicated with Φ_i, $i = 1, 2, 3$ the three chiral superfields of 4SYM and with V the vector superfield. Four more chiral superfields are necessary to fully supersymmetrize the MWT model, i.e.:

$$\left(\tilde{N}_L,\, N_L\right) \in \Lambda_1, \quad \left(\tilde{E}_L,\, E_L\right) \in \Lambda_2, \quad \left(\tilde{\bar{N}}_R,\, \bar{N}_R\right) \in N, \quad \left(\tilde{\bar{E}}_R,\, \bar{E}_R\right) \in E. \tag{18}$$

As one can see from the spectrum in Eq.(16) there is no scalar field that can be coupled to SM fermions in a gauge invariant way and play the role of the SM Higgs boson (a weak doublet with hypercharge $Y = \pm\frac{1}{2}$). We therefore (re)introduce in the theory two Higgs doublet superfields with respective charge assignment

$$H \sim \left(\mathbf{1}, \mathbf{1}, \mathbf{2}, \frac{1}{2}\right), \; H' \sim \left(\mathbf{1}, \mathbf{1}, \mathbf{2}, -\frac{1}{2}\right), \tag{19}$$

where the presence of both $Y = \pm\frac{1}{2}$ superfields is needed to give mass by gauge invariant Yukawa terms to both the upper and lower components of the weak doublets of SM fermions. With this choice it is rather natural to take the MSSM to describe the supersymmetric extension of the SM sector. All the MSSM fields are defined as singlets under $SU(2)_{TC}$. The resulting MSCT model is naturally anomaly-free, since both the MWT and the MSSM are such. We summarize in Table 1 the quantum numbers of the superfields in Eqs.(17,18,19). The renormalizable superpotential for the MSCT, allowed by gauge invariance, and which we require additionally to be $\mathcal{N} = 4$ invariant in the limit of g_{TC} much greater than the other coupling constants and to conserve baryon and lepton numbers is

$$P = P_{MSSM} + P_{TC}, \tag{20}$$

where P_{MSSM} is the MSSM superpotential, and

$$P_{TC} = -\frac{g_{TC}}{3\sqrt{2}}\epsilon_{ijk}\epsilon^{abc}\Phi_i^a\Phi_j^b\Phi_k^c + y_U\epsilon_{ij3}\Phi_i^a H_j \Phi_3^a + y_N\epsilon_{ij3}\Lambda_i H_j N + y_E\epsilon_{ij3}\Lambda_i H_j' E + y_R\Phi_3^a\Phi_3^a E. \tag{21}$$

This superpotential describes an approximately conformal theory in the limit when g_{TC} is much greater than the gauge couplings g_Y, g_L, and Yukawa couplings y_U, y_N, y_E, y_R. It is interesting

Superfield	$SU(2)_{TC}$	$SU(3)_c$	$SU(2)_L$	$U(1)_Y$
$\Phi_{1,2}$	Adj	1	$\Box$	1/2
Φ_3	Adj	1	1	-1
V	Adj	1	1	0
$\Lambda_{1,2}$	1	1	$\Box$	-3/2
N	1	1	1	1
E	1	1	1	2
H	1	1	$\Box$	1/2
H'	1	1	$\Box$	-1/2

Table 1. MSCT $\mathcal{N} = 1$ superfields

to notice that a generic Yukawa coupling constant y_{TC} in the first term has a infrared fixed point at g_{TC} [100], therefore justifying our choice of taking $y_{TC} = g_{TC}$.

The Lagrangian of the MSCT is

$$\mathcal{L} = \mathcal{L}_{MSSM} + \mathcal{L}_{TC} \ , \tag{22}$$

where the supersymmetric TC Lagrangian can be written as:

$$\mathcal{L}_{TC} = \frac{1}{2}\text{Tr}\left(W^\alpha W_\alpha|_{\theta\theta} + \bar{W}_{\dot\alpha}\bar{W}^{\dot\alpha}|_{\bar\theta\bar\theta}\right) + \Phi_f^\dagger \exp\left(2g_X V_X\right)\Phi_f|_{\theta\theta\bar\theta\bar\theta} + \left(P_{TC}|_{\theta\theta} + h.c.\right), \tag{23}$$

where

$$\Phi_f = Q, \Phi_3, \Lambda, N, E; \quad X = TC, C, L, Y \ , \tag{24}$$

with Q and Λ defined as the weak doublet superfields with components Φ_1, Φ_2, and Λ_1, Λ_2, respectively. The product $g_X V_X$ is assumed to include the gauge charge of the superfield on which it acts. The charge is Y for $U(1)_Y$, and 1 (0) for a multiplet (singlet) of a generic group $SU(N)$. The TC vector superfield V_{TC} is identified with V and its elementary field components are given in Eq.(17). The remaining vector superfields are those already defined in the MSSM [98] while the superpotential P_{TC} is given in Eq.(21). Finally, the first term on the right of Eq.(23) and its Hermitian conjugate represent the kinetic Lagrangian terms of the self-interacting techni-gluon and techni-gaugino.

4.1. The MSCT Landscape

One of the interesting general features of the MSCT spectrum is the existence of a supersymmetric fourth family of leptons. The MWT predicts the natural occurrence of a fourth family of leptons around the EW energy scale, put forward first in [7]. The physics of these fourth family of leptons has been studied in [12, 10]. From the EW point of view there is little difference between the MWT and a fourth-family extended SM at the EW scale.

Since the MSCT is a supersymmetrized version of the MWT the former now features, besides the techniquarks, a novel and natural super 4th family of leptons awaiting to be discovered at colliders, albeit with more exotic electric charges: the new electron will be doubly charged and will have a number of interesting signatures at colliders.

Several further features of the spectrum are dependent on which part of the MSCT parameter space one chooses to study. The MSCT allows model builders to investigate a large number of (perturbative and nonperturbative) inequivalent extensions of the (MS)SM. These inequivalent extensions are determined, partially, by the choice of the value of the coupling constant g_{TC} of the supersymmetric TC sector near the EW scale as well as by the vacuum choice permitted by the flat directions and, finally, by the SUSY breaking pattern. We now discuss two different limits one can take within the MSCT and sketch some of their basic features; each specific model deserves to be studied on its own and some of these models will be investigated in more detail in future publications.

The simplest case to consider is the one in which the new sector is weakly coupled at the EW scale and can be treated perturbatively. In this case the spectrum of states, which can be observed at the EW scale, is constituted by the elementary fields introduced in (17) and (18), plus the MSSM ones. However, the detailed mass spectrum will depend on the structure of the SUSY breaking terms and on the corrections induced by the EW symmetry on the supersymmetric TC sector.

The spectrum is extremely rich with several novel weakly coupled particles, such as the new techni-up and techni-down, and their respective superpartners, which can emerge at the LHC. The superpartners will be very similar to ordinary squarks but will carry TC instead of color. All the weak processes involving the production of squarks at colliders should be re-investigated to take into account the presence of these new states.

If we assume the supersymmetric TC dynamics to be strongly coupled at the EW scale, then we must use non-perturbative methods to investigate the effects of the new sector on the MSSM dynamics and vice versa. For example, we can no longer use the single particle state interpretation in terms of the underlying degrees of freedom of the supersymmetric technicolor model but rather must use an *unparticle* language given that the supersymmetric TC model is exactly conformal, before coupling it to the MSSM. The model resembles the one proposed in [101] in which, besides a TC sector, one has also coupled a natural unparticle composite sector. If no SUSY breaking terms are added directly to the 4SYM sector then conformality will be broken only via weak and hypercharge interactions. A relevant point is that one can use the machinery of the AdS/CFT correspondence to make reliable computations in the nonperturbative sector, considering the effects of the EW interactions as small perturbations.

We have introduced a very minimal supersymmetric extension of MWT and shown that one can use $\mathcal{N} = 4$ SYM as a direct extension of the SM of particle interactions.

Acknowledgments
Part of the material reviewed here is based on work done in collaboration with Matti Antola, Stefano di Chiara, Dennis D. Dietrich, Roshan Foadi, Mads T. Frandsen, Hidenori Fukano-Sakuma, Matti Järvinen, Claudio Pica, Thomas A. Ryttov and Kimmo Tuominen.

I would like to thank the organizers of the conference for having provided a vibrant scientific atmosphere.

References

[1] E. Farhi and L. Susskind, Phys. Rept. **74**, 277 (1981).
[2] S. Weinberg, Phys. Rev. D **19**, 1277 (1979).
[3] L. Susskind, Phys. Rev. D **20**, 2619 (1979).
[4] F. Sannino and K. Tuominen, Phys. Rev. D **71**, 051901 (2005) [arXiv:hep-ph/0405209].
[5] D. K. Hong, S. D. H. Hsu and F. Sannino, Phys. Lett. B **597**, 89 (2004) [arXiv:hep-ph/0406200].
[6] D. D. Dietrich, F. Sannino and K. Tuominen, Phys. Rev. D **73**, 037701 (2006) [arXiv:hep-ph/0510217].
[7] D. D. Dietrich, F. Sannino and K. Tuominen, Phys. Rev. D **72**, 055001 (2005) [arXiv:hep-ph/0505059].
[8] S. B. Gudnason, T. A. Ryttov and F. Sannino, Phys. Rev. D **76**, 015005 (2007) [arXiv:hep-ph/0612230].
[9] T. A. Ryttov and F. Sannino, Phys. Rev. D **78**, 115010 (2008) [arXiv:0809.0713 [hep-ph]].
[10] M. T. Frandsen, I. Masina and F. Sannino, Phys. Rev. D **81**, 035010 (2010) [arXiv:0905.1331 [hep-ph]].
[11] M. T. Frandsen and F. Sannino, Phys. Rev. D **81**, 097704 (2010) [arXiv:0911.1570 [hep-ph]].
[12] O. Antipin, M. Heikinheimo and K. Tuominen, JHEP **0910**, 018 (2009) [arXiv:0905.0622 [hep-ph]].

[13] R. Foadi, M. T. Frandsen, T. A. Ryttov and F. Sannino, Phys. Rev. D **76**, 055005 (2007) [arXiv:0706.1696 [hep-ph]].
[14] A. Belyaev, R. Foadi, M. T. Frandsen, M. Jarvinen, F. Sannino and A. Pukhov, Phys. Rev. D **79**, 035006 (2009) [arXiv:0809.0793 [hep-ph]].
[15] M. Antola, M. Heikinheimo, F. Sannino and K. Tuominen, JHEP **1003**, 050 (2010) [arXiv:0910.3681 [hep-ph]].
[16] O. Antipin, M. Heikinheimo and K. Tuominen, JHEP **1007**, 052 (2010) [arXiv:1002.1872 [hep-ph]].
[17] S. Nussinov, Phys. Lett. B **165**, 55 (1985).
[18] S. M. Barr, R. S. Chivukula and E. Farhi, Phys. Lett. B **241**, 387 (1990).
[19] J. Bagnasco, M. Dine and S. D. Thomas, Phys. Lett. B **320**, 99 (1994) [arXiv:hep-ph/9310290].
[20] S. B. Gudnason, C. Kouvaris and F. Sannino, Phys. Rev. D **73**, 115003 (2006) [arXiv:hep-ph/0603014].
[21] S. B. Gudnason, C. Kouvaris and F. Sannino, Phys. Rev. D **74**, 095008 (2006) [arXiv:hep-ph/0608055].
[22] K. Kainulainen, K. Tuominen and J. Virkajarvi, Phys. Rev. D **75**, 085003 (2007) [arXiv:hep-ph/0612247].
[23] C. Kouvaris, Phys. Rev. D **76**, 015011 (2007) [arXiv:hep-ph/0703266].
[24] C. Kouvaris, Phys. Rev. D **77**, 023006 (2008) [arXiv:0708.2362 [astro-ph]].
[25] M. Y. Khlopov and C. Kouvaris, Phys. Rev. D **77**, 065002 (2008) [arXiv:0710.2189 [astro-ph]].
[26] M. Y. Khlopov and C. Kouvaris, Phys. Rev. D **78**, 065040 (2008) [arXiv:0806.1191 [astro-ph]].
[27] C. Kouvaris, Phys. Rev. D **78**, 075024 (2008) [arXiv:0807.3124 [hep-ph]].
[28] K. Belotsky, M. Khlopov and C. Kouvaris, Phys. Rev. D **79**, 083520 (2009) [arXiv:0810.2022 [astro-ph]].
[29] J. M. Cline, M. Jarvinen and F. Sannino, Phys. Rev. D **78**, 075027 (2008) [arXiv:0808.1512 [hep-ph]].
[30] E. Nardi, F. Sannino and A. Strumia, JCAP **0901**, 043 (2009) [arXiv:0811.4153 [hep-ph]].
[31] R. Foadi, M. T. Frandsen and F. Sannino, Phys. Rev. D **80**, 037702 (2009) [arXiv:0812.3406 [hep-ph]].
[32] M. Jarvinen, T. A. Ryttov and F. Sannino, Phys. Lett. B **680**, 251 (2009) [arXiv:0901.0496 [hep-ph]].
[33] M. Jarvinen, C. Kouvaris and F. Sannino, Phys. Rev. D **81**, 064027 (2010) [arXiv:0911.4096 [hep-ph]].
[34] K. Kainulainen, K. Tuominen and J. Virkajarvi, JCAP **1002**, 029 (2010) [arXiv:0912.2295 [astro-ph.CO]].
[35] K. Kainulainen, K. Tuominen and J. Virkajarvi, Phys. Rev. D **82**, 043511 (2010) [arXiv:1001.4936 [astro-ph.CO]].
[36] M. T. Frandsen and S. Sarkar, Phys. Rev. Lett. **105**, 011301 (2010) [arXiv:1003.4505 [hep-ph]].
[37] C. Kouvaris and P. Tinyakov, Phys. Rev. D **82**, 063531 (2010) [arXiv:1004.0586 [astro-ph.GA]].
[38] M. Antola, S. Di Chiara, F. Sannino and K. Tuominen, arXiv:1001.2040 [hep-ph].
[39] S. Catterall and F. Sannino, Phys. Rev. D **76**, 034504 (2007) [arXiv:0705.1664 [hep-lat]].
[40] S. Catterall, J. Giedt, F. Sannino and J. Schneible, JHEP **0811**, 009 (2008) [arXiv:0807.0792 [hep-lat]].
[41] L. Del Debbio, A. Patella and C. Pica, Phys. Rev. D **81**, 094503 (2010) [arXiv:0805.2058 [hep-lat]].
[42] A. Hietanen, J. Rantaharju, K. Rummukainen and K. Tuominen, PoS **LATTICE2008**, 065 (2008) [arXiv:0810.3722 [hep-lat]].
[43] A. J. Hietanen, K. Rummukainen and K. Tuominen, Phys. Rev. D **80**, 094504 (2009) [arXiv:0904.0864 [hep-lat]].
[44] C. Pica, L. Del Debbio, B. Lucini, A. Patella and A. Rago, arXiv:0909.3178 [hep-lat].
[45] S. Catterall, J. Giedt, F. Sannino and J. Schneible, arXiv:0910.4387 [hep-lat].
[46] B. Lucini, arXiv:0911.0020 [hep-ph].
[47] F. Bursa, L. Del Debbio, L. Keegan, C. Pica and T. Pickup, Phys. Rev. D **81**, 014505 (2010) [arXiv:0910.4535 [hep-ph]].
[48] L. Del Debbio, B. Lucini, A. Patella, C. Pica and A. Rago, Phys. Rev. D **82**, 014509 (2010) [arXiv:1004.3197 [hep-lat]].
[49] L. Del Debbio, B. Lucini, A. Patella, C. Pica and A. Rago, Phys. Rev. D **82**, 014510 (2010) [arXiv:1004.3206 [hep-lat]].
[50] T. DeGrand, Phys. Rev. D **80**, 114507 (2009) [arXiv:0910.3072 [hep-lat]].
[51] T. DeGrand, Y. Shamir and B. Svetitsky, Phys. Rev. D **79**, 034501 (2009) [arXiv:0812.1427 [hep-lat]].
[52] T. DeGrand and A. Hasenfratz, Phys. Rev. D **80**, 034506 (2009) [arXiv:0906.1976 [hep-lat]].
[53] Z. Fodor, K. Holland, J. Kuti, D. Nogradi and C. Schroeder, PoS **LATTICE2008**, 058 (2008) [arXiv:0809.4888 [hep-lat]].
[54] Z. Fodor, K. Holland, J. Kuti, D. Nogradi and C. Schroeder, JHEP **0911**, 103 (2009) [arXiv:0908.2466 [hep-lat]].
[55] Z. Fodor, K. Holland, J. Kuti, D. Nogradi and C. Schroeder, JHEP **0908**, 084 (2009) [arXiv:0905.3586 [hep-lat]].
[56] J. B. Kogut and D. K. Sinclair, Phys. Rev. D **81**, 114507 (2010) [arXiv:1002.2988 [hep-lat]].
[57] T. Appelquist and R. Shrock, Phys. Lett. B **548**, 204 (2002) [arXiv:hep-ph/0204141].
[58] T. Appelquist and R. Shrock, Phys. Rev. Lett. **90**, 201801 (2003) [arXiv:hep-ph/0301108].
[59] T. Appelquist, M. Piai and R. Shrock, Phys. Rev. D **69**, 015002 (2004) [arXiv:hep-ph/0308061].

[60] F. Sannino, Acta Phys. Polon. B **40**, 3533 (2009) [arXiv:0911.0931 [hep-ph]].
[61] E. Eichten and K. D. Lane, Phys. Lett. B **90**, 125 (1980).
[62] S. Dimopoulos and L. Susskind, Nucl. Phys. B **155**, 237 (1979).
[63] E. Farhi and L. Susskind, Phys. Rev. D **20**, 3404 (1979).
[64] C. T. Hill and E. H. Simmons, Phys. Rept. **381**, 235 (2003) [Erratum-ibid. **390**, 553 (2004)] [arXiv:hep-ph/0203079].
[65] T. A. Ryttov and R. Shrock, Phys. Rev. D **82**, 055012 (2010) [arXiv:1006.5477 [hep-ph]].
[66] T. A. Ryttov and R. Shrock, Phys. Rev. D **81**, 116003 (2010) [Erratum-ibid. D **82**, 059903 (2010)] [arXiv:1006.0421 [hep-ph]].
[67] T. A. Ryttov and R. Shrock, arXiv:1005.3844 [hep-ph].
[68] T. A. Ryttov and R. Shrock, Phys. Rev. D **81**, 115013 (2010) [arXiv:1004.2075 [hep-ph]].
[69] K. Yamawaki, M. Bando and K. i. Matumoto, Phys. Rev. Lett. **56**, 1335 (1986).
[70] B. Holdom, Phys. Lett. B **150**, 301 (1985).
[71] B. Holdom, Phys. Rev. D **24**, 1441 (1981).
[72] T. W. Appelquist, D. Karabali and L. C. R. Wijewardhana, Phys. Rev. Lett. **57**, 957 (1986).
[73] H. S. Fukano and F. Sannino, Phys. Rev. D **82**, 035021 (2010) [arXiv:1005.3340 [hep-ph]].
[74] F. Sannino, arXiv:0804.0182 [hep-ph].
[75] E. Witten, Phys. Lett. B **117**, 324 (1982).
[76] D. D. Dietrich and M. Jarvinen, Phys. Rev. D **79**, 057903 (2009) [arXiv:0901.3528 [hep-ph]].
[77] D. D. Dietrich and F. Sannino, Phys. Rev. D **75**, 085018 (2007) [arXiv:hep-ph/0611341].
[78] A. Doff and A. A. Natale, Phys. Rev. D **81**, 095014 (2010) [arXiv:0912.1003 [hep-ph]].
[79] A. Doff, A. A. Natale and P. S. Rodrigues da Silva, Phys. Rev. D **80**, 055005 (2009) [arXiv:0905.2981 [hep-ph]].
[80] A. Doff and A. A. Natale, Phys. Lett. B **677**, 301 (2009) [arXiv:0902.2379 [hep-ph]].
[81] A. Doff and A. A. Natale, Int. J. Mod. Phys. E **16**, 2979 (2007).
[82] A. Doff, A. A. Natale and P. S. Rodrigues da Silva, Phys. Rev. D **77**, 075012 (2008) [arXiv:0802.1898 [hep-ph]].
[83] A. Doff and A. A. Natale, Phys. Lett. B **641**, 198 (2006) [arXiv:hep-ph/0510201].
[84] N. Evans and F. Sannino, arXiv:hep-ph/0512080.
[85] A. R. Zerwekh, Mod. Phys. Lett. A **25**, 423 (2010) [arXiv:0907.4690 [hep-ph]].
[86] E. H. Simmons, Nucl. Phys. B **312**, 253 (1989).
[87] A. Kagan and S. Samuel, Phys. Lett. B **270**, 37 (1991).
[88] C. D. Carone and E. H. Simmons, Nucl. Phys. B **397**, 591 (1993) [arXiv:hep-ph/9207273].
[89] C. D. Carone, E. H. Simmons and Y. Su, Phys. Lett. B **344**, 287 (1995) [arXiv:hep-ph/9410242].
[90] V. Hemmige and E. H. Simmons, Phys. Lett. B **518**, 72 (2001) [arXiv:hep-ph/0107117].
[91] C. D. Carone, J. Erlich and J. A. Tan, Phys. Rev. D **75**, 075005 (2007) [arXiv:hep-ph/0612242].
[92] R. S. Chivukula, A. G. Cohen and K. D. Lane, Nucl. Phys. B **343**, 554 (1990).
[93] R. Sekhar Chivukula, N. D. Christensen, B. Coleppa and E. H. Simmons, Phys. Rev. D **80**, 035011 (2009) [arXiv:0906.5567 [hep-ph]].
[94] M. Dine, W. Fischler and M. Srednicki, Nucl. Phys. B **189**, 575 (1981).
[95] B. A. Dobrescu, Nucl. Phys. B **449**, 462 (1995) [arXiv:hep-ph/9504399].
[96] H. Georgi, Phys. Rev. Lett. **98**, 221601 (2007) [arXiv:hep-ph/0703260].
[97] H. Georgi, Phys. Lett. B **650**, 275 (2007) [arXiv:0704.2457 [hep-ph]].
[98] S. P. Martin, arXiv:hep-ph/9709356.
[99] J. M. Maldacena, Adv. Theor. Math. Phys. **2**, 231 (1998) [Int. J. Theor. Phys. **38**, 1113 (1999)] [arXiv:hep-th/9711200].
[100] M. Petrini, Phys. Lett. B **404**, 66 (1997) [arXiv:hep-th/9704004].
[101] F. Sannino and R. Zwicky, Phys. Rev. D **79**, 015016 (2009) [arXiv:0810.2686 [hep-ph]].
[102] S. Catterall, JHEP **0506**, 027 (2005) [arXiv:hep-lat/0503036].
[103] J. W. Elliott, J. Giedt and G. D. Moore, Phys. Rev. D **78**, 081701 (2008) [arXiv:0806.0013 [hep-lat]].
[104] J. Giedt, R. Brower, S. Catterall, G. T. Fleming and P. Vranas, Phys. Rev. D **79**, 025015 (2009) [arXiv:0810.5746 [hep-lat]].

Constraints on Cosmological Parameters from Future Cosmic Microwave Background Experiments

Alessandro Melchiorri, Silvia Galli, Matteo Martinelli, Luca Pagano

Dipartimento di Fisica and Sezione INFN, Università di Roma "La Sapienza", Ple Aldo Moro 2, 00185, Rome, Italy

E-mail: alessandro.melchiorri@roma1.infn.it

Abstract. The Planck satellite experiment will soon let cosmologists to determine most of the cosmological parameters with unprecedented accuracy. In particular a strong improvement is expected in many parameters of interest, including neutrino mass, the amount of relativistic particles at recombination, the primordial Helium abundance and the injection of extra ionizing photon by dark matter self-annihilation. Here we review the constraints achievable by future experiments and discuss the implications for fundamental physics.

1. Introduction

The past two decades have seen dramatic improvements in measurements of the microwave background temperature fluctuations (see e.g. [1], [2] and[3]) and Planck's highly anticipated results (see [4]) will further significantly improved constraints on cosmological parameters. Moreover, on-going and planned ground-based and balloon-based experiments ([11], [12], [14], [13]) are exploring two important open frontiers: (a) the measurement of extreme ($\leq 5'$) small-scale temperature and polarization fluctuations [5] and (b) the search for primordial B-modes, the distinctive signature of gravitational waves from inflation, on large scales [6]. Proposals for a next CMB satellite as CMBPol [15] or B-POL [16] are under evaluation from american and european space agencies.

Here we briefly review the cosmological information that could come from these new datasets. We will mainly present the results obtained in a recent paper [7]. We consider a wide set of parameters focusing on those that mainly affect the "damping tail" of the CMB angular spectrum. We indeed consider additional parameters as the total neutrino mass $\sum m_\nu$ (that affects the growth of structure in the late universe), the number of extra relativistic neutrino particles N_ν^{eff} (that changes the matter-radiation epoch), and possible changes in the recombination process by changes in the fractional helium abundance Y_p, from dark matter self-annihilation processes and from variations in fundamental constants as the fine structure constant α and the Newton gravitational constant G.

As discussed in [7] we will consider 3 experimental configurations: the Planck satellite [4], the combined Planck plus ACT in polarization-sensitive detectors ACTPol [14] and, finally, the next CMBPol satellite [15]. We refer the reader to [7] for a detailed explanation of the analysis and forecast methods used.

Parameter	Planck	Planck+ACTPol	CMBPol
$\Delta(\Omega_b h^2)$	0.00013	0.000078	0.000034
$\Delta(\Omega_c h^2)$	0.0010	0.00064	0.00027
$\Delta(\theta_s)$	0.00026	0.00016	0.000052
$\Delta(\tau)$	0.0042	0.0034	0.0022
$\Delta(n_s)$	0.0031	0.0021	0.0014
$\Delta(log[10^{10}As])$	0.013	0.0086	0.0055
$\Delta(H_0)$	0.53	0.30	0.12

Table 1. 68% c.l. errors on cosmological parameters from future surveys. Standard case.

2. Constraints on a "minimal" $\Lambda - CDM$ scenario

In Table 1 we report the future constraints on the parameters of a "minimal" cosmological model. As we can see in the Table, the CMBPol experiment can provide a strong improvement (factor ~ 5) in the constraints on the baryon density, H_0 and θ_s, while the constraints on parameters as n_s are improved by a factor ~ 2. This Table will be useful in the following since it will be straightforward to identify the effect of the inclusion of an extra-parameter in the analysis.

3. Future Constraints on Neutrino Masses

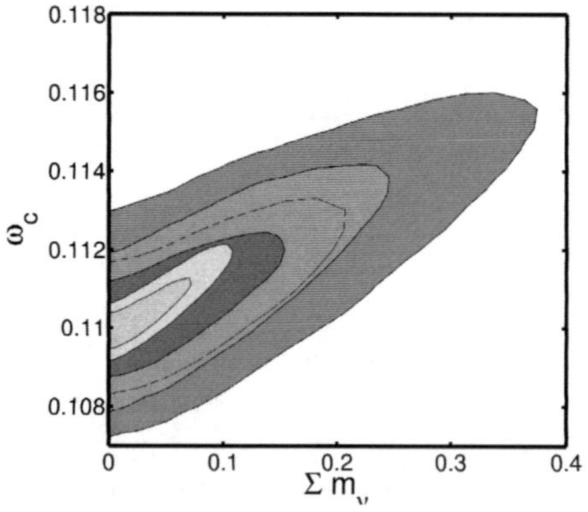

Figure 1. 68% and 95% likelihood contour plots on the $\sum m_\nu$ - ω_c plane for Planck (blu), Planck+ACTPol (red) and CMBPol (green).

Parameter	Planck	Planck+ACTPol	CMBPol
$\Delta(\Omega_b h^2)$	0.00014	0.000081	0.000033
$\Delta(\Omega_c h^2)$	0.0017	0.0010	0.00071
$\Delta(\theta_S)$	0.00028	0.00016	0.000062
$\Delta(\tau)$	0.0042	0.0034	0.0023
$\Delta(n_S)$	0.0034	0.0022	0.0016
$\Delta(\log[10^{10}A_S])$	0.013	0.0094	0.0065
$\Delta(\sum m_\nu)$	< 0.16	< 0.08	< 0.05

Table 2. 68% c.l. errors on cosmological parameters in the case of massive neutrinos.

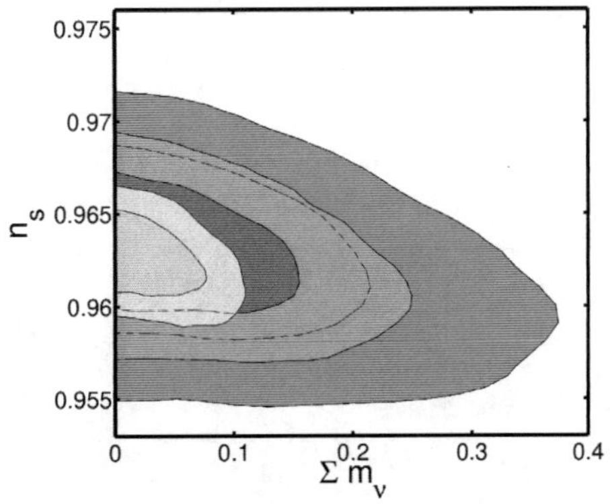

Figure 2. 68% and 95% likelihood contour plots on the $\sum m_\nu$ - ns plane for Planck (blu), Planck+ACTPol (red) and CMBPol (green).

CMB angular spectra are sensitive to a total variation in neutrino masses (see e.g. [25, 26]) defined by $\Sigma_{\nu=1,..3} m_\nu$ but can't discriminate between the mass of a single neutrino flavour (see e.g. [27])

Current oscillation experiments provide essentially two mass differences for the neutrino mass eigenstates: $\Delta m^2_{solar} \sim 8 \times 10^{-5} eV^2$ and $\Delta m^2_{atm} \sim 2.5 \times 10^{-3} eV^2$ (see e.g. [28] and references therein). An inverted hierarchy in the neutrino mass eigenstates predicts a lower limit to the total neutrino mass of about $\sum m_\nu \geq 0.10 eV$ while a direct hierarchy predicts $\sum m_\nu \geq 0.05 eV$. The goal for CMB experiments is therefore to have a sensitivity better than $\sum m_\nu \leq 0.10 eV$ for possibly ruling out the inverted hierarchy and better than $\sum m_\nu \leq 0.05 eV$ for a sure detection of neutrino mass.

As we can see from Table 2 the expected sensitivity from Planck and Planck+ACTPol fails to reach the possibility of ruling out the neutrino mass inverted hierarchy. It is however important to notice that the expected sensitivity from the KATRIN [29] beta decay experiment is of the order of $\sum m_\nu = 0.3$. Planck and Planck+ACTPol will therefore explore the same energy scale, providing a great opportunity for confirming or anticipating a mass detection from KATRIN.

Including a neutrino mass, as we can see comparing Table 2 with Table 1 as a relevant impact in the determination of the cold dark matter density ω_c that results with an uncertainty that is nearly doubled respect to the standard analysis. Moreover, also the constraints on n_s are affected. We plot in Figure 1 and Figure 2 the 2-D likelihood contour plots at 68% and 95% confidence level in the Σm_ν vs ω_c and vs n_s planes respectively. As we can see, a non negligible neutrino mass can put higher values of the cold dark matter abundance and lower values of the scalar spectral index in better agreement with observations.

4. Future Constraints on Extra Background of Relativistic Particles

An extra background of relativistic (and non-interacting) particles can be parametrized by introducing an effective number of neutrino species N_ν^{eff}. This extra-background changes the CMB anisotropies through time variations of the gravitational potential at recombination due to the presence of this non negligible relativistic component (the so-called early Integrated Sachs Wolfe effect). The main consequences is an increase in the small-scale CMB anisotropy (see e.g. [31]). The results are reported in Table 3. As we can see, comparing with the results in Table 3, the inclusion of a background of relativistic particles strongly weakens the constraints on n_s, ω_b, ω_c and θ_s. As we can see from Figures 3, 4, 5 and 6 there is indeed a strong correlation between N_ν^{eff} and these parameters. While adding ACT will improve the constraints by a factor ~ 2, CMBPol can provide constraints that could bring valuable information on the

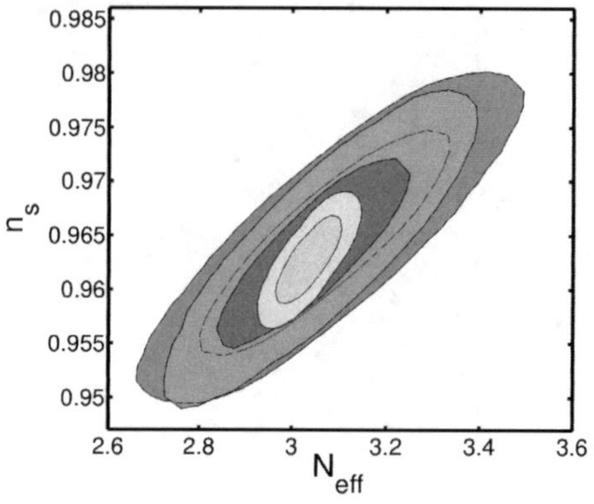

Figure 3. 68% and 95% likelihood contour plots on the N_{eff} - ns plane for Planck (blue), Planck+ACTPol (red) and CMBPol (green).

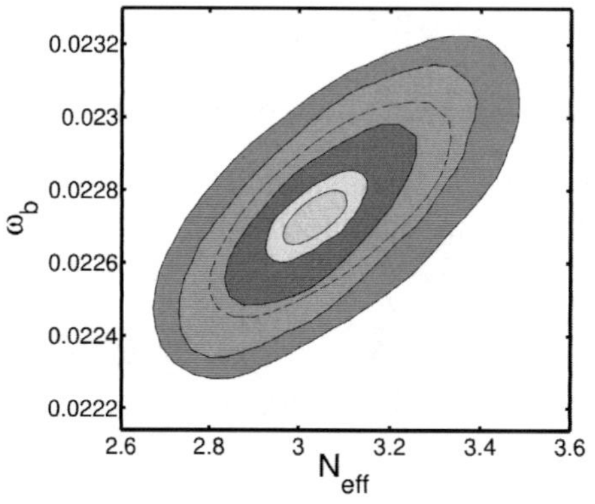

Figure 4. 68% and 95% likelihood contour plots on the N_{eff} - ω_b plane for Planck (blue), Planck+ACTPol (red) and CMBPol (green).

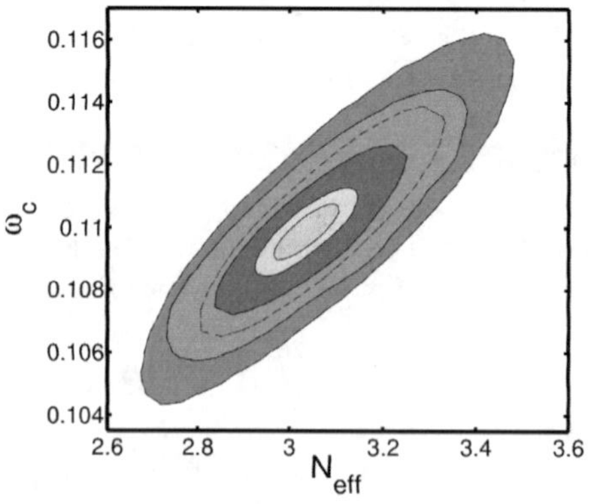

Figure 5. 68% and 95% likelihood contour plots on the N_{eff} - ω_c plane for Planck (blu), Planck+ACTPol (red) and CMBPol (green).

physics of neutrino decoupling from the photon-baryon primordial plasma. As it is well known, the standard value of neutrino parameters $N_{eff} = 3$ should be increased to $N_{eff} = 3.04$ due to an additional contribution from a partial heating of neutrinos during the electron-positron

Journal of Physics: Conference Series **259** (2010) 012004 doi:10.1088/1742-6596/259/1/012004

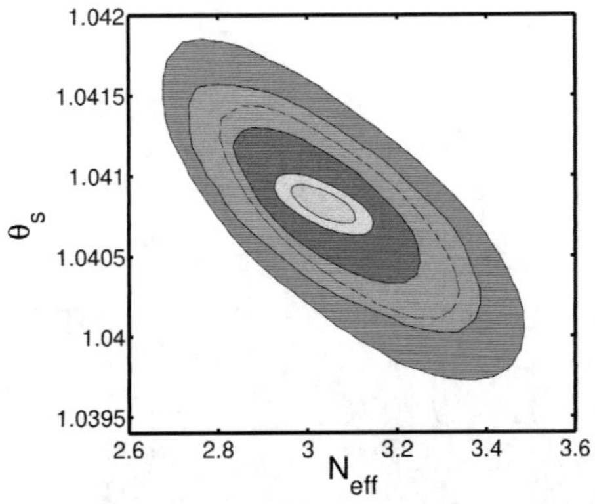

Figure 6. 68% and 95% likelihood contour plots on the N_{eff} - θ_s plane for Planck (blu), Planck+ACTPol (red) and CMBPol (green).

annihilations (see e.g. [32]). This effect, expected from standard physics, could be tested by the CMBPol experiment, albeit at just one standard deviations. However, the presence of non standard neutrino-electron interactions (NSI) may enhance the entropy transfer from electron-positron pairs into neutrinos instead of photons, up to a value of $N_{eff} = 3.12$ ([33]). This value could be discriminated by CMBPol from $N_{eff} = 3$ at ~ 3 standard deviations, shedding new light on NSI models.

	Planck	Planck+ACTPol	CMBPol
$\Delta(\Omega_b h^2)$	0.00018	0.00013	0.000051
$\Delta(\Omega_c h^2)$	0.0024	0.0015	0.00059
$\Delta(\theta_s)$	0.00042	0.00024	0.000075
$\Delta(\tau)$	0.0043	0.0035	0.0023
$\Delta(n_s)$	0.0065	0.0049	0.0026
$\Delta(log[10^{10}A_s])$	0.017	0.013	0.0077
$\Delta(N_{eff})$	0.17	0.11	0.046

Table 3. 68% c.l. errors on cosmological parameters in the case of extra background of relativistic particles N_{eff}.

5. Future Constraints on Dark Matter Self Annihilation

Annihilating particles affect the ionization history of the Universe in three main different ways. The interaction of the shower produced by the annihilation with the thermal gas can ionize it, induce Ly–α excitation of the hydrogen and heat the plasma. The first two modify the evolution of the free electron fraction x_e, the third affects the temperature of baryons [34].

The rate of energy release $\frac{dE}{dt}$ per unit volume by a relic self-annihilating dark matter particle is given by

$$\frac{dE}{dt}(z) = \rho_c^2 c^2 \Omega_{DM}^2 (1+z)^6 p_{ann} \quad (1)$$

$$p_{ann} = f \frac{<\sigma v>}{m_\chi} \quad (2)$$

with $n_{DM}(z)$ being the relic DM abundance at a given redshift z, $<\sigma v>$ is the effective self-annihilation rate and m_χ the mass of our dark matter particle, Ω_{DM} is the dark matter density parameter and ρ_c the critical density of the Universe today; the parameter f indicates the fraction of energy which is absorbed *overall* by the gas, under the approximation the energy absorption takes place locally. CMB is sensitive to the combined parameter p_{ann} only. Greater is p_{ann}, higher is the fraction of free electrons surviving after recombination, therefore widening the peak of the visibility function and dampening the peaks of the temperature and polarization angular power spectra.

Parameter	Planck	Planck+ACTPol	CMBPol
$\Delta(\Omega_b h^2)$	0.00013	0.000079	0.000032
$\Delta(\Omega_c h^2)$	0.0010	0.00063	0.00027
$\Delta(H_0)$	0.52	0.30	0.12
$\Delta(\tau)$	0.0042	0.0034	0.0023
$\Delta(n_S)$	0.0032	0.0021	0.0015
$\Delta(\log[10^{10} A_S])$	0.013	0.0085	0.0055
$\Delta(p_{ann})[m^3/s/Kg]$	$< 1.5 \cdot 10^{-7}$	$< 1.2 \cdot 10^{-7}$	$< 6.3 \cdot 10^{-8}$

Table 4. 68% c.l. errors on cosmological parameters in the case of dark matter annihilation. The upper limits on p_{ann} are at 95% c.l.

As we can see from Table 4 and comparing with the results in Table 1 the inclusion of dark matter self-annihilation doesn't affect much the constraints on the other parameters.

6. Future Constraints on Helium Abundance

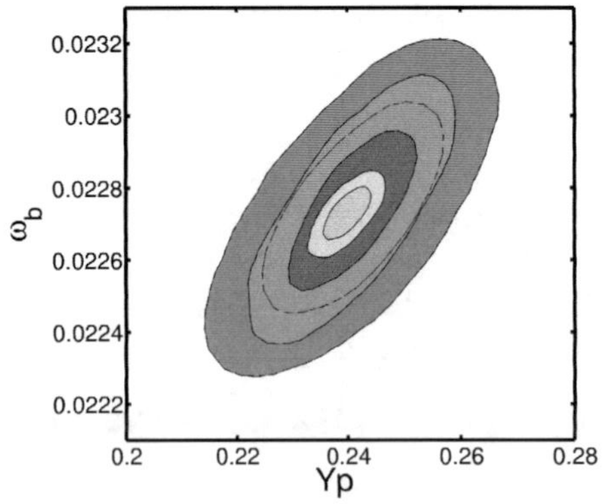

Figure 7. 68% and 95% likelihood contour plots on the Y_{He} - ω_b plane for Planck (blue), Planck+ACTPol (red) and CMBPol (green).

As recently shown by several authors ([35], [36], [37], [38]) the small scale CMB anisotropy spectrum can provide a powerful method for accurately determining the primordial Helium abundance. Current astrophysical measurements of primordial Helium converge towards a conservative aestimate of $Y_p = 0.250 \pm 0.003$ (see e.g. [39]). As we can see from Table 5 the Planck satellite mission alone will not reach such accuracy, even when combined with ACT. It is however interesting that a CMBPol-like experiment has the potential of reaching a precision comparable with current astrophysical measurements. This will open a new window of research for testing systematics in current primordial helium determinations.

Comparing the results in Table 5 with the constraints obtained in the case of a standard analysis it is easy to see that the major impact of including this parameter is on the determination

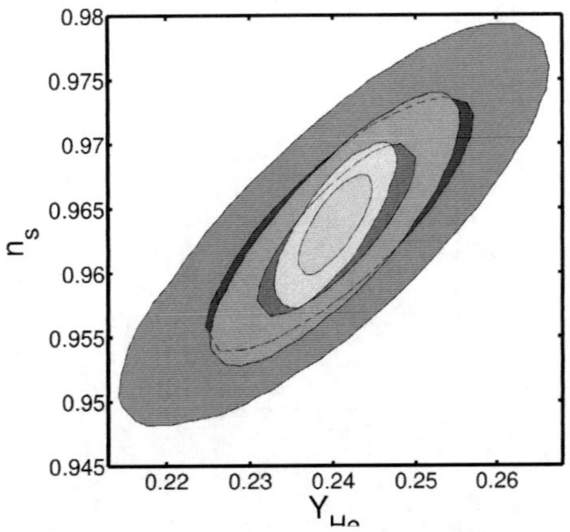

Figure 8. 68% and 95% likelihood contour plots on the Y_{He} - ns plane for Planck (blue), Planck+ACTPol (red) and CMBPol (green).

Parameter	Planck	Planck+ACTPol	CMBPol
$\Delta(\Omega_b h^2)$	0.00019	0.00013	0.000051
$\Delta(\Omega_c h^2)$	0.0010	0.00065	0.00027
$\Delta(\theta_S)$	0.00046	0.00026	0.00010
$\Delta(\tau)$	0.0043	0.0035	0.0023
$\Delta(n_S)$	0.0063	0.0043	0.0025
$\Delta(\log[10^{10}A_S])$	0.013	0.013	0.0079
$\Delta(Y_p)$	0.010	0.0061	0.0029

Table 5. 68% c.l. errors on cosmological parameters in the case of helium abundance.

of the scalar spectral index n_s and the baryon abundance, with the 1-σ c.l. increased by a factor ~ 2. In Figures 7 and 8 we plot the 2-D likelihood contours at 68% and 95% c.l. between Y_p and these parameters.

7. Future Constraints on Variations of Fundamental Constants

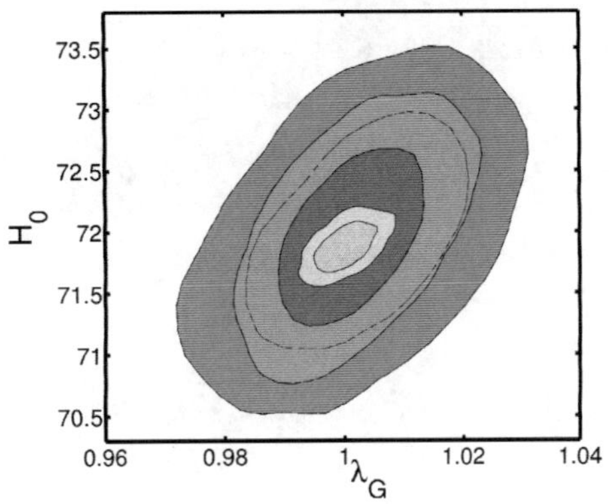

Figure 9. 68% and 95% likelihood contour plots on the λ_G - H_0 plane for Planck (blue), Planck+ACTPol (red) and CMBPol (green).

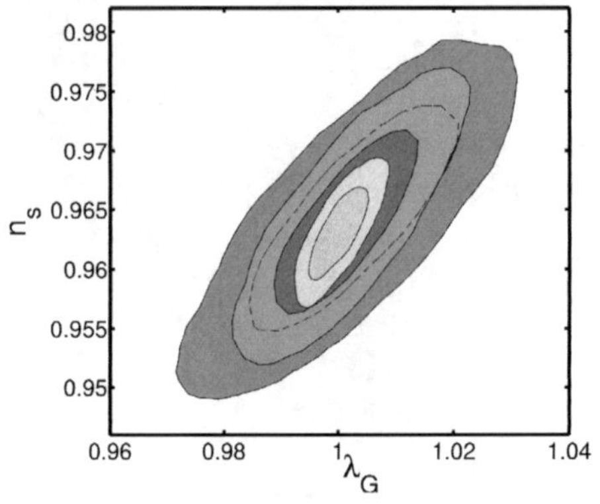

Figure 10. 68% and 95% likelihood contour plots on the λ_G - ns plane for Planck (blue), Planck+ACTPol (red) and CMBPol (green).

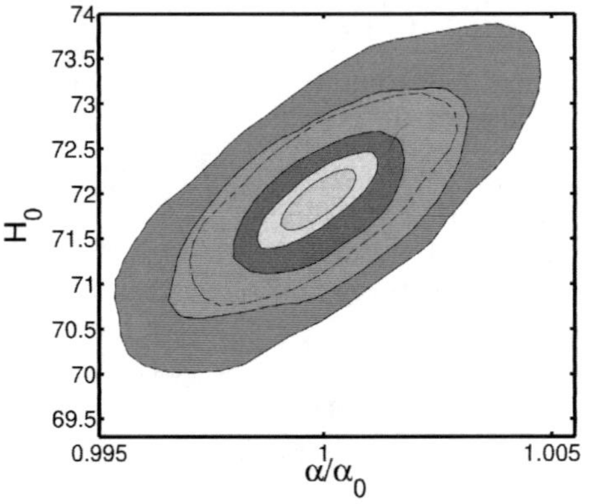

Figure 11. 68% and 95% likelihood contour plots on the α/α_0 - H_0 plane for Planck (blue), Planck+ACTPol (red) and CMBPol (green).

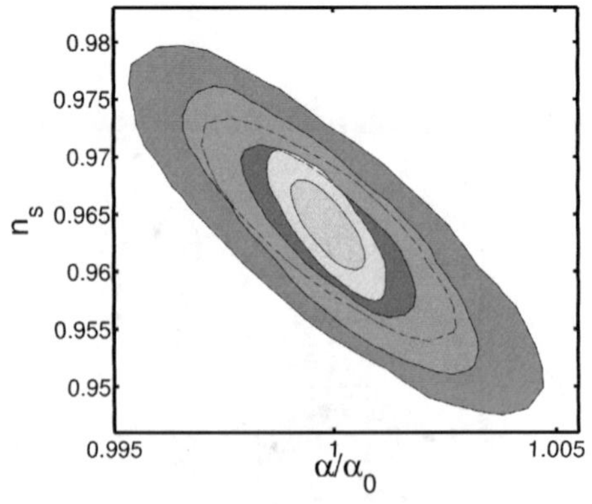

Figure 12. 68% and 95% likelihood contour plots on the α/α_0 - ns plane for Planck (blue), Planck+ACTPol (red) and CMBPol (green).

CMB anisotropies are sensitive to variations in fundamental constants as the fine structure α (see e.g. [42], [43]) or Newton's constant G ([44]) through changes in the recombination

Journal of Physics: Conference Series **259** (2010) 012004 doi:10.1088/1742-6596/259/1/012004

scenario. Varying α changes the ionization and excitation rates and could delay or accelerate recombination. Varying G does not affect recombination directly but "rescales" the expansion rate of the Universe, changing the epoch when recombination takes place.

The constraints are reported in Table 6 and Table 7 for variations in α and G respectively. In order to parametrize the variations with dimensionless quantities we have considered variations in the parameters $\Delta_\alpha = \alpha/\alpha_0$ e $\lambda_G = G/G_0$ where α_0 and G_0 are the current values of these fundamental constants, measured in laboratory[1] $\alpha_0 = 7.2973525376(50) \times 10^{-3}$ and $G_0 = 6.67428(67) \times 10^{-11} m^3 kg^{-1} s^{-2}$.

	Planck	Planck+ACTPol	CMBPol
$\Delta(\Omega_b h^2)$	0.00014	0.000089	0.000034
$\Delta(\Omega_c h^2)$	0.0012	0.00070	0.00031
$\Delta(\tau)$	0.0042	0.0034	0.0023
$\Delta(H_0)$	0.77	0.40	0.20
$\Delta(n_s)$	0.0064	0.0035	0.0025
$\Delta(log[10^{10}A_s])$	0.0086	0.011	0.0041
$\Delta(\alpha/\alpha_0)$	0.0019	0.00093	0.00051

Table 6. 68% c.l. errors on cosmological parameters from future surveys in case of a variable fine structure constant α.

	Planck	Planck+ACTPol	CMBPol
$\Delta(\Omega_b h^2)$	0.00019	0.00013	0.000048
$\Delta(\Omega_c h^2)$	0.0010	0.00068	0.00025
$\Delta(\tau)$	0.0042	0.0037	0.0022
$\Delta(H_0)$	0.60	0.40	0.13
$\Delta(n_s)$	0.0061	0.0046	0.0023
$\Delta(log[10^{10}A_s])$	0.018	0.013	0.0073
$\Delta(\lambda_G)$	0.012	0.0076	0.0030

Table 7. 68% c.l. errors on cosmological parameters from future surveys in case of a variable gravitational constant G.

As we can see from Tables 6 and 7 a variation in these fundamental constants has important effects in the determination of the scalar spectral index n_s and the Hubble costant H_0. This can also be seen in the 2-D likelihood contour plots in Figures 9, 10, 11, and 12.

8. Conclusions

Here we have briefly revived the future constraints achievable from CMB experiments on several parameters. Other than the 5 parameters of the standard Λ-CDM model we have considered new parameters mostly related to quantities that can be probed in a complementary way in laboratory and/or with astrophysical measurements. We found that CMB experiments as CMBPol could have a very important impact in the understanding of neutrino physics. CMBPol could indeed discriminate between the neutrino mass hierarchy and shed light on the physics of neutrino decoupling before BBN. Moreover, the primordial Helium abundance can be constrained with the same accuracy of current astrophysical measurements but with a much better control of systematics. Moreover, also constraints on fundamental constant can reach a level close

[1] See http://www.codata.org/

to laboratory constraints. This overlap between cosmology and other sector of physics and astronomy is definitely the most interesting aspect of future CMB research.

9. Acknowledgements

It is a great pleasure to thank our collaborators Blake Sherwin and David Spergel.

References

[1] G. F. Smoot *et al.*, Astrophys. J. **396** (1992) L1; C. L. Bennett *et al.*, Astrophys. J. **464** (1996) L1 [arXiv:astro-ph/9601067].
[2] A. D. Miller *et al.*, Astrophys. J. **524** (1999) L1 [arXiv:astro-ph/9906421]; A. Melchiorri *et al.* [Boomerang Collaboration], Astrophys. J. **536** (2000) L63 [arXiv:astro-ph/9911445]; P. de Bernardis *et al.* [Boomerang Collaboration], Nature **404** (2000) 955 [arXiv:astro-ph/0004404]. S. Hanany *et al.*, Astrophys. J. **545** (2000) L5 [arXiv:astro-ph/0005123]. N. W. Halverson *et al.*, Astrophys. J. **568** (2002) 38 [arXiv:astro-ph/0104489].
[3] D. N. Spergel *et al.* [WMAP Collaboration], Astrophys. J. Suppl. **148** (2003) 175 [arXiv:astro-ph/0302209]; C. L. Bennett *et al.* [WMAP Collaboration], Astrophys. J. Suppl. **148** (2003) 1 [arXiv:astro-ph/0302207].
[4] [Planck Collaboration], arXiv:astro-ph/0604069.
[5] M. J. White, Astrophys. J. **555** (2001) 88 [arXiv:astro-ph/0101086]; A. Cooray, arXiv:astro-ph/0203048.
[6] U. Seljak and M. Zaldarriaga, Phys. Rev. Lett. **78** (1997) 2054 [arXiv:astro-ph/9609169]; M. Kamionkowski, A. Kosowsky and A. Stebbins, Phys. Rev. D **55** (1997) 7368 [arXiv:astro-ph/9611125].
[7] S. Galli, M. Martinelli, A. Melchiorri, L. Pagano, B. D. Sherwin and D. N. Spergel, arXiv:1005.3808 [astro-ph.CO].
[8] M. Zaldarriaga and U. Seljak, Phys. Rev. D **58** (1998) 023003 [arXiv:astro-ph/9803150]; A. Lewis and A. Challinor, Phys. Rept. **429** (2006) 1 [arXiv:astro-ph/0601594].
[9] *et al.* [The ACT Collaboration], arXiv:1001.2934 [Unknown].
[10] M. Lueker *et al.*, arXiv:0912.4317 [Unknown].
[11] P. Oxley *et al.*, Proc. SPIE Int. Soc. Opt. Eng. **5543** (2004) 320 [arXiv:astro-ph/0501111].
[12] B. P. Crill *et al.*, arXiv:0807.1548 [astro-ph].
[13] J. E. Carlstrom *et al.*, arXiv:0907.4445 [astro-ph.IM].
[14] A. Kosowsky [the ACT Collaboration], New Astron. Rev. **50** (2006) 969 [arXiv:astro-ph/0608549].
[15] J. Bock *et al.* [CMBPol Collaboration], arXiv:0906.1188 [astro-ph.CO].
[16] P. de Bernardis, M. Bucher, C. Burigana, L. Piccirillo and f. P. Collaboration, Exper. Astron. **23** (2009) 5 [arXiv:0808.1881 [astro-ph]].
[17] M. Shimon, N. J. Miller, C. T. Kishimoto, C. J. Smith, G. M. Fuller and B. G. Keating, arXiv:1001.5088 [astro-ph.CO].
[18] D. Baumann *et al.* [CMBPol Study Team Collaboration], AIP Conf. Proc. **1141** (2009) 10 [arXiv:0811.3919 [astro-ph]].
[19] M. Zaldarriaga *et al.*, arXiv:0811.3918 [astro-ph].
[20] S. Dodelson *et al.*, arXiv:0902.3796 [astro-ph.CO].
[21] A. Lewis, A. Challinor and A. Lasenby, Astrophys. J. **538** (2000) 473 [arXiv:astro-ph/9911177].
[22] C. M. Hirata, U. Seljak, Phys. Rev. D **68**(2003) 083002 [arXiv:0306354 [astro-ph]];
[23] T. Okamoto, W. Hu, Phys.Rev. D **67** (2003) 083002 arXiv:astro-ph/0301031v2
[24] L. Perotto, J. Lesgourgues, S. Hannestad, H. Tu and Y. Y. Y. Wong, JCAP **0610** (2006) 013 [arXiv:astro-ph/0606227].
[25] C. P. Ma and E. Bertschinger, Astrophys. J. **455** (1995) 7 [arXiv:astro-ph/9506072].
[26] K. Ichikawa, M. Fukugita and M. Kawasaki, Phys. Rev. D **71** (2005) 043001 [arXiv:astro-ph/0409768].
[27] M. Kaplinghat, L. Knox and Y. S. Song, Phys. Rev. Lett. **91** (2003) 241301 [arXiv:astro-ph/0303344].
[28] G. L. Fogli *et al.*, Phys. Rev. D **78** (2008) 033010 [arXiv:0805.2517 [hep-ph]].
[29] Beck, M., & the Katrin collaboration 2010, Journal of Physics Conference Series, 203, 012097
[30] Monfardini, A., et al. 2006, Nuclear Instruments and Methods in Physics Research A, 559, 346
[31] R. Bowen, S. H. Hansen, A. Melchiorri, J. Silk and R. Trotta, Mon. Not. Roy. Astron. Soc. **334**, 760 (2002) [arXiv:astro-ph/0110636].
[32] G. Mangano, G. Miele, S. Pastor and M. Peloso, Phys. Lett. B **534** (2002) 8 [arXiv:astro-ph/0111408].
[33] G. Mangano, G. Miele, S. Pastor, T. Pinto, O. Pisanti and P. D. Serpico, Nucl. Phys. B **756** (2006) 100 [arXiv:hep-ph/0607267].
[34] Galli, S., Iocco, F., Bertone, G., Melchiorri, A. 2009, PRD, 80, 023505; L. Zhang, X. L. Chen, Y. A. Lei and Z. G. Si, Phys. Rev. D **74**, 103519 (2006) [arXiv:astro-ph/0603425]; L. Zhang, X. Chen, M. Kamionkowski, Z. g. Si and Z. Zheng, Phys. Rev. D **76**, 061301 (2007) [arXiv:0704.2444 [astro-ph]].

[35] R. Trotta and S. H. Hansen, Phys. Rev. D **69**, 023509 (2004) [arXiv:astro-ph/0306588].
[36] K. Ichikawa, T. Sekiguchi and T. Takahashi, Phys. Rev. D **78** (2008) 043509 [arXiv:0712.4327 [astro-ph]].
[37] Hamann, J., Lesgourgues, J., & Mangano, G. 2008, Journal of Cosmology and Astro-Particle Physics, 3, 4
[38] Komatsu, E., et al., 2010, arXiv:1001.4538
[39] F. Iocco, G. Mangano, G. Miele, O. Pisanti and P. D. Serpico, Phys. Rept. **472**, 1 (2009) [arXiv:0809.0631 [astro-ph]].
[40] R. Bean and A. Melchiorri, Phys. Rev. D **65**, 041302 (2002) [arXiv:astro-ph/0110472].
[41] W. Hu, Phys. Rev. D **65** (2002) 023003 [arXiv:astro-ph/0108090].
[42] Menegoni, E., Galli, S., Bartlett, J. G., Martins, C. J. A. P., Melchiorri, A. 2009, PRD, 80, 087302
[43] Martins, C. J. A. P., Menegoni, E., Galli, S., Mangano, G., Melchiorri, A. 2010, arXiv:1001.3418
[44] Galli, S., Melchiorri, A., Smoot, G. F., Zahn, O. 2009, PRD, 80, 023508 O. Zahn and M. Zaldarriaga, Phys. Rev. D **67** (2003) 063002 [arXiv:astro-ph/0212360].
[45] E. Komatsu *et al.* [WMAP Collaboration], Astrophys. J. Suppl. **180** (2009) 330 [arXiv:0803.0547 [astro-ph]]; J. Dunkley *et al.* [WMAP Collaboration], arXiv:0803.0586 [astro-ph].
[46] C. J. MacTavish *et al.*, Astrophys. J. **647** (2006) 799 [arXiv:astro-ph/0507503]; F. Piacentini *et al.*, Astrophys. J. **647** (2006) 833 [arXiv:astro-ph/0507507]; T. E. Montroy *et al.*, Astrophys. J. **647** (2006) 813 [arXiv:astro-ph/0507514]; Jones, W. C., et al. 2006, Astrophys. J. 647, 823
[47] C. L. Reichardt *et al.*, arXiv:0801.1491 [astro-ph].
[48] P. G. Castro *et al.* [QUaD collaboration], arXiv:0901.0810 [astro-ph.CO].
[49] J. Bock *et al.*, arXiv:0805.4207 [astro-ph].
[50] A. Lewis and S. Bridle, Phys. Rev. D **66** (2002) 103511 [arXiv:astro-ph/0205436].

16th International Symposium on Particles Strings and Cosmology (PASCOS 2010) IOP Publishing
Journal of Physics: Conference Series **259** (2010) 012005 doi:10.1088/1742-6596/259/1/012005

Self-tuning of the cosmological constant

Jihn E. Kim
Department of Physics and Astronomy and Center for Theoretical Physics,
Seoul National University, Seoul 151-747, Korea
E-mail: jekim@ctp.snu.ac.kr

Abstract. Here, I discuss the cosmological constant problems, in particular paying attention to the vanishing cosmological constant. There are three cosmological constant problems in particle physics. Hawking's idea of calculating the probability amplitude for our Universe is peaked at $\overline{\Lambda} = 0$ which I try to obtain after the initial inflationary period using a self-tuning model. I review what has been discussed on the Hawking type calculation with H^2 Lagrangian, and present a (probably) correct way to calculate the amplitude, and show that the Kim-Kyae-Lee self-tuning model allows a finite range of parameters for the $\overline{\Lambda} = 0$ to have a singularly large probability, approached from the AdS side.

1. Introduction

Recently, Nicoli commented that the Einstein equation is inconsistent from the outset [1] because the left-hand side (LHS) and the right-hand side (RHS) of the equation

$$R_{\mu\nu} - \frac{1}{2}Rg_{\mu\nu} = 8\pi G_N T_{\mu\nu} \tag{1}$$

are logically different. The LHS is exactly determined by the continuous space-time geometry. It is a classical concept on the geometry. On the other hand, the RHS is contributed by quanta of particles which have roots in the probabilistic interpretation of quantum mechanics in addition to the continuous vacuum energy if present. Any discussion in the Planck era is speculative and not yet well formulated as Kolb and Turner stress in the last Chapter of [2]. To answer this question satisfactorily, we must have a good quantum theory of gravity. So, I must admit that any solution of the cosmological constant (CC) problem may be incomplete at present, and what I present here is the CC solution idea just by replacing $4\pi G_N T_{\mu\nu}$ on the RHS of (1) by a classically interpretable $\Lambda g_{\mu\nu}$. At present, we do not know whether we are in a situation similar to classical electrodynamics in the early 20th century where the stability of an electron in Hydrogen needed a continuous distribution of positive charges, which was the Thompson model of Hydrogen. With quantum mechanics, we became to understand Hydrogen in terms of the Rutherford model. So, the CC problem may have a root in a completely different domain of physics on which we may be unfamiliar now. In this sense, we may look for any reasonable solution of the problem and it is welcome if it is not contradicted outrageously from the present perspective on the fundamental physics. In this spirit I present an idea with a specific action toward a CC solution. But there exists one nagging question on the proposal: why such an action? It may be like the stability problem of the Rutherford model with the 19th century eye of classical electrodynamics.

Journal of Physics: Conference Series **259** (2010) 012005 doi:10.1088/1742-6596/259/1/012005

The CC was introduced almost ninety-three years ago by Einstein [3]. Then, much later when the spontaneous symmetry breaking is widely discussed in the standard model, Veltman commented that the vacuum energy arising in spontaneous symmetry breaking adds to the CC [4], basically raising a question on the naturalness of setting it to zero. Even before considering the tree level CC, the loop correction to the vacuum energy was a problem since the early days of quantum mechanics. Any mode contributes $\frac{1}{2}\hbar\omega$ to the vacuum energy. In the CC discussion here, we will not rely on the probably-already-happened anthropic arguments [5]. So, we consider the CC generically, at the tree and also at loop levels unless the figures of Fig. 1 are forbidden. For example, the LHS figure of Fig. 1 corresponds to $\frac{1}{2}\hbar\omega$ per mode and the RHS figure is the two-loop vacuum energy arising from the A-terms in supergravity [6]. There can be many more diagrams contributing to the vacuum energy at the loop levels.

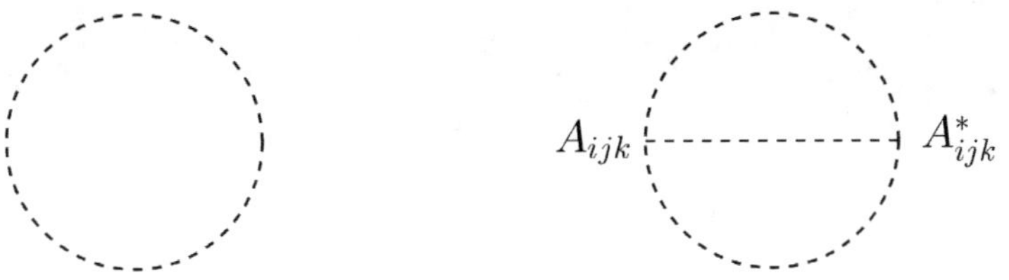

$$\simeq N\frac{m_B^2\Lambda^2}{8\pi^2} \qquad\qquad \simeq \frac{1}{(4\pi^2)^2}\sum_{ijk} A_{ijk}A^*_{ijk}\Lambda^2$$

Figure 1. Loop corrections to the CC.

If a symmetry is present in changing the CC Λ, one may try a scalar potential of Fig. 2 where the vertical axis corresponds to the CC. The vanishing CC is the point which the arrow points to and the vacuum is the point where the bullet is located. As Fig. 2 shows, in general the vanishing CC point does not corresponds to the bullet where the equation of motion is satisfied. So, a solution is not easily realizable in four space-time dimensions (4D). In addition, the CC problem must also take into account the result of spontaneous symmetry breaking [4]. The vacuum energy scale at the interesting particle physics scales are generally much greater than the current limit of $(0.002 \text{ eV})^4$. Thus, the CC is a serious fine-tuning problem. In 4D, we do not find any symmetry such that the CC is forbidden. Note, however, the other recent tries on the CC have been considered from the scale invariance and the brane statistical search [7].

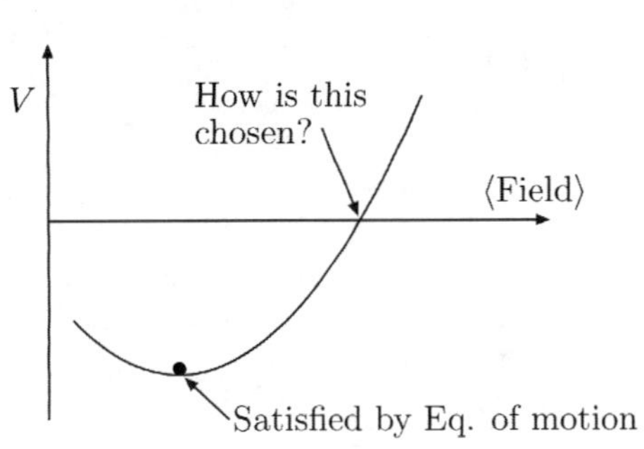

Figure 2. A form of the potential energy in terms of CC.

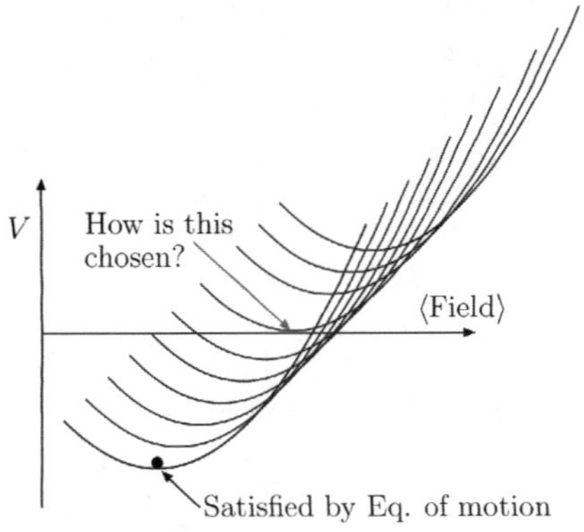

Figure 3. A form of the potential energy with multiple solutions.

Another question is at which energy scale and *temperature*, the CC is required to vanish. There exists a hierarchy of mass scales in particle physics:

- Planck scale: 2.44×10^{18} GeV
- GUT scale: $(2-3) \times 10^{16}$ GeV
- Inflation scale: $\simeq 10^{16}$ GeV, down to the EW scale
- Axion scale: $\simeq 10^{12} - 10^{11}$ GeV
- Hidden sector scale: 10^{15} GeV [8], down to 10^{6} GeV [9]
- EW scale: 100 GeV
- QCD scale: 1 GeV
- Nuclear physics scale: 10 MeV
- Electron mass scale: 0.5 MeV
- Neutrino mass scale: 0.05 eV
- Accelerating universe scale: $(0.002 \text{ eV})^4$
- Temperature of the CMB: 10^{-4} eV

For example, even though we suppose to have a CC solution at an EW scale, still 10^{-60} fine-tuning is required. It is a hierarchy among the parameters in the Lagrangian. Since the CC is dimension 4 parameter, it has the most severe fine-tuning problem in the Lagrangian. Thus, finding a solution of the CC problem is very difficult and is hence challenging. Therefore, any simple idea toward a solution is welcome at this stage if it is not obviously wrong.

2. Probability amplitude

When we consider quantum mechanics, we talk in terms of the probability amplitude: The initial state $|I\rangle$ transforming to a final state $|F\rangle$. In this spirit, Baum [10] and Hawking [11] considered the Euclidian action, only with the Ricci scalar R and the CC. The method touches upon the quantum gravity, by calculating the path integral in the Euclidian spacetime. The path integral deals with all possible paths and in the $\hbar = 0$ limit it is dominated by the classical equations of motion. If the topology change of the metric is considered, we must know the full gravity equation such as those including wormhole solutions [12], in which case an exponential of exponential function was obtained [13]. For simplicity, we do not delve into an exponential of an exponential. Without the topology change which may be a reasonable assumption much below the Planck scale, this behavior of an exponential of exponential has not been known. The Euclidian action integral which we consider first from [10, 11] has the following form

$$e^{-\tilde{I}_E} = e^{3\pi M_P^2/\Lambda}. \tag{2}$$

So, $\Lambda = 0^+$ dominates the action integral, which is interpreted as the probability for $\Lambda = 0^+$ is close to 1. But, there are questions regarding to this Baum-Hawking solution. Hawking [11] states, "My proposal requires that a variable effective CC be genarated in some manner. One possibility would be to include the values of the CC in the variables that are integrated over in the path integral." For this purpose, Ref. [11] explicitly considered in terms of $A_{\mu\nu\rho}$ (or the field strength $H_{\mu\nu\rho\sigma}$) which however is not a dynamical field in 4D.

In this scenario, the needed quantity to calculate is the action integral. Even if we understand the CC in this way, there exists another questions such as

- How do we assign the initial state?
- How does the needed primary inflation come about in this scenario?

- How does it fit to the current dark energy?

So the CC solution needs to answer the other two CC related questions also and furthermore needs an argument on, "What was the proper initial condition of our Universe?"

The existing idea of Hawking in terms of $H_{\mu\nu\rho\sigma}$ with no kinetic energy term cannot explain all the above questions. We must introduce the kinetic term with the potential shape given as that of Fig. 3 so that the point where the equation of motion is satisfied can be the point with $\Lambda = 0$. For example, $H_{\mu\nu\rho\sigma}$ can achieve this but without the kinetic energy term in 4D. If we want to use $H_{\mu\nu\rho\sigma}$ field, we must work beyond 4D.

3. A self-tuning solution with a brane

In 2000, self-tuning solutions have been tried in the RS type models in 5D. Here, I just mention the initial try and its failure. In an RS-I type model, Ref. [14] tried to show that SM fields living at the SM brane located at $x^5 = 0$ do not change, via the loop corrections at the brane, the CC solution of the bulk. Here, the bulk action is fixed with specific magnitude of coupling,

$$I_5 = \int d^5x\sqrt{-g}\left[R - \frac{4}{3}(\partial\phi)^2 - \Lambda e^{a\phi}\right] \tag{3}$$

$$+\int d^4x\sqrt{-g_4}\left[-Ve^{b\phi}\right]_{x^5=0}. \tag{4}$$

In a sense, the vacuum energy of the SM brane is cured. However, the specific form of the above bulk action is arguable for a general CC solution. Even though we allow this procedure, still it has its own problem that a singularity is present at a point y_s in the bulk. The singularity can be cured by inserting the Z_2 symmetric branes at $\pm y_s$ [15],

$$I_5 + \int d^4x\sqrt{-g_4}\left[-V_+e^{b_+\phi}\right]_{x^5=+y_s} + \int d^4x\sqrt{-g_4}\left[-V_-e^{b_-\phi}\right]_{x^5=-y_s}. \tag{5}$$

Then, to cancel the contribution of the SM brane of Eq. (4), one must fine-tune the CC contribution from the singularities of Eq. (5) [15]. Again a fine-tuning is needed: $\Lambda_{4D} = E_0 + E_+ + E_- = 0$, leading to a fine tuning between V_+, b_+, V_-, and b_-. Furthermore, there exists the no-go theorem under some plausible conditions such that one employs the usual kinetic energy term and assumes the Lorentz symmetry and the existence of 4D gravity for a large distance separation [16].

Here, we try to go beyond the above set-up. Namely, we do not specify the bulk action in terms of specific constants. Instead, we allow non-standard kinetic energy term to avoid the no-go theorem. In our discussion we will distinguish Λ's, depending where it originates, the barred ones and the rest:

$$\overline{\Lambda} = \text{obtained from } g_{\mu\nu}$$
$$\Lambda = \text{obtained from } T_{\mu\nu}.$$

In this spirit, there exists one self-tuning model by Kim, Kyae and Lee (KKL) [17]. The KKL model is worked out in the Randal-Sundrum II type model [18], with a nonstandard kinetic energy term of the antisymmetric field strength H_{MNPQ}: $\sim 1/H^2$ [17],

$$-I_E = \int d^5x_E\sqrt{g_{(5)}}\left(\tfrac{1}{2}R_{(5)} - \tfrac{2\cdot 4!}{H^2} - \Lambda_b - \Lambda_1\delta(y)\right) = \int dy\int d^4x_E$$
$$\left\{-\Psi^4\Lambda_1\delta(y) + \tfrac{1}{2}R\Psi^2 + 4\Psi^3\Psi'' + 6\Psi^2(\Psi')^2 + \tfrac{2\cdot 4!\Psi^4}{H^2} - \Psi^4\Lambda_b\right\} \tag{6}$$

where the metric is taken as $ds^2 = \beta^2(y)\eta_{\mu\nu}dx^\mu dx^\nu + dy^2$ with the signature $\eta_{\mu\nu} =$ diag.$(-1,+1,+1,+1)$. The kinetic energy term with H^2 is not developing a VEV in the low energy theory, i.e. in the long wavelength limit $(\partial_\mu A^{\nu\rho\sigma}) \to 0$. Fortunately, however, we allow the bare CC in the bulk. So, even with $\langle H^2\rangle = 0$, H^2 can be moved to the denominator,

$$\frac{1}{H^2}, \quad \text{with } \langle H^2\rangle \neq 0.$$

The field equation and the Bianchi identity are satisfied with

$$\partial_M\left(\sqrt{g}\frac{H^{MNPQ}}{H^4}\right) = 0, \quad \epsilon^{RMNPQ}[\sqrt{g_{(5)}}H_{MNPQ}] = 0\ . \tag{7}$$

With this Lagrangian, there exists a self-tuning solution

$$\beta(y) = \left(\frac{k}{a}\right)^{1/4}\frac{1}{(4k|y|+c_0)^{1/4}}\ . \tag{8}$$

But there are nearby dS and AdS solutions also [17]. It is easy to show the existence of the nearby dS and AdS solutions by trying a small $c(y)$ in $Y \equiv \beta(y)^4 = A[\text{sech}(ky+c_0)+c(y)]$. The equation satisfied by Y is

$$-\frac{1}{4}Y'' = 3\overline{\Lambda}\sqrt{Y} + \frac{2m^2k^2}{3}Y - \frac{8}{3h}Y^3 - \frac{\Lambda_1}{3}\delta(y)Y \tag{9}$$

from which one can fix for $\Lambda_b = -m^2k^2$,

$$m^2 = \frac{3}{8}, \quad \frac{16A^2}{3h}k^2\ . \tag{10}$$

So, the question is, "How does one choose the flat one?"

4. The initial state after inflation

4.1. The wave function of the universe

One way of doing quantum cosmology is to solve the Wheeler-DeWit equation [19] with an appropriate boundary condition. The obtained wave function of the Universe is independent of time. Following the terminology of Nicoli [1], it is a videotape containing all the information of the universe as cartooned in Fig. 4. There are many videotapes satisfying the Wheeler-DeWit equation. The probability to obtain a certain videotape is given by the wave function of the universe. If one obtains a videotape, he can run it in a film motion-picture projector with a certain definition of time to see how the videotape, containing all the fundamental constants of physics, evolves. Different tapes contain different fundamental constants. Of course, the CC is contained in the videotape. In this running of the videotape, he is in the classical regime where the classical Einstein equation is enough to run the film.

After Hawking proposed calculating the probability amplitude based on the Wheeler-DeWit equation [19] with the Hartle-Hawking (HH) boundary condition [20], Vilenkin suggested the outgoing wave boundary condition [21], as shown in Fig. 5. The HH wave function is real and the Vilenkin wave function is an outgoing wave whose real and impaginary parts are shown schematically in Fig. 5. This is well summarized in Page 460 of Ref. [2]. The customary Wick rotation direction is for $E > 0$ while in the Wheeler-DeWit equation the gravitational energy is negative and hence the Wick rotation has to be to the opposite direction [21], and the probability prediction becomes completely opposite to that of Hawking.

But here we follow Hawking's convention and his interpretation for the moment. Then, our result can be appropriately reinterpreted, following [21]. But our boundary condition is different from the HH or Vilenkin boundary conditions.

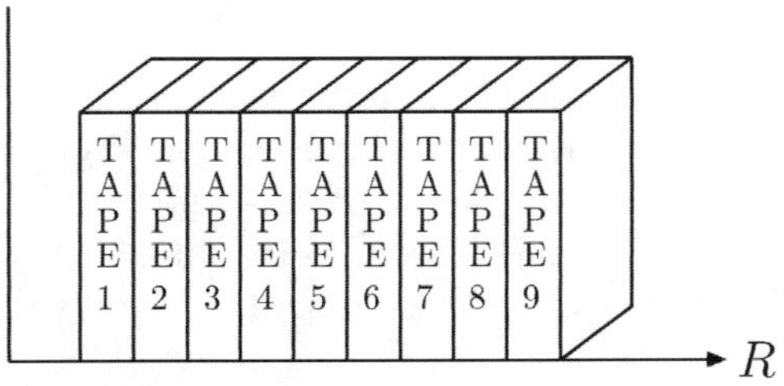

Figure 4. A stack of tapes in the video room. The picked tape is run with the classical Einstein equation. Different tapes contain different fundamental constants.

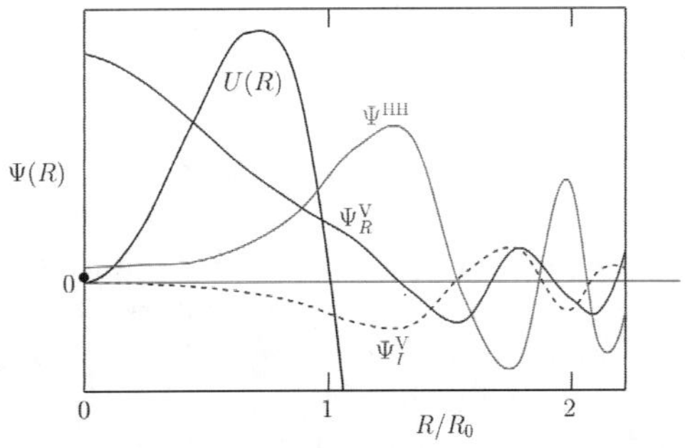

Figure 5. The universe wave functions based on the HH boundary condition and the Vilenkin boundary condition [2]. The shape of a potential $U(R)$ is also shown. The HH wave function is real while the Vilenkin wave function takes the outgoing wave only whose real and imaginary parts are shown here.

4.2. The initial state in the brane scenario

This idea of quantum cosmology must be recast with extra dimensions and branes. The KKL solution has a remarkable property as noted in [22] that the vanishing CC solution is not allowed for the parameter range of

$$|\Lambda_1| \geq \sqrt{-6\Lambda_b} \tag{11}$$

since the boundary condition at the brane $(\beta'/\beta)_{y=0^+} = -\Lambda_1/6$ cannot be satisfied. This situation is depicted in Fig. 6.

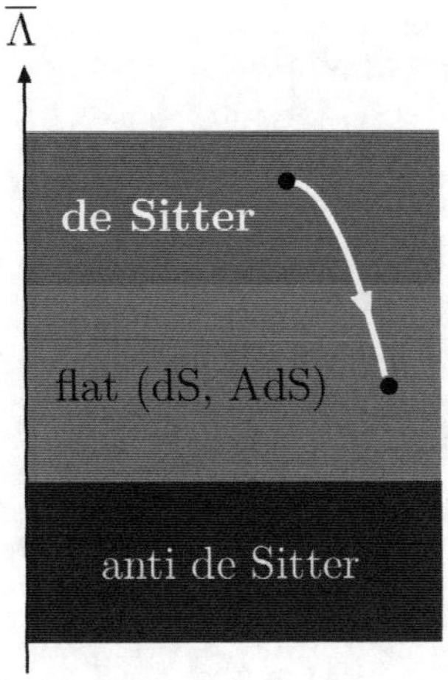

Figure 6. The initial inflation takes place for $|\Lambda_1| > \sqrt{-6\Lambda_b}$.

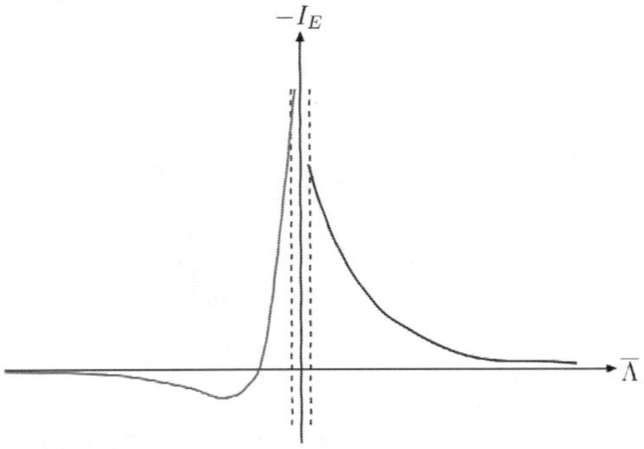

Figure 7. A schematic behavior of the probability amplitude in self-tuning models.

Therefore, we argue that for a finite range of parameter of Λ_1 satisfying (11) the inflation period continues. The particle physics action at the brane may change Λ_1 such that it does not satisfy (11); in the central region of Fig. 6 all the possibilities are open, the flat, dS and AdS solutions. Then the inflationary period might end, but the important question is what is the probability to choose the flat solution. Here comes the probability calculation. In the KKL model, we show that the probability shape is given as the RHS of Fig. 8.

The number of e-foldings during inflation is about 60, or the scale factor increases by a factor 10^{26}. So, the brane Lagrangian for inflation is tuned to satisfy this condition. Then, the initial temperature T_i drops to $10^{-26}T_i$ before the inflaton reheats up the universe. At this supercooled state before reheating, we define the initial state of the universe for the probability calculation. We require that this initial state is fuzzy enough to allow $(0.002 \text{ eV})^4$ vacuum energy, even if the probability calculation of Fig. 7 dictates the vacuum energy should be zero. Then, we require $10^{-26}T_i$ be greater than 0.002 eV, namely T_i should be greater than 2×10^{14} GeV. For example, the height of the quintessential axion of Ref. [23] can be of order $> (0.002 \text{ eV})^4$ but the fuzziness of our CC solution is of order $(0.002 \text{ eV})^4$, which is depicted in Fig. 8. So, with any inflation model where the beginning of inflation starts below 2×10^{14} GeV, the fuzziness too small and we cannot satisfy the observed acceleration [24] within our scheme of the CC solution.

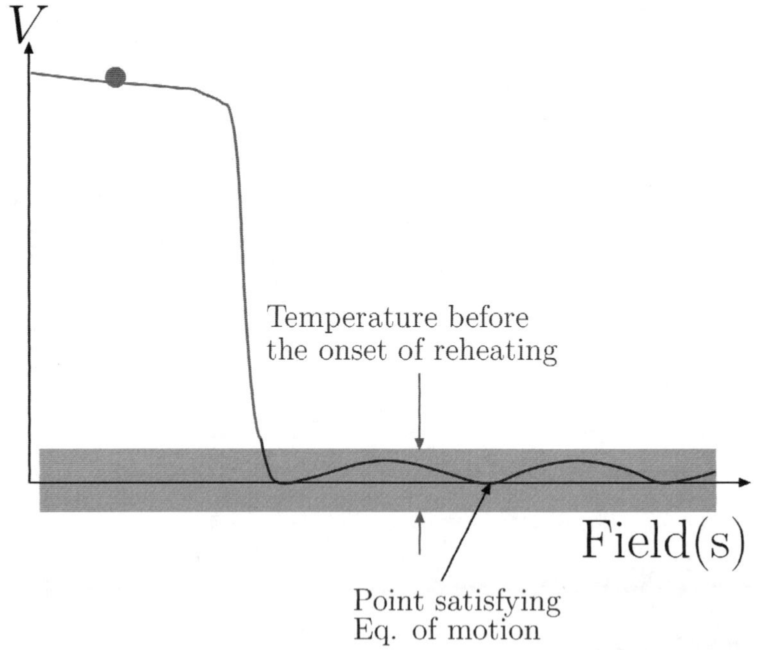

Figure 8. The initial inflation and the recent acceleration. The CC solution is swamped in the lavender band and the wave function of the universe is not exactly choosing the equilibrium point.

We take our initial condition at the time immediately after the inflation. We picked up the matter energy being in a certain region after the inflation. So, our boundary condition is a matter energy eigenstate, amounting to a kind of quantum mechanical filtering process.

5. Calculation of probability amplitude in the KKL model

Hawking calculated the probability amplitude from

$$\langle\overline{\Lambda}|I\rangle \propto \int d[g]e^{-I_E[g]} \tag{12}$$

and concluded that the volume integral is the dominant factor and concluded that the probability is the largest for $\overline{\Lambda} = 0^+$ [11]. In Hawking's case, it is not clear how the primordial inflation is taken into account.[1] In our case, the primordial inflation is achieved in the brane dynamics as explained in the previous section. Next, the initial state is given when Λ_1 rolls into the flat space region of Fig. 6. This initial state that Λ_1 sits in the flat space region is the quantum mechanical filtering process. So, to measure some state out of the state $|I\rangle$, we calculate the amplitude (12).

If particle physics Lagrangian does not contribute to (12), it is sufficient to consider the CC term and the R term only as Hawking has done. Hawking's basic argument was the size of the Euclidian volume. The dS space volume is finite, the flat space volume is infinite, and the AdS space volume is even more infinite. If we consider only the $1/\overline{\Lambda}$ term, the AdS wins in magnitude but the sign is opposite from the dS; thus the flat space is chosen. So, even if the AdS is considered, the flat space wins if we restricts only to $1/\overline{\Lambda}$ term. However, as in the self-tuning model of KKL, particle physics Lagrangian contributes to the final CC in general, and the particle physics action has the $1/\overline{\Lambda}^2$ behavior which dominates the CC term contribution in the small $\overline{\Lambda}$ region. This may change Hawking's view completely.

If we consider the sizes of volumes, the AdS wins over the flat even though both are infinite. For the flat volume, we take the $\overline{\Lambda} = 0$ limit of the dS case. For the AdS volume, we need to regularize the infinity to compare different cases of $\overline{\Lambda}$'s.

Here, I discuss what has been discussed on the H^2 Lagrangian in the Hawking type calculation, present a (probably) correct way to calculate the amplitude, and show that the KKL self-tuning model allows a finite range of parameters for which $\overline{\Lambda} = 0^-$ has the singularly large probability [25].

5.1. Hawking, Duff and Wu

Hawking presented the first calculation, using H^2 term in the Lagrangian [11]. Since then, there has been discussions on which value of the H^2 term must be used in the action integral. In this regard, a surface term $\epsilon^{\mu\nu\rho\sigma}H_{\mu\nu\rho\sigma}$ has been noted, which does not change the equation of motion. It turned out that it amounts to changing the sign of H^2 term in the action integral [26], from which Duff concluded that the probability amplitude for $\overline{\Lambda} = 0$ is least probable with $H_{\mu\nu\rho\sigma}$. Recently, Wu [27] obtained the opposite result from that of Duff. In our case also, we can consider a surface term (due to x-independent Ψ^8/H^2), following the first order formalism of [28],

$$-I_E \supset \int dy \int d^4x_E \left\{ \rho \frac{2\Psi^8}{H^2} \epsilon^{\mu\nu\rho\sigma} H_{\mu\nu\rho\sigma} - \frac{\Psi^4}{2}\rho^2 \right\}$$

which does not affect the equation of motion. If we follow Duff's method, it has the effect of changing the sign of $1/H^2$ term inside the action integral with the surface term neglected, from $\int d^5x_E\sqrt{g_{(5)}}(2\cdot 4!\Psi^4/H^2)$ to $\int d^5x_E\sqrt{g_{(5)}}(-2\cdot 4!\Psi^4/H^2)$. So, both methods satisfy the equations of motion with the action integral,

$$-I_E = \int d^5x_E \qquad \sqrt{g_{(5)}}\left[\tfrac{1}{2}R_{(5)} \underset{\uparrow\,\text{Wu}}{\overset{\text{Duff}\,\downarrow}{\pm}} \tfrac{2\cdot 4!}{H^2} - \Lambda_b - \Lambda_1\delta(y)\right] \tag{13}$$

[1] Ref. [21] tries to interpret the inflation due to the large probability amplitude for a large CC in case of the opposite sign of the Wick rotation, which is completely different from our view here.

Thus, it raises an important question, "Which method is correct?"

5.2. The α-vacuum

To our view, the confusion arises from taking a specific vacuum in their calculation [26, 27]. As in the θ-vacuum of QCD, we have the α-vacuum of antisymmetric tensor field $H_{\mu\nu\rho\sigma}$. Duff took one extremum point corresponding to $\alpha = \pi$ and Wu took another vacuum corresponding to $\alpha = 0$, and they obtained different results even though both satisfied equations of motion.

As far as the α-vacuum is concerned, the discussion is parallel whether we use H^2 or $1/H^2$ in the Lagrangian. So, for the notational brevity, we discuss α-vacuum with the H^2 kinetic energy term.

For two antisymmetric indices from μ, ν, ρ, and σ, there are six (${}_4C_2 = 6$) independent second rank antisymmetric gauge functions, for which $A_{\mu\nu\rho}$ transforms as

$$A_{\mu\nu\rho} \to A_{\mu\nu\rho} - \partial_\mu \Lambda_{\nu\rho} - \partial_\nu \Lambda_{\rho\mu} - \partial_\rho \Lambda_{\mu\nu} \; . \tag{14}$$

The gauge symmetry of the instanton solution is given by any six directions of $\Lambda_{\mu\nu}$, three for the instanton (Λ_{ij}) and three for the anti-instanton (Λ_{0i}) and the instanton action is $\int d^5x \partial_y \epsilon^{\mu\nu\rho\sigma} H_{\mu\nu\rho\sigma}$. Namely, there exist maps of

$$S_3 \to S_3. \tag{15}$$

There exists only one type of solution, i.e. only one Pontryagin index. For $\int d^4x H_{\mu\nu\rho\sigma}$ to be finite, $H_{\mu\nu\rho\sigma}$ should tend to r^{-4} for a large r. In the bulk, it arises from a 2D curl in 5D (time and the internal space y),

$$\begin{aligned} \int dy dx^0 \; \vec{\nabla} \times \vec{A} &= \int d\vec{s} \cdot \vec{A} = \int dy dx^0 [\partial_y \mathrm{A}_0 - \partial_0 \mathrm{A}_y] \\ &= \int dy dx^0 \partial_y \mathrm{A}_0 = \int dx^0 \mathrm{A}_0 = \int d^4x H_{\mu\nu\rho\sigma} \end{aligned} \tag{16}$$

where

$$\begin{aligned} \mathrm{A}_0 &= \int d^3\mathbf{x} \; \partial_0 A_{ijk} = \int d^3\mathbf{x} \; H_{\mu\nu\rho\sigma} \\ \mathrm{A}_y &= \int d^3\mathbf{x} \; \partial_y A_{ijk} = \int d^3\mathbf{x} \; H_{yijk} \; . \end{aligned} \tag{17}$$

So a gauge invariant instanton of size ρ located at x_0 takes the form

$$A_{\alpha\beta\gamma} \propto \frac{\epsilon_{\alpha\beta\gamma\mu}(x - x_0)^\mu}{(r^2 + \rho^2)^2}, \quad r = |x - x_0| \tag{18}$$

so that $A_{\mu\nu\rho}$ is proportional to r^{-3}, and $H_{\mu\nu\rho\sigma}$is proportional to r^{-4}. The 4D integral of $H_{\mu\nu\rho\sigma}$ is represented by a kind of Pontryagin integer $n = \pm 1$. Note, on the other hand, that the instanton field of nonabelian gauge groups is of pure gauge form, so that the instanton action is $\int d^4x \mathrm{Tr} F\tilde{F}$. In nonabelian gauge theories, there are many possible gauge configurations such that the irreducible instanton solution give many possible integers for the Pontryagin index. On the other hand, in our case at hand $H_{\mu\nu\rho\sigma}$ instanton gives only ± 1 for the Pontryagin index.

Now, we construct a gauge invariant α-vacuum, following the θ-vacuum construction of QCD,

$$|\alpha\rangle = \sum_{n=-\infty}^{+\infty} |n\rangle e^{in\alpha} \tag{19}$$

In the α-vacuum, after integrating out the H^2 field, what Duff chose is $\alpha = \pi$ and what Wu chose is $\alpha = 0$. However, in the α vacuum, any value of α is allowed, i.e. not restricted to $\alpha = 0$

and π. As in the θ-vacuum of QCD, any value of α is allowed in our case, and we go beyond what Duff and Wu considered.

As commented above, this α-vacuum can be defined also with the $1/H^2$ term. We calculate the action integral for $\alpha = 0$ and π with the $1/H^2$ term and for any α the action integral is between them. If one makes α a dynamical field as the QCD axion [29], then α is cosmologically settled to 0.

5.3. Parameters for the $\overline{\Lambda} = 0$ dominance

Now, we calculate the probability amplitude in the KKL model. For this, we use two Einstein equations of the bulk,

$$(\mu\nu): \quad -3\frac{\overline{\Lambda}}{\beta^2} + 3\left(\frac{\beta'}{\beta}\right)^2 + 3\frac{\beta''}{\beta} = -\Lambda_b - 2\cdot 4!\left(\frac{3}{H^2}\right) \tag{20}$$

$$(55): \quad -\frac{6\overline{\Lambda}}{\beta^2} + 6\left(\frac{\beta'}{\beta}\right)^2 = -\Lambda_b - 2\cdot 4!\left(\frac{1}{H^2}\right) \tag{21}$$

We integrate out the 4D space x and the 5th space y. In this calculation the brane tension Λ_1 is also considered. For the coefficient of $1/\overline{\Lambda}^2$ to be positive, the following condition on the parameters is required,

$$\tanh(c_0)\mathrm{sech}^2(c_0) \le \frac{k}{3}F(c_0/k, d_m) \tag{22}$$

where $F(c_0/k, d_m)$ is the result of the integration. Here, d_m is the length scale defined from the parameters of [25]. If this condition is satisfied, the action integral $-I_E$ has the behavior shown in Fig. 7, and the vanishing CC is approached from the AdS side.

In the gauge invariant α-vacuum, for the c_0 independent part with the $1/H^2$ term we obtain $(3/8k)(\pi/2)$ and $(9/2k)(\pi/2)$ for $\alpha = 0$ and π, respectively. Therefore, it seems that for the parameters satisfying Eq. (22) we obtain the $\overline{\Lambda} = 0$ dominance in the probability amplitude.

5.4. The AdS volume

Finally, we comment on our method of comparing the infinite volumes of the AdS spaces. The n-dimensional Euclidian space metric is given by

$$ds^2 = a^2\left(\frac{dr^2}{1-kr^2} + r^2 d\Omega_{n-1}^2\right) = a^2 f^2(\eta)\left(d\eta^2 + \eta^2 d\Omega_{n-1}^2\right) \tag{23}$$

where $k = 0, \pm 1$ and in the second equation the Weyl transformation is used since it is simple because of the vanishing Weyl tensor in this space,

$$\frac{dr}{\sqrt{1-kr^2}} = f(\eta)d\eta, \quad f(\eta) = \frac{r}{\eta}\ . \tag{24}$$

Now, let us specify to the AdS space of $k = -1$. Then, Eq. (24) is integrated to give $\ln \eta = -\sinh^{-1}(1/r)$, or

$$f(\eta) = \frac{r}{\eta} = \frac{2}{1-\eta^2}. \tag{25}$$

The Ricci scalar for the metric $g'_{\mu\nu} = a^2 f^2(\eta) g_{\mu\nu}$ is given by

$$R' = a^{-2}f^{-2}\left(R - 2(n-1)\nabla^2(\ln f) - (n-1)(n-2)\left(\frac{f'}{f}\right)^2\right)\ . \tag{26}$$

Noting that $R'_{\mu\nu} = \overline{\Lambda} g'_{\mu\nu}$ or $R' = n\overline{\Lambda}$, we have $\overline{\Lambda} = -(n-1)/a^2$ in the n-dimensional Euclidian AdS. Using Eq. (24), the n-dimensional Euclidian AdS volume with the metric (23) is regularized to

$$V_{\mathrm{AdS}^n} = a^n \int d^n x \left(\frac{2}{1-\eta^2}\right)^n = a^n \int_0^1 d\eta\, \eta^{n-1} V_{S^{n-1}} \left(\frac{2}{1-\eta^2}\right)^n = (2a)^n V_{S^{n-1}} \tag{27}$$

$$\cdot \frac{1}{2}\int_0^1 d\xi\, \xi^{n/2-1}(1-\xi)^{-n} = \frac{1}{2}(2a)^n V_{S^{n-1}} B(n/2, 1-n) \tag{28}$$

$$= \frac{1}{2}(2a)^n \frac{2\pi^{n/2}}{\Gamma(n/2)} \frac{\Gamma(n/2)\Gamma(1-n)}{\Gamma(1-n/2)} = \left(\frac{4(n-1)\pi}{|\overline{\Lambda}|}\right)^{n/2} \frac{\Gamma(1-n)}{\Gamma(1-n/2)}\,. \tag{29}$$

Eq. (29) factored out the diverging Gamma functions, and we can compare the $\overline{\Lambda}$ dependences. For $n = 4$, it diverges as $1/|\overline{\Lambda}|^2$ as $\overline{\Lambda}$ tends to zero.

6. Conclusion

In conclusion, we observed that (*i*) the CC problem may be understandable in higher dimensions $D > 4$, (*ii*) three CC problems should be addressed, and (*iii*) the initial state of the Universe should be defined properly after the initial inflation. A brane helps in solving the vanishing CC problem, since the loop effects of brane is not important to bulk physics. However, this idea is applicable only when there exists a self-tuning solution such as the one given in [17].

We noted that the action integral for a probability calculation is dominated from the particle physics Lagrangian, and has the amplitude proportional to $\exp[\#/\overline{\Lambda}^2]$. Near $\overline{\Lambda} = 0$, the AdS space can be made to be preferred. But slightly outside $\overline{\Lambda} = 0$, the dS space is preferred. With the parameters satisfying Eq. (22), $\overline{\Lambda} = 0$ is preferred with the Hawking's direction of the Wick rotation, which may be thought not the correct Wick rotation as commented in [21]. However, our maximum probability occurs from the negative energy side, which means that the gravitational energy is positive and our choice can be the correct direction of the Wick rotation. So, we must choose parameters such that Eq. (22) is satisfied.[2]

Also, we noted that the current acceleration should be addressed, which has not been discussed here. For this, the quintessential-axion idea may be useful [23], and we need the onset of inflation above 2×10^{14} GeV which is quite reasonable and hints a GUT inflation.

Our specific example presented here for the probability amplitude calculation uses the three index gauge field A_{MNP} in the KKL model [17]. Here, one can consider an α-vacuum. Then, α becomes a parameter in the model and any value of α between 0 and π are permitted. It is like the θ parameter of QCD. We have shown that for any value of α, there exists a finite range of parameters such that $\overline{\Lambda} = 0^-$ is chosen.

If α is made dynamical as the QCD axion, then the probability amplitude choosing $\overline{\Lambda} = 0^-$ is at $\alpha = 0$. Depending on the scale of the $H_{\mu\nu\rho\sigma}$ instanton size, there may be some cosmological interests.

Acknowledgments: I have benefitted from collaborations and discussions with B. Kyae, H. M. Lee and J.-H. Huh. I thank the organizers of PASCOS 2010, especially to C. Muñoz and J. Valle for the hospitality during the conference. This work is supported in part by the Korea Research Foundation, Grant No. KRF-2005-084-C00001.

References

[1] Nicoli H. 2010 *Talk presented as Arnold Sommerfeld Lecture* at Ludwig Maximillian Univ. Munich Germany June 28

[2] Note that if we made an error in this argument, we have a freedom to choose the opposite direction for the inequality in Eq. (22) to meet our need.

[2] Kolb E. W. and Turner M. S. 1990 *The Early Universe* [Frontiers in Phys. Vol. **69**] (Redwood City CA: Addison-Wesley Publishing Co.)
[3] Einstein A. 1917 *Sitzungsberichte der Königlich Preußischen Akad. d. Wissenschaften* (Berlin) **142** 142-152
[4] Veltman M. T. 1975 *Phys. Rev. Lett.* **34** 777
[5] Weinberg S. 1988 *Phys. Rev. Lett.* **59** 2607; Agrawal V., Barr S. M. and Donoghue J. 1998 *Phys. Rev.* D **57** 5480
[6] Choi K., Kim J. E. and Nilles H. P. 1993 *Phys. Rev. Lett.* **73** 1758
[7] Shaposhnikov M. and Zenhäusern D. 2009 *Nucl. Phys.* B **671** 162 [arXiv:0809.3406]; Diakonos F. K. and Saridakis E. N. 2009 *J. Cos. Astropart. Phys.* JCAP02(2009)030 [arXiv:0708.3143]
[8] Kim J. E. 2009 *Phys. Lett.* B **678** 197 [arXiv: 0904.1015]; Huh J.-H., Kim J. E. and Kyae B. 2009 *Phys. Rev.* D **80** 115012 [arXiv: 0904.1108]
[9] Dine M. and Nelson A. E. 1993 *Phys. Rev.* D **48** 1277
[10] Baum E. 1983 *Phys. Lett.* B **133** 185
[11] Hawking S. 1984 *Phys. Lett.* B **134** 403
[12] Giddings S. B. and Strominger A. 1988 *Nucl. Phys.* B **306** 890
[13] Coleman S. R. 1988 *Nucl. Phys.* B **310** 643
[14] Arkani-Hamed N., Dimopoulos S., Kaloper N. and Sundrum R. 2000 *Phys. Lett.* B **480** 193 [hep-th/0001197]; Kachru S., Schulz M. B. and Silverstein E. 2000 *Phys. Rev.* D **62** 045021 [hep-th/0001206]
[15] Forste S., Lalak Z., Lavignac S. and Nilles H. P. 2000 *Phys. Lett.* B **481** 360 [hep-th/0002164]
[16] Csaki C., Erlich J., Grojean C. and Hollowood T. 2000 *Nucl. Phys.* B **584** 359 [hep-th/0004133]
[17] Kim J. E., Kyae B. and Lee H. M. 2000 *Phys. Rev. Lett.* **86** 4223 [hep-th/0011118]; Kim J. E., Kyae B. and Lee H. M. 2001 *Nucl. Phys.* B **613** 306 [hep-th/0101027]
[18] Randall L. and Sundrum R. 1999 *Phys. Rev. Lett.* **83** 4690 [hep-th/9906064]
[19] DeWit B. S. 1967 *Phys. Rev.***160** 1113; Wheeler J. A. 1968 *Batelle Rencontres*, ed DeWit C. and Wheeler J. A. (New York: Benjamin)
[20] Hartle J. B. and Hawking S. W. 1983 *Phys. Rev.* D **28** 2960
[21] Vilenkin A. 1984 *Phys. Rev.* D **30** 509 and 1988 *Phys. Rev.* D **37** 888; Linde A. D. 1984 *JETP* **60** 211; Rubakov V. A. 1984 *Phys. Lett.* B **148** 280
[22] Kim J. E. 2003 *J. High Energy Phys.* JHEP01(2003)042 [hep-th/0210117]
[23] Kim J. E. and Nilles H. P. 2003 *Phys. Lett.* B **553** 1 [hep-ph/0210402]; Kim J. E. and Nilles H. P. 2009 *J. Cos. Astropart. Phys.* JCAP05(2009)010 [arXiv:0902.3610 [hep-th]]
[24] Perlmutter S. *et al.* (Supernova Cosmology Project) 1999 *Astrophys. J.* **517** 565; Riess A. G. *et al.* (Supernova Search Team) 1998 *Astron. J.* **116** 1009
[25] Kim J. E. 2010 *Phys. Rev.* D **81** 123018 [arXiv:0912.2733]
[26] Duff M. J. 1989 *Phys. Lett.* B **226** 36
[27] Wu Z. C. 2008 *Phys. Lett.* B **659** 891
[28] Aurilia A., Nicolai H. and Townsend P. K. 1980 *Nucl. Phys.* B **176** 509
[29] Callan C. G., Dashen R. F. and Gross D. J. 1976 *Phys. Lett.* B **63** 334; Jackiw R. and Rebbi C. 1976 *Phys. Rev. Lett.* **37** 172; For a review on axions, see, Kim J. E. and Carosi G. 2010 *Rev. Mod. Phys.* **82** 557

Journal of Physics: Conference Series **259** (2010) 012006 doi:10.1088/1742-6596/259/1/012006

Mixed Mediation of Supersymmetry Breaking in Models with Anomalous $U(1)$ Gauge Symmetry

Kiwoon Choi

Department of Physics, Korea Advanced Institute of Science and Technology, Daejeon 305-701, Korea

E-mail: kchoi@kaist.ac.kr

Abstract. There can be various built-in sources of supersymmetry breaking in models with anomalous $U(1)$ gauge symmetry, e.g. the $U(1)$ D-term, the F-components of the modulus superfield required for the Green-Schwarz anomaly cancellation mechanism and the chiral matter superfields required to cancel the Fayet-Iliopoulos term, and finally the supergravity auxiliary component which can be parameterized by the F-component of chiral compensator. The relative strength between these supersymmetry breaking sources depends crucially on the characteristics of D-flat direction and also on how the D-flat direction is stabilized at a vacuum with nearly vanishing cosmological constant. We examine the possible pattern of the mediation of supersymmetry breaking in models with anomalous $U(1)$ gauge symmetry, and find that various different mixed mediation scenarios can be realized, including the mirage mediation which corresponds to a mixed modulus-anomaly mediation, D-term domination giving a split sparticle spectrum, and also a mixed gauge-D-term mediation scenario.

1. Introduction

Anomalous $U(1)$ gauge symmetry, which will be referred to $U(1)_A$ in the following, appears often in 4-dimensional (4D) effective theory of string compactification [1]. It accompanies a modulus T which transforms nonlinearly under $U(1)_A$, and thereby breaks $U(1)_A$ through the Stückelberg mechanism at scales not far below the cutoff scale of 4D effective theory. The holomorphic gauge kinetic function of the model depends on T as $f_a \ni k_a T$ with k_a being a (quantized) real constant, and then the anomaly of the model is cancelled by the Green-Schwarz (GS) mechanism [2]. The nonlinear transformation of T under $U(1)_A$ induces also a modulus-dependent Fayet-Iliopoulos (FI) term [1], which might have an interesting implication to supersymmetry (SUSY) breaking. If SUSY is not broken at the scale of $U(1)_A$ breaking, this FI term should be cancelled by the D-term contribution from $U(1)_A$-charged matter fields X. Such matter fields also contribute to the $U(1)_A$ breaking, giving a contribution of the order of the FI term to the $U(1)_A$ gauge boson mass-square. In most cases, string models with $U(1)_A$ contain exotic matter fields $\Phi + \Phi^c$ which are vector-like under the standard model gauge group, while chiral under $U(1)_A$, and these exotic matter fields become massive through the Yukawa coupling to X.

Therefore, models with anomalous $U(1)$ gauge symmetry involve various built-in sources of SUSY breaking which might affect the soft terms of visible sector fields: the $U(1)_A$ D-term, F-components of the GS modulus T and the chiral matters X, and the supergravity (SUGRA) auxiliary component. Then there can be modulus mediation through the couplings between T and the visible sector gauge and matter superfields [3, 4], D-term contribution to scalar masses

[5, 6, 7, 8], gauge mediation through the loops of the exotic matter fields $\Phi + \Phi^c$ which get a SUSY breaking mass from F^X and/or D_A [9], and finally anomaly mediation through the SUGRA auxiliary component [10, 11].

Here we discuss the possible mediation pattern of SUSY breaking in models with anomalous $U(1)$ gauge symmetry [12, 13]. As we will see, the relative strength between the built-in mediation schemes crucially depends on the characteristics of the D-flat direction which corresponds to a combination of T and $\ln X$. The coefficient of D-flat combination is determined by the ratio R between the two contributions to the $U(1)_A$ gauge boson mass-square, the Stückelberg contribution associated with the non-linear $U(1)_A$ transformation of T and the conventional Higgs contribution from the vacuum values of X. Then various mixed mediation scenarios can be realized with different values of R, including the mirage mediation which corresponds to a mixed modulus-anomaly mediation [14, 15, 16, 17], D-term domination giving a split sparticle spectrum, and also a mixed gauge-D-term mediation scenario. The value of R is somewhat sensitive to the detailed form of the Kähler potential and superpotential of the model, and is determined mostly by how the D-flat direction is stabilized at a vacuum with nearly vanishing cosmological constant.

2. Models with anomalous $U(1)$

In models with anomalous $U(1)$ gauge symmetry, the quantum consistency of the model is ensured through the GS mechanism of anomaly cancellation [2]. This mechanism is implemented by a non-linear variation of the GS modulus,

$$T \to T - \frac{\delta_{\rm GS}}{2}\Lambda_A \tag{1}$$

under the $U(1)_A$ gauge transformation $V_A \to V_A - \frac{1}{2}(\Lambda_A + \Lambda_A^*)$ where V_A is the $U(1)_A$ vector superfield. In order for the anomaly cancellation to occur, holomorphic gauge kinetic function of the model should contain a T-dependent piece, $f_a \ni k_a T$ with k_a being a real constant. Under the normalization of T for which $k_a = \mathcal{O}(1)$, we have

$$\frac{\delta_{\rm GS}}{2} = \mathcal{O}\left(\frac{1}{8\pi^2}\right) \tag{2}$$

as required to cancel the anomalies due to fermion loops. Since the $U(1)_A$ gauge invariance forces the modulus Kähler potential K_0 to be a function of $t_A = T + T^* - \delta_{\rm GS} V_A$, the GS mechanism dynamically induces a FI term:[1]

$$\xi_{\rm FI} = \frac{\delta_{\rm GS}}{2} K_0', \tag{3}$$

as well as providing the Stückelberg contribution to the $U(1)_A$ vector boson mass-square:

$$\Delta M_V^2 \equiv 2g_A^2 M_{\rm GS}^2 = \frac{\delta_{\rm GS}^2}{2} g_A^2 K_0'', \tag{4}$$

where g_A is the $U(1)_A$ gauge coupling, and the prime denotes the derivative with respect to $t = T + T^*$.

To preserve SUSY at the scale where $U(1)_A$ is broken, we need $U(1)_A$-charged matter fields whose contribution to D-term cancels the FI term. Here we consider a minimal model with

[1] Unless specified, we use the convention $M_{Pl} = 1$.

single such chiral superfield X. Under the assumption that the cutoff scale Λ of 4D theory is close to M_{Pl}, the Kähler potential can be expanded as

$$K = K_0(t_A) + Z_X(t_A)e^{-2V_A}|X|^2 + \mathcal{O}(|X|^4), \tag{5}$$

where the $U(1)_A$ charge of X is normalized as $q_X = -1$. Then the $U(1)_A$ D-term is given by

$$D_A = \xi_{\rm FI} + \mathcal{Z}_X|X|^2, \tag{6}$$

where $\mathcal{Z}_X = Z_X(1 + \frac{\delta_{\rm GS}}{2}\partial_t \ln Z_X)$. Along the D-flat direction, X is triggered to acquire a vacuum expectation value as $\mathcal{Z}_X|X|^2 \simeq -\xi_{\rm FI}$, and thus participates in the breaking of $U(1)_A$. Consequently, one linear combination of T and $\ln X$ becomes the longitudinal component of V_A, providing a mass

$$M_V^2 = 2g_A^2\left(M_{\rm GS}^2 + \left(1 + \frac{\delta_{\rm GS}}{2}\partial_t \ln \mathcal{Z}_X\right)\mathcal{Z}_X|X|^2\right) \tag{7}$$

to V_A. It is thus the ratio between ξ_{FI} and M_{GS}^2,

$$R \equiv -\frac{\xi_{\rm FI}}{M_{\rm GS}^2} = -\frac{2}{\delta_{\rm GS}}\frac{K_0'}{K_0''}, \tag{8}$$

that determines which combination forms the longitudinal component of V_A. If the vacuum lies in the region with $R \gg 1$, the longitudinal component mostly comes from X. In the opposite limit with $R \ll 1$, it is mostly the GS modulus T. (Note that by definition $R \geq 0$.) It is then expected that the pattern of SUSY breaking crucially depends on which field is left at low energy scales after the breaking of $U(1)_A$, and thus on the value of R. It should be noted that the FI term is a monotonic function of $t = T + T^*$ because K_0'' is positive definite as required for sensible kinetic terms. This implies that the FI term can vanish only at a single value of t. However, regardless of the value of $\xi_{\rm FI}$, the GS mechanism always makes the $U(1)_A$ vector superfield massive through the Stückelberg mechanism.

2.1. *Supersymmetry breaking*

The chiral superfields involved in the $U(1)_A$ breaking should be stabilized eventually at a SUSY breaking minimum, and then their F-terms and the $U(1)_A$ D-term are expected to obtain nonzero vacuum expectation values. When combined with the equation of motion $D_A = -\eta^I\partial_I K$, the $U(1)_A$ gauge invariance of the superpotential W leads to the following relation:

$$m_{3/2}D_A = K_{I\bar{J}}\eta^I F^{J*}, \tag{9}$$

where $F^I = -e^{K/2}K^{I\bar{J}}(\partial_J W + W\partial_J K)^*$, $m_{3/2} = e^{K/2}W$, and η^I denotes the holomorphic Killing vector generating an infinitesimal $U(1)_A$ transformation, specifically $\eta^T = -\delta_{\rm GS}/2$ and $\eta^X = -X$. The above relation tells us that non-vanishing D-term requires a F-term SUSY breaking. For supersymmetric minimum, however, additional positive contribution to the scalar potential is required to achieve a viable de-Sitter or Minkowski vacuum. In the KKLT moduli stabilization scenario [18], such uplifting is accomplished by introducing an anti-brane stabilized at the tip of throat, yielding an uplifting potential sequestered from the bulk Calabi-Yau space including the visible sector branes. Here we assume that the uplifting sector is sequestered from the GS modulus as well as from the visible sector as in the KKLT compactification, and thus the uplifting potential takes the form [15]

$$V_{\rm lift} = e^{2K/3}\mathcal{P}, \tag{10}$$

where $\mathcal{P}$ is a positive constant which should be adjusted to make the cosmological constant to be nearly vanishing. Combining (9) with the minimization condition $\partial_I(V_F + V_D + V_{\rm lift}) = 0$, one can derive the relation [12]

$$\left(V_F + \frac{2}{3}V_{\rm lift} + 2|m_{3/2}|^2 + \frac{1}{2}M_V^2\right) D_A = -F^I F^{J*}\partial_I\partial_{\bar{J}}(\eta^L\partial_L K) + V_D\eta^I\partial_I g_A^2, \tag{11}$$

where V_F and V_D are the F-term and D-term scalar potentials in the Einstein frame, i.e.

$$V_F = K_{I\bar{J}}F^I F^{J*} - 3e^K|W|^2, \qquad V_D = \frac{g_A^2}{2}D_A^2. \tag{12}$$

Now using the relations (9) and (11) we find [13]

$$\begin{aligned}\frac{F^X}{X} &= F^T\partial_T \ln\left(\frac{K_0'}{\mathcal{Z}_X}\right) + \mathcal{O}\left(\frac{m_{3/2}D_A}{RM_{\rm GS}^2}\right), \\ g_A^2 D_A &= -\frac{R}{1+\left(1+\frac{\delta_{\rm GS}}{2}\partial_T\ln\mathcal{Z}_X\right)R}|F^T|^2\partial_T^2\ln\left(\frac{K_0'}{\mathcal{Z}_X}\right) + \mathcal{O}\left(\frac{m_{3/2}^2|F^T|^2}{M_V^2}\right),\end{aligned} \tag{13}$$

where we have used that all F and D-terms are at most of $\mathcal{O}(m_{3/2}M_{Pl})$. We note that the above relations are insensitive to the addition of sequestered uplifting potential, though in some cases the uplifting could be crucial for constructing a SUSY breaking vacuum. One can also see that $|D_A| \leq \mathcal{O}(m_{3/2}^2 M_{Pl}^2/M_V^2)$ holds at the vacuum. For models with $M_V \gg m_{3/2}$, it is found that the D-term potential gives a negligible contribution to the vacuum energy density in comparison to the F-term potential. Nonetheless, the D-term can still be an important source of soft scalar masses.

2.2. Mediation of SUSY breaking

Models with $U(1)_A$ can provide various sources of soft terms when SUSY breaking is transmitted to the visible sector. Here are the list of mediation mechanisms that can naturally appear in models with $U(1)_A$.

- *Modulus mediation*
  Modulus mediation [3, 4], which can be considered as a particular version of gravity mediation, is a natural source of soft terms in models with $U(1)_A$. For $f_a \ni k_a T$, the F-term of the GS modulus T generates gaugino masses as

$$M_a|_{\rm MM} = \frac{k_a g_a^2(\Lambda)}{2}F^T, \tag{14}$$

  at Λ close to M_{Pl}, where the gauge couplings are given by $g_a^2(\Lambda) = 1/{\rm Re}(f_a)$. If k_a are universal as suggested by the unification of gauge couplings in the MSSM, modulus-mediated gaugino masses are also universal at $M_{\rm GUT} \simeq 2\times 10^{16}$ GeV. Similarly the T-dependence of matter wave function superfields gives rise to sfermion masses $m_{\tilde{f}}^2|_{\rm MM} = \mathcal{O}(M_a^2|_{\rm MM})$. For the case that T corresponds to the dilaton or a Kähler modulus, flavor-conserving soft terms can be naturally obtained since the T-dependence is controlled by the matter modular weight. Sparticle masses induced by modulus mediation are of order the gravitino mass or less since $|F^T| \leq \mathcal{O}(m_{3/2}M_{Pl})$.

- *Anomaly mediation*
Supergravity always mediates SUSY breaking through the conformal anomaly [10, 11]. This effect can be easily understood by introducing the chiral compensator superfield C. Since the running gauge and Yukawa couplings have a logarithmic C-dependence dictated by the super-Weyl invariance, soft terms of order $F^C/8\pi^2 C \simeq m_{3/2}/8\pi^2$ are generated by radiative corrections. As the running is mostly due to gauge interactions, anomaly-mediated soft terms naturally preserve flavor. In anomaly mediation, gaugino masses are given by

$$M_a|_{\rm AM} = \frac{b_a g_a^2(Q)}{8\pi^2}\frac{F^C}{C}, \tag{15}$$

where b_a are the one-loop beta-function coefficients, and Q is the renormalization scale. Soft scalar masses are generated at two-loop order by anomaly mediation [10], so $m_{\tilde f}^2|_{\rm AM} = \mathcal{O}(M_a^2|_{\rm AM})$.
- *Gauge mediation*
In models with $U(1)_A$, soft terms can receive a gauge-mediated contribution [9] if there exist messenger fields which obtain a heavy mass by X. For messenger fields $\Phi+\Phi^c$ which couple to X through $y_\Phi X\Phi\Phi^c$ in the superpotential, the SUSY breaking by F^X is transmitted to the MSSM sector by the loops of messenger fields. In order to preserve the gauge coupling unification, the messengers are usually assumed to form a complete GUT multiplet. The resulting gauge-mediated gaugino masses at the messenger scale $\Lambda_\Phi = y_\Phi|X|$ are given by

$$M_a|_{\rm GM} = \frac{N_\Phi g_a^2(\Lambda_\Phi)}{8\pi^2}\frac{F^X}{X}, \tag{16}$$

where N_Φ is the number of messenger pairs. Gauge mediation induces soft scalar masses also, giving $m_{\tilde f}^2|_{\rm GM} = \mathcal{O}(M_a^2|_{\rm GM})$. Because the transmission of SUSY breaking occurs through the SM gauge interactions, the induced sfermion soft terms are automatically flavor-blind. Note that, if $|D_A| \geq \mathcal{O}(|F^X/X|^2)$, the usual relation $m_{\tilde f}^2 \sim M_a^2/N_\Phi$ in gauge mediation can be changed because $\Phi\Phi^c$ carries a $U(1)_A$ charge $q_\Phi + q_{\Phi^c} = 1$, and thus the gauge-mediated soft scalar masses contain the additional contribution $\delta m_{\tilde f}^2 \simeq \sum_a C_{\tilde f}^a g_a^4 \frac{N_\Phi \ln(\Lambda/\Lambda_\Phi)}{(8\pi^2)^2} g_A^2 D_A$ with $C_{\tilde f}^a$ being the quadratic Casimir of $\tilde f$ [19].

- *D-term contribution*
In the presence of $U(1)_A$, soft scalar masses receive the D-term contribution [5]

$$m_i^2\Big|_{\rm D-term} = -q_i g_A^2 D_A, \tag{17}$$

where q_i denotes the $U(1)_A$ charge of the corresponding matter field. For $D_A = \mathcal{O}(M_a^2)$ with M_a in the TeV range, we need to arrange q_i to be flavor-independent so that the D-term contributions do not cause flavor problem. If D_A is suppressed by more than the loop factor compared to M_a^2, the flavor structure is not affected by the D-term contribution and thus the matter fields are allowed to carry flavor-dependent $U(1)_A$ charges. In this case, $U(1)_A$ may be useful for generating the observed hierarchical structure of fermion masses. Whereas, for D_A much larger than M_a^2, large flavor violations can be easily avoided as sfermions get heavy mass from D_A.

Let us now discuss the relative strength between these mediation mechanisms by applying the relations (13) between the vacuum expectation values of D and F-terms. The relative size of anomaly-mediated contribution depends on the mechanism to achieve the vanishing cosmological constant. In general, anomaly mediation becomes less important if the SUSY breaking in the $U(1)_A$ sector is enough to construct a viable de-Sitter or Minkowski vacuum. For the case that the Kähler metric of X has no dependence on T, we find the F and D terms satisfy

$$\begin{aligned} F^T &\simeq -\frac{\delta_{\rm GS}}{2}R\frac{F^X}{X}, \\ g_A^2 D_A &\simeq \frac{R}{1+R}\left(1+\frac{\delta_{\rm GS}}{2}R\frac{K_0'''}{K_0''}\right)\left|\frac{F^X}{X}\right|^2. \end{aligned} \tag{18}$$

This implies

$$M_a|_{\rm MM} \sim R\, M_a|_{\rm GM}\,, \tag{19}$$

which remains valid also for large N_Φ because anomaly cancellation fixes $k_a \sim N_\Phi$. One can thus conclude that, if the vacuum arises at $R \gg 1$, gauge mediation cannot be an important source of soft terms. In contrast, for $R \ll 1$, modulus mediation would become negligible. It is also found that, for $10^{-6} \le R \le 10^3$, the D-term contribution to soft scalar masses is non-negligible compared to those from modulus and gauge mediations. In particular, for R of order unity, the D-term contribution can dominate over other mediations.

3. Stabilization of the GS modulus

Let us now explore the connection between the value of R and the stabilization of the GS modulus T. Since the vacuum lies very close to the D-flat direction, the D-term from the vacuum expectation value of X should cancel the FI term in D_A. To understand the implication of the FI term for the modulus stabilization, it is sufficient to consider the simple case that Z_X is independent of T, so we take

$$K = K_0(t_A) + |X|^2 e^{-2V_A}, \tag{20}$$

without specifying the form of the modulus Kähler potential K_0.

3.1. Non-perturbative Polonyi term

To stabilize the D-flat direction, we need a superpotential of $\{T, X\}$ providing an F-term potential for D-flat direction. An attractive example would be the non-perturbative Polonyi term induced by some stringy instanton (or hidden gaugino condensation) effects:

$$W = \omega_0 + Ae^{-aT}X, \tag{21}$$

where ω_0 is a constant of $\mathcal{O}(m_{3/2}M_{Pl}^2)$, and the gauge invariance requires $a = 2/\delta_{\rm GS} = \mathcal{O}(8\pi^2)$. By an appropriate shift of T, the coefficient A can always be made to be of order unity without loss of generality. Then the Polonyi term gives a non-trivial F-term potential with a (local) minimum at $e^{-aT}X = \mathcal{O}(\omega_0)$, while the detailed vacuum structure depends on the form of K_0.

For the Kähler potential (20) and the superpotential (21), it is easy to see that the vacuum expectation value of $\arg(e^{-aT}X)$ is simply fixed as

$$\arg(e^{-aT}X) = \begin{cases} \arg\left(\frac{\omega_0}{A}\right) + \pi, & \text{for } \; R > \frac{2-|X|^2}{1-\xi_{\rm FI}} \\ \arg\left(\frac{\omega_0}{A}\right), & \text{for } \; R < \frac{2-|X|^2}{1-\xi_{\rm FI}}, \end{cases} \tag{22}$$

and this is not affected by the addition of uplifting potential. After replacing $\arg(e^{-aT}X)$ with its expectation value, the scalar potential becomes a function of $t = T + T^*$ and $|X|$. Then the stationary conditions $\partial_{t,|X|}(V_F + V_D) = 0$ lead to

$$\begin{aligned} g_A^2 D_A &= -\frac{1}{\frac{\delta_{\rm GS}}{2}K_0''}\left(V_F' + (\partial_t \ln g_A^2)V_D\right), \\ \dot{V}_F &= \frac{2|X|^2}{\frac{\delta_{\rm GS}}{2}K_0''}\left(V_F' + (\partial_t \ln g_A^2)V_D\right), \end{aligned} \tag{23}$$

where the dot and prime denote the derivative with respect to $\ln|X|$ and t, respectively. The second relation above determines how the D-flat direction is lifted by the F-term potential. Neglecting small corrections of $\mathcal{O}(m_{3/2}^4/M_V^4)$, the condition for the above stationary solution to be a (local) minimum reads

$$\ddot{V}_F + 4\frac{|X|^4}{\frac{\delta_{\rm GS}^2}{4}K_0''^2}\left(V_F'' - \frac{K_0'''}{K_0''}V_F'\right) - 4\frac{|X|^2}{\frac{\delta_{\rm GS}}{2}K_0''}\dot{V}_F' \;>\; 0, \tag{24}$$

which corresponds to the requirement that the F-term potential should provide a positive mass-squared to the D-flat direction. Note that a supersymmetric field configuration leads to $\dot{V}_F = V_F' = 0$ and $D_A = 0$.

The field space of T can be divided into two regions depending on where $\arg(e^{-aT}X)$ is stabilized. Let us first look for a supersymmetric minimum. The equations (23) can have a supersymmetric solution only in the region with $R > 2$ where $\arg(e^{-aT}X)$ is fixed at $\arg(\omega_0/A) + \pi$. This supersymmetric solution is determined by

$$\begin{aligned} |X|^2 &= -\xi_{\rm FI}, \\ \left|\frac{A}{\omega_0}\right|\frac{e^{-at/2}}{\sqrt{-\xi_{\rm FI}}} &= \frac{1}{1-\xi_{\rm FI}}, \end{aligned} \tag{25}$$

for which the condition (24) is automatically fulfilled regardless of the detailed form of K_0. Thus, if exists, SUSY stationary point always corresponds to a local minimum of $V_F + V_D$. It is also found that there can be at most one supersymmetric solution because $\xi_{\rm FI}$ is a monotonically increasing function of t.

On the other hand, a non-supersymmetric minimum can appear in the region with $R < 2$ where $\arg(e^{-aT}X)$ is fixed at $\arg(\omega_0/A)$. Since the non-perturbative Polonyi term stabilizes the GS modulus at $at \gg 1$, it is plausible to assume that the modulus Kähler potential satisfies

$$\left|\left(\frac{\delta_{\rm GS}}{2}\right)^{n-2}\frac{\partial_t^n K_0}{K_0''}\right| \leq \frac{1}{(at)^{n-2}} \ll 1 \tag{26}$$

for $n \geq 3$ at the stationary point. With this assumption, we find that the scalar potential can have a SUSY breaking minimum with $R < 2$ at

$$\begin{aligned} |X|^2 &= -\xi_{\rm FI} + \mathcal{O}\left(\frac{m_{3/2}^2 M_{Pl}^2}{M_V^2}\right), \\ \left|\frac{A}{\omega_0}\right|\frac{e^{-at/2}}{\sqrt{-\xi_{\rm FI}}} &= \frac{1}{R} + \mathcal{O}\left(\frac{M_V^2}{M_{Pl}^2}\right), \end{aligned} \tag{27}$$

for $M_{Pl} > M_V > \sqrt{m_{3/2}M_{Pl}}$. It is straightforward to show that the above field configuration satisfies well the condition (24) and leads to

$$(V_F + V_D)_{\rm SB} \;=\; e^K|\omega_0|^2\left(-3 + \frac{1}{R}\frac{M_{\rm GS}^2}{M_{Pl}^2} + \mathcal{O}\left(\frac{M_{\rm GS}^2}{M_{Pl}^2}\right)\right), \tag{28}$$

where the second term in the brackets is the contribution from F^X. Thus the small cosmological constant requires $R = \mathcal{O}(M_{\rm GS}^2/M_{Pl}^2)$.

In order to have a SUSY breaking minimum in the region with $R > 2$, one needs a quite non-trivial modulus Kähler potential for which the assumption (26) is not valid. This can be inferred from that a non-supersymmetric solution of (23), if exists in the indicated region, is accompanied by a supersymmetric one, which is necessarily a local minimum of the potential. More than three stationary points are difficult to be obtained for the scalar potential generated by single non-perturbative superpotential term.

We now turn our attention to the dependence of the minimum on the property of the modulus Kähler potential. Written as in (25) and (27), the two minimization conditions for t and $|X|$ make it clear how the GS modulus is stabilized by the non-perturbative superpotential. One can see that the function $e^{-at/2}/\sqrt{-\xi_{\rm FI}}$ monotonically decreases in the region with $R > 1$ as t increases, while it increases monotonically in the other region. In addition, from $R' = -\frac{2}{\delta_{\rm GS}}\left(1+R\frac{\delta_{\rm GS}}{2}\frac{K_0'''}{K_0''}\right)$, it is found that R can be approximated as

$$R \;\simeq\; -\frac{2}{\delta_{\rm GS}}(t-t_*), \tag{29}$$

for $R \le \mathcal{O}(1)$ and $at_* \ge at \gg 1$, where t_* is the value of t at which the FI term vanishes. Note that, because $a = -2q_X/\delta_{\rm GS} > 0$, X can maintain the D-flatness by cancelling the FI term in the field space $t \le t_*$. Using these properties, we arrive at that the scalar potential has the following structures depending on the value of γ_*

$$\gamma_* \;=\; \left(\frac{\delta_{\rm GS}^2}{4}K_0''(t_*)\right)^{-1/2}\left|\frac{A}{\omega_0}\right|e^{-at_*/2}. \tag{30}$$

(i) For $0.26 > \gamma_*$, there arises only one supersymmetric minimum. To stabilize X at much lower than M_{Pl}, $|A/\omega_0|e^{-at_*/2}$ should be much smaller than the unity.

(ii) For $0.52 > \gamma_* > 0.26$, the scalar potential can develop two local minima, one supersymmetric and one SUSY breaking minimum. Both minima have R of order unity and $V_F + V_D < 0$.

(iii) For $\gamma_* > 0.52$, the scalar potential has only a SUSY breaking minimum. To have non-negative vacuum energy at the minimum, $|A/\omega_0|e^{-at_*/2} > \sqrt{3}$ is required.

It should be noted that the SUSY preserving minimum arises in the region $t < t_* - \delta_{\rm GS}$ where $R > 2$ is obtained, whereas the SUSY breaking minimum appears at $t_* - \delta_{\rm GS} < t < t_*$ where $R < 2$.

3.2. *Examples*

Here we provide two specific examples of the modulus Kähler potential, illustrating how the stabilization of the GS modulus is affected by the FI term. After fixing $\arg(e^{-aT}X)$ at the minimum of the potential, the two stationary conditions in (23) fix $t = T + T^*$ and $|X|$ at a minimum of $V_F + V_D$. The first condition simply corresponds to that the FI term should be cancelled by the D-term due to the vacuum expectation value of X, i.e. corresponds to the (approximate) D-flatness condition. The other condition determines where the minimum is developed along the D-flat direction. For

$$\begin{aligned} K &= K_0(T+T^*-\delta_{\rm GS}V_A) + e^{-2V_A}|X|^2, \\ W &= \omega_0 + Ae^{-aT}X, \end{aligned} \tag{31}$$

we consider two models

$$\begin{aligned} &\text{Model I} &:&\quad K_0 = -n_0\ln(T+T^*-\delta_{\rm GS}V_A), \\ &\text{Model II} &:&\quad K_0 = (T+T^*-t_*-\delta_{\rm GS}V_A)^2, \end{aligned} \tag{32}$$

with the following superpotential parameters

$$|\omega_0| = e^{-38}, \quad |A| = 1, \quad a = 4\pi^2. \tag{33}$$

In Model I, T might correspond to the dilaton or a Kähler modulus whose Kähler potential is well-approximated by the weak coupling form. On the other hand, in Model II, T might correspond to a modulus describing a singular limit of some cycle at which $\xi_{\rm FI} \propto K_0'$ vanishes. Note that K_0 is approximated by a quadratic form in the vicinity of the point with $K_0' = 0$, i.e. $t = t_*$ in Model II.

Let us summarize how the GS modulus is fixed by $V_F + V_D$ in these two examples. In Model I with $n_0 = 1$, the potential has one SUSY minimum at $t \simeq 2.04$ and $|X| \simeq 0.11 M_{Pl}$, which yields $R \simeq 80.4$. In Model II, (a) for $t_* = 2.42$, one SUSY minimum appears at $t \simeq 2.02$ and $|X| \simeq 0.14 M_{Pl}$ with $R \simeq 15.6$, (b) for $t_* = 2.14$, the potential develops two minima: one supersymmetric minimum at $t \simeq 2.06$ and $|X| \simeq 0.06 M_{Pl}$ with $R \simeq 2.87$, and one SUSY breaking minimum at $t \simeq 2.1$ and $|X| \simeq 0.04 M_{Pl}$ with $R \simeq 1.44$, (c) for $t_* = 1.82$, there is one SUSY breaking minimum at $t \simeq 1.82$ and $|X| \simeq 1.6 \times 10^{-4} M_{Pl}$ with $R \simeq 0.2 \times 10^{-4}$. These results are all in agreement with the general arguments presented in the previous subsection.

4. Mixed mediations

Here are several examples of mixed mediation that can arise naturally in models with anomalous $U(1)$ gauge symmetry [13].

- *Mirage mediation (Mixed modulus-anomaly mediation) for $R \sim \mathcal{O}(8\pi^2)$:*
  As we have discussed, a simple minded perturbative modulus Kähler potential leads to $R \sim 1/\delta_{GS} \sim 8\pi^2$. In this case, integrating out the massive $U(1)_A$ vector superfield results in an effective theory of the GS modulus T in which the matter moduli weight is shifted by the $U(1)_A$ charge q_i, through which the D-term contribution to scalar masses can be identified as the modulus-mediation. The non-perturbative superpotential in the effective theory stabilizes T at a SUSY AdS minimum, and thus an uplifting sector is needed as in the KKLT model. After uplifting, the modulus mediation becomes comparable to the loop-induced anomaly mediation [14, 15], leading to the mirage mediation pattern of soft terms [16]. Cancellation between the anomaly mediation and the radiative corrections to modulus mediation results in a significantly compressed low energy SUSY spectrum compared to other popular schemes. A characteristic feature is the mirage unification of gaugino masses at a mirage messenger scale $M_{\rm mirage}$ which is determined by the relative ratio between anomaly and modulus mediations. In mirage mediation, 1st and 2nd generation sfermion masses also show mirage unification at the same scale $M_{\rm mirage}$.

  Mirage mediation can be generalized to include contributions from gauge mediation [20, 21], which would be possible in our case if the gauge-charged messengers get masses by a $U(1)_A$ neutral singlet, not by the $U(1)_A$-charged X. Sparticle masses are then deflected from the mirage unification trajectory due to the presence of gauge mediation. The mirage unification behavior is maintained for the gauginos, but the mirage messenger scale changes from the pure mirage mediation case. For sfermion masses, mirage pattern is generically lost, and the deviation depends on the size of the gauge mediation contribution.

- *Mixed modulus-gauge-D-term mediation for $\mathcal{O}(1/(8\pi^2)^2) \leq R \leq \mathcal{O}(8\pi^2)$:*
  The situation can be quite different in models where the modulus Kähler potential yields $R \leq \mathcal{O}(8\pi^2)$. Over the range $1/(8\pi^2)^2 \leq R \leq 8\pi^2$, gaugino masses are given by a mixed modulus-gauge mediation with $M_a|_{\rm MM}/M_a|_{\rm GM} \sim F^T/(F^X/8\pi^2 X) \sim R$. On the other hand, sfermion masses are dominated mostly by the D-term contribution with a little hierarchy structure $m_{\tilde{f}} \sim \sqrt{D_A} \sim 8\pi^2 M_a/R$ for $R \geq 1$ and $m_{\tilde{f}} \sim \sqrt{D_A} \sim 8\pi^2\sqrt{R}M_a$ for $R \leq 1$.

- *Mixed gauge-anomaly-D-term mediation for* $R \sim \mathcal{O}(1/(8\pi^2)^2)$:
  Another interesting mixed mediation can arise when superpotential does not have any T-dependent nonperturbative term, but contains a Yukawa coupling $\propto X\Phi\Phi^c$ between X and the messengers $\Phi + \Phi^c$. In such case, X is stabilized at $\langle X \rangle \sim M_{GS}/8\pi^2$ by the competition between the D-term potential and the anomaly-mediated tachyonic mass, which results in $F^X/8\pi^2 X \sim F^C/8\pi^2 C \sim \sqrt{D_A}$, and thus a mixed gauge-anomaly-D-term mediation.

5. Conclusion

There can be various built-in sources of SUSY breaking in models with anomalous $U(1)$ gauge symmetry: the $U(1)$ D-term, the F-components of the Green-Schwarz modulus T and the chiral matter X introduced to cancel the Fayet-Illiopoulos term, and also the SUGRA auxiliary component parameterized by the F-component of chiral compensator C. Generically the visible sector soft masses receive modulus-mediated contribution of the order of F^T, D-term contribution of the order of $\sqrt{D_A}$, anomaly-mediated contribution of the order of $F^C/8\pi^2 C$, and finally gauge-mediated contribution of the order of $F^X/8\pi^2 X$ if there exist messengers which couple to X to get masses. We have examined the possible pattern of the mediation of SUSY breaking in such situation, and find that various mixed mediation scenarios can be realized for different values of the ratio between the FI term and the Stückelberg contribution to the $U(1)_A$ gauge boson mass-square.

Acknowledgments

We thank K. S. Jeong and M. Yamaguchi for useful discussions. This work is supported by the KRF Grants funded by the Korean Government (KRF-2008-314-C00064 and KRF-2007-341-C00010) and the KOSEF Grant funded by the Korean Government (No. 2009-0080844).

References

[1] M. Dine, N. Seiberg and E. Witten, *Nucl. Phys.* B **289**, 589 (1987)
[2] M. B. Green and J. H. Schwarz, Phys. Lett. B **149**, 117 (1984)
[3] V. S. Kaplunovsky and J. Louis, Phys. Lett. B **306**, 269 (1993)
[4] A. Brignole, L. E. Ibanez and C. Munoz, Nucl. Phys. B **422**, 125 (1994) [Erratum-ibid. B **436**, 747 (1995)]
[5] Y. Kawamura and T. Kobayashi, Phys. Lett. B **375**, 141 (1996) [Erratum-ibid. B **388**, 867 (1996)]
[6] P. Binetruy and E. Dudas, Phys. Lett. B **389**, 503 (1996)
[7] G. R. Dvali and A. Pomarol, Phys. Rev. Lett. **77**, 3728 (1996)
[8] N. Arkani-Hamed, M. Dine and S. P. Martin, Phys. Lett. B **431**, 329 (1998)
[9] G. F. Giudice and R. Rattazzi, Phys. Rept. **322**, 419 (1999)
[10] L. Randall and R. Sundrum, Nucl. Phys. B **557**, 79 (1999)
[11] G. F. Giudice, M. A. Luty, H. Murayama and R. Rattazzi, JHEP **9812**, 027 (1998)
[12] K. Choi and K. S. Jeong, JHEP **0608**, 007 (2006)
[13] K. Choi, K. S. Jeong, K. i. Okumura, C. S. Shin and M. Yamaguchi, in preparation.
[14] K. Choi, A. Falkowski, H. P. Nilles, M. Olechowski and S. Pokorski, JHEP **0411**, 076 (2004)
[15] K. Choi, A. Falkowski, H. P. Nilles and M. Olechowski, Nucl. Phys. B **718**, 113 (2005)
[16] K. Choi, K. S. Jeong and K. i. Okumura, JHEP **0509**, 039 (2005)
[17] M. Endo, M. Yamaguchi and K. Yoshioka, Phys. Rev. D **72**, 015004 (2005)
[18] S. Kachru, R. Kallosh, A. D. Linde and S. P. Trivedi, Phys. Rev. D **68**, 046005 (2003)
[19] E. Dudas, Y. Mambrini, S. Pokorski and A. Romagnoni, JHEP **0804**, 015 (2008)
[20] L. L. Everett, I. W. Kim, P. Ouyang and K. M. Zurek, Phys. Rev. Lett. **101**, 101803 (2008)
[21] K. Choi, K. S. Jeong, S. Nakamura, K. I. Okumura and M. Yamaguchi, JHEP **0904**, 107 (2009)

Journal of Physics: Conference Series **259** (2010) 012007 doi:10.1088/1742-6596/259/1/012007

Primordial Nucleosynthesis and Neutrino Physics Beyond the Standard Model

Gennaro Miele, Ofelia Pisanti, Srdjan Sarikas

Dipartimento di Scienze Fisiche, Universitá di Napoli "Federico II", Complesso Universitario di Monte S.Angelo, Via Cinthia, 80126, Napoli, Italy

E-mail: miele@na.infn.it, pisanti@na.infn.it, sarikas@na.infn.it

Abstract. The present status of standard Big Bang Nucleosynthesis (BBN) is here reviewed by comparing the theoretical predictions with the observational estimates of light nuclei abundances. In particular, the analysis reports the expected ranges for baryon fraction and effective number of neutrinos as obtained by BBN only. The comparison is also performed in case of a more detailed analysis of neutrino decoupling by assuming initial degenerate neutrino distributions and oscillation mechanism as well.

1. Introduction

Big Bang Nucleosynthesis (BBN) has a widely recognized twofold role in cosmology: it is certainly one of the observational pillars of the hot Big Bang model; at the same time, it provides one of the earliest direct cosmological probe nowadays available, constraining the properties of the universe in the age of few MeV temperature scale.

The physical framework of BBN settled down in the period between the seminal Alpher-Bethe-Gamow (known as $\alpha\beta\gamma$) paper in 1948 [1] and the essential settlement of the paradigm of the stellar nucleosynthesis of elements heavier than ^{7}Li with the B^{2}FH paper [2]. After this remarkable period — an account of which can be found in [3] — the emerged picture concerning the chemical composition of the universe was drawn where the main production of the four light elements ^{2}H, ^{3}He, ^{4}He and ^{7}Li results to be of cosmological nature, whereas all the rest of nuclides were lately produced in stars or as a consequence of stellar explosions. However, in the recent years the role played by BBN in cosmology is sensibly changed, becoming a framework where to test new cosmological models or exotic particle interactions such as lepton asymmetry, sterile neutrinos and non standard interactions, e.m. properties of neutrinos, inhomogeneities, extra degrees of freedom, extra dimensions, primordial black holes, varying coupling constants, mirror matter, scalar-tensor gravity and quintessence models (see [4] for a detailed review). Whereas in the standard scenario BBN is an extremely overconstrained framework, where the only free parameters are the baryon to photon ratio, η (equivalently, the baryon density of the universe), and the neutrino chemical potentials, ξ_{ν_α}, in the most simple extension of BBN one admits the presence of relativistic extra degrees of freedom (in addition to photons), historically parametrized in terms of the so-called *effective number of neutrinos*, $N_{\rm eff}$. This quantity, in fact, in case of no extra particle, just contains the number of active neutrinos at the time of BBN.

In the following, we will present both the analyses of BBN predictions versus observations by taking as free parameters $\Omega_B h^2$ and $N_{\rm eff}$, and assuming vanishing values of neutrino chemical potentials (not degenerate neutrinos), or alternatively by fixing $\Omega_B h^2$ at the value of WMAP and considering no extra degrees of freedom (only three active neutrinos) but assuming as free parameters the neutrino chemical potentials. In the latter case, we will follow the approach of [5] where the three flavours neutrino decoupling in case of degeneracy is computed in detail by pointing out the relevance of out of equilibrium effects.

2. Primordial abundances

We cannot observe primordial abundances directly, since stars have changed the chemical composition of the universe. For this reason, we have to look for appropriate strategies able to infer the primordial values starting from present observations. There are essentially two ways to attack this problem: *a*) to find astrophysical systems negligibly contaminated by stellar evolution, and hence with a chemical composition still equivalent to the primordial one (at least for lighter elements); or *b*) to account for galactic chemical evolution and thus to extrapolate back the present observations to the initial cosmological values. Depending on the element considered, the previous two strategies have been applied.

2.1. Deuterium

Regions negligibly evolved are certainly the ones placed at very high redshift. In the case of deuterium the astrophysical environments which seem the most appropriate are the hydrogen-rich clouds absorbing the light of background QSOs at high redshifts. In order to reliably perform the spectroscopical analysis the following requirements must be satisfied:

i) to have a neutral hydrogen column density in the range $17 < \log[N(H_I)/cm^{-2}] < 21$;

ii) a low level of metallicity [M/H] in order to reduce the chances of deuterium astration;

iii) the atoms of the clouds must have a low internal velocity dispersion, hence allowing the isotope shift of only 81.6 km/s to be resolved.

Due to such a restrictive set of requests only a few QSOs can be used for the determination of primordial deuterium, as reported in Fig. 1.

Following the approach described in [4], we derive the average value

$$\log {}^2\mathrm{H}/\mathrm{H} = -4.54 \pm 0.03 \Longrightarrow {}^2\mathrm{H}/\mathrm{H} = \left(2.87^{+0.22}_{-0.21}\right) \times 10^{-5}. \qquad (1)$$

To conclude, it is worth mentioning the recent proposal to use the fluctuations in the absorption of Cosmic Microwave Background (CMB) photons by neutral gas during the cosmic dark ages, at redshifts $z \approx 7 - 200$, to reveal the primordial deuterium abundance of the Universe. This method is based on the strength of the cross-correlation of brightness-temperature fluctuations due to resonant absorption of CMB photons in the 21-cm line of neutral

Figure 1. The nine measurements of QSAs used in the analysis of Ref. [4]. The horizontal band represents the average value of Eq. (1).

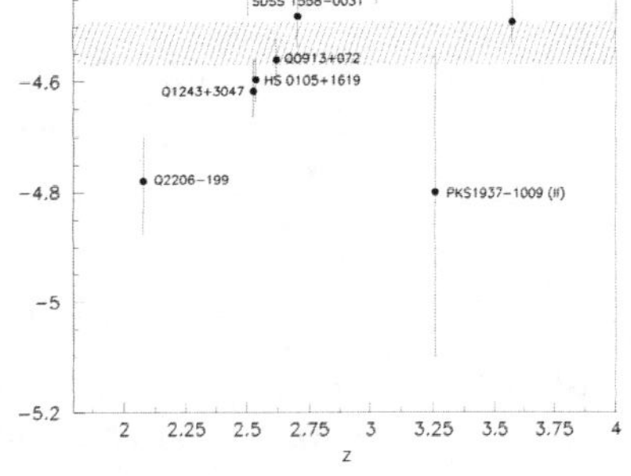

Journal of Physics: Conference Series **259** (2010) 012007 doi:10.1088/1742-6596/259/1/012007

hydrogen, with those due to resonant absorption of CMB photons in the 92-cm line of neutral deuterium. This results to be proportional to the ratio ^{2}H/H fixed during BBN. Although technically challenging, this measurement could provide the cleanest possible determination of ^{2}H/H [6]. A difficulty which has been pointed out – that may prevent the viability of the method at redshifts when the first UV sources turn on, at $z < 40$ – is that, when including Lyβ photons in the analysis, the inferred ratio ^{2}H/H would not be constant, but depend sensitively on the UV spectrum [7].

2.2. Helium - 3

Like deuterium, whose primordial yield is extremely sensitive to the baryon density parameter, ^{3}He is a crucial test of the standard BBN scenario as well. From the observational point of view, several environments are studied in order to derive its primordial abundance. Terrestrial determinations yield e.g. the ratio ^{3}He/^{4}He $\sim 10^{-6}$ from balloon measurements or $\sim 10^{-8}$ from continental rock. These observations, which show a large spread of values, confirm the idea that the terrestrial helium has no cosmological nature. In fact, most of it is ^{4}He produced by the radioactive decay of elements such as uranium and thorium. No natural radioactive decay produces ^{3}He, hence its observed terrestrial traces can be ascribed to unusual processes such as the testing of nuclear weapons or the infusion of extraterrestrial material.

Far beyond the Local InterStellar Medium (LISM), only one spectral transition allows the detection of ^{3}He, namely the 3.46 cm spin-flip transition of ^{3}He+, the analog of the widely used 21-cm line of hydrogen; this is a powerful tool for the isotope identification, as there is no corresponding transition in ^{4}He+. The emission is quite weak, hence ^{3}He has been observed outside the solar system only in a few H_{II} regions and Planetary Nebulae (PN) in the Galaxy. The values found in PN result one order of magnitude larger than proto-solar material and LISM determinations (for example, ^{3}He/H = (2-5) $\times 10^{-4}$ is measured in NGC3242 [8], confirming a net stellar production of ^{3}He in at least some stars). From the expected correlation between the metallicity of the particular galactic environment and its distance from the center of the galaxy, one would expect a gradient in ^{3}He abundance versus metallicity and/or distance. In Ref. [8] the ^{3}He/H abundance ratios are reported for the sample of simple H_{II} regions. Unfortunately, no significant correlation between the ^{3}He abundance and location (or metallicity) in the Galaxy is revealed.

The failure in observing a galactic ^{3}He dependence on metallicity, typically predicted by a chemical evolution model of the Galaxy has been referred to as the "^{3}He problem". In this scenario, by assuming a more conservative approach, the authors of Ref. [8] prefer to report an upper limit to the primordial abundance of ^{3}He by using the observations of a peculiar galactic H_{II} region,

$$^3\mathrm{He/H} < (1.1 \pm 0.2) \times 10^{-5}. \tag{2}$$

This upper bound is reported as the red band in Fig. 2. See [4] for more details.

2.3. Helium - 4

The post-BBN evolution of ^{4}He can be simply understood in terms of nuclear stellar processes which, through successive generations of stars, have burned hydrogen into ^{4}He and heavier elements, hence increasing the ^{4}He abundance above its primordial value. Since the history of stellar processing can be tagged by measuring the metallicity (Z) of the particular astrophysical environment, the primordial value of ^{4}He mass fraction, Y_p, can be derived by extrapolating the Y_p - O/H and Y_p - N/H correlations to O/H and N/H $\longrightarrow 0$. However, heavy elements like oxygen are produced by short-lived massive stars, whereas ^{4}He is essentially synthesized in all stars, so one has to minimize model-dependent evolutionary corrections. The key data for inferring ^{4}He primordial abundance are provided by observations of helium and

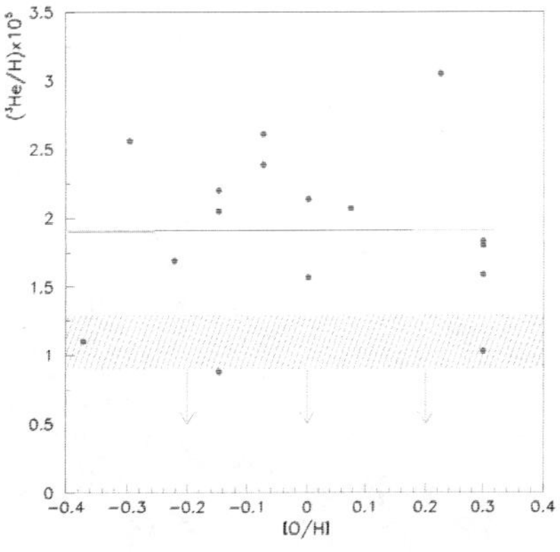

Figure 2. The data of Ref. [8] versus metallicity are reported. The black solid line stands for the linear fit of purple data, whereas the red band represents the upper bound obtained in [8].

hydrogen emission lines generated from the recombination of ionized hydrogen and helium in low-metallicity extragalactic H_{II} regions. Many attempts to determine Y_p have been made, constructing these correlations for various samples of Dwarf Irregular and Blue Compact Galaxies (BCGs). These systems are the least chemically evolved known galaxies. Plausibly, they contain very little helium synthesized in stars after the BBN, minimizing the chemical evolution problems that affect, e.g., the determination of ^{3}He.

Uncertainties in the determination of Y_p can be statistical or systematic. Statistical uncertainties can be decreased by obtaining very high signal-to-noise ratio spectra of BCGs. These BCGs are undergoing intense bursts of star formation, giving birth to high-excitation supergiant H_{II} regions and allowing an accurate determination of the helium abundance in the ionized gas through the BCG emission-line spectra. The theory of nebular emission is understood well enough not to introduce additional uncertainty. According to the standard scenario, the universe was born with zero metallicity; hence, Y_p can be determined extrapolating to $Z \to 0$, for a sample of objects, the relationship between Z and the ^{4}He abundance.

By reviewing recent determinations we adopt for the ^{4}He mass fraction (see [4] for details)

$$Y_p = 0.250 \pm 0.003. \tag{3}$$

This determination must be compared with two new analyses yielding: $Y_p = 0.2565 \pm 0.0010(\text{stat}) \pm 0.0050(\text{syst})$ [9] and $Y_p = 0.2561 \pm 0.0108$ [10], which again, pointing out a larger value for Y_p, seem to produce a serious level of tension between the data and the theoretical predictions in the standard scenario. In any case, one has to remind that the Y_p determination suffers of several problems due to different sources of systematics: i) interstellar reddening, ii) temperature of clouds, iii) electron density. These difficulties could be overcome by using more H lines.

2.4. Lithium - 7

The two stable isotopes of lithium, ^{6}Li and ^{7}Li, continue to puzzle astrophysicists and cosmologists who try to reconcile their primordial abundance, as inferred from observations, with the BBN predictions. From the astrophysical point of view the questions mainly concern the observation of lithium in cold interstellar gas and in all type of stars in which lithium lines are either detected or potentially detectable. A chance to link primordial ^{7}Li with the BBN abundance was first proposed by Spite & Spite (1982) [11], who showed that the lithium abundance in the warmest metal-poor dwarfs was independent of metallicity for $[Fe/H] < 1.5$. The constant lithium abundance, defining what is commonly called the Spite plateau, suggested that this may be the lithium abundance in pre-Galactic gas provided by the BBN.

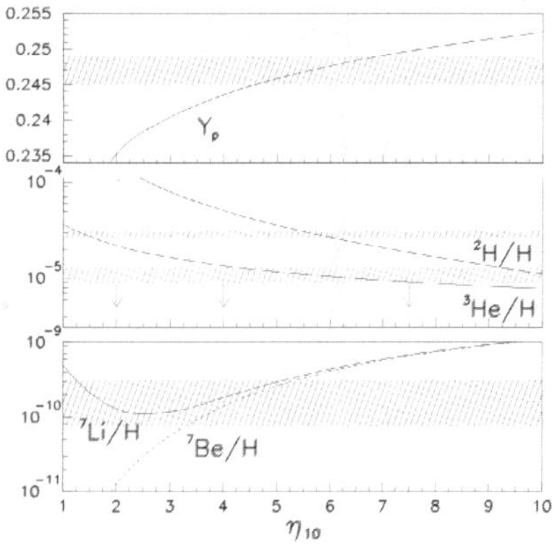

Figure 3. Values of the primordial abundances as a function of η_{10}, calculated for $\Delta N_{\rm eff} = 0$. The hatched blue bands represent the experimental determination with 1-σ statistical errors on Y_p, ^{2}H, and ^{7}Li, while the red band is the upper bound obtained in Ref. [8]. Note that for high values of η_{10} all ^{7}Li comes from ^{7}Be radioactive decay via electron capture. The vertical green band represents WMAP 5-year result, $\Omega_B h^2 = 0.02273 \pm 0.00062$ [12].

Several technical and conceptual difficulties have been responsible for quite a long tale of ^{7}Li determinations. In [4] are reviewed seven recent determinations. It is unclear how to combine the different estimates in a single number, or if the value measured is truly indicative of a primordial yield. A conservative approach (similar to the one used for ^{4}He) is to quote the simple (un-weighted) average and half-width of the above distribution of data as best estimate of the average and systematic error on ^{7}Li/H, obtaining

$$^7\text{Li/H} = \left(1.86^{+1.30}_{-1.10}\right) \times 10^{-10}, \tag{4}$$

which means an almost factor 2 of disagreement with the theoretical predictions.

3. BBN theoretical predictions versus data

The goal of a theoretical analysis of BBN is to obtain a reliable estimate of the model parameters on the basis of the experimental data on primordial abundances. In this Section we will consider first the case where the only two free parameters of BBN are the value of the baryon energy density parameter, $\Omega_B h^2$ (or equivalently the baryon to photon number density, η), and possibly, a non standard value for the relativistic energy content during BBN. The latter, after $e^\pm$ annihilation can be parameterized in terms of the effective number of neutrinos,

$$\rho_R = \left(1 + \frac{7}{8}\left(\frac{4}{11}\right)^{4/3} N_{\rm eff}\right)\rho_\gamma. \tag{5}$$

In the minimal BBN scenario the parameter set reduces to the baryon density only, since $\Delta N_{\rm eff}$ is just produced by non thermal distortions. Fig. 3 shows the dependence on $\eta_{10} \equiv \eta \cdot 10^{10}$ of the final value of the primordial yields, calculated using the nucleosynthesis code `PArthENoPE` [13], along with the experimental values of the abundances and their corresponding uncertainties, as discussed in details in Ref. [4]. To summarize, according to the analysis of [4] the inferred primordial abundances result to be

$$^2\text{H/H} = \left(2.87^{+0.22}_{-0.21}\right) \times 10^{-5}, \tag{6}$$

$$^3\text{He/H} < (1.1 \pm 0.2) \times 10^{-5}, \tag{7}$$

$$\text{Y}_p = 0.250 \pm 0.003, \tag{8}$$

Figure 4. Likelihood functions for ^{2}H/H (narrow) and Y_p (broad).

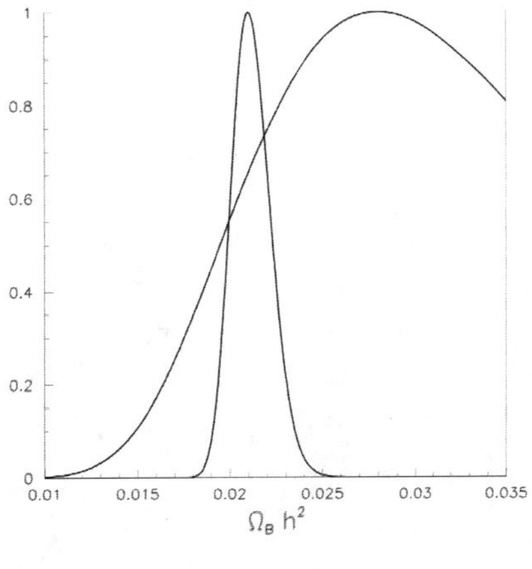

$$^7\text{Li/H} = \left(1.86^{+1.30}_{-1.10}\right) \times 10^{-10}. \tag{9}$$

To get confidence intervals for η, one constructs a likelihood function,

$$\mathcal{L}(\eta) \propto \exp\left(-\chi^2(\eta)/2\right), \tag{10}$$

with

$$\chi^2(\eta) = \sum_{ij} [X_i(\eta) - X_i^{obs}]\, W_{ij}(\eta)\, [X_j(\eta) - X_j^{obs}], \tag{11}$$

where $X_i \equiv n_i/n_b$ is the i-nuclid number density normalized with respect to the total number density of baryons and $W_{ij}(\eta)$ denotes the inverse covariance matrix,

$$W_{ij}(\eta) = [\sigma_{ij}^2 + \sigma_{i,exp}^2 \delta_{ij} + \sigma_{ij,other}^2]^{-1}, \tag{12}$$

where σ_{ij} and $\sigma_{i,exp}$ represent the nuclear rate uncertainties and experimental uncertainties of nuclide abundance X_i, respectively (we use the nuclear rate uncertainties as in Ref. [14]), while by $\sigma_{ij,other}^2$ we denote the propagated squared error matrix due to all other input parameter uncertainties (τ_n, $G_{\rm N}$, etc.). The proportionality constant in Eq. (10) can be obtained by requiring normalization to unity.

We first consider ^{2}H abundance alone, to illustrate the role of deuterium as an excellent baryometer. In this case the best fit values found are $\Omega_B h^2 = 0.021 \pm 0.001$ ($\eta_{10} = 5.7 \pm 0.3$) at 68% C.L. and $\Omega_B h^2 = 0.021 \pm 0.002$ at 95% C.L. A similar analysis can be performed using ^{4}He. In this case we get $\Omega_B h^2 = 0.028^{+0.011}_{-0.007}$ ($\eta_{10} = 7.6^{+3.0}_{-1.9}$) at 68% C.L. and $\Omega_B h^2 = 0.028^{+0.024}_{-0.012}$ at

Table 1. The theoretical values of the nuclear abundances for some value of $\Omega_B h^2$.

$\Omega_B h^2$	0.019	0.021	0.023
^{2}H/H (10^{-5})	3.36	2.87	2.48
^{3}He/H (10^{-5})	1.14	1.07	1.01
Y_p	0.2462	0.2472	0.2481
^{6}Li/H (10^{-14})	1.45	1.25	1.08
^{7}Li/H (10^{-10})	3.22	3.99	4.83
^{7}Be/H (10^{-10})	2.89	3.69	4.56

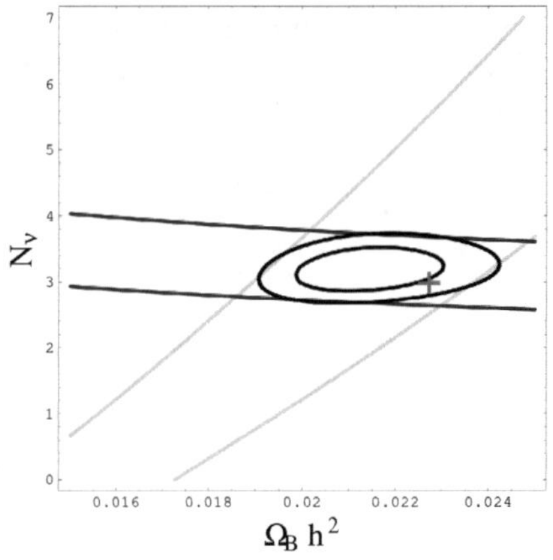

Figure 5. Contours at 68% and 95% C.L. of the total likelihood function for deuterium and ^{4}He in the plane ($\Omega_B h^2$,$N_{\rm eff}$). The bands show the 95% C.L. regions from deuterium (almost vertical) and Helium-4 (horizontal). The red cross corresponds to the standard $N_{\rm eff}$ and $\Omega_B h^2 = 0.02273$ as indicated by WMAP 5-year results.

95% C.L. Fig. 4 shows that the determination of $\Omega_B h^2$ is mainly dominated by deuterium. In any case, the result is compatible at 2-σ with WMAP 5-year result $\Omega_B h^2 = 0.02273 \pm 0.00062$ [12] and the WMAP 7-years $\Omega_B h^2 = 0.0226 \pm 0.0005$.

In Table 1 we report the values of some relevant abundances for three different baryon densities, evaluated using `PArthENoPE` [13]. Notice the very low prediction for ^{6}Li and that, for these values of baryon density, almost all ^{7}Li is produced by ^{7}Be via its eventual electron capture process.

If one relaxes the hypothesis of a standard number of relativistic degrees of freedom, it is possible to obtain bounds on the largest (or smallest) amount of radiation present at the BBN epoch, in the form of decoupled relativistic particles, or non-standard features of active neutrinos. Figure 5 displays the contour plots at 68% and 95% C.L. of the total likelihood function, in the plane ($\Omega_B h^2$, $N_{\rm eff}$). After marginalization one gets $\Omega_B h^2 = 0.021 \pm 0.001$ and $N_{\rm eff} = 3.18^{+0.22}_{-0.21}$ at 68% C.L. and $\Omega_B h^2 = 0.021 \pm 0.002$ and $N_{\rm eff} = 3.18^{+0.44}_{-0.41}$ at 95% C.L. Hence the global analysis results to be compatible with $N_{\rm eff} = 3.046$ and $\Omega_B h^2 = 0.02273$ found by WMAP at 1-σ level (see Figure 5 and [4] for details).

4. Neutrino decoupling and non thermal effects

As it is well known, $N_{\rm eff}$ can differ from zero also *via* the non thermal equilibrium terms which characterize the relic neutrino distributions. We show in Fig. 6 the asymptotic values of the flavor neutrino distribution, both without oscillations and with non-zero mixing. The dependence of the non-thermal distortions in momentum is well visible, which reflects the fact that more

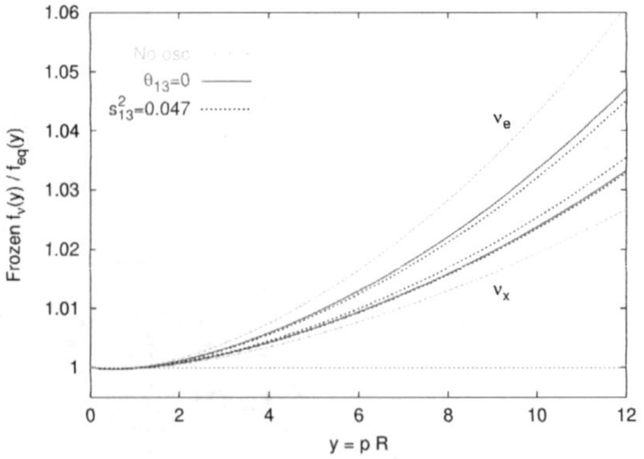

Figure 6. Frozen distortions of the flavor neutrino spectra as a function of the comoving momentum, for the best fit solar and atmospheric mixing parameters. R is the scale factor. In the case where we allow for $\theta_{13} \neq 0$ consistently with present bounds (blue dotted lines), one can distinguish the distortions for ν_μ and ν_τ (middle and lower, respectively). From [15].

Table 2. $N_{\rm eff}$ and ΔY_p obtained for different cases, with and without neutrino oscillations, as reported in [15].

Case	$N_{\rm eff}$	$10^4 \Delta Y_p$
No mix. (no QED)	3.035	1.47
No mix.	3.046	1.71
Mix., $\theta_{13} = 0$	3.046	2.07
Mix., $\sin^2(\theta_{13}) = 0.047$	3.046	2.12
Mix. Bimax., $\theta_{13} = 0$	3.045	2.13

energetic neutrinos were interacting with $e^{\pm}$ for a longer period. Moreover, the effect of neutrino oscillations is evident, reducing the difference between the flavor neutrino distortions. Fitting formulae for these distributions are available in [15].

In Table 2 we report the effect of non instantaneous neutrino decoupling on the radiation energy density, $N_{\rm eff}$, and on the ^{4}He mass fraction, Y_p. By taking also into account neutrino oscillations, one finds a global change of $\Delta Y_p \simeq 2.1 \times 10^{-4}$ which agrees with the results in [16] due to the inclusion of QED effects. Nevertheless, the net effect due to oscillations is about a factor 3 smaller than what previously estimated, due to the failure of the momentum-averaged approximation to reproduce the true distortions.

In particular, the $N_{\rm eff}$ reported in Table 2 is the contribution of neutrinos to the whole radiation energy budget, but only at the very end of neutrino decoupling. Hence, not all the $\Delta N_{\rm eff}$ there reported will be really contributing to BBN processes. In order to clarify this subtle point, we report in Table 3 the effect on all light nuclides, of the non instantaneous neutrino decoupling in the simple scenario of no neutrino oscillation, and compare this column with the simple *prescription* of adding a fix $\Delta N_{\rm eff} = 0.013$ contribution to radiation. Even though Y_p is reproduced (by construction), this is not the case for the other nuclear yields. Similar analysis have been recently presented by various groups, which might be slightly different depending on the adopted values of Y_p and/or ^{2}H/H experimental determination, see e.g. [17, 18, 19, 20, 21, 22, 23, 24, 25, 26, 27].

The situation becomes much more involved if we assume three oscillating neutrinos with an initial degeneracy, namely not vanishing chemical potentials, which can be realized by a global neutrino-antineutrino asymmetry or affect a single flavour. In any case, the presence of such an asymmetry during the neutrino decoupling can produce non thermal distortions on neutrino spectra much larger than in the non degenerate case. This effect has been recently pointed out in [5].

Applying such an approach, which means the solution of the exact Boltzmann equation for three flavour neutrino decoupling in presence of degeneracy during the BBN epoch, and taking

Table 3. Comparison of the exact BBN results with a fixed-$\Delta N_{\rm eff}$ approximation. From [15].

Nuclide	Exact (No ν-mix.)	Fixed $\Delta N_{\rm eff} = 0.013$
ΔY_p	1.71×10^{-4}	1.76×10^{-4}
$\Delta(^2{\rm H/H})$	-0.68×10^{-7}	$+0.44 \times 10^{-7}$
$\Delta(^3{\rm He/H})$	-0.11×10^{-7}	$+0.7 \times 10^{-8}$
$\Delta(^7{\rm Li/H})$	$+0.214 \times 10^{-11}$	-0.58×10^{-12}

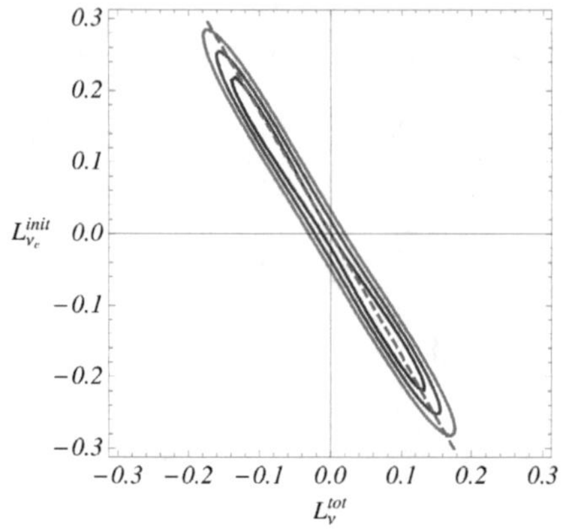

Figure 7. The regions allowed by BBN for the total neutrino asymmetry vs. initial electron neutrino asymmetry in the case $\sin^2\theta_{13} = 0$.

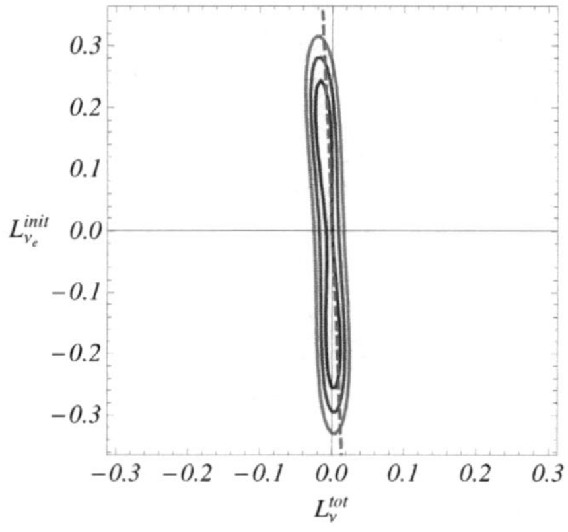

Figure 8. The regions allowed by BBN on the total neutrino asymmetry vs. initial electron neutrino asymmetry in the case $\sin^2\theta_{13} = 0.053$.

into account the corresponding distortion of the spectra in the BBN predictions one finds more interesting bounds on neutrino asymmetry than the ones previously obtained.

In particular in Fig. 7 and 8 the regions allowed by BBN in the plane total neutrino asymmetry vs initial electron neutrino asymmetry are reported, for $\sin^2\theta_{13} = 0$ and 0.053 respectively. The plots show that, in the case $\sin^2\theta_{13} = 0.053$, neutrino oscillations are able to equilibrate among the flavours and the bound on the possible total asymmetry is much strict. This is not the case for $\theta_{13} = 0$ where there is still room for a large asymmetry compatible with BBN.

5. Conclusions

The "classical parameter" constrained by BBN is the baryon to photon ratio, η, or equivalently the baryon abundance, $\Omega_B h^2$. At present, the constraint is dominated by the deuterium determination, and we find $\Omega_B h^2 = 0.021 \pm 0.001$ (1-σ). The agreement at almost 1-σ with the WMAP determination, $\Omega_B h^2 = 0.02273 \pm 0.00062$, represents a remarkable success of the Standard Cosmological Model. This agreement has been confirmed by the WMAP 7 - years data.

On the other hand, using this value as an input, a factor ~ 3 discrepancy remains with ^{7}Li determinations, which can hardly be reconciled even accounting for a conservative error budget in both observations and nuclear inputs. Even more puzzling are some detections of traces of ^{6}Li at a level far above the one expected from Standard BBN. Both nuclides indicate that either their present observations do not reflect their primordial values, and should thus be discarded for cosmological purposes, or that the early cosmology is more complicated and exciting than the Standard BBN. Neither a non-standard number of massless degrees of freedom in the plasma (parameterized via $N_{\rm eff}$) or a lepton asymmetry ξ_e (all asymmetries assumed equal) can reconcile the discrepancy.

Moreover, there are hints, coming from recent Y_p measurements, of the possible presence of one relativistic extra degree of freedom. This is not in contrast with the recent results about non thermal effects (extra radiation) in the neutrino spectra coming from an initial degeneracy and vanishing θ_{13} in the neutrino mixing matrix.

[1] R.A. Alpher, H.A. Bethe and G. Gamow, 1948. Phys. Rev., **73** 803.

[2] E.M. Burbidge, G.R. Burbidge, W.A. Fowler and F., Hoyle, 1957. Rev. Mod. Phys., **29** 547.
[3] H. Kragh, 1996. Cosmology and Controversy, Princeton Univ. Press.
[4] F. Iocco, G. Mangano, G. Miele, O. Pisanti and P.D. Serpico, 2009. Phys. Rept.**472** 1.
[5] S. Pastor, T. Pinto and G.G. Raffelt, 2009. Phys. Rev. Lett. **102**, 241302.
[6] K. Sigurdson and S.R. Furlanetto, 2006. Phys. Rev. Lett. 97 091301.
[7] L. Chuzhoy and P.R. Shapiro, 2006. Astrophys. J. 651 1.
[8] T.M. Bania et al., 2002. Nature **415** 54.
[9] Y.I. Izotov and T.X. Thuan, 2010. Astrophys. J. **710**, L67.
[10] E. Aver, K.A. Olive and E.D. Skillman,2010. JCAP **1005**, 003.
[11] F. Spite and M. Spite, 1982. Astron. Astroph. 115 357.
[12] J. Dunkley et al., 2008. [WMAP Collaboration], arXiv:0803.0586 [astro-ph].
[13] O. Pisanti et al., 2008. Comput. Phys. Commun. **178** 956.
[14] Serpico, P.D., *et al* , 2004. *JCAP* **0412** 010.
[15] G. Mangano, G. Miele, S. Pastor, T. Pinto, O. Pisanti and P.D. Serpico, 2005. Nucl. Phys. B **729** 221.
[16] S. Hannestad, 2001. Phys. Rev. D**64** 023515.
[17] E. Lisi, S. Sarkar and F. L. Villante, 1999. Phys. Rev. D **59**, 123520.
[18] S. Burles, K. M. Nollett, J. N. Truran and M. S. Turner, 1999. Phys. Rev. Lett. **82**, 4176.
[19] S. Esposito, G. Mangano, G. Miele and O. Pisanti, 2000. JHEP **0009** 038.
[20] V. Barger, J. P. Kneller, P. Langacker, D. Marfatia and G. Steigman, 2003. Phys. Lett. B **569**, 123.
[21] A. Cuoco, F. Iocco, G. Mangano, G. Miele, O. Pisanti and P. D. Serpico, 2004. Int. J. Mod. Phys. A **19**, 4431.
[22] R.H. Cyburt, B.D. Fields and K.A. Olive, 2003. Phys. Lett. B **567**, 227.
[23] R.H. Cyburt, 2004. Phys. Rev. D **70**, 023505.
[24] R.H. Cyburt, B.D. Fields, K.A. Olive and E. Skillman, 2005. Astropart. Phys. **23**, 313.
[25] S.H. Hansen, G. Mangano, A. Melchiorri, G. Miele and O. Pisanti, 2002. Phys. Rev. D **65**, 023511.
[26] G. Mangano, A. Melchiorri, O. Mena, G. Miele and A. Slosar, 2007. JCAP **0703**, 006.
[27] V. Simha and G. Steigman, 2008. JCAP **0806**, 016.

Realistic Inflation Models and Primordial Gravity Waves

Qaisar Shafi

Bartol Research Institute, Department of Physics and Astronomy, University of Delaware, Newark, DE 19716, USA

E-mail: shafi@bartol.udel.edu

Abstract. We consider a variety of realistic inflationary potentials and discuss their predictions for the tensor to scalar ratio r, a canonical measure of gravity waves generated during inflation. For a Standard Model gauge singlet inflaton with a Higgs-like potential, for example, we find that $r \gtrsim 0.02$, which may be observed by the Planck satellite. We also identify supersymmetric hybrid inflation models where r can be as high as 0.03, again within the reach of Planck. The scalar spectral index in these models lies within the WMAP one sigma bounds.

Introduction

Successful primordial inflation should:

- Explain the observed flatness and high degree of isotropy;
- Provide origin of $\frac{\delta T}{T}$;
- Offer testable predictions for n_s, r, $dn_s/d\ln k$;
- Recover Hot Big Bang Cosmology;
- Explain the observed baryon asymmetry;
- Offer plausible CDM candidate;

Need physics beyond the Standard Model (SM)?

Inflation is usually driven by some potential $V(\phi)$ and the relevant slow-roll parameters are defined as

$$\epsilon = \frac{m_p^2}{2}\left(\frac{V'}{V}\right)^2, \; \eta = m_p^2\left(\frac{V''}{V}\right).$$

Assuming slow-roll approximation (i.e. $(\epsilon, |\eta|) \ll 1$), the spectral index n_s and the tensor to scalar ratio r are given by

$$n_s \simeq 1 - 6\epsilon + 2\eta, \; r \simeq 16\epsilon.$$

The tensor to scalar ratio r is related to the energy scale of inflation by

$$V(\phi_0)^{1/4} = 3.3 \times 10^{16}\, r^{1/4}\ \mathrm{GeV}.$$

The amplitude of the curvature perturbation is given by

$$\Delta_{\mathcal{R}}^2 = \frac{1}{24\pi^2}\left(\frac{V/m_p^4}{\epsilon}\right)_{\phi=\phi_0} = 2.43 \times 10^{-9}\ (\text{WMAP7 normalization}).$$

The number of e-folds after the comoving scale $l_0 = 2\pi/k_0$ has crossed the horizon is given by

$$N_0 = \frac{1}{m_p^2}\int_{\phi_e}^{\phi_0}\left(\frac{V}{V'}\right)d\phi.$$

Inflation ends when $\max[\epsilon(\phi_e), |\eta(\phi_e)|] = 1$.

Higgs Inflation

Consider the following Higgs Potential:

$$V(\phi) = V_0\left[1 - \left(\frac{\phi}{M}\right)^2\right]^2. \qquad \longleftarrow \text{(tree level)}$$

Here ϕ is a gauge singlet field. As discussed in Refs. [1, 2], slow roll inflation may occur from

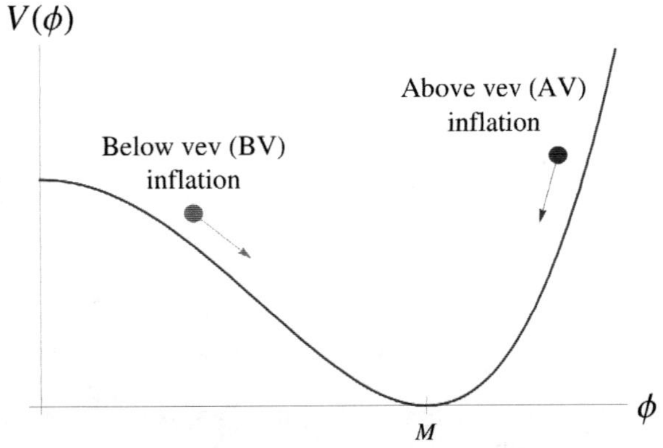

Figure 1. The Higgs potential.

either above or below the VEV M. For shorthand, we henceforth denote these regimes as BV (below VEV) and AV (above VEV) solutions. As shown in Fig. 2, WMAP data favors BV inflation, whereas, the AV branch is found to lie outside of the 1-σ bounds in Ref. [2].

Consider next the following interaction of inflaton ϕ with some GUT symmetry breaking scalar boson Φ:

$$\mathcal{L}_{int} = \frac{\lambda_\Phi^2}{2}\,\phi^2\,\Phi^2. \tag{1}$$

Including radiative corrections (quantum smearing) we have

$$V(\phi) \approx \left(\frac{m^2 M^2}{4}\right)\left[1 - \left(\frac{\phi}{M}\right)^2\right]^2 + A\,\phi^4\left[\ln\left(\frac{\phi}{M}\right) - \frac{1}{4}\right] + \frac{A\,M^4}{4}, \tag{2}$$

where $V(\phi = 0) \equiv V_0 = \frac{m^2 M^2}{4} + \frac{A\,M^4}{4}$ and $A = \frac{\mathcal{N}\lambda_\Phi^4}{32\,\pi^2}$. Note that we can use 'Minkowski space' CW corrections provided the propagating fields have masses $\gg H$ (Hubble constant).

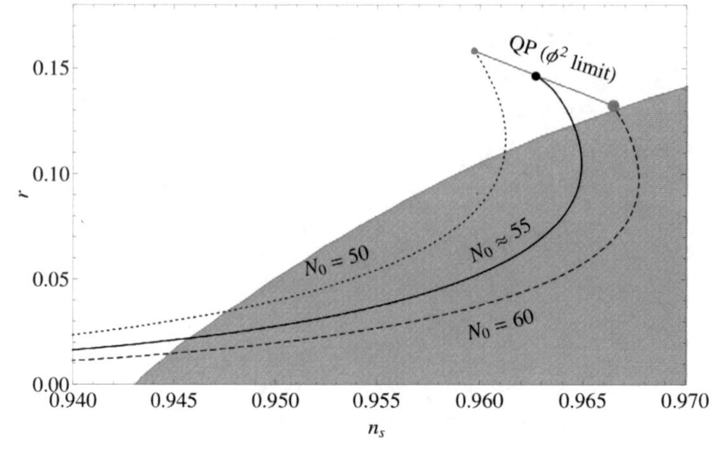

Figure 2. r vs. n_s for tree level Higgs inflation, shown together with the WMAP 1-σ (68% confidence level) bounds [3]. The blue dotted and blue dashed curves correspond to number of e-foldings $N_0 = 50$, $N_0 = 60$, respectively. Small (big) green circles correspond to the quadratic potential (QP) with $N_0 = 50$ ($N_0 = 60$).

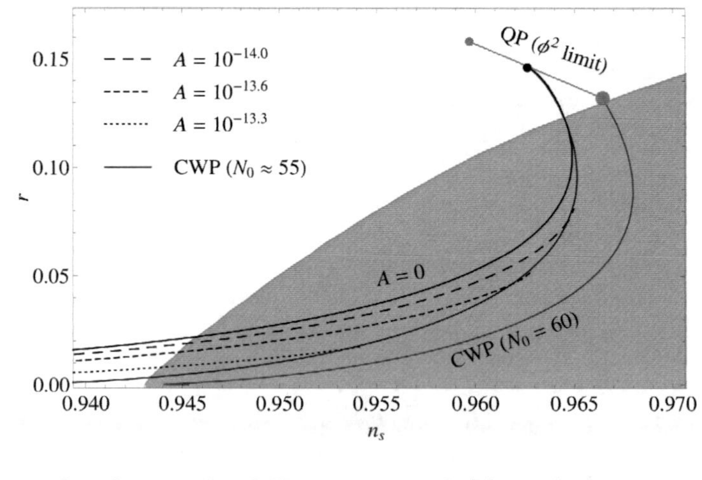

Figure 3. r vs. n_s for the tree level Higgs potential ($A = 0$) and the radiatively corrected Higgs potential ($A = 10^{-14.0}$, $10^{-13.6}$, $10^{-13.3}$). Blue and red curves represent the predictions of Coleman-Weinberg Potential (CWP) with the number of e-foldings $N_0 = 55$ and $N_0 = 60$, respectively. Small (big) green circles correspond to quadratic potential (QP) with $N_0 = 50$ ($N_0 = 60$).

The predictions of the radiatively corrected Higgs potential are shown in Fig. 3 and have been discussed in Ref. [4]. The parameter A here quantifies the radiative corrections. The CWP prediction represents the maximal smearing of the radiatively corrected Higgs inflation results.

Note that $r \gtrsim 0.02$ if $n_s \gtrsim 0.96$. Thus, Planck will test Higgs inflation soon!

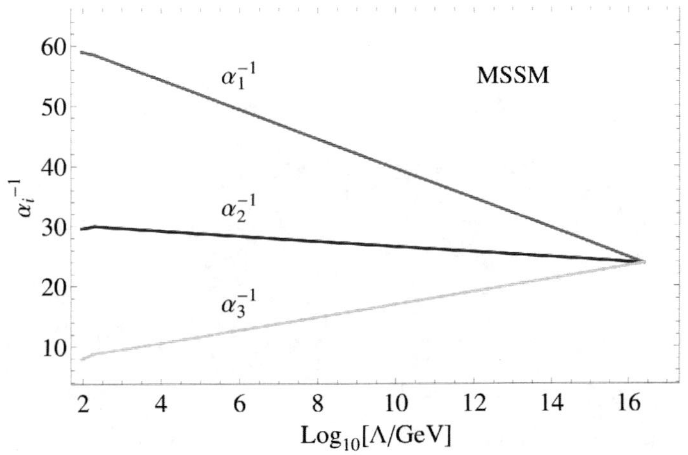

Figure 4. The gauge coupling unification in MSSM.

Susy Hybrid Inflation

Some attractive features of supersymmetry (susy) are:

- Resolution of the gauge hierarchy problem
- Unification of the SM gauge couplings at

  $M_{GUT} \sim 2 \times 10^{16}$ GeV (see Fig. 4)
- Cold dark matter candidate (LSP), typically a neutralino.

Other good reasons:

- Radiative electroweak breaking
- String theory requires susy

Leading candidate is the MSSM (Minimal Supersymmetric Standard Model).

The susy hybrid model [5, 6, 7, 8] is an attractive scenario in which inflation can be associated with symmetry breaking $G \longrightarrow H$

- Simplest inflation model is based on

$$W = \kappa\, S\,(\Phi\,\overline{\Phi} - M^2).$$

S = gauge singlet superfield, $(\Phi\,,\overline{\Phi})$ belong to suitable representation of G.

- We need $\Phi\,,\overline{\Phi}$ pair in order to preserve susy while breaking $G \longrightarrow H$ at scale $M \gg$ TeV, susy breaking scale.
- R-symmetry

$$\Phi\,\overline{\Phi} \to \Phi\,\overline{\Phi},\;\; S \to e^{i\alpha}\,S,\;\; W \to e^{i\alpha}\,W$$

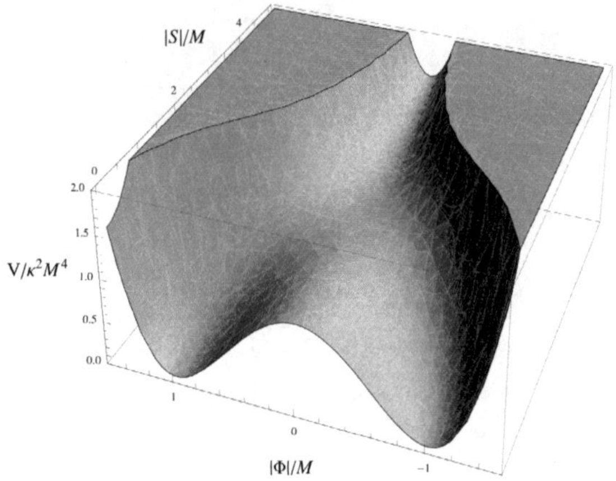

Figure 5. The supersymmetric hybrid potential.

$\Rightarrow$ W is a unique renormalizable superpotential.

- Some examples of gauge groups:

$G = U(1)_{B-L}$, (Supersymmetric superconductor)

$G = SU(5) \times U(1)$, (Φ = 10-plet of $SU(5)$), (Flipped $SU(5)$)

$G = 3_c \times 2_L \times 2_R \times 1_{B-L}$, $(\Phi = (1, 1, 2, +1))$

$G = 4_c \times 2_L \times 2_R$, $(\Phi = (\overline{4}, 1, 2))$,

$G = SO(10)$, $(\Phi = 16)$

- The susy hybrid tree level potential is

$$V_F = \kappa^2 \left(M^2 - |\Phi^2|\right)^2 + 2\kappa^2 |S|^2 |\Phi|^2,$$

with susy vacua

$$|\langle\overline{\Phi}\rangle| = |\langle\Phi\rangle| = M, \ \langle S\rangle = 0.$$

Take into account radiative corrections (because during inflation $V \neq 0$ and susy is broken by $F_S = -\kappa M^2$):

- Mass splitting in $\Phi - \overline{\Phi}$

$$m_{\pm}^2 = \kappa^2 S^2 \pm \kappa^2 M^2, \quad m_F^2 = \kappa^2 S^2$$

- One-loop radiative corrections

$$\Delta V_{\rm 1loop} = \tfrac{1}{64\pi^2} {\rm Str}[\mathcal{M}^4(S)(\ln \tfrac{\mathcal{M}^2(S)}{Q^2} - \tfrac{3}{2})]$$

- In the inflationary valley ($\Phi = 0$)

$$V \simeq \kappa^2 M^4 \left(1 + \frac{\kappa^2 \mathcal{N}}{8\pi^2} F(x)\right)$$

where $x = |S|/M$ and

$$F(x) = \frac{1}{4}\left(\left(x^4+1\right)\ln\frac{(x^4-1)}{x^4} + 2x^2 \ln\frac{x^2+1}{x^2-1} + 2\ln\frac{\kappa^2 M^2 x^2}{Q^2} - 3\right).$$

Tree Level plus radiative corrections:

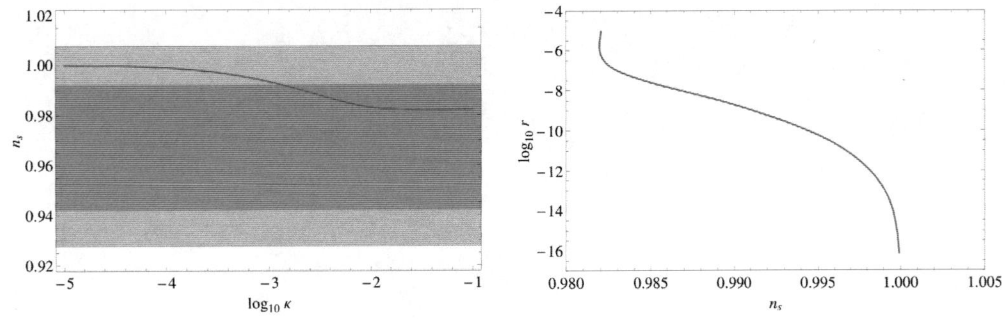

$$\delta T/T \propto (M/M_P)^2 \sim 10^{-5} \longrightarrow \text{attractive scenario } (M \sim M_G)$$

$$n_s \approx 1 - \frac{1}{N_0} \approx 0.98$$

- The minimal Kähler potential can be expanded as

$$K = |S|^2 + |\Phi|^2 + |\overline{\Phi}|^2.$$

- The sugra scalar potential is given by

$$V_F = e^{K/m_p^2}\left(K_{ij}^{-1} D_{z_i} W D_{z_j^*} W^* - 3m_p^{-2}|W|^2\right),$$

where we have defined

$$D_{z_i}W \equiv \frac{\partial W}{\partial z_i} + m_p^{-2}\frac{\partial K}{\partial z_i}W;\ K_{ij} \equiv \frac{\partial^2 K}{\partial z_i \partial z_j^*},$$

and $z_i \in \{\Phi, \overline{\Phi}, S, ...\}$.

- Take into account sugra corrections, radiative corrections and soft susy breaking terms [8]:

$$V \simeq \kappa^2 M^4 \left(1 + \left(\frac{M}{m_p}\right)^4 \frac{x^4}{2} + \frac{\kappa^2 \mathcal{N}}{8\pi^2} F(x) + a\left(\frac{m_{3/2}x}{\kappa M}\right) + \left(\frac{m_{3/2}x}{\kappa M}\right)^2\right) \quad (3)$$

where $a = 2\,|2 - A|\cos[\arg S + \arg(2 - A)]$, $x = |S|/M$ and $S \ll m_P$.

Note: No 'η problem' with minimal (canonical) Kähler potential !

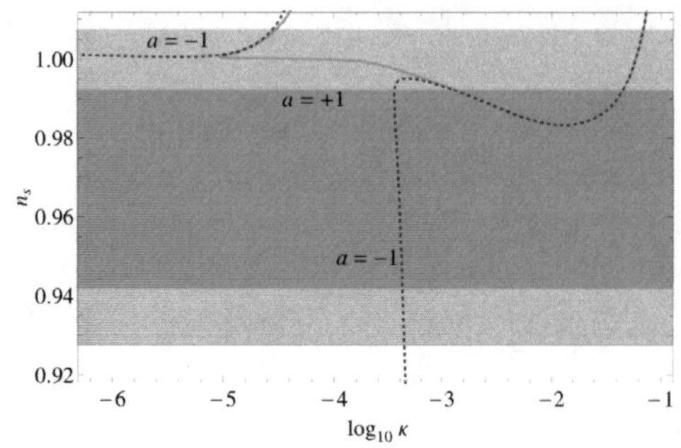

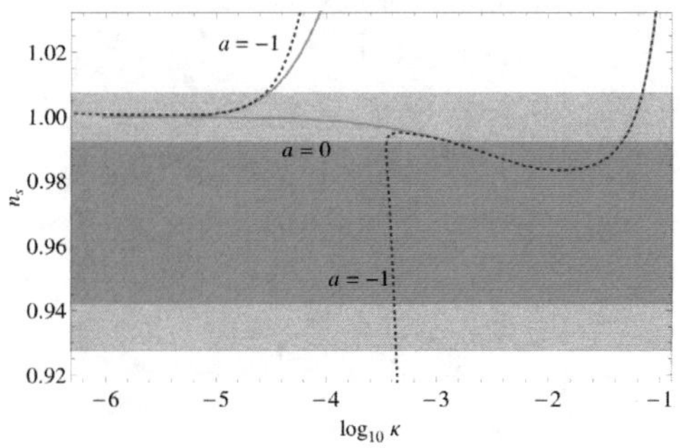

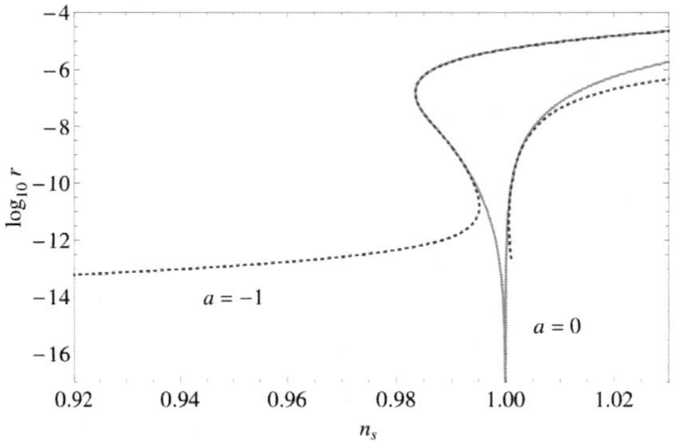

$r \lesssim 10^{-4}$ within 2-σ bounds of WMAP data (see [10])

Now consider flipped SU(5) $\equiv$ SU(5)$\times$U(1)$_X$

- Chiral superfields are arranged as

$$\mathbf{10}_1 = \begin{pmatrix} d^c & Q \\ & \nu^c \end{pmatrix}, \quad \overline{\mathbf{5}}_{-3} = \begin{pmatrix} u^c \\ L \end{pmatrix}, \quad \mathbf{1}_5 = e^c$$

- Compared to standard SU(5), these multiplets correspond to the interchange

$$u^c \longleftrightarrow d^c, \quad e^c \longleftrightarrow \nu^c$$

	Flipped SU(5) (Minimal)	SU(5) (Minimal)
Low scale susy	Yes	Yes
Doublet-triplet splitting	Yes	Fine tuning
Dimension 5 proton decay	Eliminated!	Challenging
μ problem	No	Fine tuning
Inflation	Easy	Difficult
Dimension 6 proton decay	$\tau_p \sim 10^{34}$–10^{36} yrs	$\tau_p \sim 10^{35}$–10^{36} yrs
Monopole problem	No	Yes
Seesaw mechanism	Automatic	No
Charge quantization	No	Yes
Unification of gauge couplings	Can be arranged	Yes
CDM	Yes	Yes

- Consider susy hybrid inflation in flipped SU(5), where Φ is a 10-plet

- Allowing the soft mass squared to vary, the potential appears as

$$V \simeq \kappa^2 M^4 \left(1 + \left(\frac{M}{m_P}\right)^4 \frac{x^4}{2} + \frac{\kappa^2 \mathcal{N}}{8\pi^2} F(x) + a\left(\frac{m_{3/2} x}{\kappa M}\right) + \left(\frac{M_S\, x}{\kappa M}\right)^2\right)$$

- If $M_S^2 < 0$ the soft susy breaking mass squared term drives the spectral index toward red-tilted values
- The minimal model consistent with $n_s = 0.96 - 0.97$ leads to proton lifetime predictions of order 10^{34}–10^{36} years (see Ref. [11])

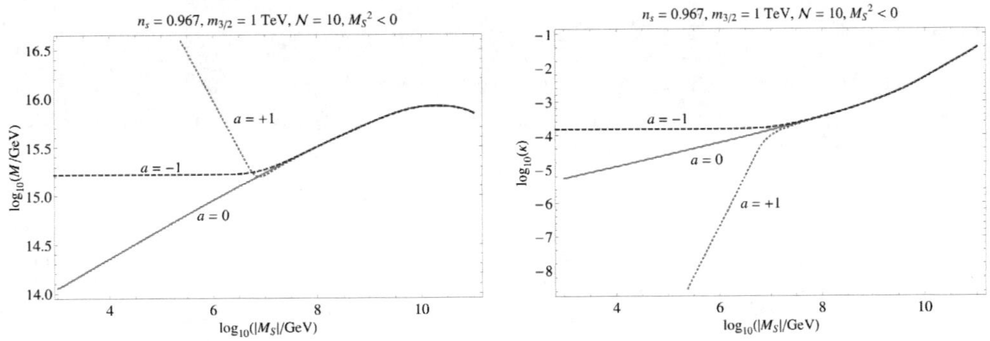

Figure 6. $\log_{10}(M/\text{GeV})$ and $\log_{10}(\kappa)$ vs. $\log_{10}(|MS|/\text{GeV})$ in the flipped $SU(5)$ model ($\mathcal{N} = 10$), with n_s fixed at the central value 0.967 (see Ref. [11]).

- The minimal susy hybrid inflation model yields r values $\lesssim 10^{-4}$;
- A more general analysis with a non-minimal Kähler potential can lead to larger r-values (see Fig. 7 and Ref. [12]).

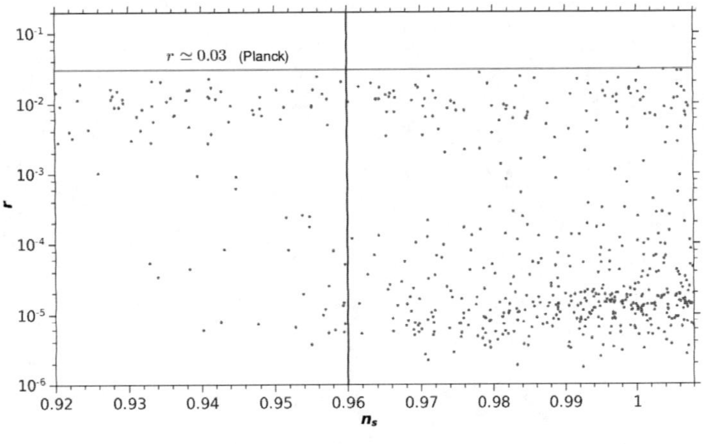

Figure 7. r vs n_s for susy hybrid model (see Ref. [12]).

Summary

- The predictions of r (primordial gravity waves) for various models of inflation are as follows:
  - Gauge Singlet Higgs Inflation:
  $$r \gtrsim 0.02 \quad \text{for} \quad n_s \geq 0.96$$
  - Susy Higgs (Hybrid) Inflation:
  $$r \lesssim 10^{-4} \text{ (minimal)}, \qquad r \lesssim 0.03 \text{ (non-minimal)}$$
- Results from PLANCK are eagerly awaited!

Acknowledgments

I would like to thank Mansoor Ur Rehman for his help in preparing this talk. The work reported here is supported in part by the Department of Energy Grant No. DE-FG02-91ER40626.

References

[1] R. Kallosh and A. D. Linde, JCAP **0704**, 017 (2007) [arXiv:0704.0647 [hep-th]].
[2] M. U. Rehman, Q. Shafi and J. R. Wickman, Phys. Rev. D **78**, 123516 (2008) [arXiv:0810.3625 [hep-ph]].
[3] E. Komatsu *et al.*, arXiv:1001.4538 [astro-ph.CO].
[4] M. U. Rehman and Q. Shafi, Phys. Rev. D **81**, 123525 (2010) [arXiv:1003.5915 [astro-ph.CO]].
[5] G. R. Dvali, Q. Shafi and R. K. Schaefer, Phys. Rev. Lett. **73**, 1886 (1994) [arXiv:hep-ph/9406319].
[6] E. J. Copeland, A. R. Liddle, D. H. Lyth, E. D. Stewart and D. Wands, Phys. Rev. D **49**, 6410 (1994) [arXiv:astro-ph/9401011].
[7] A. D. Linde and A. Riotto, Phys. Rev. D **56**, 1841 (1997) [arXiv:hep-ph/9703209].
[8] V. N. Senoguz and Q. Shafi, Phys. Rev. D **71**, 043514 (2005) [arXiv:hep-ph/0412102].
[9] M. ur Rehman, V. N. Senoguz and Q. Shafi, Phys. Rev. D **75**, 043522 (2007) [arXiv:hep-ph/0612023].
[10] M. U. Rehman, Q. Shafi and J. R. Wickman, Phys. Lett. B **683**, 191 (2010) [arXiv:0908.3896 [hep-ph]].
[11] M. U. Rehman, Q. Shafi and J. R. Wickman, Phys. Lett. B **688**, 75 (2010) [arXiv:0912.4737 [hep-ph]].
[12] Q. Shafi and J. R. Wickman, arXiv:1009.5340 [hep-ph].

16th International Symposium on Particles Strings and Cosmology (PASCOS 2010) IOP Publishing
Journal of Physics: Conference Series **259** (2010) 012009 doi:10.1088/1742-6596/259/1/012009

Neutrino Theory

S.F. King

School of Physics and Astronomy, University of Southampton, Highfield, Southampton SO17 1BJ, UK

E-mail: king@soton.ac.uk

Abstract. In this talk I give an overview of current approaches in neutrino theory and the key experimental questions which will enable progress to be made.

1. Introduction

In the three active neutrino paradigm, the lepton mixing matrix can be parameterised as in Fig.1 in terms of three angles θ_{ij}, one oscillation phase δ and (if neutrinos are Majorana particles) two Majorana phases α_i.

$$U_{MNS} = \begin{pmatrix} 1 & 0 & 0 \\ 0 & c_{23} & s_{23} \\ 0 & -s_{23} & c_{23} \end{pmatrix} \begin{pmatrix} c_{13} & 0 & s_{13}e^{-i\delta} \\ 0 & 1 & 0 \\ -s_{13}e^{i\delta} & 0 & c_{13} \end{pmatrix} \begin{pmatrix} c_{12} & s_{12} & 0 \\ -s_{12} & c_{12} & 0 \\ 0 & 0 & 1 \end{pmatrix} \begin{pmatrix} e^{i\alpha_1/2} & 0 & 0 \\ 0 & e^{i\alpha_2/2} & 0 \\ 0 & 0 & 1 \end{pmatrix}$$

Atmospheric | Reactor | Solar | Majorana

Figure 1. The lepton mixing matrix with phases factorizes into a matrix product of four matrices, associated with the physics of Atmospheric neutrino oscillations, Reactor neutrino oscillations, Solar neutrino oscillations and a Majorana phase matrix.

Ignoring the phases, the lepton mixing angles can be visualised as the Euler angles in Fig.2. The mass squared ordering is not yet determined uniquely for the atmospheric mass squared splitting, but the solar neutrino data requires $m_2^2 > m_1^2$, as shown in Fig.3. The absolute scale of neutrino masses is not fixed by oscillation data and the lightest neutrino mass may vary from about $0.0 - 0.2$ eV where the upper limit comes from cosmology. The current best fit values for the lepton angles and neutrino mass squared differences are given in Figs.4,5. Note that the reactor angle θ_{13} is not currently measured but its value is only inferred. The prospects for its future determination have been projected as in Fig.6.

2. Why go beyond the Standard Model?

It has been one of the long standing goals of theories of particle physics beyond the Standard Model (SM) to predict quark and lepton masses and mixings. With the discovery of neutrino mass and mixing, this quest has received a massive impetus. Indeed, perhaps the greatest advance in particle physics over the past decade has been the discovery of neutrino mass and

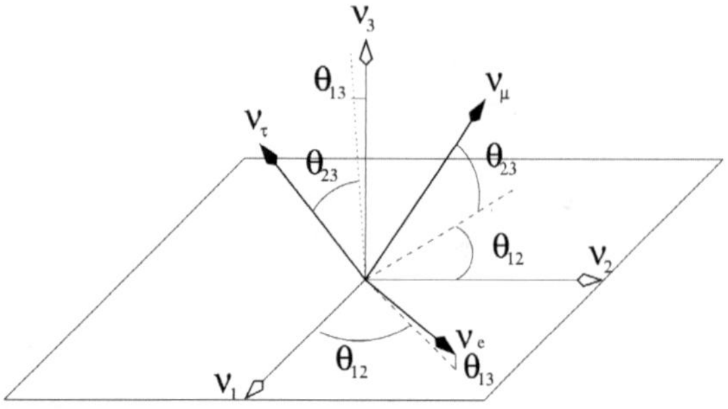

Figure 2. The relation between the neutrino weak eigenstates ν_e, ν_μ, and ν_τ and the neutrino mass eigenstates ν_1, ν_2, and ν_3 in terms of the three mixing angles θ_{12}, θ_{13}, θ_{23}. Ignoring phases, these are just the Euler angles respresenting the rotation of one orthogonal basis into another.

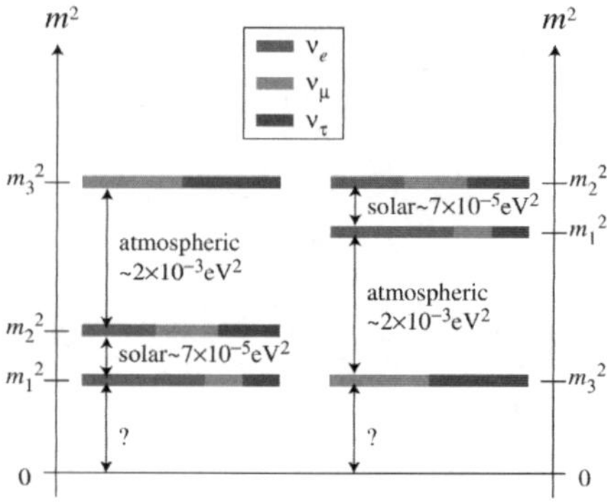

Figure 3. Alternative neutrino mass patterns that are consistent with neutrino oscillation explanations of the atmospheric and solar data. The pattern on the left (right) is called the normal (inverted) pattern. The coloured bands represent the probability of finding a particular weak eigenstate ν_e, ν_μ, and ν_τ in a particular mass eigenstate.

$$\theta_{12} = 34.4 \pm 1.0 \, \left({}^{+3.2}_{-2.9}\right)^\circ$$
$$\theta_{23} = 42.8\,{}^{+4.7}_{-2.9} \, \left({}^{+10.7}_{-7.3}\right)^\circ$$
$$\theta_{13} = 5.6\,{}^{+3.0}_{-2.7} \, (\leq 12.5)^\circ$$

Figure 4. The best fit lepton mixing angles with 1σ error (3σ error) from [1].

$$\Delta m^2_{21} = 7.59 \pm 0.20 \, \left({}^{+0.61}_{-0.69}\right) \times 10^{-5} \text{ eV}^2$$
$$\Delta m^2_{31} = \begin{cases} -2.36 \pm 0.11 \, (\pm 0.37) \times 10^{-3} \text{ eV}^2 \\ +2.46 \pm 0.12 \, (\pm 0.37) \times 10^{-3} \text{ eV}^2 \end{cases}$$

Figure 5. The best fit neutrino mass squared differences with 1σ error (and 3σ error) from [1].

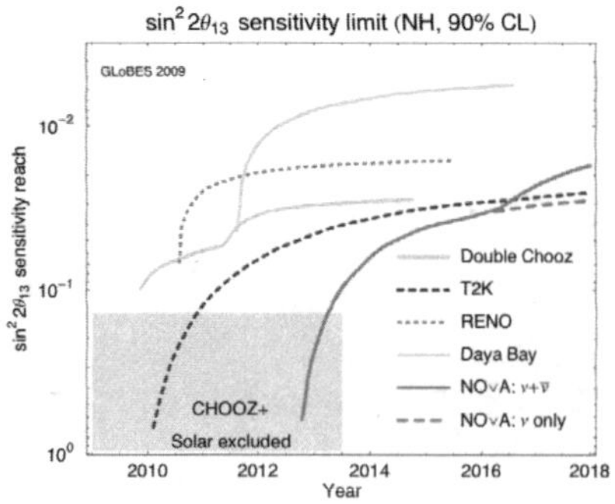

Figure 6. The future $\sin^2 2\theta_{13}$ sensitivity limit (normal hierarchy, 90% CL) from [2].

mixing involving two large mixing angles commonly known as the atmospheric angle θ_{23} and the solar angle θ_{12}, while the remaining mixing angle θ_{13}, although unmeasured, is constrained to be relatively small. The largeness of the two large lepton mixing angles contrasts sharply with the smallness of the quark mixing angles, and this observation, together with the smallness of neutrino masses, provides new and tantalizing clues in the search for the origin of quark and lepton flavour. However, before trying to address such questions, it is worth recalling why neutrino mass forces us to go beyond the SM.

Neutrino mass is zero in the SM for three independent reasons:

(i) There are no right-handed neutrinos ν_R.

(ii) There are only Higgs doublets of $SU(2)_L$.

(iii) There are only renormalizable terms.

In the SM these conditions all apply and so neutrinos are massless with ν_e, ν_μ, ν_τ distinguished by separate lepton numbers L_e, L_μ, L_τ. Neutrinos and antineutrinos are distinguished by total conserved lepton number $L = L_e + L_\mu + L_\tau$. To generate neutrino mass we must relax one or more of these conditions. For example, by adding right-handed neutrinos the Higgs mechanism of the Standard Model can give neutrinos the same type of mass as the electron mass or other charged lepton and quark masses. It is clear that the *status quo* of staying within the SM, as it is usually defined, is not an option, but in what direction should we go?

3. Neutrino Mass Models

The rest of this talk will be organized according to the plan in Fig.7. The plan in Fig.7 contains key experimental questions (in blue), leading in particular theoretical directions (in red), starting from the top left hand corner with the question "LSND?"

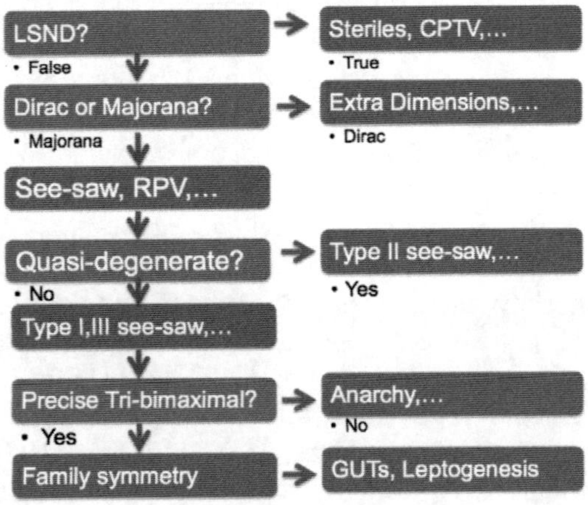

Figure 7. Plan of the talk.

3.1. LSND?

The antineutrino results from MiniBOONE may support the LSND result, but the neutrino results are consistent with the three active neutrino oscillation paradigm [3]. If LSND were correct then this could imply either sterile neutrinos and/or CPT violation, or something more exotic. For the remainder of this talk we shall focus on models based on three active neutrinos, which are well supported by atmospheric and solar neutrino experiments [4].

3.2. Dirac or Majorana?

Majorana neutrino masses are of the form $m^{\nu}_{LL}\overline{\nu_L}\nu^c_L$ where ν_L is a left-handed neutrino field and ν^c_L is the CP conjugate of a left-handed neutrino field, in other words a right-handed antineutrino field. Such Majorana masses are possible since both the neutrino and the antineutrino are electrically neutral. Such Majorana neutrino masses violate total lepton number L conservation,

Journal of Physics: Conference Series **259** (2010) 012009 doi:10.1088/1742-6596/259/1/012009

so the neutrino is equal to its own antiparticle. If we introduce right-handed neutrino fields then there are two sorts of additional neutrino mass terms that are possible. There are additional Majorana masses of the form $M^{\nu}_{RR}\overline{\nu_R}\nu^c_R$. In addition there are Dirac masses of the form $m^{\nu}_{LR}\overline{\nu_L}\nu_R$. Such Dirac mass terms conserve total lepton number L, but violate separate lepton numbers L_e, L_μ, L_τ. The question of "Dirac or Majorana?" is a key experimental question which could be decided by the experiments which measure neutrino masses directly.

3.3. Majorana neutrinos: see-saw mechanisms and R-parity violating SUSY

We have already remarked that neutrinos, being electrically neutral, allow the possibility of Majorana neutrino masses. However such masses are forbidden in the SM since neutrinos form part of a lepton doublet L, and the Higgs field also forms a doublet H, and $SU(2)_L \times U(1)_Y$ gauge invariance forbids a Yukawa interaction like HLL. So, if we want to obtain Majorana masses, we must go beyond the SM.

One possibility is to introduce Higgs triplets Δ such that a Yukawa interaction like ΔLL is allowed. However the limit from the SM ρ parameter implies that the Higgs triplet should have a VEV $\langle\Delta\rangle < 8$ GeV. One big advantage is that the Higgs triplets may be discovered at the LHC and so this mechanism of neutrino mass generation is directly testable [8].

Another possibility, originally suggested by Weinberg, is that neutrino Majorana masses originate from operators $HHLL$ involving two Higgs doublets and two lepton doublets, which, being higher order, must be suppressed by some large mass scale(s) M. When the Higgs doublets get their VEVs Majorana neutrino masses result: $m^{\nu}_{LL} = \lambda_\nu \langle H\rangle^2/M$. This is nice because the large Higgs VEV $\langle H\rangle \approx 175$ GeV can lead to small neutrino masses providing that the mass scale M is high enough. E.g. if M is equal to the GUT scale $1.75.10^{16}$ GeV then $m^{\nu}_{LL} = \lambda_\nu 1.75.10^{-3}$ eV. To obtain larger neutrino masses we need to reduce M below the GUT scale (since we cannot make λ_ν too large otherwise it becomes non-perturbative).

Typically in physics whenever we see a large mass scale M associated with a non-renormalizable operator we tend to associate it with tree level exchange of some heavy particle or particles of mass M in order to make the high energy theory renormalizable once again. This idea leads directly to the see-saw mechanism where the exchanged particles can either couple to HL, in which case they must be either fermionic singlets (right-handed neutrinos) or fermionic triplets, or they can couple to LL and HH, in which case they must be scalar triplets. These three possibilities have been called the type I [9], III [11] and II [10] see-saw mechanisms, respectively. If the coupling λ_ν is very small (for some reason) then M could even be lowered to the TeV scale and the see-saw scale could be probed at the LHC [12], however the see-saw mechanism then no longer solves the problem of the smallness of neutrino masses. Further types of see-saw mechanism have also been considered including the inverse see-saw mechanism [13] and the linear see-saw mechanism [14].

There are other ways to generate Majorana neutrino masses which lie outside of the above discussion. One possibility is to introduce additional Higgs singlets and triplets in such a way as to allow neutrino Majorana masses to be generated at either one [15] or two [16] loops. Another possibility is within the framework of R-parity violating Supersymmetry [17] in which the sneutrinos $\tilde{\nu}$ get small VEVs inducing a mixing between neutrinos and neutralinos χ leading to Majorana neutrino masses $m_{LL} \approx \langle\tilde{\nu}\rangle^2/M_\chi$, where for example $\langle\tilde{\nu}\rangle \approx$ MeV, $M_\chi \approx$ TeV leads to $m_{LL} \approx$ eV. A viable spectrum of neutrino masses and mixings can be achieved at the one loop level leading to distinctive predictions at the LHC such as [18]:

$$\tan^2\theta_{23} \approx \frac{BR(\chi^0_1 \to \mu W)}{BR(\chi^0_1 \to \tau W)}. \tag{1}$$

3.4. Quasi-degenerate?

This key experimental question may be decided by the same experiments as will also determine the nature of neutrino mass (Dirac or Majorana). Although not a theorem, it seems that a hierarchical spectrum could indicate a type I see-saw mechanism, while a (quasi) degenerate spectrum could imply a type II see-saw mechanism. It is possible that a type II see-saw mechanism could naturally explain the degenerate mass scale with the degeneracy enforced by an $SO(3)$ family symmetry, while the type I see-saw part could be responsible for the small neutrino mass splittings and the (TB) mixing [19]. An A_4 model of quasi-degenerate neutrinos with TB mixing of this kind was considered recently in [20], leading to a rather precise relationship between the neutrinoless double beta decay observable m_{ee} and the lightest neutrino mass, as shown in Fig.8. Different such relationships are predicted by the linear and inverse see-saw mechanisms [21].

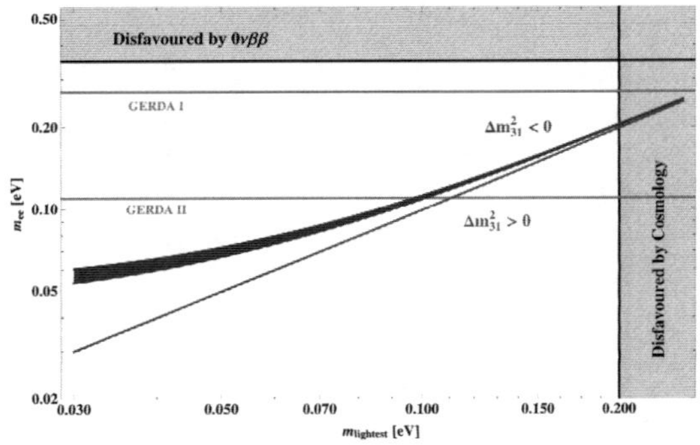

Figure 8. This plot shows the rather precise relationship between the neutrinoless double beta decay observable m_{ee} and the lightest neutrino mass, from the model in [20].

3.5. Very precise tri-bimaximal mixing?

It is a striking fact that current data on lepton mixing is (approximately) consistent with the so-called tri-bimaximal (TB) mixing pattern [22],

$$U_{TB} = \begin{pmatrix} \sqrt{\frac{2}{3}} & \frac{1}{\sqrt{3}} & 0 \\ -\frac{1}{\sqrt{6}} & \frac{1}{\sqrt{3}} & \frac{1}{\sqrt{2}} \\ \frac{1}{\sqrt{6}} & -\frac{1}{\sqrt{3}} & \frac{1}{\sqrt{2}} \end{pmatrix} P_{Maj}, \tag{2}$$

where P_{Maj} is the diagonal phase matrix involving the two observable Majorana phases. However there is no convincing reason to expect exact TB mixing, and in general we expect deviations. These deviations can be parametrized by three parameters r, s, a defined as [23]:

$$\sin\theta_{13} = \frac{r}{\sqrt{2}}, \quad \sin\theta_{12} = \frac{1}{\sqrt{3}}(1+s), \quad \sin\theta_{23} = \frac{1}{\sqrt{2}}(1+a). \tag{3}$$

The global fits of the conventional mixing angles [1] can be translated into the 1σ ranges:

$$0.07 < r < 0.21, \quad -0.05 < s < 0.003, \quad -0.09 < a < 0.04. \tag{4}$$

Clearly a non-zero value of r, if confirmed, would rule out TB mixing. However it is possible to preserve the good predictions that $s = a = 0$, by postulating a modified form of mixing matrix called tri-bimaximal-reactor (TBR) mixing [24], where only r is allowed to be non-zero.

3.6. Family Symmetry

Let us expand the neutrino mass matrix in the diagonal charged lepton basis, assuming exact TB mixing, as $M^{\nu}_{TB} = U_{TB}\mathrm{diag}(m_1, m_2, m_3)U^T_{TB}$ leading to (absorbing the Majorana phases in m_i):

$$M^{\nu}_{TB} = m_1\Phi_1\Phi_1^T + m_2\Phi_2\Phi_2^T + m_3\Phi_3\Phi_3^T \tag{5}$$

where $\Phi_1^T = \frac{1}{\sqrt{6}}(2, -1, 1)$, $\Phi_2^T = \frac{1}{\sqrt{3}}(1, 1, -1)$, $\Phi_3^T = \frac{1}{\sqrt{2}}(0, 1, 1)$, are the respective columns of U_{TB} and m_i are the physical neutrino masses. In the neutrino flavour basis (i.e. diagonal charged lepton mass basis), it has been shown that the above TB neutrino mass matrix is invariant under S, U transformations:

$$M^{\nu}_{TB} = SM^{\nu}_{TB}S^T = UM^{\nu}_{TB}U^T \ . \tag{6}$$

A very straightforward argument [25] shows that this neutrino flavour symmetry group has only four elements corresponding to Klein's four-group $Z_2^S \times Z_2^U$. By contrast the diagonal charged lepton mass matrix (in this basis) satisfies a diagonal phase symmetry T. The matrices S, T, U form the generators of the group S_4 in the triplet representation, while the A_4 subgroup is generated by S, T. Some candidate family symmetries G_f are shown in Fig.9.

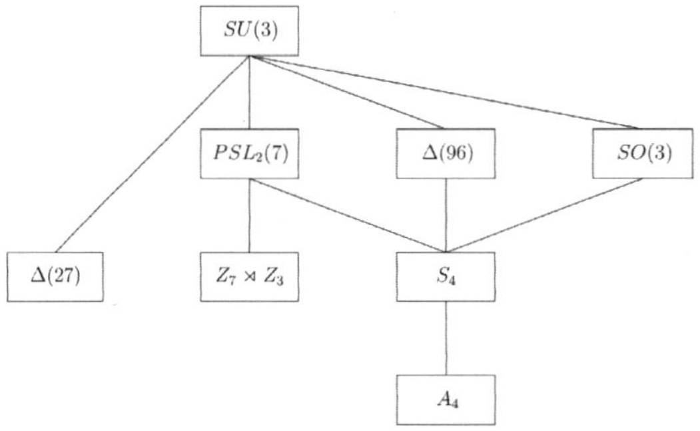

Figure 9. Some subgroups of $SU(3)$ that contain triplet representations and have been used as candidate family symmetries G_f.

3.7. Direct vs Indirect Models and Form Dominance

As discussed in [25], the flavour symmetry of the neutrino mass matrix may originate from two quite distinct classes of models. The first class of models, which we call direct models, are based on a family symmetry $G_f = S_4$, or a closely related family symmetry as discussed below, some of whose generators are directly preserved in the lepton sector and are manifested as part of the observed flavour symmetry. The second class of models, which we call indirect models, are based on some more general family symmetry G_f which is completely broken in the neutrino sector, while the observed neutrino flavour symmetry $Z_2^S \times Z_2^U$ in the neutrino flavour basis emerges as an accidental symmetry which is an indirect effect of the family symmetry G_f. In such indirect models the flavons responsible for the neutrino masses break G_f completely so that none of the generators of G_f survive in the observed flavour symmetry $Z_2^S \times Z_2^U$.

In the direct models, the symmetry of the neutrino mass matrix in the neutrino flavour basis (henceforth called the neutrino mass matrix for brevity) is a remnant of the $G_f = S_4$ symmetry of the Lagrangian, where the generators S, U are preserved in the neutrino sector, while the diagonal generator T is preserved in the charged lepton sector. For direct models, a larger family symmetry G_f which contains S_4 as a subgroup is also possible e.g. $G_f = PSL(2, 7)$ [26]. Typically direct models satisfy form dominance [27], and require flavon F-term vacuum alignment, permitting an $SU(5)$ type unification [28] typically based in A_4 family symmetry [29]. Such minimal A_4 models lead to neutrino mass sum rules between the three masses m_i, resulting in/from a simplified mass matrix in Eq.5. A_4 may result from 6D orbifold models [30] and recently an $A_4 \times SU(5)$ SUSY GUT model has been constructed in 6D [31], while a similar model in 8D enables vacuum alignment to be elegantly achieved by boundary conditions [32].

In the indirect models [25] the idea is that the three columns of U_{TB} Φ_i are promoted to new Higgs fields called "flavons" whose VEVs break the family symmetry, with the particular vacuum alignments along the directions Φ_i. In the indirect models the underlying family symmetry of the Lagrangian G_f is completely broken, and the flavour symmetry of the neutrino mass matrix $Z_2^S \times Z_2^U$ emerges entirely as an accidental symmetry, due to the presence of flavons with particular vacuum alignments proportional to the columns of U_{TB}, where such flavons only appear quadratically in effective Majorana Lagrangian [25]. Such vacuum alignments can be elegantly achieved using D-term vacuum alignment, which allows the large classes of discrete family symmetry G_f, namely the $\Delta(3n^2)$ and $\Delta(6n^2)$ groups [25]. The indirect models satisfy natural form dominance since each column of the Dirac mass matrix corresponds to a different flavon VEV. In the limit $m_1 \ll m_2 < m_3$ FD reduces to constrained sequential dominance (CSD)[33]. Examples of discrete symmetries used in the indirect approach can be found in [34].

Explicitly, the TB form of the neutrino mass matrix in Eq.5 is obtained from the see-saw mechanism in these models as follows. In the diagonal right-handed neutrino mass basis we may write $M^{\nu}_{RR} = \mathrm{diag}(M_A, M_B, M_C)$ and the Dirac mass matrix as $M^{\nu}_{LR} = (A, B, C)$ where A, B, C are three column vectors. Then the type I see-saw formula $M^{\nu} = M^{\nu}_{LR}(M^{\nu}_{RR})^{-1}(M^{\nu}_{LR})^T$ gives

$$M^{\nu} = \frac{AA^T}{M_A} + \frac{BB^T}{M_B} + \frac{CC^T}{M_C}. \tag{7}$$

By comparing Eq.7 to the TB form in Eq.5 it is clear that TB mixing will be achieved if $A \propto \Phi_3$, $B \propto \Phi_2$, $C \propto \Phi_1$, with each of $m_{3,2,1}$ originating from a particular right-handed neutrino of mass $M_{A,B,C}$, respectively. This mechanism allows a completely general neutrino mass spectrum and, since the resulting M^{ν} is form diagonalizable, it is referred to as form dominance (FD) [27]. For example, the direct A_4 see-saw models [28] satisfy FD [27], where each column corresponds to a linear combination of flavon VEVs.

A more natural possibility, called Natural FD, arises when each column arises from a separate flavon VEV, and this possibility corresponds to the case of indirect models. For example, if $m_1 \ll m_2 < m_3$ then the precise form of C becomes irrelevant, and in this case FD reduces to constrained sequential dominance (CSD)[33]. The CSD mechanism has been applied in this case to the class of indirect models with Natural FD based on the family symmetries $SO(3)$ [33, 36] and $SU(3)$ [35], and their discrete subgroups [34].

3.8. GUTs and leptogenesis

Finally we have reached the end of the decision tree, with the possibility of an all-encompassing unified theory of flavour based on GUTs and/or strings. Such theories could also include a family symmetry in order to account for the TB mixing. There are many possibilities for the choice of family symmetry and GUT symmetry. Examples include the Pati-Salam gauge group $SU(4)_{PS} \times SU(2)_L \times SU(2)_R$ in combination with $SU(3)$ [35], $SO(3)$ [33, 36], A_4 [37] or Δ_{27} [38]. Other examples are based on $SU(5)$ GUTs in combination with A_4 [39], T' [40] or S_4 [41].

In typical Family Symmetry $\otimes$ GUT models the origin of the quark mixing angles derives predominantly from the down quark sector, which in turn is closely related to the charged lepton sector. In order to reconcile the down quark and charged lepton masses, simple ansatze, such as the Georgi-Jarlskog hypothesis [42], lead to very simple approximate expectations for the charged lepton mixing angles such as $\theta^e_{12} \approx \lambda/3$, $\theta^e_{23} \approx \lambda^2$, $\theta^e_{13} \approx \lambda^3$, where $\lambda \approx 0.22$ is the Wolfenstein parameter from the quark mixing matrix. If the family symmetry enforces accurate TB mixing in the neutrino sector, then $\theta^e_{12} \approx \lambda/3$ charged lepton corrections will cause deviations from TB mixing in the physical lepton mixing angles, and lead to a sum rule relation [33, 43, 44], which can be conveniently expressed as [23] $s \approx r \cos\delta$ where $r \approx \lambda/3$ and δ is the observable CP violating oscillation phase, with RG corrections of less than one degree [45]. Such sum rules can be tested in future high precision neutrino oscillation experiments [46].

Note that in such a GUT-flavour framework, one expects the charged lepton corrections to the neutrino mixing angles to be less than of order $\theta_{12}^e/\sqrt{2}$ (where typically θ_{12}^e is a third of the Cabibbo angle) plus perhaps a further 1° from renormalization group (RG) corrections. Thus such theoretical corrections cannot account for an observed reactor angle as large as 8°, corresponding to $r = 0.2$, starting from the hypothesis of exact TB neutrino mixing.

Another interesting observation is that all models which satisfy form dominance will have zero leptogenesis [47]. This includes all models which account for tri-bimaximal mixing due to a discrete family symmetry. One way to obtain non-zero leptogenesis is to consider indirect models based on constrained sequential dominance, where the almost decoupled right-handed neutrino is the lightest one. The couplings of such a lightest right-handed neutrino are unconstrained and may allow successful leptogenesis [47].

4. Conclusion

Over the past dozen years there has been a revolution in our understanding of neutrino physics. Yet, despite this progress, it must be admitted that we still do not understand the origin or nature of neutrino mass and mixing. Neutrino mass and mixing clearly requires new physics beyond the SM, but in which direction should we go? There are many roads for model building, but we have seen that answers to key experimental questions will provide the sign posts *en route* to a unified theory of flavour. For example, if neutrinos have Majorana masses then this might signal some sort of see-saw mechanism at work. If these Majorana masses are sufficiently large then an indication might come soon from neutrinoless double beta decay experiments. In certain models such measurements would be directly related to the mass of the lightest physical neutrino mass.

One intriguing observation is that lepton mixing conforms to the tri-bimaximal pattern, at least approximately. This suggests an underlying non-Abelian discrete family symmetry that might unlock the long-standing flavour puzzle. However, even if tri-bimaximal mixing is accurately realised in the neutrino sector, due to a discrete family symmetry, when the model is embedded into a GUT theory, the charged lepton mixing corrections are typically related to quark mixing angles, leading to predicted deviations from tri-bimaximal mixing described by the sum rules $s \approx r \cos\delta$ where $r \approx \lambda/3$. Measurement of the parameters r, s, a which measure the deviation from tri-bimaximal mixing, will clearly be an important goal of the next generation of neutrino experiments.

References

[1] M. C. Gonzalez-Garcia, M. Maltoni and J. Salvado, JHEP **1004** (2010) 056 [arXiv:1001.4524 [Unknown]].
[2] P. Huber, M. Lindner, T. Schwetz and W. Winter, JHEP **0911** (2009) 044 [arXiv:0907.1896 [hep-ph]].
[3] M. Sanchez, these procedings.
[4] Y. Suzuki, these procedings.
[5] N. Arkani-Hamed, S. Dimopoulos, G. R. Dvali and J. March-Russell, Phys. Rev. D **65** (2002) 024032 [arXiv:hep-ph/9811448].
[6] S. Antusch, O. J. Eyton-Williams and S. F. King, JHEP **0508** (2005) 103 [arXiv:hep-ph/0505140].
[7] For an overview with references see e.g. S. J. Huber, arXiv:hep-ph/0211056.
[8] P. Fileviez Perez, T. Han, G. y. Huang, T. Li and K. Wang, Phys. Rev. D **78** (2008) 015018 [arXiv:0805.3536 [hep-ph]].
[9] P. Minkowski, Phys. Lett. B **67**, 421 (1977); T. Yanagida, in *Workshop on Unified Theories*, KEK report 79-18 (1979) p. 95; M. Gell-Mann, P. Ramond, R. Slansky, in *Supergravity* (North Holland, Amsterdam, 1979) eds. P. van Nieuwenhuizen, D. Freedman, p. 315; S. L. Glashow, in *1979 Cargese Summer Institute on Quarks and Leptons* (Plenum Press, New York, 1980) p. 687; R. Barbieri, D. V. Nanopoulos, G. Morchio

and F. Strocchi, Phys. Lett. B **90**, 091 (1980); R. N. Mohapatra and G. Senjanovic, Phys. Rev. Lett. **44**, 912 (1980); J. Schechter and J. W. F. Valle, Phys. Rev. D **22**, 2227 (1980).
[10] M. Magg and C. Wetterich, Phys. Lett. B **94**, 61 (1980); J. Schechter and J. W. F. Valle, Phys. Rev. D **22**, 2227 (1980); C. Wetterich, Nucl. Phys. B **187**, 343 (1981); G. Lazarides, Q. Shafi and C. Wetterich, Nucl. Phys. B **181**, 287 (1981); R. N. Mohapatra and G. Senjanovic, Phys. Rev. D **23**, 165 (1981).
[11] R. Foot, H. Lew, X. G. He and G. C. Joshi, Z. Phys. C **44**, 441 (1989).
[12] S. F. King and T. Yanagida, Prog. Theor. Phys. **114** (2006) 1035 [arXiv:hep-ph/0411030].
[13] D. Wyler and L. Wolfenstein, Nucl. Phys. B **218** (1983) 205; R. N. Mohapatra and J. W. F. Valle, Phys. Rev. D **34** (1986) 1642.
[14] M. Malinsky, J. C. Romao and J. W. F. Valle, Phys. Rev. Lett. **95** (2005) 161801 [arXiv:hep-ph/0506296].
[15] A. Zee, Phys. Lett. B **93** (1980) 389 [Erratum-ibid. B **95** (1980) 461].
[16] K. S. Babu, Phys. Lett. B **203** (1988) 132.
[17] A. S. Joshipura and M. Nowakowski, Phys. Rev. D **51** (1995) 2421 [arXiv:hep-ph/9408224].
[18] F. De Campos, O. J. P. Eboli, M. Hirsch, M. B. Magro, W. Porod, D. Restrepo and J. W. F. Valle, arXiv:1006.5075 [hep-ph].
[19] S. Antusch and S. F. King, Nucl. Phys. B **705** (2005) 239 [arXiv:hep-ph/0402121];
[20] S. Antusch, S. F. King and M. Spinrath, arXiv:1005.0708 [hep-ph].
[21] M. Hirsch, S. Morisi and J. W. F. Valle, Phys. Lett. B **679** (2009) 454 [arXiv:0905.3056 [hep-ph]].
[22] P. F. Harrison, D. H. Perkins and W. G. Scott, Phys. Lett. B **530** (2002) 167 [arXiv:hep-ph/0202074].
[23] S. F. King, Phys. Lett. B **659** (2008) 244 [arXiv:0710.0530 [hep-ph]].
[24] S. F. King, arXiv:0903.3199 [hep-ph].
[25] S. F. King and C. Luhn, JHEP **0910** (2009) 093 [arXiv:0908.1897 [hep-ph]].
[26] S. F. King and C. Luhn, Nucl. Phys. B **820** (2009) 269 [arXiv:0905.1686 [hep-ph]]; S. F. King and C. Luhn, Nucl. Phys. B **832** (2010) 414 [arXiv:0912.1344 [hep-ph]].
[27] M. C. Chen and S. F. King, JHEP **0906** (2009) 072 [arXiv:0903.0125 [hep-ph]].
[28] G. Altarelli and F. Feruglio, Nucl. Phys. B **720** (2005) 64 [arXiv:hep-ph/0504165].
[29] E. Ma and G. Rajasekaran, Phys. Rev. D **64** (2001) 113012 [arXiv:hep-ph/0106291].
[30] G. Altarelli, F. Feruglio and Y. Lin, Nucl. Phys. B **775** (2007) 31 [arXiv:hep-ph/0610165].
[31] T. J. Burrows and S. F. King, Nucl. Phys. B **835** (2010) 174 [arXiv:0909.1433 [hep-ph]].
[32] T. J. Burrows and S. F. King, arXiv:1007.2310 [Unknown].
[33] S. F. King, JHEP **0508** (2005) 105 [arXiv:hep-ph/0506297].
[34] I. de Medeiros Varzielas, S. F. King and G. G. Ross, Phys. Lett. B **644** (2007) 153 [arXiv:hep-ph/0512313].
[35] I. de Medeiros Varzielas and G. G. Ross, Nucl. Phys. B **733** (2006) 31 [arXiv:hep-ph/0507176].
[36] S. F. King and M. Malinsky, JHEP **0611** (2006) 071 [arXiv:hep-ph/0608021].
[37] S. F. King and M. Malinsky, Phys. Lett. B **645** (2007) 351 [arXiv:hep-ph/0610250].
[38] I. de Medeiros Varzielas, S. F. King and G. G. Ross, Phys. Lett. B **648** (2007) 201 [arXiv:hep-ph/0607045].
[39] G. Altarelli, F. Feruglio and C. Hagedorn, JHEP **0803** (2008) 052 [arXiv:0802.0090 [hep-ph]].
[40] M. C. Chen and K. T. Mahanthappa, Phys. Lett. B **652** (2007) 34 [arXiv:0705.0714 [hep-ph]].
[41] C. Hagedorn, S. F. King and C. Luhn, JHEP **1006** (2010) 048 [arXiv:1003.4249 [Unknown]].
[42] H. Georgi and C. Jarlskog, Phys. Lett. B **86** (1979) 297.
[43] I. Masina, Phys. Lett. B **633** (2006) 134 [arXiv:hep-ph/0508031];
[44] S. Antusch and S. F. King, Phys. Lett. B **631** (2005) 42 [arXiv:hep-ph/0508044];
[45] S. Boudjemaa and S. F. King, arXiv:0808.2782 [hep-ph].
[46] S. Antusch, P. Huber, S. F. King and T. Schwetz, JHEP **0704** (2007) 060 [arXiv:hep-ph/0702286].
[47] S. Choubey, S. F. King and M. Mitra, Phys. Rev. D **82** (2010) 033002 [arXiv:1004.3756 [hep-ph]].

16th International Symposium on Particles Strings and Cosmology (PASCOS 2010) IOP Publishing
Journal of Physics: Conference Series **259** (2010) 012010 doi:10.1088/1742-6596/259/1/012010

Lepton Flavour Violation Experiments in LHC Era

Fabrizio Cei[1]
Dipartimento di Fisica dell'Università and INFN Pisa, Largo Bruno Pontecorvo, 3, 56127 Pisa, Italy
E-mail: fabrizio.cei@pi.infn.it

Abstract. I review the present status of Lepton Flavour Violation Experiments involving muons and taus and the future perspectives of this physics field.

1. Introduction

In the minimal standard model (from now on: SM) the lepton flavour violating (from now on: LFV) processes are not allowed at all; leptons are grouped in separated doublets and the lepton flavour conservation is built in by hand assuming vanishing neutrino masses. Nevertheless, the neutrino oscillations are now established facts (for a continuously updated review see [1]) and the neutrino masses are definitely not vanishing; then, LFV in the neutral sector is an experimental reality, while until now there are no corresponding indications in the charged sector. When massive neutrinos and neutrino oscillations are introduced in the SM, LFV decays of charged leptons are predicted, but at immeasurably small levels (branching fractions $\sim 10^{-50}$ with respect to SM decays). However, Supersymmetric and expecially GUT supersymmetric theories (from now on: SUSY and SUSY-GUT) naturally accomodate finite neutrino masses and predict relatively large (and probably measurable) branching ratios (from now on: BR) for LFV processes (see for example [2], [3], [4], [5], [6]). Therefore, sizable flavour violation processes would be strong indications in favour of new physics beyond the SM.

Even if searches for charged LFV effects have, so far, yielded no results, they had a relevant impact on the particle physics development: for example, the non observation of $\mu^+ \to \mathrm{e}^+\gamma$ decay [7] established that the muon and the electron are two distinct leptons [8] and the stronger and stronger constraints on this process were basic arguments for introducing a second neutrino (ν_μ) [9]. At the beginning of the third millennium, the search for charged LFV reactions allows to explore SUSY mass scales up to $1000 \div 10000$ TeV (even out of LHC reach) and to give insights about large mass range, parity violation, number of generations ...

Figures 1 and 2 illustrates examples of recent theoretical predictions for charged LFV processes in the SUSY frame: on the left the $\mu^+ \to \mathrm{e}^+\gamma$ BR is shown as a function of $M_{1/2}$ (in GeV) for three different classes of models [5], on the right the sensitivity of some present and future experiments in the plane $(m_0, M_{1/2})$ is reported [6]. A detailed review of the mechanisms which might induce LFV processes and of the relation between LFV and other signs of new physics (like Muon Anomalous Magnetic Moment) can be found in [10].

Many experiments are under way or in preparation which would test the theoretical predictions with unprecedented levels of sensitivity.

[1] On behalf of the MEG Collaboration

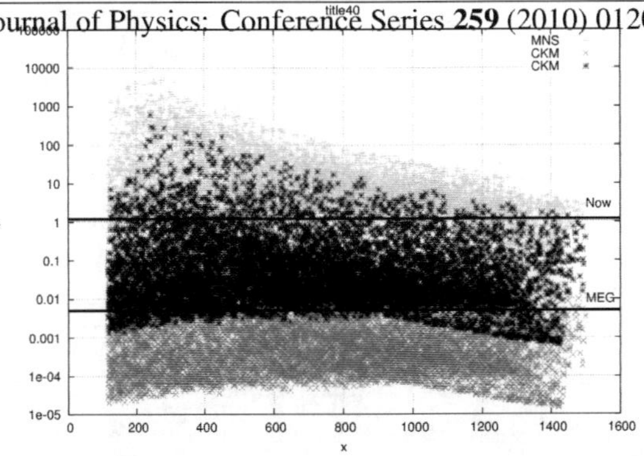

Figure 1. Branching ratio of $\mu^+ \to e^+\gamma$ decay (in units of 10^{-11}) as a function of $M_{1/2}$ (GeV) for three classes of SUSY models [5]. The horizontal line labelled "NOW" is the present experimental limit: $BR(\mu^+ \to e^+\gamma) \leq 1.2 \times 10^{-11}$ [11].

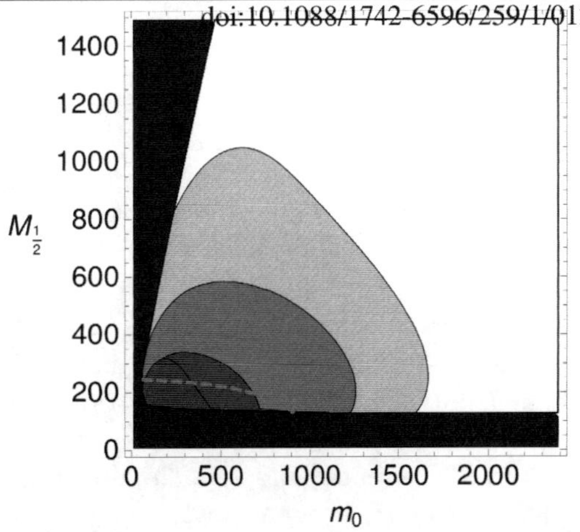

Figure 2. Sensitivity contours in the plane $(m_0, M_{1/2})$ for MEGA (Brown), BaBar + Belle (dark green), Super B (light green) and MEG (light brown). The dark brown regions are excluded by LEP data [6].

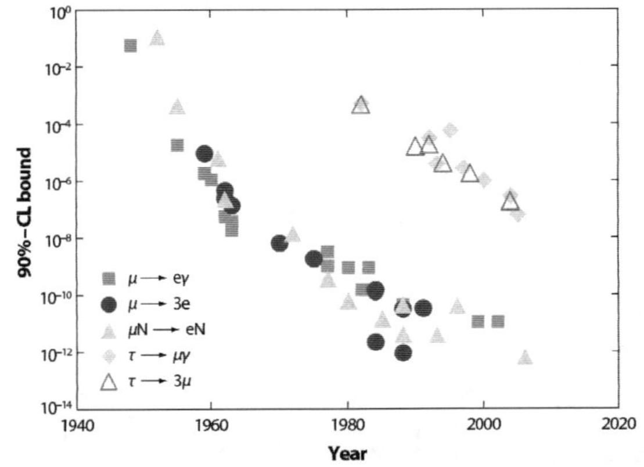

Figure 3. Improvement with time of some LFV searches (from [10]).

Note that not only positive results, but also negative results could be very significant, since they would tightly constrain the multi-dimensional SUSY parameter space. We also stress that searching for LFV processes in different channels and with different leptons is one of the most powerful tools to discriminate between different models. Figure 3 shows the improvement with time of the upper limits for some LFV processes. In this talk I discuss status and perspectives of LFV experiments involving muons and taus.

2. The muonic channel

2.1. Generalities

Muons are very sensitive probes for LFV searches since a) intense muon beams ($\sim 10^8$ μ/s) can be obtained at the meson factory machines, b) the final states are very simple, c) the LFV processes have clean signatures and d) the muon lifetime (2.2 μs) is relatively long.

The most studied muon LFV processes are the μ decay into and electron and a gamma or into three electrons and the muon to electron conversion. Other more exotic LFV reactions, like Muonium-Antimuonium conversion, rare K^0_L decays, (as $K^0_L \to e\mu$), $\mu^- A \to \mu^+ A'$, μ to τ conversion ... are not discussed here.

2.2. Search for $\mu \to e\gamma$ decay: the MEG experiment

The $\mu \to e\gamma$ decay is the historical channel where charged LFV is searched for. Positive muons (selected to avoid nuclear captures in the stopping target), coming from decay of π^+ produced in proton interactions on fixed target, are brought to stop and decay at rest, emitting simultaneously a γ and a e^+ in back-to-back directions. Since the e^+ mass is negligible, both particles carry away the same kinetic energy: $E_{e^+} = E_\gamma = m_\mu/2 = 52.83$ MeV. The signature is very simple, but, because of the finite experimental resolution, it can be mimed by two types of background: a) the *physical* or *correlated* background, due to the radiative muon decay (from now on: RMD): $\mu^+ \to e^+\bar{\nu}_\mu\nu_e\gamma$. The BR of RMD process is (1.4 ± 0.2) % of that of usual muon Michel decay $\mu^+ \to e^+\bar{\nu}_\mu\nu_e$ for $E_\gamma > 10$ MeV; b) the *accidental* or *uncorrelated* background, due to the coincidence, within the analysis window, of a e^+ coming from the usual muon decay and a γ coming from RMD, $e^+ - e^-$ annihilation in flight, e^+ bremsstrahlung in a nuclear field ...

While signal and RMD rates are proportional to the muon stopping rate R_μ, the accidental background rate is proportional to R_μ^2, since both particles come from the beam; the accidental background is dominant and sets the limiting sensitivity of a $\mu \to e\gamma$ experiment. Then, in the search for $\mu^+ \to e^+\gamma$ decay a continuous muon beam is preferred and R_μ must be carefully chosen to optimise the signal-to-noise ratio. The number of background events depends on the sizes of the signal region, which are determined (at fixed signal detection efficiency) by the experimental resolutions: the better the resolutions, the smaller the signal window and the smaller the number of background events. Physical effects which degrade the resolution, as multiple scattering and energy loss, are reduced by using "surface" muons, i.e. muons produced by pions stopped very close to the surface of π production target, which are efficiently brought to rest in thin targets. Moreover, high resolution detectors are mandatory. Table 1 shows the figures of merit obtained by previous $\mu \to e\gamma$ experiments compared with the final goals of the MEG [12] experiment; the 90 % C.L. upper limits on $\mu \to e\gamma$ BR are also reported.

Table 1. The performances of previous $\mu \to e\gamma$ experiments compared with that expected for MEG. All the quoted resolutions are FWHM. * shows an average of the numbers given in [11].

Place	Year	$\Delta E_e/E_e$	$\Delta E_\gamma/E_\gamma$	$\Delta t_{e\gamma}$	$\Delta\theta_{e\gamma}$	Upper limit	References
SIN	1977	8.7 %	9.3 %	1.4 ns	–	$< 1.0 \times 10^{-9}$	[13]
TRIUMF	1977	10 %	8.7 %	6.7 ns	–	$< 3.6 \times 10^{-9}$	[14]
LANL	1979	8.8 %	8 %	1.9 ns	37 mrad	$< 1.7 \times 10^{-10}$	[15]
LANL	1986	8 %	8 %	1.8 ns	87 mrad	$< 4.9 \times 10^{-11}$	[16]
LANL	1999	1.2 %*	4.5 %*	1.6 ns	17 mrad	$< 1.2 \times 10^{-11}$	[11]
PSI	$\approx$ 2013	0.8 %	4.0 %	0.15 ns	19 mrad	$< 1 \times 10^{-13}$	MEG

The MEG experiment [12] (Figure 4) uses the secondary $\pi E5$ muon beam line extracted from the PSI (Paul Scherrer Institute) proton cyclotron, the most powerful continuous hadronic machine in the world (maximum proton current $I = 2.2$ mA). A 3×10^7 μ^+/s beam is stopped in a 205 μm slanted polyethylene target. The e^+ momentum is measured by a magnetic spectrometer, composed by an almost solenoidal magnet (COBRA) with an axial gradient field and by a system of sixteen ultra-thin drift chambers (from now on: DC). The e^+ timing is measured by two double-layer arrays of plastic scintillators (Timing Counter, from now on: TC): the external layer is equipped with two sections of 15 scintillating bars each, the internal one with 512 scintillating fibres. The γ energy, direction and timing are measured in a $\approx$ 800 l volume liquid xenon (from now on: LXe) scintillation detector. The LXe as scintillating medium was chosen because of its large light yield (comparable with that of NaI) in the VUV region ($\lambda \approx 178$ nm), its homogeneity and the fast decay time of its scintillation light ($\approx$ 45 ns for γ's and $\approx$ 22 ns for α's) [17]. The LXe volume is viewed by 846 Hamamatsu 2″ PMTs, specially produced to be sensitive to UV light and to operate at cryogenic temperatures. Possible water or oxigen impurities in LXe are removed by circulating the liquid through a purification system.

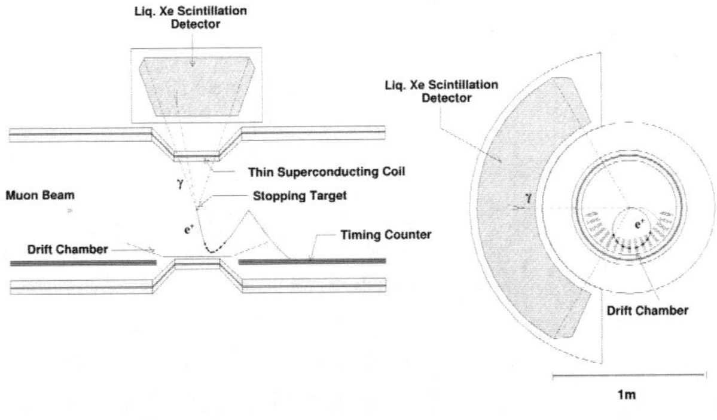

Figure 4. Layout of the MEG experiment.

A FPGA-FADC based digital trigger system was specifically developed to perform a fast estimate of the γ energy, timing and direction and of the positron timing and direction; the whole information is then combined to select events which exhibit some similarity with the $\mu \rightarrow e\gamma$ decay. The signals coming from all detectors are digitally processed by a 2 GHz custom made waveform digitizer system to identify and separate pile-up hits.

Several calibration tools (LEDs, point-like α sources deposited on wires [18], Am-Be sources, Michel decays, through going cosmic μ's, a neutron generator, 55 MeV and 83 MeV γ's from charge exchange reaction $\pi^-\mathrm{p} \rightarrow \pi^0\mathrm{n}$, γ-lines from nuclear reactions induced by a CW accelerator ...) are frequently used to measure and optimise the detector performances and to monitor their time stability. The experimental resolutions measured in summer of 2010 (the time of this conference) were: $\sigma_p/p = 0.75$ %, $\sigma_\phi = 8$ mrad and $\sigma_\theta = 11$ mrad for e^+'s, $\sigma_E/E = 2.1$ % and $\sigma_x = 5.5$ mm for γ's and $\sigma_{\Delta T} = 142$ ps for $\mathrm{e}^+ - \gamma$ relative timing. Significant improvements are expected in the following years, which should make these numbers closer to the Table 1 goals.

The data are analysed with a combination of blind and likelihood strategy. Events are pre-selected on the basis of loose cuts, requiring the presence of a track and $|\Delta T_{\mathrm{e}\gamma}| < 4$ ns. Preselected data are processed several times with improving calibrations and algorithms and events falling within a tight window ("blinding box", BB) in the $(E_\gamma, \Delta T_{\mathrm{e}\gamma})$ plane are hidden. The remaining pre-selected events fall in "sideband" regions and are used to optimise the analysis parameters, study the background and evaluate the experimental sensitivity under the zero signal hypothesis. When the optimisation procedure is completed, the BB is opened and a maximum likelihood fit is performed to the distributions of five kinematical variables (E_{e^+}, E_γ, $\Delta T_{\mathrm{e}\gamma}$, $\theta_{\mathrm{e}\gamma}$ and $\phi_{\mathrm{e}\gamma}$), in order to extract the number of Signal (S), RMD (R) and Accidental Background (B) events. Probability Distribution Functions (PDFs) are determined by using calibration measurements and MC simulations for S, theoretical formulae folded with experimental resolution for R[1] and sideband events for B. Michel positrons are used to calculate the normalization factor needed to convert an upper limit on S into an upper limit on $BR(\mu^+ \rightarrow \mathrm{e}^+\gamma)$. The analysis procedure was applied for the first time to the data collected in 2008, with reduced statistics and not optimal apparatus performances, and a first result was published [21]: $BR(\mu \rightarrow \mathrm{e}\gamma) \leq 2.8 \times 10^{-11}$ at 90 % C.L., about twice worse than the present bound [11]. In 2009 a larger and better quality data sample was collected and the analysis procedure was repeated. 370 events fell in the BB, defined as 48 MeV $< E_\gamma <$ 58 MeV and $|\Delta T_{\mathrm{e}^+\gamma}| < 0.7$ ns. Figure 5 shows the results of the maximum likelihood fit to the five kinematical variables for 2009 data. The (preliminary !) best fit result was $S = 3.0$ and $R = 35$. The analysis was repeated by different groups varying the approach (frequentistic and bayesian), the handling of sideband information and the estimated numbers of R and B in the BB; the best fit value for S ranged between 3 and 4.5 and the corresponding 90 % C.L. interval was $(0, 15)$; then, a (preliminary !) 90 % C.L. upper limit was set: $BR(\mu \rightarrow \mathrm{e}\gamma) \leq 1.5 \times 10^{-11}$, close to the current experimental limit. The experiment is expected to run at least until the end of 2012; this will produce a huge increase in statistics and, taking into account further improvements of

[1] In RMD events, the kinematical boundaries introduce a correlation between E_{e^+}, E_γ and positron-gamma relative angle which must be taken into account in the PDF.

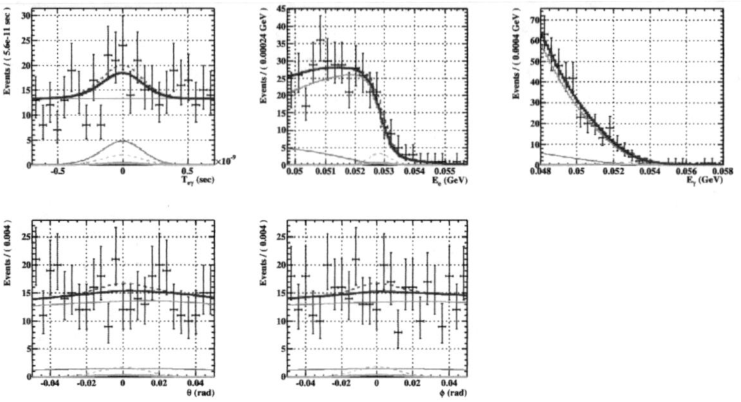

Figure 5. Results of MEG maximum likelihood analysis. From top to bottom, from left to right: $\Delta T_{e\gamma}$, E_{e^+}, E_γ, $\theta_{e\gamma}$, $\phi_{e\gamma}$. Signal PDFs are in green, RMD PDFs in red, accidental background PDFs in magenta and total PDFs in blue. The black dots represent the experimental data and the dashed lines the 90 % C.L. upper limit on the number of signal events.

detector performances, will allow to reach a sensitivity $\sim 5 \times 10^{-13}$, $(30 \div 50)$ times better than the present upper bound.

2.3. Search for $\mu \to 3e$ *decay*

In the $\mu \to 3e$ decay the final state is composed by charged particles only. In many models, for instance SUSY-GUT, its BR is related to the $\mu \to e\gamma$ BR by a factor α, since the e^+e^- pair originates from a virtual photon; then, a sensitivity of $\sim 10^{-15}$ is needed to be competitive with a $\sim 10^{-13}$ sensitivity for $\mu \to e\gamma$. The search for the $\mu \to 3e$ process is based on kinematical criteria: all possible triplets of electron tracks are formed and $\mu \to 3e$ candidate events are selected requiring a zero total momentum, an invariant mass equal to m_μ and three simultaneous tracks originating from a common vertex. The present limit $BR(\mu^+ \to e^+e^-e^+) < 10^{-12}$ at 90 % C.L. [19] dates back by 25 years, but no new $\mu \to 3e$ experiment is presently planned. However, one can ask whether this limit could be significantly improved by using the presently available beam intensity $> 10^8 \mu$/s. As for $\mu \to e\gamma$, the background for $\mu \to 3e$ decay is dominated by accidental events, coming from the coincidence of a Michel e^+ and a e^+e^- pair produced by a γ conversion in the detector or a Bhabha scattering of an other Michel e^+; the accidental event rate is proportional to R_μ^2. Since only charged particles must be detected, an electromagnetic calorimeter, with its limited resolution, is not needed and the apparatus can be based on a magnetic spectrometer only, which, however, must have a large acceptance and a relatively low momentum threshold. Morever, with intense μ beams a very high hit rate is expected in the tracking system, which causes problems of dead time, trigger and pattern recognition. With the present detector technologies a significant improvement of the limit [19] is not easy to foresee.

2.4. Search for $\mu \to e$ *conversion: Mu2E and COMET/PRISM experiments*

The $\mu \to e$ conversion is a LFV process which might take place when negative muons are captured by matter. The typical destiny of a stopped μ^- is to form a muonic atom with a nucleus; then, it can be absorbed by the nucleus or decay in orbit according to the usual Michel process. However, if the leptonic number can be violated, the muon can experience a coherent flavour change in the nuclear field, with the final state formed by a single electron only; this process is usually named $\mu \to e$ conversion. The theoretical predictions for $\mu \to e$ conversion range by some orders of magnitude, depending on the process-mediating mechanism: in SUSY frames this transition is dominated by the exchange of a virtual photon and $BR(\mu \to e) \sim 10^{-(2\div3)} BR(\mu \to e\gamma)$, but in other models more exotic schemes, like Leptoquarks, Heavy Neutrinos, a second Higgs doublet etc. are invoked. Note that some of these mechanisms can contribute to the $\mu \to e\gamma$ decay too, but some others not; therefore, the search for $\mu \to e\gamma$ decay and for $\mu \to$ e conversion provide complementary information. The present upper limit for $\mu \to$ e conversion in Au was obtained by the SINDRUM II experiment [20]: $BR(\mu \to \mathrm{e}) \leq 7 \times 10^{-13}$ at 90 % C.L.

In $\mu \to$ e conversion experiments a pulsed negative muon beam is formed from the decay of pions produced in proton collisions on fixed target and brought to stop in a layer of thin targets, where muon captures take place. The signal is a single monochromatic electron, with energy $E_\mathrm{e} = m_\mu - E_B - E_R$ where m_μ is the muon mass, E_B is the muonic atom binding energy and E_R is the nucleus recoil energy. Since E_B and E_R depend on the capturing nucleus (as a first approximation, $E_B \propto Z^2$ and $E_R \propto A^{-1}$), the E_e value is $\approx$ 105 MeV for Al and 104.3 MeV for Ti. Note that the $\mu \to$ e conversion experimental sensitivity is not limited by the accidental background, because there is only one particle in the final state. Electrons in the signal energy window can originate from a couple of beam-related sources, the muon decay in orbit (MDIO) and the radiative muon (RMC, $\mu^- A \to \nu_\mu \gamma A$) and pion (RPC, $\pi^- A \to \gamma X$) captures. The RPC background can be taken under control by reducing the π contamination in the beam ("beam purity") by means of collimators inserted within the beam line and appropriate transport systems and the MDIO by selecting a muon beam with momentum $p_\mu < 70$ MeV/c, in order to reduce the Lorentz boost of decaying electrons. Moreover, since the lifetime of muonic atoms is some hundreds of ns (for instance 860 ns in Al), one can use a pulsed beam with very short buckets ($\sim$ 100 ns), leave pions decay and search for $\mu \to$ e conversion in a delayed time window. This requires, however, that the fraction of protons arriving on the pion production target between two separate bunches is as small as possible ($\sim 10^{-9}$): this "extinction factor" is one of the key parameters in determining the final sensitivity of a $\mu \to$ e conversion experiment.

Presently there are two projects of $\mu \to$ e conversion experiments, Mu2E at Fermilab (Illinois) and COMET/PRISM at J-PARC (Japan), the latter one scheduled in two distinct stages.

Mu2E [22] (Figure 6) is derived from the original MECO project [23], which was cancelled by budget reasons in 2005. The experiment will use a 8 GeV proton beam, with 100 ns bunches, to produce a π/μ beam; then, secondary particles will be transported through a curved solenoid, arranged to single out μ^- and reject antiprotons and positive and neutral particles.

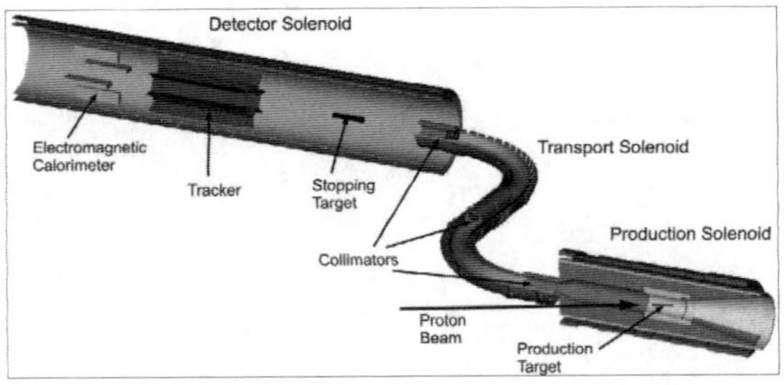

Figure 6. Layout of the Mu2E experiment.

Selected negative muons will be brought to stop in thin Al foils and electrons from muon decay or capture will be looked for by using a high resolution (900 keV FWHM at 105 MeV/c) spectrometer, with a graded magnetic field, and an electromagnetic calorimeter. The magnetic field configuration would allow to select high energy ($p > 90$ MeV/c) and recover backward going electrons.

A total number of stopped muons $N_\mu \approx 10^{18}$ is foreseen in two years of data taking; assuming $BR(\mu \to \mathrm{e}) = 10^{-15}$ and a 10^{-9} extinction factor, a 40 events signal is expected, with an estimated background < 0.5 events. On the other hand, in case of no signal Mu2E would set an upper limit: $BR(\mu \to \mathrm{e}) \leq 6 \times 10^{-17}$.

At J-PARC accelerator facility a two stage search for $\mu \to$ e conversion is planned: the goal of the first phase is to reach a sensitivity on $BR(\mu \to \mathrm{e}) < 10^{-16}$ in the COMET experiment, that of the second phase is to improve this sensitivity by two orders of magnitude in the PRISM/PRIME experiment. COMET (COherent Muon to Electron Transition, Figure 7) will use a 8 GeV pulsed proton beam with ≈ 1 μs bunch separation for two years, with a total number of stopped muons $N_\mu = 1.5 \times 10^{18}$. The experiment is conceptually similar to Mu2E, with two main differences: a) a C-shape instead of the Mu2E S-shape transport solenoid and b) a curved solenoidal spectrometer instead of the Mu2E straight solenoidal spectrometer. The C-shaped solenoid was chosen to optimise the μ momentum selection by coupling it with a suitable vertical magnetic field and the curved spectrometer to reduce the single chamber tracking rate. With an estimated background of 0.4 events, COMET would be sensitive to $\mu \rightarrow$ e $BR \geq 3 \times 10^{-17}$. In the second stage the π decay and μ transport sections of the COMET experiment will be modified and coupled with a very intense μ beam source, PRISM (Phase Rotation Intense Slow Muon source, Figure 8) [25]. A beam intensity of $10^{11 \div 12}$ μ/s is aimed, with a central momentum of 68 MeV/c. The π/μ beam will be passed through a μ storage ring, equipped with a FFAG (Fixed Field Alternating Gradient) synchrotron, where the survived pions will decay and the momentum spread will be reduced from the original ± 30 % to ± 3 %. A so small energy spread would allow to stop enough muons in very thin foils, minimising the resolution worsening due to electron interactions in the target. A final momentum resolution of 350 keV FWHM at 105 MeV/c is envisaged. The combined effect of the increased resolution and of the intense μ beam would allow to be sensitive to a $\mu \rightarrow$ e BR $< 10^{-18}$. The experimental demonstration of the phase rotation in the PRISM-FFAG ring is underway.

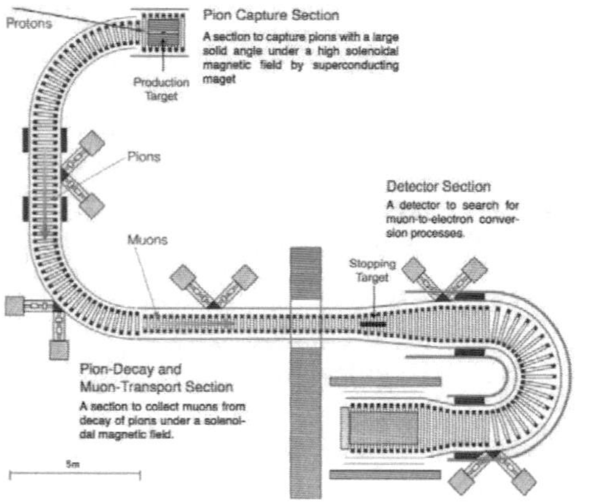

Figure 7. Layout of the COMET experiment.

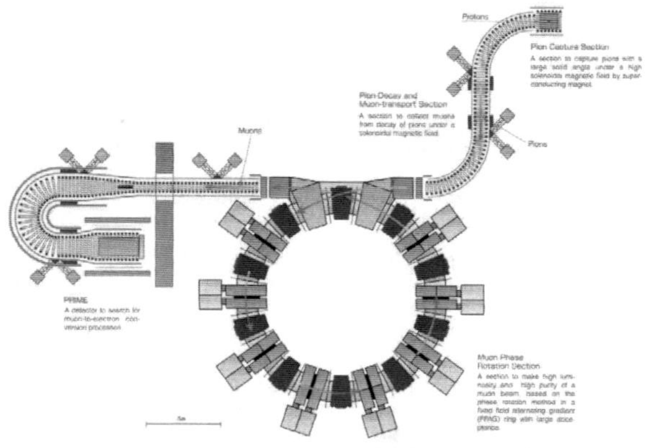

Figure 8. Layout of the PRISM/PRIME experiment.

3. The tauonic channel

3.1. Generalities

The τ lepton is in principle a very promising source of LFV decays. Thanks to the large τ mass ($m_\tau \approx 1.78$ GeV ≈ 18 m_μ) many LFV channels are open: $\tau \rightarrow \mu\gamma$, $\tau \rightarrow e\gamma$, $\tau \rightarrow 3$ leptons (from now on $\tau \rightarrow 3\ \ell$), $\tau \rightarrow$ lepton + hadron(s) (from now on $\tau \rightarrow \ell + h(s)$) .. and in many SUSY and SUSY-GUT schemes the BRs are enhanced with respect to the muon LFV decays by a factor $(m_\tau/m_\mu)^\alpha$, $\alpha \geq 3$. Therefore one expects ([26], [27]): $BR(\tau \rightarrow \mu\gamma)/BR(\mu \rightarrow e\gamma) \approx 10^{(3 \div 5)}$ and experiments searching for $\tau \rightarrow \mu\gamma$ must reach a sensitivity $\sim 10^{-(9 \div 10)}$ to be competitive with dedicated μ LFV decay experiments. The BR of other τ LFV processes are generally expected to be lower than that of $\tau \rightarrow \mu\gamma$: for instance, $\tau \rightarrow e\gamma$ is usually disfavoured because of the small coupling between first and third generation and $\tau \rightarrow 3\ \ell$ is suppressed by a factor α due to an intermediate virtual photon. However, in models with heavy Dirac neutrinos or inverted slepton hierarchy values of $BR(\tau \rightarrow e\gamma) > BR(\tau \rightarrow \mu\gamma)$ are predicted. We remind again that the complementarity between the various LFV channels is essential in the search for new physics and for a profound understanding of the flavour structure.

However, since τ lifetime is only 2.9×10^{-13} s, τ beams can not be realized and large τ samples must be obtained in intense electron or proton accelerators, operating in an energy range where the τ production cross section is large and coupled with refined detectors, able to select and reconstruct rare events. The first experiments to satisfy these requirements are that based on B factory machines, BaBar [28] at SLAC (USA) and Belle [29] at KEKB (Japan). Both experiments operate at a total Center of Mass (CM) energy at the peak of the $Y(4S)$ resonance ($\sqrt{s} = 10.54$ GeV); at this energy $\sigma\left(e^+e^- \to \tau^+\tau^-\right) \approx 0.9\,\sigma\left(e^+e^- \to b\bar{b}\right)$ so that the B factories are τ factories too. Moreover, in e^+e^- colliders the initial state is very well known and high resolution detector technologies are employed. Belle and BaBar are large central detectors, equipped with a combination of tracking devices, particle identification (PID) systems, vertex detectors and calorimeters. The main difference is related to the PID technique, based on a threshold Čerenkov counter, the time-of-flight and the tracker dE/dx for Belle and on a RICH and dE/dx in the trackers for BaBar. Several LFV decays are searched for: the gamma-leptonic $\tau \to \ell\gamma$ ($\ell = \mu, e$), the purely leptonic $\tau \to 3\,\ell$ and the leptonic-hadronic ones $\tau \to \ell + h(s)$.

3.2. Search for LFV τ decays with Belle and BaBar

The Belle and BaBar upper limits on τ LFV decays continuously improved since 2004, according to the increase of the collected and analysed data samples. In all these searches the event world is divided in two hemispheres, defined by the thrust axis: the "tag side" and the "signal side". Events with $\tau^+\tau^-$ pairs are selected by identifying in the tag side a standard SM τ decay, while possible LFV decays are searched for in the signal side. The tagging is based on the purely leptonic $\tau \to \ell\nu\bar{\nu}$ decay or on decays involving a ν and at least one prong. In the signal side LFV candidates are preselected according to the expected topology of the process under study: for $\tau \to \ell\gamma$ one single muon or electron and at least one γ; for $\tau \to 3\,\ell$ all possible triplets of charged leptons and for $\tau \to \ell + h(s)$ an isolated lepton plus the combination of tracks expected for the selected hadron(s) (for instance, for $\tau \to \ell + K^0_S$ a lepton and a $\pi^+\pi^-$ pair). Preliminary topological cuts are applied and then a blind strategy is used: the signal region is hidden and sideband data and MC simulations are used to estimate the background and optimise the selection criteria. The efficiency for LFV searches is usually $\sim 3 \div 10$ %.

In the $\tau \to \ell\gamma$ searches the main background comes from the coincidence of a γ from initial (ISR) or final state radiation and an isolated lepton from the usual τ decay. Radiative processes like $e^+e^- \to \mu^+\mu^-\gamma$ or $c\bar{c}$ pairs are also relevant background sources. It is important to note that the ISR represents an irreducible and unavoidable noise, which limits the sensitivity of these experiments to $\tau \to \ell\gamma$ decays. We will discuss later this point. The BaBar data sample analysed for $\tau \to \ell\gamma$ search corresponds to a total luminosity $\int \mathcal{L}dt > 500$ fb^{-1}; the calculated number of τ decays is $(963 \pm 7) \times 10^6$. After applying all preliminary selections, the $\tau \to \mu\gamma$ and $\tau \to e\gamma$ candidate events are studied in the $(\Delta E, M_{\ell\gamma})$ plane, where ΔE is the difference between the total energy of the $(\ell + \gamma)$ pair and the beam energy in the CM frame and $M_{\ell\gamma}$ is the $(\ell + \gamma)$ pair invariant mass. For a LFV τ decay, $\Delta E = 0$ and $M_{\ell\gamma} = M_\tau$, but because of the finite resolution one must consider a two-dimensional region. Events in a 3 σ window are blinded and the expected background is evaluated; then, the blinded region is opened and one looks at the events observed in the 2 and 3 σ windows around the nominal values. Figure 9 shows the distribution of selected events in the $(\Delta E, M_{\ell\gamma})$ plane for $\tau \to \mu\gamma$ (left) and $\tau \to e\gamma$ (right): the red dots are experimental points, the black ellipses are the 2 σ contours and the yellow and green regions contain 90 % and 50 % of MC signal events. Since the number of measured events is in agreement with the expected background, the following limits are set: $BR(\tau \to \mu\gamma) \leq 4.4 \times 10^{-8}$ and $BR(\tau \to e\gamma) \leq 3.3 \times 10^{-8}$ [30] at 90 % C.L. The Belle data sample for this search is almost equivalent (954×10^6 τ decays) and the strategy is quite similar. Figure 10 shows the distributions of Belle events in the $(M_{\ell\gamma}, \Delta E)$ plane for $\tau \to \mu\gamma$ (left) and $\tau \to e\gamma$ (right). The dashed and dotted-dashed ellipses represent the 3 σ and 2 σ contours,

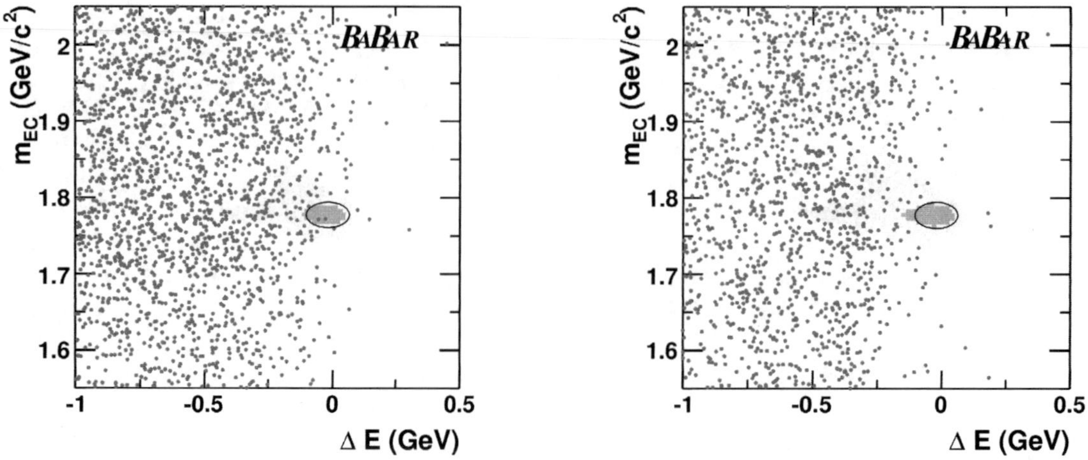

Figure 9. Results of the BaBar search for $\tau \to \mu\gamma$ (left) and $\tau \to e\gamma$ (right) LFV decays.

the diagonal dashed lines define the 2 σ band of the shorter ellipse axis and the shaded boxes indicate the signal MC events. The number of signal events was extracted by a maximum

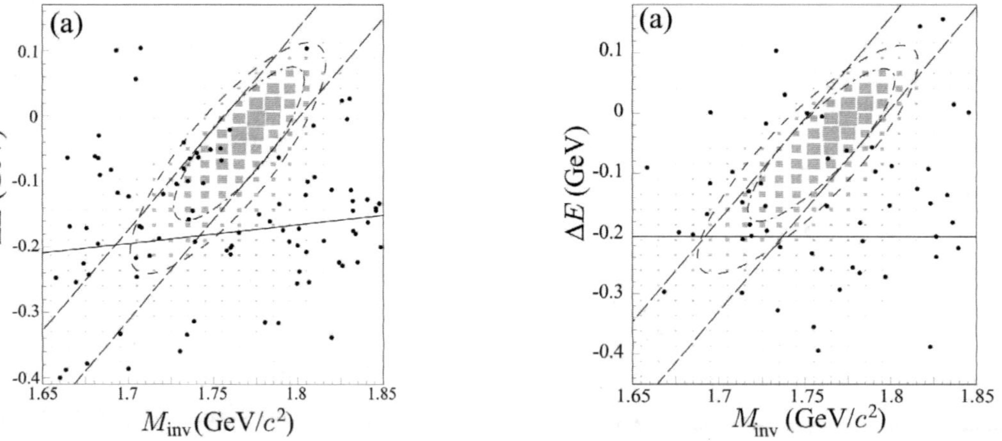

Figure 10. Results of the Belle search for $\tau \to \mu\gamma$ (left) and $\tau \to e\gamma$ (right) LFV decays.

likelihood fit obtaining $-3.9^{+3.6}_{-3.2}$ for $\tau \to \mu\gamma$ and $-0.14^{+2.18}_{-2.45}$ for $\tau \to e\gamma$; the corresponding limits are $BR(\tau \to \mu\gamma) \leq 4.5 \times 10^{-8}$ and $BR(\tau \to e\gamma) \leq 1.2 \times 10^{-7}$ [31] at 90 % C.L.

The search for $\tau \to 3\ \ell$ decay is potentially more interesting from the experimental point of view since with only charged particles in the final state the mass resolution is excellent and there are no irreducible sources of noise. The search strategy for both experiments is based on the distribution of events in the $(M_{3\ell}, \Delta E)$ plane, where $M_{3\ell}$ is the invariant mass of each possible 3-leptons combination having the required sign of charge. The main background comes from $q\bar{q}$ and Bhabha pairs, but since these events can be efficiently rejected by appropriate cuts, the residual noise in the signal region ($M_{3\ell} \approx M_\tau$ and $\Delta E \approx 0$) is very low. Table 2 shows the results of the BaBar (468 fb^{-1} data sample) [32] and Belle (782 fb^{-1} data sample) [33] searches for $\tau \to 3\ \ell$ LFV decays[2]. The 90 % C.L. upper limits on LFV decay BRs range from 1.8 to

[2] Results are shown for τ^-, but charge conjugation is implied: then, for instance $\tau^- \to e^-e^+e^-$ includes

Table 2. Efficiencies, number of expected background and observed events and 90 % C.L. BR upper limits for $\tau \to 3\ \ell$ LFV searches for BaBar and Belle (adapted from [32] and [33]).

Mode	BaBar				Belle			
	ϵ (%)	$N_{\rm bkg}$	$N_{\rm obs}$	BR^{90}	ϵ (%)	$N_{\rm bkg}$	$N_{\rm obs}$	BR^{90}
$e^-e^+e^-$	8.6 ± 0.2	0.12 ± 0.02	0	2.9×10^{-8}	6.0	0.21 ± 0.15	0	2.7×10^{-8}
$\mu^-e^+e^-$	8.8 ± 0.5	0.64 ± 0.19	0	2.2×10^{-8}	9.3	0.04 ± 0.04	0	1.8×10^{-8}
$\mu^+e^-e^-$	12.7 ± 0.7	0.34 ± 0.12	0	1.8×10^{-8}	11.5	0.01 ± 0.01	0	1.5×10^{-8}
$e^+\mu^-\mu^-$	10.2 ± 0.6	0.03 ± 0.02	0	2.6×10^{-8}	10.1	0.02 ± 0.02	0	1.7×10^{-8}
$e^-\mu^+\mu^-$	6.4 ± 0.4	0.54 ± 0.14	0	3.2×10^{-8}	6.1	0.10 ± 0.04	0	2.7×10^{-8}
$\mu^-\mu^+\mu^-$	6.6 ± 0.6	0.44 ± 0.17	0	3.3×10^{-8}	7.6	0.13 ± 0.06	0	2.1×10^{-8}

3.3×10^{-8} for BaBar and from 1.5 to 2.7×10^{-8} for Belle, depending on the individual channel. Finally, both Belle and BaBar searched for LFV $\tau \to \ell + h(s)$ decays, where a charged lepton of the same sign of the τ is emitted together with a combination of pseudo-scalar or vector hadrons (e.g. $\tau \to \ell\pi^0$, $\tau \to \ell\omega$, $\tau \to \ell\pi^+\pi^-$...). Many of these channels are very clean, without irreducible backgrounds. No evidence for LFV decays was found and the 90 % C.L. upper limits on BRs lie between 2 and 20×10^{-8} (for summaries see [34] and [35]).

Combined analyses were also performed to obtain global Belle+BaBar 90 % C.L. upper limits on LFV decays, both in bayesian and frequentistic framework. For a 2-years ago review see [10].

4. Future perspectives

4.1. Introduction

Future accelerators, like NUFACT [36] or Project X [37], are expected to deliver high intense ($\sim 10^{15}$ p/s) proton beams with energy of tens of GeV or higher; then, secondary muon beams up to $\sim 10^{14}$ μ/s could be available in the future. Moreover, two Super B projects, one in Italy [38] and one in Japan [39], which would reach luminosities $\sim 10^{35 \div 36}$ cm^2 s^{-1}, are under study and are expected to be on-line in some years. Since in both the muonic and tauonic channel one can expect very large improvements in data statistics, a spontaneous question arises: can the LFV searches take benefit from future high intensity machines ? If yes, at which level ?

4.2. Possible improvements in the muonic channel

As discussed before, the sensitivity of a $\mu \to e\gamma$ experiment is limited by the accidental background, whose rate is proportional to R_μ^2. Then, an increase in R_μ does not represent by itself an advantage, since it can cause a worsening of the signal-to-noise ratio. The $\sim 10^{-13}$ MEG projected sensitivity is a tough experimental challenge, since the required detector resolutions (Table 1) are at the limits of present technologies. Important detector improvements are needed to put the sensitivity to the $\mu \to e\gamma$ BR significantly below the MEG one. R&D studies are under way to gain an order of magnitude, taking advantage from the maximum beam intensity available at PSI ($\sim 10^8$ μ/s, possibly enhanced in the future) coupled with detector upgrades. Some proposed ideas are, for example, the use of high-resolution β spectrometers and of finely segmented targets. However, further improvements in sensitivity below 10^{-14} seem presently rather unlike. Similar considerations are valid for $\mu \to 3$e. In the $\mu \to$ e channel the situation is more promising, since the signal is an isolated electron, not affected by accidental background; then, the sensitivity is expected to increase significantly when high intensity machines will be

$\tau^+ \to e^+e^+e^-$ too.

on-line. However, very intense beams are also sources of problems, like high levels of radiation, large momentum spread etc. and several technical remarks must be taken into account. Detailed discussions of the potentialities of high intensity machines in the LFV muon decay searches and of the needed experimental solutions can be found in [40] and [41].

4.3. Potentialities of the SuperB projects

The Super B machines are projects of very intense e^+e^- machines, which would reach integrated luminosities $\int \mathcal{L}dt \approx (50 \div 75)$ ab^{-1}, ~ 50 times larger than the combined Belle+Babar data sample. With such a big increase in luminosity one could expect a sensitivity improvement on the LFV searches by two orders of magnitude. However, it is necessary to remind that the sensitivity scales as $1/\int \mathcal{L}dt$ only for a background-free experiment; otherwise, it scales only as $1/\sqrt{\int \mathcal{L}dt}$ and the expected improvement is much less significant. In the BaBar and Belle searches, the "golden" channel $\tau \to \ell\gamma$ is affected by a small, but not negligible background and the ISR represents an irreducible noise. Extrapolating the present limits on the basis of the increasing luminosity only, one obtains a predicted sensitivity on $\tau \to \ell\gamma$ decays of $\sim 2 \times 10^{-9}$, not completely satisfactory. A factor two improvement is expected by the use of polarized beams and of appropriate analysis selections and refinements in detector technologies are foreseen [42]. On the other hand, the purely leptonic channel $\tau \to 3\ \ell$ is potentially more promising, since background-free searches seem feasible; no background events were observed by Belle and BaBar and the expected number of noise events was $N_{\mathrm{bkg}} < 1$ (Table 2). Super B projects aim to reach a sensitivity $\sim 2 \times 10^{-10}$ on $\tau \to 3\ \ell$ LFV decays. Finally, the $\tau \to \ell + h(s)$ decays represent an intermediate situation, since they are almost background free but the efficiencies are largely different from channel to channel. The predicted sensitivities are in the range $(2 \div 6) \times 10^{-10}$.

4.4. LHC

τ leptons are also copiously produced in the LHC accelerator, mainly via B and D decays and, to a much lesser extent, via W and Z^0 decays. Detailed studies of possible detection of $\tau \to \mu\gamma$ decay in CMS and ATLAS were performed ([43], [44]): because of the unavoidable background, the sensitivity to this channel is not competitive with that of B factory experiments. The $\tau \to 3\mu$ channel looks more promising [45], even if only τ's from W or Z^0 decays could produce LFV processes acceptable by the trigger schemes of LHC experiments. (Dedicated trigger algorithms with improved efficiency for muons from decay of τ's originating from B or D mesons are under study.) Assuming that the backgrounds can be effectively suppressed by appropriate selection criteria, one obtains 95 % C.L. upper limits $(3.8 \div 7) \times 10^{-8}$ for an integrated luminosity of $(10 \div 30)$ fb^{-1}, comparable with the sensitivity levels reached by B factory experiments.

5. Summary and conclusions

It's an exciting era for LFV searches: a) the MEG experiment obtained a new limit on $BR(\mu^+ \to e^+\gamma)$ and is now starting a long term stable data taking to reach a projected sensitivity $\sim 5 \times 10^{-13}$; b) two $\mu \to e$ conversion experiments (Mu2E and COMET) are expected to go on-line around 2016, with expected sensitivities $\leq 10^{-16}$; c) B factory based experiments (Belle and BaBar) set first significant limits on LFV τ decay BRs, at level of few $\times 10^{-8}$; d) a factor $(10 \div 100)$ improvement in sensitivity on these searches is expected from SuperB projects.

Even more significant results could be obtained if projected high intensity accelerators will become a reality and detector technologies will be upgraded. In any case, since several theoretical predictions of LFV processes are within the reach of present or near future experiments, the discovery of LFV in the charged sector is not a dream, but a realistic hope.

5.1. Acknowledgments

First of all, I want to thank all the MEG colleagues for the long time spent in studying the detector, performing beam tests, analysing data ... I'm particularly grateful to the Pisa colleagues (A. Baldini, C. Bemporad, L. Galli, G. Gallucci, M. Grassi, D. Nicolò, A. Papa and G. Signorelli) for continuous discussions and fruitful collaborations in all activities of the experiment. Then, I want to thank some people of the Mu2E (F. Cervelli) and BaBar groups in Pisa (G. Batignani, A. Cervelli, M. Giorgi, A. Lusiani and N. Neri) for providing me with a lot of information about μ and τ physics and LFV searches at their experiments. Finally, I want to thank the organizers of this workshop and expecially Oscar Vives and Sergio Pascos for their support and for the opportunity of giving this talk in the marvellous Valencia.

References

[1] Strumia A and Vissani F, 2006 Neutrino masses and mixings and ... *Preprint* hep-ph/0606054
[2] Barbieri R and Hall L J, 1994 *Phys. Lett.* B **338** 212
[3] Barbieri R, Hall L J and Strumia A, 1995 *Nucl. Phys.* B **445** 219
[4] Hisano J *et al* , 1997 *Phys. Lett.* B **391** 341 and Erratum, 1997 *Phys. Lett.* B **391** 357
[5] Cabibbi L *et al* , 2006 *Phys. Rev.* D **74** 116002
[6] Cabibbi L *et al* , 2009 *Proc. Europh. Conf. on High Energy Physics HEP2009 (Cracow)* p 167
[7] Hincks E P and Pontecorvo B, 1948 *Phys. Rev.* **73** 257
[8] Lokanathan S and Steinberger J, 1955 *Phys. Rev.* A **98** 240
[9] Feinberg G, 1958 *Phys. Rev.* **110** 1482
[10] Marciano W J, Mori T and Roney J M, 2008 *Annu. Rev. Nucl. Part. Sci.* **58** 315
[11] Ahmed M *et al* (MEGA Collaboration), 2002 *Phys. Rev.* D **65** 112002
[12] Baldini A *et al* (MEG Collaboration), 2002 Proposal to INFN, see http://meg.psi.ch.
[13] Van der Schaaf A *et al* , 1980 *Nucl. Phys.* A **340** 249
[14] Depommier P *et al* , 1977 *Phys. Rev. Lett.* **39** 113
[15] Kinnison W W *et al* , 1982 *Phys. Rev.* D **25** 2846
[16] Bolton R D *et al* , 1988 *Phys. Rev.* D **38** 2077
[17] Baldini A *et al* , 2005 *Nucl. Instr. and Meth.* A **545** 753
[18] Baldini A *et al* , 2006 *Nucl. Instr. and Meth.* A **565** 589
[19] SINDRUM I Collaboration, 1985 *Nucl. Phys.* B **260** 1
[20] SINDRUM II Collaboration, 2003 *Proc. NOON2003 Workshop (Kanazawa)*
[21] MEG Collaboration, 2010 *Nucl. Phys.* B **834** 1
[22] Mu2E Collaboration, 2008 Proposal to Fermilab, see http://mu2e.fnal.gov/public/hep/general/proposals.html
[23] Bachman S *et al* , 1997 BNL Proposal AGS P940
[24] COMET Collaboration, 2008 Proposal to JPARC see http://j-parc.jp/NuclPart/pac_0801/pdf/Kuno.pdf
[25] PRISM Collaboration, 2008 Proposal to JPARC see http://j-parc.jp/NuclPart/Proposal_e.html, P21
[26] Rossi A, 2003 *Proc. IFAE 2003 Workshop (Lecce)*
[27] Ellis J *et al* , 2002 *Phys. Rev.* D **66** 115013.
[28] Aubert B *et al* (BaBar Collaboration), 2002 *Nucl. Instr. and Meth.* A **479** 1
[29] Abashian A *et al* (Belle Collaboration), 2002 *Nucl. Instr. and Meth.* A **479** 117
[30] Aubert B *et al* (BaBar Collaboration), 2010 *Phys. Rev. Lett.* **104** 021802
[31] Hayasaka K *et al* (Belle Collaboration), 2008 *Phys. Lett.* B **666** 16
[32] Aubert B *et al* (BaBar Collaboration), 2010 *Phys. Rev.* D **81** 111101
[33] Hayasaka K *et al* (Belle Collaboration), 2010 *Phys. Lett.* B **687** 139
[34] Hayasaka K, 2010 *Proc. Int. Conf. on High Energy Physics ICHEP2010 (Paris)*, to be published
[35] Banerjee S, 2010 *Proc. Int. Conf. on High Energy Physics ICHEP2010 (Paris)*, to be published
[36] International Design Study for the Neutrino Factory, see https://www.ids-nf.org/wiki/FrontPage
[37] see http://projectx.fnal.gov
[38] see http://www.pi.infn.it/SuperB
[39] see http://belle2.kek.jp
[40] Äystö J *et al* , 2001 CERN-TH/2001-231
[41] Albrow M G *et al* , 2005 Physics at a Fermilab proton driver *Preprint* hep-ex/0509019
[42] Giorgi M, 2010 *Proc. Int. Conf. on High Energy Physics ICHEP2010 (Paris)*, to be published
[43] Ünel N, 2005 *Proc.* 40th *Recontres de Moriond on Electroweak Interactions and Unified Theories (La Thuile)*
[44] Santinelli R, 2002 *Proc.* 7th *International Workshop on Tau Lepton Physics (TAU02) (Santa Cruz)*
[45] Raidal M *et al* , 2008 *Eur. Phys. J.* C **57** 13

Journal of Physics: Conference Series **259** (2010) 012011 doi:10.1088/1742-6596/259/1/012011

Indirect detection of dark matter, current status and recent results

Aldo Morselli
INFN Roma Tor Vergata
E-mail: aldo.morselli@roma2.infn.it

Abstract. Since its launch in the 2008, the Large Area Telescope, onboard of the Fermi Gamma-ray Space Telescope, has detected the largest amount of gamma rays, in the 20MeV 300GeV energy range and electrons + positrons in the 7 GeV- 1 TeV range. This impressive statistics allows one to perform a very sensitive indirect experimental search for dark matter. I will present the latest results on these searches.

1. The Cosmic Ray Electron spectrum

Recently the experimental information available on the Cosmic Ray Electron (CRE) spectrum has been dramatically expanded as the Fermi-LAT Collaboration [1, 2] has reported a high precision measurement of the electron spectrum from 7 GeV to 1 TeV performed with its Large Area Telescope (LAT) [3, 4]. The spectrum shows no prominent spectral features and it is significantly harder than that inferred from several previous experiments. These data together with the PAMELA data on the rise above 10 GeV of the positron fraction[5] are quite difficult to explain with just secondary production [6],[7] [9]. The temptation to claim the discovery of dark matter is strong but there are competing astrophysical sources, such as pulsars, that can give strong flux of primary positrons and electrons (see [10], [11], [12], [13] and references therein). At energies between 100 GeV and 1 TeV the electron flux reaching the Earth may be the sum of an almost homogeneous and isotropic component produced by Galactic supernova remnants and the local contribution of a few pulsars with the latter expected to contribute more and more significantly as the energy increases.

Two pulsars, Monogem, at a distance of d=290 pc and Geminga, at a distance of d=160 pc, can give a significant contribution to the high energy electron and positron flux reaching the Earth and with a set of reasonable parameters of the model of electron production we can have a nice fit of the PAMELA positron fraction[5] and Fermi data (see figures 1 and 2) , but it is true that we have a lot of freedom in the choice of these parameters because we still do not know much about these processes, so further study on high energy emission from pulsars are needed in order to confirm or reject the pulsar hypothesis.

Nevertheless a dark matter interpretation of the Fermi-LAT and of the PAMELA data is still an open possibility. In Figure 3 is shown the parameter space of particle dark matter mass versus pair-annihilation rate, for models where dark matter annihilates into monochromatic $e^{\pm}$ [13]. The preferred range for the dark matter mass lies between 400 GeV and 1-2 TeV, with larger masses increasingly constrained by the H.E.S.S. results. The required annihilation rates, when employing the dark matter density profile imply typical boost factors ranging between

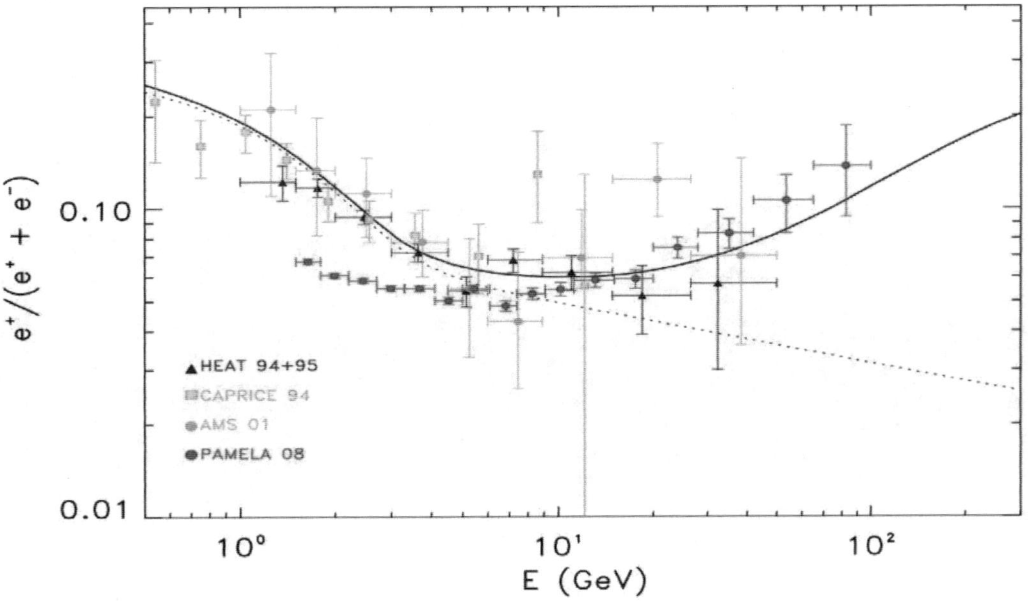

Figure 1. PAMELA data and a possible contribution from Monogem and Geminga pulsars [13]. Black-dotted line shows the background from secondary positrons in cosmic rays from GALPROP

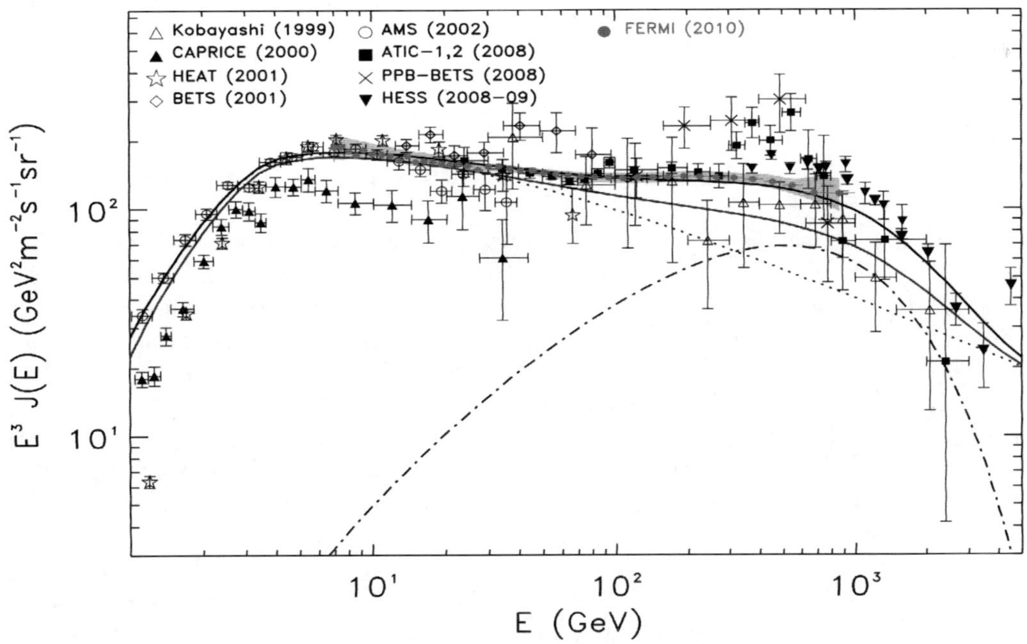

Figure 2. Electron-plus-positron spectrum (blue continuos line) for the same scenario as in figure 1. The gray band represents systematic errors on the Fermi-LAT data [3, 4]

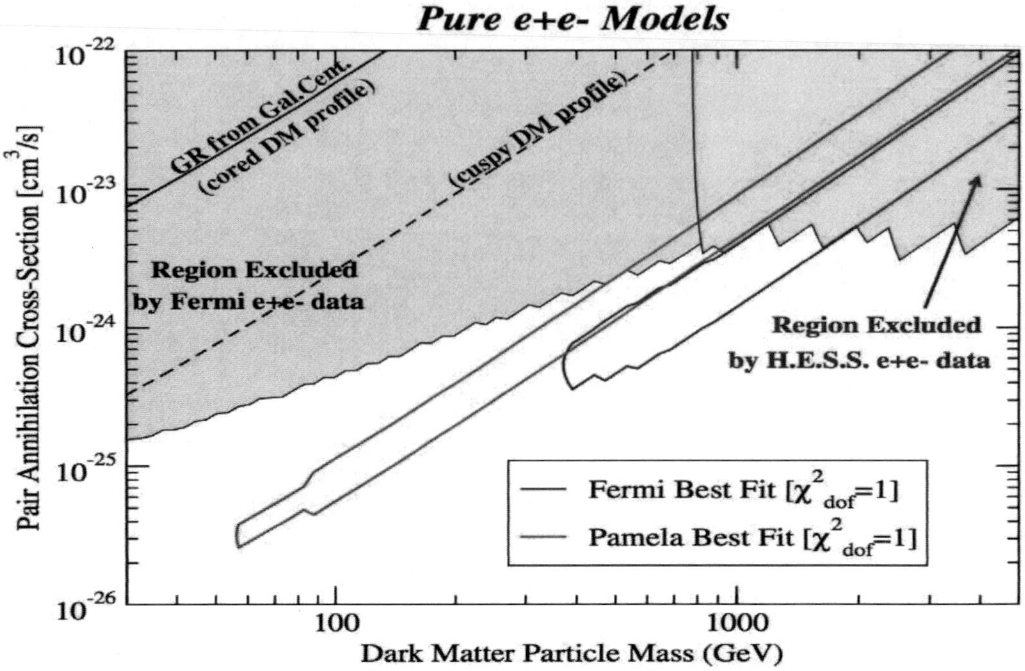

Figure 3. The parameter space of particle dark matter mass versus pair-annihilation rate, for models where dark matter annihilates into monochromatic $e^{\pm}$. Models inside the regions shaded in gray and cyan over-produce $e^{\pm}$ from dark matter annihilation with respect to the Fermi-LAT and H.E.S.S. measurements, at the 2-σ level. The red and blue contours outline the regions where the χ^2 per degree of freedom for fits to the PAMELA and Fermi-LAT data is at or below 1.

20 and 100, when compared to the value $\langle\sigma v\rangle \sim 3 \times 10^{-26}$ cm^3/sec expected for a thermally produced dark matter particle relic.

How can one distinguish between the contributions of pulsars and dark matter annihilations? Most likely, a confirmation of the dark matter signal will require a consistency between different experiments and new measurements of the reported excesses with large statistics. The observed excess in the positron fraction should be consistent with corresponding signals in absolute positron and electron fluxes in the PAMELA data and all lepton data collected by Fermi. Fermi has a large effective area and long projected lifetime, 5 years nominal with a goal 10 years mission, which makes it an excellent detector of cosmic-ray electrons up to ~1 TeV. Future Fermi measurements of the total lepton flux with large statistics will be able to distinguish a gradual change in slope with a sharp cutoff with high confidence [14]. The latter, can be an indication in favor of the dark matter hypothesis.

Another possibility is to look for anisotropies in the arrival direction of the electrons.

The Large Area Telescope on board the *Fermi* satellite (*Fermi*-LAT) detected more than 1.6 million cosmic-ray electrons/positrons with energies above 60 GeV during its first year of operation. The arrival directions of these events were searched for anisotropies of angular scale extending from ~10° up to 90°, and of minimum energy extending from 60 GeV up to 480 GeV. Upper limit for the dipole anisotropy has been set to 0.5 - 5% depending on the energy [15].

The levels of anisotropy expected for Vela-like and Monogem-like sources (i.e. sources with similar distances and ages) seem to be higher than the scale of anisotropies excluded by the results (see figure 4). However, it is worth to point out that the model results are affected by

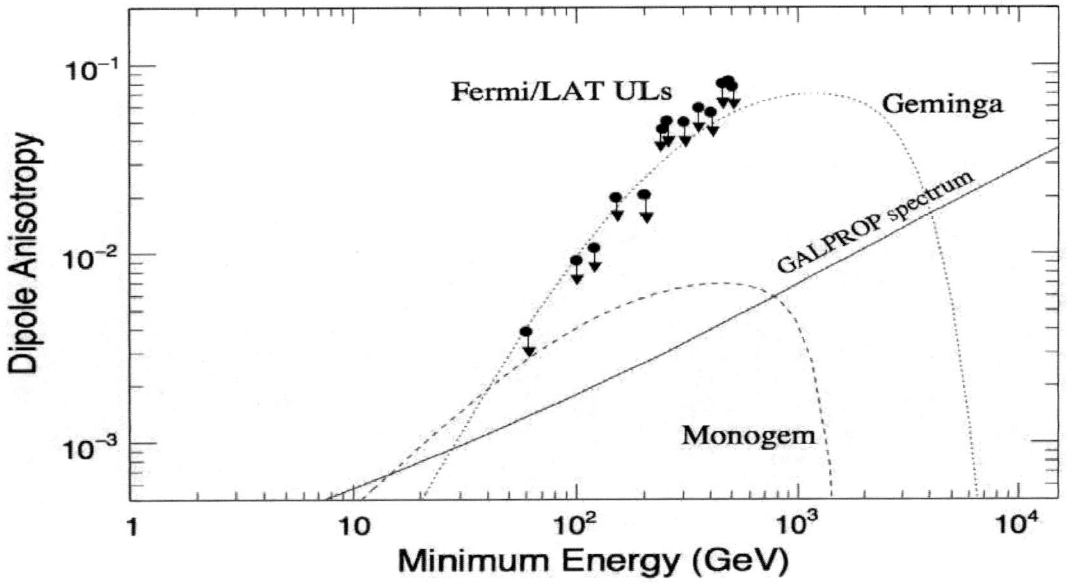

Figure 4. Dipole anisotropy δ versus the minimum energy for GALPROP (solid line), Monogem source (dashed line), and Vela source (dotted line). The 3σ confidence level from the data is also shown with circles. The solar modulation was treated using the force-field approximation with Φ=550 MV

large uncertainties related to the choice of the free parameters.

2. The gamma-ray signals

A strong leptonic signal should be accompanied by a boost in the γ-ray yield providing a distinct spectral signature detectable by Fermi.

The Galactic center (GC) is expected to be the strongest source of γ - rays from DM annihilation, due to its coincidence with the cusped part of the DM halo density prole [16, 17]. A preliminary analysis of the data, taken during the first 11 months of the Fermi satellite operations, is shown in figures 5 and 6.

The reported results were obtained with a binned likelihood analysis, performed by means of the tools developed by the Fermi/LAT collaboration (gtlike, from the Fermi analysis tools [18]). For this analysis:

- a ROI of 7°×7° was considered. This ROI was used in order to minimize the background contribution and to avoid significative leakage of the gamma ray signal under study;
- the ROI was centred at the position RA = 266.46°, Dec=-28.97°, i.e. the position of the brightest source;
- the Data taken during the first 11 months (8/2008-7/2009) have been used;
- the events were selected to have energy between 400MeV and 100GeV;
- only events classified of "diffuse" class were and which converted in the *front* part of the tracker have been selected for the analysis. The selection in energy, event classification and conversion provided us with events with very well reconstructed incoming direction and data have been binned into a 100×100bins map;
- the IRF and the events classification are those relative to the Pass6V3 version of the Fermi/LAT analysis software.

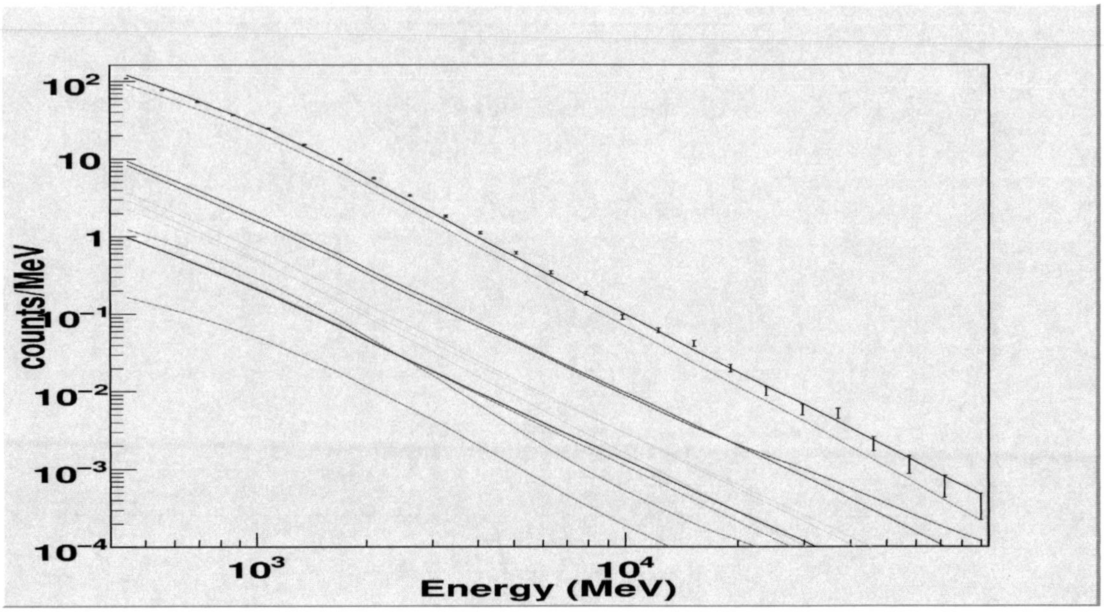

Figure 5. Spectra from the likelihood analysis of the Fermi/LAT data (number of counts vs reconstructed energy) in a 7°×7° region around the Galactic Center (number of counts vs reconstructed energy). The likelihood analysis is the standard one used with the LAT data. The main analysis steps are: (1) to select data of high quality (selection cuts on events energy, zenith angle, reconstruction and classification quality); (2) to build a emission model of the region, based on the previous knowledge and experimental evidence of new excesses with enough statistical significance; (3) to apply the likelihood analysis to the data and the considered model. For each model component a fit of the free parameters and the computation the statistical significance is obtained. Here in the plot, from above to below: -the black point are the observed data; -the black line is the sum of all the components; -the red line is the Galactic diffuse emission; - the lower black line is the isotropic extragalactic; -other components are the sources detected. These results are preliminary.

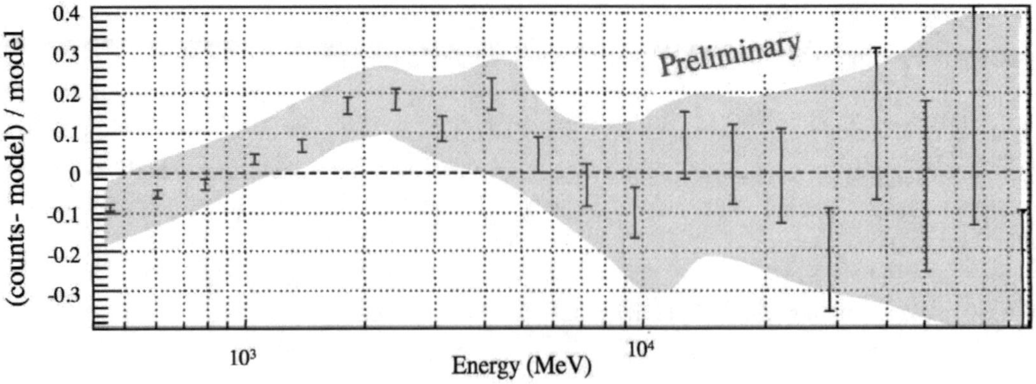

Figure 6. Residuals ((exp.data - model)/model) of the above likelihood analysis. The blue area shows the systematic errors on the effective area. These results are preliminary.

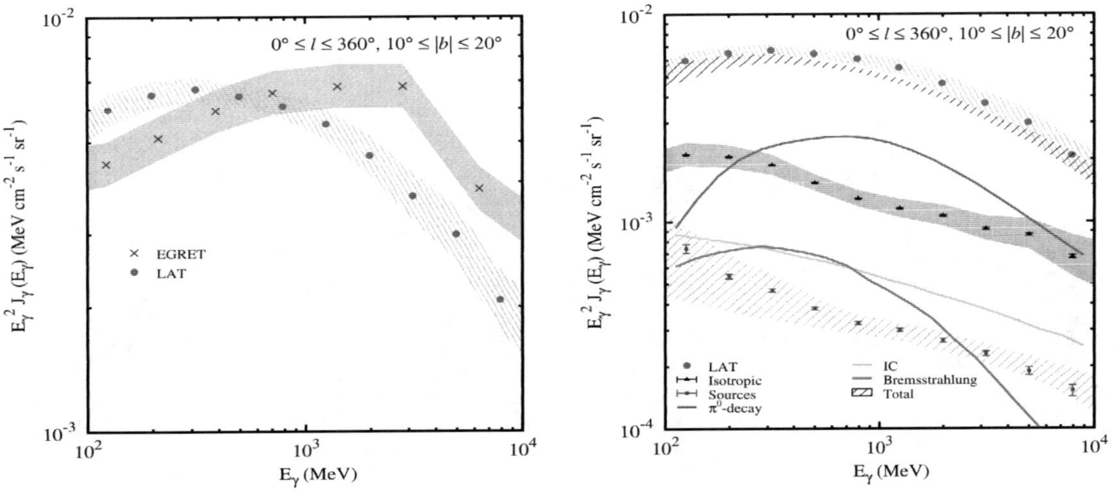

Figure 7. *Left:* Diffuse emission intensity averaged over all Galactic longitudes for latitude range $10° \leq |b| \leq 20°$. Data points: Fermi LAT, red dots; EGRET, blue crosses. Systematic uncertainties: Fermi LAT, red; EGRET, blue. *Right:* Fermi LAT data with model, source, and isotropic components for same sky region.

In order to perform the likelihood analysis for the LAT data, a model of the already known sources and the diffuse background should be built. The model in use for the presented analysis contains 11 sources in the Fermi 1 year catalog [19] which are located into or very close to the considered ROI. These sources have a point-like spatial model and a spectrum in the form of a power-law. The model also contain the diffuse gamma-ray background which is made of two components:

(i) the Galactic Diffuse gamma-ray background. The observed Galactic Diffuse emission was modelled by means of the GALPROP code [20] and [21], and the realization of the galactic emission named gll_iem_54_87Xexph7S.fit was used. During the likelihood maximization only the normalization of the GALPROP model is varied, not its components;

(ii) the Isotropic Background. This component should account for both the Extragalactic gamma-ray emission and residual charged particles. It is modelled as an isotropic emission with a template spectrum.

The diffuse gamma-ray backgrounds and discrete sources, as we know them today, can account for the large majority of the detected gamma-ray emission from the Galactic Center. Nevertheless a residual emission is left, not accounted for by the above models [22].

Improved modelling of the Galactic diffuse model as well as the potential contribution from other astrophysical sources (for instance unresolved point sources) could provide a better description of the data. Analyses are under-way to investigate these possibilities.

An excess in gamma-ray should also be seen in the Galactic diffuse spectrum. Figure 7 (left) shows the LAT data averaged over all Galactic longitudes and the latitude range $10° \leq |b| \leq 20°$. The hatched band surrounding the LAT data indicates the systematic uncertainty in the measurement due to the uncertainty in the effective area described above. Also shown on the right are the EGRET data for the same region of sky where one can see that the LAT-measured spectrum is significantly softer than the EGRET measurement [23]. Figure 7 (right) compares the LAT spectrum with the spectra of an *a priori* diffuse Galactic emission (DGE) model. While the LAT spectral shape is consistent with the DGE model used in this paper, the overall model

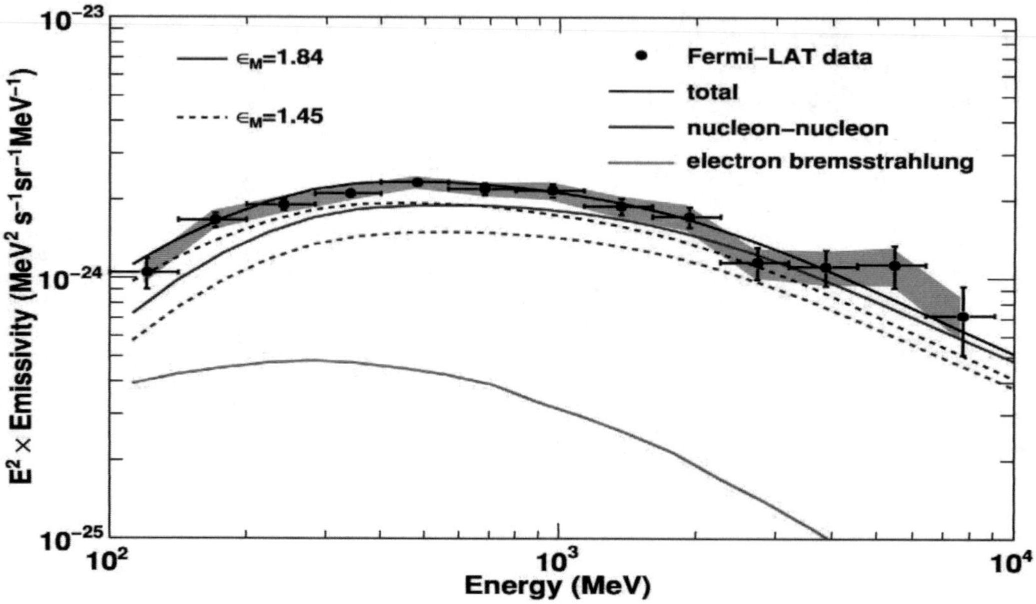

Figure 8. Differential γ-ray emissivity from the local atomic hydrogen gas compared with the calculated γ-ray production. The horizontal and vertical error bars indicate the energy ranges and 1 σ statistical errors, respectively. Estimated systematic errors of the LAT data are indicated by the shaded area. A nucleus enhancement factor ϵ_M of 1.84 is assumed for the calculation of the γ-rays from nucleon-nucleon interactions. Dotted lines indicate the emissivities for the case of $\epsilon_M = 1.45$, the lowest values in the referenced literature.

emission is too low thus giving rise to a $\sim 10 - 15\%$ excess over the energy range 100 MeV to 10 GeV. However, the DGE model is based on pre Fermi data and knowledge of the DGE. The difference between the model and data is of the same order as the uncertainty in the measured CR nuclei spectra at the relevant energies. Overall, the agreement between the LAT-measured spectrum and the model shows that the fundamental processes are consistent with our data, thus providing a solid basis for future work understanding the DGE.

Also at higher latitudes for the moment we did no observed any excess. In figure 8 it is shown the diffuse γ-rays in a mid-latitude region in the third quadrant (Galactic longitude l from 200° to 260° and latitude $|b|$ from 22° to 60°). The region contains no known large molecular cloud and most of the atomic hydrogen is within 1 kpc of the solar system. The contributions of γ-ray point sources and inverse Compton scattering are estimated and subtracted. The residual γ-ray intensity exhibits a linear correlation with the atomic gas column density in energy from 100 MeV to 10 GeV The differential emissivity from 100 MeV to 10 GeV agrees with calculations based on cosmic ray spectra consistent with those directly measured, at the 10 % level. The results obtained indicate that cosmic ray nuclei spectra within 1 kpc from the solar system in regions studied are close to the local interstellar spectra inferred from direct measurements at the Earth within $\sim$ 10 % [24].

The high-energy diffuse γ-ray emission is dominated by γ-rays produced by cosmic rays (CR) interacting with the Galactic interstellar gas and radiation fields, the so-called diffuse Galactic emission (DGE). A much fainter component, commonly designated as "extragalactic γ-ray background" (EGB) by definition has an isotropic sky distribution and is considered by many to be the superposition of contributions from unresolved extragalactic sources including active

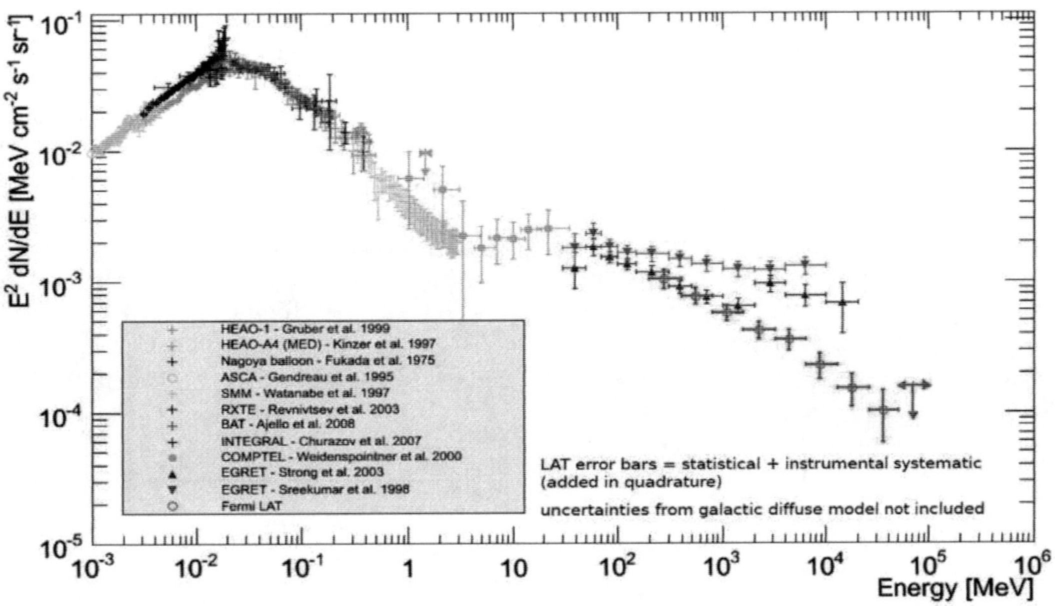

Figure 9. Fermi-LAT extragalactic intensity compared with EGRET-derived intensities. Fermi-LAT derived spectrum is compatible with a simple power-law with index $\gamma = 2.41 \pm 0.05$ and intensity $I(> 100\,\mathrm{MeV}) = (1.03 \pm 0.17) \times 10^{-5}$ cm^{-2} s^{-1} sr^{-1} where the uncertainties are systematics dominated.

galactic nuclei, starburst galaxies and γ-ray bursts and truly-diffuse emission processes.

Figure 9 shows the spectrum of the EGB above 200 MeV derived with the data taken during the first 11 months [25] , and from EGRET data [26], [27]. Our intensity extrapolated to 100 MeV based on the power-law fit, $I(> 100\,\mathrm{MeV}) = (1.03 \pm 0.17) \times 10^{-5}$ cm^{-2} s^{-1} sr^{-1}, is significantly lower than that obtained from EGRET data: $I_{\mathrm{EGRET}}(> 100\,\mathrm{MeV}) = (1.45 \pm 0.05) \times 10^{-5}$ cm^{-2} s^{-1} sr^{-1}. Furthermore, our spectrum is compatible with a featureless power law with index $\gamma = 2.41 \pm 0.05$. This is significantly softer than the EGRET spectrum with index $\gamma_{\mathrm{EGRET}} = 2.13 \pm 0.03$. To check that the different spectra are not due to the instrumental point-source sensitivities, we adopt $F(> 100\,\mathrm{MeV}) = 10^{-7}$ cm^{-2} s^{-1}, comparable to the average EGRET sensitivity, and attribute the flux of all detected LAT sources below this threshold to the EGB. We obtain an intensity $I_{res}(> 100\,\mathrm{MeV}) = (1.19 \pm 0.18) \times 10^{-5}$ cm^{-2} s^{-1} sr^{-1} and a spectrum compatible with a power-law with index $\gamma_{res} = 2.37 \pm 0.05$. Therefore, the discrepancy cannot be attributed to a lower threshold for resolving point sources. Our EGB intensity is comparable to that obtained in the EGRET re-analysis by [27] with an updated DGE model, $I_{\mathrm{SMR}}(> 100\,\mathrm{MeV}) = (1.11 \pm 0.1) \times 10^{-5}$ cm^{-2} s^{-1} sr^{-1}. However, our EGB spectrum does not show the distinctive harder spectrum above ≥ 1 GeV and peak at ~ 3 GeV found in the same EGRET reanalysis.

3. Dwarf spheroidal galaxies and Clusters of galaxies

Local Group dwarf spheroidal galaxies, the largest galactic substructures predicted by the cold dark matter scenario, are attractive targets for such indirect searches for dark matter because they are nearby and among the most extreme dark matter dominated environments. With the data taken during the first 11 months no significant γ-ray emission was detected above 100 MeV from the candidate dwarf galaxies. So we can determine upper limits to the γ-ray flux assuming both power-law spectra and representative spectra from WIMP annihilation. The resulting integral flux above 100 MeV is constrained to be at a level below around 10^{-9} photons $cm^{-2}s^{-1}$ [28] Using recent stellar kinematic data, the γ-ray flux limits can be combined with improved determinations of the dark matter density profile in 8 of the 14 candidate dwarfs to place limits on the pair annihilation cross-section of WIMPs in several widely studied extensions of the standard model, including its supersymmetric extension and other models that received recent attention. With the present data we are able to rule out large parts of the parameter space where the thermal relic density is below the observed cosmological dark matter density and WIMPs (neutralinos here) are dominantly produced non-thermally, e.g. in models where supersymmetry breaking occurs via anomaly mediation. These γ-ray limits also constrain some WIMP models proposed to explain the *Fermi* and PAMELA e^+e^- data, including low-mass wino-like neutralinos and models with TeV masses pair-annihilating into muon-antimuon pairs (see figure 10).

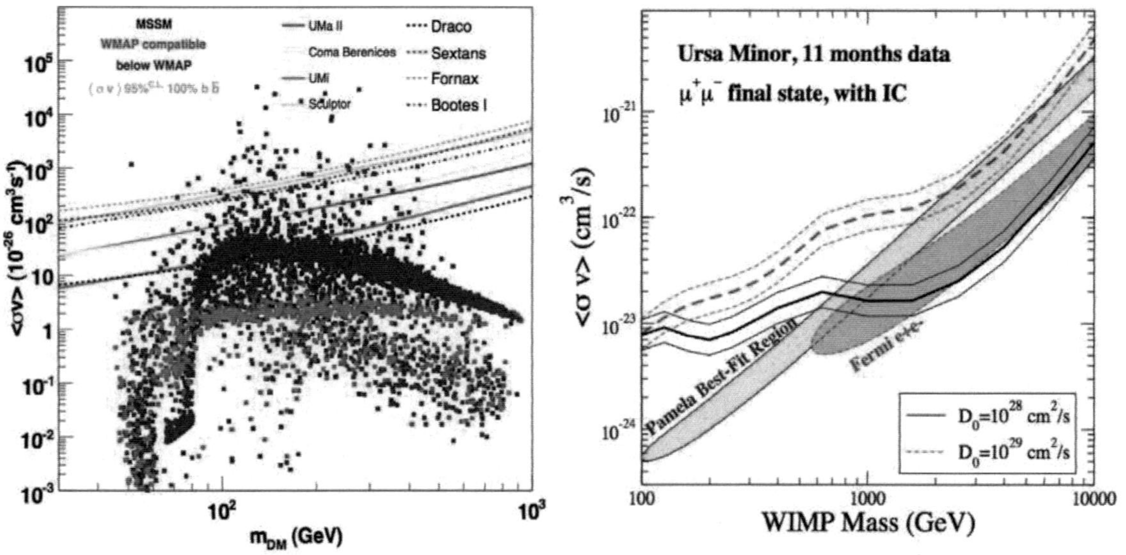

Figure 10. (Left) MSSM models in the ($m_{\rm wimp}$,$< \sigma v >$) plane consistent with all accelerator constraints. Red points have a neutralino thermal relic abundance corresponding to the inferred cosmological dark matter density (blue points have a lower thermal relic density, and we assume that neutralinos still comprise all of the dark matter in virtue of additional non-thermal production processes). The lines indicate the Fermi 95% upper limits obtained from likelihood analysis on the selected dwarfs. (Right) Constraints on the annihilation cross-section for a $\mu^+\mu^-$ final state based on the 95% confidence limits on the γ-ray flux compared to dark matter annihilation models which fit well either the PAMELA or Fermi $e^+ + e^-$ measurements. The constraints are for the Ursa Minor dwarf including both γ-ray emission from IC scattering and final state radiation. Here we consider two different diffusion coefficients and show the effect of the uncertainties in the Ursa Minor density profile.

The same kind of analysis can be done with the clusters of galaxies.

Nearby clusters and groups of galaxies are potentially bright sources of high-energy gamma-ray emission resulting from the pair-annihilation of dark matter particles. However, no significant gamma-ray emission has been detected so far from clusters in the first 11 months of observations with the Fermi Large Area Telescope. We interpret this non-detection in terms of constraints on dark matter particle properties. In particular for leptonic annihilation final states and particle

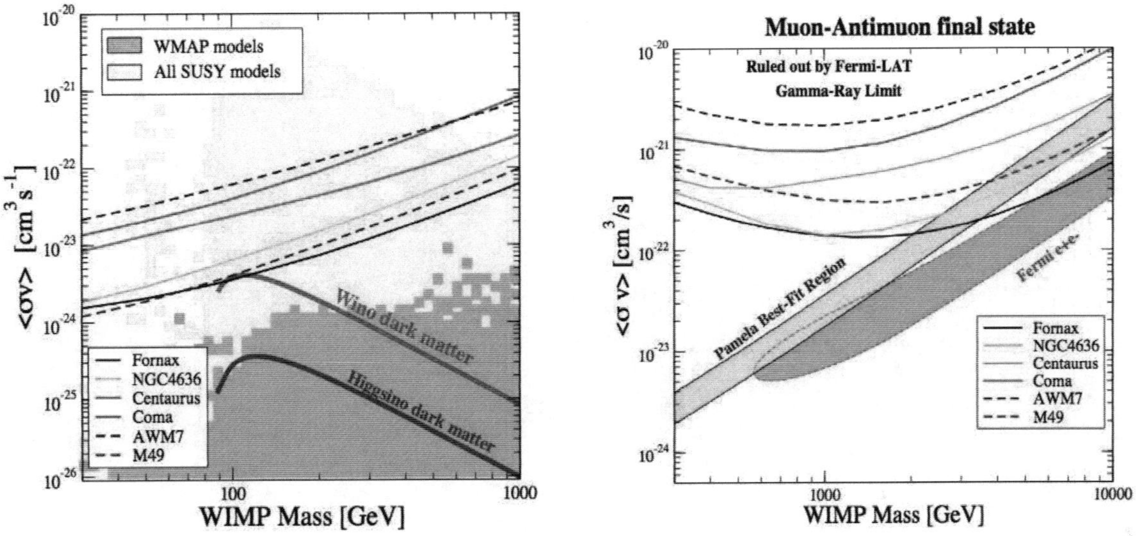

Figure 11. *Left:* The expected gamma-ray spectrum from dark matter annihilation into a $\mu^+\mu^-$ final state including both FSR and emission from IC scattering of CMB photons for different assumed particle masses. The spectral normalization is arbitrary, but the same effective J is used in all cases. *Right:* Flux upper limits as a function of particle mass for an assumed $\mu^+\mu^-$ final state, including the contributions of both FSR and IC gamma-ray emission, for the clusters. The upper limits reflect the 95% confidence level limits on the photon flux in the 100 MeV to 100 GeV energy range.

masses greater than $\sim$ 200 GeV, gamma-ray emission from inverse Compton scattering of CMB photons is expected to dominate the dark matter annihilation signal from clusters, and our gamma-ray limits exclude large regions of the parameter space that would give a good fit to the recent anomalous Pamela and Fermi-LAT electron-positron measurements [29]. For an example see figure 11.

Finally a line at the WIMP mass, due to the 2γ production channel, could be observed as a feature in the astrophysical source spectrum [14]. Such an observation is a "smoking gun" for WIMP DM as it is difficult to explain by a process other than WIMP annihilation or decay and the presence of a feature due to annihilation into γZ in addition would be even more convincing.

Up to now however no lines was observed and we obtain γ-ray line flux upper limits in the range $0.6 - 4.5 \times 10^{-9}\text{cm}^{-2}\text{s}^{-1}$ [30] and corresponding DM annihilation cross-section and decay lifetime limits shown in figure 12 and in figure 13 .

4. Conclusion

Recent accurate measurements of cosmic-ray positrons and electrons by PAMELA, and Fermi have open a new era in particle astrophysics. The CRE spectrum measured by Fermi-LAT is significantly harder than previously thought on the basis of previous data. Adopting the presence of an extra $e^{\pm}$ primary component with $\sim$ 2.4 spectral index and $E_{cut} \sim 1TeV$ allow to consistently interpret Fermi-LAT CRE data, HESS and PAMELA. Such extra-component can be originated by pulsars for a reasonable choice of relevant parameters or by annihilating dark matter for model with $M_{DM} \sim 1TeV$. Improved analysis and complementary observations (CRE anisotropy, spectrum and angular distribution of diffuse γ, DM sources search in γ) are required to possibly discriminate the right scenario. Their exotic origin has to be confirmed by

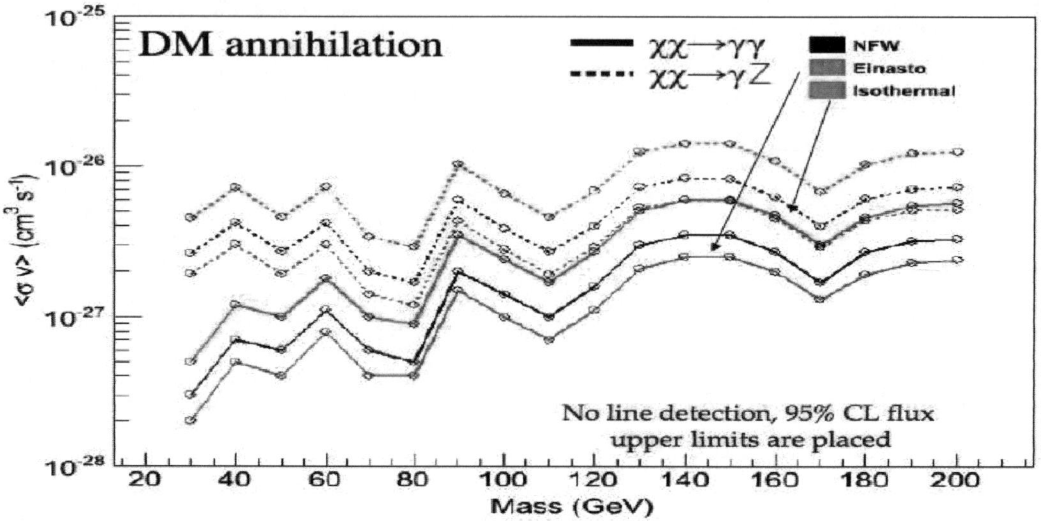

Figure 12. Cross-section limits for various dark matter halo profiles for the annihilation in monochromatic gamma-rays

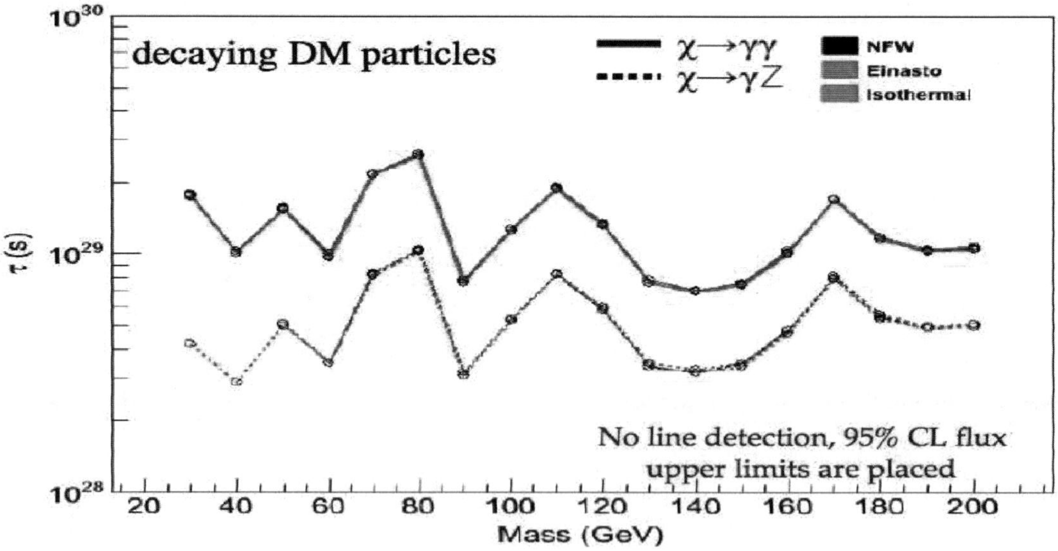

Figure 13. Lifetime limits for various dark matter halo profiles for the decay in monochromatic gamma-rays channel

complimentary findings in γ-rays by Fermi and atmospheric Cherenkov telescopes, and by LHC in the debris of high-energy proton destructions. A positive answer will be a major breakthrough and will change our understanding of the universe forever. On the other hand, if it happens to be a conventional astrophysical source of cosmic rays, it will mean a direct detection of particles accelerated at an astronomical source, again a major breakthrough. In this case we will learn a whole lot about our local Galactic environment. However, independently on the origin of these excesses, exotic or conventional, we can expect very exciting several years ahead of us.

5. References

[1] W.B. Atwood et al. [Fermi Coll.], ApJ 697 (2009) 1071-1102 [arXiv:0902.1089]
[2] A.A. Abdo et al. [Fermi Coll.] , Astroparticle Physics 32 (2009) 193-219 [arXiv:0904.2226]
[3] A.A. Abdo et al. [Fermi Coll.], Physical Review Letters 102 (2009) 181101 [arXiv:0905.0025]
[4] M. Ackermann, et al. [Fermi Coll.], Physical Review D, accepted [arXiv:1008.3999]
[5] O. Adriani et al. [PAMELA Coll.] Nature 458 (2009) 607 [arxiv:0810.4995]
[6] A.W. Strong and I.V. Moskalenko, ApJ 509 (1998) 212, ApJ 493 (1998) 694
[7] A. Lionetto, A. Morselli, and V. Zdravkovic, JCAP09 (2005) 010 [astro-ph/0502406]; V.S. Ptuskin et al., ApJ 642 (2006) 902
[8] K. Abe et al. [BESS Coll.], arXiv:0805.1754
[9] A. Morselli, I. Moskalenko, PoS(idm2008)025 [arXiv:0811.3526]
[10] A. Boulares APJ 342 (1989) 807-813
[11] F.A. Aharonian, A.M. Atoyan and H.J. Völk, Astron. Astrophys. 294 (1995) L41
[12] S. Coutu et al., Astroparticle Physics 11 (1999) 429
[13] D. Grasso et al., Astroparticle Physics 32 (2009) 140-151 [arXiv:0905.0636]
[14] E. Baltz et al. , JCAP07 (2008) 013 [arXiv:0806.2911]
[15] M. Ackermann et.al. [Fermi Coll.] , accepted for publication in Physical Review D [arXiv:1008.5119]
[16] A. Morselli at al., Nuclear Physics B (Proc. Suppl.) 113 (2002) 213
[17] A. Cesarini, F. Fucito, A. Lionetto, A. Morselli, P. Ullio, Astroparticle Physics 21 (2004) 267 [astro-ph/0305075]
[18] http://fermi.gsfc.nasa.gov/ssc/data/analysis/software/
[19] A.A. Abdo et al. [Fermi Coll.], Astrophysical Journal Supplement Series 188 (2010) 405 [arXiv:1002.2280]
[20] A. Strong et al., ApJ 613 (2004) 962S
[21] A. Strong et al., Annu. Rev. Nucl. Part. Sci. 57 (2007) 285
[22] V. Vitale and A. Morselli for the Fermi/LAT Collaboration, 2009 Fermi Symposium [arXiv:0912.3828]
[23] A. Abdo et al. [Fermi Coll.], Physical Review Letters 103 (2009) 251101 [arXiv:0912.0973]
[24] A.A. Abdo et al., [Fermi Coll.], ApJ 703 (2009) 1249-1256 [arXiv:0908.1171]
[25] A.A. Abdo et al. [Fermi Coll.], Phys.Rev.Lett.104 (2010) 101101 [arXiv:1002.3603]
[26] P. Sreekumar et al., Astrophys. J. 494 (1998) 523
[27] A.W. Strong, I.V. Moskalenko, and O. Reimer, Astrophys. J. 613 (2004) 956
[28] A.A. Abdo et al., [Fermi Coll.], ApJ 712 (2010) 147-158 [arXiv:1001.4531]
[29] M. Ackermann et al. [Fermi Coll.], JCAP 05 (2010) 025 [arXiv:1002.2239]
[30] A.A. Abdo et al., [Fermi Coll.], Physical Review Letters 104 (2010) 091302 [arXiv:1001.4836]

Journal of Physics: Conference Series **259** (2010) 012012 doi:10.1088/1742-6596/259/1/012012

The Higgs Sector and CoGeNT/DAMA-Like Dark Matter in Supersymmetric Models

John F. Gunion

Department of Physics, UC Davis, Davis CA, 95616

E-mail: `jfgunion@ucdavis.edu`

Abstract. Recent data from CoGeNT and DAMA are roughly consistent with a very light dark matter particle with $m \sim 4 - 10$ GeV and spin-independent cross section of order $\sigma_{SI} \sim (1-3)\times 10^{-4}$ pb. An important question is whether these observations are compatible with supersymmetric models obeying $\Omega h^2 \sim 0.11$ without violating existing collider constraints and precision measurements. In this talk, I review the fact the the Minimal Supersymmetric Model allows insufficient flexibility to achieve such compatibility, basically because of the highly constrained nature of the MSSM Higgs sector in relation to LEP limits on Higgs bosons. I then outline the manner in which the more flexible Higgs sectors of the Next-to-Minimal Supersymmetric Model and an Extended Next-to-Minimal Supersymmetric Model allow large σ_{SI} and $\Omega h^2 \sim 0.11$ at low LSP mass without violating LEP, Tevatron, BaBar and other experimental limits. The relationship of the required Higgs sectors to the NMSSM "ideal-Higgs" scenarios is discussed.

1. Introduction

CoGeNT [1] and DAMA [2] both have hints of dark matter detection corresponding to a very low mass particle with very large spin-independent cross section. In [3], it is claimed that a consistent explanation for both hints is provided if $\sigma_{SI} \sim (1.4 - 3.5)\times 10^{-4}$ pb, for $m_{DM} = (9-6)$ GeV. Of course, the required σ_{SI} is dependent on the assumed local relic density and for example is reduced by $\sim 60\%$ if $\rho = 0.485$ GeV/cm^3 [4] is employed rather than the usual $\rho = 0.3$ GeV/cm^3.

One would hope that CoGeNT/DAMA-like σ_{SI} at low m_{DM} could be consistent with simple supersymmetric models. However, the MSSM fails. For low $m_{\widetilde{\chi}_1^0}$, the MSSM generically predicts a value for $\Omega_{\widetilde{\chi}_1^0} h^2$ that is far above the observed value; it is only for extreme choices that one can achieve $\Omega_{\widetilde{\chi}_1^0} h^2 \sim 0.11$ [5]. For these same choices it turns out that σ_{SI} takes on its maximum possible value of $\sim 0.17\times 10^{-4}$ pb. This maximum value of σ_{SI} can be understood as follows. The cross section for $\widetilde{\chi}_1^0$-nucleon scattering is dominated by CP-even Higgs exchange and is given approximately by

$$\sigma_{SI} \approx 0.17\times 10^{-4}\ \mathrm{pb}\left(\frac{N_{13}^2}{0.1}\right)\left(\frac{\tan\beta}{50}\right)^2\left(\frac{100\mathrm{GeV}}{m_{H^0}}\right)^4 \cos^4\alpha\,, \quad (1)$$

where we have written $\widetilde{\chi}_1^0 = N_{11}\widetilde{B} + N_{12}\widetilde{W}^3 + N_{13}\widetilde{H}_d + N_{14}\widetilde{H}_u$ and referenced the most optimistic values of N_{13}^2, $\tan\beta$ and m_{H^0}. In the above, N_{13}^2 cannot be much larger than 0.1 because of limits on the Z invisible width, $\tan\beta > 50$ enters a non-perturbative Yukawa coupling domain and $m_{H^0} < 100$ GeV is not allowed by LEP limits. Further, to achieve even this most maximal value of σ_{SI} in the MSSM, one must ignore the Tevatron limit, $B(B_s \to \mu^+\mu^-) \leq 5.8 \times 10^{-8}$. Once imposed, the largest σ_{SI} for scenarios with $m_{\widetilde{\chi}_1^0}$ in the CoGeNT/DAMA region and with $\Omega_{\widetilde{\chi}_1^0}h^2 \sim 0.1$ is $\sigma_{SI} \sim 0.017 \times 10^{-4}$ pb [6], a factor of roughly 100 below the σ_{SI} needed to explain the CoGeNT/DAMA events.

Thus, it is natural to turn to the even more attractive NMSSM model. The NMSSM is defined by adding a single SM-singlet superfield $\widehat{S}$ to the MSSM and imposing a Z_3 symmetry on the superpotential, implying

$$W = \lambda\ \widehat{S}\widehat{H}_u\widehat{H}_d + \frac{\kappa}{3}\ \widehat{S}^3 \tag{2}$$

The reason for imposing the Z_3 symmetry is that then only dimensionless couplings λ, κ enter. All dimensionful parameters will then be determined by the soft-SUSY-breaking parameters. In particular, the μ problem is solved via $\mu_{\rm eff} = \lambda\langle S\rangle$ for which $\mu_{\rm eff}$ is automatically of order a TeV (as required) since $\langle S\rangle$ is of order the SUSY-breaking scale, $m_{\rm SUSY}$, which will be below a TeV.

The extra singlet field $\widehat{S}$ implies: 5 neutralinos, $\widetilde{\chi}^0_{1-5}$ with $\widetilde{\chi}^0_1 = N_{11}\widetilde{B} + N_{12}\widetilde{W}^3 + N_{13}\widetilde{H}_d + N_{14}\widetilde{H}_u + N_{15}\widetilde{S}$ being either singlet or bino, depending on M_1; 3 CP-even Higgs bosons, h_1, h_2, h_3; and 2 CP-odd Higgs bosons, a_1, a_2.

The soft-SUSY-breaking terms corresponding to the terms in W are:

$$\lambda A_\lambda S H_u H_d + \frac{\kappa}{3} A_\kappa S^3\,. \tag{3}$$

It is important to recall that when $A_\lambda, A_\kappa \to 0$, the NMSSM has an additional $U(1)_R$ symmetry, in which limit the a_1 is pure singlet and $m_{a_1} = 0$. If, $A_\lambda, A_\kappa = 0$ at M_U, RGE's give $A_\lambda \sim 100$ GeV and $A_\kappa \sim 1 - 20$ GeV, resulting in $m_{a_1} < 2m_B$ (see later) being quite natural and not fine-tuned [7]. In this situation a_1 is still primarily singlet so that $\cos\theta_A$ as defined by $a_1 = \cos\theta_A A_{MSSM} + \sin\theta_A A_S$ is typically quite small. [1] A light singlet-like a_1 also arises in the $U(1)_{PQ}$ symmetry limit of $\kappa, \kappa A_\kappa = 0$.

As is well known, the NMSSM maintains all the attractive features (especially coupling constant unification and automatic electroweak symmetry breaking from renormalization group evolution of the soft SUSY-breaking stop masses) of the MSSM while avoiding important MSSM problems. In particular, the level of finetuning is greatly reduced in "ideal Higgs" scenarios [8] in which the h_1 has SM-like WW, ZZ couplings and $m_{h_1} \lesssim 105$ GeV but escapes LEP limits via $h_1 \to a_1 a_1 \to 4\tau$ ($m_{a_1} < 2m_B$). Further, $m_{h_1} \lesssim 105$ GeV implies excellent precision electroweak consistency and suitably strong baryogenesis. In addition, the long-standing LEP excess in the $Z + b\bar{b}$ final state near $M_{b\bar{b}} \sim 100$ GeV is well fit if m_{h_1} is in the vicinity of 100 GeV and $B(h_1 \to b\bar{b}) \sim 0.1 - 0.25$, the latter being automatic when $B(h_1 \to a_1 a_1) \sim 0.75 - 0.9$.

However, if h_1 is SM-like then the arguments regarding limitations on achieving large σ_{SI} given above continue to apply. An alternative yielding much larger maximum σ_{SI} [9] is to arrange for the lightest Higgs, h_1, to have enhanced coupling to down-type quarks while it is the h_2 that couples to WW, ZZ in SM-like fashion. We term this kind of scenario an "inverted Higgs" (IH) scenario. Many large σ_{SI} scenarios have $m_{h_1} < 90$ GeV and $m_{h_2} \lesssim 110$ GeV, and are thus still pretty ideal in the sense described in the previous paragraph. We call such scenarios "inverted ideal Higgs" (IIH) scenarios. In the general NMSSM context, it is straightforward [10] to adjust m_{a_1} so as to obtain $\Omega_{\widetilde{\chi}_1^0}h^2 \sim 0.1$ (using $\widetilde{\chi}_1^0\widetilde{\chi}_1^0 \to a_1 \to X$ with m_{a_1} small). Further, in IH and IIH scenarios we found [9] that one can achieve $\sigma_{SI} \sim (0.1 - 0.2) \times 10^{-4}$ pb *without*

[1] Here, A_{MSSM} is the usual doublet pseudoscalar of the MSSM two-doublet Higgs sector and A_S is the CP-odd component of the complex scalar field residing in the singlet superfield.

violating the $B(B_s \to \mu^+\mu^-)$ bound, or any other bound. But, to get σ_{SI} as large as 1×10^{-4} requires violating $(g-2)_\mu$ quite badly, and having some enhancement of the s-quark content of the nucleon.

In a second paper [11], we showed that an extended version of the NMSSM would allow σ_{SI} as large as needed for the CoGeNT/DAMA events while maintaining consistency with all constraints, including $\Omega_{\widetilde\chi_1^0} h^2 \sim 0.11$. In particular, we considered the ENMSSM in which we only generalize the superpotential and soft-SUSY-breaking potential, keeping to just one singlet superfield. The extended superpotential is given by

$$v_0^2\hat{S} + \frac{1}{2}\mu_S\hat{S}^2 + \mu\hat{H}_u\hat{H}_d + \lambda\hat{S}\hat{H}_u\hat{H}_d + \frac{1}{3}\kappa\hat{S}^3 \,, \tag{4}$$

and the soft Lagrangian is

$$B_\mu H_u H_d + \frac{1}{2}m_S^2|S|^2 + B_S S^2 + \lambda A_\lambda S H_u H_d + \frac{1}{3}\kappa A_\kappa S^3 + H.c. \tag{5}$$

Note that the presence of explicit μ and B_μ terms. These reduce the appeal of the model somewhat, but there are string-theory-inspired sources for such explicit terms. We found that scenarios in the ENMSSM with the largest σ_{SI} are ones in which the $\widetilde\chi_1^0$ is singlino-like and the h_1 is largely singlet (rather than bino-like and mainly H_d, respectively, as in the IH NMSSM scenarios). To first approximation, $\Omega_{\widetilde\chi_1^0} h^2$ is controlled by $\widetilde\chi_1^0\widetilde\chi_1^0 \to h_1 \to X$ and σ_{SI} is determined by h_1 exchange between the $\widetilde\chi_1^0$ and the down-type quarks in the nucleon, especially s and b. We term this scenario the singlino-singlet (SS) scenario.

In a very recent paper [12], it is found that the SS type scenario can be realized in the NMSSM provided the h_1 has $m_{h_1} \lesssim 1$ GeV. They call their scenario the Dark Light Higgs (DLH) scenario. This scenario requires a considerable degree of finetuning for the couplings, but is consistent with current experimental constraints.

While we await confirmation of the CoGeNT/DAMA excesses, it is very interesting to consider the implications of all the above scenarios for Higgs physics at the LHC. That will be the topic of the remainder of this presentation.

2. The Inverted Higgs Scenarios

The largest elastic scattering cross sections arise in the case of large $\tan\beta$, significant N_{13} (the Higgsino component of the $\widetilde\chi_1^0$), and relatively light m_{H_d}, where H_d is the Higgs with enhanced coupling to down quarks, $C_{H_d d\bar{d}} \sim \tan\beta$. In this limit, the relevant scattering amplitude is

$$\frac{a_d}{m_d} \approx \frac{-g_2 g_1 N_{13} N_{11} \tan\beta}{4 m_W m_{H_d}^2} \,, \tag{6}$$

which in turn yields

$$\begin{aligned} \sigma_{SI} &\approx \frac{g_2^2\, g_1^2\, N_{13}^2 N_{11}^2\, \tan^2\beta\, m_{\widetilde\chi_1^0}^2\, m_{p,n}^4}{4\pi\, m_W^2 m_{H_d}^4\, (m_{\widetilde\chi_1^0} + m_{p,n})^2} \left[f_{T_s}^{(p,n)} + \frac{2}{27} f_{TG}^{(p,n)} \right]^2 \\ &\approx 1.7 \times 10^{-5}\ \mathrm{pb} \left(\frac{N_{13}^2}{0.10}\right) \left(\frac{\tan\beta}{50}\right)^2 \left(\frac{100\mathrm{GeV}}{m_{H_d}}\right)^4 . \end{aligned} \tag{7}$$

Constraints on the light $h_1 \sim H_d$ configuration are significant. We had to update NMHDECAY [13, 14] to include all the latest constraints. We then linked to micrOMEGAs [15] for the computation of $\Omega_{\widetilde\chi_1^0} h^2$ and σ_{SI} as in NMSSMTools [16]. The important constraints are:

(i) Constraints on the neutral Higgs sector from Zh_2 at LEP. These are important since $m_{\rm SUSY}$ should be low in order to minimize m_{h_1} and this keeps m_{h_2} low. In these cases, the h_2 can be in the "ideal" zone ($m_{h_2} < 105$ GeV) and escapes LEP detection via $h_2 \to a_1 a_1 \to 4\tau$ decays with $m_{a_1} < 2m_B$ (but close to $2m_B$ in order to avoid BaBar limits on $\Upsilon_{3S} \to \gamma a_1 \to \gamma\tau^+\tau^-$). Of course, we also require compliance with the recent ALEPH limits [17] on the $e^+e^- \to Z4\tau$ channel.

(ii) LEP constraints on $h_1 a_1$ and $h_1 a_2$. The $h_1 a_1$ cross section is $\propto maximal \times (\cos\theta_A)^2$. Thus, small $\cos\theta_A$ is desirable, which fits with both the approximate $U(1)_R$ limit mentioned earlier and the need to not have overly strong $\widetilde\chi_1^0\widetilde\chi_1^0 \to a_1^* \to X$ annihilations (as required to achieve adequate $\Omega_{\widetilde\chi_1^0}h^2$) when m_{a_1} is small and, in particular, not far from the $2m_{\widetilde\chi_1^0} \sim m_{a_1}$ resonance pole.

(iii) Tevatron direct Higgs production limits. There are two especially relevant limits given the need to focus on large $\tan\beta$ in order to achieve large σ_{SI}. The first is the limit on $b\bar b h_1$ with $h_1 \to \tau^+\tau^-$ associated production [18], which scales as $C^2_{h_1 b\bar b} \sim \tan^2\beta$, the latter being something we want to maximize. The second limits are those on $t \to h^+ b$ with $h^+ \to \tau^+\nu_\tau$ [19] (dominant at large $\tan\beta$). These are critical to include since the h^+ tends to be quite light (e.g. $\sim 120 - 140$ GeV) when h_1 is H_d-like and the h_2 is SM-like.

(iv) Limits from Υ decays. These are especially crucial for constraining scenarios with $m_{a_1} < 2m_B$. We have included the latest $\Upsilon_{3S} \to \gamma\mu^+\mu^-, \gamma\tau^+\tau^-$ limits [20, 21], whose impacts on ideal Higgs scenarios were explored in [22].

(v) B-physics constraints. The most restricting constraint arises from the very strong Tevatron limit of $B(B_s \to \mu^+\mu^-) < 5.8 \times 10^{-8}$. At large $\tan\beta$, achieving a small enough value fixes A_t as a function of $m_{\rm SUSY}$. Next comes $b \to s\gamma$. The $\mu_{\rm eff} > 0$ scenarios have roughly 1σ discrepancy with the 2σ experimental window. In contrast, the $\mu_{\rm eff} < 0$ scenarios only rarely have a $b \to s\gamma$ problem. One must also check $B^+ \to \tau^+\nu_\tau$. $\mu_{\rm eff} > 0$ scenarios pass easily, but $\mu_{\rm eff} < 0$ scenarios with the largest σ_{SI} have $1\sigma - 2\sigma$ deviations from the experimental 2σ window.

(vi) $(g-2)_\mu$. This is possibly crucial. For $\mu_{\rm eff} < 0$, the largest σ_{SI} values are achieved when $(g-2)_\mu$ is a few sigma outside the 2σ limits including theoretical uncertainties. If $(g-2)_\mu$ is strictly required to lie within the 2σ window, then the largest σ_{SI} that can be achieved for $\mu_{\rm eff} < 0$ is about a factor of 50 below that needed for the CoGeNT/DAMA events. For $\mu_{\rm eff} > 0$, the largest σ_{SI} points yield $(g-2)_\mu$ within the 2σ exp.+theor. window, but after including all other constraints the σ_{SI} values for $\mu_{\rm eff} > 0$ are not as large as those found with $\mu_{\rm eff} < 0$ (before imposing the $(g-2)_\mu$ constraint).

(vii) $\Omega_{\widetilde\chi_1^0}h^2$: Of course, we require that any accepted scenario have correct relic density within the experimental limits encoded in NMSSMTools, $0.094 \le \Omega_{\widetilde\chi_1^0}h^2 \le 0.136$.

To illustrate the impact of some of the constraints discussed above, I give three figures. In the first, Fig. 1, representative maximal values of σ_{SI} (averaged over protons and neutrons) are plotted after: a) imposing LEP limits; b) requiring $0.094 < \Omega_{\widetilde\chi_1^0}h^2 < 0.136$; c) imposing BaBar limits from Υ_{3S} decays; and d) requiring $B(B_s \to \mu^+\mu^-) < 5.8 \times 10^{-8}$. These are termed "level I" constraints. Note that $\sigma_{SI} \sim 6 \times 10^{-5}$ pb (2×10^{-5} pb) can be achieved for $\mu_{\rm eff} < 0$ ($\mu_{\rm eff} > 0$). In Fig. 2, the masses of the CP-even Higgs bosons are plotted for the $\mu_{\rm eff} > 0$ cases. Note how low all these masses are. These same plots for $\mu_{\rm eff} < 0$ would be quite similar.

The low masses imply that Tevatron limits on direct Higgs production could be important. In Fig. 3, we plot the points from Fig. 1 that are fully consistent with Tevatron limits on $b\bar b + Higgs$ with $Higgs \to \tau^+\tau^-$ and on $t \to h^+ b$ with $h^+ \to \tau^+\nu_\tau$. Typically, one finds maximal σ_{SI} values in the $(1-2) \times 10^{-5}$ pb range (a factor at least 5 below the σ_{SI} needed for CoGeNT/DAMA). Many of the points of Fig. 1 with larger σ_{SI} would survive if we relaxed these direct production

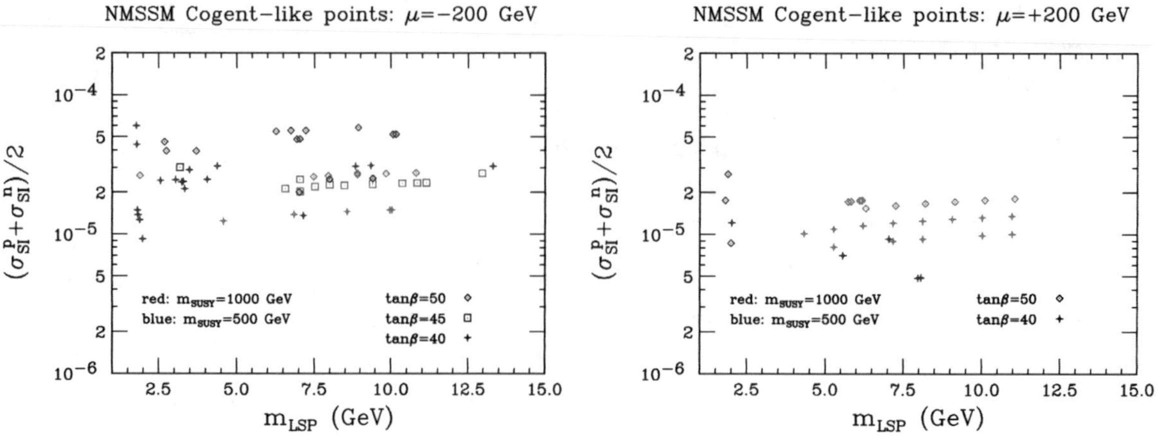

Figure 1. All points obtained after imposing level-I constraints (see text) but without imposing either Tevatron direct Higgs production limits or $(g-2)_\mu$ limits.

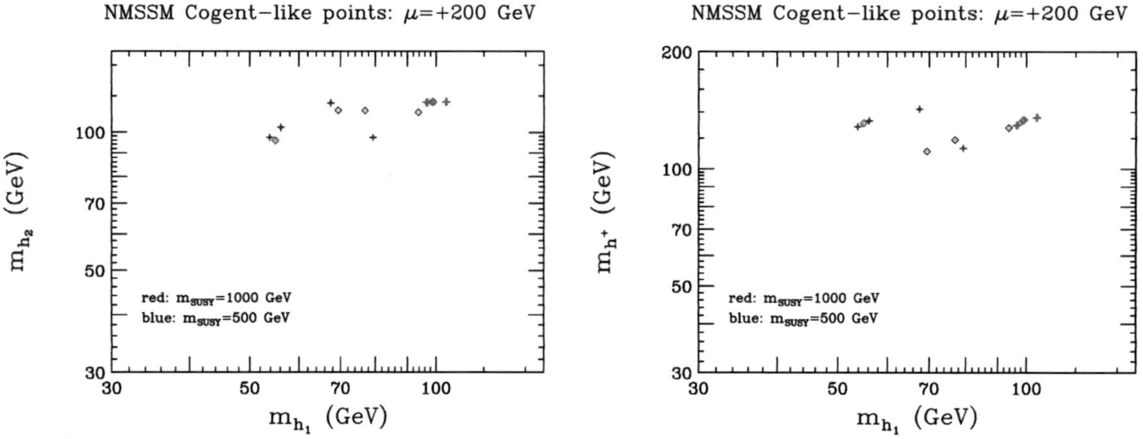

Figure 2. m_{h_2} and m_{h^+} vs. m_{h_1} for $\mu_{\rm eff} = +200$ GeV points. Only level-I (see text) constraints are imposed. There is a great amount of point overlap in these plots.

limits by 1σ (combined experimental and theoretical error). The points of the $\mu_{\rm eff} < 0$ plot in Fig. 3 would all be eliminated if the predicted $(g-2)_\mu$ is required to be within 2σ of the observed value, whereas all the $\mu_{\rm eff} > 0$ points are very consistent with the observed $(g-2)_\mu$.

To further illustrate the nature of the Higgs sector for a $\mu_{\rm eff} > 0$ scenario, I give details of a sample $\tan\beta = 40$ inverted-ideal Higgs point in Table 1. Let me make a few remarks. For this somewhat unusual point, the h_1 is fairly singlet, the h_2 and a_2 are largely H_d and it is the h_3 that is most SM-like. Nonetheless, the effective precision electroweak mass (defined by $\ln m_{eff} \equiv \sum_{i=1,2,3}[g^2_{ZZh_i}/g^2_{ZZh_{\rm SM}}]\ln m_{h_i})$ receives substantial contributions from the low mass Higgses, h_1 and h_2, and lies below 105 GeV and is thus in the range that is ideal for precision electroweak data. Next, note that Higgs decays to $\widetilde\chi^0_1\widetilde\chi^0_1$ are unimportant, but that Higgs to Higgs pair decays are often significant. Prospects for LHC detection of some of the Higgs are very good. In particular, $b\bar b h_2 + b\bar b a_2$ with $a_2, h_2 \to \tau^+\tau^-$ should be readily observable. Even a_1 discovery via $gg \to a_1 \to \mu^+\mu^-$ looks promising [23] because $C_{a_1 b\bar b} \sim 6$ and m_{a_1} is not directly under the Υ_{3S} peak. The SM-like h_3 is possibly the most difficult to detect because: a) its $\gamma\gamma$ decay mode is suppressed by $C_V(1) < 1$ and mass $m_{h_3} \sim 126$ GeV that is above the mass where the branching ratio to $\gamma\gamma$ is maximal; and, b) its decays to Higgs pairs will reduce all standard

Journal of Physics: Conference Series **259** (2010) 012012 doi:10.1088/1742-6596/259/1/012012

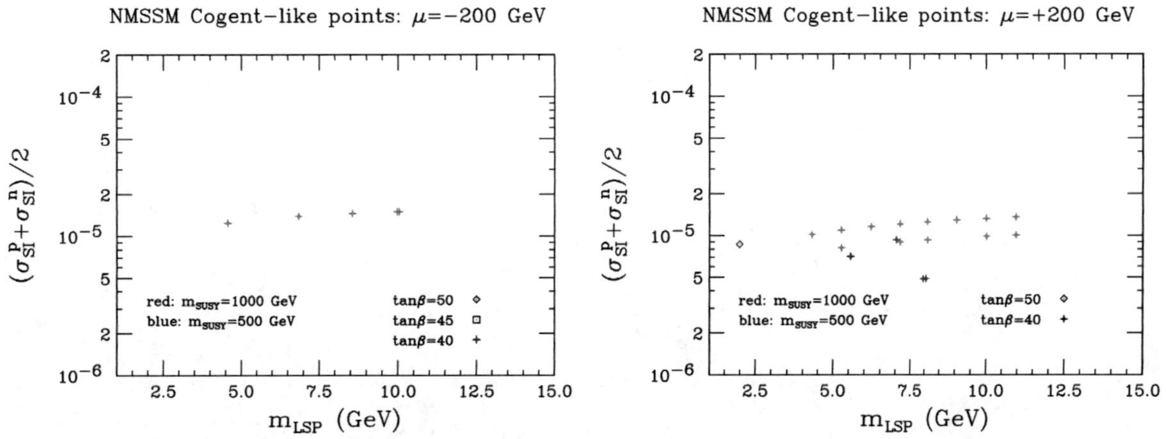

Figure 3. σ_{SI} vs. $m_{\widetilde{\chi}_1^0}$ for points fully consistent with Tevatron limits on $b\bar{b} + Higgs$ and $t \to h^+ b$. Level-I constraints are imposed. $(g-2)_\mu$ is still terrible (perfectly ok) for the surviving $\mu_{\rm eff} < 0$ ($\mu_{\rm eff} > 0$) points.

detection modes.

Finally, let me note that in the NMSSM study of [24] cross sections as large as those found here are not achieved; rather, they find $\sigma_{SI}^{Max} \sim (1-1.5) \times 10^{-6}$ pb (without enhancing the s-quark content of the nucleon). The smaller σ_{SI} is largely because they did not seek scenarios with $h_1 \sim H_d$. Ref. [24] also considers the possibility of enhancing σ_{SI} by a factor of ~ 3 by enhancing the s-quark content of the nucleon. If we adopted this same enhancement, our σ_{SI} values would approach the lowered CoGeNT/DAMA σ_{SI} range applicable if we also employed the higher local density of $\rho \sim 0.485$ GeV/cm^2 mentioned earlier. However, current analyses do not appear to allow for such a large s-quark enhancement [25].

3. The Singlino-Singlet Scenarios

In these SS scenarios, the LSP is primarily singlino and the Higgs responsible for large σ_{SI} is mainly singlet. In [11], we pursued the extended NMSSM as defined earlier and looked for scenarios of the SS type. What we found was a kind of see-saw balance between $\Omega_{\widetilde{\chi}_1^0} h^2$ and σ_{SI} such that when $\Omega_{\widetilde{\chi}_1^0} h^2 \sim 0.1$ then σ_{SI} is very naturally in the CoGeNT/DAMA preferred zone. Below, I provide a few details.

The singlino coupling to down-type quarks is given by:

$$\frac{a_d}{m_d} = \frac{g_2 \kappa N_{15}^2 \tan\beta F_s(h_1) F_d(h_1)}{8 m_W m_{h_1}^2} \tag{8}$$

where $h_1 = F_d(h_1) H_d^0 + F_u(h_1) H_u^0 + F_s(h_1) H_S^0$ and the crucial trilinear coupling that couples a singlino pair to the singlet Higgs H_S^0 is proportional to κ. This leads to

$$\sigma_{SI} \approx 2.2 \times 10^{-4}\,\text{pb} \left(\frac{\kappa}{0.6}\right)^2 \left(\frac{\tan\beta}{50}\right)^2 \left(\frac{45\,\text{GeV}}{m_{h_1}}\right)^4 \left(\frac{F_s^2(h_1)}{0.85}\right) \left(\frac{F_d^2(h_1)}{0.15}\right),$$

which is consistent with the value required by CoGeNT and DAMA/LIBRA for the indicated κ, m_{h_1} and h_1 component values. (Of course, one really sums coherently over all the CP-even Higgs bosons.) Furthermore, the large singlet fraction $F_s^2(h_1) \sim 0.85$ of the h_1 will allow it evade the constraints from LEP II and the Tevatron.

Table 1. Properties of a particularly attractive but phenomenologically complex NMSSM point with $\mu_{\rm eff} = +200$ GeV, $\tan\beta = 40$ and $m_{\rm SUSY} = 500$ GeV. All Tevatron limits ok. h_3 is the most SM-like. In the last row, the brackets give the range of B-physics predictions for this point after including theoretical errors as employed in NMHDECAY.

λ	κ	A_λ	A_κ	M_1	M_2	M_3	A_{soft}
0.081	0.01605	−36 GeV	−3.25 GeV	8 GeV	200 GeV	300 GeV	479 GeV

m_{h_1}	m_{h_2}	m_{h_3}	m_{a_1}	m_{a_2}	m_{h^+}
53.8 GeV	97.3 GeV	126.2 GeV	10.5 GeV	98.9 GeV	128.4 GeV

$C_V(h_1)$	$C_V(h_2)$	$C_V(h_3)$	m_{eff}	$C_{h_1 b\bar b}$	$C_{h_2 b\bar b}$	$C_{h_3 b\bar b}$	$C_{a_1 b\bar b}$	$C_{a_2 b\bar b}$
−0.505	0.137	0.852	101 GeV	0.24	39.7	−5.1	6.7	39.4

$m_{\widetilde\chi_1^0}$	N_{11}	N_{13}	$m_{\widetilde\chi_2^0}$	$m_{\widetilde\chi_1^\pm}$	σ_{SI}	σ_{SD}	$\Omega_{\widetilde\chi_1^0} h^2$
7 GeV	−0.976	−0.212	79.1 GeV	153 GeV	0.93×10^{-5} pb	0.45×10^{-4} pb	0.12

$B(h_1 \to a_1 a_1)$	$B(h_2 \to 2b, 2\tau)$	$B(h_3 \to 2h + 2a)$	$B(h_3 \to 2b, 2\tau)$
0.96	$0.87, 0.12$	0.3	$0.58, 0.09$

$B(a_1 \to jj)$	$B(a_1 \to 2\tau)$	$B(a_1 \to 2\mu)$	$B(a_2 \to 2b, 2\tau)$	$B(h^+ \to \tau^+ \nu)$
0.28	0.79	0.003	$0.87, 0.12$	0.97

$B(B_s \to \mu^+\mu^-)$	$B(b \to s\gamma)$	$B(h^+ \to \tau^+\nu_\tau)$	$(g-2)_\mu$
$[1.7 - 6.0] \times 10^{-9}$	$[5.8 - 12.5] \times 10^{-4}$	$[0.91 - 4.22] \times 10^{-4}$	$[4.42 - 5.53] \times 10^{-9}$

Table 2. The $\pm 2\sigma$ experimental ranges for the B physics observables tabulated in the last row of Table 1.

$B(B_s \to \mu^+\mu^-)$	$B(b \to s\gamma)$	$B(h^+ \to \tau^+\nu_\tau)$	$(g-2)_\mu$
$< 5.8 \times 10^{-8}$ (95% CL)	$[3.03 - 4.01] \times 10^{-4}$	$[0.34 - 2.3] \times 10^{-4}$	$[0.88 - 4.6] \times 10^{-9}$

Table 3. LHC Neutral Higgs Discovery Channels ($b\bar b h_2, b\bar b a_2 \to b\bar b 2\tau$ absent since $m_{h_2} \sim m_{a_2} <$ 100 GeV, the lower limit of the studies used — this should be a highly viable mode) (also $t\bar t \to b\bar t h^+ \to \tau^+\nu X$ = excellent channel at LHC)

$L = 30$ fb^{-1}	$L = 300$ fb^{-1}			
$WW \to h_3 \to 2\tau$	$b\bar b h_3 \to b\bar b 2\tau$	$gg \to h_3 \to 4\ell$	$gg \to h_3 \to 2\ell 2\nu$	$WW \to h_3 \to 2\tau$
3.8σ	2σ	1.4σ	1.1σ	14σ

Meanwhile, the thermal relic density of neutralinos is determined by the annihilation cross section and the $\widetilde\chi_1^0$ mass. In the mass range we are considering here, the dominant annihilation channel is to $b\bar b$ (or, to a lesser extent, to $\tau^+\tau^-$) through the s-channel exchange of the *same* scalar Higgs, h_1, as employed for elastic scattering, yielding:

$$\sigma_{\chi_1^0\chi_1^0} v = \frac{N_c g_2^2 \kappa^2 m_b^2 F_s^2(h_1) F_d^2(h_1)}{64\pi m_W^2 \cos^2\beta} \frac{m_{\chi_1^0}^2 (1 - m_b^2/m_{\chi_1^0}^2)^{3/2} v^2}{(4m_{\chi_1^0}^2 - m_{h_1}^2)^2 + m_{h_1}^2 \Gamma_{h_1}^2}, \tag{9}$$

where v is relative velocity between the annihilating neutralinos, $N_c = 3$ is a color factor and Γ_{h_1} is the width of the exchanged Higgs. The annihilation cross section into $\tau^+\tau^-$ is obtained by replacing $m_b \to m_\tau$ and $N_c \to 1$. This yields the thermal relic abundance of neutralinos: $\Omega_{\chi_1^0} h^2 \approx \frac{10^9}{M_{\rm Pl}} \frac{m_{\chi_1^0}}{T_{\rm FO}\sqrt{g_\star}} \frac{1}{\langle\sigma_{\chi_1^0\chi_1^0} v\rangle}$, where $g_\star$ is the number of relativistic degrees of freedom at freeze-out, $\langle\sigma_{\chi_1^0\chi_1^0} v\rangle$ is the thermally averaged annihilation cross section at freeze-out, and $T_{\rm FO}$ is the temperature at which freeze-out occurs.

Table 4. Properties of a typical ENMSSM point with $\tan\beta = 45$ and $m_{\rm SUSY} = 1000$ GeV.

λ	κ	λ_S	A_λ	A_κ	M_1	M_2	M_3	A_{soft}
0.011	0.596	-0.026 GeV	3943 GeV	17.3 GeV	150 GeV	300 GeV	900 GeV	679 GeV

B_S	μ_S	v_S^3	μ	B_μ	$\mu_{\rm eff}$	B_μ^{eff}
0	7.8 GeV	4.7 GeV	164 GeV	658 GeV	164 GeV	556 GeV

m_{h_1}	m_{h_2}	m_{h_3}	m_{a_1}	m_{a_2}	m_{h^+}
82 GeV	118 GeV	164 GeV	82 GeV	164 GeV	178 GeV

$F_S^2(h_1)$	$F_d^2(h_1)$	$F_S^2(h_2)$	$F_u^2(h_2)$	$F_S^2(h_3)$	$F_d^2(h_3)$	$F_S^2(a_1)$	$F_S^2(a_2)$
0.86	0.14	0.0	0.996	0.14	0.86	0.86	0.14

$C_V(h_1)$	$C_V(h_2)$	$C_V(h_3)$	$C_{h_1 b\bar b}$	$C_{h_2 b\bar b}$	$C_{h_3 b\bar b}$	$C_{a_1 b\bar b}$	$C_{a_2 b\bar b}$
-0.0096	0.999	-0.041	16.8	2.9	41.7	-16.9	41.7

$m_{\widetilde\chi_1^0}$	N_{11}^2	$N_{13}^2+M_{14}^2$	N_{15}^2	σ_{SI}	$\Omega_{\widetilde\chi_1^0}h^2$
4.9 GeV	0.0	0.0	1.0	2.0×10^{-4} pb	0.105

$B(h_1\to\widetilde\chi_1^0\widetilde\chi_1^0)$	$B(h_1\to 2b,2\tau)$	$B(h_2\to\widetilde\chi_1^0\widetilde\chi_1^0)$	$B(h_2\to 2b,2\tau)$	$B(h^+\to\tau^+\nu)$
0.64	0.33, 0.03	0.003	0.88, 0.092	0.97

$B(a_1\to\widetilde\chi_1^0\widetilde\chi_1^0)$	$B(a_1\to 2b,2\tau)$	$B(a_2,h_3\to\widetilde\chi_1^0\widetilde\chi_1^0)$	$B(a_2,h_3\to 2b,2\tau)$
0.64	0.33, 0.03	0.05	0.85, 0.095

For the range of masses and cross sections considered here, we find $m_{\chi_1^0}/T_{\rm FO}\approx 20$, yielding a thermal relic abundance of

$$\Omega_{\chi_1^0}h^2 \approx 0.11\left(\frac{0.6}{\kappa}\right)^2\left(\frac{50}{\tan\beta}\right)^2\left(\frac{m_{h_1}}{45\,{\rm GeV}}\right)^4\left(\frac{7\,{\rm GeV}}{m_{\chi_1^0}}\right)^2\left(\frac{0.85}{F_s^2(h_1)}\right)\left(\frac{0.15}{F_d^2(h_1)}\right), \tag{10}$$

i.e. naturally close to the measured dark matter density, $\Omega_{\rm CDM}h^2 = 0.1131\pm0.0042$ *for the same choices for* κ, m_{h_1} *and composition fractions as give CoGeNT/DAMA-like* σ_{SI}. The only question is can we achieve the above situation without violating LEP and other constraints. Basically, one wants a certain level of decoupling between the singlet sectors and the MSSM sectors, but not too much. To find out, we performed parameter scans with an extended version of NMHDECAY and micrOMEGAs that includes both the non-NMSSM parameters of Eqs. (4) and (5) as well as the latest B-physics and Tevatron constraints. We find points for $15<\tan\beta<45$ that are consistent (within the usual $\pm2\sigma$ combined theory plus experimental windows – excursions in $b\to s\gamma$ and $b\bar b h, h\to\tau^+\tau^-$ that fall slightly outside this window are present at high $\tan\beta$) with all collider and B-physics constraints having the appropriate thermal relic density and σ_{SI} as large as *few* $\times 10^{-4}$ pb.

The complete framework has contributions to σ_{SI} and $\Omega_{\chi_1^0}h^2$ beyond Eqs. (9) and (10) and high-σ_{SI} points typically have large contributions from the non-singlet Higgses. I confine myself to discussing one 'typical' point that does the job. Its properties are tabulated in Table 4.

Let us note the following regarding this particular point.

(i) What you see is that the h_1, a_1 have separated off from something that is close to an MSSM-like Higgs sector with $h_2\sim h^0$ being SM-like and $h_3\sim H^0$, $a_2\sim A^0$ and $h^+\sim H^+$.

(ii) Detection of the h_2 would be possible via the usual SM-like detection modes planned for the MSSM h^0.

(iii) There are some $h_2, a_2\to\widetilde\chi_1^0\widetilde\chi_1^0$ decays, but at such a low branching ratio level that detection of these invisible decay modes would be unlikely, even if very interesting.

(iv) Decays to pairs of Higgs of any of the heavier Higgs bosons are not of importance. Of course, by choosing $m_{\rm SUSY}=1000$ GeV so that $m_{h_2}>114$ GeV (beyond the LEP limits), we have not forced the issue. It will be interesting to look for SS scenarios that are ideal-Higgs-like with $m_{h_2}<110$ GeV.

(v) One sees that h_1 and a_1 decay primarily to $\widetilde{\chi}_1^0\widetilde{\chi}_1^0$ but that there also decays to $b\bar{b}$ and $\tau^+\tau^-$ with reduced branching ratios of 0.33 and 0.03 compared to the normal $B(b\bar{b}) \sim 0.85$ and $B(\tau^+\tau^-) \sim 0.12$.

(vi) h_1 and a_1 do have somewhat enhanced couplings to $b\bar{b}$ (in this example $C_{h_1 b\bar{b}}, C_{a_1 b\bar{b}} \sim \sqrt{F_d^2(h_1, a_1)}\tan\beta \sim 17$) and so the rates for $gg \to b\bar{b}h_1 + gg \to b\bar{b}a_1$ will be quite substantial. However, the reduced $B(h_1, a_1 \to \tau^+\tau^-) \sim 0.03$ implies that detection of such production in the $b\bar{b} + \tau^+\tau^-$ final state might prove challenging, probably requiring very high L at the LHC.

(vii) Further work is needed to quantify discovery prospects in the $gg \to b\bar{b} + (h_1, a_1) \to b\bar{b} + \not{E}_T$ channel.

(viii) At this large $\tan\beta$, detection of the h_3 and a_2 would certainly be possible in $gg \to b\bar{b}h_3 + b\bar{b}a_2$ in the $h_3, a_2 \to \tau^+\tau^-$ decay channel.

(ix) For this sample case, the charged Higgs is *just* too heavy to allow $t \to h^+ b$ decays and so one would have to turn to $gg \to \bar{t}bh^+ + t\bar{b}h^-$ with detection of the charged Higgs in the $\tau\nu_\tau$ final state. Further investigation is needed to assess the feasibility of such detection, but at least the cross section is very enhanced by virtue of the large $\tan\beta$ value.

A few final notes regarding this scenario. First, it is the very large value of A_λ and the very small λ that keep the singlet and MSSM Higgs sectors fairly separate. Second, the new parameters of the ENMSSM, μ and B_μ must be substantial. This is generally the case if you desire an SS scenario with $m_{h_1} > few$ GeV.

Table 5. Properties of the SS DLH NMSSM point with $\tan\beta = 13.77$, $m_{\widetilde{q}} = 1000$ GeV and $m_{\widetilde{\ell}} = 200$ GeV.

λ	κ	λs	A_λ	A_κ	M_1	M_2	M_3	A_{soft}
0.1205	0.00272	168 GeV	2661 GeV	−24.03 GeV	100 GeV	200 GeV	660 GeV	750 GeV

m_{h_1}	m_{h_2}	m_{h_3}	m_{a_1}	m_{a_2}	m_{h^+}
0.811 GeV	116 GeV	244 GeV	16.7 GeV	244 GeV	244 GeV

$F_S^2(h_1)$	$F_d^2(h_1)$	$F_S^2(h_2)$	$F_u^2(h_2)$	$F_S^2(h_3)$	$F_d^2(h_3)$	$F_S^2(a_1)$	$F_S^2(a_2)$
0.997	0.00017	0.0036	0.99	0.0	0.994	1.00	0.00

$C_V(h_1)$	$C_V(h_2)$	$C_V(h_3)$	$C_{h_1 b\bar{b}}$	$C_{h_2 b\bar{b}}$	$C_{h_3 b\bar{b}}$	$C_{a_1 b\bar{b}}$	$C_{a_2 b\bar{b}}$
0.06	0.998	0.0	0.183	0.994	13.77	−0.12	13.77

$m_{\widetilde{\chi}_1^0}$	N_{11}^2	$N_{13}^2 + N_{14}^2$	N_{15}^2	σ_{SI}	$\Omega_{\widetilde{\chi}_1^0} h^2$
7.2 GeV	0.0036	0.017	0.98	2.34×10^{-4} pb	0.112

$B(h_1 \to \widetilde{\chi}_1^0\widetilde{\chi}_1^0)$	$B(h_1 \to 2s, 2g, 2\mu)$	$B(h_2 \to \widetilde{\chi}_1^0\widetilde{\chi}_1^0)$	$B(h_2 \to \widetilde{\chi}_1^0\widetilde{\chi}_2^0)$	$B(h_2 \to 2b, 2\tau)$
0.027	0.833, 0.14, 0.027	0.05	0.45	0.37, 0.038

$B(h^+ \to t\bar{b})$	$B(h^+ \to \widetilde{\chi}_{1,2}^+\widetilde{\chi}_{1,2,3,4,5}^0)$
0.138	0.80

$B(a_1 \to \widetilde{\chi}_1^0\widetilde{\chi}_1^0)$	$B(a_1 \to 2b, 2\tau, 2\mu)$
0.25	0.70, 0.042, 0.00015

$B(a_2, h_3 \to \widetilde{\chi}_1^0\widetilde{\chi}_1^0)$	$B(a_2, h_3 \to 2t, 2b, 2\tau)$	$B(a_2, h_3 \to \widetilde{\chi}_{1,2,3,4,5}^0\widetilde{\chi}_{1,2,3,4,5}^0)$	$B(a_2, h_3 \to \widetilde{\chi}_{1,2}^+\widetilde{\chi}_{1,2}^-)$
0.00	0.013, 0.126, 0.023	0.32	0.48

As noted earlier, in [12] an alternative SS scenario can be realized in the strict NMSSM, but only if $m_{h_1} \lesssim 1$ GeV. The properties of their representative point are tabulated in Table 5. Some observations regarding this scenario are the following.

(i) The h_1 is very light and very singlet. It is so weakly coupled to the down and up quarks that it can probably only be detected directly via $\Upsilon_{3S} \to \gamma h_1$ with $h_1 \to \mu^+\mu^-$. For

Journal of Physics: Conference Series **259** (2010) 012012 doi:10.1088/1742-6596/259/1/012012

current data from BaBar and using $B(h_1 \to \mu^+\mu^-) \sim 0.027$ (see the Table), the limit from $\Upsilon_{3S} \to \gamma h_1 \to \gamma\mu^+\mu^-$ is $C_{h_1 b\bar{b}} \sim 0.2 - 0.3$ for $m_{h_1} \sim 1$ GeV (the limit fluctuates very rapidly). For this scenario the value of $C_{h_1 b\bar{b}} = 0.183$ (see the Table) is thus just below the BaBar limit. This indicates that increased statistics could very well reveal the light h_1 since $C_{h_1 b\bar{b}}$ cannot be much below this value and still provide a large enough σ_{SI} to explain the CoGeNT/DAMA events.

(ii) Meanwhile, the h_2 is completely SM-like and its discovery at the LHC or Tevatron would be possible in the usual channels for a SM Higgs of the same mass.

(iii) The a_1 has a very small branching ratio to $\mu^+\mu^-$ (since $m_{a_1} > 2m_B$) and would have to be searched for in the $b\bar{b}$ or $\tau^+\tau^-$ decay mode. Since the a_1 is very singlet its production cross sections would be so small that this would likely be an impossible task.

(iv) The h_3, a_2, h^+ form a decoupled degenerate doublet with common mass of around 244 GeV. The $b\bar{b}h_3$ and $b\bar{b}a_2$ couplings are both enhanced by a factor of $\tan\beta = 13.77$. The most promising LHC signal would be $gg \to b\bar{b}h_3 + b\bar{b}a_2$ with decay $h_3, a_2 \to \tau^+\tau^-$. Of course, $B(h_3, a_2 \to \tau^+\tau^-) \sim 0.023$ is uncomfortably small and even this signal would be quite weak (and does not emerge in the NMHDECAY LHC estimates as viable).

4. Conclusions

The CoGeNT/DAMA data suggests a large spin-independent cross section for dark matter scattering on nucleons, $\sigma_{SI} \sim (1-3)\times 10^{-4}$ pb, at low dark matter mass, $m_{DM} \sim 4-9$ GeV. This cannot be achieved for the lightest neutralino in the minimal supersymmetric model (MSSM) after imposing all constraints, including, in particular, $\Omega_{\widetilde{\chi}_1^0} h^2 \sim 0.11$ and the Tevatron limit of $B(B_s \to \mu^+\mu^-) < 5.8 \times 10^{-8}$. It is then natural to ask if the simplest extension of the MSSM obtained by adding a singlet superfield to the MSSM (the NMSSM and ENMSSM models) can allow simultaneous compatibility between CoGeNT/DAMA events and all other constraints, or must one turn to more exotic supersymmetric or other models. As reviewed here, the NMSSM and ENMSSM *can* achieve large σ_{SI} at low $m_{\widetilde{\chi}_1^0}$ while satisfying all constraints, but only if the Higgs sector has the appropriate structure and properties. Indeed, given the LEP, Tevatron, BaBar and other constraints, only a limited number of possibilities within the NMSSM and ENMSSM have been delimited to date. These include:

(i) The "inverted-Higgs" (IH) scenarios where the lightest CP-even Higgs, h_1, is H_d-like while the h_2 has SM-like couplings to WW, ZZ, but might decay via $h_2 \to a_1 a_1$.
After imposing $\Omega_{\widetilde{\chi}_1^0} h^2 \sim 0.11$, and all other constraints, the value of σ_{SI} that can be achieved falls short by something like a factor of $5 - 10$ (assuming a reasonable $(g-2)_\mu$ is required) in comparison to the value of $\sigma_{SI} \sim (1-3) \times 10^{-4}$ pb that is apparently required by CoGeNT/DAMA. However, this kind of scenario could become interesting if: a) the relevant σ_{SI} values at small $m_{\widetilde{\chi}_1^0}$ turn out to be somewhat smaller (*e.g.* new data or increased local density ρ); b) the $(g-2)_\mu$ restrictions employed turn out to be incorrect; and/or c) the s-quark content of the nucleons has been underestimated.
All Higgs bosons in the IH scenarios are quite light and discovery prospects are good. The $\widetilde{\chi}_1^0$ is primarily bino. One can even have an "inverted-ideal-Higgs" (IIH) scenario in which the ZZ coupling squared weighted Higgs mass, m_{eff}, is below 105 GeV and is thus in the ideal range for precision electroweak data, finetuning and electroweak baryogenesis. We focused on describing such an IIH scenario in our discussion.

(ii) The singlet-singlino (SS) scenarios in which the h_1 and $\widetilde{\chi}_1^0$ are primarily singlet and singlino, respectively, with m_{h_1} fairly small ($m_{h_1} \sim 40 - 70$ GeV) in the ENMSSM case and very small ($m_{h_1} \lesssim 1$ GeV) in the NMSSM DLH case.

In both these cases, one can achieve the required $\sigma_{SI} \sim 2 \times 10^{-4}$ pb while maintaining $\Omega_{\widetilde{\chi}_1^0} h^2 \sim 0.11$ and obeying all constraints.
In the ENMSSM case, detecting the h_1 directly at the LHC in $gg \to b\bar{b}h_1$ with $h_1 \to \tau^+\tau^-$ might prove possible.
In the DLH case, detection of the h_1 would require a significant (but not enormous) increase in statistics relative to current BaBar data for $\Upsilon_{3S} \to \gamma\mu^+\mu^-$.
In both cases, most of the other Higgs bosons would be readily detectable.

In general, there is an intimate connection between achieving large σ_{SI} at small $m_{\widetilde{\chi}_1^0}$ and a relatively unusual Higgs sector structure. In the SS cases, detection of the singlet h_1 will be highly non-trivial, but absolutely necessary if we are to understand the source of the large σ_{SI} and achieve a quantitative understanding of $\Omega_{\widetilde{\chi}_1^0} h^2$. If at least some Higgs bosons are discovered and their properties measured and if the CoGeNT/DAMA events are confirmed as dark matter detection, then it is also possible that the NMSSM and ENMSSM supersymmetric models will be ruled out, requiring that more exotic possibilities be considered.

5. Acknowledgments

This work was supported by US DOE grant DE-FG03-91ER40674. I wish to thank my collaborators for their contributions to our joint work. I also wish to thank the organizers of PASCOS 2010 for their generous support and the Aspen Center for Physics for support during the course of this project. I also wish to thank C. Wagner for several useful communications.

References

[1] C. E. Aalseth *et al.* [The CoGeNT Collaboration], arXiv:1002.4703 [astro-ph.CO].
[2] R. Bernabei *et al.*, arXiv:1002.1028 [astro-ph.GA].
[3] D. Hooper, J. I. Collar, J. Hall and D. McKinsey, arXiv:1007.1005 [hep-ph].
[4] M. Pato, O. Agertz, G. Bertone, B. Moore and R. Teyssier, Phys. Rev. D **82**, 023531 (2010) [arXiv:1006.1322 [astro-ph.HE]].
[5] E. Komatsu *et al.* [WMAP Collaboration], Astrophys. J. Suppl. **180**, 330 (2009) [arXiv:0803.0547 [astro-ph]].
[6] D. Feldman, Z. Liu and P. Nath, arXiv:1003.0437 [hep-ph].
[7] R. Dermisek and J. F. Gunion, Phys. Rev. D **75**, 075019 (2007) [arXiv:hep-ph/0611142].
[8] R. Dermisek and J. F. Gunion, Phys. Rev. Lett. **95**, 041801 (2005) [arXiv:hep-ph/0502105].
[9] J. F. Gunion, A. V. Belikov and D. Hooper, arXiv:1009.2555 [hep-ph].
[10] J. F. Gunion, D. Hooper and B. McElrath, Phys. Rev. D **73**, 015011 (2006) [arXiv:hep-ph/0509024];
[11] A. V. Belikov, J. F. Gunion, D. Hooper and T. M. P. Tait, arXiv:1009.0549 [hep-ph].
[12] P. Draper, T. Liu, C. E. M. Wagner, L. T. M. Wang and H. Zhang, arXiv:1009.3963 [hep-ph].
[13] U. Ellwanger, J. F. Gunion and C. Hugonie, JHEP **0502**, 066 (2005) [arXiv:hep-ph/0406215].
[14] U. Ellwanger and C. Hugonie, Comput. Phys. Commun. **175**, 290 (2006) [arXiv:hep-ph/0508022].
[15] G. Belanger, F. Boudjema, A. Pukhov and A. Semenov, Comput. Phys. Commun. **176**, 367 (2007) [arXiv:hep-ph/0607059].
[16] U. Ellwanger and C. Hugonie, http://www.th.u-psud.fr/NMHDECAY/nmssmtools.html.
[17] S. Schael *et al.* [ALEPH Collaboration], JHEP **1005**, 049 (2010) [arXiv:1003.0705 [hep-ex]].
[18] D. Benjamin *et al.* [Tevatron New Phenomena & Higgs Working Group], arXiv:1003.3363 [hep-ex].
[19] V. M. Abazov *et al.* [D0 Collaboration], Phys. Lett. B **682**, 278 (2009) [arXiv:0908.1811 [hep-ex]].
[20] B. Aubert *et al.* [BABAR Collaboration], Phys. Rev. Lett. **103**, 081803 (2009) [arXiv:0905.4539 [hep-ex]].
[21] B. Aubert *et al.* [BABAR Collaboration], Phys. Rev. Lett. **103**, 181801 (2009) [arXiv:0906.2219 [hep-ex]].
[22] R. Dermisek and J. F. Gunion, Phys. Rev. D **81**, 075003 (2010) [arXiv:1002.1971 [hep-ph]].
[23] R. Dermisek and J. F. Gunion, Phys. Rev. D **81**, 055001 (2010) [arXiv:0911.2460 [hep-ph]].
[24] D. Das and U. Ellwanger, arXiv:1007.1151 [hep-ph].
[25] J. Giedt, A. W. Thomas and R. D. Young, Phys. Rev. Lett. **103**, 201802 (2009) [arXiv:0907.4177 [hep-ph]]. We wish to thank H. Leutwyler for very useful communications regarding these issues.

Journal of Physics: Conference Series **259** (2010) 012013 doi:10.1088/1742-6596/259/1/012013

Physics from extra dimensions

I. Antoniadis[1]

Department of Physics
CERN - Theory Unit
1211 Geneva 23, Switzerland

E-mail: ignatios.antoniadis@cern.ch

Abstract. Lowering the string scale in the TeV region provides a theoretical framework for solving the mass hierarchy problem and unifying all interactions. The apparent weakness of gravity can then be accounted by the existence of large internal dimensions, in the submillimeter region, and transverse to a braneworld where our universe must be confined. I review the main properties of this scenario and its implications for observations at both particle colliders, and in non-accelerator gravity experiments.

1. Introduction

During the last few decades, physics beyond the Standard Model (SM) was guided from the problem of mass hierarchy. This can be formulated as the question of why gravity appears to us so weak compared to the other three known fundamental interactions corresponding to the electromagnetic, weak and strong nuclear forces. Indeed, gravitational interactions are suppressed by a very high energy scale, the Planck mass $M_P \sim 10^{19}$ GeV, associated to a length $l_P \sim 10^{-35}$ m, where they are expected to become important. In a quantum theory, the hierarchy implies a severe fine tuning of the fundamental parameters in more than 30 decimal places in order to keep the masses of elementary particles at their observed values. The reason is that quantum radiative corrections to all masses generated by the Higgs vacuum expectation value (VEV) are proportional to the ultraviolet cutoff which in the presence of gravity is fixed by the Planck mass. As a result, all masses are "attracted" to become about 10^{16} times heavier than their observed values.

Besides compositeness, there are two main theories that have been proposed and studied extensively during the last years, corresponding to different approaches of dealing with the mass hierarchy problem. (1) Low energy supersymmetry with all superparticle masses in the TeV region. Indeed, in the limit of exact supersymmetry, quadratically divergent corrections to the Higgs self-energy are exactly cancelled, while in the softly broken case, they are cutoff by the supersymmetry breaking mass splittings. (2) TeV scale strings, in which quadratic divergences are cutoff by the string scale and low energy supersymmetry is not needed. Both ideas are experimentally testable at high-energy particle colliders and in particular at LHC. Below, I discuss their implementation in string theory.

[1] On leave from CPHT (UMR CNRS 7644) Ecole Polytechnique, F-91128 Palaiseau.

Journal of Physics: Conference Series **259** (2010) 012013 doi:10.1088/1742-6596/259/1/012013

2. Strings and extra dimensions

The appropriate and most convenient framework for low energy supersymmetry and grand unification is the perturbative heterotic string. Indeed, in this theory, gravity and gauge interactions have the same origin, as massless modes of the closed heterotic string, and they are unified at the string scale M_s. As a result, the Planck mass M_P is predicted to be proportional to M_s:

$$M_P = M_s/g\,, \qquad (1)$$

where g is the gauge coupling. In the simplest constructions all gauge couplings are the same at the string scale, given by the four-dimensional (4d) string coupling, and thus no grand unified group is needed for unification. In our conventions $\alpha_{\rm GUT} = g^2 \simeq 0.04$, leading to a discrepancy between the string and grand unification scale $M_{\rm GUT}$ by almost two orders of magnitude. Explaining this gap introduces in general new parameters or a new scale, and the predictive power is essentially lost. This is the main defect of this framework, which remains though an open and interesting possibility [1].

The other other perturbative framework that has been studied extensively in the more recent years is type I string theory with D-branes. Unlike in the heterotic string, gauge and gravitational interactions have now different origin. The latter are described again by closed strings, while the former emerge as excitations of open strings with endpoints confined on D-branes [2]. This leads to a braneworld description of our universe, which should be localized on a hypersurface, i.e. a membrane extended in p spatial dimensions, called p-brane (see Fig. 1). Closed strings propagate in all nine dimensions of string theory: in those extended along the p-brane, called parallel, as well as in the transverse ones. On the contrary, open strings are attached on the p-brane. Obviously, our p-brane world must have at least the three known dimensions of space.

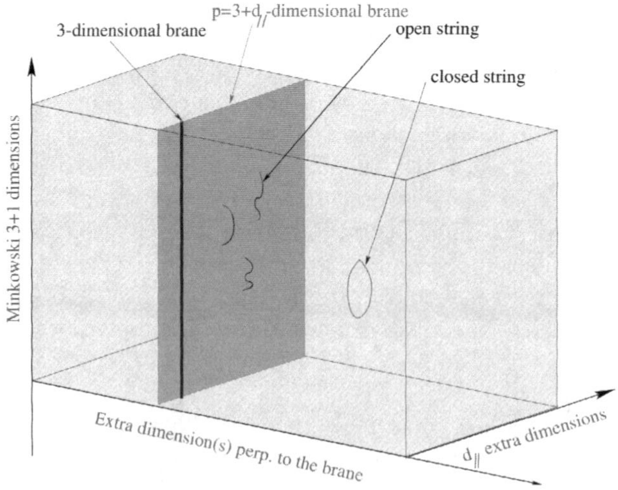

Figure 1. In type I string framework, our Universe contains, besides our 3 spatial dimensions (denoted by the blue line), some extra dimensions ($d_\parallel = p-3$) parallel to our world p-brane (green plane) where endpoints of open strings are confined, as well as some transverse dimensions (yellow space) where only gravity (described by closed strings) propagate.

But it may contain more: the extra $d_\parallel = p-3$ parallel dimensions must have a finite size, in order to be unobservable at present energies, and can be as large as TeV$^{-1} \sim 10^{-18}$ m [3]. On the other hand, transverse dimensions interact with us only gravitationally and experimental bounds are much weaker: their size should be less than about 0.1 mm [4]. In the following, I review the main properties and experimental signatures of low string scale models [5, 6].

2.1. Framework of low scale strings

In type I theory, the different origin of gauge and gravitational interactions implies that the relation between the Planck and string scales is not linear as (1) of the heterotic string. The requirement that string theory should be weakly coupled, constrain the size of all parallel dimensions to be of order of the string length, while transverse dimensions remain unrestricted. Assuming an isotropic transverse space of $n = 9 - p$ compact dimensions of common radius $R_\perp$, one finds:

$$M_P^2 = \frac{1}{g_s^2} M_s^{2+n} R_\perp^n \,, \qquad g_s \simeq g^2 \,. \tag{2}$$

where g_s is the string coupling. It follows that the type I string scale can be chosen hierarchically smaller than the Planck mass [7, 5] at the expense of introducing extra large transverse dimensions felt only by gravity, while keeping the string coupling small [5]. The weakness of 4d gravity compared to gauge interactions (ratio M_W/M_P) is then attributed to the largeness of the transverse space $R_\perp$ compared to the string length $l_s = M_s^{-1}$.

An important property of these models is that gravity becomes effectively $(4+n)$-dimensional with a strength comparable to those of gauge interactions at the string scale. The first relation of Eq. (2) can be understood as a consequence of the $(4+n)$-dimensional Gauss law for gravity, with

$$M_*^{(4+n)} = M_s^{2+n}/g^4 \tag{3}$$

the effective scale of gravity in $4+n$ dimensions. Taking $M_s \simeq 1$ TeV, one finds a size for the extra dimensions $R_\perp$ varying from 10^8 km, .1 mm, down to a Fermi for $n = 1, 2$, or 6 large dimensions, respectively. This shows that while $n = 1$ is excluded, $n \geq 2$ is allowed by present experimental bounds on gravitational forces [4, 8]. Thus, in these models, gravity appears to us very weak at macroscopic scales because its intensity is spread in the "hidden" extra dimensions. At distances shorter than $R_\perp$, it should deviate from Newton's law, which may be possible to explore in laboratory experiments (see Fig. 2).

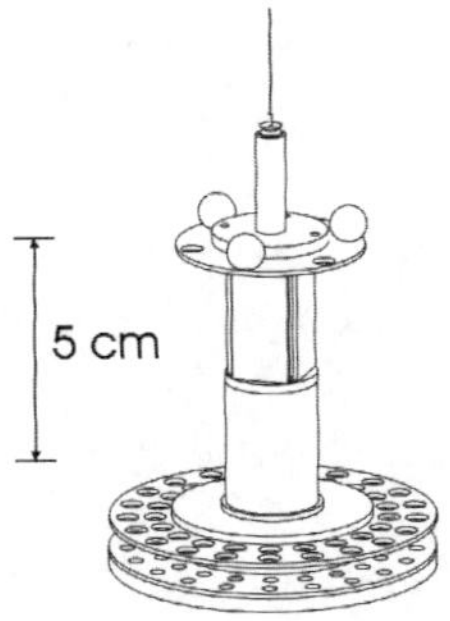

Figure 2. Torsion pendulum that tested Newton's law at 55 μm.

3. Large number of species

Here, we open a parenthesis to describe that low scale gravity with large extra dimensions is actually a particular case of a more general framework, where the ultraviolate (UV) cutoff is lower than the Panck scale due to the existence of a large number of particle species coupled to gravity [9]. Indeed, it was shown that the effective UV cutoff M_* is given by

$$M_*^2 = M_P^2/N\,, \tag{4}$$

where the counting of independent species N takes into account all particles which are not broad resonances, having a width less than their mass. The derivation is based on black hole evaporation but here we present a shorter argument using quantum information storage [10]. Consider a pixel of size L containing N species storing information. The energy required to localize N wave functions is then given by N/L, associated to a Schwarzschild radius $R_s = N/LM_P^2$. The latter must be less than the pixel size in order to avoid the collapse of such a system to a black hole, $R_s \leq L$, implying a minimum size $L \geq L_{min}$ with $L_{min} = \sqrt{N}/M_P$ associated precisely to the effective UV cutoff $M_* = L_{min}$ given in eq. (4). Imposing $M_* \simeq 1$ TeV, one should then have $N \sim 10^{32}$ particle species below about the TeV scale!

In the string theory context, there are two ways of realizing such a large number a particle species by lowering the string scale at a TeV (without changing the behavior of gravity at large distances):

(i) In large volume compactifications with the Standard Model (SM) localized on D-brane stacks, as described in the previous section. The particle species are then the Kaluza-Klein (KK) excitations of the graviton (and other possible bulk modes) associated to the large extra dimensions, given by

$$N = R_\perp^n l_s^n\,, \tag{5}$$

up to energies of order $M_* \simeq M_s$.

(ii) By introducing an infinitesimal string coupling $g_s \simeq 10^{-16}$ with the SM localized on Neveu-Schwarz NS5-branes in the framework of little strings [11]. In this case, the particle species are the effective number of string modes that contribute to the black hole bound [12]:

$$N = 1/g_s^2\,. \tag{6}$$

and gravity does not become strong at $M_s \sim \mathcal{O}(\text{TeV})$.

Note the both TeV string realizations above are compatible with the general expression (2), but in the second case there is no relation between the string and gauge couplings.

4. Experimental implications in accelerators

We now turn to the experimental predictions of TeV scale strings. Their main implications in particle accelerators are of three types, in correspondence with the three different sectors that are generally present: (i) new compactified parallel dimensions, (ii) new extra large transverse dimensions and low scale quantum gravity, and (iii) genuine string and quantum gravity effects. On the other hand, there exist interesting implications in non accelerator table-top experiments due to the exchange of gravitons or other possible states living in the bulk.

4.1. World-brane extra dimensions

In this case $RM_s \gtrsim 1$, and the associated compactification scale $R_\parallel^{-1}$ would be the first scale of new physics that should be found increasing the beam energy [3, 13, 14]. The main consequence is the existence of KK excitations for all SM particles that propagate along the extra parallel dimensions. Their masses are given by:

$$M_m^2 = M_0^2 + \frac{m^2}{R_\parallel^2} \quad ; \quad m = 0, \pm1, \pm2, \ldots \tag{7}$$

where we used $d_{\parallel} = 1$, and M_0 is the higher dimensional mass. The zero-mode $m = 0$ is identified with the 4d state, while the higher modes have the same quantum numbers with the lowest one, except for their mass given in (7). There are two types of experimental signatures of such dimensions [13, 15, 16]: (i) virtual exchange of KK excitations, leading to deviations in cross-sections compared to the SM prediction, that can be used to extract bounds on the compactification scale; (ii) direct production of KK modes.

On general grounds, there can be two different kinds of models with qualitatively different signatures depending on the localization properties of matter fermion fields. If the latter are localized in 3d brane intersections, they do not have excitations and KK momentum is not conserved because of the breaking of translation invariance in the extra dimension(s). KK modes of gauge bosons are then singly produced giving rise to generally strong bounds on the compactification scale and new resonances that can be observed in experiments. Otherwise, they can be produced only in pairs due to the KK momentum conservation, making the bounds weaker but the resonances difficult to observe.

In addition to virtual effects, KK excitations can be produced on-shell at LHC as new resonances [15] (see Fig. 3). There are two different channels, neutral Drell–Yan processes

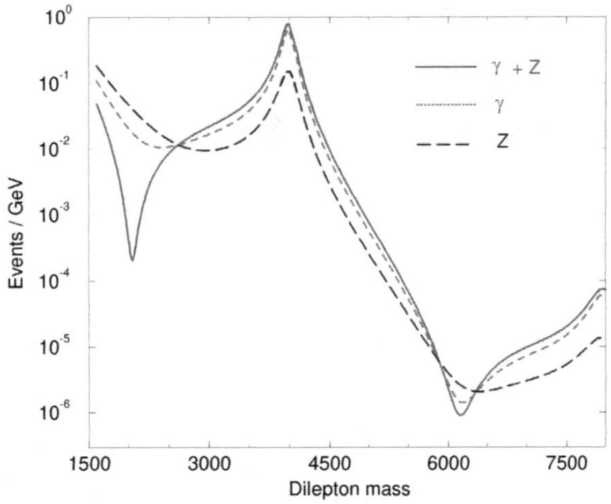

Figure 3. Production of the first KK modes of the photon and of the Z boson at LHC, decaying to electron-positron pairs. The number of expected events is plotted as a function of the energy of the pair in GeV.

$pp \to l^+l^-X$ and the charged channel $l^{\pm}\nu$, corresponding to the production of the KK modes $\gamma^{(1)}, Z^{(1)}$ and $W_{\pm}^{(1)}$, respectively. The discovery limits are about 6 TeV, while the exclusion bounds 15 TeV.

On the other hand, if all SM particles propagate in the extra dimension (called universal), KK modes can only be produced in pairs and the lower bound on the compactification scale becomes weaker, of order of 300-500 GeV. Moreover, no resonances can be observed at LHC, so that this scenario appears very similar to low energy supersymmetry. In fact, KK parity can even play the role of R-parity, implying that the lightest KK mode is stable and can be a dark matter candidate in analogy to the LSP [17].

Table 1. Limits on $R_\perp$ in mm.

Experiment	$n=2$	$n=4$	$n=6$
Collider bounds			
LEP 2	5×10^{-1}	2×10^{-8}	7×10^{-11}
Tevatron	5×10^{-1}	10^{-8}	4×10^{-11}
LHC	4×10^{-3}	6×10^{-10}	3×10^{-12}
NLC	10^{-2}	10^{-9}	6×10^{-12}
Present non-collider bounds			
SN1987A	3×10^{-4}	10^{-8}	6×10^{-10}
COMPTEL	5×10^{-5}	-	-

4.2. Extra large transverse dimensions

The main experimental signal is gravitational radiation in the bulk from any physical process on the world-brane. In fact, the very existence of branes breaks translation invariance in the transverse dimensions and gravitons can be emitted from the brane into the bulk. During a collision of center of mass energy $\sqrt{s}$, there are $\sim(\sqrt{s}R_\perp)^n$ KK excitations of gravitons with tiny masses, that can be emitted. Each of these states looks from the 4d point of view as a massive, quasi-stable, extremely weakly coupled (s/M_P^2 suppressed) particle that escapes from the detector. The total effect is a missing-energy cross-section roughly of order:

$$\frac{(\sqrt{s}R_\perp)^n}{M_P^2}\sim\frac{1}{s}\left(\frac{\sqrt{s}}{M_s}\right)^{n+2}. \tag{8}$$

Explicit computation of these effects leads to the bounds given in Table 1.

Fig. 4 shows the cross-section for graviton emission in the bulk, corresponding to the process $pp\to jet+graviton$ at LHC, together with the SM background [18]. There is a particular energy

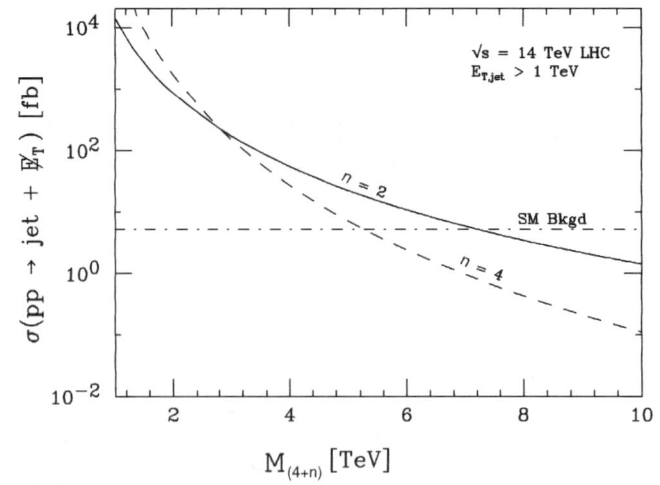

Figure 4. Missing energy due to graviton emission at LHC, as a function of the higher-dimensional gravity scale M_*, produced together with a hadronic jet. The expected cross-section is shown for $n=2$ and $n=4$ extra dimensions, together with the SM background.

and angular distribution of the produced gravitons that arise from the distribution in mass of KK states of spin-2. This can be contrasted to other sources of missing energy and might be a smoking gun for the extra dimensional nature of such a signal.

In Table 1, there are also included astrophysical and cosmological bounds. Astrophysical bounds [19, 20] arise from the requirement that the radiation of gravitons should not carry on too much of the gravitational binding energy released during core collapse of supernovae. The best cosmological bound [21] is obtained from requiring that decay of bulk gravitons to photons do not generate a spike in the energy spectrum of the photon background measured by the COMPTEL instrument. Bulk gravitons are expected to be produced just before nucleosynthesis due to thermal radiation from the brane. The limits assume that the temperature was at most 1 MeV as nucleosynthesis begins, and become stronger if temperature is increased.

4.3. String effects

At low energies, the interaction of light (string) states is described by an effective field theory. Their exchange generates in particular four-fermion operators that can be used to extract independent bounds on the string scale. In analogy with the bounds on longitudinal extra dimensions, there are two cases depending on the localization properties of matter fermions. If they come from open strings with both ends on the same stack of branes, exchange of massive open string modes gives rise to dimension eight effective operators, involving four fermions and two space-time derivatives [22, 23]. The corresponding bounds on the string scale are then around 500 GeV. On the other hand, if matter fermions are localized on non-trivial brane intersections, one obtains dimension six four-fermion operators and the bounds become stronger: $M_s \gtrsim 2-3$ TeV [23, 6]. At energies higher than the string scale, new spectacular phenomena are expected to occur, related to string physics and quantum gravity effects, such as possible micro-black hole production [24, 25, 26]. Particle accelerators would then become the best tools for studying quantum gravity and string theory.

Direct production of string resonances in hadron colliders leads generically to a universal deviation from Standard Model in jet distribution [27]. In particular, the first Regge excitation of the gluon has spin 2 and a width an order of magnitude lower than the string scale, leading to a characteristic peak in dijet production; similarly, the first excitations of quarks have spin 3/2. The dijet cross-section is shown in Fig. 5 for LHC energies. The reason for the universal behavior is that tree N-point open superstring amplitudes involving at most two fermions and gluons are completely model independent from the details of the compactification, including the number of supersymmetries that are left unbroken in four dimensions (even if all are broken). Such tree-level amplitudes do not receive contributions from KK, string winding or closed string graviton modes, but are given as a universal sum over exchanges of open string Regge excitations lying in Regge trajectories with masses

$$M_n^2 = M_s^2 n \quad ; \quad n = 0, 1, \ldots \tag{9}$$

and maximal spin $n+1$.

The relevant partonic cross-sections for dijet production, involving at most two quarks are $|\mathcal{M}(gg \to gg)|^2$, $|\mathcal{M}(gg \to q\bar{q})|^2$, $|\mathcal{M}(q\bar{q} \to gg)|^2$ and $|\mathcal{M}(qg \to qg)|^2$, where g and q denote gluons and quarks, respectively. They can be obtained from the first two, up to crossing symmetries, at the full (tree) string level and they are model independent [28]. The low energy expansion of these amplitudes reproduce the usual QCD expressions, while higher order terms describe the string corrections due to the exchange of Regge excitations. Besides these amplitudes, there are also those involving four quarks, such as $|\mathcal{M}(q\bar{q} \to q\bar{q})|^2$ and $|\mathcal{M}(qq \to qq)|^2$. These are model dependent, because the details of the compactification do not decouple. The reason is that they involve four vertices containing twist fields that describe quark states arising from open strings stretched in brane intersections. However, taking into account

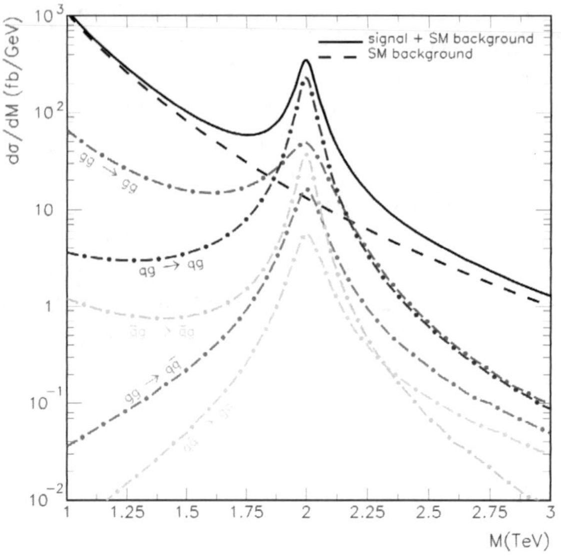

Figure 5. Production of the first Regge excitations at LHC in the dijet channel, for $M_s = 2$ TeV. The cross-section is plotted as a function of the dijet invariant mass M.

QCD color factors, their contribution is suppressed because parton luminosities in proton-proton collisions at TeV energies favor at least one gluon in the initial state. As a result, the dominant contribution comes from the model independent cross-sections described above, leading to the effect of Fig. 5.

We finish this section with some comments related to the possible micro-black hole production. Independently on the unresolved issue of the convergence of string perturbation theory in the kinematic region relevant to micro-black hole formation, there is a simple argument showing that at least within the perturbative TeV string framework, the energy threshold for black hole production is far above the LHC reach. Indeed, a string size black hole has a horizon radius $r_H \sim 1$ in string units, while the d-dimensional Newton's constant behaves as $G_N \sim g_s^2$. It follows that the mass of a d-dimensional black hole is [29]:

$$M_{\rm BH} \sim r_H^{d/2-1}/G_N \simeq 1/g_s^2 \,. \tag{10}$$

Thus, for a weakly coupled theory, this energy threshold is much higher than the string and the higher dimensional Planck scales M_s and M_* of eq. (3). Comparing this energy threshold with the mass of Regge excitations (9), one finds $n \sim 1/g_s^4$ which is actually compatible with the relation one obtains by identifying the black hole entropy $S_{\rm BH} \sim 1/G_N \sim 1/g_s^2$ with the perturbative string entropy $S_{\rm string} \sim \sqrt{n}$. Using now relation (2), and the value of the Standard Model gauge couplings $g_s \simeq g^2 \sim 0.1$, one finds that the energy threshold $M_{\rm BH}$ of micro-black hole production is about four orders of magnitude higher than the string scale, implying that one would produce 10^4 string states before reaching $M_{\rm BH}$.

5. Supersymmetry in the bulk and short range forces

5.1. Sub-millimeter forces

Besides the spectacular predictions in accelerators, there are also modifications of gravitation in the sub-millimeter range, which can be tested in "table-top" experiments that measure gravity

at short distances. There are three categories of such predictions:
(i) Deviations from the Newton's law $1/r^2$ behavior to $1/r^{2+n}$, which can be observable for $n = 2$ large transverse dimensions of sub-millimeter size.
(ii) New scalar forces in the sub-millimeter range, related to the mechanism of supersymmetry breaking, and mediated by light scalar fields φ with masses [30, 5]:

$$m_\varphi \simeq \frac{m_{susy}^2}{M_P} \simeq 10^{-4} - 10^{-6}\ \text{eV}\,, \tag{11}$$

for a supersymmetry breaking scale $m_{susy} \simeq 1 - 10$ TeV. They correspond to Compton wavelengths of 1 mm to 10 μm. m_{susy} can be either $1/R_\parallel$ if supersymmetry is broken by compactification [30], or the string scale if it is broken "maximally" on our world-brane [5]. A universal attractive scalar force is mediated by the radion modulus $\varphi \equiv M_P \ln R$, with R the radius of the longitudinal or transverse dimension(s). For $n = 2$, there may be an enhancement factor of the radion mass by $\ln R_\perp M_s \simeq 30$ decreasing its wavelength by an order of magnitude [31].

The coupling of the radius modulus to matter relative to gravity can be easily computed and is given by:

$$\sqrt{\alpha_\varphi} = \frac{1}{M}\frac{\partial M}{\partial \varphi}\ ;\ \ \alpha_\varphi = \begin{cases} \frac{\partial \ln \Lambda_{QCD}}{\partial \ln R} \simeq \frac{1}{3} \ \text{ for } R_\parallel \\ \\ \frac{2n}{n+2} = 1 - 1.5 \text{ for } R_\perp \end{cases} \tag{12}$$

where M denotes a generic physical mass. Such a force can be tested in microgravity experiments and should be contrasted with the change of Newton's law due the presence of extra dimensions that is observable only for $n = 2$ [4, 8]. The resulting bounds from an analysis of the radion effects are [32]: $M_* \gtrsim 6\,\text{TeV}$.
(iii) Non universal repulsive forces much stronger than gravity, mediated by possible abelian gauge fields in the bulk [19, 33]. Such fields acquire tiny masses of the order of M_s^2/M_P, as in (11), due to brane localized anomalies [33]. Although their gauge coupling is infinitesimally small, $g_A \sim M_s/M_P \simeq 10^{-16}$, it is still bigger that the gravitational coupling E/M_P for typical energies $E \sim 1$ GeV, and the strength of the new force would be $10^6 - 10^8$ stronger than gravity.

In Fig. 6 we depict the actual information from previous, present and upcoming experiments [8, 31]. The solid lines indicate the present limits from the experiments indicated. The excluded regions lie above these solid lines. Measuring gravitational strength forces at short distances is challenging. The horizontal lines correspond to theoretical predictions, in particular for the graviton in the case $n = 2$ and for the radion in the transverse case. These limits are compared to those obtained from particle accelerator experiments in Table 1.

6. Standard Model on D-branes

The gauge group closest to the Standard Model one can easily obtain with D-branes is $U(3) \times U(2) \times U(1)$. The first factor arises from three coincident "color" D-branes. An open string with one end on them is a triplet under $SU(3)$ and carries the same $U(1)$ charge for all three components. Thus, the $U(1)$ factor of $U(3)$ has to be identified with *gauged* baryon number. Similarly, $U(2)$ arises from two coincident "weak" D-branes and the corresponding abelian factor is identified with *gauged* weak-doublet number. Finally, an extra $U(1)$ D-brane is necessary in order to accommodate the Standard Model without breaking the baryon number [34].

It turns out that there are two possible ways of embedding the Standard Model particle spectrum on these stacks of branes [34], which are shown pictorially in Fig. 7. The quark doublet Q corresponds necessarily to a massless excitation of an open string with its two ends on the two different collections of branes (color and weak). As seen from the figure, a fourth brane stack is needed for a complete embedding, which is chosen to be a $U(1)_b$ extended in

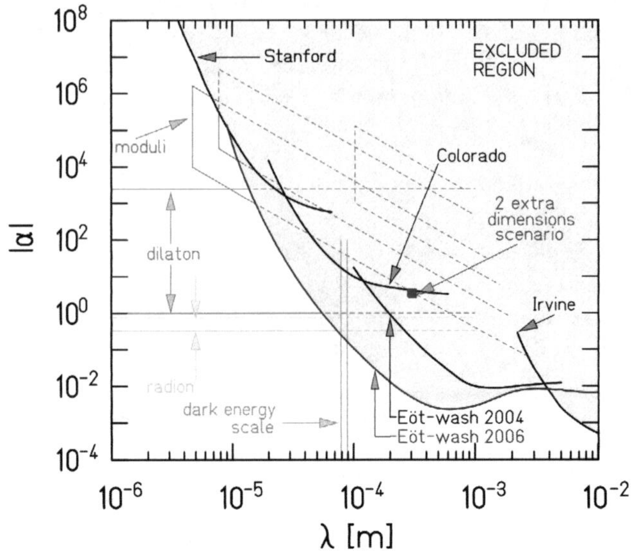

Figure 6. Present limits on new short-range forces (yellow regions), as a function of their range λ and their strength relative to gravity α. The limits are compared to new forces mediated by the graviton in the case of two large extra dimensions, and by the radion.

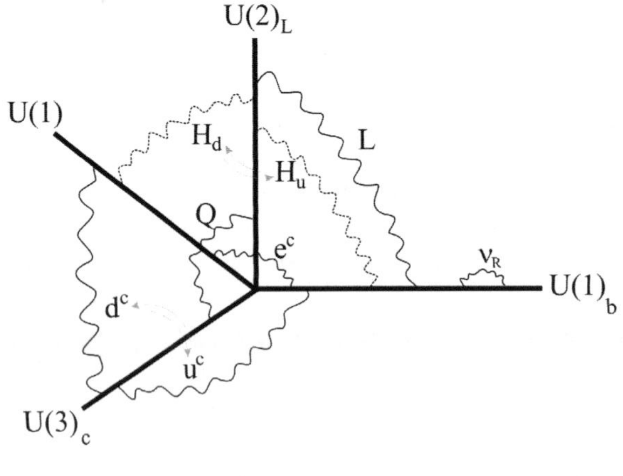

Figure 7. A minimal Standard Model embedding on D-branes.

the bulk. This is welcome since one can accommodate right handed neutrinos as open string states on the bulk with sufficiently small Yukawa couplings suppressed by the large volume of the bulk [35]. The two models are obtained by an exchange of the up and down antiquarks, u^c and d^c, which correspond to open strings with one end on the color branes and the other either on the $U(1)$ brane, or on the $U(1)_b$ in the bulk. The lepton doublet L arises from an open string stretched between the weak branes and $U(1)_b$, while the antilepton l^c corresponds to a string with one end on the $U(1)$ brane and the other in the bulk. For completeness, we also show the two possible Higgs states H_u and H_d that are both necessary in order to give tree-level masses to all quarks and leptons of the heaviest generation.

Acknowledgments
Work supported in part by the European Commission under the ERC Advanced Grant 226371 and the contract PITN-GA-2009-237920, and in part by the CNRS grant GRC APIC PICS 3747.

References

[1] Fore a review, see e.g. K. R. Dienes, Phys. Rept. **287** (1997) 447 [arXiv:hep-th/9602045]; and references therein.
[2] C. Angelantonj and A. Sagnotti, *Phys. Rept.* **371** (2002) 1 [Erratum-ibid. **376** (2003) 339] [arXiv:hep-th/0204089].
[3] I. Antoniadis, *Phys. Lett.* B **246** (1990) 377.
[4] D. J. Kapner, T. S. Cook, E. G. Adelberger, J. H. Gundlach, B. R. Heckel, C. D. Hoyle and H. E. Swanson, Phys. Rev. Lett. **98** (2007) 021101.
[5] N. Arkani-Hamed, S. Dimopoulos and G. R. Dvali, *Phys. Lett.* B **429** (1998) 263 [arXiv:hep-ph/9803315]; I. Antoniadis, N. Arkani-Hamed, S. Dimopoulos and G. R. Dvali, *Phys. Lett.* B **436** (1998) 257 [arXiv:hep-ph/9804398].
[6] For a review see e.g. I. Antoniadis, *Prepared for NATO Advanced Study Institute and EC Summer School on Progress in String, Field and Particle Theory, Cargese, Corsica, France (2002)*; and references therein.
[7] J. D. Lykken, *Phys. Rev.* D **54** (1996) 3693 [arXiv:hep-th/9603133].
[8] J. C. Long and J. C. Price, *Comptes Rendus Physique* **4** (2003) 337; R. S. Decca, D. Lopez, H. B. Chan, E. Fischbach, D. E. Krause and C. R. Jamell, *Phys. Rev. Lett.* **94** (2005) 240401; R. S. Decca et al., arXiv:0706.3283 [hep-ph]; S. J. Smullin, A. A. Geraci, D. M. Weld, J. Chiaverini, S. Holmes and A. Kapitulnik, arXiv:hep-ph/0508204; H. Abele, S. Haeßler and A. Westphal, in 271th WE-Heraeus-Seminar, Bad Honnef (2002).
[9] G. Dvali, arXiv:0706.2050 [hep-th]; Int. J. Mod. Phys. A **25** (2010) 602 [arXiv:0806.3801 [hep-th]]; G. Dvali and M. Redi, Phys. Rev. D **77** (2008) 045027 [arXiv:0710.4344 [hep-th]]; R. Brustein, G. Dvali and G. Veneziano, JHEP **0910** (2009) 085 [arXiv:0907.5516 [hep-th]].
[10] G. Dvali and C. Gomez, Phys. Lett. B **674** (2009) 303 [arXiv:0812.1940 [hep-th]].
[11] I. Antoniadis and B. Pioline, *Nucl. Phys.* B **550** (1999) 41 [arXiv:hep-th/9902055]; I. Antoniadis, S. Dimopoulos and A. Giveon, JHEP **0105** (2001) 055 [arXiv:hep-th/0103033].
[12] G. Dvali and D. Lust, arXiv:0912.3167 [hep-th]; G. Dvali and C. Gomez, arXiv:1004.3744 [hep-th].
[13] I. Antoniadis and K. Benakli, *Phys. Lett.* B **326** (1994) 69.
[14] K. R. Dienes, E. Dudas and T. Gherghetta, Phys. Lett. B **436** (1998) 55 [arXiv:hep-ph/9803466]; Nucl. Phys. B **537** (1999) 47 [arXiv:hep-ph/9806292].
[15] I. Antoniadis, K. Benakli and M. Quirós, *Phys. Lett.* B **331** (1994) 313 and *Phys. Lett.* B **460** (1999) 176; P. Nath, Y. Yamada and M. Yamaguchi, *Phys. Lett.* B **466** (1999) 100 T. G. Rizzo and J. D. Wells, *Phys. Rev.* D **61** (2000) 016007; T. G. Rizzo, *Phys. Rev.* D **61** (2000) 055005; A. De Rujula, A. Donini, M. B. Gavela and S. Rigolin, *Phys. Lett.* B **482** (2000) 195;
[16] E. Accomando, I. Antoniadis and K. Benakli, *Nucl. Phys.* B **579** (2000) 3.
[17] G. Servant and T. M. P. Tait, *Nucl. Phys.* B **650** (2003) 391.
[18] G. F. Giudice, R. Rattazzi and J. D. Wells, *Nucl. Phys.* B **544** (1999) 3; E. A. Mirabelli, M. Perelstein and M. E. Peskin, *Phys. Rev. Lett.* **82** (1999) 2236; T. Han, J. D. Lykken and R. Zhang, *Phys. Rev.* D **59** (1999) 105006; K. Cheung and W.-Y. Keung, *Phys. Rev.* D **60** (1999) 112003; C. Balázs *et al.*, *Phys. Rev. Lett.* **83** (1999) 2112; L3 Collaboration (M. Acciarri *et al.*), *Phys. Lett.* B **464** (1999) 135 and **470** (1999) 281: J. L. Hewett, *Phys. Rev. Lett.* **82** (1999) 4765.
[19] N. Arkani-Hamed, S. Dimopoulos and G. Dvali, *Phys. Rev.* D **59** (1999) 086004.
[20] S. Cullen and M. Perelstein, *Phys. Rev. Lett.* **83** (1999) 268; V. Barger, T. Han, C. Kao and R. J. Zhang, *Phys. Lett.* B **461** (1999) 34.
[21] K. Benakli and S. Davidson, *Phys. Rev.* D **60** (1999) 025004; L. J. Hall and D. Smith, *Phys. Rev.* D **60** (1999) 085008.
[22] E. Dudas and J. Mourad, Nucl. Phys. B **575** (2000) 3 [arXiv:hep-th/9911019]; S. Cullen, M. Perelstein and M. E. Peskin, *Phys. Rev.* D **62** (2000) 055012; D. Bourilkov, *Phys. Rev.* D **62** (2000) 076005; L3 Collaboration (M. Acciarri *et al.*), *Phys. Lett.* B **489** (2000) 81.
[23] I. Antoniadis, K. Benakli and A. Laugier, *JHEP* **0105** (2001) 044.
[24] P. C. Argyres, S. Dimopoulos and J. March-Russell, Phys. Lett. B **441** (1998) 96 [arXiv:hep-th/9808138]; T. Banks and W. Fischler, arXiv:hep-th/9906038.

[25] S. B. Giddings and S. Thomas, *Phys. Rev.* D **65** (2002) 056010 [arXiv:hep-ph/0106219]; S. Dimopoulos and G. Landsberg, *Phys. Rev. Lett.* **87** (2001) 161602 [arXiv:hep-ph/0106295].
[26] P. Meade and L. Randall, arXiv:0708.3017 [hep-ph].
[27] L. A. Anchordoqui, H. Goldberg, D. Lust, S. Nawata, S. Stieberger and T. R. Taylor, Phys. Rev. Lett. **101** (2008) 241803 [arXiv:0808.0497 [hep-ph]].
[28] D. Lust, S. Stieberger and T. R. Taylor, Nucl. Phys. B **808** (2009) 1 [arXiv:0807.3333 [hep-th]].
[29] G. T. Horowitz and J. Polchinski, Phys. Rev. D **55** (1997) 6189 [arXiv:hep-th/9612146].
[30] I. Antoniadis, S. Dimopoulos and G. Dvali, *Nucl. Phys.* B **516** (1998) 70; S. Ferrara, C. Kounnas and F. Zwirner, *Nucl. Phys.* B **429** (1994) 589.
[31] I. Antoniadis, K. Benakli, A. Laugier and T. Maillard, *Nucl. Phys.* B **662** (2003) 40 [arXiv:hep-ph/0211409].
[32] E. G. Adelberger, B. R. Heckel, S. Hoedl, C. D. Hoyle, D. J. Kapner and A. Upadhye, Phys. Rev. Lett. **98** (2007) 131104.
[33] I. Antoniadis, E. Kiritsis and J. Rizos, *Nucl. Phys.* B **637** (2002) 92.
[34] I. Antoniadis, E. Kiritsis and T. N. Tomaras, *Phys. Lett.* B **486** (2000) 186; I. Antoniadis, E. Kiritsis, J. Rizos and T. N. Tomaras, *Nucl. Phys.* B **660** (2003) 81.
[35] K. R. Dienes, E. Dudas and T. Gherghetta, *Nucl. Phys.* B **557** (1999) 25 [arXiv:hep-ph/9811428]; N. Arkani-Hamed, S. Dimopoulos, G. R. Dvali and J. March-Russell, *Phys. Rev.* D **65** (2002) 024032 [arXiv:hep-ph/9811448]; G. R. Dvali and A. Y. Smirnov, *Nucl. Phys.* B **563** (1999) 63.

16th International Symposium on Particles Strings and Cosmology (PASCOS 2010) IOP Publishing
Journal of Physics: Conference Series **259** (2010) 012014 doi:10.1088/1742-6596/259/1/012014

String Theory Phenomenology

Angel Uranga
Instituto de Física Teórica UAM/CSIC, 28049 Madrid, Spain
E-mail: angel.uranga@uam.es

Abstract. We review the qualitative features of string theory compactifications with low energy effective theory close to (supersymmetric versions of) the Standard Model of Particle Physics. We cover Calabi-Yau compactifications of heterotic strings, intersecting brane models, and the recent setup of F-theory models.

1. String theory

1.1. Generalities

String theory is the theoretically best motivated proposal to provide a unified description of gravitational and gauge interactions, in a framework consistent at the quantum level. The main idea underlying string theory is to include additional degrees of freedom at high-energies to get rid of the ultraviolet divergences arising in the standard quantization of gravity. In this sense, it is very analogous to the role of electroweak theory in solving the ultraviolet problems of the Fermi theory of weak interactions, by the addition of new high-energy degrees of freedom, the W and Z bosons. The new degrees of freedom introduced by string theory however show up at a new scale, the string scale M_s. Although in particular constructions it might be close to the familiar TeV scale, it is in general expected to be much higher, in the fantastically high regime close to the GUT scale of 10^{16} GeV or the Planck scale $M_p \sim 10^{19}$ GeV.

The main novelty of string theory is the nature of the proposed new degrees of freedom. They are not given by a few new massive particles. String theory rather proposes that all particles are actually not point-like, but extended one-dimensional objects, *strings*, with characteristic size $L_s \sim 1/M_s$. Different particles are just different oscillation modes of a single kind of object, the string, endowing the theory with a great unification power. Even though there is an infinite number of such oscillation modes, the mass of the corresponding particle increases with the oscillation excitation number, with steps of order M_s. Hence at familiar energies $E \ll M_s$, only the string groundstates are observable. In this regime there is not enough resolution to detect the extended nature of the string, and these particles look like point-like. At higher energies however, the additional excited states of the string become accessible, leading to a tower of new particles, whose role is to control the ultraviolet properties of the theory.

The light particle spectrum of string theories is model dependent, as we will precise later on, but there are some general features, for both closed and open strings. A general property of string theory is that the massless spectrum of closed strings contains a spin-2 particle which behaves as a graviton. Thus string theory naturally includes Einstein's General Relativity description of gravitational interactions. Moreover, the light spectrum of open strings produces (possibly non-abelian) gauge bosons, and matter fields charged under them, like spin-1/2 matter fermions, or

Journal of Physics: Conference Series **259** (2010) 012014 doi:10.1088/1742-6596/259/1/012014

spin-0 bosons. It is remarkable that such particles are of the same kind as those encountered in Nature, and this underlies the potential application of string theory as a unification framework for all particles and interactions.

1.2. Compactification, 4d physics and the landscape

The nice ultraviolet properties of string theory require a high-energy oscillation spectrum corresponding to vibration modes in 10 dimensions. In fact there are five perturbatively different 10d string theories with these well-behaved properties, known as the type IIA, type IIB, type I, $SO(32)$ heterotic and $E_8 \times E_8$ heterotic string theories. They are however closely related in the regime beyond perturbation theory. Thus the structure of string theory when formulated in 10d is a beautiful construct, with many gists of new symmetries, known as string dualities, and a yet mysterious unique underlying structure, often referred to as M-theory.

Although it is remarkable that a physical theory can fix the number of spacetime dimensions in which it must be formulated, the result may seem disappointing. This however does not imply that the number of dimensions observable at low energies $E \ll M_s$ is 10. There is a standard mechanism, known as compactification, by which the number of observable dimensions can be reduced to four. The idea is to consider the 10d spacetime to be $M_4 \times \mathbf{X_6}$, the product of 4d Minkowski space and a 6d internal manifold, which should be compact, i.e. have a finite volume. Introducing a characteristic size L, which we take $L \gg 1/M_s$, the theory at energies $E \ll 1/L \ll M_s$ appear to have only four dimensions, since there is not enough resolution to explore the extra dimensions.

As we describe below, the geometry and other ingredients in the internal dimensions however encode many physical properties of the resulting 4d theory, like the gauge group, matter content, and quantitative data like the values of couplings constants, etc. This implies that, since there are many many many different ways of compactifying the 10d theory, there is a multitude of possible 4d string vacua, each leading to completely different physical theories. This is the recently popularized view of the landscape of vacua in string theory, whose interpretation and potential usefulness is still subject to much controversy and leads to not few coffee time discussions. Although the situation may seem to spoil the uniqueness and beauty of the theory, for our present purposes we may regard it as not dissimilar to General Relativity, where despite the uniqueness of the theory leads to a multitude of possible solutions. In General Relativity, the crucial point is to focus on the relevant physical properties to pick the right solution to describe a given system, e.g. spherical symmetry to describe the Solar System. In a similar spirit, our point is to focus on particular string constructions containing the basic building blocks of Particle Physics.

1.3. Purpose of String Phenomenology

If string theory is realized in Nature, it should be able to describe not just vague sectors of gauge interaction, but a very precise one: the Standard Model, or some of its promising extensions at the TeV scale. The aim of string phenomenology is thus first the study of general classes of constructions with at least a chance to lead to the Standard Model. By this we mean the appearance of non-abelian gauge interactions, replicated fermions charged under them, scalar fields coupling to them, possibly supersymmetry, etc. Subsequently, and within each class, it pursues the construction of explicit models as close as possible to the Standard Model, with the hope of learning more about its high energy regime when realized in string theory. The field is relatively old, dating back to the mid-eighties, yet there is continuous progress, with recent insights into the questions of compactifications with fluxes, and the inclusion of non-perturbative effects.

For the purposes of this talk, we illustrate the approach of string phenomenology with several classes of constructions. We start with heterotic string compactifications on Calabi-Yau spaces in section 2. Next we move on to brane world models in section 3, realized in terms of intersecting or magnetized D-branes. We conclude with some recent developments on moduli stabilization in section 4, and their application to supersymmetry breaking at energies accessible to the LHC, and the computation of the spectrum of hopefully soon-to-be-observed sparticles.

2. Heterotic string Calabi-Yau compactifications

Heterotic string theory in 10d contains a set of massless fields given by the graviton (and some supersymmetric partners), gauge bosons of the (very large) groups $E_8 \times E_8$ or $SO(32)$, and their supersymmetric gaugino partners in the adjoint representation. In the following we review the construction in [1] of compactifications of the $E_8 \times E_8$ theory, on 6d spaces preserving 4d $\mathcal{N} = 1$ supersymmetry, known as Calabi-Yau compactifications.

As described above we choose the 10d spacetime to have a geometry $M_4 \times \mathbf{X}_6$, with compact $\mathbf{X}_6$. Consistency of the compactification requires turning on non-trivial magnetic fields F_a^{mn} for the internal components $m, n = 4, \ldots, 9$ of some of the 10d gauge bosons. This leads to a breaking of the gauge group in the compactification, so that the observable 4d gauge group is given as the subgroup commuting with the magnetic fields turned on in $\mathbf{X}_6$. Choosing the magnetic fields along suitable generators of e.g. one of the E_8 gauge factors, it is possible to construct explicit models with a quite realistic 4d gauge group, like $SU(5)$ or $SU(3) \times SU(2) \times U(1)$ (possibly times additional $U(1)$ factors, and additional decoupled non-abelian sectors, known as 'hidden sectors'). In the following we denote this sector as the Standard Model (SM) sector of these compactifications.

The introduction of magnetic fields has a second interesting effect. Since the 10d spin-1/2 fermions (the gauginos) transform in the adjoint representation of the 10d group, they couple to these magnetic fields. This modifies the spectrum of light particles they give rise to in the 4d theory. In fact they behave as charged particles in a magnetic field, with a spectrum of Landau levels. The number of groundstates of such system corresponds to a multiplicity for the corresponding massless charged fermions in the 4d theory, in other words, the number of families of the resulting SM sector. Although there is no universal prediction for this number, in particular no hint for a special role of 3-family models, it is remarkable that heterotic string theory leads naturally to the appearance of replicated sets of charged fermions. This natural replication is also a general feature in other realistic constructions in coming sections.

There are several constructions leading to models very close to the SM, but their detailed discussion requires heavy mathematical tools, well beyond this overview. For a flavor of these examples, some references describing recently constructed models are [2, 3]. We instead turn to a back-of-the-envelope computation of the value of the string scale M_s in the general class of heterotic compactifications.

The structure of the 10d action for gravitational and gauge interactions is sketchily

$$S_{\text{het},10d} = \int d^{10}x \, \frac{M_s^8}{g_s^2} \, R_{(10d)} + \int d^{10}x \, \frac{M_s^6}{g_s^2} \, (\operatorname{tr} \mathcal{F}_{\mu\nu} \mathcal{F}^{\mu\nu})_{(10d)} \tag{1}$$

where $R_{(10d)}$ and $\mathcal{F}$ are the 10d curvature scalar and non-abelian gauge field strength. The M_s prefactors are included for dimensional reasons, while the g_s dependence arises because both gravitational and gauge interactions arise from closed string in heterotic string theory.

Compactification on $\mathbf{X}_6$ can be implemented by simply performing the integral over the six internal dimensions, and truncating the 10d dynamical fields to the massless 4d fields. This leads to additional factors of the $\mathbf{X}_6$ volume V_6

$$S_{\text{het},4d} = \int d^4x \, \frac{M_s^8 V_6}{g_s^2} \, R_{(4d)} + \int d^4x \, \frac{M_s^6 V_6}{g_s^2} \, (\operatorname{tr} \mathcal{F}_{\mu\nu} \mathcal{F}^{\mu\nu})_{(4d)} \tag{2}$$

where now $R_{(4d)}$ and $\mathcal{F}$ are the 4d curvature scalar and non-abelian gauge field strength. The coefficients of this 4d action can be compared with the observed values of the strength of gravitational and gauge interactions, i.e. the 4d Planck mass M_p and SM gauge coupling constants (which for simplicity we consider unified at some value $g_{\rm SM} \sim \mathcal{O}(10^{-1})$, as suggested by renormalization group extrapolation of observed couplings to high scales). Namely

$$M_p^2 = \frac{M_s^8 V_6}{g_s^2} \sim 10^{18}\,\text{GeV} \quad , \quad \frac{1}{g_{\rm SM}^2} = \frac{M_s^6 V_6}{g_s^2} \sim \mathcal{O}(10^{-1}) \tag{3}$$

This leads to the equality

$$M_s = M_p\, g_{\rm SM} \sim 10^{17}\,\text{GeV} \tag{4}$$

The string scale is therefore fantastically high, compared with any experimental Particle Physics scale, but tantalizingly close to other theoretical scales, like the GUT scale of the Planck scale. The electroweak hierarchy problem is usually addressed by invoking 4d $\mathcal{N} = 1$ supersymmetry, as we had anticipated with hindsight. The question of supersymmetry breaking and its scale is postponed until section 4.

3. D-brane models and their generalizations

In this section we turn to an alternative and very popular model building setup in string theory, based on IIA and IIB string theories. Although these theories do not contain gauge bosons in 10d, they arise when considering the theory in the presence of a new kind of object, the D-branes. The models hence lead to the localization of gauge sectors on lower-dimensional subspaces of spacetime, the volume of D-branes, while gravity still propagates in 10d spacetime. This is the so-called brane world picture.

3.1. D-branes

Some of the most successful setups to realize the Standard Model in string theory are based on D-branes. For the purposes of this talk, Dp-branes are higher-dimensional planes, with p space dimensions and propagating in time, and on which open strings can have their endpoints. Consequently, open string oscillation modes correspond to gauge and matter particles which must propagate on the volume of these D-branes, while closed strings lead to the graviton still propagating in full 10d spacetime, see Figure 1.

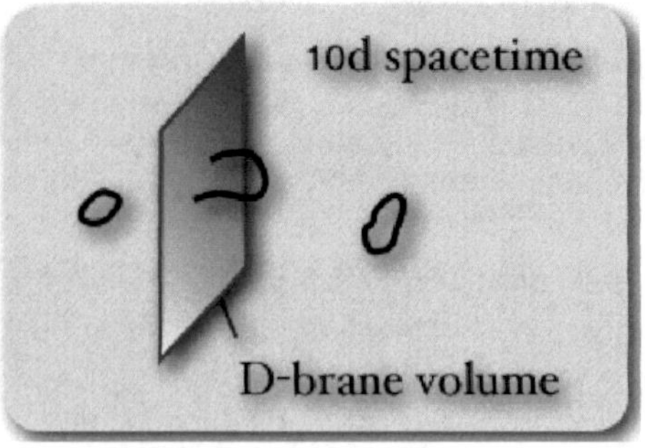

Figure 1. In the presence of a D-brane, open strings localized on its volume, while closed string still propagate over full 10d spacetime.

This has several important consequences. For instance, it implies a modification of the above argument to compute the value of the string scale. In this new setup, the structure of the 10d action for gravitational and gauge interactions is sketchily

$$S_{\mathrm{D}p,10d} = \int d^{10}x \, \frac{M_s^8}{g_s^2} \, R_{(10d)} + \int d^{p+1}x \, \frac{M_s^{p-3}}{g_s} \, (\mathrm{tr}\, \mathcal{F}_{\mu\nu}\mathcal{F}^{\mu\nu})_{(p+1)d} \tag{5}$$

where $R_{(10d)}$ is the 10d curvature scalar and $\mathcal{F}$ is the non-abelian gauge field strength on the $(p+1)$-dimensional volume of the Dp-brane. The M_s prefactors arise dimensional reasons, while the g_s dependences follow because in this setup gravitational and gauge interactions arise from closed and open string sectors, respectively.

Upon compactification on $\mathbf{X}_6$ to 4d, the gravitational term picks up a factors of $V_{\mathbf{X}_6}$, while the gauge term get a factor V_{p-3}, the volume of the $(p-3)$ dimensions of $\mathbf{X}_6$ actually wrapped by the D-brane volume.

$$S_{\mathrm{D}p,4d} = \int d^4x \, \frac{M_s^8 V_6}{g_s^2} \, R_{(4d)} + \int d^4x \, \frac{M_s^{p-3} V_{p-3}}{g_s} \, (\mathrm{tr}\, \mathcal{F}_{\mu\nu}\mathcal{F}^{\mu\nu})_{4d} \tag{6}$$

This leads to the expression of the 4d Planck mass and gauge couplings as

$$M_p^2 = \frac{M_s^8 V_6}{g_s^2} \sim 10^{18}\,\mathrm{GeV} \quad , \quad \frac{1}{g_{\mathrm{YM}}^2} = \frac{M_s^{p-3} V_{p-3}}{g_s} \sim \mathcal{O}(10^{-1}) \tag{7}$$

Assuming for simplicity $\mathbf{X}_6$ to factorize into dimensions along and transverse to the D-brane, with volumes satisfying $V_6 = V_{p-3} V_\perp$, we obtain

$$M_p^2 g_{\mathrm{SM}}^2 = \frac{M_s^{11-p} V_\perp}{g_s} \sim 10^{17}\,\mathrm{GeV} \tag{8}$$

This richer formula leads to several possibilities for the string scale. It can be assumed to be very large, as in heterotic models, invoking 4d $\mathcal{N}=1$ supersymmetry to address the electroweak hierarchy problem. Alternatively, the string scale can be lowered, even to the TeV range, by allowing the transverse dimensions to have a very large volume $V_\perp$. Since the latter cannot be explored by gauge sector particles, but only by the graviton, their size is only very mildly constrained by experiment, and can be as large as a fraction of a millimiter. This geometric generation of a large 4d Planck scale from a low fundamental scale can provide an alternative explanation for the hierarchy between the electroweak and Planck scales [4].

A second important property of D-brane models is that they provide a geometrization of many features of the gauge sectors. For instance, the appearance of non-abelian $SU(n)$ gauge interactions from D-brane open string sectors is achieved by considering a stack of n overlapping D-branes. There are n^2 possible open string sectors, defined by the n possible D-branes on which the open strings can start, times those on which they can end, see figure 2. This leads to a natural $n \times n$ matrix structure, signaling the existence of a $U(n)$ gauge symmetry, whose gauge bosons are the massless oscillation states of the corresponding open strings (in the following and for simplicity, we are somewhat cavalier in our treatment of $U(1)$ factors). Geometrization of other features will be manifest in the explicit model building setups, to which we turn in next sections.

3.2. Intersecting D-brane models

In this section we overview intersecting brane models in string theory, see [5, 6] for original reference or e.g. [7] for a pedagogical review.

A natural generalization of configurations of overlapping D-branes is to consider D-branes which overlap on the 4d Minkowski directions, but intersect non-trivially in the internal

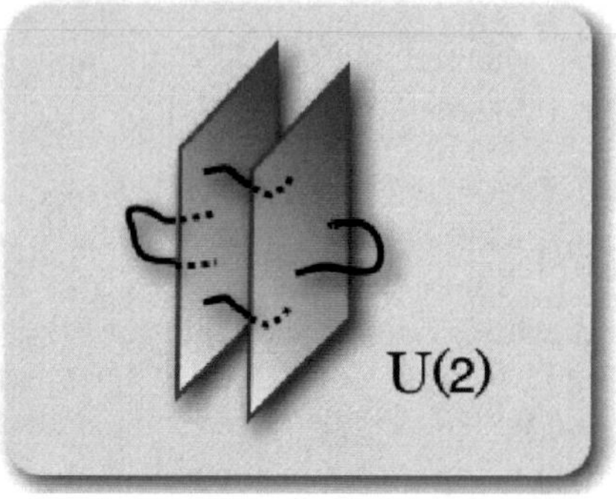

Figure 2. Stacks of n overlapping D-branes lead to n^2 possible open string sectors, and lead to non-abelian $U(n)$ gauge symmetries. The figure illustrates the case of $n = 2$.

directions of $\mathbf{X}_6$. As shown in [8], there are new sectors of open strings stretching between the two intersecting D-brane stacks. Remarkably, for particular configurations, like D6-branes intersecting at points in $\mathbf{X}_6$, the corresponding oscillation state spectrum contains massless 4d chiral fermions, which moreover couple to the two gauge symmetry factors carried by the D-branes. For instance, the intersection of two stacks with 3 and 2 D-branes, respectively, leads to a non-abelian gauge group $SU(3) \times SU(2)$, and chiral fermions charged under these 'color' and 'weak isospin' interactions (in the correct $(\overline{3}, \overline{2})$ representation).

Interestingly, a general feature of the internal geometry is that several D-brane in general intersect at several points in $\mathbf{X}_6$, see figure 3. This results in a multiplicity of 4d fermions with precisely the same gauge quantum numbers, i.e. a replication of fermion families. In other words, in the present string theory setup the existence of several fermion families is natural (although admittedly there is no preference for 3-family models).

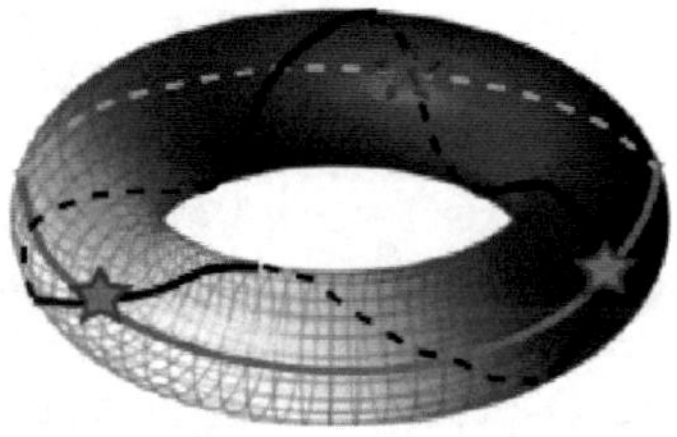

Figure 3. Pictorial representation of D-branes wrapped in the extra dimensions and intersecting at three points, leading to a 4d theory with three families of charged fermions.

This turns the problem of embedding the SM gauge group and fermion content in string string theory into a question about the geometry of D-branes in the internal dimensions and the multiplicity of their intersections. There are indeed concrete proposed patterns of intersections which accommodate the SM gauge group and all SM matter fermions and the Higgs multiplet, with correct non-abelian and hypercharge quantum numbers, and explicit string compactifications realizing them [9], see figure 4.

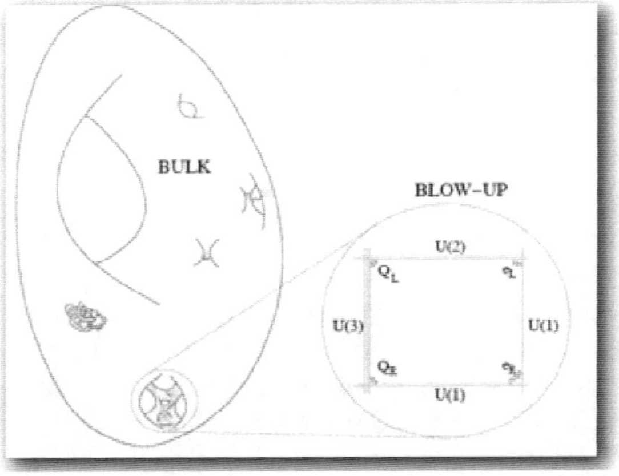

Figure 4. Pictorial representation of intersecting D-branes in the extra dimensions, producing the gauge group and matter content of the SM. For simplicity the picture displays one intersection of each kind instead of the three required for SM family replication.

3.3. Magnetized branes and F-theory

The duality symmetries of string theory in 10 dimensions persist through compactification, and lead to equivalent realization of the same 4d model in terms of different geometrical structure in the internal space. For instance, the intersecting D-brane models of previous section admit a dual realization in terms of D-branes with non-trivial magnetic fields on their volumes.

This picture, known as magnetized D-brane models [10, 5], is interesting in its own. In analogy with heterotic models in section 2, the 4d gauge group and spectrum of massless fermions are determined by properties of the magnetic fluxes turned on the (D-brane volume subspace of) the internal space $\mathbf{X}_6$. An important difference with respect to heterotic models is that the original gauge group is not E_8, and can be chosen to be closer to the SM one. This is particularly helpful to get rid of unwanted massless matter and hence to construct more economical Particle Physics models. A popular choice is to consider stacks of five overlapping D7-branes, leading to an $SU(5)$ gauge group, subsequently broken in the compactification by non-trivial magnetic flux along the $U(1)$ hypercharge generator. This provides a string theory generalization of 4d GUT models, in which the breaking exploits the presence of the extra dimensions and does not require the presence of large GUT multiplets in the 4d theory. This approach has been intensely pursued in a non-perturbative generalization of D7-brane systems, going under the name of F-theory. We refer the reader to e.g. [11] for further details on the construction of these F-theory GUTS, and move on to consider some general phenomenological properties of Particle Physics models in string theory.

3.4. Some phenomenological properties

In the following we sketch the status of several phenomenologically interesting questions in the above string compactifications. Lack of space prevents a systematic discussion in the different classes of string models, and we switch rather freely among them to describe the insights from the most successful such models. Hence, although many properties are actually realized in explicit string models, it is fair to say that there is no example displaying all of them simultaneously. Still, the interest of explicit examples lies, as already mentioned, not in themselves but rather in the general lessons they can teach.

Gauge coupling unification

Journal of Physics: Conference Series **259** (2010) 012014 doi:10.1088/1742-6596/259/1/012014

In many string models the different gauge factors of the SM arise from a single gauge factor in higher dimensions. In these models gauge coupling unification occurs at the string scale, so if taken high enough (around $10^{16} - 10^{17}$ GeV, reproduces in a qualitatively correct way the picture of gauge coupling unification suggested by extrapolation of the observed couplings to high energies.

Yukawa couplings

In intersecting brane models, Yukawa couplings are mediated by so-called 'open worldsheet instantons', morally open strings propagating among the intersections corresponding to the fields in the coupling of interest. This process is exponentially suppressed in the area swept out by such open strings. Since different SM families lie at different intersections, the above process suggests a tantalizing mechanism to generate hierarchically different fermion masses. This idea has been pursued further in the dual magnetized brane picture (or rather, its F-theory generalization), in which Yukawa couplings arise as overlap integrals of the wavefunctions of the Landau groundstates corresponding to the fields in the coupling of interest. These wavefunctions have Gaussian profiles, with different SM families localized at points in the internal space. Hence, the overlap integrals for the different SM families have hierarchically different values. These ideas have led to quite explicit proposals to generate interesting fermion mass textures.

String scale

As already mentioned, the value of the string scale M_s is necessarily high $\sim 10^{17}$ GeV in heterotic models. In D-brane models, it can also be chosen to be large. In these models with a high fundamental scale, the usual proposed solution to the electroweak hierarchy problem is supersymmetry.

In D-brane models there is the alternative possibility to allow for a lower string scale. The extreme case is to lower it down to the TeV scale. An in-between possibility is to consider the string scale around the so-called intermediate scale $\sim 10^{11}$ GeV. The latter is interesting for an appealing mechanism of supersymmetry breaking.

Proton decay

An important issue in any extension of the SM is proton decay, in particular in models with not too high fundamental scale, where higher dimensional operators violating baryon number might not be sufficiently suppressed. In D-brane models leading to the SM gauge group already at the string scale, baryon number is often a global symmetry unbroken in perturbation theory. The only baryon number violating processes are thus non-perturbative, and hence exponentially suppressed, leading to negligible proton decay rates. In models with underlying GUT groups at the string scale (including certain D-brane models, as well as F-theory GUTs and heterotic models), baryon number violating operators are possible, and it is necessary to engineer the configuration to have no such dimension-5 couplings. For a choice of high string scale $M_s \sim 10^{16}$ GeV, natural for such models, the resulting proton decay rates are close to the present experimental bounds.

4. Flux compactifications

In the discussion of previous models we have purposely overlooked a phenomenologically very relevant question. String compactifications as we have described them lead to 4d theories with typically hundreds of massless neutral scalars coupling to matter with gravitational strength. These fields, known as moduli, have actually a flat potential, and hence can have arbitrary vevs at no energetic cost. These moduli vevs parametrize the quantitative values of parameters in the compactification. For instance, the parameters determining the size and shape of the compactification space, the value of the string coupling constant, etc. These moduli fields are troublesome from the phenomenological viewpoint, since they generically lead to unobserved fifth forces, or potentially to problems in the cosmological evolution of the Universe.

There is by now a quite general mechanism for moduli stabilization, namely for the construction of string models where these vevs are fixed dynamically by a moduli potential is known as the moduli stabilization. It is based on the use of 'flux compactifications' [12, 13], in which one turns on magnetic fluxes for the internal components of generalized gauge fields, described by antisymmetric tensors with p indices, and which are related to the graviton by 10d supersymmetry. These are analogous to the magnetic fluxes on the D-brane volumes in magnetized brane setups in section (3.3), with the difference that they belong to the gravitational sector of the theory, rather than being localized on the D-branes.

The introduction of these fluxes produces a non-trivial potential energy for the compactification moduli as follows. Change in the moduli vevs correspond to changes in the compactification geometry, for instance a change in the overall size of the internal space $\mathbf{X}_6$. Such changes in the geometry modify the vacuum energy of the configuration, since they change the magnetic flux density, e.g. it dilutes when $\mathbf{X}_6$ gets larger, and becomes denser when $\mathbf{X}_6$ gets smaller. A vacuum energy depending on the moduli vev is precisely a potential for the moduli fields, which fixes their vevs to the corresponding minimum. Hence the geometry of the compactification is fixed by a dynamical process of energy minimization.

Flux compactifications lead to another interesting effect. Fluxes gravitate and in compactifications with large flux density in localized regions of the internal space, they can lead to gravitational potential wells in the extra dimensions. This produces a gravitational redshift of 4d energy scales, depending on the position in the extra dimensions. These models, known as warped compactifications, thus reproduce the physics of the warped extra dimension or Randall-Sundrum scenario [14], which can potentially address the generation of the electroweak hierarchy.

Finally, flux compactification lead to an appealing mechanism of supersymmetry breaking and its mediation to the SM [15, 16, 17]. Consider as starting point any 4d $\mathcal{N}=1$ supersymmetric compactification of the kind studied in previous sections, e.g. the D-brane models in section 3, leading to a realistic SM sector. The model requires the introduction of fluxes in order to trigger moduli stabilization, and these fluxes in general may not be supersymmetry preserving. This leads to a breaking of supersymmetry of order M_s in the gravitational sector of the theory, which is eventually transmitted to the D-brane SM sector. The order of magnitude of the supersymmetry breaking scale in the SM sector is generically

$$M_{\rm SUSY} = \frac{M_s^2}{M_p} \tag{9}$$

where the M_p denominator arises from the gravitational strength of the mediation mechanism. In general this favors models of intermediate string scale $M_s \sim 10^{10} - 10^{12}$ GeV, since they lead naturally to TeV scale supersymmetry breaking in the visible SM sector. However, particular models with other choices of string scale may still yield the correct supersymmetry breaking scale if they produce additional factors in the analog of (9).

Concrete models allow a precise computation of the MSSM soft terms at the string scale, and allow the computation of their renormalization group evolution to low energies. This provides precise determination of the superparticle spectrum for the models at hand. We refer the reader to the literature, e.g. [19, 18] for such detailed analysis.

5. What is it good for?

A clear conclusion from the above overview of string phenomenology is that there are many string models leading to semi-realistic Particle Physics models, yet none reproduces all features of the SM to the last quantitative details.

Given the large set of string vacua, and even of these semi-realistic string vacua, it would be naive to expect to find the correct one by simple inspection, i.e. by directly constructing an explicit example. This may seem disappointing. However, as already emphasized in the introduction, the main purpose of building explicit SM-like examples in string phenomenology is learning about new properties and mechanisms of interest for Particle Physics, which may be valid more generally than in the particular example where they are uncovered.

Along these lines, string theory has proven extremely useful in motivating and realizing, in a ultraviolet complete setup, several scenarios which provide a fruitful arena for phenomenological model building of Physics beyond the SM. For instance, extra dimensions, the brane world scenario, compactifications with warped dimensions, to name a few.

String theory moreover provides well defined patterns for the scales and ranges of parameters for these scenarios, which may very well be subject to experimental test in the near future. Although such finding would not prove string theory right, it would conform important supporting evidence for it. An example of such precise patterns is given by the structure of soft terms in supersymmetric models with flux supersymmetry breaking, explained in section 4

Finally, certain string theory scenarios lead to smoking gun signatures, which if observed would be hard to explain in any other theory. An example in (admittedly contrived for some tastes) TeV scale string models is the existence of string resonances (i.e massive particles corresponding to excited oscillation states of strings) at experimentally accessible energies.

All in all, string theory and its potential application to Particle Physics is a very rewarding enterprise. In particular string phenomenology is a field reaching its mature age, and for which LHC data will undoubtedly pave the way towards a second youth.

Acknowledgments

Acknowledgments I thank the organizers of the PASCOS '10 conference for building up such a nice event, and my collaborators in the last years for their insights and important lessons. My work is supported by grant FPA2009-07908 from the Spanish Ministerio de Ciencia e Innovación, grant HEPHACOS ESP-1473 from Comunidad de Madrid, and Marie Curie ITN PITN-GA-2009-237920.

References

[1] P. Candelas, G. T. Horowitz, A. Strominger and E. Witten, 'Vacuum Configurations For Superstrings' Nucl. Phys. B **258** (1985) 46.

[2] V. Braun, Y.-H. He, B. Ovrut, T. Pantev, 'The Exact MSSM Spectrum from String Theory', JHEP0605:043,2006.

[3] V. Bouchard, R. Donagi, 'An SU(5) Heterotic Standard Model', arXiv:hep-th/0512149.

[4] N. Arkani-Hamed, S. Dimopoulos and G. R. Dvali, 'The hierarchy problem and new dimensions at a millimeter' Phys. Lett. B **429** (1998) 263
I. Antoniadis, N. Arkani-Hamed, S. Dimopoulos and G. R. Dvali, 'New dimensions at a millimeter to a Fermi and superstrings at a TeV', Phys. Lett. B **436** (1998) 257

[5] R. Blumenhagen, L. Goerlich, B. Kors and D. Lust, 'Noncommutative compactifications of type I strings on tori with magnetic background flux', JHEP **0010** (2000) 006.

[6] G. Aldazabal, S. Franco, L. E. Ibanez, R. Rabadan and A. M. Uranga, 'D = 4 chiral string compactifications from intersecting branes' J. Math. Phys. **42** (2001) 3103
G. Aldazabal, S. Franco, L. E. Ibanez, R. Rabadan and A. M. Uranga, 'Intersecting brane worlds' JHEP **0102** (2001) 047.

[7] A. M. Uranga, 'Chiral four-dimensional string compactifications with intersecting D-branes', Class. Quant. Grav. **20** (2003) S373.

[8] M. Berkooz, M. R. Douglas and R. G. Leigh, 'Branes intersecting at angles' Nucl. Phys. B **480** (1996) 265.

[9] L. E. Ibanez, F. Marchesano and R. Rabadan, 'Getting just the standard model at intersecting branes', JHEP **0111** (2001) 002.

[10] C. Angelantonj, I. Antoniadis, E. Dudas and A. Sagnotti, 'Type-I strings on magnetised orbifolds and brane transmutation', Phys. Lett. B **489** (2000) 223.

[11] J. J. Heckman and C. Vafa, 'From F-theory GUTs to the LHC', arXiv:0809.3452 [hep-ph].

[12] K. Dasgupta, G. Rajesh and S. Sethi, 'M theory, orientifolds and G-flux', JHEP **9908** (1999) 023.

[13] S. B. Giddings, S. Kachru and J. Polchinski, 'Hierarchies from fluxes in string compactifications', Phys. Rev. D **66** (2002) 106006.

[14] L. Randall and R. Sundrum, 'A large mass hierarchy from a small extra dimension', Phys. Rev. Lett. **83** (1999) 3370.

[15] M. Grana, 'MSSM parameters from supergravity backgrounds', Phys. Rev. D **67** (2003) 066006.

[16] P. G. Camara, L. E. Ibanez and A. M. Uranga, 'Flux-induced SUSY-breaking soft terms', Nucl. Phys. B **689** (2004) 195.
P. G. Camara, L. E. Ibanez and A. M. Uranga, 'Flux-induced SUSY-breaking soft terms on D7-D3 brane systems' Nucl. Phys. B **708** (2005) 268.

[17] M. Grana, T. W. Grimm, H. Jockers and J. Louis, 'Soft Supersymmetry Breaking in Calabi-Yau Orientifolds with D-branes and Fluxes', Nucl. Phys. B **690** (2004) 2.

[18] L. Aparicio, D. G. Cerdeno and L. E. Ibanez, 'Modulus-dominated SUSY-breaking soft terms in F-theory and their test at LHC', JHEP **0807** (2008) 099.

[19] J. P. Conlon, C. H. Kom, K. Suruliz, B. C. Allanach and F. Quevedo, 'Sparticle Spectra and LHC Signatures for Large Volume String Compactifications' JHEP **0708** (2007) 061.

16th International Symposium on Particles Strings and Cosmology (PASCOS 2010) IOP Publishing
Journal of Physics: Conference Series **259** (2010) 012015 doi:10.1088/1742-6596/259/1/012015

Holographic Principle and the Surface of Last Scatter

Paul H. Frampton

Department of Physics and Astronomy, University of North Carolina, Chapel Hill, NC 27599-3255.
Institute for the Physics and Mathematics of the Universe, University of Tokyo, Kashiwa, Chiba 277-8568, Japan

E-mail: `frampton@physics.unc.edu, paul.frampton@ipmu.jp`

Abstract. I discuss a solution to the dark energy problem, which arises when the visible universe is approximated by a black hole, in a quasi-static asymptotically-flat approximation. Using data, provided by WMAP7, I calculate the Schwarzschild radius r_S and compare to the measured physical radius of the visible universe, bounded by the surface of last scatter. The ratio, $\epsilon(t_0) = r/r_S$ is found to be comparable to $\epsilon = 1$, as allowed by the holographic principle. The measurement of a shift parameter, σ, introduced by Bond, Efstathiou and Tegmark in 1997, plays an important role in the accuracy of the calculation. The approximation leads to a surprisingly small discrepancy, presumably explicable by the de Sitter, and expanding, nature of the actual universe.

1. Introduction

In fundamental theoretical physics, there was, at the beginning of the twenty-first century, an impossible seeming problem. The problem is the dark energy in cosmology, comprising some seventy percent of the universe.

Another cosmological problem, closely entwined with the dark energy problem, is the question well posed, now almost eight decades ago, by Tolman[1] as to whether one can construct a consistent cyclic model, given the seemingly contradictory constraint imposed by the second law of thermodynamics. The most developed solution of this Tolman conundrum is that suggested in [2].

The most important observational advance in cosmology since the early studies of cosmic expansion in the 1920s was the dramatic and, at that time, surprising discovery in the waning years of the twentieth century that the expansion is accelerating. This was first announced in February 1998, and it was based on the concordance of two groups' data on Supernovae Type 1A [3, 4]

Many subsequent experiments concerning the Cosmic Microwave Background (CMB), Large Scale Structure (LSS) and other measurements have all confirmed the 1998 claim. I therefore adopt the position that the accelerated expansion rate is an observed fact.

Assuming general relativity, together with the cosmological principle of homogeneity and isotropy, the scale factor $a(t)$ in the FRW metric satisfies [5, 6] the Friedmann-Lemaître equation, with different energy density components subsumed into ρ:

$$H(t)^2 = \left(\frac{\dot{a}}{a}\right)^2 = \left(\frac{8\pi G}{3}\right)\rho. \tag{1}$$

I normalize $a(t_0) = 1$ at the present, time $t = t_0$, and ρ is an energy density source which drives the expansion of the universe. Two established contributions to ρ are ρ_m from matter (including dark matter) and ρ_γ radiation, so that

$$\rho \supseteq \rho_m + \rho_\gamma \tag{2}$$

with $\rho_m(t) = \rho_m(t_0)a(t)^{-3}$ and $\rho_\gamma(t) = \rho_\gamma(t_0)a(t)^{-4}$.

For the observed accelerated expansion, a phenomenological approach is to add to the sources, in Eq.(1), a dark energy term $\rho_{DE}(t)$ with

$$\rho_{DE}(t) = \rho_{DE}(t_0)a(t)^{-3(1+\omega)}, \tag{3}$$

where $\omega = p/\rho$ is the equation of state. For the case $\omega = -1$, as for a cosmological constant Λ, and if one discards the matter and radiation terms which are smaller than the dark energy term, one can easily integrate the Friedmann-Lemaître equation to find

$$a(t) = a(t_0)\ e^{Ht} \tag{4}$$

where $\sqrt{3}H = \sqrt{\Lambda} = \sqrt{8\pi G\rho_{DE}}$.

By differentiation of Eq. (4) with respect to time p times, one obtains for the pth derivative

$$\frac{d^p}{dt^p}a(t)|_{t=0} = (H)^p \tag{5}$$

Therefore, if Λ is positive, as in a De Sitter geometry, not only is the acceleration ($p = 2$) positive and non-zero, but so are the jerk ($p = 3$), the snap ($p = 4$), the crackle ($p = 5$), the pop ($p = 6$) and all derivatives for $p \geq 7$.

The insertion of the dark energy term Eq. (3) in Eq. (1) works very well as a part of the ΛCDM model. However, it is an *ad hoc* procedure which gives no insight into what dark energy is.

2. Dark Energy Problem

With this background, I shall now move to a different explanation for the accelerated expansion which obviates any dark energy, including any need for a cosmological constant.

The following considerations may initially appear to be trivial, circular, tautological or some combination thereof. It appears, nevertheless, by hindsight that the 1998 discovery of cosmic acceleration could have been far less surprising theoretically had one previously thought of the $\Lambda = 0$ universe as a black hole.

Journal of Physics: Conference Series **259** (2010) 012015 doi:10.1088/1742-6596/259/1/012015

I now adopt this different approach, with no dark energy, where instead the central role is played by the assumption of the holographic principle[7, 8] (see also [9]) and by the overriding concept of entropy.

The essential assumption is the aforementioned holographic principle, by which I understand that all the information about the universe is encoded on its two-dimensional surface. What this implies is, however unlikely it seems and however contrary to everyday experience that the three-dimensional world I apparently observe is somehow an illusion. This can lead to a reinterpretation of the cosmic acceleration, and possibly the most dramatic new insight into gravity in over three centuries.

Consider the Schwarzschild radius (r_s), and the physical radius (R), of the Sun ($\odot$). They are $(r_s)_\odot = 3km$ and $R_\odot = 800,000km$. Their ratio is $(\rho)_\odot \equiv (R/r_s)_\odot = 2.7 \times 10^5$. One can readily check that for the Earth or for the Milky Way the ratio $\rho = (R/r_s)$ is likewise much larger than one: $\rho >> 1$. Such objects are nowhere close to being black hole. Now consider the visible universe (VU) up to the surface of last scatter. As we shall disucss later, with mass $M_{VU} \sim 5.5 \times 10^{23} M_\odot$, it has $(r_s)_{VU}$ surprisingly close t $(r)_{VU} \sim 14Gpc$, even with the approximations of a quasi-static and asymptotically-flat universe. The visible universe, within which we all live, is well approximated by a black hole. The following addressing of the dark energy problem follows from this observation.

At the horizon, there is a PBH temperature [10, 11, 12], T_β, which I can estimate as

$$T_\beta = \frac{\hbar}{k_B} \frac{H}{2\pi} \sim 3 \times 10^{-30} K. \tag{6}$$

This temperature of the horizon information screen leads to a concomitant FDU acceleration [13, 14, 15] $a_{Horizon}$, outward, of the horizon given by the relationship

$$a_{Horizon} = \left(\frac{2\pi c k_B T_\beta}{\hbar} \right) = cH \sim 10^{-9}\, m/s^2\,. \tag{7}$$

When T_β is used in Eq. (7), I arrive at a cosmic acceleration which is essentially in agreement with the observations[3, 4].

From this viewpoint, the dark energy is non-existent. Instead there is a consequence of the second law of thermodynamics, acting to create the appearance of a dark energy component of the driving density on the right-hand-side of the Friedman-Lemaître equation, Eq.(1).

I have discussed a theory underlying the accelerated expansion of the universe based on entropy. This approach provides a physical understanding of the acceleration phenomenon which was lacking in the description as dark energy.

The entropy of the universe has received some recent attention [16], in part because it relates to the feasibility of constructing a consistent cyclic model. For example, the cyclic model in [2], assuming its internal consistency will indeed be fully confirmed, provides the solution to a difficult entropy question originally posed, seventy-five years earlier, by Tolman [1]. The accelerated expansion rate is no longer surprising. It is the inevitable consequence of information storage on the surface of the visible universe.

Journal of Physics: Conference Series **259** (2010) 012015 doi:10.1088/1742-6596/259/1/012015

This solution of the dark energy problem not only solves a cosmological problem but also casts a completely new light on the nature of the gravitational force [17, 18, 19, 20]. Since the expansion of the universe, including the acceleration thereof, can only be a gravitational phenomenon, I arrive at the viewpoint that gravity is a classical result of the second law of thermodynamics. This means that gravity cannot be regarded as on the same footing with the electroweak and strong interactions. Only if one assumes higher spatial dimensions, it can be argued[21] that non-gravitational interactions are also be emergent.

Although this can be the most radical change in gravity theory for three centuries, it is worth emphasizing that general relativity and its classical tests remain unscathed, as does the prediction of gravitational waves.

The result calls into question almost all of the work done on quantum gravity, since the discovery of quantum mechanics. For gravity, there is no longer necessity for a graviton. In the case of string theory, the principal motivation[22, 23] for the profound and historical suggestion by Scherk and Schwarz that string theory be reinterpreted, not as a theory of the strong interaction, but instead as a theory of the gravitational interaction, came from the natural appearance of a massless graviton in the closed string sector.

This is not saying that string theory is dead. What it is saying is, that string theory cannot be a theory of the fundamental gravitational interaction, since there is no fundamental gravitational interaction.

The way this new insight emerged, as a solution of the dark energy problem itself, was as a natural line of thought, following the discovery of a cyclic model in [2], and the subsequent investigations [24, 25, 26, 27, 28] of the entropy of the universe, including a possible candidate for dark matter[26, 29]. It is also strongly overlapping with the discussions in [17, 18].

Another ramification, of such a solution of the dark energy, problem is the status, fundamental versus emergent, of the three spatial dimensions, that we all observe every day. Because the solution assumes the holographic principle[7, 8], at least one spatial dimension appears as emergent. Regarding the visible universe as a sphere, with radius of about 14 Gpc, the emergent space dimension is then, in spherical polar coordinates, the radial coordinate, while the other two coordinates, the polar and azimuthal angles, remain fundamental. Physical intuition, related to the isotropy of space, may suggest that, if one space dimension is emergent, then so must be all three. This merits further investigation, and may require a generalization of the holographic principle in [7, 8]. On the other hand, a fundamental time coordinate is useful in dynamics.

Of course, this present discussion of cosmic acceleration, is merely one small step towards the ultimate goal, of a cyclic model, in which time never begins or ends.

3. Holographic Principle

An interesting and profound idea about the degrees of freedom describing gravity, is the holographic principle [7, 8].

For the case of a sphere, with mass M, of radius r, where r will be the co-moving radius for the expanding universe, a form, of the holographic principle, states that the ration $\epsilon = r_S/r$, has an upper limit equal to that of a black hole, *i.e.*

$$\epsilon = \left(\frac{r_S}{r}\right) \leq 1. \tag{8}$$

With G as Newton's constant, $r_S = 2GM$ is the Schwarzschild radius. It is interesting, from the viewpoint of the physical understanding of the visible universe, to use accurate observational data to check, whether the simplied, and non-covariant, Eq.(8) is satisfied at the present time, $t = t_0$, and in the past, cognizant that, with dark energy, if r sufficiently increases, Eq.(8) will eventually be violated, if the universe expands exponentially quickly, and if one assumes the approximations used here. Note that, at the time of [7], before dark energy, if Eq.(8) is now satisfied, one might expect it to remain so.

The holographic principle is supported, by string theory. The AdS/CFT correspondence [30] is an explicit realization of Eq.(8), and so, apart from the non-trivial subtlety that our universe is dS, not AdS, from the viewpoint of string theory, there is every reason to believe the holographic principle, and to wish to check Eq.(8). It is related to recent considerations of the entropy of the universe [17, 27, 28].

However, physics is an empirical science, and therefore the scientific method dictates that we should find a physical example, in which Eq.(8) can be calculated. The result, reported here, is that a detailed and accurate check of Eq.(8), as applied to the visible universe, fails, by a statistically-significant amount, although in the past, a few billion years ago, it was satisfied.

I should define, precisely, what is meant by the visible universe. It is the sphere centered for convenience at the Earth with radius $d_A(Z^*) = 14.0 \pm 0.1Gpc$. The value of $d_A(Z^*)$ is the particle horizon corresponding to the recombination red shift $Z^* = 1090 \pm 1$, and is measured directly by WMAP7 [31], without needing the details of the expansion history. Thus, "visible" means with respect to electromagnetic radiation.

4. The Visible Universe

The motion is that the visible universe, so defined, is a physical object which should be subject to the holographic principle. It is an expanding, rather than a static, object, yet my understanding is that the principle, at least in its covariant form, is still expected to be valid.

I shall use the notation employed by the WMAP7 paper [31], from which all observational data are taken.

The present age, t_0, of the universe is measured to be

$$t_0 = 13.75 \pm 0.13Gy \tag{9}$$

The comoving radius, $d_A(Z^*)$, of the visible universe, is, likewise, measured to one percent accuracy:

$$d_A(Z^*) \equiv (1 + Z^*)D_A(Z^*) = c\int_{t^*}^{t_0} \frac{dt}{a(t)} = 14.0 \pm 0.1Gpc, \tag{10}$$

where it is noted that the measurement, of $d_A(Z^*)$, does not require knowledge, of the expansion history, $a(t)$, for $t^* \leq t \leq t_0$.

Journal of Physics: Conference Series **259** (2010) 012015 doi:10.1088/1742-6596/259/1/012015

The critical density, ρ_c, is provided by the formula

$$\rho_c = \left(\frac{3H_0^2}{8\pi G}\right), \tag{11}$$

whose value depends on H_0, as does the total, baryonic plus dark, matter density, ρ_m

$$\rho_m \equiv \Omega_m \rho_c. \tag{12}$$

Because the error on the Hubble parameter, H_0, is several per cent, it is best to avoid H_0, in checking the holographic principle.

The mass of the matter, M(Z*), contained in the visible universe, is first estimated, to be augmented later, as

$$M(Z^*) = \frac{4\pi}{3} d_A(Z^*)^3 \rho_m \tag{13}$$

which gives $M(Z^*) \sim 5.5 \times 10^{23} M_\odot$, consistent with the oft-used estimates of $\sim 10^{11}$ galaxies, each of mass $\sim 10^{12} M_\odot$ including dark matter. The Schwarzschild radius, $r_S(Z^*)$, is given by $R_S(Z^*) \equiv 2GM(Z^*)$.

Collecting results enables the desired accurate check of the simplified holographic principle, which compares the radius, r, for the visible universe with that for a black hole, r_S of the same mass.

A shift parameter, σ, was defined by Bond, Efstathiou and Tegmark (BET) in [32], as

$$\sigma = \frac{\sqrt{\Omega_m H_0^2}}{c}(1 + Z^*) D_A(Z^*), \tag{14}$$

which was, with great prescience, introduced by BET, as a dimensionless quantity, to be measured, accurately, by CMB observations.

This BET shift parameter, σ, of Eq. (14), is given in [31], as (Note that σ is designated R in [31])

$$\sigma = 1.725 \pm 0.018. \tag{15}$$

A little algebra shows that the BET shift parameter σ provides the most accurate available check, of the holographic principle, by virtue of the result

$$\epsilon(t_0) = \left[\frac{r_S}{r}\right] \equiv \sigma^2 = 2.976 \pm 0.062 \tag{16}$$

which is surprisingly close to the saturation of Eq.(8). The derivation of Eq.(16) is as follows:

$$\begin{aligned}
\epsilon &= \left(\frac{r_S}{r}\right) && (17)\\
&= \left(\frac{1}{(1+Z^*)D_A(Z^*)}\right)\left(2G.\frac{4\pi}{3}(1+Z^*)^3 D_A(Z^*)^3 \Omega_m \frac{3H_0^2}{8\pi G}\right) && (18)\\
&= \Omega_m H_0^2 (1+Z^*)^2 D_A(Z^*)^2 && (19)\\
&= \sigma^2 = 2.976 \pm 0.062 \qquad Q.E.D. && (20)
\end{aligned}$$

I used the definition in Eq.(14) for this calculation. It is very fortunate that the square of σ, defined in Eq. (14), agrees with the calcualtion of ϵ, in Eq. (16).

This is suggestive that the expression, Eq.(13), overestimates somewhat the mass because the density is smaller at the sphere's center, where the observer is. More accurately, one can be to take into account this variation of density with redshift by using instead

$$\begin{aligned}
M(Z^*) &= \int_0^{d_A(Z^*)} dr 4\pi r^2 \rho_m \left(\frac{r}{d_A(Z^*)}\right)^3 \\
&= \frac{1}{2}\frac{4\pi d_A(Z^*)^3 \rho_m}{3}, && (21)
\end{aligned}$$

which goves $M(Z^*)m \sim 2.7 \times 10^{23} M_\odot$ and reduces the estimate of $\epsilon(t_0)$ by a factor two to

$$\epsilon(t_0) = 1.48 \pm 0.03, \qquad (22)$$

with a discrepancy of only 48 percent which is surprisingly small compared to the size of either of the radii, both of which are $\sim 10^{61}$ in natural (Planck) units. The mass of the visible universe in Eq.(21) for the volume in asymptotically-flat space is a phenomenological estimate, but the best available.

5. Discussion

To my knowledge, the visible universe is, at present, the only physical object, for which it is possible to calculate, and compare with experiment, or observation, the simplified holographic principle.

The result Eq. (22) is surprisingly close to $\epsilon = 1$. The departure from Eq.(8) can presumably be ascribed to the two assumptions of (i) asymptotic flatness and (ii) a quasi-static universe. It is therefore a very interesting challenge to relax these assumptions and further to use the precision WMAP7 data to attempt to confirm that the holographic principle, Eq. (8), is precisly saturated by the observable universe.

Such a calculational check is under way. For example, one may try replacing the Schwarzschild metric by the McVittie metric[33] which accommodates both a cosmological constant and an expanding universe. Whether this can confirm even more accurately the holographic principle, only time will tell.

Acknowledgements

I am grateful to Professor Dam Son, for visiting the University of Tokyo. The idea, for how to address the dark energy problem, occurred during Son's three lectures, about the holographic principle, at Hongo campus, on February 6, 2010. Useful discussions with Damien Easson, Hirosi Ooguri, George Smoot and Shigeki Sugimoto are acknowledged. This work was supported in part by the World Premier International Research Center Initiative (WPI initiative), MEXT, Japan and by U.S. Department of Energy Grant No. DE-FG02-05ER41418.

[1] R. C. Tolman, Phys. Rev. **38,** 1758 (1931).
[2] L. Baum and P.H. Frampton, Phys. Rev. Lett. **98,** 071301 (2007). hep-th/0610213.
[3] S. Perlmutter et al., Supernova Cosmology Project, Astrophys. J. **517,** 565 (1998). astro-ph/9812133.
[4] A.G. Reiss et al., Supernova Search Team, Astrn. J. **116,** 1009 (1998). astro-ph/9805201.
[5] A. Friedmann, Z.Phys. **10,** 377 (1922).
[6] G. Lemaître, Ann. Soc. Sci. Bruxelles,**A47,** 49 (1927).
[7] G. 't Hooft, in Salamfestschrift, editors: A. Ali, J. Ellis and S. Randjbar-Daemi, World Scienti?c (1994), pg. 284 arXiv:gr-qc/9310026.
[8] L. Susskind, J. Math. Phys. **36,** 6377 (1995). hep-th/9409089.
[9] R. Bousso, JHEP, **9907:**004 (1999). hep-th/9905177.
R. Bousso, Rev. Mod. Phys. **74,** 825 (2002). hep-th/0203101.
[10] L. Parker, Phys. Rev. **183,** 1057 (1969).
[11] J. D. Bekenstein, Phys. Rev. **D7,** 2333 (1973).
[12] S.W. Hawking, Commun. Math. Phys. **43,** 199 (1975).
[13] S.A. Fulling, Phys. Rev. **D7,** 2850 (1973).
[14] P.C.W. Davies, J. Phys. **A8,** 609 (1975).
[15] W.G. Unruh, Phys. Rev. **D14,** 870 (1976).
[16] C.A. Egan and C.H. Lineweaver, Astrophys. J. **710,** 1825 (2010). arXiv:0909.3983[astro-ph.CO].
[17] E. Verlinde, arXiv:1001.0785[hep-th].
[18] T. Jacobson, Phys. Rev. Lett. **74,** 1260 (1995). gr-qc/9504004.
[19] T. Padmanaban, Mod. Phys. Lett. A25, 1129 (2010). arXiv: 0912.3165 [gr-qc].
[20] C.M. Ho, D. Minic and Y.J. Ng, arXiv: 1005.3537 [hep-th].
[21] P.G.O. Freund. arXiv: 1008.4147 [hep-th].
[22] T.Yoneya, Prog.Theor.Phys. **51,** 1907 (1974).
[23] J. Scherk and J.H. Schwarz, Nucl. Phys. **B81,** 118 (1974).
[24] P.H. Frampton and T.W. Kephart, JCAP, **06:**008 (2008). arXiv:0711.0193[gr-qc].
[25] P.H. Frampton, S.D.H. Hsu, T.W. Kephart and D. Reeb, Class. Quant. Grav. **26,** 145005 (2009). arXiv:0801.1847 [hep-th].
[26] P.H. Frampton, JCAP **10:** 016 (2009). arXiv:0905.3632[hep-th].
[27] D.A. Easson, P.H. Frampton and G.F. Smoot. arXiv:1002.4278[hep-th].
[28] D.A. Easson, P.H. Frampton and G.F. Smoot. arXiv:1003.1528[hep-th].
[29] P.H. Frampton, K. Kawasaki, F. Takahashi and T. Yanagida, JCAP **04:**023 (2010). IPMU-09-0157. arXiv:1001.2308[hep-ph].
[30] J. Maldacena, Adv. Theor. Math. Phys. **2,** 231 (1998). hep-th/9711200.
[31] E. Komatsu, *et al.*, *Seven-Year Wilkinson Microwave Anisotropy Probe (WMAP) Observations: Cosmological Interpretation.* arXiv:1001.4538 [astro-ph.CO] (2010).
[32] J.R. Bond, G. Efstathiou and M. Tegmark, Mon.Not.Roy.Astron.Soc. **291,** L33-L41 (1997). astro-ph/9702100.
[33] G.C. McVittie, Mon. Not. Royal Astron. Soc. **93,** 326 (1933).

Particles in astrophysics and cosmology: a dark connection

Nicolao Fornengo

Department of Theoretical Physics, University of Torino and INFN/Torino, via P. Giuria 1, 10125 Torino, Italy

E-mail: nicolao.fornengo@unito.it, fornengo@to.infn.it

Abstract. The particle physics interpretation of the missing–mass, or dark–matter, problem of cosmological and astrophysical nature is going to be posed under deep scrutiny in the next years. From the particle physics side, accelerators will deeply test theoretical ideas of new physics beyond the Standard Model, where particle candidates of dark matter are predicted. From the astrophysical side, many probes are already providing a great deal of independent information on the foreseen signals which can be produced by the galactic or extra-galactic dark matter. The ultimate hope is in fact the emergence of dark matter signals from the various sources of backgrounds and the rise of a coherent picture of new physics from the accelerator physics, astrophysics and cosmology sides. A very ambitious and far–reaching project, which will bring to a deeper level our understanding of the fundamental laws which rule the Universe.

1. Introduction

The presence of dark matter has been established on very different cosmological and astrophysical scales by a large number of experimental observations, most notably from the dynamics of galaxy clusters, from the flatness of the rotational curves of galaxies and from the observation of weak lensing phenomena, as well as by the theoretical understanding of structure formation. A significant amount of cold, collision–less and dissipation–less dark matter is therefore needed in clustered systems: much more matter is present in these systems than luminous matter and sizably more than baryonic matter, as the comparison of the amount of dark matter in the Universe with the amount of baryons from primordial nucleosynthesis clearly shows.

Non–baryonic cold dark matter is therefore needed, and this fact poses challenges not only to Cosmology and Astrophysics, but to Fundamental Physics as well, since no viable Dark Matter (DM) candidate is present in the Standard Model of particle physics. Extensions like supersymmetric theories or theories of extra–dimensions typically accomodate succesfull DM candidates, like neutralinos or sneutrinos is Supersymmetry (SUSY) or Kaluza–Klein (KK) excitations in theories of extra–dimensions. These particles may be present in our current Universe and act as DM if they have been produced during the early phases of the evolution of the Universe and then remain as relics from that early stage. They obviously need to be stable on cosmological time scales and to be produced in the early Universe in the right amount to form the whole amount of DM we see today in cosmic structures. Even though these DM particles have very faint interaction with ordinary matter, hopefully they could have the chance to be revealed through some process: these sought–after signals are of astrophysical origin, since they are produced in galaxies or galaxy clusters, but they have a particle physics nature as well, since they must rely on the very particle physics properties of DM. At the same time, if DM is a new kind of elementary particle, a natural way to search for its existence and to determine its properties is at accelerators, which have the capabilities to directly discover new physics and to pinpoint the properties of new particles.

The studies to identify the nature of dark matter therefore stand at the intersection of three major fields of research: Particle Physics, Cosmology and Astrophysics. Fundamental particle physics has the responsibility to provide a suitable DM candidate. From the theoretical point of view, we need to identify and develop models of physics beyond the Standard Model, able to incorporate one (or more) suitable DM candidates. From the experimental side, accelerator searches are the main tools of investigation and discovery of new fundamental states of matter. Cosmology describes the environment where DM particle particles are formed in the early Universe and the evolution history of these DM particles along the structure formation history. The occurrence of the right amount of a particle DM candidate is a delicate balance between cosmological evolution, production mechanisms (thermal, non thermal) and particle physics properties. The ability of a particle DM candidate to form the correct structures in the Universe depends both on the DM properties and on the gravitational evolution in the expanding Universe. Astrophysics describes the environment where the signals of the presence of DM (galaxies or galaxy clusters) are hopefully produced: the size and properties of these signals depend on the astrophysical context where they originate and the ability to observe them is limited by (typically overwhelming) backgrounds of astrophysical origin.

The "particle dark matter crossroad" is therefore a unique place where to put under deep scrutiny our ideas on the fundamental laws which rule the Universe. In these notes I will briefly report on the current status of some subjects which stands at this crossroad, starting from direct and indirect astrophysical signals of DM, and then moving to some brief comment of the connection between accelerator searches and cosmology. We will concentrate on the case of WIMP (Weakly Interacting Massive Particle) DM.

2. Multichannel search of WIMP dark matter

Galactic DM may be searched for in many ways: by looking at the recoil energy directly deposited in a low–background detector (direct detection) or by looking for annihilation products which are produced in the galactic environment (antimatter, gamma–rays, neutrinos) or in the Earth and Sun (neutrinos). Some of these annihilation products may further produce secondary signals: it is the case for electrons and positrons, which can produce secondary gamma-rays from inverse Compton scattering on radiation fields in the galaxy (CMB, infrared and starlight), or radio emission from synchrotron emission on galactic magnetic fields. An additional interesting effect on the CMB is represented by the possibility to distort the CMB spectrum again by means of Compton scattering on the electron/positron gas produced by DM annihilation in galaxy clusters. This version of the Sunyaev–Zeldovich effect is nevertheless small and very difficult to access [1, 2].

All these astrophysical signals depend on the way the non–relativistic DM particles are distributed in the galactic (or extra–galactic) environment. Moreover, charged signal (positrons, antiprotons, antideuterons) and those secondary effects due to these antimatter signals, are affected by their propagation in the galactic environment. Large uncertainties in the theoretical prediction arise from the modelling of the DM distribution and galactic propagation. Some examples will be given in the next subsections.

3. Direct searches

Direct detection of DM relies on the ability to detect the recoil of a nucleus induced by the elastic (or even inelastic) scattering of a DM particle. The expected recoil spectrum is an–almost exponential function of the nuclear recoil energy and the main source of background

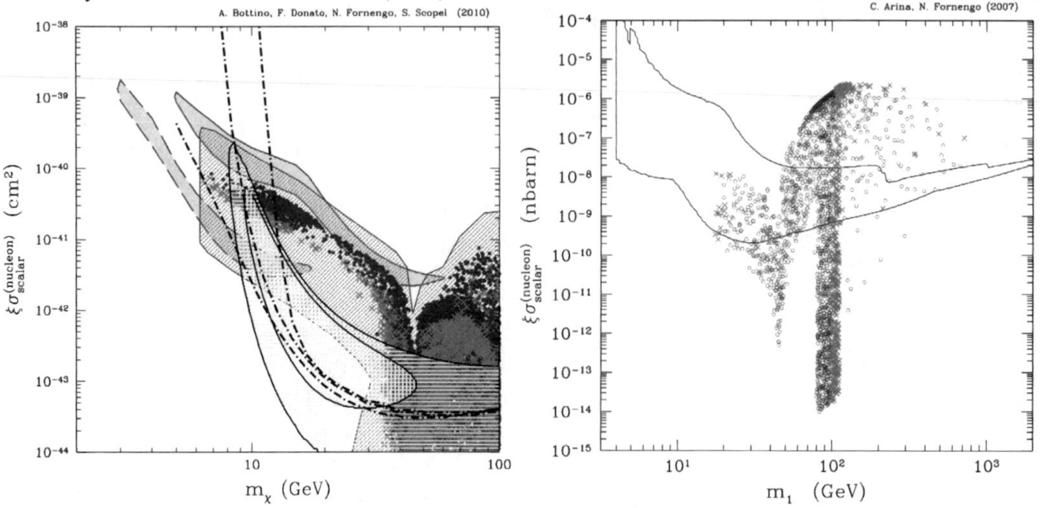

Figure 1. LEFT PANEL: DM–nucleon scattering cross section $\xi\sigma_{\rm scalar}^{\rm (nucleon)}$ as a function of the WIMP mass. The [green] shaded regions denote the DAMA/LIBRA [3] annual modulation regions, under the hypothesis that the effect is due to a WIMP with a coherent interaction with nuclei; the region delimitated by the solid line refers to the case where the channeling effect is not included, the one with a dashed contour to the case where the channeling effect is included [4]. The [violet] band displays the region related to the two CDMS candidate events [5], obtained under the hypothesis of a backgound contribution as in Ref. [6], normalized to 0.8 events in the whole energy window of CDMS II. The contours refer to 68% and 85% C.L. The dot–dashed lines show the bounds from XENON100, [7] under three different choices of the scintillation efficiency. The scatter plot represents supersymmetric configurations calculated in a Minimal Supersymmetric Standard Model (MSSM) where gaugino universality is not assumed. The [red] crosses denote configurations with a neutralino relic abundance which matches the WMAP cold dark matter amount ($0.098 \leq \Omega_\chi h^2 \leq 0.122$), while the [blue] dots refer to configurations where the neutralino is subdominant ($\Omega_\chi h^2 < 0.098$). The region covered by a [blue] slant hatching denotes the extension of the scatter plot upwards and downwards, the hadronic uncertainties in the scattering coherent cross–section are included. RIGHT PANEL: Sneutrino–nucleon scattering cross section $\xi\sigma_{\rm nucleon}^{\rm (scalar)}$ as a function of the sneutrino mass m_1, for a full scan of the supersymmetric parameter space [8]. [Red] crosses refer to models with sneutrino relic abundance in the cosmologically relevant range; [blue] open circles refer to cosmologically subdominant sneutrinos. The solid curve shows the DAMA/NaI region, compatible with the annual modulation effect observed by the experiment [9].

arises from neutron interactions. The motions of the Earth with respect to the galactic system induces typical signatures in the recoil, with peculiar features which are not expected to be shared by the neutron background. The motion of the local group inside the Galaxy would induce a backward/forward asymmetry in the direction of the recoils (directionality), the motion of the Earth around the Sun induces an annual modulation of the rate, of the order of a few percent and energy–dependent, while the rotation of the Earth around its axis would induce an even smaller diurnal effect.

Currently annual modulation is studied by the DAMA Collaboration. A clear evidence for an annual dependence of the rate is observed along 13 annual cycles, with a cumulative exposure of 427050 Kg×day [9]. Modulation is observed only on single–hit events in the signal energy-window, while multiple–hit events and detector stability parameters do not modulate. The evidence has a 8.9σ C.L. and the modulation amplitude in the 2–6 keV energy window is $S_m = (0.0116 \pm 0.0013)$ cpd/kg/keV, the modulation phase is 146 ± 7 days (compatible with 152 days, the expected phase for an isotropic halo – notice that this value can change if the halo is not isotropic [10, 9]) and the period of modulation is (0.999 ± 0.002) years. When interpreted as due to DM, the DAMA annual modulation effect defines compatibility regions in the plane DM mass vs DM–nucleon scattering cross section, as shown in Fig. 1. In the same figure, the DAMA regions are confronted with theoretical predictions in models where light neutralinos arise as a consequence of gaugino non–universality [4]. The DAMA regions shown in Fig. 1 refer to a specific galactic halo model. In case of different halo models, the regions are accordingly modified [9]. The right panel of the same figure shows the comparison of the DAMA region with sneutrino DM.

Currently, all other direct DM detector are not directly testing the presence of annual modulation in their recoil rate. Instead, they rely on sophisticated techniques of background rejection and/or interpretation. Recently the CDMS Collaboration reported the results of their final exposure (612 Kg×day) with 2 events which pass all cuts (for an expected background of 0.8 events) [11]. The statistics is definitely extremely small and the CDMS data should cautiously be used to set bounds, but it is interesting to notice that these two events would be compatible with the DAMA result for low–mass DM, as shown in Fig. 1. Just after CDMS presented their final exposure result, CoGeNT (exposure of 18.48 Kg×day) reported an irreducible excess of bulk–like events at very low recoil energies, which also are compatible with DAMA in the 10 GeV DM mass range [12]. Also CRESST (exposure of 333 Kg×day) has recently reported an excess of events over the expected background [13]. Since these events refer to interaction with Oxygen, they would be compatible with DAMA and CoGeNT for the same DM masses and similar scattering cross sections. Even though these results from CDMS, CoGeNT and CRESST, with a low statistic, cannot be currently considered as possible evidences, it is nevertheless interesting that with the current sensitivities direct detection detectors start showing some excesses, which appear to be compatible among themselves and with the long–standing and much more strong DAMA effect, in the case of WIMP DM in the 10 GeV range.

This year also XENON100 reported its first results (exposure of 170 Kg×day) [7]: in this case no event passed all cuts. The 90% C.L. bound posed by XENON100 is also shown in Fig. 1. The actual bound from the XENON detector is currently limited by the knowledge of the scintillation efficiency at low energies ($L_{\rm eff}$), a quantity for which there are currently different determination with a large spread among them [7]. Fig. 1 shows the impact of this large uncertainty for the interesting low DM mass range. A precise determination of $L_{\rm eff}$ at low recoil energies for the Xenon experiments will then be very relevant in assessing the bound (or the compatibility region, in case of detection of a signal) from this type of detectors [14].

Directional detectors are currently under study, while diurnal modulation may be accessible to the DAMA setup, if the mass of the detector is increased in the ton range.

4. Neutrinos as dark matter messengers

DM captured and accumulated inside bodies like the Earth and the Sun may annihilate and produce a neutrino flux which can escape the body. Spectral and angular features may be exploited to desentangle the signal from the atmospheric neutrino background [15, 16]. Neutrino oscillation, and transport in the Sun, have been shown to be relevant effects, which cannot be neglected [15, 16]. The typical signal relies in the search for a muon neutrino flux, which induces upgoing muons in the neutrino telescope. An explicit example of a signal coming from the Sun, and relevant for the same light DM neutralinos discussed in the previous Section, is shown in Fig. 2, where predictions for stopping muons in the SK detector are shown for different DM halo properties [17].

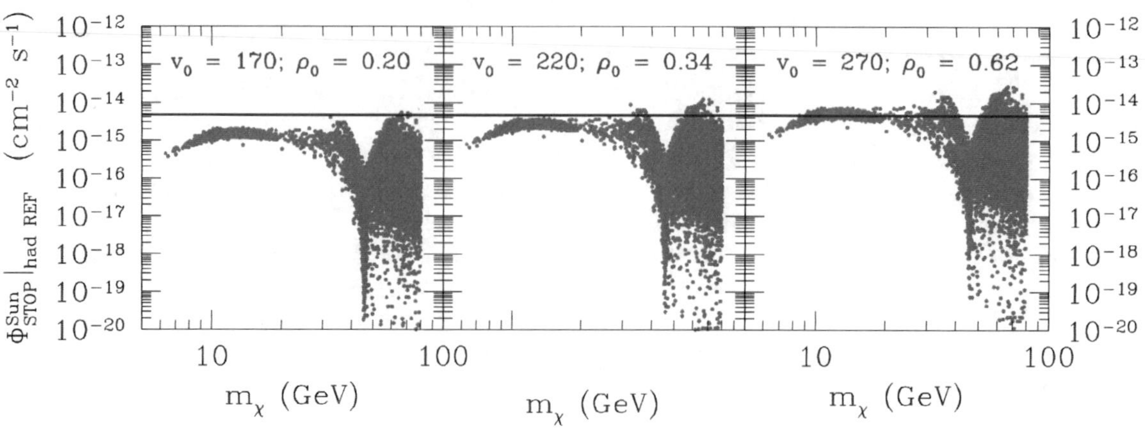

Figure 2. Stopping–muons flux, generated by light neutralino pair-annihilation inside the Sun [17]. The three panels show the dependence of the muon flux on the local rotational velocity v_0 (in units of Km s^{-1}) and the local density ρ_0 (in units of GeV cm^{-3}) of the dark matter halo. The horizontal line represents the experimental limit on through-going muons from the Earth obtained using the SK data [18].

5. Antiprotons

Annihilation in the galactic environment may produce antimatter, thus adding an exotic contribution to cosmic rays. The antiproton signal at low–energies has a mild feature and when compared to the background the capability to clearly disentagle a signal from the background is hard, expecially when considering that astrophysical uncertainties will still be a major component in the theoretical determination of the signal [19]. This can be seen in Fig. 3, where predictions for the differential flux both for the secondary production (the background) and for the signal for some representative DM masses is shown. For heavy DM, the spectral feature could allow discrimination against the background, but this requires pretty strong boost factors, which appear to be disfavoured by recent studies [31]. Special annihilation mechanism, like the Sommerfeld enhancement [32], could prevent the necessity of large astrophysical boost factors.

Current data from PAMELA on the $\bar{p}/p$ ratio show no deviation from expectations on the secondary component: this implies that antiprotons may be used to set relevant constraints on the presence of DM and on its properties, once theoretical uncertainties are properly taken into account [19, 33, 26]. Fig. 3 shows a scan of the SUSY parameter space of a low–energy realization of the Minimal Supersymmetric Standard Model (MSSM) where neutralino is the DM candidate [19, 26]. Theoretical uncertainties of astrophysical origin are sizeable [19]. E.g., in the right panel of Fig. 3 the scatter plot can be shifted upward or downward by about a factor of 6–10 [19], due to uncertainties in galactic propagation.

6. Antideuterons

Antideuterons as a DM indirect signal have been proposed in Ref. [34]. Recently a reanalysis has been developed, where also theoretical uncertainties have been quantified [35]. Some results are reported in Fig. 4, where it is shown that the low–energy spectrum offers a unique opportunity to desentangle a signal from the background, since at kinetic energies below 1–3 GeV the production of secondary antideuteros (the background) suffers from a kinematical suppression. Therefore low kinetic energies are the place where experimental efforts should concentrate. Antideuterons appear to offer the best possibility to detect a signal, even in the absence of a boost factor.

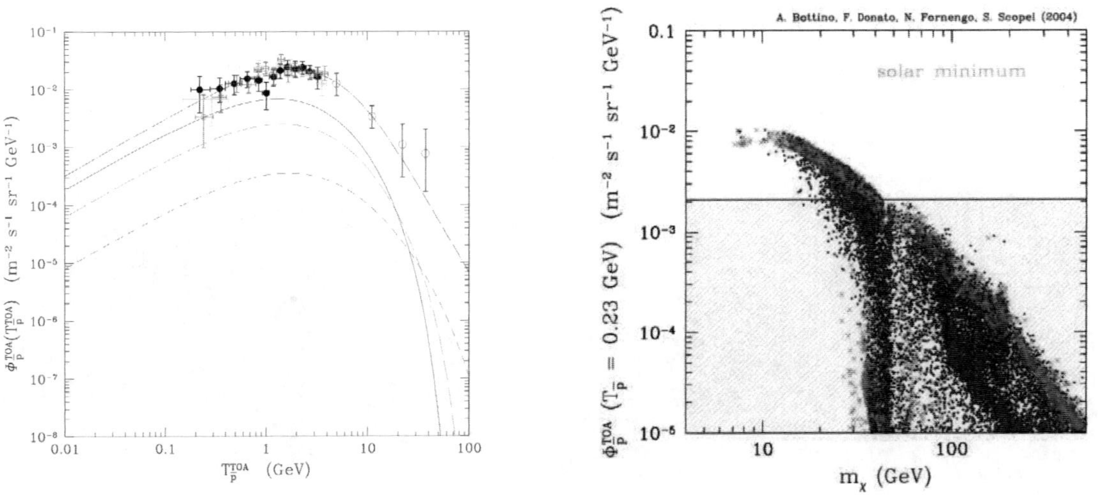

Figure 3. LEFT PANEL: Primary Top-Of-Atmosphere (TOA) antiproton fluxes as a function of the antiproton kinetic energy, for some representative spectra from neutralino annihilation [19]: the solid, long–dashed, short–dashed, dotted lines refer to $m_\chi = 60, 100, 300, 500$ GeV, respectively. The astrophysical parameters correspond to the median choice. Solar modulation is for minimal solar activity. The upper dot–dashed curve corresponds to the antiproton secondary flux [20, 21]. Full circles, open squares, stars and empty circles show the data from BESS 1995-97 [22], BESS 1998 [23], AMS [24] and CAPRICE [25]. RIGHT PANEL: Antiproton flux at $T_{\bar{p}} = 0.23$ GeV vs. the neutralino mass, at solar minimum and for the best fit set for the astrophysical parameters [26]. A spherical isothermal DM density profile has been used. The scatter plots are derived by a full scan of the parameter space of non–universal gaugino models which predict low–mass neutralinos [27, 28, 29]. Crosses [red] and dots [blue] denote neutralino configurations with $0.095 \le \Omega_\chi h^2 \le 0.131$ and $\Omega_\chi h^2 < 0.095$, respectively. The shaded region denotes the amount of primary antiprotons which can be accommodated at $T_{\bar{p}} = 0.23$ GeV without entering in conflict with the experimental BESS data [22, 23] and secondary antiproton calculations [30]. The scatter plot has an astrophysical uncertainty which can shift it either up or down by one order of magnitude.

Foreseen experiment (GAPS, AMS) will have a unique chance to probe this signature directly in the next decade [35]. An example of the capability to probe the SUSY parameter space with a future experimental mission (GAPS) is shown in the rigth panel of the same figure. Neutralino consifurations with masses up to a few hundreds of GeV may be probed, and signals as large as 100 events are possible.

7. Positrons and electrons

Positrons offer a very interesting possibility and have recently gained a lot of attention as a consequence of the release of the PAMELA data on the positron fraction [45] first and then of FERMI [55], ATIC [54] and HESS [56] on the sum of electron and positrons.

Similary to the case of the other indirect detection signal, astrophysical uncertainties largely affect also the positron flux [41], as well as the the background flux of secondary origin [42], and they have to be taken into consideration when comparison with data is attempted. When theoretical and experimental uncertainties are considered [42], theoretical determinations agree with available data on the single positron flux [42], which is available up to a few tens of GeV.

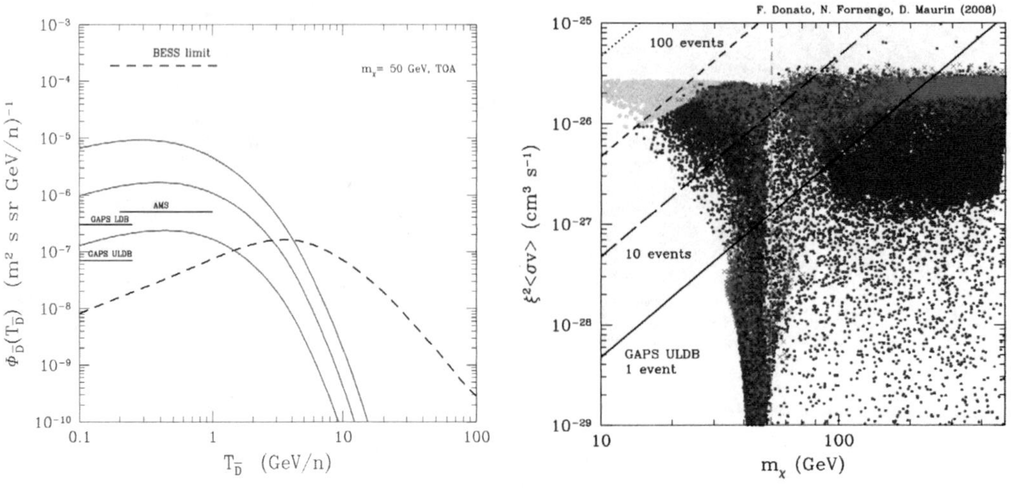

Figure 4. LEFT PANEL: Top Of Atmosphere (TOA) primary (solid lines) and secondary (dashed line) antideuteron fluxes, modulated at solar minimum, for a Weakly Interacting Massive Particle (WIMP) with $m_\chi = 50$ GeV and for the three propagation models which encompass astrophysical uncertainties [35]. The secondary flux is shown for the median propagation model. The upper dashed horizontal line shows the current BESS upper limit on the search for cosmic antideuterons [36]. The three horizontal solid [blue] lines are the estimated sensitivities for (from top to bottom): AMS–02 [37], GAPS on a long (LDB) and ultra–long (ULDB) duration balloon flights [38, 39, 40]. RIGHT PANEL: GAPS ULDB reach compared to predictions for neutralino DM in low–energy supersymmetric models, shown in the plane effective annihilation cross section $\xi^2\langle\sigma_{\rm ann}v\rangle_0$ vs. neutralino mass m_χ [35]. The solid, long–dashed and short–dashed lines show the estimate of the capability of GAPS ULDB of measuring 1, 10 and 100 events, respectively, for the median propagation model. The scatter plot reports the quantity $\xi^2\langle\sigma_{\rm ann}v\rangle_0$ calculated in a low–energy MSSM (for masses above the vertical [green] dashed line) and in non–universal gaugino models which predict low–mass neutralinos [27, 28, 29]. [Red] Crosses refer to cosmologically dominant neutralinos, while [blue] dots stand for subdominant neutralinos. Grey point are excluded by antiproton searches.

At higher energies, both the positron fraction (shown by PAMELA up to 100 GeV) and the total lepton flux (shown by FERMI, ATIC and HESS up to few TeV) clearly show an excess over the pure secondary background [51, 59]. The lepton fluxes at high energies clearly indicate the presence of local sources of electrons/positrons, which show up in the fluxes above few tens of GeV. A natural explanation of this effect comes from astrophysical sources (pulsars, supernova remnants) which have been shown to be able to reproduce the experimental data in a natural way [51, 59, 60]. Fig. 5 shows the agreement between pure astrophysical contributions (secondary leptons and sources) with the FERMI and HESS data. Other astrophysical mechanisms able to explain the rise of the positron flux at high energies have also been discussed [61].

A contribution to the positron (and electron) fluxes may also be originated by DM annihilation. The positron flux from DM annihilation may possess spectral features, depending on the final state of the particle DM annihilation [41]. Fig. 5 shows the case of a DM particle of 100 GeV mass which annihilates dominantly in pure final states. Hard spectra, like the annihilation into e^+e^- or into gauge bosons or also to $\tau^+\tau^-$ are potentially able to reproduce

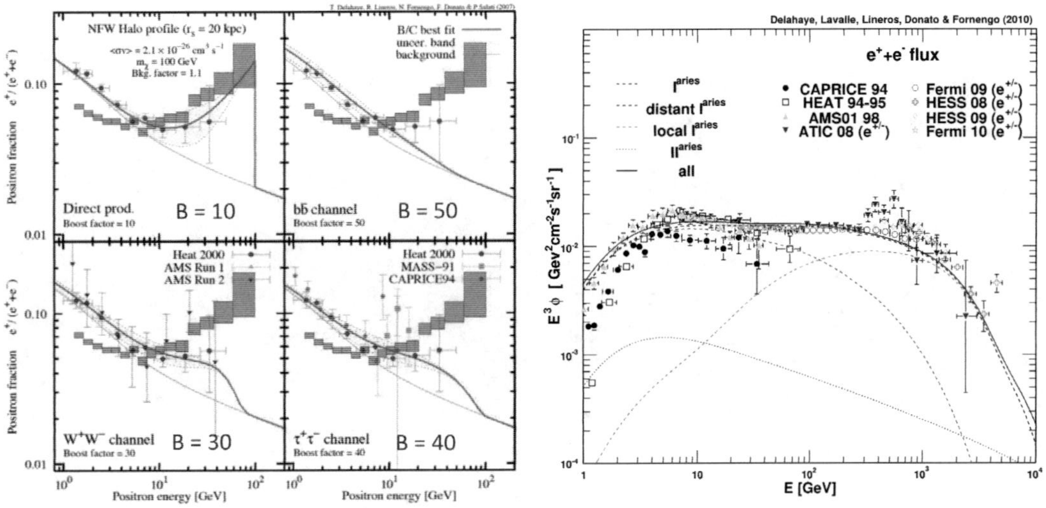

Figure 5. LEFT PANEL: Positron fraction $e^+/(e^- + e^+)$ versus the energy E for a DM particle with a mass of 100 GeV and for a Navarro-Frenk-White (NFW) profile [41]. The four cases refer to different annihilation final states: direct e^+e^- production (top left), $b\bar{b}$ (top right), W^+W^- (bottom left) and $\tau^+\tau^-$ (bottom right). In each panel, the thick solid [red] curve refers to the best–fit choice of the astrophysical parameters. The colored [yellow] area features the total uncertainty band arising from positron propagation. In each panel, the thin [brown] solid line stands for the background of Ref. [43, 44]. The [red] boxes denote the PAMELA data [45]. Experimental data from HEAT [46], AMS01 [47, 48], CAPRICE [49] and MASS [50] are also plotted. RIGHT PANEL: Electron + positron flux vs the energy E. The solid line shows a template flux of pure astrophysical origin, while the various dashed and dotted curves break up the total flux into contributions from local and distant astrophysical sources (pulsars supernova remnants and secondaries) [51]. Data are taken from CAPRICE [49], HEAT [52], AMS [53], ATIC [54], FERMI [55] and HESS [56].

the PAMELA result [41, 62], while instead hadronic final states produce softer spectra. Typically, however, the positron signal requires sizable boost factors in order to prevail over the background: this implies that if also hadrons are produced, strong bounds come from antiproton searches [63, 62]. Analyses of the PAMELA data, in combination with other types of indirect searches, show that large boost factors are needed and the DM needs to produce mostly leptons (and not hadrons) unless it is very heavy, with masses above few TeV [62]

From the point of view of particle physics, a WIMP DM which annihilates rather dominantly into leptons is typically difficult to realize: the most natural candidates, like neutralinos or sneutrinos, do not produce sizeable amounts of leptons, and it is therefore necessary to resort to different types of candidates. A great deal of theoretical activity has been devoted to the identification of suitable DM candidates to explain the leptonic "anomaly".

Another difficulty with the DM interpretation of the PAMELA and FERMI data stands in the necessity of large boost factors: the signal has to be largely boosted over the standard production: this may be realized by DM overdensities (which are very unlikely and basically excluded [31]), by peculiar effects in the low–velocity annihilation cross section (Sommerfeld enhancement) [32]) or by a non–standard cosmology at the time of DM production in the early Universe [57]. The whole host of indirect detection signals may be used to set bounds on these

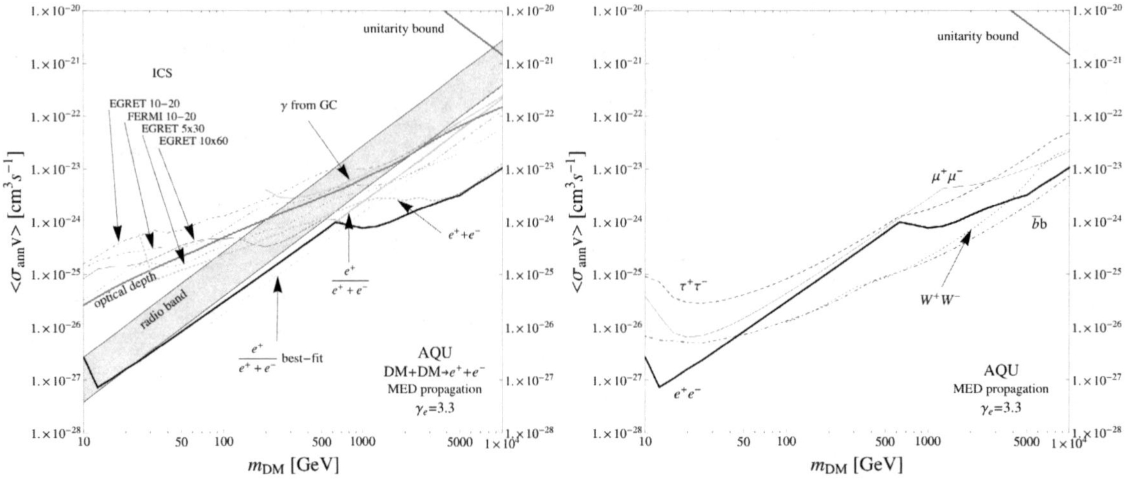

Figure 6. LEFT PANEL: Upper limits on the DM annihilation cross–section (versus the DM mass) coming from different astrophysical observations, for the Aquarius DM distribution and for DM annihilations into e^+e^- [57]. The MED propagation model for cosmic rays has been used. The region above the thick black line is excluded by the convolution of all the implemented constraints. The shaded band labelled as "radio band" denotes the uncertainty on the radio constraint. The dashed line labelled as "$e^+/(e^+ + e^-)$ best–fit" denotes the values of the DM annihilation cross–section required to explain the PAMELA data on the positron fraction. The unitarity bound assuming s-wave annihilations [58] is also shown. RIGHT PANEL: Summary of the astrophysical bounds on the DM annihilation cross–section vs. the DM mass, for the Aquarius DM distribution and for different DM annihilation channels: e^+e^-, $\mu^+\mu^-$, $\tau^+\tau^-$, W^+W^- and $\bar{b}b$. Cosmic rays are propagated in the MED model.

possibilities. This may be rephrased into bounds on the annihilation cross section of the DM particle: if the boost factor is due to the Sommerfeld enhancement or to an enhanced annihilation cross section due to alternative cosmologies the bounds are directly on the cross section; if the boost is due to DM overdensities, the bound is actually on the product of the annihilation cross section times the astrophysical boost factor. Fig. 6 shows one of these analyses [57], where a whole set of different bounds are applied.

8. Gamma rays

Gamma rays are another important tool for studying DM annihilation in the Galaxy and to probe regions of the galactic environment which are partly different from those explored by charged cosmic rays. Spectral features of the gamma–ray signal are not typically very strong, except for the case of direct annihilation into a gamma–line, which instead would be a striking signature of DM annihilation. The gamma line is typically strongly suppressed for suitable DM candidates, and therefore very hard to be probed. The gamma–ray signal typically requires (sizable) boost factors in order to be observable on the top of the astrophysical gamma rays. FERMI will be in the next years a unique laboratory to study gamma rays and it will provide valuable insight also on the DM problem [64]. Currently FERMI is providing an interesting bound on the extra–galactic (or high–latitude) gamma–rays flux [65], which starts to set interesting bounds on the amount of gamma–rays produced by DM annihilation [66].

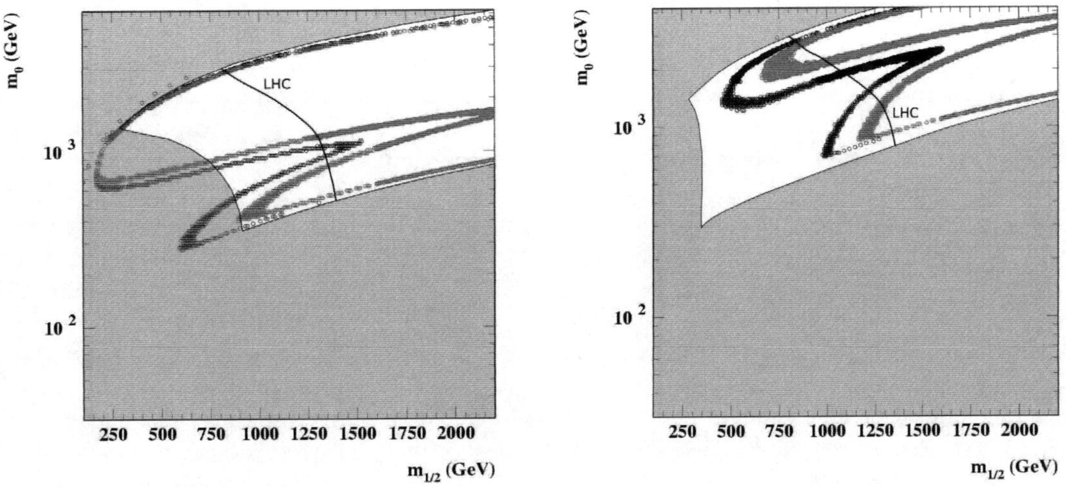

Figure 7. Minimal SUGRA parameter space m_0 (universal soft scalar mass) vs $m_{1/2}$ (gaugino mass) for $tan\beta = 45$ (left panel) and $tan\beta = 53$ (right panel) and common trilinear coupling $A_0 = 0$. The shaded areas are excluded by bounds on supersymmetry searches and supersymmetry contribution to rare processes. The dark [black] circles show the region of parameter space where the neutralino relic abundance matches the WMAP range for cold dark matter in standard cosmology. The light [red] points refer to the same situation in a scalar–tensor cosmology. From Ref. [67]. The solid line denotes the expected reach of LHC.

9. Accelerator physics and cosmology

Dark matter candidates are potentially present in almost any extension of the Standard Model of particle physics. In supersymmetric theories with R–parity conservation, both neutralinos and sneutrinos are successful cold dark matter candidates, although other possibilities are present, like e.g. gravitinos. In the next years LHC, and hopefully in the future the ILC, will probe these new physics models and a quite intriguing interconnection between high–energy physics studies, astophysics and cosmology will be posed under deep scrutiny. An example of this interplay is depicted in Fig. 7, where a section of the minimal SUGRA parameter space is shown, together with the expected reach of the LHC. A fraction of this parameter space is already excluded by LEP, Tevatron and studies of rare processes. In the allowed region, Fig. 7 shows the sector which is compatible with a relic neutralino able to explain the dark matter content of the Universe, a sector which is just a small fraction of the relevant parameter space. The same figure also shows the effect induced by the thermal history of the Universe: alternative cosmologies, different from the FRW cosmology, imply a modified decoupling epoch and an ensuing different relic abundance: therefore, the cosmologically relevant regions in parameter space are shifted. The example shown in Fig. 7 refers to scalar–tensor cosmologies. Reconstruction of the particle physics properties of dark matter and the underlying particle physics model represent a window also on the early Universe fisical properties [57, 67, 68].

10. Conclusions

In these notes, we have briefly discussed the study of particle dark matter, which stands at the intersection of the fields of Particle Physics, Cosmology and Astrophysics. Rich theoretical and experimental activities are currently undertaken, with the ultimate goal to discover the nature of dark matter and the new physics which underlies it.

We have seen that astrophysical DM searches may be proficiently used to set constraints on the properties of particle DM, as long as a signal is not observed. If a signal is detected, this may guide us toward the identification of some of the properties of the DM candidate (and to some extent of the underlying new physics model): the most direct information will likely be on the mass of the DM particle and on the size of its interactions. A fine structure of the DM particle properties will be difficult, and to this aim more than one type of astrophysical signal (if accessible) would be likely to be necessary.

Different detection signals probe different properties of the DM particle and feel different features of the galactic environment. To identify a DM signal it is therefore of greatest importance to exploit specific and typical signatures of the various types of signals, like annual modulation or directionality in direct detection, or the low–energy signal–to–background behavior of antideuterons in cosmic rays. In any case, a better knowledge of the astrophysical environment and of the astrophysical backgrounds will be very important.

Cosmological properties and astrophysical signals of particle DM candidates can either guide or complement accelerator physics searches. LHC has recently started its operations and will be fully operational in the next years. A great wealth of data will then be available and in case of detection of a signal from new physics, a first indication on the nature of the underlying model may hopefully arise. These results will necessarily have to be analyzed also in the context of the presence of a suitable DM candidate. Accelerators, with their unique capability of identifying (at least part of the) new physics particles and their properties, will allow to shape out the predictions also for DM signals. The two approaches are therefore both fundamental and complementary in the study of the DM hypothesis: only accelerators can prove the existence of new physics beyond the Standard Model and directly discover the new physical states, but only astrophysical DM searches can prove that the new physical states explain the DM puzzle and explicitly identify the DM presence in the astrophysical environment. The interplay between the two approaches may even be able in the future to tell us something on the cosmological evolution of the early Universe. The "particle dark matter crossroad" is therefore an interesting place to visit in the next years.

Work supported by research grants funded by Ministero dell'Istruzione, della Università e della Ricerca (MIUR) under contract PRIN 2008NR3EBK, Università di Torino (UniTO), Istituto Nazionale di Fisica Nucleare (INFN) under project FA51, Italian Space Agency (ASI) under contract N° ASI-INAF I/088/06/0 and the spanish MICINNs Consolider–Ingenio 2010 Programme under grant MULTIDARK CSD2009-00064.

References

[1] Colafrancesco S *Astron. Astrophys.* 422(2004)L23
[2] Lavalle J 2010 *Preprint arXiv:1008.5124 [astro-ph.HE]*
[3] Bernabei R et al 2008 *Eur. Phys. J.* C 56(2008)333
[4] Bottino A, Donato F, Fornengo N and Scopel S 2008 *Phys. Rev.* D78(2008)083520
[5] Ahmed Z et al (CDMS Collaboration) 2009 *Preprint arXiv:0912.3592*
[6] Kopp J,Schwetz T and Zupan J 2010 JCAP1002(2010)014
[7] XENON100 Collaboration 2010 *Preprint arXiv:1005.0380v2 [astro-ph.CO]* ; XENON100 Collaboration 2010 *Preprint arXiv:1005.2615v1 [astro-ph.CO]*
[8] Arina C and Fornengo N 2007 JHEP0711(2007)029
[9] Bernabei R et al 2003 *Riv. Nuovo Cim.* 26N1(2003)1; Bernabei R et al 2004 *Int. J. Mod. Phys.* D13(2004)2127
[10] Fornengo N and Scopel S 2003 *Phys. Lett.* B576(2003)189
[11] Bernabei R et al. (DAMA Collabration) 2010 *Preprint arXiv:1002.1028 [astro-ph.GA]*
[12] Aalseth et al (CoGeNT Collaboration) 2010 *Preprint arXiv:1002.4703 [astro-ph.CO]*
[13] Seidel 2010 WONDER Workshop, LNGS, March 2010; Jochum, GGI, Florence, May 2010
[14] Savage C, Gelmini G, Gondolo P and Freese K 2010 *Preprint arXiv:1006.0972 [astro-ph.CO]*; Collar J

2010 *Preprint arXiv:1006.2031 [astro.ph-CO]*; Hooper D, Collar J, Hall J and McKinsey D 2010 *Preprint arXiv:1007.1005 [hep-ph]*; Sorensen P 2010 JCAP09(2010)033
[15] Cirelli M et al 2005 *Nucl. Phys.* B727(2005)99; Erratum-ibid B790(2008)338
[16] Blennow M, Edsjo J and Ohlsson T 2008 JCAP0801(2008)021
[17] Niro N, Bottino A, Fornengo N and Scopel S 2009 *Phys. Rev.* D80(2009)095019
[18] Ashie Y et al 2005 *Phys. Rev.* D71(2005)112005
[19] Donato F et al 2004 *Phys. Rev.* D69(2004)063501
[20] Donato et al 2001*Ap. J.* 563(2001)172
[21] Maurin et al 2002 *Preprint astro-ph/0212111*
[22] Orito S et al. (BESS Collaboration) 2000 *Phys. Rev. Lett.* 84(2000)1078
[23] Maeno T et al (BESS Collaboration) 2001 *Astropart. Phys.* 16(2001)121
[24] Aguilar M et al (AMS Collaboration) 2002 *Phys. Rep.* 366(2002)331
[25] Boezio M et al (CAPRICE Collaboration) 2001 *Astrophys. J.* 561(2001)787
[26] Bottino A, Donato F, Fornengo N, and Scopel S 2004 *Phys. Rev.* D70(2004)015005
[27] Bottino A, Fornengo N, Polesello G, Scopel S 2008 *Phys. Rev.* D77(2008)115026
[28] Bottino A, Donato F, Fornengo N, and Scopel S 2003 *Phys. Rev.* D68(2003)043506
[29] Bottino A, Fornengo N, and Scopel S 2003 *Phys. Rev.* D67(2003)063519
[30] Donato F et al 2001 *Ap. J.* 563(2001)172
[31] Lavalle J, Yuan Q, Maurin D and Bi X 2008 *Astron. Astrophys.* 479(2008)427
[32] Lattanzi M and Silk J 2009 *Phys.Rev.* D79(2009)083523
[33] Bottino A, Donato F, Fornengo N and Salati P 2003 *Phys. Rev.* D72(2003)083518
[34] Donato F, Fornengo N, Salati P 2000 *Phys. Rev.* D62(2000)043003
[35] Donato F, Fornengo N, Maurin D 2008 *Phys. Rev.* D78(2008)043506
[36] Fuke H et al 2005 *Phys. Rev. Lett.* 95(2005)081101
[37] Choutko V and Giovacchini F (AMS Collaboration) 2008 Proc. ICRC 2007, Vol. 4 (2008) p 765
[38] Hailey C et al 2004 *Nucl. Inst. and Meth. in Phys. Res.* B214(2004)122
[39] Hailey C et al 2006 JCAP1(2006)7
[40] Koglin J 2008 Proc. of TAUP 2007, JOP Conf. Series 120 (2008)
[41] Delahaye T et al 2008 *Phys. Rev.* D77(2008)063527
[42] Delahaye T et al 2009 *Astron. Astrophys* 501(2009)821
[43] Baltz E and Edsjo J 1999 *Phys. Rev.* D59(1999)023511
[44] Moskalenko I and Strong A 1998 *Ap. J.* 493(1998)694
[45] Adriani O et al 2009 *Nature* 458(2009)607
[46] Barwick S et al 1997 *Ap. J.* 482(1997)L191
[47] Aguilar M et al 2007 *Phys. Lett.* B646(2007)145
[48] Alcaraz J et al 2000 *Phys. Lett.* B484(2000)10
[49] Boezio M et al 2000 *Ap. J.* 532(2000)653
[50] Grimani C et al 2002 *Astron. Astrophys.* 392(2002)287
[51] Delahaye T, Lavalle J, Lineros R, Donato F and Fornengo N 2010 *Preprint arXiv:1002.1910 [astro-ph.HE]*
[52] DuVernois et al. 2001 *Ap. J.* 559(2001)296
[53] Alcaraz J et al. 2000 *Phys. Lett.* B484(2000)10
[54] Chang J et al. 2008 *Nature* 456(2008)362
[55] Abdo A et al. 2009 *Phys. Rev. Lett.* 102(2009)181101
[56] Aharonian F et al 2008 *Phys. Rev. Lett.* 101(2008)261104; Aharonian F et al 2009 *Astron. Astrophys* 508(2009)561
[57] Catena R, Fornengo N, Pato M, Pieri L and Masiero A 2010 *Phys. Rev.* D81(2010)123522
[58] Beacom F, Bell N and Mack G 2007 *Phys. Rev. Lett.* 99(2007)231301
[59] Grasso D et al (FERMI-LAT Collaboration) 2009 *Astropart. Phys.* 32(2009)140
[60] Profumo S 2008 *Preprint arXiv:0812.4457 [astro-ph]*
[61] Blasi P 2009 *Phys. Rev. Lett.* 103(2009)051104
[62] Cirelli M, Kadastik M, Raidal M and Strumia A 2009 *Nucl. Phys.* B813(2009)1
[63] Donato F, Maurin D, Brun P, Delahaye T and Salati P 2009 *Phys. Rev. Lett.* 102(2009)071301
[64] Alvarez et al 2007 *Preprin arXiv:0712.1548*
[65] Abdo A et al 2009 *Phys. Rev. Lett.* 103(2009)251101
[66] Abdo A et al 2010 JCAP04(2010)014
[67] Catena R, Fornengo N, Masiero A, Pietroni M and Schelke M 2008 JHEP0810(2008)003
[68] Donato D, Fornengo N and Schelke M 2007 JCAP0703(2007)021; Schelke M, Catena R, Fornengo N, Masiero A and Pietroni M 2006 *Phys. Rev.* D74(2006)083505

Journal of Physics: Conference Series **259** (2010) 012017 doi:10.1088/1742-6596/259/1/012017

Present status and future prospects for a Higgs boson discovery at the Tevatron and LHC

Howard E. Haber

Santa Cruz Institute for Particle Physcs
University of California, Santa Cruz, CA 95064 USA

E-mail: haber@scipp.ucsc.edu

Abstract. Discovering the Higgs boson is one of the primary goals of both the Tevatron and the Large Hadron Collider (LHC). The present status of the Higgs search is reviewed and future prospects for discovery at the Tevatron and LHC are considered. This talk focuses primarily on the Higgs boson of the Standard Model and its minimal supersymmetric extension. Theoretical expectations for the Higgs boson and its phenomenological consequences are examined.

1. Introduction

The origin of electroweak symmetry breaking and the mechanism that generates the masses of the known fundamental particles is one of the central challenges of particle physics. The Higgs mechanism [1] in its most general form can be used to explain the observed masses of the $W^{\pm}$ and Z bosons as a consequence of three Goldstone bosons ($G^{\pm}$ and G^0) that end up as the longitudinal components of the massive gauge bosons. These Goldstone bosons are generated by the underlying dynamics responsible for electroweak symmetry breaking. However, the fundamental nature of this dynamics is still unknown. Two broad classes of electroweak symmetry breaking mechanisms have been pursued theoretically. In one class of theories, the electroweak symmetry breaking dynamics is weakly-coupled, while in the second class of theories the dynamics is strongly-coupled [2].

The electroweak symmetry breaking dynamics of the Standard Model (SM) employs a self-interacting complex doublet of scalar fields, which consists of four real degrees of freedom [3]. Elementary scalar dynamics yields a non-zero scalar field vacuum expectation value (vev), which leads to the breaking of the electroweak symmetry. Consequently, three massless Goldstone bosons are generated, while the fourth scalar degree of freedom that remains in the physical spectrum is the CP-even neutral Higgs boson of the Standard Model. Precision electroweak data favors a Higgs mass below 200 GeV, in which case the scalar self-interactions are weak [4, 5]. In the weakly-coupled approach to electroweak symmetry breaking, the Standard Model is very likely embedded in a supersymmetric theory [6] in order to stabilize the large gap between the electroweak and the Planck scales in a natural way [7]. The minimal supersymmetric extension of the Standard Model (MSSM) employs two complex Higgs doublets, resulting in five physical scalar degrees of freedom. In a large range of the MSSM parameter space, the properties of the lightest scalar of the MSSM are nearly indistinguishable from those of the SM Higgs boson.

It is possible that the dynamics responsible for electroweak symmetry breaking is strongly-coupled [8]. Precision electroweak data disfavors the simplest classes of models based on this approach. Nevertheless, it remains possible that the physics of electroweak symmetry breaking is more complicated than suggested by the Standard Model or its supersymmetric extensions. However, in this talk I will not consider these theoretical alternatives nor their phenomenological consequences.

2. Theory of weakly-coupled Higgs bosons

2.1. The Standard Model Higgs Boson

In the Standard Model, the Higgs mass is given by: $m_h^2 = \frac{1}{2}\lambda v^2$, where λ is the Higgs self-coupling parameter. Since λ is unknown at present, the value of the Standard Model Higgs mass is not predicted. In contrast, the Higgs couplings to fermions [bosons] are predicted by the theory to be proportional to the corresponding particle masses [squared-masses]. In particular, the SM Higgs boson is a CP-even scalar, and its couplings to gauge bosons, Higgs bosons and fermions are given by:

$$g_{hf\bar{f}} = \frac{m_f}{v}\,, \qquad g_{hVV} = \frac{2m_V^2}{v}\,, \qquad g_{hhVV} = \frac{2m_V^2}{v^2}\,,$$
$$g_{hhh} = \tfrac{3}{2}\lambda v = \frac{3m_h^2}{v}\,, \qquad g_{hhhh} = \tfrac{3}{2}\lambda = \frac{3m_h^2}{v^2}\,, \tag{1}$$

where $V = W$ or Z and $v = 2m_W/g = 246$ GeV. In Higgs production and decay processes, the dominant mechanisms involve the coupling of the Higgs boson to the $W^\pm$, Z and/or the third generation quarks and leptons. Note that a h^0gg coupling (g=gluon) is induced by virtue of a one-loop graph in which the Higgs boson couples dominantly to a virtual $t\bar{t}$ pair. Likewise, a $h^0\gamma\gamma$ coupling is generated, although in this case the one-loop graph in which the Higgs boson couples to a virtual W^+W^- pair is the dominant contribution. Further details of the SM Higgs boson properties are given in ref. [3]. Reviews of the SM Higgs properties and its phenomenology, with an emphasis on the impact of loop corrections to the Higgs decay rates and cross-sections can be found in Refs. [9, 10, 11].

2.2. Extended Higgs sectors

For an arbitrary Higgs sector, the tree-level ρ-parameter is given by [12]

$$\rho_0 \equiv \frac{m_W^2}{m_Z^2\cos^2\theta_W} = 1 \quad \Longleftrightarrow \quad (2T+1)^2 - 3Y^2 = 1\,, \tag{2}$$

independently of the Higgs vevs, where T and Y specify the weak-isospin and the hypercharge of the Higgs representation to which it belongs. Y is normalized such that the electric charge of the scalar field is $Q = T_3 + Y/2$. The simplest solutions are Higgs singlets $(T,Y) = (0,0)$ and hypercharge-one complex Higgs doublets $(T,Y) = (\frac{1}{2},1)$. Thus, we shall consider non-minimal Higgs sectors consisting of multiple Higgs doublets (and perhaps Higgs singlets), but no higher Higgs representations, to avoid the fine-tuning of Higgs vevs.

The two-Higgs-doublet model (2HDM) consists of two hypercharge-one scalar doublets. Of the eight initial degrees of freedom, three correspond to the Goldstone bosons and five are physical: a charged Higgs pair, $H^\pm$ and three neutral scalars. In contrast to the SM where the Higgs-sector is CP-conserving, the most general 2HDM allows for Higgs-mediated CP-violation (see ref. [13] for a complete specification of the 2HDM tree-level interactions). If CP is conserved, the Higgs spectrum contains two CP-even scalars, h^0 and H^0 and a CP-odd scalar A^0. Thus, new features of the extended Higgs sector include: (i) charged Higgs bosons; (ii) a CP-odd Higgs boson (if CP is conserved in the Higgs sector); (iii) possible Higgs-mediated CP-violation and neutral Higgs states of indefinite CP. More exotic Higgs sectors allow for doubly-charged Higgs bosons, etc. Further details of extended Higgs sectors can be found in ref. [3].

2.3. The Higgs sector of the MSSM

The Higgs sector of the MSSM is a 2HDM, whose Yukawa couplings and Higgs potential are constrained by supersymmetry (SUSY). Minimizing the Higgs potential, the neutral components of the Higgs fields acquire vevs v_u and v_d, where $v^2 \equiv v_d^2 + v_u^2 = 4m_W^2/g^2 = (246\text{ GeV})^2$. The ratio of the two vevs is an important parameter of the model, $\tan\beta \equiv v_u/v_d$. The five physical Higgs particles consist of a charged Higgs pair $H^\pm$, one CP-odd scalar A^0, and two CP-even scalars h^0, H^0, obtained by diagonalizing the 2×2 CP-even Higgs squared-mass matrix. All tree-level Higgs masses and couplings and the angle α (which parameterizes the diagonalization of the 2×2 CP-even Higgs squared-mass matrix) can be expressed in terms of two Higgs sector parameters, usually chosen to be m_A and $\tan\beta$. See refs. [10, 14] for further details.

At tree level, $m_h \leq m_Z|\cos 2\beta| \leq m_Z$, which is ruled out by LEP data. But, this inequality receives quantum corrections. The tree-level Higgs mass is shifted due to an incomplete cancellation of particles and their superpartners (which would be an exact cancellation if supersymmetry were unbroken) that contribute at one-loop to the Higgs self-energy function. The Higgs mass upper bound is then modified to [15]

$$m_h^2 \lesssim m_Z^2 + \frac{3g^2m_t^4}{8\pi^2 m_W^2}\left[\ln\left(\frac{M_S^2}{m_t^2}\right) + \frac{X_t^2}{M_S^2}\left(1 - \frac{X_t^2}{12M_S^2}\right)\right], \tag{3}$$

where $X_t \equiv A_t - \mu\cot\beta$ governs stop mixing and M_S^2 is the average top-squark squared-mass. The state-of-the-art computation includes the full one-loop result, all the significant two-loop contributions, some of the leading three-loop terms, and renormalization-group improvements [16]. The final conclusion is that $m_h \lesssim 130$ GeV [assuming that the top-squark mass is no heavier than about 2 TeV]. This upper bound is reached when $\tan\beta \gg 1$ and $m_A \gg m_Z$ in the so-called *maximal mixing scenario*, which corresponds to choosing $X_t/M_S \sim 2$ so that the predicted value of m_h is maximal.

Tree-level couplings of the MSSM Higgs bosons with gauge bosons are often suppressed by an angle factor, either $\cos(\beta-\alpha)$ or $\sin(\beta-\alpha)$, as shown in the table below.

$\cos(\beta-\alpha)$	$\sin(\beta-\alpha)$	angle-independent
$H^0W^+W^-$	$h^0W^+W^-$	—
H^0ZZ	h^0ZZ	—
ZA^0h^0	ZA^0H^0	ZH^+H^- , γH^+H^-
$W^\pm H^\mp h^0$	$W^\pm H^\mp H^0$	$W^\pm H^\mp A^0$

Tree-level Higgs-fermion couplings may be either suppressed or enhanced with respect to the SM value, $gm_f/2m_W$. The charged Higgs boson couplings to fermion pairs, with all particles pointing into the vertex, are:

$$g_{H^-t\bar{b}} = \frac{g}{\sqrt{2}m_W}\Big[m_t\cot\beta\, P_R + m_b\tan\beta\, P_L\Big],$$
$$g_{H^-\tau^+\nu} = \frac{g}{\sqrt{2}m_W}\Big[m_\tau\tan\beta\, P_L\Big],$$

where $P_{R,L} \equiv \frac{1}{2}(1 \pm \gamma_5)$, and the neutral Higgs boson couplings are:

$$
\begin{aligned}
h^0 b\bar{b} \quad (\text{or } h^0\tau^+\tau^-): &\quad -\frac{\sin\alpha}{\cos\beta} = \sin(\beta-\alpha) - \tan\beta\cos(\beta-\alpha)\,, \\
h^0 t\bar{t}: &\quad \frac{\cos\alpha}{\sin\beta} = \sin(\beta-\alpha) + \cot\beta\cos(\beta-\alpha)\,, \\
H^0 b\bar{b} \quad (\text{or } H^0\tau^+\tau^-): &\quad \frac{\cos\alpha}{\cos\beta} = \cos(\beta-\alpha) + \tan\beta\sin(\beta-\alpha)\,, \\
H^0 t\bar{t}: &\quad \frac{\sin\alpha}{\sin\beta} = \cos(\beta-\alpha) - \cot\beta\sin(\beta-\alpha)\,, \\
A^0 b\bar{b} \quad (\text{or } A^0\tau^+\tau^-): &\quad \gamma_5\,\tan\beta\,, \\
A^0 t\bar{t}: &\quad \gamma_5\cot\beta\,,
\end{aligned}
$$

where the γ_5 indicates a pseudoscalar coupling. Especially noteworthy is the possible $\tan\beta$-enhancement of certain Higgs-fermion couplings. Typically, $1 \lesssim \tan\beta \lesssim m_t/m_b$ in most MSSM models.

2.4. The decoupling limit

In many models with extended Higgs sectors, a parameter regime exists in which one Higgs boson is light (of order m_Z) and all other Higgs scalars are very heavy ($\gg m_Z$). In this case, one can formally integrate out the heavy scalar states. The effective low-energy Higgs theory is precisely that of the SM Higgs boson. This is called the decoupling limit [17]. The MSSM exhibits the decoupling limit for $m_A \gg m_Z$. In this case, it is easy to verify that $m_A \simeq m_H \simeq m_{H^\pm}$, up to corrections of $\mathcal{O}(m_Z^2/m_A)$, and $\cos(\beta-\alpha) = 0$ up to corrections of $\mathcal{O}(m_Z^2/m_A^2)$. Examining the Higgs couplings listed above, one can check that in the limit of $\cos(\beta-\alpha) \to 0$, all h^0 couplings to SM particles approach their SM limits, as expected.

3. Present status of the Higgs boson

3.1. Direct Higgs mass bounds

From 1989–2000, experiments at LEP searched for $e^+e^- \to Z \to h^0 Z$ (where one of the Z-bosons is on-shell and one is off-shell). No significant evidence was found leading to a lower bound on the Higgs mass, $m_h > 114.4$ GeV at 95% CL [18]. Recent data from the Tevatron extends the disallowed Higgs mass region by excluding 158 GeV $< m_h <$ 175 GeV at 95% CL [19].

The MSSM Higgs mass bounds are more complicated, since they depend on a variety of MSSM parameters. Since supersymmetric radiative corrections to the lightest MSSM Higgs mass must be significant to avoid exclusion, the MSSM Higgs mass bounds depend sensitively on the multi-dimensional MSSM parameter space. In the maximal-mixing scenario, the LEP Higgs search rules out (at 95% CL) charged Higgs bosons with $m_{H^\pm} <$ 79.3 GeV and neutral Higgs masses with $m_h <$ 92.9 GeV and $m_A <$ 93.4 GeV [20]. In certain other scenarios, it is possible to significantly relax these bounds. For example, in the CPX scenario [21], supersymmetric radiative corrections introduce CP-violating phenomena into the MSSM Higgs sector, in which case an arbitrarily light Higgs boson is still allowed for $3 \lesssim \tan\beta \lesssim 10$ [22].

3.2. Indirect Higgs mass bounds

Precision electroweak data puts indirect bounds on the Higgs mass. One can fit a plethora of electroweak observables, based on the predictions of the SM, in which the Higgs mass is allowed to float. The global fit of precision electroweak data yields a χ^2 distribution for the goodness of fit as a function of the Higgs mass. If the direct Higgs search bounds are omitted, the favored central value of the Higgs mass obtained by the LEP Electroweak Working group is 89 GeV, with a one-sigma experimental uncertainty of +35 and −26 GeV [4]. The corresponding one-sided 95% CL upper bound excludes $m_h < 158$ GeV. Including the LEP Higgs search data, the upper bound increases to 185 GeV. Similar results are obtained by the GFITTER collaboration, which quotes [5]:

$$m_h = 120.6^{+17.0[+34.3]}_{-5.2[-6.2]}\ \text{GeV}\,, \tag{4}$$

based on all direct and indirect data, where both 1σ and 2σ error bars are given (the latter in the square brackets). The minimum χ^2 associated with the GFITTER fit is $\chi^2_{\rm min} = 17.82$ for 14 degrees of freedom.

3.3. Can a light Higgs boson be avoided?

If new physics beyond the Standard Model (SM) exists, then one expects shifts in the W and Z masses due to new contributions to the one-loop corrections. In many cases, these effects can be parameterized in terms of two quantities, S and T introduced by Peskin and Takeuchi [23]:

$$\overline{\alpha}T \equiv \frac{\Pi^{\rm new}_{WW}(0)}{m_W^2} - \frac{\Pi^{\rm new}_{ZZ}(0)}{m_Z^2}\,, \tag{5}$$

$$\frac{\overline{\alpha}}{4\overline{s}_Z^2\overline{c}_Z^2}\,S \equiv \frac{\Pi^{\rm new}_{ZZ}(m_Z^2)-\Pi^{\rm new}_{ZZ}(0)}{m_Z^2} - \left(\frac{\overline{c}_Z^2-\overline{s}_Z^2}{\overline{c}_Z\overline{s}_Z}\right)\frac{\Pi^{\rm new}_{Z\gamma}(m_Z^2)}{m_Z^2} - \frac{\Pi^{\rm new}_{\gamma\gamma}(m_Z^2)}{m_Z^2}\,, \tag{6}$$

where $s \equiv \sin\theta_W$, $c \equiv \cos\theta_W$, and barred quantities are defined in the $\overline{\rm MS}$ scheme evaluated at m_Z. The $\Pi^{\rm new}_{V_aV_b}(q^2)$ are the new physics contributions to the one-loop gauge boson vacuum polarization functions. The region in the S–T plane consistent with precision electroweak data is shown in Fig. 1(a).

In order to avoid the conclusion of a light Higgs boson, new physics beyond the SM must enter at an energy scale between 100 GeV and 1 TeV. This new physics can be detected at the LHC

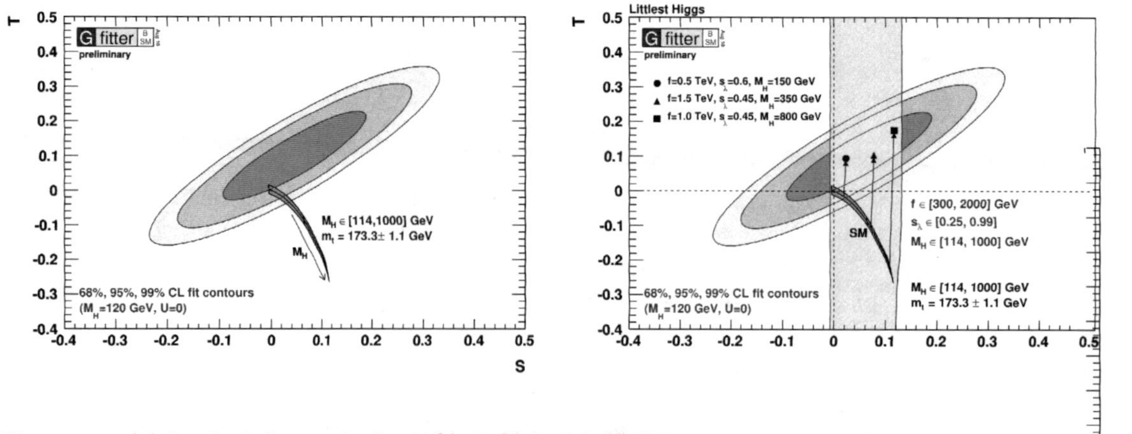

Figure 1. (a) In the left panel, the 68%, 95% and 99% CL contours in the S vs. T plane based on the GFITTER global Standard Model fit to electroweak precision data. The point $(S,T) = (0,0)$ corresponds to $m_t = 173.1$ GeV and $m_h = 120$ GeV. (b) In the right panel, the GFITTER fit to electroweak precision data is exhibited for the littlest Higgs model of ref. [24] (with T-parity) for the parameters shown. Heavier Higgs masses are allowed, if the effects of the Higgs bosons to the gauge boson vacuum polarization functions are compensated by the effects due to new particles and interactions contained in the littlest Higgs model. Taken from ref. [5].

either through direct observation of new physics beyond the SM or by improved precision measurements at future colliders that can detect small deviations from SM predictions. Although the precision electroweak data is suggestive of a light Higgs boson, one cannot definitively rule out a significantly heavier Higgs boson if the new physics conspires to yield a total contribution to S and T that is consistent with the precision electroweak data [e.g., see Fig. 1(b)].

4. Prospects for Higgs discovery at the Tevatron

The Tevatron will continue to take data through the end of 2011. In addition to an increased integrated luminosity, there is still room for some improvements in the Higgs search analysis. The possibility of extending the Tevatron run by another three years is currently under discussion. To facilitate this discussion, a collaboration of Fermilab theorists and experimentalists produced a white paper making the case for an extended Run III of the Tevatron from the end of 2011 through 2014 [25]. The projected improvement of the Tevatron Higgs search is shown in Fig. 2.

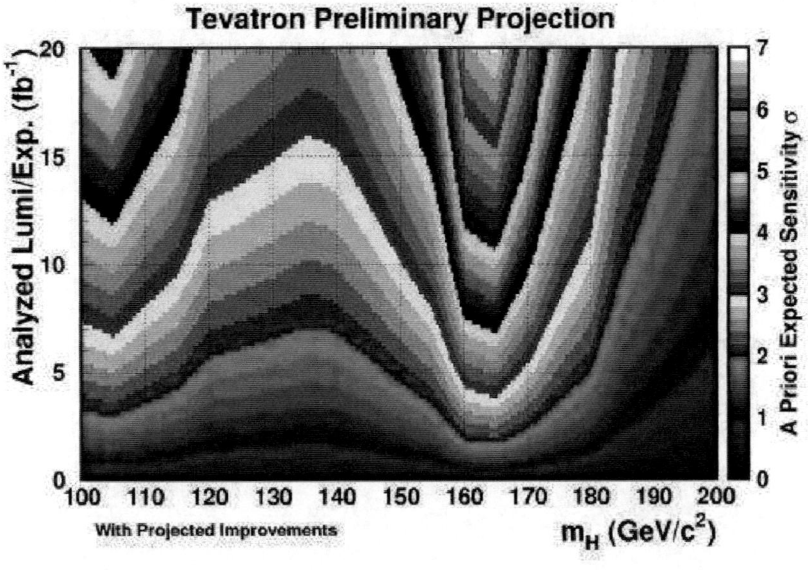

Figure 2. Higgs Boson sensitivity with projected improvements per experiment [25].

In Table 1, the expected sensitivity of the Tevatron Higgs search for Higgs masses of 115 GeV, 130 GeV and 145 GeV as a function of the analyzable luminosity per experiment is given.

Analyzable Lum/Exp	115 GeV	130 GeV	145 GeV
5 fb^{-1}	2.2 σ	1.7 σ	1.9 σ
10 fb^{-1}	3.1 σ	2.5 σ	2.7 σ
15 fb^{-1}	3.8 σ	3.0 σ	3.2 σ
20 fb^{-1}	4.4 σ	3.5 σ	3.7 σ

Table 1. Sensitivity to the Standard Model Higgs Boson combining all modes. The low mass ≤ 130 GeV mode is principally $q\bar{q} \to (W, Z) + (h \to b\bar{b})$; the higher mass ≥ 130 GeV mode is principally $gg \to h \to WW^*$. Taken from ref. [25].

5. Higgs phenomenology at the LHC

Although it is possible that evidence for the Higgs boson may emerge from future Tevatron running, the discovery of the Higgs boson and the identification of its properties are expected to take place at the LHC. Once the Higgs boson is discovered, a program of Higgs physics at the LHC must address the following important questions:

- How many Higgs states are there?
- Assuming one Higgs-like state is discovered,
  - is it a Higgs boson?
  - is it *the* SM Higgs boson?

The measurement of Higgs boson properties will be critical in order to answer the last two questions:

- mass, width, CP-quantum numbers (is the Higgs sector CP-violating?);
- branching ratios and Higgs couplings;
- reconstructing the Higgs potential (Higgs self-couplings).

5.1. Mechanisms for SM Higgs production at the LHC

At hadron colliders, the relevant Higgs production processes are shown in Fig. 3.

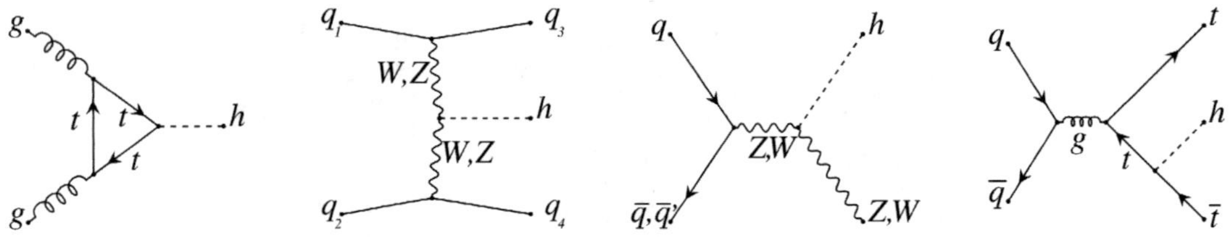

Figure 3. SM Higgs production mechanisms at hadron colliders

The Higgs boson is then detected via its decay products. The dominant channels for observing the Higgs boson at the LHC are:

$$\begin{aligned}
&gg \to h^0\,, \quad h^0 \to \gamma\gamma\,, VV^{(*)}\,,\\
&qq \to qqV^{(*)}V^{(*)} \to qqh^0, \quad h^0 \to \gamma\gamma,\, \tau^+\tau^-,\, VV^{(*)}\,,\\
&q\bar{q}^{(\prime)} \to V^{(*)} \to Vh^0\,, \quad h^0 \to b\bar{b}\,, WW^{(*)}\,,\\
&gg, q\bar{q} \to t\bar{t}h^0, \quad h^0 \to b\bar{b},\, \gamma\gamma,\, WW^{(*)}\,.
\end{aligned}$$

where $V = W$ or Z is a physical gauge boson and V^* is a virtual (off-mass-shell) gauge boson.

5.2. SM Higgs cross sections and branching ratios

The SM Higgs production cross-sections for $\sqrt{s} = 14$ TeV at the LHC are shown in Fig. 4. The discovery channels for Higgs production depend critically on the Higgs branching ratios. The branching ratios and total width of the SM Higgs boson are shown in Fig. 5.

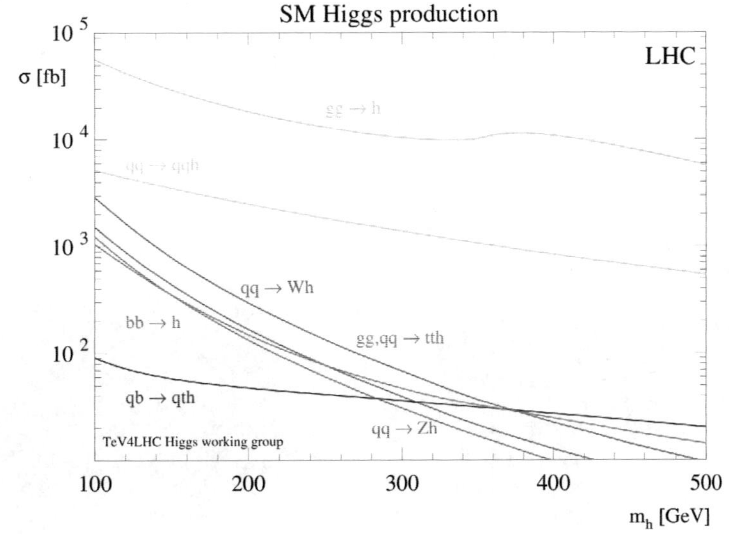

Figure 4. SM Higgs production cross-sections for $\sqrt{s} = 14$ TeV at the LHC (taken from ref. [26]).

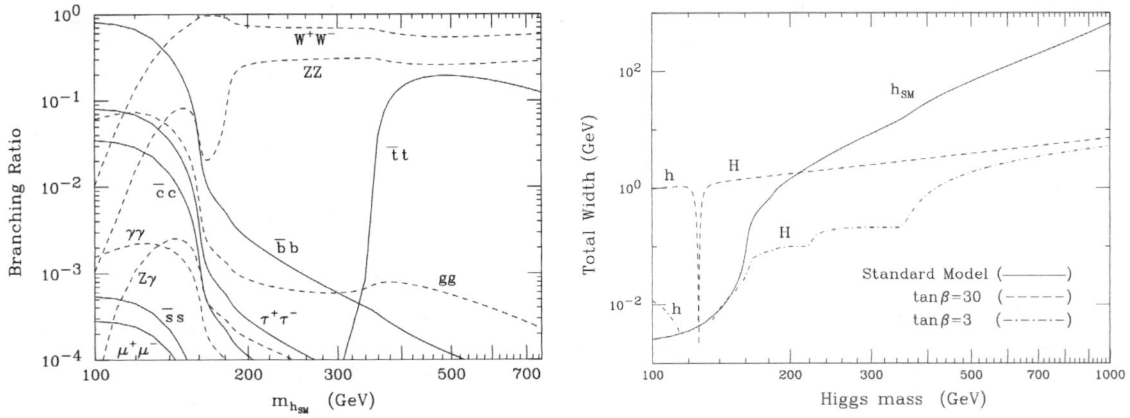

Figure 5. (a) In the left panel, the branching ratios of the SM Higgs boson are shown as a function of the Higgs mass. Two-boson [fermion-antifermion] final states are exhibited by solid [dashed] lines. (b) In the right panel, the total width of the Standard Model Higgs boson (denoted by $h_{\rm SM}$) is shown as a function of its mass. For comparison, the widths of the two CP-even scalars, h^0 and H^0 of the MSSM are exhibited for two different choices of MSSM parameters ($\tan\beta = 3$ and 30 in the maximal mixing scenario; the onset of the $H^0 \to h^0 h^0$ and $H^0 \to t\bar{t}$ thresholds in the $\tan\beta = 3$ curve are clearly evident). Taken from ref. [10].

5.3. LHC prospects for SM Higgs discovery

An examination of Fig. 5 indicates that for $m_h < 135$ GeV, the decay $h \to b\bar{b}$ is dominant, whereas for $m_h > 135$ GeV, the decay $h \to WW^{(*)}$ is dominant (where one of the W bosons must be virtual if $m_h < 2m_W$). These two Higgs mass regimes require different search strategies. For the lower mass Higgs scenario, gluon-gluon fusion to the Higgs boson followed by $h \to b\bar{b}$ cannot be detected as this signal is overwhelmed by QCD two-jet backgrounds. Instead, the Tevatron

Higgs search employs associated production of the Higgs boson with a W or Z followed by $h \to b\bar{b}$. This process is more difficult at the LHC, where the signal-to-background discrimination is more severe. Instead, the LHC relies on the rare decay mode, $h \to \gamma\gamma$, which has a branching ratio of about 2×10^{-3} in the Higgs mass regime of interest. In the higher Higgs mass regime, both the Tevatron and the LHC rely primarily on $h \to WW^{(*)}$ for Higgs masses below 200 GeV. Should the Higgs mass be significantly larger than the current expectations based on precision electroweak data, then the LHC can discover the Higgs boson through the so-called golden channel, $h \to ZZ \to \ell^+\ell^-\ell^+\ell^-$.

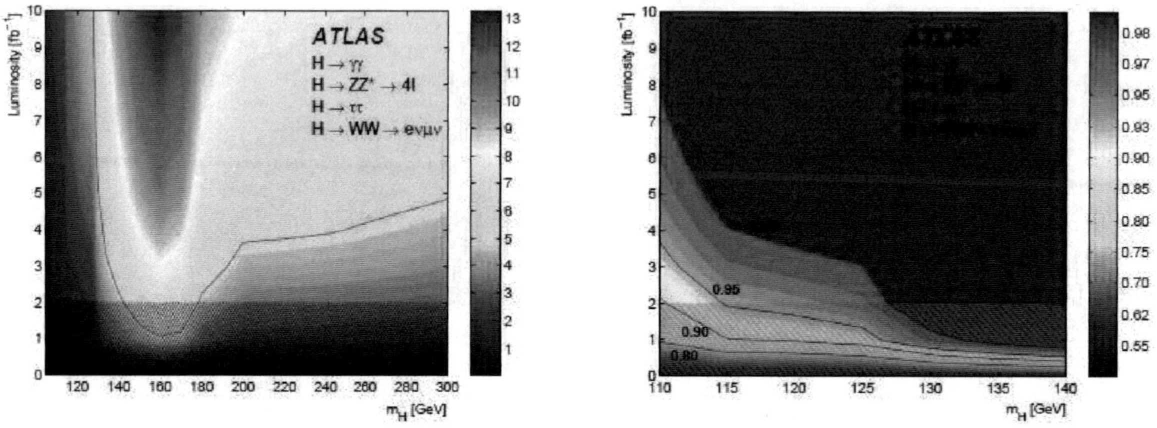

(a) (b)

Figure 6. (a) The left panel depicts the significance contours for different SM Higgs masses and integrated luminosities. The thick curve represents the 5σ discovery contour. The median significance is shown with a color according to the legend. (b) The right panel depicts the expected luminosity required to exclude a Higgs boson with a mass m_H at a confidence level given by the corresponding color. In both panels, the hatched area below 2 fb^{-1} indicates the region where the approximations used in the combination are not accurate, although they are expected to be conservative. Taken from ref. [27].

The ATLAS projections [27] exhibited in Fig. 6 suggest that in the Higgs mass range of 114 GeV $\lesssim m_h \lesssim$ 130 GeV, the discovery of the Higgs (primarily through the rare $h \to \gamma\gamma$ decay mode) will require an integrated luminosity larger than 10 fb^{-1} at $\sqrt{s} = 14$ TeV. The CMS projections [28] shown in Fig. 7 are slightly more optimistic in the low mass Higgs regime based on an "optimized" neural net analysis of the $h \to \gamma\gamma$ mode.

Currently, the LHC is running at half the design energy at significantly lower luminosities than those displayed in Figs. 6 and 7. Current projections suggest that at the end of 2011, the LHC running at $\sqrt{s} = 7$ TeV will accumulate a total integrated luminosity per experiment of about 1 fb^{-1}. Based on these projections, the ATLAS and CMS Higgs searches will yield a 5σ discovery in only a limited Higgs mass range, with a possible 95% exlusion over a somewhat larger Higgs mass range [29, 30]. As an example, Fig. 8 depicts the 95% Higgs mass exclusion region anticipated at the end of the 2011 run [29]. These results would likely match or even exceed the corresponding high mass ($m_h \geq 130$ GeV) exclusion range expected from the Tevatron.

After the Higgs boson discovery, one must check that its properties are consistent with the theoretical expectations. Measurements of the spin, parity and couplings to gauge bosons and fermions typically need a large data sample (which requires tens to hundreds of fb^{-1}). However, in some special circumstances, one can discriminate between different hypotheses for spin and

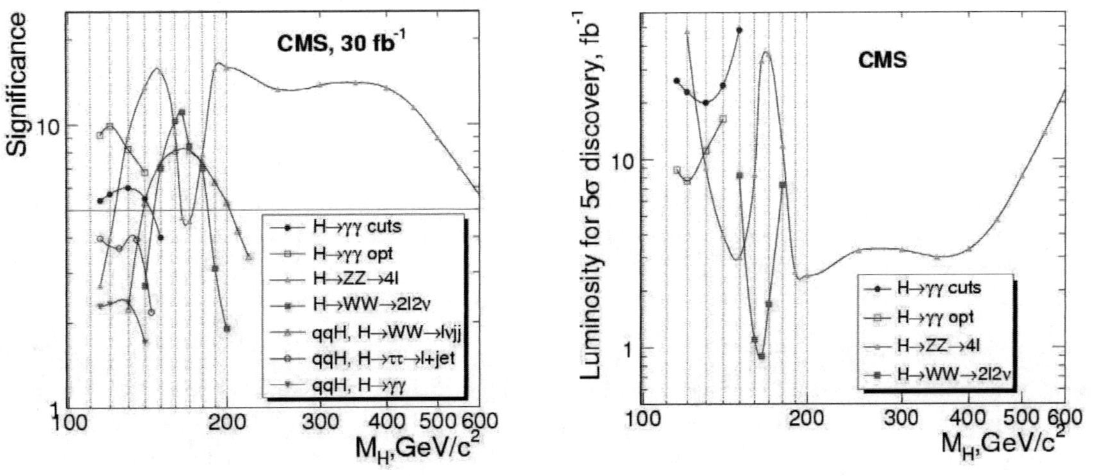

Figure 7. (a) The left panel depicts the signal significance as a function of the SM Higgs boson mass for 30 fb^{-1} at $\sqrt{s} = 14$ TeV, for the different Higgs boson production and decay channels. (b) The right panel depicts the integrated luminosity required for a 5σ discovery of the inclusive Higgs boson production, $pp \to h + X$, with the Higgs boson decay modes $h \to \gamma\gamma$, $h \to ZZ \to \ell^+\ell^-\ell^+\ell^-$ and $h \to W^+W^- \to \ell^+\nu_\ell\,\ell^-\bar{\nu}_\ell$. Taken from ref. [28].

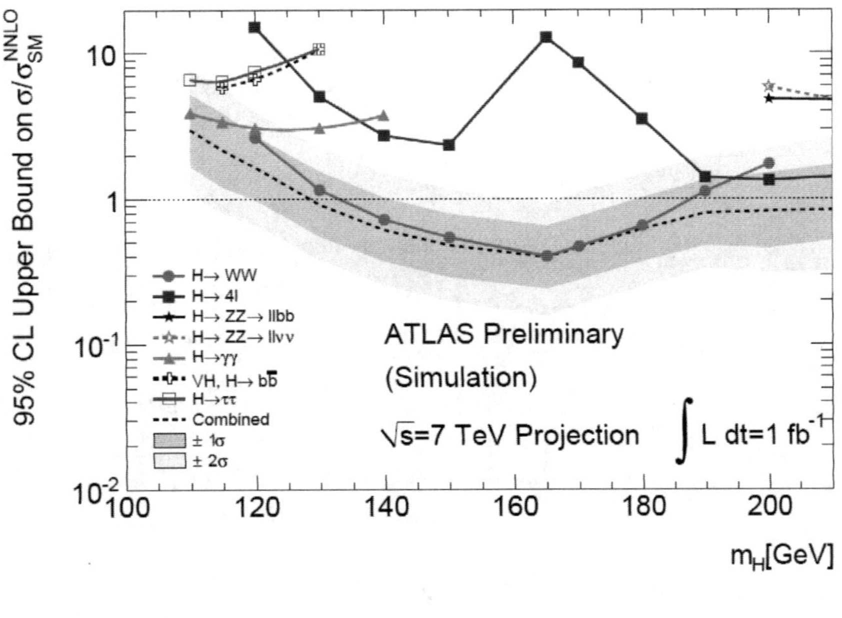

Figure 8. The multiple of the cross-section of a SM Higgs boson that can be excluded using 1 fb^{-1} of data at $\sqrt{s} = 7$ TeV. At each mass, every channel giving reporting on it is used. The green and yellow bands indicate the range in which the limit is expected to lie, depending on the data. Taken from ref. [29]

$\mathbb{H}_0 \Downarrow \mathbb{H}_1 \Rightarrow$	0^+	0^-	1^-	1^+
0^+	–	17	12	16
0^-	14	–	11	17
1^-	11	11	–	35
1^+	17	18	34	–

$\mathbb{H}_0 \Downarrow \mathbb{H}_1 \Rightarrow$	0^+	0^-	1^-	1^+
0^+	–	52	37	50
0^-	44	–	34	54
1^-	33	32	–	112
1^+	54	55	109	–

Table 2. Minimum number of observed events such that the median significance for rejecting the null background-only hypothesis $\mathbb{H}_0$ in favor of the standard Higgs signal plus background hypothesis $\mathbb{H}_1$ (assuming $\mathbb{H}_1$ is right) exceeds 3σ and 5σ, respectively, with m_H=145 GeV. Based on an analysis of the $H \to ZZ^*$ decay mode. Taken from ref. [31].

parity with a small data sample. For example, for Higgs boson masses above about 130 GeV, it will be possible to isolate a data sample of $h \to ZZ^{(*)} \to \ell^+\ell^-\ell^+\ell^-$. By analyzing the distributions and correlations of the five relevant angular variables that describe the $\ell^+\ell^-\ell^+\ell^-$ events, one can achieve a median expected discrimination significance of 3σ with as few as 19 events for $m_h = 200$ GeV and even better discrimination for the off-shell decays of an $m_h = 145$ GeV Higgs boson [31], as shown in Table 2. Thus, if a Higgs boson is discovered at the LHC with 1 fb^{-1}, then it may already be possible to confirm a preference for the expected 0^+ spin-parity assignment with a small sample of Higgs events.

5.4. LHC prospects for MSSM Higgs discovery

In the decoupling limit, the properties of the lightest CP-even Higgs boson of the MSSM (denoted by h^0) are nearly identical to that of the SM Higgs boson. Thus, all the SM Higgs search techniques apply without modification in the MSSM. Moreover, the other MSSM Higgs bosons, $H^\pm$, H^0 and A^0, which are significantly heavier than m_h (and m_Z), are roughly degenerate in mass (cf. Section 2.4). In the opposite limit (far from the decoupling limit), all physical MSSM Higgs bosons are typically below 200 GeV in mass and no single Higgs state possesses couplings that precisely match those of the SM Higgs boson [32].

Cross-sections and branching ratios for the MSSM Higgs boson (which depend on $\tan\beta$ as well as the corresponding Higgs mass) can be found in refs. [10, 14]. Here, we simply note that:

- gluon-gluon fusion can produce both neutral CP-even and CP-odd Higgs bosons.
- VV fusion ($V = W$ or Z) can produce only CP-even Higgs bosons (at tree-level). Moreover, in the decoupling limit, the heavy CP-even Higgs boson is nearly decoupled from the VV channel.
- Neutral Higgs bosons can be produced in association with $b\bar{b}$ and with $t\bar{t}$ in gluon-gluon scattering.
- Charged Higgs bosons can be produced in association with $t\bar{b}$ in gluon-gluon scattering.
- If $m_{H^\pm} < m_t - m_b$, then $t \to bH^-$ is an allowed decay, and the dominant $H^\pm$ production mechanism is via $t\bar{t}$ production.
- Higgs bosons can be produced in pairs (e.g., H^+H^-, $H^\pm h^0$, $h^0 A^0$).
- Higgs bosons can be produced in cascade decays of SUSY particles.
- The total width of any one of the MSSM Higgs bosons is always $\lesssim 1\%$ of its mass [cf. Fig. 5(b)].
- Higgs search strategies depend on the location of the model in the m_A–$\tan\beta$ plane.

Simulations by the ATLAS and CMS collaborations shown in Fig. 9 exhibit which regions of the m_A–$\tan\beta$ plane are sensitive to the MSSM Higgs search. In particular, for large values

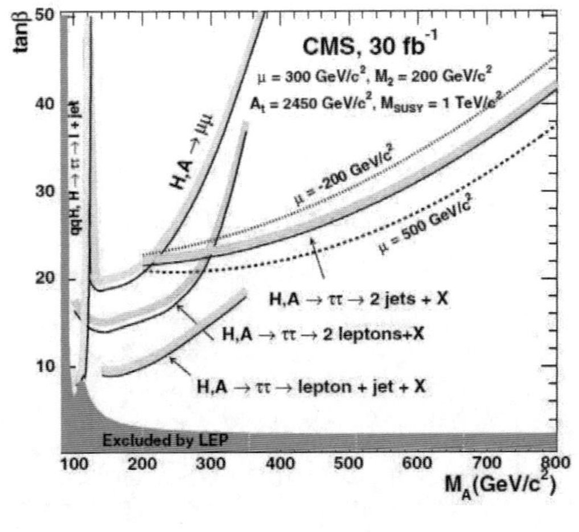

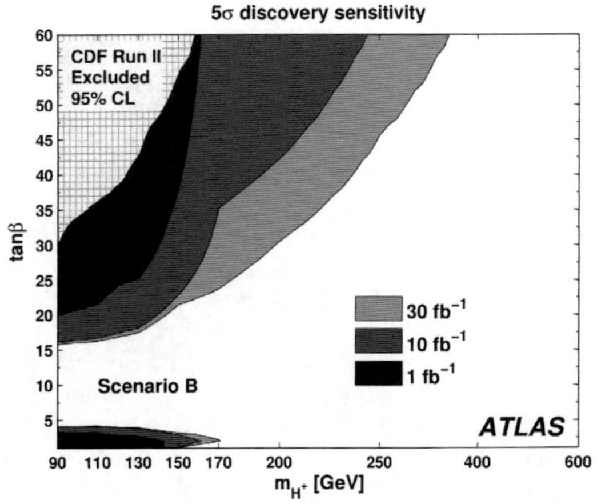

Figure 9. In the left panel, the 5σ discovery potential in the maximal mixing scenario for the heavy H^0 and A^0 scalars at CMS is shown for $\sqrt{s} = 14$ TeV and 30 fb^{-1} of data (taken from ref. [33]). In the right panel, the 5σ discovery sensitivity for the charged Higgs mass is shown as a function of $m_{H^\pm}$ and $\tan\beta$ (taken from ref. [27]).

of $m_A \gtrsim 200$ GeV (corresponding to the decoupling limit), there is no sensitivity for the discovery of the heavy MSSM Higgs states, $H^\pm$, H^0 and A^0, for moderate values of $\tan\beta$. This is the infamous "LHC wedge" region, in which the h^0 of the MSSM (whose properties are nearly indistinguishable from the SM Higgs boson) is the only MSSM Higgs boson that can be discovered at the LHC. Further details can be found in refs. [10, 14].

Acknowledgments

I would like to thank Jose Valle for providing me with the opportunity to visit Valencia and to speak at the PASCOS-2010 conference. I am especially grateful to Jose and his colleagues at the Universitat de València for their kind hospitality during my visit. I also appreciate a number of useful comments and suggestions from Sebastian Grab and Lorenzo Ubaldi, who read the draft version of this manuscript. This work is supported in part by the U.S. Department of Energy, under grant number DE-FG02-04ER41268.

References

[1] P.W. Higgs, Phys. Rev. Lett. **12** (1964) 132; Phys. Rev. **145** (1966) 1156; F. Englert and R. Brout, Phys. Rev. Lett. **13** (1964) 321; G.S. Guralnik, C.R. Hagen, and T.W.B. Kibble, Phys. Rev. Lett. **13** (1964) 585.
[2] S. Weinberg, Phys. Rev. D **13** (1976) 974; ibid. **19** (1979) 1277; L. Susskind, Phys. Rev. D **20** (1979) 2619.
[3] J.F. Gunion, H.E. Haber, G.L. Kane and S. Dawson, *The Higgs Hunter's Guide* (Westview Press, Boulder, CO, 2000).
[4] The ALEPH, CDF, D0, DELPHI, L3, OPAL, SLD Collaborations, the LEP Electroweak Working Group, the Tevatron Electroweak Working Group, and the SLD Electroweak and Heavy Flavor groups, CERN-PH-EP/2009-023 (13 November, 2009); the most recent updated results can be found at `http://lepewwg.web.cern.ch/LEPEWWG/`.
[5] H. Flacher, M. Goebel, J. Haller *et al.*, Eur. Phys. J. C **60** (2009) 543. [arXiv:0811.0009 [hep-ph]]; the most recent updated results can be found at `http://gfitter.desy.de/`.
[6] H.P. Nilles, Phys. Rept. **110** (1984) 1; H.E. Haber and G.L. Kane, Phys. Rept. **117** (1985) 75; S.P. Martin, "A Supersymmetry Primer," in *Perspectives on Supersymmetry II* , edited by G.L. Kane (World Scientific, Singapore, 2010) pp. 1–153 [arXiv:hep-ph/9709356v5].

[7] E. Witten, Nucl. Phys. B **188** (1981) 513; L. Susskind, Phys. Rept. **104** (1984) 181.
[8] For a recent review, see e.g., C.T. Hill, E. H. Simmons, Phys. Rept. **381** (2003) 235. [hep-ph/0203079].
[9] B.A. Kniehl, Phys. Rept. **240** (1994) 211.
[10] M. Carena and H.E. Haber, Prog. Part. Nucl. Phys. **50** (2003) 63 [arXiv:hep-ph/0208209].
[11] A. Djouadi, Phys. Rept. **457** (2008) 1 [arXiv:hep-ph/0503172].
[12] H.-S. Tsao, in *Proceedings of the 1980 Guangzhou Conference on Theoretical Particle Physics*, edited by H. Ning and T. Hung-yuan (Science Press, Beijing, 1980) p. 1240.
[13] H.E. Haber and D. O'Neil, Phys. Rev. D **74** (2006) 015018 [arXiv:hep-ph/0602242]; Erratum–ibid. **74** (2006) 059905.
[14] A. Djouadi, Phys. Rept. **459** (2008) 1 [arXiv:hep-ph/0503173].
[15] H.E. Haber and R. Hempfling, Phys. Rev. Lett. **66** (1991) 1815; Y. Okada, M. Yamaguchi and T. Yanagida, Prog. Theor. Phys. **85** (1991) 1; J.R. Ellis, G. Ridolfi and F. Zwirner, Phys. Lett. B **257** (1991) 83.
[16] See e.g., G. Degrassi, S. Heinemeyer, W. Hollik, P. Slavich and G. Weiglein, Eur. Phys. J. C **28** (2003) 133 [arXiv:hep-ph/0212020]; B.C. Allanach, A. Djouadi, J.L. Kneur, W. Porod and P. Slavich, JHEP **0409** (2004) 044 [arXiv:hep-ph/0406166]; S.P. Martin, Phys. Rev. D **75** (2007) 055005 [arXiv:hep-ph/0701051]; P. Kant, R.V. Harlander, L. Mihaila and M. Steinhauser, JHEP **1008** (2010) 104 [arXiv:1005.5709 [hep-ph]].
[17] H.E. Haber and Y. Nir, Nucl. Phys. B **335** (1990) 363; J.F. Gunion and H.E. Haber, Phys. Rev. D **67** (2003) 075019 [arXiv:hep-ph/0207010].
[18] R. Barate *et al.* [LEP Working Group for Higgs boson searches and ALEPH and DELPHI and L3 and OPAL Collaborations], Phys. Lett. B **565** (2003) 61. [hep-ex/0306033].
[19] The TEVNPH Working Group for the CDF and D0 Collaborations, FERMILAB-CONF-10-257-E (July 28, 2010) [arXiv:1007.4587 [hep-ex]].
[20] The ALEPH, DELPHI, L3 and OPAL Collaborations, and the LEP Higgs Working Group, LHWG Note 2001-04 (July 9, 2001); LHWG Note 2001-05 (July 4, 2001). Updates can be found at: `http://lephiggs.web.cern.ch/LEPHIGGS/www/Welcome.html`.
[21] D.A. Demir, Phys. Rev. D **60** (1999) 095007 [arXiv:hep-ph/9905571]; A. Pilaftsis and C.E.M. Wagner, Nucl. Phys. B **553** (1999) 3 [arXiv:hep-ph/9902371]; S.Y. Choi, M. Drees and J.S. Lee, Phys. Lett. B **481** (2000) 57 [arXiv:hep-ph/0002287]; M. Carena, J.R. Ellis, A. Pilaftsis and C.E.M. Wagner, Nucl. Phys. B **586** (2000) 92 [arXiv:hep-ph/0003180]; S. Heinemeyer, Eur. Phys. J. C **22** (2001) 521 [arXiv:hep-ph/0108059]; T. Ibrahim and P. Nath, Phys. Rev. D **63** (2001) 035009 [arXiv:hep-ph/0008237]; ibid. **66** (2002) 015005 [arXiv:hep-ph/0204092].
[22] S. Schael *et al.* [The ALEPH, DELPHI, L3 and OPAL Collaborations, and the LEP Higgs Working Group], Eur. Phys. J. C **47** (2006) 547 [arXiv:hep-ex/0602042].
[23] M.E. Peskin and T. Takeuchi, Phys. Rev. Lett. **65** (1990) 964; Phys. Rev. D **46** (1992) 381.
[24] N. Arkani-Hamed, A.G. Cohen, E. Katz and A.E. Nelson, JHEP **0207** (2002) 034 [arXiv:hep-ph/0206021]; H.C. Cheng and I. Low, JHEP **0408** (2004) 061 [arXiv:hep-ph/0405243]; J. Hubisz, P. Meade, A. Noble and M. Perelstein, JHEP **0601** (2006) 135 [arXiv:hep-ph/0506042].
[25] M. Carena *et al.*, "Run III: Continued Running of the Tevatron Collider Beyond 2011" (May 26, 2010), available from `http://beamdocs.fnal.gov/AD-public/DocDB/ShowDocument?docid=3617`.
[26] U. Aglietti *et al.* [TeV4LHC Higgs Working Group], "Tevatron-for-LHC Report: Higgs," FERMILAB-CONF-06-467-E-T [arXiv:hep-ph/0612172].
[27] G. Aad *et al.* [The ATLAS Collaboration], *Expected Performance of the ATLAS Experiment: Detector, Trigger and Physics*, CERN-OPEN-2008-020 (December, 2008) [arXiv:0901.0512 [hep-ex]].
[28] *CMS Physics Technical Design Report*, Volume II: Physics Performance, edited by A. De Roeck, M. Grünwald, J. Mnich, L. Pape and M. Spiropulu, CERN/LHCC 2006-021, CMS TDR 8.2 (26 June 2006).
[29] The ATLAS Collaboration, *ATLAS Sensitivity Prospects for the Higgs Boson Production at the LHC Running at 7, 8 or 9 TeV*, ATLAS NOTE ATL-PHYS-PUB-2010-015 (October 27, 2010).
[30] The CMS Collaboration, *The CMS physics reach for searches at 7 TeV*, CMS NOTE-2010/008 (July 20, 2010).
[31] A. De Rujula, J. Lykken, M. Pierini, C. Rogan and M. Spiropulu, Phys. Rev. D **82** (2010) 013003 [arXiv:1001.5300 [hep-ph]].
[32] E. Boos, A. Djouadi, M. Muhlleitner and A. Vologdin, Phys. Rev. D **66** (2002) 055004 [arXiv:hep-ph/0205160]; E. Boos, A. Djouadi and A. Nikitenko, Phys. Lett. B **578** (2004) 384 [arXiv:hep-ph/0307079].
[33] S. Abdullin *et al.*, Eur. Phys. J. C **39**, s2 (2005) s41-s61.

Journal of Physics: Conference Series **259** (2010) 012018 doi:10.1088/1742-6596/259/1/012018

B-Physics at the Tevatron

John Ellison
(for the CDF and DØ Collaborations)
Department of Physics and Astronomy, University of California, Riverside CA 92521 USA
E-mail: john.ellison@ucr.edu

Abstract. Recent B-Physics results from the CDF and DØ experiments at the Tevatron are described, with emphasis on rare decays and searches for CP violation.

1. Introduction

We report on recent B-Physics results from the Tevatron. The topics covered include measurement of the polarization amplitudes in $B_s^0 \to \phi\phi$, the search for rare flavor-changing neutral-current decays, CP violation in $B_s^0 \to J/\psi\phi$ and semileptonic B_s^0 decays, and a new measurement of the like-sign asymmetry in dimuon events.

The data used for the results described here are based on approximately 2.9 – 6.1 fb^{-1} of data collected by the CDF and DØ experiments at the Fermilab Tevatron, a proton-antiproton collider operating at a center-of-mass energy of 1.96 TeV. The DØ and CDF detectors are described in detail in [1] and [2] respectively.

2. Polarization Amplitudes in $B_s^0 \to \phi\phi$

The decay $B_s^0 \to \phi\phi$ is an example of a pseudoscalar meson decaying to two vector mesons and is described by three polarization amplitudes: $A_0, A_\parallel$, and $A_\perp$ corresponding to the three polarization configurations of the final state decay products. Due to the $V - A$ nature of the weak interaction and helicity conservation in QCD, the naive expectation is that the longitudinal polarization amplitude dominates in B decays to two light vector mesons: $|A_0|^2 \gg |A_\parallel|^2 \approx |A_\perp|^2$. This has been confirmed in tree-level $b \to u$ decays and there is evidence for this in $d \to d$ penguin decays. However, in the decay $B \to \phi K^*$, a $b \to s$ decay, it has recently been found that $|A_0|^2 \approx |A_\parallel|^2 \approx |A_\perp|^2$. Since the $B_s^0 \to \phi\phi$ decay also proceeds through a penguin decay, it can be used to check this result. Also, with sufficient statistics, this decay channel will be useful for a comparison of CP violation in $B_s^0 \to \phi\phi$ and $B_s^0 \to J/\psi\phi$ decays.

CDF have studied the $B_s^0 \to \phi\phi$ decay, where both ϕ mesons decay via $\phi \to K^+K^-$, using 300 events reconstructed in 2.9 fb^{-1} of data [3]. The polarization amplitudes are measured using a fit to the mass and three decay angles of the B_s^0 decay products. The fit results are:

$$
\begin{aligned}
|A_0|^2 &= 0.348 \pm 0.041(\text{stat}) \pm 0.021(\text{syst}) \\
|A_\parallel|^2 &= 0.287 \pm 0.043(\text{stat}) \pm 0.011(\text{syst}) \\
|A_\perp|^2 &= 0.365 \pm 0.044(\text{stat}) \pm 0.027(\text{syst}) \\
\cos\delta_\parallel &= -0.91\,^{+0.15}_{-0.13}(\text{stat}) \pm 0.09(\text{syst})
\end{aligned}
$$

where $\cos\delta_{\parallel}$ is the relative strong phase. These results show that $|A_0|^2 \approx |A_{\parallel}|^2 \approx |A_{\perp}|^2$ and therefore the naive prediction is disfavored. The results are consistent within the uncertainties with the theoretical expectations of QCD factorization calculations (see [4] for example), while they do not agree with simple perturbative QCD [5].

3. Rare Decays: $B_{s,d} \to \mu^+\mu^-$

The decays $B_{s,d} \to \mu^+\mu^-$ are flavor changing neutral current processes that proceed via GIM-suppressed penguin diagrams, and consequently they have small branching fractions in the SM:

$$\begin{aligned} B(B_s^0 \to \mu^+\mu^-) &= (3.6 \pm 0.3) \times 10^{-9} \\ B(B^0 \to \mu^+\mu^-) &= (1.1 \pm 0.1) \times 10^{-10} \end{aligned}$$

The SM signal is currently beyond the sensitivity of CDF and DØ at the Tevatron, but new physics can significantly enhance the branching fractions. Examples are the MSSM ($B \sim \tan^6\beta$), SUSY R-parity violating models, and GUTs. Therefore, the observation of these decays with the current data would necessarily indicate new physics.

DØ have reported a new result in the search for $B_s^0 \to \mu^+\mu^-$ based on 6.1 fb^{-1} of data [6]. Events are selected using a Bayesian neural network based on five variables: the B_s^0 transverse momentum and pointing angle; the vertex χ^2 and transverse decay length significance; and the smallest impact parameter significance of the two muons. A control sample of $B^+ \to J/\psi[\to \mu^+\mu^-]K^+$ is used for normalization. Figure 1 shows the $\mu\mu$ invariant mass distribution for events in the signal region of the neural network discriminant β.

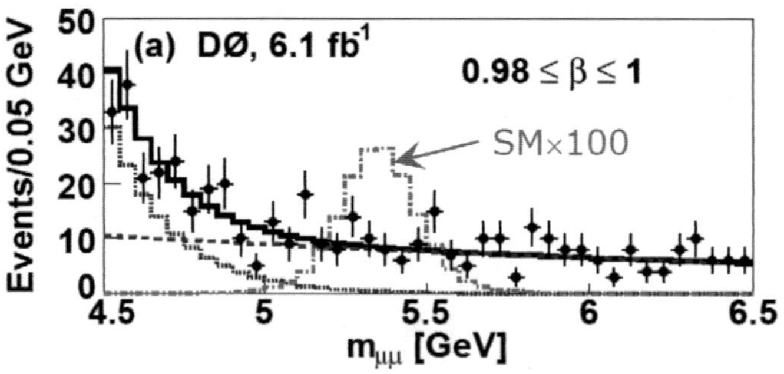

Figure 1. The $\mu\mu$ invariant mass distribution for events in the signal region of the neural network discriminant β, showing the data (dots with uncertainties), expected background distribution (solid line), and the SM signal distribution multiplied by a factor of 100. The dimuon background contributions from double semileptonic decays $B(D) \to \mu^+\nu X$ (dashed line) and sequential semileptonic decays $B \to \mu^+\nu\bar{D}, \bar{D} \to \mu^-\bar{\nu}X$ (dotted line) are also shown.

No evidence for a signal is observed and a limit is set on the branching fraction at the 95% C.L.:

$$B(B_s^0 \to \mu^+\mu^-) < 5.1 \times 10^{-8}$$

with an expected limit of 3.8×10^{-8}.

CDF have searched for $B_s^0 \to \mu^+\mu^-$ and $B^0 \to \mu^+\mu^-$ in a data set of 3.7 fb^{-1} and set the following 95% C.L. limits [7]:

$$\begin{aligned} B(B_s^0 \to \mu^+\mu^-) &< 4.3 \times 10^{-8} \\ B(B^0 \to \mu^+\mu^-) &< 7.6 \times 10^{-9} \end{aligned}$$

with expected limits of 3.3×10^{-8} and 9.1×10^{-9}, respectively.

4. CP Violation in $B_s^0 \to J/\psi\phi$

The $B_s^0 \to J/\psi\ \phi$ decay involves a final state that is a mixture of CP-even and CP-odd states. CP violation can occur via the interference between direct and mixed decays and is described by the phase β_s. CDF have updated their analysis of this mode using a 5.2 fb^{-1} data set [8]. In order to separate the CP-even and CP-odd states, CDF use a maximum likelihood fit to the mass, lifetime, and time-dependent angular distributions of the $B_s^0 \to J/\psi(\to \mu^+\mu^-)\ \phi(K^+K^-)$ decay. The fit yields the CP-violating phase β_s and the width difference $\Delta\Gamma_s \equiv \Gamma_L - \Gamma_H$, where Γ_L and Γ_H are the widths of the light and heavy eigenstates.

The B_s^0 flavor at production is determined using a combined opposite-side plus sameside tagging algorithm. The initial state flavor tag improves the sensitivity to the CP-violating phase and removes a sign ambiguity on β_s for a given $\Delta\Gamma_s$. In the fit, ΔM_s is constrained to its measured value. The number of signal events reconstructed is 6500. The average lifetime and width difference resulting form the fit are:

$$\begin{aligned} \bar{\tau}_s &= 1.53 \pm 0.025(\text{stat}) \pm 0.012(\text{syst})\ \text{ps} \\ \Delta\Gamma_s &= 0.075 \pm 0.035(\text{stat}) \pm 0.01(\text{syst})\ \text{ps}^{-1} \end{aligned}$$

The likelihood is coverage-corrected to account for non-Gaussian tails and systematic uncertainties. Confidence level contours in the $\phi_s - \Delta\Gamma_s$ plane are shown in Fig. 2(a). The likelihood as a function of β_s is shown in Fig. 2(b). The resulting 95% C.L. limits on the CP-violating phase β_s are:

$$\beta_s = [-\pi/2, -1.44] \cup [-0.13, 0.68] \cup [0.89, \pi/2] \quad 95\%\ \text{C.L.}$$

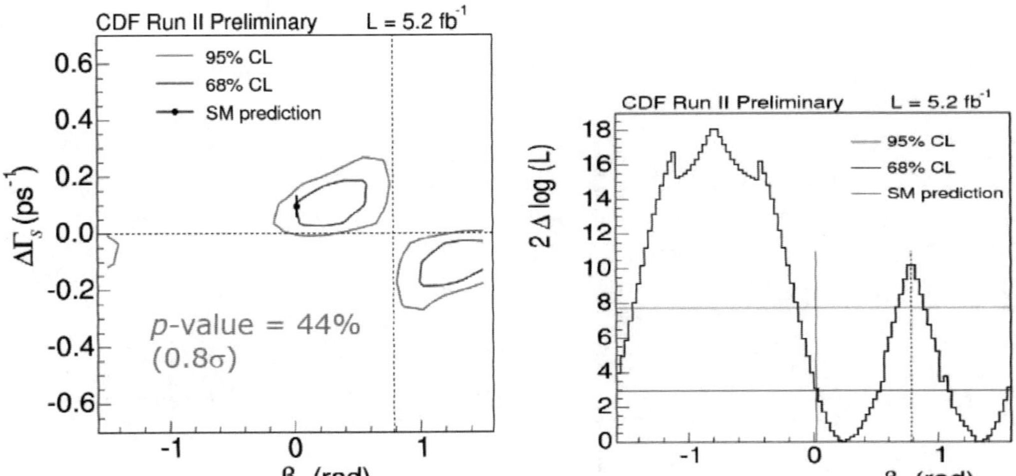

Figure 2. (a) 65% and 95% confidence level regions in the $\beta_s - \Delta\Gamma_s$ plane for the CDF $B_s^0 \to J/\psi\phi$ analysis based on 5.2 fb^{-1} of data. (b) Coverage-corrected profile likelihood as a function of β_s.

The p-value for the fit is 44%, corresponding to a deviation from the standard model prediction of 0.8 σ.

5. Search for CP Violation in Semileptonic B_s^0 Decays

In this section we report a new search for CP violation in the decay $B_s^0 \to D_s^- \mu^+ \nu X$ by measurement of the flavor-specific asymmetry using a time-dependent analysis with flavor tagging. Two final state samples were reconstructed: $D_s^- \to \phi\pi^-$, $\phi \to K^+K^-$ and $D_s^- \to K^{*0}K^-$, $K^{*0} \to K^+\pi^-$.

The flavor-specific asymmetry is defined as

$$a_{fs}^s = \frac{\Gamma\left(\bar{B}_s^0(t) \to f\right) - \Gamma\left(B_s^0(t) \to \bar{f}\right)}{\Gamma\left(\bar{B}_s^0(t) \to f\right) + \Gamma\left(B_s^0(t) \to \bar{f}\right)}$$

The SM prediction is very small, $a_{fs}^s = (-2.06 \pm 0.57) \times 10^{-5}$, and any deviation from this prediction would be evidence for CP-violating new physics.

The analysis uses a data set of integrated luminosity 5 fb^{-1} [9]. The technique used is similar to that used in the DØ B_s^0 oscillation analysis. A fit to the invariant $KK\pi$ mass of the selected data is shown in Fig. 3.

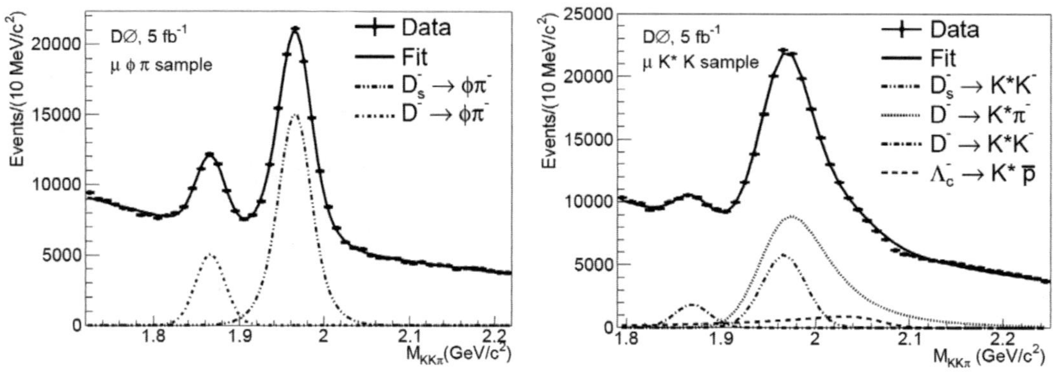

Figure 3. The $KK\pi$ invariant mass distributions in the $\mu\phi\pi$ and $\mu K^{*0}K$ samples together with the fit results.

An unbinned likelihood fit is used to extract the asymmetry. The systematic uncertainties are mainly due to uncertainties in the $c\bar{c}$ contribution, uncertainties in the efficiency vs visible proper decay length, and uncertainties in the $B_s^0 \to D_s^{(*)}\mu\nu$ branching fractions. Accounting for these yields the final result for the flavor-specific asymmetry:

$$a_{fs}^s = -0.0017 \pm 0.0091(\text{stat})\,^{+0.0014}_{-0.0015}(\text{syst})$$

This result is consistent with the SM prediction and is the most precise measurement to date.

6. Anomalous Like-Sign Dimuon Asymmetry

The final result reported here is the new DØ measurement of the like-sign dimuon asymmetry [10]. The dimuon charge asymmetry of semileptonic B decays is defined by

$$A_{sl}^b \equiv \frac{N_b^{++} - N_b^{--}}{N_b^{++} + N_b^{--}}$$

and has contributions from B_d and B_s^0 decays:

$$A_{sl}^b = (0.506 \pm 0.043)a_{sl}^d + (0.494 \pm 0.043)a_{sl}^s$$

Journal of Physics: Conference Series **259** (2010) 012018 doi:10.1088/1742-6596/259/1/012018

where

$$a^q_{sl} = \frac{\Gamma(\bar{B}^0_q \to \mu^+ X) - \Gamma(B^0_q \to \mu^- X)}{\Gamma(\bar{B}^0_q \to \mu^+ X) + \Gamma(B^0_q \to \mu^- X)}, \qquad q = d, s$$

The SM prediction for the semileptonic charge asymmetry is very small, $A^b_{sl} = (-2.3\,^{+0.5}_{-0.6}) \times 10^{-4}$, and new physics could change its value if new CP-violating phases are introduced.

The "raw" asymmetries measured by DØ are the like-sign dimuon asymmetry

$$A \equiv \frac{N(\mu^+\mu^+) - N(\mu^-\mu^-)}{N(\mu^+\mu^+) + N(\mu^-\mu^-)} = (0.564 \pm 0.053)\%$$

and the inclusive muon charge asymmetry

$$a \equiv \frac{N(\mu^+) - N(\mu^-)}{N(\mu^+) + N(\mu^-)} = (0.955 \pm 0.003)\%$$

These measurements are obtained from 3.7 million events in the like-sign dimuon sample and 1.5 billion events in the inclusive muon sample, respectively. The relation of A^b_{sl} to these raw asymmetries is as follows:

$$\begin{aligned} A &= KA^b_{sl} + A_{\text{bkg}} \\ a &= kA^b_{sl} + a_{\text{bkg}} \end{aligned}$$

After accounting for non-symmetric background processes from prompt/physics sources (A_{bkg} and a_{bkg}) and dilution due to other processes that do not contribute to asymmetry (K and k), the final result is:

$$A^b_{sl} = (-0.957 \pm 0.251(\text{stat}) \pm 0.146(\text{syst}))\%$$

which represents a 3.2 σ deviation from the theoretical prediction of the SM. A comparison with other measurements is shown in Fig. 4. As can be seen, the result is consistent with other measurements with the uncertainties.

7. Conclusions

The high luminosity of the Tevatron coupled with the relatively large b production cross section allows many measurements to be made that provide tests of the SM and allow searches for physics beyond the SM, especially in the area of CP violation and rare decays. The measurements are ongoing and are complementary to results obtained at the B-factories. Future measurements at the Tevatron and at the LHC will further explore the rich field of flavor physics.

Journal of Physics: Conference Series **259** (2010) 012018 doi:10.1088/1742-6596/259/1/012018

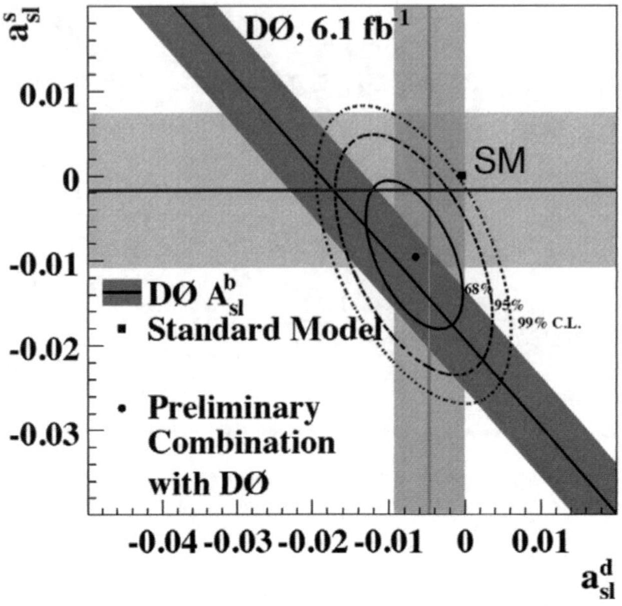

Figure 4. a^d_{sl} vs. a^s_{sl} showing the DØ measurement of A^b_{sl} (black line with red error band), the world average measurement of a^d_{sl} (green line with green error band), and the DØ direct measurement of a^s_{sl} (blue line with blue error band). Also shown are the 68%, 95%, and 99% C.L. regions for the combination of the three results.

References

[1] V.M. Abazov et al., (The DØ Collaboration), "The Upgraded D0 Detector," Nucl. Instr. and Methods A **565**, 463, (2006).
[2] R. Blair et al., (The CDF Collaboration), "The CDF II Detector Technical Design Report", Fermilab-Pub-96/390-E.
[3] CDF Note 10064, (The CDF Collaboration), 2010.
[4] M. Beneke, J. Rohrer, and D. Yang, Nuclear Physics B, **774**, 64, (2007).
[5] A. Ali et al., Phys. Rev. D **6**, 074018, (2007)
[6] V.M. Abazov et al., (The DØ Collaboration), arXiv.org:1006:3469, to appear in Phys. Lett. B.
[7] CDF Note 9892, (The CDF Collaboration), 2009.
[8] CDF Note 10206, (The CDF Collaboration), 2010.
[9] V.M. Abazov et al., (The DØ Collaboration), Phys. Rev. D **82**, 012003, (2010).
[10] V.M. Abazov et al., (The DØ Collaboration), Phys. Rev. Lett. **105**, 081801 (2010); V.M. Abazov et al., (The DØ Collaboration), Phys. Rev. D **82**, 032001, (2010).

16th International Symposium on Particles Strings and Cosmology (PASCOS 2010) IOP Publishing
Journal of Physics: Conference Series **259** (2010) 012019 doi:10.1088/1742-6596/259/1/012019

Top quark and Electroweak measurements at the Tevatron

Lucio Cerrito

Department of Physics, Queen Mary University of London, London, UK
on behalf of the CDF and D0 Collaborations

E-mail: l.cerrito@qmul.ac.uk

Abstract. We present recent preliminary measurements at the Tevatron of $t\bar{t}$ and single top production cross section, top quark mass and width, top pair spin correlations and forward-backward asymmetry. In the electroweak sector, we present the Tevatron average of the W boson width, and preliminary measurements of the W and Z forward-backward asymmetries and WZ, ZZ diboson production cross sections. All measurements are based on larger amount of collision data than previously used and are in agreement with the standard model.

1. Introduction

The Tevatron accelerator complex produces $p\bar{p}$ collisions with a center of mass energy of $\sqrt{s} = 1.96$ TeV, and the CDF and D0 experiments have so far recorded ~ 8 fb^{-1} of collision data. Just a fraction of this data allowed establishing a successful experimental programme of physics of the top quark and of the electroweak sector. In the following, we summarise the main measurements and report on preliminary results based on the latest data analyses.

2. Top quark pair and single top production

Top quarks are produced and observed at the Tevatron in their pair and singlet modes. Measurements of the cross section allow testing of the production and decay mechanisms of the top and its coupling in the standard model (SM). The pair production of top and anti-top in $p\bar{p}$ collisions has been measured with increasing precision since its discovery in 1995 [1, 2] at the Tevatron. Figure 1 shows the most recent measurements using data collected by the D0 experiment [3] in various decay channels. As top quarks decay almost exclusively to $W + b$, the final states of $t\bar{t}$ events are labelled as *dilepton*, *lepton plus jets* and *all-hadronic* depending on whether a leptonic decay has occurred in both, one only or none of the two final-state W bosons respectively. All results are in agreement with each other and in agreement with the next-to-leading order (NLO) calculation for $t\bar{t}$ production in $p\bar{p}$ collisions at $\sqrt{s} = 1.96$ TeV of 7.5$\pm$0.8 pb [4, 5, 6]. Preliminary measurements are also shown in figure 1, performed with 4.3–5.3 fb^{-1} of collision data, and yield individually a precision of $\sim 11\%$, similar to the theoretical uncertainty. Since a large component (about 6%) of the experimental uncertainty on the $t\bar{t}$ cross section is due to the uncertainty on the luminosity determination, higher precision can be obtained by measuring the ratio of $t\bar{t}$ to Z-boson cross sections σ_{tt}/σ_Z [9] and using the theoretical Z boson cross section calculation as input. With such an approach, the uncertainty on the luminosity largely cancels out and the most accurate preliminary determination of the $t\bar{t}$

cross section, combining samples in all decay channels seen with the CDF II detector [7, 8] yields: $\sigma_{t\bar{t}} = 7.50 \pm 0.31(\text{stat.}) \pm 0.34(\text{syst.}) \pm 0.15(\text{lumi.})$ pb (for a top mass of $m_t = 172.5\ GeV/c^2$), in excellent agreement with NLO calculations and with an experimental uncertainty of only $\pm 6.4\%$.

The electroweak production of single top quarks has been established at the Tevatron in 2009 [10, 11]. A preliminary measurement using 4.8 fb^{-1} of collision data recorded with the D0 detector has now been completed in the τ+jets decay channel, measuring $\sigma_{tbX+tqbX} = 3.4^{+2.0}_{-1.8}$ pb. Figure 2 shows the current best measurements of single top cross section (s-channel and t-channel combined) and the preliminary Tevatron combined cross section of $2.76^{+0.58}_{-0.47}$ pb, in agreement with the NLO theoretical prediction of 3.4 ± 0.2 pb [12, 13, 14] (for $m_t = 170\ GeV/c^2$). This also allows extraction of the Cabibbo-Kobayashi-Maskawa matrix element $|V_{tb}| = 0.88 \pm 0.07$ with a 95% confidence level (C.L.) lower limit of $|V_{tb}| > 0.77$.

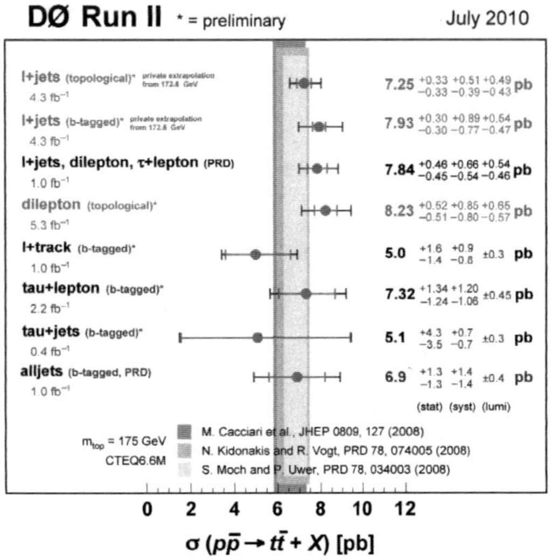

Figure 1. Summary of $t\bar{t}$ production cross section measurements using samples in different decay channels recorded with the D0 detector.

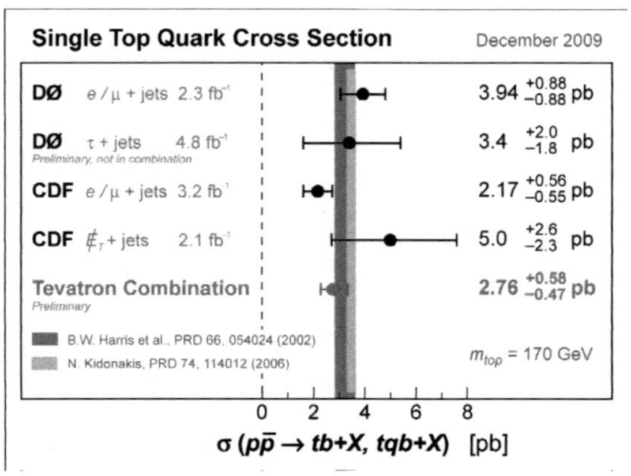

Figure 2. Measurements of single top cross section performed with data collected by the CDF and D0 collaborations. Also indicated the preliminary Tevatron combination.

3. Top quark mass and properties

The mass of the top quark is of interest both because the top is the heaviest known fundamental particle and because a precise measurement helps constraining the mass of the Higgs boson. We present two preliminary measurements using events in the dilepton and lepton plus jets final states reconstructed with the CDF II detector. The analyses were based respectively on 4.8 fb^{-1} and 5.6 fb^{-1} of collision data and yield experimental top mass values of $m_t = 170.6 \pm 2.2(\text{stat.}) \pm 3.1(\text{syst.})$ GeV/c^2 and $m_t = 173.0 \pm 0.6(\text{stat.}) \pm 1.1(\text{syst.})$ GeV/c^2. Figure 3 shows the CDF combination of top mass measurements, $m_t = 173.1 \pm 0.7(\text{stat.}) \pm 0.9(\text{syst.})$ GeV/c^2, which with an accuracy of 0.67% is more precise than the previous world average [15]. Moreover, measurements derived using all the $t\bar{t}$ decay channels are consistent with each other, and a preliminary determination of the top and anti-top mass difference using 5.6 fb^{-1} of collision data recorded with the CDF II detector gives $m_t - m_{\bar{t}} = -3.3 \pm 1.4(\text{stat.}) \pm 1.0(\text{syst.})$ GeV/c^2. Taking into account the preliminary results quoted above, the new Tevatron combined experimental top mass determination is $m_t = 173.3 \pm 1.1$ GeV/c^2 [16]. Projections indicate that the Tevatron experiments should reach individually a precision of ± 0.8–1.0 GeV/c^2 with ~ 10 fb^{-1} of data.

The experimental study of physics of top quarks at the Tevatron includes the determination of a variety of properties in addition to the production cross section and the top quark mass. Measurement of properties such as the production mechanism, the forward-backward asymmetry and differential cross sections for example are only recently approaching the accuracy needed to test the SM predictions. We present two preliminary results on the width of the top quark. In the first, using 4.3 fb^{-1} of Tevatron's $p\bar{p}$ collision data collected by the CDF II detector, we exploit the reconstructed top quark mass to set an upper limit on the top quark width (Γ_{top}) of $\Gamma_{top} < 7.6$ GeV/c^2 at 95% C.L. We also set a central limit of 0.3 GeV/$c^2 < \Gamma_{top} < 4.4$ GeV/c^2 at 68% C.L. In the second result, we extract the total width of the top quark from the partial decay width $\Gamma(t \to Wb)$ and the branching fraction $BR(t \to Wb)$. $\Gamma(t \to Wb)$ is obtained from the measured t-channel cross section for single top quark production in 2.3 fb^{-1} of $p\bar{p}$ data from the D0 collaboration, and $BR(t \to Wb)$ is extracted from a measurement of the ratio $R = BR(t \to Wb)/BR(t \to Wq)$ in $t\bar{t}$ events in 1 fb^{-1} of integrated luminosity. Assuming $BR(t \to Wq) = 1$, where q includes any kinematically accessible quark, the result is: $\Gamma_t = 2.1 \pm 0.6$ GeV/c^2 which corresponds to a top quark lifetime of $\tau_t = (3 \pm 1) \cdot 10^{-25}$ s.

A preliminary measurement of the top quark charge, based on 2.7 fb^{-1} of collision data collected with the CDF II detector and using soft electron and muon b-tagging, now excludes a top quark charge of $-4e/3$ with 95% C.L, extending the significance of a previous result. Measurements of the dynamics of production and decay of the top quark are of importance because they are sensitive to new production and decay mechanisms. For example, SM top pair production induces a characteristic spin correlation which can be modified by new production mechanisms such as Z' bosons or Kaluza-Klein gluons [17]. We report on a preliminary measurement of $t\bar{t}$ helicity fractions and spin correlation in 5.3 fb^{-1} of reconstructed lepton+jets data with the CDF II detector. Choosing the direction of the top quark momentum as the spin quantization axis (the *helicity basis*), we find a spin correlation coefficient $\kappa_{hel} = 0.48 \pm 0.48(\text{stat.}) \pm 0.22(\text{syst.})$, in agreement with the NLO expectation of $\kappa \simeq 0.35$ [18]. We also report a preliminary measurement of the forward-backward asymmetry of pair produced top quarks in 5.3 fb^{-1} of data collected with the CDF II detector. A small asymmetry of 0.038 ± 0.006 in the laboratory rest frame is expected in Quantum Chromo Dynamics at NLO [19] and large positive asymmetries with large uncertainties were measured at CDF and D0 [20, 21]. We find $A_{lab} = 0.15 \pm 0.05(\text{stat.}) \pm 0.02(\text{syst.})$, in agreement with the SM.

4. *W* boson mass and width

A precise measurement of the W boson mass (m_W) provides a measurement of quantum loop corrections, and in the SM these corrections to m_W are contributed predominantly by the top quark and the Higgs boson loops. Therefore, m_W in conjunction with the top quark mass constrains the mass m_H of the undiscovered Higgs boson. The most recent combination of m_W measurements at the Tevatron is reported in detail elsewhere [22], and when combined with measurements at the LEP collider yields the value $m_W = 80,399 \pm 23$ MeV/c^2 [23]. Overall, precision electroweak measurements, including the preliminary value presented in this article of $m_t = 173.3 \pm 1.1$ GeV/c^2, indicate the preferred Higgs boson mass of $m_H = 89^{+35}_{-26}$ GeV/c^2, or $m_H < 185$ GeV/c^2 at 95% C.L.

In addition to the W boson mass, the decay width of the W is also of importance towards a stringent test of the SM and can be used to constrain other SM parameters, such as the V_{cs} Cabibbo-Kobayashi-Maskawa matrix element. A recent combination of the CDF and D0 Run I and Run II measurements, which are based on fits to tails of the W transverse mass [24] distributions, yields $\Gamma_W = 2,046 \pm 49$ MeV/c^2 [25], and a preliminary world average including both the Tevatron and LEP2 gives $\Gamma_W = 2,085 \pm 42$ MeV/c^2.

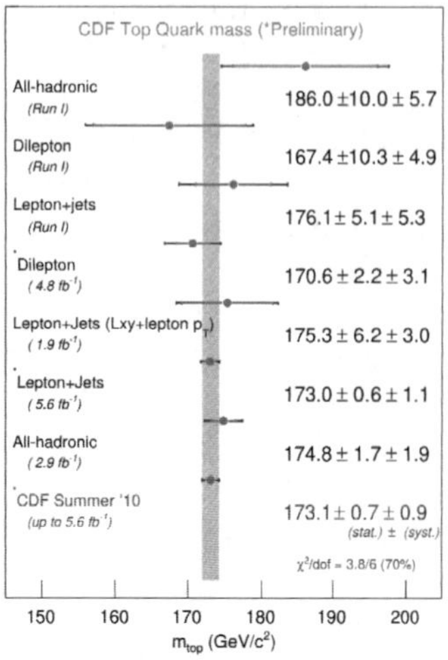

Figure 3. Preliminary and published best measurements of the mass of the top quark in each decay channel, based on data from Run I (1992-1996) and Run II (2001-present) analysed by the CDF collaboration. Shown also is the average of those measurements.

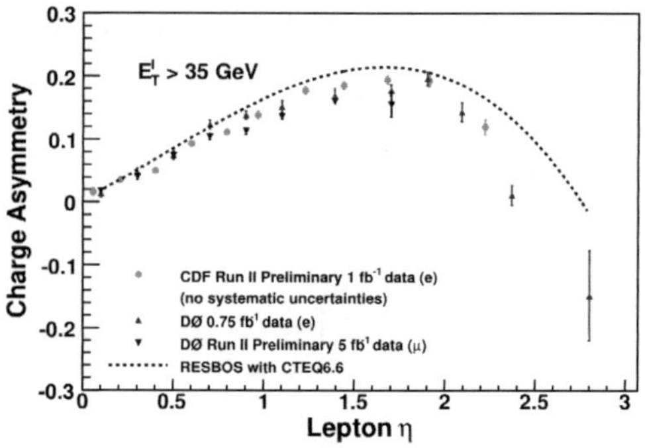

Figure 4. Comparison of the W boson charge asymmetry measured by the CDF and D0 collaborations as a function of lepton $|\eta|$, for lepton's transverse energies $E_T^\ell > 35$ GeV. The data are compared to a simulation based on `RESBOS` with `CTEQ6.6` PDFs.

5. *W* and *Z* charge asymmetries

The production of W bosons at the Tevatron is subject to a well known charge asymmetry, due to the different fraction of the proton momentum carried on average by u and d quarks:

$$A(y_W) = \frac{d\sigma^+/dy_W - d\sigma^-/dy_W}{d\sigma^+/dy_W + d\sigma^-/dy_W} \tag{1}$$

where y_W is the W boson rapidity [26] and $d\sigma^\pm/dy_W$ is the differential cross section for W^+ or W^- boson production. Previous measurements [27, 28] have been published by the D0 and CDF collaborations, based on candidate $W \to e\nu$ events in 0.75 fb^{-1} and 1 fb^{-1} of collision data respectively, in which comparisons to Monte Carlo (MC) simulations of W production and decay showed generally good agreement. We present a direct comparison of these analyses with an additional preliminary measurement based on candidate $W \to \mu\nu$ events in 4.9 fb^{-1} of collision data collected with the D0 detector. The data are compared to MC event simulation from `RESBOS` [29] with `CTEQ6.6` [30] parton distribution functions (PDFs), as a function of the lepton pseudo-rapidity η, in different ranges of the lepton's transverse energy (E_T^ℓ). While in the low E_T^ℓ the data are well reproduced by the MC, for $E_T^\ell > 35$ GeV discrepancies start to emerge and the MC is now seen to overestimate the asymmetry at large $|\eta|$ (see figure 4) with respect to both CDF and D0 experimental data.

The presence of both vector and axial-vector couplings in $q\bar{q} \to Z/\gamma^* \to \ell^+\ell^-$ gives rise to an asymmetry also for Z bosons in the polar angle of the negatively charged lepton momentum relative to the incoming quark direction in the rest frame of the lepton pair. We present a preliminary measurement of such forward-backward asymmetry in the kinematic region where the invariant mass (M_{ee}) of the lepton pair is between 50 and 600 GeV/c^2. The data sample consists of 4.1 fb^{-1} of $p\bar{p}$ collisions recorded by the CDF II detector. Figure 5 shows the data compared with `PYTHIA v6.216` [31], indicating very good agreement across the entire M_{ee} spectrum. The forward-backward asymmetry is sensitive to $\sin^2\theta_W^{eff}$, which is an effective parameter that includes higher order corrections. The current world average value of $\sin^2\theta_W^{eff}$ is 0.23145±0.00013 [32], the D0 collaboration measured $\sin^2\theta_W^{eff} = 0.2326 \pm 0.0019$ from an earlier Z forward-backward asymmetry measurement [33] based on 1.1 fb^{-1} of collision data and we plan to soon report $\sin^2\theta_W^{eff}$ from the CDF preliminary measurement presented here.

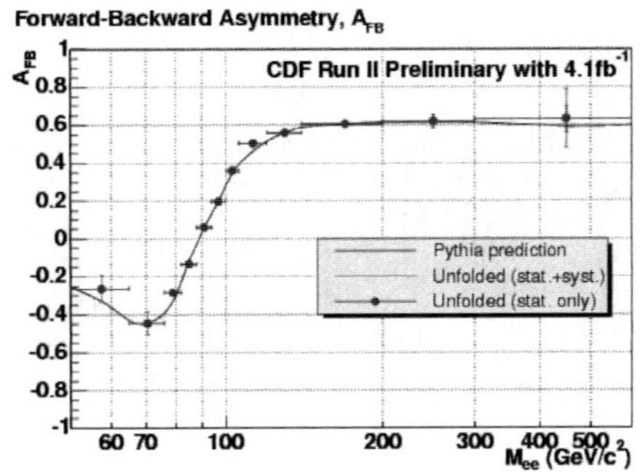

Figure 5. Forward-backward charge asymmetry of high mass Drell-Yan dielectron pairs.

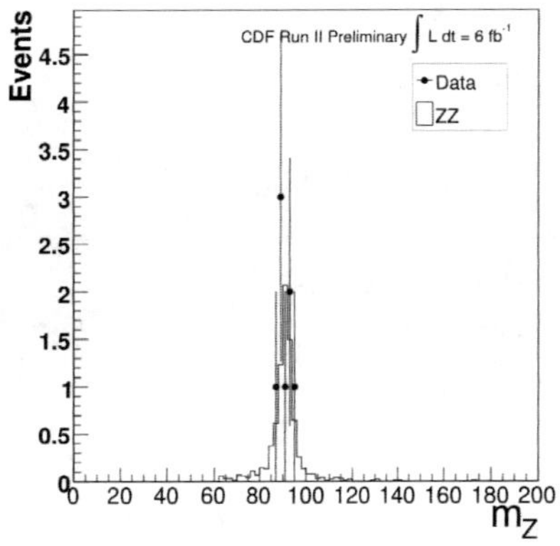

Figure 6. Invariant mass for the Z candidates reconstructed in the 4 ZZ candidate events.

6. WW, WZ, ZZ diboson production

The measurement of heavy diboson production in $p\bar{p}$ collisions is a way of probing the structure of the electroweak sector, as the presence of anomalous triple gauge bosons couplings for instance could indicate new physics at a higher mass scale. Moreover, as these processes are backgrounds to some Higgs boson decay signatures, their measurement provides milestones on the way to sensitivity to a Higgs boson production measurement. Heavy diboson production has been established at the Tevatron in several final states using 1–3 fb^{-1} [34, 35]. As the collision data available increases, the precision of these measurements is being progressively improved. We report on a preliminary measurement of WZ and ZZ production in three- and four-lepton channels, using $\sim$ 6 fb^{-1} of data collected by the CDF experiment. From 50 $WZ \to \ell\nu$ candidate events and 4 $ZZ \to \ell\ell$ candidate events, we measure $\sigma(p\bar{p} \to WZ) = 4.1 \pm 0.7$ pb and $\sigma(p\bar{p} \to ZZ) = 1.7^{+1.2}_{-0.7}(\text{stat.}) \pm 0.2(\text{syst.})$ pb. The systematic uncertainties are minimised by measuring the experimental ratio to the inclusive $\sigma(p\bar{p} \to Z)$ cross section and using the theoretical value of $\sigma(Z \to \ell\ell)$ production cross section. Figure 6 shows the distribution of invariant mass for the Z candidates reconstructed in the 4 ZZ candidate events. The measurements are in agreement with the LO theoretical prediction of $\sigma(p\bar{p} \to WZ) = 3.46 \pm 0.21$ pb and $\sigma(p\bar{p} \to ZZ) = 1.4 \pm 0.1$ pb, and in agreement with earlier measurements at the Tevatron [36, 37]. We also report on a preliminary measurement of combined $WW + WZ$ production cross section in the channel with an identified electron or muon, large missing transverse energy and two jets in 4.6 fb^{-1} of CDF II data. The analysis employs a matrix element technique which calculates event probability densities for signal and background hypotheses. The probabilities are used to form a discriminant variable which is evaluated for signal and background events. The resulting template distributions are fit to the data using a binned likelihood approach, and yield a cross section $\sigma_{WW+WZ} = 16.5^{+3.3}_{-3.0}$ pb. The measurement increases the precision of previous determinations [38, 39] and is in agreement with the theory prediction of 16.1±0.9 pb [40].

7. Conclusions

We have presented recent preliminary measurements at Tevatron of top pair and single production cross section, top quark mass and width, top pair spin correlations and forward-backward asymmetry. In the electroweak sector, we have presented a new Tevatron average of the W boson width, and preliminary measurements of the W and Z forward-backward asymmetries and WZ, ZZ diboson production cross sections. All measurements are based on larger amount of collision data than previously used and are in agreement with the SM.

References

[1] F. Abe *et al.* (CDF Collaboration), *Phys. Rev. Lett.* **74** 2626 (1995).
[2] S. Abachi *et al.* (D0 Collaboration), *Phys. Rev. Lett.* **72** 2632 (1995).
[3] V.M. Abazov *et al.* (D0 Collaboration), *Nucl. Instrum. Methods Phys. Res.* A **565** 463 (2006).
[4] M. Cacciari *et al.*, *Journal of High Energy Physics* **09** 127 (2008).
[5] N. Kidonakis and R. Vogt, *Phys. Rev.* D **78** 074005 (2008).
[6] U. Langenfeld, S. Moch and P. Uwer, *Phys. Rev.* D **80** 054009 (2009).
[7] The CDF II Detector Technical Design Report No. Fermilab-Pub-96/390-E;
[8] D. Acosta *et al.*, (CDF Collaboration) *Phys. Rev.* D **71**, 052003 (2005).
[9] T. Aaltonen *et al.* (CDF Collaboration), *Phys. Rev. Lett.* **105** 012001 (2010).
[10] T. Aaltonen *et al.* (CDF Collaboration), *Phys. Rev. Lett.* **103** 092009 (2009).
[11] V.M. Abazov *et al.* (D0 Collaboration), *Phys. Rev. Lett.* **103** 092001 (2009).
[12] B.W. Harris, E. Laenen, L. Phaf, Z. Sullivan, and S. Weinzier. *Phys. Rev.* D **66** 054024 (2002).
[13] N. Kidonakis, *Phys. Rev.* D **74** 114012 (2006).
[14] J. Campbell, R.K. Ellis, F. Tramontano, *Phys. Rev.* D **70** 094012 (2004).
[15] The Tevatron Electroweak Working Group, Fermilab-TM-2427-E, arXiv:0903.2503v1 [hep-ex] (2009).
[16] The Tevatron Electroweak Working Group, Fermilab-TM-2466-E, arXiv:1007.3178v1 [hep-ex] (2009).
[17] M. Arai, N. Okada, K. Smolek, V. Simak, *Phys. Rev.* D **75** 095008 (2007).
[18] W. Bernreuther, A. Brandenburg, Z.G. Si, P. Uwer, CERN-PH-TH/2004-206, arXiv:0410197 [hep-ph] (2004).
[19] O. Antunano, J.H. Kuhn and G. Rodrigo, *Phys. Rev.* D **77** 014003 (2008).
[20] T. Aaltonen *et al.*, (CDF Collaboration) *Phys. Rev. Lett.* **101** 202001 (2008).
[21] V.M. Abazov *et al.*, (D0 Collaboration) *Phys. Rev. Lett.* **100** 142002 (2008).
[22] The Tevatron Electroweak Working Group, Fermilab-TM-2439-E, arXiv:0908.1374v1 [hep-ex] (2009).
[23] The LEP and Tevatron Electroweak Working Groups, CERN-PH-EP/2009-023, arXiv:0911.2604v2 [hep-ex] (2009).
[24] The W transverse mass is defined as $M_T = \sqrt{2(p_T^\ell p_T^\nu - \vec{p}_T^\ell \cdot \vec{p}_T^\nu)}$, where $\vec{p}_T^\ell$ and $\vec{p}_T^\nu$ are the measured transverse momentum of the charged lepton and the transverse momentum of the neutrino respectively.
[25] The Tevatron Electroweak Working Group, Fermilab-TM-2460-E, arXiv:1003.2826v1 [hep-ex] (2010).
[26] We assume (z, ϕ, θ) coordinate system with the z axis in the direction of the proton beam; ϕ and θ are the azimuthal and polar angle, respectively. The rapidity (y_W) and the pseudorapidity (η) are defined as $y_W = ln\frac{E+P_z}{E-P_z}$ and $\eta = -\text{lntan}\left(\frac{\theta}{2}\right)$.
[27] V.M. Abazov *et al.* (D0 Collaboration) *Phys. Rev. Lett.* **101** 211801 (2008).
[28] T. Aaltonen *et al.* (CDF Collaboration) *Phys. Rev. Lett.* **102** 181801 (2009).
[29] C. Balazs and C.P. Yuan, *Phys. Rev.* D **56**, 5558 (1997).
[30] P.M. Nadolsky *et al.*, Phys. Rev. D **78** 013004 (2008).
[31] T. Sjostrand *et al.*, *Comput. Phys. Commun.* **135**, 328 (2001).
[32] K. Nakamura *et al.* (Particle Data Group), *J. Phys.* G **37**, 075021 (2010).
[33] V.M. Abazov *et al.* (D0 Collaboration) *Phys. Rev. Lett.* **101** 191801 (2008).
[34] V.M. Abazov *et al.* (D0 Collaboration) *Phys. Rev. Lett.* **94** 151801 (2005); V.M. Abazov *et al.* (D0 Collaboration) *Phys. Rev.* D **76** 111104 (2007); V.M. Abazov *et al.* (D0 Collaboration) *Phys. Rev. Lett.* **101** 171803 (2008).
[35] D.E. Acosta *et al* (CDF Collaboration) *Phys. Rev. Lett.* **94** 211801 (2005); A. Abulencia *et al.* (CDF Collaboration) *Phys. Rev. Lett.* **98** 161801 (2007); T. Aaltonen *et al.* (CDF Collaboration) *Phys. Rev. Lett.* **100** 201801 (2008).
[36] T. Aaltonen *et al.* (CDF Collaboration) *Phys. Rev. Lett.* **98** 161801 (2007).
[37] V.M. Abazov *et al.* (D0 Collaboration) arXiv:1006.0761v1 [hep-ph] (2010).
[38] T. Aaltonen *et al.* (CDF Collaboration) *Phys. Rev. Lett.* **104** 101801 (2010).
[39] V.M. Abazov *et al.* (D0 Collaboration) *Phys. Rev. Lett.* **102** 161801 (2009).
[40] J. M. Campbell and R. K. Ellis, *Phys. Rev.* D **60**, 113006 (1999).

Journal of Physics: Conference Series **259** (2010) 012020 doi:10.1088/1742-6596/259/1/012020

SUSY Searches at the Tevatron

Jared Yamaoka

Research Associate, Duke University Department of Physics, Physics Bldg Science Drive, Durham, NC 27708 USA

E-mail: yamaoka@cern.ch

Abstract. We present the results of several searches for Supersymmetric (SUSY) particles at the CDF and D0 experiments at the Tevatron at Fermilab. Using data sets as large as 6.3 fb^{-1}, a wide variety of SUSY models are explored. Since no excesses are observed in data above the Standard Model background, limits on the production cross section times branching ratio are set. Where appropriate these limits are interpreted within specific SUSY models to calculate mass exclusion limits.

1. Introduction

Supersymmetry (SUSY) [1] is one of the most popular extensions of the Standard Model (SM). SUSY can lessen the hierarchy problem and allows the unification of the gauge couplings which the SM suffers from. Each SM particle has a corresponding SUSY particle (sparticle) that contributes to the Higgs mass squared with opposite sign relative to the contributions of SM particles. This protects the weak mass scale from divergences. SUSY is a broken symmetry since the sparticles do not have the same mass as their SM partners. If R-parity (R_p) is conserved the lightest SUSY particle (LSP) is stable and provides a candidate for cosmological dark matter. This paper presents several results which are interpreted within different SUSY models.

2. mSUGRA

The minimal super gravity (mSUGRA) model of SUSY breaking [2] is described by five independent parameters: the unified scalar and gaugino masses m_0 and $m_{1/2}$, the ratio of the vacuum expectation values of the two Higgs doublets, $\tan\beta$, the unified trilinear coupling A_0, and the sign of the Higgs mass parameter μ. These parameters define all the properties (masses, cross sections, etc.) of the model. In this model the LSP is weakly interacting and typically identified via missing transverse energy ($\not{E}_T$).

2.1. Gaugino Pair Production in Trilepton Final State

The classical signature for SUSY is chargino($\tilde{\chi}_1^{\pm}$) and neutralino ($\tilde{\chi}_2^0$) decaying to leptons and LSPs. This leads to a final state with three leptons and $\not{E}_T$ from a neutrino and LSPs which escape undetected. The advantage of this *trilepton* channel is that three isolated leptons with large $\not{E}_T$has a very low expected background from SM processes. The key disadvantage is the leptons can be relatively low p_T so identifying them is more difficult. This process also has a low cross section times branching ratio.

In the D0 analysis [3], we define four different channels distinguished by the lepton content of the final state. For channels without τ, one *low-*p_T and one *high-*p_T selection is designed to

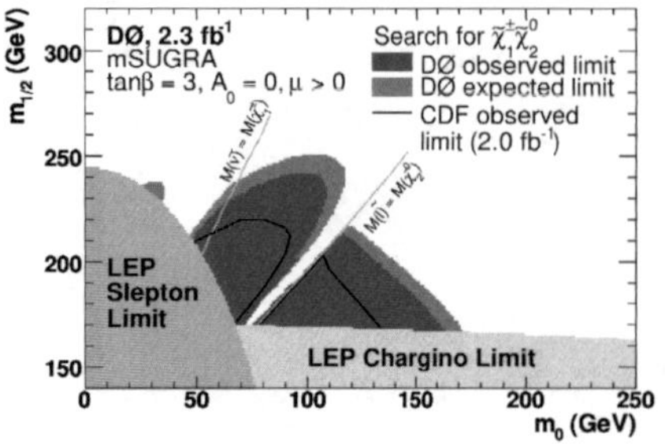

Figure 1. D0 exclusion limit in the $m_{1/2}$-m_0 plane.

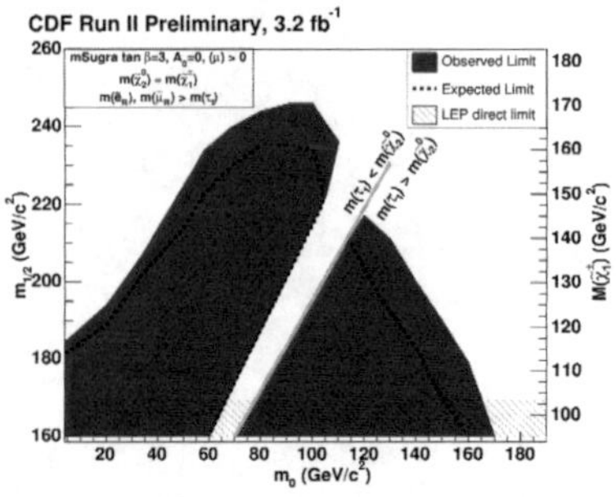

Figure 2. CDF exclusion limit in the $m_{1/2}$-m_0 plane.

exploit the different kinematic properties for different mass ranges. Primarily these selections tune the p_T cuts of the leading leptons to maximize the expected sensitivity. The channels which contain τ use only one set of cuts similar to the low-p_T selection. This analysis is performed with data collected until June, 2007, totaling 2.3 fb^{-1}. The analysis using hadronic τ decays is based on 1.0 fb^{-1}. We see agreement with SM expectations so we set an exclusion limit. Assuming a parameter space (m_0 = 60 GeV, $\tan\beta$= 3, A_0 = 0, and μ > 0) we constrain the chargino ($\tilde{\chi}_1^{\pm}$) mass to be greater than 155 GeV, see Figure 1.

In the CDF analysis [4], we select events and separate them, based on lepton type (excluding τ), into exclusive samples based on expected purity. This channel independence allows easy statistical combination of the results. We create two control samples to test our background estimates against data. These trilepton events are chosen in a kinematic region where we do not expect any signal. Once we are satisfied with the agreement in the control regions, we apply SUSY specific cuts and look at signal region data to compare against background. This analysis is performed with data collected until 1 July, 2008, totaling 3.2 fb^{-1}. We see agreement with SM expectations so we set an exclusion limit. Using the parameter space mentioned above we constrain the $\tilde{\chi}_1^{\pm}$mass to be greater than 164 GeV, see Figure:2.

2.2. Squarks and Guinos in $\not{E}_T$+ Jets

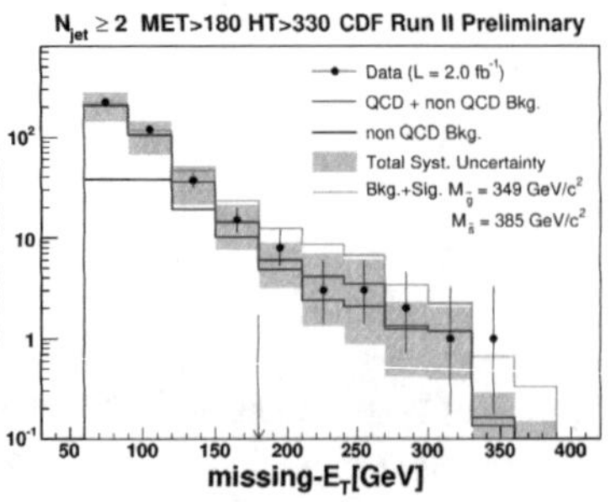

Figure 3. CDF $\not{E}_T$ distribution for two jet events.

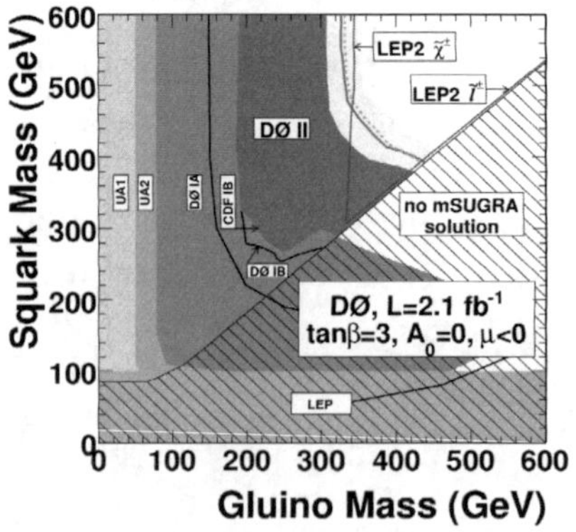

Figure 4. D0 mass exclusion in the $m_{\tilde{q}}$-$m_{\tilde{g}}$ plane.

Since R-parity is conserved in the mSUGRA model, supersymmetric particles are produced in pairs. At the Tevatron, squarks ($\tilde{q}$) and gluinos ($\tilde{g}$), the superpartners of quarks and gluons, would be produced, if sufficiently light, by the strong interaction. Their decay then leads to SM particles and to the LSP. These decays lead to final states containing jets and $\not{E}_T$.

Using 2.1 fb^{-1} and 2.0 fb^{-1} respectively, the D0 analysis [5] and the CDF analysis [6] perform similar searches looking for excess events in two, three, or four jets plus $\not{E}_T$. Both find good agreement between the data and SM expectations. Figure 3 shows this agreement for the CDF analysis. Since there is no excess we can set a limit on the $\tilde{q}(\tilde{g})$ mass, see Figure 4. Within the mSUGRA model assuming, $m_{\tilde{q}} \sim m_{\tilde{g}}$, the D0 analysis and the CDF analysis constrain $m_{\tilde{q}(\tilde{g})}$ to be greater then 390 GeV.

3. MSSM

Similar to mSUGRA, we can have a generic minimal supersymmetric extension of the SM (MSSM). In these MSSM R-parity conserving models [7], sparticles are produced in pairs and the lightest supersymmetric particle (LSP) is stable and denoted $\tilde{\chi}_1^0$.

3.1. $\tilde{b}\bar{\tilde{b}}$ Production

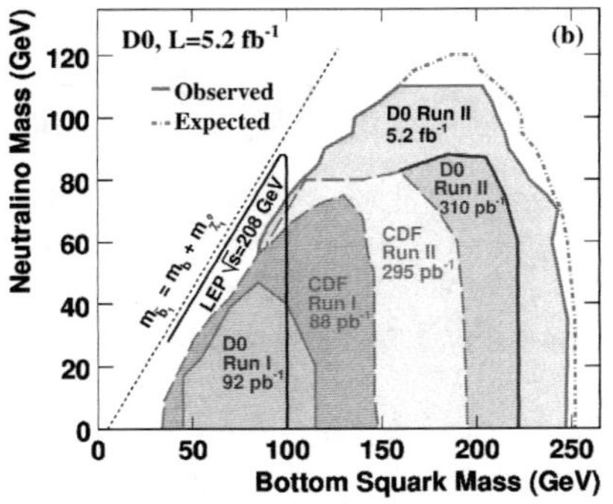

Figure 5. D0 exclusion plot in the $m_{\tilde{\chi}_1^0}$-$m_{\tilde{b}}$ plane.

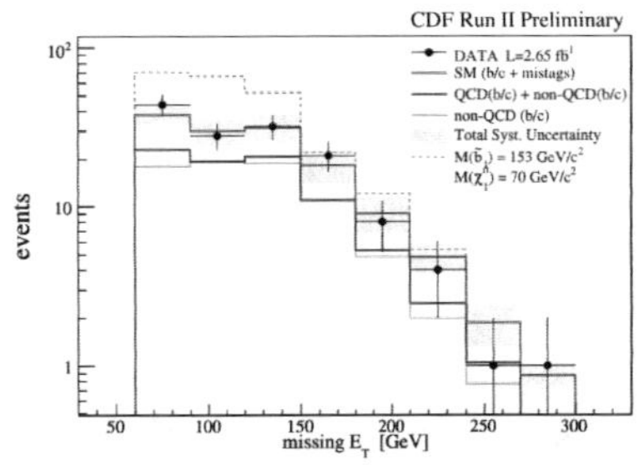

Figure 6. CDF $\not{E}_T$distribution.

Assuming a SUSY particle mass hierarchy such that the bottom squark ($\tilde{b}$) decays exclusively as $\tilde{b} \to b\tilde{\chi}_1^0$, the expected signal for direct $\tilde{b}\bar{\tilde{b}}$ production is characterized by the presence of two energetic jets from the hadronization of the bottom quarks and large $\not{E}_T$from the two LSPs in the final state.

In the D0 analysis [8], final states with two or three jets are selected. Additional requirements are that the two leading jets are identified b jets and large $\not{E}_T$are imposed. An artificial neural network (NN) is used to identify jet associated with b quarks. This result is based on 5.2 fb^{-1} of integrated luminosity.

In the CDF analysis [9], final state with two jets, with at least one identified as a b quark are selected. b quarks are identified via displaced vertices. Further optimization is done in two signal regions: $m_{\tilde{b}} - m_{\tilde{\chi}_1^0} < 90$ GeV and $m_{\tilde{b}} - m_{\tilde{\chi}_1^0} > 90$ GeV. This result is based on 2.65 fb^{-1} of integrated luminosity.

Figure 6 shows the agreement of the data and SM expectation in the CDF analysis. Both analysis see agreement with SM expectations. Figure 5, shows the mass exclusion in the $m_{\tilde{\chi}_1^0}$-$m_{\tilde{b}}$ plane. Assuming a massless $\tilde{\chi}_1^0$, the D0 analysis constrains $m_{\tilde{b}}$ to be greater than 247 GeV. Assuming $m_{\tilde{\chi}_1^0} < 80$ GeV, the CDF analysis constrains $m_{\tilde{b}}$ to be greater than 230 GeV.

3.2. $\tilde{t}\bar{\tilde{t}}$ Production

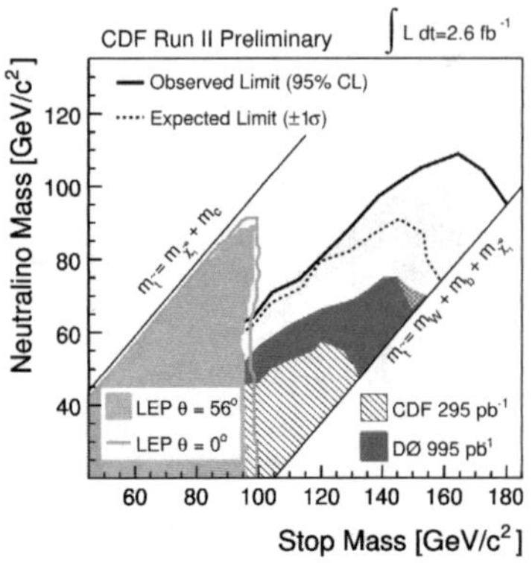

Figure 7. CDF exclusion in the $m_{\tilde{\chi}_1^0}$-$m_{\tilde{t}}$ plane.

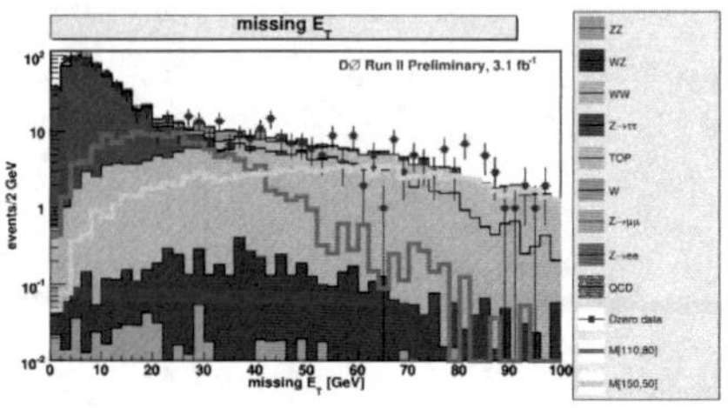

Figure 8. D0 $\not{E}_T$ distribution.

In the MSSM, the mixing between the chiral states of the scalar partner particles of the SM fermions is greatest for the top quark due to its large mass. Thus, it is possible that the light stop ($\tilde{t}$) is the lightest squark and has the largest production cross section at the Tevatron. If R-parity is conserved, then the top squarks would be produced in pairs with the dominant diagrams being quark/anti-quark annihilation and gluon/gluon fusion. If the two body decays $\tilde{t} \to b\tilde{\chi}_1^0$ and $\tilde{t} \to t\tilde{\chi}_1^0$ are kinematically forbidden, then the likely decay modes are the two body $\tilde{t} \to c\tilde{\chi}_1^0$, the three body $\tilde{t} \to b\tilde{\nu}^0 l^+$, and the four body $\tilde{t} \to b\tilde{\chi}_1^0 f\bar{f}$.

In the D0 analysis [10], we search for the three body decays of top squark pairs in the in the $b\bar{b}e^{\pm}\mu^{\mp}\tilde{\nu}\nu$ final state. The analysis also assumes $\tilde{t}$ decay via the three body decay mode with a 100% branching fraction, and the sneutrino ($\tilde{\nu}^0$) is the LSP, Events with opposite charged e and μ are selected with no requirement on the number of jets. Then selection is further optimized for large and small differences between $m_{\tilde{t}}$ and $m_{\tilde{\nu}}$. Using 3.1 fb^{-1} of integrated luminosity, we see agreement with SM expectations. Figure 8 shows the agreement in the $\not{E}_T$for data and SM expectation. For $m_{\tilde{\nu}} < 80$ GeV, the D0 analysis constrains $m_{\tilde{t}}$ to be greater than 200 GeV.

In the CDF analysis [11], we search for the two body decay $\tilde{t} \to c\tilde{\chi}_1^0$. Here we make the assumption that the two body branching ratio is 100%, and the $\tilde{\chi}_1^0$ is the LSP. A NN based flavor separator is used to identify jets coming from c quarks. Events with 2 jets and $\not{E}_T$are selected. Additionally, one of the jet is required to be identified as a c quark. The sensitivity is futher optimized with a NN to separate the signal like events from background like events. Using 2.6 fb^{-1} of integrated luminosity, we see agreement with SM expectations. Figure 7 shows the limit in the $m_{\tilde{\chi}_1^0}$-$m_{\tilde{t}}$ plane. Assuming $m_{\tilde{\chi}_1^0} = 80$ GeV the CDF analysis constrains $m_{\tilde{t}}$ to be greater than 180 GeV.

4. GMSB Di-photon

In gauge mediated supersymmetry breaking (GMSB) models, the masses of the particles arise from SM gauge interactions and are proportional to the effective SUSY breaking scale Λ. As the gravitino ($\tilde{G}$) does not participate in SM gauge interactions, it has a small mass [12] and is the LSP. Assuming R-parity conservation, the SUSY process with the largest cross section at the Tevatron would be chargino and neutralino production, followed by decay chains to the next-to-lightest SUSY particle (NLSP).We consider the case when the lightest neutralino ($\tilde{\chi}_1^0$)

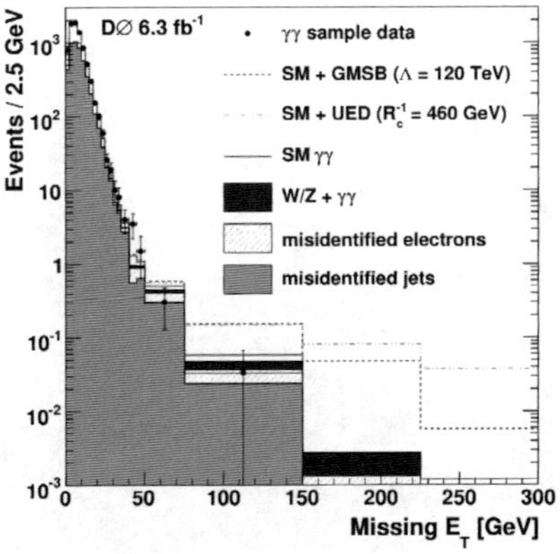

Figure 9. D0 $\not\!\!E_T$distribution.

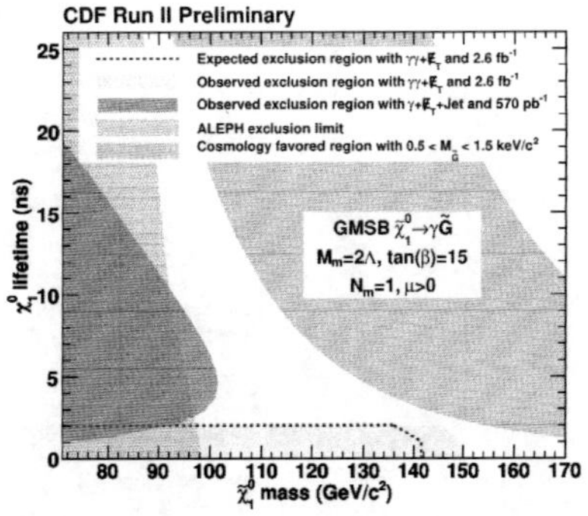

Figure 10. CDF exclusion limit in the $\tilde{\chi}_1^0$ *lifetime*-$m_{\tilde{\chi}_1^0}$ plane.

is the NLSP, which decays promptly with the dominant branching fraction yielding a photon (γ) and an essentially massless gravitino. The two gravitinos escape detection, resulting in the final state $\gamma\gamma + \not\!\!E_T + X$ where X denotes leptons and jets produced in the decay chains.

Using 6.3 fb^{-1} and 2.6 fb^{-1} of data respectively, D0 [13] analysis and CDF [14] analysis see agreement with SM expectations. Figure 9 shows the data and SM expectation for $\not\!\!E_T$in the D0 data. Figure 10 show the exclusion limit in the $\tilde{\chi}_1^0$ *lifetime*-$m_{\tilde{\chi}_1^0}$ plane. The D0 analysis constrains $m_{\tilde{\chi}_1^0}$ to be greater than 175 GeV and CDF analysis constrains $m_{\tilde{\chi}_1^0}$ to be greater than 149 GeV.

5. Hidden Valley Dark Photon

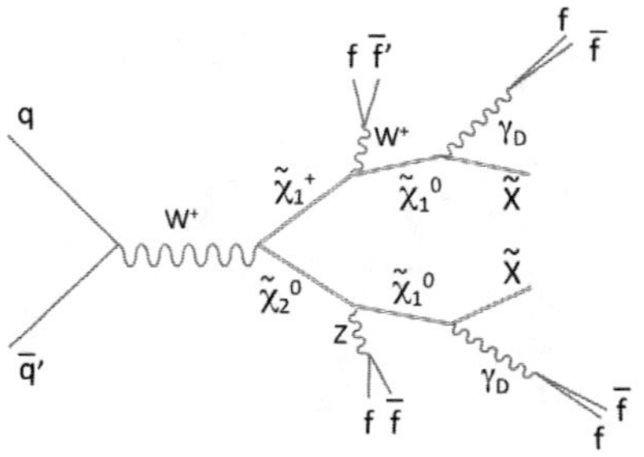

Figure 11. Feynmann diagram for γ_D decaying to lepton jets.

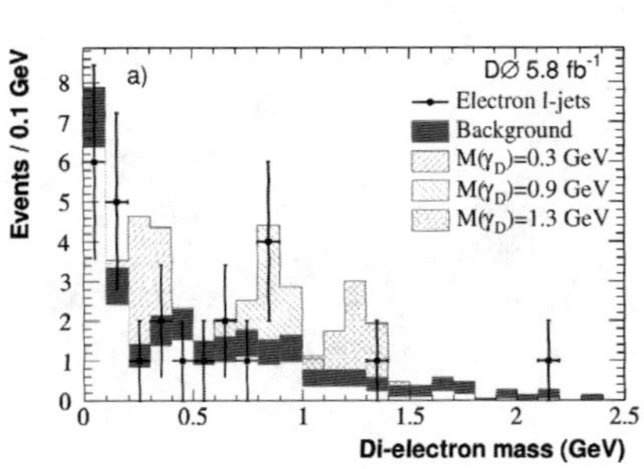

Figure 12. D0 *ee* invariant mass.

Hidden-valley models [15] contain a hidden sector that is very weakly coupled to SM particles. The force carrier in the hidden sector, the dark-photon (γ_D), must have a mass below ~2 GeV to agree with various astrophysical experimental reults, and generally decays into SM charged-fermion pairs. In many models, γ_D has a short lifetime, and does not travel an observable distance before decaying. Assuming SUSY, there will be partners for both the SM and the

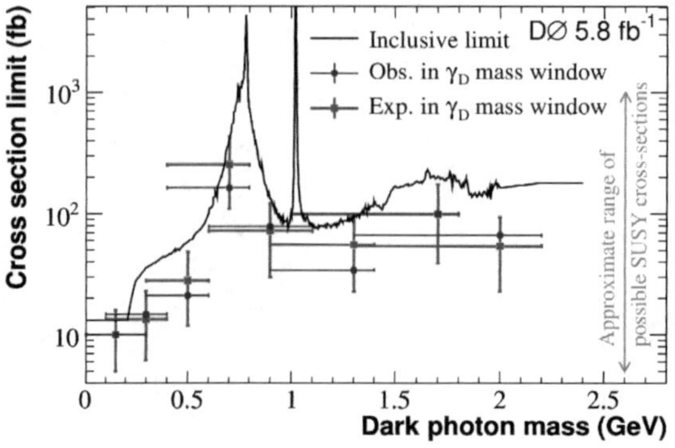

Figure 13. D0 Cross section limit as a function of γ_D mass.

hidden sector particles, and this can lead to decay chains as shown in Figure 11 Since hidden-sector particles have small mass and they are produced with high velocities, their decays through the hidden sector can produce jets of tightly collimated particles from decays of γ_D. If the γ_D is less than twice the pion mass the jets will consist only of charged leptons. Even for larger m_{γ_D}, the lepton content of these jets will be high, and we therefore refer to them as leptonic jets (l-jets).

Using 5.8 fb^{-1} of data in the D0 analysis, we uses l-jets in association with large $\not{E}_T$ to search for γ_D [16]. Figure 12 shows the invariant mass of the electron l-jets. While no evidence of signal was found, we are able to set cross section limits on SUSY events decaying to two l-jets and $\not{E}_T$, see Figure 13.

6. R_p Violating (RPV) Tau Sneutrino ($\tilde{\nu}_\tau$)

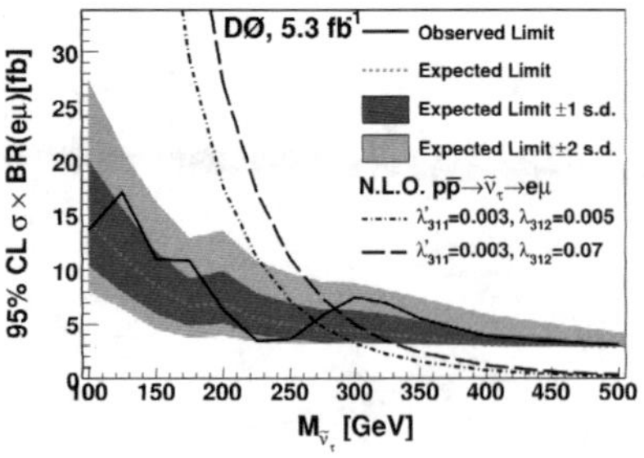

Figure 14. D0 $\tilde{\nu}_\tau$ exclusion.

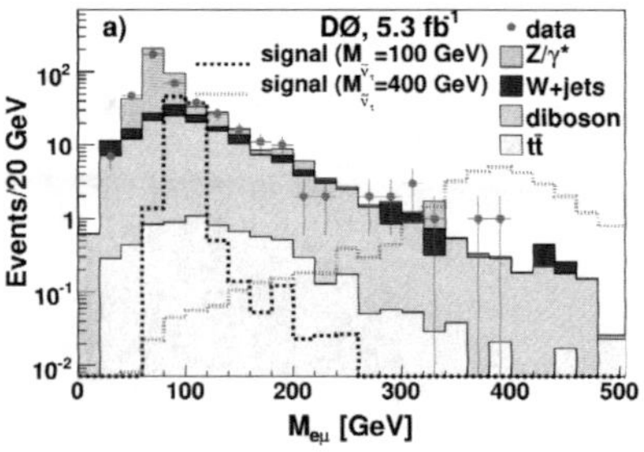

Figure 15. D0 $e\mu$ invariant mass.

In some SUSY models [17], the R-parity which differentiates SM particles from their supersymmetric partners could be violated. Such models can introduce lepton favor violation and provide non-zero neutrino masses and angles consistent with neutrino-oscillation data.

In the D0 analysis [18], events with one e and one μ are selected. Further requirement, such as removing events with jets, are used to suppress SM backgrounds. Figure 15 shows the agreement between data and SM backgrounds. Since no excess is observed, we set a limit on the cross section times branching ratio, see Figure 14.

In the CDF analysis [19] all combinations of different flavor leptons are used ($e\mu$, $e\tau$, $\mu\tau$). Using an integrated luminosity of 1 fb^{-1} we sees agreement with the SM expectation. Assuming a certain set of coupling parameters, we set a lower limit on the mass of the $\tilde{\nu}_\tau$ of 441 GeV.

7. Summary

The Tevatron is collecting high quality data at an unprecedented rate and both CDF and D0 are probing many new models. While there is currently no evidence of new physics, we are able to set limits on several particle masses within a variety of SUSY models.

References

[1] Wess J and Zumino B 1974 *Nucl. Phys.* **B70** 39–50
[2] Cremmer E, Fayet P and Girardello L 1983 *Phys. Lett.* **B122** 41
[3] Abazov V M *et al.* (D0) 2009 *Phys. Lett.* **B680** 34–43 (*Preprint* 0901.0646)
[4] Aaltonen T *et al.* (CDF Collaboration) 2008 *Phys. Rev. Lett.* **101** 251801
[5] Abazov V M *et al.* (D0) 2008 *Phys. Lett.* **B660** 449–457 (*Preprint* 0712.3805)
[6] Aaltonen T *et al.* (CDF Collaboration) 2009 *Phys. Rev. Lett.* **102** 121801
[7] Haber H E and Kane G L 1985 *Phys. Rept.* **117** 75–263
[8] Abazov V M *et al.* (D0) 2010 *Phys. Lett.* **B693** 95–101 (*Preprint* 1005.2222)
[9] Aaltonen T *et al.* (CDF) 2010 *Phys. Rev. Lett.* **105** 081802 (*Preprint* 1005.3600)
[10] Abazov V M *et al.* (D0) 2010 *D0 Public Note* URL http://www-d0.fnal.gov/Run2Physics/WWW/results/final/NP/N10H/N10H.pdf
[11] Aaltonen T *et al.* (CDF Collaboration) 2009 *CDF Public Note* URL http://www-cdf.fnal.gov/physics/exotic/r2a/20090709.stop_charm/public_note/publicnote_stop_jul09_v02.ps
[12] Dine M, Leigh R and Kagan A 1993 *Phys. Rev. D* **48** 4269–4274
[13] Abazov V M *et al.* (D0) 2008 *Phys. Lett.* **B659** 856–863 (*Preprint* 0710.3946)
[14] Aaltonen T *et al.* 2010 *Phys. Rev. Lett.* **104** 011801
[15] Strassler M J and Zurek K M 2007 *Phys. Lett.* **B651** 374–379 (*Preprint* hep-ph/0604261)
[16] Abazov V M *et al.* (D0) 2010 (*Preprint* 1008.3356)
[17] Weinberg S 1982 *Phys. Rev. D* **26** 287–302
[18] Abazov V M *et al.* (D0) 2010 (*Preprint* 1007.4835)
[19] Aaltonen T *et al.* (The CDF) 2010 (*Preprint* 1004.3042)

The latest results from the ATLAS experiment

B. H. Samset, on behalf or the ATLAS collaboration
Department of Physics, University of Oslo, Norway
E-mail: b.h.samset@fys.uio.no

Abstract. The ATLAS detector at the LHC is a general purpose experiment for studying $p+p$ collisions at TeV energies. This talk presents early ATLAS physics measurements, based on O(10 nb^{-1}) of data from the first half year of LHC running. Topics include minimum bias and soft QCD physics, W and Z observation, and various measurements of detector performance.

1. Introduction

This talk presents the latest results[1] of the ATLAS experiment at the Large Hadron Collider (LHC). The LHC started operations in December 2009, and by the time of this conference had collided protons at center–of–mass energies of 0.9, 2.36 and 7.0 TeV.

The ATLAS detector is described in detail elsewhere [1], as are its expected capabilities for physics discoveries [2]. In brief, ATLAS is a general purpose detector for LHC physics, of barrel plus endcap design. From the collision point and outwards it consists of an inner detector system for precision tracking, followed by electromagnetic and hadronic calorimeters, and finally a muon spectrometer. Barrel (inner) and toroidal (outer) magnetic fields provide bending for particle momentum determination.

ATLAS has been online and operational for all three collision energies delivered by the LHC. As of the PASCOS conference, the LHC had delivered an integrated luminosity at 7.0 TeV of 287 nb^{-1}, of which ATLAS had recorded 270 nb^{-1}. The results to be presented here are based on O(10 nb^{-1}) of data at 7.0 TeV. While a dataset of this size is not enough for significant discoveries, it is sufficient to give us a detailed understanding of the performance of the ATLAS detector and rediscover the Standard Model of particle physics in a new energy regime

Topics to be covered in this proceeding include the early performance of various parts of the detector, measurements of heavy resonances like J/ψ, $W^{\pm}$ and Z^0, and a detailed study of charged particle production in minimum bias collisions at 7.0 TeV.

2. ATLAS detector performance

To be able to measure cross sections, knowledge of the absolute luminosity delivered by the LHC at the ATLAS interaction point is required. ATLAS measures luminosity using a combination of independent detectors and LHC van der Meer scans, and estimate an instantaneous luminosity uncertainly of 11%. This number is applied for the analyses presented in sections 3 and 4. Details of the ATLAS luminosity measurements can be found in [3].

Figures 1 and 2 show examples of the early performance of the ATLAS inner detector for tracking and reconstruction of resonances. Figure 1 shows the invariant mass peak of oppositely

[1] Status as of the PASCOS 2010 conference, 19th July 2010.

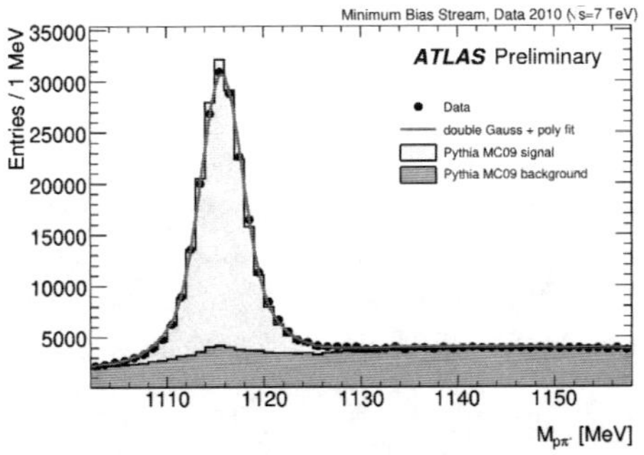

Figure 1. Comparison of measured and predicted Lambda mass spectrum, from the ATLAS barrel detector region (both tracks satisfy $|\eta| < 1.2$). The black circles are data, while the histograms show Monte Carlo simulation (normalised to data). The red line is a line-shape function fitted to data. See [4] for details.

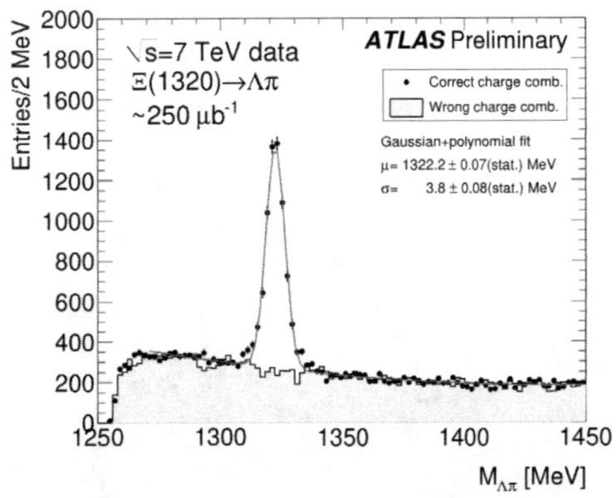

Figure 2. The invariant masses of the reconstructed Ξ cascade decay candidates. The distributions for correct charge combinations and wrong charge combinations are overlaid. The fit (red curve) is described in [5].

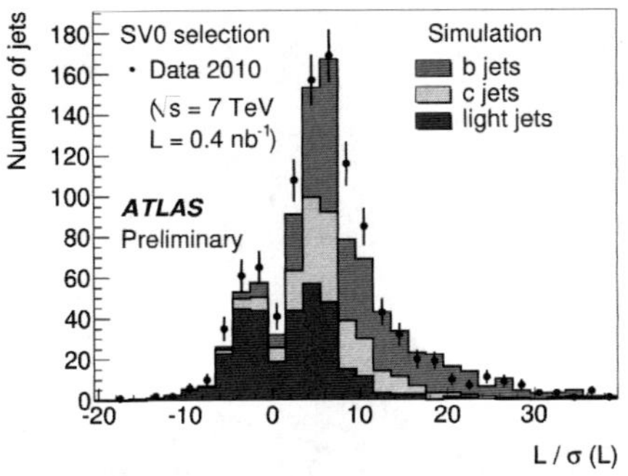

Figure 3. ATLAS b–tagging performance. The three-dimensional decay length significance, signed with respect to the calorimeter jet axis, for all secondary vertices reconstructed in data events (markers). The expectation from simulated events (histogram) for non-diffractive minimum bias events, normalized to the number of jets in the data, is superimposed. See [7] for details.

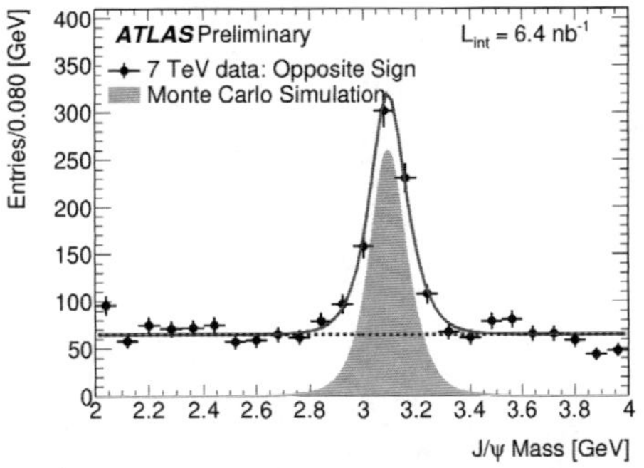

Figure 4. Invariant mass distribution of reconstructed J/ψ candidates. The points with error bars are data. The solid line is the result of maximum likelihood unbinned fit to all di-muon pairs in the mass window 2-4 GeV$/c^2$ and the dashed line is the result for the background of the same fit. The result of the fit to the prompt J/ψ is represented by solid area. From [8].

charged track pairs subject to requirements on the position of a secondary vertex position and pointing angle, around the known mass of the Λ^0 baryon. Both data (black points), MonteCarlo (histograms) and a fit to the signal plus background(red line) are shown. Figure 2 shows a similar analysis for the decay of Ξ(1320), where the previously reconstructed Λ^0s are also used. In both cases, mass values and resonance widths are consistent with the values quoted by the Particle Data Group [6]. See Refs [4] and [5] for details.

The performance of both the ATLAS inner detector tracking and the calorimeter based jet reconstruction is indicated by figure 3, which shows b–tagging results using the decay length

Journal of Physics: Conference Series **259** (2010) 012021 doi:10.1088/1742-6596/259/1/012021

significance method. See Ref. [7] for details. The figure shows a region $L/\sigma(L) \geq 10$ where jets from b–jets dominate, providing a clear handle on b quark production. It can also be seen that to reproduce the observed data (black circles), both b–jets, c–jets and light jet flavour production must be taken into account.

Figure 4 indicates the performance of the ATLAS muon system, here used to reconstruct the $J/\psi \to \mu^+\mu^-$ decay. Oppositely charged pairs of well reconstructed muons are selected and their invariant mass fitted in the window of 2-4 GeV/c^2. The reconstructed mass value for J/ψ is close to the Particle Data Group [6] value and the background level is as expected from MonteCarlo simulations. See [8] for details.

3. Observation of W and Z candidates

Production of $W^{\pm}$ and Z^0 bosons is of great interest at the LHC, both for precision studies of the Standard Model and for determination of backgrounds for searches for new physics. Within the data sample analysed by ATLAS by the time of this conference, 23.1 ± 1.2 (stat) ± 1.7 (syst) ± 4.6 (lumi) $W \to e\nu$ decays and 28.7 ± 0.5 (stat) ± 3.9 (syst) ± 5.7 (lumi) $W \to \mu\nu$ decays are expected. Only 1.6 ± 0.1 (syst) ± 0.3 (lumi) $Z \to ee$ and 3.2 ± 0.7 (syst) ± 0.6 (lumi) $Z \to \mu\mu$ events are predicted by the Standard Model.

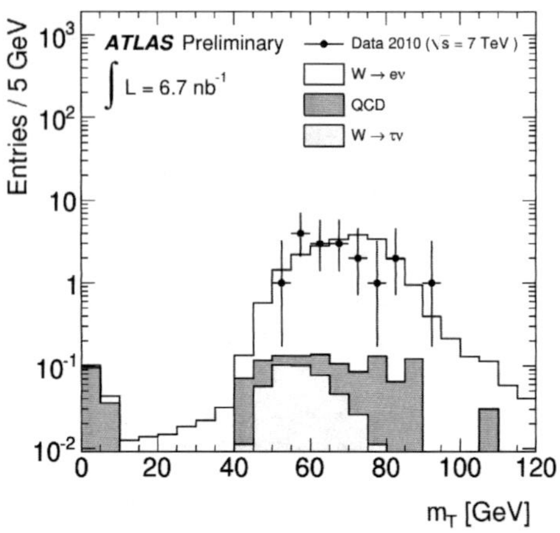

Figure 5. m_T of a selected electron–E_T^{miss} system after requiring $E_T^{miss} > 25$ GeV. A 6.7 nb^{-1} data sample (black circles) is imposed on a MonteCarlo simulation of signal and expected backgrounds (histograms). From [9].

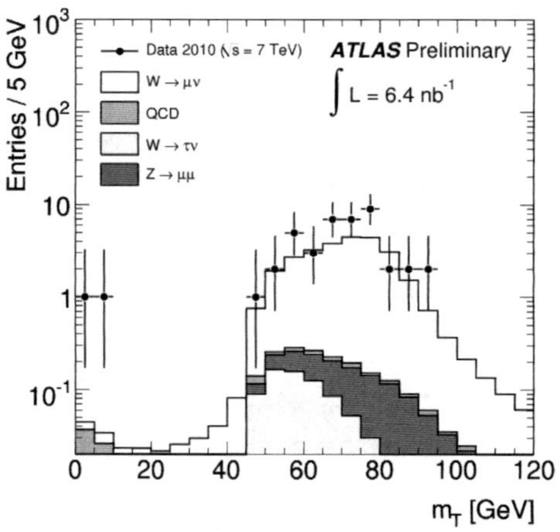

Figure 6. m_T of a selected muon–E_T^{miss} system after requiring $E_T^{miss} > 25$ GeV. A 6.4 nb^{-1} data sample (black circles) is imposed on a MonteCarlo simulation of signal and expected backgrounds (histograms). From [9].

Figures 5 and 6 show the final transverse mass $m_T = \sqrt{(2p_T^l p_T^\nu(1 - cos(\phi_l - \phi_\nu))}$ distributions for the electron and muon channels respectively. Here p_T^l and ϕ^l are the transverse momentum and angle in the transverse plane of a selected lepton, while the values p_T^ν and ϕ^ν are taken to be the magnitude and direction of the missing transverse energy (E_T^{miss}) vector as measured by ATLAS.

Well reconstructed and isolated leptons with high transverse momenta are selected, and combined with a requirement of $E_T^{miss} > 25$ GeV. For the electron events, an inner detector track pointing to an energy deposit in the electromagnetic calorimeter is required, as well as $E_T^{el} > 20$ GeV. For the muons a match between an inner detector track and a muon system track is required, as well as $p_T^\mu > 20$ GeV.

Journal of Physics: Conference Series **259** (2010) 012021 doi:10.1088/1742-6596/259/1/012021

The data points (black crosses) are well described by the expected Monte Carlo based sum of signal events and backgrounds from QCD and W/Z decay events in other channels. The total number of W candidates observed is 17 in the electron channel and 40 in the muon channel, consistent with the Standard Model predictions listed above. See [9] for details.

Z bosons decaying into e^+e^- and $\mu^+\mu^-$ are also searched for, with similar analysis cuts on the e and μ candidates as described above. A single $Z \to ee$ candidate is left after all cuts, with an invariant mass of $m_{ee} = 91.4$ GeV/c^2. Similarly two events survive the cuts on the $Z \to \mu\mu$ channel, with invariant masses of $m_{\mu\mu} = 81.6$ GeV/c^2 and 80.2 GeV/c^2 respectively. In both cases this is consistent with the Standard Model predictions. See [9] for details, and the ATLAS public webpage [14] for example event displays of these and other events.

4. Minimum–bias charged particle production

An early measurable at the LHC is the charged particle density in minimum bias collisions, as function of transverse momentum of the leading track, pseudorapidity and other variables. ATLAS reported on this for 900 GeV data in [10], and has presently extended this analysis to 7 TeV. A similar analysis for the middle energy of 2.36 TeV, which as also been delivered by the LHC, is underway.

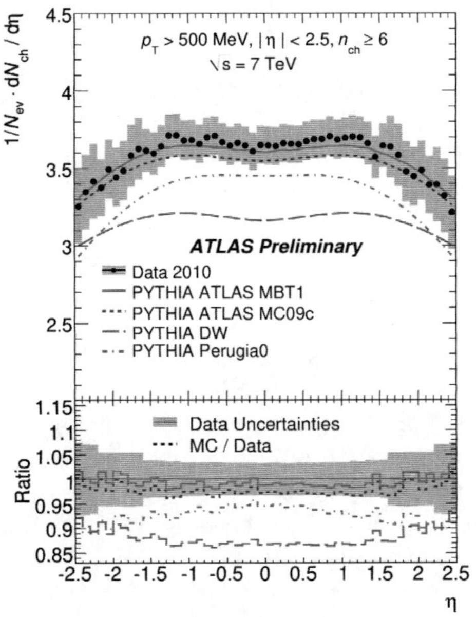

Figure 7. Charged particle density as function of pseudorapidity in minimum bias collisions at 7 TeV with $n_{ch} \geq 6$, compared to the new tune AMBT1 (red solid line), the MC09c tune (blue dashed line), Perugia0 tune (magenta dash-dotted line) and DW tune (brown long dashed line). From [11].

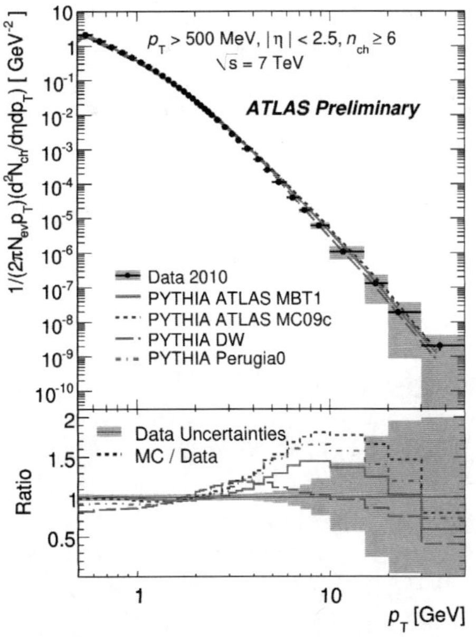

Figure 8. Charged particle density as function of transverse momentum in minimum bias collisions at 7 TeV with $n_{ch} \geq 6$, compared to the new tune AMBT1 (red solid line), the MC09c tune (blue dashed line), Perugia0 tune (magenta dash-dotted line) and DW tune (brown long dashed line). From [11].

Figures 7 and 8 shows preliminary ATLAS results on charged particle production at 7 TeV as function of pseudorapidity and transverse momentum of the leading track respectively. Soft collisions, or interactions yielding a low charged particle density ($n_{charged}$), are less well understood theoretically than hard interactions, so to minimize the model dependence a reduced phase space has been studied. The final analysis cuts are $p_T > 500$ MeV and $n_{charged} \geq 6$. For details of the analysis, see [11].

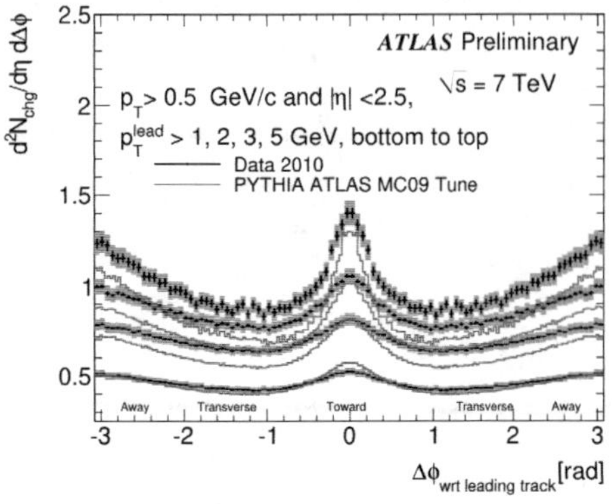

Figure 9. ϕ distribution of track densities for $p_T > 0.5$ GeV and $|\eta| < 2.5$ with respect to the leading track ϕ, compared to ATLAS Pythia MC09 tune predictions. The distributions obtained by restricting the minimum leading track p_T to different values are overlaid. The plots were symmetrized by reflecting them about $\phi = 0$. From [13].

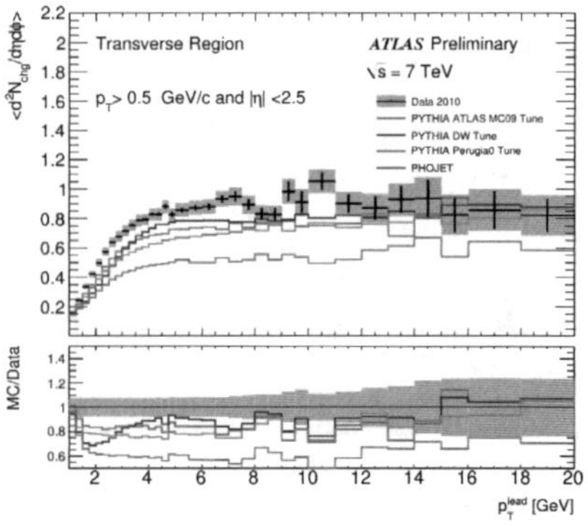

Figure 10. Density of charged particles with $p_T > 0.5$ GeV and $|\eta| < 2.5$ as a function of the leading track p_T, for the transverse region defined by the leading track and compared with Pythia ATLAS MC09, DW and Perugia0 tunes and Phojet predictions. Phojet predictions are only shown in the region where sufficient statistics are available. See [13] for details.

The two main features when comparing the results to model calculations based on PYTHIA [12] is that all the most common model tunes, including Perugia0 and DW, underestimate the total multiplicity by 10-15% and also predict too hard p_T spectra. With the aim of providing an improved PYTHIA tuning for LHC data, ATLAS has performed a study of the underlying event (UE) in soft collisions. A thorough knowledge of the UE is required e.g. for doing lepton isolation in high–p_T analyses.

The present analysis selects tracks in regions towards, away from, and transverse to the leading track, and attempts to isolate the effects of the underlying event by studying these regions separately. Figure 9 shows the azimuthal angle between a selected track and the leading track for various cuts on the leading track p_T, compared to the previous ATLAS MonteCarlo-based PYTHIA tune MC09. The activity in the towards and away regions is readily seen, while there is still a plateau of activity observed in the transverse region. Figure 10 shows, as an example, the $\langle d^2N_{ch}/dp_T d\phi \rangle$ as function of leading track p_T for the transverse region, compared to model predictions. See [13] for details.

Based on the above figures, and similar ones for other measurables such as the mean transverse momentum as function of charged particle density, ATLAS has performed a re–tuning of the PYTHIA parameters for multi–parton interactions and color reconnections,and released an LHC tune called AMBT1. It uses as input ATLAS UE data at center–of–mass energies 0.9 and 7 TeV, as well as data from CDF and D0 at the TeVatron. The resulting MonteCarlo fits to data, which can be seen in figures 7 and 8, provide a better description of both the overall multiplicities and spectral shapes, though it is clear that the model description is still far from perfect even for these overall variables.

5. Conclusions

After the LHC produced its first record energy collisions in 2010, ATLAS has promptly delivered high quality results covering a wide spectrum of physics processes. Analyses have been performed at all levels, from detector commissioning and calibration, through reconstruction of stable objects and resonances, to early searches for new physics. This talk has presented results from a small number of these, while yet more will be shown at various conferences during Summer 2010.

At the same time the LHC continues to deliver 7 TeV beams with ever increasing luminosities. This talk has demonstrated that ATLAS can efficiently reconstruct both high and low p_T objects, and perform accurate physics analyses based on the ever increasing integrated luminosity.

For further and more recent results, please see the ATLAS public results webpage [14].

References

[1] G. Aad *et al.* [The ATLAS Collaboration], JINST **3** (2008) S08003.
[2] G. Aad *et al.* [The ATLAS Collaboration], arXiv:0901.0512 [hep-ex].
[3] The ATLAS Collaboration, ATLAS-CONF-2010-060.
[4] The ATLAS Collaboration, ATLAS-CONF-2010-033.
[5] The ATLAS Collaboration, ATLAS-CONF-2010-032.
[6] K. Nakamura et al. (Particle Data Group), J. Phys. G 37, 075021 (2010)
[7] The ATLAS Collaboration, ATLAS-CONF-2010-042.
[8] The ATLAS Collaboration, ATLAS-CONF-2010-045.
[9] The ATLAS Collaboration, ATLAS-CONF-2010-044.
[10] G. Aad *et al.* [ATLAS Collaboration], Phys. Lett. B **688** (2010) 21 [arXiv:1003.3124 [hep-ex]].
[11] The ATLAS Collaboration, ATLAS-CONF-2010-031.
[12] T. Sjostrand, S. Mrenna and P. Z. Skands, JHEP **0605** (2006) 026 [arXiv:hep-ph/0603175].
[13] The ATLAS Collaboration, ATLAS-CONF-2010-029.
[14] The ATLAS public results webpage, https://twiki.cern.ch/twiki/bin/view/Atlas/AtlasResults

Journal of Physics: Conference Series **259** (2010) 012022 doi:10.1088/1742-6596/259/1/012022

LHCb: first results and prospects for the 2010-11 run

J P Palacios on behalf of the LHCb Collaboration

Universitaet Zuerich, Physik-Institut, Winterthurerstrasse 190, 8057 Zuerich, Switzerland

E-mail: juan.palacios@cern.ch

Abstract. The LHCb physics programme is now underway. The data collected in 2009 and in the first period of the 2010 run have allowed for a range of interesting minimum bias physics measurements, including those of prompt K_S^0, Lambda, anti-Lambda, proton and anti-proton production. Heavy flavour signals have also been collected and detailed studies of J/ψ production, both prompt and from B-decays, are underway. The status of these measurements will be reviewed, and the prospects will be given for the core LHCb CP-violation and rare decay programme in the remainder of the 2010-11 run.

1. Introduction

Beyond the standard model physics (BSM) predicts the existence of new particles at the TeV scale. These can be detected indirectly via appearance as virtual particles in loop and penguin diagrams contributing to electro-weak decays of heavy hadrons. This approach allows for the detection of such particles at energy scales lower than the kinematic limit for their direct production. It also provides knowledge of the phases in their couplings, giving information on the flavour structure of BSM physics. The LHCb detector [1] at the LHC collider is designed to search for new physics by precisely measuring elements of the CKM matrix as well as investigating rare flavour changing neutral current (FCNC) processes. In both approaches, the presence of new physics would manifest itself in deviations from the behaviour predicted by the standard model of particle physics (SM). This manuscript concentrates on the most important LHCb measurements for the first part of the 2010 run, based on 14 nb^{-1} of integrated luminosity.

2. Key programme for 2010-2011 run

The core program of LHCb has been presented in detail in [2]. It consists of precise measurements of CKM parameters in CP violating decays, searches for FCNC processes that are extremely rare in the SM, measurements of production asymmetries in rare FCNC decays, and the measurement of anomalous polarisation in radiative B meson decays. With the 1 $\mathrm{fb}^{-1}$of data predicted for the 2010-2011 run, it will be possible to make significant impact in all but the latter of these measurements. Furthermore, the early LHC running has provided an excellent opportunity to study CP violation effects in charm decays. Hence charm physics has become one of the key measurements for the 2010-2011 run.

2.1. Early data particle production

The emphasis during early running has been to understand the detector and trigger. This involved the measurement of the produiction rates well known hadrons, such as the K_s^0, Λ, D

and π^0 mesons to calibrate the momentum, energy scales and particle identification. Studies of di-μ final states, particularly the production of J/Ψ mesons [3], set the stage for studies of rare decays involving two muons in the final state. The production cross-section of K_s^0 has been measured and published in [4]. The measurement of $\bar{\Lambda}/\Lambda$ production ratios at the high η range of LHCb is important for the tuning of particle generators [5]. Its study is well underway. Figure 1 shows the $\bar{\Lambda}/\bar{\Lambda}$ production ratio at 0.9 and 7 TeV centre of mass energy. The ratio can be seen to be consistently lower than expected for $\sqrt{s} = 0.9$ TeV.

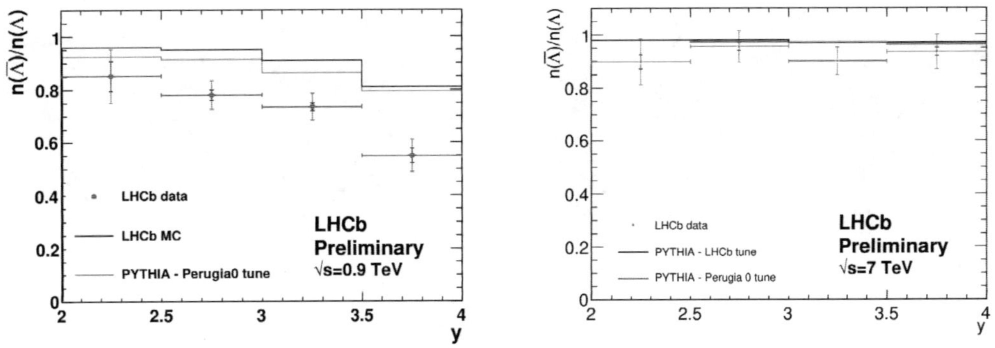

Figure 1. Λ and $\bar{\Lambda}$ production ratio as a function of rapidity y at 0.9 (left) and 7 TeV (right) centre of mass energy

2.2. Search for rare decays $B_s^0 \to \mu^+\mu^-$, $D^0 \to \mu^+\mu^-$

The decay channel $B_s^0 \to \mu^+\mu^-$ has a branching ratio of $\mathcal{B} = (3.35 \pm 0.32) \times 10^{-9}$ in the SM [6]. This rare decay mode can be greatly enhanced by physics beyond the standard model. For example, the minimal sypersymmetric standard model (MSSM) predicts an enhancement as the sixth power if tanβ. The current experimental limit from CDF, based on 3.7 fb^{-1} of data, is $\mathcal{B} < 3.6 \times 10^{-8}$ at 90% CL [7].

LHCb will select events such that the branching ratio can be estimated from data, without reliance on simulations. Three uncorrelated observables - the B_s^0 invariant mass, muon likelihood, and geometrical likelihood - are used for selection. These can be calibrated with different control samples. The two-body invariant mass has been calibrated in early data using control channels $K_S^0 \to \pi^+\pi^-$ and $J/\psi \to \mu^+\mu^-$ decays. It is planned to calibrate the mass resolution at the mass scale of the B_s^0 using the decay channel $B_s^0 \to K^+K^-$. Simulation studies show an expected resolution of around 20 MeV.

The branching ratio will be determined using control channels $B^+ \to J/\psi K^+$ and $B_d^0 \to K\pi$. This requires an understanding of the ratio of B_s^0 production over that of light B mesons, B_d^0 and B_u^+, which can be obtained through a measurement of the relative rates of $B_s^0 \to D_s^-\pi^+$ and $B_d^0 \to D^-K^+$ [8].

LHCb expects a 90% confidence limit of 16×10^{-9} with 0.2 fb^{-1} of data. This compares favourably with the Tevatron expected limit for 8 fb^{-1}, as can be seen in Figure 2. With 1.0 fb^{-1} the LHCb exclusion limit will be down to $\mathcal{B} = 7 \times 10^{-9}$. Alternatively, the signal observation limit is expected to be for $\mathcal{B}$ 3.5 times the SM prediction.

The decay channel $D^0 \to \mu^+\mu^-$ presents similar characteristics to its B_s^0 counterpart, with a SM branching fraction $\mathcal{B} \approx 3 \times 10^{-13}$. The Belle collaboration has measured $\mathcal{B} < 1.4 \times 10^{-7}$ at 90% CL[9]. LHCb can improve on that limit, expecting to set an upper limit fo $\mathcal{B} < 4 \times 10^{-8}$ with only 0.1 fb^{-1}.

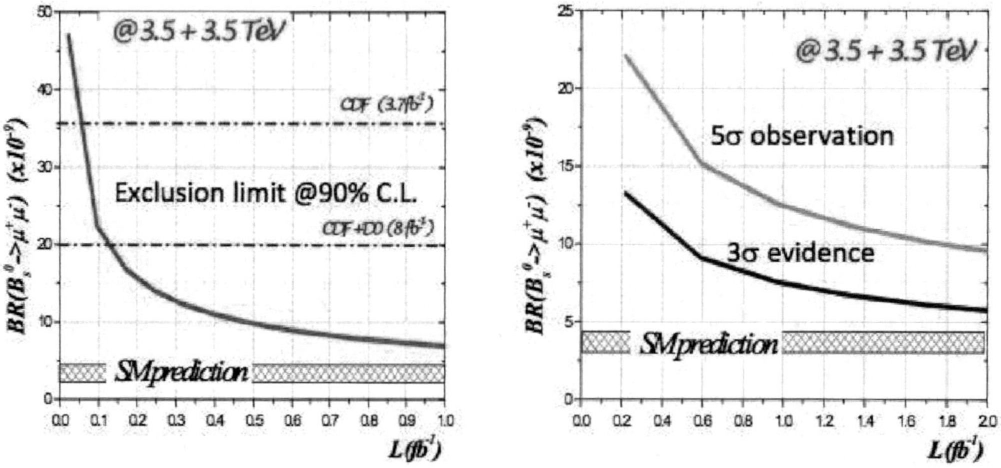

Figure 2. Left: the expected 90% $B_s^0 \to \mu^+\mu^-$ branching fraction exclusion limit at LHCb as a function of integrated luminosity. The current and expected limits from the Tevatron are shown. Right: the expected 3σ observation and 5σ discovery branching fraction as a function of integrated luminosity. The SM prediction is shown.

2.3. Angular asymmetries in $B_d^0 \to K^*\mu^+\mu^-$

The $B_d^0 \to K^*\mu^+\mu^-$ decay has a branching fraction of 1.0×10^{-6} in the SM. The angular distributions of the final state particles contain information on the couplings of virtual particles in loops contributing to the decay, and are well constrained within the SM. Therefore the angular distributions provide a probe for new physics.

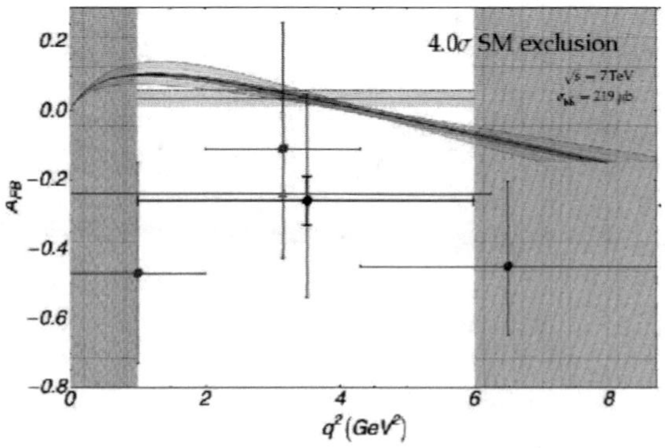

Figure 3. A_{FB} in $B_d^0 \to K^*\mu^+\mu^-$ as a function of q^2. Standard model prediction (black line with red and green error bands) and average (purple line) for $A_{FB}(q^2)$, Belle data (blue points), BaBar data (red point), and predicted LHCb result (black point) assuming Belle measured value and 1.0 fb^{-1} of integrated luminosity.

The zero crossing point of the forward-backward asymmetry of the $\mu^+\mu^-$ as a function of the q^2 of the muons, $A_{FB}(q^2)$, has minimal hadronic uncertainties and high sensitivity to new couplings, and therefore is an optimal experimental observable. The standard model prediction, together with the current values from Belle and BaBar can be seen in Figure 3. These show an intriguing hint of a divergence from the SM. LHCb expects to find 1400 events in the first 1.0 fb^{-1} of data, leading to a sensitivity of $\sigma(A_{FB}) = 0.07$.

2.4. B_s mixing phase ϕ_s

The decay $B_s^0 \to J/\Psi\phi$ is sensitive to the mixing phase ϕ_s through the interference between mixing and decay amplitudes. The contribution from penguin diagrams to ϕ_s is small, leading to a small theoretical error on the standard model prediction of $\phi_s = (-36.0^{+1.6}_{-2.0})$mrad [10]. New particles present in the box diagrams that give rise to B_s^0 mixing can modify this phase, making it an ideal probe for new physics. The extraction of ϕ_s from $B_s^0 \to J/\psi\phi$ decays requires measuring the time dependent B_s^0 and $\bar{B}_s^0$ decay rates, thus requiring the tagging of the initial flavour of the B_s^0. Simulations show that LHCb has a mis-tag rate of 33% and an effective tagging efficiency of 6% in this channel. The fast oscillations of the B_s^0 require a very high proper time resolution. LHCb expects 40 fs from simulation. This resolution can be measuredfrom the negative edge of the proper time distribution of easily identifiable prompt decays, such as $J/\psi \to \mu^+\mu^-$. The J/ψ and ϕ are both vector mesons. This means that they can be either in a CP-odd or CP-even final state, each of which has a different angular distribution for the final state particles. The extraction of the ϕ_s phase requires that these components be extracted statistically. The procedure to do so shall be cross checked against $B_d^0 \to J/\psi K^*$ decays, which share the same helicity structure and have been measured accurately in the B factories. Special attention has to be paid to angular biases introduced by trigger and selection cuts. The analysis can be complemented with B_s^0 decays into final states such as $J/\psi f0$ and $J/\psi\eta'$, which are CP eigenstates. These have lower yields but do not require an angular analysis. LHCb expects some 30×10^3 $B_s^0 \to J/\psi(\mu^+\mu^-)\phi(K^+K^-)$ events in 1.0fb^{-}1, resulting in a sensitivity of $\sigma(\phi_s)$ = 0.07.

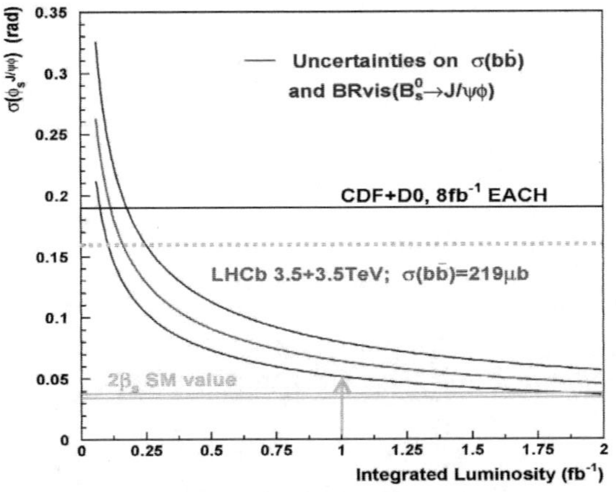

Figure 4. Sensitivity for ϕ_s from $B_s^0 \to J/\psi\phi$ as a function of integrated luminosity. The standard model prediction and Tevatron expected sensitivity for 8 fb^{-1} are shown. The expected LHCb sensitivity for the 2010-2011 run, corresponding to 1.0 fb^{-1}, is $\sigma(\phi_s) \approx$ 0.07

2.5. Charm physics

LHCb expects to collect of the order of 10^8 D mesons with 1 fb^{-1}, allowing for precise measurements of CP violation in $D^0\bar{D}^0$ mixing and direct CP violation in decays of charged D mesons, both of which are small in the Standard Model. With only 0.1 fb^{-1}, 17 million tagged $D^0 \to K\pi$ and 1.3 million tagged $D^0 \to K^+K^-$ are expected, allowing to surpass the sensitivity on CP violation in D mixing currently available from B factories. Several million single Cabibbo suppressed $D^+ \to K^+K^-\pi^+$ decays are expected, allowing for sensitive searches for direct CP violation. In both aspects, LHCb expects to surpass the results of the B factories with the data obtained in the 2010-2011 run.

The production of D^0 mesons in channels with a well known branching fraction can be used to measure the $b\bar{b}$ production cross section in pp collisions. In one such study, $b \to D^0X\mu^-\bar{\nu}$ decays have been identified and resolved from prompt D^0 by using the impact parameter of the

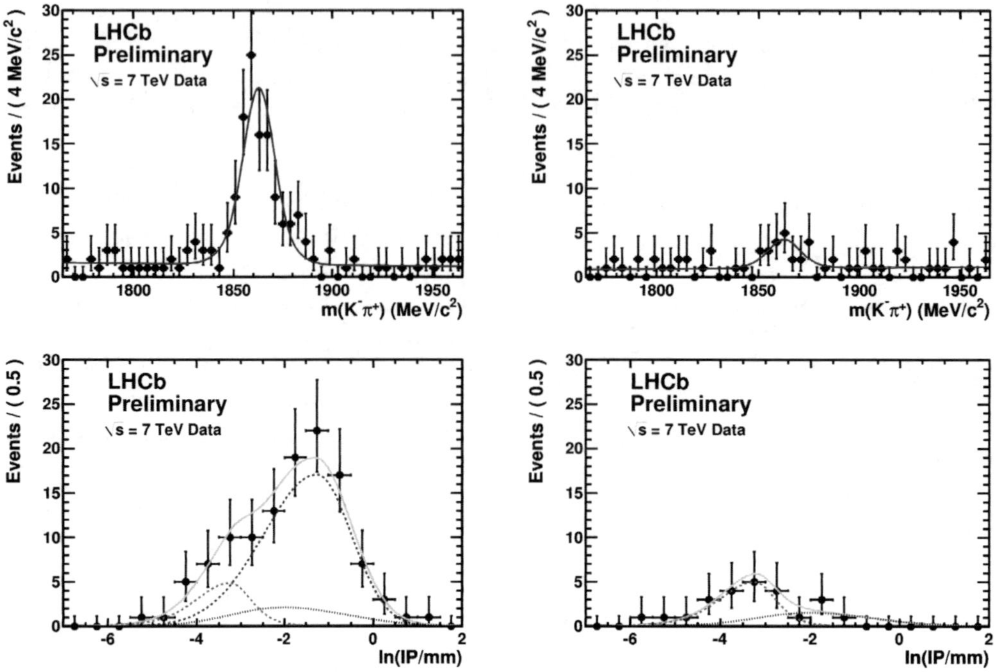

Figure 5. $D^0 \to K^-\pi^+$ invariant mass (top) and logarithm of the D^0 impact parameter with respect to the primary vertex in millimeters (bottom) for $D^0\mu$ candidates with 'right sign' (left) and 'wrong sign' (right) correlation, in 3 nb^{-1} of data at $\sqrt{s} = 7$ TeV. Fit results are superimposed as curves. In the bottom plots, the black dotted curve represents the non-D^0 background estimated from the mass sidebands, and the blue and red dotted curves represent the D^0 signal from b-hadron decays and from prompt production.

decay products with respect to the primary vertex. The charge of the μ and the decay products of the D^0 has been used to determine the combinatorial background component. Both aspects are illustrated in Figure 5 This analysis has been extended and published in [11].

3. Conclusions

The LHCb detector has successfully initiated its heavy flavour programme. The production and decays of well known particles has been used to understand the detector response and key elements of the core physics programme. We expect to improve on current experimental constraints on new physics in the flavour sector with 0.2 fb^{-1}. With the 1.0 fb^{-1} of integrated luminosity expected for the 2010-2011 run LHCb will be in a good position to find physics beyond the SM.

4. Bibliography

[1] Alves A *et al.* [LHCb Collaboration] 2008 JINST 3 S08005.
[2] Adeva B *et al.* [LHCb Collaboration] arXiv:0912.4179v2 [hep-ex]
[3] LHCb Collaboration 2010 "Measurement of the J/ψ production cross section at $\sqrt{s} = 7$ TeV in LHCb" LHCb-CONF-2010-010
[4] Aaij R *et al.* [LHCb Collaboration] 2010 "Prompt K^0_S production in *pp* collisions at $\sqrt{s} = 7$ TeV" *Phys. Lett.* B **693** 69-80
[5] LHCb Collaboration 2010 "Measurement of prompt $\bar{\Lambda}/\Lambda$ and $\bar{\Lambda}/K^0_s$ production ratios in inelastic non-diffractive *pp* collisions at $\sqrt{s} = 0.9$ anf 7 TeV" LHCb-CONF-2010-011

[6] Blanke B, Buras A, Guardagnoli D and Tarantino C 2006 JHEP0610:003
[7] CDF Collaboration, CDF Public Note 9892.
[8] Fleisher R, Serra N and Tuning N 2010 "New strategy for B_s branching ratio measurements and the search for new physics in $B_s^0 \to \mu^+\mu^-$" *Phy. Rev.* D **82** 034038
[9] Petric M for the Belle Collaboration, arXiv:1005.5445 [hep-ex]
[10] Charles J *et al.* [CKMfitter Group] 2005 *Eur. Phys. J.* C **41**, 1-131 (updated results available at: http://ckmfitter.in2p3.fr)
[11] Aaij R *et al.* arXiv:1009.2731v2 [hep-ex]

Searches for New Physics at the NA62 experiment

Spasimir Balev

European Organization for Nuclear Research, CERN CH-1211, Geneve 23, Switzerland

E-mail: `Spasimir.Balev@cern.ch`

Abstract. The experiment NA62 at CERN collected a large sample of $K^{\pm}$ leptonic decays in order to perform precise test of lepton universality, by measuring the helicity suppressed ratio $R_K = \Gamma(K^{\pm} \to e^{\pm}\nu)/\Gamma(K^{\pm} \to \mu^{\pm}\nu)$. The final result of the analysis based on 59,963 $K^{+} \to e^{+}\nu$ candidates ($\sim$ 40% of the total data set) is $R_K = (2.486 \pm 0.013) \times 10^{-5}$ and consistent with the predictions of the Standard Model. The aim for the future program of the experiment is to measure the branching ratio of the ultra-rare decay $K^{+} \to \pi^{+}\nu\nu$ with 10% precision.

1. The present of NA62: Search for lepton flavour violation

1.1. Introduction

The ratio of kaon leptonic decay rates $R_K = \Gamma(K^{\pm}_{e2})/\Gamma(K^{\pm}_{\mu 2})$ is known in the Standard Model (SM) with excellent precision due to cancellation of the hadronic effects: $R_K^{SM} = (m_e/m_\mu)^2 \left(\frac{m_K^2-m_e^2}{m_K^2-m_\mu^2}\right)^2 (1+\delta R_{QED}) = (2.477 \pm 0.001) \times 10^{-5}$ [1], where $\delta_{QED} = (-3.78 \pm 0.04)\%$ is a correction due to the inner bremsstrahlung (IB) $K_{l2\gamma}$ process which is included by definition into R_K [1]. Being helicity suppressed due to $V-A$ structure of the charged weak current, R_K is sensitive to non-SM effects. In particular MSSM allows non-vanishing $e-\tau$ mixing, mediated by H^{+}, which can lead to few percent enhancement of R_K [2].

The present world average of $R_K = (2.493 \pm 0.031) \times 10^{-5}$ [3] is dominated by the recent KLOE final result [4]. The NA62 experiment collected data during 2007 and 2008 aiming to reach accuracy of $\sim$ 0.4%. The final result on partial data set is presented here.

1.2. Experimental setup

The NA62 experiment used the NA48/2 beam line [5] and detector setup [6] with optimisations for K_{e2} data collection. The beam line of NA48/2 experiment is designed to deliver simultaneously K^{+} and K^{-}, produced on a beryllium target from SPS primary protons. The beams of (74±2) GeV/c momentum are selected by a system of magnetic elements. After cleaning, shaping and focusing, the beams enter 114 m long vacuum decay volume. The momenta of the charged decay products are measured by magnetic spectrometer consisting of four drift chambers (DCHs) and a dipole magnet. The resolution of the spectrometer is $\sigma(p)/p = 0.5\% \oplus 0.009\%p$ (p in GeV/c). A scintillator hodoscope (HOD), located after the spectrometer, sends fast trigger signals from charged particles and measures their time with a resolution of 150 ps. The electromagnetic energy of particles is measured by a liquid krypton

[1] Unlike the structure dependent (SD) $K_{l2\gamma}$.

calorimeter (LKr), a quasi-homogeneous ionization chamber with an active volume of 10 m^3, 27 X_0 deep and segmented transversely into 13,248 cells (2×2 cm^2 each). The energy resolution is $\sigma(E)/E = 0.032/\sqrt{E} \oplus 0.09/E \oplus 0.0042$ and the spatial resolution in the transverse coordinates x and y for a single electromagnetic shower is $\sigma_x = \sigma_y = 0.42/\sqrt{E} \oplus 0.06$ cm (E in GeV). A beam pipe traversing the centers of the detectors allows undecayed beam particles and muons from decays of beam pions to continue their path in vacuum.

The K_{e2} decays are selected by trigger requiring coincidence of hits in the HOD planes (Q_1 signal) together with sufficient energy deposit (> 10 GeV) in LKr. The $K_{\mu 2}$ events are selected by Q_1 signal, downscaled by a factor 150. Both triggers also use loose requirement on DCH multiplicity.

1.3. Analysis strategy

Due to the topological similarity of K_{e2} and $K_{\mu 2}$ decays a large part of the selection conditions are common for both decays. We require the presence of single reconstructed charged track with momentum 13 GeV/$c < p <$ 65 GeV/c (the lower limit is due to the 10 GeV LKr energy deposit requirement in K_{e2} trigger). The track extrapolated to DCH, LKr and HOD should be within their geometrical acceptances. The CDA between the charged track and the nominal kaon beam axis should be less than 1.5 cm. The event is rejected if a cluster in the LKr with energy larger than 2 GeV and not associated with track is present, in order to suppress the background from other kaon decays.

A kinematical separation between K_{e2} and $K_{\mu 2}$ for low track momenta is possible, based on the reconstructed missing mass, assuming the track to be an electron or a muon: $M^2_{\text{miss}}(l) = (P_K - P_l)^2$, where P_l ($l = e, \mu$) is the four-momentum of the lepton. Since the kaon four-momentum P_K is not measured directly in every event, its average is monitored in each SPS spill with fully reconstructed $K^\pm \to 3\pi^\pm$ decays. A cut $M_1^2 < M^2_{\text{miss}}(e) < M_2^2$ is applied to select K_{e2} candidates, and $M_1^2 < M^2_{\text{miss}}(\mu) < M_2^2$ for $K_{\mu 2}$ ones, where M_1^2 and M_2^2 vary from 0.010 to 0.016 $(\text{GeV}/c^2)^2$ for different track momenta, depending on M_{miss} resolution. Particle identification is based on the ratio E/p of track energy deposit in the LKr to its momentum measured by the spectrometer. Particles with $0.95 < E/p < 1.1$ for $p > 25$ GeV/c and $0.90 < E/p < 1.1$ otherwise[2], are identified as electrons, while particles with $E/p < 0.85$ – as muons.

The analysis is based on counting the number of reconstructed K_{e2} and $K_{\mu 2}$ candidates with the selection described above. Since the decays are collected simultaneously, the result does not depend on kaon flux measurement and the systematic effects due to the detector efficiency cancel to first order. To take into account the momentum dependence of signal acceptance and background level, the measurement is performed independently in bins of reconstructed lepton momentum. The ratio R_K in each bin is computed as

$$R_K = \frac{1}{D} \cdot \frac{N(K_{e2}) - N_B(K_{e2})}{N(K_{\mu 2}) - N_B(K_{\mu 2})} \cdot \frac{f_\mu \times A(K_{\mu 2}) \times \epsilon(K_{\mu 2})}{f_e \times A(K_{e2}) \times \epsilon(K_{e2})} \cdot \frac{1}{f_{\text{LKr}}}, \tag{1}$$

where $N(K_{l2})$ are the numbers of selected K_{l2} candidates ($l = e, \mu$), $N_B(K_{l2})$ are numbers of background events, f_l are efficiencies of electron and muon identification criteria, $A(K_{l2})$ are geometrical acceptances, $\epsilon(K_{l2})$ are trigger efficiencies, f_{LKr} is the global efficiency of the LKr readout, and $D = 150$ is the downscaling factor of the $K_{\mu 2}$ trigger. In order to compute $A(K_{l2})$, a detailed Geant3-based Monte-Carlo simulation is employed.

1.4. Backgrounds

$N_B(K_{e2})$ in (1) is dominated by $K_{\mu 2}$ events with track misidentified as electron, mainly in case of high energetic bremsstrahlung after the magnetic spectrometer, when the photon takes more

[2] The background to K_{e2} is concentrated in the region $p > 25$ GeV/c, hence the need of tighter electron ID.

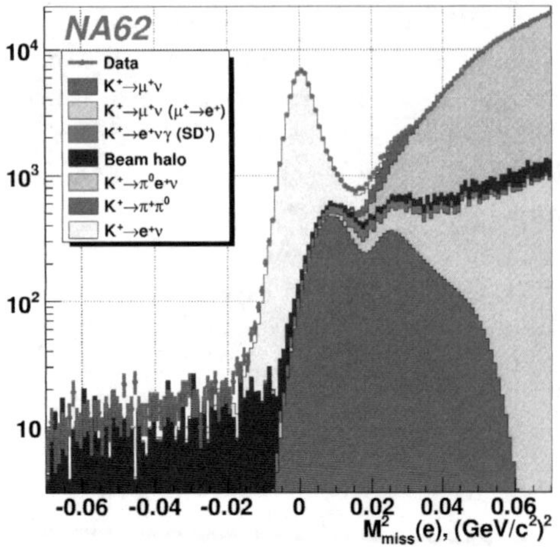

Figure 1. $M^2_{\mathrm{miss}}(e)$ distributions for K_{e2} candidates and for various backgrounds.

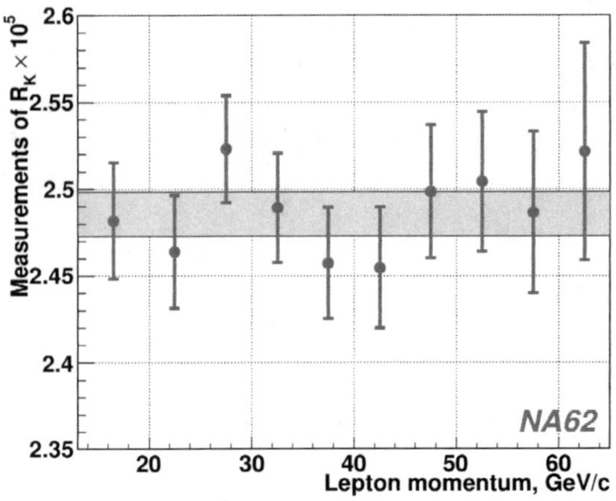

Figure 2. R_K in track momentum bins.

than 95% of muon's energy. The probability for such process is measured directly by clean sample of muons passing $\sim 10X_0$ of lead (Pb) before hitting the LKr. A Geant4 simulation is used to evaluate the Pb correction to the probability for muon misidentification which occurs via two principal mechanisms: 1) muon energy loss in Pb by ionization, dominating at low momenta; 2) bremsstrahlung in the last radiation lengths of Pb increasing the probability for high track momenta. The background is evaluated to be $(6.10 \pm 0.22)\%$.

Since the incoming kaon is not tracked and the signature of K_{l2} decays is a single reconstructed track, the background from beam halo should be considered. The performance of muon sweeping system results in lower background in K^+_{e2} sample ($\sim 1\%$) than in K^-_{e2} sample ($\sim 20\%$), therefore $\sim 90\%$ of data were collected with the K^+ beam only, and small fractions were recorded with simultaneous beams and K^- beam only. The halo background in K^+_{e2} was measured to be $(1.14 \pm 0.06)\%$ directly by using data, collected when no K^+ beam is present.

The other backgrounds considered are: $(0.27 \pm 0.04)\%$ from $K_{\mu 2}$ with subsequent $\mu \to e$ decay[3]; $(1.15 \pm 0.17)\%$ from $K_{e2\gamma}(\mathrm{SD})$[4]; 0.06% for both K_{e3} and $K_{2\pi}$ decays.

The number of K_{e2} candidates is 59,963 before background subtraction. The $M^2_{\mathrm{miss}}(e)$ distribution of data events and backgrounds are presented in Fig. 1.

1.5. Systematic uncertainties and results

The electron identification efficiency is measured directly as a function of track momentum and its impact point at LKr using electrons from K_{e3} decays. The average f_e is $(99.27 \pm 0.05)\%$ (f_μ is negligible). The geometric acceptance correction $A(K_{\mu 2})/A(K_{e2})$ is known with permille precision and depends on the radiative $K_{e2\gamma}(\mathrm{IB})$ decays, simulated following [7] with higher order corrections according to [8]. The trigger efficiency correction $\epsilon(K_{e2})/\epsilon(K_{\mu 2}) \approx 99.6\%$ is significant only in the first analysis bin $13 < p < 20$ GeV/c and accounts for the difference in the trigger conditions, namely the requirement of $E > 10$ GeV energy deposited in LKr for K_{e2} only. Additional small systematic uncertainty arises due to the global LKr readout efficiency, measured to be $(99.80 \pm 0.01)\%$.

[3] This background is suppressed according to Michel distribution, as muons from $K_{\mu 2}$ decays are fully polarized.
[4] The recent KLOE measurement [4] is used for this estimation. The uncertainty on the background is dominated by R_K stability with respect to the level of supression of events with extra activity in LKr.

The independent measurements of R_K in track momentum bins are presented in Fig. 2. The final NA62 result, based on 40% of the accumulated statistics is $R_K = (2.486 \pm 0.011_{\text{stat}} \pm 0.007_{\text{syst}}) \times 10^{-5} = (2.486 \pm 0.013) \times 10^{-5}$, consistent with SM expectation. The analysis of the whole data set will allow to reach uncertainty of 0.4%. The combined new world average is $(2.487 \pm 0.012) \times 10^{-5}$.

2. The future of NA62: Measurement of $BR(K^+ \to \pi^+ \nu\nu)$

2.1. Introduction

The future program of the experiment NA62 at CERN SPS is currently in an advanced stage of development. The main goal of the experiment is to measure the branching ratio of the very rare $K^+ \to \pi^+ \nu\bar{\nu}$ decay, by detecting approximately 100 events with 10% background.

The $K^+ \to \pi^+ \nu\bar{\nu}$ and $K_L \to \pi^0 \nu\bar{\nu}$ decays are exceptionally clean modes, dominated by short distance dynamics due to power-like GIM mechanism, and therefore are excellent probes in flavour physics. At the quark level the process $d \to s\nu\nu$ is realised by combination of Z^0 penguin and double W exchange. The leading SM contribution to the matrix element is generated by top quark loops and can be computed with negligible theoretical uncertainty. In case of $K^+ \to \pi^+ \nu\bar{\nu}$ decay there is a small contribution from charm quark, while the contribution from up quark is negligible in both decay modes. The hadronic matrix element can be extracted from the well measured $K \to \pi e \nu$ decays rates with negligible theoretical uncertainty. The current estimations of the branching ratios for the the $K \to \pi\nu\bar{\nu}$ decays within the SM are $BR(K^+ \to \pi^+ \nu\bar{\nu}) = (8.22 \pm 0.84) \cdot 10^{-11}$ and $BR(K_L \to \pi^0 \nu\bar{\nu}) = (2.76 \pm 0.40) \cdot 10^{-11}$ [9].

Measurement of the branching ratios of both $K^+ \to \pi^+ \nu\bar{\nu}$ and $K_L \to \pi^0 \nu\bar{\nu}$ decay modes leads to the determination of the V_{td} element in the CKM matrix, i.e. of the Wolfenstein parameters (ρ, η) that define the unitarity triangle, independently of the results from B-physics.

The strong suppression of $K \to \pi\nu\bar{\nu}$ decays within the SM follows from the absence of tree-level contributions and the hard GIM mechanism. However, this also leads to high sensitivity to possible new-physics effects. Since the cleanness of these decays modes remains valid in all realistic extensions of SM, a precise measurement of their branching ratios provide sensitive test of the flavour structure of any model beyond SM. Evidence of new physics can be seen in $K \to \pi\nu\bar{\nu}$ decays even without significant signals in B-decays or without particles within LHC reach [10].

Only three $K^+ \to \pi^+ \nu\bar{\nu}$ events were discovered by BNL E949 Collaboration. The measured value of the branching ratio is compatible with the SM prediction: $(1.47^{1.30}_{-0.89}) \cdot 10^{-10}$ [11].

The NA62 experiment [12] will use kaon decays in-flight technique, based on the NA48 apparatus and infrastructure, and the same CERN-SPS beam line which produced the kaon beam for all NA48 experiments.

2.2. Backgrounds

The experimental signature of the studied $K^+ \to \pi^+ \nu\bar{\nu}$ decays is a single reconstructed track in the detector downstream the decay volume in time coincidence with a kaon measured by the upstream beam tracker. The kinematics of a one-track event can be fully described by the variable $m^2_{miss} = (P_K - P_\pi)^2$, where P_K and P_π are the four-momenta of the kaon and pion, respectively. Two categories of backgrounds can be considered:

- **Kinematically constrained background.** The decays in this category correspond to $\sim$95% of the total K^+ branching fraction. The $K^+ \to \pi^+ \pi^0$ decay splits the kinematical region of the signal in two parts, called region I and region II. $K^+ \to \mu^+ \nu$ and $K^+ \to \pi\pi\pi$ decays sit on the opposite sides of these regions where cuts must be applied as well. A rejection factor larger than 10^{12} can be achieved only if efficient photon rejection and particle identification complements the kinematical rejection.

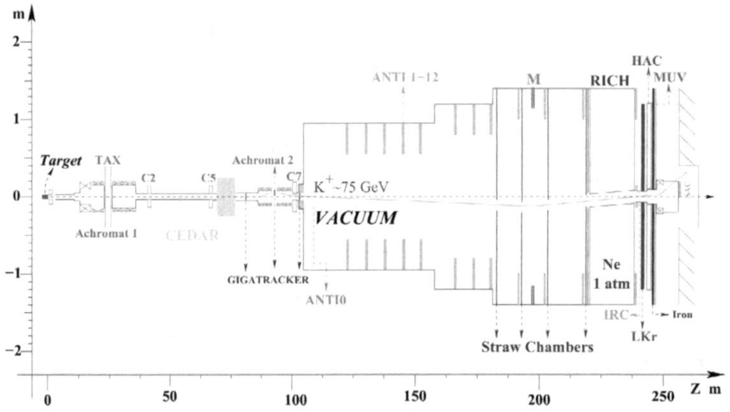

Figure 3. Layout of NA62 experimental apparatus.

- **Kinematically not constrained background.** The rejection of such background profits from relatively small branching fractions. However, photon veto system and particle identification are the only experimental tools available to reduce such a background.

The expected level of background is $\sim 13.7\%$. The signal acceptance is found to be 14.4%. With a flux of about $4.8 \cdot 10^{12}$ kaon per year of data taking, the expected number of signal events is 55 events/year.

2.3. The Experimental set-up
The layout of the experimental apparatus is presented in Fig. 3.

The K^+ beam will be produced from SPS protons hitting a beryllium target with momentum 400 GeV/c and intensity $1.1 \cdot 10^{12}$ protons/s. The nominal momentum of the K^+ beam will be $P_K = 75$ GeV/c with $\Delta P_K/P_K = 1\%$. The expected fraction of K^+ in the beam is 6%. After the first achromat a differential Cerenkov counter (CEDAR) existing at CERN, will be used after its upgrade for new experimental conditions for kaon tagging, in order to keep the beam background under control.

The beam spectrometer (Gigatracker) will be placed in the second achromat station. It consists of thin silicon micro-pixel detectors ($< 0.5\%$ X_0 per station) for redundant momentum measurement of the incoming beam with 200 ps time resolution, necessary to provide a tight coincidence between incoming kaon and outgoing pion. The integrated beam rate at the Gigatracker will be 750 MHz. A set of ring anti-counters (CHANTI) will be placed after the last Gigatracker station and a large one around the beginning of the decay volume, in order to veto secondary charged particles produced in the collimator.

After the 80 m long decay volume, a magnetic spectrometer, operating in vacuum, will be used to measure the momentum of the out-going pion. The spectrometer is designed with four straw chambers with four coordinate views each. Chambers should introduce a small material contribution (0.5% X_0 per chamber) and have a good spatial resolution (130 μm per view). Full-size prototypes were built and tested in beam during 2007 run of NA62 experiment and in 2010.

After the magnetic spectrometer a gas Ring Imaging Cerenkov counter (RICH) will be placed, providing muon/pion separation in the momentum interval (15–35) GeV/c. The detector will be used also for precise measurement of the pion crossing time with resolution ~100 ps, sending signals to the trigger system. In addition the RICH will provide a redundant measurement of velocity of the charged particles. The detector consists of a 18 m long tube with a diameter of 2.8 m, filled with Neon at atmospheric pressure. The Cerenkov radiation will be collected by two mirrors with 17 m focal length. A full-length, 60 cm in diameter prototype has been integrated in the NA62 set-up and tested during the 2007 run.

A full hermeticity for photons up to 50 mrad will be provided by four subdetector systems. The decay volume and the region after the straw spectrometer will be surrounded by 12 sets of ring-shaped lead glass anti-counters (Large Angle Veto - LAV). The construction phase is advanced, with two of the stations built and tested. The Liquid Krypton Calorimeter, built for the NA48 experiment, will be used as part of the photon veto system in the forward region. During test run in 2006 it was shown that inefficiency below 10^{-5} can be reached. A program of consolidation and update of the readout electronics of the LKr is under way. The photon veto system is completed by the Intermediate Ring Calorimeter (IRC), placed at the entrance of LKr, and Small Angle Calorimeter (SAC) at the very end of the detector system, after the muon deflecting magnet, both covering the angular regions around and in the beam. A SAC prototype was tested during 2006 and an upper limit on its inefficiency was found to be $6.4 \cdot 10^{-5}$.

The muon veto system of the experiment will consist of three parts - the first two modules (MUV1 and MUV2) follow directly the LKr and work as hadronic calorimeters for measurement of deposited energies and shower shapes of incident particles. The third station will be used in the on-line selection.

The trigger system of NA62 experiment will be at two levels. The first one (L0) is a hardware trigger, which will decrease the events rate from $\sim$10 MHz to $\sim$1 MHz, employing signals from the RICH, the photon veto system and the muon detector. The second level (L1/2) is software and its aim is to decrease the event rate to kHz level.

The R&D program of the NA62 experiment is near completion with some of the detectors already in construction phase. The data taking is foreseen for 2013 and 2014. Other physics opportunities, such as searches for lepton flavour violation and new low mass particles are being considered as part of the program of the experiment.

References

[1] Cirigliano V and Rosell I, 2007 *Phys. Rev. Lett.* **99** 231801
[2] Masiero A, Paradisi P and Petronzio R, 2006 *Phys. Rev.* **D74** 011701; 2008 *JHEP* **0811** 42
[3] Nakamura K *et al.*, 2010 *J. Phys. G* **37** 075021
[4] Ambrosino F *et al.*, 2009 *Eur. Phys. J.* **C64** 627; Erratum-ibid. 2010 **C65** 703
[5] Batley J R *et al.*, 2007 *Eur. Phys. J.* **C52** 875
[6] Fanti V *et al.*, 2007 *Nucl. Inst. Methods* **A574** 433
[7] Bijnens J, Ecker G and Gasser J, 1993 *Nucl. Phys.* **B396** 81
[8] Weinberg S, 1965 *Phys. Rev.* **140** B516; Gatti C, 2006 *Eur. Phys. J.* **C45** 417
[9] The *Flavianet* Kaon WG [http://www.lnf.infn.it/vus/]
[10] Buras A J *et al.*, 2005 *Nucl. Phys. B* **714** 103; Isidori G *et al.*, 2006 *JHEP* **0608** 064; Blanke M*et al.*, 2007 *JHEP* **0701** 066; Buras A J *et al.*, 2006 *JHEP* **0611** 062; Bryman D *et al.*, 2006 *Int. J. Mod. Phys. A* **21** 487; Grinstein B *et al.*, 2007 *Nucl. Phys. B* **763** 35
[11] Anisimovsky V V *et al.*, 2004 *Phys. Rev. Lett.* **93** 031801
[12] Proposal to measure the Rare Decay at the CERN SPS, CERN-SPSC-2005-013, 11/6/2005

Journal of Physics: Conference Series **259** (2010) 012024 doi:10.1088/1742-6596/259/1/012024

Muon performance and related ATLAS first physics

María Moreno Llácer for the ATLAS collaboration

IFIC - Instituto de Física Corpuscular (CSIC-UV), Edificio Institutos de Investigación, Apdo. Correos 22085, E-46071, Valencia - España

E-mail: mamolla@cern.ch

Abstract. Muon final states provide clean signatures for many physics processes at the LHC. Therefore, an excellent identification of muons and momentum resolution in a very wide energy range is required. The performance of the ATLAS muon reconstruction and identification is here studied with the cosmic rays data recorded during the commissioning period and the first LHC proton-proton collision data at $\sqrt{s}$= 7 TeV. Measured resolutions, efficiencies, muon isolation and residual distributions of reconstructed muon tracks are well reproduced by the Monte-Carlo simulation. In addition, the first observation in the ATLAS experiment of W and Z bosons in the muon channel at $\sqrt{s}$= 7 TeV collisions is described. The selection yields to 40 $W \rightarrow \mu\nu$ and 2 $Z \rightarrow \mu\mu$ candidates in a total integrated luminosity of approximately of 6.4 nb^{-1} and 7.9 nb^{-1} respectively, in agreement with Monte-Carlo expectations.

1. Introduction

ATLAS (A Toroidal LHC Apparatus) [1] is one of the four major experiments at the LHC (Large Hadron Collider), in which protons are colliding at a center of mass energy of 7 TeV. It consists of three main sub-systems: the inner detector (ID), the calorimeters (electromagnetic and hadronic) and the muon spectrometer (MS). It uses a superconducting magnet system with a central solenoid around the inner detector and large air-core toroid magnets for the muon spectrometer. The commissioning of the ATLAS detector with physics data already started while the detector was being mechanically and electrically completed by collecting cosmic rays with the parts of the detectors that were becoming available. Several global cosmic rays runs with near all sub-detectors operating at full coverage and with different magnetic field configurations were undertaken. Around 500 million events were recorded in 2008 and 2009 providing the first opportunity to study the performance of the combined reconstruction algorithms with real data. First collisions at 7 TeV took place in March 2010 and near 300 nb^{-1} were collected by ATLAS at the time of the conference. These data have been used to study the detector performance in detail and re-discover the Standard Model. Some examples of this wide spectrum of physics results are presented in this report.

2. Muon reconstruction algorithms

LHC physics requires an excellent muon identification and a very precise momentum measurement in a very wide energy range. For that, four kinds of muon candidates are distinguished in ATLAS depending on the way they are reconstructed: *stand-alone muons*, *combined muons*, *segment tagged muons* and *calorimeter tagged muons*. The highest quality category of muons are combined muons, i.e. muons that are formed by combining an inner

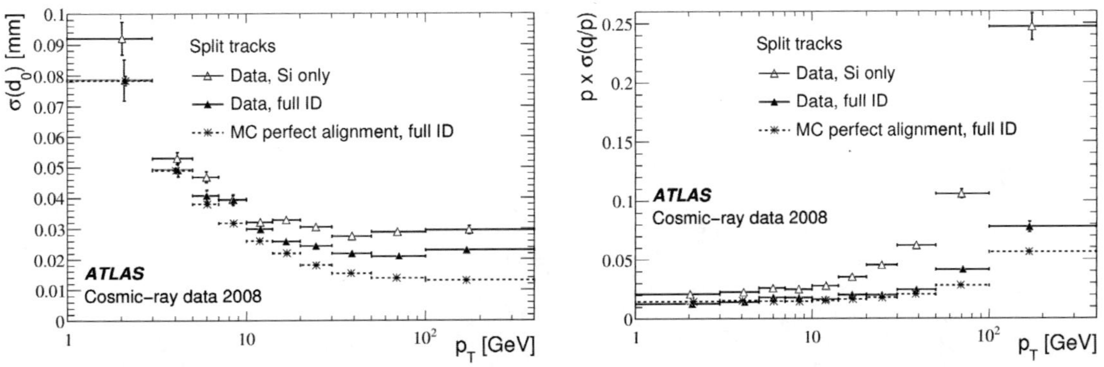

Figure 1: *Transverse impact parameter (left) and momentum (right) resolution determined from data as a function of transverse momentum. Resolutions of full inner detector (solid triangles) and silicon-only (open triangles) tracks are compared to those from full tracks in Monte-Carlo simulation (stars).*

detector track with a muon spectrometer track. Unless stated, we will only consider combined muons in this article.

3. Performance of the muon reconstruction with cosmic ray data

In the absence of LHC beams, the only source of physics signals in the ATLAS detectors are cosmic rays interactions in the atmosphere, which produce muons that can penetrate to the ATLAS cavern and deposit energy in all detector sub-systems. Cosmic rays have been used as part of the commissioning effort throughout the detector installation procedure, starting in 2005. Several global runs with all the detectors operating near full coverage were taken in 2008 and 2009 and more than 500 million cosmic rays were recorded.
Cosmic rays in ATLAS come mostly from above, and arrive mainly via two large access shafts used for the detector installation. Muons crossing the central region sometimes pass near the nominal interaction point. By selecting on track impact parameter with respect to the interaction point, it is possible to identify a sample of projective muons similar to collision-like events. Several analysis have been performed with these data for the inner detector [2] and the muon spectrometer [3].

For the inner detector, the track parameters resolutions have been estimated by splitting the full track into two tracks, an upper and a lower one and comparing their track parameters. On the left plot of Figure 1 one can see the transverse impact parameter resolution as a function of the muon momentum for data and Monte-Carlo. The resolution is dominated by multiple scattering at low p_T and by the intrinsic resolution in the intermediate and high p_T regions.
On the right plot of Figure 1 the relative momentum resolution is shown. The resolution is degraded at high momentum, as expected since particles at high p_T bend less in the magnetic field and make the determination of the curvature more difficult. A fractional momentum resolution of less than 2% is reached for muons of $p_T < 50$ GeV.

In Figure 2 the relative momentum resolution for muons reconstructed in the muon spectrometer is shown. The resolution function can be fitted with the sum in quadrature of three terms, the uncertainty on the energy loss corrections p_0, the multiple scattering contribution p_1 and the intrinsic resolution p_2:

$$\frac{\sigma_{p_T}(MuonSpectrometer)}{p_T} = \frac{p_0}{p_T} \oplus p_1 \oplus p_2 p_T \tag{1}$$

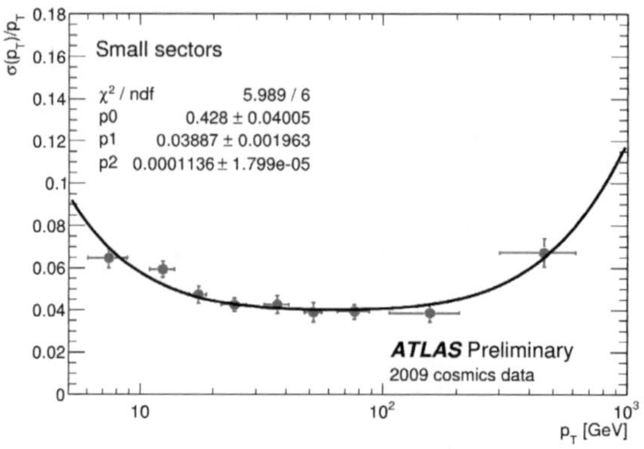

Figure 2: *Transverse momentum resolution for stand-alone tracks in the muon spectrometer as a function of the muon transverse momentum.*

The results of the fit are in agreement with the expected values but for the intrinsic term, which is sensitive to the alignments which are still in progress. The resolution is close to the nominal one for $p_T < 50$ GeV.

4. Performance of the muon reconstruction with collision data at 7 TeV

Already with the first LHC proton-proton collisions (March 2010), ATLAS devoted a substantial amount of work to understand the detector performance to assure solid physics measurements [4], [5]. In this section, a few examples of performance studies based on a fraction of the processed data at the time of the conference are presented. The data are compared to expectations based on Monte-Carlo simulations. The latter samples were generated at $\sqrt{s}$= 7 TeV with PYTHIA [6] using MRST LO [7] PDF, simulated with GEANT4 [8] and fully reconstructed with the same reconstruction algorithms used for data.

4.1. Inclusive muon spectrum

At a center of mass energy of 7 TeV, the main source of muons are light mesons, heavy mesons and Gauge bosons. The left plot of Figure 3 shows the transverse momentum distribution of the reconstructed muons compared to the Monte-Carlo predictions. For the Monte-Carlo, different components in the muon spectrum are distinguished: muons originating from pions or kaons decays which, due to their long lifetime, may cross large part of the detector before decaying and that are reconstructed in the inner detector (green), prompt muons from decays of short-lived particles (red) and others (white). According to the simulation, the spectrum is dominated by light meson decays at low p_t, while the contribution from prompt muons becomes more important at high transverse momentum.

4.2. Momentum resolution

The right plot of Figure 3 shows the relative difference of the momentum measured in the inner detector and in the muon spectrometer. The distribution has a narrow core and a tail to positive values. The shape of the distribution is similar in the Monte-Carlo simulation. According to the simulation, the tail of the distribution to positive values is caused by muons from pion and kaon decays-in-flight. There is a larger tail in the measured data than in the simulated distribution, the cause of which was found to be remaining misalignments in one of the endcaps of the inner detector, which has already been improved at the time of writing.

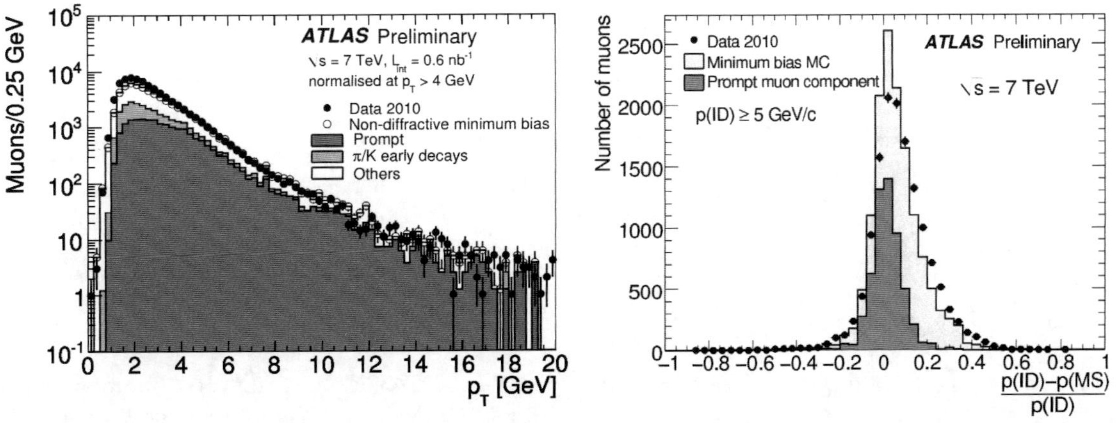

Figure 3: *Left: p_T spectrum of the reconstructed combined muons. Right: Relative difference between the inner detector and stand-alone muon momentum for both data and simulation.*

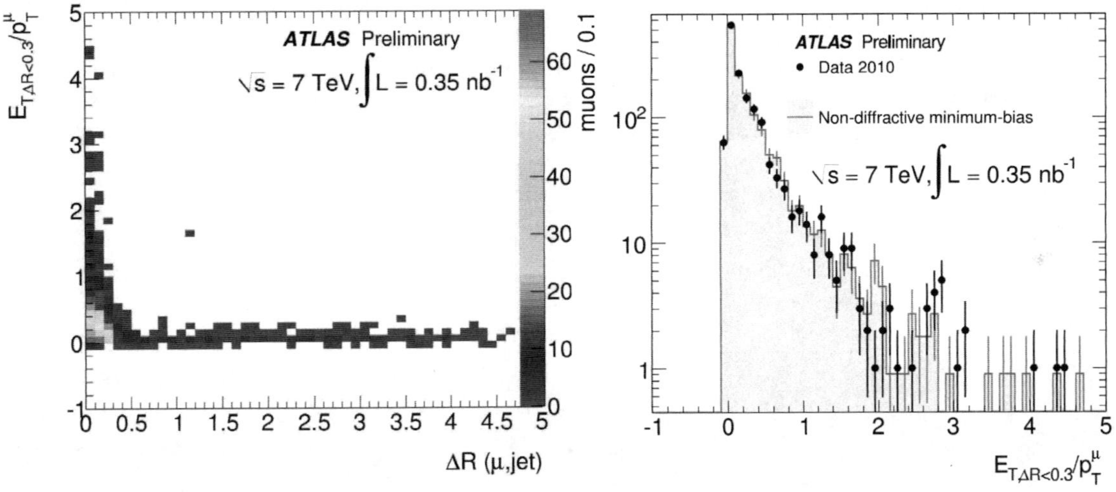

Figure 4: *Left: Dependence of the measured transverse energy deposited around a muon in a cone of $\Delta R < 0.3$ normalized to the transverse momentum of the muon on the distance to the closest jet. Right: Measured sum of the transverse momentum of tracks around a combined muon in a cone of $\Delta R < 0.3$.*

4.3. Validation of muon energy deposits in the calorimeters

Muons inside jets tend to be produced by hadron decays, therefore a powerful tool for selecting prompt muons is the requirement that the muon is isolated. The isolation can be performed in two ways: by applying an upper limit on the energy deposited in a cone around the muon (subtracting the energy deposited by the muon itself), or by cutting on the sum of the transverse momentum of the tracks in a cone around the muon. The size of the cone is typically $0.2 \leq \Delta R \leq 0.4$. The left plot of Figure 4 shows the distribution of the transverse energy deposited in a cone of size $\Delta R = 0.3$ around the muon normalized to the transverse muon momentum for muons with $p_T > 6$ GeV. One can see that muons inside jets have much larger

Table 1: *Number of observed and expected events from W and Z bosons decay in the muon channel.*

	$W \rightarrow \mu\nu$ **channel**	$Z \rightarrow \mu\mu$ **channel**
Observed	40	2
Expected	$28.7 \pm 0.5(\text{stat}) \pm 3.9(\text{syst}) \pm 5.7(\text{lumi})$	$3.2 \pm 0.7(\text{syst}) \pm 0.6(\text{lumi})$
Signal	$25.9 \pm 3.6(\text{syst}) \pm 5.2(\text{lumi})$	$3.2 \pm 0.7(\text{syst}) \pm 0.6(\text{lumi})$
Background	$2.8 \pm 0.5(\text{stat}) \pm 0.8(\text{syst}) \pm 0.6(\text{lumi})$	$(2.1 \pm 0.8(\text{syst}) \pm 0.4(\text{lumi}))10^{-4}$

energy deposits around them. The right plot shows the sum of the transverse momentum of the tracks in a cone of $\Delta R = 0.3$ around the jet for both data and Monte-Carlo. An excellent agreement between the measured and the simulated distributions is achieved.

5. First physics results with muons at 7 TeV collisions

The W and Z bosons are produced abundantly at the LHC, and for the first time in proton-proton collisions. The well known properties of these bosons provide significant constraints in the determination of the ATLAS detector performance, powerful tools to constrain the Parton Density Function (PDF) inside the proton and understanding of these processes as backgrounds to new physics ones. At the time of the conference, the ATLAS experiment had observed 40 $W \rightarrow \mu\nu$ and 2 $Z \rightarrow \mu\mu$ candidates [9], from a total integrated luminosity of approximately 6.4 nb^{-1} and 7.9 nb^{-1} respectively.

5.1. *Observation of W production*

First of all, collision candidate events are selected by requiring a muon trigger and a primary vertex with at least three tracks. In addition, the muons are required to have $p_T > 20$ GeV, the E_T^{miss} is required to be above 25 GeV and the transverse mass of the W candidate events is required to be $\mathrm{m}_T = \sqrt{2p_T^l p_T^\nu(1 - cos(\phi^l - \phi^\nu))} > 40$ GeV. The m_T distribution in the selected events is shown in Figure 5 before and after the E_T^{miss} cut. All events beyond $m_T > 40$ GeV are finally selected as W candidate events. The data are compared to expectations based on Monte-Carlo simulations and both distributions agree quite well. In Table 1, the number of observed data events and the Monte-Carlo predictions are summarized giving the statistical, systematic and luminosity uncertainty in the number of expected signal and background events. The QCD component of the background is estimated from data, while the electroweak component is obtained from Monte-Carlo.

5.2. *Observation of Z production*

For the Z boson search, candidates are required to contain two isolated muons of opposite charge and $p_T > 15$ GeV. The invariant mass is required to be $80 < m_Z < 100$ GeV. With an integrated luminosity of 7.9 nb^{-1}, two Z boson candidates with invariant masses of 80.2 GeV and 87.6 GeV have been observed in the muon channel.

6. Conclusions

The performance of the muon reconstruction algorithms in ATLAS has been studied with the first data available: cosmic rays and first proton-proton collisions at $\sqrt{s}$= 7 TeV. The ATLAS detector has delivered high quality results covering a wide spectrum of physics processes. With

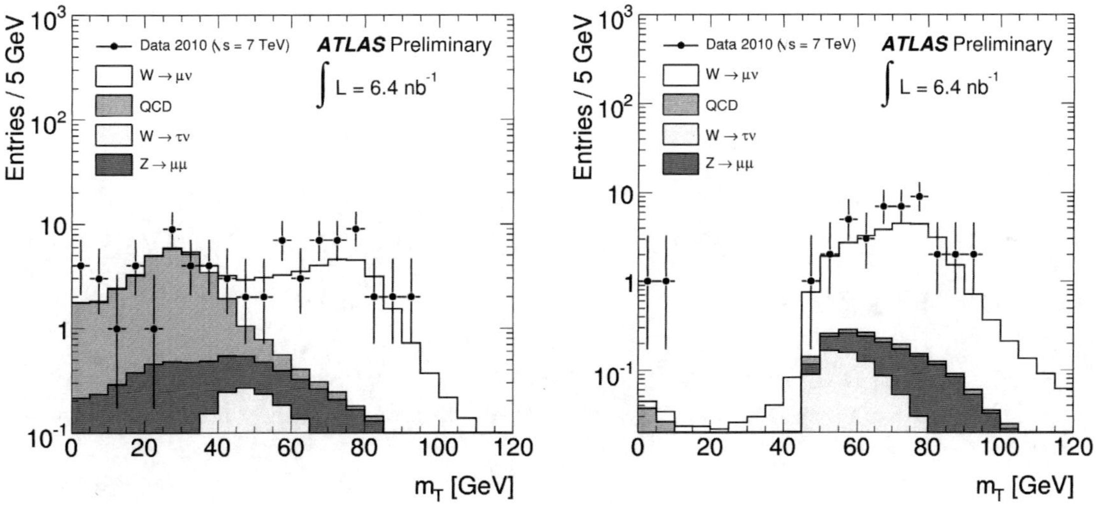

Figure 5: *Transverse invariant mass of W candidates in the muon channel before (left) and after (right) the $E_T^{miss} > 25$ GeV cut.*

the integrated luminosity collected at the time of the conference, ATLAS has scrutinized the detector performance and reconstructed stable objects. Data and Monte-Carlo comparisons have been performed and have allowed to validate the simulations. The measured track parameters and resolutions and the energy deposition in the calorimeters are well predicted by the Monte-Carlo simulation. In addition, the production of W and Z bosons has been observed.

References

[1] ATLAS Collaboration, *"Expected performance of the ATLAS Experiment - Detector, Trigger and Physics"*, `arXiv:0901.0512 [hep-ex]`.

[2] ATLAS Collaboration, *"The ATLAS inner detector commissioning and calibration"*, `arXiv:1004.5293 [hep-ex]`

[3] ATLAS Collaboration, *"Commissioning of the ATLAS Muon Spectrometer with Cosmic Rays"*, `arXiv:1006.4384 [hep-ex]`

[4] ATLAS Collaboration, *"Preliminary studies for the measurement of the inclusive muon spectrum in pp collisions at $\sqrt{s} = 7$ TeV with the ATLAS detector"*, ATLAS-CONF-2010-035

[5] ATLAS Collaboration, *"Muon Performance in Minimum Bias pp Collision Data at $\sqrt{s} = 7$ TeV with ATLAS"*, ATLAS-CONF-2010-036

[6] T. Sjostrand et al, *"PYTHIA 6.4 physics and manual"*, JHEP(05) 026 (2006)

[7] A. D. Martin et al, *"Parton distributions for the LHC"*, Eur. Phys. J. C 63, 189-285 (2009)

[8] S. Agostinelli et al, *"GEANT4-A simulation toolkit"*, Nuclear Instr. Methods A 506, 250 (2003)

[9] ATLAS Collaboration, *"Observation of $W \to l\nu$ and $Z \to ll$ production in proton-proton collisions at $\sqrt{s} = 7$ TeV with the ATLAS detector"*, ATLAS-CONF-2010-044

High precision spectra at large redshift for dynamical DE cosmologies

S.A. Bonometto, L. Casarini

Department of Physics G. Occhialini – Milano–Bicocca University, Piazza della Scienza 3, 20126 Milano, Italy & I.N.F.N., Sezione di Milano

A.V. Macciò

Max-Planck-Institut für Astronomie, Königstuhl 17, 69117 Heidelberg

G. Stinson

Jeremiah Horrocks Institute, University of Central Lancashire, Presto PR1 2HE

Abstract. The next generation mass probes will investigate DE nature by measuring non–linear power spectra at various z, and comparing them with high precision simulations. Producing a complete set of them, taking into account baryon physics and for any DE state equation $w(z)$, would really be numerically expensive. Regularities reducing such duty are essential. This paper presents further n–body tests of a relation we found, linking models with DE state parameter $w(z)$ to const.–w models, and also tests the relation in hydro simulations.

1. Introduction

The background metrics of space–time can be written in the forms

$$ds^2 = c^2dt^2 - a^2(t)\,d\ell^2 = a^2(\tau)\,(d\tau^2 - d\ell^2)$$

defining ordinary or conformal time, t or τ. In systems which abandoned cosmic expansion, the former expression yields the Minkowskian metrics: the no longer growing scale factor can be inglobated in spatial coordinates. In such systems t is the natural time coordinate; e.g., many cyclic motions become (quasi–)periodical in respect to t. On the contrary, when a system resent the overall expansion, the natural time coordinate is τ. For instance, sonic waves in the pre–recombination plasma, are (quasi–)periodical in respect to τ. This periodicity yields the setting of maxima and minima in CMB spectra, as well as the scales of BAO's in matter spectra.

Let us consider then the comoving distance between today's observer and the LSB (last scattering band), $L = \int_{rec}^{0} d\ell$. It coincides with the conformal time $T = \int_{rec}^{0} d\tau$ a photon takes to reach the observer from the LSB. Accordingly, if two models have the same comoving distance from the LSB, we expect them to share those features which are determined by the cosmological dynamics. As a matter of fact, it has been known since several years [1] that, if they share $\Omega_{c,b}$, h and n_s (density parameters, Hubble parameter in units of 100 (km/s)/Mpc, primeval scalar index), their linear spectra are quite close. In particular, models with the same conformal age, differing just because of DE state equation, exhibit almost coincident linear spectra.

When non–linear effects are taken into account, the dynamics is no longer simply ruled by τ. Non linear spectral deformations arise when matter halos form, abandoning the overall expansion. The point is whether they are so widely spread to yield a significantly discrepant spectral behavior at a given k–scale. Greater k's correspond to scales entering a non–linear regime earlier; therefore, we expect greater discrepancies at greater k's.

A test of these expectations can only be performed in a numerical way, by running model simulations. Let us then outline that forthcoming tomographic lensing experiments (see, e.g., [2]) will be able to measure spectra with a precision $\mathcal{O}(1\,\%)$ [3], so that this is the limit above which spectral discrepancies can no longer be disregarded. In turn, such experiments are expected to test spectra up to $k \simeq 10\,h\mathrm{Mpc}^{-1}$, corresponding to length scales $\mathcal{O}(0.5\,h^{-1}\mathrm{Mpc})$, well inside galaxy clusters. To test spectral regularities, then, hydro effects cannot be disregarded.

A number of authors have considered these problems in some detail. The k value above which hydrodynamics causes spectral distorsions $> 1\,\%$ has been found to be ~ 2–$3\,h\,\mathrm{Mpc}^{-1}$ [4]. Models with an assigned polynomial DE state equation $w(z)$ (models $\mathcal{A}$) were then compared with models with constant w (models $\mathcal{W}$, or *auxiliary* models), selected so to have the same L and the same σ_8 (mean square density fluctuation at $8\,h^{-1}\mathrm{Mpc}$), besides of $\Omega_{b,c}$ and h [5]. Spectral discrepancies, within $k \sim 2$–$3\,h\,\mathrm{Mpc}^{-1}$ were found to keep within $1\,\%$.

A critical feature of forthcoming mass probes is their capacity to explore the Universe at $z > 0$. Models with the same conformal age at $z = 0$, and different DE state equation, have different ages at $z > 0$ and spectral discrepancies soon exceed $1\,\%$. In principle, then, one should find an auxiliary model $\mathcal{W}(z)$ of $\mathcal{A}$, at any z. But equal H at $z \neq 0$ means different H at $z = 0$. Accordingly, simulations performed in boxes of identical size in units of $h_o^{-1}\mathrm{Mpc}$, can no longer be compared (the $_o$ suffix is used just here, to outline that h is taken at $z = 0$). If boxes of suitable different sizes are then used, sample variance is the dominant effect.

This difficulty was overcome by abandoning the request that $\mathcal{A}$ and $W(z)$ have the same $H(z)$, and requiring them to share just the reduced density parameters $\omega_{c,b} = \Omega_{c,b}h^2$ *(weak requirement)* [7]. The definition of ρ_{cr} (critical density) however implies that, at any z,

$$\Omega_{c,b} \times H^2 = \Omega_{c,b} \times (8\pi G/3)\rho_{cr}~,$$

so that the $\omega_{c,b} \propto \Omega_{c,b}H^2$ scales as $\Omega_{c,b}\rho_{cr} = \rho_{c,b} \propto a^{-3}$, indipendently of DE nature. Models with equal $\omega_{c,b}$ at $z = 0$ will then share them at any z. The $\mathcal{W}(z)$ model expected to approach $\mathcal{A}$ at z, then depends on z just because of its different *constant* DE state parameter $\tilde{w}(z)$. No constraint holds on $H(z)$, any more; thus, we can take the same H value in all models at $z = 0$.

Notice that the $\tilde{w}(z)$ dependence so defined is quite different from $w(z)$, the DE state equation in $\mathcal{A}$. This outlines a significant experimental risk, as observers fitting data against constant w models, would find $\tilde{w}(z)$, instead of the actual DE state equation $w(z)$.

Here we extend the test in [7] to a set of models, sampling cosmologies consistent with data [8, 9], and confirm that the weak requirement works up to the required k (see also [7]). We also test the requirement including hydrodynamics, confirming spectral regularities up to $k \sim 10\,h\mathrm{Mpc}^{-1}$ (see also [10]). Further hydrodynamical tests are in progress.

A side result we shall outline concerns the rate of star formation. The size of the box used allows to follow such process in a rather approximate way. We however tested that our simulations produce an amount of star which is not unreasonable. Previous tests in boxes of comparable side admittedly found more substantial difficulties [11]. The point is then that, in spite of the proximity of the cosmologies treated, we find a significantly different star production. If this property is confirmed by simulations in boxes apt to follow in detail galaxy formation, it could envisage a new important feature to trace DE nature observationally.

2. N–body tests for cosmologies consistent with current data

Using available data, the WMAP team [8, 9] tried to constrain the coefficients w_0 and w_a in a polynomial expression $w(a) = w_0 + (1 - a)\,w_a$ for the DE state parameter. However, the

Journal of Physics: Conference Series **259** (2010) 012025 doi:10.1088/1742-6596/259/1/012025

likelihood ellipses (Figure 1) have significantly shifted from WMAP5 to WMAP7. The setting

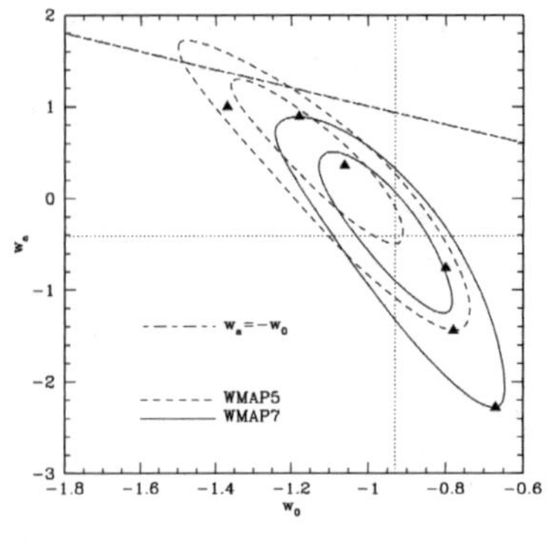

Figure 1. Likelihood ellipses on the w_o–w_a plane from WMAP5 and WMAP7 data releases. Black triangles indicate the $\mathcal{A}$ models considered. State equations beyond the $w_o = -w_a$ line should be modified at high–z. The dotted lines cross on the WMAP7 best fit.

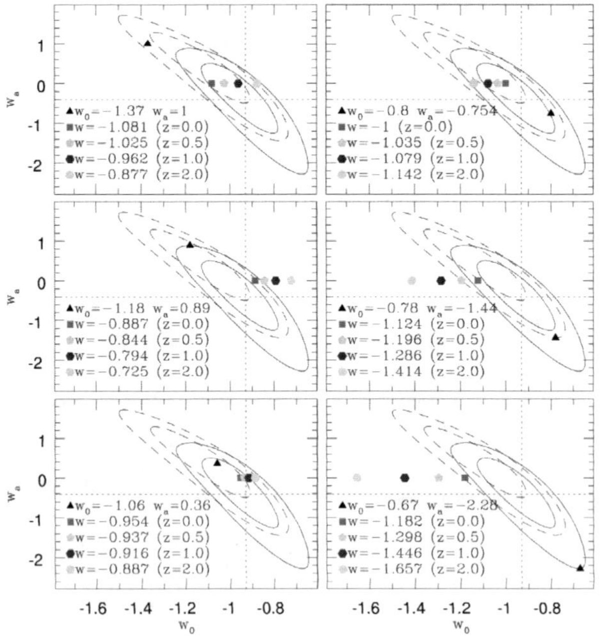

Figure 2. Each box refers here to a single $\mathcal{A}$ model in Figure 1 (black triangle) and related $\mathcal{W}(z)$ models (color polygones). Notice that, in most cases, distances between colored polygones are smaller than their distance from the black triangle. This outlines the possibility of a serious bias in data analysis, if data are tested by assuming constant–w cosmologies.

of the six $\mathcal{A}$ models considered is shown in the Figure. The other parameters are consistent with both WMAP5 and WMAP7 data: $\Omega_m = 0.274$, $h = 0.7$, $\sigma_8 = 0.81$ and $n_s = 0.96$.

We test the spectra of $\mathcal{A}$ models *vs.* the corresponding $\mathcal{W}(z)$. All simulations start at $z = 24$, from realizations fixed by an identical random seed. We use the PKDGRAV code [6], modified to deal with any variable $w(a)$, as in [7]. In a box of side $L_{box} = 256h^{-1}$Mpc we set $N = 256^3$ particles and use a gravitational softening $\epsilon = 25h^{-1}$kpc. We have 4 auxiliary models $\mathcal{W}(z)$ for

$z = 0,\ 0.5,\ 1,\ 2$ (see Fig. 2), for each $\mathcal{A}$ model, run just down to the redshift where they are tested. For all models, mass functions agree with Sheth & Tormen [12] predictions.

The first $\mathcal{A}$ cosmology considered, consistent with WMAP5, is outside the 2–σ curve for WMAP7. The significant shift of the ellipses is not due to the fresh CMB inputs. Rather, omitting the distance prior, as well as the whole SDSS data [8, 9], surely had an impact on it.

Figure 2 shows that the distance between $\mathcal{W}$ models, on the $w_a = 0$ line by definition, is mostly smaller than their distance from $\mathcal{A}$. Distances depend on w_o and are smaller for the central w_o values. There is therefore a possible observational problem, as outlined in the Introduction.

Model spectra are worked out by FFT–ing the matter density field, computed on a regular grid $N_G \times N_G \times N_G$ (with $N_G = 2048$) via a Cloud in Cell algorithm.

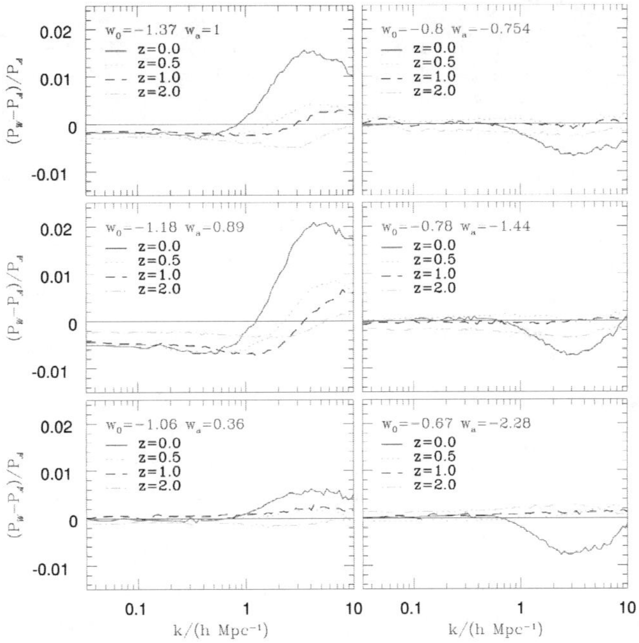

Figure 3. Spectral discrepancies at the redshift indicated by line colors. Each box refers to an $\mathcal{A}$ model with the same ordering as in Fig. 2. Top discrepancies at $z = 0$; only there they approach 2 % for some model with $w_o \ll -1$ and for $k \sim 4$–$5\,h\,\mathrm{Mpc}^{-1}$ (here however gas dynamics should not be ignored). Within ~ 2–$3\,h\,\mathrm{Mpc}^{-1}$, discrepancies keep $< 1\,\%$.

Figure 3 then shows spectral discrepancies. Within $k = 3\,h\,\mathrm{Mpc}^{-1}$ the top discrepancy occurs for the two models with lowest w_o (< -1), at $z = 0$. Other models exhibit a nicer behavior, allowing to guess that the greatest discrepancies, between $\mathcal{A}$ and $\mathcal{W}(z = 0)$ evolutionary histories, occur for models with $w_o \ll -1$. As expected, discrepancies decrease at higher z, where non–linearities had no time to affect low k's. At $z = 0.5$ they are in the permil range, for all models.

3. Hydrodynamical tests

We then compare $\mathcal{A}$ and $\mathcal{W}(z)$ models, also including baryon physics. Hydro simulations are much more time demanding and here we present results for one model: $w_o = -0.8$, $w_a = -0.754$; top right in Figures 2 and 3. The assigned Ω_m value comes from $\Omega_b = 0.046$, $\Omega_c = 0.228$.

The program used is GASOLINE, a multi-stepping, parallel Tree–smoothed particle hydro (SPH) n–body code [13]. It includes radiative and compton cooling. The star formation, based on [14], allows gas particles above constant density and temperature thresholds in convergent

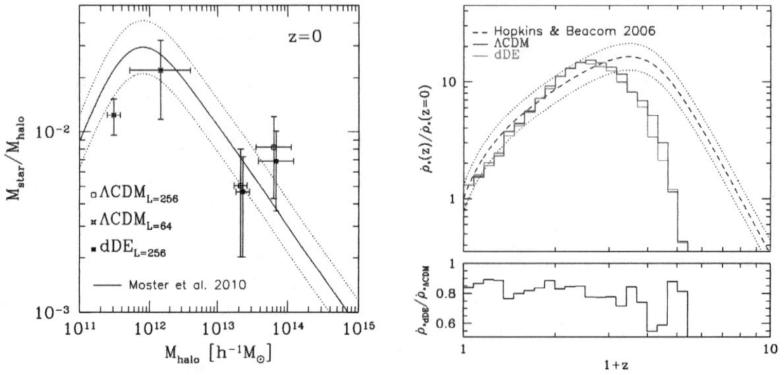

Figure 4. *L.h.s.*: Ratio between stellar and total mass in haloes. Empty and solid squares show the results for the 256 box; crosses for the 64 box. Results are compared with data in [18]; the 40% scatter yields $\sim \pm 2\sigma$. *R.h.s.*: Star formation evolution in ΛCDM and dDE in the 256 box; the dashed (black) line are data fits [19], assuming a Salpeter IMF; curves are normalized to the $z = 0$ value. The lower panel shows the ratio between star productions in dDE and ΛCDM .

flows to spawn star particles at a rate proportional to the local dynamical time [15]. The program includes SN feedback [15] and a UV background, following [16]. We applied GASOLINE the same changes made in PKDGRAV to handle dDE. Further details can be found in [7, 17].

All simulations use $(2\times)256^3$ particles and most are run in a box of $256\,h^{-1}$Mpc. CDM (gas) particles have mass $m_c h/M_{\odot} = 6.33 \times 10^{10}$ ($m_b h/M_{\odot} = 1.28 \times 10^{10}$). The force resolution ϵ (softening) is 1/40 of the intra-particle separation. For $256\,h^{-1}$Mpc this yields $\epsilon \simeq 25\ h^{-1}$kpc (wavenumber $\kappa = 2\pi/\epsilon \simeq 150\,h\,\mathrm{Mpc}^{-1}$). All simulations start at $z = 24$. The $\mathcal{A}$ model is selected so that $\mathcal{W}(z = 0)$ is ΛCDM. This ΛCDM was also run in a box of $64\,h^{-1}$Mpc, still with $(2\times)256^3$ particles. All simulation parameters scale accordingly.

Figure 4 (l.h.s.) shows the ratios M_{star}/M_{halo} *vs.* M_{halo} in both 64 and $256\,h^{-1}$Mpc boxes, compared with predictions in [18]. The Figure shows that star formation generally follows the halo occupation distribution trend. Slightly too many stars form in haloes in the top M_{halo} bin, due to overcooling. This might be fixed by including AGN effects.

The r.h.s. of Fig. 4 shows star formation histories in $256\,h^{-1}$Mpc boxes. Data [19] are also shown, with $\pm 20\,\%$ error bars, approximating 2-σ's. Our outputs are shifted upwards by a factor 1.5, so to account for an overall deficiency in star formation. Furthermore, observations and simulations peak for $z \simeq 2.3$ and $\simeq 1.5$: stars form later in simulations. These discrepancies arise because stars do not form in halos with < 200 gas particles [20], corresponding to a mass $\sim 1.3 \times 10^{12}\,h^{-1}M_{\odot}$ in our simulations. In spite of that, at our resolution level star formation seems described well enough to allow spectral comparison on the scales relevant to us.

The bottom frame shows the ratio between star formation in $\mathcal{A}$ and ΛCDM. The decrease by $\sim 20\%$ arises from a change of $w(z)$, yelding a different evolutionary history, within an identical conformal age. Future work will seek the cause of this unexpectedly large shift.

Let us now come to spectral comparisons between $\mathcal{A}$ and $\mathcal{W}(z)$ models at $z = 0$ and 0.5, 1, 2 . They are shown in Figures 5. For the model considered, regularities persist when gas dynamics is important, although residual discrepancies (mostly of the order of a few permils) are greater than in n–body simulations. The largest discrepancy is found in the gas spectra, even if the gas spectrum distortion at $z = 0$, attaining $\sim 5\,\%$, apparently does not imply a significant distortion in the global spectrum.

Altogether we conclude that even the inclusion of hydrodynamics keeps the discrepancies between auxiliary and true model in the permil range, for the model considered.

Journal of Physics: Conference Series **259** (2010) 012025 doi:10.1088/1742-6596/259/1/012025

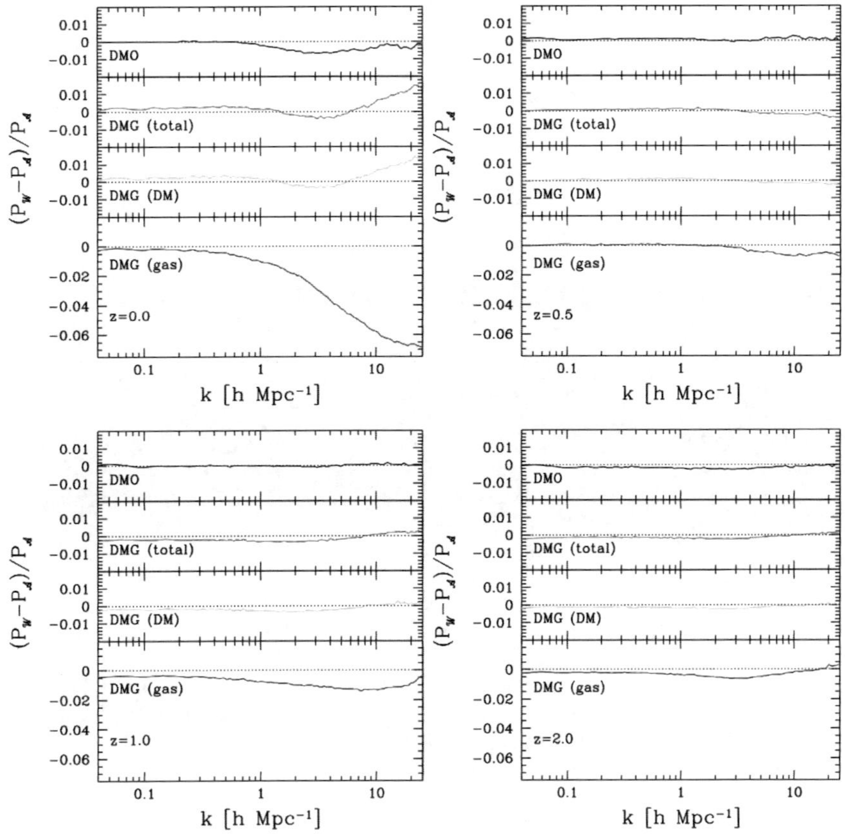

Figure 5. Comparison between $\mathcal{A}$ and $\mathcal{W}(z)$ spectra, for n–body (DMO) and hydro (DMG) simulations (various components shown). Top panels are for $z = 0$ (l.h.s.) and $z = 0.5$ (r.h.s.); bottom panels for $z = 1$ and $z = 2$.

4. References

[1] Linder E V and Jenkins A, 2003 *Mon. Not. R. Astron. Soc.* **346** 573
[2] A. Refregier *et al.* , 2010, arXiv:1001.0061
[3] Huterer D & Takada M, 2005 *Astrophys. J.* **23** 369
[4] Y.P. Jing, P. Zhang, W.P. Lin, L. Gao, V. Springel, 2006, APJ, L119
[5] Francis M J, Lewis G F and Linder E V, 2007
[6] Stadel J. G., 2001, PhD Thesis
[7] Casarini L, Macciò A V & Bonometto S A, 2009 *J. Cosmol. Astropart. Phys.* **3** 14
[8] Komatsu E et al., 2009 *Astrophys. J. Suppl.* **180** 330
[9] Komatsu E et al., arXiv:1001.4538
[10] Casarini L, Macciò A V, Bonometto S A, Stinson G., 2010, arXiv:1005.4683
[11] Rudd D. H., Zentner A. R., Kravtsov A. V., 2008, *Astrophys. J.* , 672, 19
[12] Sheth R. K., Tormen G., 2002, *Mon. Not. R. Astron. Soc.* , 329, 61
[13] Wadsley J. W., Stadel J., Quinn T., 2004, NewA, 9, 137
[14] Katz N., 1992, *Astrophys. J.* , 391, 502
[15] Stinson G., Seth A., Katz N., Wadsley J., Governato F., Quinn T., 2006, MNRAS, 373, 1074
[16] Haardt F., Madau P., 1996, *Astrophys. J.* , 461, 20
[17] Governato F., Willman B., Mayer L., Brooks, A., Stinson G., Valenzuela O., Wadsley J., Quinn T., 2007, *Mon. Not. R. Astron. Soc.* , 374, 1479
[18] Moster B. P., Somerville R. S., Maulbetsch C., van den Bosch F. C., Macciò A. V., Naab T., Oser L., 2010, *Astrophys. J.* , 710, 903
[19] Hopkins A. M., Beacom J. F., 2006, *Astrophys. J.* , 651, 142
[20] Christensen C., Quinn T., Stinson G., Bellovary J., Wadsley J., 2010, arXiv:1005.1681

Journal of Physics: Conference Series **259** (2010) 012026 doi:10.1088/1742-6596/259/1/012026

The impact of renormalization on the observable predictions of inflation

Jose Navarro-Salas

Departamento de Fisica Teorica and IFIC, Centro Mixto Universidad de Valencia-CSIC. Facultad de Fisica, Universidad de Valencia, Burjassot-46100, Valencia, Spain

E-mail: jnavarro@ific.uv.es

Abstract. We analyze the generation of primordial perturbations in a (single-field) slow-roll inflationary universe. In momentum space, these (Gaussian) perturbations are characterized by a zero mean and a non-zero variance $\Delta^2(k,t)$. However, in position space the variance diverges in the ultraviolet. The requirement of a finite variance in position space forces one to regularize $\Delta^2(k,t)$. This can be achieved by proper renormalization in an expanding universe in a unique way. This affects the predicted scalar and tensorial power spectra (evaluated when the modes acquire classical properties) for wavelengths that today are at observable scales. As a consequence, the imprint of slow-roll inflation on the CMB anisotropies is significantly altered.

1. Introduction

A sufficiently long period of accelerated expansion in the very early universe is able to solve the questions raised by the standard big bang cosmology [1]. The hot big bang cosmology is an extremely successful theory. It explains the existence of the cosmic microwave background (CMB) and its thermal nature, the observed expansion of the universe, the abundance of light elements and the astrophysical fits for the age of the universe. However, it leaves without answer why our universe appears so homogeneous and nearly flat at large scales. Inflation offers a natural answer to these questions and, at the same time, provides a predictive mechanism to account for the small observed inhomogeneities [2] responsible for the structure formation in the universe and the anisotropies present in the cosmic microwave background (CMB), as first detected by the COBE satellite and further analyzed by the WMAP satellite [3]. Inflation predicts production of primordial density perturbations and relic gravitational waves as amplifications of vacuum fluctuations together with a quantum-to-classical transition at the scale of Hubble sphere crossing. Primordial perturbations leave an imprint in the CMB anisotropies, which are, therefore, of major importance for understanding our universe and its origin. The potential-energy density of a scalar (inflaton) field is assumed to cause the inflationary expansion, and the amplification of its quantum fluctuations and those of the metric are inevitable consequences in an expanding universe [4]. The metric fluctuations provide the initial conditions for the acoustic oscillations of the plasma at the onset of the subsequent radiation-dominated epoch. The detection of the effects of primordial gravitational waves in future high-precision measurements of the CMB anisotropies, as for instance in the PLANCK satellite mission [5], will serve as a highly non-trivial test for inflation. Therefore, it is particularly important to scrutinize, from all points of view, the standard predictions of inflation (see, for instance, [6]) to be tested empirically.

In this respect, it was pointed out in [7] (see also [8]) that quantum field renormalization significantly modifies the amplitude of quantum fluctuations, and hence the corresponding power spectra, in de Sitter inflation. The analysis was further developed in [9, 10] to understand how the fundamental testable predictions of (single-field) slow-roll inflation could be affected by renormalization. We summarize here our results and the logic of our approach.

2. Reexamining the calculation of the spectrum of perturbations

Let us assume that $\varphi(\vec{x},t)$ represents a perturbation obeying a free field wave-equation on the inflationary background $ds^2 = -dt^2 + a^2(t)d\vec{x}^2$, where $a(t)$ is a quasi-exponential expansion factor ($a(t) \sim e^{Ht}$). At the quantum level, this field is expanded as

$$\varphi(\vec{x},t) = \frac{1}{(2\pi)^{3/2}} \int d^3k [\varphi_k(t) a_{\vec{k}} e^{i\vec{k}\vec{x}} + \varphi_k^*(t) a_{\vec{k}}^\dagger e^{-i\vec{k}\vec{x}}] , \tag{1}$$

where the creation and annihilation operators satisfy the canonical commutation relation $[a_{\vec{k}}, a_{\vec{k}'}^\dagger] = \delta^3(\vec{k} - \vec{k}')$. The mode functions $\varphi_k(t)$ are required to satisfy the adiabatic condition (see, for instance, [11]). The power spectrum for this perturbation, $\Delta_\varphi^2(k,t)$, is usually defined in terms of the Fourier transform of the variance of the field

$$\langle \hat{\varphi}_{\vec{k}}(t) \hat{\varphi}_{\vec{k}'}^\dagger(t) \rangle = \delta^3(\vec{k} - \vec{k}') \frac{2\pi^2}{k^3} \Delta_\varphi^2(k,t) , \tag{2}$$

where $\hat{\varphi}_{\vec{k}}(t) \equiv \varphi_k(t) a_{\vec{k}}$. These modes describe a perturbation field characterized, in momentum space, by a zero mean $\langle \hat{\varphi}_{\vec{k}}(t) \rangle = 0$ and the variance (2). The advantage of working in momentum space resides in the fact that different modes fluctuate independently of each other, as explicitly displayed by the presence of the delta function in (2). This way, the quantum field is regarded as an infinite collection of oscillators, each with a different value of $\vec{k}$. In position space the perturbation is also characterized by a zero mean $\langle \varphi(\vec{x},t) \rangle = 0$ and a variance (or dispersion)

$$\langle \varphi^2(\vec{x},t) \rangle = \frac{1}{(2\pi)^3} \int d^3k d^3k' \langle \hat{\varphi}_{\vec{k}}(t) \hat{\varphi}_{\vec{k}'}^\dagger(t) \rangle e^{i(\vec{k}-\vec{k}')\vec{x}} , \tag{3}$$

which, due to spatial homogeneity, turns out to be independent of $\vec{x}$. This variance is formally related to the power spectrum by

$$\langle \varphi^2(\vec{x},t) \rangle = \int_0^\infty \frac{dk}{k} \Delta_\varphi^2(k,t) . \tag{4}$$

As is well-known in quantum field theory, the above expectation value quadratic in the field φ is divergent. It suffers from quadratic and logarithmic ultraviolet divergences

$$\langle \varphi^2(\vec{x},t) \rangle \sim \frac{1}{4\pi^2} \int_0^\infty dk \left(\frac{k}{a^2} + \frac{\dot{a}^2}{a^2 k} + \ldots\right) . \tag{5}$$

The first term corresponds to the usual contribution from vacuum fluctuations in Minkowski space. This contribution can be eliminated by standard renormalization in flat space. The logarithmic divergence, however, is characteristic of vacuum fluctuations in a curved background. Because the different k-modes fluctuate independently of each other, one could be tempted to get rid of this logarithmic ultraviolet divergence by simply eliminating the modes with $k > aH$ and leaving the rest unaffected (see, for instance, [12]). If one eliminates this divergence using a window function in this way, as is usual for random fields, then one obtains $\Delta_\varphi^2(k) \approx H^2/4\pi^2$, where $\Delta_\varphi^2(k)$ is defined by the quantity $\Delta_\varphi^2(k,t)$ evaluated a few Hubble times after the "horizon

Journal of Physics: Conference Series **259** (2010) 012026 doi:10.1088/1742-6596/259/1/012026

crossing time" t_k $(a(t_k)/k = H(t_k))$, since this is the time scale at which the modes behave as classical perturbations. However, one should take into account that the field fluctuations are quantum in nature and, therefore, one should consider the subtle points of quantum field theory (QFT) regarding the ultraviolet divergences.

Even though free quantum field theory is usually regarded as an infinite set of independent harmonic oscillators (one for each k-mode), there are fundamental holistic aspects of QFT that can not be properly understood in terms of independent modes. Renormalization is the hallmark of the holistic aspect of QFT. This is clear in the fact that, although the renormalization schemes in QFT in curved spacetimes are based on the ultraviolet behavior of the theory, the infrared sector is also affected by renormalization, leading potentially to observable consequences. This can be explicitly displayed by considering, for instance, the Casimir effect. The energy density between the two conducting plates obtained by proper renormalization provides the well-known and experimentally tested expression. However, a naive subtraction obtained by introducing a high-frequency cut-off in the integrals in momentum space (i.e., treating the k-modes as being independent) produces a quite different result (see the discussion of [13]).

Taking this into account, we see that the logarithmic divergence in (5) should be dealt with by renormalization and one can not rule out, a priori, the possibility that the treatment of the divergences at very high values of k may produce some impact at lower momenta. Therefore, we propose that in the standard definitions of the spectrum $\Delta^2_\varphi(k,t)$, as given in (2) and (4), one should replace the unrenormalized $\langle\varphi^2(\vec{x},t)\rangle$ by the renormalized variance, $\langle\varphi^2(\vec{x},t)\rangle_{ren}$. Writing $\tilde{\Delta}^2_\varphi(k,t)$ for the spectrum defined in this way, the definition in (4) (and similarly in (2)) is replaced by the corresponding renormalized expression

$$\langle\varphi^2(\vec{x},t)\rangle_{ren} = \int_0^\infty \frac{dk}{k}\tilde{\Delta}^2_\varphi(k,t) \ . \tag{6}$$

This way, the physical variance $\langle\varphi^2(\vec{x},t)\rangle_{ren}$ remains a well-defined quantity, in the same way as one could obtain a finite expression for the expectation values of the quantum stress-energy tensor. To complete the physical consistency of this approach, it would be desirable to define a unique expression for the necessary subtractions required to produce a consistent $\langle\varphi^2(\vec{x},t)\rangle_{ren}$. Since the power spectrum is defined in momentum space, the natural scheme is renormalization in momentum space, so we define

$$\langle\varphi^2(\vec{x},t)\rangle_{ren} = \frac{4\pi}{(2\pi)^3}\int_0^\infty k^2dk(|\varphi_k(t)|^2 - C_k(t)) \ , \tag{7}$$

where $C_k(t)$ represents the expected counterterms. Therefore, we have to apply well established renormalization techniques in QFT in curved spacetimes to properly remove these quantum divergences. As we will see all schemes of renormalization leads to the same counterterms.

3. Renormalization in momentum space

While in particle physics experiments the absolute value of the vacuum energy is unobservable, in the presence of gravity the absolute value of the vacuum energy matters. This is why normal ordering suffices to extract relevant physical quantities for free fields in Minkowski space but requires a refinement when the spacetime background departs from Minkowski space. A particularly intuitive method of renormalization was proposed by Bunch-Parker [14] in the late seventies. It was proposed aiming at extending to curved space the standard momentum-space methods of perturbation theory for interacting fields in Minkowsky space. This renormalization scheme introduce a local momentum-space representation of two-point functions which allows to remove the ultraviolet divergences of quantized fields in a general curved background. This way the standard Minkowskian propagator of a scalar free field in momentum space

$$\frac{1}{k^2+m^2} , \tag{8}$$

is replaced by a series with leading terms

$$\frac{1}{k^2+m^2} + \frac{(\frac{1}{6}-\xi)R}{(k^2+m^2)^2} + \dots , \tag{9}$$

where m is the mass and ξ is the coupling curvature parameter of the corresponding wave equation

$$[\nabla_\mu \nabla^\mu - m^2 - \xi R]\varphi(x) = 0 . \tag{10}$$

The Fourier transform leading to momentum space is crucially performed with respect to normal Riemann coordinates at a given point, which constitutes the best possible approximation in curved space to the inertial coordinates of Minkowski space. The presence of the curvature introduce additional divergent terms which are well counted by the above corrections to the Minkowskian propagator. For a massive field the above expansion, when translated to configuration space, turns out to be equivalent [14] to the DeWitt-Schwinger propertime representation of two-point functions [15, 16] (see also [11, 17]). However, in contrast to the DeWitt-Schwinger renormalization scheme, which requires $m \neq 0$ since it is based on an expansion in inverse power of the mass, the above momentum-space series can be extended to the massless limit. Following [14], the renormalized two-point function is given, in general, by the following subtraction

$$\langle\varphi(x)\varphi(x')\rangle_{ren} = \langle\varphi(x)\varphi(x')\rangle - i\frac{g^{-1/4}(x)}{(2\pi)^4}\int d^nke^{iky}[\frac{1}{p^2+m^2} + \frac{(\frac{1}{6}-\xi)R}{(p^2+m^2)^2}] , \tag{11}$$

where y^μ are the Riemann normal coordinates around the reference point x'.

We will now restrict now attention to a massless field and a cosmological metric of the form $ds^2 = -dt^2 + a^2(t)d\vec{x}^2$. At $t = t'$, and for a generic expansion factor $a(t)$, we have $y^0 = 0$ and $\vec{y} = a(t)(\vec{x} - \vec{x}')$. Performing then the k_0 integration we get for the subtraction term

$$-i\frac{g^{-1/4}(x)}{(2\pi)^4}\int d^4ke^{iky}[\frac{1}{p^2} + \frac{(\frac{1}{6}-\xi)R}{p^4}] = \frac{1}{2(2\pi)^3a^3}\int d^3\vec{k}\left[\frac{1}{w_k} + \frac{(\frac{1}{6}-\xi)R}{2w_k^3}\right]e^{i\vec{k}(\vec{x}-\vec{x}')} , \tag{12}$$

where $w_k = k/a(t)$. A further angular integration yields to the following subtraction counterterm

$$\frac{1}{4\pi^2a^3}\int_0^\infty dkk^2\left(\frac{a}{k} + (1-6\xi)\frac{a^3}{2k^3}\left[\left(\frac{\dot{a}}{a}\right)^2 + \frac{\ddot{a}}{a}\right]\right)\frac{\sin k|\vec{x}-\vec{x}'|}{k|\vec{x}-\vec{x}'|} . \tag{13}$$

At coincidence $\vec{x} \to \vec{x}'$ we have

$$(4\pi^2a(t)^3)^{-1}\int_0^\infty k^2dk[w_k(t)^{-1} + (W_k(t)^{-1})^{(2)}] , \tag{14}$$

where the counterterms $w_k(t)^{-1} + (W_k(t)^{-1})^{(2)}$ are

$$w_k^{-1} + (W_k^{-1})^{(2)} = \frac{a}{k} + (1-6\xi)\frac{a^3}{2k^3}\left[\left(\frac{\dot{a}}{a}\right)^2 + \frac{\ddot{a}}{a}\right] . \tag{15}$$

As pointed out in [10] the same counterterms are exactly obtained, in the massless limit, by the method of adiabatic renormalization [18], and this is why we have used the notation of adiabatic renormalization for naming the momentum-space counterterms. The normalized modes are usually expressed in momentum space as $(2(2\pi a(t))^3)^{-1/2}e^{i\vec{k}\vec{x}}\bar{\varphi}_k(t)$. Then the renormalized variance $\langle\varphi^2(\vec{x},t)\rangle_{ren}$ is finally given by the expression

$$\langle\varphi^2(\vec{x},t)\rangle_{ren} = (4\pi^2 a(t)^3)^{-1}\int_0^\infty k^2 dk[|\bar{\varphi}_k(t)|^2 - w_k(t)^{-1} - (W_k(t)^{-1})^{(2)}] \ . \quad (16)$$

Therefore, for the field perturbations arising from inflation the counterterms $C_k(t)$ in (7) are univocally constructed thus defining a unique expression for the spectrum $\tilde{\Delta}^2_\varphi(k,t)$. The holistic nature of QFT is then explicitly realized through (6). Although the counterterms are fully determined by the ultraviolet behavior of the modes, the long wavelength sector, and hence the new $\tilde{\Delta}^2_\varphi(k,t)$, is significantly affected by the subtractions.

4. New predictions for the power spectra

To evaluate the physical power spectrum we have to specify when the classical perturbations are created. It happens when the decaying mode becomes very small relative to the dominant one [19] and this corresponds a few n efolds after the Hubble radius exit time t_k (we note that n is related to the squeezing parameter). In the slow-roll scenario, when H slowly decreases with time, the effects of renormalization have a non-trivial impact on $\tilde{\Delta}^2_\varphi(k,t)$ when this quantity is evaluated a few Hubble times after the time of horizon crossing t_k. We then obtain for the tensorial and scalar spectra

$$\begin{aligned} \tilde{\Delta}^2_t(k,n) &\approx \frac{2}{M_P^2}\left(\frac{H(t_k)}{2\pi}\right)^2 \epsilon(t_k)(2n-3/2) \\ \tilde{\Delta}^2_s(k,n) &\approx \frac{1}{2M_P^2\epsilon(t_k)}\left(\frac{H(t_k)}{2\pi}\right)^2 (3\epsilon(t_k)-\eta(t_k))(2n-3/2) \ , \end{aligned} \quad (17)$$

where ϵ, η are the standard slow-roll parameters. Note that the parameter n enters in the power spectra parameterizing the (unknown) time at which the modes exhibit classical behavior. However, since both tensorial and scalar spectra have the same dependence on n, the tensor-to-scalar ratio r is not sensitive to the unknown parameter n, in the same way as it is insensitive to the scale of inflation $H(t_k)$. As a consequence of (17), the imprint of slow-roll inflation on the CMB anisotropies is significantly altered. Note that, in contrast with the standard (unrenormalized) approach, the tensorial amplitude does not uniquely depend on the scale of inflation. Moreover, one can also obtain a non-trivial change in the consistency condition that relates the tensor-to-scalar ratio r to the spectral indices [10] for single-field inflation

$$r = 4(1-n_s-n_t) + \frac{4n_t'}{n_t^2-2n_t'}\left(1-n_s-\sqrt{2n_t'+(1-n_s)^2-n_t^2}\right) \ . \quad (18)$$

The tensor-to-scalar ratio is now related with the spectral indices n_t, $(1-n_s)$, and also $n_t' \equiv dn_t/d\ln k$. This modification has far reaching consequences. For instance, since the observations from WMAP [3](with BAO+SN) strongly suggest that $(1-n_s) \approx 0.030\pm0.015$ (with $r < 0.22$), expression (18) allows for an exact scale invariant tensorial power spectrum, $n_t = 0$, while being compatible with a non-zero ratio $r \approx 0.12 \pm 0.06$, which is forbidden by the standard prediction ($r = -8n_t$). We also note that the new predictions remain in agreement with observation for the simplest forms of inflation (ϕ^2 and ϕ^4 potentials). The influence of relic gravitational waves on the CMB may soon come within the range of planned measurements, offering a non-trivial test of the new predictions.

Acknowledgments
This work has been partially supported by the spanish grant FIS2008-06078-C03-02. The author thanks I. Agullo, G.J. Olmo and L. Parker for collaboration on the topic of this talk.

References
[1] Guth A 1981 *Phys. Rev.* D**23** 347
Starobinsky A A 1980 *Phys. Lett.* B**91** 99
Linde, A D 1982 *Phys. Lett.* B**108**, 389 (1982); 1983 *Phys. Lett.* B**129** 177
Albrecht A and Steinhardt P J 1982 *Phys. Rev. Lett.* **48** 1220
Sato K 1981 *Mon. Not. Roy. Astron. Soc.* **195** 467
[2] Mukhanov V F and Chibisov G V 1981 *JETP Letters* **33** 532
Hawking S W 1982 *Phys. Lett.* B**115** 295
Guth A and Pi S-Y 1982 *Phys. Rev. Lett.* **49** 1110
Starobinsky A A 1982 *Phys. Lett.* B**117** 175
Bardeen J M, Steinhardt P J and Turner, M S 1983 *Phys. Rev.*D**28** 679
[3] Komatsu E 2009 *et al., J. Suppl. Ser.* **180** 330
[4] Parker L 1968 *Phys. Rev. Lett.* **21** 562; 1969 *Phys. Rev.* **183** 1057
[5] Efstathiou G., Lawrence C. and Tauber J., ESA-SCI(2005)1
[6] Liddle A R and Lyth D H 2000 *Cosmological inflation and large-scale structure*, Cambridge University Press
Dodelson S 2003 *Modern Cosmology*, Academic Press
Weinberg S 2008 *Cosmology*, Oxford University Press
[7] Parker L 2007 Amplitude of perturbations from inflation, hep-th/0702216.
[8] Agulló I, Navarro-Salas J, Olmo G J and Parker L 2008 *Phys. Rev. Lett.* **101**, 171301
[9] Agulló I, Navarro-Salas J, Olmo G J and Parker L 2009 *Phys. Rev. Lett.* **103**, 061301
Agulló I, Navarro-Salas J, Olmo G J and Parker L 2009 *Gen. Rel. Grav.* **41**, 2301
[10] Agulló I, Navarro-Salas J, Olmo G J and Parker L 2010 *Phys. Rev. D* **81** 043514
[11] Parker L and Toms D J 2009 *Quantum field theory in curved spacetime: quantized fields and gravity*, Cambridge University Press
[12] Lyth D H and Liddle A R 2009 *The primordial density perturbation*, Cambridge University Press
[13] Hollands C and Wald R 2004 *Quantum field theory is not merely quantum mechanics applied to the low energy effective degrees of freedom*, arXiv:gr-qc/0405082v1
[14] Bunch T S and Parker L 1979 *Phys. Rev.* D**20** 2499
[15] DeWitt B S 1975 *Phys. Rep.* **19C** 296
[16] Christensen S M 1976 *Phys. Rev.* D**14** 2490
[17] Birrel N D and Davies P C W 1982 *Quantum fields in curved space*, Cambridge University Press
[18] Parker L and Fulling S A 1974 *Phys. Rev. D* **9** 341
Fulling S A and Parker L 1974 *Ann. Phys. (N.Y.)* **87** 176
Fulling S A, Parker L and Hu B L 1974 *Phys. Rev. D* **10** 3905
[19] Kiefer C, Lohmar I, Polarski D and Starobinsky A A 2007 *Class. Quantum Grav.* **24** 1699
Polarski D and Starobinsky A A 1996 *Class. Quantum Grav.* **13** 377
Albrecht A, Ferreira P, Joyce M and Prokopec T 1994 *Phys. Rev. D* **50** 4807.

16th International Symposium on Particles Strings and Cosmology (PASCOS 2010) IOP Publishing
Journal of Physics: Conference Series **259** (2010) 012027 doi:10.1088/1742-6596/259/1/012027

Nongaussian Features from Inflationary Particle Production

Neil Barnaby

School of Physics and Astronomy, 116 Church Street S.E., Minneapolis, MN, 55455 USA

E-mail: barnaby@physics.umn.edu

Abstract. The inflaton field can be expected to couple to a number of additional fields whose energy density does not play any significant role in driving inflation. Such couplings may lead to isolated bursts of particle production during inflation, for example via parametric resonance or a phase transition, and leave observable imprints in the cosmological fluctuations. I illustrate this effect for a simple prototype interaction $g^2(\phi - \phi_0)^2\chi$ between the inflaton, ϕ, and iso-inflaton, χ. Using both classical lattice simulations and analytical quantum field theory computations, I show that this mechanism generates localized bump-like features in the power spectrum and also a completely new type of nongaussianity. Observations are consistent with relatively large features of this type and the nongaussianity from particle production may be observable in future missions.

1. Introduction

The inflationary paradigm has become a cornerstone of modern cosmology. As measurements of the Cosmic Microwave Background (CMB) radiation grow increasingly precise, it has become topical to look beyond the simplest single-field, slow-roll inflationary scenario. In particular, it is interesting to determine the extent to which non-minimal signatures, such as features in the primordial power spectrum or observable nongaussianities, can be accommodated by microscopically sensible inflation models. Efforts in this direction are valuable because they allow us to test our theoretical prejudices and they provide observers with well-motivated templates for departures from the standard scenario. If detected, such signatures might open an observational window into fundamental particle physics at extremely high energy scales.

The motion of the inflaton may trigger the production of some non-inflaton (iso-curvature) particle *during* inflation [1, 2, 3]. Inflationary particle production is a generic feature of models from string theory (*eg* - brane/axion monodromy [4]) and also supersymmetric field theory [5]. Examples have been studied where particle production occurs via parametric resonance [1, 2, 3, 6, 7, 8, 9], as a result of a phase transition [10, 11], or otherwise [12]. These constructions have attracted interest recently for a number of reasons, including the possibility to exploit the dissipative effect of particle production to slow the motion of the inflaton on a steep potential [9, 12] (see also [13]).

In this talk, which is based [1, 2, 3], I will discuss the observational signatures of inflationary particle production. This scenario provides a simple, microscopically well-motivated mechanism to generate features in the primordial power spectrum and also observably large nongaussianity. The bispectrum from particle production is particularly novel: we find a completely new kind of nongaussian signature which has been overlooked in previous literature. For reasonable values of model parameters, this should be detectable in future missions.

2. Particle Production During Inflation

To illustrate the basic physics of inflationary particle production, we consider the following prototype action

$$S = \int d^4x\sqrt{-g}\left[-\frac{1}{2}(\partial\phi)^2 - V(\phi) - \frac{1}{2}(\partial\chi)^2 - \frac{\mu^2}{2}\chi^2 - \frac{g^2}{2}(\phi-\phi_0)^2\chi^2\right] \tag{1}$$

which is understood to be minimally coupled to gravity. Here ϕ is the inflaton, χ is the iso-inflaton and we leave the potential $V(\phi)$ unspecified expect to assume that it is sufficiently flat to drive a long period of inflation with $H \equiv \dot{a}/a \cong$ const where $a(t)$ is scale factor. Note that we have included a "bare" mass μ^2 for the iso-inflaton; even if χ is classically massless at $\phi = \phi_0$, then such a term will be generated by radiative corrections.

The scenario we have in mind is the following. Inflation starts with $\phi \gg \phi_0$ so that the iso-inflaton is extremely massive $m_\chi^2 \equiv \mu^2 + g^2(\phi-\phi_0)^2 \gg H^2$ and stays pinned in the vacuum, $\chi = 0$. At the moment when $\phi = \phi_0$ the iso-inflaton mass varies non-adiabatically $|\dot{m}_\chi|/m_\chi^2 \gtrsim 1$ and χ particles are produced quantum mechanically. To describe this burst of particle production one must solve the following equation for the χ-particle mode functions:

$$\ddot{\chi}_k + 3H\dot{\chi}_k + \left[\frac{k^2}{a^2} + \mu^2 + k_\star^4 t^2\right]\chi_k = 0 \tag{2}$$

Here we have approximated $\phi \cong \phi_0 + vt$ and introduced the scale $k_\star \equiv \sqrt{g|v|} = \sqrt{g/(2\pi\mathcal{P}_\zeta^{1/2})}H$ where $\mathcal{P}_\zeta^{1/2} = 5\times 10^{-5}$ is the amplitude of the vacuum fluctuations from inflation. For reasonable values of the coupling $g^2 \gtrsim 10^{-7}$ we have $k_\star > H$ and particle production is rapid as compared to the expansion time. In this regime one can solve (2) very accurately for the occupation number of the created particles [14, 15]

$$n_k = e^{-\pi(\mu^2+k^2)/k_\star^2} \tag{3}$$

Clearly particle production effects may be suppressed if μ^2 is very large. For string theory or supersymmetric models, one naturally has $\mu^2 \sim H^2 \ll k_\star^2$ and there is no suppression.

Following the initial burst of particle production, two distinct physical effects take place. First, the energetic cost of producing a gas of non-equilibrium χ particles drains energy from the inflaton, forcing $\dot{\phi}(t)$ to drop abruptly. This velocity dip is a *backreaction* effect and contributes a negligible "ringing" pattern to the power spectrum [8]. The second physical effect, which yields the dominant contribution to the cosmological fluctuations, is called *rescattering* [1] and will be the subject of the remainder of this talk.

3. Rescattering and Infra-Red Cascading

The importance of rescattering effects for the observable cosmological fluctuations in models with particle production was first recognized in [1]. Fig. 1 illustrates the key process: bremsstrahlung emission of long-wavelength $\delta\phi$ fluctuations from rescattering of the produced χ particles off the condensate $\phi(t)$. The time scale for such processes, $k_\star^{-1}$, is fast as compared to the expansion time, H^{-1}. Moreover, the production of inflaton fluctuations $\delta\phi$ deep in the IR is extremely energetically inexpensive, since the inflaton is nearly massless. The combination of the short time scale for rescattering and the energetic cheapness of radiating IR $\delta\phi$ leads to a rapid build-up of power in long wavelength inflaton modes: infra-red (IR) cascading. These long-wavelength inflaton modes freeze outside of the horizon and lead to a bump-like feature in the primordial power spectrum.

In [1] the model (1) was studied using lattice field theory simulations, without neglecting any physical processes (that is to say that full nonlinear structure of the theory, including backreaction and rescattering effects, was accounted for consistently). However, this same dynamics can be understood analytically [3] by solving the equation for the inflaton fluctuations $\delta\phi$ in the approximation that all interactions are neglected, except for the diagram Fig. 1:

$$\delta\ddot{\phi} + 3H\delta\dot{\phi} - \frac{\vec{\nabla}^2}{a^2}\delta\phi + V_{,\phi\phi}\delta\phi \cong -g^2\left[\phi(t) - \phi_0\right]\chi^2 \qquad (4)$$

The solution of (4) may be split into two parts: the solution of the homogeneous equation and the particular solution which is due to the source term. Schematically we have

$$\delta\phi(t,\mathbf{x}) = \underbrace{\delta\phi_{\rm vac}(t,\mathbf{x})}_{\rm homogeneous} + \underbrace{\delta\phi_{\rm resc}(t,\mathbf{x})}_{\rm particular} \qquad (5)$$

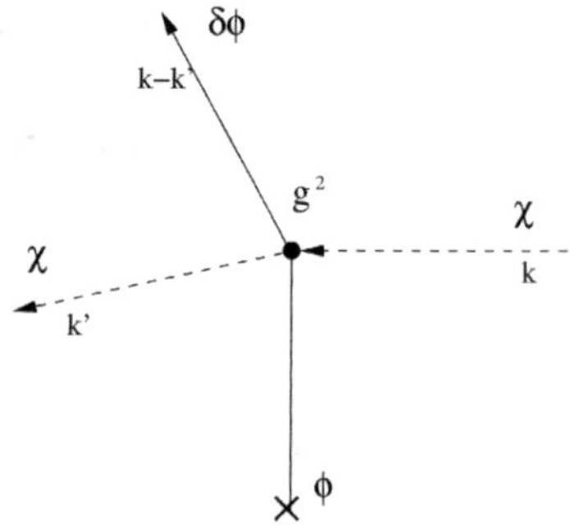

Figure 1. Rescattering diagram

The former contribution is the homogeneous solution which behaves as $\delta\phi_{\rm vac} \sim H/(2\pi)$ on large scales and, physically, corresponds to the usual scale invariant vacuum fluctuations from inflation. The particular solution, $\delta\phi_{\rm resc}$, corresponds physically to inflaton fluctuations which are generated by rescattering. The abrupt growth of χ inhomogeneities at $t = 0$ sources the particular solution $\delta\phi_{\rm resc}$, leading to the production of inflaton fluctuations which subsequently cross the horizon and become frozen. We have studied the dynamics of rescattering and IR cascading in the model (1) using both fully nonlinear lattice simulations and also analytical quantum field theory computations. We have found remarkable agreement between these two independent approaches, for a wide range of model parameters [1, 3].

Notice that the particular solution $\delta\phi_{\rm resc}$ of equation (4) is bi-linear in the gaussian field χ, suggesting that the fluctuations from rescattering will be highly nongaussian. We have explored the nongaussianity of the inflaton modes from rescattering in two different ways: by analytically computing the bispectrum and also by numerical evaluation of the Probability Distribution Function (PDF) [3]. We define the PDF, $P(\delta\phi)$, as the probability that the inflaton field has a fluctuation of size $\delta\phi = \phi - \langle\phi\rangle$. This is plotted in the left panel of Fig. 2 for several time steps during the evolution. In order to make the physics of inflationary particle production clear, we have subtracted off the usual vacuum fluctuations of the inflaton. That is, the PDF in the left panel of Fig. 2 is associated only with the contribution $\delta\phi_{\rm resc}$ in (5).

We can understand physically the behaviour of PDF plotted in the left panel of Fig. 2. Shortly after the initial burst of particle production the inflaton perturbations $\delta\phi$ are extremely nongaussian, due to the sudden appearance of the source term $J \propto \chi^2$ in the equation of motion (4). Very quickly, in less than an e-folding, nonlinear interactions begin to drive the system towards gaussianity. A very similar behaviour has been observed in lattice simulations of out-of-equilibrium interacting scalar fields during preheating [16]. In the case of rescattering during preheating, the system will eventually become gaussian when the fields thermalize. However, in our case the universe is still inflating during the process of rescattering and IR cascading. As a result, nongaussian inflaton fluctuations generated by rescattering are stretched out by the quasi-de Sitter expansion and must freeze once their wavelength crosses the Hubble scale. Hence, at late times the PDF does not become completely gaussian, but rather freezes-in with some non-trivial skewness. Within a few e-foldings from the moment of particle production the time evolution of the PDF has become completely negligible.

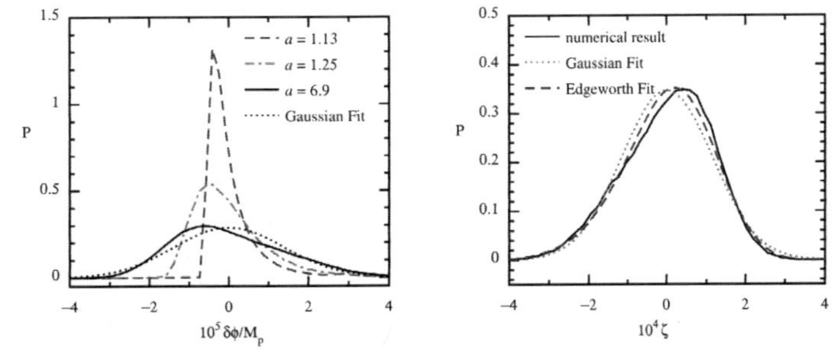

Figure 2. The left panel show the PDF of the inflaton fluctuations generated by rescattering and IR cascading, at a series of different values of the scale factor, a. The dotted black curve shows a Gaussian fit at late times and we have normalized the scale factor so that $a = 1$ at the moment when particle production occurs. The right panel shows the PDF of the *total* curvature fluctuation ζ, evaluated at late times (well after all relevant modes have crossed the horizon and frozen). The solid black curve is the exact result and the dotted red curve is a gaussian fit. We have also plotted the leading correction to the gaussian result in the Edgeworth expansion. For illustration, we have chosen $g^2 = 0.1$ and a standard chaotic inflation potential $V(\phi) = m^2\phi^2/2$ in both panels.

4. Features and Nongaussianity

Let us now consider the observational signatures associated with the dynamics described in the last section. The primordial power spectrum in the model (1) is well approximated by [2]

$$P_\zeta(k) = A_s\left(\frac{k}{k_0}\right)^{n_s-1} + A_{\rm IR}\left(\frac{\pi e}{3}\right)^{3/2}\left(\frac{k}{k_{\rm IR}}\right)^3 e^{-\frac{\pi}{2}\left(\frac{k}{k_{\rm IR}}\right)^2} \tag{6}$$

The first term corresponds to the usual vacuum fluctuations from inflation (with amplitude A_s and spectral index n_s). The second term in (6) corresponds to the bump-like feature from particle production with amplitude $A_{\rm IR}$ (that depends on g^2) and location $k_{\rm IR}$ (that depends on ϕ_0). Current observational data are compatible with features as large as $A_{\rm IR}/A_s \sim 0.1$ for $k_{\rm IR}$ corresponding to CMB scales, whereas even larger features are allowed on smaller scales. In Fig. 3 we plot the power spectrum (6) for a representative choice of parameters.

Nongaussian statistics have attracted a considerable amount of interest recently, owing to their potential as a tool for observationally discriminating between the plethora of inflationary models in the literature. Nongaussianity is often characterized using the *bispectrum* [17], defined as the 3-point correlation function of the fourier transform of the curvature fluctuation:

$$\langle \zeta_{\mathbf{k_1}}\zeta_{\mathbf{k_2}}\zeta_{\mathbf{k_3}}\rangle = (2\pi)^3\delta^{(3)}(\mathbf{k_1}+\mathbf{k_2}+\mathbf{k_3})B_\zeta(k_i) \tag{7}$$

Translational invariance ensures that $B_\zeta(k_i)$ depends on three wave-numbers which form a triangle: $\mathbf{k_1}+\mathbf{k_2}+\mathbf{k_3} = 0$. A general bispectrum $B_\zeta(k_i)$ may be characterized by specifying its size (amplitude of B_ζ), shape (whether B_ζ peaks on squeezed, equilateral or flattened triangles) and running (the dependence of B_ζ on the size of the triangle) [18].

As remarked above, the fluctuations generated by rescattering and IR cascading are highly nongaussian. This nongaussianity is very different from other models, such as the local, equilateral or enfolded shapes, which have been studied in the literature [3]. The bispectrum $B_\zeta(k_i)$ peaks strongly for triangles with a characteristic size $\sim k_{\rm IR}$, corresponding to the location of the bump in the power spectrum (6), and is therefore very far from scale invariant. The shape of $B_\zeta(k_i)$ is also novel [3].

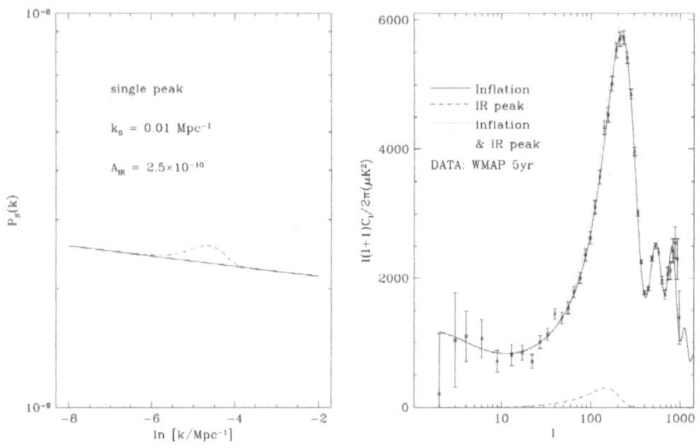

Figure 3. The left panel shows a sample bump in the power spectrum with amplitude $A_{\rm IR} = 2.5 \times 10^{-10}$, which corresponds to a realistic coupling $g^2 \sim 0.01$, and location $k_{\rm IR} = 0.01\,{\rm Mpc}^{-1}$. This example represents a distortion of $\mathcal{O}(10\%)$ as compared to the usual vacuum fluctuations and is consistent with the data at 2σ. The right panel shows the corresponding CMB angular TT power spectrum.

The unusual shape/running of the bispectrum from particle production makes it difficult to compare the magnitude of the nongaussianity in our scenario to more familiar models, such as the local template. In order to quantify the nongaussianity from particle production we find it useful to compute the cummulants of the Probability Distribution Function (PDF). In the right panel of Fig. 2 we plotted the PDF of the inflaton fluctuations from rescattering. However, the quantity that is relevant for observations is the PDF of the *total* curvature perturbation, ζ, including *both* the contributions in (5). In the right panel of Fig. 2 we plot this quantity, evaluated at very late times, well after all relevant modes have crossed the horizon and become frozen.

Let us define the central moments of the PDF as

$$\langle \zeta^n \rangle \equiv \int \zeta^n P(\zeta) d\zeta \tag{8}$$

These moments carry information about the correlation functions of ζ integrated over all wave-numbers, and therefore provide a useful tool to compare models with very different shape/running properties [19]. The dimensionless skewness, $\hat{\kappa}_3 \equiv \langle\zeta^3\rangle/\langle\zeta^2\rangle^{3/2}$, and kurtosis, $\hat{\kappa}_4 \equiv \langle\zeta^4\rangle/\langle\zeta^2\rangle^2 - 3$, encode departures from gaussianity. These are summarized in Table 1. We have also computed an "equivalent $f_{NL}^{\rm local}$" which, for a given g^2, is the magnitude of f_{NL} necessary to reproduce the skewness $\hat{\kappa}_3$ with a local ansatz $\zeta = \zeta_g + \frac{3}{5} f_{NL}\left[\zeta_g^2 - \langle\zeta_g^2\rangle\right]$.

Table 1. Moments of the Probability Distribution Function

g^2	skewness ($\hat{\kappa}_3$)	kurtosis ($\hat{\kappa}_4$)	"equivalent" $f_{NL}^{\rm local}$
1	−0.51	0.2	−4500
0.1	−0.49	−0.1	−4300
0.01	−0.006	$< \mathcal{O}(10^{-3})$	−53

The coupling, g^2, controls both the nongaussianity and also the magnitude of the bump-like feature in (6). A key question is whether nongaussian effects can be observable in a regime where the feature is small enough to have evaded detection. The answer seems to be affirmative: for $g^2 = 0.01$ the spectrum (6) fits the data at 2σ [2] while $\hat{\kappa}_3 = -0.006$. This level of skewness would be produced by a local model with $f_{NL} \sim -53$, which is comparable to current observational bounds and is well within the accuracy of future missions.

5. Conclusions

Particle production during inflation is a simple and microscopically well-motivated mechanism that generates localized features in the primordial power spectrum and also significant nongaussianities. Such signatures provide a novel example of non-decoupling,[1] suggesting a possibility to probe extremely high scale physics with cosmology. The new type of nongaussianity that we have discovered is phenomenologically interesting and can be large even without tine-tuning the inflationary trajectory or appealing to re-summation of an infinite series of high dimension operators.

The key process that generates cosmological perturbations in our model, IR cascading, is interesting it its own right: it is qualitatively different from other mechanisms in the literature (in that we do not rely on the quantum vacuum fluctuations of some light iso-curvature fields) and the underlying dynamics are relevant also for preheating, moduli trapping and non-equilibrium quantum field theory more generally.

I am grateful to the organizers of PASCOS 2010 for providing me with the opportunity to present this material.

[1] N. Barnaby, Z. Huang, L. Kofman and D. Pogosyan, Phys. Rev. D **80**, 043501 (2009) [arXiv:0902.0615 [hep-th]].
[2] N. Barnaby and Z. Huang, arXiv:0909.0751 [astro-ph.CO].
[3] N. Barnaby, arXiv:1006.4615 [astro-ph.CO].
[4] E. Silverstein and A. Westphal, Phys. Rev. D **78**, 106003 (2008) [arXiv:0803.3085 [hep-th]]. L. McAllister, E. Silverstein and A. Westphal, arXiv:0808.0706 [hep-th].
[5] A. Berera and T. W. Kephart, Phys. Rev. Lett. **83**, 1084 (1999) [arXiv:hep-ph/9904410].
[6] D. J. H. Chung, E. W. Kolb, A. Riotto and I. I. Tkachev, Phys. Rev. D **62**, 043508 (2000) [arXiv:hep-ph/9910437].
[7] O. Elgaroy, S. Hannestad and T. Haugboelle, JCAP **0309**, 008 (2003) [arXiv:astro-ph/0306229].
[8] A. E. Romano and M. Sasaki, Phys. Rev. D **78**, 103522 (2008) [arXiv:0809.5142 [gr-qc]].
[9] D. Green, B. Horn, L. Senatore and E. Silverstein, Phys. Rev. D **80**, 063533 (2009) [arXiv:0902.1006 [hep-th]].
[10] N. Barnaby and J. M. Cline, Phys. Rev. D **73**, 106012 (2006) [arXiv:astro-ph/0601481]. N. Barnaby and J. M. Cline, Phys. Rev. D **75**, 086004 (2007) [arXiv:astro-ph/0611750].
[11] L. A. Kofman and A. D. Linde, Nucl. Phys. B **282**, 555 (1987). L. A. Kofman and D. Y. Pogosian, Phys. Lett. B **214**, 508 (1988). D. S. Salopek, J. R. Bond and J. M. Bardeen, Phys. Rev. D **40**, 1753 (1989). J. A. Adams, B. Cresswell and R. Easther, Phys. Rev. D **64**, 123514 (2001) [arXiv:astro-ph/0102236].
[12] M. M. Anber and L. Sorbo, arXiv:0908.4089 [hep-th].
[13] A. Berera, Phys. Rev. Lett. **75**, 3218 (1995) [arXiv:astro-ph/9509049].
[14] L. Kofman, A. D. Linde and A. A. Starobinsky, Phys. Rev. D **56**, 3258 (1997) [arXiv:hep-ph/9704452].
[15] L. Kofman, A. D. Linde, X. Liu, A. Maloney, L. McAllister and E. Silverstein, JHEP **0405**, 030 (2004) [arXiv:hep-th/0403001].
[16] G. N. Felder and L. Kofman, Phys. Rev. D **63**, 103503 (2001) [arXiv:hep-ph/0011160].
[17] V. Acquaviva, N. Bartolo, S. Matarrese and A. Riotto, Nucl. Phys. B **667**, 119 (2003) [arXiv:astro-ph/0209156]. J. M. Maldacena, JHEP **0305**, 013 (2003) [arXiv:astro-ph/0210603]. D. Seery and J. E. Lidsey, JCAP **0506**, 003 (2005) [arXiv:astro-ph/0503692].
[18] X. Chen, M. x. Huang, S. Kachru and G. Shiu, JCAP **0701**, 002 (2007) [arXiv:hep-th/0605045].
[19] M. LoVerde, A. Miller, S. Shandera and L. Verde, JCAP **0804**, 014 (2008) [arXiv:0711.4126 [astro-ph]].

[1] The produced χ particles are extremely massive for (almost) the entire history of the universe, however, their effect cannot be integrated out due to the non-adiabatic time dependence of the χ modes.

Journal of Physics: Conference Series **259** (2010) 012028 doi:10.1088/1742-6596/259/1/012028

Inflation with racetrack superpotential and matter field

Marek Olechowski

Institute of Theoretical Physics, University of Warsaw, ul. Hoa 69, PL-00-681 Warsaw, Poland

E-mail: Marek.Olechowski@fuw.edu.pl

Abstract. Two classes of racetrack inflation models, saddle point and inflection point ones, are constructed in a fully supersymmetric framework with the matter field F-term as a source of supersymmetry breaking and uplifting. Two kinds of the matter sector are considered: the Polonyi model and the quantum corrected O'Raifeartaigh model. The inflaton is dominated by the matter field in the former case, and by the volume modulus in the latter. Fine-tuning of the parameters is relaxed as compared to the corresponding racetrack models with non-supersymmetric uplifting. A low scale of supersymmetry breaking may be obtained in some cases.

1. Introduction

Many string inspired models of inflation are based on the KKLT mechanism of moduli stabilization [1]. In the simplest case, the effective superpotential for the volume modulus contains a constant term originating from fluxes and one non-perturbative exponential term. With two such non-perturbative terms, and for appropriate choice of parameters, the resulting potential is suitable for slow roll inflation. There are two classes of these so-called racetrack models. Inflation may be realized in the vicinity of a flat saddle point of the potential [2] or in the vicinity of a flat inflection point [3]. The inflaton is dominated by the imaginary component of the volume modulus in the former case and by its real component in the latter. In both classes of models the vacuum is obtained by appropriate uplifting of a (deep) AdS minimum. Unfortunately, the uplifting term in the KKLT mechanism comes from $\overline{\text{D3}}$-branes and breaks supersymmetry (SUSY) explicitly. This is the main problem of the KKLT approach (supersymmetric formalism used for a non-supersymmetric model). There were many proposals to overcome this difficulty. Uplifting and spontaneous SUSY breaking due to F-terms and/or D-terms for matter fields were used to obtain consistent versions of the KKLT mechanism for moduli stabilization. However, very few inflation models with spontaneous SUSY breaking were proposed so far [4, 5]. We show below that saddle point and inflection point racetrack models can be constructed in a fully supersymmetric framework using the matter field F-terms to brake SUSY and uplift the potential.

2. Constraints on Kähler potential

The F-term contribution to the supergravity potential,

$$V = e^G \left(G^{i\bar{j}} G_i G_{\bar{j}} - 3 \right) , \tag{1}$$

depends on the superpotential W and the Kähler potential K via the combination

$$G(\Phi_i, \Phi_i^\dagger) = K(\Phi_i, \Phi_i^\dagger) + \log W(\Phi_i) + \log \overline{W}(\Phi_i^\dagger) \tag{2}$$

(in standard notation: $G_i \equiv \partial G/\partial\Phi_i$, $G_{i\bar{j}} \equiv \partial^2 G/\partial\Phi_i\partial\Phi_j^\dagger$, etc. and $G^{i\bar{j}}$ is the matrix inverse to $G_{i\bar{j}}$). In general, this potential is a complicated function of the fields Φ_i and it is difficult to analyze its structure, e.g. properties of stationary points. However, in all phenomenologically reasonable models, some relatively simple constraints on the Kähler potential can be found. The necessary condition for a stable dS vacuum is given by [6]:

$$R(f) < \frac{2}{\widehat{G}^2}, \tag{3}$$

where $\widehat{G}^2 \equiv G^i G_i = 3 + e^{-G}V$ and $R(f) \equiv R_{i\bar{j}p\bar{q}} f^i f^{\bar{j}} f^p f^{\bar{q}}$ is the sectional curvature along the Goldstino (i.e. SUSY breaking) direction defined by the unit vector f with the components $f_i = G_i/\widehat{G}$. It is important that the sectional curvature $R(f)$ depends only on the Kähler potential and not on the superpotential (because it involves only mixed holomorphic-antiholomorphic derivatives of G). The right hand side of (3) depends on the potential but for any positive vacuum energy, $\widehat{G}^2$ is bigger than 3. Therefore, somewhat weaker condition $R(f) < 2/3$ do not depend on the superpotential and must be fulfilled in any model with a dS vacuum.

Almost the same necessary condition as (3) must be fulfilled [7, 8] in models of slow roll inflation. There are two differences. First: in a minimum all first derivatives of the potential vanish. In an inflationary region the derivative in the inflaton direction is non-zero but it must be very small in order to fulfill the slow roll conditions. Second: in the case of a minimum all directions in the field space must be stable i.e. all eigenvalues of the matrix of the second derivatives of the potential must be positive. In the case of inflation, one of these eigenvalues may be negative but its absolute value must be very small. So, formally, the necessary condition for an inflationary region is slightly weaker than (3). However, in practice, it is more difficult to find a model allowing for inflation than a model with a stable dS minimum. The reason is that the value of $\widehat{G}^2$ in the inflationary region is bigger (usually substantially bigger) than in the dS minimum with a very small cosmological constant.

To illustrate the importance of condition (3) we consider supergravity models with the no-scale Kähler potential $K = -3\ln(T + \overline{T} - |\Phi|^2)$. The Kähler field space in such models has constant curvature so, irrespectively of the details of SUSY breaking, the sectional curvature $R(f) = 2/3$ and the necessary condition (3) is violated. So, no matter what superpotential we choose, the no-scale models have neither stable dS minima nor regions suitable for inflation. The same is true for no-scale models without matter fields for which $K = -3\ln(T + \overline{T})$. The last statement may look surprising because there are no-scale racetrack inflation models (e.g. [2, 3]). This apparent discrepancy appears because the conditions discussed above are valid for supersymmetric models while the uplifting terms in most inflation models break SUSY explicitly (see [7]).

3. Racetrack inflation with matter F-term uplifting

We investigate viability of inflationary models based on the racetrack superpotential with uplifting and spontaneous SUSY breaking provided by a hidden sector matter field. As we will see later, it is enough to consider quite simple models in which both the superpotential and the Kähler potential are sums of contributions depending on the volume modulus, $T = t + i\tau$, and one complex matter field, $\Phi = \phi + i\theta$:

$$K = -3\ln(T + \overline{T}) + K^{(\Phi)}(\Phi, \overline{\Phi}), \tag{4}$$

$$W = A + Ce^{-cT} + De^{-dT} + W^{(\Phi)}(\Phi)\,. \tag{5}$$

The matter field superpotential also may be very simple:

$$W^{(\Phi)} = c_0 - \mu^2\Phi\,. \tag{6}$$

We consider two forms of the matter field Kähler potential: the canonical one

$$K_{\mathrm{P}}^{(\Phi)} = \Phi\overline{\Phi}\,, \tag{7}$$

and

$$K_{\mathrm{O}}^{(\Phi)} = \Phi\overline{\Phi} - \frac{(\Phi\overline{\Phi})^2}{\Lambda^2}\,. \tag{8}$$

The subscripts P and O were used because (7) and (6) describe the Polonyi model of SUSY breaking while (8) and (6) with $c_0 = 0$ correspond to an effective quantum corrected O'Raifeartaigh model with the superpotential $W = mXY + \lambda\Phi X^2 - \mu^2\Phi$ after integrating out heavy fields X and Y at the energy scale of order Λ [9].

For the separable Kähler potential (4) the necessary condition (3) simplifies to

$$\frac{2}{3}\Theta_T^4 + R_\Phi\Theta_\Phi^4 < \frac{2}{\widehat{G}^2}\,, \tag{9}$$

where R_Φ is the curvature scalar of the one dimensional submanifold associated with the matter field. The analogous curvature in the modulus direction, R_T, was replaced by its value equal 2/3. The spherical coordinates parameterizing SUSY breaking $\Theta_i^2 \equiv G_{i\bar{i}}f^if^{\bar{i}}$ (no summation over i or $\bar{i}$) satisfy the normalization condition $\Theta_T^2 + \Theta_\Phi^2 = 1$.

In the following we discuss briefly four models of inflation with uplifting from the matter filed. For both uplifting mechanisms, Polonyi and O'Raifeartaigh, we consider saddle point and inflection point inflationary scenarios.

4. Polonyi uplifting

The Kähler potential of the Polonyi model is canonical so the sectional curvature in the matter field subspace vanishes: $R_\Phi = 0$. In such a case the necessary condition for a dS minimum and the inflationary region (9) simplifies to:

$$\Theta_T^4 < \frac{3}{\widehat{G}^2}\,. \tag{10}$$

It is clear that the matter field contributing to SUSY breaking helps to get a stable dS vacuum and slow-roll inflation. In the limit $\Theta_T \to 0$, when the matter field very strongly dominates SUSY breaking, the necessary condition (10) may be satisfied for arbitrary value of $\widehat{G}^2$.

In the Polonyi model with superpotential (6) and the canonical Kähler potential (7), and for appropriate choice of the parameters, SUSY is broken in a Minkowski minimum located at

$$\phi_{\min} = \sqrt{3} - 1, \qquad \theta_{\min} = 0\,. \tag{11}$$

The position of this minimum does not change substantially after coupling the Polonyi model to the volume modulus. However, the existence of a near Minkowski (dS with a very small curvature) vacuum is not enough for a given model to accommodate inflation. There must be also a region in the field space where the potential is suitable for inflation. In addition, such region must be connected to the vacuum by a trajectory along which the inflaton may evolve after the slow roll period of inflation.

We have shown [10] that all these conditions may be fulfilled in the model with the racetrack superpotential for the volume modulus coupled to the Polonyi matter sector:

$$K = -3\ln(T+\overline{T}) + \Phi\overline{\Phi}\,, \qquad W = W_0 + Ce^{-cT} + De^{-dT} - \mu^2\Phi\,. \tag{12}$$

There are regions in the parameter space for which inflation may be realized in the vicinity of a saddle point or an inflection point of the potential. In both cases our models with the Polonyi uplifting differ significantly from the corresponding models with non-supersymmetric uplifting.

In the inflection point scenario, the inflationary dynamics is effectively described by the real parts of the fields, t and ϕ, because the axions, τ and θ, vanish during and after inflation. An example of the potential suitable for inflation in such case is shown in figure 1. The trajectory of the inflaton (left panel of figure 1) shows that inflation proceeds along the valley (mainly) in the ϕ-direction. This valley becomes unstable in the t direction for big enough values of ϕ (see the right panel of figure 1). So, after violation of the slow-roll conditions, the inflaton rolls down (first mainly in t-direction) towards the (near) Minkowski vacuum.

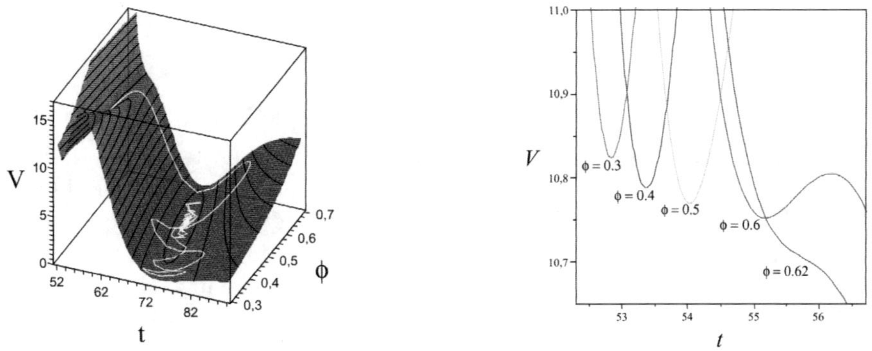

Figure 1. Inflection point model with the Polonyi uplifting for the parameters: $W_0 = -5.393\cdot10^{-5}$, $C=-2/5$, $D=2/50$, $\mu^2=1.097814\cdot10^{-4}$, $c=\pi/20$, $d=\pi/30$. Left panel: The potential multiplied by 10^{15} for $\tau=\theta=0$. The white curve represents the inflaton trajectory. Right panel: Several $\phi=$ const sections of the inflationary valley.

As in most inflationary models, some fine tuning of the parameters is necessary to obtain a long enough period of inflation. We checked how precisely the constant term in the superpotential (12), W_0, must be tuned when all other parameters are fixed (except μ^2 which is always adjusted to cancel the cosmological constant in the vacuum). As a measure of the tuning we used $\Delta \equiv |(W_{0_{\max}} - W_{0_{\min}})/\frac{1}{2}(W_{0_{\max}} + W_{0_{\min}})|$ where $W_{0_{\min}}$ ($W_{0_{\max}}$) is the minimal (maximal) value of W_0 giving at least 60 e-folds of inflation and the spectral index consistent with WMAP data. The necessary fine tuning in our model is at the level of $10^{-4}\div10^{-3}$, and is about $4\div5$ orders of magnitude weaker than in the analogous model with non-SUSY uplifting. Such big difference is related to the overshooting problem which is very severe [11] in the original racetrack inflection point model [3] but much more easy to overcome in the presence of Polonyi uplifting.

It is more difficult to analyze inflation from (a vicinity of) a saddle point because all four real fields are involved. We constructed models in which SUSY breaking is dominated by the matter field, with the volume modulus contribution, Θ_T^2, at the level of 10^{-1} at the saddle point and 10^{-2} in the vacuum. The inflaton is a mixture of axions with a bigger contribution from the matter field axion θ. The necessary fine tuning is again reduced when compared to a corresponding model with non-SUSY uplifting: this time Δ grows from 10^{-4} to 10^{-3}.

Both models with Polonyi uplifting have some common features. SUSY breaking and the inflaton are dominated by the matter field. Fields ϕ and θ are somewhat lighter than t and τ but all of them are heavier than the Hubble scale during inflation so no substantial isocurvature perturbations are produced. The tensor to scalar perturbations ratio and running of the spectral index are negligible from the observational point of view, with typical values $r \approx 10^{-7}$ and $(\mathrm{d}n_s/\mathrm{d}\ln k) \approx 10^{-3}$. In each case, the spectral index is very well (with precision better than one percent) approximated by analytic formulae derived for one-field inflation from a saddle or an inflection point [3]. The last is true also for other models discussed below.

5. O'uplifting

Now we discuss the model in which the volume modulus with the racetrack superpotential is coupled to the matter field of the radiatively corrected O'Raifeartaigh model:

$$K = -3\ln(T+\overline{T}) + \Phi\overline{\Phi} - \frac{(\Phi\overline{\Phi})^2}{\Lambda^2}\,, \qquad\qquad W = W_0 + Ce^{-cT} + De^{-dT} - \mu^2\Phi\,. \tag{13}$$

This model formally approaches the Polonyi one in the limit $\Lambda \to \infty$. However, the consistency of the used approximations requires $|\Phi| \ll \Lambda \ll 1$. The sectional curvature in the matter subspace equals $R_\Phi = -4/[\Lambda^2(1-4|\Phi|^2/\Lambda^2)]$. It is negative and, for (necessarily) small values of $|\Phi|$ and Λ, it has big absolute value. Hence, it is much more easy to fulfill the necessary condition (9) using the O'uplifting.

We have checked that both, inflection point and saddle point, inflation may be realized in the racetrack models with O'uplifting. Examples of appropriate potentials and inflaton trajectories are shown in figures 2 and 3. In all such models, SUSY breaking during inflation and in the vacuum is dominated by the matter field. An important difference, as compared to the Polonyi uplifting, is the composition of the inflaton which now is dominated by the volume modulus. More precisely, the real component t in the inflection point model and the imaginary component τ in the saddle point model. The fine tuning necessary to realize phenomenologically successful inflation is similar to that in the case with non-SUSY uplifting.

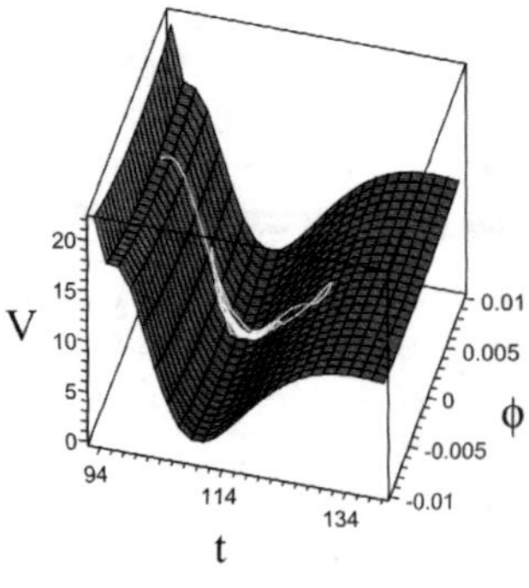

Figure 2. The potential multiplied by 10^{21} and the inflaton trajectory (white curve) for the inflection point racetrack model with O'uplifting for parameters: $\Lambda = 0.1$, $W_0 = -3.426786982 \cdot 10^{-7}$, $C = -2/5$, $D = 1/200$, $\mu^2 = 5.168125 \cdot 10^{-7}$, $c = \pi/20$, $d = \pi/30$. The axion fields τ and θ are fixed at 0.

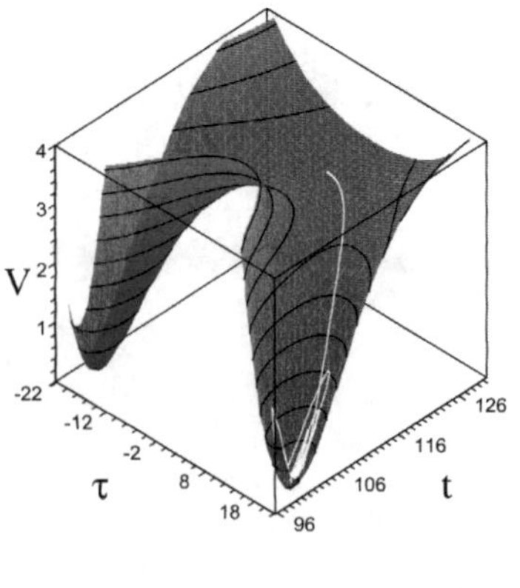

Figure 3. The potential multiplied by 10^{16} and the inflaton trajectory (white curve) for the saddle point racetrack model with O'uplifting for parameters: $\Lambda = 0.1$, $W_0 = -1/13005$, $C = 1/20$, $D = -7/80$, $\mu^2 = 1.073049 \cdot 10^{-4}$, $c = \pi/50$, $d = \pi/45$. The matter field is fixed at $\phi = \phi_{\rm vac}$, $\theta = \theta_{\rm vac}$.

6. Models with low scale of SUSY breaking

In most string inspired models the scale of SUSY breaking is at least as high as the Hubble scale during inflation, which is unsatisfactory from the phenomenological point of view. A low scale of SUSY breaking may be obtained if inflation ends in a (near) SUSY Minkowski minimum [12]. However, in such case the (non-SUSY) uplifting is very small and the no-scale Kähler potential violates necessary condition (3), so inflation cannot be realized [7]. It was shown in [11] that racetrack inflation models with low scale SUSY breaking can be constructed if α' and/or string loop corrections to the Kähler potential are big enough. It is interesting to ask whether similar models can be realized when the volume modulus is coupled to the matter field and no string corrections to the Kähler potential are included. The answer to this last question is positive. However, some conditions must be met. The superpotential for the matter field must be more complicated than the linear one given in (6). The simplest model we found is defined by

$$K = -3\ln(T+\overline{T}) + \Phi\overline{\Phi}\,, \qquad W = W_0 + Ce^{-cT} + De^{-dT} + \mu^2(\Phi - \Phi_0)^2\,. \qquad (14)$$

The inflation takes place in the vicinity of an inflection point and is very strongly dominated by the matter field (non-negligible contribution from the modulus appears only for negative c and d). More detailed analysis of such models may be found in [10].

7. Conclusions

We have shown that racetrack inflation models can be constructed using the fully supersymmetric framework with the matter F-terms as sources of uplifting and SUSY breaking. Using Polonyi or O'Raifeartaigh matter sector one can get the potential with an inflection point or a saddle point suitable for inflation. The inflaton is dominated by the matter field (volume modulus) for Polonyi (O'Raifeartaigh) uplifting. For the Polonyi matter sector, the fine tuning necessary to get long enough inflation and a correct value of the spectral index is weaker than in the corresponding models with non-SUSY uplifting (especially for the inflection point model). In the case of O'uplifting, the matter field almost decouples from the inflationary dynamics which is quite similar to corresponding models with non-SUSY uplifting. Inflation with a low scale of SUSY breaking may be obtained in some cases with Polonyi uplifting.

Acknowledgments

The author would like to thank the Organizers of PASCOS 2010 for their hospitality. Work partially supported by the EC 7th Framework Project PITN-GA-2009-237920 "Unification in the LHC era", 6th Framework Project MRTN-CT-2006-035863 "The Origin of Our Universe: Seeking Links between Fundamental Physics and Cosmology" and by Polish MNiSzW scientific research grant N N202 103838 (2010 - 2012).

References

[1] Kachru S, Kallosh R, Linde A D and Trivedi S P 2003 *Phys. Rev.* D **68** 046005
[2] Blanco-Pillado J J *et al.* 2004 *JHEP* **0411** 063
[3] Linde A D and Westphal A 2008 *JCAP* **0803** 005
[4] Brax P, Davis A C, Davis S C, Jeannerot R and Postma M, 2008 *JCAP* **0801**, 008 (2008)
[5] Westphal A 2005 *JCAP* **0511** 003
[6] Covi L, Gomez-Reino M, Gross C, Louis J, Palma G A and Scrucca C A, 2008 *JHEP* **0806** 057
[7] Badziak M and Olechowski M 2008 *JCAP* **0807** 021
[8] Covi L, Gomez-Reino M, Gross C, Louis J, Palma G A and Scrucca C A, 2008 *JHEP* **0808** 055
[9] Kallosh R and Linde A D 2007 *JHEP* **0702** 002
[10] Badziak M and Olechowski M 2010 *JCAP* **1002** 026
[11] Badziak M and Olechowski M 2009 *JCAP* **0902** 010
[12] Kallosh R and Linde A D 2004 *JHEP* **0412** 004

16th International Symposium on Particles Strings and Cosmology (PASCOS 2010) IOP Publishing
Journal of Physics: Conference Series **259** (2010) 012029 doi:10.1088/1742-6596/259/1/012029

Dynamical solutions in the Nishino-Salam-Sezgin model

Masato Minamitsuji

Department of Physics, Graduate School of Science and Technology, Kwansei-Gakuin University, Gakuen 2-1, Sanda, 669-1337 Japan

E-mail: masato.minamitsuji@kwansei.ac.jp

Abstract. We investigate the dynamical 3-form flux compactifications and their implications for cosmology in our brane world in the six-dimensional Nishino-Salam-Sezgin model, which is usually referred to as the Salam-Sezgin model. We take the background of the 3-form field acting on the internal space and timelike dimensions without the $U(1)$ gauge field strength. Our solutions are obtained by exchanging the roles of the radial coordinate and the time coordinate from the assumptions made in deriving the previously known static solutions. At the center of the internal space, there is a conical singularity which may be interpreted as our 3-brane world, but it is difficult to realize a warped structure in the direction of the ordinary 3-space. Except for the oscillating solutions, our dynamical evolutions lead to the ever expanding ordinary three-space. However, its expansion rate is not enough to account for the present acceleration of our Universe.

1. Introduction

Six-dimensional models have been recognized as the important playground to study cosmology and gravity with extra dimensions compactified by fluxes of antisymmetric tensor fields. The warped, flux compactification solutions were reported in the context of Nishino-Salam-Sezgin (NSS) six-dimensional supergravity [1, 2]. Note that this model is usually referred to as the Salam-Sezgin model especially when compactification is discussed. However, the compactification on S^2, which was found by Salam and Sezgin, was constructed in the theory first proposed by Nishino and Sezgin. Therefore, it is more suitable for us to refer to the model as the NSS model.

In order to explore the cosmology, one useful approach is to use the dynamical solutions. In this direction, exact dynamical solutions were obtained in [3, 4, 5, 6]. The field components of the bosonic part of the NSS model are the gravity, the $U(1)$-gauge field, Kalb-Ramond 2-form (i.e., 3-form field strength) and the dilaton. In Refs. [3, 4], warped time dependent solutions with the $U(1)$-field strength, without the Kalb-Ramond 3-form field strength in the NSS model, were explored. The solutions have all six dimensions evolving in time. There are three types of the exact time dependent solutions: One of them is the scaling solution, in which both the scale factors of the external and internal spaces have the linear dependence on the proper time. The other two solutions have more complicated time dependence. In the earlier times, these solutions can be seen as the generalizations of the solutions obtained in Ref. [7], but in the later times both approach the above scaling solution. The scaling solution is the generalization of the static solution obtained in [1] and becomes an attractor. In this sense, the static brane solution in the 2-form background is always unstable. An extension of the time dependent 2-form solution in the NSS model to more general Einstein-Maxwell-dilaton models was considered in Ref. [5]. The exact wave solution was also obtained in [6].

The purpose of this work is to present the other classes of time dependent solutions in the NSS model in the presence of the nonvanishing 3-form field strength but in the absence of the $U(1)$ field strength. Our solutions are obtained by exchanging the roles of the radial coordinate and the time coordinate in the static solutions obtained recently in Ref. [8]. For more details, see our recent publications [9].

2. Nishino-Salam-Sezgin model

The action of the bosonic part of the NSS supergravity [2] is given by

$$I = \int d^6x\sqrt{-g}\left[\frac{1}{2\kappa^2}(R-(\partial\phi)^2) - \frac{1}{4}e^{-\phi}F_{\mu\nu}^2 - \frac{1}{12}e^{-2\phi}H_{\mu\nu\rho}^2 - \frac{g^2}{2}e^{\phi}\right], \tag{1}$$

where ϕ is the dilaton, $H = dB$ is the field strength for the Kalb-Ramond field B and $F = dA$ is the field strength for the $U(1)$ gauge field A. The parameters g and κ are gauge and gravitational coupling constants, respectively. Most of the preceding works have discussed the $U(1)$ field acting on the internal space dimensions, $F \neq 0$, with the vanishing 3-form $H = 0$. Instead, we consider the nonvanishing 3-form field acting on the internal space and timelike dimensions $H \neq 0$, with the vanishing $U(1)$ field $F = 0$.

3. Warped static solutions

3.1. Solutions

In this section, we summarize the properties of the warped static solutions with the following metric

$$ds_6^2 = -e^{2u_0(y)}dt^2 + e^{2u_1(y)}\sum_{i=1}^{3}(dx^i)^2 + e^{2v(y)}dy^2 + e^{2w(y)}d\theta^2, \tag{2}$$

where the coordinates t and x^i describe our four-dimensional world, and the remainings y and θ are transverse to it. The metric components u_0, u_1, v, w and the dilaton ϕ are assumed to be functions of y. We also define $U := u_0 + 3u_1 - v + w$. We assume that for the 3-form field strength has the form

$$H = E'(y)\, dt \wedge dy \wedge d\theta. \tag{3}$$

The type of solutions depends on the signature of

$$f_2^2 = 2f_1^2 - 4(G_\phi - G_w)^2 - \frac{32}{3}G_\phi^2, \tag{4}$$

where f_1, G_ϕ and G_w are constants, which appear in integrating the Einstein equations and the dilaton equation of motion. For $f_2^2 > 0$

$$\begin{aligned} ds^2 &= \frac{e^{-2(2G_\phi - G_w)y}}{\cosh^{\frac{1}{2}}[f_1(y-y_1)]|\sinh[f_2(y-y_2)]|^{\frac{1}{2}}}\left[-dt^2 + |\sinh[f_2(y-y_2)]|e^{2\left(\frac{7}{3}G_\phi - G_w\right)y}\sum_{i=1}^{3}(dx^i)^2\right] \\ &+ \frac{e^{-2G_\phi y}|\sinh[f_2(y-y_2)]|^{\frac{1}{2}}}{\cosh^{\frac{5}{2}}[f_1(y-y_1)]}\left[\frac{2c^2f_1^4}{\kappa^2g^4f_2^2}dy^2 + e^{-2(G_w - G_\phi)y}\frac{\cosh^2[f_1(y-y_1)]}{\sinh[f_2(y-y_2)]}d\theta^2\right], \\ e^\phi &= \frac{g^2f_2^2}{2f_1^2c^2}e^{2G_\phi y}\frac{\cosh^{\frac{1}{2}}[f_1(y-y_1)]}{|\sinh[f_2(y-y_2)]|^{\frac{1}{2}}}, \\ H &= \frac{\sqrt{2}f_2^3g^2}{4\kappa f_1^2c^2}\frac{1}{\sinh^2[f_2(y-y_2)]}dt\wedge dy\wedge d\theta, \end{aligned} \tag{5}$$

where c, y_1 and y_2 are the other integration constants. For $f_2^2 < 0$

$$\begin{aligned}
ds^2 &= \frac{e^{-2(2G_\phi - G_w)y}}{\cosh^{\frac{1}{2}}[f_1(y-y_1)]|\sin[|f_2|(y-y_2)]|^{\frac{1}{2}}}\Big[-dt^2 + |\sin[|f_2|(y-y_2)]|e^{2(\frac{7}{3}G_\phi - G_w)y}\sum_{i=1}^{3}(dx^i)^2\Big] \\
&+ \frac{e^{-2G_\phi y}|\sin[|f_2|(y-y_2)]|^{\frac{1}{2}}}{\cosh^{\frac{5}{2}}[f_1(y-y_1)]}\Big[\frac{2c^2 f_1^4}{\kappa^2 g^4 |f_2|^2}dy^2 + e^{-2(G_w - G_\phi)y}\frac{\cosh^2[f_1(y-y_1)]}{|\sin[|f_2|(y-y_2)]|}d\theta^2\Big], \\
e^\phi &= \frac{g^2|f_2|^2}{2f_1^2c^2}e^{2G_\phi y}\frac{\cosh^{\frac{1}{2}}[f_1(y-y_1)]}{|\sin[|f_2|(y-y_2)]|^{\frac{1}{2}}}, \\
H &= \frac{\sqrt{2}|f_2|^3 g^2}{4\kappa f_1^2 c^2}\frac{1}{\sin^2[|f_2|(y-y_2)]}dt\wedge dy\wedge d\theta.
\end{aligned} \tag{6}$$

For $f_2^2 = 0$

$$\begin{aligned}
ds^2 &= \frac{e^{-2(2G_\phi - G_w)y}}{\cosh^{\frac{1}{2}}[f_1(y-y_1)]|y-y_2|^{\frac{1}{2}}}\Big[-dt^2 + |y-y_2|e^{2(\frac{7}{3}G_\phi - G_w)y}\sum_{i=1}^{3}(dx^i)^2\Big] \\
&+ \frac{e^{-2G_\phi y|y-y_2|^{1/2}}}{\cosh^{\frac{5}{2}}[f_1(y-y_1)]}\Big[\frac{2c^2 f_1^4}{\kappa^2 g^4}dy^2 + e^{-2(G_w - G_\phi)y}\frac{\cosh^{\frac{1}{2}}[f_1(y-y_1)]}{|y-y_2|}d\theta^2\Big], \\
e^\phi &= \frac{g^2}{2c^2 f_1^2}e^{2G_\phi y}\frac{\cosh^{\frac{1}{2}}[f_1(y-y_1)]}{|y-y_2|^{\frac{1}{2}}}, \\
H &= \frac{\sqrt{2}g^2}{4\kappa f_1^2 c^2}\frac{1}{|y-y_2|^2}dt\wedge dy\wedge d\theta.
\end{aligned} \tag{7}$$

3.2. Conditions to realize the Lorentz symmetry in four dimensions

In these solutions, the warp functions in the time and three spatial directions are different and in general the Lorentz symmetry is broken. One way to recover this symmetry is to put a five-dimensional brane with a Kaluza-Klein circle at a certain place in y direction and impose the symmetry locally. Here we are interested in the region of $y > 0$ with the assumptions $y_1 > 0$ and $y_2 > 0$. First, the original spacetime is cut at $y = y_0 > \max(y_1, y_2)$ and the singular part of $y < y_0$ is removed. Then, an identical copy of the remaining piece is glued to the original one, at the codimension-one boundary $y = y_0$. The boundary at $y = y_0$ contains the ordinary four-dimensional spacetime with a Kaluza-Klein circle. The induced metric on the brane is given by

$$ds_5^2 = e^{2w(y_0)}d\theta^2 - e^{2u_0(y_0)}dt^2 + e^{2u_1(y_0)}\sum_{i=1}^{3}(dx^i)^2. \tag{8}$$

To support the brane we need to consider brane matter with anisotropic stress. We also assume that the extra circular direction of θ is compact enough.

Since the oscillating solutions of $f_2^2 < 0$ have singularities other than $y = y_2$, we exclude them. Our discussion here is applied to solutions of $f_2^2 \geq 0$. The requirement to obtain the Lorentz symmetry at some $y = y_0$ is given by $u_1' = u_0'$, which leads to

$$\coth[f_2(y_0 - y_2)] = \frac{2}{f_2}(G_w - \frac{7}{3}G_\phi) = \frac{2(g_w - \frac{7}{3}g_\phi)}{\sqrt{2 - 4(g_\phi - g_w)^2 - \frac{32}{3}g_\phi^2}}, \tag{9}$$

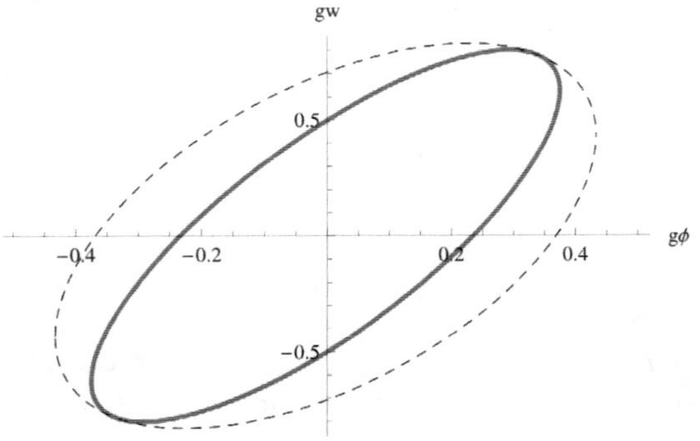

Figure 1. The dashed (black) and solid (red) curves show Eqs. (10) and (11), respectively. On the other hand, outside the solid curve, the right hand side of Eq. (9) is always positive. Also, inside the dashed curve, the solutions of $f_2^2 \geq 0$ can exist. Therefore, the regions where the brane world with the Lorentz symmetry is realized are upper and lower ones surrounded by these two curves.

where $g_\phi := \frac{G_\phi}{f_1}$ and $g_w := \frac{G_w}{f_1}$. Without loss of the generality, we may assume $f_1 > 0$. The quantity in the square root vanishes for

$$g_w = g_\phi \pm \sqrt{\frac{1}{2} - \frac{8}{3}g_\phi^2}. \tag{10}$$

In addition, the right hand side of Eq. (9) becomes ± 1 for

$$g_w = \frac{5}{3}g_\phi \pm \sqrt{\frac{1}{4} - \frac{16}{9}g_\phi^2}. \tag{11}$$

Noting that $|\coth x| > 1$ and Eq. (10), the allowed region for the recovery of the Lorentz symmetry at $y = y_0$ in the (g_w, g_ϕ) plane is shown in Fig. 1. The dashed (black) and solid (red) curves show Eqs. (10) and (11), respectively. First, outside the solid curve, the right hand side of Eq. (9) is always positive. Second, inside the dashed curve, the solutions of $f_2^2 \geq 0$ exist. Therefore, the regions where the four-dimensional world with the Lorentz symmetry is realized are those surrounded by these two different curves.

4. Time dependent solutions

4.1. Solutions

Assuming the same metric ansatz Eq. (2), we investigate the time dependent solutions by exchanging the roles of the t and y as well as those of u_0 with v. We also assume that the 3-form field strength has the form of

$$H = \dot{F}(t)\, dt \wedge dy \wedge d\theta. \tag{12}$$

After integration, we obtain the following solutions. For a positive f_2^2, where

$$f_2^2 = 2f_1^2 - 4(G_w - G_\phi)^2 - \frac{32}{3}G_\phi^2, \tag{13}$$

the solution is given by

$$ds^2 = \frac{e^{-2G_\phi t}\cosh^{\frac{1}{2}}[f_2(t-t_2)]}{|\sinh[f_1(t-t_1)]|^{\frac{5}{2}}}\Big[-\frac{2c^2 f_1^4}{\kappa^2 g^4 f_2^2}dt^2 + e^{\frac{8}{3}G_\phi t}\sinh^2[f_1(t-t_1)]\sum_{i=1}^{3}(dx^i)^2\Big]$$

$$
\begin{aligned}
&+ \frac{e^{-2(2G_\phi - G_w)t}}{|\sinh[f_1(t-t_1)]|^{\frac{1}{2}}\cosh^{\frac{1}{2}}[f_2(t-t_2)]}\left[dy^2 + e^{4(G_\phi - G_w)t}d\theta^2\right],\\
e^\phi &= \frac{f_2^2 g^2}{2c^2 f_1^2}\frac{|\sinh[f_1(t-t_1)]|^{\frac{1}{2}}}{\cosh^{\frac{1}{2}}[f_2(t-t_2)]}e^{2G_\phi t},\\
H &= \frac{\sqrt{2}f_2^3 g^2}{4\kappa f_1^2 c^2}\frac{1}{\cosh^2[f_2(t-t_2)]}dt\wedge dy\wedge d\theta. \qquad (14)
\end{aligned}
$$

There are the other type of solutions with the oscillating metric components. Now, we are interested in the cosmological evolutions and focus only on the above type solutions.

4.2. Cosmological evolutions

Let us discuss the cosmological behavior of our solutions. Therefore, we investigate the late-time behavior $t \gg t_1, t_2$. Then, we obtain the asymptotic form of the metric at the later times:

$$ds^2 \approx -dT^2 + T^{2q_1}\sum_{i=1}^{3}(dx^i)^2 + T^{2q_v}dy^2 + T^{2q_w}d\theta^2, \qquad (15)$$

where T is the cosmic proper time, and q_1, q_v and q_w are the powers of the cosmic expansion in the 3-space and two-dimensional internal space, defined by

$$q_1 := \frac{f_2 - f_1 + \frac{4}{3}G_\phi}{f_2 - 5f_1 - 4G_\phi}, \quad q_v := \frac{-f_2 - f_1 - 4(2G_\phi - G_w)}{f_2 - 5f_1 - 4G_\phi}, \quad q_w := \frac{-f_2 - f_1 - 4G_w}{f_2 - 5f_1 - 4G_\phi}, \qquad (16)$$

respectively. Here, we omit unimportant numerical coefficients. In Fig. 2, we show q_1 as the functions of $g_w := \frac{G_w}{|f_1|}$ and $g_\phi := \frac{G_\phi}{|f_1|}$. The power of the fastest three-space expansion is given by $q_{1,\max} = \frac{9+4\sqrt{3}}{33} \approx 0.483$ for $g_w = g_\phi = -\frac{\sqrt{3}}{4}$, which is less than the radiation-dominated universe $\frac{1}{2}$ For the same value of $g_w = g_\phi$, $q_v = q_w$ takes the minimal value $q_{v,\min} = q_{w,\min} = \frac{1-2\sqrt{3}}{11} \approx -0.224$. These powers of $(q_{1,\max}, q_{(v,w),\min})$ are the same as in the time dependent solution in the 2-form backgrond of NSS model [7]. Thus, it is not enough to explain the present cosmic acceleration. In Fig. 3, The solid (black) curve shows the region where solutions of $f_2^2 > 0$ exist. The thick (red) curve shows the points below which $q_1 > q_v$, while in the left side to the dashed (blue) curve, $q_1 > q_w$. The dotted line shows $g_w = g_\phi$. Therefore, in the region surrounded by these three curves, q_1 is positive and greater than q_v and q_w. Thus, the ordinary three space can expand faster than the y- and θ-directions in the small region of the negative g_w and g_ϕ which is surrounded by three different curves. Although our solutions are not suitable to explain the present acceleration of our Universe, this is an interesting feature.

5. Conclusions

We derived the time dependent ones in the 3-form background in the six-dimensional Nishino-Salam-Sezgin model and discussed their properties, after reviewing the warped static solutions obtained in [8]. The time dependent solutions are obtained, by exchanging the role of space and time from the above static solutions. Focusing on the solutions with the non-oscillating metric components with time, we investigated the cosmological evolutions obtained from the late-time behaviors. Although in a region of the parameter space the ordinary three-dimensional space can expand faster than both the internal directions, the power of the fastest expansion is approximately 0.483 and is not enough to explain the acceleration of the universe.

Other than the solutions discussed in this article, we have also investigated the dynamical 1-brane solutions in the NSS model. Readers who are interested in these solutions should refer to Refs. [9, 10] for more details.

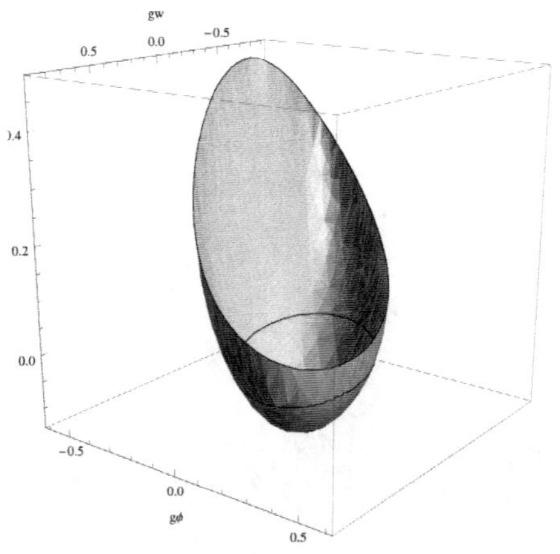

Figure 2. q_1 is shown as the function of g_w and g_ϕ. The solid curve shows the level of $q_1 = 0$.

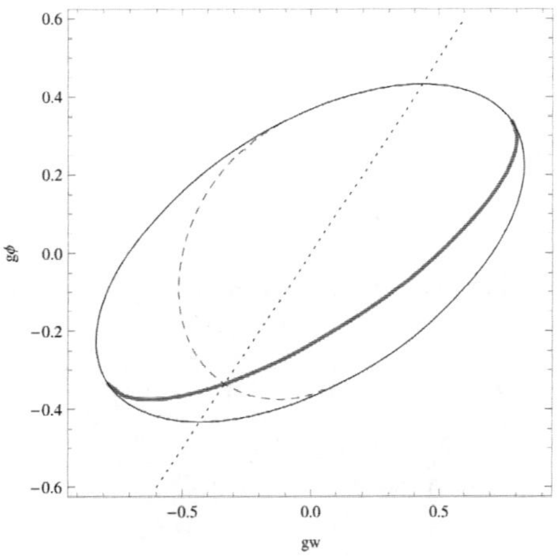

Figure 3. The solid (black) curve shows the region where solutions of $f_2^2 > 0$ exist. The thick (red) curve shows the points below which $q_1 > q_v$, while in the left side to the dashed (blue) curve, $q_1 > q_w$. The dotted line shows $g_w = g_\phi$.

Acknowledgments
The author would like to thank K. i. Maeda, N. Ohta and K. Uzawa for discussions in completing these works.

References

[1] Y. Aghababaie *et al.*, JHEP **0309**, 037 (2003) [arXiv:hep-th/0308064]; G. W. Gibbons, R. Guven and C. N. Pope, Phys. Lett. B **595**, 498 (2004) [arXiv:hep-th/0307238]; C. P. Burgess, F. Quevedo, G. Tasinato and I. Zavala, JHEP **0411**, 069 (2004) [arXiv:hep-th/0408109].
[2] H. Nishino and E. Sezgin, Phys. Lett. B **144**, 187 (1984); A. Salam and E. Sezgin, Phys. Lett. B **147**, 47 (1984); H. Nishino and E. Sezgin, Nucl. Phys. B **278**, 353 (1986).
[3] A. J. Tolley, C. P. Burgess, C. de Rham and D. Hoover, New J. Phys. **8**, 324 (2006) [arXiv:hep-th/0608083];
[4] E. J. Copeland and O. Seto, JHEP **0708**, 001 (2007) [arXiv:0705.4169 [hep-th]].
[5] T. Kobayashi and M. Minamitsuji, JCAP **0707**, 016 (2007) [arXiv:0705.3500 [hep-th]].
[6] A. J. Tolley, C. P. Burgess, C. de Rham and D. Hoover, JHEP **0807**, 075 (2008) [arXiv:0710.3769 [hep-th]].
[7] K. i. Maeda and H. Nishino, Phys. Lett. B **154**, 358 (1985); K. i. Maeda and H. Nishino, Phys. Lett. B **158**, 381 (1985).
[8] H. R. Afshar and S. Parvizi, JHEP **0910**, 086 (2009) [arXiv:0906.0707 [hep-th]].
[9] M. Minamitsuji, N. Ohta and K. Uzawa, Phys. Rev. D **81**, 126005 (2010) [arXiv:1003.5967 [hep-th]].
[10] K. i. Maeda, M. Minamitsuji, N. Ohta and K. Uzawa, Phys. Rev. D **82**, 046007 (2010) [arXiv:1006.2306 [hep-th]].

16th International Symposium on Particles Strings and Cosmology (PASCOS 2010) IOP Publishing
Journal of Physics: Conference Series **259** (2010) 012030 doi:10.1088/1742-6596/259/1/012030

Torsion in extra-dimensions

Kameshwar C. Wali[1]
Department of Physics, Syracuse University, Syracuse, NY-13244, USA
E-mail: `wali@physics.syr.edu`

Abstract. We consider a variant of the 5 dimensional Kaluza-Klein theory within the framework of Einstein-Cartan formalism. By imposing a set of constraints on torsion and Ricci rotation coefficients, we show that the torsion components are completely expressed in terms of the metric. and the Ricci tensor in 5D corresponds exactly to what one would obtain from torsion-free general relativity on a 4D hypersurface. The contributions of the scalar and vector fields of the standard K-K theory to the Ricci tensor and the affine connections are completely nullified by the contributions from the torsion. As a consequence, geodesic motions do not distinguish the torsion free 4D space-time from a hypersurface of 5D space-time with torsion satisfying the constraints. Since torsion is not an independent dynamical variable in this formalism, the modified Einstein equations are different from those in the general Einstein-Cartan theory. This leads to important cosmological consequences such as the emergence of cosmic acceleration.

1. Introductory remarks

Torsion, a geometrical property of space-time, represents spin degrees of freedom. In analogy to Einstein equations where the energy-momentum of matter fields is coupled to the geometrical Riemannian metric, we can couple angular momentum and spin of matter fields to the geometrical torsion The resulting theory of gravity, known as the Einstein-Cartan theory[1] treats the metric and torsion as two independent geometrical characteristics of space-time [2]. This geometrical Property of space-time, *torsion*arises in the definition of covariant derivative operator of vector and tensor fields, forming an anti-symmetric part of the affine connection coefficients (ACCs). In the absence of torsion, metric compatibility leads the determination of the ACCs in terms of the metric components and their derivatives, known as Christoffel symbols, which are symmetric.

Historically, beginning with the Kaluza-Klein (KK) theory, there has been a great interest in introducing extra dimensions of space-time to unify gravity with elementary particle interactions. In the KK theory, the scalar and vector fields, which are the extra dimensional components of the metric tensor contribute to the affine connection and the Ricci tensor and hence modify their values from the corresponding values in 4D space-time[3]. The contribution of these fields to the Einstein tensor are normally interpreted as gravity induced matter.

In the present work, we incorporate torsion into 5D KK theory[4].In general torsion introduces new parameters other than the metric. But with the choice of a set of minimal conditions on torsion, so as to restrict torsion to the extra dimension,we determine all its components in terms of the metric. Interestingly, the imposed conditions lead to a complete cancellation between the

[1] Work done in collaboration with Karthik H.Shankar, *Mod Phys Lett A.* **25** (2010) 2121

modifications induced by the extra dimensional metric components and the contributions from the torsion. Thus the Ricci tensor in 5D space-time in the resulting formalism is exactly the same as the Ricci tensor in a torsion free 4D space-time. We conclude by briefly discussing some ensuing physical consequences.

2. Basic Formalism

We begin by defining and collecting together from reference [5], the relevant standard relations we need in both the coordinate and the inertial frames. Let $(i,j,k,...)$ and $(\mathrm{A},\mathrm{B},...)$ denote coordinate and inertial frame indices respectively and $\hat{\mathbf{e}}_i = \partial_i$ and $\hat{\theta}^i = dx^i$ be the basis of the tangent and dual spaces at each point in space-time. We define the corresponding inertial basis to be $\hat{\mathbf{e}}_{\mathrm{A}} = \mathrm{e}^{i}_{\cdot\mathrm{A}}\hat{\mathbf{e}}_i$ and $\hat{\theta}^{\mathrm{A}} = \mathrm{e}^{\cdot\mathrm{A}}_{i}\hat{\theta}^i$, where the vielbeins $\mathrm{e}^{i}_{\cdot\mathrm{A}}$ and $\mathrm{e}^{\cdot\mathrm{B}}_{j}$ satisfy the orthonormality conditions,

$$\mathrm{e}^{\cdot\mathrm{A}}_{i}\mathrm{e}^{j}_{\cdot\mathrm{A}} = \delta^j_i; \qquad \mathrm{e}^{\cdot\mathrm{A}}_{i}\mathrm{e}^{i}_{\cdot\mathrm{B}} = \delta^{\mathrm{A}}_{\mathrm{B}}. \tag{1}$$

By definition, the metric in the inertial frame is Minkowskian η_{AB}, and the metric tensor in the coordinate system is $\mathbf{g}_{ij} = \mathrm{e}^{\cdot\mathrm{A}}_{i}\mathrm{e}^{\cdot\mathrm{B}}_{j}\eta_{AB}$ and $\mathbf{g}^{ij} = \mathrm{e}^{i}_{\cdot\mathrm{A}}\mathrm{e}^{j}_{\cdot\mathrm{B}}\eta^{AB}$. The covariant derivative operator $(\tilde{\nabla})$ can be defined in terms of the coordinate basis, or equivalently in terms of the inertial frame basis,

$$\tilde{\nabla}_{\hat{\mathbf{e}}_i}\hat{\mathbf{e}}_j = \tilde{\Gamma}^{k}_{\cdot ij}\hat{\mathbf{e}}_k, \qquad \tilde{\nabla}_{\hat{\mathbf{e}}_{\mathrm{A}}}\hat{\mathbf{e}}_{\mathrm{B}} = \omega^{\mathrm{C}}_{\cdot\mathrm{AB}}\hat{\mathbf{e}}_{\mathrm{C}}, \tag{2}$$

where $\tilde{\Gamma}^{i}_{\cdot jk}$ and $\omega^{\mathrm{A}}_{\cdot\mathrm{BC}}$ are the affine and the Ricci rotation coefficients respectively. The relationship between these two quantities follows from the transformation laws between the coordinate frame$(\hat{\mathbf{e}}_i)$ and inertial frame$(\hat{\mathbf{e}}_{\mathrm{A}})$,

$$\omega^{\mathrm{A}}_{\cdot\mathrm{BC}} = \mathrm{e}^{i}_{\cdot\mathrm{B}}(\tilde{\nabla}_{\hat{\mathbf{e}}_i}\mathrm{e}^{j}_{\cdot\mathrm{C}})\mathrm{e}^{\cdot\mathrm{A}}_{j}. \tag{3}$$

As is well known, the affine connection by itself is not a tensor, but its antisymmetric part, the torsion, is a tensor.

$$\mathrm{T}^{i}_{\cdot jk} = \tilde{\Gamma}^{i}_{\cdot jk} - \tilde{\Gamma}^{i}_{\cdot kj}. \tag{4}$$

Again, using the transformation laws between the coordinate and the inertial frames, we have

$$\mathrm{T}^{i}_{\cdot jk} = \mathrm{e}^{\cdot\mathrm{B}}_{j}\mathrm{e}^{\cdot\mathrm{C}}_{k}\mathrm{e}^{i}_{\cdot\mathrm{A}}\mathrm{T}^{\mathrm{A}}_{\cdot\mathrm{BC}}. \tag{5}$$

Furthermore, with the standard assumption of metric compatibility, namely $\nabla_{\hat{\mathbf{e}}_i}g_{jk} = 0$, we obtain

$$\tilde{\Gamma}^{i}_{\cdot jk} = \hat{\Gamma}^{i}_{\cdot jk} + \mathrm{K}^{i}_{\cdot jk}, \tag{6}$$

where $\hat{\Gamma}^{i}_{\cdot jk}$ is the Christoffel connection,

$$\hat{\Gamma}^{i}_{\cdot jk} = \left\{ {}_{j}{}^{i}{}_{k} \right\} = \frac{1}{2}\mathbf{g}^{im}[\partial_j\mathbf{g}_{km} + \partial_k\mathbf{g}_{jm} - \partial_m\mathbf{g}_{jk}], \tag{7}$$

and $\mathrm{K}^{i}_{\cdot jk}$ is the contorsion tensor.

$$\mathrm{K}^{i}_{\cdot jk} = \frac{1}{2}\left[\mathrm{T}^{i}_{\cdot jk} + \mathrm{T}_{j\cdot k}^{\cdot i} + \mathrm{T}_{k\cdot j}^{\cdot i}\right]. \tag{8}$$

3. 5D space-time Metric

Consider a foliation of the 5D space-time in terms of a family of 4D hypersurfaces, which are parametrized by the coordinate system $\{x^\mu\}$, where $(\mu, \nu, ...)$ denote the coordinate indices on these hypersurfaces. Let x^5 denote the parametrization of the family, and 5 denote the corresponding coordinate index. The hypersurface coordinates $\{x^\mu\}$ together with x^5 will then span the entire 5D space-time. Let the metric and its inverse on each of the hypersurface be $\mathrm{g}_{\mu\nu}$ and $\mathrm{g}^{\mu\nu}$ respectively, which can in principle depend on the x^5 coordinate. Let $\{\mathrm{e}^{\mu}_{\cdot a}, \mathrm{e}^{\cdot a}_{\mu}\}$ denote the tetrad system on these hypersurfaces satisfying the orthonormality relations $\mathrm{e}^{\mu}_{\cdot a}\mathrm{e}^{\cdot b}_{\mu} = \delta^a_b$ and $\mathrm{e}^{\mu}_{\cdot a}\mathrm{e}^{a}_{\cdot\nu} = \delta^\mu_\nu$. Here $(a, b, ...)$ denote the tetrad indices on these 4D hypersurfaces. The metric on the hypersurface is then given by $\mathrm{g}_{\mu\nu} = \mathrm{e}^{\cdot a}_{\mu}\mathrm{e}^{\cdot b}_{\nu}\eta_{ab}$ and its inverse $\mathrm{g}^{\mu\nu} = \mathrm{e}^{\mu}_{\cdot a}\mathrm{e}^{\nu}_{\cdot b}\eta^{ab}$.

We will now construct the vielbiens in the 5D space-time by extending the tetrad system on the 4D hypersurfaces. We take the components of the 5D vielbeins to be $\mathrm{e}^{i}_{\cdot\mathrm{A}} = (\mathrm{e}^{\mu}_{\cdot a}, \mathrm{e}^{\mu}_{\cdot\dot{5}}, \mathrm{e}^{5}_{\cdot a}, \mathrm{e}^{5}_{\cdot\dot{5}})$ and $\mathrm{e}^{\cdot\mathrm{A}}_{i} = (\mathrm{e}^{\cdot a}_{\mu}, \mathrm{e}^{\cdot a}_{5}, \mathrm{e}^{\cdot\dot{5}}_{\mu}, \mathrm{e}^{\cdot\dot{5}}_{5})$. The index $\dot{5}$ corresponds to the fifth dimension of the inertial frame. The orthonormality relations in 5D (eq. 1) immediately leads to [2]

$$\begin{aligned} &\mathrm{e}^{\mu}_{\cdot\dot{5}} = 0,\ \mathrm{e}^{5}_{\cdot a} = -\mathrm{e}^{\mu}_{\cdot a}\mathrm{A}_\mu,\ \mathrm{e}^{5}_{\cdot\dot{5}} = \Phi^{-1}, \\ &\quad \mathrm{e}^{\cdot a}_{5} = 0,\ \mathrm{e}^{\cdot\dot{5}}_{\mu} = \mathrm{A}_\mu\Phi,\ \mathrm{e}^{\cdot\dot{5}}_{5} = \Phi. \end{aligned} \tag{9}$$

The metric in the 5D is then given by $\mathbf{g}_{ij} = \mathrm{e}^{\cdot\mathrm{A}}_{i}\mathrm{e}^{\cdot\mathrm{B}}_{j}\eta_{AB}$ and $\mathbf{g}^{ij} = \mathrm{e}^{i}_{\cdot\mathrm{A}}\mathrm{e}^{j}_{\cdot\mathrm{B}}\eta^{AB}$.

$$\begin{aligned} \mathbf{g}_{\mu\nu} &= \mathrm{g}_{\mu\nu} + \epsilon\mathrm{A}_\mu\mathrm{A}_\nu\Phi^2,\ \mathbf{g}_{\mu5} = \epsilon\mathrm{A}_\mu\Phi^2,\ \mathbf{g}_{55} = \epsilon\Phi^2, \\ \mathbf{g}^{\mu\nu} &= \mathrm{g}^{\mu\nu},\ \mathbf{g}^{\mu5} = -\mathrm{A}^\mu,\ \mathbf{g}^{55} = \mathrm{A}_\lambda\mathrm{A}^\lambda + \epsilon\Phi^{-2}. \end{aligned} \tag{10}$$

The raising and lowering of indices on the vector field is done w.r.t the 4D metric $\mathrm{g}_{\mu\nu}$. The parameter $\epsilon = \pm1$ denotes whether the extra dimension is space-like or time-like. Note that the induced metric on the hypersurfaces (induced by the 5D geometry), $\mathrm{g}_{\mu\nu} + \epsilon\mathrm{A}_\mu\mathrm{A}_\nu\Phi^2$ is different from the 4D metric $\mathrm{g}_{\mu\nu}$ on them, but quite evidently related by a gauge transformation.

We will now impose a set of constraints on torsion and the Ricci rotation coefficients consistent with Cartan's first structure equation [5] that relates torsion and connection coefficients. With a minimal modification of the standard general relativity in mind, we chose the set of constraints [7] so that the torsion components on the 4D hypersurface vanish (see also [8]), yet leaving non-vanishing torsion components completely determined in terms of the metric components.

$$\textbf{Condition 1}: \mathrm{T}^{a}_{\cdot\mathrm{BC}} = 0$$

Using eq. 5 and noting that $\mathrm{e}^{\mu}_{\cdot\dot{5}} = 0$, we find $\mathrm{T}^{\mu}_{\cdot ik} = 0$. This implies that the only nonzero components of torsion are $\mathrm{T}^{5}_{\cdot ik}$. To determine these, we impose the following condition on the Ricci rotation coefficients,

$$\textbf{Condition 2}: \omega^{\dot{5}}_{\cdot\mathrm{BC}} = 0$$

Condition 2 along with metric compatibility implies $\omega^{\mathrm{A}}_{\cdot\mathrm{B}\dot{5}} = 0$. We can now use eq. 3 to write,

$$\omega^{\mathrm{A}}_{\cdot\mathrm{B}\dot{5}} = \mathrm{e}^{i}_{\cdot\mathrm{B}}(\tilde{\nabla}_{\hat{\mathrm{e}}_i}\mathrm{e}^{j}_{\cdot\dot{5}})\mathrm{e}^{\cdot\mathrm{A}}_{j} = \mathrm{e}^{i}_{\cdot\mathrm{B}}(\partial_i\mathrm{e}^{j}_{\cdot\dot{5}} + \tilde{\Gamma}^{j}_{\cdot ik}\mathrm{e}^{k}_{\cdot\dot{5}})\mathrm{e}^{\cdot\mathrm{A}}_{j} = 0 \tag{11}$$

Since $\mathrm{e}^{\mu}_{\cdot\dot{5}} = \mathrm{e}^{a}_{\cdot 5} = 0$, the above equation implies,

$$\tilde{\Gamma}^{\mu}_{\cdot i5} = 0, \qquad \tilde{\Gamma}^{5}_{\cdot i5} = -\mathrm{e}^{\cdot\dot{5}}_{5}\partial_i\mathrm{e}^{5}_{\cdot\dot{5}} \tag{12}$$

[2] There exists another class of vielbeins that satisfy the orthonormality relations. But they are essentially related to eq. 9 by gauge. Here, we choose to work with the vielbeins given by eq. 9 to make the formalism readily comparable to the existing Kaluza-Klein literature.

Using the above equations along with eq.6, we can express the contorsion in terms of the Christoffel symbols $\hat{\Gamma}$. and the vielbeins,

$$\begin{aligned} \mathrm{K}^{\mu}_{\cdot i5} &= -\hat{\Gamma}^{\mu}_{\cdot i5}, \\ \mathrm{K}^{5}_{\cdot i5} &= -\left(\hat{\Gamma}^{5}_{\cdot i5} + \mathrm{e}^{\dot{5}}_{5}\partial_i \mathrm{e}^{5}_{\cdot\dot{5}}\right) = -\hat{\Gamma}^{5}_{\cdot i5} + \mathrm{J}_i, \end{aligned} \quad (13)$$

where $\mathrm{J}_i \equiv \Phi^{-1}\partial_i\Phi$.

The above equations are sufficient to determine all the components of torsion in terms of the metric components

$$\begin{aligned} \mathrm{T}^{\mu}_{\cdot ij} &= \mathrm{T}^{5}_{\cdot 55} = 0, \\ \mathrm{T}^{5}_{\cdot \mu\nu} &= 2\partial_{[\mu}\mathrm{A}_{\nu]} + 2\mathrm{J}_{[\mu}\mathrm{A}_{\nu]}, \\ \mathrm{T}^{5}_{\cdot \mu 5} &= \mathrm{J}_{\mu} - \partial_5\mathrm{A}_{\mu} - \mathrm{A}_{\mu}\mathrm{J}_5, \end{aligned} \quad (14)$$

A remarkable result that follows is that in addition to yielding the non-vanishing components of torsion in terms of the metric components, the solution to eq. 13 also yields the following condition on the metric on the 4D hypersurfaces.

$$\partial_5 \mathrm{g}_{\mu\nu} = 0. \quad (15)$$

This implies all the hypersurfaces in the foliating family have the same 4D metric. To place things in perspective, we observe that in the standard Kaluza-Klein theory, the assumption of cylindrical condition makes all the quantities, namely $\mathrm{g}_{\mu\nu}$, A_{μ} and Φ independent of x^5. Whereas in our formulation, the constraints automatically imply $\mathrm{g}_{\mu\nu}$ is independent of x^5, while A_{μ} and Φ could still depend on x^5.

With all the $\mathrm{K}^{i}_{\cdot jk}$ determined from eqns. 14 and 8, we now use eq. 6 to calculate all the affine connection coefficients.

$$\begin{aligned} \tilde{\Gamma}^{\lambda}_{\cdot 55} &= \tilde{\Gamma}^{\lambda}_{\cdot \nu 5} = \tilde{\Gamma}^{\lambda}_{\cdot 5\nu} = 0, \\ \tilde{\Gamma}^{5}_{\cdot \mu\nu} &= \nabla_{\mu}\mathrm{A}_{\nu} + \mathrm{J}_{\mu}\mathrm{A}_{\nu}, \\ \tilde{\Gamma}^{5}_{\cdot 5\mu} &= \partial_5\mathrm{A}_{\mu} + \mathrm{J}_5\mathrm{A}_{\mu}, \\ \tilde{\Gamma}^{5}_{\cdot \mu 5} &= \mathrm{J}_{\mu},\ \tilde{\Gamma}^{5}_{\cdot 55} = \mathrm{J}_5,\ \tilde{\Gamma}^{\lambda}_{\cdot \mu\nu} = \Gamma^{\lambda}_{\cdot \mu\nu}. \end{aligned} \quad (16)$$

Here $\Gamma^{\lambda}_{\cdot \mu\nu}$ corresponds to the Christoffel symbols (analogous to eq. 7) obtained from torsion free 4D space-time with metric $\mathrm{g}_{\mu\nu}$. Note that the components of 5D Christoffel symbols along the hypersurface coordinates is not equal to the Christoffel symbols calculated on 4D spacetime, that is, $\hat{\Gamma}^{\lambda}_{\cdot \mu\nu} \neq \Gamma^{\lambda}_{\cdot \mu\nu}$. The symbol ∇_{μ} corresponds to the covariant derivative operator on the torsion-free 4D space-time, where the Christoffel symbols are exactly the affine connection coefficients.

Substituting the above connection coefficients in the Ricci tensor defined by

$$\tilde{R}_{ik} = \partial_k\tilde{\Gamma}^{j}_{\cdot ji} - \partial_j\tilde{\Gamma}^{j}_{\cdot ki} + \tilde{\Gamma}^{j}_{\cdot km}\tilde{\Gamma}^{m}_{\cdot ji} - \tilde{\Gamma}^{j}_{\cdot jm}\tilde{\Gamma}^{m}_{\cdot ki}, \quad (17)$$

we find

$$\tilde{R}_{\mu\nu} = R_{\mu\nu},\ \tilde{R}_{\mu 5} = \tilde{R}_{5\mu} = \tilde{R}_{55} = 0. \quad (18)$$

Here $R_{\mu\nu}$ represents the Ricci tensor on the torsion-free 4D space-time. It also follows that the 5D Ricci scalar is exactly the same as the Ricci scalar in the torsion free 4D space time, that is $\tilde{R} = R$. We also note, in general, the presence of torsion makes the Ricci tensor non-symmetric, but the constraints we have imposed on the torsion leaves the Ricci tensor symmetric.

It is straightforward to see that the formalism and the results obtained thus far are not specific to 4 and 5 dimensions, they can be generalized to any arbitrary D and D+1 dimensions. We shall now consider some implications of the formalism with respect to geodesic equations and solutions to Einstein equations.

4. Geodesic Equations

The 5D geodesic equations split into

$$\begin{aligned}\ddot{x}^5 &+\tilde{\Gamma}^5_{\cdot\mu\nu}\dot{x}^\mu\dot{x}^\nu + \left(\tilde{\Gamma}^5_{\cdot\mu 5} + \tilde{\Gamma}^5_{\cdot 5\mu}\right)\dot{x}^\mu\dot{x}^5 + \tilde{\Gamma}^5_{\cdot 55}\left(\dot{x}^5\right)^2 = 0,\\ \ddot{x}^\lambda &+\Gamma^\lambda_{\cdot\mu\nu}\dot{x}^\mu\dot{x}^\nu = 0 \end{aligned} \qquad (19)$$

We note that the components of the geodesic equations along the hypersurface are exactly the same as the geodesic equations in the torsion free 4D space-time. This is in contrast with the conventional Kaluza Klein theory where the 4D geodesic equations are modified by the presence of the fields A_μ and Φ. In this formalism, the presence of torsion completely nullifies the effect of these fields in the 4D geodesic equations. Furthermore, it is worth noting that the geodesic of a particle can be confined to a 4D hypersurface by requiring $\dot{x}^5 = 0$. From eq. 19, we see that this requires the additional condition,

$$\tilde{\Gamma}^5_{\cdot\mu\nu} = \nabla_\mu A_\nu + J_\mu A_\nu = 0. \qquad (20)$$

If this condition can be satisfied, then there will be no observable difference, as far as a test particle is concerned, whether we live in a torsion free 4D space-time or on a hypersurface within the 5D space-time with torsion. However, it is apparent that this condition is a strong constraint requiring the vector and scalar fields satisfy the above equation for a given 4D metric. It is conceivable that for some 4D metrics, no choice of vector and scalar fields would satisfy the above constraint. In such cases, the particle would be free to move in the x^5 direction unless constrained by an external force or if the fifth co-ordinate is compact and small as usually assumed in most adaptations of the Kaluza-Klein theory.

5. Einstein's equations

To obtain the field equations, we need to vary the action with respect to the independent dynamical variables of the theory. In Einstein-Cartan theory, the independent variables are the metric and the torsion. By imposing the geometrical constraints using Cartan structure equations, we have expressed torsion in terms of the 5D metric. Hence, taking the Lagrangian density to be the Ricci scalar and varying the action with respect to the 5D metric, we obtain the following modified Einstein equations.

$$R_\mu^{\ \nu} - \frac{1}{2}R\delta_\mu^{\ \nu} + H_\mu^{\ \nu} = \Sigma_\mu^{\ \nu} \qquad (21)$$

$$-A^\alpha R_{\mu\alpha} - A^\alpha H_{\mu\alpha} = \Sigma_\mu^{\ 5} \qquad (22)$$

$$-\frac{1}{2}R = \Sigma_5^{\ 5} \qquad (23)$$

where

$$H_{\mu\nu} = \nabla_{(\mu}B_{\nu)} - (\nabla\cdot B)g_{\mu\nu} + J_{(\mu}B_{\nu)} - (J\cdot B)g_{\mu\nu} \qquad (24)$$

and $B_\mu = T^5_{\cdot\mu 5}$ is a vector in 4D torsion free space time.

In the above equations Σ is the stress tensor that one would obtain when matter fields are included in the Lagrangian prior to variation of action. Its 4D components will be the observed stress energy tensor and can be identified with the stress energy tensor of the usual 4D torsion free general relativity. Its conservation requires

$$\nabla_\nu H_\mu^{\ \nu} = 0. \qquad (25)$$

Note that when $H_{\mu\nu} = 0$, eq. 21 reduces to the torsion free 4D Einstein equations. One could set $\Sigma_\mu^{\ 5}$ to zero and use eqns. 22, 25 to solve for the fields A_μ and Φ.

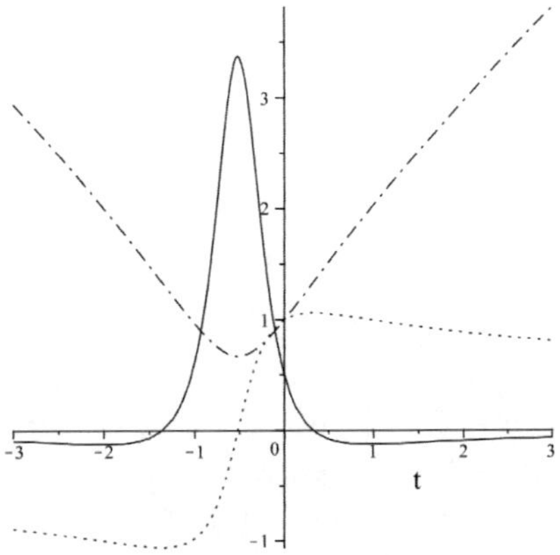

Figure 1. The dashed-dot curve denotes $a(t)$, the dotted curve denotes $\dot{a}(t)$, and the solid curve denotes $\ddot{a}(t)$. The time axis is in units of H_o^{-1}.

6. Robertson-Walker cosmology:

To illustrate an application of the formalism, let us consider spatially flat homogenous and isotropic universe with metric

$$\mathrm{d}s^2 = -dt^2 + a^2(t)\left(dx^2 + dy^2 + dz^2\right). \tag{26}$$

The assumption of homogeneity and isotropy of the 4D geometry requires that $\mathrm{A}_\mu = (\mathrm{A}_t, 0, 0, 0)$ and A_t and Φ are functions of t and not the spatial coordinates. To further simplify, we shall also assume that the fields do not depend on x^5. From eq. 14, these constraints imply that $\mathrm{B}_\mu = \mathrm{J}_\mu$ and the only non vanishing component of J_μ is J_t, and of A_μ is A_t. Applying the conservation equation (eq. 25) yields

$$(i)\ \dot{\Phi} = 0, \mathrm{J}_t = 0, \ \text{or} \qquad (ii)\ \Phi = \dot{a}(t), \mathrm{J}_t = \ddot{a}/\dot{a}. \tag{27}$$

Case (i) yields $\mathrm{H}_{\mu\nu} = 0$, and eq. 21 yields the usual Friedman equation along with matter conservation.

$$(\dot{a}/a)^2 = \frac{8\pi}{3}\rho.$$

Case (ii) yields non vanishing $\mathrm{H}_{\mu\nu}$, which when applied to eq. 21 gives the modified Friedman equation

$$(\dot{a}/a)^2 + (\ddot{a}/a) = \frac{8\pi}{3}\rho, \tag{28}$$

along with matter conservation which implies $\rho a^3 = \rho_o$ is a constant in a matter dominated universe.

To solve the equation, we specify initial conditions at the current instant of time, $a = 1$, $\dot{a} = \mathrm{H}_o$, the Hubble constant and $\ddot{a} = -q_o\mathrm{H}_o^2$, where q_o is the current deceleration parameter. These conditions can be used to calculate ρ_o, the current matter density (including dark matter). Taking $q_o = -0.5$, which is consistent with the current observations [6], the solution to eq. 28 is plotted in figure 1. From the dashed-dot curve (scale factor), note that the universe started expanding at $t = -0.518\,\mathrm{H}_o^{-1}$, from a size of $a = 2/3$, prior to which it was in a contracting phase. This is in contrast to the solution of the usual Friedmann equations which yields a big bang

($a = 0$) at $t = -0.667\,\mathrm{H}_o^{-1}$. From the solid curve, note that the acceleration is currently positive but decreasing and would become negative beyond $t = 0.319\,\mathrm{H}_o^{-1}$. This is qualitatively consistent with the analysis of observed data in [9], but is in sharp contrast with the standard ΛCDM model which predicts that the acceleration would continue to increase for ever. Clearly, more detailed studies are needed to understand the full implications of this formalism on cosmology.

7. Concluding remarks

Torsion is an integral, geometrical property of space time. The model we have explored has a very general mathematical result pertaining to affine connection and Ricci tensor in Einstein-Cartan theory in higher dimensions. In higher dimensional theories, it provides an alternative way to confine gravity in a torsion free lower dimension, yet modifying it in a significant way to be relevant to astrophysics and cosmology. We have explored some of the interesting consequences of the model and in the process of further study.

Acknowledgments

This work was supported in part (K. C. Wali) by the U. S. Department of Energy (DOE) under contract no. DE-FG02-85ER40237.

References

[1] F. W. Hehl, P. v.d. Hyde, G. D. Kerlick and J. M. Nester, Rev. Mod. Phys. **48**, 393 (1976). For a historical review of concept of torsion and its role in constructing a non-Riemanninan structure of space-time.

[2] I. L. Shapiro, Phys. Rept. **357**, 113 (2002). The report reviews the physical aspects of torsion including both classical and quantum aspects and the necessity of torsion to couple the spin of matter fields to space-time.

[3] J. M. Overduin and P. S. Wesson, Phys. Rept. **283**, 303 (1997). For various aspects and recent developments of torsion free Kaluza-Klein theory.

[4] M. W. Kalinowski, Int. J. Theo. Phys. **20**, 563 (1981). For an analysis of Kaluza-Klein theory in the presence of torsion.

[5] M. Nakahara, sections 7.8.1 and 7.8.4, *Geometry, Topology and Physics*, Second edition (IOP Publishing Ltd., London, 2003).

[6] R. A. Daly and S. G. Djorgovski,CP937, *Supernova 1987A: 20 Years After*, edited by S. Immler, K. Weiler, and R. McCray (American Institute of Physics 978-0-7354-0448-9, 2007).

[7] N. A. Viet and K. C. Wali, Phys. Rev. **D67**, 124029 (2003). These minimal conditions on torsion were utilized in Viet, Wali to obtain a fully covariant action involving all the metric fields in a discretized Kaluza-Klein theory. See also N. A. Viet, arxiv: hep-th/0303128.

[8] C. G. Bohmer and L. Fabbri, Mod. Phys. Lett. **A22**, 2727 (2007).

[9] A. Shafieloo, V. Sahni and A. A. Starobinsky, arXiv: astro-ph/0903.5141.

Journal of Physics: Conference Series **259** (2010) 012031 doi:10.1088/1742-6596/259/1/012031

Matrix Inflation and its String Theory Origins

A. Ashoorioon[1], H. Firouzjahi[2] & M. M. Sheikh-Jabbari[2]

[1] Institutionen för fysik och astronomi Uppsala Universitet, Box 803, SE-751 08 Uppsala, Sweden

[2] School of Physics, Institute for Research in Fundamental Sciences (IPM), P .O. Box 19395-5531, Tehran, Iran

E-mail: amjad.ashoorioon@fysast.uu.se

Abstract. Motivated by the dynamics of N coincident D3 branes in some specific flux compactifications, we construct an inflationary model in which inflation is driven by three $N \times N$ hermitian matrices $\Phi_i, i = 1, 2, 3$, hence the name Matrix Inflation, or M-flation for short. We show that one can consistently restrict the classical dynamics to a sector in which the Φ_i are proportional to the $N \times N$ irreducible representation of SU(2). In this sector our model behaves like an effective inflaton field, which takes super-Planckian field values, and $3N^2 - 1$ isocurvature fields. These may have the observational effects such as production of iso-curvature perturbations on cosmic microwave background. Moreover, the existence of these extra scalars provides us with a natural preheating mechanism and exit from inflation. Due to the super-Planckian excursions of the canonical effective inflaton, the model is capable of producing a considerable amount of gravity waves that can be probed by future CMB polarization experiments. Furthermore, the fine-tunings associated with unnaturally small couplings in the chaotic type inflationary scenarios are removed. We also show that even if the cutoff of the theory is lowered by the square of number of species, one can still use the effective field theory approach to justify the absence of higher dimensional operators.

1. Introduction

The inflationary paradigm, the idea that the early Universe has undergone a nearly exponential expansion phase, has appeared as the leading candidate for explaining the recent cosmological observations data [1]. The simplest and still successful model of inflation is a massive, free scalar field minimally coupled to the Einstein gravity. Nonetheless, motivated by various beyond the Standard Model particle physics or supergravity and string theory settings, many models of inflation have been constructed by introducing more non-trivial potentials for the scalar field and/or the addition of other scalar fields to the model. The latter are called multi-field models.

Cosmological perturbation in theories of multi-field inflation, in which one deals with more than one scalar field, have been studied [2]. In the multiple field inflationary models one can perform a rotation in the field space of scalar fields where the inflaton field is evolving along a trajectory while the remaining fields are orthogonal to it. These extra fields, like the inflaton itself, have quantum fluctuations which once stretched to super-Hubble scales can become classical and can therefore contribute to the power spectrum of iso-curvature as well as curvature perturbations, the details of which depends on the post inflationary dynamics and the reheating scenario.

In [3,4], inspired by the potential derived from a stringy setup in which a stack of N D3-branes is exposed to an appropriate flux in a specific background, we formulated an inflationary model in which the inflatons are $N \times N$ hermitian matrices, hence the name "Matrix" inflation, or M-flation. As we will see, our model has a simple but still rich dynamics for a specific truncation of the theory which is known as SU(2) truncation. We argue that Matrix inflation can solve the fine-tunings associated with the standard chaotic inflationary scenarios. Furthermore, there will be a tower of iso-curvature perturbations that except for one, whose amplitude at the Hubble scale might be observable, the rest are quite tiny and probably unobservable. The existence of the isocurvature mode can have significant observational consequences for the CMB observations [5]. Moreover, we argue that our model has an embedded efficient preheating mechanism and also observable gravity waves at the horizon scale that can be probed by future experiments. We also show that one can still use the effective field theory approach even if the UV cutoff of the theory is lowered. We should emphasize here that in calculating the potentials from the dynamics of coincident branes in an appropriate flux compactification, we did not take into account the back-reactions of the compactifications and the moduli stabilization effects on the potential.

2. String Theory Model

In the context of string theory, the world-volume theory of N coincident p-branes is described by a (supersymmetric) $U(N)$ gauge theory. In this system, the transverse positions of the branes, Φ_I, $I = p+1, ..., 9$, which from the world-volume theory are scalars in the adjoint representation of $U(N)$, are hence $N \times N$ matrices. For the case of our interest, $p = 3$, there are six such scalars. The DBI action for the system of N coincident D3-branes in the background RR six form flux (sourced by a distribution of D5-branes) is given by

$$S = \frac{1}{(2\pi)^3 l_s^4 g_s} \int d^4x \, \mathrm{STr}\left(1 - \sqrt{-|g_{ab}|}\sqrt{|Q^I_J|} + \frac{i g_s}{2 \cdot 2\pi l_s^2}[X^I, X^J] C^{(6)}_{I\,J\,0123}\right). \tag{1}$$

Here l_s is the string scale and g_s is the perturbative string coupling. The operator STr on a product of matrices is the trace of their symmetrized product. The induced metric on branes, g_{ab}, is given by $g_{ab} = G_{MN}\partial_a X^M \partial_b X^N$, where X^M indicates the ten-dimensional positions of the branes and G_{MN} is the ten dimensional background metric. Here the indices $I, J = 4, 5, \cdots, 9$ represent the coordinates perpendicular to the branes world-volume, the indices $a, b = 0, 1, 2, 3$ represent the brane world-volume coordinates and the capital letters $M, N = 0, 1, \cdots, 9$ indicate the ten-dimensional coordinates. The matrix $Q^I_J = \delta^{IJ} + \frac{i}{2\pi l_s^2}[X^I, X^J]$ is due to non-commutativity properties of the system [6] and $C^{(6)}_{I\,J\,0123}$ is a rank-6 antisymmetric Ramond-Ramond (RR) field which has two legs along the direction transverse to the D3-brane.

We consider the ten-dimensional IIB supergravity background

$$ds^2 = -2dx^+dx^- - \hat{m}^2 \sum_{i=1}^{3}(x^i)^2(dx^+)^2 + \sum_{I=1}^{8} dx_I dx_I, \tag{2}$$

$$C_{+123ij} = \frac{2\hat{\kappa}}{3}\epsilon_{ijk}x^k \tag{3}$$

where i, j indices, which are ranging over $1, 2, 3$, parameterize three out of six transverse directions to D3-brane and x^I include three spatial directions along the brane and five of the transverse directions to D3-branes. With

$$\hat{m}^2 = 4g_s^2\hat{\kappa}^2/9 \tag{4}$$

the above background, with constant dilaton, is a solution to the supergravity equations of motion. If we turn-off fluctuations along the directions transverse to the branes and the x^i directions (this may be done if we compactify six dimensions on a $CY3$ with two 3-cycles, one small and one large), fix the light-cone gauge on the D3-branes, expand the action and keep up to the fourth order four X^I. We obtain

$$\begin{aligned} S &= \frac{1}{(2\pi)^3 l_s^4 g_s} \int d^4x \; \mathrm{Tr}[\frac{-1}{2}\partial_\mu X_i \partial^\mu X_i - V(X)] \\ V &= -\frac{1}{4\cdot(2\pi l_s^2)^2}\,[X_i,X_j][X_i,X_j] + \frac{i g_s \hat{\kappa}}{3\cdot 2\pi l_s^2}\epsilon^{ijk} X_i[X_j,X_k]X_i + \frac{1}{2}\hat{m}^2 X_i^2 \; . \end{aligned} \tag{5}$$

If we redefine

$$X_i = \sqrt{(2\pi)^3 g_s}\; l_s^2 \Phi_i \; , \qquad \lambda = 2\pi g_s \; , \qquad \hat{\kappa} = \frac{\kappa}{g_s\cdot\sqrt{2\pi g_s}} \; , \qquad \hat{m}^2 = m^2 \; , \tag{6}$$

The potential takes the form

$$V = \mathrm{Tr}\left(-\frac{\lambda}{4}[\Phi_i,\Phi_j][\Phi_i,\Phi_j] + \frac{i\kappa}{3}\epsilon_{jkl}[\Phi_k,\Phi_l]\Phi_j + \frac{m^2}{2}\Phi_i^2\right) . \tag{7}$$

It is possible to show that one can consistently restrict the classical dynamics to a sector in which we are effectively dealing with a single scalar field $\hat{\phi}$. This sector, which will be called the $SU(2)$ sector, is obtained for matrix configurations of the form

$$\Phi_i = \hat{\phi}(t) J_i \; , \qquad i = 1,2,3, \tag{8}$$

where J_i are the basis for the N dimensional irreducible representation of the $SU(2)$ algebra

$$[J_i, J_j] = i\,\epsilon_{ijk} J_k \; , \qquad \mathrm{Tr}(J_i\,J_j) = \frac{N}{12}(N^2-1)\,\delta_{ij}\, . \tag{9}$$

Since both Φ_i and J_i are hermitian, we conclude that $\hat{\phi}$ is a real scalar field.

Plugging these into the action (5) and adding the four-dimensional Einstein gravity, we obtain

$$S = \int d^4x\sqrt{-g}\left[\frac{M_P^2}{2}R + \mathrm{Tr}J^2\left(-\frac{1}{2}\partial_\mu\hat{\phi}\partial^\mu\hat{\phi} - \frac{\lambda}{2}\hat{\phi}^4 + \frac{2\kappa}{3}\hat{\phi}^3 - \frac{m^2}{2}\hat{\phi}^2\right)\right] , \tag{10}$$

where $\mathrm{Tr}J^2 = \sum_{i=1}^3 \mathrm{Tr}(J_i^2) = N(N^2-1)/4$. Interestingly enough, this represents the action of chaotic inflationary models with a non-standard kinetic energy. Upon the field redefinition

$$\hat{\phi} = \left(\mathrm{Tr}J^2\right)^{-1/2}\phi = \left[\frac{N}{4}(N^2-1)\right]^{-1/2}\phi\, , \tag{11}$$

the kinetic energy for the new field ϕ becomes standard, while the potential takes the form of a symmetry-breaking potential, taking into account the condition (4)

$$V_0(\phi) = \frac{\lambda_{eff}}{4}\phi^4 - \frac{2\kappa_{eff}}{3}\phi^3 + \frac{m^2}{2}\phi^2 = \frac{\lambda_{eff}}{4}\phi^2\,(\phi-\mu)^2 \tag{12}$$

where

$$\lambda_{eff} = \frac{2\lambda}{\mathrm{Tr}J^2} = \frac{8\lambda}{N(N^2-1)}, \quad \kappa_{eff} = \frac{\kappa}{\sqrt{\mathrm{Tr}J^2}} = \frac{2\kappa}{\sqrt{N(N^2-1)}}, \quad \mu \equiv \sqrt{2}m/\sqrt{\lambda_{eff}} \tag{13}$$

The minimum $\phi = \mu$ for the potential (12) corresponds to the supersymmetric background where N D3-branes blow-up into a giant D5-brane [7]. It is worth noting that in the brane theory setting the $U(N)$ symmetry appears as a gauge symmetry.

As a result of motion of D3-branes in the background $C^{(6)}$ flux, two of the directions transverse to D3-branes blow-up into (fuzzy) two sphere, which in the large N limit behaves as a D5-brane with world-volume $R^4 \times S^2$. In this geometric picture our inflaton field ϕ is nothing but the radius of this two-sphere.

3. Mass spectrum of Ψ_i modes in Matrix Inflation

We notice that there are $3N^2 - 1$ fields denoted as Ψ_i perpendicular to the SU(2) direction that even though do not contribute to the classical inflationary dynamics, do have quantum fluctuations and will hence affect the cosmological perturbation analysis. To compute these effects we need to have the mass spectrum of the $3N^2 - 1$ modes coming from the Ψ_i.

Assuming $\Psi_i = \Phi_i - \hat{\phi} J_i$ where $\mathrm{Tr}(\Psi_i J_i) = 0$, we expand the action up to the second order in Ψ_i. The kinetic term readily takes the standard form $\frac{1}{2}\mathrm{Tr}(\partial_\mu \Psi_i \partial^\mu \Psi_i)$. After a slightly lengthy but straightforward computation the potential to second order in Ψ_i is obtained as

$$V_{(2)} = \mathrm{Tr}\left[\frac{\lambda}{2}\hat{\phi}^2\ \Omega_i\Omega_i + \frac{m^2}{2}\ \Psi_i\Psi_i + \left(-\frac{\lambda}{2}\hat{\phi}^2 + \kappa\hat{\phi}\right)\Psi_i\Omega_i\ \right] \tag{14}$$

where

$$\Omega_k \equiv i\epsilon_{ijk}[J_i, \Psi_j]\ . \tag{15}$$

From the above form we see that if we have the eigenvectors (eigen-matrices) of the Ω_i we can compute the spectrum of Ψ_i in terms of $\hat{\phi}$-field (to be viewed as the inflaton). Finding the eigenvectors of Ω_i is mathematically the same problem as finding the vector spherical harmonics. (For a detailed discussion see e.g. [8], section 5.2.) If we denote the Ω eigenvalues by ω, i.e. $\Omega_i = \omega\Psi_i$, we obtain

$$V_{(2)} = \left(\frac{\lambda_{eff}}{4}\phi^2(\omega^2 - \omega) + \kappa_{eff}\,\omega\,\phi + \frac{m^2}{2}\right)\ \mathrm{Tr}\,\Psi_i\Psi_i\ . \tag{16}$$

If we have the possible values of ω we can read off the effective (ϕ-dependent) mass of the Ψ_i modes

$$M^2 = \frac{\lambda_{eff}}{2}\phi^2(\omega^2 - \omega) + 2\kappa_{eff}\,\omega\phi + m^2 \tag{17}$$

Following the analysis of [8], we find that ω can take three values:

- *" The zero modes"* $\omega = -1$. This happens for modes of the form

  $$\Psi_i = [J_i, \Lambda],$$

  with Λ being an arbitrary matrix. Since U(N) symmetry is a gauge theory, zero modes can be removed by the gauge transformations. Noting that Λ is an arbitrary matrix there are N^2 of such modes, all with the same mass.
- *"The α modes"*: $\omega = -(l+1)$, $l \in \mathbb{Z}$, $0 \le l < N$. Each mode for a given l has a multiplicity of $2l + 1$. Therefore, there are N^2 of α-modes.
- *"The β modes"*: $\omega = l$, $l \in \mathbb{Z}$, $0 < l < N$. Each mode for a given l has a multiplicity of $2l + 1$. Therefore, there are $N^2 - 1$ of β-modes.

As expected, there are altogether $3N^2 - 1$ zero, α, and β modes.

4. Power spectra in the presence of Ψ_i modes

With the mass spectrum for Ψ_i modes computed in the previous section we can compute the power spectra of the adiabatic and the iso-curvature perturbations.

There are two distinguishable branches for the model to inflate, $\phi > \mu$ and $\mu/2 < \phi < \mu$, that we will analyze them separately here.

(a) $\phi > \mu$

To match the observational constraints from WMAP, the above parameters have to take the following values:

$$\lambda_{eff} \simeq 4.91 \times 10^{-14}, \quad m \simeq 4.074 \times 10^{-6} M_P, \quad \kappa_{eff} \simeq 9.574 \times 10^{-13} M_P \tag{18}$$

$$\phi_i \simeq 43.57 M_P \quad , \quad \phi_f \simeq 27.07 M_P \quad , \quad \mu \simeq 26 M_P, \tag{19}$$

where ϕ_f and ϕ_i are respectively the values of the inflaton at the end of inflation and when the current Hubble scale left the horizon. The scalar spectral index for the adiabatic perturbations is $n_{\mathcal{R}} \simeq 0.959$.

The lowest mass in the tower of $\Psi_{r,lm}$ iso-curvature modes belongs to the the $l = 0$ $\alpha-$mode, whose mass is $M^2_{\alpha,0}(\phi) = \lambda_{eff}\phi^2 - 2\kappa_{eff}\phi + m^2$. The amplitudes and spectral index of the iso-curvature spectrum are respectively $P_{S_{\alpha,0m}} \simeq 1.162 \times 10^{-11}$ and $n_{\Psi_{\alpha,0m}} \simeq 0.981$. The next ones in the tower of iso-curvature modes have negligible amplitudes at Hubble scale (i.e. $\gtrsim 10^{-15}$) and therefore their contributions could be ignored. The amplitude of tensor spectrum at Hubble scale is $P_T(k_{60}) \simeq 4.84 \times 10^{-10}$, $r \simeq 0.2$ with the spectral index $n_T \simeq -0.025$. Planck should be able to verify this model.

(b) $\mu/2 < \phi < \mu$

Here to satisfy the constraints from the amplitude and spectral index from WMAP5, one has to adjust the the parameters as below:

$$\phi_i \simeq 23.5 M_P \quad , \quad \phi_f \simeq 35.03 M_P \quad , \quad \mu \simeq 36 M_P. \tag{20}$$

$$\lambda_{eff} \simeq 7.187 \times 10^{-14}, \quad m \simeq 6.824 \times 10^{-6} M_P, \quad \kappa_{eff} \simeq 1.940 \times 10^{-12} M_P. \tag{21}$$

The index of the adiabatic spectrum for such values of parameters is $n_{\mathcal{R}} \simeq 0.961$.

Again the least massive iso-curvature mode is the $l = 0$ $\alpha-$mode. Its amplitude and spectral index are, respectively, $P_{S_{\alpha,0m}} \simeq 1.46 \times 10^{-11}$ and $n_{\Psi_{\alpha,0m}} \simeq 0.987$. Next iso-curvature modes have larger masses and therefore their amplitudes are negligible in comparison with the adiabatic one ($\gtrsim 10^{-12}$). The amplitude of tensor spectrum at Hubble scale is $P_T(k_0) \simeq 1.307 \times 10^{-11}$, i.e. $r \simeq 0.048$ with the spectral index $n_T \simeq -0.006$. Such gravity wave spectrum could be detected by CMBPOL [9]. The tensor spectrum is very close to being scale-invariant in this case.

5. Particle creation and preheating scenario

Even though one can argue that during the slow-roll inflation particle creation and hence its back-reaction on the dynamics of the ϕ field is not large, it becomes important when ϵ, η are of order one. To show this schematically, we approximate the potential with a quartic potential and follow the analysis performed in [10] for massless preheating, using the potential

$$V(\phi, \chi) = \frac{\lambda}{4}\phi^4 + \frac{1}{2}g\phi^2\chi^2 \tag{22}$$

As shown in [10], the ϕ equation of motion toward the end of inflation is an (unharmonic) oscillator around $\phi = 0$, the amplitude of the solution is decreasing as $t^{-1/2}$. The equation for the Ψ modes, can be also solved explicitly in this case. One can see that the equation for all three zero, α and β modes is of the form of $g^2/\lambda = n(n+1)/2$ (with $n = 1, l, l-1$ respectively for zero, α and β modes). As discussed in [10] for these specific values of g^2/λ we have the significant property that there is an enhancement in the parametric resonance leading to considerable creation of zero, α and β modes. For the odd n (i.e. for our zero modes, odd l α-mode, and even l β-mode) the particle creation is peaked around zero momentum k modes. For the even n modes, however, the particle creations is peaked around momenta $k^2 = \frac{3}{2}H^2_{inf}\epsilon\sqrt{\frac{g^2}{2\lambda}}$, where ϵ is computed at the beginning of the slow-roll inflation and H_{inf} is the Hubble during inflation. For low n, the Floquet index $\mu_k \propto \ln n_k$ (n_k is the number density of the produced particles at momentum k) is around 0.15 for odd n and around 0.5 for even n. Therefore, among the low n modes the main contribution to preheating is coming from odd n [10]. As discussed in [10] the bigger k is, the more energy can be transferred from the inflationary sector to the Ψ

sector and a more efficient preheat mechanism. This means that α and β modes with large l, l of order N, make the biggest contribution. Of course, one should note that zero modes are not physical in the stringy motivated model matrix inflation, where they are gauged. All in all, due to the existence of the large l modes and for large N in our model, we expect to have a very efficient preheating model.

6. UV behavior

In in a theory with many particles the scale where quantum gravity effects become large is lowered to

$$\Lambda^2 = \frac{M_P^2}{N_{\rm cl}}, \tag{23}$$

where $N_{\rm cl}$ counts all the species with mass below the cutoff Λ [11], here in our case $N_{\rm cl} = 3N^2$. Here we check that the amount of excursions for the effective inflaton ϕ is less than this cutoff for our model. This will guarantee that the higher dimensional operators do not affect the dynamics. We first focus on the branch in which the setup inflates in the region $\phi > \mu$. To find the number of species in our case, we assume that $\lambda = 1$. To obtain the value of $\lambda = 4.91 \times 10^{-14}$, one needs $N = 54618$ D3 branes. The amount of excursion in the physical field space, $\Delta\hat{\phi}$, is $2(\phi_f - \phi_i)/\sqrt{(N(N^2-1))} = 2.58 \times 10^{-6} M_P$. The value of UV cutoff could also be calculated from (23) which turns out to be $1.05 \times 10^{-5} M_p$. Hence the higher dimensional operators are suppressed at least by the factors of $\left(\frac{\Delta\hat{\phi}}{\Lambda}\right)^2 \simeq 0.06$ for the mass term and $\left(\frac{\Delta\hat{\phi}}{\Lambda}\right)^3 \simeq 0.014$ (and smaller) for the cubic (and quartic) term. The other interesting thing that is worthwhile mentioning is that one obtains the mass parameter for the model to be close to the cutoff, $m \simeq 4.074 \times 10^{-6} M_P$, and one would only need a tuning of order $\simeq 50\%$ to achieve the value required to fit the observational data. Repeating the same analysis for the branch in which $\mu/2 < \phi < \mu$, one realizes that that the higher dimensional operators are suppressed at least by factors of $\left(\frac{\Delta\hat{\phi}}{\Lambda}\right)^2 \simeq 0.02$, therefore the UV behavior is under much better control. One can conclude that at least the stringy picture of matrix inflation is safe from the UV criticism of [12].

References

[1] E. Komatsu *et al.* , "Five-Year Wilkinson Microwave Anisotropy Probe Observations:Cosmological Interpretation," arXiv:0803.0547 [astro-ph].
[2] C. Gordon, D. Wands, B. A. Bassett and R. Maartens, "Adiabatic and entropy perturbations from inflation," Phys. Rev. D **63**, 023506 (2001), [arXiv:astro-ph/0009131];
[3] A. Ashoorioon, H. Firouzjahi and M. M. Sheikh-Jabbari, JCAP **0906**, 018 (2009) [arXiv:0903.1481 [hep-th]].
[4] A. Ashoorioon, H. Firouzjahi and M. M. Sheikh-Jabbari, JCAP **1005**, 002 (2010) [arXiv:0911.4284 [hep-th]].
[5] A. Mangilli, L. Verde and M. Beltran, arXiv:1006.3806 [astro-ph.CO].
[6] R. C. Myers, "Dielectric-branes," JHEP **9912**, 022 (1999) [arXiv:hep-th/9910053].
[7] O. DeWolfe, S. Kachru and H. L. Verlinde, "The giant inflaton," JHEP **0405**, 017 (2004) [arXiv:hep-th/0403123].
[8] K. Dasgupta, M. M. Sheikh-Jabbari and M. Van Raamsdonk, "Matrix perturbation theory for M-theory on a PP-wave," JHEP **0205**, 056 (2002) [arXiv:hep-th/0205185].
[9] D. Baumann *et al.* [CMBPol Study Team Collaboration], "CMBPol Mission Concept Study: Probing Inflation with CMB Polarization," arXiv:0811.3919 [astro-ph].
[10] P. B. Greene, L. Kofman, A. D. Linde and A. A. Starobinsky, "Structure of resonance in preheating after inflation," Phys. Rev. D **56**, 6175 (1997) [arXiv:hep-ph/9705347].
[11] G. Dvali, arXiv:0706.2050 [hep-th].
[12] M. Postma, arXiv:1009.0421 [astro-ph.CO].

16th International Symposium on Particles Strings and Cosmology (PASCOS 2010) IOP Publishing
Journal of Physics: Conference Series **259** (2010) 012032 doi:10.1088/1742-6596/259/1/012032

Cosmological intersecting brane solutions in string theory

Kunihito Uzawa

Department of Physics, Kindai University, Higashi-Osaka, Osaka 577-8502, Japan

E-mail: uzawa@phys.kindai.ac.jp

Abstract. The recent discovery of an explicit dynamical description of p-branes makes it possible to investigate the existence of intersection of such objects. We generalize the solutions depending on the overall transverse space coordinates and time to those which depend also on the relative transverse space and satisfy new intersection rules. We give classification of these dynamical intersecting brane solutions involving two branes, and discuss the application of these solutions to cosmology and show that these give Friedmann-Lemaître-Robertson-Walker cosmological solutions.

1. Introduction

The dynamics of the brane world model in five or six dimensions have been much explored (see [1, 2] and references therein) because of the possible cosmological and phenomenological interests. Although some results have recently emerged on the applications of the solutions in higher-dimensional supergravity to the brane world cosmology, e.g., in [1], the construction of the cosmological model in string theory is much less extensive. One motivation for the present work is to improve this situation. For this purpose, it is first necessary to construct dynamical brane solutions depending on the time as well as space coordinates.

It has already been known [3, 4] that dynamical brane solutions arise when the gravity is coupled not only to a single gauge field but also to several combinations of scalars and forms, as generalization of the static intersecting brane solutions in the supergravity. Here we construct dynamical brane solutions by generalizing these static solutions to a dynamical one. The first class of dynamical solutions we study has the dependence on the time as well as overall transverse space coordinates in the metric and obeys the well-known intersection rules. However, it has also been known for some time that some static intersecting brane solutions may not follow these intersection rules. These intersecting brane solutions are derived for the case when the branes depend on the relative transverse directions of the intersecting branes [5].

Our goal is to exhaust and classify all two intersecting brane solutions which depend on the time and (relative) transverse dimensions and to study their applications to the cosmological evolutions, in particular in the ten-dimensional string theory and eleven-dimensional supergravity theory. We first find cosmological solutions for possible intersections including the above exceptional cases for two intersecting branes by extending the similar solutions obeying the usual intersection rules [3, 4]. Our results on the dynamical branes are given for general cases of arbitrary dimensions and forms, but in their applications to cosmology, we mainly focus on the dynamical branes in ten- and eleven-dimensional supergravities because these are the most important low-energy effective theories of superstrings.

2. Intersecting brane solutions in D-dimensional theory

We consider a D-dimensional theory composed of the metric g_{MN}, dilaton ϕ, and the antisymmetric tensor fields of rank (p_r+2) and (p_s+2):

$$\begin{aligned} S &= \frac{1}{2\kappa^2}\int\left[R*\mathbf{1}-\frac{1}{2}d\phi\wedge *d\phi-\frac{1}{2}\frac{1}{(p_r+2)!}\mathrm{e}^{\epsilon_r c_r\phi}F_{(p_r+2)}\wedge *F_{(p_r+2)}\right. \\ &\quad \left.-\frac{1}{2}\frac{1}{(p_s+2)!}\mathrm{e}^{\epsilon_s c_s\phi}F_{(p_s+2)}\wedge *F_{(p_s+2)}\right], \end{aligned} \tag{1}$$

where κ^2 is the D-dimensional gravitational constant, $*$ is the Hodge operator in the D-dimensional spacetime, $F_{(n)}$ is an n-form field strength, and c_I, ϵ_I ($I=r,\ s$) are constants given by

$$c_I^2=4-\frac{2(p_I+1)(D-p_I-3)}{D-2},\qquad \epsilon_I=\begin{cases}+ & \text{if } p_I-\text{brane is electric}\\ - & \text{if } p_I-\text{brane is magnetic}.\end{cases} \tag{2}$$

To solve these field equations, we assume that the D-dimensional metric takes the form

$$ds^2=h_r^\alpha h_s^\beta\left[h_r^{-1}h_s^{-1}q_{\mu\nu}(\mathrm{X})dx^\mu dx^\nu+h_s^{-1}\gamma_{ij}(\mathrm{Y}_1)dy^idy^j+h_r^{-1}w_{mn}(\mathrm{Y}_2)dv^mdv^n+u_{ab}(\mathrm{Z})dz^adz^b\right], \tag{3}$$

where $q_{\mu\nu}$, γ_{ij}, w_{mn}, and u_{ab} are the metrics depending only on x^μ, y^i, v^m, and z^a coordinates of dimensions $(p+1)$, (p_s-p), (p_r-p), and $(D+p-p_r-p_s-1)$, respectively. The parameters α and β in the metric (3) are given as

$$\alpha=\frac{p_r+1}{D-2},\qquad \beta=\frac{p_s+1}{D-2}. \tag{4}$$

Here we suppose that $p_s(p_r)$-brane extends along X and Y_1 (Y_2) spaces.

The D-dimensional metric (3) implies that the p-brane solutions are characterized by a function which depends on the coordinates transverse to the brane as well as the worldvolume coordinate. For the configurations of two branes, we should sort the coordinates in three sets and the powers of harmonic functions are different for each set of coordinates according to the intersection rules. One set of the coordinates is the overall worldvolume coordinates, which are common to the two branes. The others are overall transverse coordinates and the last are the relative transverse coordinates, which are transverse to only one of the two branes.

The field equations of intersecting branes allow for the following three kinds of possibilities on p_r- and p_s-branes in D dimensions [5]:

(I) Both h_r and h_s depend on the overall transverse coordinates: $h_r=h_r(x,z)$, $h_s=h_s(x,z)$.

(II) Only h_s depends on the overall transverse coordinates, but the other h_r does on the corresponding relative coordinates: $h_r=h_r(x,y)$, $h_s=h_s(x,z)$.

(III) Each of h_r and h_s depends on the corresponding relative coordinates: $h_r=h_r(x,y)$, $h_s=h_s(x,v)$.

For completeness, let us first consider case (I) though this has been already discussed in [4]. We assume that the the scalar field ϕ and the gauge field strengths are given as

$$\mathrm{e}^\phi = h_r^{\epsilon_r c_r/2}h_s^{\epsilon_s c_s/2}, \tag{5}$$

$$F_{(p_r+2)} = d\left[h_r^{-1}(x,z)\right]\wedge\Omega(\mathrm{X})\wedge\Omega(\mathrm{Y}_2),\qquad F_{(p_s+2)}=d\left[h_s^{-1}(x,z)\right]\wedge\Omega(\mathrm{X})\wedge\Omega(\mathrm{Y}_1), \tag{6}$$

where $\Omega(\mathrm{X})$, $\Omega(\mathrm{Y}_1)$, and $\Omega(\mathrm{Y}_2)$ denote the volume forms of dimensions $(p+1)$, (p_s-p), and (p_r-p), respectively:

$$\Omega(\mathrm{X}) = \sqrt{-q}\, dx^0 \wedge \cdots \wedge dx^p, \quad \Omega(\mathrm{Y}_1) = \sqrt{\gamma}\, dy^1 \wedge \cdots \wedge dy^{p_s-p}, \quad \Omega(\mathrm{Y}_2) = \sqrt{w}\, dv^1 \wedge \cdots \wedge dv^{p_r-p}. \tag{7}$$

Here, q, γ, and w are the determinant of the metric $q_{\mu\nu}$, γ_{ij}, and w_{mn}, respectively. With the assumptions of fields, it is easy to show that the field equations reduce to

$$R_{\mu\nu}(\mathrm{X}) = 0, \quad R_{ij}(\mathrm{Y}_1) = 0, \quad R_{mn}(\mathrm{Y}_2) = 0, \quad R_{ab}(\mathrm{Z}) = 0, \tag{8}$$

$$h_r = h_0(x) + h_1(z), \quad h_s = k_0(x) + k_1(z), \tag{9}$$

$$D_\mu D_\nu h_0 = 0, \quad \triangle_{\mathrm{Z}} h_1 = 0, \quad \triangle_{\mathrm{Z}} h_s = 0, \quad \text{for } \partial_\mu h_s = 0, \tag{10}$$

$$D_\mu D_\nu k_0 = 0, \quad \triangle_{\mathrm{Z}} k_1 = 0, \quad \triangle_{\mathrm{Z}} h_r = 0, \quad \text{for } \partial_\mu h_r = 0, \tag{11}$$

where D_μ is the covariant derivative with respect to the metric $q_{\mu\nu}$, and $\triangle_{\mathrm{Z}}$ is the Laplace operators on the space of Z, and $R_{\mu\nu}(\mathrm{X})$, $R_{ij}(\mathrm{Y}_1)$, $R_{mn}(\mathrm{Y}_2)$, and $R_{ab}(\mathrm{Z})$ are the Ricci tensors of the metrics $q_{\mu\nu}(\mathrm{X})$, $\gamma_{ij}(\mathrm{Y}_1)$, $w_{mn}(\mathrm{Y}_2)$, and $u_{ab}(\mathrm{Z})$, respectively, and we used the intersection rule $\chi = 0$:

$$\chi \equiv p + 1 - \frac{(p_r+1)(p_s+1)}{D-2} + \frac{1}{2}\epsilon_r \epsilon_s c_r c_s. \tag{12}$$

If $F_{(p_r+2)} \neq 0$ and $F_{(p_s+2)} \neq 0$, the functions h_1 and k_1 are nontrivial.

As a special example, let us consider the case

$$q_{\mu\nu} = \eta_{\mu\nu}, \quad \gamma_{ij} = \delta_{ij}, \quad w_{mn} = \delta_{mn}, \quad u_{ab} = \delta_{ab}, \tag{13}$$

where $\eta_{\mu\nu}$ is the $(p+1)$-dimensional Minkowski metric and δ_{ij}, δ_{mn}, δ_{ab} are the (p_s-p)-, (p_r-p)-, and $(D+p-p_r-p_s-1)$-dimensional Euclidean metrics, respectively. For $\partial_\mu h_s = 0$, the solution for h_r and h_s can be obtained explicitly as

$$h_r(x,z) = A_\mu x^\mu + B + \sum_l M_l |z^a - z_l^a|^{-D-p+p_r+p_s+3}, \quad h_s(z) = C + \sum_c M_c |z^a - z_c^a|^{-D-p+p_r+p_s+3}, \tag{14}$$

where A_μ, B, C, M_l, and M_c are constant parameters, and z_l^a and z_c^a are constants representing the positions of the branes. We can also choose the solution in which the p_s-brane part depends on both t and z.

Now we discuss the intersecting brane solutions in eleven-dimensional supergravity and in ten-dimensional string theories. For the M-branes in eleven-dimensional supergravity, there is 4-form field strength without dilaton, so the intersection rule $\chi = 0$ gives

$$p = \frac{(p_r+1)(p_s+1)}{9} - 1, \tag{15}$$

where p denotes the number of overlapping dimensions of the p_r and p_s branes. Then we get the intersections involving the M2 and M5-branes [5]

$$\mathrm{M2} \cap \mathrm{M2} = 0, \quad \mathrm{M2} \cap \mathrm{M5} = 1, \quad \mathrm{M5} \cap \mathrm{M5} = 3. \tag{16}$$

For the ten-dimensional string theories, the couplings to dilaton for the RR-charged D-branes are given by

$$\epsilon_r c_r = \frac{1}{2}(3 - p_r), \quad \epsilon_s c_s = \frac{1}{2}(3 - p_s). \tag{17}$$

The condition $\chi = 0$ then gives

$$p = \frac{1}{2}(p_r + p_s - 4). \tag{18}$$

The intersections for the D-branes are thus given by [5]

$$\mathrm{D}p_r \cap \mathrm{D}p_s = \frac{1}{2}(p_r + p_s) - 2. \tag{19}$$

We finally consider the intersections for NS-branes. The parameters c_r for fundamental string (F1) and solitonic 5-brane are $\epsilon_1 c_1 = -1$ (for F1) and $\epsilon_5 c_5 = 1$ (for NS5), respectively. Then the intersections involving the F1 and NS5-branes are [5]

$$\mathrm{F1} \cap \mathrm{NS5} = 1, \quad \mathrm{NS5} \cap \mathrm{NS5} = 3, \quad \mathrm{F1} \cap \mathrm{D}\bar{p} = 0, \quad \mathrm{D}\bar{p} \cap \mathrm{NS5} = \bar{p} - 1, \quad (1 \le \bar{p} \le 6)\,. \tag{20}$$

There is no solution for the F1-F1 and D0-NS5 intersecting brane systems because the numbers of space dimensions for each pairwise overlap are negative by the intersection rule.

We next consider case (II). We also take the following ansatz for the scalar field ϕ and the gauge field strengths:

$$\mathrm{e}^{\phi} = h_r^{\epsilon_r c_r/2}\, h_s^{\epsilon_s c_s/2}, \tag{21}$$

$$F_{(p_r+2)} = d\left[h_r^{-1}(x,y)\right] \wedge \Omega(\mathrm{X}) \wedge \Omega(\mathrm{Y}_2), \quad F_{(p_s+2)} = d\left[h_s^{-1}(x,z)\right] \wedge \Omega(\mathrm{X}) \wedge \Omega(\mathrm{Y}_1), \tag{22}$$

where $\Omega(\mathrm{X})$, $\Omega(\mathrm{Y}_1)$, and $\Omega(\mathrm{Y}_2)$ are defined in Eq. (7). Since we use the same procedure as in the case (I), we can derive the intersection rule $\chi = 0$ from the field equations. For $\chi = 0$ and $\partial_\mu h_s = 0$, it is easy to show that the field equations reduce to

$$R_{\mu\nu}(\mathrm{X}) = 0, \quad R_{ij}(\mathrm{Y}_1) = 0, \quad R_{mn}(\mathrm{Y}_2) = 0, \quad R_{ab}(\mathrm{Z}) = 0, \tag{23}$$

$$h_r = h_0(x) + h_1(y), \quad h_s = h_s(z), \quad D_\mu D_\nu h_0 = 0, \quad \triangle_{\mathrm{Y}_1} h_1 = 0, \quad \triangle_{\mathrm{Z}} h_s = 0, \tag{24}$$

where $\triangle_{\mathrm{Y}_1}$ is the Laplace operators on the space of Y_1. If $F_{(p_r+2)} \neq 0$ and $F_{(p_s+2)} \neq 0$, the functions h_1 and k_1 are nontrivial.

Let us consider the following case Eq. (13) in more detail. For $\partial_\mu h_s = 0$, the solution for h_r and h_s can be obtained explicitly as

$$h_r(x,y) = A_\mu x^\mu + B + \sum_l M_l |y^i - y^i_l|^{-p_s+p+2}, \quad h_s(z) = C + \sum_c M_c |z^a - z^a_c|^{-D-p+p_r+p_s+3}, \tag{25}$$

where A_μ, B, C, y^i_l, z^a_c, M_l, and M_c are constant parameters.

One can easily get the solution for $\partial_\mu h_r = 0$ and $\partial_\mu h_s \neq 0$ if the roles of Y_1 and Y_2 are exchanged. Since the dynamical solution (25) obeys the same intersection rule $\chi = 0$, the intersections of M-branes in eleven-dimensional supergravity and D-branes in ten-dimensional string theories are given as (16), (19), and (20).

Finally, we consider case (III). We also assume that the scalar field ϕ and the gauge field strengths are given as

$$\mathrm{e}^{\phi} = h_r^{\epsilon_r c_r/2}\, h_s^{\epsilon_s c_s/2}, \tag{26}$$

$$F_{(p_r+2)} = h_s\, d\left[h_r^{-1}(x,y)\right] \wedge \Omega(\mathrm{X}) \wedge \Omega(\mathrm{Y}_2), \quad F_{(p_s+2)} = h_r\, d\left[h_s^{-1}(x,v)\right] \wedge \Omega(\mathrm{X}) \wedge \Omega(\mathrm{Y}_1), \tag{27}$$

where $\Omega(\mathrm{X})$, $\Omega(\mathrm{Y}_1)$, and $\Omega(\mathrm{Y}_2)$ denote the volume $(p+1)$-, $(p_s - p)$-, and $(p_r - p)$-forms, respectively.

Under the assumption, the field equations give the intersection rule $\chi = -2$. This is different from the usual rule applicable to the cases (I) and (II). Upon using the intersection rule $\chi = -2$, it is easy to show that the field equations reduce to

$$R_{\mu\nu}(\mathrm{X}) = 0, \quad R_{ij}(\mathrm{Y}_1) = 0, \quad R_{mn}(\mathrm{Y}_2) = 0, \quad R_{ab}(\mathrm{Z}) = 0, \tag{28}$$

$$h_r = h_0(x) + h_1(y), \quad h_s = k_0(x) + k_1(v), \tag{29}$$

$$D_\mu D_\nu h_0 = 0, \quad \triangle_{\mathrm{Y}_1} h_1 = 0, \quad \triangle_{\mathrm{Y}_2} h_s = 0, \quad \text{for } \partial_\mu h_s = 0, \tag{30}$$

$$D_\mu D_\nu k_0 = 0, \quad \triangle_{\mathrm{Y}_1} h_r = 0, \quad \triangle_{\mathrm{Y}_2} k_1 = 0, \quad \text{for } \partial_\mu h_r = 0, \tag{31}$$

where $\triangle_{Y_1}$ and $\triangle_{Y_2}$ are the Laplace operators on the spaces of Y_1 and Y_2, respectively. The functions h_1 and k_1 are nontrivial for $F_{(p_r+2)} \neq 0$ and $F_{(p_s+2)} \neq 0$.

Now we consider the case in Eq. (13). For $\partial_\mu h_s = 0$, the solution for h_r and h_s can be obtained explicitly as

$$h_r(x, y) = A_\mu x^\mu + B + \sum_l M_l |y^i - y^i_l|^{-p_s+p+2}, \quad h_s(v) = C + \sum_c M_c |v^m - v^m_c|^{-p_r+p+2}, \quad (32)$$

where A_μ, B, C, y^i_l, v^m_c, M_l, and M_c are constant parameters. We can also get the solution in which the function h_s depends on both t and v.

Let us consider the intersecting brane solutions in eleven-dimensional supergravity and in ten-dimensional string theories. We first discuss the intersections of M-branes in eleven-dimensional supergravity. The intersection rule $\chi = -2$ leads to

$$p = \frac{(p_r+1)(p_s+1)}{9} - 3. \quad (33)$$

Then we get the intersection involving the M5-brane

$$\mathrm{M5} \cap \mathrm{M5} = 1. \quad (34)$$

Equation (33) tells us that the numbers of intersection for M2-M2 and M2-M5 branes are negative, which means that there is no intersecting solution of these brane systems.

Next we consider the intersection in the ten-dimensional string theory. The couplings to dilaton for the RR-charged D-branes are

$$\epsilon_r c_r = \frac{1}{2}(3 - p_r), \quad \epsilon_s c_s = \frac{1}{2}(3 - p_s), \quad (35)$$

and the condition $\chi = -2$ is expressed as $p = \frac{1}{2}(p_r + p_s - 8)$. The intersections for the RR-charged D-branes are thus given by

$$\mathrm{D}p_r \cap \mathrm{D}p_s = \frac{1}{2}(p_r + p_s) - 4. \quad (36)$$

We finally consider the intersections for NS-branes. The parameters c_r for fundamental string (F1) and solitonic 5-brane are $\epsilon_1 c_1 = -1$ for F1 and $\epsilon_5 c_5 = 1$ for NS5, respectively. Then the intersection with F1-brane is forbidden by the intersection rule. The intersections involving the NS5-branes are

$$\mathrm{NS5} \cap \mathrm{NS5} = 1, \quad \mathrm{D}\bar{p} \cap \mathrm{NS5} = \bar{p} - 3, \quad (3 \le \bar{p} \le 8). \quad (37)$$

There is no brane solution involving other intersections because the numbers of space dimensions for each pairwise overlap become negative by the intersection rule.

We list the Friedmann-Lemaître-Robertson-Walker cosmological solutions with an isotropic and homogeneous three-space for the solutions (14), (25), (32) in [5] for M-branes, D-branes, F1 and NS5-branes. Since the time dependence in the metric comes from only one brane in the intersections, the obtained expansion law is simple. In order to find an expanding universe, one may have to compactify the vacuum bulk space as well as the brane worldvolume. If we have introduced the cosmic time τ, we find that the fastest expanding case in the Jordan frame has the power $a \propto \tau^{7/15}$, which is too small to give a realistic expansion law like that in the matter dominated era ($a \propto \tau^{2/3}$) or that in the radiation dominated era ($a \propto \tau^{1/2}$), where a is the scale factor of our Universe. When we compactify the extra dimensions and go to the four-dimensional Einstein frame, the power exponents are different depending on how we compactify the extra dimensions even within one solution. We give the power exponent of the fastest expansion of our four-dimensional universe in the Einstein frame in [5]. We again see that the expansion is too small. Hence, we have to conclude that in order to find a realistic expansion of the universe in this type of models, one has to include additional "matter" fields on the brane.

3. Discussions

We have derived intersecting dynamical brane solutions and discussed their dynamics in the ten- and eleven-dimensional supergravity models. These solutions are obtained by replacing a constant A in the warp factor $h = A + h_1(y)$ of a supersymmetric solution by a function $h_0(x)$ of the coordinates x^μ [3, 4, 6]. Our solutions can contain only one function depending on both time as well as overall or relative transverse space coordinates. In particular, the solutions in Sec. 2 tell us that the brane which depends on overall transverse coordinate can be extended to the time dependent case. It is possible to get the dynamical intersecting brane solutions which obey the intersection rule $\chi = -2$ different from the usual one, as we have discussed in Sec. 2.

We have used the intersection rules to find the cosmological solution because it is not so easy to find it analytically without their rules. The intersection rules have led to the functions h_r and h_s which can be written by linear combinations of terms depending on both coordinates of worldvolume and transverse space. This feature is expected to be shared by a wide class of supersymmetric solutions beyond the examples considered in the present paper, because the result has been obtained by analyzing the general structure of solutions for warped compactification with field strength of the ten- or eleven-dimensional supergravities under ansatz that is natural to include supersymmetric solutions as a special case. We have found that these solutions give a Friedmann-Lemaître-Robertson-Walker universe if we regard the homogeneous and isotropic part of the brane worldvolumes as our spacetime [5]. Unfortunately, the power of the scale factor is so small that the solutions of field equations cannot gives a realistic expansion law. This means that we have to consider additional matter on the brane in order to get a realistic expanding universe. As the number p_r or p_s increases, the power of the scale factor becomes large. We find that the intersection with D6-brane in ten-dimensional theory give the fastest expansion of our universe because the three-dimensional spatial space of our universe stays in the transverse space to the D6-brane. Though the power of the scale factor for the transverse space in solutions with the D7- or D8-branes is larger than those with the D6-brane, the number of the transverse space to these branes is less than three. Hence, these solutions cannot provide the isotropic universe if we assume that the transverse space to the brane is the part of our universe [5].

The solutions we have obtained may give some moduli instabilities because of the flat direction of the moduli potential in the lower-dimensional effective theories after compactifications [1, 4, 6, 7]. Such instability will grow unless the global or local minimum of the potential can be produced by some correction in the effective theory.

The dynamical solutions contain only one function depending on both time and transverse space coordinates. One possible reason for this is that the ansatz concerning the structure of the D-dimensional metric is too restrictive. It is interesting to examine if our solutions can be extended to more general solutions by relaxing the assumptions of the field ansatz.

References

[1] M. Minamitsuji, N. Ohta and K. Uzawa, Phys. Rev. D **81** (2010) 126005 [arXiv:1003.5967 [hep-th]].
[2] K. i. Maeda, M. Minamitsuji, N. Ohta and K. Uzawa, Phys. Rev. D **82** (2010) 046007 [arXiv:1006.2306 [hep-th]].
[3] P. Binetruy, M. Sasaki and K. Uzawa, Phys. Rev. D **80** (2009) 026001 [arXiv:0712.3615 [hep-th]].
[4] K. i. Maeda, N. Ohta and K. Uzawa, JHEP **0906** (2009) 051 [arXiv:0903.5483 [hep-th]].
[5] M. Minamitsuji, N. Ohta and K. Uzawa, [arXiv:1007.1762 [hep-th]], to appear in PRD.
[6] H. Kodama and K. Uzawa, JHEP **0507** (2005) 061 [arXiv:hep-th/0504193].
[7] H. Kodama and K. Uzawa, JHEP **0603** (2006) 053 [arXiv:hep-th/0512104].

Journal of Physics: Conference Series **259** (2010) 012033 doi:10.1088/1742-6596/259/1/012033

The good, the bad and the ugly .... of Hořava gravity

Antonio Padilla

School of Physics & Astronomy, University of Nottingham, Nottingham NG7 2RD, UK

E-mail: antonio.padilla@nottingham.ac.uk

Abstract. I review the good, the bad and the ugly of the non-projectable versions of Hořava gravity. I explain how this non-relativistic theory was constructed and why it was touted with such excitement as a quantum theory of gravity. I then review some of the issues facing the theory, explaining how strong coupling occurs and why this is such a problem for both phenomenology and the question of renormalisability. Finally I comment on possible violations of Equivalence Principle, and explain why these could be an issue for Blas *et al*'s "healthy extension".

1. Introduction

Let us begin with a few words regarding the title. A couple of days after I gave this talk in Valencia, Blas *et al* submitted the following paper to the arXiv: *Models of non-relativistic quantum gravity: the good, the bad and the healthy* [1]. At the time I did not know whether the similarity with the title of my talk was deliberate, or mere coincidence. My initial suspicion was that it had to be deliberate since what they call healthy, I have called ugly! However, I have since been assured by the authors of [1] that it was nothing more than a rather amusing coincidence.

By now Hořava gravity comes in many different forms [2, 3, 4]. From now on, when I refer to "Hořava gravity" I will mean the so-called *non-projectable* theories for which the lapse function can depend on space $N = N(x, t)$. I will begin by explaining what is so good about Hořava's original idea, and why his toy model was touted with such excitement as a renormalisable theory of quantum gravity. The key idea is to break the isotropic scaling of space and time, allowing us to improve the UV behaviour of the theory without introducing ghosts. To do this we must break diffeomorphism invariance, and this is where the trouble starts. By breaking diffeomorphism invariance we introduce an additional scalar degree of freedom. What happens to this scalar mode as we approach the classical GR limit in the infra-red? Does it decouple or does it become strongly coupled? Unfortunately it is the latter [5, 6]. This is the bad of Hořava gravity. Filling a conspicuous gap in the existing literature, we will explain precisely why this strong coupling is so bad for both the renormalisability of the theory and its phenomenology.

Following these developments, Blas *et al* proposed an extension of Hořava's theory which they claimed did not run into problems with strong coupling, and should be a phenomenologically viable toy model of quantum gravity [3, 7]. This is what they refer to as the "healthy" extension, and I have referred to as "ugly". Perhaps "ugly" is a bit harsh. What I find unappealing is that they introduce a large dimensionless parameter ($\sim 10^7$) by hand, through a very low scale of Lorentz violation. This large parameter is absolutely crucial in order to solve the problems described in the previous paragraph[3, 7]. As we will explain, the situation may even be worse than this since the low scale of Lorentz violation might lead to unacceptably large violations of the Equivalence Principle [8]. Establishing whether or not this is the case is a work in progress [9]. Should we find that Blas *et al*'s model [3] can solve the strong coupling problem without any additional phenomenological problems, then I think the large parameter is a small price to pay for a phenomenologically consistent toy model of quantum gravity. Beauty, as they say, is in the eye of the beholder! For now, we should be patient and wait for the results of [9].

2. The Good

To understand what is so good about Hořava's original idea, we need to understand the trouble with gravity. Classical gravity is well described by General Relativity. At the quantum level, GR is non-renormalisable, essentially because the coupling constant has negative mass dimension, $[G_N] = -2$, and the graviton propagator scales like $1/k^2$. To see ways in which we might get round this, let us consider the following toy model,

$$\mathcal{L} = -\frac{1}{2}(\partial\phi)^2 + \lambda\phi^6 \tag{1}$$

As with GR, the propagator scales like $1/k^2$, and the coupling constant has mass dimension $[\lambda] = -2$, so the theory looks like it might well be non-renormalisable. One way to fix this might be to add relativistic higher derivative corrections to improve the UV behaviour of the propagator. Schematically,

$$\frac{1}{k^2} \to \frac{1}{k^2} + \frac{1}{k^2}\lambda k^4 \frac{1}{k^2} + \frac{1}{k^2}\lambda k^4 \frac{1}{k^2}\lambda k^4 \frac{1}{k^2} + \ldots = \frac{1}{k^2 - \lambda k^4} \tag{2}$$

After canonically normalising in the UV, $\phi \to \hat{\phi}/\sqrt{|\lambda|}$, we get a new coupling constant $\hat{\lambda} = \lambda^{-2}$ whose mass dimension is non-negative. The theory appears to be power counting renormalisable, but it has come at a price – the higher derivatives introduce additional ghost-like degrees of freedom. This is easily seen by rewriting the propagator as

$$\frac{1}{k^2 - \lambda k^4} = \frac{1}{k^2} - \frac{1}{k^2 - \lambda^{-1}} \tag{3}$$

and is patently unacceptable. Note that the problem really lies with higher order time derivatives as opposed to space derivatives. This observation suggests a possible resolution. Let us abandon Lorentz invariance and introduce higher order spatial derivatives without introducing any higher order time derivatives.The former should improve the UV behaviour of the propagator, whereas the latter guarantees the absence of ghosts. To this end, we modify the kinetic term

$$-\frac{1}{2}(\partial\phi)^2 \to \frac{1}{2}\dot{\phi}^2 - \frac{1}{2}\phi(-\Delta)^z\phi, \tag{4}$$

where Δ is the spatial Laplacian. We now have non-relativistic dispersion relation $w^2 \propto p^{2z}$, or in other words, time and space scale differently, $[x] = -1$, $[t] = -z$. For large enough z, it follows that the coupling constant has a non-negative scaling dimension, $[\lambda] = 4z - 6$. As an example, consider the $z = 3$ theory: we expect it to be power counting renormalisable, but we do not expect ghosts. Of course, the price we have paid is to have broken Lorentz invariance, which is well tested at low energies. However, we can cope with this by adding a relevant operator of the form $\mathcal{L}_{rel} = \frac{1}{2}c^2\phi\Delta\phi$. This leaves the good UV physics unaffected, but allows Lorentz invariance to be restored as an emergent symmetry in the IR, with an emergent speed of light c.

Journal of Physics: Conference Series **259** (2010) 012033 doi:10.1088/1742-6596/259/1/012033

Hořava's nice idea is to apply these methods to gravity [2]. First we must abandon Lorentz invariance, which means choosing a preferred time, t and making an ADM split

$$ds^2 = -N(x,t)^2 c^2 dt^2 + g_{ij}(x,t)(dx^i + N^i(x,t)dt)(dx^j + N^j(x,t)dt) \tag{5}$$

As usual, g_{ij} is the spatial metric, N^i is the shift vector, and N is the lapse function. By allowing N to depend on space we note that we are dealing with the *non-projectable* version of the theory. Now we see that we no longer have full diffeomorphism invariance. It has been replaced by "foliation preserving diffeomorphisms"

$$x^i \to \tilde{x}^i = \tilde{x}^i(x,t), \qquad t \to \tilde{t} = \tilde{t}(t) \tag{6}$$

The action is constructed from the following covariant objects: the spatial metric g_{ij} and the extrinsic curvature $K_{ij} = \frac{1}{2N}(\dot{g}_{ij} - 2\nabla_{(i}N_{j)})$, where ∇_i is the spatial covariant derivative. In principle, one could also include terms with $\nabla_i \log N$ [3], but let us come back to that in section 4. To build the gravitational analogue of the action (4) we first replace the kinetic term

$$\frac{1}{2}\dot{\phi}^2 \to \frac{1}{\kappa}\sqrt{g}N(K_{ij}K^{ij} - \lambda K^2) \tag{7}$$

where κ is the gravitational coupling with scaling dimension $[\kappa] = z - 3$, and λ is a dimensionless parameter that also runs with scale. Clearly, for the $z = 3$ theory the gravitational coupling constant is dimensionless and we might expect the theory to be power counting renormalisable! Sticking with $z = 3$, we build the remaining part of the UV action as follows:

$$-\frac{1}{2}\phi(-\Delta)^3\phi \to \beta\kappa\sqrt{g}N\nabla_k R_{ij}\nabla^k R^{ij} + \dots \tag{8}$$

where β is a dimensionless parameter and R_{ij} is the spatial Ricci tensor. Here "..." denotes any of the other possible dimension 6 operators that one might wish to include in the potential. Just as we needed to restore Lorentz invariance for the low energy scalar, now we need to be able to restore GR at low energies in the gravitational theory. To do this we add a relevant operator

$$\mathcal{L}_{rel} = \frac{c^2}{\kappa}\sqrt{g}NR \tag{9}$$

The full $z = 3$ gravitational theory is now given by

$$S = \int dtd^3x\ \frac{1}{\kappa}\sqrt{g}N(K_{ij}K^{ij} - \lambda K^2 + c^2 R) + \text{higher order spatial derivatives} \tag{10}$$

If we neglect the higher order spatial derivatives (as we would in the IR), we are left with something that looks very similar to the action for GR,

$$S_{GR} = \frac{c^4}{16\pi G_N}\int dtd^3x\ \sqrt{g}N\left(\frac{K_{ij}K^{ij} - K^2}{c^2} + R\right) \tag{11}$$

The claim then is that λ flows to 1 at low energies, such that we recover GR with an emergent speed of light, c, and an emergent Newton's constant, $G_N = \kappa c^2/16\pi$. It would seem that we have a theory of gravity that is well behaved at high energies, and recovers GR at low energies!

Journal of Physics: Conference Series **259** (2010) 012033 doi:10.1088/1742-6596/259/1/012033

3. The Bad

As the cynic might say, if something is too good to be true, it is probably a lie. For Hořava gravity, it turns out that there are some concerns of varying degrees of seriousness. At the not-so-serious end we have to contend with the enormous number of possible terms to be included in the potential – in other words, the "..." present in equation (8). Hořava tried to reduce the number by borrowing the principle of *detailed balance* from Condensed Matter Theory [2], although this has since been ruled out phenomenologically [5]. Whilst this is merely an aesthetic consideration, there is a more serious issue involving light cones and fine tuning. We have argued that Lorentz invariance is not exact, and emerges in the IR due to the presence of relevant operators. However, there is no reason to expect different particle species to see the same emergent light-cones. For this to happen we need to fine-tune the coefficients of the relevant operators in each case. A third concern relates to the formal structure of the theory as opposed to phenomenology – the constraint algebra is dynamically inconsistent. This manifests itself through the lapse vanishing asymptotically for generic solutions to the constraint equations[11].

The most serious issue of all stems from the breaking of diffeomorphism invariance. To understand why, we briefly return to our scalar example. Here we broke Lorentz invariance, which is just a symmetry of the *background* and does not affect the number of degrees of freedom. In contrast, for our gravity theory we broke diffeomorphism invariance, which is the *dynamical* symmetry of General Relativity. We therefore expect new dynamical degrees of freedom to appear. Now, the critical question is: what happens to these additional degrees of freedom as we approach the so-called "GR limit" at low energies? I can think of just two possibilities: either they decouple (which is good), or they become strongly coupled (which is bad).

In Hořava gravity it turns out to be the latter [5, 6]. Fluctuations about Minkowski space suggest that there exists an additional scalar mode that becomes strongly coupled at an energy scale of around $\Lambda_{naive} \sim \sqrt{\frac{c^3|1-\lambda|}{\kappa}} \sim M_{pl}c^2\sqrt{|1-\lambda|}$ [5]. **This means that the extra mode is strongly coupled on all scales in the "GR limit" where** $\lambda \to 1$. In fact, the situation is even worse than this. A more careful analysis [6] including fluctuations on general curved backgrounds yields a strong coupling scale of the order $\Lambda_{strong} \sim \Lambda_{naive}^{\frac{3}{4}}(c/L)^{\frac{1}{4}}$, where L measures the characteristic length scale of the background[1] . This means that strong coupling kicks in at all scales not just in the "GR limit" ($\lambda \to 1$), but also in the Minkowski limit ($L \to \infty$).

In each case, the dynamics of the extra mode was revealed using the Stuckelberg trick [12]. Let us remind ourselves of the spirit of this trick by means of a simpler example. The massless photon is described by a Lagrangian $\mathcal{L} = -\frac{1}{4}F_{\mu\nu}^2$ and has just two degrees of freedom due to gauge invariance, $A_\mu \to A_\mu + \partial_\mu\psi$. Now consider a theory of massive photons (with some interactions)

$$\mathcal{L} = -\frac{1}{4}F_{\mu\nu}^2 - \frac{m^2}{2}A_\mu^2 + A_\mu J^\mu \tag{12}$$

As we no longer have gauge invariance, we allow $\partial_\mu J^\mu \neq 0$ and note that the massive photon picks up an extra degree of freedom. What happens to this extra mode as $m \to 0$? To reveal this we perform a Stuckelberg trick, artificially restoring gauge invariance by means of the field redefinition $A_\mu = \tilde{A}_\mu + \frac{1}{m}\partial_\mu\phi$. The new action is given by

$$\mathcal{L} = -\frac{1}{4}\tilde{F}_{\mu\nu}^2 - \frac{1}{2}(\partial\phi)^2 + J^\mu\left(\tilde{A}_\mu + \frac{1}{m}\partial_\mu\phi\right) + \mathcal{O}(m) \tag{13}$$

[1] The background spatial curvature $\bar{R}_{ij} \sim 1/L^2$ and the background extrinsic curvature $\bar{K}_{ij} \sim c/L$.

Journal of Physics: Conference Series **259** (2010) 012033 doi:10.1088/1742-6596/259/1/012033

and is manifestly invariant under $\tilde{A}_\mu \to \tilde{A}_\mu + \partial_\mu \psi$, $\phi \to \phi - m\psi$. As $m \to 0$, the current-scalar interaction diverges, or in other words, the Stuckelberg field, ϕ, becomes strongly coupled in the massless limit!

The Stuckelberg trick in Hořava gravity works along the same lines, by artificially restoring missing gauge invariance [5, 6]. Indeed, we restore diffeomorphism invariance by redefining the ADM slicing in terms of a Stuckelberg field, $\phi(x,t)$. That is the slices go from $t =$ constant $\to$ $\phi(x,t) =$ constant. It is this Stuckelberg field that becomes strongly coupled at the scale Λ_{strong}.

What is so bad about strong coupling? Is it bad for renormalisability? Not necessarily. It is well known that QED becomes strongly coupled in the UV due to the presence of a Landau pole, and yet the theory is still renormalisable. However, Hořava gravity is not so well behaved. To understand this, consider why we believed the theory to be renormalisable in the first place. We inferred a schematic form of the action in terms of perturbative degree of freedom, simply by identifying curvatures with derivatives of the graviton, e.g. $R_{ij} \rightsquigarrow \nabla^2 h_{ij}$. The problem is that this completely ignores the Stuckelberg field. Furthermore, the Stuckelberg action looks nothing like the renormalisable actions studied in [10]. Crucially, the dispersion relation for the Stuckelberg field does *not* scale like $z = 3$ in the UV [7, 8] so we no longer expect our theory to be renormalisable.

Is strong coupling bad for phenomenology? Again, not necessarily. In DGP gravity, for example, strong coupling is linked to the so-called Vainshtein effect which helps to screen an extra scalar on solar system scales [13]. The problem with Hořava gravity is that strong coupling occurs over nearly all scales close to the "GR limit". Strong coupling can be associated with a breakdown of perturbation theory, and so perturbation theory will hardly ever apply in Hořava gravity. What this means is that the effective perturbative degrees of freedom of GR are no longer applicable –the true degrees of freedom will be bound states of the graviton and Stuckelberg fields. Without knowing the precise form of those bounds states, we lose some predictive power, but is the theory ruled out? I suspect it is. The reason is that perturbative GR is actually very well tested by the binary pulsar observations of Hulse and Taylor [14]. It is difficult to see how these can be reproduced given that a perturbative description hardly ever applies.

I conclude this section with the following orthogonal observation: Hořava gravity may suffer from violations of the Equivalence Principle [8]. There are strict bounds [15] on the size of these violations which one should be able to use to rule out regions of parameter space. To understand how the violations may arise, we need to understand how to couple matter in these theories. In the relativistic Stuckelberg picture, the matter action $S_m[\Psi_n; \gamma_{\mu\nu}, \phi]$ depends on the matter fields, Ψ_n, the spacetime metric, $\gamma_{\mu\nu}$, and the Stuckelberg field, ϕ. In the absence of full diffeomorphism invariance, we no longer have the usual conservation of energy-momentum. Instead, using the foliation preserving diffeomorphisms, we can show that [8]

$$h_{\alpha\nu}\nabla_\mu T^{\mu\nu} = 0, \qquad \frac{1}{\sqrt{-\gamma}}\frac{\delta S_m}{\delta\phi} = -\frac{n_\nu \nabla_\mu T^{\mu\nu}}{\sqrt{-(\nabla\phi)^2}} \tag{14}$$

where $n_\mu = \partial_\mu\phi/\sqrt{-(\nabla\phi)^2}$ is the unit normal to the spatial slices, and $h_{\mu\nu} = \gamma_{\mu\nu} + n_\mu n_\nu$ is the spatial projector. Thus, non-conserved sources can carry a form of Stuckelberg "charge", $\Gamma \sim \nabla T^{\mu\nu}/T^{\mu\nu}$.

With this in mind, we considered the field due to a slowly varying point mass with $T^{\mu\nu} = M\exp(-t\Gamma_{\text{source}})\delta^3(\vec{x})\text{diag}(1,0,0,0)$ [8]. This has Stuckelberg charge Γ_{source}. Probe particles also carrying Stuckelberg charge feel an extra Stuckelberg force, that dominates beyond a critical radius, $r_c \sim (\lambda - 1)/\sqrt{\Gamma_{\text{probe}}\Gamma_{\text{source}}}$. In this Stuckelberg region, different probes with different Γs will fall at different rates, in violation of the Equivalence Principle. Such violations are measured by the Eotvos parameter [15], $\eta \sim \frac{\Gamma_1 - \Gamma_2}{\Gamma_1 + \Gamma_2}$, which can be large even for small probe charges Γ_1 and Γ_2. To avoid these problems, we need to push the critical radius out to large values so that the Stuckelberg force is subdominant on astrophysical scales. This means taking large λ or small Γs. With large λ we deviate too much from GR and face problems with Lorentz violation [1]. We therefore require small Stuckelberg charges. In fact, one might be tempted to assume that all sources are conserved so that all Stuckelberg charges vanish. Whilst this is fine classically, we would expect quantum corrections to introduce some Stuckelberg charge as there is no symmetry to prevent it. The typical value of Γ ought to be non-zero, but suppressed by some power of the Lorentz symmetry breaking scale M_{UV}.

4. The Ugly (or the Beautiful?)

Up until now our discussion has focussed on the non-projectable version of Hořava gravity, in its original manifestation [2]. In an attempt to alleviate some of the more serious problems I have discussed (such as dynamical inconsistency and strong coupling), Blas *et al* proposed an extension of this model, including terms in the Lagrangian of the form $a_i = \nabla_i \log N$ [3]. I refer the reader to [16] for a discussion on the constraint algebra and dynamical consistency, and focus instead on explaining the proposed resolution of the strong coupling problem. Recall that the main issue with this was the fact that the Stuckelberg action looks nothing like the renormalisable actions studied in [10], and that crucially, the dispersion relation for the Stuckelberg field does *not* scale like $z = 3$ in the UV [7, 8]. By adding terms to the action like

$$\frac{\alpha}{\kappa}\sqrt{g}Na_i a^i, \qquad A\sqrt{g}Na_i a_j R^{ij}, \qquad B\kappa a_i a^i a_j a_k R^{jk} \tag{15}$$

the dispersion relation is modified so that it has the correct UV scaling ($w^2 \propto p^6$). Specifically,

$$w^2 = c_s^2 p^2 + \frac{p^4}{M_A^2} + \frac{p^6}{M_B^4}, \tag{16}$$

where $c_s^2 = \frac{\lambda-1}{\alpha}$, $M_A \sim \left(\frac{\alpha}{A}\right)^{1/2} M_{pl}$, $M_B \sim \left(\frac{\alpha}{B}\right)^{1/4} M_{pl}$. For $M_B \sim M_{pl}$ strong coupling is still a problem, kicking in at a scale $\Lambda_{strong} \sim (\lambda-1)^{3/4}\alpha^{-1/4}M_{pl}c^2$ [17, 8]. This comes as no surprise – to avoid strong coupling problems we need to introduce some new physics below the would be strong coupling scale. In fact, we need to have $M_B \ll \sqrt{\alpha}M_{pl}$ in order that the $z = 3$ scaling kicks in before strong coupling [7]. In some ways this proposal is reminiscent of string theory in which the string scale is introduced below the Planck scale where strong coupling would otherwise occur.

The reason we have labelled this version of theory as ugly is that the numbers involved are quite large. To see this note that the condition $M_B \ll \sqrt{\alpha}M_{pl}$ corresponds to $B \gg \alpha^{-1}$. As tests of Lorentz violation require $\alpha \lesssim 10^{-7}$, we are forced to take $B \gg 10^7$ [7]. Of course, as we stated in the introduction, this would be a small price to pay for a consistent quantum theory of gravity! We should also point out that it really is just a question of aesthetics – the choice of M_B parametrically below M_{pl} is not technically unnatural. The point is that M_B sets the cut off and so neither M_B nor M_{pl} receive large corrections.

So, do we have a consistent quantum theory of gravity? Not necessarily – it is too early to say. Whilst Blas *et al*'s model might cure strong coupling issues, it could do so in such a way that it runs into real trouble with tests of Equivalence Principle. Recall that we expect the typical value of the Stuckelberg charge to be non-zero, but suppressed by some power of the Lorentz symmetry breaking scale M_{UV}. The Stuckelberg charges must therefore be relatively large since we need a low scale of Lorentz violation $M_{UV} \sim M_B \ll M_{pl}$ to avoid strong coupling. As λ cannot be made arbitrarily large, it is clear that Blas *et al*'s model could face a challenge from EP tests as described at the end of the previous section. It all boils down to the tension in choosing the value of M_B: choose too high a value and you get strong coupling, too low a value and you fail EP tests. We are currently working out the details to see if an acceptable choice for M_B really exists [9].

Finally, I would like to highlight a very recent result in which it was claimed that one must take $\lambda < 1/3$ in order to have a stable vacuum in the quantum version of the theory [18]. It is difficult to see how this can be compatible with the phenomenological requirement that $|\lambda - 1| < 10^{-7}$ at low energies. We should not suppose that λ runs from $\lambda < 1/3$ in the UV to $\lambda = 1$ in the IR, since $\lambda = 1/3$ is expected to be a fixed point [2], and the Stuckelberg field is a ghost whenever $1/3 < \lambda < 1$ [3].

Acknowledgments
I would like to thank my collaborators, both past and present, including Paul Saffin, Gustavo Niz, Christos Charmousis, Thomas Sotiriou and especially Ian Kimpton. I would also like to thank Antonios Papazoglou and Oriol Pujolas for useful discussions. I am funded by a Royal Society University Research Fellowship.

References
[1] D. Blas, O. Pujolas and S. Sibiryakov, arXiv:1007.3503 [hep-th].
[2] P. Horava, Phys. Rev. D **79**, 084008 (2009) [arXiv:0901.3775 [hep-th]].
[3] D. Blas, O. Pujolas and S. Sibiryakov, Phys. Rev. Lett. **104**, 181302 (2010) [arXiv:0909.3525 [hep-th]].
[4] P. Horava and C. M. Melby-Thompson, arXiv:1007.2410 [hep-th].
[5] C. Charmousis, G. Niz, A. Padilla and P. M. Saffin, JHEP **0908** (2009) 070 [arXiv:0905.2579 [hep-th]].
[6] D. Blas, O. Pujolas and S. Sibiryakov, JHEP **0910**, 029 (2009) [arXiv:0906.3046 [hep-th]].
[7] D. Blas, O. Pujolas and S. Sibiryakov, Phys. Lett. B **688**, 350 (2010) [arXiv:0912.0550 [hep-th]].
[8] I. Kimpton and A. Padilla, JHEP **1007** (2010) 014 [arXiv:1003.5666 [hep-th]].
[9] I. Kimpton, A. Padilla and T. Sotiriou, *Work in progress.*
[10] M. Visser, Phys. Rev. D **80** (2009) 025011 [arXiv:0902.0590 [hep-th]]. M. Visser, arXiv:0912.4757 [hep-th].
[11] M. Henneaux, A. Kleinschmidt and G. L. Gomez, Phys. Rev. D **81** (2010) 064002 [arXiv:0912.0399 [hep-th]].
[12] H. Ruegg and M. Ruiz-Altaba, Int. J. Mod. Phys. A **19** (2004) 3265 [arXiv:hep-th/0304245].
[13] G. R. Dvali, G. Gabadadze and M. Porrati, Phys. Lett. B **485**, 208 (2000) [arXiv:hep-th/0005016]. C. Deffayet, G. R. Dvali, G. Gabadadze and A. I. Vainshtein, Phys. Rev. D **65**, 044026 (2002) [arXiv:hep-th/0106001]. G. Dvali, New J. Phys. **8**, 326 (2006) [arXiv:hep-th/0610013].
[14] R. A. Hulse and J. H. Taylor, Astrophys. J. **195**, L51 (1975).
[15] S. Schlamminger, K. Y. Choi, T. A. Wagner, J. H. Gundlach and E. G. Adelberger, Phys. Rev. Lett. **100** (2008) 041101 [arXiv:0712.0607 [gr-qc]]. S. Baessler, B. R. Heckel, E. G. Adelberger, J. H. Gundlach, U. Schmidt and H. E. Swanson, Phys. Rev. Lett. **83** (1999) 3585.
[16] J. Kluson, JHEP **1007**, 038 (2010) [arXiv:1004.3428 [hep-th]].
[17] A. Papazoglou and T. P. Sotiriou, Phys. Lett. B **685** (2010) 197 [arXiv:0911.1299 [hep-th]].
[18] F. W. Shu, arXiv:1009.3677 [hep-th].

Journal of Physics: Conference Series **259** (2010) 012034 doi:10.1088/1742-6596/259/1/012034

Stability of gravity-scalar systems for domain-wall models with a soft wall

Damien P. George

Nikhef Theory Group, Science Park 105, 1098 XG Amsterdam, The Netherlands

E-mail: dpgeorge@nikhef.nl

Abstract. We show that it is possible to create an RS soft-wall model, a model with a compact extra dimension, without using fundamental branes. All that is required are bulk scalar fields minimally coupled to gravity. Of crucial importance is the stability of the size of the extra dimension. Without branes, one cannot easily implement the Goldberger-Wise mechanism, and instead it must be shown that the scalar configuration is stable in its own right. We use the superpotential approach for generating solutions, the so called 'fake supergravity' scenario, and show that configurations generated in such a way are always free of tachyonic modes. Furthermore, we show that the model is also free of zero modes (in the spin-0 sector) if all the scalars have odd parity. We discuss the hierarchy problem in soft-wall models, and applications of our analysis to the AdS/QCD correspondence.

1. Introduction

Extra dimensions are a plausible extension to the standard model of particle physics. In particular, much attention has been given to the type I Randall-Sundrum (RS) model [1], where the electroweak hierarchy can be naturally generated by the warping of a compact extra dimension. In their original realisation, these models have fundamental branes of negative tension, or hard-walls, at the boundary of the extra dimension. Recently there has been interest in a new type of RS-like warped spacetime: a compact spacetime where the negative tension brane is replaced with a physical singularity. These soft-wall models were originally designed to yield linear Regge trajectories in the context of the AdS/CFT correspondence [2], but have since been the basis of actual models beyond the standard model [3, 4, 5, 6, 7, 8, 9, 10], and also provide a holographic dual description of unparticle models [11, 12].

Our aim is take the soft-wall model and go one step further by removing the final, positive-tension brane at the origin, and replacing it with a suitable scalar field profile, such as a domain wall. Such a model will then describe a compact extra dimension without the use of fundamental dynamical branes. This is interesting because it is a purely field theoretical construction and requires no appeal to string theory. Figure 1 depicts the progression from RS1 models to domain-wall soft-wall models. In what follows, we analyse the stability of such domain-wall models, and show that they are capable of solve the hierarchy problem. See reference [13] for details.

2. Warped Background Configuration

We work in the framework of 5D general relativity, with N scalar fields Φ_i minimally coupled to gravity. We also allow the presence of brane terms for completeness. The action for such a

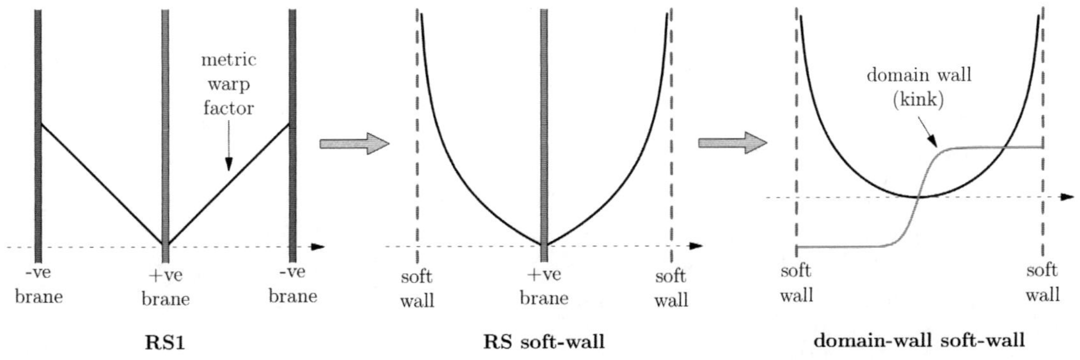

Figure 1. From RS1 to RS soft-wall to domain-wall soft-wall. The domain-wall set-up provides a purely field theoretic construction of a compact extra dimension.

set-up is given by

$$\mathcal{S} = \int d^4x\, dy \left[\sqrt{-g}\left(M^3 R + \mathcal{L}_{\text{matter}}\right) - \sqrt{-g_4}\lambda\right] , \tag{1}$$

where M is the 5D Planck mass and the matter and brane terms are

$$\mathcal{L}_{\text{matter}} = -\frac{1}{2}\sum_{i=1}^{N} g^{PQ}\partial_P\Phi_i\partial_Q\Phi_i - V(\{\Phi_i\}) , \tag{2}$$

$$\lambda = \lambda(\{\Phi_i\}) = \sum_\alpha \lambda_\alpha(\{\Phi_i\})\delta(y - y_\alpha) . \tag{3}$$

Here, $V(\{\Phi_i\})$ is the scalar potential and $\lambda_\alpha(\{\Phi_i\})$ the brane localised potential (including brane tension) for the brane at y_α. Our aim is to study the general stability conditions for non-trivial background configurations of this class of models.

We specialise to backgrounds that depend only on the extra dimension y, and denote the scalar background solutions $\phi_i(y)$. The RS background metric ansatz is

$$ds^2 = e^{-2\sigma(y)}\eta_{\mu\nu}dx^\mu dx^\nu + dy^2 , \tag{4}$$

where μ, ν index the 4D subspace. Einstein's equations are $G_{MN} = \frac{1}{2M^3}T_{MN}$ and yield, together with the Euler-Lagrange equations,

$$6M^3\sigma'' = \sum_i \phi_i'^2 + \lambda(\{\phi_j\}) , \tag{5}$$

$$6M^3\left(4\sigma'^2 - \sigma''\right) = -2V\left(\{\phi_j\}\right) - \lambda\left(\{\phi_j\}\right) , \tag{6}$$

$$\phi_i'' - 4\sigma'\phi_i' - V_i(\{\phi_j\}) - \lambda_i(\{\phi_j\}) = 0 . \tag{7}$$

The notation $V(\{\phi_i\})$ means that the potential is to be evaluated with the background fields ϕ_i. A subscript i on V or λ denotes partial differentiation with respect to the field Φ_i.

Given a particular V and λ, the space of background configurations is parameterised by the integration constants of the system. For a particular choice of integration constants, that is, a particular background, we want to know if such a choice is stable within the space of configurations. We shall study local, perturbative stability by adding small fluctuations to the background, turning the problem into an eigenvalue problem. For the initial stages our analysis will be for an arbitrary scalar potential V. Later on we will need to specialise to the fake supergravity approach.

3. Perturbations

The general ansatz which takes into account both spin-0 and spin-2 perturbations is

$$ds^2 = e^{-2\sigma(y)}\left[(1-2F(x^\mu,y))\eta_{\mu\nu} + h_{\mu\nu}(x^\mu,y)\right]dx^\mu dx^\nu + \left[1+G(x^\mu,y)\right]^2 dy^2\,, \tag{8}$$

$$\Phi_i(x^\mu,y) = \phi_i(y) + \varphi_i(x^\mu,y)\,. \tag{9}$$

Here we work in the axial gauge, $h_{\mu 5} = 0$, with transverse traceless part $\partial^\mu h_{\mu\nu} = \eta^{\mu\nu}h_{\mu\nu} = 0$. The (ij) Einstein's equations (off-diagonal spatial) enforce $G = 2F$ which we take from now on. The rest of Einstein's equations correspond to $(\mu\nu)$, $(\mu 5)$ and (55), and taking the 4-trace of the $(\mu\nu)$ equations shows that the spin-2 and spin-0 perturbations decouple from one another.

3.1. Spin-2 Perturbations

The spin-2 perturbation $h_{\mu\nu}$ decouples from F and φ_i. The equation for $h_{\mu\nu}$ is

$$-e^{2\sigma}\Box h_{\mu\nu} - h''_{\mu\nu} + 4\sigma' h'_{\mu\nu} = 0\,. \tag{10}$$

There is always a zero mode, $h'_{\mu\nu} = 0$, which is normalisable. This is the well-known 4D massless graviton from the RS model [14]. In conformal coordinates z defined by $dy = e^{-\sigma}dz$ with rescaled $h_{\mu\nu} = e^{3\sigma/2}\tilde{h}_{\mu\nu}$ we can write this equation as a Schrödinger-like equation in a self-adjoint form:

$$(\partial_z - \frac{3}{2}\sigma')(-\partial_z - \frac{3}{2}\sigma')\tilde{h}_{\mu\nu} = \Box\tilde{h}_{\mu\nu}\,. \tag{11}$$

Applying supersymmetric quantum mechanics we find that there are no tachyonic modes in the spin-2 sector. Thus the spin-2 fluctuations do not destabilise the configuration.

3.2. Spin-0 Perturbations

The spin-0 sector is significantly more complicated than the spin-2 sector, since physical modes are mixtures of F and φ_i. The equations for these perturbations consist of two of Einstein's equations and the Euler-Lagrange equations (sum over repeated scalar indices):

$$6M^3(F' - 2\sigma' F) = \phi'_i\varphi_i\,, \tag{12}$$

$$6M^3(-e^{2\sigma}\Box F - 2\sigma' F' + F'') = 2\phi'_i\varphi'_i + 2\,\lambda|_{\rm bg}\,F + \lambda_i|_{\rm bg}\,\varphi_i\,. \tag{13}$$

$$e^{2\sigma}\Box\varphi_i + \varphi''_i - 4\sigma'\varphi'_i - 6F'\phi'_i - (4V_i + 2\lambda_i)|_{\rm bg}\,F - (V_{ij} + \lambda_{ij})|_{\rm bg}\,\varphi_j = 0\,. \tag{14}$$

To simplify these equations we go to conformal coordinates z as above and rescale the fields by $F = e^{3\sigma/2}\chi(z)/\sqrt{12}$ and $\varphi_i = M^{3/2}e^{3\sigma/2}\psi_i(z)$. Equations (12) through (14) then become a set of coupled Schrödinger-like equations:

$$\begin{aligned} -\chi'' + (\mathcal{V}_{00} + \mathcal{B}_{00})\chi + (\mathcal{V}_{0i} + \mathcal{B}_{0i})\psi_i &= \Box\chi\,, \\ -\psi''_i + (\mathcal{V}_{ij} + \mathcal{B}_{ij})\psi_j + (\mathcal{V}_{0i} + \mathcal{B}_{0i})\chi &= \Box\psi_i\,, \end{aligned} \tag{15}$$

where the effective potential $\mathcal{V}$ has components

$$\mathcal{V}_{00} = \frac{9}{4}\sigma'^2 + \frac{5}{2}\sigma''\,, \quad \mathcal{V}_{0i} = \frac{2\phi''_i}{\sqrt{3M^3}}\,, \quad \mathcal{V}_{ij} = \left(\frac{9}{4}\sigma'^2 - \frac{3}{2}\sigma''\right)\delta_{ij} + \frac{\phi'_i\phi'_j}{M^3} + e^{-2\sigma}\,V_{ij}|_{\rm bg}\,, \tag{16}$$

and the brane terms are

$$\mathcal{B}_{00} = \frac{1}{3M^3}e^{-2\sigma}\,\lambda|_{\rm bg}\,, \qquad \mathcal{B}_{0i} = \frac{1}{\sqrt{3M^3}}e^{-2\sigma}\,\lambda_i|_{\rm bg}\,, \qquad \mathcal{B}_{ij} = e^{-2\sigma}\,\lambda_{ij}|_{\rm bg}\,. \tag{17}$$

Equation (15) defines the physical spin-0 spectrum. The symmetry of the cross-coupling ensures that the eigenvalues of this spectrum are real. To demonstrate stability of a given background configuration, we must also show that all eigenvalues are positive.

4. Fake supergravity

Our aim is to find a class of models which have strictly positive eigenvalues in the spin-0 sector, ensuring the stability of the background configuration. We specialise to the case without brane terms ($\lambda = 0$) and where the scalar potential is generated from a superpotential $W(\{\Phi_i\})$ using the fake supergravity approach [15, 16]

$$V(\{\Phi_i\}) = \sum_i \frac{1}{2}\left[W_i(\{\Phi_i\})\right]^2 - \frac{1}{3M^3}\left[W(\{\Phi_i\})\right]^2 , \tag{18}$$

where $W_i \equiv dW/d\Phi_i$. For such a potential, equations (5) through (7) simplify to

$$\sigma'(y) = W(\{\phi_i\})/6M^3 , \qquad \phi_i'(y) = W_i(\{\phi_i\}) . \tag{19}$$

This is a set of *first order* equations; W encodes for both V and some of the integration constants. Without loss of generality we can take $\sigma(y_0) = 0$, leaving N integration constants for the boundary values of ϕ_i. So, given a particular W, the set $\{\phi_i(y_0)\}$ will then uniquely define a configuration. We want to know if such a configuration is stable.

Using the fake supergravity approach, one can write the spin-0 effective potential as $\mathcal{V} = S^2 + S'$ where

$$S = e^{-\sigma}\left.\begin{pmatrix} \frac{1}{12M^3}W & \frac{1}{\sqrt{3M^3}}W_j \\ \frac{1}{\sqrt{3M^3}}W_i & \frac{-1}{4M^3}\delta_{ij}W + W_{ij}\end{pmatrix}\right|_{\rm bg} . \tag{20}$$

Then equation (15) takes the self-adjoint form

$$(\partial_z + S^\dagger)(-\partial_z + S)\Psi = \Box\Psi , \tag{21}$$

where $\Psi = (\chi, \psi_i)$. For a warped metric the relevant boundary terms vanish and supersymmetric quantum mechanics tells us that the eigenvalues of the spin-0 sector are non-negative. The background is thus free of tachyonic modes. To ensure complete stability, we must still analyse the existence of zero mode solutions. Such zero modes can, for example, correspond to changes in the size of the compact extra dimension, rendering the set-up unstable.

4.1. Zero modes with $N = 1$

With $N = 1$ scalar we can use the Einstein constraint equation to eliminate ψ_1 in terms of χ. Then let $\chi = S_{01}g$, with S_{01} defined by equation (20). The Schrödinger-like equation for g is

$$(\partial_z - S_{11})(-\partial_z - S_{11})g + S_{01}^2 g = \Box g . \tag{22}$$

Multiplying this equation from the left by g^* and integrating over the extra dimension yields

$$\int |(-\partial_z - S_{11})g|^2\, dz + \int |S_{01}g|^2 + (\text{boundary terms}) = \Box \int |g|^2 dz . \tag{23}$$

For a warped metric the boundary terms vanish, and the existence of a zero mode with $\Box g = 0$ requires both $(-\partial_z - S_{11})g = 0$ and $S_{01}g = 0$. Thus $g = 0$ and we can conclude that systems with $N = 1$ scalar using the fake supergravity approach do not have a zero mode.

4.2. Zero modes with $N \geq 2$

Set-ups with more than one scalar field can in general posses a zero mode. Nevertheless, we can formulate a simple criterion that can be used to find models which do not have a zero mode: for a system of definite parity, the number of independent normalisable zero modes is at most equal

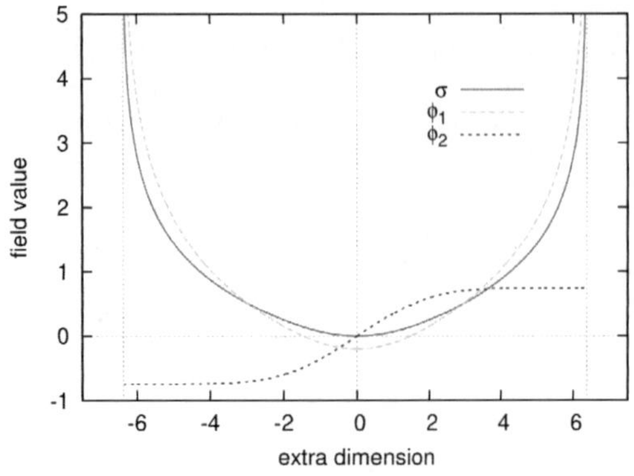

Figure 2. Background solutions for Example 1. Parameters are $\nu = 1.4$, $a = 0.5$, $b = 0.3$. The size of the extra dimension is not stabilised and depends on the value $\phi_1(0)$.

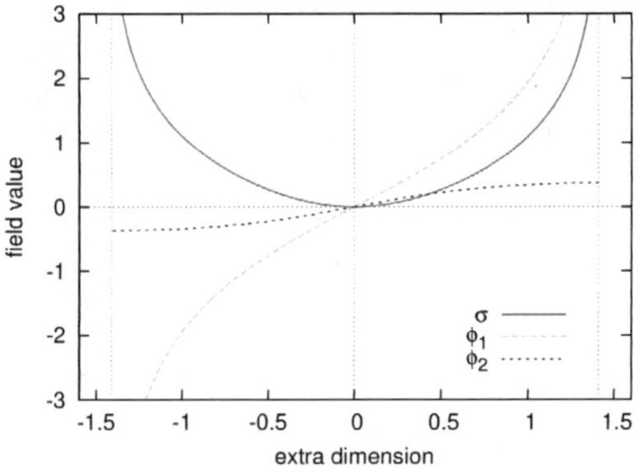

Figure 3. Background solutions for Example 2. Parameters are $\alpha = 1$, $\nu = 1.4$, $a = 0.5$, $b = 0.3$. The configuration is stable and is an example of a domain-wall soft-wall model.

to the number of even scalar fields. This is essentially a statement about integration constants. In the fake supergravity approach, the background configuration is given by the solutions to the first order equations (19). The restriction to models with parity eliminates those integration constants associated with odd-parity scalars, since their field value must vanish at $y = 0$. Unique solutions to the fake supergravity equations are then parameterised by the integration constants of the even-parity fields. The final point is that zero modes move us continuously through this space of solutions, so there cannot be more zero modes than the number of even fields.

5. Applications

We shall look at two example models with $N = 2$ scalar fields to illustrate the above criterion about the existence of zero modes. The second model is an example of a stable domain-wall soft-wall set-up. Following this we make some comments about AdS/QCD.

5.1. Example 1 — unstable even system

Consider the superpotential

$$W(\Phi_1, \Phi_2) = e^{\nu\Phi_1}\left(a\,\Phi_2 - b\,\Phi_2^3\right), \tag{24}$$

where Φ_1 and Φ_2 are the dilaton and kink fields respectively. We choose Φ_1 to have even and Φ_2 to have odd parity in y. See Figure 2 for representative background solutions (the plots use units where $6M^3 = 1$). In these solutions there is a choice of integration constant for Φ_1 and our criterion tells us that a corresponding spin-0 zero mode can exist. Such a mode in fact does exist, and is given by $(\chi, \psi_1, \psi_2) = N(-\nu, \sqrt{2}, 0)e^{-3\sigma/2}\eta(x^\mu)$ where N is a normalisation constant and η is the 4D mode. This zero mode physically corresponds to changes in the size of the extra dimension, so the background configuration is not stable.

5.2. Example 2 — stable odd system

The second model has the superpotential

$$W(\Phi_1, \Phi_2) = \alpha\,\sinh(\nu\,\Phi_1) + \left(a\,\Phi_2 - b\,\Phi_2^3\right), \tag{25}$$

where Φ_1 and Φ_2 are now both taken to have odd parity. As a consequence of this parity choice, there are no integration constants to choose, and, by our criterion, there cannot exist any zero

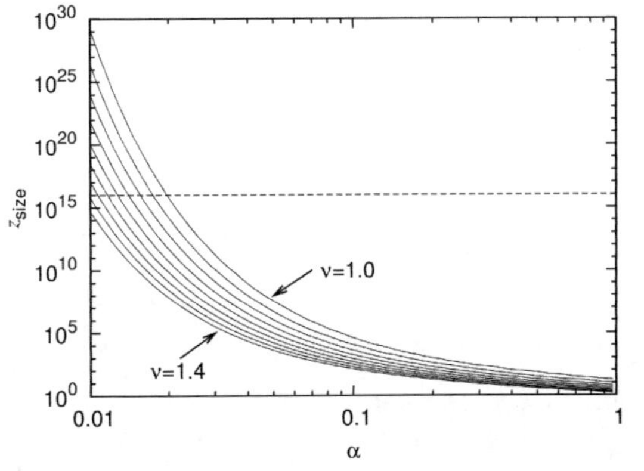

Figure 4. Solving the hierarchy problem using the domain-wall soft-wall model of Example 2. The size of the compact extra dimension in conformal coordinates, z_{size}, sets the characteristic mass scale. Bulk fields have a KK mass scale $m_{\text{KK}} \sim z_{\text{size}}^{-1}$ and the hierarchy problem is solved if $z_{\text{size}} \sim 10^{16}$ can be obtained for $\mathcal{O}(1)$ model parameters. The plot has $a = b = 1$ in equation (25) and shows z_{size} versus α for $\nu = 1.0 \dots 1.4$. To obtain the relevant hierarchy one should take $\alpha \simeq 0.02$ and $\nu \simeq 1.0$.

modes. The background solution is unique and stabilised, and the size of the extra dimension is fixed by the parameters in W; see Figure 3. The field Φ_2 is in a kink configuration and replaces the brane in usual RS soft-wall models, providing an example of a domain-wall model with a soft wall. Figure 4 shows how the electroweak hierarchy can be generated in such a model.

5.3. Application to AdS/QCD

Soft-wall models were originally motivated by attempts to construct 5D AdS models of QCD [2]. Having a soft wall in the infrared (large y) yields linear Regge trajectories; the meson excitations in the QCD dual theory have masses which scale like $m_n^2 \sim n$, where n is the excitation number. Scalar fluctuations in the AdS picture correspond to glueball and scalar meson excitations in the dual QCD theory. Since the fake supergravity approach is common in the literature, and it seems that more than one scalar is required to dynamically generate the background (see for example [3]), our Schrödinger-like equation for the spin-0 eigenvalues, equations (20) and (21), can be used to compute the scalar spectrum with multiple background fields.

Acknowledgments

This research was done in collaboration with S. M. Aybat. It was supported by the Netherlands Foundation for Fundamental Research of Matter (FOM) and the National Organisation for Scientific Research (NWO).

References

[1] Randall L and Sundrum R 1999 *Phys. Rev. Lett.* **83** 3370–3373 (*Preprint* `hep-ph/9905221`)
[2] Karch A, Katz E, Son D T and Stephanov M A 2006 *Phys. Rev.* **D74** 015005 (*Preprint* `hep-ph/0602229`)
[3] Batell B and Gherghetta T 2008 *Phys. Rev.* **D78** 026002 (*Preprint* `0801.4383`)
[4] Falkowski A and Perez-Victoria M 2008 *JHEP* **12** 107 (*Preprint* `0806.1737`)
[5] Batell B, Gherghetta T and Sword D 2008 *Phys. Rev.* **D78** 116011 (*Preprint* `0808.3977`)
[6] Delgado A and Diego D 2009 *Phys. Rev.* **D80** 024030 (*Preprint* `0905.1095`)
[7] Mert Aybat S and Santiago J 2009 *Phys. Rev.* **D80** 035005 (*Preprint* `0905.3032`)
[8] Gherghetta T and Sword D 2009 *Phys. Rev.* **D80** 065015 (*Preprint* `0907.3523`)
[9] Cabrer J A, von Gersdorff G and Quiros M 2009 (*Preprint* `0907.5361`)
[10] von Gersdorff G 2010 (*Preprint* `1005.5134`)
[11] Cacciapaglia G, Marandella G and Terning J 2009 *JHEP* **02** 049 (*Preprint* `0804.0424`)
[12] Falkowski A and Perez-Victoria M 2009 *Phys. Rev.* **D79** 035005 (*Preprint* `0810.4940`)
[13] Aybat S M and George D P 2010 *JHEP* **09** 010 (*Preprint* `1006.2827`)
[14] Randall L and Sundrum R 1999 *Phys. Rev. Lett.* **83** 4690–4693 (*Preprint* `hep-th/9906064`)
[15] DeWolfe O, Freedman D Z, Gubser S S and Karch A 2000 *Phys. Rev.* **D62** 046008 (*Preprint* `hep-th/9909134`)
[16] Freedman D Z, Nunez C, Schnabl M and Skenderis K 2004 *Phys. Rev.* **D69** 104027 (*Preprint* `hep-th/0312055`)

Plasma graviton production in TeV-scale gravity

E Yu Melkumova

Department of Physics, Moscow State University, 119899, Moscow, Russia

E-mail: elenamelk@srd.sinp.msu.ru

Abstract. We develop the theory of interaction of classical plasma with Kaluza-Klein (KK) gravitons in the ADD model of TeV-scale gravity. Plasma is described within the kinetic approach as the system of charged particles and Maxwell field both confined on the brane. Interaction with multidimensional gravity living in the bulk with n compact extra dimensions is introduced within the linearized theory. The KK gravitons emission rates are computed taking into account plasma collective effects through the two-point correlation functions of the fluctuations of the plasma energy-momentum tensor. Apart from known mechanisms (such as bremsstrahlung and gravi-Primakoff effect) we find essentially collective channels such as the coalescence of plasma waves into gravitons which may be manifest in turbulent plasmas. Our results indicate that commonly used rates of the KK gravitons production in stars and supernovae may be underestimated.

1. Introduction

TeV-scale gravity proposal due to Arkani-Hamed, Dimopoulos and Dvali (ADD) [1] suggests that the standard model particles live in the four-dimensional subspace (the brane) of the D-dimensional bulk with $n = D - 4$ extra dimensions compactified on a torus which are inhabited only by gravity. The D-dimensional Planck mass M_* is supposed to be TeV-scale, and the large extra dimensions (LED) to have sub-millimeter size. In this scenario there is an infinite tower of massive KK gravitons [2] whose existence may be detected using table-top and collider experiments, astrophysical and cosmological observations. It has been pointed out that one of the strongest bounds on the parameters come from the supernova SN1987A data [1]-[10]. Processes contributing to energy loss due to graviton emission from SN1987A, the red giants and the Sun include photon-photon annihilation, electron-positron annihilation, nucleon-nucleon gravitational bremsstrahlung, gravi-Compton-Primakoff scattering [3]-[10].

Previous calculations of the KK graviton emission rates were performed using one-particle Feynman rules with subsequent averaging over Bose-Einstein and Maxwell-Boltzmann distributions to get the results reliable for finite temperature T [4]. These calculations, however, did not take into account plasma collective effects such as Debye screening, interaction of charged particles with plasma waves etc, which may be important in astrophysical conditions. This is the goal of the present contribution. We generalize kinetic theory of interaction of plasmas with gravitational waves developed earlier in 3 + 1 dimensions [11] to the ADD model and present calculation of gravi-bremsstrahlung rate from non-relativistic thermal isotropic electron-ion plasma.

2. Kinetic approach

Consider collisionless plasma consisting of charged particles of types α with parameters e_α, m_α described by microscopic distribution function [12, 13]

$$\mathcal{F}_\alpha(x, \mathbf{p}) = \sum_{i=1}^{N_\alpha} \delta(\mathbf{r} - \mathbf{r}_i(t))\delta(\mathbf{p} - \mathbf{p}_i(t)), \tag{1}$$

normalized as $\int \mathcal{F}_\alpha(x, \mathbf{p}) d\mathbf{p}\; d^3x = N_\alpha$, satisfying the kinetic equation

$$\frac{\partial \mathcal{F}_\alpha}{\partial t} + \mathbf{v}\frac{\partial \mathcal{F}_\alpha}{\partial \mathbf{r}} + \mathbf{e}_\alpha(\mathbf{E} + [\mathbf{vB}])\frac{\partial \mathcal{F}_\alpha}{\partial \mathbf{p}} = \mathbf{0}. \tag{2}$$

They interact via the electromagnetic field $\mathbf{E}$, $\mathbf{B}$ satisfying Maxwell equations

$$\mathrm{div}\mathbf{E} = 4\pi\rho, \quad \mathrm{rot}\mathbf{E} = -\frac{\partial \mathbf{B}}{\partial t}, \quad \mathrm{div}\mathbf{B} = 0, \quad \mathrm{rot}\mathbf{B} = 4\pi\mathbf{j} + \frac{\partial \mathbf{E}}{\partial \mathbf{t}}, \tag{3}$$

with the source terms

$$\rho(x) = \sum_\alpha e_\alpha \int \mathcal{F}_\alpha(x, \mathbf{p}) d\mathbf{p}, \quad \mathbf{j}(x) = \sum_\alpha e_\alpha \int \frac{\mathbf{p}}{p_\alpha^0} \mathcal{F}_\alpha(x, \mathbf{p}) d\mathbf{p}. \tag{4}$$

To describe gravitational radiation we will need the 3-spatial components of the plasma energy momentum tensor

$$T_{ij} = {}_mT_{ij} + {}_FT_{ij},$$

where

$${}_mT_{ij} = \sum_\alpha \int \frac{p_i p_j}{p_\alpha^0} \mathcal{F}_\alpha(x, \mathbf{p})\, d\mathbf{p}, \quad p_\alpha^0 = \sqrt{\mathbf{p}^2 + m_\alpha^2}, \quad x \equiv \{t, \mathbf{r}\}, \tag{5}$$

$${}_FT_{ij} = -\frac{1}{4\pi}\left(E_i E_j + B_i B_j - \delta_{ij}\frac{E^2 + B^2}{2}\right). \tag{6}$$

To solve the system of equations (2-4) we use perturbation theory in terms of the electric charges. First we separate fluctuations from the average distribution using the approach of [11, 12]:

$$\mathcal{F}(t, \mathbf{r}, \mathbf{p}) = f_\alpha^0 + \delta f_\alpha^0 + \delta f_\alpha, \tag{7}$$

where $f^0 \equiv < \mathcal{F}_\alpha(x, \mathbf{p}) >$ - the equilibrium Maxwell distribution function:

$$f_\alpha^0 = \frac{N_{0\alpha}}{(2\pi v_{T_\alpha}^2)^{\frac{3}{2}}} e^{-\frac{1}{2}\frac{v^2}{v_{T_\alpha}^2}}, \quad v_{T_\alpha} = \sqrt{\frac{T_\alpha}{m_\alpha}}, \quad v_{T_\alpha} << 1, \tag{8}$$

δf_α^0- represents fluctuations due to chaotic particle motion ("zero" fluctuations), while $\delta f_\alpha(t, \mathbf{r}, \mathbf{p})$ - stands for fluctuations due to electromagnetic interaction of the particles. Using the fact that f_α^0 satisfies the free equation (for zero charges), we obtain:

$$\frac{\partial \delta f_\alpha}{\partial t} + \mathbf{v}\frac{\partial \delta f_\alpha}{\partial \mathbf{r}} + e_\alpha(\mathbf{E} + [\mathbf{vB}])\frac{\partial(f_\alpha^0 + \delta f_\alpha^0 + \delta f_\alpha)}{\partial \mathbf{p}} = 0. \tag{9}$$

We then further expand the fluctuation δf_α in power series in terms of charges: $\delta f_\alpha = \delta f^1_\alpha + \delta f^2_\alpha + ...$, and use the Fourier-transformation

$$f(x) = \frac{1}{(2\pi)^4}\int f(k)e^{-ik_\mu x^\mu}d^4k$$

to get in the first two orders

$$\delta f^1_\alpha(k,\mathbf{p}) = -\frac{ie_\alpha}{(\omega-\mathbf{kv})}\mathbf{F}(k)\frac{\partial f^0_\alpha}{\partial \mathbf{p}}, \qquad \mathbf{F}\equiv\mathbf{E}+[\mathbf{vB}], \tag{10}$$

$$\delta f^2_\alpha(k,\mathbf{p}) = -\frac{ie_\alpha}{(2\pi)^4(\omega-\mathbf{kv})}\int \mathbf{F}(k-k_1)\frac{\partial}{\partial\mathbf{p}}(\delta f^0_\alpha(k_1,\mathbf{p})+\delta f^1_\alpha(k_1,\mathbf{p}))d^4k_1. \tag{11}$$

The corresponding expansion of the charged particles energy-momentum tensor will read:

$$_mT_{ik} = {_mT^0_{ik}} + \delta_m T^0_{ik} + \sum_{l=1}^{\infty}\sum_\alpha \int \frac{p_ip_j}{p^0_\alpha}\delta f^l_\alpha(x,\mathbf{p})\,d\mathbf{p}. \tag{12}$$

First non-zero contribution to gravitational radiation comes from terms of the second order in the fields **E**, **B**, namely $\delta T_{ik} = \delta\, {_mT^2_{ik}} + \delta\, {_FT_{ik}}$.

3. Gravitational radiation

In ADD model one considers [2] the linearized $D = 4+n$ dimensional Einstein equations

$$\Box\psi_{MN} = \varkappa^2_D T_{MN},$$

where $\psi_{MN} = h_{MN} - \eta_{MN}h^P_P/2,\ \ h_{MN} = g_{MN} - \eta_{MN}$, the indices M, N run over the brane μ, $\nu = 0,1,2,3$ and n directions on the torus, and the harmonic gauge $\partial_N\psi^{MN} = 0$ is understood. Formula for the total (integrated over space and time) energy loss on gravitational radiation in ADD model was recently derived in [14]:

$$\mathcal{E} = \frac{\varkappa^2_D}{16\pi^3 V_d}\sum_{\mathbf{N}\in\mathbb{Z}^d}\int d\mathbf{k}T_{SN}(k)T^*_{LR}(k)\tilde\Lambda^{SNLR}\Big|_{k^0=\sqrt{|\mathbf{k}|^2+(2\pi\mathbf{N}/L)^2}}, \qquad V_d = (2\pi L)^{D-4}, \tag{13}$$

where $T_{SN}(k)$ is the four-dimensional Fourier-transform, and

$$\tilde\Lambda^{SNLR} = \frac{1}{2}\left[\eta^{SL}\eta^{NR} + \eta^{SR}\eta^{NL}\right] - \frac{1}{D-2}\eta^{SN}\eta^{LR} \tag{14}$$

is the polarization projection operator. Since the stress tensor has only brane indices $\mu = 0, i$ and it satisfies the conservation equation $k_\mu T^{\mu\nu}(k) = 0$, one can eliminate its components $T^{0\nu}$ in favor of T^{ij}. The resulting expression will contain the sum of various order terms from which the lowest non-zero contribution comes from the product of the second order terms. We also have to normalize the energy loss of the stationary plasma per unit volume and per unit time, which is achieved by introducing the two-point correlation function according to the relation

$$< T_{\alpha\beta}(k)T^*_{\alpha'\beta'}(k+\Delta k) > = < T_{\alpha\beta}T^*_{\alpha'\beta'} >_k \delta^4(\Delta k).$$

The final expression for the gravitational emission rate to all graviton modes per unit plasma volume per unit time will read

$$P = \frac{\varkappa_D^2}{16\pi^3 V_d} \sum_{\mathbf{N}\in\mathbb{Z}^d} \int d\mathbf{k} < \delta T_{ik}\delta T^*_{i'k'} >_k \tilde{\Lambda}^{iki'k'} \Big|_{k^0=\sqrt{|\mathbf{k}|^2+(2\pi\mathbf{N}/L)^2}}, \quad (15)$$

$$\tilde{\Lambda}^{iki'k'} = \frac{1}{2}\left[\Delta^{ii'}\Delta^{kk'} + \Delta^{ik'}\Delta^{i'k}\right] - \frac{1}{D-2}\Delta^{ik}\Delta^{i'k'}, \quad \Delta_{ik} = \delta_{ik} - n_i n_k, \quad \mathbf{n} = \frac{\mathbf{k}}{k^0}. \quad (16)$$

In our approach we did not use explicit decomposition of the full multidimensional metric perturbation h_{MN} in massless and massive modes, these correspond to $\mathbf{N} = 0$ and $\mathbf{N} \neq 0$ terms in the sum respectively. For massless modes the three-dimensional vector $\mathbf{n}$ is the unit vector, but not for massive modes. Note that h_{MN} has bulk components due to the trace term in the wave equation.

To calculate gravitational bremsstrahlung from the non-relativistic plasma we need the longitudinal field and the related fluctuation function. Keeping as the source in Maxwell equations terms of the zero and the first order

$$\mathrm{div}\mathbf{E} = 4\pi e \int d\mathbf{v}(\delta f^0 + \delta f^1), \quad \delta f^1 = -i\frac{e}{m}\frac{1}{\omega - \mathbf{kv}}\mathbf{E}_k\frac{\partial f^0}{\partial \mathbf{v}}, \quad (17)$$

one can express the first order electric field through zero fluctuations:

$$\mathbf{E}_k = -4\pi i \frac{\mathbf{k}}{|\mathbf{k}|^2}\frac{\rho^0_{\mathbf{k}\omega}}{\varepsilon_L(k)}, \qquad \rho^0_k = \sum e \int d\mathbf{v}\delta f^0_k(\mathbf{v}), \quad (18)$$

$$\varepsilon_L(k) = 1 + \frac{4\pi e^2}{m\omega|\mathbf{k}|^2}\int d\mathbf{p}\frac{(\mathbf{kv})}{(\omega - \mathbf{kv})}\mathbf{k}\frac{\partial f_0(\mathbf{v})}{\partial \mathbf{v}}, \quad (19)$$

the last line being the longitudinal permittivity of the isotropic plasma. Using this, one then finds the two-point correlation functions defined above [13].‡ Correlation function for zero-order particle density $< \delta n^2 >^0_k$ is calculable for the Maxwell distribution (8) in the finite form, while the field correlation functions can be then expressed through this quantity. Using the equations (17) and (18) one find the following relations between $< \rho^2 >_k$, $< E^2 >_k$ and $< \rho^2 >^0_k$:

$$< \rho^2 >_k = \frac{< \rho^2 >^0_k}{|\varepsilon_L(k)|^2}, \quad < E^2 >_k = \frac{16\pi^2}{|\mathbf{k}|^2}\frac{< \rho^2 >^0_k}{|\varepsilon_L(k)|^2}, \quad < \delta\rho^2 >^0_k = \sqrt{2\pi}\frac{n_0 e_\alpha}{|\mathbf{k}|v_T}\mathrm{e}^{-\frac{\omega^2}{2|\mathbf{k}|^2 v_T^2}}. \quad (20)$$

Substituting this to (fluctuations of) the energy-momentum tensors and making use of the formula (15), one obtains after some rearrangements:

$$< \delta T_{ik}\delta T^*_{i'k'} >_k \tilde{\Lambda}^{iki'k'} =$$
$$\int d^4k_1 \left\{ A(k_1,k)\frac{T}{m}\frac{< \delta n^2 >^0_{k_2}}{|\varepsilon_L(k_2)|^2} < \delta n^2 >^0_{k_1} + B(k_1,k)\frac{< \delta n^2 >^0_{k_1}}{|\varepsilon_L(k_1)|^2}\frac{< \delta n^2 >^0_{k_2}}{|\varepsilon_L(k_2)|^2} \right\}. \quad (21)$$

According to the relations (20), the first term here is the product of the field and particle density correlation functions, while the second is quadratic in the field correlators. The following frequency regions has to be distinguished:

‡ Density fluctuations can be characterized by correlation functions whose Fourier-transforms exhibits homogeneity and isotropy: $< \delta n(\mathbf{r}_1,t_1)\delta n(\mathbf{r}_2,t_2) > \equiv < \delta n^2 >_{\mathbf{r},t}$, $\mathbf{r} = \mathbf{r}_1 - \mathbf{r}_2, t = t_1 - t_2$. The space-time Fourier transformation will give the spectral fluctuation density: $< \delta n^2 >_k = \int d\mathbf{r} \int dt e^{-i\mathbf{kr}+i\omega\mathbf{r}} < \delta n^2 >_{\mathbf{r},t}$, $< \delta n(k)\delta n(k') > = (2\pi)^4 < \delta n^2 >_k \delta^4(k-k')$.

i) Zero real part of the complex permittivity $Re\varepsilon_L(\omega, \mathbf{k}) = 0$ in the denominators corresponds to propagation of the longitudinal Langmuir waves with the frequency

$$\omega \simeq \omega_L, \quad \omega_L = \sqrt{\frac{4\pi^2 e^2 N_0}{m}}$$

and the long wavelength: $|\mathbf{k}| < \frac{1}{r_D}$, where $r_{D_e} = \frac{v_{T_e}}{\omega_L}$ is the Debye shielding radius.

ii) For $\omega > \omega_L$, $|\mathbf{k}| v_{T_e} \gg 1$, then $\varepsilon_L \simeq 1$ which corresponds to electromagnetic intaraction switched off.

iii) For $\omega < \omega_L$ one has $\varepsilon_L(k) \simeq 1 + (|\mathbf{k}| r_D)^{-2}$ which corresponds to the Debye shielding.

Correspondingly, we have to distinguish radiation losses due to gravitational bremsstrahlung (denominators are non-zero, both terms being taken into account), and the coalescence of two Langmuir plasmons into graviton (the second term with both denominators at Langmuir frequencies).

3.1. Gravitational bremsstrahlung

This corresponds to $Re\varepsilon_L \neq 0$. The main contribution to the integral comes from the region $\omega < |\mathbf{k}_1| v_{T_e}$, so for $k_0 > \omega_L$, $|\mathbf{k}_1| r_D > 1$, then $\varepsilon_L \simeq 1$. In opposite limit $k_0 < \omega_L$, Debye shielding becomes manifest, then $\varepsilon_L(k_1) \simeq 1 + (|\mathbf{k}_1| r_D)^{-2}$. Therefore we obtain in the two cases:

$$P \simeq \frac{128}{\sqrt{\pi}}(1+\sqrt{2}) e^4 N_0^2 v_{T_e} \frac{T^{n+1}}{M_*^{n+2}} \Big\{ I_1(n) \ln\Lambda + \frac{8}{3} I_2(n) \Big\}, \quad k^0 > \omega_L, \tag{22}$$

$$P \simeq \frac{128}{\sqrt{\pi}}(1+\sqrt{2}) e^4 N_0^2 v_{T_e} \frac{\omega_L^{n+1}}{M_*^{n+2}} I_1(n) \ln\Lambda, \quad k^0 < \omega_L, \tag{23}$$

where $\ln\Lambda$ is the Coulomb logarithm, $\Lambda = r_{D_e}/\hat{r}$, $r_{D_e} = \frac{v_{T_e}}{\omega_L}$ – is the electron Debye radius, $\hat{r}$ is determined from the condition $e^2/\hat{r} = T$ and $M_* = \frac{1}{\varkappa_D^2}$ is the D-dimensional Planck mass. The coefficients $I_1(n)$, $I_2(n)$ depend on the number of extra dimensions n, $2 \leq n \leq 7$ as follows: $(I_1(n) = 1.1,\ 1.07\ 0.97\ 0.83,\ 0.67,\quad I_2(n) = 4.8,\ 6.35,\ 7.3,\ 7.5,\ 7.1$. If $n = 0$ $(I_1(n) = 1,\ I_2(n) = 1)$ these results coincide with those - in $3+1$ dimensions [11]. Without collective effects, the result in $3+1$ dimensions corresponds to the quadruple formula calculation with consequent thermal averaging [15].

3.2. Coalescence of two Langmuir plasmons into graviton

For coalescence of two Langmuir plasmons into the graviton the field correlation function can be presented as

$$< E^2 >_k = 4\pi^2 T(\delta(\omega - \omega_L) + \delta(\omega + \omega_L)) \tag{24}$$

The graviton frequency will be twice the Langmuir frequency and the rate of emission will be

$$P^0 = 2e^4 N_e^4 v_{T_e} \frac{(2\omega_L)^{n+1}}{M_*^{n+2}} I_n, \tag{25}$$

with $I_n = 1.4,\ 1.9,\ 2.2,\ 2.3,\ 2.2,$ for $n = 2, \ldots, 6$.

Compare with the result of the ref. [4] for the bremsstrahlung losses which was obtained using the one-particle approach:

$$P \simeq \sum_j \frac{\Gamma(\frac{5}{2}+n) n_e n_j Z_j^2 \alpha^2 I_{GB}(n)}{\Gamma(\frac{3}{2})} \frac{T^{n+1}}{M_*^{n+2}}, \tag{26}$$

where n_e and n_j are the electrons and ions densities, Z_j is the ions charge, $\alpha = e^2$, and the numerical integral $I_{GB}(n)$ is presented for $n = 2, 3: \quad I_{GB}(n) = 0.7, 0.3$. Being applied to red giants with the electron temperature $T \sim 8.6$ keV and the electron density $n_e = 3.0 \cdot 10^{29} cm^{-3}$, assuming the conservative upper limit on the energy-loss rates of red giants $\dot{\varepsilon} \sim 100 erg\, g^{-1}\, sec^{-1}$, the authors of [4] obtained the upper limit for Planck mass M_* to be 10^3 for $n = 2$ and $1.7 \cdot 10^{-5}$ for $n = 3$. Our result (23) for the bremsstrahlung losses is about three time greater than (26). Moreover, an additional enhancement comes from the in coalescence process (25).

The work was supported by RFBR project 08-02-01398-a.

References

[1] Arkani-Hamed N, Dimopoulos S and Dvali G 1998 *Phys. Lett* **B 429** 263; 1999 *Phys. Rev* **D 59** 086004

[2] Han T, Lykken J D, and Zhang R-J 1999 *Phys. Rev.* **D59** 105006

[3] Cullen S and Perelstein M 1999 *Phys. Rev. Lett.* **83** 268

[4] Barger V D, Han T, Kao C and Zhang R-J 1999 *Phys. Lett.* **B461** 34

[5] Hanhart C, Phillips D R, Reddy S, and Savage M J 2001 *Nucl. Phys.* **B595** 335

[6] Hanhart C, Pons J A, Phillips D R and Reddy S 20001 *Phys. Lett.* **B509** 1

[7] Hannestad S and Raffelt G *Phys. Rev. Lett.* **87** 051301

[8] Hannestad S and Raffelt G 2003 *Phys. Rev.* **D67**, 125008 (2003); [Erratum-ibid. 2004 **D69** 029901]

[9] Hanhart C, Phillips D R, and Reddy S 2001 *Phys.Lett.* **B499**, 9

[10] Satheeshkrumar V S and Suresh, P K arXiv:0805.3429 [astro-ph]

[11] Galtsov D V, Grats Yu V and Melkumova E Yu 1983 *Sov.Phys.J.* **26** 455

[12] Sitenko A D 1967 *Electromagnetic fluctuations in plasmas* New York, Acad. Press 255 p.

[13] Sitenko A D 1973 *Phys.Scr.* **7** 193

[14] Gal'tsov D, Kofinas G, Spirin P, and Tomaras T N (in preparation)

[15] Weinberg S 1965 *Phys.Rev.* **B140** B516

M(atrix) model interaction with 11D supergravity

Igor A. Bandos

Department of Theoretical Physics, University of the Basque Country UPV/EHU, P.O. Box 644, 48080 Bilbao, Spain and IKERBASQUE, Basque Foundation for Science, Bilbao, Spain

E-mail: igor_bandos@ehu.es, bandos@ific.uv.es

Abstract. We present the equations of motion for multiple M0–brane (mM0) system in an arbitrary curved supergravity superspace which generalizes the M(atrix) model equations for the case of arbitrary supergravity background. Although these were obtained in the frame of superembedding approach to mM0, we do not make a review of this approach in this contribution but concentrate discussion on the structure of the equations.

1. Introduction

M(atrix) model conjecture [1] attracted - and still attracts- much attention as a relatively simple approach to a nonperturbative description of String/M-theory, the most promising candidate on the rôle of quantum theory of gravity and of the unified theory of all fundamental interactions. The basic Lagrangian used to study that 'M(atrix) theory' was the one obtained by dimensional reduction of D=10 $U(N)$ supersymmetric Yang–Mills theory (SYM) down to d=1. This is treated as a (very low energy and gauge fixing) description of the dynamical system of N D0–branes (D-particles or Dirichlet particles) moving in flat target $D = 10$ type IIA superspace.

D0-brane is the simplest representative of the family of Dp-branes (Dirichlet p–branes with p=1 for string, p=2 for membrane etc.) defined as hypersurfaces where string can have its ends [2, 3]. As String theory is a theory of gravity, a 'frozen' hypersurface cannot exist in its frame so that Dp–branes are dynamical supersymmetric extended objects. Their equations of motion can be obtained from actions defined on their $(p+1)$ dimensional worldvolumes W^{p+1} embedded in target D=10 type II superspace $\Sigma^{(10|32)}$ with ten bosonic and 32=2×16 fermionic dimensions.

In general, the interaction between constituents of a multiple Dp (mDp) system is due to the strings stretched between different Dp–branes. At low energy the massive string excitations can be ignored and the string is described by its massless modes which are the fields of SYM multiplet. When a set of N nearly coincident Dp-branes is considered, the strings scratched between different branes give rise to the fields of $U(N)$ SYM multiplet [4]. This explains why the degrees of freedom of the multiple Dp-brane system are the ones of the maximally supersymmetric (16 supersymmetries) $U(N)$ SYM multiplet. However, being the objects of the type II String Theory, multiple Dp-brane system should actually possess 32 component supersymmetry, the 16 of which are realized linearly (and preserved by the ground state of mDp system and in the SYM description) while the other 16 supersymmetries are realized nonlinearly (spontanousely broken by the mDp system). The mechanism of enhancement of 10D $\mathcal{N} = 1$ supersymmetry (with 16 generators) till 10D type II supersymmetry (with 32 generators) by passing from the SYM to nonlinear action containing Dirac–Born–Infeld and Wess–Zumino terms is well understood for the case of a single Dp-brane (see [5] and [6, 7] for more references).

Thus the $d = 1\ \mathcal{N} = 16\ U(N)$ SYM action provides a very-low-energy description of the nearly coincident mD0 system in flat target type II superspace and just this Lagrangian was used to describe the M(atrix) theory of [1]. The restoration of the 11D Lorentz invariance starting from $1d\ \mathcal{N} = 16$ SYM Lagrangian was also considered in [1] and allowed to treat this as the Matrix model Lagrangian in flat 11D supergravity background. However, till very recent time, the matrix model equations were known for very few particular non-flat supergravity backgrounds including pp-waves [8] and the light–like linear dilaton background [9].

The matrix model equations for an arbitrary supergravity background, which we will discuss here, have been obtained in [11] by developing superembedding approach to the system of multiple M0-branes. Having said that, we have to explain briefly what is M0-brane and what is the superembedding approach to super-p-branes (although we will not review this in the present contribution but just present the results obtained in its frame).

Single M0–brane is just the eleven dimensional massless superparticle [5]. It is the simplest representative of the family of 11D M-branes also including supermembrane (M2-brane) [12, 13] and M-theory 5-brane (M5-brane) [14, 15, 16]. M0-brane can also be called M-wave; the name multiple gravitons was used in [17] for the (bosonic limit of the) multiple M0 (mM0) system.

The superembedding approach to supersymmetric extended objects, (super-)p-branes [13, 18, 14, 6, 7], following the so–called STV (Sorokin–Tkach–Volkov) approach to superparticles and superstrings [19] (see [6] for review and more refs) describes p-branes in terms of embedding of its *worldvolume superspace* $\mathcal{W}$ into the *target superspace* Σ. For the case of M0-brane these are the d=1 $\mathcal{N}$=16 superspace $\mathcal{W}^{(1|16)}$ and the D=11 $\mathcal{N}$=1 superspace $\Sigma^{(11|32)}$, respectively. The embedding of $\mathcal{W}$ into Σ obeys the so-called *superembedding equation* which, in the case of higher dimensional p-branes (with also sufficiently high co-dimension D-p) specifies completely the p–brane dynamics, *i.e.* contains p–brane equations of motion among its consequences. This on-shell nature of the superembedding equation, which holds, in particular, for all the above mentioned M-branes, makes possible to use the superembedding approach to derive equations of motion for new branes (as it was with M5-brane [14]) and also for multi–brane systems (although this presently is developed for multi-particles and multi-wave cases only [20, 10, 11])[1].

2. Basic variables and Abelian ($U(1)$) part of the Matrix model equations

As we have already said, the basic action staying beyond the M(atrix) theory of [1] is the dimensional reduction of the maximal, D=10 $\mathcal{N}$=1, $U(N)$ SYM theory down to $d = 1$. In such a dimensional reduction the spacial components of the $u(N)$ valued SYM potential give rise to nanoplet of anti-hermitian $N \times N$ matrices of scalar fields $\hat{\mathbf{X}}^i$, while the D=10 Majorana–Weyl fermionic field of the 10D SYM gives rise to 16 $u(N)$ valued fermionic matrices $\hat{\mathbf{\Psi}}_q$, $q = 1, \ldots, 16$.

One can extract the trace part of the above matrix fields thus separating the variables describing the center of mass (center of energy) motion and the relative motion of the constituents of the multiple D0-brane (mD0) system,

$$\hat{\mathbf{X}}^i = \hat{x}^i(\tau)\, I_{N\times N} + \mathbf{X}^i(\tau)\,, \qquad i = 1, ..., 9\,, \qquad tr(\mathbf{X}^i) = 0\,, \tag{1}$$

$$\hat{\mathbf{\Psi}}_q = \hat{\psi}_q(\tau) I_{N\times N} + \mathbf{\Psi}_q(\tau)\,, \qquad q = 1, ..., 16\,, \qquad tr(\mathbf{\Psi}_q) = 0\,. \tag{2}$$

The Abelian center of energy variables $\hat{x}^i(\tau)$, $\hat{\psi}_q(\tau)$ are the same as describing a single D0 after fixing the so–called static gauge. The same variables are used also for the gauge fixed description of M0-brane, while before the gauge fixing, the Lorentz (and/or diffeomorphism) covariant description of a single M0 brane is given in terms of the $\Sigma^{(11|32)}$ coordinate functions

$$\hat{Z}^M(\tau) = (\hat{x}^\mu(\tau)\,, \hat{\theta}^{\check{\alpha}}(\tau))\,, \qquad \mu = 0, 1, \ldots, 9, 10\,, \qquad \check{\alpha} = 1, \ldots, 32\,. \tag{3}$$

[1] The boundary fermion approach [27], describing mDp systems on a 'minus one-quantization level', also uses a version of superembedding approach, but with nonstandard superspace including boundary fermion directions.

The number of bosonic coordinate functions in the covariant description of D0-brane is 10 (*versus* 11 for M0, Eq. (3)). The similarity of the variable used for the gauge fixed description is due to that the D0–brane, which is the massive 10D $\mathcal{N}=2$ superparticle (see [21] for the first appearance of its D=6 counterpart) can be obtained by dimensional reduction of the 11D massless superparticle in $D=11$, which is to say of the M0-brane [5].

However, the gauge fixed description of M0-brain has a higher residual symmetry: the SO(9) residual bosonic symmetry of the gauge fixed description of D0–brane is enlarged in this case up to SO(1,1)⊗SO(9) group[2] so that the basic fields and parameters are also characterized by their SO(1,1) weight. In particular, a parameter of the κ–symmetry [21, 22] (which can be identified with worldline supersymmetry [19] and reflects a part of spacetime supersymmetry which is preserved by M0-brane ground state) has SO(1,1) weight 1 which we mark by sign superscript in its notation: ϵ^{+q} [10, 11]. Clearly the SO(1,10) spinorial parameter $\varepsilon^{\alpha}(x)$ (and $\varepsilon^{\alpha}(Z)$) of the 11D spacetime (superspace) supersymmetry is inert under the above SO(1,1), so that when describing the spacetime supersymmetry preserved by M0-brane (*i.e.* linearly realized in the M0–brane model), $\hat{\varepsilon}^{\alpha}:=\varepsilon^{\alpha}(\hat{Z}(\tau))$, it is expressed through the above κ–symmetry parameter ϵ^{+q} by using the 16×32 matrix $v_q^{-\alpha}$ which carries the $SO(1,1)$ weight -1,

$$\hat{\varepsilon}^{\alpha}=\epsilon^{+q}v_q^{-\alpha}\,, \qquad \alpha=1,\ldots,32\,, \qquad q=1,\ldots,16\,. \tag{4}$$

This rectangular *spinor moving frame* matrix carrying the $Spin(1,10)$ index $\alpha=1,\ldots,32$ and $Spin(9)$ index $q=1,\ldots,16$ is constrained to obey

$$v_q^{-}\Gamma_a v_p^{-}=u_a^{=}\delta_{qp}\,, \qquad 2v_{\alpha q}^{-}v_{\beta q}^{-}=\Gamma^a_{\alpha\beta}u_a^{=}\,, \tag{5}$$

where $\Gamma^a_{\alpha\beta}:=\Gamma^{a}{}_{\alpha}{}^{\gamma}C_{\gamma\beta}$ are 11D Dirac matrices contracted with 11D charge conjugation matrix. The vector $u_a^{=}(\tau)$ in (5) is light-like (as a consequence of (5)). Actually, it is proportional to the momentum of the M0-brane, the fact which can be expressed by stating that

$$\hat{E}^a_{\#}:=D_{\#}\hat{Z}^M(\tau)E^a_M(\hat{Z})=u_a^{=}(\tau)\,, \tag{6}$$

where $D_{\#}$ is covariant derivative on the worldline, $D_{\#}\hat{Z}^M=e^{\tau}_{\#}\partial_\tau\hat{Z}^M(\tau)$ and $E^a_M(Z)$ are the component of the target superspace supervielbein form $E^a=dZ^ME^a_M(Z)$. (Notice that in our notation the subscript plus index is equivalent to superscript minus one and *vice versa*). The supervielbein form of the curved 11D superspace $E^A=(E^a,E^{\alpha})=dZ^ME^A_M(Z)$ carries degrees of freedom of the 11D supergravity supermultiplet if it obeys the superspace supergravity constraints [24, 25]. The most important of these are collected in $T^a:=DE^a=-iE^{\alpha}\wedge E^{\beta}\Gamma^a_{\alpha\beta}$.

One can consider $u_a^{=}(\tau)$ as a part of *moving frame* attached to the M0 worldline,

$$\left(\frac{u_a^{\#}+u_a^{=}}{2}\,,\,u_a^i\,,\,\frac{u_a^{\#}-u_a^{=}}{2}\right)\in SO(1,D-1)\Leftrightarrow\begin{cases}u_a^{=}u^{a\,=}=0\,, & u_a^{=}u^{a\,i}=0\,, \quad u_a^{=}u^{a\,\#}=2\,,\\ u_a^{\#}u^{a\,\#}=0\,, & u_a^{\#}u^{ai}=0\,, \quad u_a^iu^{aj}=-\delta^{ij}\,.\end{cases} \tag{7}$$

The M0 equations of motion can be conveniently written with the use of the space-like moving frame vector u_a^i (in (7)) and the set of spinors $v_{\alpha p}{}^{-}$ ('square roots' of $u_a^{=}$ in the sense of (5)) as

$$D_{\#}\hat{E}_{\#}{}^a\,u_a{}^i=0\,, \qquad \hat{E}_{\#}{}^{\alpha}v_{\alpha p}{}^{-}:=D_{\#}\hat{Z}^M(\tau)E_M{}^{\alpha}(\hat{Z})v_{\alpha p}{}^{-}=0\,. \tag{8}$$

These equations of motion also imply that $v_{\alpha q}^{-}$ and $u_a^{=}$ are covariantly constant, $Dv_{\alpha q}^{-}=0$ and $Du_a^{=}=0$ (see [10, 11] for more detail).

[2] Actually, the residual symmetry of the gauge fixed M0-brane is $[SO(1,1)\otimes SO(9)]\otimes K_9$, see [23, 7] and refs therein, but the K_9 part is inessential for our discussion here.

The moving frame variables (7) can be also used to extract, in the covariant manner, different projections of the supergravity fluxes to the worldline. The following projections of the 4-form field strength ($F_4 = dC_3$), gravitino field strength $T_{ab}{}^{\alpha} = T_{[ab]}{}^{\alpha}$ and Riemann tensor of supergravity $R_{dc\, ba} = R_{[dc]\, [ba]}$

$$\hat{F}_{\#ijk} := F^{abcd}(\hat{Z}) u_a^{=} u_b{}^i u_c{}^j u_d{}^k\,, \qquad \hat{R}_{\#ij\#} := R_{dc\, ba}(\hat{Z}) u^{d=} u^{ci} u^{bj} u^{a=}\,, \\ \hat{T}_{\#\, i+q} := T_{ab}{}^{\alpha}(\hat{Z})\, v_{\alpha q}^{-}\, u_a^{=} u_b^i \tag{9}$$

play a special rôle: only they influence the closure of the worldline supersymmetry algebra[3].

In our superembedding approach [10, 11] *the center of energy motion of the multiple M0 system is described by the same set of equations (8) and characterized by the same supersymmetry properties as a single M0*. Then it is natural to expect that the supergravity fluxes can enter the equations of relative motion of the mM0 system only through (9). This is indeed the case.

3. Non-Abelian ($SU(N)$) part of the Matrix model equations in an arbitrary $11D$ supergravity background

In the light of relation between M0 and D0 brane, it is natural to expect that the relative motion of the constituents of the mM0 system is described by the nanoplet of bosonic traceless matrices $\mathbf{X}^i$ and 16-plet of fermionic traceless matrices $\mathbf{\Psi}_q$ introduced in (1) for the case of D0-branes. The superembedding approach of [10, 11] imposes on these fields a set of dynamical equations which includes the following fermionic equation

$$\dot{\Psi}_q = -\tfrac{1}{4}\gamma^i_{qp}\, [\mathbf{X}^i\,,\, \Psi_p] + \tfrac{1}{24}\hat{F}_{\#ijk}\gamma^{ijk}_{qr}\Psi_r - \tfrac{1}{4}\mathbf{X}^i\hat{T}_{\#i\,+q}\,, \tag{10}$$

as well as Gauss constraint

$$\left[\mathbf{X}^i\,,\, \dot{\mathbf{X}}^i\right] = 4i\,\{\Psi_q\,,\, \Psi_q\} \tag{11}$$

and proper bosonic equation of motion

$$\ddot{\mathbf{X}}^i = \tfrac{1}{16}\, [\mathbf{X}^j, [\mathbf{X}^j, \mathbf{X}^i]] + i\gamma^i_{qp}\, \{\Psi_q\,,\, \Psi_p\} + \tfrac{1}{4}\mathbf{X}^j\hat{R}_{\#j\,\#i} + \tfrac{1}{8}\hat{F}_{\#ijk}\left[\mathbf{X}^j,\, \mathbf{X}^k\right] - 2i\Psi_q\hat{T}_{\#i\,+q}\,. \tag{12}$$

Here, for the safe of simplicity, we have used the dot notation for the covariant time derivative, so that $\dot{\Psi}_q = D_{\#}\Psi_q$, $\dot{\mathbf{X}}^i := D_{\#}\mathbf{X}^i$, $\ddot{\mathbf{X}}^i := D_{\#}D_{\#}\mathbf{X}^i$.

The third term in the *r.h.s.* of the bosonic equation (12) shows that, generically, the field $\mathbf{X}^i$ is massless with the mass matrix generated by the special projection of Riemann tensor in (9). The last, fifth term describes the contribution of the fermionic flux. The most interesting forth term, $\hat{F}_{\#ijk}[\mathbf{X}^j, \mathbf{X}^k]$, describes the 'dielectric coupling' characteristic for the Emparan-Myers 'dielectric brane effect' [29, 30]. It is essentially non-Abelian, which is also true for the first and second terms in the *r.h.s.* of (12), although these two are also present in the case of flat background without fluxes and in 1d dimensional reduction of 10D SYM.

In this flat 11D superspace case Eqs. (10), (11) and (12) also coincide with the equations describing mD0 system in flat 10D type IIA superspace [20][4]. However, when discussing mM0 system we have to define also the $SO(1,1)$ weight of these matrix valued fields (the notion which did not appear when mD0 system is considered). Superembedding approach [10] suggests to choose the weight of the bosonic and fermionic matrix field to be (-2) and (-3), respectively,

$$\mathbf{X}^i = \mathbf{X}^{=i} \equiv \mathbf{X}^{--i} \qquad \mathbf{\Psi}_q := \mathbf{\Psi}_q^{=-} \equiv \mathbf{\Psi}_q^{---}\,. \tag{13}$$

[3] This is tantamount to saying that their superspace generalizations enter the description of the geometry of the worldline superspace $\mathcal{W}^{(1|16)}$ and normal bundle over it; see [11].

[4] The dimensional reduction of the curved superspace mM0 equations was not studied yet; it might provide a simple way to reproduce the coupling of mD0 to type IIA supergravity fluxes.

Also the covariant time derivative $D_{\#} = D_{++} \equiv D^{--}$ carries the negative $SO(1,1)$ weith (-2). As it was noticed in [11], these observations, considered together with that only the projections of fluxes presented in (9) may influence the dynamics, makes the structure of the equations for the relative motion of the m$M0$ constituents quite rigid: very few terms are allowed by $SO(1,1) \times SO(9)$ symmetry in addition to the ones which were found as a result of calculations. This restricts possible deformations/nonlinear generalizations of Eqs. (10), (11) and (12), at least when the center of mass motion of the mM0 system is not deformed.

4. Conclusion and discussion

In this contribution we have presented the Matrix model equations in an arbitrary D=11 supergravity background which were obtained from the superembedding approach to multiple M0-brane (mM0) system [10, 11]. The set of these equations splits naturally on an Abelian ('$U(1)$') part, describing the center of energy motion of the mM0 system, and a non-Abelian ('$SU(N)$') part containing equations for anti-hermitian traceless $N \times N$ matrices $\mathbf{X}^i$, describing the relative motion of the mM0 constituents, and their superpartners $\mathbf{\Psi}_q$. The equations describing the center of energy motion of the mM0 system are the same as equations of motion for a single M0-brane, which implies that the center of energy moves on light–like geodesic in a (generically curved) 11D spacetime. The flat target superspace limit of the equations describing the relative motion of the mM0 constituents can be also obtained by making dimensional reduction of the D=10 $SU(N)$ SYM model down to d=1. However, our equations describe the mM0 system in an arbitrary supergravity superspace thus generalizing the Matrix model equations for an arbitrary supergravity background. The right hand sides of these equations contain the contributions from the supergravity fluxes, this is to say, form the antisymmetric tensor field strength $F_{abcd}(x)$, gravitino field strength $T_{ab}{}^{\alpha}(x)$ and Riemann tensor $R_{ab\,cd}(x)$ of the 11D supergravity. The natural application is to use our general equations to obtain the Matrix model in physically interesting backgrounds such as $AdS_4 \times S^7$ and $AdS_7 \times S^4$ superspaces.

Notice that all the terms with background contributions to the *r.h.s.*'s of our mM0 equations are linear in fluxes. This is in disagreement with expectations based on the study of the Myers-type actions [30, 17]. Although the Myers action is purely bosonic and resisted all the attempts of its straightforward supersymmetric and Lorentz covariant generalization eleven years (except for the cases of lower dimensional and lower co-dimensional Dp-branes [26]) a particular progress in this direction was reached recently, in the frame of the boundary fermion approach [27] which gives a supersymmetric and Lorentz covariant description of multiple Dp-brane system but on a 'minus one-quantization' level (the quantization of boundary fermion sector is needed to be done to arrive at 'classical' action). Taking this and also evidences from the string amplitude calculations into account, we do not exclude the possibility that the above mentioned discrepancy implies that our approach gives only an approximate description of the Matrix model interaction with supergravity fluxes (but, if so, it is Lorentz covariant, supersymmetric and going beyond the $U(N)$ SYM approximation).

If this is the case, a way to search for a more general interaction lays through modification of the basic equations of the superembedding approach of [10, 11], namely the superembedding equation, defining the embedding of the mM0 center of energy superspace $\mathcal{W}^{(1|16)}$ into the target $D = 11$ supergravity superspace $\Sigma^{(11|32)}$, and the basic constraints of the $d = 1$, $\mathcal{N} = 16$ SYM model on the center of energy superspace. (Notice that the modification of the basic superembedding-type equations in the boundary fermion approach were suggested recently in [28]). The problem of the deformation of the basic constraint determining the equations for the relative motion of mM0 constituents is the curved superspace generalization of the studies in [31, 32, 28]. However, unfortunately, if we allow consistent deformations of the basic equations of our superembedding approach, although most probably these exist, they would certainly make the equations very complicated up to being unpractical.

Interestingly enough, if we do not deform the superembedding equation, but allow for a deformation of the $d = 1\mathcal{N} = 16$ $SU(N)$ SYM constraints on the center of mass superspace (see [10, 11]), the situation seems to be much more under control due to the rigid structure of the mM0 equations [11]. In this case the center of mass motion and supersymmetry of the corresponding superspace is influenced only by the projections (9) of the supergravity fluxes, so that it is reasonable to assume that only these fluxes can enter the equations of relative motion of the mM0 constituents. Then, as we have already noticed, the requirement of SO(1,1)×SO(9) symmetry leaves very few possibilities to add the new terms to the ones already present in the *r.h.s.*'s of the equations (10)–(12). In particular, the only possible nonlinear contribution to the *r.h.s.* of the bosonic equations (12) is $\mathbf{X}^j \hat{F}_{\#jkl}\hat{F}_{\#ikl}$ describing the contribution proportional to the second power of the 4-form flux to the mass matrix of the $su(N)$-valued fields $\mathbf{X}^j$.

Acknowledgments
The author is grateful to Y. Lozano for useful discussions/communications. This work was supported in part by the research grants FIS2008-1980 from the MICINN of Spain and the Basque Government Research Group Grant ITT559-10.

References

[1] Banks T, Fischler W, Shenker S H and Susskind L 1997, *Phys. Rev.* D **55**, 5112-5128.
[2] Sagnotti A 1987 Open strings and their symmetry groups *Preprint* ROM2F-87-25 available as hep-th/0208020.
[3] Horava P 1989 Nucl. Phys. B **327**, 461; *Phys. Lett.* B **231**, 251.
[4] Witten E 1996 *Nucl. Phys.* B **460**, 335 (*Preprint* hep-th/9510135).
[5] Bergshoeff E and Townsend P K 1997 Super-*D*-branes *Nucl.Phys.* B **490**, 145–162 (*Preprint* hep-th/9611173).
[6] Sorokin D P 2000 Superbranes and superembeddings, *Phys. Rept.* **329**, 1-101 (*Preprint* hep-th/9906142).
[7] Bandos I A 2009 Superembedding approach to Dp-branes, M-branes and multiple D(0)-brane systems *Preprint* arXiv:0912.2530 [hep-th].
[8] Berenstein D E, Maldacena J M and Nastase H S JHEP04(2002)013 (*Preprint* hep-th/0202021).
[9] Craps B, Sethi S and Verlinde E P, JHEP10(2005)005 (*Preprint* hep-th/0506180).
[10] Bandos I A 2010 *Phys. Lett.* B **687**, 258-263 (*arXiv:*0912.5125[hep-th]).
[11] Bandos I A 2010 *Phys. Rev. Lett.* **105**, 071602 (*arXiv*:1003.0399 [hep-th]); Multiple M0-brane system in an arbitrary eleven dimensional supergravity background, *Preprint* 1009.3459 [hep-th].
[12] Bergshoeff E, Sezgin E and Townsend P K 1987 *Phys. Lett.* B **189**, 75.
[13] Bandos I, Sorokin D, Tonin, Pasti P and Volkov D V 1995 *Nucl. Phys.* B **446**, 79 (*Preprint* hep-th/9501113).
[14] Howe P S and Sezgin E 1996 D = 11, p = 5 *Phys. Lett.* B **394**, 62 (*Preprint* hep-th/9611008).
[15] Bandos I A, Lechner K, Nurmagambetov A, Pasti P, Sorokin D P and Tonin M 1997 Covariant action for the super-five-brane of M-theory, *Phys. Rev. Lett.* **78**, 4332 (*Preprint* hep-th/9701149).
[16] Aganagic M, Park J, Popescu C and Schwarz 1997 *Nucl. Phys.* B **496**, 191-214 (*Preprint* hep-th/9701166).
[17] Janssen B and Lozano Y 2002 *Nucl. Phys.* B **643**, 399-430; B **658**, 281 (*Preprint* hep-th/0205254, 0207199).
[18] Howe P S and Sezgin E 1996 Superbranes *Phys. Lett.* B **390**, 133 (*Preprint* hep-th/9607227).
[19] Sorokin D P, Tkach V I and Volkov D V 1989 *Mod. Phys. Lett.* A **4** 901-908.
[20] Bandos I A 2009 *Phys. Lett.* B **680**, 267–273 (2009) (*Preprint* arXiv:0907.4681 [hep-th]).
[21] de Azcárraga J A and Lukierski J 1982 *Phys. Lett.* B **B113**, 170–174.
[22] Siegel W 1983 *Phys. Lett.* B **128**, 397.
[23] Bandos I A 2008 *Phys. Lett.* B **659**, 388 *Nucl. Phys.* B **796**, 360 (*Preprint* 0707.2336, 0710.4342 [hep-th]).
[24] Cremmer E and Ferrara S 1980 *Phys. Lett.* B **91**, 61.
[25] Brink L and Howe P S, *Phys. Lett.* B **91**, 384.
[26] Sorokin D P JHEP08(2001)022; Drummond J M, Howe P S and Lindstrom U 2002 *Class. Quant. Grav.* **19**, 6477; Panda S and Sorokin D JHEP02(2003)055.
[27] Howe P S, Lindstrom U and Wulff L JHEP08(2005)041; JHEP02(2007)070.
[28] Howe P S, Lindstrom U and Wulff L 2010 JHEP07(2010)028 (*Preprint* arXiv:1004.3466 [hep-th]).
[29] Emparan R 1998 Born-Infeld strings tunneling to D-branes, *Phys. Lett.* B **423**, 71 (*Preprint* hep-th/9711106).
[30] Myers R C 1999 Dielectric-branes JHEP12(1999)022 (*Preprint* hep-th/9910053).
[31] Cederwall M, Nilsson B E W and Tsimpis D 2001 JHEP07(2001)042 (*Preprint* hep-th/0104236).
[32] Movshev M and Schwarz A 2009 Supersymmetric Deformations of Maximally Supersymmetric Gauge Theories. I *Preprint* arXiv:0910.0620 [hep-th].

Journal of Physics: Conference Series **259** (2010) 012037 doi:10.1088/1742-6596/259/1/012037

Latest results of the direct dark matter search with the EDELWEISS-2 experiment

Valentin Kozlov for the EDELWEISS collaboration

Karlsruhe Institute of Technology, Institut für Kernphysik, Postfach 3640, 76021 Karlsruhe, Germany

E-mail: Valentin.Kozlov@kit.edu

Abstract. EDELWEISS-2 is a Ge-bolometer experiment located in the underground laboratory Laboratoire Souterrain de Modane (LSM, France). For the second phase of the experiment, the collaboration has developed new cryogenic detectors with an improved background rejection (interleaved electrode design) [1]. A continuous operation of ten of these bolometers at LSM together with an active muon veto shielding has been achieved. First results based on an effective exposure of 144 kg·d taken in 2009 have been published recently [2], the acquired data set has since then been doubled. The already published data correspond to an improvement in sensitivity of about 15 compared to EDELWEISS-1 [3]. We present and discuss the latest bolometer data including the identification of muon-induced background events and special measurements of muon-induced neutrons in LSM.

1. Introduction

A large number of cosmological observations, e.g. rotation curves of spiral galaxies, mass-to-light ratios of galaxy clusters measured by the use of gravitational lensing, anisotropy in cosmic microwave background (WMAP), can be consistently explained by introducing a non-visible matter in the Universe or so-called dark matter (DM). The required properties of a corresponding DM particle lead to the generic name WIMP, weakly interacting massive particle. An excellent candidate for this DM arises from the models of supersymmetry (SUSY), being the lightest SUSY particle, LSP. These particles can be searched for in a terrestrial detector by means of nuclear recoils originating from a WIMP-nucleon elastic scattering. The recoil energy, which has to be measured in such a direct detection experiment, can be calculated within the typically accepted models and it is in the range of 1 to 100 keV while the expected interaction rate is below 0.01 event per day and per kilogram of the target material. In order to be sensitive to this low count rate at such energies, experiments on direct DM search are commonly located deep underground to reduce the influence of cosmic rays, select low-radioactivity materials to be used, develop methods and the detectors themselves to have a very powerful background rejection, study general background conditions of the experiment.

2. EDELWEISS experiment

EDELWEISS-2 is a direct DM search experiment (Fig. 1) located in the Laboratoire Souterrain de Modane (LSM), an underground lab in the Fréjus French-Italian road tunnel in the Alps. This laboratory has a shielding of 4850 m.w.e., which reduces the muon flux down to about

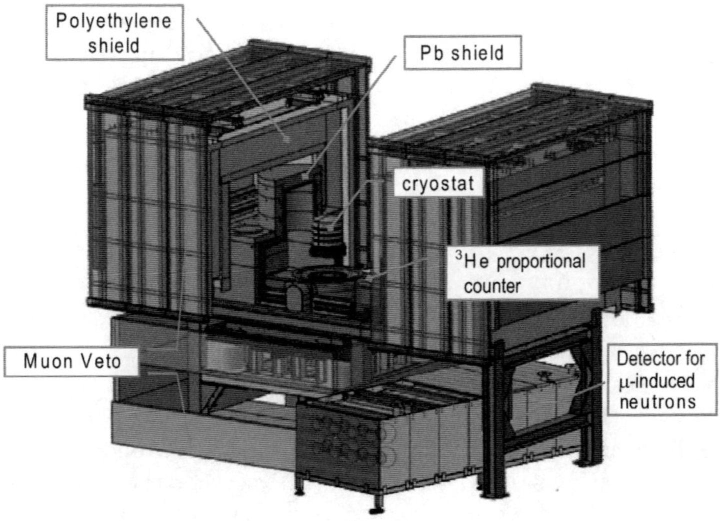

Figure 1. General layout of the EDELWEISS-2 experimental set-up with additional ^{3}He proportional counter for thermal neutrons and liquid scintillator detector to measure muon-induced neutrons.

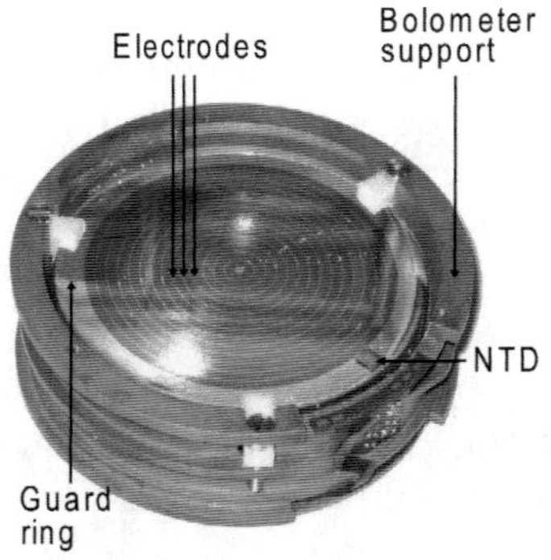

Figure 2. 400 g Ge bolometer with the interleaved electrode design.

5 $\mu/m^2/$day. Bolometers of pure natural Ge are used in the experiment both as the detectors and the target material. When these detectors are cooled down to about 20 mK, a deposited energy of a particle can be measured by means of ionization and phonon excitation signals. Aluminum electrodes evaporated onto the top and the bottom surfaces of the Ge crystal are used to collect electrons and holes. The temperature increase is measured by the use of neutron-transmutation-doped germanium (NTD) as temperature sensor (Fig. 2). The ratio of the two signals, so-called Q-value, is different for nuclear and electron recoils with nuclear recoils having Q$\sim$0.3 when normalized to Q=1 for electron recoils. This separation in Q allows a powerful γ/β-background rejection. A general overview of the set-up is shown in Fig. 1: The central part of the experiment is a dilution cryostat which can host up to 40 kg of detectors. A lead layer of 20 cm shields the bolometers against an external γ/β-background while 50 cm of polyethylene is used to moderate neutrons. A muon veto system (>98% coverage) consisting of 100 m^2 of plastic scintillator to tag remaining cosmic muons completes the installation. In addition, a continuous control of the Rn level is performed near the cryostat, a ^{3}He-gas detector is installed inside the shields to monitor the thermal neutron flux. A neutron counter system based on 1 m^3 of Gd-loaded liquid scintillator [11] and coupled electronically with the muon veto allows to study muon-induced neutron background. The scientific goal of EDELWEISS-2 is to reach a sensitivity of 10^{-8} pb for the WIMP-nucleon spin-independent (SI) cross-section in 2011 and, with an upgrade of the set-up, to achieve 5×10^{-9} pb by 2013.

3. Bolometers with interleaved electrodes

Although the background conditions in EDELWEISS-2 are improved in comparison to EDELWEISS-1 by a factor of 3 for γ's and a factor of 2 in contamination of ^{210}Pb, bolometers with higher rejection power than used in EDELWEISS-1 had to be developed in order to be able to achieve the aimed sensitivity for DM search. This especially concerns the rejection of so-called surface events [1], which have Q-values in the region-of-interest (ROI) due to an incomplete charge collection near the detector surface. The EDELWEISS collaboration thus has developed and successfully tested new bolometers with interleaved electrode design (Fig. 2)[1]. A main principle of these detectors, named inter-digitized electrode detectors, or ID's, is to have concentric ring electrodes for charge collection instead of one planar electrode. By tuning the electric potentials applied on the neighboring electrodes the electrical field is modified near the surface such that depending on an interaction position, i.e. in the bulk or near the surface, a created charge is accumulated either on a set of electrodes with the highest potential, called *fiducial* electrodes, or in addition on low potential electrodes, called correspondently *veto* electrodes. This allows to reliably reject the near-surface events [1]. A first bolometer of this kind was tested for a first time in November 2007 having 200 g mass with 50% as fiducial volume, in January 2009 ten of the 400 g detectors were installed in LSM for the direct DM search.

4. One year of WIMP search with ID detectors

After commissioning of the new detectors, they were continuously operated starting from April 2009 to May 2010 constituting 418 days in total, out of which 322 days were under stable running conditions and 305 days have been considered for the WIMP search, the remaining 17 days being used for various calibrations. During a physics run, a regular energy calibration with ^{133}Ba source was performed as well as every day regeneration of the detectors in order to avoid the formation of space charge. After the first 6 months of operation, the data were analyzed offline with 2 independent analysis chains [2]. Data of 9 out of 10 detectors were used in the analysis. One detector had one deficient guard and one missing veto electrode and in the absence of reliable proof that information from these two electrodes can be efficiently compensated by the remaining channels, this detector was left out of the analysis. The WIMP search threshold was fixed *a priori* to 20 keV in order to ensure that the data set is compatible with a nearly full efficiency for the chosen energy threshold. The data with appropriate baseline resolution and online heat threshold were selected on an hour-by-hour basis. This requirement reduced the effective exposure by about 20%. A reduced χ^2 cut was applied on the pulse fit for heat and fiducial signals of each event. The efficiency of this cut is controlled with the γ-ray background and its typical value is 97%. The fiducial mass was measured individually for each detector using the low-energy peaks ($\sim$10 keV) associated with the cosmogenic activation of germanium, resulting in a uniform contamination by the isotopes of ^{65}Zn and ^{68}Ge. An average fiducial mass of 166$\pm$6 g was deduced for 400 g detectors and a conservative value of 160 g is used in the calculation of the total exposure. Both analyses agreed in the final result of the six-month run: one nuclear recoil candidate was observed in an effective exposure of 144 kg·d. Two further events were in coincidence with either an other bolometer or the muon veto and thus excluded as a WIMP candidate. This result was intepreted in terms of limits on the spin-independent cross-section of WIMP scattering off nucleons, e.g. a cross-section of 1.0×10^{-7} pb is excluded at 90% CL for a WIMP mass of 80 GeV/c^2 (Fig. 3(*right*, EDELWEISS-II 2009)). This result corresponds to an improvement in sensitivity of about 15 compared to EDELWEISS-1 [3] and demonstrates the very high rejection capabilities of ID detectors in an actual WIMP search. Detailed description of the six-month analysis can be found in Ref. [2].

In May 2010 the WIMP search run was stopped and the full data set was analyzed by blindly applying the same cuts as for the first six months. Preliminary results of one of the two analysis chains (second one is still ongoing as for the time of writing) are shown in Fig. 3. In an effective exposure of 322 kg·d there are in total 4 events located in the nuclear recoil band which passed all the cuts: three are close to the threshold and one is at 175 keV (Fig. 3(*left*)). Using an unbiased method to determine the energy window of optimal sensitivity [4] leads to an upper limit on the WIMP-nucleon elastic scattering cross-section of 5×10^{-8} pb (90% CL) for a WIMP mass of 80 GeV/c^2 (Fig. 3(*right*, EDELWEISS-II (prelim.))). The fact that there is no event between 23 and 175 keV results in an almost linear improvement of sensitivity for WIMP masses above 50 GeV, i.e. an improvement of $\sim$2 compared to the data in the 6-months analysis published

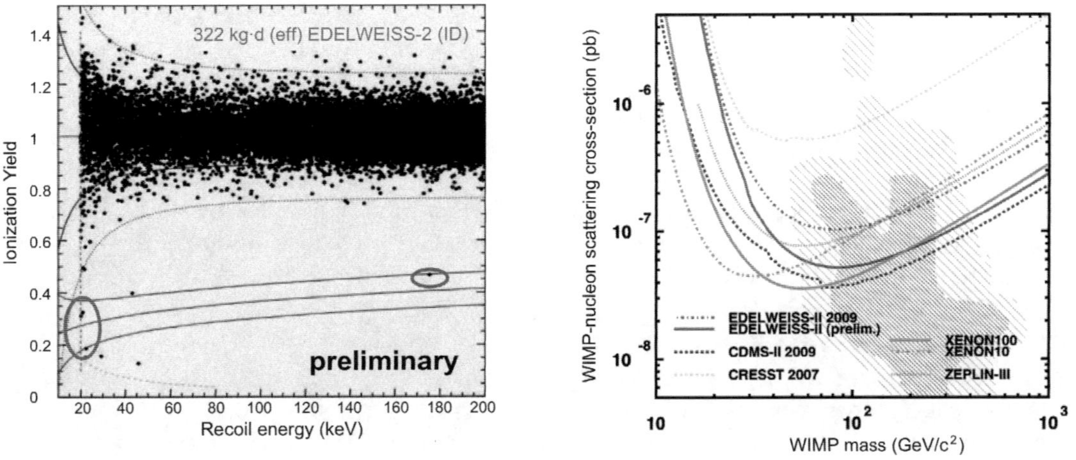

Figure 3. Recent results of the Edelweiss-2 experiment: one-year data obtained in the physics run with new ID bolometers, 4 events passed all the cuts are encircled (*left*); the upper limits on the WIMP-nucleon spin-independent cross-section as a function of WIMP mass (*right*).

earlier [2]. The 4 observed events are to be compared with expectations for various background sources discussed in the following.

5. Background studies

The present estimate of an upper limit of the total background of known origin is 1.6 events (90% CL) in the corresponding exposure of 322 kg·d. The different sources being studied at present are the residual gamma, surface events caused mainly by the remaining beta activity, neutron background from radioactive decays and the muon-induced neutrons. The fact that in the WIMP run 3 events are observed very close to the threshold points towards the need to better understand the detector performance and more detailed studies to determine the acceptance of events in this region.

Regular γ-calibrations with ^{133}Ba source allow to study the rejection power of all 10 detectors together against the γ-background. The number of events found in the nuclear recoil band in ROI during these calibrations compared to the total number of γ-events in the detector bulk yields the γ-rejection factor, which is then used to estimate the corresponding contribution in the WIMP search run. One can still improve the γ-rejection power by optimizing the polarization voltages of the bolometers, implement 2 NTD sensors for heat measurement, introduce different electrode combinations for segmentation. A faster readout electronics can help to avoid pile-up events. The number of unrejected surface events is estimated by multiplying the number of observed low-ionization yield events in the WIMP search run before the rejection of surface events by the rejection power measured previously in Ref. [1]. The surface event background of ID detectors can be further decreased with a better surface treatment, optimization of the mass-to-surface ratio and by improving an energy resolution. The contribution of neutrons from radioactive decays has been simulated using GEANT3 in Ref. [5] and studied in Ref. [6, 7]. The presence of U/Th traces in the lead shield was recently measured as well as the efficiency of EDELWEISS neutron shields was checked with an external strong AmBe source ($\sim 10^5$ neutrons/s). The number of neutrons from interactions of untagged muons is estimated from the number of observed coincidences between the bolometers and the muon veto, multiplied by the muon veto efficiency. The effect of neutrons can be further reduced by an additional polyethylene layer between the lead shield and the cryostat. Comprehensive GEANT4 simulations were developed [8] within the collaboration in order to understand the muon-induced background in detail. In the simulations it was shown, for instance, that the muon-induced bolometer events with an energy deposition of E≥50 keV are dominantly the electron recoils while events with E≤50 keV are neutron recoils. Analysis of real data measured in coincidence between the muon veto and bolometers highlights this fact experimentally [9, 10]. However, the analysis of data suffers from the low coincidence rate, e.g. only several events are detected in 100 live days. This is why a special neutron counter based on a liquid Gd-loaded scintillator was designed and installed. It routinely acquires data since end of 2008 and the count rate of muon-induced neutrons in the order of 1 neutron/day has been achieved [11].

6. Conclusion and outlook

EDELWEISS-2 experiment successfully runs since end of 2007 aiming to reach sensitivity on WIMP-nucleon SI cross-section of 10^{-8} pb in 2011 and 5×10^{-9} pb by 2013. New bolometers with the interleaved electrode design have been developed for direct DM search. Ten of these detectors were continuously operated in LSM during 418 days together with an active muon veto shielding. Preliminary data analysis of the full effective exposure of 322 kg·d shows an improvement in sensitivity of about 30 compared to EDELWEISS-1, e.g. a cross-section of 5×10^{-8} pb is excluded at 90% CL for a WIMP mass of 80 GeV/c^2. The detector development continues towards further improvement and 4 of new *fully interdigitized electrode* bolometers, each with a total mass of 800 g and 80% fiducial volume, were installed in LSM in July 2010. This development together with the detailed studies of the background conditions in LSM, in particular, muon induced neutrons give us the confidence to reach aimed sensitivity and put a good base for a DM experiment of next generation, e.g. EURECA [12], a 1-ton cryogenic detector array.

Acknowledgments

This work is supported in part by the German Research Foundation (DFG) through its collaborative research center SFB-TR27 ("Neutrinos and Beyond"), by the French Agence Nationale pour la Recherche and the Russian Foundation for Basic Research (grant No. 07-02-00355-a).

References

[1] Broniatowski A et al. 2009 *Phys. Lett.* B **681** 305 and references there in (*Preprint* arXiv:0905.0753v1 [astro-ph.IM]).
[2] Armengaud E et al. 2010 *Phys. Lett.* B **687** 294 (*Preprint* arXiv:0912.0805v1 [astro-ph.CO]).
[3] Sanglard V et al. 2005 *Phys. Rev.* D **71** 122002.
[4] Yellin S 2002 *Phys. Rev.* D **66** 032005.
[5] Chabert L PhD thesis (Université Claude Bernard Lyon 1) tel-00007093.
[6] Chardin G, Gerbier G 2003 *Proceedings of the 4th International Workshop on Identification of Dark Matter* (World Scientific, Singapore) Eds N.J. Spooner, V. Kudryavtsev p. 470.
[7] Rozov S et al. *Preprint* arXiv:1001.4383v1 [astro-ph.IM].
[8] Horn M 2007 *Simulations of the muon-induced neutron background of the EDELWEISS-II experiment for Dark Matter search* Ph.D. thesis (Universität Karlsruhe (TH)) and *FZKA scientific report* 7391 (http://bibliothek.fzk.de/zb/berichte/FZKA7391.pdf)
[9] Chantelauze A for the Edelweiss Collaboration 2007 *SUSY2007* Karlsruhe Germany (*Preprint* arXiv:0710.5849v1 [astro-ph]).
[10] Chantelauze A 2009 *Investigation of the muon-induced background of the EDELWEISS-II experiment* Ph.D. thesis (Université Blaise Pascal) Clermont-Ferrand.
[11] Kozlov V et al. 2010 *Astropart. Phys.* **34** 97 (*Preprint* arXiv:1006.3098v1 [astro-ph.IM]).
[12] Kraus H et al. 2007 *Nucl. Phys.* B (Proc. Suppl.) **173** 168.

Journal of Physics: Conference Series **259** (2010) 012038 doi:10.1088/1742-6596/259/1/012038

Prospects of Measuring Lepton CP-violation with LBNE at DUSEL

Jelena Maricic for LBNE Collaboration

Department of Physics, Drexel University, 3141 Chestnut St, Philadelphia, PA, USA

E-mail: jelena@physics.drexel.edu

Abstract. Excellent measurement of the neutrino oscillation parameters achieved in recent years has set the scene for probing the size of the leptonic CP-violation angle. The Long Baseline Neutrino Experiment (LBNE) will have an unprecedented sensitivity to CP-violation angle and a range of other physics questions. LBNE will be a massive neutrino detector located at the Deep Underground Science and Engineering Laboratory (DUSEL) in the Homestake mine in the town of Lead, South Dakota, USA. Two independent detector technologies have been utilized for detector design: water Cherenkov and liquid argon time projection chamber (LArTPC) type of detector and both technologies have similar sensitivity to the main physics questions to be answered. The experiment will measure the value of CP-violation phase δ in lepton sector, ordering of neutrino masses and value of the angle θ_{13}. In addition, the LBNE will be able to search for proton decay, get a detailed energy spectrum in the case of galactic supernovae, detect solar and atmospheric neutrinos, possibly geoneutrinos and even measure the relic supernovae neutrino flux. In order to address most of the aforementioned physics questions, the detector will be placed at large depth of 1480 m (WC). The scientific goals require a minimal size of 2×100 kton equivalent water Cherenkov fiducial volume or 2×17 kton LArTPC, or the combination of the both. The LBNE will obtain 3σ C.L. measurement of θ_{13} at the 0.001 level, for any value of CP-δ phase. In addition LBNE will resolve the neutrino mass hierarchy at 3σ C.L. measurement of the neutrino mass hierarchy if $\sin^2 2\theta_{13} \geq 0.01$ for any value of CP-δ phase and measure about 50% of all CP-δ phases with 3σ C.L. for $\sin^2 2\theta_{13} \geq 0.01$. The experiment will make these measurements using an electron neutrino appearance signal in the muon neutrino beam sent to LBNE from Fermilab, 1300 km away. The beam will be 700 kW and then 2 MW. The experiment will run in both neutrino and anti-neutrino mode. In addition to detectors at DUSEL, a near detector complex at Fermilab is foreseen for beam normalization measurement. The report on the physics reach, design status and current undergoing research and development toward construction of the LBNE.

1. Introduction

There are several important open questions in the neutrino physics today: the value of charge parity (CP) violation phase δ in lepton sector; the ordering of the three neutrino masses (normal mass hierarchy ($m_1 \leq m_2 \leq m_3$ or inverted mass hierarchy $m_3 \leq m_1 \leq m_2$); and the value of the last non-measured neutrino mixing angle θ_{13}. The Long Baseline Neutrino Experiment (LBNE) will have the potential to address these questions and maintain a broad physics program at the same time - search for the proton decay, galactic supernovae detailed neutrino spectrum measurement, relic supernovae neutrinos, solar and atmospheric neutrinos and possibly geoneutrinos. The LBNE detectors will be located at the Deep Underground Science and Engineering Laboratory (DUSEL) in the Homestake mine in the town of Lead,

South Dakota, USA. Figure 1 shows the overview of the water Cherenkov detector. Strong neutrino beam will be sent from Fermi National Laboratory, 1300 km away (see Fig. 2). The experiment will measure the flux of electron (anti-)neutrino appearance signal in the muon (anti-)neutrino beam in order to search for CP-δ phase, θ_{13} and mass hierarchy.

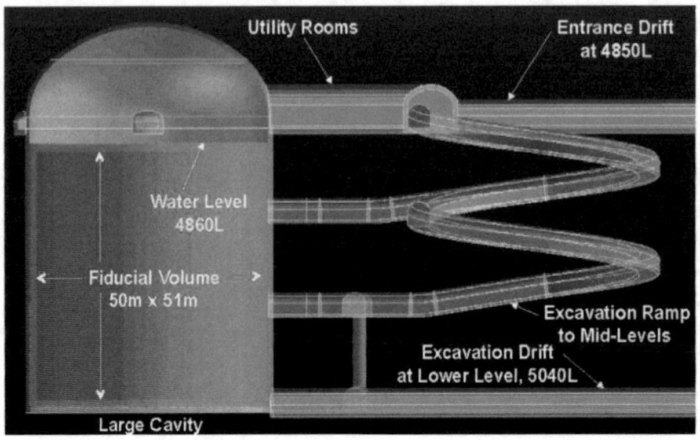

Figure 1. Outline of a 100 kton water Cerenkov detector module. Up to three modules are planned to be installed at DUSEL.

Figure 2. Muon neutrino beam will be sent to LBNE from Fermi National Laboratory, 1300 km away.

2. Scientific potential

2.1. Beam neutrino physics

Using a strong muon neutrino beam from Fermi National Laboratory, appearance search for electron neutrinos will be conducted at the neutrino detector in Homestake mine, to search for the value of CP violation phase, ordering of neutrino masses and value of the the neutrino oscillation angle θ_{13}. Evaluation of the physics prospects has been conducted for a wide band beam with initial proton beam energy 60 to 120 GeV (with 700 kW and 2 MW power) that produces neutrinos in 1-10 GeV range. Wide band beam, at the 1300 km baseline allows observation of two oscillation maxima at 0.8 GeV and 2.4 GeV as shown in Fig. . The physics studies have been conducted by the LBNE physics working group [1] for 700 kW and 2 MW beam. Fig. 4 shows muon neutrino charged current signal in the case of normal and inverted hierarchy. Weaker signal in the case of inverted hierarchy drives the beam intensity requirements. The studies of the sensitivity to δ-CP, mass hierarchy and θ_{13} have been performed for two different

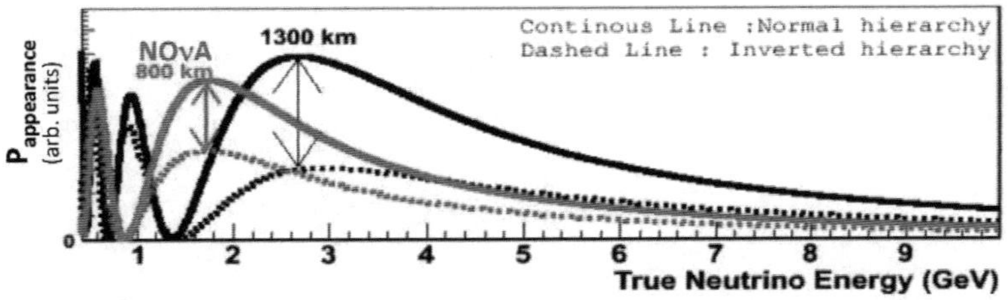

Figure 3. LBNE will utilize a wide band beam from Fermilab. A long baseline allows observation of two distinct oscillation maxima in comparison to the shorter baselines (signal at a distance of NOνA experiment shown).

beam configuration in which the latter has a slightly narrower peak shifted to higher energy and both beam options give similar sensitivity.. Results are shown in Fig. 6, Fig. 5 and Fig. 7.

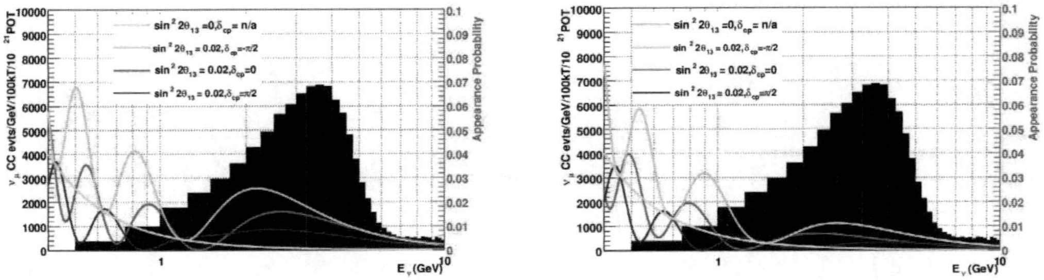

Figure 4. Muon neutrino spectrum at LBNE baseline for different neutrino mixing parameter values. The left figure corresponds to normal hierarchy and $\Delta m_{32} = 2.5 \times 10{-}3$ eV^2, while the right figure corresponds to inverted hierarchy case and $\Delta m_{32} = -2.5 \times 10{-}3$ eV^2. The original muon neutrino spectrum sent from Fermilab is shown as black histogram.

2.2. Nucleon decay

Grand Unified theory predicts proton decay, but this process has not been observed yet. A number of theoretical models predict proton decay with a lifetime in the range of 10^{35-36} years. There are a lot of possible branches for the proton decay with the modes $p \rightarrow e^+ + \pi^0$ and $p \rightarrow K^+ + \nu$ being common benchmarks. Based on the experience from SuperKamiokande detector, 20 years of running with 300 kton water Cherenkov detector will result in 10^{35} years lifetime limit at 90% C.L. for $p \rightarrow e^+ + \pi^0$ decay mode. The atmospheric neutrinos represent concurrent background to kaon decay at rest, important in the mode $p \rightarrow K^+ + \nu$. As a result, for a 20 year long run, expected sensitivity is in the range $1 - 2 \times 10^{34}$ years long proton lifetime in this decay mode.

2.3. Cosmological supernovae

Universe is permeated by the diffuse neutrino flux from all previous supernovae and can be detected on Earth with neutrino detectors. Learning more about this flux and spectrum of relic supernovae neutrinos will provide insight in the star formation mechanism and era when they were formed. Supernovae relic neutrinos are detected in water Cherenkov via inverse beta decay interaction on protons, resulting in positron and neutron capture on proton in the energy

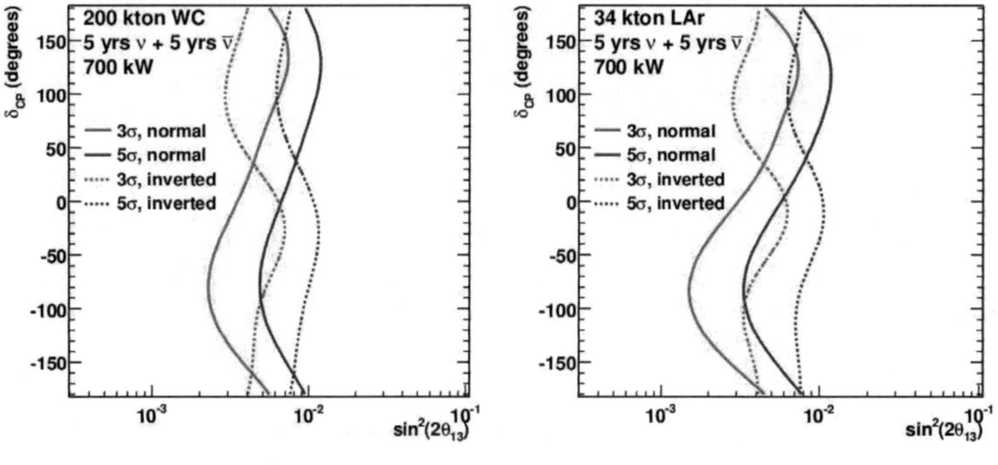

Figure 5. 3σ (red) and 5σ (blue) sensitivity of LBNE to $\sin\theta_{13}$ as a function of δ-CP for 200 kt of WC (left) and 34 kt of LAr (right). This assumes 5+5 years of and running in a 700 kW beam. Curves are shown for both normal (solid) and inverted (dashed) mass hierarchies.

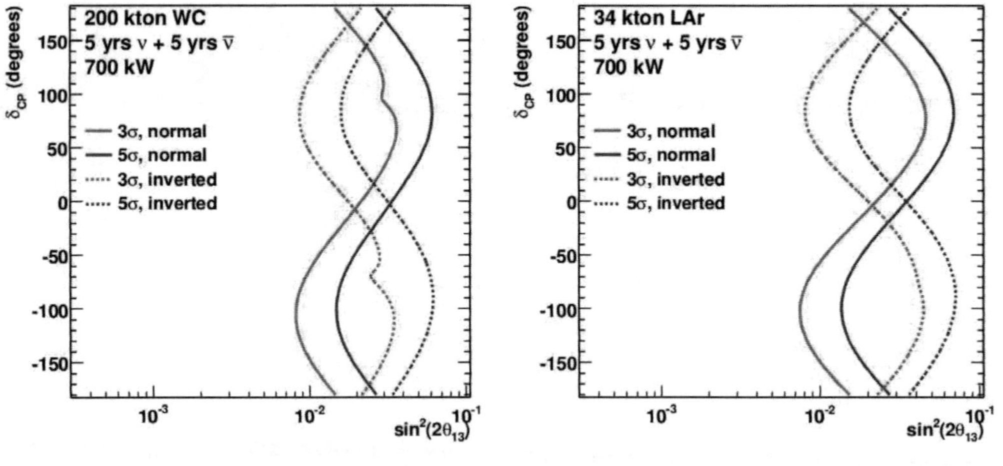

Figure 6. Resolution of the mass hierarchy for 200 kt of WC (left) and 34 kt of LAr (right). This assumes 5+5 years of ν and $\bar{\nu}$ running in a 700 kW beam. To the right of the curves, the normal (solid) or inverted (dashed) mass hierarchy can be excluded at the 3σ (red) or 5σ (blue) level for the indicated values of true $\sin^2 2\theta_{13}$ and δ-CP.

range between 10 MeV (above reactor neutrino energy) and below 100 MeV (atmospheric anti-neutrino background). Above 20 MeV, the dominant background is due to atmospheric neutrino interactions. Water Cherenkov detector has no means of tagging γ photons from the neutron capture due to its low energy of 2.2 MeV. However, if the the detector is loaded with Gd, then the γ's emitted in neutron capture are at the level of 8 MeV and can be tagged successfully. A 200 kton Gd loaded water Cherenkov detector should be able to unambiguously detect relic supernovae neutrino flux.

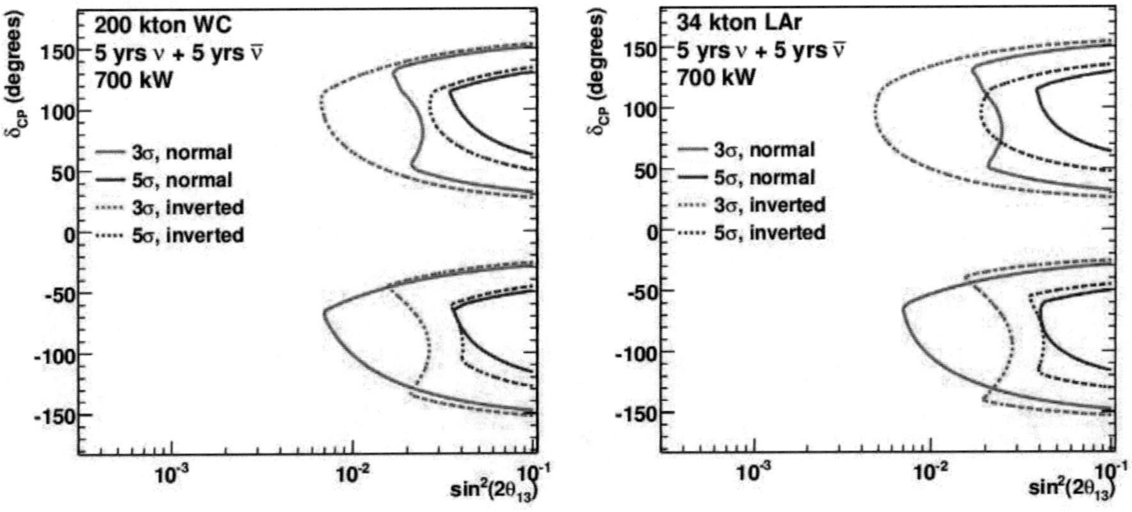

Figure 7. 3σ (red) and 5σ (blue) sensitivity of LBNE to CP violation for 200 kt of WC (left) and 34 kt of LAr (right). This assumes 5+5 years of ν and $\bar{\nu}$ running in a 700 kW beam. Curves are shown for both normal (solid) and inverted (dashed) mass hierarchies.

2.4. Galactic Supernovae and Solar neutrinos

Galactic supernova will produce a strong signal in the LBNE neutrino detector and provide a detailed temporal and spectral profile of detected neutrinos. This information will lead to better understanding of core collapse, accretion, neutron star cooling and possible transition to quark matter or black hole, tests of exotic physics and measurement of neutrino oscillation parameters including mass hierarchy and θ_{13}. Such a large neutrino detector will collect the largest number of solar and atmospheric neutrinos in the energy range from 5 M eV to 1 T eV.

3. Experimental design

The LBNE consists of several integral components that are required to reach the physics goals and these are: neutrino detector at DUSEL inside the Homestake mine, neutrino beam that will be built at Fermi lab and near detector that will be built adjacent to the neutrino beam to measure detailed beam profile, normalize neutrino flux and monitor neutrino beam stability. In addition, near detector will perform a whole plethora of short baseline neutrino studies and accuretly measure neutrino interaction cross-sections for various channels. Two complementary technologies are planned for neutrino detector in South Dakota. The first one is well understood and proven to work, water Cherenkov detector technology (as used in SuperKamiokande and SNO experiments) and the second one is Liquid Argonne detector technology that is promising, but requires a lot of research and development before a large scale detector (of the order of 20 kton) can be built. Water Cherenkov detector (Fig. 1) will consist of two 100 kton fiducial volume modules and they are envisioned as right cylinders 54 m high and 53 m in diameter. Approximately 50,000 high quantum ten inch PMTs are needed per module to get 30% photocoverage and achieve high sensitivity to the wide LBNE physics program.

4. Status and Prospects

Cavern excavation studies and cost estimates have been performed to prepare a cavern at 1479 m depth for the water Cherenkov and liquid Argonne detectors. A vigorous research and development effort in the USA is underway to produce conceptual design for LBNE by the end of 2010, followed by the full technical design. The plan is to start building the detector by 2013-2014 fiscal year. Construction time is foreseen to last 5 years for data taking to start in 2018-2019. LBNE has a strong discovery potential and will provide answers to a number of open questions in particle and astroparticle physics.

5. References

[1] A. Beck *et al* (LBNE Physics Working Group), presented at INT workshop, Seattle, USA, August 9-11, 2010

Journal of Physics: Conference Series **259** (2010) 012039 doi:10.1088/1742-6596/259/1/012039

Search for neutrino-less double beta decay with EXO

Razvan Gornea on behalf of the EXO collaboration

LHEP, Albert Einstein Center for Fundamental Physics, Bern University, Switzerland

E-mail: gornea@gmail.com

Abstract. Neutrino oscillation experiments have shown that neutrinos have very small but non vanishing masses. These experiments however are not able to determine neither the absolute mass scale of neutrinos nor whether they are two-component Majorana particles, i.e. their own antiparticles. Neutrino-less double beta decay can only occur if the neutrinos are Majorana particles, a preferred scenario in most possible schemes leading to finite masses. Among several viable candidate isotopes, EXO has chosen Xe-136 to search for this decay. Its main advantage is that the final state, i.e. the barium ion, can be tagged using optical spectroscopy. The detection of the double beta decay daughter nucleus can be the key to a background free measurement of such a rare process. An intermediate size detector (EXO-200) of 200 kg enriched xenon (80% Xe-136) is about to take data at the WIPP underground site in New Mexico. A ton-scale experiment is being designed with Ba ion tagging capability. EXO-full will detect, in addition to the two electrons, the coincident appearance of a barium ion. This improved event signature is expected to provide total elimination of the background from radioactive impurities.

1. Introduction

The double beta decay has the potential to reveal new information about the nature and properties of neutrinos [1]. This rare nuclear transition can proceed through multiple channels: with the emission of two electrons and two anti-neutrinos ($2\nu\beta\beta$, allowed in the standard model), without the release of neutrinos ($0\nu\beta\beta$) or accompanied by the radiation of a light neutral fermion. The half-life of the neutrino-less decay $0\nu\beta\beta$ depends on the effective neutrino mass, a linear combination of the three neutrino masses, and therefore its measurement provides additional information about the neutrinos masses which cannot be established from neutrino oscillation experiments [2, 3]. The observation of this exotic process would indicate that the neutrino is a Majorana particle. The spectrum of the energy carried by the two electrons is very different for the various decay channels. For $0\nu\beta\beta$ this is simply a peak at the Q value of the decay. Therefore, events from each channel can be discriminated even when using detectors which are not sensitive to the topology of electron tracks.

The EXO R&D program aims for the development of a ton-scale, ultra low background TPC filled with xenon enriched in Xe-136 [4]. The collaboration has procured 200 kg of 80% enriched xenon that will be employed in a cryogenic liquid TPC equipped with both charge and scintillation light readout for a first phase called EXO-200. A high-pressure gaseous TPC is still under consideration for the next phase because it offers better spatial resolution facilitating efficient event selection using the topology of electron tracks. Future detection schemes will involve identification of the final state (i.e. Ba^+ tagging) which enables complete rejection of the radioactive background.

The EXO-200 project will test the feasibility of the liquid option. Considerable efforts have been made to design and build a detector with very low radioactive background. All the materials and components used for EXO-200 have been screened. The design priority was minimizing the amount of the materials surrounding the TPC. EXO-200 does not include final state tagging as this technique is still under development. EXO-200 will also provide interesting physics results. It will explore the quasi-degenerate neutrino mass scheme. The two-neutrino double beta decay has never been observed in xenon, the best limit gives $T_{1/2} > 10^{22}$ years [5]. EXO-200 should be able to measure this channel if current theoretical predictions, which anticipate half-live values not far away from the experimental limit, are accurate. Meanwhile, the collaboration is intensively investigating various techniques for the transport and tagging of the barium ion.

The expected performance of EXO-200 prototype is given in table 1 along with the projections for the future phases for which final state tagging is expected to provide complete radioactive background rejection.

Table 1. Projected performance for EXO-200 and following phases. The detection efficiency is mostly dominated by the fiducial cut. For EXO-200 the background is dominated by the contribution of the residual activity. For future phases, the expected background is exclusively from the two-neutrino channel and is a function of the energy resolution. Calculations by [6, 7] used for nuclear matrix elements.

Detector name	Mass [ton]	Efficiency [%]	Run time [year]	$\Delta E(Q)$ [%]	Background [events]	$T_{1/2}$ limit [year]	$m_{\beta\beta}$ limit [meV]
EXO-200	0.2	70	2	1.6	40	6.4×10^{25}	130
EXO-full	1	70	5	1.6	0.5	2.0×10^{27}	24
EXO-full	10	70	10	1.0	0.7	4.1×10^{28}	5.3

2. EXO-200 preparation

The reduction of the background due to the residual radioactivity (especially from K-40 and the isotopes of the U and Th chains) is necessary for the observation of the double beta decay. Ultimately, the allowed channel becomes a background for the $0\nu\beta\beta$ and therefore improving the energy resolution of the detector is an important goal.

2.1. Radioactive background survey

The collaboration set up an intensive measurement campaign to determine the residual radioactive contamination of many materials and components considered for the detector construction [8]. The K, U and Th concentrations of more than 350 materials have been measured. A database containing the results for 225 interesting candidates has been made available to the community developing detectors with similar low background requirements. Various methods have been employed for this survey: standard mass spectrometry (MS), glow discharge MS (GD-MS), inductively coupled plasma MS (ICP-MS), neutron activation analysis (NAA), alpha and gamma counting. To reach optimal sensitivity, each method imposes particular constraints on sample preparation. Direct gamma counting offers the best sensitivity to cost ratio for large mass samples whereas ICP-MS performs the best for small mass samples that are chemically compatible with acid based preconcentration methods. A Monte Carlo simulation that includes the geometry of the EXO-200 detector, the measured activities and event selection algorithms has been developed. It enabled us to predict the background induced by the residual contamination.

2.2. Energy resolution study

The collaboration has studied the energy resolution achievable in liquid xenon using a small test cell equipped with a UV sensitive PMT and a charge readout system [9]. The measurements with a thin Bi-207 source indicate that ionization and scintillation signals in liquid xenon are anti-correlated as illustrated in figure 1. The cell has been operated at multiple drift voltages and the results point out that this effect is due to the dependence of the scintillation yield on the amount of charge recombination. Therefore, simultaneous measurements of light and charge provide better energy resolution than any single approach as it is shown in figure 2 (σ_{min} is the minor axis of the 2-D peak seen in figure 1). The energy resolution at the Q value (2458 keV [10]) for the decay of Xe-136 is expected to be 1.6%. Careful considerations have been applied during EXO-200 design to insure optimal scintillation and charge collection readout.

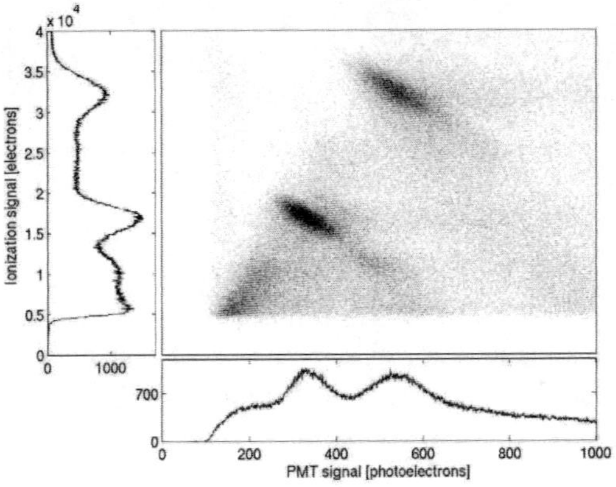

Figure 1. Two-dimensional scintillation and ionization spectrum (4 kV/cm drift field).

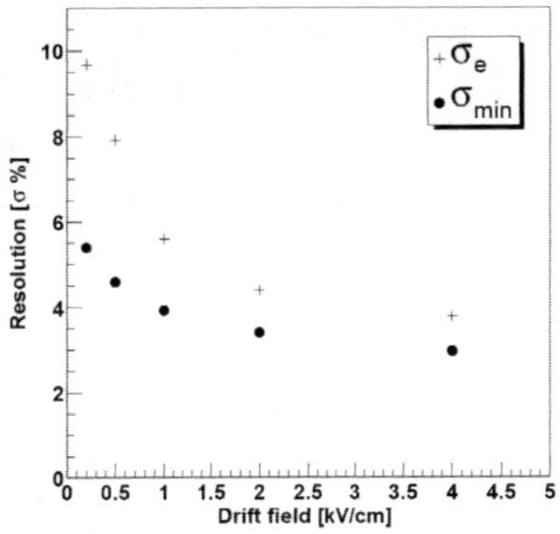

Figure 2. Energy resolution for the ionization channel alone compared to the minimal value obtained from a fit to the two-dimensional spectrum (570 keV peak).

3. EXO-200 detector

3.1. EXO-200 chamber

The TPC has a cylindrical shape (40 cm diameter and 35 cm length) and is segmented in two zones by a central photo-etched cathode made of phosphor bronze. Both regions are equipped with induction and charge collection wire grids for ionization measurement and tracking, and with LAAPD (Large-Area Avalanche Photodiode) planes for the collection of scintillation light (peaked at 174 nm). Each LAAPD plane contains 250 UV sensors (with $QE > 1$ at 174 nm) which have an active diameter of 1.6 cm and are operated at a voltage around 1400 V for a gain of 150×. Lateral Teflon sheets serve as ultraviolet reflectors. Figure 3 shows a diagram and a 3-D drawing of the TPC. The chamber is made from ultra-low radioactivity copper by clean welding (i.e. e-beam and TIG welding) of 1.5 mm thick rings, performed in a controlled environment.

3.2. EXO-200 cryostat

The TPC is immersed in a refrigeration cryostat filled with 4.2 tons of high purity heat transfer fluid. Many candidates have been evaluated and HFE-7000 was chosen primarily because it has the lowest residual radioactive contamination of all [8]. This fluid serves as the inner gamma ray and neutron shield and as a thermal bath that maintains the chamber temperature uniform.

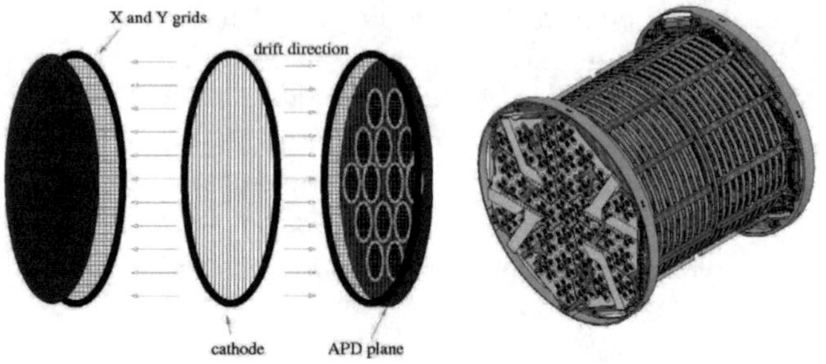

Figure 3. TPC diagram and 3-D drawing.

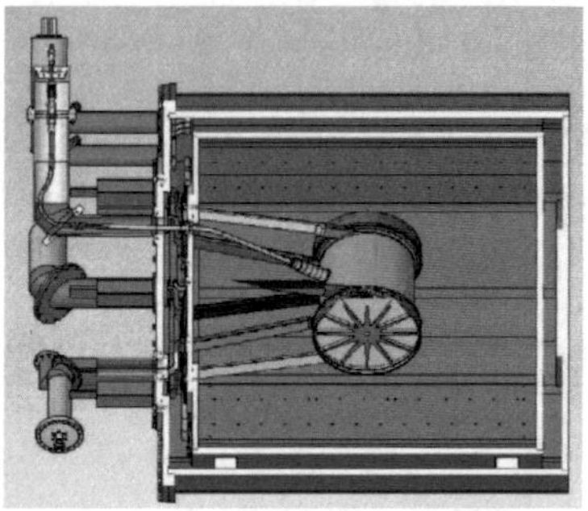

Figure 4. Diagram of the EXO-200 cryostat.

The cryostat has a cylindrical shape (1.5 m diameter and 1.5 m length) and it is also made of ultra-low activity copper. A 3-D drawing is shown in figure 4. This cryogenic scheme has been fully commissioned including the xenon and HFE handling systems.

3.3. EXO-200 installation

The WIPP underground experimental area is located in a salt mine at a depth of 665 m. EXO has been allocated a vast space provided with utility services. The collaboration installed a series of clean room modules supported by adjustable pillars in which the detector and its support infrastructure are enclosed. Figure 5 shows pictures of the EXO-200 underground setup. A muon veto composed of plastic scintillator panels (20 units of 65×315 cm^2 and 11 units of 65×375 cm^2) covers the module that hosts the detector. Monte Carlo simulations have been used to optimize the configuration of these panels to obtain a high tagging efficiency (i.e. 99.7% resulting in the reduction of muon related background by a factor of 20×). Data taking using natural xenon will start in November 2010. These test runs will commission the detector and its data acquisition system. Operation with enriched xenon is expected for the first half of 2011.

Figure 5. EXO-200 underground experimental facility. 180° panoramic view inside the clean room (top) and outside in the drift (bottom).

4. Final state R&D

Precise ion location in the drift volume is important for an efficient final state detection. Therefore, the barium ion drift properties in liquid (for the baseline option) and in gas (for the gaseous option) xenon must be investigated. The reliable ion extraction and transport to the tagging apparatus is essential for a practical device. The Ba^+ ion can be tagged in a RF trap using resonant light scattering. A 493.41 nm and a 649.69 nm laser are both required for resonant light scattering. The readout is performed in the blue frequency range and the red laser is used as a switch. In-situ tagging techniques are also explored.

4.1. Barium tagging

The double beta decay of xenon produces Ba^{++} ions and, for the liquid phase, charge reduction to Ba^+ is expected. For the gas phase, various additives can be employed with xenon to obtain singly ionized barium. Laser based Ba^+ tagging has been accomplished in a RF trap using low pressure helium cooling ($P = 10^{-3}$ torr) [11, 12]. Figure 6 illustrates the excellent resolving power of this method for counting ions.

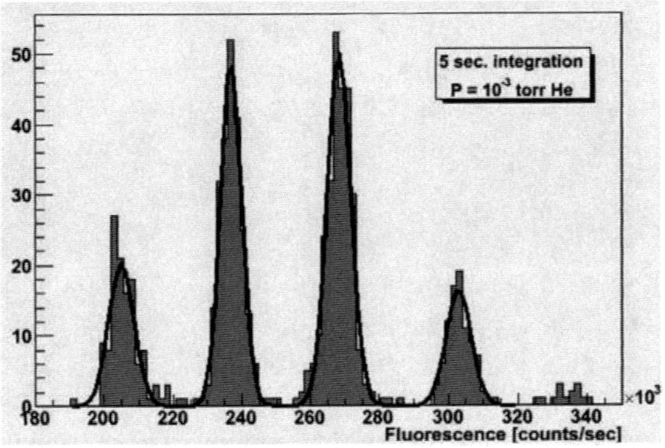

Figure 6. Histogram of the light scattered from barium ions (5 s time slices). The first peak is associated with the background and the following peaks correspond respectively to 1, 2 and 3 ions trapped.

4.2. Barium ion extraction

A cryostat able to cool 100 kg of liquid xenon has been developed to test the insertion of the collection probe and to study its effects on the TPC performance. The test cryostat remains to be commissioned for operation with argon and xenon. COMSOL based simulations have been conducted to determine the design of the cathode and the tip. A mechanical displacement device and the adapted TPC will be built and then integrated with the cryostat. Measurements with gamma sources and a custom barium ion source [14] will establish the optimal procedure for barium ion extraction. Later, this setup will be combined with either a cryogenic collection tip or a tip providing a surface with high ablation resistance and the efficiency for collecting barium ions will be measured.

4.3. Ion collection probe

Several techniques are being investigated. The concept of a cryogenic collection probe revolves around the idea of trapping the barium ion in a few layers of xenon to secure its transport and delivery to the analysis chamber [13]. Such a tip requires liquid He cooling to reach a sufficiently low temperature ($\approx$ 40 K) to inhibit xenon sublimation in vacuum. It is also possible to adsorb the ion on a clean surface with the disadvantage of neutralization. A promising technique, resonant ionization spectroscopy (RIS), has been explored for pyrolytic graphite and silicon. In this case, a desorption laser is used in combination with RIS lasers of specific frequency that selectively ionize the barium atom. The 553.5 nm and 389.7 nm lasers are tuned to barium transitions and are able to pump the atom to a highly excited state from which it decays to an ionized state. Figure 7 shows the time of flight spectrum of ions ejected by a desorption laser from a silicon surface with a thin barium layer produced by sputtering. When the RIS lasers are also employed, barium ions greatly dominate the spectrum. Presently, an efficiency of 0.1% has been reached and a significant improvement is expected for the next generation setup targeting single ion detection.

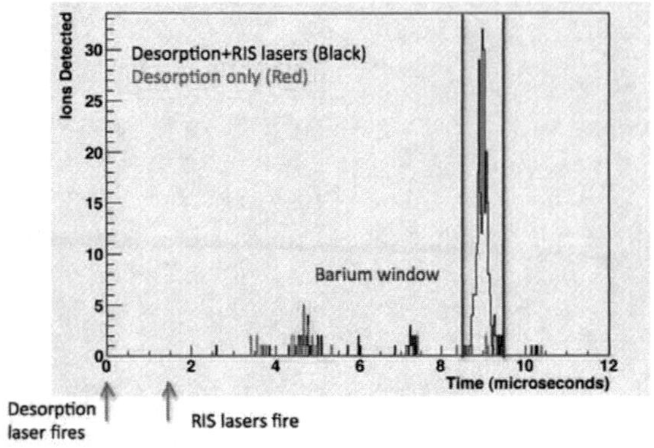

Figure 7. Time of flight spectrum for desorbed ions from a silicon surface with a thin layer of barium. The red curve illustrates the signal when only the desorption laser is employed. When RIS lasers, tuned to barium transitions, are also used then a large number of barium ions is observed as indicated by the black curve.

4.4. In-situ tagging

The collaboration is also exploring the feasibility of in-situ tagging by combining the cryogenic probe concept with in-situ laser irradiation. Spectral broadening effects occur in matter and therefore only a weak fluorescence signal is obtained (weak in respect to resonant light scattering in vacuum). In liquid, tests demonstrated sensitivity to 10^5 barium atoms. Large improvements in the laser intensity and in the efficiency of the fluorescence light collection are needed to reach single atom detection. In gas, EXO is also investigating the possibility of ion transport through a combination of electric drift fields and RF carpets.

5. Conclusion

EXO has finalized the commissioning of EXO-200 and data taking with natural xenon will start in November 2010. The enriched xenon will be used in 2011 for physics runs. The EXO-200 detector will provide valuable data for future phases and hopefully interesting physics results. Various techniques for final state tagging are explored in preparation for the ton-scale detector.

References

[1] P. Vogel 2007 *Nucl. Phys. B (Proc. Suppl.)* **168** 23
[2] Giunti C 2000 *Nucl. Phys.* B **87** 1-3 324-6
[3] Amsler C et al. 2002 *Phys. Lett.* B **545** 57-61
[4] Akimov D et al. 2005 *Nucl. Phys.* B **138** 224-6
[5] R. Bernabei et al. 2002 *Phys. Lett.* B **546** 23-28
[6] Rodin V A et al. 2007 *Nucl. Phys.* A **793** 1-4 213-5
[7] Caurier E et al. 2008 *Phys. Rev. Lett.* **100** 052503
[8] Leonard D S et al. 2008 *Nucl. Instrum. Meth.* A **591** 490-509
[9] Conti E et al. 2003 *Phys. Rev.* B **68** 054201
[10] Redshaw M et al. 2007 *Phys. Rev. Lett.* **98** 053003
[11] Green M et al. 2007 *Phys. Rev.* A **76** 023404
[12] Flatt B et al. 2007 *Nucl. Instrum. Meth.* A **578** 399-408
[13] Fierlinger P et al. 2008 *Rev. Sci. Instrum.* **79** 045101
[14] Montero Diez M et al. 2010 *to be published* **arXiv** 1008.3422

Journal of Physics: Conference Series **259** (2010) 012040 doi:10.1088/1742-6596/259/1/012040

The NEXT experiment: neutrinoless double beta decay searches at the LSC

Justo Martín-Albo, on behalf of the NEXT Collaboration
Instituto de Física Corpuscular (IFIC), CSIC & Univ. de Valencia, Valencia, Spain
E-mail: justo.martin-albo@ific.uv.es

Abstract. NEXT is a proposed 100-kg high-pressure xenon gas TPC that will search for neutrinoless double beta decay in ^{136}Xe. Such a detector, thanks to its excellent energy resolution and its powerful background rejection provided by the distinct double beta decay topological signature, may become one of the leading experiments of the field. The project is proceeding through a fast research and development phase. The first prototypes, containing about 1 kg of pressurized xenon will be taking data in 2010, and the final detector, NEXT-100, is planned to run in late 2013 at the Laboratorio Subterráneo de Canfranc (LSC), Spain.

1. Introduction

Neutrinos, unlike the other Standard Model fermions, could be Majorana particles, that is, truly neutral particles indistinguishable from their antiparticles. The existence of Majorana neutrinos may have very deep physics implications. For example, they provide a natural explanation for the smallness of neutrino masses, the so-called *seesaw mechanism* [1–4]. Furthermore, Majorana neutrinos, through *leptogenesis*, could have induced the baryon asymmetry of the Universe [5].

The only practical way to establish experimentally that neutrinos are their own antiparticles is the detection of neutrinoless double beta decay ($\beta\beta0\nu$). This is a hypothetical, very slow nuclear transition in which a nucleus with Z protons decays into a nucleus with $Z+2$ protons and the same mass number A, emitting two electrons that carry essentially all the energy released ($Q_{\beta\beta}$). Such a process occurs if and only if neutrinos are massive, Majorana particles [6,7]. Also, $\beta\beta0\nu$ violates lepton number conservation, and is therefore forbidden in the Standard Model.

Any source of lepton number violation (LNV) can induce $\beta\beta0\nu$ and contribute to its amplitude. If we assume, nevertheless, that the dominant LNV at low energies is the Majorana mass term for the neutrino, the rate of $\beta\beta0\nu$ is proportional to the *effective Majorana mass* of the ν_e:

$$m_{\beta\beta} = \Big|\sum_i m_i\, U_{ei}^2\Big|, \tag{1}$$

where m_i are the neutrino mass eigenstates and U_{ei} are elements of the neutrino mixing matrix. Therefore, a measurement of the $\beta\beta0\nu$ decay rate would provide direct information on neutrino masses [8].

The most sensitive limit to date was set by the Heidelberg-Moscow (HM) experiment [9]: $T^{0\nu}_{1/2}(^{76}\mathrm{Ge}) \geq 1.9\times10^{25}$ years. A subset of this collaboration claimed to observe evidence for a $\beta\beta0\nu$ signal, with a best value for the half-life of 1.5×10^{25} years (95% CL), resulting in an

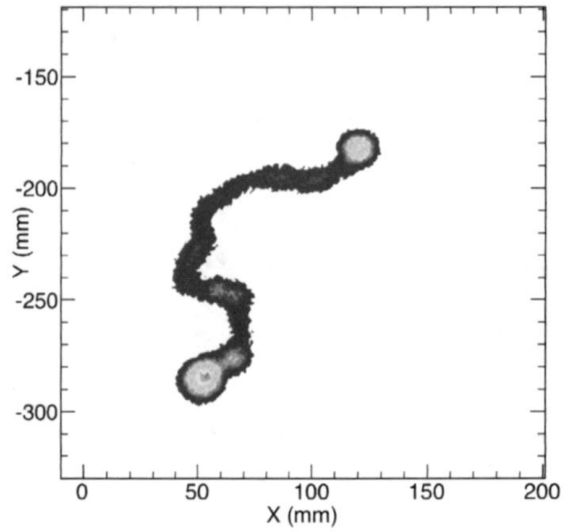

Figure 1. Simulation of a neutrinoless double beta decay event in HPXe (10 bar). The ionization track, tortuous due to the multiple scattering, ends in two *blobs*. This signature can be exploited to reject backgrounds.

effective Majorana mass of 0.39 eV [10]. The claim has been subject of severe criticism [11]. Nevertheless, this controversial result, together with the observation of neutrino oscillations [12] (which proves that neutrinos have a non-null mass, an essential condition for $\beta\beta0\nu$ to exist) have revived the interest in neutrinoless double beta decay searches, and several new experiments with sensitivity to $m_{\beta\beta}$ up to ~100 meV are currently under design and construction.

2. High-pressure xenon for $\beta\beta0\nu$ searches

Double beta decay experiments are, in general, calorimeters searching for a small peak at $Q_{\beta\beta}$ in their energy spectrum. This is a challenging task: backgrounds from natural radioactivity are abundant in this region of energies and can easily overwhelm the signal peak. Good energy resolution is therefore very important, because increases the signal-to-noise ratio, but it is not enough by itself. Indeed, experiments based on germanium diodes (such as HM), devices with excellent energy resolution, have been background-limited. Consequently, all new-generation experiments try to exploit other signatures, such as event reconstruction, to further discriminate between signal and background.

A high-pressure xenon gas (HPXe) TPC can provide both good energy resolution and event topological information for $\beta\beta0\nu$ searches. Double beta decay events have a distinctive topological signature in HPXe that can be used to reject backgrounds: a ionization track, of about 30 cm long at 10 bar, tortuous due to the multiple scattering, an with larger deposition (*blobs*) in both ends (see Figure 1). The Gotthard experiment [13], consisting in a small xenon TPC (~5 kg) operated at 5 bars, proved the utility of this signature, achieving an impressively low background rate of about 0.01 counts/keV/kg/year. However, the detector suffered of a modest energy resolution, 6.6% FWHM at $Q_{\beta\beta}$, probably due to the the use of conventional avalanche amplification in a wire plane, and to the addition of methane (4%) to the xenon (in order to increase the drift velocity and to suppress diffusion), that quenched the primary signals.

Measurements in other small HPXe systems [14, 15] have shown that optimal energy resolution, <0.5% FWHM at $Q_{\beta\beta}$, is possible using *electroluminescence* (EL) for the amplification of the signals: the charges from primary ionization are accelerated by a moderate electric field (~3–4 kV/cm/bar), producing a proportional emission of UV light with sub-

Journal of Physics: Conference Series **259** (2010) 012040 doi:10.1088/1742-6596/259/1/012040

poissonian fluctuations. This performance seems independent of the gas pressure below some 50 bar.

Additionally, good energy resolution can also be obtained at low pressure using new-generation avalanche amplification devices such as microbulk Micromegas [16, 17].

Xenon is the only noble gas that has a $\beta\beta$-decaying isotope, ^{136}Xe. Its $Q_{\beta\beta}$-value is high enough (2458 keV) to be used in a $\beta\beta0\nu$ experiment. The natural abundance of the isotope is 9%, but it can be enriched by centrifugation at a reasonable cost. In addition, xenon does not have any other long-lived radioactive isotopes, and being a noble gas, it can be easily purified.

3. The NEXT experiment

The Neutrino Experiment with a Xenon TPC (NEXT) [18] will search for $\beta\beta0\nu$ in ^{136}Xe using a 100 kg HPXe electroluminescent TPC. The project is approved for operation in the Canfranc Underground Laboratory (LSC), Spain. Following ideas in [19], the TPC will have separated readout systems for calorimetry and tracking to facilitate both measurements.

The detection process is as follows. Particles interacting in the HPXe transfer their energy to the medium through ionization and excitation. The excitation energy is manifested in the prompt emission of VUV ($\sim$175 nm) scintillation light. The ionization tracks (positive ions and free electrons) left behind by the particle are prevented from recombination by a strong electric field. Negative charge carriers drift then toward the TPC anode, entering a region with an even more intense electric field. There, further VUV photons are generated isotropically by the EL process. Therefore, both scintillation and ionization produce an optical signal, to be detected with a photosensors array (PMTs) located behind the cathode or in the chamber sides. The detection of the primary scintillation light constitutes the start-of-event (t_0), whereas the detection of EL light provides an energy measurement. EL light can be used also for tracking by detecting it with an array of photosensors (MPPCs, for example) located behind the anode. An alternative to this may be the use of microbulk Micromegas for both the energy measurement and the tracking.

The experiment is proceeding through a fast research and development phase. Currently, several prototypes of moderate size are being analyzed and will provide us with a better understanding of the technology. By early 2011, the NEXT Collaboration will decide on the definitive design of the NEXT-100 detector, to be installed in the Laboratorio Subterráneo de Canfranc (Spain) in 2013.

4. Backgrounds and sensitivity of NEXT

The importance of a background source in NEXT depends on the energy resolution and also on the track identification capabilities of the detector. Besides, NEXT is a fully active detector: background events with charged particles entering the active volume can be rejected defining a fiducial volume few centimeters away from the chamber walls. The reliability of such a veto depends on the accuracy of the t_0 measurement. Only high-energy photons can create signal-like tracks (through Compton interactions, pair-creation and photoelectric absorption) far from the detector walls .

The $\beta\beta0\nu$ peak of ^{136}Xe is located in the energy region of the naturally-occurring radioactive processes. Although the half-life of the parents of the natural decay chains is comparable to the age of the universe, it is very short compared to the desired half-life sensitivity of the new $\beta\beta0\nu$ experiments (10^{26} years). For that reason, even a small quantity of these nuclides creates significant event rates. Only those processes with energy around or above $Q_{\beta\beta}$ and able to mimic a signal track can become a background. In our case, the dangerous radioactive isotopes are ^{208}Tl and ^{214}Bi, from the thorium and uranium series, respectively.

The daughter of ^{208}Tl emits a de-excitation photon of 2614 keV with a 100% intensity. The Compton edge of this gamma is at 2382 keV, well below $Q_{\beta\beta}$. However, the scattered gamma

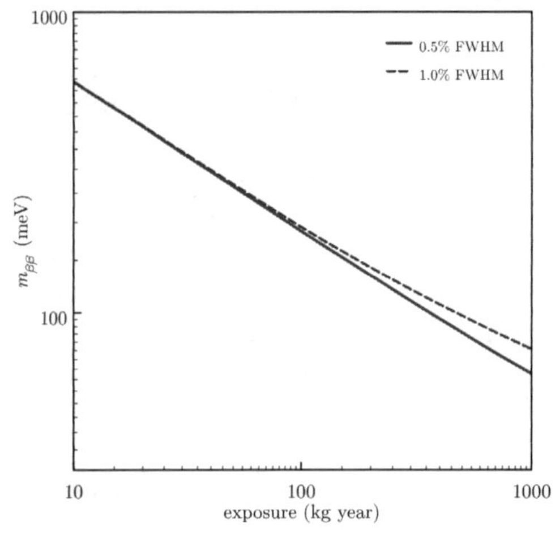

Figure 2. Sensitivity of NEXT-100 to $m_{\beta\beta}$ for an estimated level of background in the ROI of 0.4 counts/keV/ton/year.

can interact and produce other electron tracks close enough to the initial Compton-electron so they are reconstructed as a single object falling in the energy Region of Interest (ROI). Pair-creation events are not able to produce single-track events in the ROI. Photoelectric electrons are produced above our ROI but can loose energy via bremsstrahlung and populate the window, in case the emitted photons escape out of the detector.

After the decay of ^{214}Bi, its daughter emits a number of de-excitation gammas with energies above 2.3 MeV. The gamma line at 2447 keV (intensity: 1.57%) is very close to $Q_{\beta\beta}$. Energy resolution is essential to separate it from the signal peak. The gamma lines above $Q_{\beta\beta}$ have low intensity (below 0.1%), but their Compton spectra can produce background tracks in the ROI.

All materials contain ^{208}Tl and ^{214}Bi impurities in a given amount. Careful selection of radio-pure components —some of them with activities as low as 1 μBq/kg — and purification of the xenon is mandatory.

In addition, one has to worry about radon gas: either ^{222}Rn (half-life of 3.8 d) from the ^{238}U chain or ^{220}Rn (half-life of 55 s) from the ^{232}Th chain. As a gas, it diffuses into the air and produces background in the detector proximities. This contamination can be translated into ^{214}Bi, for the ^{222}Rn, or into ^{208}Tl, for the ^{220}Rn, their only decay products affecting the experiment. Radon may be eliminated from the TPC volume by recirculation of the xenon through appropriate filters. Also, some underground laboratories have installed charcoal scrubbers into the airstream to remove radon from the detector proximities.

Cosmic particles can also affect our experiment by producing high energy photons or activating materials. This is the reason why double beta decay experiments are conducted deep underground. At these depths, muons are the only surviving cosmic ray particles, but their interactions with the rock produce neutrons and electromagnetic showers. Active or passive shieldings will be used to minimize this background contribution. The NEXT Collaboration is considering two possible approaches for the shielding: a lead castle or a water tank.

Taking into account background estimates, signal detection efficiency and background rejection factors, we have evaluated the sensitivity of the NEXT experiment to $m_{\beta\beta}$. The result is shown in Figure 2. In a few years of data-taking, NEXT will be able to fully explore the degenerate hierarchy of neutrino masses, possibly outperforming most of the existing proposals.

Acknowledgments
This work was supported by the Spanish Ministerio de Ciencia e Innovación under grants Consolider-Ingenio 2010 CSD2008-0037 (CUP) and FPA2009-13697-C04.

References
[1] Minkowski P 1977 *Phys. Lett.* **B67** 421
[2] Gell-Mann M, Ramond P and Slansky R 1979 (*Preprint* `Print-80-0576 (CERN)`)
[3] Yanagida T 1979 *Proc. Works. on the Baryon Number of the Universe and Unified Theories* (Tsukuba, Japan)
[4] Mohapatra R N and Senjanovic G 1980 *Phys. Rev. Lett.* **44** 912
[5] Fukugita M and Yanagida T 1986 *Phys. Lett.* **B174** 45
[6] Schechter J and Valle J W F 1982 *Phys. Rev.* **D25** 2951
[7] Hirsch M, Kovalenko S and Schmidt I 2006 *Phys. Lett.* **B642** 106–110 (*Preprint* `arXiv:hep-ph/0608207`)
[8] Bilenky S M, Faessler A and Simkovic F 2004 *Phys. Rev.* **D70** 033003 (*Preprint* `arXiv:hep-ph/0402250`)
[9] Baudis L *et al.* 1999 *Phys. Rev. Lett.* **83** 41–44 (*Preprint* `arXiv:hep-ex/9902014`)
[10] Klapdor-Kleingrothaus H V, Dietz A, Harney H L and Krivosheina I V 2001 *Mod. Phys. Lett.* **A16** 2409–2420 (*Preprint* `arXiv:hep-ph/0201231`)
[11] Aalseth C E *et al.* 2002 *Mod. Phys. Lett.* **A17** 1475–1478 (*Preprint* `arXiv:hep-ex/0202018`)
[12] Gonzalez-Garcia M C and Maltoni M 2008 *Phys. Rept.* **460** 1–129 (*Preprint* `arXiv:0704.1800 [hep-ph]`)
[13] Luscher R *et al.* 1998 *Phys. Lett.* **B434** 407–414
[14] Akimov D Y *et al.* 1996 *Nucl. Instrum. Meth.* **A379** 484–487
[15] Fernandes L M P *et al.* 2010 *JINST* **5** P09006 (*Preprint* `arXiv:1009.2719 [astro-ph.IM]`)
[16] Balan C *et al.* 2010 (*Preprint* `arXiv:1009.2960 [physics.ins-det]`)
[17] Cebrian S *et al.* 2010 (*Preprint* `arXiv: 1009.1827 [physics.ins-det]`)
[18] Grañena F *et al.* (NEXT Collaboration) 2009 (*Preprint* `arXiv:0907.4054 [hep-ex]`)
[19] Nygren D 2009 *Nucl. Instrum. Meth.* **A603** 337–348

16th International Symposium on Particles Strings and Cosmology (PASCOS 2010) IOP Publishing
Journal of Physics: Conference Series **259** (2010) 012041 doi:10.1088/1742-6596/259/1/012041

Search for the θ_{13} mixing angle with the Double Chooz experiment

Inés Gil-Botella [1]

CIEMAT, Basic Research Department, Avenida Complutense, 22, 28040 Madrid, Spain

E-mail: ines.gil@ciemat.es

Abstract. The Double Chooz reactor neutrino experiment will search for a non vanishing θ_{13} mixing angle with unprecedented sensitivity, which might open the way to unveiling CP violation in the leptonic sector. The measurement of this angle will be based in a precise comparison of the antineutrino spectrum at two identical detectors located at different distances from the Chooz nuclear reactor cores in France. Double Chooz is particularly attractive because of its capability to measure $\sin^2(2\theta_{13})$ to 3σ if $\sin^2(2\theta_{13}) > 0.05$ or to exclude $\sin^2(2\theta_{13})$ down to 0.03 at 90% C.L. for $\Delta m^2_{31} = 2.5 \times 10^{-3}$ eV^2 in three years of data taking with both detectors. The installation of the far detector at 1 km from the reactors is almost completed and the first neutrino interactions are expected by end 2010.

1. Introduction: The search for the θ_{13} mixing angle

During the last years, experiments with solar, atmospheric, reactor and beam neutrinos have provided clear evidence of the existence of neutrino oscillations driven by non-zero neutrino masses and neutrino mixing. The two mass differences which govern the oscillation probabilities have been measured with good precision and the θ_{12} and θ_{23} mixing angles are known to be large (almost maximal for θ_{23}), dominating the solar and atmospheric oscillations respectively [1]. However, the θ_{13} angle (contributing to the effective decoupling between solar and atmospheric oscillations), the sign of Δm^2_{31} (mass hierarchy) and the δ_{CP} violating phase are still unknown.

In particular, only an upper limit on the value of θ_{13} has been established indicating that the angle is very small compared to the other mixing angles. A three-flavor global analysis of the existing data provides a constraint on θ_{13} being $\sin^2(2\theta_{13}) < 0.12$ at 90% C.L [2]. This limit is essentially dominated by the result obtained by the CHOOZ reactor experiment [3] in France. This experiment measured the fraction of $\bar{\nu}_e$'s surviving at a distance of 1.05 km from the reactor cores to be R = 1.01 ± 2.8% (stat) ± 2.7% (syst). This result was mainly limited by the systematic uncertainties induced by the imperfect knowledge of the neutrino production and interaction.

The main goal of the present generation of neutrino oscillation experiments is the measurement of the last unknown mixing angle. This is of fundamental interest not only to understand the leptonic mixing but also because it determines the possibilities to observe CP violation in the leptonic sector with the forthcoming neutrino experiments.

[1] On behalf of the Double Chooz collaboration

2. Reactor neutrinos

The information on the θ_{13} mixing angle can be essentially obtained from accelerator or nuclear reactor neutrino experiments. The long baseline accelerator experiments measure the appearance of ν_e's in a ν_μ beam generated at long distance from the detector. The $\nu_\mu \to \nu_e$ transition depends on several oscillation parameters like the CP phase and the sign of Δm^2_{31}, in addition to θ_{13}. Moreover, they can also be sensitive to matter effects due to the long baselines. Therefore, the measurement of θ_{13} will be affected by correlations and degeneracies between parameters and the sensitivity of the accelerator experiments to this parameter will be reduced.

On the other hand, reactor experiments are unique to provide an unambiguous determination of θ_{13}. Nuclear reactors are very intense sources of $\bar{\nu}_e$'s coming from the β-decay of the neutron-rich fission fragments of fuel elements (^{238}U, ^{235}U, ^{239}Pu and ^{241}Pu). The mean energy released by fission is about 200 MeV and about 2 neutrinos above 1.8 MeV are emitted per fission. Thus, the expected neutrino flux is of the order of 10^{20} $\bar{\nu}_e$/s in 4π for a typical 3 GW$_{th}$ reactor.

Reactor neutrino experiments will look for the disappearance of $\bar{\nu}_e$'s with energies extending up to 10 MeV over distances of the order of kilometers (short baselines) to maximize the disappearance probability. Due to the low energy range of the emitted electron antineutrinos, only disappearance measurements can be performed. The survival probability of $\bar{\nu}_e$'s emitted by a nuclear power station can be written as:

$$\begin{aligned} P(\bar{\nu}_e \to \bar{\nu}_e) &= 1 - \sin^2(2\theta_{13})\sin^2\frac{\Delta m^2_{31}L}{4E} - \cos^4\theta_{13}\sin^2(2\theta_{12})\sin^2\frac{\Delta m^2_{21}L}{4E} \\ &\quad + 2\sin^2\theta_{13}\cos^2\theta_{13}\sin^2\theta_{12}\left(\cos\frac{(\Delta m^2_{31}-\Delta m^2_{21})L}{2E} - \cos\frac{\Delta m^2_{31}L}{2E}\right) \end{aligned} \quad (1)$$

where E is the antineutrino energy and L the distance from the source to the detector. Only the first term is relevant for short baselines. The oscillation amplitude is proportional to $\sin^2(2\theta_{13})$ and independent of the CP phase and the mass hierarchy. In addition, the combination of the MeV range neutrino energies and the short baseline implies that the modification of the oscillation probability induced by matter effects can be neglected to first approximation. Therefore, reactor neutrino experiments can perform a clean measurement of θ_{13}, not affected by degeneracies or correlations between different oscillation parameters.

2.1. Neutrino signal and background in reactor experiments

Reactor antineutrinos are detected through the inverse beta decay $\bar{\nu}_e + \mathrm{p} \to e^+ + \mathrm{n}$. The signature of the neutrino interaction is the coincidence of a prompt signal from the positron energy loss and the resultant annihilation γ-rays and a delayed signal from the neutron capture on a nucleus. Many liquid scintillator $\bar{\nu}$ experiments use scintillator loaded with Gadolinium in their fiducial volume because of its large neutron capture cross section and high total γ yield of 7-8 MeV. The time of coincidence in this case is reduced to 30 μs ($\sim$200 μs in the case of capture in H). The kinematics of the inverse beta decay reaction implies an energy threshold of 1.8 MeV. Above this limit the antineutrino energy is accurately reconstructed from the prompt signal.

The signature of a neutrino interaction can be mimicked by two kind of background events: accidentals and correlated. The accidental background corresponds to the coincidence of a positron-like signal coming from natural radioactivity of the surrounding environment or of the detector materials, with a neutron induced by cosmic muon spallation in the surrounding rock and captured in the detector. The accidental background rate can be estimated quite accurately by measuring a fraction of the energy deposit which do not occur in temporal proximity to any other energy deposit. The second type of background are correlated events that mimic both parts of the coincidence signal. The positron-like and neutron-like signals come either from fast neutrons induced by cosmic muons, which can produce proton-recoils in the target scintillator (misidentified as e^+) and subsequently are captured after thermalisation, or from the β - n decay of long-lived (~100 ms) cosmogenic radioisotopes (^9Li, ^8He) produced by cosmic muons passing through the detector.

3. The Double Chooz experiment

Some of the largest systematic uncertainties of the CHOOZ reactor experiment were related to the accuracy to which the original neutrino flux and spectrum were known. In order to improve the CHOOZ sensitivity to the θ_{13} mixing angle, a relative comparison between two or more identical detectors located at different distances from the power plant is required. The first one, located at few hundred meters from the nuclear cores, monitors the neutrino flux and spectrum shape before neutrinos oscillate. The second detector, located 1-2 km away from the cores, searches for a departure from the global solid angle effect of the neutrino energy spectrum. At the same time, the statistical error can also be reduced by increasing the exposure and the fiducial volume of the detector. Moreover, backgrounds can be further reduced with a better detector design, using veto detectors and external shields against muons and external radioactivity.

Based on the previous strategy, the Double Chooz experiment [4] will try to improve our knowledge on the θ_{13} mixing angle within a competitive time scale and for a modest cost. The Double Chooz collaboration involves institutes from Brazil, France, Germany, Japan, Russia, Spain, UK and USA.

The experiment is being installed in the Chooz-B nuclear power plant in the Northeast of France. The maximum operating thermal power of the two cores is 8.54 GW_{th}. The far detector is located at 1050 m distance from the cores in the same underground laboratory used by the CHOOZ experiment, saving cost and time. It provides a quickly-prepared and well-shielded (300 m.w.e.) site with near-maximal oscillation effect and valuable knowledge about backgrounds from CHOOZ. A second identical detector (near detector) will be installed at 400 m away from the reactor cores to cancel the lack of knowledge of the neutrino spectrum and reduce the systematic errors related to the detector. For the near site, a 85 m air ramp plus a 115 m tunnel will be excavated under a small natural hill (overburden 115 m.w.e.) and a near lab will be equipped.

3.1. The detector design

The CHOOZ detector design can be optimized in order to reduce backgrounds. The Double Chooz detectors (Fig. 1) consist of concentric cylinders and an outer plastic scintillator muon veto. The innermost volume ("target") contains about 10 tons of Gd-loaded liquid scintillator (0.1% Gd) within a transparent acrylic vessel. This will be the volume for neutrino interactions. It is surrounded by a 55 cm thick layer of unloaded scintillator ("gamma-catcher") contained in a second acrylic vessel. This scintillating volume is necessary to fully contain the energy deposition of gamma rays from the neutron capture on Gd as well as the positron annihilation gamma rays inside the central region. It also improves the rejection of the fast neutron background.

Surrounding the gamma-catcher, a 105 cm thick region contains non-scintillating oil inside a stainless steel "buffer" vessel. This volume allows to reduce by two orders of magnitude, with respect to CHOOZ, the level of accidental backgrounds coming mainly from the radioactivity of the photomultiplier tubes (PMTs). 390 low background 10" PMTs are installed on the inner wall and lids of the tank to collect the light from the central scintillating volumes, providing about 13% photocathode coverage.

The central detector is encapsulated within a "inner muon veto" tank, 50 cm thick, filled with scintillating organic liquid and instrumented with 78 8" PMTs. It allows the identification of muons passing near the active detector that can create spallation neutrons and backgrounds coming from outside. Because of space constraint, the 70 cm sand shielding of CHOOZ is

Figure 1. The Double Chooz detector design.

replaced by a 15 cm iron layer to protect the detector from rock radioactivity and to increase the target volume. An "outer muon veto" made of plastic scintillator strips covers the top of the main system and provides additional rejection power for cosmic-induced events. It can be used for constant mutual efficiency monitoring with the inner veto.

Calibration systems will be deployed periodically into the detector allowing to check the stability of the system and measure the detection efficiency and the detector energy response for positrons, gammas and neutrons. The deployment of radioactive sources must be performed in a clean environment under a dry nitrogen atmosphere. A glove box interface with an associated clean room will be installed on the top of the detector.

3.2. Improvements with respect to CHOOZ

The CHOOZ experiment took data only during few months due to loss of transparency of the target liquid scintillator, arriving to a statistical error of 2.8%. Double Chooz expects to reduce this error by the enlargement of the target volume up to 10.3 m^3. Moreover, a new high-quality liquid scintillator has been developed with an improved stability of at least five years. The statistical error should therefore go down to ~0.5% after three years of data taking.

Many systematic uncertainties that affected CHOOZ and all previous single-baseline reactor neutrino experiments are greatly reduced by having both near and far detectors. Table 1 summarizes the systematic uncertainties in the measurement of the antineutrino flux comparing both CHOOZ and Double Chooz detectors.

Though an uncertainty from the neutrino contribution of spent fuel pools remains, it is negligible for Double Chooz. The neutrino rates are proportional to the number of free protons inside the target volumes, which will be experimentally determined by a weighing procedure at the filling with a precision of 0.2%. Both detectors will be filled with the same batch of liquid scintillator. In addition, a comprehensive calibration system consisting of radioactive sources deployed in different detector regions, laser light flashes and LED pulses, will be enforced to correct the unavoidable differences between the two detector responses. The optimization of the Double Chooz detector design allows to simplify the analysis and to reduce the detection

Table 1. Systematic uncertainties in CHOOZ and Double Chooz reactor experiments.

	CHOOZ	**Double Chooz**
Reactor fuel cross section	1.9%	–
Reactor power	0.7%	–
Energy per fission	0.6%	–
Number of protons	0.8%	0.2%
Detection efficiency	1.5%	0.5%
TOTAL	2.7%	0.6%

efficiency systematic errors up to 0.5% while keeping high statistics. Therefore, the dominant detector related systematic error is expected to be kept below 0.6%.

The selection of high pure materials for detector construction and passive shielding around the active region provide an efficient protection against accidental background events. The inner and outer veto systems and the inner detector muon electronics are designed to address the correlated background. The total background rate is estimated to be 3-4 events per day in the far detector to be compared with a neutrino rate of ~50 per day. The error originated by the background subtraction in both detectors is expected to be less than 1%.

3.3. Expected sensitivity

The Double Chooz experiment will consist of two different phases. In a first phase, the far detector will start taking data alone. In a few months the previous CHOOZ limit will be surpassed. After ~1.5 years of data taking with one detector, Double Chooz will be sensitive to $\sin^2(2\theta_{13}) > 0.06$ (Fig. 2), a factor 3 better than CHOOZ.

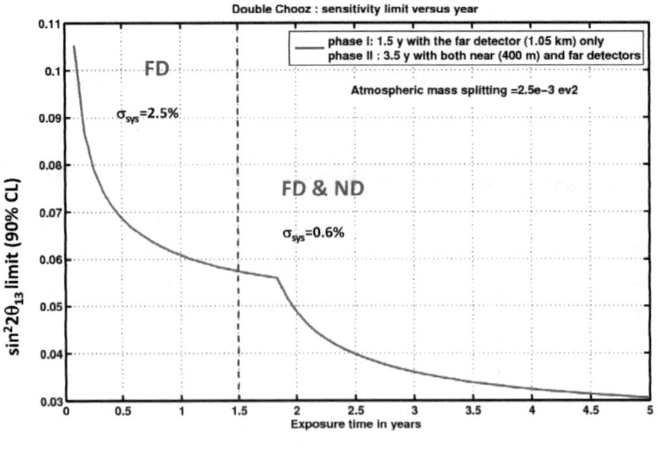

Figure 2. Double Chooz expected sensitivity limit (90% C.L.) to $\sin^2(2\theta_{13})$ as a function of time for $\Delta m^2_{31} = 2.5 \times 10^{-3}$ eV^2, assuming that the near detector is built 1.5 years after the start of the far detector operation (end of 2010).

In a second phase, with both near and far detectors running simultaneously, the systematic errors can be reduced up to 0.6%. Double Chooz can measure $\sin^2(2\theta_{13})$ up to 0.05 at 3σ or exclude $\sin^2(2\theta_{13})$ down to 0.03 at 90% C.L. after three years of operation if no oscillation is seen.

3.4. Current status

The Double Chooz experiment is finishing the integration phase of the far detector in the existing underground laboratory at Chooz. The installation started in May 2008 with the integration of the external shield and followed by the assembly of the inner veto tank. Then, the buffer vessel was completed and the 390 10" PMTs successfully mounted in summer 2009. Finally, the acrylic gamma-catcher and target vessels were installed inside the buffer (Fig. 3) and the detector was closed. All the liquid scintillators have been produced and delivered, the DAQ and electronics systems are ready and the experiment is currently about to start the filling stage. First neutrino data are expected by end 2010.

Figure 3. View of the acrylic vessels inside the buffer tank with the photomultipliers oriented towards the center of the detector.

The tender process for the civil work of the near lab will start in October 2010 and the near site will be available by end 2011 to start with the near detector installation. The goal is to have both, near and far detectors, operative by mid 2012.

4. Conclusions

Double Chooz is a new generation reactor neutrino experiment using two identical detectors at different distances to measure the still unknown θ_{13} mixing angle. Double Chooz has almost completed the installation of the far detector and it is expected to start taking data at the end of 2010. The near detector will be ready by mid 2012 for the second phase of the Double Chooz experiment. After three years of operation with both detectors, Double Chooz will be able to measure $\sin^2(2\theta_{13})$ to 3σ if $\sin^2(2\theta_{13}) > 0.05$ or exclude the mixing angle down to $\sin^2(2\theta_{13}) > 0.03$ at 90% C.L. if no oscillation is observed.

References

[1] See for example Gonzalez-García M C, Maltoni M and Salvado J 2010 *J. High Energy Phys.* JHEP04(2010)056 (*Preprint* arXiv:1001.4524 [hep-ph])
[2] Mezzetto M and Schwetz T 2010 *Preprint* arXiv:1003.5800 [hep-ph]
[3] Apollonio M *et al.* [CHOOZ Collaboration] 2003 *Eur. Phys. J.* C **27** 331-374
[4] Ardellier F *et al.* [Double Chooz Collaboration] 2006 Double Chooz: A search for the neutrino mixing angle theta(13) *Preprint* arXiv:hep-ex/0606025

16th International Symposium on Particles Strings and Cosmology (PASCOS 2010) IOP Publishing
Journal of Physics: Conference Series **259** (2010) 012042 doi:10.1088/1742-6596/259/1/012042

New MINOS Oscillation Results

Jeffrey K. de Jong* **for the MINOS Collaboration**
Department of Physics, University of Oxford, Oxford, OX1 3RH, United Kingdom
E-mail: jeffrey.dejong@physics.ox.ac.uk

Abstract. The MINOS experiment (Main Injector Neutrino Oscillation Search) is a two detector long-baseline neutrino oscillation experiment. Oscillation parameters are determined by comparing the spectrum and composition of the neutrino beam measured at the far detector with that measured at the near detector. This paper summarises the four main beam neutrino oscillation results: muon neutrino and muon anti-neutrino disappearance, electron neutrino appearance and oscillations to sterile neutrinos, derived from the data collected during MINOS' first four years of operation.

1. The MINOS experiment and the Neutrino Beam

The MINOS experiment [1] was designed to accurately measure the atmospheric neutrino oscillation parameters Δm^2_{32} and $\sin^2(2\theta_{23})$ using the Neutrinos at the Main Injection(NuMI) beam at Fermi National Accelerator Laboratory. It consists of two functionally identical, steel/scintillator sampling calorimeters. The 0.98 kton near detector is located approximately 1 km from the NuMI target, the 5.4 kton far detector is located 734 km further downstream in the Soudan Mine Underground Lab in northern Minnesota. The detectors consist of alternating steel and scintillating planes. The scintillating planes are comprised of strips measuring 4.1 cm wide and 1.0 cm thick. Each scintillating plane is attached to a 2.54 cm thick toroidally magnetised (1.3 T) steel plane. Neighbouring scintillating layers are rotated by 90° to allow for three dimensional track reconstruction, and the magnetic field allows particle charge and momentum determination.

The NuMI neutrino beam is generated by the decay of focused πs and Ks produced when 120 GeV protons strike the NuMI target. The neutrino energy can be controlled by the location of the target with respect to the focusing horns. Changing the polarity of the focusing horns allows MINOS to either produce a predominantly muon neutrino beam, or a beam with a larger anti-neutrino component. In the standard configuration, the Low Energy(LE) Forward Horn Current(FHC) mode of operation the beam is comprised of 91.7% ν_μ, 7.0% $\overline{\nu}_\mu$ and 1.3% (ν_e+$\overline{\nu}_e$) and peaks around a few GeV. With the horn polarity reversed(RHC) the beam is 39.9% ν_μ, 58.1% $\overline{\nu}_\mu$ and 2.0% (ν_e+$\overline{\nu}_e$).

2. Oscillation Analyses

Each of the MINOS oscillation analyses are performed utilizing a different neutrino interaction. The $\nu_\mu/\overline{\nu}_\mu$ disappearance analysis identifies charged current (CC) interactions. These interactions are characterized by a hadronic shower at the event vertex and a long (anti-)muon track. Similarly the ν_e appearance analysis looks for CC events which possess both a hadronic and electromagnetic shower. By summing the hadronic and leptonic energy components of the event the original neutrino energy can be reconstructed. The Neutral Current interactions, which contain no flavour information on the incident neutrino, are characterised by short diffuse hadronic showers. The scattered neutrino carries off some of the original neutrino energy so the NC analysis is performed using the visible shower energy.

The unoscillated neutrino energy spectra are first measured at the near detector and then measured further downstream at the far detector. Comparing the shape and composition of the spectra allow the oscillation parameters to be measured and the similarity of the two detectors minimises the systematic errors relating to beam, and neutrino interaction uncertainties.

The muon neutrino disappearance, electron neutrino appearance and oscillations to sterile neutrinos results presented below are determined using 7.0×10^{20} protons-on-target in the standard low energy(LE) forward horn current(FHC) running scenario. The muon anti-neutrino disappearance result is from 1.7×10^{20} POTs delivered in the low energy configuration with both the focussing horn polarity and detector magnetic field polarities reversed.

2.1. ν_μ *Disappearance*

The MINOS experiment measures the atmospheric oscillation parameters Δm^2_{32} and $\sin^2(2\theta_{23})$ by measuring the probability that a given muon neutrino will oscillate while it travels between the near and far detector. In the simple two-flavour approximation, the survival probability of a ν_μ with an energy E [GeV] travelling a distance L [km] is given by:

$$P(\nu_\mu \rightarrow \nu_\mu) = 1 - sin^2(2\theta)sin^2(\frac{1.27\Delta m^2 L}{E}) \tag{1}$$

where Δm^2 is the mass splitting and θ is the mixing angle. Figure 1 shows the selected charged current ν_μ spectrum at the far detector. The unoscillated far detector spectrum is predicted by first measuring the neutrino spectrum at the near detector. The beam Monte-Carlo is then tuned to match the observed near detector spectrum. This tuning determines the kinematics of the hadrons at the production point, which then can be used to extrapolate the spectrum at the far detector.

The ν_μ charged currents events are characterized by a hadronic shower, and a long muon track. The neutral current background becomes non-negligible at low neutrino energies where the muon track does not extend outside the shower. A total of 1986 events are observed, where as in the absence of neutrino oscillations are 2451 are expected. The best fit to the far detector spectrum places $|\Delta m^2_{23}|=2.35^{+0.11}_{-0.08}\times10^{-3}\text{eV}^2$, and constrains $\sin^2(2\theta_{23}) > 0.91$ at 90% C.L. The contours for the best fit to the data in Figure 1 can be found in Fig 2. Alternative neutrino disappearance models where also fit; including pure neutrino decay[8], disfavoured at $>6\sigma$; and neutrino decoherence [9], disfavoured at $>8\sigma$.

Further details of this analysis can be found in [2, 3].

2.2. $\overline{\nu}_\mu$ *Disappearance*

A similar study to that discussed in the previous section was performed on the anti-neutrino data set. A total of 1.7×10^{20} protons on target were accumulated in the reversed horn current running period between September 2009 and March 2010. Despite the large muon-neutrino contamination in the RHC running, the magnetic field facilitates the separation of μ^+ from μ^- resulting in a high anti-neutrino purity in the final data set, >98% for $E_\nu < 6$GeV.

Figure 3 plots the reconstructed energy spectrum of $\overline{\nu}_\mu$ CC events at the far detector. A total of 97 events are observed with the no-oscillation prediction of 155 events[1]. This result disfavours the null oscillation hypothesis at 6.3σ. The best fit oscillation parameters to the anti-neutrino

[1] For an equivalent number POTs we observe fewer anti-neutrino events due to their lower interaction cross-section and their lower production rate.

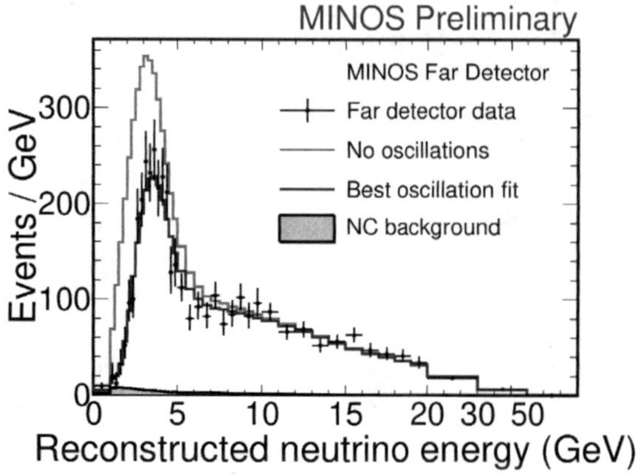

Figure 1. The reconstructed energy spectrum of selected ν_μ CC events at the far detector. The red line in this prediction in the case of no oscillations, the blue line is the best fit to oscillations.

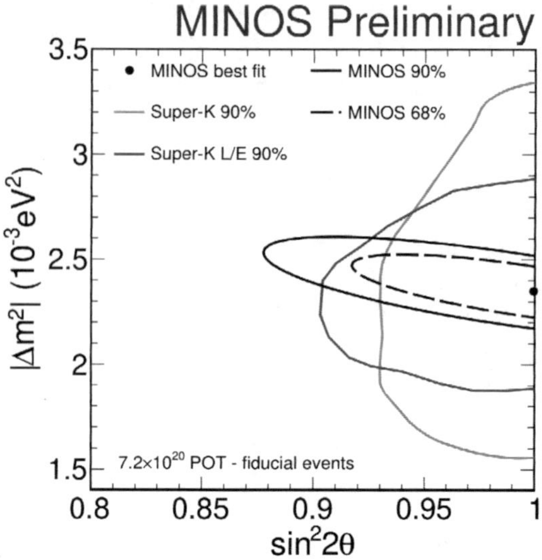

Figure 2. Contours for the oscillation fit to the data in Fig.1. Previous experimental limits are also shown.

data in Figure 3 are $|\overline{\Delta m^2_{32}}|=3.36^{+0.45}_{-0.40}\times10^{-3}$ eV^2 and $\sin^2(2\overline{\theta_{23}})=0.81\pm0.11$. The contours for the neutrino and anti-neutrino oscillation figures are shown in Figure 4. While the anti-neutrino data favours a slightly higher $|\Delta m^2|$ than the neutrino data, which would violate CPT, the results are consistent at 2σ. MINOS plans to double the anti-neutrino data set through another period of RHC running in early 2011 to investigate this further.

2.3. ν_e Appearance

In the instance where θ_{13} is non-zero electron neutrinos are expected to be produced through the sub-dominant oscillation channel $\nu_\mu \rightarrow \nu_e$. The MINOS detectors where optimised for muon identification; the width of the scintillating strips is on the order of a Moliere radius making electromagnetic shower identification difficult. However, an 11 variable artificial neural network (ANN) was developed relying mostly on characterisations of the longitudinal and transverse energy deposition profiles. The ANN selection variable measured and predicted at the far detector is plotted in Figure 5.

As the data is expected to be background dominant, at the CHOOZ limit MINOS expects a signal to background ratio of 1:2, a data driven approach was used to determine the exact composition of the background at the MINOS near detector. The NC, CC and beam ν_e are then extrapolated separately to the far detector. A total of 49.1 $\pm$ 7.0(stat)$\pm$2.7(syst) events are expected, and 54 events are observed. This corresponds to a 0.7σ excess, consistent with no ν_e appearance.

Using this data we are then able to place limits on the θ_{13} parameter. Further more, due to matter effects MINOS is sensitive to the neutrino mass hierarchy. In the scenario where Δm^2_{32}=2.43$\times10^{-3}$eV^2, $\sin^2(2\theta_{23})=1$ and $\delta_{CP}=0$ we constrain $\sin^2(2\theta_{13}) < 0.12$ in the normal mass hierarchy and $\sin^2(2\theta_{13}) < 0.2$ in the inverted mass hierarchy at 90% C.L. The 90% contours as a function of δ_{CP} are shown in Figure 6. Further details of this analysis can be found in [4, 5].

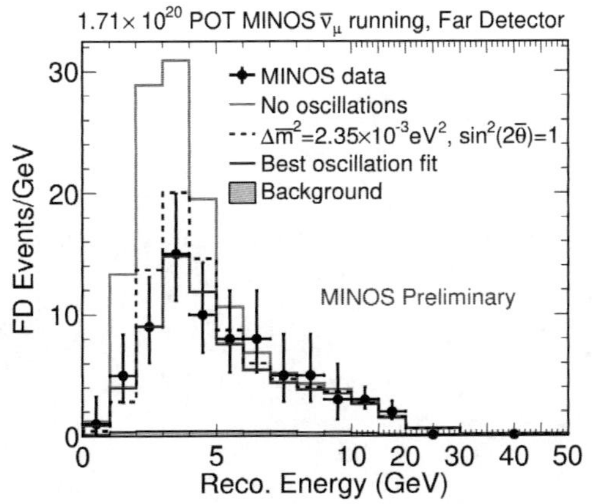

Figure 3. The reconstructed energy spectrum of selected $\overline{\nu}_\mu$ CC events at the far detector. The red line in this prediction in the case of no oscillations, the dashed line is the prediction using the fit to neutrino data, the blue line is the best fit to oscillations.

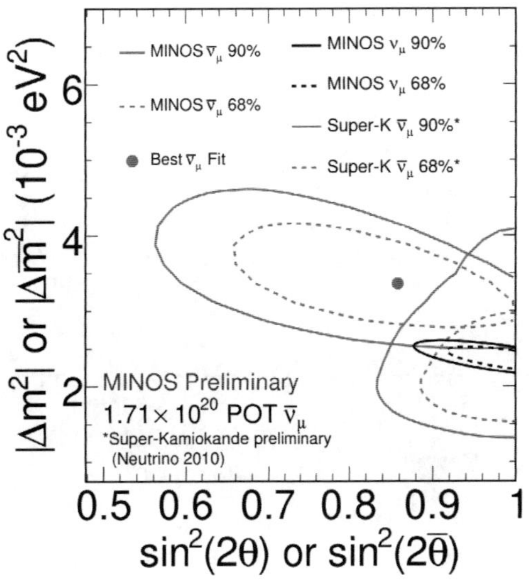

Figure 4. Contours for the oscillation fit to the anti-neutrino data in Figure 3(in red). Previous experimental limits are also shown(in green). The MINOS neutrino disappearance contours from Figure 2 are also shown.

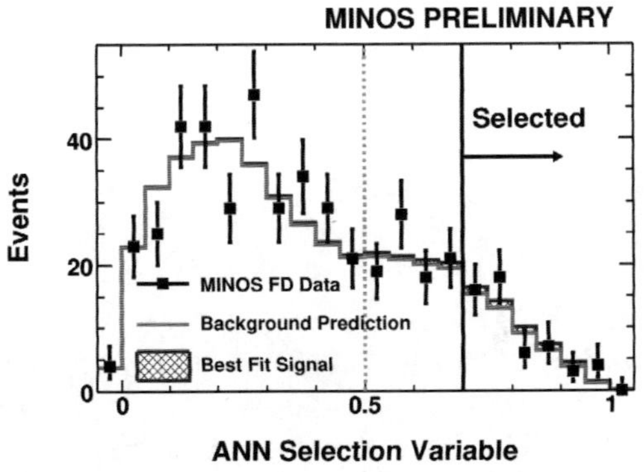

Figure 5. The reconstructed PID of all selected events at the far detector. Events with values a value of 0.7 are ν_e-like. The solid line is the background prediction in the case of no ν_e appearance.

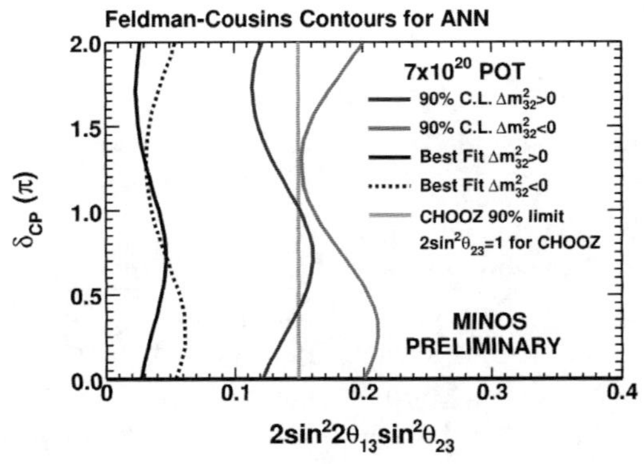

Figure 6. θ_{13} contours, the green vertical line is the CHOOZ limit for a Δm^2=2.43×10^{-3}eV^2

2.4. Neutral Current Interactions

The Neutral Current event rate should not change in the standard three active flavour oscillation model. A deficit in the Neutral Current event rate could indicate mixing to sterile neutrinos. Presented here are the results assuming a single sterile neutrino with mass splitting equivalent to the atmospheric mass scale.

The selection of neutral current events relies mainly on the exclusion of a long muon track, low energy ν_μ CC interactions where the muon track does not extend past the hadronic shower represent an irreducible background to the NC signal. Furthermore, the prediction of the event rate at the far detector is dependent on the assumed value of θ_{13} as a non-zero value will produce an excess of ν_e CC events at the far detector which, by virtue of not having a track will be identified as NC.

Figure 7 plots the visible energy of the selected NC events at the far detector. In total we expect 757 events, where as 802 events are observed. This result is consistent with no deficit of neutral current events.

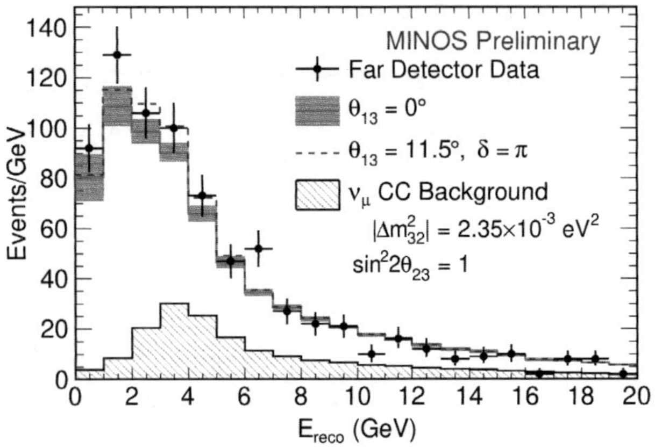

Figure 7. The visible energy spectrum of the selected far detector NC event sample, including the predictions at differing θ_{13} and the CC ν_μ background.

Defining the fraction of neutrinos which oscillate to sterile flavours as

$$f_s = \frac{P_{\nu_\mu \to \nu_s}}{1 - P_{\nu_\mu \to \nu_\mu}} \quad (2)$$

we find that $f_s < 0.22(0.40)$ at 90% C.L. without(with) ν_e appearance.

3. Conclusions

MINOS has measured neutrino oscillation parameters with great precision; it currently has the best measurement of Δm^2_{32}, $\Delta\overline{m^2_{32}}$, and the worlds best constraints on the θ_{13} mixing angle and the sterile neutrino fraction f_s. The results presented here are based on 7×10^{20} POTs in the FHC running scenario and 1.7×10^{20} POTs RHC. The MINOS experiment continues to accumulate data and will further improve all the measurements presented here.

4. Acknowledgements

This work was supported by the US DOE, the UK STFC, the US NSF, the State and University of Minnesota, the University of Athens, Greece and Brazil's FAPESP and CNPq. We are grateful to the Minnesota Department of Natural Resources, the crew of Soudan Underground Laboratory, and the staff of Fermilab for their contributions to this effort.

References

[1] Michael, D.G*et. al.* [MINOS Collaboration] 2008 The magnetized steel and scintillator calorimeters of the MINOS experiment *Nucl. Instrum. Method.* **A596** 190-228

[2] Michael, D.G*et. al.* [MINOS Collaboration] 2006 Observation of muon neutrino disappearance with the MINOS detectors and the NuMI neutrino beam *Phys.Rev.Lett* **97** 191801

[3] Adamson, P*et. al.* [MINOS Collaboration] 2008 Measurement of Neutrino Oscillations with the MINOS Detectors in the NuMI beam*Phys.Rev.Lett* **101** 131802
[4] Adamson, P*et. al.* [MINOS Collaboration] 2009 Search for muon-neutrino to electron-neutrino transitions in MINOS *Phys.Rev.Lett* **103** 261802
[5] Adamson, P*et. al.* [MINOS Collaboration] 2010 New constraints on the muon-neutrino to electron-neutrino transitions in MINOS *Phys.Rev.* **D82** 051102
[6] Adamson, P*et. al.* [MINOS Collaboration] 2008 Search for active neutrino disappearance using neutral-current interactions in the MINOS long base-line experiment. *Phys.Rev.Lett* **103** 261802
[7] Adamson, P*et. al.* [MINOS Collaboration] 2010 Search for sterile neutrino mixing in the MINOS long baseline experiment. *Phys.Rev.* **D81** 052004
[8] Fogli, G.L.*et. al.* 2003 *Phys.Rev.* **D67** 093006
[9] Barger, V.*et. al.* 1999 *Phys.Rev.Lett.* **82** 2640
[10] Apollonio, M. *et. al.* 2003 *Eur.Phys.Journ.* **C2** 331

Journal of Physics: Conference Series **259** (2010) 012043 doi:10.1088/1742-6596/259/1/012043

The degenerate gravitino scenario

Lotfi Boubekeur

Departament de Física Teòrica, Universitat de València, E-46100, Burjassot, Spain.

Instituto de Física Corpuscular (IFIC), Universitat de València-CSIC,
Edificio de Institutos de Paterna, Apt. 22085, E-46071, Valencia, Spain.

E-mail: lotfi.boubekeur@uv.es

Abstract. We propose an alternative solution to the gravitino problem in supersymmetric theories. In the "degenerate gravitino" scenario the mass difference between the gravitino and the LOSP (lightest ordinary MSSM particle) is much smaller than the gravitino mass itself. As a consequence, the energy released in the decay of the next to lightest sypersymmetric particle (NLSP) is reduced. The cosmological and astrophysical constraints on the gravitino abundance, and hence on the reheating temperature, become softer than in the usual case. We explore the parameter space of this scenario and confront it with the above mentioned constraints. In the case of neutralinos LOSP, mass splittings of order 10^{-2} GeV are possible, while for stau LOSP the splitting can be of order tens of GeVs. In both cases, maximal reheating temperatures, compatible with thermal leptogenesis are reached.

1. Introduction

The gravitino is notoriously known to cause a lot of trouble in cosmology [1, 2]. Even after inflation was discovered and recognized as a solution to the relic problem of standard cosmology, it was soon realized that gravitinos are re-created at reheating [3]. As their number density depends on the reheating temperature, this last cannot be too high [4]. This is at the origin of the well-known tension on the reheating temperature in thermal leptogenesis. In the case of a decaying gravitino, the problem comes from the energetic decay products that can potentially destroy the light elements. On the other hand, if the gravitino is stable, the danger comes from overclosure of the Universe. Eventhough gravitinos produced at reheating can make out all of dark matter, there can be other production mechanisms that produce gravitinos, like decay of heavier MSSM particles or the inflaton. These considerations make the gravitino abundance severely constrained. Several mechanisms have been devised to address this problem. One approach is to make the gravitinos decay before big bang nucleosynthesis (BBN) rendering them harmless [2]. The other approach is to make the gravitino stable in which case it can play the role of cold dark matter [5]. In [6], we aimed to investigate alternative solutions to the gravitino problem that allow for high reheating temperatures compatible with thermal leptogenesis. As the gravitino couplings are fixed by supergravity, this does not leave much freedom and the only free parameter available is the gravitino mass itself $m_{3/2}$. We were led to a specific scenario where the gravitino is almost degenerate with the LOSP, which as we will see allows to address the cosmological gravitino problem.

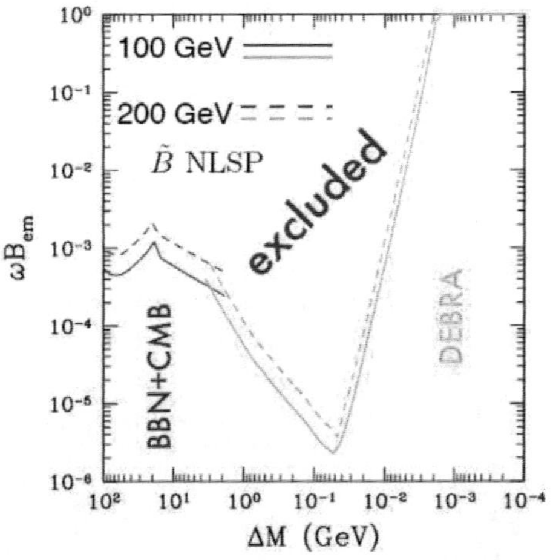

Figure 1. Combined constraints on the degenerate gravitino scenario for the case of gravitino LSP with Bino NLSP.

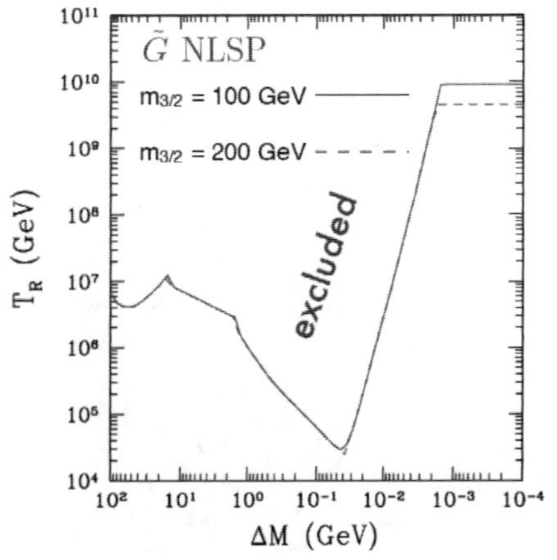

Figure 2. Constraints on the maximal reheating temperature after inflation for the Gravitino NLSP with Bino LSP.

2. The degenerate gravitino scenario

The degenerate gravitino scenario proposed in [6] is defined through the requirement that the mass splitting between the gravitino and the LOSP is smaller than the gravitino mass itself i.e. $\Delta M = |m_{3/2} - m_{\rm LOSP}| \equiv \delta\, m_{3/2} \ll m_{3/2}$. Therefore, effectively, we are left with only the gravitino and the LOSP as all remaining MSSM particles are assumed to be heavier and the gravitino can be either the LSP or the NLSP. Furthermore, the smallness of the mass splitting has two main consequences. First the decay products are much softer and they do not cause trouble with BBN. Second, the NLSP lifetime increases and its decay occurs much later, so the gravitino is almost stable and can play the role of cold dark matter or part of it. As the gravitinos in our case decay much later than usual, it is not enough to check that the gravitinos do not conflict with BBN. Gravitino decay will affect different eras and thus will be subject to different constraints.

3. Constraints

Let us begin with the most straightforward constraint. The relic density of cold dark matter will have two components: a thermal one produced through scatterings or freeze-out and a non thermal one produced through the decay of the NLSP. The sum of these two should match the observed CDM relic abundance. This requirement is expressed as $\Omega_{\rm CDM}\, h^2 = \Omega_{\rm LSP}^{\rm TP}\, h^2 + \Omega_{\rm LSP}^{\rm NTP}\, h^2 \simeq 0.11$. It also is useful to define a new parameter $\omega \equiv Y_{\rm NLSP}/Y_{\rm CDM}$, which quantifies the amount of present cold dark matter coming from the NLSP decay. In this equation $Y_{\rm NLSP}$ refers to the NLSP yield just before its decay. In the following, the various constraints will apply on δ and ω according to the NLSP lifetime. Next, we consider BBN constraints. At temperatures of order $T \sim 1$ MeV, the light nuclei are synthesized in the primordial plasma. These temperatures corresponds to times between 1 sec and 10^3 sec. The obtained abundances in standard BBN calculations are in striking agreement with observation. However, the injection of energetic particles in the primordial plasma at BBN or later can disrupt the standard BBN processes [7, 8, 9, 10], leading to a disagreement between theory and observation. In our scenario, we need to worry only about 2-body electromagnetic decays [1] as it is the only ones allowed by the small mass splitting.

For longer lifetimes, we should consider CMB constraints. In addition to BBN constraints, for long lifetimes $\tau_{\rm NLSP} \gtrsim 10^7$ sec., there are strong bounds from the shape of the CMB black-body spectrum. As pointed out in [4], the late injection of electromagnetic energy may distort the frequency dependence of the CMB spectrum from its observed blackbody shape.

After recombination, at the cosmic time around 10^{13} sec, the number density of free electrons drops quickly and the photons are almost free from the interactions. Therefore the photons from the decaying particles can reach us now and contribute to the Diffuse Electromagnetic Background RAdiation (DEBRA).

The present photon flux from two-body decay can be written as

$$\frac{d\Phi}{dE_\gamma} = \frac{c}{4\pi}\int_{t_i}^{t_0} \frac{dt}{\tau_{\rm NLSP}} \frac{\rho_c\,\Omega_{\rm WMAP}\,\omega\,B_{\rm em}}{m_{\rm NLSP}} e^{-t/\tau_{\rm NLSP}}\delta(E_\gamma - aE_{\rm em}), \tag{1}$$

where $E_{\rm em} \simeq m_{\rm LSP}\delta$ is the energy of the photon at production, $\tau_{\rm NLSP}$ and $m_{\rm NLSP}$ are the lifetime and mass of NLSP, $\rho_c = 3H_0^2/8\pi G_N = 8.0992h^2 \times 10^{-47}\,{\rm GeV}^4$ and $a = a(t)$ is the time-dependent scale factor with $a(t_0) = 1$ at present time t_0.

The expected flux Eq. (1), has to be compared with bounds from SPI, COMPTEL and EGRET, [11]. Fig. 1 shows combined CMB, BBN and DEBRA constraints in the case of a Bino NLSP. As one can see, for splittings of order 100 GeV, $B_{\rm em}\omega$ is below order 10^{-3}. As ΔM decreases to 10^{-1} GeV, $\omega B_{\rm em}$ is bound to be below 10^{-6}. Starting from $\Delta M \simeq 10^{-1}$ GeV, higher values of $\omega B_{\rm em}$ are allowed, until reaching $\Delta M \simeq 10^{-1}$ GeV, where all values of $\omega B_{\rm em}$ are allowed. What about the allowed reheating temperature T_R after inflation? One can translate the previous combined bounds on bounds on T_R. Fig. 2 shows such a bound for a scenario with gravitino NLSP and Bino LSP. On the other hand, for the opposite case where the gravitino is the LSP, maximal reheating temperatures of order 10^9 GeV, compatible with thermal leptogenesis are always possible provided the relic density of NLSPs is suppressed ($\omega \ll 1$).

Up to now, we discussed the situation where the gravitino is almost degenerate with the neutralino. In the MSSM, the LSP can also be the lightest stau. This region is usually discarded because in this case cold dark matter is charged, which is excluded by heavy water data [12]. However in our case, there is no reason to exclude it since the stau can be the NLSP, with the gravitino as the LSP. In addition to the previous bounds that apply equally, the so-called Catalyzed BBN (CBBN) [13] bounds on the staus abundance ($Y_{\tilde\tau} < Y_{\rm CBBN}$) apply. These are shown in Fig. 3. Applying the CBBN constraint $Y_{\rm CBBN} = 2 \times 10^{-16}$ derived in [14, 15] results in a very stringent constraints that eliminates almost all of parameter space. On the other hand, applying the most conservative CBBN constraint $Y_{\rm CBBN} = 10^{-14} - 10^{-15}$ [16, 17] still allows some parameter space to survive.

What about the degenerate gravitino in the simplest framework of weak scale supersymmetry? the so-called CMSSM (Constrained MSSM) which is one of its simplest and most popular realizations for phenomenological studies. It is defined in terms of only five free parameters: a common scalar (m_0), gaugino ($m_{1/2}$) and tri–linear (A_0) mass parameters (all specified at the GUT scale) plus the ratio of Higgs vacuum expectation values $\tan\beta$ and ${\rm sign}(\mu)$, where μ is the Higgs/higgsino mass parameter whose square is computed from the conditions of radiative electroweak symmetry breaking. Furthermore, the gravitino mass $m_{3/2}$ is kept as a free parameter. We have analyzed the degenerate gravitino scenario in the context of the CMSSM framework . In addition to the previously mentioned constraints, we applied low energy constraints to obtain the highest reheating temperature that can be reached in the CMSSM

[1] This means that we will consider only decays NLSP$\rightarrow$LSP+ X ($X = \gamma, \tau$) as we will see later, and $B_{\rm em} \simeq 1$ hereafter, where $B_{\rm em}$ stands for the electromagnetic branching ratio.

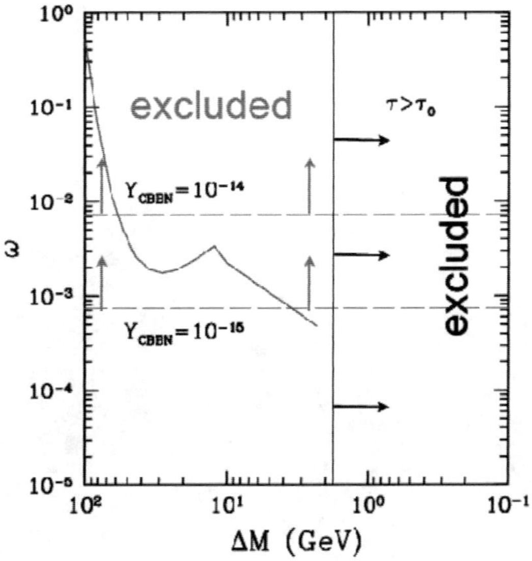

Figure 3. CBBN bounds on the degenerate gravitino scenario with stau NLSP. The region on the right of the vertical line is ruled-out by heavy water data.

fulfilling the WMAP constraint on the CDM abundance and the relevant collider bounds; namely, direct SUSY searches, $(g-2)_\mu$ using ($e^+e^- \to$hadrons) data and the $BR(B \to X_s\gamma)$. The results are shown in Fig. 4 for two representative sets of CMSSM parameters. Note that the highest reheating temperatures are always reached around regions of the parameter space where the neutralino relic density is suppressed, like for example the stau-neutralino coannihilation region for low $\tan\beta$ and "focus point" or "hyperbolical branch" for large $\tan\beta$.

4. Conclusions

We proposed an alternative solution that addresses the gravitino problem in thermal leptogenesis by making the gravitino degenerate with the LOSP. This has the direct consequence that the injected energy is suppressed, making the decay products less dangerous for BBN. Due to the small mass splitting, the gravitino and the LOSP are typically long-lived. We analyzed this scenario (the "degenerate gravitino" scenario) by confronting it to cosmological and astrophysical constraints. Since the NLSP decays at or after BBN, we considered in addition to BBN, constraints from CMB spectral distortions and diffuse gamma rays observations. First we performed a model-independent analysis by considering a generic NLSP-LSP degenerate scenario where the NLSP decays through NLSP$\to$LSP+ X ($X = \gamma, \tau$). Since the final cold dark matter relic density is the sum of both thermal and non-thermal contributions, we required that the total cold dark matter is consistent with cosmological observation. Then, using the results of this analysis, we studied the degenerate gravitino scenario in the context of the CMSSM where three types of spectra arise, they are: gravitino NLSP with neutralino LSP and gravitino LSP with neutralino or stau NLSP. Each of these cases has been analyzed in this framework defined by the usual high energy parameters (m_0, $m_{1/2}$, A_0, $\tan\beta$, sign(μ)) and the gravitino mass $m_{3/2}$, and confronted with low-energy observables. We find that high reheating temperatures consistent with thermal leptogenesis can be found in all three scenarios if a sizable part of cold dark matter comes from gravitinos produced at reheating. In this case, depending on $\tan\beta$, we are led to regions in the parameter space where the relic density of stau and neutralinos are

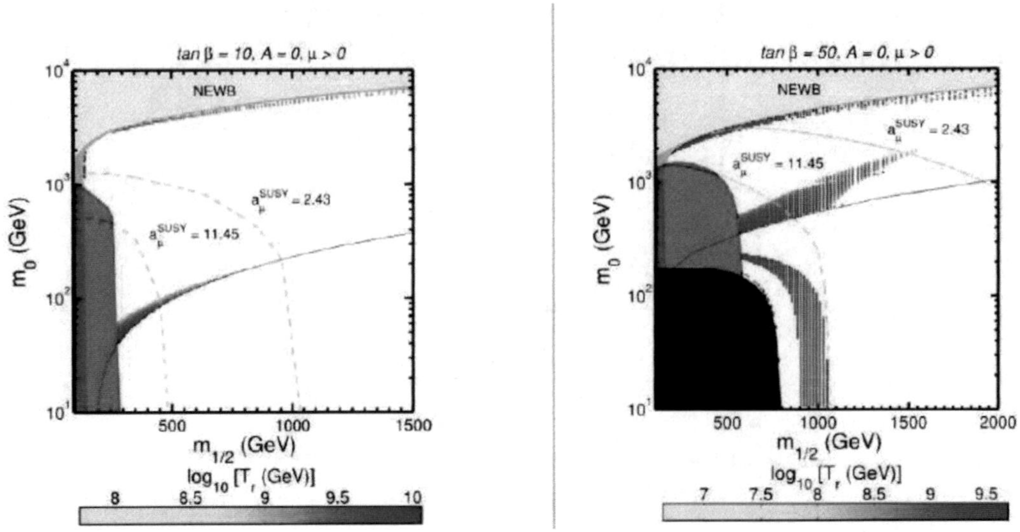

Figure 4. The reheating temperature in the CMSSM in the degenerate gravitino scenario for two sets of parameters. In both panels the red (medium grey) region is forbidden by the Higgs bound from LEP and in the very light grey region no correct electroweak symmetry breaking is obtained. The light brown (light grey) band below the NEWB area corresponds to the region forbidden by the LEP chargino bound. The regions below the dashed green lines satisfy the $(g-2)_\mu$ constraint at the 2 σ ($a_\mu^{\rm SUSY} > 2.43 \times 10^{-10}$) or 3 σ level ($a_\mu^{\rm SUSY} > 11.45 \times 10^{-10}$). In both plots, we used $m_t = 173.1$ GeV.

somewhat suppressed. In general to satisfy all the constraints, the mass splitting between the NLSP and LSP should be $\Delta M \simeq 10^{-2}$ GeV for the degenerate neutralino-gravitino scenario, which implies very long-lived NLSPs which are beginning to decay at present. On the other hand, in the gravitino-stau scenario, splittings in the range 10 GeV $\lesssim \Delta M \lesssim$ 90 GeV are still consistent with reheating temperatures of the order of 10^9 GeV if we consider the conservative CBBN constraint.

Let us comment on the required degeneracy in the "degenerate gravitino" scenario. Although a degeneracy of the order of $\Delta M \simeq 10^{-2}$ GeV certainly implies a certain amount of fine-tuning, this tuning is only two orders of magnitude stronger than the usual tuning required in the coannihilation or funnel regions to obtain the right relic density in the MSSM. On the other hand, notice also that the fine tuning in our scenario is much softer that the tuning required in other scenarios like inelastic dark matter [18].

Finally, it is also important to consider the phenomenological consequences of this scenario in colliders. In the case of neutralino LSP or NLSP, the only indirect signal of this scenario will be that the relic density of neutralinos inferred from the measurements of supersymmetric masses and couplings at LHC, will not match the observed cold dark matter abundance and will be smaller. However, the measurements at direct detection experiments will agree with the cross sections obtained from colliders. On the other hand, for stau NLSP, the collider signatures would be spectacular, as the staus would be completely stable at collider scales and slow charged tracks will appear in the detector [19]. In this case direct detection experiments will give a null results as all the dark matter at present times is made of gravitinos. Therefore, the "degenerate gravitino" scenario will be probed at colliders and direct detection experiments if SUSY is discovered at LHC.

Acknowledgments
I thank Ki Young Choi, Roberto Ruiz de Austri and Oscar Vives for the enjoyable collaboration. I acknowledge financial support from spanish MEC and FEDER (EC) under grant FPA2008-02878 and Generalitat Valenciana under the grant PROMETEO/2008/004.

References
[1] H. Pagels and J. R. Primack, Phys. Rev. Lett. **48** (1982) 223.
[2] S. Weinberg, Phys. Rev. Lett. **48** (1982) 1303.
[3] J. R. Ellis, J. S. Hagelin, D. V. Nanopoulos, K. A. Olive and M. Srednicki, Nucl. Phys. B **238** (1984) 453.
[4] J. R. Ellis, J. E. Kim and D. V. Nanopoulos, Phys. Lett. B **145** (1984) 181.
[5] M. Bolz, A. Brandenburg and W. Buchmuller, Nucl. Phys. B **606** (2001) 518 [Erratum-ibid. B **790** (2008) 336] [arXiv:hep-ph/0012052].
[6] L. Boubekeur, K. Y. Choi, R. Ruiz de Austri and O. Vives, JCAP **1004**, 005 (2010) [arXiv:1002.0340 [hep-ph]].
[7] R. H. Cyburt, J. R. Ellis, B. D. Fields and K. A. Olive, Phys. Rev. D **67** (2003) 103521 [arXiv:astro-ph/0211258].
[8] K. Jedamzik, Phys. Rev. D **70** (2004) 063524 [arXiv:astro-ph/0402344].
[9] M. Kawasaki, K. Kohri and T. Moroi, Phys. Lett. B **625** (2005) 7 [arXiv:astro-ph/0402490] and M. Kawasaki, K. Kohri and T. Moroi, Phys. Rev. D **71** (2005) 083502 [arXiv:astro-ph/0408426].
[10] K. Jedamzik, Phys. Rev. D **74** (2006) 103509 [arXiv:hep-ph/0604251].
[11] H. Yuksel and M. D. Kistler, Phys. Rev. D **78** (2008) 023502 [arXiv:0711.2906 [astro-ph]].
[12] T. Yamagata, Y. Takamori and H. Utsunomiya, Phys. Rev. D **47**, 1231 (1993).
[13] M. Pospelov, Phys. Rev. Lett. **98** (2007) 231301 [arXiv:hep-ph/0605215]
K. Kohri and F. Takayama, Phys. Rev. D **76** (2007) 063507 [arXiv:hep-ph/0605243]
M. Kaplinghat and A. Rajaraman, Phys. Rev. D **74** (2006) 103004 [arXiv:astro-ph/0606209].
[14] K. Hamaguchi, T. Hatsuda, M. Kamimura, Y. Kino and T. T. Yanagida, Phys. Lett. B **650** (2007) 268 [arXiv:hep-ph/0702274].
[15] J. Pradler and F. D. Steffen, Phys. Lett. B **666** (2008) 181 [arXiv:0710.2213 [hep-ph]].
[16] K. Jedamzik, JCAP **0803** (2008) 008 [arXiv:0710.5153 [hep-ph]].
[17] K. Jedamzik and M. Pospelov, arXiv:0906.2087 [hep-ph].
[18] D. Tucker-Smith and N. Weiner, Phys. Rev. D **64** (2001) 043502 [arXiv:hep-ph/0101138].
[19] S. Tarem, and S. Bressler, E. Duchovni and L. Levinson, Atlas note ATL-PHYS-PUB-2005-022, 2005;
S. Tarem, S. Bressler, H. Nomoto and A. Dimattia, Atlas note ATL-PHYS-PUB-2008-001, 2008;
J. R. Ellis, A. R. Raklev and O. K. Oye, Atlas note ATL-PHYS-PUB-2007-016, 2007;
K. Ishiwata, T. Ito and T. Moroi, Phys. Lett. B **669** (2008) 28 [arXiv:0807.0975 [hep-ph]];
S. Asai, K. Hamaguchi and S. Shirai, Phys. Rev. Lett. **103**, 141803 (2009) [arXiv:0902.3754 [hep-ph]].

Journal of Physics: Conference Series **259** (2010) 012044 doi:10.1088/1742-6596/259/1/012044

SFT based cosmological models

Alexey S. Koshelev[1]

Theoretische Natuurkunde, Vrije Universiteit Brussel and The International Solvay Institutes, Pleinlaan 2, B-1050 Brussels, Belgium

E-mail: `alexey.koshelev@vub.ac.be`

Abstract. We consider the appearance of multiple scalar fields in SFT inspired non-local models with a single scalar field at late times. In this regime all the scalar fields are free. This system minimally coupled to gravity is mainly analyzed in this note. We build one exact solution to the equations of motion. We consider an exactly solvable model which obeys a simple exact solution in the cosmological context for the Friedmann equations and that reproduces the behavior expected from SFT in the asymptotic regime.

1. Introduction

It is known that String Field Theory (SFT) [1] and p-adic string theory [2] tend to be UV-complete. One can therefore expect that the resulting (effective) models should be free of pathologies. Furthermore, models originating from SFT exhibit one general non-standard property: they have terms with infinitely many derivatives, i.e. non-local terms. Higher derivative terms usually produce the well known Ostrogradski instability. However, the Ostrogradski result is related to derivatives order higher than two but finite. In the case of infinitely many derivatives it is possible that instabilities do not appear.

Contemporary cosmological observational data [3] strongly support that the present Universe exhibits an accelerated expansion providing thereby an evidence for a dominating DE component. Recent results of WMAP together with the data on Ia supernovae give the following bounds for the DE state parameter $w_{\rm DE} = -1.02^{+0.14}_{-0.16}$. Note that the present cosmological observations do not exclude an evolving DE state parameter w. Non-local models of the type obtained from SFT may have effective phantom behavior and can therefore be interesting for the present cosmology. To construct a stable model with $w < -1$ one should construct from the fundamental theory, which is stable and admits quantization, an effective theory with the Null Energy Condition (NEC) violation. This is a hint towards SFT inspired cosmological models.

In this note we briefly review a new class of cosmological models based on string field theory (SFT) and p-adic string theory. We build one exact solution to the equations of motion and introduce an exactly solvable model ([4]) obeying a simple analytic solution which reproduce the behavior of the initial model in the asymptotic regime (see [5, 6, 7, 8] and references therein for a more detailed analysis on the subject).

[1] Postdoctoral researcher of FWO-Vlaanderen

Journal of Physics: Conference Series **259** (2010) 012044 doi:10.1088/1742-6596/259/1/012044

2. Model setup

We are interested in non-local models arising from SFT in large-time regime . Our starting point is the action

$$S = \int d^4x\sqrt{-g}\left(\frac{R}{16\pi G_N} + \frac{1}{g_o^2}\left(-\frac{1}{2}\partial_\mu T\partial^\mu T + \frac{1}{2\alpha'}T^2 - \frac{1}{\alpha'}v(\bar{T})\right) - \Lambda'\right). \quad (1)$$

We work in $1+3$ dimensions with signature $(-,+,+,+)$, the coordinates are denoted by Greek indices $\mu,\nu,\ldots$, running from 0 to 3. G_N is the Newtonian constant, $8\pi G_N = 1/M_P^2$, where M_P is the Planck mass, α' is the string length squared (we do not assume $\alpha' = 1/M_P^2$), g_o is the open string coupling constant and Λ' is a constant term. $\bar{T} = \mathcal{G}(\alpha'\Box)T$, with $\Box = D^\mu\partial_\mu = \frac{1}{\sqrt{-g}}\partial_\mu\sqrt{-g}g^{\mu\nu}\partial_\nu$ and D_μ is the covariant derivative, T is a scalar field primarily associated with the open string tachyon. The function $\mathcal{G}(\alpha'\Box)$ can be non-polynomial, thus producing a manifest non-locality. Fields are dimensionless while $[g_o] =$ length. $v(\bar{T})$ is the open string tachyon self-interaction. It does not have a term quadratic in T. Factor $1/\alpha'$ in front of the tachyon potential looks unusual and can be easily removed by a rescaling of fields. For our purposes it is convenient to keep all the fields dimensionless.

Such a four-dimensional action is motivated by string field theory. In SFT one has $\mathcal{G}(\alpha'\Box) = e^{-\frac{\beta}{2}\alpha'\Box}$, where β is a parameter determined exclusively by the conformal field theory of the string, but we keep this function general. We stress once again that appearance of non-localities is a general feature of SFT based models and it is exactly the feature that we are going to explore here[2].

Introducing the field $T_b = \bar{T}$ and dimensionless coordinates $\bar{x}_\mu = x_\mu/\sqrt{\alpha'}$, we can rewrite the above action as

$$S = \int d^4x\sqrt{-g}\left(\frac{R}{16\pi G_N} + \frac{1}{g_o^2}\left(-\frac{1}{2}\partial_\mu\tilde{T}_b\partial^\mu\tilde{T}_b + \frac{1}{2}\tilde{T}_b^2 - v(T_b)\right) - \Lambda'\right). \quad (2)$$

where we omit bars for simplicity and set hereafter $\alpha' = 1$. We emphasize that the potential of the field T_b is $V = -\frac{1}{2}T_b^2 + v(T_b)$. Assuming an extremum of the potential V exists, one can linearize the theory around it using $T_b = T_0 - \tau$. As a result one obtains

$$V = -\frac{1}{2}\tau^2 + \frac{v(T_0)''}{2}\tau^2 + V(T_0).$$

The action (2), linearized around the extremum of the potential, can be written as

$$S = \int d^4x\sqrt{-g}\left(\frac{R}{16\pi G_N} + \frac{1}{2g_o^2}\tau\mathcal{F}(\Box)\tau - \Lambda\right), \quad (3)$$

where $\mathcal{F}(\Box) = (\Box+1)\mathcal{G}^{-2}(\Box) - m^2$ with $m^2 \equiv v(T_0)''$ and $\Lambda = \Lambda' + \frac{V(T_0)}{g_o^2}$. In SFT one would have $\mathcal{F}_{\mathrm{SFT}}(\Box) = (\Box+1)e^{\beta\Box} - m^2$. $\mathcal{F}$ is in fact the inverse propagator and it is natural to expect $\beta < 0$ in the SFT case corresponding to convergent propagator at large momenta. From this SFT example we can draw a very important lesson. Assume $m = 0$. Then $\mathcal{F}_{\mathrm{SFT}} = (\Box+1)e^{\beta\Box}$ and the non-locality does not show up at all. Indeed, the poles of the propagator do not feel the exponent. Another way of thinking is that the exponential factor can be eliminated by a field redefinition. The situation is dramatically different for $m \neq 0$. The propagator $1/\mathcal{F}_{\mathrm{SFT}}$ has infinitely many poles. This is a manifestation of the non-locality. Also note that the physics

[2] The appearance of higher derivatives is not an exclusive feature of this theory. Non-commutative theories, for instance, also have higher derivative, but these non-local structures are very different.

would be totally different for full function $\mathcal{F}$ and its truncated series expansion because the pole structure may get modified considerably.

A very important role is played by the spectrum of the theory, determined by the equation

$$\mathcal{F}(J) = (J+1)\mathcal{G}^{-2}(J) - m^2 = 0. \tag{4}$$

We call it *characteristic* equation. It can be an algebraic or a transcendental. We consider $\mathcal{F}(\Box)$ of general form, with the only assumption that all roots are simple. The analyticity of the function $\mathcal{F}$ on the complex plane is also important for representing $\mathcal{F}$ by the convergent series expansion:

$$\mathcal{F} = \sum_{n=0}^{\infty} f_n \Box^n, \quad f_n \in \mathbb{R}. \tag{5}$$

Reality of coefficients is required by the hermiticity of the Lagrangian. Strictly speaking even the analyticity requirement can be weakened, but in this case one has to be careful with poles and the corresponding convergence domain of the series.

Classical solutions of the equation of motion were studied and analyzed in [6]. The key point in the analysis is the fact that action (3) is fully equivalent to the action

$$S_{local} = \int d^4x\sqrt{-g}\left(\frac{R}{16\pi G_N} - \frac{1}{g_o^2}\sum_i \frac{\mathcal{F}'(J_i)}{2}\left(g^{\mu\nu}\partial_\mu\tau_i\partial_\nu\tau_i + J_i\tau_i^2\right) - \Lambda\right) \tag{6}$$

with many free local scalar fields. Here J_i are roots of the characteristic equation (4) and there are as many scalar fields as roots of the characteristic equation.

It is important that the roots J_i as well as the coefficients $\mathcal{F}'(J_i)$ can be complex. This do not represent a problem, since all the local scalar fields are just mathematical functions without physical interpretation. What is important is that the original field τ must be real since it represents a physical excitation. This is simple to achieve. The roots J_i are either real or complex conjugate. Complex conjugate J_i would yield complex conjugate (up to an overall constant factor) solutions for τ_i. Thus, making the linear combination of all τ_i real is just a matter of choosing the integration constants appropriately. Finally, note that the coefficients $\mathcal{F}'(J_i)$ can be eliminated by a field rescaling.

In [8] it was proven that cosmological perturbations in the free theory with one non-local scalar field (3) and in the corresponding local theory (6) with many scalar field are equivalent as well. Moreover, several examples of evolution and perturbations were developed. The difficulty here, however, is that not many exactly solvable models with multiple scalar fields are known. Studying perturbations without having an exact solution is not an easy task. Of course, there is no need to have infinitely many scalar fields. On the contrary, on can set almost all of them to zero by choosing trivial integration constants. The point is that the models for which exact solutions are known are mostly models with one single field. Reducing the action (6) to a single local field completely misses the rich and non-trivial structure coming from SFT. The most intriguing case of complex masses requires at least two fields. In fact, we need the original function τ to be real and thus one cannot keep only one complex field without its complex conjugate.

3. Exactly solution with Cold Dark Matter (CDM)

We specialize to the case of spatially flat Friedmann–Robertson–Walker (FRW) Universe, with metric

$$ds^2 = -dt^2 + a^2(t)\left(dx_1^2 + dx_2^2 + dx_3^2\right), \tag{7}$$

where $a(t)$ is the scale factor and t is the cosmic time. The Hubble parameter, as usual, is $H = \dot{a}/a$ and the dot here and hereafter in this paper denotes a derivative with respect to t.

Background solutions for τ are therefore assumed to be spacially homogeneous as well. The local action (6) admits an exact solution to its equations of motion, which are

$$\begin{aligned} 3H^2 &= 4\pi G\left(\tilde{\sum_i}\mathcal{F}'(J_i)\left(\dot{\tau}_i^2+J_i\tau_i^2\right)\right)+8\pi G_N\Lambda\,,\\ \dot{H} &= -4\pi G\tilde{\sum_i}\mathcal{F}'(J_i)\dot{\tau}_i^2\,, \end{aligned} \tag{8}$$

and

$$\ddot{\tau}_i+3H\dot{\tau}_i+J_i\tau_i=0,\quad \text{for all } i\,. \tag{9}$$

One can explicitly check that there is the following solution

$$\begin{aligned} \tau_i &= \tau_{i0}\cosh(H_0t)^{-\frac{3}{2}\beta}P^n_{\frac{3}{2}\beta-1}(\tanh(H_0t))\,,\\ H &= H_0\beta\tanh(H_0t) \end{aligned} \tag{10}$$

and one more scalar field $\tau_{(0)}$ corresponding to $J_0=0$. where $\tau_{i0}=\sqrt{2\frac{\Gamma(\frac{3}{2}\beta-n)}{\Gamma(\frac{3}{2}\beta+n)}}$, $\frac{3}{2}\beta-1$ is not an integer and P^μ_ν are the associated Legendre functions of the first kind.

This solution is an exact solution provided we add some amount of the CDM in the system which can be naturally attributed to the brane to which our initial string is attached. To check the validity of this solution one has to use the "addition theorem" for the Legendre functions. Asymptotic behavior of this solution is exactly one expected in [6].

4. Exactly solvable model

An exact analytic solution for equations of motion following from the action (6) (and furthermore (3)) is not known. However, it is much more transparent to work with exact solutions rather than with asymptotics when one wants to study cosmological perturbations. There is a chance to modify the potential such that: first, the model becomes exactly solvable and, second, all the new terms vanish rapidly in the regime of interest, so that the previous picture is restored. Furthermore an exactly solvable model in cosmology has its own value just because not so many exactly solvable models are known. In our particular case we deal with many scalar fields and this complicates the problem. Moreover we have complex coefficients in the Lagrangian and this is an unexplored problem.

We consider the following modified action

$$\begin{aligned} S=\int d^4x\sqrt{-g}\Bigg[\frac{R}{16\pi G_N}+\frac{1}{g_o^2}\Bigg(-\sum_i\frac{\mathcal{F}'(J_i)}{2}\Big(g^{\mu\nu}(\partial_\mu\tau_{i+}\partial_\nu\tau_{i+}+\partial_\mu\tau_{i-}\partial_\nu\tau_{i-})+\\ +J_i(\tau_{i+}^2+\tau_{i-}^2)\Big)-\frac{3\pi G}{2}\left(\sum_i\mathcal{F}'(J_i)(\alpha_{i+}\tau_{i+}^2+\alpha_{i-}\tau_{i-}^2)\right)^2+c.c.\Bigg)-\Lambda\Bigg], \end{aligned} \tag{11}$$

where $H_0=\sqrt{\frac{8\pi G_N\Lambda}{3}}$, $G\equiv G_N/g_o^2$ is a dimensionless analog of the Newton constant and $\alpha_{i\pm}$ are the two solutions of $J_i=-\alpha_i(\alpha_i+3H_0)$. To simplify the notation, we introduce new indeces $P,Q,R\dots$ taking the degeneracy in α_i into account. Namely, we consider $P,Q,R\dots=i_+,i_-,j_+\dots$, bearing in mind that one can now have $J_P=J_Q$ for some values of P and Q. With this piece of notation the modified action (11) takes the form

$$S=\int d^4x\sqrt{-g}\left[\frac{R}{16\pi G_N}+\frac{1}{g_o^2}\left(-\tilde{\sum_P}\frac{\mathcal{F}'(J_P)}{2}\left(g^{\mu\nu}\partial_\mu\tau_P\partial_\nu\tau_P+J_P\tau_P^2\right)-\frac{3\pi G}{2}\left(\tilde{\sum_P}\mathcal{F}'(J_P)\alpha_P\tau_P^2\right)^2\right)-\Lambda\right] \tag{12}$$

where $\tilde{\sum}_P$ indicates that complex conjugate quantities are included in the sum[3].

The local action (12) admits an exact solution to its equations of motion, which are

$$\begin{aligned} 3H^2 &= 4\pi G\left(\tilde{\sum_P}\mathcal{F}'(J_P)\left(\dot{\tau}_P^2 + J_P\tau_P^2\right) + 3\pi G\left(\tilde{\sum_P}\mathcal{F}'(J_P)\alpha_P\tau_P^2\right)^2\right) + 8\pi G_N\Lambda\,, \\ \dot{H} &= -4\pi G\tilde{\sum_P}\mathcal{F}'(J_P)\dot{\tau}_P^2\,, \end{aligned} \tag{13}$$

and

$$\ddot{\tau}_P + 3H\dot{\tau}_P + J_P\tau_P + 6\pi G\alpha_P\tau_P\tilde{\sum_Q}\mathcal{F}'(J_Q)\alpha_Q\tau_Q^2 = 0, \quad \text{for all } P\,. \tag{14}$$

One can explicitly check that there is the following solution

$$\begin{aligned} \tau_P &= \tau_{P0}e^{\alpha_P t}\,, \\ H &= H_0 - 2\pi G\tilde{\sum_P}\mathcal{F}'(J_P)\alpha_P\tau_{P0}^2 e^{2\alpha_P t}\,. \end{aligned} \tag{15}$$

This solution is valid for any number of fields (including single field model) and for any values of parameters α_P (i.e. real, complex, etc.). Moreover, we see that if $\mathrm{Re}(\alpha_P) < 0$ for all P then the quartic term in the scalar fields potential vanishes and we are left with free fields. Thus for large times the model (6) is restored and we can speak about the asymptotic regime of SFT based models. Stability of the solution deserves a deeper analysis.

Acknowledgments

The author would like to thank the organizers of the "PASCOS 2010" conference for the opportunity to present this work and for creating a very stimulating environment for scientific discussions. The author is grateful to I.Ya. Aref'eva, F. Bezrukov, B. Craps, B. Dragovich, G. Dvali, A. Mazumdar, V. Mukhanov, and S.Yu. Vernov for useful comments and stimulating discussions. This work is supported in part by the Belgian Federal Science Policy Office through the Interuniversity Attraction Poles IAP VI/11, the European Commission FP6 RTN programme MRTN-CT-2004-005104 and by FWO-Vlaanderen through the project G.0428.06, RFBR grant 08-01-00798 and state contract of Russian Federal Agency for Science and Innovations 02.740.11.5057.

References

[1] K. Ohmori 2001 *Preprint* hep-th/0102085;
I.Ya. Aref'eva, D.M. Belov, A.A. Giryavets, A.S. Koshelev and P.B. Medvedev 2001 *Preprint* hep-th/0111208;
W. Taylor 2003 *Preprint* hep-th/0301094.

[2] L. Brekke, P.G.O. Freund, M. Olson, E. Witten 1988 *Nucl. Phys.* B **302** 365;
P.H. Frampton, Ya. Okada 1988 *Phys. Rev.* D **37** 3077;
V.S. Vladimirov, I.V. Volovich, E.I. Zelenov 1994 *p-adic Analysis and Mathematical Physics* (WSP, Singapore);
B. Dragovich, A. Yu. Khrennikov, S. V. Kozyrev, I. V. Volovich 2009 *Anal. Appl.* **1** 1 (*Preprint* arXiv:0904.4205).

[3] E. Komatsu 2009 *Astrophys. J. Suppl.* **180** 330 (*Preprint* arXiv:0803.0547);
M. Kilbinger et al. 2008 *Preprint* arXiv:0810.5129.

[4] A.S. Koshelev 2009 *Preprint* (arXiv:0912.5457);
Federico Galli, Alexey S. Koshelev 2010 *Theor. Math.Phys.* **164** 1169 (*Preprint* arXiv:1010.1773).

[3] For example $\tilde{\sum}_P J_P\tau_P^2 = \sum_P\left(J_P\tau_P^2 + J_P^*\tau_P^{2*}\right)$.

[5] I.Ya. Aref'eva 2006 *AIP Conf. Proc.* **826** 301 (*Preprint* astro-ph/0410443);
I.Ya. Aref'eva 2007 *AIP Conf. Proc.* **957** 297 (*Preprint* arXiv:0710.3017).
[6] A.S. Koshelev 2007 *J. High Energy Phys.* JHEP **0704** 029 (*Preprint* hep-th/0701103).
[7] I.Ya. Aref'eva and A.S. Koshelev 2008 *J. High Energy Phys.* JHEP **0809** 068, (*Preprint* arXiv:0804.3570).
[8] A.S. Koshelev and S.Yu. Vernov 2009 *Preprint* (arXiv:0903.5176);
A.S. Koshelev and S.Yu. Vernov 2010 *Preprint* (arXiv:1009.0746).

Journal of Physics: Conference Series **259** (2010) 012045 doi:10.1088/1742-6596/259/1/012045

Galileon Fifth Forces

Clare Burrage
Départment de Physique Théorique, Université de Genève,
24 Quai E. Ansermet, CH-1211, Genève, Switzerland

E-mail: clare.burrage@unige.ch

Abstract. The Galileon is a proposal to explain the current accelerated expansion of the universe through a scalar field with non-linear dynamics. The form of these non-linearities is protected by the Galileon symmetry and the theory is ghost free. I review the way in which the Galileon scalar field hides from experimental probes of gravity dynamically, showing that the theory is less constrained than previously supposed. I discuss the implications of this for quantum strong coupling scale.

1. Introduction

The past decade has provided compelling evidence that the expansion of the universe is accelerating. The simplest explanation for this is the presence of a cosmological constant term in the Einstein equations governing the evolution of the universe

$$R_{\mu\nu} - \frac{1}{2}Rg_{\mu\nu} = 8\pi G T_{\mu\nu} + \Lambda g_{\mu\nu}, \tag{1}$$

where the effective cosmological constant which drives the accelerated expansion of the universe is the sum of a bare cosmological constant and a contribution from the energy density of the vacuum

$$\Lambda_{eff} = \Lambda + 8\pi G\langle\rho_{vac}\rangle. \tag{2}$$

Field theory estimates of the vacuum energy density indicate that it scales as the fourth power of the energy cut-off of the theory. This leads to the cosmological constant problem: the scale predicted for Λ is far larger than the observed value $\Lambda \sim 10^{-3}$ eV.

It is therefore worth considering whether there is an alternative explanation for the accelerated expansion of the universe. In particular whether our failure to explain observations of the current expansion of the universe indicates a failure of Einstein gravity on the largest scales in the universe. One example of such an alternative explanation is the DGP model [1] where the universe is confined to a 3-brane embedded in a 5D Minkowski bulk. In this scenario gravity appears four-dimensional below a length scale L, and five-dimensional at larger distances. The DGP model can describe an FRW universe which gets accelerated at late times by the leaking of gravity into the fifth dimension without the need for a cosmological constant (or a dynamical dark energy field) on the brane. However it has been shown that the self accelerating branch of the DGP model contains negative energy classical solutions known as ghosts and is therefore unstable [2].

Are there other modifications of gravity which could describe the accelerated expansion of the universe, without the problems encountered by the DGP model? Such a theory would have to describe significant modifications of gravity on the largest observable scales in the universe but negligible modifications of gravity within the solar system to be in agreement with observations. By necessity therefore, it must be non linear. To avoid the problems of the DGP model it must also be free of negative energy solutions - which requires that the equations of motion are second order. These two requirements led to the construction of the Galileon model by Nicolis, Rattazzi and Trincherini [3], which we now proceed to describe.

2. The Galileon

Consider modifications of general relativity where locally, and for weak fields, the deviations can be written in terms of a light scalar field, π. The Galileon model assumes that the theory obeys a covariant generalisation of the Galileon shift symmetry

$$\pi(x) \to \pi(x) + b_\mu x^\mu + c. \tag{3}$$

The coupling between π and other matter fields is assumed to be linear, and the back reaction of the energy density of the scalar field is neglected.

There are then only five possible Lagrangians which obey the Galileon symmetry and give rise to second order equations of motion [3]:

$$\mathcal{L}_1 = \pi, \tag{4}$$

$$\mathcal{L}_2 = -\frac{1}{2}\partial\pi\cdot\partial\pi, \tag{5}$$

$$\mathcal{L}_3 = -\frac{1}{2}(\Box\pi)\partial\pi\cdot\partial\pi, \tag{6}$$

$$\mathcal{L}_4 = -\frac{1}{4}[(\Box\pi)^2\partial\pi\cdot\partial\pi - 2(\Box\pi)\partial\pi\cdot\Pi\cdot\partial\pi - (\Pi\cdot\Pi)(\partial\pi\cdot\partial\pi) + 2\partial\pi\cdot\Pi\cdot\Pi\cdot\partial\pi], \tag{7}$$

$$\begin{aligned}\mathcal{L}_5 = &-\frac{1}{5}[(\Box\pi)^3\partial\pi\cdot\partial\pi - 3(\Box\pi)^2\partial\pi\cdot\Pi\cdot\partial\pi - 3\Box\pi(\Pi\cdot\Pi)(\partial\pi\cdot\partial\pi) + 6(\Box\pi)\partial\pi\cdot\Pi\cdot\Pi\cdot\partial\pi \\ &+2(\Pi\cdot\Pi\cdot\Pi)(\partial\pi\cdot\partial\pi) + 3(\Pi\cdot\Pi)\partial\pi\cdot\Pi\cdot\partial\pi - 6\partial\pi\cdot\Pi\cdot\Pi\cdot\Pi\cdot\partial\pi],\end{aligned} \tag{8}$$

where $\Pi^\mu_\nu = \partial^\mu\partial_\nu\pi$ and $\cdot$ indicates a contraction of indices. The most general Galileon Lagrangian is a linear combination of the above with arbitrary constant coefficients

$$\mathcal{L}_\pi = \sum_{i=1}^{5} c_i\mathcal{L}_i. \tag{9}$$

The Galileon symmetry prevents other non-linear terms from being generated by quantum fluctuations. A five dimensional theory has been constructed which gives rise to the Galileon model in an appropriate four-dimensional limit [4]. Then the constant coefficients c_i are related to brane tensions, and the coefficients of the curvature invariants in the higher dimensional Lagrangian. Previously it has been suggested that observational constraints force all of the c_i to be at the Planck scale. Here we show that this is not the case, broadening the spectrum of Galileon phenomenology.

The Galileon theory can describe the current accelerated expansion of the universe as there is a solution to the Galileon equations of motion which corresponds to de Sitter space, even if all other sources are absent [3]

$$\pi_{dS}(x) = -\frac{1}{4}H^2 x_\mu x^\mu. \tag{10}$$

This is a self accelerating solution, analogous to those found for the DGP model. But unlike the DGP model the Galileon scenario is free of ghost-like instabilities. The existence of this self acceleration solution imposes one algebraic constraint on the five coefficients c_i:

$$c_1 - 2c_2H^2 + 3c_3H^4 - 3c_4H^6 + \frac{3}{2}c_5H^8 = 0. \tag{11}$$

3. Galileon Forces

The existence of new light scalar fields as an explanation for the accelerated expansion of the universe has long been known to be problematic; such fields typically mediate new long range fifth forces. Laboratory and solar system tests of gravity tightly constrain the existence of such forces, and a light canonical scalar field can only exist if it couples to matter with a strength many orders of magnitude weaker than the gravitational force. However non-linearities were included in the Galileon model precisely to try to avoid this problem. We now turn to how these non-linearities help to screen the effects of the scalar field within the solar system.

In the presence of matter π is perturbed from its background value $\pi \to \pi_{dS} + \pi$, and it is these perturbations which potentially mediate strong forces between massive objects. The force due to the Galileon field can be shown to be proportional to the gradient of the scalar field profile around a massive object [5], therefore in order to know the strength of the Galileon force, we must first determine the field profile generated by massive objects.

The spherically symmetric, static equation of motion for the Galileon field near an object of mass M is

$$d_2\left(\frac{\pi'}{r}\right) + 2d_3\left(\frac{\pi'}{r}\right)^2 + 2d_4\left(\frac{\pi'}{r}\right)^3 = \frac{M}{4\pi r^3}, \tag{12}$$

where the d_i are linear combinations of the coefficients c_i. Stable solutions to this equation exist if $d_2 > 0$, $d_4 \geq 0$, $d_3 \geq \sqrt{(3/2)d_2d_4}$, $d_5 < 0$ [3]. If the non-linearities are absent $d_3 = d_4 = d_5 = 0$ we recover the linear solution

$$\pi(r) = \pi_0 - \frac{M}{4\pi d_2 r}, \tag{13}$$

and the ratio of the Galileon force to the Newtonian gravitational one is

$$\frac{F_\pi}{F_N} = \frac{2M_P^2}{d_2}. \tag{14}$$

As previously discussed, we see that the scalar force can only be suppressed by tuning the parameter d_2 to be much larger than the Planck scale. However in the full Galileon theory the force can be made substantially weaker when the non-linearities become important.

The equation of motion (13) can be rewritten

$$g + \left(\frac{R_1}{r}\right)^3 g^2 + \left(\frac{R_2}{r}\right)^6 g^3 = 1 \tag{15}$$

where $\pi'/r = (M/4\pi r^3 d_2)g(r)$ and $R_1^3 = d3M/2\pi d_2^2$, $R_2^6 = M^2d_4/8\pi^2d_2^3$. Then the Galileon force is dynamically suppressed whenever $g(r) < 1$. It's clear that whenever d_3 or d_4 is non zero g is forced to tend to zero in the interior of any massive object. The radius within which the non-linearities become important and g becomes less than unity is known as the Vainshtein radius

$$r_\star \sim \max\{R_1, R_2\}. \tag{16}$$

Unlike in the original massive gravity scenario discussed by Vainshtein, or the DGP model, the Vainshtein radius of an object is not fixed in the Galileon model. There remains enough freedom in the model to make the Vainshtein radius of any object arbitrarily large or small.

3.1. Within the Vainshtein Radius

There are two regimes of behaviour of the field within the Vainstein radius, depending on which of the nonlinearities dominate in (13), forming concentric shells around the source object. We call these region A and region B.

In region A: $\alpha R_2 \ll r \ll R_1$, where $\alpha + (R_2/R_1)^3 < 1$. Here the ratio of the Galileon to gravitational force is

$$\frac{F_\pi}{F_N} = \frac{M_P^2}{4\pi d_2}\left(\frac{r}{R_1}\right)^{3/2} = \frac{M_P^2}{2}\left(\frac{1}{2\pi d_3 M}\right)^{1/2} r^{3/2}. \tag{17}$$

In region B: $0 < r \ll \alpha R_2$. Here the ratio of the Galileon to gravitational force is

$$\frac{F_\pi}{F_N} = \frac{M_P^2}{4\pi d_2}\left(\frac{r}{R_2}\right)^{2} = \frac{M_P^2}{2}\left(\frac{1}{\pi d_4 M^2}\right)^{1/3} r^{2}. \tag{18}$$

The force inside the Vainshtein radius is always independent of d_2 and therefore we can vary the position of the Vainshtein radius and the strength of the Galileon force inside that Vainshtein radius independently. In particular note that, as the position of the Vainshtein radius is arbitrary, we do not need to impose that the Galileon force is weaker than gravity everywhere inside the Vainshtein radius in order to be in agreement with observational constraints.

To ensure that the Galileon field satisfies the observational constraints from any gravitational experiment dynamically, and without fine tuning of the parameters, we must follow a two step procedure. Firstly we must ensure that a gravitational experiment lies within its own Vainshtein radius, so that the scalar force is dynamically suppressed within the experiment. This imposes a constraint on either R_1 or R_2. Secondly we must impose that the Galileon force within the experiment is sufficiently weak to evade detection. This imposes a constraint on d_3 or d_4 respectively.

Determination of the full set of constraints imposed on the Galileon model by gravitational experiments remains for future work. Here we make only the first step along this path by requiring that gravitationally bound objects should not feel significant deviations from general relativity. This requires that the Vainshtein radius of a gravitationally bound object must be bigger than its size, and that within the bound radius the Galileon force is suppressed compared to the gravitational one. This imposes two constraints on the parameters of the model:

$$d_2 \lesssim M_P^2, \tag{19}$$
$$d_3 \gtrsim 10^{118}. \tag{20}$$

4. The Strong Interaction Scale

As well as determining the strength of the Galileon force the couplings c_i, or equivalently d_i, also control the energy scale at which quantum fluctuations of the Galileon field becomes strongly self interacting. If this scale is too low, within reach of current experiments, we might worry that we do not have sufficient control over the theory to predict whether the theory really is in agreement with observations.

The Lagrangian for a canonically normalised perturbation ϕ about the de Sitter background $\pi \to \pi_{dS} + \phi$ can be written schematically as

$$\mathcal{L}_\phi = -(\partial\phi\cdot\partial\phi)\left(\frac{1}{2} + \frac{1}{2}\frac{d_3}{d_2^{3/2}}[P] + \frac{1}{4}\frac{d_4}{d_2^2}[P]^2 + \frac{1}{5}\frac{d_5}{d_2^{5/2}}[P]^3\right) \tag{21}$$

where $[P]$ denotes polynomials in $\partial\partial\phi$ with the required contractions of indices. Inspection of this Lagrangian might lead us to conclude that strong interactions become important at potentially low scales, for example $d_2^{1/2}/d_3^{1/3}$. However this would be misleading as no experiment can ever be done without the presence of matter. Instead we measure fluctuations in the neighbourhood of massive objects and we should compute the strong coupling scale within an object's Vainstein radius.

Within the Vainshtein radius renormalisations from the background alter the strong coupling scale. For example if $d_4 = d_5 = 0$ we find that the Lagrangian for a perturbation is

$$\mathcal{L}_\phi = Z_{\mu\nu}\partial^\mu\phi\partial^\nu\phi - \frac{d_3}{2}(\Box\phi)\partial\phi\cdot\partial\phi, \tag{22}$$

where $Z_{\mu\nu}$ can be diagonalised with all eigenvalues at the scale

$$Z \sim -\frac{d_2}{2}\left[1 - \frac{1}{2}\left(\frac{R_1}{r}\right)^{3/2}\right]. \tag{23}$$

Combining this with the previously found constraints on d_2 and d_3, (20) and (21), we find that the quantum cut off at the surface of the earth is

$$\tilde{\Lambda}(x) \sim \frac{Z^{1/2}}{d_3^{1/3}} \lesssim \frac{1}{1\ \text{cm}} \tag{24}$$

This is problematic for the Galileon model as gravitational experiments at the surface of the earth have probed gravitational forces on distance scales shorter than this would allow. However this problem is alleviated if the other non-linearities also play a role. For example, allowing all the d_i to be non-zero, if the surface of the earth lies within region B, defined above, then the strong coupling scale is again determined by a diagonalisible matrix $Z_{\mu\nu}$, where the eigenvalues are now

$$Z_r \sim 3d_2\left(\frac{R_2}{r}\right)^2, \tag{25}$$

$$Z_t \sim 3d_2\left(\frac{R_2}{r}\right)^2 - \frac{d_5 d_2^{3/2}}{\sqrt{2}d_4^{3/2}}, \tag{26}$$

$$Z_\Omega \sim -d_2 + \frac{2d_3^2}{3d_4}, \tag{27}$$

which can all be made sufficiently large that the quantum strong interaction scale can be put well beyond the reach of experiments.

5. Conclusions

The Galileon is a scalar field which couples to the fields of the Standard Model. It has a background self accelerating solution which explains observations of the current accelerated expansion of the universe without the need for a cosmological constant. Its dynamics are non-linear but protected by the Galileon symmetry, and the equations of motion for the field are second order so that the theory is ghost free.

The non-linearities of the theory allow the field to hide from fifth-force experiments without fine tuning the parameters of the theory. The Galileon field effectively becomes weakly coupled in the presence of matter, and so hides from gravitational experiments dynamically. The size of the Vainshtein radius, which controls when the non-linearities become important, is arbitrary in the Galileon model, and remains to be constrained by experiment. In contrast to previous assumptions this means that we are not required to set all the parameters of the theory at the

Planck scale. It also means that the strong coupling scale of the theory can be made arbitrarily high, and therefore placed out of reach of experiments.

In summary the Galileon model is less constrained by experiment than was previously supposed, and much of its phenomenology remains to be explored!

Acknowledgments
The research described here was done in collaboration with David Seery and published as [5]. CB was supported by the German Science Foundation (DFG) under the Collaborative Research Centre (SFB) 676.

References
[1] Dvali G R, Gabadadze G and Porrati M 2000 *Phys. Lett.* **B485** 208–214 (*Preprint* `hep-th/0005016`)
[2] Luty M A, Porrati M and Rattazzi R 2003 *JHEP* **09** 029 (*Preprint* `hep-th/0303116`)
[3] Nicolis A, Rattazzi R and Trincherini E 2009 *Phys. Rev.* **D79** 064036 (*Preprint* `0811.2197`)
[4] de Rham C and Tolley A J 2010 *JCAP* **1005** 015 (*Preprint* `1003.5917`)
[5] Burrage C and Seery D 2010 *JCAP* **1008** 011 (*Preprint* `1005.1927`)

Journal of Physics: Conference Series **259** (2010) 012046 doi:10.1088/1742-6596/259/1/012046

Gauge non-singlet (GNS) inflation in SUSY GUTs

Jochen P Baumann

Max-Planck-Institut für Physik (Werner-Heisenberg-Institut), Föhringer Ring 6, 80805 München, Germany

E-mail: `jbaumann@mppmu.mpg.de`

Abstract. In this talk based on reference [1] we explore the novel possibility that the inflaton responsible for cosmological inflation is a gauge non-singlet matter field in supersymmetric (SUSY) Grand Unified Theories (GUTs). We consider hybrid-like inflation models in SUSY (more specifically tribrid inflation models [2]), where we show that the scalar components of gauge non-singlet superfields, together with fields in conjugate representations, may form a D-flat direction suitable for inflation.

We apply these ideas to SUSY models with an Abelian gauge group, a Pati-Salam gauge group and finally Grand Unified Theories based on $SO(10)$. Here, the scalar components of the matter superfields in the **16**s may combine with a single $\overline{\mathbf{16}}$ to form the inflaton. Thus in some sense this "unified particle" can act as the "mother of the universe", as it contains the inflaton responsible for inflation as well as all the matter present today.

1. Introduction

The inflationary paradigm remains extremely successful in solving the horizon and flatness problems of the standard Big Bang cosmology, and at the same time in explaining the origin of structure of the observable Universe and the absence of relic topological defects [3] (for a textbook review, see e.g. [4]).

A long standing question in inflation models is: Who is the inflaton? We are still far from answering this question. Indeed it is still unclear whether the inflaton, the (presumed) scalar field responsible for inflation, should originate from the observable (matter) sector or the hidden (e.g. moduli) sector of the theory. However the connection between inflation and particle physics is rather difficult to achieve in the observable sector due to the lack of understanding of physics beyond the Standard Model (SM) and in the hidden sector due to the lack of understanding of the string vacuum.

However over the past dozen years there has been a revolution in particle physics due to the experimental discovery of neutrino mass and mixing (for a review, see e.g. [5]), and this improves the prospects for finding the inflaton in the observable sector. Indeed, if the SM is extended to include the seesaw mechanism [6–11] and SUSY, the right-handed sneutrinos, the superpartners of the right-handed neutrinos, become excellent inflaton candidates.

Despite the unknown identity of the inflaton, conventional wisdom dictates that it must be a gauge singlet since otherwise radiative corrections would spoil the required flatness of the inflaton potential. For example in SUSY models scalar components of gauge non-singlet superfields have quartic terms in their potential, due to the D-terms, leading to violations of the slow-roll conditions which are inconsistent with recent observations by WMAP. In addition, gauge non-singlet inflatons would be subject to one-loop Coleman-Weinberg corrections from loops with gauge fields which could easily lead to large radiative corrections that induce an unacceptably large slope of the inflaton potential. Furthermore a charged inflaton is in general also subject to two-loop corrections to its mass which can easily be larger than the Hubble scale [12]. Such a contribution is in principle large enough to spoil inflation for any gauge non-singlet scalar field, leading to a sort of "gauge η-problem".

In this talk we shall argue that, contrary to conventional wisdom, the inflaton may in fact be a gauge *non*-singlet (GNS). For definiteness we shall confine ourselves here to examples of SUSY hybrid inflation and show that the scalar components of gauge non-singlet superfields, together with fields in conjugate representations, may form a D-flat direction suitable for inflation. Along this D-flat trajectory the usual F-term contributes the large vacuum energy. We apply these ideas first to a simple Abelian gauge group $G = U(1)$, then to a realistic SUSY Pati-Salam model, then to $SO(10)$ SUSY GUTs, where the scalar components of the matter superfields in the **16**s may combine with a single $\overline{\mathbf{16}}$ to form the inflaton, with the right-handed sneutrino direction providing a possible viable trajectory for inflation.

Indeed, assuming the sneutrino inflationary trajectory, we calculate the one-loop Coleman-Weinberg corrections and the two-loop corrections usually giving rise to the "gauge η-problem" and show that both corrections do not spoil inflation. In addition we show that the monopole problem [13] can be resolved. The usual η-problem arising from SUGRA [14] may be resolved using a Heisenberg symmetry [15] with stabilized modulus [16].

2. SUSY Hybrid Inflation with a GNS Inflaton

SUSY hybrid inflation is typically based on the superpotential [17]

$$W_0 = \kappa\, S \left(H\bar{H} - M^2\right) \tag{1}$$

where the superfield S is a singlet under some gauge group G, while the superfields H and $\bar{H}$ reside in conjugate representations (reps) of G. The F-term of S provides the vacuum energy to drive inflation, the scalar component of the singlet S is identified as the slowly rolling inflaton, and the scalar components of H and $\bar{H}$ are waterfall fields which take zero values during inflation but are switched on when the inflaton reaches some critical value, ending inflation and breaking the gauge group G at their global minimum $\langle H\rangle = \langle\bar{H}\rangle = M$. Typically G is identified as a GUT group and H, $\bar{H}$ are the Higgs which break that group..

Consider the following simple extension of the superpotential in Eq. (1),

$$W = W_0 + \frac{\zeta}{\Lambda}\left(\phi\,\bar{\phi}\right)\left(H\bar{H}\right) \tag{2}$$

where we have included an additional pair of GNS superfields ϕ and $\bar{\phi}$ in conjugate reps of G which couple to the Higgs superfields via a non-renormalizable coupling controlled by a dimensionless coupling constant ζ and a scale Λ. At first glance, we might expect the presence of the effective operator in Eq. (2), that we have added to the superpotential W_0 in Eq. (1), to not perturb the usual SUSY hybrid inflation scenario described above. However its presence allows the new possibility that inflation is realized via slowly rolling scalar fields contained in the superfields ϕ and $\bar{\phi}$ with the singlet field S staying fixed at zero during (and after) inflation. In a SUGRA framework, non-canonical terms for S in the Kähler potential can readily provide a large mass for S such that it quickly settles at $S = 0$. On the other hand, large SUGRA mass contributions can be avoided for ϕ and $\bar{\phi}$ using a Heisenberg symmetry [16].

While the singlet S field is held at a zero value by SUGRA corrections, the scalar components of ϕ, $\bar{\phi}$, having no such SUGRA corrections, are free to take non-zero values during the inflationary epoch. The non-zero ϕ, $\bar{\phi}$ field values provide positive mass squared contributions to all components of the waterfall fields H and $\bar{H}$ during inflation, thus stabilizing them at

Journal of Physics: Conference Series **259** (2010) 012046 doi:10.1088/1742-6596/259/1/012046

zero by the F-term potential from the second term in Eq. (2). As in standard SUSY hybrid inflation, the F-term of S, arising from W_0 in Eq. (1), yields a large vacuum energy density $V_0 = \kappa^2 M^4$ which drives inflation and breaks SUSY. Since ϕ, $\bar{\phi}$ are the only fields which are allowed to take non-zero values during inflation, they may be identified as inflaton(s) provided that their potential is sufficiently flat. Since both ϕ and $\bar{\phi}$ carry gauge charges under G, their vacuum expectation values (VEVs) break G already during inflation, thus, although ϕ and $\bar{\phi}$ are GNS fields under the original gauge group G, they are clearly gauge singlets under the surviving subgroup of $G' \subset G$ respected by inflation. This trivial observation will help to protect the ϕ and $\bar{\phi}$ masses against large radiative corrections. Another key feature is that the quartic term in the ϕ and $\bar{\phi}$ potential arising from D-term gauge interactions may be avoided in a D-flat valley in which the conjugate fields ϕ and $\bar{\phi}$ take equal VEVs.

3. Explicit Example with $G = U(1)$

Let us now explicitly calculate the full global SUSY potential for the model in Eq. (2), assuming an Abelian gauge group $G = U(1)$. For $G = U(1)$ and equal charge for ϕ and H we obtain a D-term contribution (setting a possible Fayet-Iliopoulos term to zero)

$$V_D = \frac{g^2}{2}\left(|\phi|^2 - |\bar{\phi}|^2 + |H|^2 - |\bar{H}|^2\right)^2 , \tag{3}$$

which on the inflationary trajectory $\langle H\rangle = \langle \bar{H}^*\rangle = 0$ obviously has a D-flat direction $\langle\phi\rangle = \langle\bar{\phi}^*\rangle$. Under the assumption that the D-term potential Eq. (3) has already stabilized the fields in the D-flat valley, the remaining potential is generated from the F-term part which can be calculated with the equations of motion $F^{*i} = -\partial W/\partial\phi_i$. Plugging in the D-flatness condition $\langle\phi\rangle = \langle\bar{\phi}^*\rangle$ and setting $S = 0$, the F-term potential reduces to

$$V_F = \left|\kappa^2\left(M^2 - H\bar{H}\right)\right|^2 + 2\frac{|\zeta|^2}{\Lambda^2}|\phi|^2|H|^2|\bar{H}|^2 + \frac{|\zeta|^2}{\Lambda^2}|\phi|^4|H|^2 + \frac{|\zeta|^2}{\Lambda^2}|\phi|^4|\bar{H}|^2 , \tag{4}$$

which is of the typical hybrid-form, cf. Fig. 1.

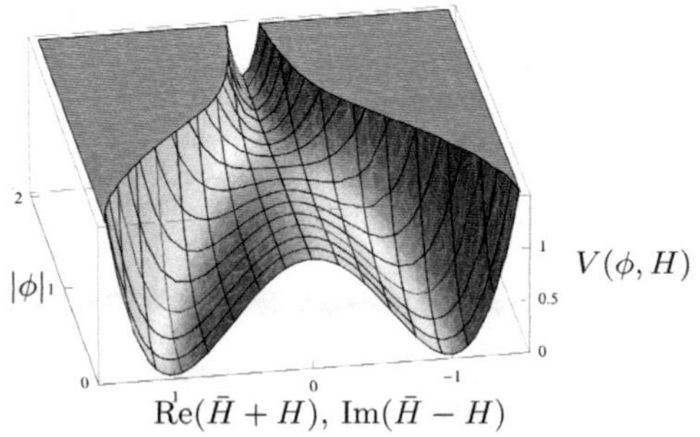

Figure 1: Plot of the F-term hybrid inflation potential in the D-flat valleys $\phi = \bar{\phi}^*$, $H = \bar{H}^*$.

Indeed, in the inflationary valley $S = H = \bar{H} = 0$ it has a flat inflaton direction $|\phi|$ and a tachyonic waterfall direction below some critical value $|\phi_c|$.

The tree-level flat direction is only lifted radiatively due to inflaton-dependent, SUSY breaking waterfall masses. Diagonalizing the mass matrices in the $(H, \bar{H})$-basis, the eigenvalues calculated from Eqs. (2) and (4) are 1 Dirac fermion with the squared mass $m_F^2 = |\zeta|^2|\phi|^4/\Lambda^2$ and 2 complex scalars with squared masses $m_S^2 = |\zeta|^2|\phi|^4/\Lambda^2 \pm |\kappa|^2 M^2$. Plugging these values into the formula for the one-loop Coleman-Weinberg potential and integrating back the equation of motion of the inflaton field to the point 60 e-folds before the end of inflation we obtain the following predictions for the model at hand

$$n_s \simeq 0.98 \quad , \quad r \lesssim 0.01 \quad , \quad M \simeq 7\cdot 10^{15}\,\text{GeV} . \tag{5}$$

4. Sneutrino Inflation in SUSY Pati-Salam

In this section we discuss a fully realistic example of SUSY hybrid inflation with a GNS inflaton where G is identified with the SUSY Pati-Salam gauge group. Following the general ideas presented in the previous section, in the model under construction inflation will proceed along a trajectory in field space where the D-term contribution vanishes and the F-term contribution dominates the vacuum energy. In addition to that we want to associate the inflaton field to the "matter sector" of the theory so that the model is closely related to low energy particle physics.

Typically if there are only matter fields in the (CP conjugated) right-handed Pati-Salam reps R_i^c this would lead to large D-term contributions incompatible with inflation. Therefore, in addition to the matter fields R_i^c we also introduce another field $\bar{R}^c$ in the conjugate rep of the gauge group. For simplicity, we will discuss here the case where $i = 1, \ldots, 4$ and where there is only one $\bar{R}^c$. The introduction of $\bar{R}^c$ is also necessary in order to keep all the waterfall directions stabilized during inflation. After inflation, one linear combination of the fields R_i^c will pair up with $\bar{R}^c$ and become heavy, while three other combinations remain light and contain the three generations of SM fields. In addition, the superfields containing the right-handed neutrinos of the seesaw mechanism will obtain their large masses after inflation.

More specifically, the (minimalist) field content of our model reads

$$
\begin{aligned}
R_i^c &= (\bar{\mathbf{4}}, \mathbf{1}, \bar{\mathbf{2}}) = \begin{pmatrix} u_i^c & u_i^c & u_i^c & \nu_i^c \\ d_i^c & d_i^c & d_i^c & e_i^c \end{pmatrix}, \ \bar{R}^c = (\mathbf{4}, \mathbf{1}, \mathbf{2}) = \begin{pmatrix} \bar{u}^c & \bar{u}^c & \bar{u}^c & \bar{\nu}^c \\ \bar{d}^c & \bar{d}^c & \bar{d}^c & \bar{e}^c \end{pmatrix} \\
H^c &= (\bar{\mathbf{4}}, \mathbf{1}, \bar{\mathbf{2}}) = \begin{pmatrix} u_H^c & u_H^c & u_H^c & \nu_H^c \\ d_H^c & d_H^c & d_H^c & e_H^c \end{pmatrix}, \ \bar{H}^c = (\mathbf{4}, \mathbf{1}, \mathbf{2}) = \begin{pmatrix} \bar{u}_H^c & \bar{u}_H^c & \bar{u}_H^c & \bar{\nu}_H^c \\ \bar{d}_H^c & \bar{d}_H^c & \bar{d}_H^c & \bar{e}_H^c \end{pmatrix}
\end{aligned}
\tag{6}
$$

where $H^c, \bar{H}^c$ are the waterfall Higgs superfields breaking PS to the SM after inflation and in addition we introduce two further gauge singlet fields, namely S and X. The introduction of the spurion X, which aquires a VEV before the inflationary epoch, is due to two additional symmetries, namely a R-symmetry and a discrete $\mathbb{Z}_{10}$ symmetry, which allow us to restrict the allowed terms in the superpotential, which reads as follows

$$
\begin{aligned}
W = &\kappa\, S \left(\frac{\langle X \rangle}{\Lambda} H^c \bar{H}^c - M^2 \right) \\
&+ \frac{\lambda_{ij}}{\Lambda} (R_i^c \bar{H}^c)(R_j^c \bar{H}^c) + \frac{\gamma}{\Lambda} (\bar{R}^c H^c)(\bar{R}^c H^c) + \frac{\zeta_i}{\Lambda} (R_i^c \bar{R}^c)(H^c \bar{H}^c) + \frac{\xi_i}{\Lambda} (R_i^c \bar{H}^c)(\bar{R}^c H^c) \,.
\end{aligned}
\tag{7}
$$

Here, two multiplets enclosed in brackets are contracted with their respective $SU(4)_C$ and $SU(2)_R$ indices.

We would like to remark at this point that the symmetries and charge assignments we have chosen are not unique and should mainly illustrate that it is possible to obtain the desired form of the superpotential by symmetry.

It turns out that the model has several tree-level flat directions in $R_i^c, \bar{R}^c$ field space and in principle inflation can proceed along any of them. However, let us now focus on the inflationary dynamics when the inflaton fields acquire VEVs along the sneutrino direction. As an explicit example we consider the simplest case where only one of the $R^c = R_1^c \neq 0$ as well as $\bar{R}^c \neq 0$ are slow-rolling, while all the others remain at zero $R_{i\neq 1}^c = 0$.

Computing the D-term potential in the case at hand, we find that it has a flat direction for $|\nu^c| = |\bar{\nu}^c|$. From now on, we assume that inflation occurs in this D-flat valley. Therefore the scalar potential during inflation has to be calculated with the D-flatness condition $|\nu^c| = |\bar{\nu}^c|$ imposed.

Looking at the resulting scalar potential we find that due to large F-term contributions to their masses from the VEVs of the inflaton fields, the waterfall fields are fixed at zero during inflation (S, as always, is fixed at 0 during and after inflation by SUGRA corrections). However,

Journal of Physics: Conference Series **259** (2010) 012046 doi:10.1088/1742-6596/259/1/012046

as the inflaton fields slowly roll to smaller values, the masses of the waterfall fields decrease and finally one direction in field space becomes tachyonic. The $H^c, \bar{H}^c$ fields now quickly roll to their true minima and inflation ends by the "waterfall".

One can calculate the critical values at which the system gets destabilized by setting the dynamical masses to zero. We find

$$|\tilde{\nu}^c_{\text{crit}}| = \sqrt{\frac{2\,|\kappa|\,M}{|\zeta|}}\,, \tag{8}$$

for the $\text{Re}(u^c_H + \bar{u}^c_H)$, $\text{Im}(\bar{u}^c_H - u^c_H)$, ... directions and the real, positive solutions

$$|\tilde{\nu}^c_{\text{crit}}| = \sqrt{\frac{2\,|\kappa|\,M}{|\zeta + \xi + 2\gamma|}}\,, \qquad |\tilde{\nu}^c_{\text{crit}}| = \sqrt{\frac{2\,|\kappa|\,M}{|\zeta + \xi - 2\gamma|}}\,, \tag{9}$$

for the $\text{Re}(\bar{\nu}^c_H - \nu^c_H)$- and $\text{Im}(\bar{\nu}^c_H - \nu^c_H)$-directions.

For generic non-zero values of γ (and for example small ξ), either the $\text{Re}(\bar{\nu}^c_H - \nu^c_H)$- or the $\text{Im}(\bar{\nu}^c_H - \nu^c_H)$-direction will become tachyonic for larger values of the inflaton VEV than the $\text{Re}(u^c_H + \bar{u}^c_H)$, ... directions. Consequently, it destabilizes first and the waterfall will happen in this direction in field space, breaking G_{PS} down to the Standard Model. Furthermore, since the waterfall happens in a preferred direction, the production of monopoles is avoided.

Also, since the gauge symmetry is already broken by the inflaton VEVs during inflation, the inflaton direction basically decouples from the gauge interactions. Therefore the potentially dangerous one- and two-loop corrections to the inflaton potential are under control and pose no threat to the realization of slow-roll inflation.

After calculating the full mass spectrum on the inflationary trajectory and plugging it into the formula for the one-loop Coleman-Weinberg potential, one can again derive the inflationary predictions, which are very similar to those for the case $G = U(1)$ presented in the last section.

5. $SO(10)$ SUSY GUTs

In order to embed our model of the previous section into a $SO(10)$ framework we first have to make it explicitly left-right-symmetric by adding left-charged matter-supermultiplets as well as left-charged Higgs-supermultiplets to the theory.

At this stage the model contains two copies of the inflaton sector discussed in the previous section, one charged under $SU(2)_R$ and one charged under $SU(2)_L$, as well as additional couplings between the two sectors. In the absence of a discrete left-right symmetry we would expect the couplings in the left and right sector to be not exactly equal. With two potential sectors for inflation, inflation may happen in both of them with the respective sneutrinos playing the role of the inflaton. Thus we might have an "inflaton race" between the two sectors. Once the waterfall happens in one of the two sectors (with different couplings in each sector we do not expect this to happen simultaneously), inflation ends since the vacuum energy given by the F_S-term vanishes. At the same time the masses of the matter fields get fixed by the VEVs of the waterfall fields and the couplings between the left and the right sector.

In terms of the PS framework considered in the preceding section, each of the left- and right-charged leptoquark superfields as well as the Higgs superfields are unified into **16** reps and their conjugate counterparts into $\overline{\mathbf{16}}$ reps as

$$\begin{aligned} \mathbf{16} &= (\mathbf{4},\mathbf{2},\mathbf{1}) \oplus (\bar{\mathbf{4}},\mathbf{1},\bar{\mathbf{2}})\,, \\ \overline{\mathbf{16}} &= (\bar{\mathbf{4}},\bar{\mathbf{2}},\mathbf{1}) \oplus (\mathbf{4},\mathbf{1},\mathbf{2})\,. \end{aligned} \tag{10}$$

Choosing again proper symmetries to forbid unwanted operators, we arrive at the following superpotential relevant for inflation

$$W_{\text{inf}} = \kappa\, S\left(H\bar{H} - M^2\right) + \frac{\lambda_{ij}}{\Lambda} F_i F_j \bar{H}\bar{H} + \frac{\gamma}{\Lambda} \bar{F}\bar{F} H H + \frac{\zeta_i}{\Lambda} F_i \bar{F} H \bar{H} + \dots\ , \tag{11}$$

where F_i and $\bar{F}$ now contain the inflaton as well as the SM matter fields.

We assume that $SO(10)$ is broken to G_{PS} before inflation and then inflation, as well as the waterfall after inflation are realized as discussed in the previous section.

Again, we would like to emphasize at this point that the minimalist field content and the choice of symmetries mainly serves the purpose of giving a proof of existence that GNS inflation can be realized in $SO(10)$ GUTs. In a fully realistic model, which e.g. may also contain a full flavor sector, different symmetries may have to be chosen and the field content may have to be extended.

6. Conclusions

We have forged a connection between inflation and particle physics by identifying the inflaton with a unified matter particle in a SUSY GUT. When this is done the inflaton carries a gauge charge. This can lead to dangerous contributions to the inflaton potential from the D-terms and quantum corrections at the one- and two-loop level that are potentially big enough to violate the slow-roll conditions and render (hybrid) inflation impossible.

We have first presented our general setup - which relies on a "tribrid-like" superpotential together with a Heisenberg symmetry with stabilized modulus in the Kähler potential to solve the SUGRA η-problem - in the simplest case of an inflaton charged under $U(1)$.

We have then extended our model to the Pati-Salam gauge group, where we have explicitly calculated the inflaton potential up to one-loop order and also estimated the two-loop corrections for the case of inflation along the D-flat right-handed sneutrino trajectory. We have shown that all corrections are under control and that furthermore no topological defects are produced in the waterfall phase ending inflation.

Finally, we have sketched how the model can be embedded into a $SO(10)$ GUT, where now the inflaton resides in the 16-dim spinor representation of $SO(10)$, thus unifying the inflaton and the SM matter fields in one particle.

References

[1] Antusch S *et al.* 2010 *JHEP* **08** 100 (*Preprint* 1003.3233)
[2] Antusch S, Dutta K and Kostka P M 2010 *AIP Conf. Proc.* **1200** 1007–1010 (*Preprint* 0908.1694)
[3] Guth A H 1981 *Phys. Rev.* **D23** 347–356
[4] Liddle A R and Lyth D H ISBN-13-9780521828499
[5] King S F 2004 *Rept. Prog. Phys.* **67** 107–158 (*Preprint* hep-ph/0310204)
[6] Minkowski P 1977 *Phys. Lett.* **B67** 421
[7] Gell-Mann M, Ramond P and Slansky R 1979 In Supergravity, P. van Nieuwenhuizen & D.Z. Freedman (eds.), North Holland Publ. Co.
[8] Yanagida T 1979 In Proceedings of the Workshop on the Baryon Number of the Universe and Unified Theories, Tsukuba, Japan, 13-14 Feb 1979
[9] Glashow S L 1979 Lectures given at the Cargese Summer School, Cargese, Corsica
[10] Mohapatra R N and Senjanovic G 1980 *Phys. Rev. Lett.* **44** 912
[11] Schechter J and Valle J W F 1982 *Phys. Rev.* **D25** 774
[12] Dvali G R 1995 *Phys. Lett.* **B355** 78–84 (*Preprint* hep-ph/9503375)
[13] Kibble T W B 1976 *J. Phys.* **A9** 1387–1398
[14] Copeland E J, Liddle A R, Lyth D H, Stewart E D and Wands D 1994 *Phys. Rev.* **D49** 6410–6433 (*Preprint* astro-ph/9401011)
[15] Binetruy P and Gaillard M K 1987 *Phys. Lett.* **B195** 382
[16] Antusch S, Bastero-Gil M, Dutta K, King S F and Kostka P M 2009 *JCAP* **0901** 040 (*Preprint* 0808.2425)
[17] Dvali G R, Shafi Q and Schaefer R K 1994 *Phys. Rev. Lett.* **73** 1886–1889 (*Preprint* hep-ph/9406319)

Cosmological effects of multivacua theories: gravitational waves from first order transitions during inflation

Diego Chialva
Nordita Institute, Stockholm, Sweden
E-mail: chialva@nordita.org

Abstract. We study the gravitational waves emitted from the physical processes related to first order phase transitions during inflation. The spectrum shows features that differ both from the case of waves generated via quantum vacuum oscillations of the metric and both from the case of transitions occurring outside inflation. We also study the detectability of the waves.

1. Introduction

Primordial gravitational waves generated during inflation can open a window of opportunities for research, as they decouple quite immediately from their surrounding and carry valuable information about the physics that produced them.

Many important and interesting sources arise when first order phase transitions occur: in this short communication we will deal with the emission, features and detectability of gravitational waves produced by this kind of transitions if they occur during inflation. We will mainly follow the analysis and results of [1].

This scientific investigation has different motivations. Most notably, the discovery of many metastable minima in the effective theories of high-energy physics and quantum gravity [2] raises the interest in studying scenarios where phase transitions occur at early times, such as during inflation. It has also been recently fully investigated a model where inflation is entirely driven by a series of first order phase transitions: chain inflation [3].

Gravitational waves generated by first order transitions have been a field of intense research, documented in the literature, but only for what concerns transitions at reheating or at the electroweak and QCD scales [4].

2. Setup and approach

Our scenario is a period of inflation where some first order phase transitions occur. The underlying theoretical model can be very complicated, with a potential exhibiting many metastable minima, multiple fields and various dynamical phases while the fields pass through the minima. There are different possibilities for inflation: for example it can take place via a mechanism such as chain inflation or also in other ways, for example by slow-roll, depending on the behaviour of the fields of the model.

The first order phase transitions occur via nucleation of bubbles of the new phases within the old ones. The latent heat is released by the collisions of the bubbles and the decays of their walls into a radiation-dominated fluid. Many sources of gravitational waves appear thanks to this dynamics.

We simplify the description of this setup by describing it through a set of physical parameters. The emission and features of the gravitational waves can also be studied using these parameters and avoiding the complicated field theory description. The analysis has also the advantage of obtaining general results, which can then be adapted to the various specific models by computing the physical parameters from first principles in the model of interest. In this way the model can be tested.

3. Analysis of the setup

We list here the physical parameters we will use in our analysis of wave production. We also investigate and discuss the most important bounds on these parameters, which ensure the success of the transitions, reaching percolation and large scale thermalization, and the efficiency of inflation (small backreaction).

Background. For our purposes, we can describe the background evolution with only two parameters: the Hubble parameter H and a parameters $\varepsilon = -\frac{\dot{H}}{H^2}$ yielding its time evolution. We consider a quasi-de Sitter evolution, for which $\varepsilon < 1$.

Transitions. As we will see, only a few features of the transitions suffices for studying the wave emission during inflation. The relevant physical parameters are[1]: the nucleation rate per unit time $\widetilde{\Gamma}_n$, the transition time-scale β_n^{-1}, the nucleation radius of bubbles r_n, the energy density $\Delta\epsilon_n$ released by the transition.

These parameters are constrained by the requirement of self-consistency of the scenario. In particular, the transitions must be successful in reaching percolation and large scale thermalization without backreacting too heavily on the background. This forces the ratio of the typical transition scale and the Hubble scale to be bounded as

$$10 < \frac{\beta_n}{H} < \left(\frac{\pi^2}{S_E^{(n)}}\right)^{\frac{1}{5}} 10^{\frac{8}{5}} . \tag{1}$$

Radiation fluid. A radiation component of the Universe necessarily appears as a consequence of the collisions and decay of the bubble walls. Although it is (and must be) a subdominant component compared to the vacuum energy driving inflation, the radiation is in general relevant when discussing perturbations, see for example [3].

In order to be compatible with a FRW description of inflation, at least at certain scales, the fluid must rapidly thermalize[2]. Its most important physical parameter will then be its temperature T. Furthermore, as in general the fluid constituents will be charge carriers, the (gauge) coupling(s) g will be important parameter(s).

The fluid temperature must be bounded if we want backreaction not to stop inflation. In [1], using the value of the spectrum of density perturbations measured by WMAP [7] to relate the Hubble scale to the Planck mass, it was found

$$T < 10^{-2}\kappa^{-\frac{1}{4}}\varepsilon^{\frac{1}{2}} M_{Planck} , \tag{2}$$

where $\kappa = \kappa(T)$ counts the number of relativistic degrees of freedom at temperature T. It was also found that $\frac{T}{r_n} < 1$, which indicates that bubble nucleation via vacuum tunneling (describable in terms of some order parameter) dominates over thermal nucleation [6].

[1] We allow for more than one transition to occur, therefore indicating each of them with a suffix n, $1 \le n \le N$, N kept generic.

[2] As this depends on the specific features of the model and investigating it would go beyond the scope of our analysis, we assume that thermalization occurs sufficiently rapidly.

4. Gravitational wave solutions

Gravitational waves are solution of the linearized Einstein equation (in a suitable gauge)

$$h_{ij}^{''} + 2\mathcal{H}h_{ij}^{'} - \nabla^2 h_{ij} = 16\pi G a^2 \pi_{ij}^T\,. \tag{3}$$

where $\mathcal{H} = aH$ and π_{ij}^T is the traceless symmetric transverse tensor part of the anisotropy stress tensor, obtained by projection from the total energy-momentum tensor. Here[3], $h(k,\eta)$ is the mode function of the graviton field $h(\vec{x},\eta)$ expanded in eigenfunctions of the Laplace-Beltrami differential operator with eigenvalues $-k^2$. The wavenumber k is related to the physical momentum by $k = a(t)p$. The $'(\dot{})$ indicates derivative by $\eta\,(t)$.

During inflation the solution for a generic source is

$$\begin{aligned} h = & \left(c_1 + i\frac{\pi}{4}\int^{\eta} d\eta'(\eta')^{1-\nu^T} H_{\nu^T}^{(2)}(-k\eta')S(k,\eta')\right)(-\eta)^{\nu^T} H_{\nu^T}^{(1)}(-k\eta) \\ & + \left(c_2 - i\frac{\pi}{4}\int^{\eta} d\eta'(\eta')^{1-\nu^T} H_{\nu^T}^{(1)}(-k\eta')S(k,\eta')\right)(-\eta)^{\nu^T} H_{\nu^T}^{(2)}(-k\eta)\,, \end{aligned} \tag{4}$$

where $H_{\nu^T}^{(1,2)}$ are Hankel functions, $\nu^T = \frac{3}{2} + \varepsilon \sim \frac{3}{2}$ and $S(k,\eta) = 16\pi G a^2\, \pi^T(k,\eta)\,.$

We will also need the solution during matter or radiation domination for modes within the horizon ($k > \mathcal{H}$):

$$h = \frac{A_+(k)}{a\sqrt{k}}e^{-ik\eta} + \frac{A_-(k)}{a\sqrt{k}}e^{ik\eta}\,. \tag{5}$$

The two quantities that are relevant when discussing the production, features and detectability of gravitational waves are the spectrum

$$\langle h^*(k,t)h(k',t)\rangle = \delta(k-k')\frac{2\pi^2}{k^3}P_h(k,t) \tag{6}$$

and the energy density radiated per octave

$$k\frac{d\rho_h(k,t)}{dk} = k^3\frac{\langle \dot{h}^*(k,t)\dot{h}(k',t)\rangle}{8\pi G} = k^3\frac{\langle h^{*'}(k,\eta)h'(k',\eta)\rangle}{8\pi G a^2}\,. \tag{7}$$

5. Sources and emission

In presence of first order phase transitions many different sources of gravitational waves can appear.

5.1. Transitions and bubble collisions

The energy density released by a transition goes partly in the acceleration of the bubble walls and partly into the velocity spectrum of the fluid coming from the collision of bubbles of a preceding transition. The features of the gravitational waves emission are different depending on how the energy is divided into these two channels.

The fraction of energy going into the fluid velocity can be estimated looking for the solutions of the continuity equations for energy and momentum across the bubble wall. It is found that there are no stationary solutions compatible with inflation [1]. This means that the energy released by the transition goes predominantly into the acceleration of the wall (near-vacuum description).

The collisions of bubbles are complicated non-equilibrium processes which must be studied with the help of numerical simulations. The best simulations available today are performed in

[3] We will avoid writing polarizations when not strictly necessary, to avoid cluttering of formulas.

the setup of static universe [7]. They show that only two quantities are effectively important for the spectrum of gravitational waves: the overall released energy and the completion time β_n^{-1} for the phase transition, which sets the peak frequency of emission. The complicate small scale dynamics add up incoherently and are sub-dominant.

The emission from phase transitions occurring during inflation exhibits three important differences: *i*) the static quality of the background is an approximation valid for scales shorter than the horizon, *ii*) the total released energy is

$$\Delta\epsilon_n \sim -\frac{d\rho}{dt}\beta_n^{-1} \simeq 6H^3 M_{\rm Planck}^2 \beta_n^{-1}\varepsilon\,, \tag{8}$$

smaller by the factor $2\frac{H}{\beta_n}\varepsilon$ than the total energy density ρ dominated by the vacuum component, *iii*) the evolution of the waves during inflation is peculiar as modes can exit the horizon and the wave amplitude freezes (as we can see from (4)).

The spectrum of superhorizon modes at time t after the emission time t_n is

$$P_h^{\rm sup} = \sum_{n=1}^{N} \frac{9}{2\pi^4}\left(\frac{H}{\beta_n}\right)^6 \varepsilon^2\chi_n^2 \begin{cases} \frac{f_{\beta_n}}{p} & f_{H_n} < p < f_{\beta_n} \\ \left(\frac{f_{\beta_n}}{p}\right)^5 & p > f_{\beta_n} \end{cases} \tag{9}$$

$$= \sum_{n=1}^{N} \frac{2}{\pi^2} A_n \left(\frac{H}{M_{\rm Planck}}\right)^2 \begin{cases} \frac{\tilde\beta_n}{k} & a(t_n)H < k < \tilde\beta_n \\ \left(\frac{\tilde\beta_n}{k}\right)^5 & k > \tilde\beta_n \end{cases}\,, \tag{10}$$

where we have taken into account the redshift of the waves between the emission time t_n and the time $t_{\rm ex}$ of exit from horizon for the mode k

$$\frac{a(t_n)}{a(t_{\rm ex})} = \chi_n \frac{\beta_n}{p_n} \qquad \chi_n = \frac{H}{\beta_n}. \tag{11}$$

Superhorizon modes are the most interesting ones as they are not suppressed after they exit the horizon. Note how (9) differs both from the spectrum of tensor modes generated by quantum oscillations of the metric in vacuum (in particular, no scale invariance), and both from the spectrum sourced by first order transitions occurring *outside* inflation (different frequency dependence). Finally, the backreaction of (9) is strong only for a huge number of transitions ($N \sim 10^{12}$).

5.2. From turbulence, (hyper)magnetic[4] fields, viscosity

There are other sources related to first order phase transitions which could generate a sizable spectrum of gravitational waves [8]. However, we will show that during inflation they appear to be very suppressed or even not to occur.

Turbulence

Turbulence can be an important source of waves. However it is possible to prove that it cannot occur for the range of values of the physical parameters that are in accordance with inflation (see section 3). Indeed, turbulence is signalled by large Reynold numbers, but from (1), (2), we find

$$Re = \frac{v_f^{(\beta_n)}\beta_n^{-1}}{\nu} \sim v_f^{(\beta_n)}\frac{T}{\beta_n} g^4 \log(g^{-1}) \leq 1 \tag{12}$$

where $v_f^{(\beta_n)}$ is the characteristic velocity of the fluid flow at the injection scale, which is the bubble size at collision $\sim \beta_n^{-1}$.

[4] With the attribute *hyper* we denote generically the magnetic fields associated with some U(1) gauge symmetry at high energy, here sourced by the phase transitions.

Plasma physics and gauge fields

In the dynamical range interesting for our study, the relativistic fluid of charge carriers can be described as a plasma. Gravitational waves could be efficiently sourced by hypermagnetic fields generated by the local currents arising because of the bubble dynamics and collisions. The hypermagnetic fields are generated at the typical scale of the bubbles at collision and then amplified by hypermagnetic turbulence. However, the latter does not occur when the physical parameters lie within the bounds we have found to be consistent with inflation (see section 3). Indeed, the magnetic Reynold number is

$$Re_\mu \sim v_f^{(\beta_n)} \frac{T}{\beta_n g^2 \log(g^{-1})}, \tag{13}$$

which, even for reasonably small couplings ($g \sim 0.1$), is much less than 100.

Fluid viscosity

The anisotropic stress tensor generated by viscosity of a fluid with velocity $\vec{u}$ is [9]

$$\pi_{ij}^{(\mathrm{visc})} = -\varsigma \left(\partial_j u_i^{(f)} + \partial_i u_j^{(f)} - \frac{2}{3} \nabla \cdot \vec{u}^{(f)} \delta_{ij} \right) - \zeta \nabla \cdot \vec{u}^{(f)} \delta_{ij}. \tag{14}$$

The bulk viscosity ζ is vanishing for a relativistic fluid. The shear viscosity ς could instead be large, as it is given by $\varsigma = \varsigma_0 \frac{T^3}{g^4 \log(g^{-1})}$ for a weakly coupled fluid [10].

One can however calculate $\Omega^{(\mathrm{visc})}$, the energy density radiated per octave normalized to the total energy density, at the time of emission, by using (7) and normalizing it. However, for scales compatible with the plasma description ($k > \mathcal{H}$), one finds that this radiated energy is much smaller than the one emitted by the collision of bubbles at the same (hydrodynamical) scales:

$$\frac{\Omega^{(\mathrm{visc})}}{\Omega^{(\mathrm{coll})}} \lesssim g^4 \log(g^{-1}) \mathcal{V}_f^2 \Big|_{k=\lambda_f^{-1}} \ll 1. \tag{15}$$

6. Detection

We will discuss the possible detection of the gravitational waves via the temperature anisotropy of the CMBR or directly by interferometers.

6.1. CMBR

.

We choose to concentrate on the temperature anisotropies and not the polarization, although both are affected by gravitational waves emission. In particular, we focus our discussion on the scalar-to-tensor ratio r.

For illustrative reasons, we now make the simplifying assumption that $\frac{\beta_n}{H(t_n)}$ does not depend too much on n, and we consider only multipoles $\ell_k = \frac{2\tilde{k}}{H_0}$ for wavenumbers $k_\beta = a_n \beta_n$ and $k_\mathcal{H} \sim a(t_n) H$, as they are respectively the typical comoving scale of the transitions and the Hubble scale. Using (9), the bounds in section 3 and the typical scalar density spectrum, we obtain

$$c_s\, 10^{-4}\, N\, \varepsilon^2 \left(\frac{S_E}{\pi^2} \right)^{\frac{8}{5}} \frac{\ell_{\tilde{\beta}}}{\ell_*} \leq r \leq c_s\, 10 \frac{9}{4\pi^2} N\, \varepsilon^2\, \frac{\ell_{\tilde{\beta}}}{\ell_*}\,. \tag{16}$$

A value of $r \sim 0.07$ is detectable. We see that $N \sim 3 \cdot 10$ transitions at the lowest allowed rate $\frac{\beta}{H} \sim 10$ would leave detectable marks at multipoles $\sim \ell_{k_\mathcal{H}}$. The signal however rapidly weakens for higher multipoles and/or faster transitions.

Journal of Physics: Conference Series **259** (2010) 012047 doi:10.1088/1742-6596/259/1/012047

6.2. Interferometers

To discuss direct detection of these primordial waves by interferometers such as LIGO, LISA, DECIGO, we need to evolve them after horizon re-entering. Frequencies falling within the sensitivity of those three interferometers must have been within the horizon before matter-radiation equality and nucleosynthesis. From (5) we see that h evolves as $\sim a^{-1}$ and therefore we can compute the *strain* produced by the waves and compare it with the sensitivity of the interferometers at the relevant frequencies.

The detailed results in [1] show that the detectability of the emitted waves depends on the number of transitions, their timescales and the number of e-foldings that must have occurred to sufficiently redshift their frequencies after the waves were emitted.

To give an example, we can specialize to the case of chain inflation, which is characterized by fast transitions. One can find that in the case of many transitions ($N \gg \frac{\beta}{H_{\rm infl}}$, where $H_{\rm infl}$ is the Hubble parameter during inflation), the waves would be detectable at DECIGO for all allowed values of $\frac{\beta}{H_{\rm infl}}$, detectable at LISA for $\frac{\beta}{H_{\rm infl}} < 11$, but not detectable at LIGO.

Note that waves that re-enter the horizon before nucleosynthesis must not interfere with the latter. There is therefore a constraint on the wave amplitude (or equivalently the strain) [11]. In [1] it was shown that the constraint is easily satisfied by the waves we discuss, thanks to the bounds in section 3.

7. Conclusion

We have studied the emission and features of gravitational waves from first order phase transitions during inflation. On the positive side, we have shown that the spectrum of waves presents features that are very different from those of the waves generated by vacuum quantum oscillations of the metric or by transitions occurring outside inflation, allowing to clearly distinguish them. On the negative side, the intensity of the sources is highly constrained by the requests of small backreaction, homogeneity and isotropy at large scales, and this makes the amplitude of the spectrum of a single transition quite suppressed.

However, the accumulation due to a moderate number of slow but successful transitions (which could occur in settings such as the string landscape) would leave a detectable imprint both in the CMBR and at interferometers. For faster transitions, the signal rapidly weakens.

References

[1] D. Chialva, arXiv:1004.2051 [astro-ph.CO].
[2] U. H. Danielsson, N. Johansson and M. Larfors, JHEP **0703** (2007) 080; D. Chialva *et al.*, JHEP **0802** (2008) 016.
[3] D. Chialva and U. H. Danielsson, JCAP **0810**, 012 (2008); JCAP **0903**, 007 (2009).
[4] A. Kosowsky, M. S. Turner and R. Watkins, Phys. Rev. D **45**, 4514 (1992); A. Kosowsky, A. Mack and T. Kahniashvili, Phys. Rev. D **66**, 024030 (2002); C. Grojean and G. Servant, Phys. Rev. D **75**, 043507 (2007); C. Caprini and R. Durrer, Phys. Rev. D **74**, 063521 (2006); T. Kahniashvili, et al., Phys. Rev. D **78**, 043003 (2008); R. Easther, J. T. Giblin and E. A. Lim, Phys. Rev. Lett. **99**, 221301 (2007); J. Garcia-Bellido, D. G. Figueroa and A. Sastre, Phys. Rev. D **77**, 043517 (2008); Dufaux J.F. *et al.*, Uzan J.P., Phys. Rev. D **76**, 123517 (2007); M. S. Turner and F. Wilczek, Phys. Rev. Lett. **65**, 3080 (1990).
[5] G. Hinshaw *et al.*, WMAP Collaboration, arXiv: 0803.0732.
[6] A. D. Linde, Chur, Switzerland: Harwood (1990) 362 p. (Contemporary concepts in physics, 5).
[7] A. Kosowsky and M. S. Turner, Phys. Rev. D **47**, 4372 (1993); M. Kamionkowski, A. Kosowsky and M. S. Turner, Phys. Rev. D **49**, 2837 (1994); S. J. Huber and T. Konstandin, JCAP **0809** (2008) 022; C. Caprini *et al.*, Phys. Rev. D **79** (2009) 083519;
[8] T. Kahniashvili, L. Kisslinger and T. Stevens, Phys. Rev. D **81**, 023004 (2010); C. Caprini, R. Durrer and G. Servant, JCAP **0912**, 024 (2009);
[9] Steven Weinberg, *Gravitation and Cosmology*, Wiley & Sons, 1972, ISBN 0471925675
[10] P. B. Arnold, G. D. Moore and L. G. Yaffe, JHEP **0011** (2000) 001; Le Bellac, "Thermal field theory", Cambridge University press, ISBN-13: 9780521654777; S. Weinberg, Astrophys. J. **168** (1971) 175.
[11] M. Maggiore, Phys. Rept. **331** (2000) 283.

Journal of Physics: Conference Series **259** (2010) 012048 doi:10.1088/1742-6596/259/1/012048

Hydrodynamics of the electroweak phase transition in an extension of the Standard Model with dimension-6 interactions

M Sopena[1], **SJ Huber**[1, 2]

[1] Department of Physics and Astronomy, Sussex University, Brighton, East Sussex BN1 9QE, UK

E-mail: `m.sopena@sussex.ac.uk`

Abstract. Extensions of the Standard Model are being considered as viable settings for a first-order electroweak phase transition which would satisfy Sakharov's three conditions for the generation of the baryon asymmetry of the Universe. These extensions would provide a sufficiently strong phase transition and remove the main obstacles which appear in the context of the Standard Model: A far-too-high lower bound on the Higgs mass, immediate wipeout of the newly-created baryon asymmetry and insufficient CP violation. We apply standard semiclassical treatments of the hydrodynamics of a first-order phase transition to the case of a recently-introduced dimension-6 extension of the Standard Model which (within the present bounds on the Higgs mass) could produce the observed baryon asymmetry of the Universe. We express the friction term in the hydrodynamic equations in terms of the particle content of the model and produce predictions for the velocity of the expanding bubble wall in the stationary regime.

1. Introduction

A 1st-order electroweak phase transition at the ~ 100 GeV scale could satisfy Sakharov's three conditions for electroweak baryogenesis: Deviation from thermal equilibrium, CP violation, and baryon number violation. In the Standard Model, B-violation proceeds through finite-temperature 'sphaleron' transitions in the presence of external W fields. These are Boltzmann-suppressed and become relevant at $T \gtrsim 100$ GeV, with $E_{Sph}/T \sim v/T$, where v is the Higgs VEV. In a first-order transition, bubbles of the new phase nucleate and grow, with B-violation taking place *outside* the growing bubbles, in the symmetric phase, and the baryon asymmetry then being transported across the bubble wall into the new phase where it must survive washout. However we now know that within the Standard Model the washout-preventing condition $v/T > 1$ is extremely hard to satisfy, not to mention that the electroweak phase transition is not even first-order for $m_h \gtrsim 75$ GeV [1]. This is the origin of the current interest in extended settings (including supersymmetry) in which a first-order electroweak phase transition which avoided washout of the newly-created baryon asymmetry could take place within existing experimental bounds. Such a setting could even be constrained through signatures detectable in the near future, for instance at the LHC or through measurements of an electric dipole moment for the neutron.

[2] DPhil supervisor

We present the results of applying the usual WKB-approximation hydrodynamic analysis to one such (non-supersymmetric) extension of the Standard Model analysed in [2]. The model features dimension-6 interactions regulated by a cut-off scale M. The model is fully parametrised by M and the Higgs mass m_h and provides a broad enough region in parameter space where a baryon-asymmetry-preserving, first-order phase transition may take place.

2. The dimension-6 potential

The finite-temperature effective potential for the dimension-6 model is written as [2]

$$\begin{aligned} V(\phi, T) = & \frac{1}{2}(-\mu^2 + (\frac{1}{2}\lambda + \frac{3}{16}g_2^2 + \frac{1}{16}g_1^2 + \frac{1}{4}y_t^2)T^2)\phi^2 \\ & -\frac{g_2^3}{16\pi}T\phi^3 + \frac{\lambda}{4}\phi^4 + \frac{3}{64\pi^2}y_t^4\phi^4 \ln(\frac{Q^2}{c_F T^2}) \\ & +\frac{1}{8M^2}(\phi^6 + 2\phi^4T^2 + \phi^2T^4) \end{aligned} \tag{1}$$

where $Q \equiv m_{top} = 178GeV$, $c_F \approx 13.94$, and μ and λ are determined via the conditions

$$\frac{\partial V(\phi,0)}{\partial \phi}|_{\phi=v_0} = 0, \qquad \frac{\partial^2 V(\phi,0)}{\partial \phi^2}|_{\phi=v_0} = m_h^2 \tag{2}$$

where $v_0 = 246$ GeV. Dimension-6 non-renormalisable operators can parametrise the effects of new physics beyond the cut-off M. In order for these to be relevant at the weak scale we require $M \lesssim 1$ TeV.

3. Nucleation temperature. The hydrodynamic equations.

For a first-order transition to take place, the expansion of the effective potential at the relevant temperature scale, which has a global minimum at zero Higgs VEV, must develop a *second* local minimum at a nonzero VEV as T decreases (see eg [3]). As T decreases further the value of V at the second minimum approaches the value at zero until both become degenerate at the so-called *critical temperature.* For $T < T_c$ nucleation of bubbles of the broken symmetry phase becomes possible. As T falls further below T_c, production of bubbles of the broken symmetry phase accelerates and existing bubbles grow, eventually filling all space. The *nucleation temperature* $T_n < T_c$ is that at which the integrated probability of bubble nucleation in the horizon volume reaches unity,

$$P(T = T_n) = \int_{T_n}^{T_c} dP = \int_{T_n}^{T_c} (\Gamma/Vol) \cdot V_H \cdot dt = \int_{T_n}^{T_c} \frac{T^4}{H^4} e^{-F_c/T} \frac{dT}{T} = 1 \tag{3}$$

where we have taken $\Gamma/Vol = \Lambda^4(T)e^{-F_c/T} \approx T^4 e^{-F_c/T}$, F_c being the free energy of the *critical bubble* (large enough to spontaneously grow) at the temperature T.

The phase transition ends when the fraction of the horizon volume taken over by the new phase becomes one.

The usual treatment of the hydrodynamical problem models the early Universe plasma as a perfect relativistic fluid with a conserved energy-momentum tensor,

$$\partial_\mu T^{\mu\nu} = \partial_\mu (T^{\mu\nu}_{field} + T^{\mu\nu}_{fluid}) =$$
$$= \partial_\mu \Big(\partial^\mu \phi \partial^\nu \phi - g^{\mu\nu} (\frac{1}{2} \partial_\alpha \phi \partial^\alpha \phi) +$$
$$+(\rho + P) u^\mu u^\nu - P g^{\mu\nu} \Big) = 0 \tag{4}$$

which, applying the relevant thermodynamic relations and introducing a *friction term* [4], results, in the rest frame of the advancing bubble wall, working in a single spatial dimension (perpendicular to the planar wall, along the direction of advance), and assuming a stationary situation, in the system

$$\frac{d^2\phi(x)}{dx^2} = \frac{\partial V(\phi, T)}{\partial \phi} + \eta \frac{\phi^2}{T_{s1}} v\gamma \frac{d\phi(x)}{dx} \tag{5}$$
$$(4aT^4 - T\frac{\partial V(\phi, T)}{\partial T})\gamma^2 v = C_1 \tag{6}$$
$$(4aT^4 - T\frac{\partial V(\phi, T)}{\partial T})\gamma^2 v^2 + P_r - V(\phi, T) + \frac{1}{2}(\frac{d\phi}{dx})^2 = C_2 \tag{7}$$

where v (the fluid velocity), T and the Higgs VEV ϕ depend exclusively on the spatial coordinate and γ is the usual relativistic factor $(1 - v^2)^{-1/2}$. η is our friction parameter and T_{s1} the plasma temperature in the symmetric phase ahead of the wall. We solve this system with vanishing ϕ derivative at both extremes of the integration interval and vanishing Higgs VEV at the extreme in the symmetric phase and we obtain the shape of the wall (fig 1). Solutions to the hydrodynamic equations are usually described as either *subsonic* (deflagration) or *supersonic* (detonation) relative to the velocity of sound in the plasma [4] [5] [6]. Subsonic solutions are preceded by a shock front which heats up and accelerates the plasma at rest. A supersonic wall is followed by a rarefaction wave which brings the plasma back to rest. Stationary solutions of the hydrodynamic equations are similarity solutions, maintaining their shape as the bubble grows. For a deflagration we calculate the whole bubble profile by solving the conservation equation in spherical coordinates in the region between the bubble wall and the shock front [6] and the leap in v, T across the shock front [4]. The results of the calculation for a choice of model parameters is shown in figure 2.

4. The friction parameter

The friction term introduced in a phenomenological way above can be calculated explicitly from the particle populations in the plasma [7]. The equation of motion for the Higgs field can be written covariantly as

$$\Box\phi + \frac{\partial V_{eff}(\phi, T)}{\partial \phi} + \Sigma \frac{dm^2}{d\phi} \int \frac{d^3p}{(2\pi)^3 2E} \delta f(p, x) = 0 \tag{8}$$

with the mass dependence on the Higgs VEV $m^2 = \frac{y^2\phi^2}{2}$ for fermions, $m^2 = \frac{g^2\phi^2}{4}$ for bosons, and δf expressing the *deviation from equilibrium* of particle populations responsible for the friction (the sum is over all relevant species). In the WKB approximation distributions evolve following a Boltzmann equation

$$\frac{df}{dt} = \partial_t f + \dot{\vec{x}} \cdot \partial_{\vec{x}} f + \dot{\vec{p}} \cdot \partial_{\vec{p}} f = -C[f] \tag{9}$$

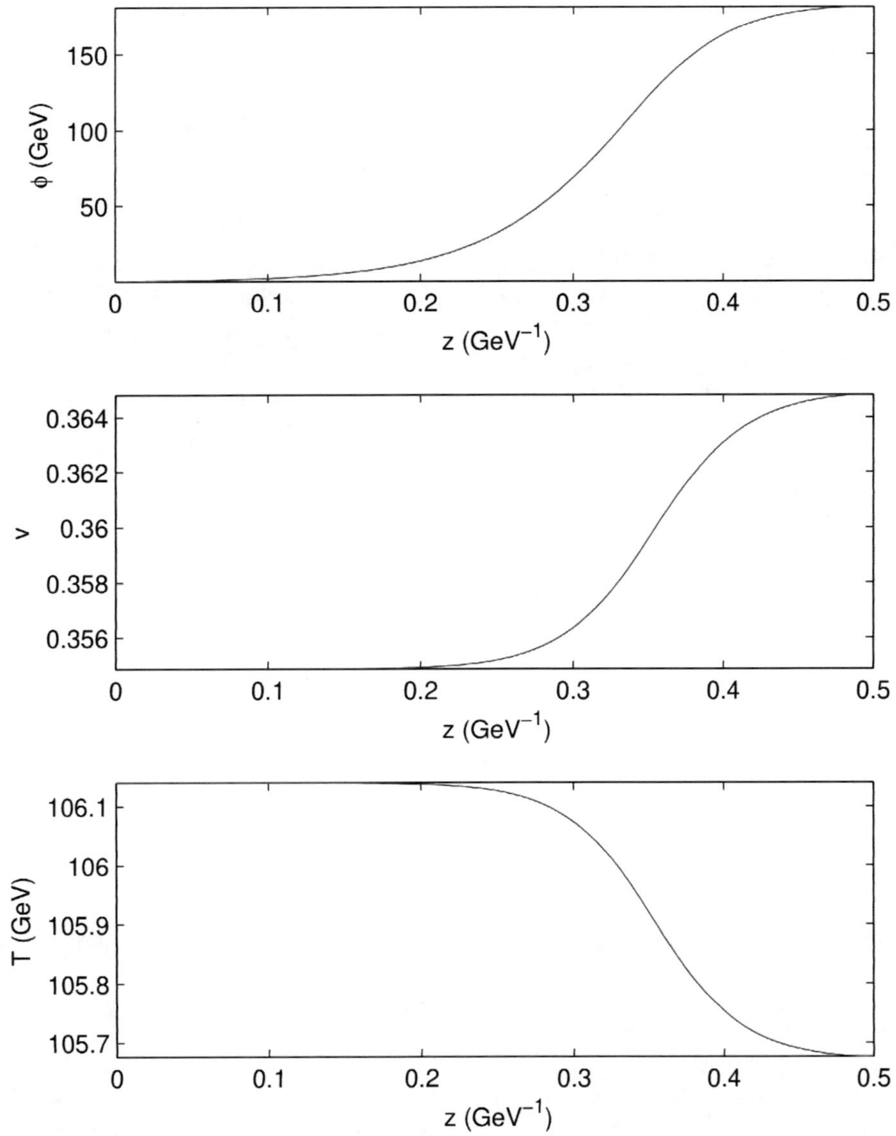

Figure 1. Higgs VEV, fluid velocity and temperature profiles across a subsonic bubble wall in the case $M = 800$, $m_h = 115$, $\eta = 0.398$. Quantities are in the rest frame of the wall. The symmetric phase is to the left [9].

Journal of Physics: Conference Series **259** (2010) 012048 doi:10.1088/1742-6596/259/1/012048

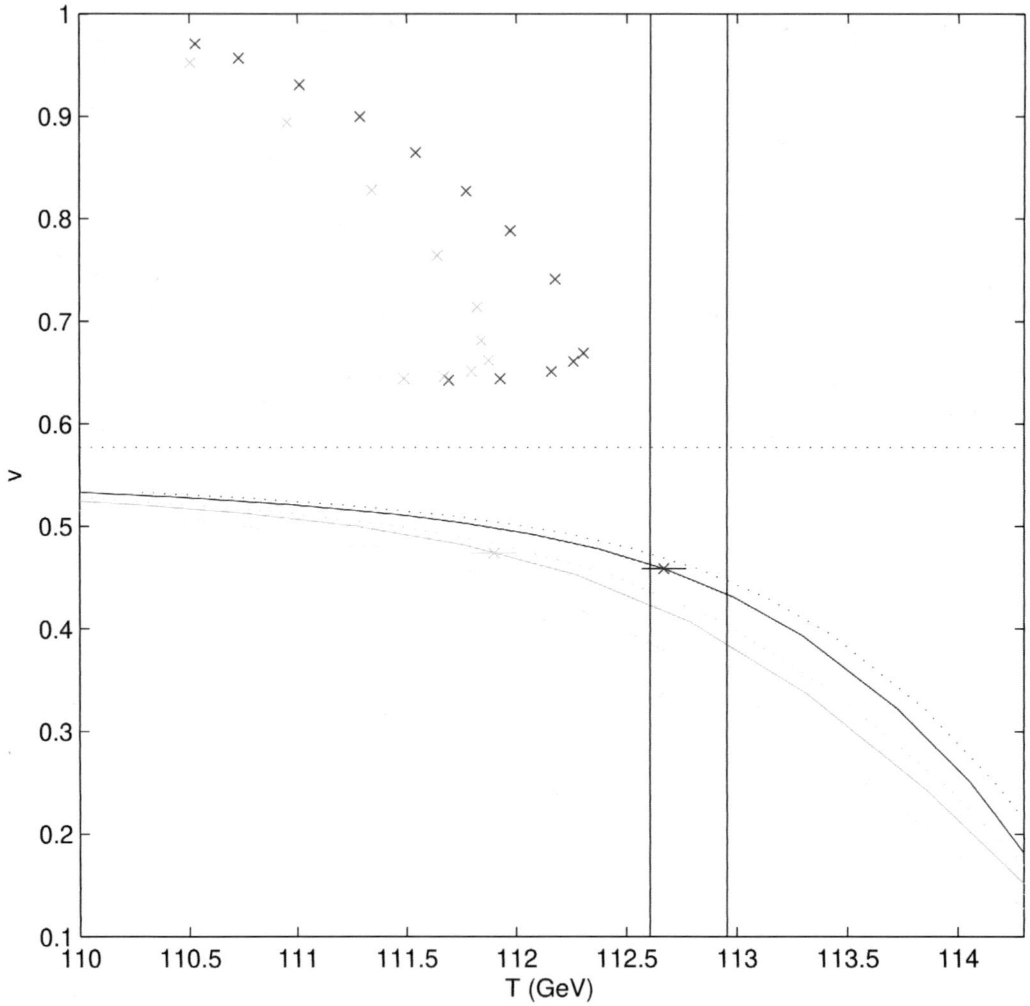

Figure 2. Wall velocity vs temperature of the universe for values of the friction coefficient η 0.3 (blue), 0,4 (green) and 0.5 (yellow), $M = 800$ GeV, $m_h = 120$ GeV. The horizontal dotted line marks the speed of sound in the plasma. The continuous lines below are subsonic solutions. For comparison, dotted lines to the right of them mark solutions found when neglecting the sphericity of the bubble. The horizontal error bars mark the point where the stability criterium for subsonic solutions [8] changes sign (the stable region lies to the right of the error bars). The crosses above the velocity of sound mark two branches of supersonic solutions, the lower one being unphysical [5]. The two vertical lines mark the nucleation and finalisation temperatures for the phase transition for this choice of parameters. In this example and for these values of η stationary supersonic solutions would be excluded for the duration of the phase transition [9].

Table 1. Fitted ηs and velocities for different values of M for two slices at $m_h = 115, 150$ GeV [9].

m_h	M	T_n	$\xi = \phi_0/T_n$	η	v_w
115	900	115.92	1.26	0.477	0.34
	800	105.49	1.74	0.398	0.38
	700	88.86	2.47	0.305	0.45
	650	75.10	3.14	0.240	0.53
	630	67.00	3.56	0.207	?
	610	54.70	4.44	0.153	0.74
	600	45.65	5.35	0.110	0.83
150	700	144.64	0.92	0.539	0.35
	650	136.62	1.19	0.490	0.36
	600	126.46	1.48	0.438	0.39
	550	112.71	1.87	0.380	0.43
	500	91.61	2.53	0.298	0.50

Following [7] we adopt the *relaxation time approximation* to the collision integral $C[f] = -\frac{\delta f}{\tau}$. In this way we can reexpress δf and write the friction term (for one particle species) as

$$\phi^2\phi'\tau\beta\gamma v \int \frac{d^3p}{(2\pi)^3 4E^2} \frac{e^{\beta\gamma(E-vp_z)}}{(e^{\beta\gamma(E-vp_z)} \pm 1)^2} \tag{10}$$

The sum of these contributions is equivalent to $\frac{\eta}{T}\phi^2\phi'\gamma v$ (the equilibrium distribution for fermions/bosons boosted to a frame moving with velocity v in the z direction is $f_0 = \frac{1}{e^{\beta\gamma(E-vP_z)} \pm 1}$ with $\beta = \frac{1}{T}$, $\gamma = \frac{1}{\sqrt{1-v^2}}$). In this way we can calculate the friction parameter (and thus the wall velocity) for any Standard Model-like situation as a function of the strength of the phase transition given by $\xi_n \equiv \frac{\phi_0}{T_n}$ (here ϕ_0 is the Higgs VEV in the broken symmetry phase). The relevant prefactors to the momentum integral can be calibrated by reproducing the Standard Model wall velocities in [7].

Proceeding in an iterative fashion for each choice of model parameters we arrive at the results in table 1. Hydrodynamic equations become increasingly hard to solve numerically for wall velocities close to the speed of sound, as in the case $m_h = 115$ GeV, $M = 630$ GeV. The slice $m_h = 115$ features a jump for a strength of the phase transition $\xi \sim 3.5$ from subsonic solutions (usually assumed in standard baryogenesis scenarios) to supersonic ones (the standard assumption for production of gravitational waves).

In summary, we apply standard hydrodynamic treatments to the electoweak phase transition in an extension of the Standard Model with dimension-6 interactions. We calculate the friction coefficient in the hydrodynamic equations as a function of the particle content of the cosmic fluid and produce predictions for the bubble wall velocity for a range of model parameters.

References

[1] Kajantie K, Laine M, Rummukainen K and Shaposhnikov ME 1996 *Phys. Rev. Lett.* **77** 2887-90
[2] Bodeker D, Fromme L, Huber SJ and Seniuch M 2005 JHEP **0502** 026 [arXiv:hep-ph/0412366].
[3] Anderson GW and Hall LJ 1992 *Phys. Rev. D* **45** 2685.
[4] Ignatius J, Kajantie K, Kurki-Suonio H and Laine M 1994 *Phys. Rev. D* **49** 3854 [arXiv:astro-ph/9309059].
[5] Megevand A and Sanchez AD 2009 *Nucl. Phys. B* **820** 47 [arXiv:0904.1753 [hep-ph]].
[6] Espinosa JR, Konstandin T, No JM and Servant G 2010 *JCAP* **1006** 028 [arXiv:1004.4187 [hep-ph]].
[7] Moore GD and Prokopec T 1995 *Phys. Rev. D* **52** 7182 [arXiv:hep-ph/9506475].
[8] Huet PY, Kajantie K, Leigh RG, Liu BH and McLerran LD 1993 *Phys. Rev. D* **48** 2477 [arXiv:hep-ph/9212224].
[9] Huber SJ and Sopena M, to appear 2010

Journal of Physics: Conference Series **259** (2010) 012049 doi:10.1088/1742-6596/259/1/012049

Oscillations in the primordial bispectrum

P. Daniel Meerburg

Astronomical Institute "Anton Pannekoek", University of Amsterdam, Amsterdam 1098 SJ, The Netherlands

E-mail: p.d.meerburg@uva.nl

Abstract. There exist several models of inflation that produce primordial bispectra that contain a large number of oscillations. In this paper we discuss these models, and aim at finding a method of detecting such bispectra in the data. We explain how the recently proposed method of mode expansion of bispectra might be able to reconstruct these spectra from separable base functions. Extracting these base functions from the data might then lead to observational constraints on these models.

1. Introduction

The aim of this review is to summarize the work done in [1], as presented at PASCOS 2010, Valencia in July of this year. In this work we investigate the method of mode expansion [2, 3] for primordial bispectra that contain (a large number of) oscillations. The discussed inflationary models/secanrios include features in the inflaton potential [4, 5], resonant models [5, 6, 7], and models with a slight modification of the initial vacuum state [8, 9].

Although these models generally predict quite large primordial bispectral amplitude (i.e. f_{NL}), they are hard to constrain because their shapes are highly oscillatory (typically they also predict a large number of oscillations on top of the baryonic acoustic collations in the power spectrum, see e.g. [6, 10]) A way to inquire unconstrained types of bispectra is to compare them with constrained types, such as local [11, 12], equilateral [13] and orthogonal [14] type bispectra. The comparison can be qualified using the correlation function between two shapes in comoving momentum space defined as [2, 15]

$$F_X \star F_Y \equiv \int_{\Delta_k} dk_1 dk_2 dk_3 w_k S_X S_Y, \tag{1}$$

where S is the shape of the bispectrum and w_k is a weight function, which was chosen as $w_k = 1/k_t$ in [3] to increase resemblance with the Fisher matrix (correlation) found in multipole space. The integral runs over the tetrahedral domain which is set by the triangle constraints together with the maximal observable k_{max}:

$$k_a \leq k_b + k_c \text{ for } k_a \geq k_b, k_c$$
$$k_a, k_b, k_c \leq k_{max},$$

where $a, b, c = \{1, 2, 3\}$, $a \neq b \neq c$. Note that for optimal estimates one should in fact consider the correlation function in multipole space, however this is computationally challenging and as a first estimate the above correlator is time efficient and quite precise [3].

Most constrained bispectra have a thing in common: they are relatively smooth. Consequently when comparing these to highly oscillating bispectra one finds that resulting constraints become relatively weak. In order to improve the analysis of these bispectra and the resulting bounds on associated physical parameters, one should directly constrain these models in the data. Unfortunately constraining bispectra from the data is very time consuming and computationally challanging, in particular since these models have two additional free parameters beside their amplitude: the phase (γ) and the frequency (ω_x). In order to search for signatures in the data, one should scan through a range of possible frequencies and phases since it is possible to completely miss the signal if one searches only for one specific/fixed frequency and phase. This compromises the analysis for these types of spectra to the extend that one should to try to consider other means.

In a recent paper [2, 3], the authors propose a very elegant method of constraining any type of bispectra directly from the data. The idea is to expand a given (theoretical) bispectrum into a set of orthogonal basis functions R_n, i.e.

$$S(x_1, x_2, x_3) \simeq \sum_{n=0}^{N} \alpha_n R_n(x_1, x_2, x_3). \tag{2}$$

where $x_i = k_i/k_{max}$, with k_{max} the largest observable k in the CMB, while $S(x_1, x_2, x_3)$, known as the (model-dependent) shape of the bispectrum, is defined via the primordial three point correlation function:

$$\langle \zeta_{\vec{k}_1} \zeta_{\vec{k}_2} \zeta_{\vec{k}_3} \rangle = (2\pi)^7 f_{NL} \Delta^2 \delta^K \left(\sum_{i=1}^{3} k_i \right) \frac{S(k_1, k_2, k_3)}{k_1^2 k_2^2 k_3^2}. \tag{3}$$

Here ζ is the gauge invariant curvature perturbation ($\zeta = -H\delta\phi/\dot{\phi}_0$) which is constant after horizon exit, Δ is the amplitude of the primordial power spectrum (i.e. for single field slow-roll $\Delta = H^2/8\pi\epsilon$, where H is the Hubble rate at the end of inflation and ϵ the slow-roll parameter). f_{NL} is the amplitude of the primordial bispectrum and its value depend on the model under consideration. If the set of R_n is optimally constructed and S is relatively smooth, the number of modes necessary to reconstruct the spectrum is low. For example to reconstruct the bispectrum predicted by Dirac-Born-Infeld models of inflation (equilateral type) only $n = 5$ modes lead to a 99% correlation between the original and the reconstructed shape (in k space) [1, 3]. The advantage of this method of direct measurement is obvious. The constructed mode functions are aimed at being able to reconstruct a large number of bispectra. Measuring these modes in the data could therefore constrain a large number of models at once (see [16]). As such, mode expansion can be considered an optimal way of extracting the bispectrum from the data, without focussing on specific shape (e.g. local and equilateral).

In the following (sec. 2) we will use the method of mode expansion to a class of primordial bispectra that oscillate rapidly. We compare the existing set of mode functions, based on polynomials to a new set of base functions, based on Fourier functions. We show that the latter basis is more efficient in expanding oscillatory type bispectra. We will discuss (sec. 3) some other related observations concerning the expansion, and explain how these could be useful in future data analysis. We will conclude in section 4.

2. Models and their expansion

In [1] we investigated the ability to reconstruct inflationary bispectra that are not smooth using R_n bispectral basis functions. We investigated three different shapes predicted by different physics:

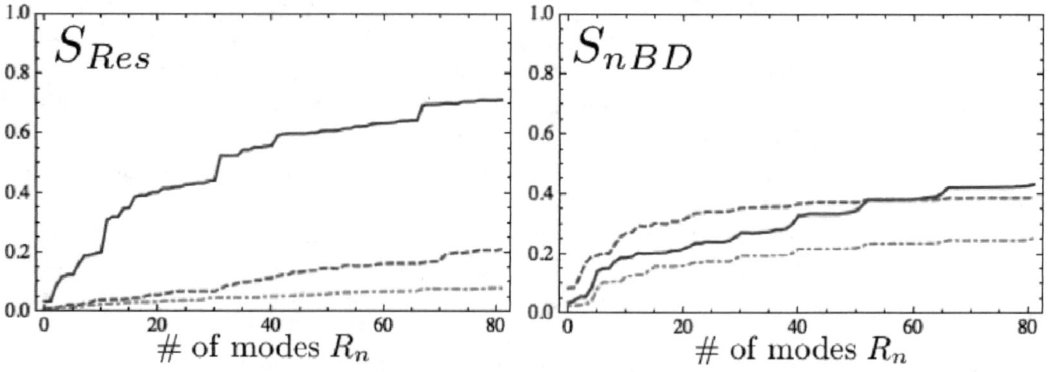

Figure 1: The correlation of S_{Res} and S_{nBD} with their respective reconstruction using R_n basis functions. The correlation is shown for three different frequencies: $\omega_{r,v} = 20$ (solid), 40 (dashed) and 60 (dot-dashed).

$$S_{Feat} = k_{max}^{-6} \sin(\omega_f x_t + \gamma_1), \tag{4}$$

$$S_{Res} = k_{max}^{-6} \sin(\omega_r \ln x_t + \gamma_2), \tag{5}$$

$$S_{nBD} = \frac{\omega_v k_{max}^{-6}}{x_1 x_2 x_3} \sum_j \frac{1}{x_j^3} \left(\frac{1}{2} \frac{\cos\left(\omega_v \frac{x_{j+1}+x_{j+2}}{x_j} + \gamma_3\right)}{\omega_v \left(\frac{x_{j+1}+x_{j+2}}{x_j} - 1\right)} - \frac{\sin\omega_v \left(\omega_v \frac{x_{j+1}+x_{j+2}}{x_j} + \gamma_3\right)}{\omega_v^2 \left(\frac{x_{j+1}+x_{j+2}}{x_j} - 1\right)^2} \right.$$
$$\left. \frac{\cos\delta - \cos\left(\omega_v \frac{x_{j+1}+x_{j+2}}{x_j} + \gamma_3\right)}{\omega_v^3 \left(\frac{x_{j+1}+x_{j+2}}{x_j} - 1\right)^3} \right). \tag{6}$$

For details explaining how to compute these bispectra and the non-Gaussian amplitude f_{NL} for each of these models we would like to refer to [1, 4, 5, 6, 7, 8]. Generally, the various frequencies ω are model dependent as is the phase γ. Typically $\omega \gg \mathcal{O}(1)$.

As a first attempt we take the R_n basis functions generated to be orthonormal on the tetrahedral domain. The details of R_n mode construction on the this domain can be found in [3, 16]. As expected, it requires many modes to establish significant correlation between the reconstructed shape S^r and the predicted shape S^p. In figure 1 we show the various reconstruction attempts of eq. (5) and eq. (6) for several frequencies.

The R_n basis functions are based on polynomials, e.g. $R_3 \propto x_1^3 + x_2^3 + x_3^3 + c$ and higher n correspond to larger powers of x_i. Such basis functions are not particularly optimized for oscillatory bispectra. In [1] we considered another set of mode functions, denoted $\mathcal{F}_n$, based on $e^{-i\omega x}$ (Fourier reconstruction). It must be noted that S_{Feat} is already of this form and therefore we did not consider this model in the expansion. To further investigate the reconstructive power of $\mathcal{F}_n$ we considered three additional toy model shapes:

$$S_1 = \left(\sin\frac{\omega_1}{x_1+1} + \sin\frac{\omega_1}{x_2+1} + \sin\frac{\omega_1}{x_3+1}\right), \tag{7}$$

$$S_2 = \sin\omega_2 x_1 x_2 x_3, \tag{8}$$

$$S_3 = \left(\sin\frac{\omega_3 x_t}{x_1+1} + \sin\frac{\omega_3 x_t}{x_2+1} + \sin\frac{\omega_3 x_t}{x_3+1}\right). \tag{9}$$

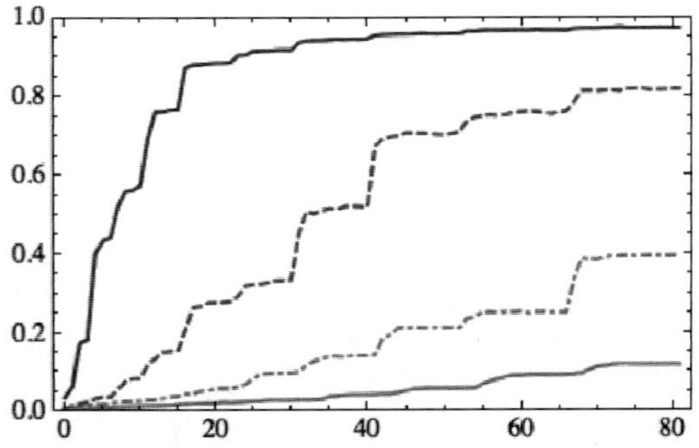

(a) Mode expansion for resonant type non-Gaussianity (eq. (5)) using the $\mathcal{F}_n$ basis functions. We considered the following frequecnies: frequencies $\omega_r = 20$ (blue, solid), 40 (purple, dashed), 60 (yellow, dot-dashed) and 80 (green, solid).

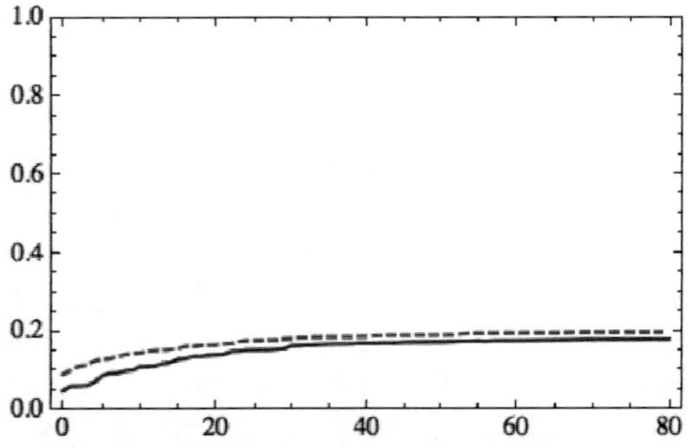

(b) Mode expansion for S_{nBD} (eq.(6)) using the same mode functions as in 2a. Here we only considered two frequencies: $\omega_v = 20$ (blue, solid) and 40 (purple, dashed). The correlation seems to saturate at around 20%.

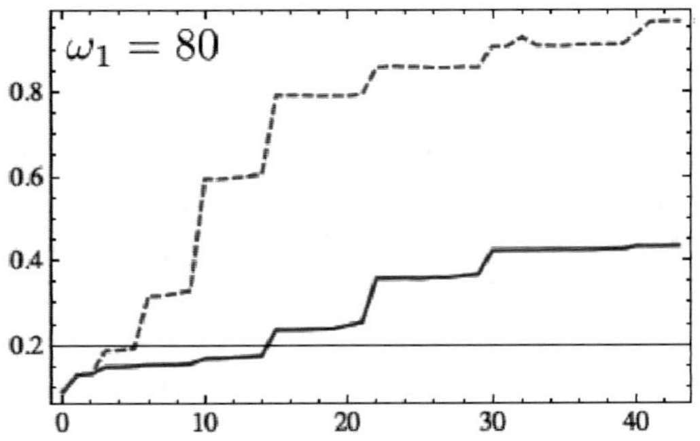

(c) Mode expansion comparison between $\mathcal{F}_n$ (top) and R_n (bottom) with a fixed frequency for S_1.

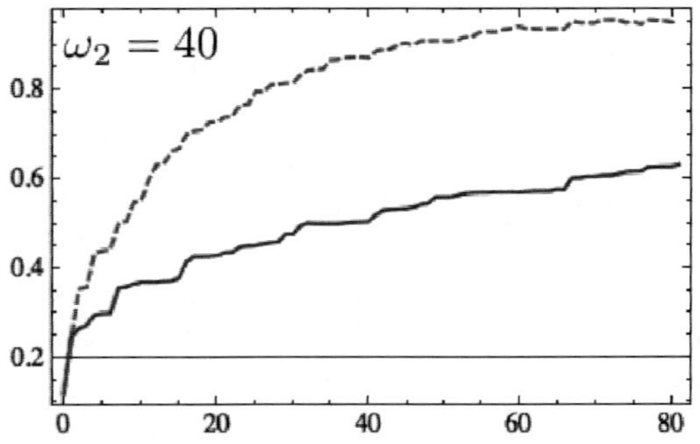

(d) Mode expansion comparison between $\mathcal{F}_n$ (top) and R_n (bottom) with a fixed frequency for S_2.

Figure 2: Mode expansion for various models.

We have plotted the number of modes versus the total correlation between the original model and the expansion for S_{Res}, S_{nBD}, S_1 and S_2 in figure 2. As can be seen, for most of these bispectra reconstruction using $\mathcal{F}_n$ versus R_n is much more effective, reducing the number of modes to reach similar correlation by a factor of 5. However, for nS_{nBD}, $\mathcal{F}_n$ reconstruction is not improved. The possible explanation why Fourier expansion is even worse than polynomial expansion for this type of bispectrum, seems to be related to the rapid change in frequency in a fixed direction. Fourier expansion is optimized for scale invariant frequencies. The polynomial expansion is simply optimized in reproducing as many different shapes as possible, explaining the observation that it is able to slowly increase correlation with the addition of modes while Fourier expansion seems to saturate around 20%. Given the large enhancement of the amplitude f_{NL}^{nBD} of this type of non-Gaussianity (which scales as ω_v^3, see e.g. [8], one might still be able to extract some information from that data even with such small correlations [17].

3. Discussion

In general the expansion of the oscillatory primordial bispectra becomes unavailing for really large frequencies, both using $\mathcal{F}_n$ and R_n. There are however a number of interesting observations which could make constraining and expanding oscillating bispectra much more viable. First of all the expansion in mode functions of the resonant bispectrum has a very discrete character; if you consider fig. 2a and fig 2c only few modes actually contribute significantly to the convergence of the correlation. One could only try to expand the spectrum only in those modes, which could significantly reduce the number of modes necessary. Since the important modes seem to be related to the direction of propagation of the oscillation, we find that this conclusion is independent of the phase. In other words, only the value of the weights (α_n) will differ, not the mode numbers (n) that are relevant for the expansion, e.g. the resonant shape S_{Res} appears to propegate in the k_t direction. As such only those mode numbers have a non-zero α that corresponds to base mode functions that contain $\mathcal{F}_n = \{1, e^{ink_t}\}$.

Secondly, given the discreteness of the correlation it is (obviously) not necessary to constrain all mode functions in the CMB data to get an indication of there is an oscillatory three point signal and what the possible frequency of this signal might be. For resonant non-Gaussianities we only need to consider those modes that have a significantly large α, and the measured value of the weights α would be a direct measure of the frequency. If one could extract the multipole projected Fourier modes that are responsible for most of the weight, this could in principle provide signatures of primordial bispectra with frequencies much larger than $\omega_r = 80$ shown in fig. 2a. Measuring modes up to e.g. $n = 100$ would not only provide information about the frequency of the signal, but could also hint on the type of primordial bispectrum.

4. Conclusion

We have investigated the viability of mode expansion for bispectra that contain oscillations. The motivation for investigating such features and their mode expansion, is that recently it has been shown that several scenarios or mechanisms can produce such features not only in the power spectrum, but also in the bispectrum. The appearance of oscillations in the bispectrum makes comparison with existing bispectral constraints, based on smooth bispectra, very inefficient and there exists substantial room for improvement. Polynomial expansion has been proposed to achieve factorization of a given theoretical bispectrum and we have investigated this for three different models. As expected, the larger the frequency of the primordial bispectrum, the more modes it requires to establish a reasonable approximation of the original spectrum. Fortunately, both the resonant and non-BD bispectrum have an amplitudes that scale with their frequency. Therefore, a small improvement in correlation could lead to a significant improvement in the ability to constrain the model by measuring these modes in the data and reconstructing the primordial signal.

Complementarily, we have proposed a different basis expansion, based on Fourier functions instead of polynomials. Such expansion is more relevant for resonant and non-BD scenario, since the feature bispectrum can already be transformed into Fourier modes analytically, using identities. We have shown that Fourier modes are much more efficient for the resonant bispectrum, reducing the number of modes necessary to establish the same correlation as polynomial modes by at least a factor of 5. For the non-BD bispectrum both Fourier expansion and polynomial expansion are difficult. In the case of Fourier expansion correlation increases fast with the addition of modes, but quickly saturates to maximum of about 20%. We believe that this is due to the exact form of the bispectrum, which has many small features near the edges of the tetrahedral domain. One might hope that some of these very small features are washed out when you compute the multipole equivalent. We hope to investigate this in a future attempt. In addition we have investigated three toy-spectra, not based on any particular model, which have a different oscillating orientation compared to the three theoretical models. Expanding these in Fourier modes show similar improvement (only two have been shown) compared to polynomial expansion as the resonant bispectrum. In general, we therefore belief that Fourier expansion is much more effective in the expansion of oscillatory spectra compared to polynomial basis expansion.

We showed that for resonant inflation only a limited number of modes contribute significantly in reproducing the original bispectrum. This would allow to consider (observationally) only those modes that contribute substantially. This holds independent of the phase and frequency of the signal and is due to the specific form of this bispectrum, which oscillates (primarily) in the k_t direction. Because the modes that are important for the reconstruction of the original bispectrum are independent of the frequency, this also implies that when one would observe these modes in the data one could in fact find evidence for much larger frequencies than discussed here, simply because for larger frequencies these modes will also matter but their respective weight (α) will be smaller. Despite the fact that we could not optimally expand the non-BD bispectrum using Fourier modes, we did look into the three toy-sepctra. We found that other modes are important. Moreover, the modes that are important directly represent the orientation of the oscillating spectrum and could therefore discriminate between different bispectra quite effectively. If this conclusion holds after forward projection into multipole space, measuring a number of Fourier mode functions in the CMB data would present an efficient way of deducing whether oscillations are present in the data and could give both an indication of the frequency and the shape of the primordial bispectrum.

Acknowledgments

He would also like to thank the organizers for allowing to present this recent work. The author was supported by the Netherlands Organization for Scientific Research (NWO), NWO-toptalent grant 021.001.040. The author received additional fundings for attending this conference from the Leids Kerkhoven-Bosscha Foundation (LKBF).

References

[1] Meerburg P D 2010, "Oscillations in the Primordial Bispectrum: Mode Expansion" *Phys. Rev.* D **82** 063517
[2] Fergusson J R and Shellard E P S 2009 "The shape of primordial non-Gaussianity and the CMB bispectrum" *Phys. Rev.* D **80** 043510
[3] Fergusson J R Liguori M and Shellard E P S 2010, "General CMB and Primordial Bispectrum Estimation I: Mode Expansion, Map-Making and Measures of f_{NL}" arXiv:0912.5516
[4] Chen X Easther R and Lim E A 2007 "Large non-Gaussianities in single field inflation" *JCAP* **0706** 023
[5] Chen X Easther R and Lim E A 2008, "Generation and Characterization of Large Non-Gaussianities in Single Field Inflation" *JCAP* **0804** 010
[6] Flauger R McAllister L Pajer E Westphal A and Xu G 2009, "Oscillations in the CMB from Axion Monodromy Inflation" *JCAP* **1006** 009
[7] Flauger R and Pajer E 2010, "Resonant Non-Gaussianity" arXiv:1002.0833
[8] Meerburg P D van der Schaar J P and Corasaniti P S 2009 "Signatures of Initial State Modifications on Bispectrum Statistics" *JCAP* **0905** 018
[9] Meerburg P D van der Schaar J P and Jackson M G 2009 "Bispectrum signatures of a modified vacuum in single field inflation with a small speed of sound," *JCAP* **1002** 001 (2010)
[10] Greene B Schalm K van der Schaar J P and Shiu G 2004, "Extracting new physics from the CMB" *In the Proceedings of 22nd Texas Symposium on Relativistic Astrophysics at Stanford University, Stanford, California, 13-17 Dec 2004, pp 0001*
[11] A. Gangui, F. Lucchin, S. Matarrese and S. Mollerach, "The Three Point Correlation Function Of The Cosmic Microwave Background In Inflationary Models," *Astrophys. J.* **430**, 447 (1994).
[12] E. Komatsu and D. N. Spergel, "Acoustic signatures in the primary microwave background bispectrum," *Phys. Rev.* D **63**, 063002 (2001).
[13] P. Creminelli, A. Nicolis, L. Senatore, M. Tegmark and M. Zaldarriaga, "Limits on non-Gaussianities from WMAP data," *JCAP* **0605**, 004 (2006).
[14] L. Senatore, K. M. Smith and M. Zaldarriaga, "Non-Gaussianities in Single Field Inflation and their Optimal Limits from the WMAP 5-year Data," *JCAP* **1001**, 028 (2010)
[15] Babich D Creminelli P and Zaldarriaga M 2004, "The shape of non-Gaussianities" *JCAP* **0408** 009
[16] Fergusson J R Liguori M and Shellard E P S 2010, "The CMB Bispectrum" arXiv:1006.1642
[17] Meerburg P D and van der Schaar J P 2010, "Minimal cut-off vacuum state constraints from CMB bispectrum statistics" arXiv:1009.5660

16th International Symposium on Particles Strings and Cosmology (PASCOS 2010) IOP Publishing
Journal of Physics: Conference Series **259** (2010) 012050 doi:10.1088/1742-6596/259/1/012050

Black Holes in 5D Hořava Lifshitz Theory of Gravity

George Koutsoumbas, Eleftherios Papantonopoulos,
Pavlos Pasipoularides, Minas Tsoukalas

Department of Physics, National Technical University of Athens,
Zografou Campus GR 157 73, Athens, Greece.

E-mail: kutsubas@central.ntua.gr, lpapa@central.ntua.gr, paul@central.ntua.gr, minasts@central.ntua.gr,

Abstract. In this talk we discuss black hole solutions in five dimensions in the context of Hořava-Lifshtz gravity. There are several classes of solutions which are characterized by an AdS_5, dS_5 or flat large distance asymptotic behaviour, plus the standard $1/r^2$ tail of the usual five-dimensional Schwarzschild black holes. In addition there are solutions with an unconventional short or large distance behaviour, and for a special range of the coupling parameters solutions which coincide with black hole solutions of conventional relativistic five-dimensional Gauss-Bonnet gravity.

1. Introduction

A lot of attention has been addressed lately to Hořava-Lifshtz model of gravity [1],[2]. The model is based on an anisotropy between space and time coordinates, which is expressed via the scalings $t \to b^z t$ and $x \to bx$, where z is a dynamical critical exponent. In the particular model the four-dimensional diffeomorphism invariance of general relativity is sacrificed in order to achieve power-counting renormalizability. Instead we have foliation preserving diffeomorphisms

$$x^i \to x'^i = x'^i(x^i, t), \quad t \to t' = t'(t). \tag{1.1}$$

The action contains a kinetic plus a potential term, which both respect a restricted $(3+1)$ diffeomorphism invariance. Due to (1.1) we can add higher order spatial derivative terms, which improve the UV behavior, rendering the model power-counting renormalizable. However the absence of full diffeomorphism invariance introduces an additional scalar mode which can lead to strong coupling problems or instabilities [5], [6], [7], [8], [9].

Originally the potential term, [1] was formed under the so called "detailed balance principle". An alternative way for constructing an action is to include all possible operators which are compatible with the renormalizability of the model. This implies that all operators with dimension less or equal to six are allowed in the action (for the exact form of the action see [3],[4]). There are two versions which are known as projectable and non-projectable. In the projectable version the lapse function N^2 depends only on the time coordinate, while in the non–projectable version N^2 is a function of both the space and time coordinates.

Issues connected with broken Lorentz invariance were studied in the spherically symmetric solutions of 4D HL gravity. In the case of detailed balance, such spherically symmetric solutions were found [10], but they exhibited an unconventional large distance asymptotic behaviour. The correct Schwarzchild-flat asymptotic behaviour can be recovered if the detailed balance action is modified in the IR by a term proportional to Ricci scalar, and the cosmological constant term is considered to be zero [11]. A similar study, in the case of non-vanishing cosmological constant, has also been carried out [12]. A generalization to topological black holes was obtained in [13]. Finally, a systematic study of static spherically symmetric solutions of 4D HL gravity was presented in [14],[17] where the most general spherically symmetric solution for $\lambda \neq 1$ and general coupling parameters was obtained.

In this talk we present a full study of spherically symmetric solutions in the non-projectable version of the five-dimensional Horava-Lifshitz gravity, for $z = 4$, [16]. For the construction of the 5D action we do not use the "detailed balance principle", but we include all the terms which are compatible with the renormalizability of the model. In particular, we can include all spatial curvature terms with dimension less than or equal to eight. However, the large number of possible terms, which are allowed in the action, leads to equation of motion of great complexity. For this reason we restrict our study only to terms of up to second order in the curvature. Also, we suppose that in the IR limit 5D the HL gravity reduces to the 5D General Relativity plus a bulk cosmological constant. A class of spherically symmetric solutions of the 5D HL gravity has been considered previously [15], but only for a very specific choice of the couplings.

2. Action and equations of motion

The 5D action of the model is constructed from a kinetic plus a potential term according to the equation

$$S = \frac{1}{16\pi G_5 c} \int dt d^d x \sqrt{|g|} N \left\{ \mathcal{L}_K + \mathcal{L}_V \right\} \tag{2.1}$$

in which d ($D = d+1 = 5$) is the spatial dimension and G_5 is the five dimensional Newton constant. The kinetic part in the above Lagrangian of Eq. (2.1) can be expressed via the extrinsic curvature as:

$$\mathcal{L}_K = (K^{ij}K_{ij} - \lambda K^2), \quad K_{ij} = \frac{1}{2N}\left\{ -\partial_t g_{ij} + \nabla_i N_j + \nabla_j N_i \right\}, \quad i,j = 1,2,3,4 \tag{2.2}$$

For the construction of the potential term we will follow the approach, according to which the potential term is constructed by including all possible renormalizable operators, that have dimension smaller or equal to eight, hence we write

$$\mathcal{L}_V = \mathcal{L}_R + \mathcal{L}_{R^2} + \mathcal{L}_{R^3} + \mathcal{L}_{\Delta R^2} + \mathcal{L}_{R^4} + \mathcal{L}_{\Delta R^3} + \mathcal{L}_{\Delta^2 R^2} \ . \tag{2.3}$$

where the symbol Δ is defined as $\Delta = \partial^i \partial_i$ ($i = 1,2,3,4$).

The dimensions of the various terms in the Lagrangian read

$$[R] = 2, \ [R^2] = 4, \ [R^3] = [\Delta R^2] = 6, [R^4] = [\Delta R^3] = [\Delta^2 R^2] = 8 \ . \tag{2.4}$$

In this work we are mainly interested in the lowest order operator $\mathcal{L}_R$ and the operator $\mathcal{L}_{R^2}$, which contains contributions of second order in the curvature:

$$\mathcal{L}_R = \eta_{0a} + \eta_{1a} R, \quad \mathcal{L}_{R^2} = \eta_{2a} R^2 + \eta_{2b} R^{ij} R_{ij} + \eta_{2c} R^{ijkl} R_{ijkl} \ . \tag{2.5}$$

Note that in the case of three spatial dimensions the term $R^{ijkl}R_{ijkl}$ is absent, as the Weyl tensor in three dimensions automatically vanishes. However, in four spatial dimensions this term cannot be omitted from the action.

The first term $\mathcal{L}_R$ is necessary in order to recover 5D general relativity with a cosmological constant in the IR limit. The second term $\mathcal{L}_{R^2}$, includes all possible quadratic terms in curvature, and becomes important in the short distance regime of the theory. Moreover, η_{0a} plays the role of the cosmological constant, while η_{1a}, η_{2a}, η_{2b}, and η_{2c} are dimensionful coupling constants with dimensions

$$[\eta_{1a}] = 6, \quad [\eta_{2a}] = [\eta_{2b}] = [\eta_{2c}] = 4 \ . \tag{2.6}$$

In the present analysis we ignore higher order Lagrangian terms, of dimension six and eight.

If this model is to make sense, it is necessary that the 5D general relativity (with a cosmological constant in our case) is recovered in the IR limit. Although there is no theoretical proof for this difficult question, we will assume that the renormalization group flow towards the IR leads the parameter λ to the value one ($\lambda = 1$), hence 5D general relativity is recovered. Also, to obtain the Einstein-Hilbert action

$$S_{EH} = \frac{1}{16\pi G_5} \int dx^0 d^4x \sqrt{g} N \left(\tilde{K}_{ij} \tilde{K}^{ij} - \tilde{K}^2 + R + \eta_{0a} \right) \ , \tag{2.7}$$

we have to set $\eta_{1a} = c^2$, and

$$\tilde{K}_{ij} = \frac{1}{2N} \left\{ -\partial_0 g_{ij} + \nabla_i \left(\frac{N_j}{c} \right) + \nabla_j \left(\frac{N_i}{c} \right) \right\} \ . \tag{2.8}$$

where the time-like coordinate x_0 is defined as $x_0 = ct$.

We will present 5D spherically symmetric solutions of the Horava-type gravity model. We use the following ansatz for the metric

$$ds^2 = -N(r)^2 dt^2 + f^{-1}(r)\, dr^2 + r^2 d\Omega_k^2 \ , \tag{2.9}$$

in which r is a radius coordinate that corresponds to the extra dimension, and $d\Omega_k^2$ is the metric of a 3D maximally symmetric space, where k is the spatial curvature of 3D hypersurfaces and for $k = 1, -1, 0$ we have a sphere, hyperboloid or 3D torus topology correspondingly. In what follows it is convenient to perform the transformation

$$f(r) = k + r^2 Z(r) \ . \tag{2.10}$$

Then the action of the model to second order in curvature terms is

$$S = \frac{1}{16\pi G_5} \int dt d^d x \sqrt{|g|} N \left(K^{ij} K_{ij} - \lambda K^2 + \eta_{0a} + \eta_{1a} R + \eta_{2a} R^2 + \eta_{2b} R^{ij} R_{ij} + \eta_{2c} R^{ijkl} R_{ijkl} \right) \tag{2.11}$$

can be put into the form

$$S\left[N(r), Z(r), \frac{dZ(r)}{dr}\right] = \int_0^{+\infty} dr\ L\left[N(r), Z(r), \frac{dZ(r)}{dr}\right] \ , \tag{2.12}$$

after we integrate out the angular coordinates, where

$$L\left[N, Z, \frac{dZ}{dr}\right] \sim r^3 \sqrt{\frac{N^2}{f}} \left(P \left(r\frac{dZ}{dr} \right)^2 + M(Z) \left(r\frac{dZ}{dr} \right) + Q(Z) \right) \tag{2.13}$$

and

$$\begin{aligned} P &= 3(3\eta_{2a} + \eta_{2b} + \eta_{2c}) \ , \\ M(Z) &= 6(12\eta_{2a} + 3\eta_{2b} + 2\eta_{2c})Z - 3\eta_{1a} \ , \\ Q(Z) &= 12(12\eta_{2a} + 3\eta_{2b} + 2\eta_{2c})Z^2 - 12\eta_{1a} Z + \eta_{0a} \ . \end{aligned} \tag{2.14}$$

The solutions come from the Euler-Lagrange equations of the aforementioned Lagrangian.

3. Solutions

3.1. No quadratic terms, $\eta = 0$ and $\varrho = 0$
If we set $\eta = \varrho = 0$ we find the simple solution

$$f(r) = k + r^2 Z = k + \frac{\eta_{0a}}{12} r^2 + \frac{\tilde{C}_\mu}{3r^2} , \tag{3.1}$$

where $\tilde{C}_\mu$ is a constant of integration and $N(r)^2 = f(r)$. If we set

$$\Lambda_{eff} = -\eta_{0a} , \quad \mu = -\frac{\tilde{C}_\mu}{3} , \tag{3.2}$$

the above equation takes the well-known form

$$f(r) = k - \frac{\Lambda_{eff}}{12} r^2 - \frac{\mu}{r^2} , \tag{3.3}$$

which is the standard AdS_5 (for $\Lambda_{eff} < 0$) or dS_5 (for $\Lambda_{eff} > 0$) or asymptotically flat (for $\Lambda_{eff} = 0$) Schwarzschild black hole solution of 5D general relativity with a cosmological constant.

3.2. $\eta = 0$ and $\varrho \neq 0$
If $\eta = 0$ then $P = 0$, then the equation for Z reads as

$$r\frac{dZ}{dr} = -\frac{\eta_{0a} - 12Z + 12\varrho Z^2}{-3 + 6\varrho Z} . \tag{3.4}$$

Integration of this equation yields:

$$3\varrho Z^2 - 3Z + \frac{\eta_{0a}}{4} + \frac{\tilde{C}_\mu}{r^4} = 0 , \tag{3.5}$$

where $\tilde{C}_\mu$ is an integration constant which is related to the mass of the black hole. The algebraic equation (3.5) gives two solutions

$$Z(r) = \frac{1}{2\varrho} + \sigma \frac{\sqrt{3(3 - \varrho\eta_{0a})r^4 - 12\varrho\tilde{C}_\mu}}{6\varrho\, r^2} , \tag{3.6}$$

where σ is a sign ($\sigma = \pm 1$), so for the function $f(r) = k + r^2 Z$ we obtain

$$f(r) = k + \frac{r^2}{2\varrho}\left[1 + \sigma\sqrt{\left(1 - \frac{\varrho\eta_{0a}}{3}\right) - \frac{4\varrho\tilde{C}_\mu}{3r^4}}\right] . \tag{3.7}$$

In what follows we will assume that $\varrho\tilde{C}_\mu < 0$, because for $\varrho\tilde{C}_\mu > 0$ the range of radius r has a lower bound ($r > r_{min}$).

The Euler-Lagrange equations for $N(r)$ yield

$$N(r)^2 = f(r) . \tag{3.8}$$

This solution is in direct connection with the solutions for black holes in Gauss-Bonnet theory which take the following form

$$f(r) = k + \frac{r^2}{4\hat{a}}\left[1 + \sigma\sqrt{1 - 8\hat{a}n^2 + 8\frac{a\mu_g}{r^4}}\right] , \quad \sigma = \pm 1 , \tag{3.9}$$

in which μ_g is a constant of integration which is related with the mass of the black hole, and the parameter $n^2 = -2\Lambda$ corresponds to a negative bulk cosmological constant.

If we replace

$$\hat{a} \to \frac{\varrho}{2},\ n^2 \to \frac{\eta_{0a}}{12},\ \mu_g \to -\frac{C_\mu}{3} \tag{3.10}$$

in equation (3.9), we recover the black hole solution of Eq. (3.7), for the specific case $\eta = 0$ and $\varrho \neq 0$ of the previous section.

Note, that the condition $\eta = 3\eta_{2a}+\eta_{2b}+\eta_{2c} = 0$ is satisfied in the case of GB coefficient $\eta_{2a} = \hat{a}$, $\eta_{2b} = -4\hat{a}$ and $\eta_{2c} = \hat{a}$, but there are other different combinations of the coupling parameters η_{2a}, η_{2b}, η_{2c} which give $\eta = 0$ and $\varrho \neq 0$. This is a very interesting result which merits further investigation. Note also that the relation $1 - \frac{\varrho\eta_{0a}}{3} = 0$ corresponds to the Chern-Simons limit of GB gravity.

3.3. $\varrho = 0$ and $\eta \neq 0$

Substituting $\varrho = 0$ and $\eta \neq 0$ we find for the function $f(r) = k + r^2 Z(r)$

$$f(r) = k + \frac{\eta_{0a}}{12} r^2 + \frac{r^2}{16\eta}\left(W_L^2\left(\frac{\tilde{C}_\mu}{r^4}\right) + 2W_L\left(\frac{\tilde{C}_\mu}{r^4}\right)\right) . \tag{3.11}$$

where $W_L(x)$ is the Lambert function, which is defined as the real solution of the equation $e^{W_L(x)}W_L(x) = x$. The large r asymptotic behaviour of Eq. (3.11) is found to be

$$f(r) = k + \frac{\eta_{0a}}{12} r^2 + \frac{\tilde{C}_\mu}{8\eta r^2} + O\left(\frac{1}{r^6}\right) . \tag{3.12}$$

The function $N(r)^2$ can be expressed in the following closed form

$$N(r)^2 = f(r)\tilde{N}(Z(r))^2 = \frac{\tilde{C}_\mu^2 f(r)}{r^8\left(W_L^2\left(\frac{\tilde{C}_\mu}{r^4}\right) + W_L\left(\frac{\tilde{C}_\mu}{r^4}\right)\right)^2} . \tag{3.13}$$

In the large r regime we find, from the above equation, that:

$$N(r)^2 = f(r)\left(1 + \frac{\tilde{C}_\mu^2}{r^8} + O\left(\frac{\tilde{C}_\mu^3}{r^{12}}\right)\right) , \tag{3.14}$$

hence in the large distance limit we recover the standard asymptotic behavior $N(r)^2 \simeq f(r)$.

3.4. Static solutions in the generic case ($\eta \neq 0$ and $\varrho \neq 0$)

For the generic case where $\eta \neq 0$ and $\varrho \neq 0$ we will expose the main features of the solutions. For more details can be found in [16]. First we define the following constants

$$A \equiv -3\eta_{0a}\eta + \frac{9\eta}{\varrho} , \quad B \equiv \varrho(\varrho - 4\eta) . \tag{3.15}$$

Case (i): $B \geq 0$ and $|\sqrt{B}/\varrho| < 1$, $A > 0$

In this case we have in general two branches of solutions for function $f(r)$ where both of them have AdS_5, dS_5 or flat large distance behavior. In the case of $k = 0$ one of the solutions exhibits a naked singularity. On the other hand we can see that the large distance behavior of $N(r)^2$, is identical with that of $f(r)$. In the case where $A < 0$, the radius r must have an upper bound and the function $f(r)$ has no large distance limit. This case lacks of physical interest.

Case (ii): $B \geq 0$ and $|\sqrt{B}/\varrho| > 1$, $A > 0$

Here we have only one branch for the function $f(r)$. However there is an even more significant difference, as this class of solutions does not have regular AdS_5, dS_5 or flat asymptotic behavior. Furthermore we see that $N(r)^2$ vanishes for large values of r. For $A < 0$ the solutions behave similarly.

Case (iii): $B > 0$ and $A = 0$

Setting $A = 0$, we obtain for the functions $f(r)$ and $\hat{N}(r)$:

$$f(r) = k + \frac{r^2}{2\varrho} - \frac{C_0}{3} r^{2\frac{\sqrt{B}+\varrho}{\sqrt{B}-\varrho}} \tag{3.16}$$

and

$$\hat{N}(r) = \frac{\bar{C}_N}{C_0^2\sqrt{B}|\varrho + \sqrt{B}|} r^{-4\frac{\sqrt{B}+\varrho}{\sqrt{B}-\varrho}} \ . \tag{3.17}$$

As we see the above solutions exhibit an unconventional asymptotic behaviour which is not of the type AdS_5, dS_5 or flat 5D Schwarzschild form.

Case (iv): $B < 0$ and $A > 0$

Here if $B < 0$ then A must always be positive. Function $f(r)$ has several branches of solutions. In contrast with the previous case ($B > 0$) the range of radius r terminates at a lower non-zero bound. Solutions for function $f(r)$ have normal AdS_5, dS_5 large distance behavior. Again there are solutions where naked singularities occur.

Case (v): $B = 0$

Finally for $B = 0$ we have the following solutions

$$f(r) = k + \left(\frac{1}{2\varrho} + \frac{\sigma\sqrt{A}}{3\varrho}\right) r^2 \pm \frac{\tilde{C}_\mu}{3\varrho r^2} \ . \tag{3.18}$$

This has the standard form of a AdS_5, dS_5 or flat solution. For $N(r)^2$ we get,

$$N(r)^2 = \frac{1}{f(r)} \ . \tag{3.19}$$

4. Conclusions

We presented static spherically symmetric solutions in the framework of the 5D Horava-Lifshitz gravity. We considered an action consisting of terms up to second order in the curvature and we solve the theory with a non-projectable spherically symmetric ansatz for the metric. The black hole spectrum we found is controlled by three parameters η, ϱ and η_{0a}, where η_{0a} is a cosmological constant.

More specifically, there are three main sets of solutions: the two special cases ($\eta = 0$, $\varrho \neq 0$), and ($\eta \neq 0$, $\varrho = 0$) and the generic case ($\eta \neq 0$, $\varrho \neq 0$). In all cases we obtained analytic black hole solutions which have the standard AdS_5, dS_5 of flat asymptotic behaviour, plus the well-known $1/r^2$ tail. However, we also obtained solutions with an unconventional short and large distance asymptotic behaviour. Also, in many cases we obtained solutions with a naked singularity.

We also found static solutions which, after a proper identification of coupling parameters, coincide with static black hole solutions of relativistic gravity theories with quadratic curvature correction terms. One class of these solutions consists of the Schwarzschild-AdS black hole solutions of five-dimensional Lanczos-Lovelock gravity theories. Another class of solutions contains the well-known Gauss-Bonnet black hole solutions. The interesting result we obtained in our investigation is that the non-relativistic solutions of the HL gravity corresponding to the Gauss-Bonnet solutions can be obtained for various combinations of the coupling parameters η and ϱ and not just the standard Gauss-Bonnet combination. This may be attributed to the fact that the HL static solutions are insensitive to the coupling parameter λ, so they hold even if $\lambda \neq 1$ (the value which signals the breaking of Lorentz invariance).

Acknowledgments

M.T. would like to thank the organizers of PASCOS 2010 for giving him the opportunity to give this talk.

References

[1] Horava P 2009, *JHEP* **0903** 020 (*Preprint* 0812.4287 [hep-th])
[2] Horava P 2009, *Phys. Rev.* D **79**, 084008 (*Preprint* 0901.3775 [hep-th])
[3] Sotiriou T P, Visser M and Weinfurtner S 2009, *JHEP* **0910**, 033 (*Preprint*0905.2798 [hep-th])
[4] Sotiriou T P, Visser M and Weinfurtner S 2009, *Phys. Rev. Lett.* **102** 251601 (*Preprint* 0904.4464 [hep-th])
[5] Charmousis C, Niz G, Padilla A and Saffin P M 2009, *JHEP* **0908** 070 (*Preprint* 0905.2579 [hep-th])
[6] Blas D, Pujolas O and Sibiryakov S 2010, *Phys. Rev. Lett.* **104** 181302 (*Preprint* 0909.3525 [hep-th])
[7] Papazoglou A and Sotiriou T P 2010, *Phys. Lett.* B **685** 197 (*Preprint* 0911.1299 [hep-th])
[8] Kimpton I and Padilla A 2010, *JHEP* **1007**, 014 (*Preprint* 1003.5666 [hep-th])
[9] Padilla A 2010, (*Preprint* 1009.4074 [hep-th])
[10] Lu H, Mei J and Pope C N 2009, *Phys. Rev. Lett.* **103**, 091301 (*Preprint* 0904.1595 [hep-th])
[11] Kehagias A and Sfetsos K 2009, *Phys. Lett.* B **678** 123 (*Preprint* 0905.0477 [hep-th])
[12] Park M i 2009, *JHEP* **0909** 123 [arXiv:0905.4480 [hep-th]].
[13] R. G. Cai, L. M. Cao and N. Ohta, Phys. Rev. D **80** (2009) 024003 (*Preprint* 0904.3670 [hep-th])
[14] Kiritsis E and Kofinas G 2010, *JHEP* **1001** 122 (*Preprint* 0910.5487 [hep-th])
[15] Cai R G, Liu Y and Sun Y W 2009, *JHEP* **0906**, 010 (*Preprint* 0904.4104 [hep-th]) [arXiv:0904.4104 [hep-th]].
[16] Koutsoumbas G, Papantonopoulos E, Pasipoularides P and Tsoukalas M 2010, *Phys. Rev.* D **81** 124014 (*Preprint* 1004.2289 [hep-th])
[17] Koutsoumbas G and Pasipoularides P 2010, *Phys. Rev.* D **82** 044046 (*Preprint* 1006.3199 [hep-th])

Constraints on D Dimensional Warped Spaces.

Paul R. Archer

[a] *Department of Physics & Astronomy, University of Sussex, Brighton BN1 9QH, UK*

E-mail: p.archer@sussex.ac.uk

Abstract. In order to investigate the phenomenological implications of allowing gauge fields to propagate in warped spaces of more than five dimensions, we consider a toy model of a space warped by the presence of a anisotropic bulk cosmological constant. After solving the Einstein equation, three classes of solutions are found, those in which the additional ($D > 5$) dimensions are growing, shrinking or remaining constant. It is found that gauge fields propagating in these spaces have a significantly different Kaluza Klein (KK) mass spectrum and couplings from that of the Randall and Sundrum model. This leads to a greatly reduced lower bound on the KK scale, arising from electroweak constraints, for spaces growing towards the IR brane.

1. Introduction

The Randall and Sundrum (RS) model [1] has proved a popular extension to the standard model for a number of reasons. Firstly it offers a non supersymmetric resolution to the gauge hierarchy problem and secondly it provides a natural model with which to describe flavour physics [2, 3, 4]. However allowing the standard model gauge fields to propagate in the bulk gives rise to corrections to the electroweak observables (EWO's) at tree level, which can naively force the lower bound on the Kaluza Klein (KK) mass scale to be $M_{KK} \gtrsim 11$ TeV, corresponding to a mass of the first gauge boson excitation of about 27 TeV [5]. This bound can be reduced to $M_{KK} \gtrsim 2-3$ TeV by allowing the fermions to propagate in the bulk and localising them towards the UV brane and at the same time introducing either large brane kinetic terms [6] or a bulk $SU(2)_R \times SU(2)_L \times U(1)$ gauge symmetry [7].

In spite of this, before the RS model can be considered a full resolution to the hierarchy problem it requires a UV completion. There has been considerable work on providing a UV completion via string theory, particularly with regard to the AdS/CFT correspondence, for example [8, 9, 10], but this work necessarily requires models of more than five dimensions. Although the RS model is often seen as the low energy effective theory of $AdS_5 \times S^5$, it is quite possible that the $D > 5$ additional dimensions are also warped [11, 12].

With this in mind we first look generally at what really determines the size of these EW corrections, before taking a 'bottom up' approach to extending the RS model to more than 5D, by considering a space with a anisotropic bulk cosmological constant.

2. General Warped Extra Dimensions

In this section we consider a $4+1+\delta$ dimensional spaced warped with respect to (w.r.t) a single 'preferred' direction r, described by

$$ds^2 = a^2(r)\eta_{\mu\nu}dx^\mu dx^\nu - b^2(r)dr^2 - c^2(r)d\Omega_\delta^2, \tag{1}$$

where $\eta_{\mu\nu} = \text{diag}(+ - - -)$ is the 4D Minkowski metric and $d\Omega_\delta^2 = \gamma_{ij} d\phi^i d\phi^j$, with i, j running from $1 \ldots \delta$. As in the RS model the space is bounded by two branes at $r = r_{\text{ir}}$ and $r = r_{\text{uv}}$.

2.1. Resolving the Hierarchy Problem

From a phenomenological perspective, one of the central ideas of extra dimensional models is that the 4D bare parameters of a given theory are not fundamental but have been scaled down from their higher dimensional values. Hence to resolve the hierarchy problem it is required that the dimensionful parameters of the higher dimensional theory be all at the same order of magnitude or in particular that the fundamental Higgs mass ($m_{\text{fund}}^{\text{Higgs}}$) be similar to the fundamental Planck mass (M_{fund}). In computing the 4D effective parameters there are essentially two effects at work. Firstly the bare parameters are scaled by the volume of the extra dimension. So for example the fundamental Planck mass is scaled by

$$M_{\text{P}}^2 \sim \int d^{\delta+1}x \, a^2 bc^\delta \sqrt{\gamma} \, M_{\text{Fund}}^{\delta+3}. \tag{2}$$

This is the mechanism used in the ADD model [13, 14]. Secondly fields localised throughout the space would have masses suppressed by gravitational red shifting or warping. So the Higgs, localised on the IR brane, would have a 4D effective mass

$$m_{\text{4D}}^2 = a^2(r_{\text{ir}}) m_{\text{fund}}^2. \tag{3}$$

In the RS model the volume effects are $\mathcal{O}(1)$ and hence a warp factor of $a(r_{\text{ir}})^{-1} \equiv \Omega \sim 10^{15}$ is required in order to resolve the hierarchy problem [1]. Although clearly these mechanisms are not mutually exclusive and the fundamental scale of nature could be neither the Planck scale nor the EW scale but some intermediate scale.

2.2. Electroweak Corrections

Likewise when one computes the EW corrections it is found that they are largely determined by the extent to which the 4D gauge couplings have been scaled. To be more precise, after making the usual KK decomposition

$$A_\mu = \sum_n A_\mu^{(n)}(x^\mu) f_n(r) \Theta_n(\phi_1, \ldots, \phi_\delta) \text{ such that } \int d^{1+\delta}x \, bc^\delta \sqrt{\gamma} f_n f_m \Theta_n \Theta_m = \delta_{nm}, \tag{4}$$

where the profiles are given by

$$f_n'' + \frac{(a^2 b^{-1} c^\delta)'}{(a^2 b^{-1} c^\delta)} f_n' - \frac{b^2}{c^2} \alpha_n f_n + \frac{b^2}{a^2} m_n^2 f_n = 0, \quad -\frac{1}{\sqrt{\gamma}} \partial_{\phi_i} (\sqrt{\gamma} \gamma^{ij} \partial_{\phi_j} \Theta_n) = \alpha_n \Theta_n, \tag{5}$$

and $'$ denotes derivative w.r.t r. Then the tree level corrections to the EWO's are determined by three parameters, the coupling of the KK gauge fields to the Higgs relative to that of the zero mode,

$$F_n \equiv \frac{f_n(r_{\text{ir}})}{f_0(r_{\text{ir}})} \quad \text{or} \quad F_n \equiv \frac{f_n(r_{\text{ir}}) \Theta_n(\phi_{\text{ir}})}{f_0(r_{\text{ir}}) \Theta_0(\phi_{\text{ir}})}, \tag{6}$$

the coupling of the KK gauge fields to the fermions relative to that of the zero mode

$$F_\psi^{(n)} \equiv \frac{f_\psi^{(0,n,0)}}{f_\psi^{(0,0,0)}} \quad \text{where} \quad f_\psi^{(l,n,m)} \equiv \int d^{\delta+1}x \, ba^3 c^\delta \sqrt{\gamma} \, f_L^{(l)} \Theta_L^{(l)} \, f_n \Theta_n \, f_L^{(m)} \Theta_L^{(m)}. \tag{7}$$

and the mass of the KK gauge fields m_n^2. Here $f_L^{(n)}$ and $\Theta_L^{(n)}$ are the KK profiles of the fermions, analogous to (4). The two definitions of the gauge-Higgs coupling are related to whether the Higgs is localised on a 3 brane or just w.r.t r as explained in [15].

Often the EW corrections are parametrised in terms of the Peskin-Takeuchi parameters [16], however the approach taken in [15] was to fix input parameters by comparison with $\hat{\alpha}(M_Z)^{-1}$, G_f and $\hat{M}_Z$ and then compute corrections to eight EWO's. In practice the resulting constraints from the two methods do not differ significantly. It is found that typically the tightest constraints come from the weak mixing angle (s_Z^2), which when there is a bulk $SU(2)\times U(1)$ gauge symmetry is given by

$$s_Z^2 \approx s_p^2 \left(1 - \frac{c_p^2}{c_p^2 - s_p^2} \sum_{n=1} \left[\frac{m_z^2 F_n^2}{m_n^2} - \frac{m_w^2 \left(F_n - F_\psi^{(n)}\right)^2}{m_n^2} \right] + \mathcal{O}(m_n^{-4}) \right). \tag{8}$$

While if there is a bulk $SU(2)_R \times SU(2)_L \times U(1)$ gauge symmetry it is given by

$$s_Z^2 \approx s_p^2 \left(1 + \frac{c_p^2}{c_p^2 - s_p^2} \sum_{n=1} \left[\frac{m_w^2 F_\psi^{(n)\,2}}{m_n^2} - \frac{2m_w^2 F_n F_\psi^{(n)}}{m_n^2} + s'^2 m_w^2 \left(\frac{\tilde{F}_n^2}{\tilde{m}_n^2} - \frac{F_n^2}{m_n^2} \right) + \mathcal{O}(m_n^{-4}) \right] \right), \tag{9}$$

where s' is the mixing angle associated with the breaking of $SU(2)_R \times U(1) \to U(1)$, $\tilde{F}_n^2$ and $\tilde{m}_n^2$ are the equivalent couplings and masses for the $SU(2)_R$ gauge fields, $s_p^2 = \frac{1}{2}\left(1 - \sqrt{1 - \frac{4\pi\alpha}{\sqrt{2}G_f \hat{M}_Z^2}}\right)$ and $c_p^2 = 1 - s_p^2$. Note that when $F_n \sim \tilde{F}_n$ and $m_n \sim \tilde{m}_n$ then the fourth term of (9) cancels and the constraints become linearly dependent on F_n. This term is essentially the remnant of the Peskin-Takeuchi T parameter.

2.3. Reducing the Gauge-Higgs Coupling in 5D

Typically F_ψ can always be reduced by localising the fermions towards the UV brane away from the KK gauge fields and hence the lower bound on the KK mass scale is dominated by the value of F_n. For the RS model (with $\Omega = 10^{15}$) $F_n \approx 8.3$, while for universal extra dimensions $F_n = \sqrt{2}$. One can then ask the question, whether a space can be found in which the KK gauge fields effectively decouple from the Higgs? In five dimensions the gauge Higgs coupling is given by

$$F_n = \frac{\sqrt{\int b\, dr}\, f_n(r_{\text{ir}})}{\sqrt{\int b f_n^2\, dr}}. \tag{10}$$

However it can be shown that the warp factor is related to the gauge profile by

$$\Omega^2 = \frac{b^2(r_{\text{uv}}) f_n(r_{\text{uv}}) f_n''(r_{\text{ir}})}{b^2(r_{\text{ir}}) f_n(r_{\text{ir}}) f_n''(r_{\text{uv}})}. \tag{11}$$

Hence if one requires a large warp factor to resolve the hierarchy problem, then one would need a very unusually shaped profile, in order to reduce the gauge-Higgs coupling. So in agreement with [17], one can conclude that five dimensional spaces that resolve the hierarchy problem via warping, will typically have large EW corrections.

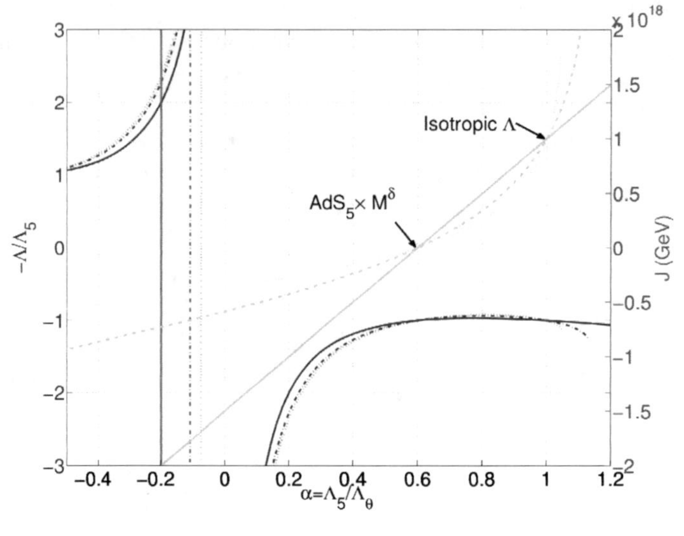

Figure 1. The solutions to the Einstein equations for 6D (solid lines), 8D(dash-dot lines) and 10D (dot-dot lines). J is plotted in green and $-\frac{\Lambda}{\Lambda_5}$ in blue. The red lines correspond to $2k+\delta J=0$ and hence the spaces to the left of them will have exponentially suppressed fundamental Planck masses. Here we have fixed $\Omega\equiv e^{kR}=10^{15}$ and $M_{KK}\equiv\frac{k}{\Omega}=1$ TeV.

3. A D dimensional Extension of the RS Model

We now consider a $4+1+\delta$ dimensional space, bounded by two 3 branes, described by

$$S=\int d^{5+\delta}x\sqrt{G}\left[\Lambda-\frac{1}{2}M^{3+\delta}R+\mathcal{L}_{\text{bulk}}\right]+\int d^4x\sqrt{g_{\text{ir}}}\left[\mathcal{L}_{\text{ir}}+V_{\text{ir}}\right]$$
$$+\int d^4x\sqrt{g_{\text{uv}}}\left[\mathcal{L}_{\text{uv}}+V_{\text{uv}}\right],\qquad(12)$$

where the co-ordinates run over $(x_\mu,r,\theta_1\ldots\theta_\delta)$, $r\in[0,R]$ and $\theta_i\in[0,R_\theta]$. We allow for an anisotropic bulk cosmological constant of the form $\Lambda=\text{diag}\,(\Lambda\eta_{\mu\nu},\Lambda_5,\Lambda_\theta,\ldots,\Lambda_\theta)$. We also introduce the parameter $\alpha\equiv\frac{\Lambda_5}{\Lambda_\theta}$. It can then be shown that the Einstein equations admit solutions (although admittedly not unique solutions) of the form

$$ds^2=e^{-2kr}\eta_{\mu\nu}dx^\mu dx^\nu-dr^2-\sum_{i=1}^{\delta}e^{-2Jr}d\theta_i^2.\qquad(13)$$

These solutions are plotted in figure 1. There are basically three classes of solution. When $\alpha>\frac{3}{5}$, $J>0$ and the warping of the δ additional dimensions are shrinking towards the IR brane. When $\alpha=\frac{3}{5}$ the δ additional dimensions are not warped. This scenario has been studied in [18, 19]. Finally when $\alpha<\frac{3}{5}$, $J<0$ and the warping of the δ additional dimensions are growing towards the IR brane. Spaces to the left of the vertical (red) lines in figure 1, would have volume suppressed Planck masses according to (2) and hence risk introducing an additional hierarchy between the AdS curvature and the Planck mass.

Moving now to consider gauge fields propagating in these spaces. The gauge KK profiles are given by (5),

$$f_n''-(2k+\delta J)f_n'-\sum_{i=1}^{\delta}e^{2Jr}\frac{l_i^2}{R_\theta^2}f_n+e^{2kr}m_n^2f_n=0,\qquad(14)$$

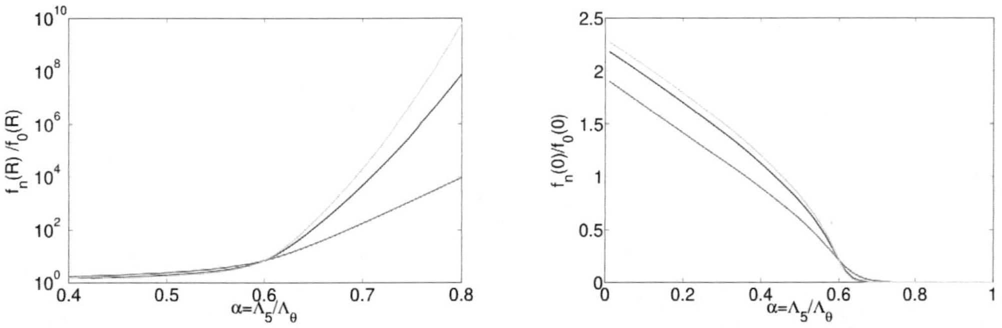

Figure 2. The relative coupling of the $l_i = 0$ KK gauge modes with a IR localised Higgs / fermion (on the left) and a UV localised fermion (on the right) in 6D (red), 8D (blue) and 10D (green). Here $\Omega \equiv e^{kR} = 10^{15}$ and $M_{KK} \equiv \frac{k}{\Omega} = 1\text{TeV}$.

where $\partial_i^2\Theta_n = -\frac{l_i^2}{R_\theta^2}\Theta_n$. When $l_i \neq 0$, (14) has no analytical solutions, although to get a feel for the solutions one can make the substitution $x^2 = \left(e^{2kr}m_n^2 - e^{2Jr}\sum_{i=1}^{\delta}\frac{l_i^2}{R_\theta^2}\right)/\gamma^2$ such that $x' = \gamma x$. So now when $\gamma \approx$ constant, (14) can be solved to give

$$f(x) \approx N x^{\frac{1}{2}\frac{2k+\delta J}{\gamma}}\left(\mathbf{J}_{-\frac{1}{2}\frac{2k+\delta J}{\gamma}}(x) + \beta\mathbf{Y}_{-\frac{1}{2}\frac{2k+\delta J}{\gamma}}(x)\right). \tag{15}$$

Typically this approximation is valid in the IR region where the gauge fields are localised and also (14) can be solved numerically to check these results [20]. As in the RS model, the KK mass eigenvalues are then determined by the zeros of the Bessel functions, hence

$$m_n \sim X_n\frac{\sqrt{\gamma^2 + \frac{e^{2JR}}{R_\theta^2}\sum_i^\delta l_i^2}}{e^{kR}}, \tag{16}$$

where $X_n \sim \mathcal{O}(1)$. Now returning to the three classes of solutions found in the previous section, when $J > 0$ (i.e. $\alpha > \frac{3}{5}$) then $\frac{e^{JR}}{R_\theta} \gg \gamma \sim k$ and the $l_i \neq 0$ KK modes would gain masses far larger than $M_{KK} = k/\Omega$ and essentially decouple from the low energy theory. On the other hand if $J \sim 0$ then $\frac{e^{JR}}{R_\theta} \sim \gamma$ and the $l_i \neq 0$ KK modes would have masses of $\mathcal{O}(M_{KK})$. However of potential interest to LHC phenomenologists, is the case when $J < 0$, $\frac{e^{JR}}{R_\theta} \ll \gamma$ and a hyperfine splitting in the KK mass spectrum would be introduced.

Moving on to consider the relative gauge couplings, plotted in figure 2, one can see the combined effects of warping and volume enhancement. Due to warping the KK modes will be localised towards the IR brane, while the zero mode is flat. So when the additional dimensions are warped ($J \neq 0$), the volume of those extra dimensions will scale the KK modes very differently to that of the zero mode. The result is a significant enhancement or suppression of the relative gauge couplings. The effect of this on the EW constraints is plotted in figure 3. The constraints for $\alpha > \frac{3}{5}$ ($J > 0$) are not plotted since the gauge fields would become strongly coupled and hence the tree level analysis would not be valid. Although one could assume the constraints would be large. There also appears to be an overall lower bound, corresponding to $M_{KK} \gtrsim 2$ TeV, that arises from the difficulty in getting F_n^2 or $F_nF_\psi < 1$.

4. Conclusion

The model presented in the second half of these proceedings can only really be considered as a toy model, whose sole purpose was to demonstrate the phenomenological implications of having

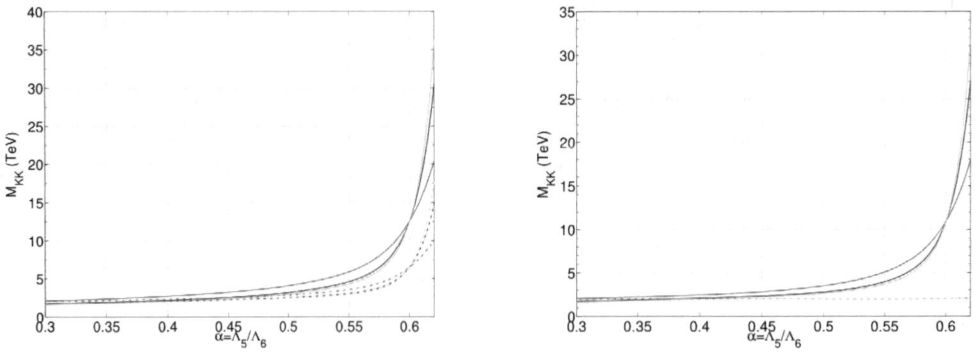

Figure 3. The lower bound on $M_{KK} \equiv \frac{k}{\Omega}$ arising from the EW observable s_Z^2. With a bulk $SU(2) \times U(1)$ gauge symmetry (top) and a bulk $SU(2)_R \times SU(2)_L \times U(1)$ custodial symmetry (bottom). Here the fermions are localised to the IR brane (solid line) or the UV brane (dot-dash line) while the gauge fields propagate in 6D (red), 8D (blue) and 10D (green). The Higgs is localised w.r.t r such that $\Omega = 10^{15}$.

additional ($D > 5$) warped dimensions. Having said that, the volume scaling of the relative coupling as well as the additional KK modes, are generic effects that appear in all spaces looked at, including solutions from string theory such as [12][15]. If one is to speculate that all strongly coupled Yang Mills theories have an AdS dual space then restricting ones phenomenological studies to that of AdS_5 may lead to misinterpretation of signals at the LHC. Here it has been demonstrated that the EW constraints, on models with warped extra dimensions, can be reduced with or without a custodial symmetry or UV localised fermions, by the inclusion of an additional dimension warped in the opposite direction to that of the four large dimensions. It has also been demonstrated that gauge fields propagating in spaces with additional warped dimensions would have a very different KK spectrum, couplings and hence phenomenology to that of the RS model.

[1] Randall L and Sundrum R 1999 *Phys. Rev. Lett.* **83** 3370–3373 (*Preprint* hep-ph/9905221)
[2] Grossman Y and Neubert M 2000 *Phys. Lett.* **B474** 361–371 (*Preprint* hep-ph/9912408)
[3] Gherghetta T and Pomarol A 2000 *Nucl. Phys.* **B586** 141–162 (*Preprint* hep-ph/0003129)
[4] Huber S J and Shafi Q 2001 *Phys. Lett.* **B498** 256–262 (*Preprint* hep-ph/0010195)
[5] Csaki C, Erlich J and Terning J 2002 *Phys. Rev.* **D66** 064021 (*Preprint* hep-ph/0203034)
[6] Carena M S, Ponton E, Tait T M P and Wagner C E M 2003 *Phys. Rev.* **D67** 096006 (*Preprint* hep-ph/0212307)
[7] Agashe K, Delgado A, May M J and Sundrum R 2003 *JHEP* **08** 050 (*Preprint* hep-ph/0308036)
[8] Arkani-Hamed N, Porrati M and Randall L 2001 *JHEP* **08** 017 (*Preprint* hep-th/0012148)
[9] Perez-Victoria M 2001 *JHEP* **05** 064 (*Preprint* hep-th/0105048)
[10] Reece M and Wang L T 2010 (*Preprint* 1003.5669)
[11] Bao R and Lykken J D 2005 *Phys. Rev. Lett.* **95** 261601 (*Preprint* hep-ph/0509137)
[12] Klebanov I R and Strassler M J 2000 *JHEP* **08** 052 (*Preprint* hep-th/0007191)
[13] Arkani-Hamed N, Dimopoulos S and Dvali G R 1998 *Phys. Lett.* **B429** 263–272 (*Preprint* hep-ph/9803315)
[14] Antoniadis I, Arkani-Hamed N, Dimopoulos S and Dvali G R 1998 *Phys. Lett.* **B436** 257–263 (*Preprint* hep-ph/9804398)
[15] Archer P R and Huber S J 2010 (*Preprint* 1004.1159)
[16] Peskin M E and Takeuchi T 1992 *Phys. Rev.* **D46** 381–409
[17] Delgado A and Falkowski A 2007 *JHEP* **05** 097 (*Preprint* hep-ph/0702234)
[18] Davoudiasl H, Hewett J L and Rizzo T G 2003 *JHEP* **04** 001 (*Preprint* hep-ph/0211377)
[19] Davoudiasl H and Rizzo T G 2008 *JHEP* **11** 013 (*Preprint* 0809.4440)
[20] Archer P R and Huber S J To appear

Journal of Physics: Conference Series **259** (2010) 012052 doi:10.1088/1742-6596/259/1/012052

Technicolor in hot and cold phases

Kimmo Tuominen

Department of Physics, P.O.Box 35 (YFL),
FI-40014 University of Jyväskylä, Finland,
and
Helsinki Institute of Physics, P.O. Box 64,
FI-00014 University of Helsinki, Finland

E-mail: kimmo.i.tuominen@jyu.fi

Abstract. We review recent developments on models of dynamical electroweak symmetry breaking. We first consider determination of the conformal window, i.e. the phase diagram, of cold gauge theory and how this can be applied to classify phenomenologically viable models of walking technicolor type. In the second part of the talk we consider finite temperature phase diagram of a generic walking type theory.

1. Introduction

Recently models of dynamical electroweak symmetry breaking [1, 2, 3, 4], i.e. technicolor, have received much attention. A general property of phenomenologically viable technicolor theories is that they be of walking type. This assumes that there are two widely different energy scales, $\Lambda_{\rm T}$ and $\Lambda_{\rm ET} \sim 10^3\Lambda_{\rm T}$, between which the coupling constant of the theory evolves very slowly; the theory is almost conformal towards the infrared. Below $\Lambda_{\rm T}$ and above $\Lambda_{\rm ET}$ the coupling runs similarly to asymptotically free theories (like QCD). In (Extended) Technicolor theories $\Lambda_{\rm T} = \Lambda_{\rm TC} \approx 246$ GeV and $\Lambda_{\rm ET} = \Lambda_{\rm ETC}$. The need for walking behavior comes from the requirement that the successes of the Standard Model should not be spoiled: the contributions from new physics to the electroweak precision parameters should be small and the contributions to flavor changing neutral current interactions should be suppressed.

The breakthrough with respect to earlier models has been the realization that one can achieve near conformal dynamics with a very small number of flavors using matter in higher dimensional representations [5, 6, 7, 8]. These results were obtained by considering the phase diagram of cold gauge theory as a function of number of colors and flavors and fermion representations [5], which will be the first topic of this talk; after that we will discuss finite temperature phase diagram of a generic walking technicolor theory using gauge/gravity duality.

2. Phase diagrams at $T = 0$

Let us first review how zero temperature phase diagrams of gauge theories can be constructed and how they help to identify plausible candidates for walking technicolor; we consider only non-supersymmetric theories. Generally, the evolution of a non-abelian gauge coupling is governed by two complementary effects: By the anti-screening of gauge bosons and by the screening of matter fields. If the former dominates the theory is in an asymptotically free QCD-like phase with confinement and chiral symmetry breaking in the vacuum, while if the latter dominates

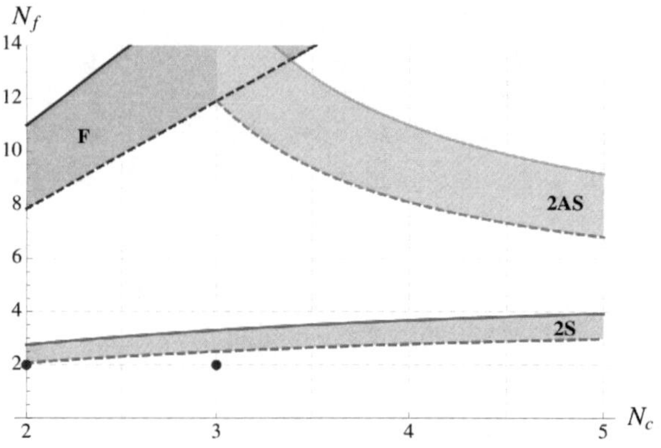

Figure 1. A phase diagram of a nonsupersymmetric SU(N_c) gauge theory as a function of number of colors, N_c and flavors N_f. The shaded bands are the conformal windows corresponding to different fermion representations. The symbols F and 2(A)S stand for fundamental and 2-index (anti)symmetric representations.

the theory is in QED-like non-abelian Coulomb phase. By increasing the amount of matter charges in the theory, one can change the evolution of the coupling from asymptotically free to the QED-like running. In the intermediate case between these two behaviors, the theory is governed in the infrared by a stable fixed point.

Consider the two-loop beta function, $\beta(g) = -\frac{\beta_0}{16\pi^2}g^3 - \frac{\beta_1}{(16\pi^2)^2}g^5$. Asymptotic freedom is lost when β_0 becomes negative and this signals the phase transition into non-abelian Coulomb phase. The loss of asymptotic freedom defines the upper boundary of the conformal window. On the other hand, if $g^{*2} = -16\pi^2\beta_0/\beta_1$ is positive, i.e. $\beta_1 < 0$ while $\beta_0 > 0$, the theory has a fixed point, $\beta(g^*) = 0$ and the theory is in the infrared conformal phase, i.e. inside the conformal window. Finally, if $\beta_1 > 0$, then the theory is asymptotically free and in a QCD-like phase with confinement and chiral symmetry breaking. However, in addition to the existence of a fixed point in the two-loop beta function one needs to take into account that at sufficiently strong coupling, chiral symmetry breaking is triggered and the theory cannot flow into the fixed point. Approximate solutions to Schwinger-Dyson equation for the fermion two-point function [9] suggest that the critical coupling for chiral symmetry breaking is $g_c^2 = 4\pi^2/(3C_2(R))$, where R denotes the fermion representation under investigation. With g_c^2 and g^{*2} defined as above, the condition $g_c^2 = g^{*2}$ determines an estimate for the lower boundary of the conformal window. For walking one wants to be slightly below the lower boundary of the conformal window so that $g_c^2 \lesssim g^*$.

The resulting phase diagram for different fermion representations, fundamental and two-index (anti)symmetric ones, is shown in Fig. 1 as a function of numbers of colors and flavors. The shaded regions show the conformal window. For adjoint representation this would be a horizontal band coinciding with the two-index symmetric one at $N_c = 2$, and we chose not to plot it explicitly to maintain clarity of the figure. These four representations exhaust the interesting cases: yet higher representations lead to loss of asymptotic freedom already for very low values of N_f [10], and in particular do not admit asymptotically free $N_c \to \infty$ limit. We stress the fact that the lower boundary of the conformal window shown in Fig. 1 is a semiquantitative estimate. Various other methods than the one we have used here can be applied [11, 12, 13, 14], and the results are quantitatively similar. Recently there has been also a lot of developments towards determining the conformal window for $N_c = 2$ or $N_c = 3$ and various fermion representations using lattice simulations [15, 16, 17, 18, 19, 20, 21, 22, 23].

Since phenomenological constraints, most notably the precision S-parameter [24], require walking technicolor with as few flavors as possible, Fig. 1 immediately implies two candidate theories: $N_f = 2$ in the two index symmetric representation of either SU(2) or SU(3). The former was proposed as a minimal walking technicolor model [5] , and its phenomenological consequences have been investigated in detail both for colliders [25, 26, 27] and for cosmology [28, 29, 30]. A particularly important feature which was established in [6], was that the composite Higgs particle can be parametrically light in comparison to other hadronic states. This is due to the fact that the Higgs scalar is the order parameter of the quantum phase transition to the conformal phase and hence its mass is expected to vanish as $M_\sigma^2 \sim (N_f^{(c)} - N_f)\Lambda^2$, where Λ is the intrinsic scale of the strong dynamics, as the scaling field N_f approaches the critical value $N_f^{(c)}$. This result was recently reconsidered in [31].

3. Hot technimatter

Let us then turn to another aspect: thermodynamics of a generic walking theory and gauge/gravity duality which has been proposed to describe various aspects of SU(N_c) gauge theories. Most of the studies have concentrated on QCD thermodynamics [32, 33, 34], but here we will apply them to walking technicolor.

The model [32, 33] starts from a metric ansatz

$$ds^2 = b^2(z)\left[-f(z)dt^2 + d\mathbf{x}^2 + \frac{dz^2}{f(z)}\right] \tag{1}$$

plus a scalar field $\phi(z) = \log\lambda(z)$. The three functions $b(z), f(z)$ in the metric and the scalar field $\phi(z)$ are determined from the Einstein equations following from the gravity action (in standard notation)

$$S = \frac{1}{16\pi G_5}\int d^5x\,\sqrt{-g}\left[R - \tfrac{4}{3}\,(\partial_\mu\phi)^2 + V(\phi)\right]. \tag{2}$$

For the details of these equations and the numerical method of solution, see [33, 35]. An advantage of this gauge/gravity duality approach is that the information on the boundary theory we will need is a knowledge of the beta function of its gauge coupling. We will consider an ansatz for a walking type beta function dependent on few parameters; in principle the parameter values can be obtained if the dynamics of the underlying theory is known.

The beta function ansatz we shall use is

$$\beta(\lambda) = -c\lambda^2\frac{(1-\lambda)^2 + e}{1 + a\lambda^3}, \qquad \lambda = N_c g^2, \tag{3}$$

which is tuned to asymptotic freedom in the UV ($\lambda \to 0$) and to walking near $\lambda = 1$ if $e > 0$ is small. The case $e = 0$ corresponds to a theory with an infrared fixed point (IRFP). A beta function, of course, is scheme dependent while the consequences we derive, an equation of state and associated phase structure, are entirely physical. One can thus say that our model defines the regularisation scheme leading to the coupling constant and its beta function we start from. For the solution of the Einstein equations the potential V must be known, and to begin with, we do not know V but only the beta function. However, we can construct a potential which reproduces the desired beta function, (3), on the boundary as $T \to 0$, i.e. $f \to 1$, and implies confinement at strong coupling. As discussed in [36], a condition for confinement is that the equation $\beta(\lambda) + \frac{3}{2}\,\lambda = 0$ have a solution; for details of the construction of the potential, see [35]

To illustrate the resulting thermodynamics, we consider $c = 3a/2 = 10$, $e = 0.1$ One has two first order transitions for large enough c. Decreasing T from very large values one encounters a first order transition at $T = T_{\rm ET}$ with the structure shown in Fig. 2. Below $T_{\rm ET}$ there follows

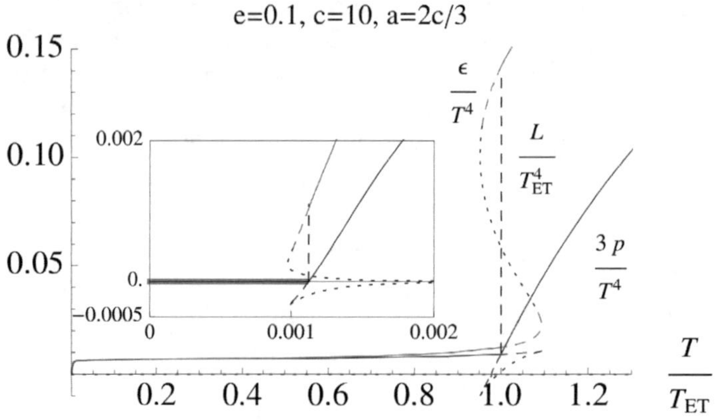

Figure 2. Energy density and $3p$ scaled by T^4 plotted vs $T/T_{\rm ET}$ for the beta function (3) with $e = 0.1$, $c = 3a/2 = 10$. The normalisation is such that at $T \gg T_{\rm ET}$ both ϵ/T^4 and $3p/T^4$ approach $\pi^2/15$. There is a first order transition at $T = T_{\rm ET}$, metastable branches are dashed and the unstable branch is dotted. There is a second 1st order phase transition at $T = T_{\rm T} \approx 0.0011 T_{\rm ET}$, the details are shown in the inset. In between, for $T_{\rm T} < T < T_{\rm ET}$ there is a quasiconformal phase with nearly constant p/T^4. The thick line below $T_{\rm T}$ corresponds to the non-black hole low T phase with $p = 0$.

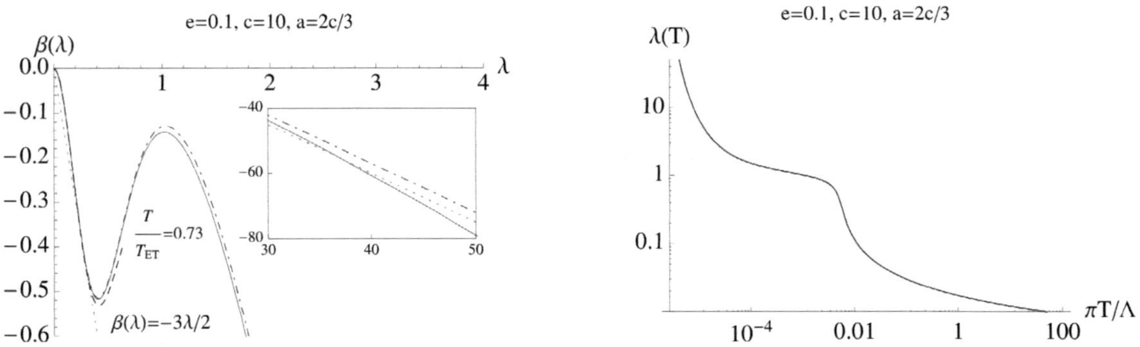

Figure 3. Left: The input beta function (parametrisation (3), dotdashed line) and the output beta function, computed from the numerical solution at $T = 0.73 T_{\rm ET}$ (dashed black line) and at the very low T value $T = T_{\rm T}$ (continuous line), for parameter values shown in the figure. The inset shows how the output beta function at $T = T_{\rm T}$ crosses the (dotted) confinement line [36], $-\frac{3}{2}\lambda$, at the position of the quasi conformal $\rightarrow$ confined transition. Right: Evolution of the coupling corresponding to the low T output beta function.

a wide *quasiconformal* phase in which p/T^4 is almost constant. For the IRFP case, $e = 0$ this would extend down to $T = 0$; now at about $T \approx 0.01 T_{\rm ET}$ pressure starts decreasing and crosses $p = 0$ at some $T = T_{\rm T} = 0.0011 T_{\rm ET}$. Below this temperature the quasi conformal phase becomes metastable (see inset in Fig. 2) and the vacuum phase with $p = 0$ is the stable one. This is a first order transition from a quasiconformal phase to a confining phase, the energy density ϵ dropping suddenly to zero.

Retaining c and a fixed to their above values, we next study how the phase diagram depends on the parameter e which controls the departure from the limit of exactly infrared conformal theory. See Fig. 4. At the origin, $T = e = 0$, the line of first order transitions, $T_{\rm T}(e)$, terminates at a second order quantum transition, in which the vacuum theory becomes infrared conformal.

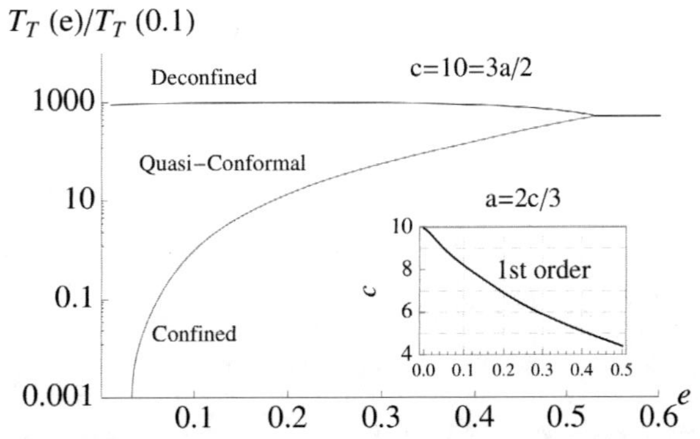

Figure 4. The (T, e) phase diagram for $c = 3a/2 = 10$. The upper curve shows $T_{\rm ET}(e)$, the lower $T_{\rm T}(e)$, both normalised by $T_{\rm T}(0.1)$. The inset shows for what values of c for various e the upper ET transition is of first order, for c below the line the transition is a cross-over. The line to the right of the triple point is the deconfinement transition line.

At finite values of $e \lesssim 0.5$ there are, as a function of temperature, three phases corresponding to confined, quasi-conformal and deconfined matter. When e increases, the upper transition temperature $T_{\rm ET}(e)$ decreases slowly, while the lower transition temperature $T_{\rm T}(e)$ increases rapidly. At some critical e_c these lines merge and the quasi- conformal phase disappears; for $e > e_c$ the phase structure will consist of a single transition line separating deconfined and confined phases. In the case depicted in the figure all transition lines correspond to a first order transition and hence the intersection is a triple point.

The inset in Fig. 4 shows for what values of c the transition changes from a crossover to a 1st order one, for various e.

4. Conclusions

We considered phase diagrams related to technicolor both at zero and finite temperature. At zero temperature the role of the phase diagrams of gauge theories is to help identify plausible models for walking technicolor, i.e. theories which are naturally near the conformal window in the (N_c, N_f)-plane with modest number of new matter fields. The results obtained using approximative analytic methods provide important input which can be further tested using lattice simulations.

Once such theories, like minimal walking technicolor, are identified their properties can be studied further. Here we considered in particular finite temperature phase diagrams using a method based on gauge/gravity duality treating the strong dynamics in isolation. A natural next step would be to embed the obtained phase structure into the full electroweak theory and study the implications for early universe and its evolution history.

Acknowledgments

I would like to thank J. Alanen, D. D. Dietrich, K. Kajantie and F. Sannino for collaboration on the topics presented here.

References

[1] Weinberg S 1979 *Phys. Rev.* D **19** 1277
[2] Susskind L 1979 *Phys. Rev.* D **20** 2619
[3] Hill C T and Simmons E H 2003 *Phys. Rept.* **381** 235 [Erratum-ibid. **390** 553].

[4] Sannino F 2008 Dynamical stabilization of the fermi scale: phase diagram of strongly coupled theories for (minimal) walking technicolor and unparticles *preprint* 0804.0182 [hep-ph]
[5] Sannino F and Tuominen K 2005 *Phys. Rev.* D **71** 051901
[6] Dietrich D D, Sannino F and Tuominen K 2005 *Phys. Rev.* D **72** 055001
[7] Dietrich D D, Sannino F and Tuominen K 2006 *Phys. Rev.* D **73** 037701
[8] Dietrich D D and Sannino F 2007 *Phys. Rev.* D **75** 085018
[9] Cohen A G and Georgi H 1989 *Nucl. Phys.* B **314** 7
[10] Eichten E, Kang K and Koh I G 1982 *J. Math. Phys.* **23** 2529
[11] Ryttov T A, and Sannino F 2008 *Phys. Rev.* D **78** 065001
[12] Antipin O and Tuominen K 2010 *Phys. Rev.* D **81** 076011
[13] Poppitz E and Unsal M 2009 *J. High Enenrgy Phys.* JHEP 0909 050
[14] Frandsen M T, Pickup T and Teper M 2010 Delineating the conformal window *eprint* 1007.1614 [hep-ph]
[15] Catterall S and Sannino F 2007 *Phys. Rev.* D **76** 034504
[16] Catterall S, Giedt J, Sannino F and Schneible J 2008 *J. High Energy Phys.* JHEP 0811 009
[17] Hietanen A J, Rantaharju J, Rummukainen K and Tuominen K 2009 *J. High Energy Phys.* JHEP 0905 025
[18] Hietanen A J, Rummukainen K and Tuominen K 2009 *Phys. Rev.* D **80** 094504
[19] Shamir Y, Svetitsky B and DeGrand T 2008 *Phys. Rev.* D **78** 031502
[20] DeGrand T, Shamir Y and Svetitsky B 2009 *Phys. Rev.* D **79** 034501
[21] Del Debbio L, Patella A and Pica C 2010 *Phys. Rev.* D **81** 094503
[22] Fodor Z, Holland K, Kuti J, Nogradi D and Schroeder C 2009 *J. High Energy Phys.* JHEP 0911 103
[23] Kogut J B and Sinclair D K 2010 *Phys. Rev.* D **81** 114507
[24] Peskin M E and Takeuchi T 1992 *Phys. Rev.* D **46** 381
[25] Foadi R, Frandsen M T, Ryttov T A and Sannino F 2007 *Phys. Rev.* D **76** 055005
[26] Antipin O, Heikinheimo M and Tuominen K 2009 *J. High Energy Phys.* JHEP 0910 018
[27] Frandsen M T, Masina I and Sannino F 2010 *Phys. Rev.* D **81** 035010
[28] Gudnason S B, Kouvaris C and Sannino F 2006 *Phys. Rev.* D **73** 115003
[29] Kainulainen K, Tuominen K and Virkajarvi J 2007 *Phys. Rev.* D **75**, 085003
[30] Cline J M, Jarvinen M and Sannino F *Phys. Rev.* D **78** 075027
[31] Appelquist T and Bai Y 2010 A light dilaton in walking gauge theories *eprint* 1006.4375 [hep-ph]
[32] Gursoy U, Kiritsis E, Mazzanti L and Nitti F 2009 *J. High Energy Phys.* JHEP 0905 033
[33] Gursoy U, Kiritsis E, Mazzanti L and Nitti F 2009 *Nucl. Phys.* B **820** 148
[34] Alanen J, Kajantie K and Suur-Uski V 2009 *Phys. Rev.* D **80** 126008
[35] Alanen J, Kajantie K and Tuominen K 2010 *Phys. Rev.* D **82** 055024
[36] Gursoy U, Kiritsis E and Nitti F 2008 *J. High Energy Phys.* JHEP 0802 019

Journal of Physics: Conference Series **259** (2010) 012053 doi:10.1088/1742-6596/259/1/012053

Electroweak phase transition in technicolor

Matti Järvinen[1]
CP^3-Origins, Campusvej 55, 5230 Odense M, Denmark
E-mail: mjarvine@physics.uoc.gr

Abstract.
Several phenomenologically viable walking technicolor models have been proposed recently. I demonstrate that these models can have first order electroweak phase transitions, which are sufficiently strong for electroweak baryogenesis. Strong dynamics can also lead to several separate transitions at the electroweak scale, with the possibility of a temporary restoration and an extra breaking of the electroweak symmetry. First order phase transitions will produce gravitational waves, which may be detectable at future experiments.

1. Introduction

Progress in the understanding of the phase diagram of generic asymptotically free gauge theories has led to a renewed interest in technicolor models (for a recent review see [1]). Explicit examples of technicolor models, not in conflict with electroweak precision tests, have been put forward in [2, 3, 4, 5].

I review the recent work [6, 8, 7, 9] on the finite temperature electroweak phase transition in technicolor. The standard model has a cross-over electroweak transition [10]. It is interesting to see if first order phase transitions are possible in technicolor [11] or other [12, 13, 14, 15] extensions of the standard model, because this would remove an obstacle for realizing electroweak baryogenesis (see, e.g., [16]).

This work concentrates on Minimal Walking Technicolor (MWT) [2] and Ultra Minimal Technicolor (UMT) [5]. These two models can be seen as templates for two qualitatively different cases of generic strongly interacting extensions of the standard model. I use effective field theory to describe the strong dynamics of the strongly interacting technicolor sector. The parameter space of the effective theory covers a variety of different possible strong dynamics. The fact that UMT has two matter sectors and therefore in general two chiral phase transitions makes it essentially different from MWT.

In MWT, the standard model Higgs is replaced by two (Dirac) techniquarks which transform in the adjoint representation of the technicolor gauge group $SU(2)_{\rm TC}$. MWT as such suffers from the Witten topological anomaly [17], which is cured by adding an extra generation of massive leptons. UMT has two sectors of matter transforming under the fundamental and adjoint representations of $SU(2)_{\rm TC}$, leading typically to two chiral phase transitions. The fundamental sector contains two Dirac techniquarks, and the adjoint sector has two Weyl techniquarks.

[1] Present address: Crete Center for Theoretical Physics, Department of Physics, University of Crete, 71003 Heraklion, Greece

2. Tools

2.1. Effective theory

The relevant degrees of freedom in the technicolor theory at the electroweak scale are the (lightest) composite technihadrons. Their dynamics cannot be solved from the underlying strongly interacting theory, but it can be described by using an effective field theory, which implements the chiral symmetry breaking pattern ($SU(4) \to SO(4)$ for MWT) [18]. In this study I include the lightest scalar states.

For MWT, the simplest effective Lagrangian of the scalar sector reads

$$\mathcal{L}_{\rm eff} = \frac{1}{2}{\rm Tr}\left[D_\mu M D^\mu M^\dagger\right] - \mathcal{V}(M) + \mathcal{L}_{\rm ETC} \tag{1}$$

where

$$\begin{aligned} \mathcal{V}(M) &= -\frac{m^2}{2}{\rm Tr}[MM^\dagger] + \frac{\lambda}{4}{\rm Tr}\left[MM^\dagger\right]^2 + \lambda'{\rm Tr}\left[MM^\dagger MM^\dagger\right] \\ &- 2\lambda''\left[{\rm Det}(M) + {\rm Det}(M^\dagger)\right] , \end{aligned} \tag{2}$$

and D_μ is the electroweak covariant derivative. The matrix M contains 20 scalar states, including the pions which are eaten by the Z and W bosons in the electroweak transition, a set of extra Goldstone modes, and a composite Higgs. Consistency requires us to include also the pseudoscalar partner Θ of the Higgs and a scalar partner triplet A of the eaten pions. The masses of these states may be close to the electroweak scale, and therefore they play an important role in the electroweak phase transition. I also need to add masses to the extra Goldstone modes. These masses are contained in the term $\mathcal{L}_{\rm ETC}$ above. They may arise through the coupling to the electroweak sector [19] and/or through extended technicolor interactions. The couplings λ, λ' and λ'' are free parameters in this study, and they can be traded for a set of masses of the composite particles. In the future, these parameters may be constrained by lattice studies or by experimental data.

The chiral symmetry of UMT is $SU(4) \times SU(2) \times U(1)$, including the separate $SU(4)$ and $SU(2)$ transformations of the two sectors of matter and one anomaly free $U(1)$ transformation. The symmetry is expected to break spontaneously to $Sp(4) \times SO(2) \times Z_2$. The dynamics of the two matter sectors of UMT are described by Lagrangians analogous to (1). The chiral symmetry also allows terms which mix the two sectors [5, 8]. At the level of elementary particles, these terms can be understood to arise from technigluon exchange between the two types of matter.

2.2. Effective potential

I study the phase transition by using the standard effective potential:

$$V(\phi, T) = V_{\rm tree}(\phi) + V_{1-{\rm loop}}(\phi, T = 0) + V_{1-{\rm loop}}(\phi, T) \tag{3}$$

where ϕ is the vacuum expectation value of the composite Higgs. I use the effective Lagrangian described above to describe the dynamics of technihadrons. For UMT the potential depends on two Higgs VEVs, corresponding to its two sectors. The one-loop temperature dependent term includes ring resummation [20]. The top and the electroweak gauge bosons are the only relevant standard model particles. I also use the method of [21] to interpolate between the high and low temperature asymptotics of the temperature-dependent one-loop term.

At high temperatures the effective potential is minimized at $\phi = 0$, and the electroweak symmetry is intact. First order phase transition is possible if the effective potential admits another minimum at nonzero $\phi = \phi_{\rm br}$ at low temperatures, separated from the symmetric one by a potential barrier. Below the critical temperature T_c, the symmetric vacuum becomes

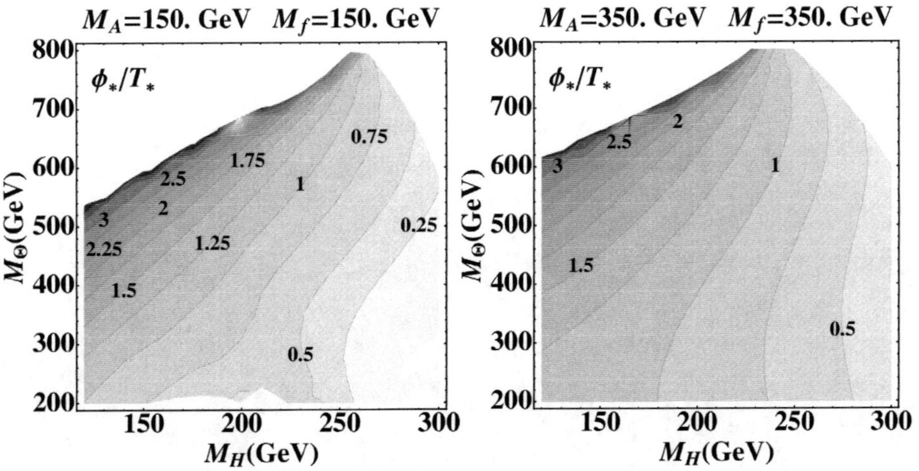

Figure 1. The strength ϕ_*/T_* of the electroweak transition in MWT for various parameter values.

unstable: $V(0,T) > V(\phi_{\rm br},T)$. First order phase transition proceeds via bubble nucleation at some temperature T_* ($T_* < T_c$). The bubble profile for the above effective potential can be solved numerically, and it can be used to estimate the bubble nucleation rate. T_* is then defined as the temperature where the nucleation rate per Hubble volume and time becomes equal to one. The parameter which characterizes the strength of the transition is ϕ_*/T_*, where ϕ_* is the Higgs expectation value in the broken phase vacuum at the nucleation temperature. Electroweak baryogenesis requires

$$\frac{\phi_*}{T_*} \gtrsim 1 \,. \tag{4}$$

3. Results

3.1. Strength of the electroweak transition in MWT

The plot the ratio ϕ_*/T_* for MWT in Fig. 3.1 [6, 9]. In these plots I assumed that the masses of the extra Goldstone modes of MWT, which arise from $\mathcal{L}_{\rm ETC}$, are much higher than the electroweak scale so that the contributions from these Goldstones can be neglected. On the horizontal axis I have the mass of the composite Higgs, and the vertical axis has the mass of the pseudoscalar partner Θ of the Higgs. The additional parameters are the mass of the scalar partners A of the eaten pions, and the mass M_f of the fourth family leptons. The transition is strong enough for electroweak baryogenesis ($\phi_*/T_* \gtrsim 1$) in a vast region of parameter space.

3.2. Strength of the electroweak transition in UMT

The parameter ϕ_*/T_* is shown in Fig. 3.2 for both the transitions of UMT in the case where coupling between two sectors of matter is absent [8, 9]. The horizontal axes have the corresponding composite Higgs masses, and the vertical axes have the masses of the scalar partners of the Goldstone bosons. The right hand plot presents the strength of the transition related to the techniquarks which transform under the fundamental representation of $SU(2)_{\rm TC}$ and are gauged under the electroweak. Therefore, this transition triggers the electroweak phase transition. The effect of mixing between the two matter sectors on the strength of the transitions was analyzed in [8]. It was typically seen to weaken the transitions.

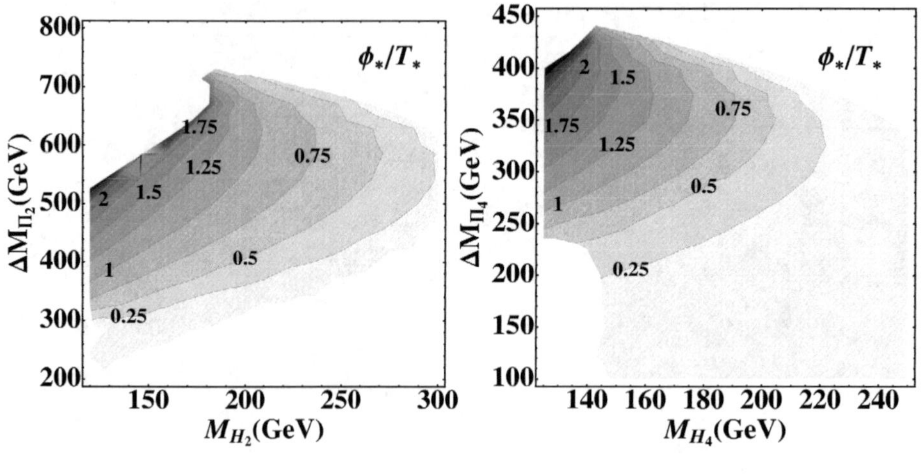

Figure 2. The strength ϕ_*/T_* of the phase transitions of the two sectors in UMT. Left: transition of matter transforming under the adjoint of $SU(2)_{\rm TC}$. Right: transition of matter transforming under the fundamental of $SU(2)_{\rm TC}$.

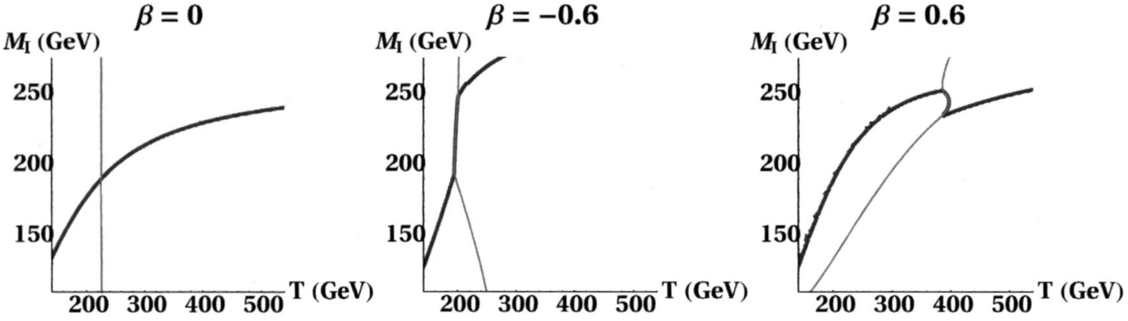

Figure 3. Phase diagrams for UMT in the $T - M_{H_4}$ plane. There is the possibility to have a simultaneous transition of both the sectors (middle), and also for an extra electroweak phase transition (right).

3.3. Extra electroweak transition in UMT

The two matter sectors of UMT interact strongly, which may lead to nontrivially structured phase transitions [7, 8]. Three qualitatively different phase diagrams on the $T - M_I$ plane, where $M_I = M_{H_4}$ is the Higgs mass of the sector charged under electroweak, are shown in Fig. 3.3. The thick blue line is the transition of the matter which transforms under the fundamental of $SU(2)_{\rm TC}$, i.e., the electroweak transition, and the thin red line is the transition of the sector containing matter in the adjoint of $SU(2)_{\rm TC}$. Since the Higgs mass M_I is (unknown but) fixed by the underlying theory, the diagrams are probed along horizontal lines from right to left as the universe cools down.

The left plot shows a phase diagram in the absence of interactions between the two matter sectors (the mixing angle β between the two Higgses is zero). In the other plots the interactions are turned on ($\beta \neq 0$). The middle plot shows a case (for well chosen parameter values [7]) where it is possible to have a simultaneous transition of both the sectors (blue and red transitions overlap). The right hand plot shows an even more interesting case where electroweak symmetry (the blue line) is broken twice and restored once for M_I around 240 GeV, which potentially

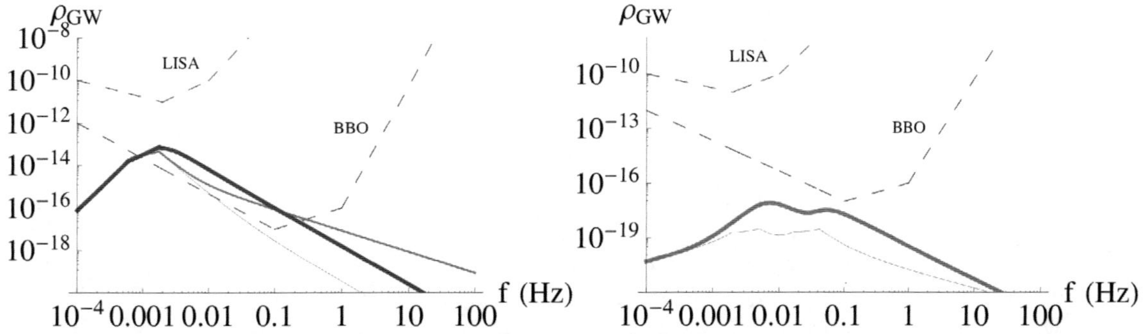

Figure 4. The energy density of the gravitational waves produced by the phase transitions for specific choices of parameter values. Left: the spectrum for MWT compared to the expected reach of LISA and BBO. Right: estimates for UMT.

enhances baryogenesis.

3.4. Production of gravitational waves

The bubble profile can also be used to estimate the strength of the gravitational waves produced at the phase transition(s). I plot the spectrum of the produced waves for rather optimal, but not fine-tuned choice of parameters both for MWT and for UMT in Fig. 3.4 [9]. The various solid lines follow different estimates for the production of gravitational waves from bubble collisions and from the turbulent motion of the plasma [22, 23]. The dashed lines are the estimated sensitivities of the LISA [24] and the Big Bang Observer [25] experiments. For MWT (left hand plot) the produced waves can possibly be detected at the Big Bang Observer, while for UMT (right hand plot) the signal appears to be too weak. Notice however that the energy spectrum of UMT has several distinct peaks produced by the transitions of the two matter sectors and by the different wave production mechanisms. This demonstrates the fact that the structure of the phase transition may be probed through the observed spectrum of gravitational waves.

4. Conclusion

I analyzed the electroweak phase transition in two technicolor models, MWT and UMT, by using effective field theories. These two examples demonstrate that strong first order electroweak phase transition is possible in strongly interacting extensions of the standard model. Therefore, such extensions have the potential to drive electroweak baryogenesis. I pointed out that strong interactions at the electroweak scale may lead to nontrivial phase structures, and possibly to extra electroweak phase transitions. First order phase transitions also produce gravitational waves, which can potentially be detected at future experiments.

After this talk was presented, I finished a related work in collaboration with F. Sannino [26], which analyzes chiral phase transitions in technicolor theories by using chiral perturbation theory. This approach suggests that the critical temperatures specifically in MWT and UMT are higher than the electroweak scale, so that $\phi_*/T_* < 1$ or the transitions are of second order. In any case, partially gauged technicolor models [3, 4] remain as good candidates for producing strong first order electroweak phase transitions. In these models the transitions are expected to be strengthened thanks to the larger number of techniflavors.

Acknowledgments

This work was done in collaboration with J. M. Cline, C. Kouvaris, T. A. Ryttov, and F. Sannino. I was partially supported by the Marie Curie Excellence Grant under contract MEXT-CT-2004-013510, by the Villum Kann Rasmussen foundation, by Regional Potential program of the E.U. FP7-REGPOT-2008-1: CreteHEPCosmo-228644, and by Marie Curie contract PIRG06-GA-2009-256487.

References

[1] F. Sannino, arXiv:0911.0931 [hep-ph].

[2] F. Sannino and K. Tuominen, Phys. Rev. D **71**, 051901 (2005) [arXiv:hep-ph/0405209].

[3] D. D. Dietrich, F. Sannino and K. Tuominen, Phys. Rev. D **72**, 055001 (2005) [arXiv:hep-ph/0505059].

[4] D. D. Dietrich and F. Sannino, Phys. Rev. D **75**, 085018 (2007) [arXiv:hep-ph/0611341].

[5] T. A. Ryttov and F. Sannino, Phys. Rev. D **78**, 115010 (2008) [arXiv:0809.0713 [hep-ph]].

[6] J. M. Cline, M. Järvinen and F. Sannino, Phys. Rev. D **78**, 075027 (2008) [arXiv:0808.1512 [hep-ph]].

[7] M. Järvinen, T. A. Ryttov and F. Sannino, Phys. Lett. B **680**, 251 (2009) [arXiv:0901.0496 [hep-ph]].

[8] M. Järvinen, T. A. Ryttov and F. Sannino, Phys. Rev. D **79**, 095008 (2009) [arXiv:0903.3115 [hep-ph]].

[9] M. Järvinen, C. Kouvaris and F. Sannino, Phys. Rev. D **81**, 064027 (2010) [arXiv:0911.4096 [hep-ph]].

[10] K. Kajantie, M. Laine, K. Rummukainen and M. E. Shaposhnikov, Nucl. Phys. B **466**, 189 (1996) [arXiv:hep-lat/9510020]. K. Kajantie, M. Laine, K. Rummukainen and M. E. Shaposhnikov, Phys. Rev. Lett. **77**, 2887 (1996) [arXiv:hep-ph/9605288]. K. Rummukainen, M. Tsypin, K. Kajantie, M. Laine and M. E. Shaposhnikov, Nucl. Phys. B **532**, 283 (1998) [arXiv:hep-lat/9805013].

[11] Y. Kikukawa, M. Kohda and J. Yasuda, Phys. Rev. D **77**, 015014 (2008) [arXiv:0709.2221 [hep-ph]].

[12] M. Carena, M. Quiros and C. E. M. Wagner, Phys. Lett. B **380**, 81 (1996) [arXiv:hep-ph/9603420].

[13] J. M. Cline and K. Kainulainen, Nucl. Phys. B **482**, 73 (1996) [arXiv:hep-ph/9605235]. M. Laine, Nucl. Phys. B **481**, 43 (1996) [Erratum-ibid. B **548**, 637 (1999)] [arXiv:hep-ph/9605283]. M. Losada, Phys. Rev. D **56**, 2893 (1997) [arXiv:hep-ph/9605266].

[14] J. M. Cline and G. D. Moore, Phys. Rev. Lett. **81**, 3315 (1998) [arXiv:hep-ph/9806354].

[15] M. Laine and K. Rummukainen, Phys. Rev. Lett. **80**, 5259 (1998) [arXiv:hep-ph/9804255]; Nucl. Phys. B **535**, 423 (1998) [arXiv:hep-lat/9804019].

[16] J. M. Cline, arXiv:hep-ph/0609145.

[17] E. Witten, Phys. Lett. B **117**, 324 (1982).

[18] T. Appelquist, P. S. Rodrigues da Silva and F. Sannino, Phys. Rev. D **60**, 116007 (1999) [arXiv:hep-ph/9906555]. S. B. Gudnason, C. Kouvaris and F. Sannino, Phys. Rev. D **73**, 115003 (2006) [arXiv:hep-ph/0603014]. R. Foadi, M. T. Frandsen, T. A. Ryttov and F. Sannino, Phys. Rev. D **76**, 055005 (2007) [arXiv:0706.1696 [hep-ph]].

[19] D. D. Dietrich and M. Järvinen, Phys. Rev. D **79**, 057903 (2009) [arXiv:0901.3528 [hep-ph]].

[20] P. Arnold and O. Espinosa, Phys. Rev. D **47**, 3546 (1993) [Erratum-ibid. D **50**, 6662 (1994)] [arXiv:hep-ph/9212235].

[21] J. M. Cline and P. A. Lemieux, Phys. Rev. D **55**, 3873 (1997) [arXiv:hep-ph/9609240].

[22] M. Kamionkowski, A. Kosowsky and M. S. Turner, Phys. Rev. D **49**, 2837 (1994) [arXiv:astro-ph/9310044]. C. Caprini, R. Durrer and G. Servant, Phys. Rev. D **77**, 124015 (2008) [arXiv:0711.2593 [astro-ph]]. S. J. Huber and T. Konstandin, JCAP **0809**, 022 (2008) [arXiv:0806.1828 [hep-ph]].

[23] G. Gogoberidze, T. Kahniashvili and A. Kosowsky, Phys. Rev. D **76**, 083002 (2007) [arXiv:0705.1733 [astro-ph]]. C. Caprini, R. Durrer and R. Sturani, Phys. Rev. D **74**, 127501 (2006) [arXiv:astro-ph/0607651]. T. Kahniashvili, A. Kosowsky, G. Gogoberidze and Y. Maravin, Phys. Rev. D **78**, 043003 (2008) [arXiv:0806.0293 [astro-ph]].

[24] P. L. Bender, et al., *LISA Pre-Phase A Report; Second Edition*, MPQ 233 (1998).

[25] G. M. Harry, P. Fritschel, D. A. Shaddock, W. Folkner and E. S. Phinney, Class. Quant. Grav. **23**, 4887 (2006) [Erratum-ibid. **23**, 7361 (2006)].

[26] M. Järvinen and F. Sannino, arXiv:1009.5380 [hep-ph].

Perturbative Minimal Superconformal Technicolor

M Antola[1], K Tuominen[2 3], F Sannino[2] and S Di Chiara[2]

[1] Department of Physics and Helsinki Institute of Physics, P.O.Box 64, FI-00014, Finland
[2] CP³-Origins, Campusvej 55, DK-5230 Odense M, Denmark
[3] Helsinki Institute of Physics, P.O.Box 64, FI-00014, Finland and University of Jyvaskyla, P.O. Box 35, FI-40014, Finland

E-mail: matti.antola@helsinki.fi

Abstract. We investigate the perturbative regime of the Minimal Supersymmetric Conformal Technicolor. The model allows for a stable vacuum correctly breaking the electroweak symmetry. The particle spectrum features several light particles stemming out from the new $\mathcal{N} = 4$ sector of the theory. We find that the Tevatron and the LHC can rule out a significant portion of the parameter space of this model.

1. Introduction

In technicolor models [1, 2] one gives a mass to the electroweak gauge bosons via new strong dynamics. These models must be extended in order to give mass to the standard model (SM) fermions [3, 4]. In typical extensions one expects large flavor changing neutral current (FCNC) processes. Using near conformal dynamics alleviates the FCNC problem [5], but the price for the early models was a seemingly inconsolable tension with the electroweak (EW) precision data [6].

For over a decade it was hoped that near conformal dynamics might anyways eliminate the tension with precision data even if one had a large number of technidoublets gauged under the electroweak (EW) symmetry. However recently in [7] it has been shown that to be phenomenologically viable the near conformal models should contain the minimal number of flavors gauged under the EW symmetry.

The simplest models of this type which are shown to pass the precision tests, or have the smallest deviation from the precision data, while still providing a (near) conformal behavior were put forward in [8]. Among these, the Minimal Walking Technicolor (MWT) features the most economical particle content. In MWT the gauge group is $SU(2)_{TC} \times SU(3)_C \times SU(2)_L \times U(1)_Y$ and the field content of the technicolor sector is constituted by two flavors of techni-fermions and one techni-gluon all in the adjoint representation of $SU(2)_{TC}$. The model features also a pair of Dirac leptons, whose left-handed components are assembled in a weak doublet, necessary to cancel the Witten anomaly [9] arising when gauging the new technifermions with respect to the weak interactions.

The model requires, however, still additional ingredients in order to give mass to the SM fermions. For example, one may postulate the existence of an Extended Technicolor (ETC) sector, traditionally featuring new gauge interactions linking the SM fermions to the techniquarks, which can generate mass terms for the SM fermions (as well as for the techni-mesons and -baryons) via a new dynamical mechanism. Another alternative is to reintroduce new bosons (*bosonic technicolor*) [10, 11] able to give masses to the SM fermions using standard Yukawa interactions. Eventhough these models are phenomenologically viable, they suffer from a SM-like fine tuning and are therefore unnatural. Supersymmetric technicolor has been considered [12, 13] as a way to naturalize bosonic technicolor. Another possibility would be to imagine the new scalars also to be composite of some new strong dynamics.

In [14] we made the observation that the techni-fermions and techni-gluons of the Minimal Walking Technicolor fit perfectly in an $\mathcal{N}=4$ supermultiplet, provided that we also include three scalar superpartners. In fact the $SU(4)$ global symmetry of MWT is nothing but the well known $SU(4)_R$ R symmetry of the $\mathcal{N}=4$ Super Yang Mills theory. This is the global quantum symmetry that does not commute with the supersymmetry transformations.

Supersymmetrizing MWT in this way leads to an approximate $\mathcal{N}=4$ supersymmetry of the technicolor sector that is broken to $\mathcal{N}=1$ only by EW gauge and Yukawa interactions. Due to approximate $\mathcal{N}=4$ invariance the beta function of the technicolor gauge coupling is zero at one loop, i.e. the associated technicolor model is approximately conformal. We called this model Minimal Supersymmetric Conformal Technicolor (MSCT).

This model can also be viewed as the first extension of the SM exhibiting maximal supersymmetry in four dimensions. MSCT constitutes an interesting theoretical as well as phenomenological model to explore since it naturally allows to investigate different regimes according to how strongly coupled the maximally supersymmetric Yang-Mills theory is taken to be. Here we will assume the technicolor coupling is perturbative.

2. The Model

The fermionic particle content of the MWT is given explicitly as

$$Q_L^a = \begin{pmatrix} U^a \\ D^a \end{pmatrix}_L , \qquad U_R^a , \quad D_R^a, \quad a=1,2,3 \; ; \quad L_L = \begin{pmatrix} N \\ E \end{pmatrix}_L , \qquad N_R \, , \; E_R \, , \tag{1}$$

where U and D are techni-fermions in the adjoint representation of $SU(2)_{TC}$, whose left-handed components form a doublet under $SU(2)_L$, and the chiral leptons required to cancel the Witten anomaly are denoted by N and E. The following generic hypercharge assignment is free from gauge anomalies:

$$Y(Q_L) = \frac{y}{2} \qquad Y(U_R, D_R) = \left(\frac{y+1}{2}, \frac{y-1}{2}\right) \tag{2}$$

$$Y(L_L) = -3\frac{y}{2} \qquad (N_R, E_R) = \left(\frac{-3y+1}{2}, \frac{-3y-1}{2}\right) \tag{3}$$

The global symmetry of the technicolor sector is $SU(4)$ which breaks explicitly to $SU(2)_L \times U(1)_Y$ by the natural choice of the electroweak embedding [8, 15]. Electroweak symmetry breaking is triggered by a fermion bilinear condensate. The vacuum is stable against the SM quantum corrections [16].

To build MSCT we set $y=1$ in Eqs. (2) so that $\bar{D}_R^a$ is a singlet under EW symmetry and can play the role of the techni-gaugino. We define the $\mathcal{N}=4$ supermultiplet in terms $\mathcal{N}=1$ superfields, whose scalar and fermionic components are expressed by:

$$\left(\tilde{U}_L,\, U_L\right) \in \Phi_1, \quad \left(\tilde{D}_L,\, D_L\right) \in \Phi_2, \quad \left(\tilde{\bar{U}}_R,\, \bar{U}_R\right) \in \Phi_3, \quad (G,\, \bar{D}_R) \in V, \tag{4}$$

where we used a tilde to label the scalar superpartner of each fermion. We indicated with Φ_i, $i=1,2,3$ the three chiral superfields of 4SYM and with V the vector superfield. The superfields associated with the remaining MWT fermions N and E are given by:

$$\left(\tilde{N}_L,\, N_L\right) \in \Lambda_1, \quad \left(\tilde{E}_L,\, E_L\right) \in \Lambda_2, \quad \left(\tilde{\bar{N}}_R,\, \bar{N}_R\right) \in N, \quad \left(\tilde{\bar{E}}_R,\, \bar{E}_R\right) \in E. \tag{5}$$

Table 1. MSCT $\mathcal{N}=1$ superfields and transformation properties. Adj denotes adjoint and F fundamental.

Superfield	$SU(2)_{TC}$	$SU(2)_L$	$U(1)_Y$
$\Phi_{1,2}$	Adj	F	1/2
Φ_3	Adj	1	-1
V	Adj	1	0
$\Lambda_{1,2}$	1	F	-3/2
N	1	1	1
E	1	1	2
H	1	F	1/2
H'	1	F	-1/2

The quantum numbers of the superfields in Eqs.(4,5) and of those labeled by H and H', which contain each a Higgs scalar weak doublet, are given in Table 1.

The renormalizable lepton and baryon number conserving superpotential for the MSCT is

$$P = P_{MSSM} + P_{TC}, \tag{6}$$

where P_{MSSM} is the minimal supersymmetric standard model (MSSM) superpotential, and

$$P_{TC} = -\frac{g_{TC}}{3\sqrt{2}}\epsilon_{ijk}\epsilon^{abc}\Phi_i^a\Phi_j^b\Phi_k^c + y_U\epsilon_{ij}\Phi_i^a H_j \Phi_3^a + y_N\epsilon_{ij}\Lambda_i H_j N + y_E\epsilon_{ij}\Lambda_i H_j' E + y_R E\Phi_3^a\Phi_3^a. \tag{7}$$

In the last equation $\Phi_i^a = Q_i^a$, $i = 1,2$, with a the technicolor index. Gauge invariance and $\mathcal{N}=1$ supersymmetry does not ensure the Yukawa coupling of the first term to be equal to g_{TC}, however, setting it to this value amounts to the $\mathcal{N}=4$ limit. We have also investigated the running of a general Yukawa coupling and shown that it tends towards g_{TC} at low energies [17]. This result justifies our choice to set it equal to the technicolor gauge coupling itself.

3. Electroweak Symmetry Breaking

In perturbative MSCT we assume SUSY is broken softly and electroweak symmetry breaking is driven by the negative mass squared of $\tilde{D}_L$, $\tilde{H}$ and $\tilde{H}'$[1]. Electroweak symmetry breaking follows the pattern $SU(2)_L \times U(1)_Y \times SU(2)_{TC} \to U(1)_{EM} \times U(1)_{TC}$. The phenomenological constraints on a new $U(1)$ massless gauge boson were studied in [18]. The lower limits on the scale of dimension six operators with SM fields and the new massless photon are in the TeV range.

We choose the vacuum expectation value (vev) of the techni scalar to be aligned in the third direction of the $SU(2)_{TC}$ gauge space. We define the vevs:

$$\left\langle \tilde{D}_L^3 \right\rangle = \frac{v_{TC}}{\sqrt{2}}, \quad \left\langle \tilde{H}_0 \right\rangle = s_\beta \frac{v_H}{\sqrt{2}}, \quad \left\langle \tilde{H}_0' \right\rangle = c_\beta \frac{v_H}{\sqrt{2}}, \tag{8}$$

where all the vevs are chosen to be real, $s_\beta = \sin\beta$, and $c_\beta = \cos\beta$.

After EWSB the mass terms of gauge bosons are written as a function of the mass eigenstates as:

$$-\mathcal{L}_{g\text{-}mass} = g_{TC}^2 v_{TC}^2 G_\mu^+ G^{-\mu} + \frac{g_L^2}{2}\left(v_{TC}^2 + v_H^2\right) W_\mu^+ W^{-\mu} + \frac{g_L^2 + g_Y^2}{4}\left(v_{TC}^2 + v_H^2\right) Z_\mu Z^\mu \tag{9}$$

[1] Notice that we indicate the scalar component of each weak doublet superfield with a tilde.

Journal of Physics: Conference Series **259** (2010) 012054 doi:10.1088/1742-6596/259/1/012054

where

$$G^{\pm}_{\mu} = \frac{1}{\sqrt{2}}\left(G^1_{\mu} \mp i\,G^2_{\mu}\right)\ ,\ W^{\pm}_{\mu} = \frac{1}{\sqrt{2}}\left(W^1_{\mu} \mp i\,W^2_{\mu}\right)\ ,\ Z_{\mu} = c_w W^3_{\mu} - s_w B\ ,\ t_w = \frac{g_Y}{g_L}. \tag{10}$$

The $\pm$ exponent of the techni-gluon refers to the $U(1)_{TC}$ charge, while the $\pm$ exponent on the EW gauge bosons refer to the usual EM charge. We see the electroweak scale is set by

$$\sqrt{v_H^2 + v_{TC}^2} = 246\ \text{GeV}. \tag{11}$$

The remaining, massless states are the techni-photon and the EW photon:

$$G_{\mu} = G^3_{\mu}\ , \qquad A_{\mu} = s_w W^3_{\mu} + c_w B \tag{12}$$

4. Spectrum and Constraints

We can write the gauge boson, fermion, and scalar squared mass matrices in block diagonal form in the basis of EM- and TC-charges and L and B numbers [17]. The mass matrices of all the SM fermions and their superpartners assume the same form, in terms of the Higgs vevs, as of those obtained in the MSSM and can be found for example in [19]. The EW gauginos, Higgs scalar doublets and their superpartners mix with the $\mathcal{N} = 4$ technicolor sector. Finally the fields $N_L, \bar{N}_R$, and their scalar superpartners will not mix at tree level with other SM fields with EM charge $Q_{EM} = 1$ (where we defined $Q_{EM} = T^3_L + Y$).

We also calculate the one-loop contributions to the CP-even and -odd neutral (both under $U(1)_{EM}$ and $U(1)_{TC}$) scalars by numerically evaluating the derivatives of the mass eigenvalues with respect to the fields evaluated on the vevs [20]. In this first estimate we compute to the neutral Higgs masses neglecting the contributions from top-stop mass splitting. We consider the fields given in Table 1, plus the W and B bosons and their superpartners. It is seen that the masses of the lightest physical states, h^0_0 and A_0, which are massless at tree level, gain a mass at 1-loop with a strong dependence on the size of the Yukawa couplings.

The lower bounds on the mass of the lightest neutralino and chargino are [21]:

$$m_{\chi^0_0} > 46\ \text{GeV}\,,\ m_{\chi^{\pm}_0} > 94\ \text{GeV}\,. \tag{13}$$

These limits refer to the MSSM, but are rather general, since they are extracted mostly from the Z decay to neutralino-antineutralino pair the former, and from photo-production of a chargino-antichargino pair at LEPII the latter. We can therefore assume these limits to hold also for the MSCT. Because of their generality and independence from the coupling strength (as long as it is not negligible), we use the lower bound on the chargino mass also for the mass of the doubly-charged chargino E. Note that the presence of the term proportional to y_R in the superpotential, Eq.(7) allows it to decay into singly charged ordinary particles, thereby escaping cosmological constraints on charged stable particles.

The techni-gaugino $\bar{D}^3_R$ is an EW singlet fermion and therefore is a right-handed neutrino, which can be very light. Because of this, and because the mass of the lightest techineutralino is a decreasing function of the mass of $\bar{D}^3_R$, we assume the soft SUSY breaking mass of the technigaugino to be small.

Other useful limits on the parameters are obtained by using the fact that the smallest eigenvalue of a semi-positive definite square matrix is smaller or equal to any eigenvalue of the principal submatrices. By using the neutralino, chargino, and doubly charged chargino mass matrices, we get for example

$$v_{TC} > 2\frac{46\ \text{GeV}}{\sqrt{g_L^2 + g_Y^2}} = 124\ \text{GeV}\,,\ v_H < 213\ \text{GeV}\,,\ \frac{y_E c_\beta v_H}{\sqrt{2}} > 94\ \text{GeV}\,,\ m_t = \frac{y_t}{y_E} t_\beta m_{cc} \tag{14}$$

where the subscript t refers to the top quark and m_{cc} is the doubly charged lepton mass. One of the most important inequalities is

$$y_t > \frac{173}{213}\sqrt{\frac{1}{\frac{1}{2}-\frac{94^2}{y_E^2 213^2}}}\,. \tag{15}$$

This last bound is plotted in Figure 1, where the shaded area shows the values of y_t and y_E excluded by the experiment: it is evident from the plot in Figure 1 that either y_t or y_E is constrained to be larger than about 1.3. Such large Yukawa couplings imply a Landau pole at some intermediate scale and we cannot discuss unification of the gauge couplings.

By using the renormalization group equations [17] we find that a phenomenologically reasonable compromise for the values of y_t and y_E allowing perturbativity at the energy scale of a few TeVs is, respectively, 1.65 and 2.2. The value of the lightest chargino mass is controlled mainly by y_U, which we take equal to y_t, while $y_{TC} = y_N = 1.1$, since they are less constrained to be large. In Figure 2 are plotted $y_{TC}, y_U, y_t, y_N, y_E$ as a function of the renormalization scale M: the couplings are normalized for $M = m_Z$ to $y_t = y_U = 1.65$, $y_E = 2.2$, $y_N = y_{TC} = 1.1$.

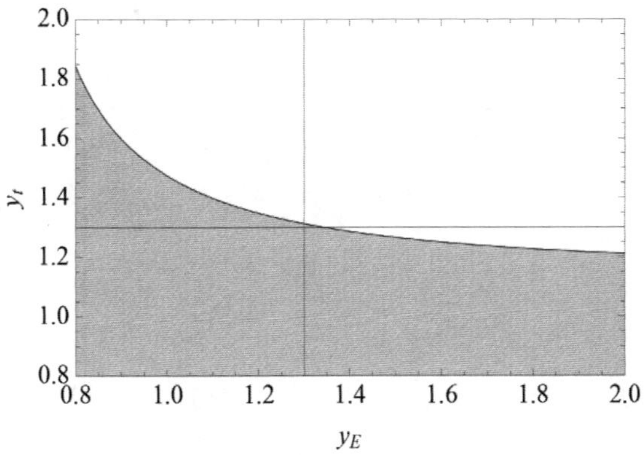

Figure 1. Shaded area shows experimentally excluded values of the Yukawa couplings y_t and y_E.

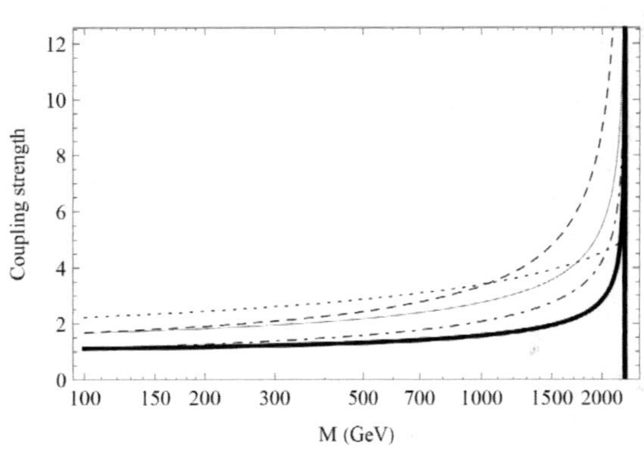

Figure 2. Plot of $y_{TC}, y_U, y_t, y_N, y_E$ as a function of the renormalization scale M: the couplings are normalized for $M = m_Z$ to $y_t = y_U = 1.65$, $y_E = 2.2$, $y_N = y_{TC} = 1.2$.

By maximizing the minimum eigenvalue of $\mathcal{M}_n \cdot \mathcal{M}_n^\dagger$ on the parameter space allowed by the bounds (14,15) with $y_t = 1.65$ and $y_E = 2.2$, we obtain

$$m_{\chi_0^0}^{max} = 46 \text{ GeV}\,,\; m_{\chi_0^\pm} = 60 \text{ GeV}, \tag{16}$$

where $m_{\chi_0^\pm}$ is calculated at the point in parameter space that maximizes $m_{\chi_0^0}$. With these chosen values of the Yukawa couplings at the EW scale, a Landau pole arises around 2.3 TeV and the chargino mass is light compared to the experimental bounds. A scan of the allowed parameter space, with the same values of the Yukawa couplings shows that it is not possible to satisfy both constraints in Eq.(13) simultaneously. However, by further lowering the Landau pole scale we can achieve a heavy enough spectrum:

$$m_{\chi_0^0}^{max} = 47 \text{ GeV}\,,\; m_{\chi_0^\pm}^{max} = 96 \text{ GeV}\,,\; m_{h_0^0} = 95 \text{ GeV}\,,\; m_{A_0} = 32 \text{ GeV}\,,\; M_{pole} = 400 \text{ GeV} \tag{17}$$

The results correspond to having chosen at the EW scale the values $y_N = 1.8$, $y_t = g_{TC} = y_U = 2.3$, $y_E = 2.4$.

Journal of Physics: Conference Series **259** (2010) 012054 doi:10.1088/1742-6596/259/1/012054

5. Conclusions

Our preliminary results on the spectrum of the MSCT indicate that the Tevatron and the LHC can rule out a significant portion of the parameter space of this model. Part of the particle spectrum is very similar to the one of the MSSM, however, MSCT also features several new light states, with respect to the EW scale, such as a doubly charged particle. Therefore an interesting experimental signature would be the discovery of a doubly charged particle together with a very light chargino and/or neutralino. Finally, since the Yukawa couplings are larger than the SM ones we expect 100% increase of several production cross sections such as the Higgs scalar (h_0^0) one via the gluon-gluon fusion process.

Another characteristic of the model relevant for collider experiments is that due to the presence of one extra Higgs-type particle, coming from the supertechnicolor sector, the spectrum features scalars and pseudoscalars lighter than in the MSSM case. These states will also yield interesting signatures at collider experiments.

Since our model features, at the EW scale, a new $\mathcal{N} = 4$ supertechnicolor sector, collider experiments have the possibility to explore directly string theory. This is so since the new scalars coming from this sector can be directly identified with the *extra* six space coordinates of ten dimensional supergravity. This link is even more clear when considering the present supertechnicolor sector in the nonperturbative regime which can be investigated using AdS/CFT techniques.

References

[1] Weinberg S 1979 *Phys. Rev.* D **19** 1277
[2] Susskind L 1979 *Phys. Rev.* D **20** 2619
[3] Dimopoulos S and Susskind L 1979 *Nucl. Phys.* B **155** 237
[4] Eichten E and Lane K D 1980 *Phys. Lett.* B **90** 125
[5] Holdom B 1981 *Phys. Rev.* D **24** 1441
[6] Peskin M E and Takeuchi T 1990 *Phys. Rev. Lett.* **65** 964
[7] Sannino F *Preprint* arXiv:1006.0207 [hep-lat]
[8] Sannino F and Tuominen K 2005 *Phys. Rev.* D **71** 051901 (*Preprint* arXiv:hep-ph/0405209)
[9] Witten E 1982 *Phys. Lett.* B **117** 324
[10] Simmons E H 1989 *Nucl. Phys.* B **312** 253
[11] Antola M, Heikinheimo M, Sannino F and Tuominen K 2010 *JHEP* **1003** 050 (*Preprint* arXiv:0910.3681 [hep-ph])
[12] Dine M, Fischler W and Srednicki M 1981 *Nucl. Phys.* B **189** 575
[13] Dobrescu B A 1995 *Nucl. Phys.* B **449** 462 (*Preprint* arXiv:hep-ph/9504399)
[14] Antola M, Di Chiara S, Sannino F and Tuominen K, *Preprint* arXiv:1001.2040 [hep-ph]
[15] Dietrich D D, Sannino F and Tuominen K 2005 *Phys. Rev.* D **72** 055001 (*Preprint* arXiv:hep-ph/0505059)
[16] Dietrich D D and Jarvinen M 2009 *Phys. Rev.* D **79** 057903 (*Preprint* arXiv:0901.3528 [hep-ph])
[17] Antola M, Di Chiara S, Sannino F and Tuominen K *Preprint* arXiv:1009.1624 [hep-ph]
[18] Dobrescu B A 2005 *Phys. Rev. Lett.* **94** 151802 (*Preprint* arXiv:hep-ph/0411004)
[19] Martin S P *Preprint* arXiv:hep-ph/9709356
[20] Elliott T, King S F and White P L 1994 *Phys. Rev.* D **49** 2435
[21] Amsler C *et al.* [Particle Data Group] *Phys. Lett.* B **667**, 1 (2008)

Journal of Physics: Conference Series **259** (2010) 012055 doi:10.1088/1742-6596/259/1/012055

Tuned Two-Higgs-Doublet Model

B. Grzadkowski[1], **P. Osland**[2]

[1] Institute of Theoretical Physics, University of Warsaw, Hoża 69, PL-00-681 Warsaw, Poland
[2] Department of Physics and Technology, University of Bergen, Postboks 7803, N-5020 Bergen, Norway

E-mail: [1] Bohdan.Grzadkowski@fuw.edu.pl

E-mail: [2] per.osland@ift.uib.no

Abstract. We consider a Two-Higgs-Doublet Model (2HDM) constrained by the condition that assures cancellation of quadratic divergences up to the leading two-loop order. Regions in the parameter space consistent with existing experimental constraints and with the cancellation condition are determined. The possibility for CP violation in the scalar potential is discussed and regions of $\tan\beta - M_{H^\pm}$ with substantial amount of CP violation are found. The model allows to ameliorate the little hierarchy problem by lifting the minimal scalar Higgs boson mass and by suppressing the quadratic corrections to scalar masses. The cutoff originating from the naturality arguments is therefore lifted from ~ 0.6 TeV in the Standard Model to $\gtrsim$2.5 TeV in the 2HDM, depending on the mass of the lightest scalar.

1. Introduction

This project aims at extending the Standard Model (SM) in such a way that there would be no quadratic divergences up to the leading order at the two-loop level of the perturbation expansion. The quadratic divergences were first discussed within the SM by Veltman [1], who, adopting dimensional reduction [2], found the following quadratically divergent one-loop contribution to the Higgs boson (h) mass

$$\delta^{(\mathrm{SM})} m_h^2 = \frac{\Lambda^2}{\pi^2 v^2}\left[\frac{3}{2}m_t^2 - \frac{1}{8}\left(6m_W^2 + 3m_Z^2\right) - \frac{3}{8}m_h^2\right], \quad (1)$$

where Λ is a UV cutoff and $v \simeq 246$ GeV denotes the vacuum expectation value of the scalar doublet. The issue of quadratic divergences was then investigated further within other regularization schemes (e.g. point splitting [3]) and also in [4] without reference to any regularization scheme.

Within the SM precision measurements require a light Higgs boson, therefore the correction (1) exceeds the mass itself even for small values of Λ, e.g. for $m_h = 130$ GeV one obtains $\delta^{(\mathrm{SM})} m_h^2 \simeq m_h^2$ already for $\Lambda \simeq 600$ GeV. On the other hand, if we assume that the scale of new physics is widely separated from the electro-weak scale, then constraints that emerge from analysis of operators of dimension 6 require $\Lambda \gtrsim$ a few TeV. The lesson from this observation is that regardless of what physics lies beyond the SM, some amount of fine tuning is necessary; either we tune to lift the cutoff above $\Lambda \simeq 600$ GeV, or we tune when precision observables measured at LEP are fitted. Tuning both in corrections to the Higgs mass and in LEP physics is, of course, also a viable alternative which we are going to explore below. So, we will look for new physics in the TeV range which will allow to lift the cutoff implied by quadratic corrections to m_h^2 to the multi-TeV range *and* which will be consistent with all the experimental constraints—both require some amount of tuning. Note that within the SM the requirement $\delta^{(\rm SM)} m_h^2 = 0$ implies an unrealistic value of the Higgs boson mass $m_h \simeq 310$ GeV.

Here we are going to argue that the Two-Higgs-Doublet Model (2HDM) in certain region of its parameter space can soften the little hierarchy problem both by suppressing quadratic corrections to scalar masses *and* it allows to lift the central value for the lightest Higgs mass up to a value which is well above the LEP limit.

2. The Two-Higgs-Doublet Model

In order to accommodate CP violation we consider here a 2HDM with softly broken $_2$ symmetry which acts as $\Phi_1 \to -\Phi_1$ and $u_R \to -u_R$ (all other fields are neutral). The scalar potential then reads

$$\begin{aligned} V(\phi_1,\phi_2) &= -\frac{1}{2}\left\{m_{11}^2\phi_1^\dagger\phi_1 + m_{22}^2\phi_2^\dagger\phi_2 + \left[m_{12}^2\phi_1^\dagger\phi_2 + \text{H.c.}\right]\right\} + \frac{1}{2}\lambda_1(\phi_1^\dagger\phi_1)^2 + \frac{1}{2}\lambda_2(\phi_2^\dagger\phi_2)^2 \\ &\quad +\lambda_3(\phi_1^\dagger\phi_1)(\phi_2^\dagger\phi_2) + \lambda_4(\phi_1^\dagger\phi_2)(\phi_2^\dagger\phi_1) + \frac{1}{2}\left[\lambda_5(\phi_1^\dagger\phi_2)^2 + \text{H.c.}\right] \end{aligned} \tag{2}$$

The minimization conditions at $\langle\phi_1^0\rangle = v_1/\sqrt{2}$ and $\langle\phi_2^0\rangle = v_2/\sqrt{2}$ can be formulated as follows:

$$\begin{aligned} m_{11}^2 &= v_1^2\lambda_1 + v_2^2(\lambda_{345} - 2\nu), \\ m_{22}^2 &= v_2^2\lambda_2 + v_1^2(\lambda_{345} - 2\nu), \end{aligned} \tag{3}$$

where $\lambda_{345} \equiv \lambda_3 + \lambda_4 + \text{Re}\,\lambda_5$ and $\nu \equiv \text{Re}\, m_{12}^2/(2v_1v_2)$.

We assume that ϕ_1 and ϕ_2 couple to down- and up-type quarks, respectively (the so-called 2HDM II).

2.1. Quadratic divergences

At the one-loop level the cancellation of quadratic divergences for the scalar Green's functions at zero external momenta (Γ_i, $i = 1,2$) in the 2HDM type II model implies [5]

$$\Gamma_1 \equiv \frac{3}{2}m_W^2 + \frac{3}{4}m_Z^2 + \frac{v^2}{2}\left(\frac{3}{2}\lambda_1 + \lambda_3 + \frac{1}{2}\lambda_4\right) - 3\frac{m_b^2}{c_\beta^2} = 0, \tag{4}$$

$$\Gamma_2 \equiv \frac{3}{2}m_W^2 + \frac{3}{4}m_Z^2 + \frac{v^2}{2}\left(\frac{3}{2}\lambda_2 + \lambda_3 + \frac{1}{2}\lambda_4\right) - 3\frac{m_t^2}{s_\beta^2} = 0, \tag{5}$$

where $v^2 \equiv v_1^2 + v_2^2$, $\tan\beta \equiv v_2/v_1$ and we use the notation: $s_\theta \equiv \sin\theta$ and $c_\theta \equiv \cos\theta$. Note that when $\tan\beta$ is large, the two quark contributions can be comparable. In the type II model the mixed, $\phi_1 - \phi_2$, Green's function is not quadratically divergent.

The quartic couplings λ_i can be expressed in terms of the mass parameters and elements of the rotation matrix needed for diagonalization of the scalar masses (see, for example, Eqs. (3.1)–(3.5) of [6]). Therefore, for a given choice of α_i's, the squared neutral-Higgs masses M_1^2, M_2^2 and M_3^2 can be determined from the cancellation conditions (4)–(5) in terms of $\tan\beta$, μ^2 and $M_{H^\pm}^2$. It is worth noticing that scalar masses resulting from a scan over α_i, $M_{H^\pm}$ and $\tan\beta$ exhibit a striking mass degeneracy in the case of large $\tan\beta$: $M_1 \simeq M_2 \simeq M_3 \simeq \mu^2 + 4m_b^2$.

At the two-loop level the leading contributions to quadratic divergences are of the form of $\Lambda^2 \ln\Lambda$. They could be determined adopting a method noticed by Einhorn and Jones [4], so that the cancellation conditions for quadratic divergences up to the leading two-loop order read:

$$\Gamma_1 + \delta\Gamma_1 = 0 \qquad \text{and} \qquad \Gamma_2 + \delta\Gamma_2 = 0 \tag{6}$$

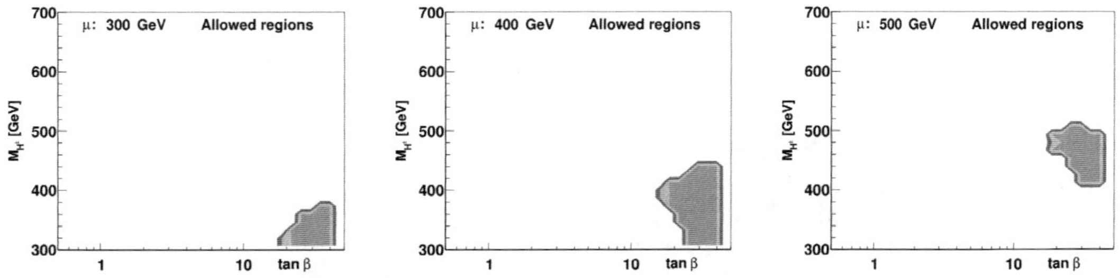

Figure 1. Two-loop allowed regions in the $\tan\beta$–$M_{H^\pm}$ plane, for $\Lambda = 2.5$ TeV, for $\mu = 300, 400, 500$ GeV (as indicated). Red: positivity is satisfied; yellow: positivity and unitarity both satisfied; green: also experimental constraints satisfied at the 95% C.L., as specified in the text.

with

$$\delta\Gamma_1 = \frac{v^2}{8}[9g_2\beta_{g_2} + 3g_1\beta_{g_1} + 6\beta_{\lambda_1} + 4\beta_{\lambda_3} + 2\beta_{\lambda_4}]\ln\left(\frac{\Lambda}{\bar{\mu}}\right) \tag{7}$$

$$\delta\Gamma_2 = \frac{v^2}{8}[9g_2\beta_{g_2} + 3g_1\beta_{g_1} + 6\beta_{\lambda_2} + 4\beta_{\lambda_3} + 2\beta_{\lambda_4} - 24g_t\beta_{g_t}]\ln\left(\frac{\Lambda}{\bar{\mu}}\right) \tag{8}$$

where β's are the appropriate beta functions while $\bar{\mu}$ is the renormalization scale. Hereafter we will be solving the conditions (6) for the scalar masses M_i^2 for a given set of α_i's, $\tan\beta$, μ^2 and $M_{H^\pm}^2$. For the renormalization scale we will adopt v, so $\bar{\mu} = v$. Then those masses together with the corresponding coupling constants, will be used to find predictions of the model for various observables which then can be checked against experimental data.

2.2. Allowed regions

In order to find phenomenologically acceptable regions in the parameter space we impose the following experimental constraints: the oblique parameters T and S, $B_0 - \bar{B}_0$ mixing, $B \to X_s\gamma$, $B \to \tau\bar{\nu}_\tau X$, $B \to D\tau\bar{\nu}_\tau$, LEP2 Higgs-boson non-discovery, R_b, the muon anomalous magnetic moment and the electron electric dipole moment (for details concerning the experimental constraints, see refs. [7, 6, 8]). Subject to all these constraints, we find allowed solutions of (6). For instance, imposing all the experimental constraints we find allowed regions in the $\tan\beta$–$M_{H^\pm}$ plane as illustrated by the red domains in the $\tan\beta$–$M_{H^\pm}$ plane, see Fig. 1 for fixed values of μ. The allowed regions were obtained scanning over the mixing angles α_i and solving the two-loop cancellation conditions (6). Imposing also unitarity in the Higgs-Higgs-scattering sector [9, 10, 11] (yellow regions), the allowed regions are only slightly reduced. Requiring that also experimental constraints are satisfied the green regions are obtained.

For parameters that are consistent with unitarity, positivity, experimental constraints and the two-loop cancellation conditions (6), we show in Fig. 2 scalar masses resulting from a scan over α_i, $M_{H^\pm}$ and $\tan\beta$. As we have noticed for the one-loop spectrum, large $\tan\beta$ implies similar scalar masses. This is indeed what is being observed in Fig. 2 also for the two-loop case. The allowed solutions "peak" around $M_{H^\pm} \sim \mu$ with $20 \lesssim \tan\beta \lesssim 50$.

2.3. CP violation

Here we will verify the possibility of having CP violation in the scalar potential (2), subject to the two-loop cancellation of quadratic divergences (6). In order to parametrize the magnitude of CP violation we adopt the $U(2)$-invariants introduced by Lavoura and Silva [12] (see also

Journal of Physics: Conference Series **259** (2010) 012055 doi:10.1088/1742-6596/259/1/012055

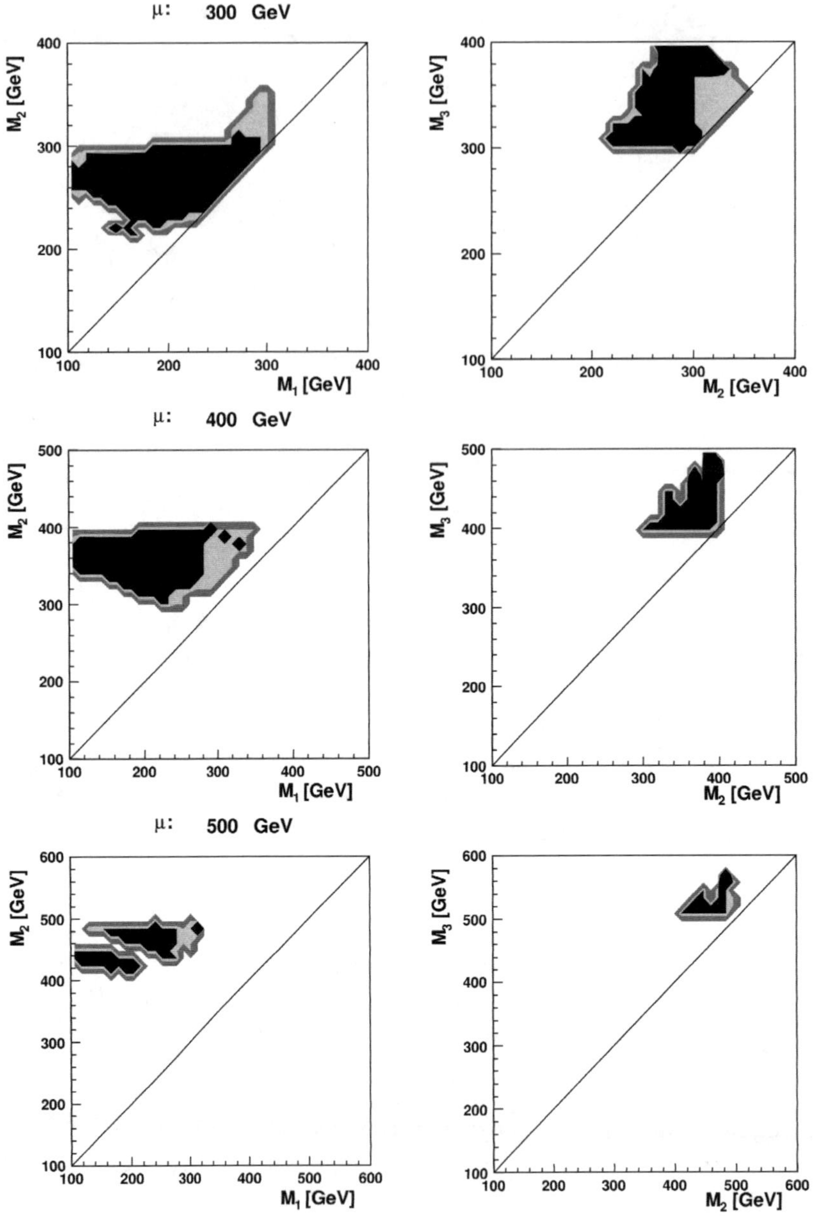

Figure 2. Two-loop distributions of allowed masses M_2 vs M_1 (left panels) and M_3 vs M_2 (right) for $\Lambda = 2.5$ TeV, resulting from a scan over the full range of α_i, $\tan\beta \in (0.5, 50)$ and $M_{H^\pm} \in (300, 700)$ GeV, for $\mu = 300, 400, 500$ GeV. Red: Positivity is satisfied; yellow: positivity and unitarity both satisfied; green: also experimental constraints satisfied at the 95% C.L., as specified in the text.

[13]). However here we use the basis-invariant formulation of these invariants J_1, J_2 and J_3 as proposed by Gunion and Haber [14]. As is proven there (theorem #4) the Higgs sector is

CP-conserving if and only if $\mathrm{Im}\, J_i = 0$ for all i. In the basis adopted here the invariants read [7]:

$$\mathrm{Im}\, J_1 = -\frac{v_1^2 v_2^2}{v^4}(\lambda_1 - \lambda_2)\mathrm{Im}\, \lambda_5, \tag{9}$$

$$\begin{aligned}\mathrm{Im}\, J_2 = & -\frac{v_1^2 v_2^2}{v^8}\left[\left((\lambda_1 - \lambda_3 - \lambda_4)^2 - |\lambda_5|^2\right) v_1^4 + 2(\lambda_1 - \lambda_2)\mathrm{Re}\, \lambda_5 v_1^2 v_2^2 \right. \\ & \left. - \left((\lambda_2 - \lambda_3 - \lambda_4)^2 - |\lambda_5|^2\right) v_2^4\right] \mathrm{Im}\, \lambda_5,\end{aligned} \tag{10}$$

$$\mathrm{Im}\, J_3 = \frac{v_1^2 v_2^2}{v^4}(\lambda_1 - \lambda_2)(\lambda_1 + \lambda_2 + 2\lambda_4)\mathrm{Im}\, \lambda_5. \tag{11}$$

It is seen that there is no CP violation when $\mathrm{Im}\, \lambda_5 = 0$, see [7] for more details.

As we have noted earlier, $\tan\beta$ above ~ 40 implies approximate degeneracy of scalar masses. That could jeopardize the CP violation in the potential since it is well known that the exact degeneracy $M_1 = M_2 = M_3$ results in vanishing invariants $\mathrm{Im}\, J_i$ and no CP violation (exact degeneracy implies $\mathrm{Im}\, \lambda_5 = 0$). Using the one-loop conditions (4)–(5) one immediately finds that $\lambda_1 - \lambda_2 = 4(m_b^2/c_\beta^2 - m_t^2/s_\beta^2)/v^2$, which implies

$$\mathrm{Im}\, J_1 = 4\, \mathrm{Im}\, \lambda_5 \frac{c_\beta^2 m_t^2 - s_\beta^2 m_b^2}{v^2} = -4\, \mathrm{Im}\, \lambda_5 \left(\frac{m_b}{v}\right)^2 + \mathcal{O}\left(\frac{\mathrm{Im}\, \lambda_5}{\tan\beta^2}\right) \tag{12}$$

Note that if $\tan\beta$ is large then $\mathrm{Im}\, J_1$ is suppressed not only by $\mathrm{Im}\, \lambda_5 \simeq 0$ (as caused by $M_1 \simeq M_2 \simeq M_3$) but also by the factor (m_b^2/v^2), as implied by the cancellation conditions (4)–(5). The same suppression factor appears for $\mathrm{Im}\, J_3$. The case of $\mathrm{Im}\, J_2$ is more involved, however when m_b^2/v^2 is neglected all the invariants (9)–(11) have the same simple asymptotic behavior for large $\tan\beta$:

$$\mathrm{Im}\, J_i \sim \frac{\mathrm{Im}\, \lambda_5}{\tan^2\beta} \tag{13}$$

Those conclusions qualitatively remain also at the two-loop level. For a quantitative illustration we plot in Fig. 3 maximal values of the invariants in the $\tan\beta$–$M_{H^\pm}$ plane with all the necessary constraints imposed, looking for regions which still allow for substantial CP violation. At high values of $\tan\beta$ these invariants are of the order of 10^{-3}, in qualitative agreement with the discussion above. It is worth noticing that the corresponding invariant in the SM; $\mathrm{Im}\, Q = \mathrm{Im}\,(V_{ud}V_{cb}V_{ub}^\star V_{cd}^\star)$ [15] is of the order of $\sim 2 \times 10^{-5} \sin\delta_{KM}$ (V_{ij} and δ_{KM} are elements of the CKM matrix and CP-violating phase, respectively). Therefore the model considered here allows for CP violation at least two orders of magnitude larger than in the SM.

3. Summary

The goal of this project was to build a minimal realistic model which would ameliorate the little hierarchy problem through suppression of the quadratic divergences in scalar boson mass corrections and through lifting the mass of the lightest Higgs boson. It has been shown that it could be accomplished within the Two-Higgs-Doublet Model type II. Phenomenological consequences of requiring no quadratic divergences in corrections to scalar masses were discussed. The 2HDM type II was analyzed taking into account the relevant existing experimental constraints. Allowed regions in the parameter space were determined. An interesting scalar mass degeneracy was noticed for $\tan\beta \gtrsim 40$. The issue of possible CP violation in the scalar potential was discussed and regions of $\tan\beta - M_{H^\pm}$ with substantial strength of CP violation were identified. The cutoff implied by the naturality arguments is lifted from ~ 600 GeV in the SM up to at least $\gtrsim 2.5$ TeV, depending on the mass of the lightest scalar. In order to accommodate a possibility for dark matter a scalar gauge singlet should be added to the model.

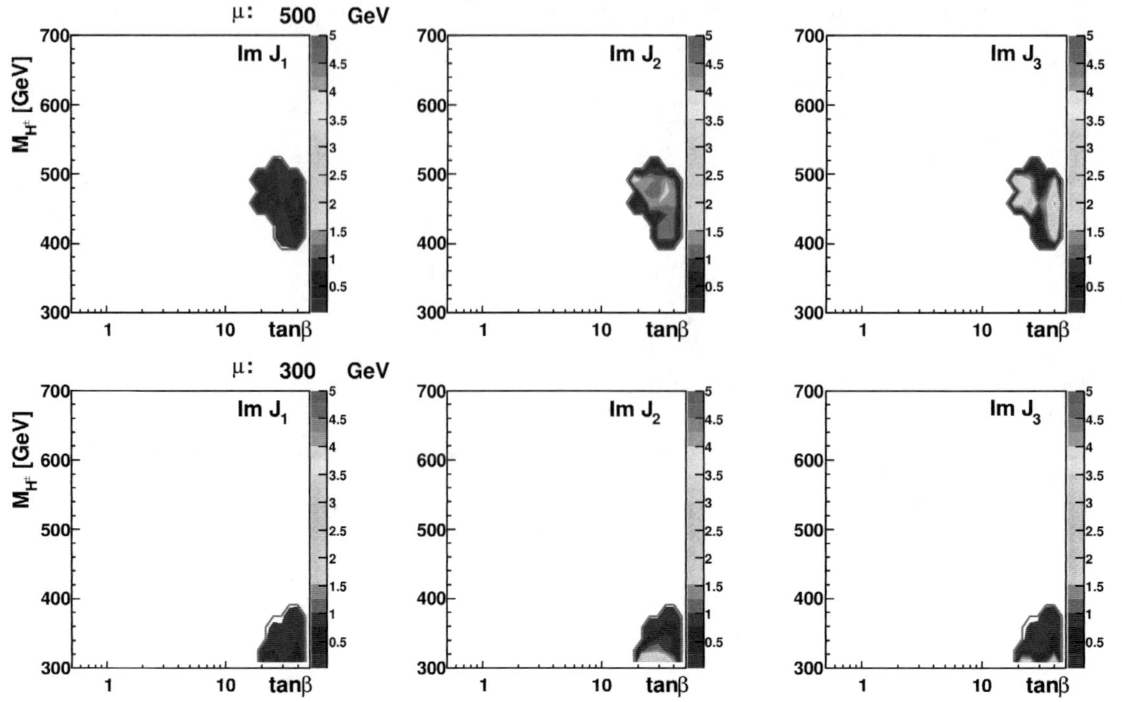

Figure 3. Absolute values of the imaginary parts of the $U(2)$-invariants $|\text{Im } J_i|$ at the two-loop level for $\Lambda = 2.5$ TeV, for $\mu = 500$ GeV (top) and $\mu = 300$ GeV (bottom). The color coding in units 10^{-3} is given along the right vertical axis.

Acknowledgments

This work is supported in part by the Ministry of Science and Higher Education (Poland) as research project N N202 006334 (2008-11). B. G. acknowledges support of the European Community within the Marie Curie Research & Training Networks: "HEPTOOLS" (MRTN-CT-2006-035505) and "UniverseNet" (MRTN-CT-2006-035863). The research of P. O. has been supported by the Research Council of Norway.

References

[1] Veltman M J G 1981 *Acta Phys. Polon* B **12** 437
[2] Siegel W 1979 *Phys. Lett.* B **84** 193 Capper D M , Jones D R T and van Nieuwenhuizen P 1980 *Nucl. Phys.* B **167** 479
[3] Osland P and Wu T T 1992 *Z. Phys.* C **55** 569 Osland and T. T. Wu 1992 *Z. Phys.* C **55** 585
[4] Einhorn M B and Jones D R T 1992 *Phys. Rev.* D **46** 5206
[5] Newton C and Wu T T 1994 *Z. Phys.* C **62** 253
[6] El Kaffas A W, Osland P and Ogreid O M 2007 *Nonlin. Phenom. Complex Syst.* **10** 347
[7] Grzadkowski B, Ogreid O M and Osland P, *Phys. Rev.* D **80** 055013
[8] El Kaffas A W, Osland P and Ogreid O M 2007 *Phys. Rev.* D **76** 095001
[9] Kanemura S, Kubota T and Takasugi E 1993 *Phys. Lett.* B **313** 155
[10] Akeroyd A G, Arhrib A and Naimi E M 2000 *Phys. Lett.* B **490** 119, Arhrib A arXiv:hep-ph/0012353
[11] Ginzburg I F and Ivanov I P arXiv:hep-ph/0312374, 2005 *Phys. Rev.* D **72** 115010 [arXiv:hep-ph/0508020].
[12] Lavoura L and Silva J P 1994 *Phys. Rev.* D **50** 4619
[13] Branco G C, Rebelo M N and Silva-Marcos J I 2005 *Phys. Lett.* B **614** 187
[14] Gunion J F and Haber H E 2005 *Phys. Rev.* D **72** 095002
[15] Bernreuther W 2002 *Lect. Notes Phys.* **591** 237

Journal of Physics: Conference Series **259** (2010) 012056 doi:10.1088/1742-6596/259/1/012056

Latest results from the CMS experiment

Andrea Giammanco, on behalf of the CMS collaboration
Université Catholique de Louvain, Louvain-la-Neuve, Belgium
E-mail: andrea.giammanco@cern.ch

Abstract. The Compact Muon Solenoid (CMS) experiment at CERN's Large Hadron Collider (LHC) has started to collect and analyze proton-proton collisions at $\sqrt{s} = 7$ TeV. At the time of this conference, CMS had collected ≈ 300 nb^{-1} of collision data, including small samples at lower energies of 0.9 TeV and 2.36 TeV. These samples allowed some precise studies of QCD and heavy flavour physics, the first cross section measurements of the W and Z bosons at 7 TeV, the first hints of reobservation of the top quark, and even the start of the exploration of the allowed parameter space of some new physics models. The results presented here demonstrate good detector performance and illustrate well understood physics signatures that may lead in the near future to discovery physics at the LHC.

1. Introduction

At the time of this conference, the Large Hadron Collider (LHC) at CERN [1] had delivered proton-proton collisions at the record-breaking energy of 7 TeV for four months; the CMS experiment has collected an integrated luminosity of 300 nb^{-1}.

We have analysed the data for validating Standard Model physics at a higher center-of-mass energy than ever before and have searched for physics beyond the Standard Model. This sudden flood of new results imposed on us some difficult choices in the selection of topics to be covered in 20 minutes. In particular, I chose to over-represent the very latest results on "hard probes" like search for heavy quarks and electroweak bosons, and the very first public results on searches for new physics.

2. LHC and CMS Status

The LHC restarted operation in late November 2009, colliding protons at $\sqrt{s} = 0.9$ TeV and through December accumulated 10 μb^{-1} at this energy and 0.4 μb^{-1} at 2.36 TeV, followed by collisions at 7 TeV at the end of March 2010. Since the initial collision, luminosity has increased during the commissioning phase for the accelerator and experiments.

The Compact Muon Solenoid (CMS) experiment [2] is a general purpose detector designed for high-energy and high-rate hadron and heavy-ion collision. From the interaction point outward, CMS includes silicon pixel and strip tracking detectors with acceptance in $|\eta| < 2.4$; a lead-tungstate crystal electromagnetic calorimeter (ECAL) covering $|\eta| < 3$ for photon and electron identification; a brass-scintillator hadron calorimeter (HCAL) covering $|\eta| < 3$ and a quartz fiber forward calorimeter (HF) covering $3 < |\eta| < 5$; a superconducting solenoid operated at 3.8 Tesla; and a muon detection system consisting of drift tubes and resistive plate chambers covering $|\eta| < 1.2$, and cathode strip and resistive plate chambers covering $0.9 < |\eta| < 2.4$.

Through 21 July 2010, the day of this presentation, CMS recorded an integrated luminosity of 303 nb^{-1}, out of 346 nb^{-1} delivered by the LHC for an efficiency of almost 90%. The peak instantaneous luminosity achieved at the end of this period was of the order of $10^{30}cm^{-2}s^{-1}$, and at the time of writing this proceeding the LHC commissioning is quickly progressing towards an instantaneous luminosity goal of $10^{32}cm^{-2}s^{-1}$, which the experiment plans to maintain through the end of 2011, hoping to achieve an integrated luminosity goal of 1000 pb^{-1} by the end of 2011.

3. Hadron Physics with the Inner Tracking System

Several important QCD studies could be performed with the limited data set and the exclusive use of the inner tracking system. Figure 1 shows the measurement of the charged particle multiplicity and spectrum in pp collisions at 0.9, 2.36 and 7 TeV [3, 4]. It is observed that the growth with center-of-mass energy doesn't scale as foreseen from previous fits (see also Ref. [5] for an extensive comparison of data to several generator settings for $\langle p_T \rangle$ versus track multiplicity).

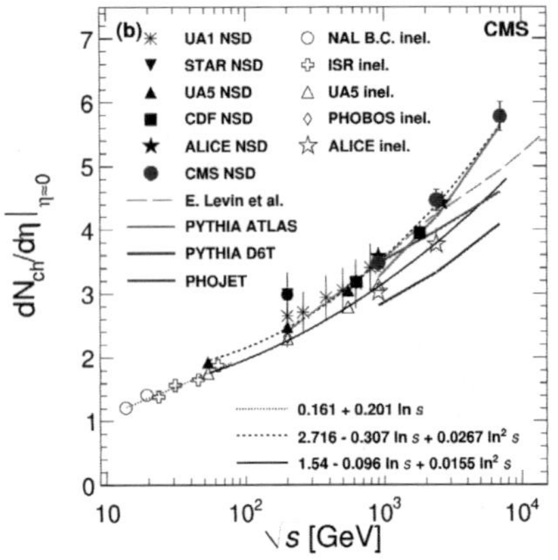

Figure 1. Growth of particle multiplicity with center-of-mass energy in pp and $p\bar{p}$ collisions.

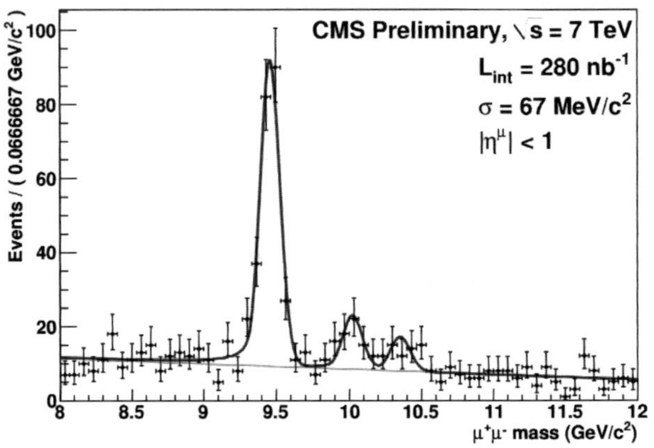

Figure 2. Reconstruction of the Υ resonances decaying into muons, with both muons within $|\eta| < 1$.

The vertex reconstruction capability has been demonstrated by the reconstruction of several unstable hadrons, some identified with the help of lifetime cuts (e.g., $K_S \to \pi^+\pi^-$, $\Lambda \to p\pi$, $\Xi \to \Lambda\pi$, $\Omega \to \Lambda K$, and several D and B mesons, see Sec. 4), and some which decay too quickly to be discerned from the primary vertex (e.g., $K^* \to K_S\pi$, $\Sigma \to \Lambda\pi$, $\Xi^0 \to \Xi^\pm\pi^\mp$, $\phi \to K^+K^-$). In the latter case the track sample has been enriched in kaons by exploiting the dE/dx measurement from the microstrips, see Sec. 6 [6, 7, 8, 9].

An extensive program of Monte Carlo tuning has already started with the inclusive and exclusive measurements performed, and comparisons between the peak positions and the known mass values of all the observed resonances are being used to validate or correct the tracker alignment, the magnetic field mapping, the material budget of the detector, and to validate particle identification. For example, the kinematic properties of the $\Lambda \to p\pi$ decay allow to know which of the two tracks is the proton track and which one is the pion track, providing a validation of dE/dx-based particle-id, and the $K_S \to \pi^+\pi^-$ and $\phi \to K^+K^-$ decays allow to study the probability of pions and kaons to be identified as muons.

4. Heavy Quarks (c, b, t)

The muon system of the CMS detector allows for a high-resolution measurement of the di-muonic resonances ranging in mass from the ρ and ω to the Z. With 280 nb^{-1} of data, well resolved peaks were observed for the Υ resonances up to 3S [10], see Figure 2, and the J/ψ yield was such as to allow for the first differential cross sections to be meaningfully compared to the existing models [11].

The D and B mesons, copiously produced at 7 TeV pp collisions, can travel distances of the order of the millimeter before decaying; variables directly or indirectly related to the flight distance are thus used for the identification of c or b flavours. A direct decay length cut has been used for identification of several D mesons [12] with very early data at $\sqrt{s} = 7$ TeV, and for the $B^{\pm} \rightarrow J/\psi K^{\pm}$ [13] and $B_S \rightarrow J/\psi\phi$ [14] decays with less than 300 nb^{-1}.

Several techniques are being used to measure the $b\bar{b}$ cross section. For example one analysis exploits the fact that $\approx 20\%$ of the b decay chains contain muons [15], while another public result at the time of the conference makes use of the so called b tagging algorithms [16], which takes advantage of the high precision in resolving secondary vertices from b hadrons. Many algorithms have been developed in CMS, and their agreement with simulation is quite impressive even for very limited early data [17]. As an example, figure 3 shows a particularly complex discriminator built by combining the likelihoods of the measured 3D impact parameters of all tracks in a jet, overweighting the four most displaced tracks (four being the average number of tracks from a b-hadron decay).

Figure 4 shows a particularly convincing di-muon candidate event with two b-tagged jets at $\sqrt{s} = 7$ TeV data presented at this conference. The confidence in the event as a golden top quark candidate arises from very good understanding of b-tagging using our silicon system.

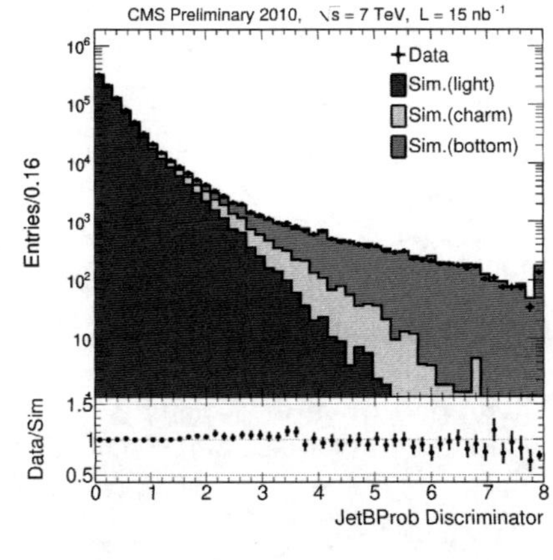

Figure 3. One of the jet b-tagging discriminators used in CMS, see definition in the text.

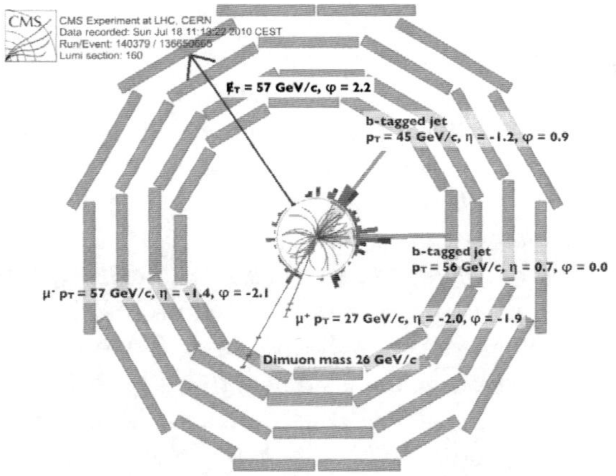

Figure 4. A "golden" top quark candidate, interpreted as $t\bar{t} \rightarrow \mu^+\mu^-\nu\bar{\nu}b\bar{b}$.

5. W and Z Bosons

The leptonic decays of W and Z bosons provide particularly distinct signatures at hadron colliders. Such clean events play a major role in commissioning the lepton-id in CMS and in the understanding of the first data.

Figure 5 shows the distinctive "Jacobian peak" in transverse mass distribution of W reconstructed from its decay into a muon and missing transverse energy. Figure 6 shows the invariant mass of Z decaying into pairs of high-p_T electrons passing tight quality selections combining the tracking and calorimetric informations. Figure 7 shows the first cross sections at 7 TeV with 198 nb^{-1} [18], yielding values in agreement with expectations. An extensive program of understanding physics with W and Z bosons is already ongoing; for example, W asymmetry study, Z forward-backward asymmetry, and studies of vector bosons associated with jets. The first two of these studies will be important to constrain the parton distribution functions and the last one to characterize one of the main backgrounds to top quark measurements and various SUSY searches.

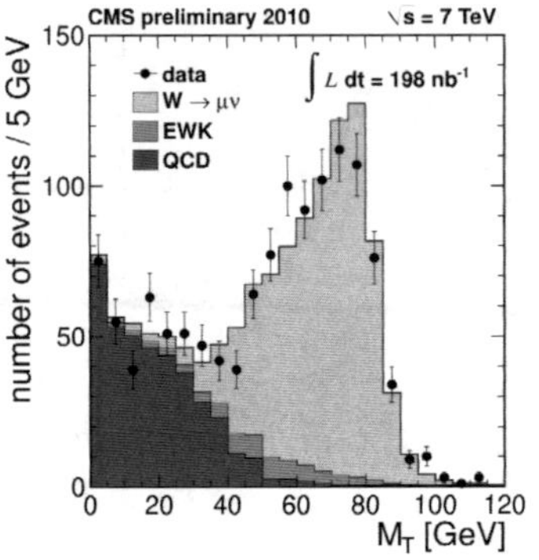

Figure 5. Transverse mass distribution for $W \to \mu\nu$ candidates.

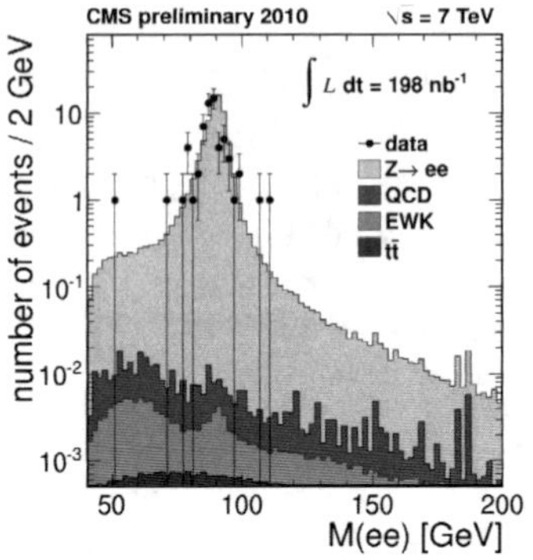

Figure 6. Invariant mass distribution for $Z \to e^+e^-$ candidates.

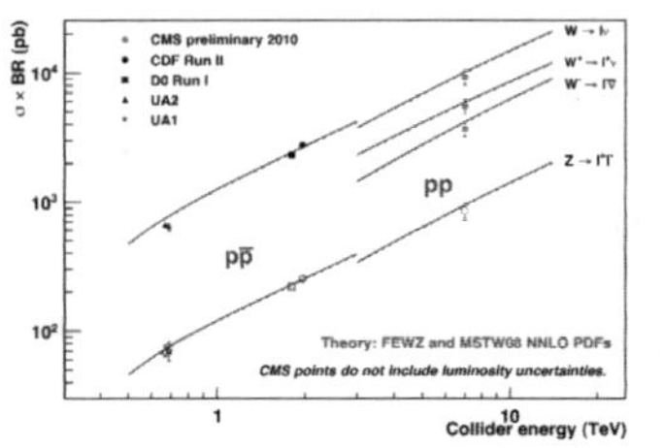

Figure 7. W and Z cross sections times branching ratio for their decay into electon and muon channels, versus center-of-mass energy.

6. The First CMS Searches

The LHC is, first of all, a machine meant to discover new physics. Nevertheless, it came as a surprise that so early a few analyses reached a sensitivity comparable with previous experiments. Searches for new particles with striking signatures could reach a very good sensitivity even with an incomplete understanding of all the detector subsystems; moreover, several new physics models predict massive particles with direct coupling to gluons, which implies the possibility to produce them by gluon-gluon fusion and therefore a big advantage of LHC over previous accelerators, in analogy with top quark production.

Figure 8 shows the reach of a search for resonances in the di-jet final state [20]. From the smoothness of the di-jet mass distribution generic cross section limits have been derived, which in turn have been used to constrain models of new physics. The first 120 nb^{-1} of data were sufficient to exclude at 95% C.L. string resonances with mass less than 1.67 TeV, excited quarks with mass less than 0.59 TeV, and axigluons and colorons with mass less than 0.52 TeV.

A signature-based search is performed for heavy stable charged particles (HSCPs) [21] by combining momentum and ionization energy loss measurements, exploiting the property of slowly moving particles to release much more energy by ionization than ultra-relativistic ones. Two selections are applied, one where candidate inner tracks are matched to muon-like tracks, which gives a very high signal-to-background ratio for models where the HSCP is lepton-like (e.g., stable staus or charginos), and one entirely based on the inner tracking system, which is thus sensitive to the so called R-hadrons, i.e., bound states of colored heavy particles (e.g., stable stops or gluinos) which have a very high probability of interacting in the calorimeters. Figure 9 shows the distribution of reconstructed masses (from the inversion of the dE/dx versus P relation) for selected tracks passing also the muon identification requirements. With only 198 nb^{-1} of data, lower limits at 95% C.L. on the mass of stable gluinos were set at 284 GeV (271 GeV) from analysis with (without) muon identification.

Complementary to this analysis, using dedicated calorimeter triggers, CMS is also searching for long-lived gluinos which may stop in the calorimeters and decay during no-beams periods [22]. In a dataset with a peak instantaneous luminosity of $1.3 \times 10^{30} cm^{-2}s^{-1}$, an integrated luminosity of 203 - 232 nb^{-1} depending on the gluino lifetime, and a search interval corresponding to 115 hours of LHC operation, no significant excess above background was observed, allowing to set stringent limits on gluino pair production over 14 orders of magnitude of gluino lifetime, extending existing limits from the Tevatron [23].

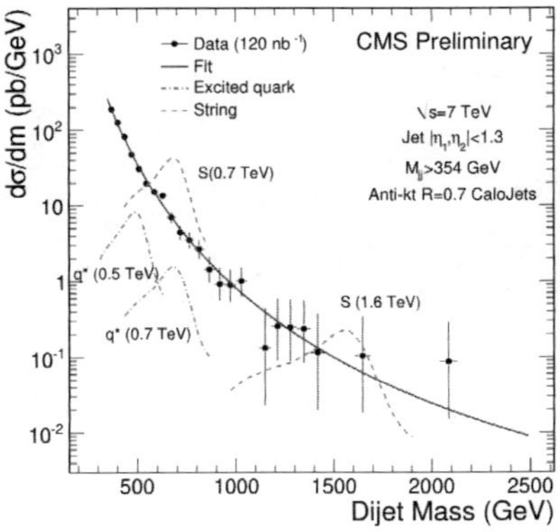

Figure 8. Distribution of di-jet invariant mass, and predictions for some new physics models.

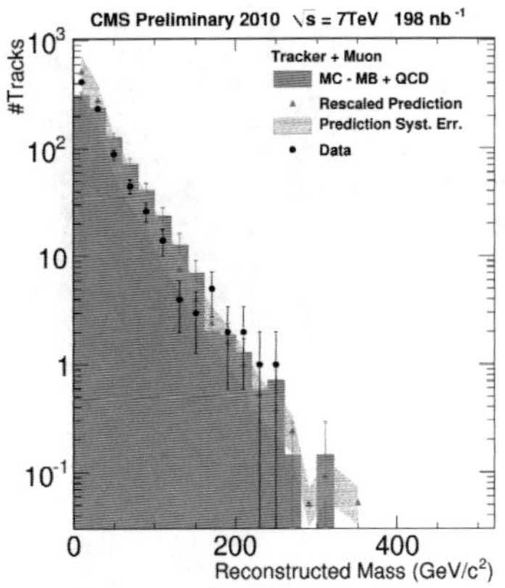

Figure 9. Distribution of reconstructed masses for the muon-matched HSCP search.

7. Outlook and Conclusion

The CMS detector is performing well. It has already demonstrated excellent reconstruction of the many ingredients expected in many searches for new physics: missing transverse energy, b-tagging, leptons, and jets. All the known particles have been reobserved, and many "known unknowns" are being measured; some searches based on simple topologies or unusual signatures have already started, in some cases already improving over the existing limits, thanks to the steep dependence of the cross sections on the collision energy. Some of the success can be credited to the long commissioning campaign with cosmic rays, and the overall quality of the first results is impressive and bodes well for the future.

References

[1] Evans L and Bryant P editors 2008 *JINST* **3** S08001.
[2] The CMS Collaboration 2008 *JINST* **3** S08004.
[3] The CMS Collaboration 2010 *JHEP* **02** 041.
[4] The CMS Collaboration 2010 *Phys. Rev. Lett.* **105** 022002.
[5] The CMS Collaboration 2010 *Physics Analysis Summary* QCD-10-006.
[6] The CMS Collaboration 2010 *Physics Analysis Summary* TRK-10-001 (*Preprint* arXiv:1007.1988).
[7] The CMS Collaboration 2010 *Physics Analysis Summary* QCD-10-007.
[8] The CMS Collaboration 2010 *Detector Performance Summary* CMS-DP-2010-004.
[9] The CMS Collaboration 2010 *Detector Performance Summary* CMS-DP-2010-001.
[10] The CMS Collaboration 2010 *Physics Analysis Summary* BPH-10-003.
[11] The CMS Collaboration 2010 *Physics Analysis Summary* BPH-10-002.
[12] The CMS Collaboration 2010 *Webpage* https://twiki.cern.ch/twiki/bin/view/CMSPublic/PhysicsResultsTRK
[13] The CMS Collaboration 2010 *Detector Performance Summary* CMS-DP-2010-033.
[14] The CMS Collaboration 2010 *Detector Performance Summary* CMS-DP-2010-024.
[15] The CMS Collaboration 2010 *Physics Analysis Summary* BPH-10-007.

Journal of Physics: Conference Series **259** (2010) 012056 doi:10.1088/1742-6596/259/1/012056

[16] The CMS Collaboration 2010 *Physics Analysis Summary* BPH-10-009.
[17] The CMS Collaboration 2010 *Physics Analysis Summary* BTV-10-001.
[18] The CMS Collaboration 2010 *Physics Analysis Summary* EWK-10-002.
[19] The CMS Collaboration 2010 *Physics Analysis Summary* TOP-10-004.
[20] The CMS Collaboration 2010 *Physics Analysis Summary* EXO-10-001, updated in EXO-10-010.
[21] The CMS Collaboration 2010 *Physics Analysis Summary* EXO-10-004.
[22] The CMS Collaboration 2010 *Physics Analysis Summary* EXO-10-003.
[23] The D0 collaboration 2007 *Phys. Rev. Lett.* **99** 131801.

16th International Symposium on Particles Strings and Cosmology (PASCOS 2010) IOP Publishing
Journal of Physics: Conference Series **259** (2010) 012057 doi:10.1088/1742-6596/259/1/012057

Implications of CP violating 2HDM in B physics

Anjan S Joshipura, Bhavik P Kodrani

Physical Research Laboratory, Navarangpura, Ahmedabad 380009, India

E-mail: anjan@prl.res.in, bhavik@prl.res.in

Abstract. The charged fermion mass matrices are invariant under $U(1)^3$ symmetry linked to the fermion number transformation. Under the condition that the definition of this symmetry in arbitrary weak basis does not depend upon Higgs parameters such as ratio of vacuum expectation values, a class of two Higgs doublet models (2HDM) can be identified in which tree level flavor changing neutral currents normally present in 2HDM are absent. However unlike the type I or type II Higgs doublet models, the charged Higgs couplings in these models contain additional flavor dependent CP violating phases. These phases can account for the recent hints of the beyond standard model CP violation in the B_d and B_s mixing. In particular, there is a range of parameters in which new phases do not contribute to the K meson CP violation but give identical new physics phases in the B_d and B_s meson mixing.

1. Introduction

Cabibbo Kobayashi Maskawa (CKM) mechanism of CP violation has been established as the dominant source of CP violation by observations at B-factories and Tevatron. These experiments have also shown some hints of physics beyond standard model [1–5]. One of them is CP violating observable $S_{J/\psi K_s}$ of $B_d^0 - \bar{B}_d^0$ mixing which is measured in time dependent CP asymmetry of $B_d^0 \to J/\psi K_s$ by BABAR experiment at the Stanford linear accelerator center (SLAC) and by Belle experiment at KEK [6, 7]. Value of $S_{J/\psi K_s}$ obtained from fit using $\epsilon_k, \Delta m_q, |V_{cb}|, \alpha, \gamma$ as inputs shows deviations from experimentally determined value at about 2σ [8]. Another hint for new physics comes from determination of CP violating phase ϕ_s of $B_s^0 - \bar{B}_s^0$ mixing. ϕ_s is determined from analysis of time dependent angular distribution of decay products in flavor tagged decay $B_s^0 \to J/\psi\phi$ by CDF and D0 experiments at Fermilab Tevatron collider through decay chain $B_s^0 \to J/\psi\phi$, $J/\psi \to \mu^+\mu^-$, $\phi \to K^+K^-$ [9, 10]. SM predicted value for ϕ_s is much small compared to the experimentally determined value. UTfit group has combined all the available experimental constraints on B_s mixing and performed a model independent analysis of NP contribution to $B_s^0 - \bar{B}_s^0$ mixing. From this analysis UTfit group has concluded that phase ϕ_s of $B_s^0 - \bar{B}_s^0$ mixing determined from their analysis deviates from the SM prediction by about 3σ [1, 11]. A similar analysis performed by CKMfitter group shows that the deviation from SM prediction is about 2.5 σ [12].

Inconsistencies between SM prediction and the experimental determination in $B_d^0 - \bar{B}_d^0$ and $B_s^0 - \bar{B}_s^0$ mixing discussed here can be resolved if there is some new physics beyond SM which can give extra contribution to above mentioned CP violating observables. Here we explore the possibility of explaining these deviations in two Higgs doublet model (2HDM). The most general 2HDM generates large flavor changing neutral currents (FCNC) for moderate Higgs mass. To avoid FCNC an additional discrete symmetry is imposed which results in type-I or type II

2HDM. In these models the charged Higgs couplings do not provide any additional source of CP violation. Therefore they cannot explain possibly large CP violating phases in the neutral B mixings if confirmed in future. Hence we need to go beyond the type-I and type-II 2HDM to explain new CP violating phases. One way is to restrict the structure of FCNC couplings using flavor symmetries rather than eliminating them altogether [13–17] . This leads to models with suppressed FCNC and additional phases whose phenomenological consequences have been studied in [18–20]. A different aproach based on shared flavour symmetry is given in [21]. An intersting possibility is to have models without tree level FCNC but containing additional phases in the charged Higgs couplings. Here we wish to discuss such models motivated by the studies of flavor symmetries of mass matrix in 2HDM.

Flavor symmetries are often applied to restrict the structure of Yukawa couplings all of which cannot be directly determined from experiments. The relation between the structures of mass matrices and symmetries was studied in [22–25]. It was shown by Lam that a symmetry can always be found under which an arbitrary neutrino mass matrix M_ν remains invariant [25]. Grimus, Lavoura and Ludl [22] showed that any Hermitian mass matrix $M_f M_f^\dagger$ obtained form a fermion mass matrix M_f always possesses a symmetry $G_f = U(1) \times U(1) \times U(1)$ and the corresponding G for the mass matrix of the Majorana neutrinos is $Z_2 \times Z_2 \times Z_2$. Here we generalize this study to the non hermitian mass matrices in 2HDM.

In the next section we will describe general 2HDM and the structure of FCNC and charged higgs interaction. Study of the symmetries of mass matrices in 2HDM and their consequences on FCNC and charged higgs interaction will also be given. In sction 3. phenomenology of this model in neutral mesons mixing will be given. In section 4. we will obtain constraints on NP paramateres in present case from the neumerical analysis of $B_s^0 - \bar{B}_s^0$ mixing and charge asymmetry of like sign dimuon events. Last section will present a summary.

2. Mass matrix symmetries and 2HDM

Yukawa couplings for 2HDM are given as

$$-\mathcal{L}_Y = \bar{Q}'_L(\Gamma_{1d}\phi_1 + \Gamma_{2d}\phi_2)d'_R + \bar{Q}'_L(\Gamma_{1u}\tilde{\phi}_1 + \Gamma_{2u}\tilde{\phi}_2)u'_R + \text{H.c.}\ , \quad (1)$$

where Γ_{iq} $(i = 1, 2; q = u, d)$ are matrices in the generation space. $\phi_{1,2}$ denote Higgs doublets and $\tilde{\phi}_i = i\tau_2\phi_i^*$. Q'_{iL} refer to three generations of doublet quarks and primed fields in the above equation refer to various quark fields in the weak basis. The neutral component of a specific linear combination of the Higgs fields $\phi \equiv \cos\beta\phi_1 + \sin\beta e^{-i\theta}\phi_2$ is responsible for the mass generation

$$M_q = v(\cos\beta\Gamma_{1q} + \sin\beta\Gamma_{2q}e^{i\theta_q}) = V_{qL}D_qV_{qR}^\dagger\ , \quad (2)$$

where $\langle\phi_1^0\rangle = v\cos\beta$; $\langle\phi_2^0\rangle = v\sin\beta e^{i\theta}$ with $v \sim 174$ GeV and $\theta_d = -\theta_u = \theta$. The matrices $V_{qL,R}$ diagonalize M_q. We define the matrices $S_{qL,R}$ as

$$S_{qL,R} = V_{qL,R}P_qV_{qL,R}^\dagger\ , \quad (3)$$

It can be seen that the mass matrices have invariance given by following equation.

$$S_{qL}^\dagger M_q S_{qR} = M_q\ . \quad (4)$$

Where $P_q = \text{diag.}(e^{i\alpha_{1q}}, e^{i\alpha_{2q}}, e^{i\alpha_{3q}})$. Here $S_{qL,R}$ define two different $U(1) \times U(1) \times U(1)$ symmetries G_u and G_d for up and down quarks. We put a mild requirement on possible $S_{qL,R}$ namely that the form of $S_{qL,R}$ be independent of the parameters $\tan\beta$ and θ which are determined entirely in the Higgs sector. With this requirement and from eq.(2) of mass matrices, we get

$$S_{qL}^\dagger\Gamma_{iq}S_{qR} = \Gamma_{iq} \qquad i = 1, 2\ . \quad (5)$$

This shows that individual Yukawa couplings should also respect the symmetry. Let us parametrize Γ_{iq} as

$$\Gamma_{iq} \equiv V_{qL} \tilde{\Gamma}_{iq} V_{qR}^\dagger \ . \tag{6}$$

Eqs. (3,5) then imply

$$P_q^\dagger \tilde{\Gamma}_{iq} P_q = \tilde{\Gamma}_{iq} \ . \tag{7}$$

If G_u, G_d refer to the full $U(1) \times U(1) \times U(1)$ symmetry with totally independent α_{iq} then the only non-trivial solution of eq.(7) is a diagonal $\tilde{\Gamma}_{iq}$ for every i and q. Yukawa couplings are then given as

$$\Gamma_{iq} = V_{qL} \gamma_{iq} V_{qR}^\dagger \ , \tag{8}$$

where γ_{iq} are diagonal matrices with complex entries. More general forms for $\tilde{\Gamma}_{iq}$ are allowed if one demands invariance with respect to subgroups of $G_u \times G_d$. All the flavor violations are induced by Higgs combination $\phi_F \equiv -\sin\beta\phi_1 + \cos\beta\phi_2 e^{-i\theta}$. The couplings of the neutral component ϕ_F^0 are given as

$$-\mathcal{L}_Y^0 = \bar{q}_L F_q q_R \phi_F^0 + \text{H.c.} \tag{9}$$

with

$$\begin{aligned} F_q &\equiv V_{qL}^\dagger (-\sin\beta\Gamma_{1q} + \cos\beta\Gamma_{2q} e^{i\theta_q}) V_{qR} \\ &= (-\sin\beta\gamma_{1q} + \cos\beta\gamma_{2q} e^{i\theta_q}) \ , \end{aligned} \tag{10}$$

where we have used eq.(8) to obtain the second line. It is seen that the FCNC matrix F_q become diagonal along with the mass matrices and the tree level FCNC are absent. But the phases of $(F_q)_{ii}$ cannot be removed in the process of making the quark masses real and remain as physical parameters. The charged component H^+ of ϕ_F correspond to the physical charged Higgs field and its couplings in our case are given by

$$-\mathcal{L}_{H^+} = \frac{H^+}{v} (\bar{u}_{iL} V_{ij} (F_d)_{jj} d_{jR} - \bar{u}_{iR} V_{ij} (F_u^*)_{ii} d_{jL}) + \text{H.c.} \tag{11}$$

The above couplings are similar to the charged Higgs couplings in 2HDM of type-I and II. In those models, $(F_q)_{ii}$ are proportional to the corresponding quark masses m_{iq} and are real. Here $(F_q)_{ii}$ are general complex numbers which can provide new phases in the $B_{d,s} - \bar{B}_{d,s}$ mixing.

An interesting class of 2HDM without the tree level FCNC have been obtained from general 2HDM by assuming that two Yukawa couplings Γ_{1q} and Γ_{2q} are proportional to each other [26]. In the present case, the two Yukawa coupling matrices are not proportional to each other but the tree level FCNC are still absent. The phases of $(F_q)_{ii}$ in the charged Higgs couplings are dependent on the flavour index i unlike in models of [26] which are characterized by universal phases one for the up and the other for the down quarks. If the diagonal matrices γ_{1q} and γ_{2q} in eq.(8) are proportional to each other then the present class of models reduce to the one in [26].

If $S_{uL} \neq S_{dL}$ then neither the Yukawa interactions given in eq.(1) nor the charged current weak interactions remain invariant under symmetries of the mass matrices. This means that radiative corrections will not preserve [27] the structure implied by eq.(8). However as in case of the 2HDM of type-I and type-II as well as the aligned models of [26], the full Lagrangian of the present model is formally invariant under the fermion number transformation $q_{iL,R} \to e^{i\alpha_{qi}} q_{iL,R}$ accompanied by the change in the CKM matrix elements $V_{ij} \to e^{i\alpha_{iu}} V_{ij} e^{-i\alpha_{jd}}$. As a consequence of this all the radiative corrections in the model would display structure similar to the one obtained in the Minimal Flavor Violating [28] models.

3. Neutral Mesons mixings

In this section we discuss phenomenological application of the model in new contribution to neutral meson mixing induced by the charged Higgs couplings in eq.(11). The $(F_d)_{ii}$ and $(F_u)_{ii}$ entering the H^+ couplings are determined by the diagonal Yukawa couplings γ_{iq} which also determine corresponding quark masses, see eq.(8). We make a simplifying assumption that the first two generation quark masses and the corresponding $(F_q)_{ii}$ are small compared to the third generation masses and $(F_q)_{33}$. Hence eq.(11) reduces to

$$-\mathcal{L}_{H^+} \approx \frac{H^+}{v}\left(\bar{u}_{iL}V_{i3}(F_d)_{33}b_R - \bar{t}_R V_{3j}(F_u^*)_{33}d_{jL}\right) + \text{H.C.} \tag{12}$$

The charged Higgs contribution to the $K^0 - \bar{K}^0$ mixing arise only from the second term. The phase in $(F_u)_{33}$ can be absorbed in the definition of H^+. As a result the above Lagrangian does not generate any new CP violating phases in the K meson mixing as long as $(F_q)_{jj}$ are neglected for $j = 1, 2$. However it can lead to non-trivial phases in the $B_q - \bar{B}_q$ mixing since the charged Higgs exchanges in this case involve both $(F_d)_{33}$ and $(F_u)_{33}$ and their phases can not be simultaneously removed. More interestingly, the interaction in eq.(12) distinguishes between the d and s quarks only through the CKM factor and not through additional phases. This results in strong correlations among the CP violation in B_s and B_d system. This can be seen as follows.

$B_q^0 - \bar{B}_q^0$ $(q = d, s)$ mixing amplitude can be parameterized in the presence of new physics contribution as

$$\langle B_q|\mathcal{H}^{SM} + \mathcal{H}^{NP}|\bar{B}_q\rangle \equiv \langle B_q|\mathcal{H}^{SM}|\bar{B}_q\rangle\,(1 + \kappa_q e^{i\phi_q^{NP}}) \equiv |\langle B_q|\mathcal{H}^{SM}|\bar{B}_q\rangle|\,\rho_q e^{-2i(\beta_q+\phi_q)} , \tag{13}$$

where β_q represent the relevant phase in case of the SM and ϕ_q are the charged Higgs induced phases. The box diagrams involving WH and HH lead to new physics contribution involving the charged Higgs H. The interaction in eq.(12) lead to the following effective Hamiltonian at the weak scale:

$$\mathcal{H}^{NP} = C_1 \bar{q}_L\gamma^\mu b_L \bar{q}_L\gamma_\mu b_L + C_2 \bar{q}_R b_L \bar{q}_R b_L , \tag{14}$$

where $C_{1,2}$ are the Wilson coefficients. Taking matrix element of the above equation and comparing with the SM result leads to

$$\kappa_q e^{i\phi_q^{NP}} = \frac{4\pi^2}{G_F^2 M_W^2 \eta_B S_0(x_t)}\left[C_1 - 5/24C_2\left(\frac{M_{B_q}}{m_b + m_q}\right)^2 \frac{B_{2q}}{B_{1q}}\right] , \tag{15}$$

where G_F, M_W respectively denote the Fermi coupling constant and the W boson mass. $S_0(x_t) \approx 2.3$ for $m_t \sim 161$ GeV. $\eta_B \approx 0.55$ refers to the QCD correction to the Wilson operator in the SM, $B_{1q,2q}$ are the bag factors which enter the operator matrix elements in eq.(14) and M_{B_q}, m_b, m_q respectively denote the masses for the B_q mesons for $q = d, s$, b quark and the d, s quarks. The Wilson coefficients $C_{1,2}$ are independent of the flavour $q = d, s$ of the light quark in B_q. Mild dependence of κ_q on q arise from the operator matrix element multiplying C_2 in eq.(15). This leads to two predictions: To a good approximation, (i) $\kappa_d \approx \kappa_s$ and (ii) $\phi_d^{NP} \approx \phi_s^{NP}$. This implies from eq.(13) that

$$\frac{\Delta M_d}{\Delta M_s} \approx \frac{\Delta M_d^{SM}}{\Delta M_s^{SM}} , \tag{16}$$

where ΔM_q denote the values of the mass difference between heavy and light mass eigenstates in B_q system in presence of new physics. Equality of κ_d and κ_s as well as ϕ_d^{NP} and ϕ_s^{NP} along with eq.(13) also imply

$$\phi_d \approx \phi_s \tag{17}$$

The detailed phenomenological consequences of this prediction are already discussed in [29] in a model independent manner. It appears to be in the right direction for explaining the CP violating anomalies. In case of B_d, the unitarity triangle angle β

$$\sin 2\beta = 0.84 \pm 0.09$$

as determined [8] using the information from V_{cb}, ϵ_K and $\frac{\Delta M_d}{\Delta M_s}$ is found to be higher than the value

$$\sin 2\beta = 0.681 \pm 0.025$$

obtained from the mixing induced asymmetry in $B \to J/\psi K_S$ decay. Since the latter measures $\beta + \phi_d$, the above information can be reconciled [29] with a negative $\phi_d \approx -10°$. $\phi_d \approx \phi_s$ then implies a sizable asymmetry [29] $S_{\psi\phi} = \sin 2(\beta_s - \phi_s) \sim 0.4$ in B_s decay as indicated by the analysis of UTfit [11] or the CKMfitter [12] group.

In a recent study global fit to various obseravbles was carried out in different scenarios [30]. One of the scenario has the relation $\kappa_d \equiv \kappa_s$ and $\phi_d^{NP} \equiv \phi_s^{NP}$. The resulting fit in this scenario is better than the SM fit. Results from this fit are a comman NP phase $2\phi_d = 2\phi_s = -14.4^{+6.7}_{-4.2}$ and $\sin 2\beta = 0.83 \pm 0.05$ at 2σ. These results agree with our results obtained here.

4. Numerical analysis

In this section we discuss our numerical analysis in which we obtain constraints on model parameters from study of $B_s^0 - \bar{B}_s^0$ mixing. New physics contribution to $B_s^0 - \bar{B}_s^0$ mixing is parametrized by UTfit group as [11]

$$C_{Bs} e^{-2i\phi_{B_s}} = \left(1 + \frac{\Delta M_s^{NP}}{\Delta M_s^{SM}}\right) \tag{18}$$

Also from eq.(13), $C_{B_s} = |1 + \kappa_s e^{-2i(\phi_s^{NP} - \beta_s)}|$ and $\phi_{B_S} = -\frac{1}{2}\text{Arg}[1 + \kappa_s e^{-2i(\phi_s^{NP} - \beta_s)}]$. UTfit collaboration has obtained the allowed range of the parameters C_{B_S} and ϕ_{B_s} as [11]

$$\begin{aligned} C_{B_s} &= [0.68, 1.51] \ , \\ \phi_{B_s} &= [-30.5, -9.9] \cup [-77.8, -58.2] \end{aligned} \tag{19}$$

We calculate C_{B_s} and ϕ_{B_s} in the present model neglecting first and second generation masses and corresponding F_{qii}. The Charged higgs mass M_h is allowed to vary in the range $100 - 500$ GeV while β and couplings F_{q33} are varied in the allowed range. Values of C_{B_s} and ϕ_{B_s} which can be obtained in this case is shown in Figure(1).

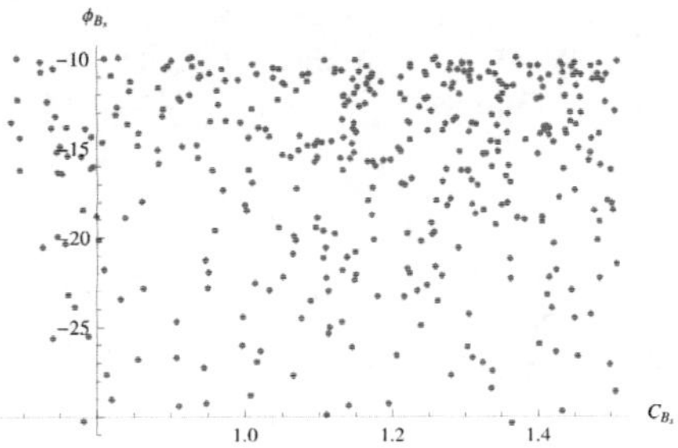

Figure 1. Values of NP parameter C_{B_s} and ϕ_{B_s} which can be obtained in limit when first two generation masses and F_{qii} , where $q = u, d$ and $i = 1, 2$, can be neglected

Journal of Physics: Conference Series **259** (2010) 012057 doi:10.1088/1742-6596/259/1/012057

We have also checked that our model is compatible with the recent determination of like-sign dimuon charge asymmetry of semileptonic b-hadron decays by D0 collaboration [31]. The model discussed here can give extra contributation required to explain the deviation of the experimental value of like-sign dimuon charge asymmetry of semileptonic b-hadron decays from SM prediction.

5. Summary

Using the flavor symmetries which does not depend on the higgs parameter $\tan\beta$ and θ, we obtained a class of 2HDM in which FCNCs can be eliminated without imposing discrete symmetries. Unlike type - I and type-II 2HDM, charged higgs interaction in this model contains new phases which are flavor dependant. In the limit of vanishing first and second generation masses and corresponding couplings F_{qii} where $q = u, d$ and $i = 1, 2$, $K^0 - \bar{K}^0$ mixing does not get any new CP phase while the $B_d^0 - \bar{B}_d^0$ and $B_s^0 - \bar{B}_s^0$ mixing gets new phases. New contribution to CP violation in this model in $B_d^0 - \bar{B}_d^0$ and $B_s^0 - \bar{B}_s^0$ mixing are correlated and it is possible to explain CP violating anomalies in B_d and B_s systems. The model parameters can also generate like sign dimuon charge asymmetry required by the experimental data.

References

[1] Bona M *et al.* 2009 (*Preprint* 0906.0953)
[2] Botella F J, Branco G C and Nebot M 2007 *Nucl. Phys.* **B768** 1–20 (*Preprint* hep-ph/0608100)
[3] Lenz A 2008 *Nucl. Phys. Proc. Suppl.* **177-178** 81–86 (*Preprint* 0705.3802)
[4] Lenz A and Nierste U 2007 *JHEP* **06** 072 (*Preprint* hep-ph/0612167)
[5] Nierste U 2007 *Nucl. Phys. Proc. Suppl.* **170** 135–140 (*Preprint* hep-ph/0612310)
[6] Aubert B *et al.* (BABAR) 2009 *Phys. Rev.* **D79** 072009 (*Preprint* 0902.1708)
[7] Chen K F *et al.* (Belle) 2007 *Phys. Rev. Lett.* **98** 031802 (*Preprint* hep-ex/0608039)
[8] Lunghi E and Soni A 2009 *JHEP* **08** 051 (*Preprint* 0903.5059)
[9] Aaltonen T *et al.* (CDF) 2008 *Phys. Rev. Lett.* **100** 161802 (*Preprint* 0712.2397)
[10] Abazov V M *et al.* (D0) 2008 *Phys. Rev. Lett.* **101** 241801 (*Preprint* 0802.2255)
[11] Bona M *et al.* (UTfit) 2009 *PMC Phys.* **A3** 6 (*Preprint* 0803.0659)
[12] Deschamps O 2008 (*Preprint* 0810.3139)
[13] Antaramian A, Hall L J and Rasin A 1992 *Phys. Rev. Lett.* **69** 1871–1873 (*Preprint* hep-ph/9206205)
[14] Cheng T P and Sher M 1987 *Phys. Rev.* **D35** 3484
[15] Branco G C, Grimus W and Lavoura L 1996 *Phys. Lett.* **B380** 119–126 (*Preprint* hep-ph/9601383)
[16] Hall L J and Weinberg S 1993 *Phys. Rev.* **D48** 979–983 (*Preprint* hep-ph/9303241)
[17] Joshipura A S 1991 *Mod. Phys. Lett.* **A6** 1693–1700
[18] Joshipura A S and Kodrani B P 2010 *Phys. Rev.* **D81** 035013 (*Preprint* 0909.0863)
[19] Joshipura A S and Kodrani B P 2009 *Phys. Lett.* **B670** 369–373 (*Preprint* 0706.0953)
[20] Joshipura A S and Kodrani B P 2008 *Phys. Rev.* **D77** 096003 (*Preprint* 0710.3020)
[21] Datta A 2008 *Phys. Rev.* **D78** 095004 (*Preprint* 0807.0795)
[22] Grimus W, Lavoura L and Ludl P O 2009 *J. Phys.* **G36** 115007 (*Preprint* 0906.2689)
[23] Lam C S 2008 *Phys. Rev. Lett.* **101** 121602 (*Preprint* 0804.2622)
[24] Lam C S 2008 *Phys. Rev.* **D78** 073015 (*Preprint* 0809.1185)
[25] Lam C S 2006 *Phys. Rev.* **D74** 113004 (*Preprint* hep-ph/0611017)
[26] Pich A and Tuzon P 2009 *Phys. Rev.* **D80** 091702 (*Preprint* 0908.1554)
[27] Ferreira P M, Lavoura L and Silva J P 2010 *Phys. Lett.* **B688** 341–344 (*Preprint* 1001.2561)
[28] D'Ambrosio G, Giudice G F, Isidori G and Strumia A 2002 *Nucl. Phys.* **B645** 155–187 (*Preprint* hep-ph/0207036)
[29] Buras A J and Guadagnoli D 2008 *Phys. Rev.* **D78** 033005 (*Preprint* 0805.3887)
[30] Lenz A *et al.* 2010 (*Preprint* 1008.1593)
[31] Abazov V M *et al.* (D0) 2010 (*Preprint* 1005.2757)

16th International Symposium on Particles Strings and Cosmology (PASCOS 2010) IOP Publishing
Journal of Physics: Conference Series **259** (2010) 012058 doi:10.1088/1742-6596/259/1/012058

How light can the lightest neutralino be?

Olaf Kittel

Departamento de Física Teórica y del Cosmos and CAFPE,
Universidad de Granada, E-18071 Granada, Spain

E-mail: kittel@th.physik.uni-bonn.de

Abstract. In this talk we summarize previous work on mass bounds of a light neutralino in the Minimal Supersymmetric Standard Model. We show that without the GUT relation between the gaugino mass parameters M_1 and M_2, the mass of the lightest neutralino is essentially unconstrained by collider bounds and precision observables. We conclude by considering also the astrophysics and cosmology of a light neutralino.

1. Introduction

The lightest supersymmetric particle, the LSP, plays a special role in the search for Supersymmetry (SUSY) [1] at colliders. For conserved R-parity or proton hexality [2, 3], the LSP is stable and thus the end product of cascade decays of any produced SUSY particle. Thus the nature of the LSP is decisive for all supersymmetric signatures at the LHC and ILC. Here we ask the question 'How light can the lightest neutralino be?', and discuss bounds from collider physics and precision observables, to summarize previous works [4–9]. Note that over the last decade there has been tremendous interest to derive bounds on the neutralino mass mainly from its relic density to explain the dark matter of the universe [10–13].

2. Neutralino framework

In the Minimal Supersymmetric Standard Model (MSSM) [1], the masses and mixings of the neutralinos and charginos are given by their mass matrices [1, 14]

$$\mathcal{M}_0 = M_Z \begin{pmatrix} M_1/M_Z & 0 & -s_\theta c_\beta & s_\theta s_\beta \\ 0 & M_2/M_Z & c_\theta c_\beta & -c_\theta s_\beta \\ -s_\theta c_\beta & c_\theta c_\beta & 0 & -\mu/M_Z \\ s_\theta s_\beta & -c_\theta s_\beta & -\mu/M_Z & 0 \end{pmatrix}, \quad \mathcal{M}_\pm = M_W \begin{pmatrix} M_2/M_W & \sqrt{2} s_\beta \\ \sqrt{2} c_\beta & \mu/M_W \end{pmatrix}, \quad (1)$$

respectively, with $c_\beta = \cos\beta$, $s_\beta = \sin\beta$ and $c_\theta = \cos\theta_w$, $s_\theta = \sin\theta_w$, and the weak mixing angle θ_w. Besides the masses of the W and Z boson, M_W and M_Z, respectively, the neutralino and chargino sectors at tree level only depend on the $U(1)_Y$ and $SU(2)_L$ gaugino masses M_1 and M_2, respectively, the higgsino mass parameter μ, and the ratio $\tan\beta = v_2/v_1$ of the vacuum expectation values of the two Higgs fields. The neutralino (chargino) masses are the square roots of the eigenvalues of $\mathcal{M}_0\mathcal{M}_0^\dagger$ ($\mathcal{M}_\pm\mathcal{M}_\pm^\dagger$) [14].

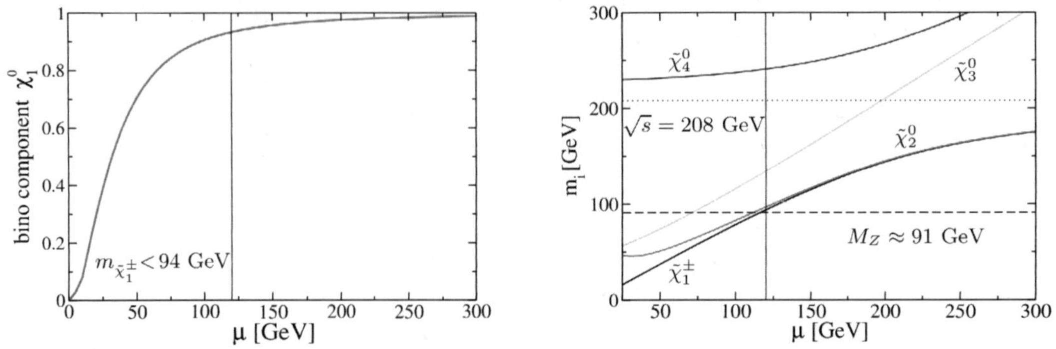

Figure 1. Bino admixture of $\tilde\chi_1^0$ (left plot) and masses of charginos and neutralinos (right plot) for $M_2 = 200$ GeV, $\tan\beta = 10$, and M_1 as given in Eq. (3), such that $m_{\tilde\chi_1^0} = 0$ GeV [8]. Left to the vertical lines at $\mu \approx 135$ GeV, the chargino mass is $m_{\tilde\chi_1^\pm} < 94$ GeV. In the right panel, the dotted line indicates the reach of LEP2 ($\sqrt{s} = 208$ GeV) for $e^+e^- \to \tilde\chi_1^0\tilde\chi_i^0$ production, and the dashed line indicates the mass of the Z boson, $M_Z \approx 91$ GeV.

The PDG cites as the laboratory bound on the lightest neutralino mass [14]

$$m_{\tilde\chi_1^0} > 46 \text{ GeV} \tag{2}$$

at 95% C.L., which is based on the chargino searches at LEP, $m_{\tilde\chi_1^\pm} \gtrsim 100$ GeV [14]. These yield lower limits on $M_2, |\mu| \gtrsim 100$ GeV. Furthermore, this bound assumes an underlying SUSY GUT, i.e, $M_1 = 5/3\tan^2(\theta_w)M_2 \approx 0.5\,M_2$. The experimental bound on M_2 then implies the lower bound on M_1, which give rise to the lower bound in Eq. (2).

However, if one drops the GUT relation, M_1 is an independent parameter, allowing to tune the neutralino mass determined from the lowest-order mass matrix $\mathcal{M}_0$ freely [4,8,10,15]. This choice can be made stable against radiative corrections [8]. The neutralino mass is identically zero for [15]

$$\det(\mathcal{M}_0) = 0 \quad \Rightarrow M_1 = \frac{M_Z^2 M_2 \sin^2\theta_w \sin(2\beta)}{\mu M_2 - M_Z^2\cos^2\theta_w \sin(2\beta)} \approx 0.05\frac{M_Z^2}{\mu} = \mathcal{O}(1\,\text{GeV}). \tag{3}$$

For $M_1 \ll M_2, |\mu|$, the neutralino $\tilde\chi_1^0$ is mainly a bino, see Fig. 1, i.e., it couples to hypercharge. This will automatically reduce the contribution to the invisible Z width, $Z \to \tilde\chi_1^0\tilde\chi_1^0$. The masses of the other neutralinos and charginos are then of the order of M_2 and $|\mu|$, see Fig. 1. In the following, we discuss bounds on the neutralino mass from production at LEP and from precision observables. Finally, we summarize bounds from astrophysics and cosmology.

3. Collider bounds

Neutralino production at LEP: If we assume $m_{\tilde\chi_1^0} = 0$, the associated production $e^+e^- \to \tilde\chi_1^0\tilde\chi_2^0$ would be accessible at LEP up to the kinematical limit of $\sqrt{s} = m_{\tilde\chi_2^0} = 208$ GeV. In order to compare with the results of the LEP searches we make use of the model-independent upper bounds on the topological neutralino production cross section obtained by OPAL with $\sqrt{s} = 208$ GeV [16],

$$\sigma(e^+e^- \to \tilde\chi_1^0\tilde\chi_2^0) \times \text{BR}(\tilde\chi_2^0 \to Z\tilde\chi_1^0) \times \text{BR}(Z \to q\bar q). \tag{4}$$

Taking into account $\text{BR}(Z \to q\bar q) \approx 70\%$, one can roughly read off from the OPAL[1] plots [16],

$$\sigma(e^+e^- \to \tilde\chi_1^0\tilde\chi_2^0) \times \text{BR}(\tilde\chi_2^0 \to Z\tilde\chi_1^0) < 70\,\text{fb}. \tag{5}$$

[1] We analyze this bound assuming conservatively that $\text{BR}(\tilde\chi_2^0 \to Z\tilde\chi_1^0) = 1$. Note that OPAL has only considered the hadronic Z decay channel, $Z \to q\bar q$. If other leptonic neutralino decays open, for example 2-body (or 3-body) decays via sleptons, see the dot-dashed line in Fig. 2(b), this would lead to a reduction of the hadronic signal OPAL searched for, and thus would allow for higher neutralino production cross sections. In that sense our bounds on these production cross sections are conservative.

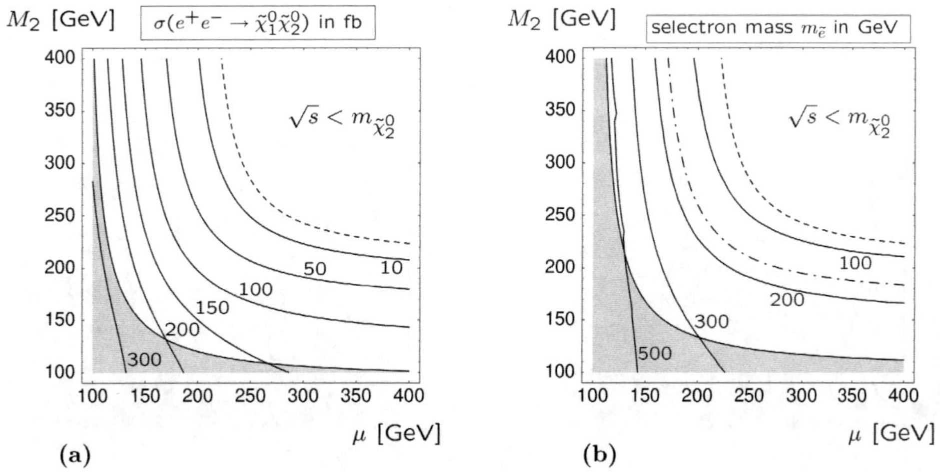

Figure 2. **(a)** Contour lines in the μ–M_2 plane of the neutralino production cross section $\sigma(e^+e^- \to \tilde{\chi}_1^0\tilde{\chi}_2^0)$ with $\tan\beta = 10$, and $m_{\tilde{e}_R} = m_{\tilde{e}_L} = m_{\tilde{e}} = 200$ GeV, at $\sqrt{s} = 208$ GeV. At each point, M_1 is chosen such that $m_{\tilde{\chi}_1^0} = 0$. **(b)** Contour lines in the μ–M_2 plane of the lower bounds on the selectron mass $m_{\tilde{e}_R} = m_{\tilde{e}_L} = m_{\tilde{e}}$, such that $\sigma(e^+e^- \to \tilde{\chi}_1^0\tilde{\chi}_2^0) = 70$ fb for $m_{\tilde{\chi}_1^0} = 0$ with $\tan\beta = 10$. In **(a)**, **(b)**, the dashed lines indicate the kinematical limit $m_{\tilde{\chi}_2^0} = \sqrt{s} = 208$ GeV, in the gray shaded areas the chargino mass is $m_{\tilde{\chi}_1^\pm} < 94$ GeV. Along the dot-dashed contour in **(b)** the relation $m_{\tilde{e}} = m_{\tilde{\chi}_2^0}$ holds.

This is already a very tight bound, since typical neutralino production cross sections can be of the order of 100 fb, see Fig. 2(a). For bino-like neutralinos, the main contribution to the cross section is due to $\tilde{e}_R$ exchange. Thus, the bound on the neutralino production cross section can be translated into lower bounds on the selectron mass $m_{\tilde{e}}$, for $m_{\tilde{\chi}_1^0} = 0$. In Fig. 2(b), we show lower bounds of the selectron mass, such that along the contours the bound $\sigma(e^+e^- \to \tilde{\chi}_1^0\tilde{\chi}_2^0) = 70$ fb is fulfilled.

Radiative neutralino production: Another search channel at LEP is radiative neutralino production, $e^+e^- \to \tilde{\chi}_1^0\tilde{\chi}_1^0\gamma$. However, due to the large SM background from radiative neutrino production $e^+e^- \to \nu\bar{\nu}\gamma$, we find that the significance is always $S < 0.1$ for $\mathcal{L} = 100$ pb^{-1} and $\sqrt{s} = 208$ GeV [6, 7]. Cuts on the photon energy or angle do not help, due to similar distributions of signal and background. At the ILC however, radiative neutralino production will be measurable, due to a higher luminosity and the option of polarized beams [6, 7, 17].

4. Bounds from precision observables and rare decays

In the following we study the impact of a light or massless neutralino on electroweak precision physics. As an example, we focus on the invisible Z width, $\Gamma_{\rm inv}$, which is potentially very sensitive to a light or massless neutralino, due to the contribution $Z \to \tilde{\chi}_1^0\tilde{\chi}_1^0$, which involves the higgsino contribution of the neutralino. However, a light neutralino is mainly bino-like for $|\mu| \gtrsim 125$ GeV, see Fig. 1. In Fig. 3, we show the difference $\delta\Gamma = (\Gamma_{\rm inv} - \Gamma_{\rm inv}^{\rm exp})/\Delta\Gamma$ from the measured invisible width $\Gamma_{\rm inv}^{\rm exp} = 499.0 \pm 1.5$ MeV [14, 18], in units of the experimental error $\Delta\Gamma = 1.5$ MeV, to the theoretical prediction $\Gamma_{\rm inv}$. The calculations of $\Gamma_{\rm inv}$ include the full $\mathcal{O}(\alpha)$ SM and MSSM contributions, supplemented with leading higher-order terms [19]. The deviation from the measured width $\Gamma_{\rm inv}^{\rm exp}$ is larger than 5σ only for $|\mu| \lesssim 125$ GeV, where an increasing higgsino admixture leads to a non-negligible neutralino coupling to the Z. However those parts of the μ-M_2 planes are mostly already excluded by direct chargino searches at LEP. Note also that already the SM contribution to $\Gamma_{\rm inv}$ is more than 1σ larger than the experimental value $\Gamma_{\rm inv}^{\rm exp}$ [18, 19].

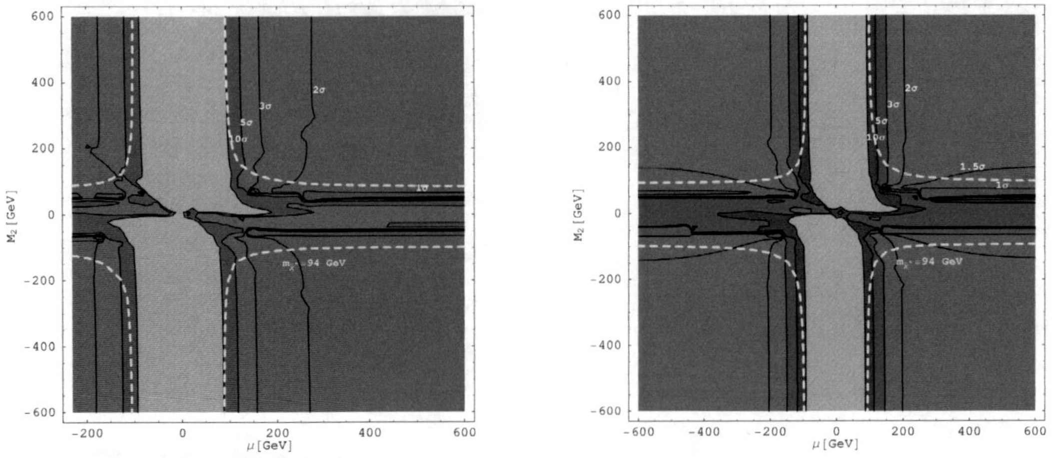

(a) (b)

Figure 3. Contour lines in the μ–M_2 plane for the difference $\delta\Gamma = (\Gamma_{\rm inv} - \Gamma_{\rm inv}^{\rm exp})/\Delta\Gamma$ of theory prediction and experimental value of the invisible Z width in units of the experimental error $\Delta\Gamma = 1.5$ MeV, for $m_{\tilde\chi_1^0} = 0$ GeV, $\tan\beta = 10$, and **(a)** $A_\tau = A_t = A_b = m_{\tilde g} = M_A = 2M_{\tilde f} = 500$ GeV, **(b)** $A_\tau = A_t = A_b = m_{\tilde g} = M_A = M_{\tilde f} = 600$ GeV. Along the dashed line $m_{\tilde\chi_1^\pm} = 94$ GeV.

A massless or light neutralino has low impact on the W boson mass, the effective leptonic weak mixing angle $\sin^2\theta_{\rm eff}$, the electric dipole moments of the electron, neutron and mercury, and the anomalous magnetic moment of the muon $(g-2)_\mu$. We thus refer the reader to the original paper [8]. Rare meson decays into a light bino-like neutralino have also been analyzed [20], but no constraints on the neutralino mass could be set.

5. Bounds from astrophysics and cosmology

Supernova cooling: Light neutralinos of masses of order 100 MeV could be thermally produced inside a Supernova. If their mean free path is of the order of the Supernova core size or lager, the neutralinos escape freely and lead to an additional cooling of the Supernova [5, 21, 22]. To be in agreement with observations of the Kamiokande and IMB Collaborations from SN 1987A, see Ref. [5], the cooling must not shorten the neutrino signal. The energy that is emitted by the neutralinos is much smaller than that emitted by the neutrinos if $m_{\tilde\chi_1^0} \gtrsim 200$ MeV [5], with $m_{\tilde e} = 500$ GeV. For heavy sleptons, $m_{\tilde e} \gtrsim 1200$ GeV, however, no bound on the neutralino mass can be set [5].

Hot dark matter: We consider the case of a nearly massless neutralino, $m_{\tilde\chi_1^0} \lesssim \mathcal{O}(1\,{\rm eV})$. Since the very light bino contributes to the hot dark matter of the universe, we assume here implicitly that the cold dark matter originates from another source. The bino relic energy density, $\rho_{\widetilde B}$, divided by the critical energy density of the universe, ρ_c, is given by [23]

$$\Omega_{\widetilde B} \equiv \frac{\rho_{\widetilde B}}{\rho_c} = \frac{43}{11}\zeta(3)\,\frac{8\pi G_N}{3H_0^2}\,\frac{g_{\rm eff}(\widetilde B)}{g_{*S}(T)}\,T_\gamma^3\, m_{\widetilde B}\,. \tag{6}$$

In order for the bino hot dark matter not to disturb the large structure formation, we assume its contribution to be less than the upper bound on the energy density of the neutrinos, as determined by the WMAP data [24]

$$\Omega_{\widetilde B} h^2 \le [\Omega_\nu h^2]_{\rm max} = 0.0076\,. \tag{7}$$

From Eqs. (6) and (7), we find the conservative upper bound

$$m_{\widetilde B} \le 0.7\ \text{eV}\,. \tag{8}$$

Thus a very light bino with mass below about 1 eV is consistent with structure formation. This line of argument was originally used by Gershtein and Zel'dovich [25] and Cowsik and McClelland [26] to derive a neutrino upper mass bound, by requiring $\Omega_\nu \leq 1$. We have here obtained an upper mass bound for a hot dark matter bino.

Cold dark matter: The impact of a light neutralino on its thermal relic density has widely been studied [10]. If the neutralino accounts for the dark matter, its mass has to be $m_{\tilde{\chi}_1^0} > 3 \ldots 20$ GeV. Although seeming theoretically unmotivated, those bounds could in principle be evaded by allowing a small amount of R-parity violation [4], and/or additional dark matter candidates.

Note that many authors have revisited the case of a light WIMP in the sub 10 GeV mass range, to explain recent results from the DAMA/LIBRA, CDMS and/or CoGeNT experiments [11]. In the MSSM, to ensure their effective annihilation, such particles must exchange a light $\mathcal{O}$(GeV) pseudoscalar Higgs at large values of $\tan\beta$. Since this is however ruled out by recent TEVATRON results on SUSY Higgs searches, the authors of Ref. [12] recently concluded that a light MSSM neutralino of $m_{\tilde{\chi}_1^0} < 15$ GeV should be excluded. In Ref. [13] it was pointed out that also improved measurements on $B_s \to \mu\mu$ exclude neutralinos with such light masses to accommodate the CoGeNT preferred region.

Acknowledgments

I would like to thank Herbi Dreiner, Sven Heinemeyer, Ulrich Langenfeld, and Georg Weiglein for the collaborations underlying this work. This work was partially supported by SFB TR-33 The Dark Universe.

References

[1] H. E. Haber and G. L. Kane, Phys. Rept. **117**, 75 (1985).
[2] H. Dreiner, C. Luhn and M. Thormeier, *Phys. Rev.* D **73** (2006) 075007 [arXiv:hep-ph/0512163].
[3] H. Dreiner, et al. *Nucl. Phys.* **B 795** (2008) 172 [arXiv:0708.0989 [hep-ph]].
[4] D. Choudhury, H. K. Dreiner, P. Richardson and S. Sarkar, Phys. Rev. D **61**, 095009 (2000); A. Dedes, H. K. Dreiner and P. Richardson, Phys. Rev. D **65**, 015001 (2001) [arXiv:hep-ph/0106199].
[5] H. K. Dreiner, C. Hanhart, U. Langenfeld and D. R. Phillips, Phys. Rev. D **68**, 055004 (2003) [arXiv:hep-ph/0304289].
[6] H. K. Dreiner, O. Kittel and U. Langenfeld, Phys. Rev. D **74** (2006) 115010 [arXiv:hep-ph/0610020].
[7] H. K. Dreiner, O. Kittel and U. Langenfeld, *Eur. Phys. J.* **C 54** (2008) 277 [arXiv:hep-ph/0703009].
[8] H. K. Dreiner, S. Heinemeyer, O. Kittel, U. Langenfeld, A. M. Weber and G. Weiglein, Eur. Phys. J. C **62**, 547 (2009) [arXiv:0901.3485 [hep-ph]]; arXiv:0707.1425 [hep-ph].
[9] H. K. Dreiner, AIP Conf. Proc. **1200**, 73 (2010) [arXiv:0910.1509 [hep-ph]].
[10] D. Hooper and T. Plehn, Phys. Lett. B **562**, 18 (2003) [arXiv:hep-ph/0212226];
G. Belanger, F. Boudjema, A. Pukhov and S. Rosier-Lees, arXiv:hep-ph/0212227;
A. Bottino, N. Fornengo and S. Scopel, Phys. Rev. D **67**, 063519 (2003) [arXiv:hep-ph/0212379];
A. Bottino, F. Donato, N. Fornengo and S. Scopel, Phys. Rev. D **68**, 043506 (2003) [arXiv:hep-ph/0304080]; Phys. Rev. D **70**, 015005 (2004) [arXiv:hep-ph/0401186];
G. Belanger, F. Boudjema, A. Cottrant, A. Pukhov and S. Rosier-Lees, JHEP **0403**, 012 (2004) [arXiv:hep-ph/0310037];
J. S. Lee and S. Scopel, Phys. Rev. D **75**, 075001 (2007) [arXiv:hep-ph/0701221];
D. Hooper, T. Plehn and A. Vallinotto, Phys. Rev. D **77**, 095014 (2008) [arXiv:0801.2539 [hep-ph]].
[11] A. Bottino, F. Donato, N. Fornengo, S. Scopel, Phys. Rev. D **81** (2010) 107302 [arXiv:0912.4025 [hep-ph]];
E. Kuflik, A. Pierce and K. M. Zurek, Phys. Rev. D **81**, 111701 (2010) [arXiv:1003.0682 [hep-ph]];
A. L. Fitzpatrick, D. Hooper and K. M. Zurek, Phys. Rev. D **81**, 115005 (2010) [arXiv:1003.0014 [hep-ph]];
J. Hisano, K. Nakayama and M. Yamanaka, Phys. Lett. B **684**, 246 (2010) [arXiv:0912.4701 [hep-ph]];
M. Asano, S. Matsumoto, M. Senami and H. Sugiyama, JHEP **1007**, 013 (2010) [arXiv:0912.5361 [hep-ph]].
[12] D. A. Vasquez, G. Belanger, C. Boehm, A. Pukhov and J. Silk, arXiv:1009.4380 [hep-ph].
[13] D. Feldman, Z. Liu and P. Nath, Phys. Rev. D **81** (2010) 117701 [arXiv:1003.0437 [hep-ph]].
[14] C. Amsler et al. [Particle Data Group], *Phys. Lett.* **B 667** (2008) 1.

[15] I. Gogoladze, J. D. Lykken, C. Macesanu and S. Nandi, Phys. Rev. D **68**, 073004 (2003).
[16] G. Abbiendi *et al.* [OPAL Collaboration], Eur. Phys. J. C **35** (2004) 1 [arXiv:hep-ex/0401026].
[17] G. A. Moortgat-Pick *et al.*, arXiv:hep-ph/0507011.
[18] The ALEPH, DELPHI, L3, OPAL, SLD Collaborations, the LEP Electroweak Working Group, the SLD Electroweak and Heavy Flavour Groups, hep-ex/0509008; hep-ex/0612034; LEP Electroweak Working Group, see: `http://lepewwg.web.cern.ch/LEPEWWG/Welcome.html`.
[19] S. Heinemeyer, W. Hollik, A. M. Weber and G. Weiglein, JHEP **0804** (2008) 039 [arXiv:0710.2972 [hep-ph]].
[20] H. K. Dreiner, S. Grab, D. Koschade, M. Kramer, B. O'Leary and U. Langenfeld, Phys. Rev. D **80**, 035018 (2009) [arXiv:0905.2051 [hep-ph]].
[21] J. Grifols, E. Masso and S. Peris, *Phys. Lett.* **B 220** (1989) 591.
[22] J. Ellis, K. Olive, S. Sarkar and D. Sciama, *Phys. Lett.* **B 215**, 404 (1988).
[23] E. Kolb and M. Turner "*The Early Universe*" (Westview Press 1990).
[24] J. Dunkley et al. [WMAP Collaboration], arXiv:0803.0586 [astro-ph].
[25] S. Gershtein and Y. Zeldovich, *JETP Lett.* **4** (1966) 120 [*Pisma Zh. Eksp. Teor. Fiz.* **4** (1966) 174].
[26] R. Cowsik and J. McClelland, Phys. Rev. Lett. **29** (1972) 669.

Journal of Physics: Conference Series **259** (2010) 012059 doi:10.1088/1742-6596/259/1/012059

Minimal matter at the LHC

Roberto Franceschini
ITPP, EPFL, CH-1015, Lausanne, Switzerland
E-mail: roberto.franceschini@epfl.ch

Abstract. A systematic approach for a model independent description of phenomenology at the LHC is presented. The most general linear interaction of scalars and fermions with pairs of standard model fields is considered at the renormalizable level. Searches for the new states are devised and the reach of TeVatron and LHC is discussed. Extensions to more general cases with more complicated spectra and interactions mediated by higher dimensional operators are briefly discussed.

1. Introduction

The experiment at the LHC are taking collision data since 2009 and the accelerator schedule foresees an integrated luminosity of 1/fb collected by the end of 2011. The data from the LHC will come out collisions at the highest energies reached so far in accelerators.

The collisions of the LHC are probing physics at energy scales in the multi-TeV range and are likely to produce the Higgs boson or other particles that appear in beyond the standard model physics scenarios.

The unprecedented complexity of the final states of the collisions and absence of an outstanding model for BSM physics requires to be open minded in terms of the possible observations of the LHC.

Motivated models are a very useful guidance to investigate possible signature for the collider, however it might well be that our model building experience is biasing our analysis in such a way that we are missing some signature of potential interest for an observation at the LHC. Thus we want to investigate what is the generic phenomenology that can be studied at the LHC and we want to do that in a model independent fashion.

The model independent treatment of the LHC phenomenology will give us a rationale to understand the reach of LHC in terms of discovery of different kinds of new physics. Furthermore we will produce a list of the possible signatures for the LHC under mild assumptions on the interaction of the new physics. This will allow to judge how much the model inspired signatures studies for the LHC have actually covered the space of possible signatures. Finally we will discuss how much of the phenomenology can be discussed without specifying a model, that is to say we will address the problem of the model independent characterization of the new physics.

To formulate a model of BSM physics requires to specify the particle content of the model, that is to say the quantum numbers of the states of the theory, and a set of symmetries that we want to impose on top of the Lorentz and gauge invariance of the SM.

The quantum numbers are restricting the possible interactions by the enforcing of the symmetries imposed on the theory. Additional symmetry can involve flavor and thus specify the flavor structure of the interactions. The actual interactions are the result of the combination of these ingredients.

In this contribution we will describe how the simplest realizations of BSM physics result in set of simple and distinctive signatures for the collider. More complicated scenarios and motivated models of new physics beyond the standard model can in many cases be, at least coarsely, captured by the simple examples that we will discuss. The simplest model that we can imagine has no additional symmetry with respect of the SM and has only one new light state. For concreteness we can assume that the decay of the new state is dominated by renormalizable interactions with bilinears of SM fields. Unless the new physics is very close to strong coupling with the SM the higher dimensional operators that can potentially change the decay phenomenology of the new states are expected to be of little relevance.

Under the specified assumptions our new physics state will have an interaction

$$\mathcal{L} = \lambda \Phi_{new} \phi_{SM} \phi_{SM} \,, \tag{1}$$

where λ is a generic coupling of the new physics to the SM sector. The coupling λ can be rather large in the cases in which the interaction is not constrained by current experiments. However we shall consider the case of nearly vanishing coupling λ in which all the effects on physics at scales lighter than the mass of the new particle Φ_{new} can be neglected. Due to the absence of competing interaction terms the decay of Φ_{new} will be dominated by the coupling λ.

The production of new states Φ_{new} with mass M will occur through pair production via the gauge interactions. Direct production of single resonances through the coupling λ is also possible, but in the cases where λ is taken small this is bringing only a small contribution. The single production can be sizable in the cases where the coupling λ is not constrained to be overly small, however this should be studied in a case-by-case fashion and we decide to deal only with pair production, which will serve as a guaranteed starting point for every analysis.

2. States of minimal matter

Here we shall consider what are the possible signatures of new particles present in BSM scenario. From eq. (1) we can derive the complete list of one particle states that can interact linearly with the SM. The full list of the fermionic and scalar states can be found in [1]. The case of vector is effectively covered, although in a different spirit, in [2] [3].

The list of all the possible fermions that can couple to the SM contains more than ten states and in principle each of them is a separate case for a study of the phenomenology for the collider. Indeed the differences in the quantum numbers of each state are potentially observable in the details of the final states of the collisions. Nonetheless the coarsest features of the final state of the collisions in which these new states are produced are determined basically just by the $SU(3)$ representation of the new object and by its coupling to heavy flavors. As a matter of fact the new fermions that can couple according to eq. (1) are either colored or not colored, i.e. they are *heavy quarks* that decay into a jet and a vector of the SM or are *heavy leptons* that decay to a lepton and a vector of the SM

- heavy quarks $Q \to jh, jW, jZ$
- heavy leptons $L \to \ell h,\ \ell W,\ \ell Z$

where by ℓ we mean either a charged SM lepton or a neutrino, by j we mean a light quark jet and the vectors are longitudinally polarized. The list of states contains some exotic charge as leptons of charge ++ or quarks of charge 5/3 or 4/3.

In case the new object has a preference for third generation couplings the possible decays are

- heavy quarks $Q^{u,d} \to t, b\, h$ or $t, b\, Z$ or $b, t\, W$
- exotic charged heavy quarks $Q^u_{5/3} \to tW^+, Q^d_{4/3} \to \bar{b}W^+$

- heavy leptons $L^+, L^0 \to \tau, \nu_\tau h$ or $\tau, \nu_\tau W$ or $\tau, \nu_\tau Z$

For the case of scalars the classes of states that can be used to describe the coarser phenomenology for the LHC are heavy quarks

- heavy quarks that decay into di-jet resonances, $S_c \to jj$
- heavy leptons that decay into leptons $L \to \ell\ell$
- leptoquarks that decay to mixed jet+leptons final states $LQ \to j\ell$

3. Collider phenomenology

Despite the large number of possible combinations of quantum numbers the huge list of states that we can couple to the SM reduces to a rather short list of items to be studied at the collider. Focusing on the signatures without heavy flavors all the possibilities for the pair production of scalars and fermions are

- scalar di-jet resonances that result in $4j$ signatures
- scalar di-lepton resonances that result in 4ℓ signatures
- heavy quarks that result in $2j2V$ final states
- heavy leptons that result in $2\ell 2V$ final states
- leptoquarks that result in $jj\ell\ell$ signatures

where $V = h, Z, W$.

Detailed searches for each of these final states have been devised in [1] here we will limit to say that interesting signatures emerge in final states with four light jet, four charged leptons, $l^+l^+l^-jj$mET, $4jl^-l^+$.

Indeed these are the final states that we find to be accessible for a discovery at the LHC in the 2009-2011 run at 7 TeV. The potential of the LHC for the discovery of resonances in these channels is summarized in figure 1. The potential is expressed in the plane of the mass of the new resonance M and the integrated luminosity of the LHC and compared with the discovery reach of TeVatron under the assumption that 10/fb will be available for the analysis at each experiment of TeVatron. The labels on the vertical dashed lines indicate the number of events produced at TeVatron, while the label on the solid curves corresponds to the number of events produced at the LHC. The region in red corresponds to masses and integrated luminosities for which the TeVatron have more events than the LHC or the LHC has less than one event. The yellow areas correspond to the mass and integrated luminosities where the LHC have more events than TeVatron, in some case in a number sufficient to make a discovery at both the machines. Finally the green area is the region where the TeVatron has a limited reach due to the lack of produced events and the LHC instead can make a discovery.

In general we can see that the LHC must reach 1/fb of integrated luminosity to be competitive with the TeVatron on all the mentioned searches. Furthermore we notice that the discovery regions for the LHC start with luminosity of order a fraction of 1/fb and that the signatures that can be observed over the SM background are either very clean, as the $l^+l^+l^-l^-$ signature, or rather dirty and "jettish" as the $4j$ signature.

4. Conclusions

In this contribution we discussed the model-independent characterization of new physics in simple set-ups that can capture the coarsest features of the actual model of new physics that the LHC will explore soon. Ignoring the possible flavor structure of the interactions of the new particles and assuming that their decay is mediated by renormalizable interactions we discussed the complete list of final states that can arise in the pair production of the new states by gauge interactions. Our result can simply be generalized to include heavy flavors in the final state.

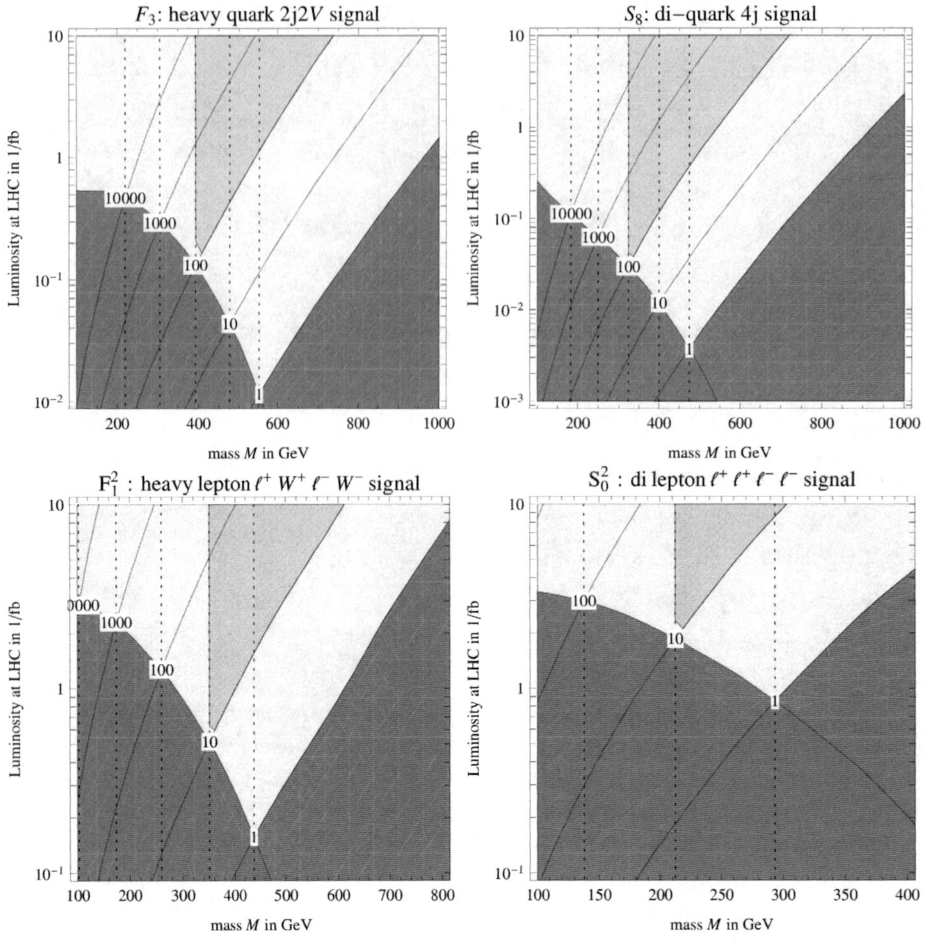

Figure 1. Discovery reach of LHC at 7 TeV compared with TeVatron with 10/fb of integrated luminosity. The comparison is made as a function of the mass of the resonance and of the integrated luminosity collected at LHC.

The inclusion of higher dimensional operators is less trivial and can give important differences with respect to the scenario discussed here. Indeed higher dimensional operators can arise when the new physics sector experiences some strong coupling. For instance the heavy quarks could have a higher dimensional operator of magnetic moment type

$$\frac{1}{\Lambda}\bar{Q}_{L,R}\sigma^{\mu\nu}V_{\mu\nu}q_{R,L}\,,$$

that can alter significantly the branching fractions and the set of available final states for the decay of the heavy quark Q. In other cases the higher dimensional operators can embody the result of having integrated out some heavy state as is for instance the case of four fermions operators

$$\frac{1}{\Lambda^2}\bar{Q}\Gamma^{(i)}q\,\bar{q}\Gamma^{(i)}q\,,$$

for the decay of a heavy quark $Q \to jjj$. The cases in which we can integrate out states and obtain effective operators for the decay of the lightest states represent a simple case in which a spectrum with several levels can be discussed in terms of a finite simple set of possibilities that can be categorized in terms of the properties of the objects appearing in the interaction.

More complicated spectra, as those that admit long decay chains, can result in a signature that is mainly determined by the properties of the lowest lying state of the chain rather than by the properties of the originating objects. In this cases a systematic discussion of the phenomenology of the lightest states is still doable, but a complete classification of the phenomenology of the parents seems overly rich to be enclosed in few representative cases. The full discussion of the phenomenology resulting from spectra with more than one state requires a more complex discussion that we leave for further works.

References

[1] E. Del Nobile, R. Franceschini, D. Pappadopulo and A. Strumia, Nucl. Phys. B **826**, 217 (2010) [arXiv:0908.1567 [hep-ph]].
[2] C. Kilic, T. Okui and R. Sundrum, JHEP **1002**, 018 (2010) [arXiv:0906.0577 [hep-ph]].
[3] C. Kilic and T. Okui, JHEP **1004**, 128 (2010) [arXiv:1001.4526 [hep-ph]].

Journal of Physics: Conference Series **259** (2010) 012060 doi:10.1088/1742-6596/259/1/012060

The search for a light CP-odd Higgs and light dark matter at colliders

Miguel-Angel Sanchis-Lozano

IFIC, Centro Mixto Universitat de València-CSIC, 46100 Burjassot, Spain

E-mail: mas@ific.uv.es

Abstract. The possible existence of a light CP-odd Higgs boson in the NMSSM, mixing with η_b resonances, can yield sizable effects on the bottomonium spectroscopy and decays to be observed at (Super) B factories. On the other hand, recent findings from astroparticle and cosmology favoring a light dark matter constituent with mass about 10 GeV or less cast a new interest into the search for invisible decays of Υ resonances performed ar B factories.

1. Introduction

As is well-known, certain symmetries in the field Lagrangian describing a possible new physics scenario can contain exceptionally light physical particles whereas the scale of the theory may be much greater. In particular, the Next-to-Minimal Supersymmetric Standard Model (NMSSM, see [1] and references therein), can yield a light CP-odd scalar A_1 of mass m_{A_1} about 10 GeV in the Higgs spectrum while evading LEP and B physics bounds (see e.g. Fig.3 of Ref.[2]).

In the simplest version of the NMSSM, the scale invariant superpotential and associated soft trilinear couplings read

$$W = \lambda S H_u H_d + \frac{1}{3}\kappa S^3 + \ldots,\ V_{soft} = (\lambda A_\lambda S H_u H_d + \frac{1}{3}\kappa A_\kappa S^3) + h.c. + \ldots \tag{1}$$

A vev of the singlet field $s \equiv \langle S \rangle$ generates an effective $\mu_{eff} = \lambda s$ term, and it is convenient to define also an effective $B_{eff} = A_\lambda + \kappa s$ term. The Higgs sector of the NMSSM contains six independent parameters, which can be chosen as λ, κ, A_λ, A_κ, $\tan\beta$, μ_{eff}.

On the other hand, two physical pseudoscalar states appear in the Higgs spectrum of the NMSSM, as superpositions of the MSSM-like state A_{MSSM} and the singlet-like state A_S,

$$A_1 = \cos\theta_A A_{MSSM} + \sin\theta_A A_S\ , \tag{2}$$

where θ_A is the mixing angle.

To a good approximation (for moderate A_λ, small A_κ and large $\tan\beta$), the mass of the lightest CP-odd Higgs boson and $\cos\theta_A$ can be written as [3]

$$m_{A_1}^2 \simeq 3\kappa s \left(\frac{3\lambda^2 v^2 A_\lambda \sin 2\beta}{2\mu_{eff}B_{eff} - 3\kappa s A_\kappa \sin 2\beta} - A_\kappa \right)\ ;\ \cos\theta_A \simeq -\frac{\lambda v (A_\lambda - 2\kappa s)\sin 2\beta}{2\mu_{eff}B_{eff} + 3\kappa s A_\kappa \sin 2\beta}\ . \tag{3}$$

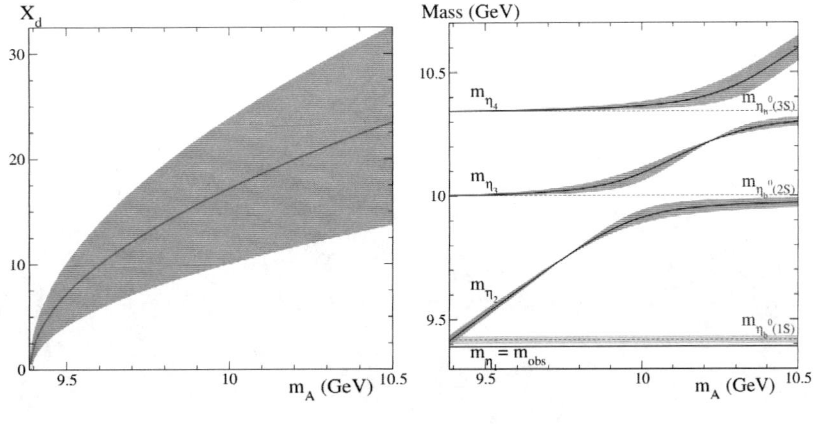

Figure 1. Left: X_d as a function of m_A (in GeV) such that one eigenvalue of $\mathcal{M}^2$ coincides with the BABAR result ([25]). Right: The masses of all eigenstates as function of m_A. Note the striking possibility of a negative hyperfine splittings (i.e. the pseudoscalar η_i level would lie higher than the corresponding vector state due to the mixing) absolutely unaccountable by quark potential models of hadrons.

The reduced coupling X_d of the light physical A_1 Higgs boson to down-type quarks and leptons (normalized with respect to the coupling of the CP-even Higgs boson of the Standard Model) is given by

$$X_d = \cos\theta_A \tan\beta \ . \tag{4}$$

At first sight the r.h.s. expression in Eq.(3) seems to imply (from $\sin 2\beta \sim 2/\tan\beta$ for large $\tan\beta$) that $\cos\theta_A$ decreases indeed with $\tan\beta$ – this would be the case in the PQ-symmetry-limit ($\kappa \to 0$) or R-symmetry-limit (A_κ, $A_\lambda \to 0$), where the second term in the denominator tends to zero. On the other hand, it follows from the minimization equations of the scalar potential of the NMSSM (as in the MSSM), for fixed soft Higgs mass terms and μ_{eff}, that $\tan\beta$ is proportional to $1/|\mu_{eff}B_{eff}|$ for large $\tan\beta$ [4], hence large values of $\tan\beta$ are associated to small values of $|B_{eff}|$ (since $|\mu_{eff}| > 100$ GeV from the lower bound on chargino masses). It is useful to replace $|B_{eff}|$ by the parameter

$$M_A^2 = \frac{2\mu_{eff}B_{eff}}{\sin 2\beta} \ , \tag{5}$$

which sets the scale for the masses of the complete $SU(2)$ multiplet of Higgs states including a scalar, a pseudoscalar and a charged Higgs as in the MSSM (in our case, the corresponding pseudoscalar is the heavier one A_2). In terms of M_A^2, X_d can be written approximately as

$$X_d \simeq -\frac{\lambda v(A_\lambda - 2\kappa s)}{M_A^2 + 3\kappa s A_\kappa} \times \tan\beta \ , \tag{6}$$

and it is reasonable to examine the large $\tan\beta$ region keeping M_A fixed.

It also follows from the l.h.s expression in Eq.(3) that there exist always values of A_κ of the same sign as A_λ (typically both negative) where m_{A_1} is small [5], while $\cos\theta_A \simeq 0.1 - 0.6$ and hence X_d is not suppressed. This requires a moderate fine-tuning of A_κ (or M_A). To summarize, the following conditions can be fulfilled simultaneously in the NMSSM, which yield possibly observable effects in Υ decays:

- $m_{A_1} \lesssim 10.5$ GeV from, e.g., appropriate values of A_κ;
- a large value of X_d, if $\tan\beta$ is large while M_A in the denominator of (6) remains moderate.

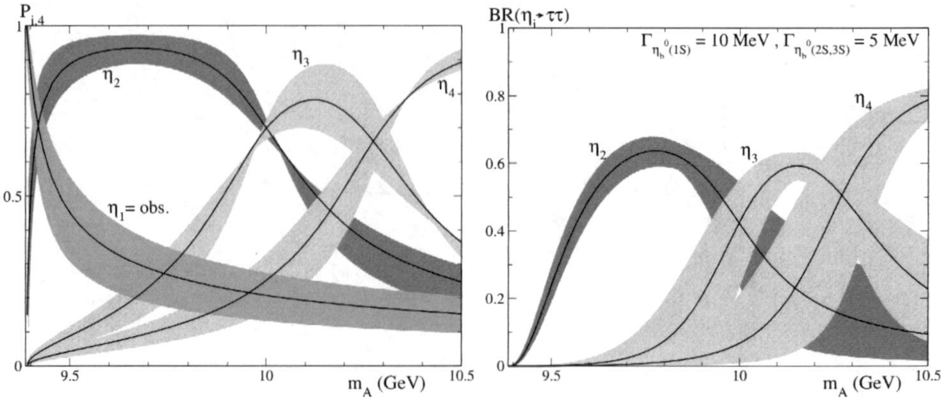

Figure 2. Left: The A-components $|P_{i,4}|$ for all 4 eigenstates as functions of m_A. Right: The branching ratios into $\tau^+\tau^-$ for the eigenstates η_2, η_3 and η_4 as functions of m_A.

2. Mixing of A_1 with the $\eta_b(nS)$ resonances

In the presence of a pseudoscalar Higgs boson with a mass close to one of the different $\eta_b(nS)$ resonances, a significant mixing between these states can occur [6]. For the $\eta_b^0(1S) - \eta_b^0(2S) - \eta_b^0(3S) - A$ mass matrix we make the ansatz

$$\mathcal{M}^2 = \begin{pmatrix} m^2_{\eta_b^0(1S)} & 0 & 0 & \delta m_1^2 \\ 0 & m^2_{\eta_b^0(2S)} & 0 & \delta m_2^2 \\ 0 & 0 & m^2_{\eta_b^0(3S)} & \delta m_3^2 \\ \delta m_1^2 & \delta m_2^2 & \delta m_3^2 & m_A^2 \end{pmatrix} . \tag{7}$$

The diagonal elements $m^2_{\eta_b^0(nS)}$ are assumed to be known (within errors) from QCD with for $m_{\eta_b^0(1S)}$. The results depend only weakly on the $\eta_b^0(2S)$ and $\eta_b^0(3S)$ masses, for which we take $m_{\eta_b^0(2S)} = 10002$ MeV, $m_{\eta_b^0(3S)} = 10343$ MeV. The off-diagonal elements δm_n^2 can be computed in the framework of a non-relativistic quark model in terms of the radial wave functions at the origin [6, 7].

In order that one eigenvalue of $\mathcal{M}^2$ coincides with the BABAR result ([25]) subsequently denoted as m_{obs}, m_A in (7) has to satisfy

$$m_A^2 = m_{obs}^2 + \frac{\delta m_1^4}{m^2_{\eta_b^0(1S)} - m^2_{obs}} + \frac{\delta m_2^4}{m^2_{\eta_b^0(2S)} - m^2_{obs}} + \frac{\delta m_3^4}{m^2_{\eta_b^0(3S)} - m^2_{obs}} . \tag{8}$$

Once m_A is expressed in terms of X_d, X_d remains the only unknown parameter in $\mathcal{M}^2$. Varying m_{obs} $m_{\eta_b^0(1S)}$ and δm_i^2, we obtain for X_d as a function of m_A the result shown in Fig. 1.

Now the masses of all 4 eigenstates of $\mathcal{M}^2$ can be computed, which are shown together with the error bands from m_{obs}, $m_{\eta_b^0(1S)}$ and δm_i^2 (in orange/grey) in Fig. 1 as functions of m_A. Henceforth we denote the 4 eigenstates of $\mathcal{M}^2$ by η_i, $i = 1 \dots 4$ where, by construction, $m_{\eta_1} \equiv m_{obs}$. For clarity we have indicated in Fig. 1 our assumed values for $m_{\eta_b^0(nS)}$ as horizontal dashed lines. For m_A not far above 9.4 GeV (where X_d is relatively small) the effects of the mixing on the states $\eta_b^0(2S)$ and $\eta_b^0(3S)$ are negligible, but for larger m_A the spectrum can differ considerably from the one expected without the presence of A!

	$\mathcal{B}(e^+e^-)$	$\mathcal{B}(\mu^+\mu^-)$	$\mathcal{B}(\tau^+\tau^-)$	$R_{\tau/e}(nS)$	$R_{\tau/\mu}(nS)$
$\Upsilon(1S)$	2.38 ± 0.11	2.48 ± 0.05	2.60 ± 0.10	0.09 ± 0.06	0.05 ± 0.04
$\Upsilon(2S)$	1.91 ± 0.16	1.93 ± 0.17	2.00 ± 0.21	0.05 ± 0.14	0.04 ± 0.06
$\Upsilon(3S)$	2.18 ± 0.21	2.18 ± 0.21	2.29 ± 0.30	0.05 ± 0.16	0.05 ± 0.16

Table 1. Measured leptonic branching ratios $\mathcal{B}(\Upsilon(nS) \to \ell\ell)$ (in %) and error bars (summed in quadrature) of $\Upsilon(1S)$, $\Upsilon(2S)$, and $\Upsilon(3S)$ resonances.

Next we consider the branching ratios of the eigenstates into $\tau^+\tau^-$, which are induced by their A-components in the decomposition of the eigenstates into the states before mixing, as

$$\eta_i = P_{i,1}\,\eta_b^0(1S) + P_{i,2}\,\eta_b^0(2S) + P_{i,3}\,\eta_b^0(3S) + P_{i,4}\,A\,. \tag{9}$$

where analytic expressions for the coefficients $P_{i,j}$ can be found in [8]. In Fig. 2 we show our results for the A-components $P_{i,4}$ for all 4 eigenstates together with the error bands from m_{obs}, $m_{\eta_b^0(1S)}$ and δm_i^2, and the corresponding branching fractions into taus induced by the mixing.

In view of (9) let us finally remark that what should be understood as a Higgs state is, to some extent, a matter of convention. It seems natural to call "Higgs" the mass eigenstate with the largest A_1-component (P_{i4}), implying non-trivial consequences from the experimental point of view, sometimes overlooked [9].

3. Possible lepton universality breaking

At present, a direct detection of a CP-odd Higgs with a mass in the particularly interesting region 9.4 GeV $\lesssim m_{A_1} \lesssim$ 10.5 GeV via a peak in the photon spectrum seems to be quite difficult. As pointed out in [10], one manifestation of the existence of such a light CP-odd Higgs boson of mass could be a breakdown of LU (lepton universality) in Υ decays, if the (not necessarily soft) radiated photon escapes undetected in the experiment, or simply is not specifically searched for in the analysis of events. (The leptonic width is, in fact, an inclusive quantity with a sum over an infinite number of photons.)

Higgs-mediated Υ decays would lead to an excess of its tauonic branching ratio (BR), which can be assessed through the ratio

$$\mathcal{R}_{\tau/\ell} = \frac{\mathcal{B}_{\tau\tau} - \mathcal{B}_{\ell\ell}}{\mathcal{B}_{\ell\ell}} = \frac{\mathcal{B}_{\tau\tau}}{\mathcal{B}_{\ell\ell}} - 1 \tag{10}$$

where $\mathcal{B}_{\tau\tau}$ denotes the tauonic, and $\mathcal{B}_{\ell\ell}$ the electronic ($\ell = e$) or muonic ($\ell = \mu$) branching ratios of the Υ resonance, respectively. A statistically significant non-zero value of $\mathcal{R}_{\tau/\ell}$ would be a strong argument in favour of a pseudoscalar Higgs boson mediating the process.

In Table 1 we summarize the current situation of LU obtained from [11]. As already mentionned in the introduction, a $\sim 1\,\sigma$ effect seems visible in most cases, leading to an overall (positive) $\sim 2\,\sigma$ effect.

We have estimated the possible amount of LU breaking in the NMSSM with m_{A_1} in the 9.4-10.5 GeV range and large X_d in order to verify whether it can be assessed experimentally [7]. In the set of plots of Fig.3, $\mathcal{R}_{\tau/\ell}$ is shown for $\Upsilon(1S)$, $\Upsilon(2S)$ and $\Upsilon(3S)$, as a function of $m_{A_{10}}$. We have considered both resonant and non-resonant production:

$$\mathcal{R}_{\tau/\ell} = \mathcal{R}^{\eta_b}_{\tau/\ell} + \mathcal{R}^{A_1}_{\tau/\ell} \tag{11}$$

through a dipole transition yielding an η_b resonance or via Wilczek mechanism, respectively [7].

Journal of Physics: Conference Series **259** (2010) 012060 doi:10.1088/1742-6596/259/1/012060

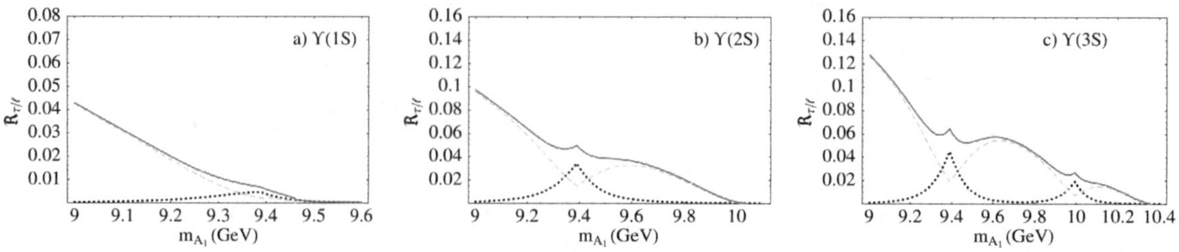

Figure 3. $\mathcal{R}_{\tau/\ell}$ versus the pseudoscalar Higgs mass for *a)* $\Upsilon(1S)$, *b)* $\Upsilon(2S)$, and *c)* $\Upsilon(3S)$ decays using $X_d = 12$, $m_{\eta_{b0}(1S)} = 9.389$ GeV [25], assuming $m_{\eta_{b0}(2S,3S)} = 9.997, 10.32$ GeV respectively, and $\Gamma_{\eta_{b0}(1S,2S,3S)} = 5$ MeV. The contributions from $\mathcal{R}^{A_1}_{\tau/\ell}$ are indicated as dashed green lines, the contributions from $\mathcal{R}^{\eta_b}_{\tau/\ell}$ as dotted black lines, and their sum $\mathcal{R}_{\tau/\ell}$ as solid red lines.

4. Invisible decays of Υ resonances

Measurements of invisible (meaning undetected) decay rates of ψ and Υ resonances at Charm and B factories can discover or place strong constraints on dark matter scenarios where candidate dark matter constituents are lighter than the b-quark [12].

According to the SM, invisible decays of the $\Upsilon(1S)$ can proceed via $b\bar{b}$ annihilation into a $\nu\bar{\nu}$ pair with a branching fraction $B(\Upsilon(1S) \to \nu\bar{\nu}) \simeq 10^{-5}$ [13], well below current experimental sensitivity. However, models containing low-mass dark matter components might enhance such invisible decay modes up to observable rates [14, 15]. Interestingly, from the astroparticle and cosmological side DAMA and CoGeNT experiments have recently reported on the observation of events compatible with a light dark matter candidate with a mass inside the interval $[5, 10]$ GeV [16]. Such a mass range for Light Dark Matter (LDM) constituents is especially attractive in certain cosmological scenarios, e.g., the Asymmetric Dark Matter model where the cosmological dark matter density arises from the baryon asymmetry of the universe and is expected to have a mass of order $[1-10]$ GeV (see [17] and references therein).

Searches for the $\Upsilon(1S) \to$ invisible mode have been carried out by Argus [19], CLEO [20], Belle [21] and BaBar [22], making use of the cascade decay: $\Upsilon(2S,3S) \to \pi^+\pi^-\Upsilon(1S)$, where the dipion provides the tagging of the $\Upsilon(1S)$ decaying into an invisible mode. The most recent experimental data from BaBar [22] provide the following upper limit at the 90% CL:

$$B[\Upsilon(1S) \to \text{invisible}] \ < \ 3 \times 10^{-4} \tag{12}$$

which is only one order of magnitude above the SM expectation. Notice that this bound only applies to LDM candidates with mass less than half the $\Upsilon(1S)$ mass.

However, as stressed in [23] the above invisible decay mode induced by scalar or pseudoscalar mediators (like CP-odd or CP-even Higgs bosons) actually vanishes no matter the character of the dark matter candidate (either scalar, Dirac or Majorana fermion) inasmuch the decaying resonance is a $J^{PC} = 1^{--}$ state. Actually, in order to get a non-vanishing decay rate, a new vector U-boson associated to the gauging of an extra $U(1)$ symmetry would be required [23]. Hence a different type of search has been perfomed at B factories by looking at the decay $\Upsilon \to \gamma +$ invisible which can proceed via a light scalar or pseudoscalar Higgs mediator decaying into a LDM pair. Let us remark the interest of this quest in connection with the possibility of a light CP-odd Higgs boson A_1 which could naturally become the mediator of the decay into undetected dark matter particles: $\Upsilon \to \gamma\ A_1(\to$ invisible$)$. The result for $\Upsilon(3s)$ decays [24] is

$$B[\Upsilon(3S) \to \gamma + \text{invisible}] \ < \ (0.7-30) \times 10^{-6} \ \text{ at 90\% CL}, \ \ s^{1/2}_{inv} < 7.8 \text{ GeV} \tag{13}$$

where s_{inv} denotes the invariant-mass squared of the hypothetical LDM pair.

Improving limits from searches of this type at higher mass encounters experimental systematic challenges because the energy of the final state photon gets progressively lower, where resolution , fake photon candidates, and noise are all typically less favorable than at higher photon energies. Nevertheless, such improvements, if achieved, would provide important constraints on theoretical possibilities by reaching higher LDM masses.

5. Conclusions and outlook

We have emphasized that the interval 9.4 GeV $\lesssim m_{A_1} \lesssim$ 10.5 GeV can have an effect on the $\eta_b(1S)$ mass as measured by Babar via mixing, and explain the possibly excessive $\Upsilon(1S) - \eta_b(1S)$ hyperfine splitting. In addition, violation of lepton universality in inclusive radiative Υ decays can be a signal for an additional CP-odd Higg which could (should) be tested in current or forthcoming experiments at (Super) B factories [26, 27].

Finally, recent findings from astroparticle and cosmology favoring a light dark matter constituent with mass about 10 GeV or less add a new interest to the search for invisible decays of Υ resonances performed at (Super) B factories.

Acknowledgments
This work was supported by grants FPA2008-02878 and GVPROMETEO2010-056.

References

[1] U. Ellwanger, C. Hugonie and A. M. Teixeira, arXiv:0910.1785 [hep-ph].
[2] F. Domingo, U. Ellwanger, E. Fullana, C. Hugonie and M. A. Sanchis-Lozano, JHEP **0901** (2009) 061 [arXiv:0810.4736 [hep-ph]].
[3] R. Dermisek and J. F. Gunion, Phys. Rev. D **75** (2007) 075019 [arXiv:hep-ph/0611142].
[4] B. Ananthanarayan and P. N. Pandita, Int. J. Mod. Phys. A **12** (1997) 2321 [arXiv:hep-ph/9601372].
[5] R. Dermisek and J. F. Gunion, Phys. Rev. D **76** (2007) 095006 [arXiv:0705.4387 [hep-ph]].
[6] M. Drees and K.-I. Hikasa, Phys. Rev. D **41** (1990) 1547.
[7] E. Fullana and M. A. Sanchis-Lozano, Phys. Lett. B **653** (2007) 67 [arXiv:hep-ph/0702190].
[8] F. Domingo, U. Ellwanger and M. A. Sanchis-Lozano, Phys. Rev. Lett. **103** (2009) 111802 [arXiv:0907.0348 [hep-ph]].
[9] M. A. Sanchis-Lozano, arXiv:1003.0312 [hep-ph].
[10] M. A. Sanchis-Lozano, Mod. Phys. Lett. A **17** (2002) 2265 [arXiv:hep-ph/0206156];
M. A. Sanchis-Lozano, Int. J. Mod. Phys. A **19** (2004) 2183 [arXiv:hep-ph/0307313];
M. A. Sanchis-Lozano, PoS **HEP2005** (2006) 334 [arXiv:hep-ph/0510374].
[11] C. Amsler *et al.* [Particle Data Group], Phys. Lett. B **667** (2008) 1.
[12] B. McElrath, Phys. Rev. D **72** (2005) 103508 [arXiv:hep-ph/0506151].
[13] L. N. Chang, O. Lebedev and J. N. Ng, Phys. Lett. B **441** (1998) 419 [arXiv:hep-ph/9806487].
[14] P. Fayet, Phys. Rev. D **74** (2006) 054034 [arXiv:hep-ph/0607318].
[15] G. K. Yeghiyan, Phys. Rev. D **80** (2009) 115019 [arXiv:0909.4919 [hep-ph]].
[16] A. L. Fitzpatrick, D. Hooper and K. M. Zurek, Phys. Rev. D **81** (2010) 115005
[17] D. E. Kaplan, M. A. Luty and K. M. Zurek, Phys. Rev. D **79** (2009) 115016 [arXiv:0901.4117 [hep-ph]].
[18] T. Cohen, D. J. Phalen, A. Pierce and K. M. Zurek, Phys. Rev. D **82** (2010) 056001 [arXiv:1005.1655 [hep-ph]].
[19] H. Albrecht *et al.* [ARGUS Collaboration], Phys. Lett. B **179** (1986) 403.
[20] P. Rubin *et al.* [CLEO Collaboration], Phys. Rev. D **75** (2007) 031104 [arXiv:hep-ex/0612051].
[21] O. Tajima *et al.* [Belle Collaboration], Phys. Rev. Lett. **98** (2007) 132001 [arXiv:hep-ex/0611041].
[22] B. Aubert *et al.* [BABAR Collaboration], Phys. Rev. Lett. **103** (2009) 251801 [arXiv:0908.2840 [hep-ex]].
[23] P. Fayet, Phys. Rev. D **81**, 054025 (2010) [arXiv:0910.2587 [hep-ph]].
[24] B. Aubert *et al.* [BaBar Collaboration], arXiv:0808.0017 [hep-ex].
[25] B. Aubert *et al.* [BABAR Collaboration], Phys. Rev. Lett. **101** (2008) 071801 [arXiv: 0807.1086 [hep-ex]].
[26] M. Bona *et al.*, arXiv:0709.0451 [hep-ex].
[27] D. G. Hitlin *et al.*, arXiv:0810.1312 [hep-ph].

Journal of Physics: Conference Series **259** (2010) 012061 doi:10.1088/1742-6596/259/1/012061

New GUT predictions for m_τ/m_b and m_μ/m_s

Stefan Antusch

Max-Planck-Institut für Physik (Werner-Heisenberg-Institut),
Föhringer Ring 6, D-80805 München, Germany

E-mail: `antusch@mppmu.mpg.de`

Abstract. Group theoretical factors from GUT symmetry breaking can lead to predictions for the ratios of quark and lepton masses (or Yukawa couplings) at the unification scale. Due to supersymmetric (SUSY) threshold corrections the viability of such predictions can depend strongly on the SUSY parameters. For three common minimal SUSY breaking scenarios with anomaly, gauge and gravity mediation we discuss which GUT scale ratios m_e/m_d, m_μ/m_s, y_τ/y_b and y_t/y_b are allowed when phenomenological constraints from electroweak precision observables, B physics, $(g-2)_\mu$, mass-limits on sparticles from direct searches as well as, optionally, constraints from the observed dark matter density are taken into account. We derive possible new predictions for the GUT scale mass ratios and compare them with the phenomenologically allowed ranges. We find that new GUT scale predictions such as $m_\mu/m_s = 9/2$ or 6 and $y_\tau/y_b = 3/2$ or 2 are often favoured compared to the ubiquitous relations $m_\mu/m_s = 3$ or $y_\tau/y_b = 1$. The talk is based on the results obtained in Refs. [1, 2].

1. Introduction

In order to construct successful GUT models of flavour, it is desirable to know the approximate GUT scale values of the quark and lepton masses and mixing parameters. The experimental data on the masses of the strange quark and the muon, for example, extrapolated to the GUT scale by means of renormalisation group (RG) running within the SM, gave rise to the so-called Georgi Jarlskog (GJ) relations [3] of $m_\mu/m_s = 3$ and $m_e/m_d = 1/3$ at the GUT scale, which can be realised from a Clebsch-Gordan factor after GUT symmetry breaking. The GJ relations have become a popular building block in many classes of unified flavour models.

In SUSY GUTs, another intriguing possibility emerges, which is the unification of all third family Yukawa couplings, i.e. of y_t, y_b, y_τ and furthermore, in the context of the seesaw mechanism, of y_ν at the GUT scale. To investigate whether this relation can be realised in a given model of low energy SUSY, it is well known that a careful inclusion of SUSY threshold corrections is required [4, 5, 6, 7]. These threshold effects are particularly important in the case of large $\tan\beta$, where $y_t = y_b = y_\tau$ seems achievable. Their relevance for realising GUT scale predictions for ratios of quark and lepton masses, in particular the GJ relation, has recently been discussed in [8, 2].

In this talk we discuss the impact of SUSY threshold corrections on quark and charged lepton Yukawa couplings and on the ratios m_μ/m_s, m_e/m_d, y_τ/y_b and y_t/y_b, which are important input parameters for GUT model building. We discuss recently proposed new predictions [1] for the GUT scale mass ratios and compare them with the phenomenologically allowed ranges. We find that new GUT scale predictions such as $m_\mu/m_s = 9/2$ or 6 and $y_\tau/y_b = 3/2$ or 2 are often favoured compared to the ubiquitous relations $m_\mu/m_s = 3$ or $y_\tau/y_b = 1$.

2. GUT predictions for mass ratios

In GUTs joined representations contain different kinds of fermions. GUT invariant operators, which generate the low energy Yukawa couplings after symmetry breaking, therefore typically relate some of the Yukawa couplings to each other. An operator like $\mathbf{5}_F\overline{\mathbf{10}}_F\langle\bar{\mathbf{5}}_H\rangle$ in SU(5), where $\mathbf{5}_F$ stands for the five dimensional matter, $\overline{\mathbf{10}}_F$ for the ten dimensional matter and $\langle\bar{\mathbf{5}}_H\rangle$ for the vacuum expectation value (vev) of the five dimensional Higgs representation, generates after SU(5) breaking identical GUT scale Yukawa couplings for down type quarks and charged leptons. This kind of operator generates b-τ-unification of Yukawa couplings. If the vev of the five dimensional representation is replaced by the vev of a 45 dimensional representation, the ratio is changed to $y_\mu/y_s = -3$. This relation is the well-known Georgi Jarlskog (GJ) relation [3].

In [1] Yukawa coupling ratios predicted from dimension four and five operators in SU(5) and Pati-Salam (PS) gauge theory are studied systematically. For this the following three assumptions have been made:

(i) The Yukawa matrices in the flavour basis are hierarchical and dominated by the diagonal elements. This is reasonable for the second and third generation and in this way the ratio of the masses directly corresponds to the ratio of the (2,2) resp. (3,3) entries of the Yukawa matrices, which can be predicted from the underlying group structure.

(ii) The (2,2) and (3,3) elements are dominated by one single operator. Otherwise different ratios mix up and the predicitons are obscured.

(iii) All the fields are included within common SO(10) models, so that an embedding in SO(10) is possible.

In the conclusions we will comment on which relations are viable within our setup.

3. SUSY threshold corrections

To relate GUT scale Yukawa coupling ratios to low energy fermion masses within a SUSY theory it is very important to include SUSY threshold corrections in the renormalization group running, especially for large $\tan\beta$. The matching formula for the down type quarks and charged leptons is

$$y_i^{MSSM} = \frac{y_i^{SM}}{\cos\beta(1+\epsilon_i\tan\beta)},$$

where i stands for d, s, b, e, μ and τ and ϵ_i is a one loop factor. Explicit expressions for ϵ_i can be found in [2], where the importance of SUSY threshold corrections for the GUT scale Yukawa coupling ratios in the electroweak unbroken phase are analyzed.

Table 1. Ranges for the GUT scale ratios m_μ/m_s, m_e/m_d, y_τ/y_b and y_t/y_b corresponding to example ranges of SUSY parameters g_+, g_- and a defined in [2] including $A_t = 0$ for $\tan\beta = 30$. Case 0 refers to the case without SUSY threshold corrections. The present experimental errors for the quark masses have been included.

Ratio	Case 0	Case g_+	Case g_-	Case a
m_e/m_d	[0.28, 0.67]	[0.30, 0.86]	[0.21, 0.61]	[0.20, 0.62]
m_μ/m_s	[3.39, 6.07]	[3.73, 7.79]	[2.54, 5.49]	[2.40, 5.63]
y_τ/y_b	[1.27, 1.38]	[1.20, 2.02]	[0.71, 1.43]	[0.60, 1.39]
y_t/y_b	[2.56, 3.02]	[2.36, 4.19]	[1.50, 3.28]	[1.14, 2.87]

In table 1 some exemplary results are shown. For case g_+, which is inspired by unification of gaugino masses with positive μ the ratios are shifted to larger values compared to case 0, where no thresholds are included. For case g_- and a, which are inspired by unification of gaugino masses with negative μ and anomaly mediated SUSY breaking scenarios, the ratios are shifted to smaller values. So the SUSY spectrum resp. the SUSY breaking parameters and especially their signs have a strong impact on the GUT scale mass ratios via SUSY threshold corrections.

4. Experimental constraints

For the calculation of the complete SUSY spectrum a modified version of SoftSUSY 2.0.18 [10] was used, where the full SUSY threshold corrections and left right mixing of sfermions in the electroweak unbroken phase for the first two generations were included. Via the SLHA [12] interface the spectra are imported to micrOMEGAs 2.2 CPC [11] for the calculation of further observables, which served as constraints.

For the results shown in figure 1 we have used the following constraints:

- Direct detection: Masses of $\chi^\pm$, $\tilde{l}$ and $\tilde{\nu}$ heavier than LEP bounds
- Electroweak precision observables:
  - $M_W = 80.429 \pm 0.039$ GeV
  - $\sin^2\theta_{eff} = 0.23153 \pm 0.00016$
- B-physics:
  - $BR(b \to s\gamma) = \left(3.55^{+0.36}_{-0.37}\right) \times 10^{-4}$
  - $BR(B_s \to \mu^+\mu^-) \leq 5.8 \times 10^{-8}$
- Anomalous magnetic moment of the muon: $0 \leq a_\mu \leq 35.9 \times 10^{-10}$

For a detailed discussion of these constraints and the discussion of an additional dark matter constraint see [1].

5. Summary and Conclusions

Table 2. Summary of the results. A checkmark (red cross) denotes (in)compatibility within the experimental constraints.

Prediction	Gauge group	mAMSB	mGMSB	CMSSM
$m_\mu/m_s = 3$	SU(5), PS	$\surd$	$\times$	$\times$
$y_t = y_b = y_\tau$	SU(5), PS	$\times$	$\times$	$\times$
$y_\tau/y_b = 1$	SU(5), PS	$\surd$	$\times^1$	$\times^1$
$m_\mu/m_s = 6$	SU(5)	$\times$	$\surd$	$\surd$
$m_\mu/m_s = 9/2$	SU(5)	$\surd$	$\surd$	$\surd$
$y_\tau/y_b = 3/2$	SU(5)	$\times$	$\surd$	$\surd$
$2y_t = 2y_b = y_\tau$	PS	$\times$	$\surd$	$\surd$
$y_t = 2y_b = 2y_\tau$	PS	$\surd$	$\times$	$\times$

The first thing to notice about our results summarized in table 2 is, that the common GUT assumptions of the GJ relation and complete third family Yukawa coupling unification are challenged in our setup. Whereas the GJ relation is possible only for the mAMSB scenario for $\tan\beta \sim 30$, complete third family Yukawa coupling unification is incompatible with our results in all three scenarios (see also figure 1). In the mAMSB case, where the corrections point in the right direction to achieve this relation without fine tuning, unification cannot be achieved because of non successful electroweak symmetry breaking and tachyons in the spectrum.

Nevertheless we find valid alternatives to the standard assumptions. For the second generation we find the SU(5) relations $y_\mu/y_s = 9/2$ for $\tan\beta \sim 20$ and $y_\mu/y_s = 6$ for $\tan\beta \sim 50$ (not for mAMSB). The first of these is even valid for all three SUSY breaking scenarios (and smaller $\tan\beta$), which is simply due to the fact, that it emerges when the threshold corrections are small and therefore the dependence on the spectrum is comparatively weak. The relation $y_\mu/y_s = 9/2$ has recently been applied to a SU(5) GUT model in [13].

For the third generation the SU(5) and PS relation $y_\tau/y_b = 1$ can be realised in mAMSB for $\tan\beta \sim 30$.[1] The SU(5) relation $y_\tau/y_b = 3/2$ appears in mGMSB for $\tan\beta \sim$ 30-45 and in CMSSM for $\tan\beta \sim$ 20-50. There are also alternatives in PS models, namely the relations $2y_t = 2y_b = y_\tau$ for $\tan\beta \sim 60$ in mGMSB and CMSSM and $y_t = 2y_b = 2y_\tau$ for $\tan\beta \sim 30$ in mAMSB. These relations are however generated from dimension six operators and have been obtained in [9], where a slightly different setup has been used.

In summary we found that in our setup new GUT predictions for quark and lepton mass ratios such as $m_\mu/m_s = 9/2$ or 6 and $y_\tau/y_b = 3/2$ or 2 are often favoured compared to the ubiquitous relations $m_\mu/m_s = 3$ and $y_\tau/y_b = 1$. These alternative relations point to characteristic SUSY spectra and breaking mechanisms which can be tested at the CERN LHC and future colliders.

Acknowledgments

This work is partially supported by the DFG cluster of excellence "Origin and Structure of the Universe".

References

[1] S. Antusch and M. Spinrath, Phys. Rev. D **79** (2009) 095004 [arXiv:0902.4644 [hep-ph]].
[2] S. Antusch and M. Spinrath, Phys. Rev. D **78** (2008) 075020 [arXiv:0804.0717 [hep-ph]].
[3] H. Georgi and C. Jarlskog, Phys. Lett. B **86** (1979) 297.
[4] L. J. Hall, R. Rattazzi and U. Sarid, Phys. Rev. D **50** (1994) 7048.
[5] M. S. Carena, M. Olechowski, S. Pokorski and C. E. M. Wagner, Nucl. Phys. B **426** (1994) 269.
[6] R. Hempfling, Phys. Rev. D **49** (1994) 6168.
[7] T. Blazek, S. Raby and S. Pokorski, Phys. Rev. D **52** (1995) 4151.
[8] G. Ross and M. Serna, Phys. Lett. B **664** (2008) 97 [arXiv:0704.1248 [hep-ph]].
[9] B. C. Allanach, S. F. King, G. K. Leontaris and S. Lola, Phys. Rev. D **56** (1997) 2632 [arXiv:hep-ph/9610517].
[10] B. C. Allanach, Comput. Phys. Commun. **143** (2002) 305 [arXiv:hep-ph/0104145].
[11] G. Belanger, F. Boudjema, A. Pukhov and A. Semenov, Comput. Phys. Commun. **149** (2002) 103 [arXiv:hep-ph/0112278]; *Ibid.* **174** (2006) 577 [arXiv:hep-ph/0405253]; *Ibid.* **176** (2007) 367 [arXiv:hep-ph/0607059]; *Ibid.* **180** (2009) 747 [arXiv:0803.2360 [hep-ph]].
[12] P. Skands *et al.*, JHEP **0407** (2004) 036 [arXiv:hep-ph/0311123].
[13] S. Antusch, S. F. King and M. Spinrath, arXiv:1005.0708 [hep-ph].

[1] In our study we have focused on $\tan\beta$ between 20 and 60. We note that there can also be b-τ-unification of Yukawa couplings for $\tan\beta \sim 1$ (see e. g. [8]).

Journal of Physics: Conference Series **259** (2010) 012061 doi:10.1088/1742-6596/259/1/012061

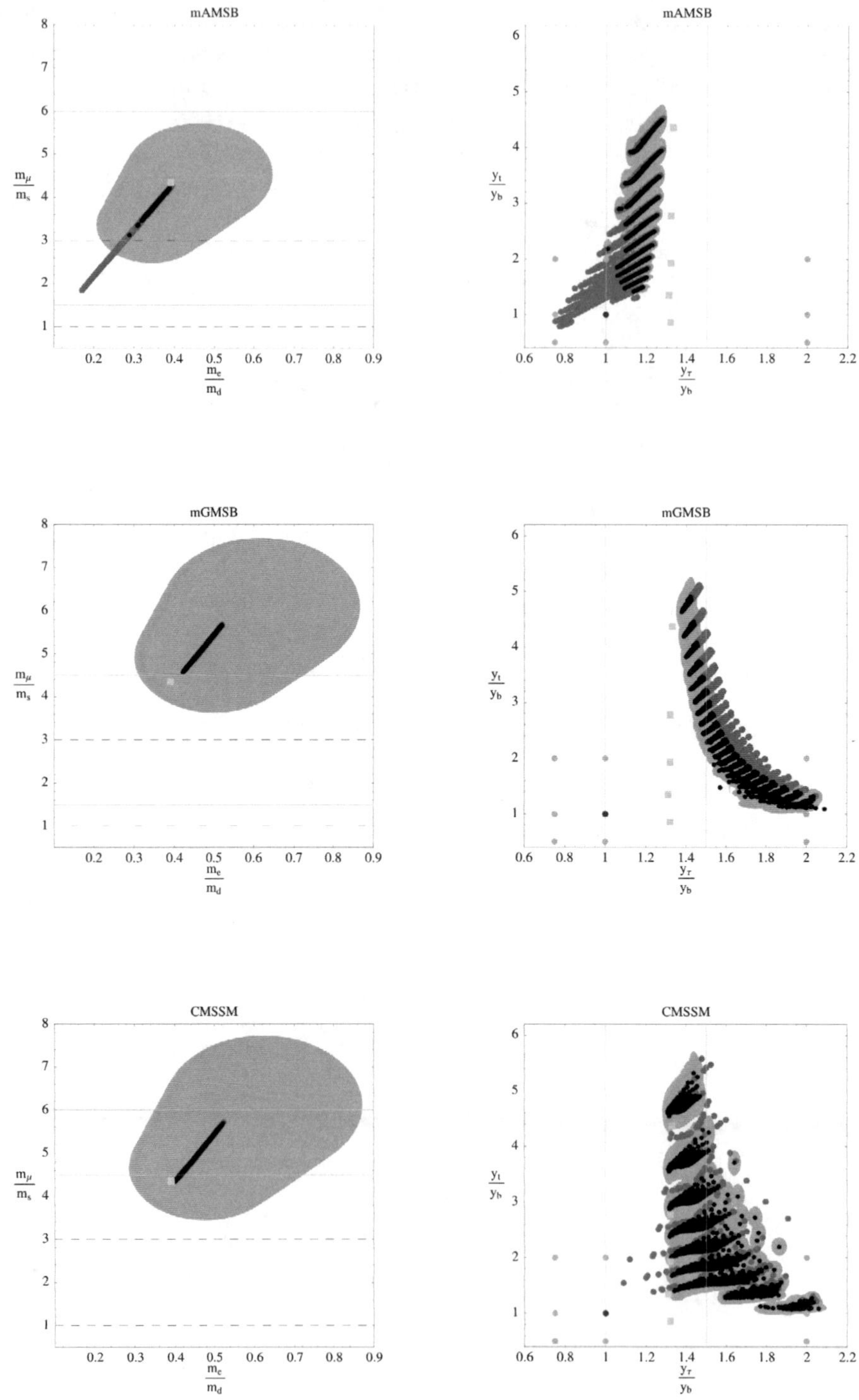

Figure 1. Final results for mAMSB, mGMSB and CMSSM. The (red) black points are the (excluded) allowed points after applying the constraints. The grey regions indicate the uncertainties from experimental quark mass errors. The green lines are predictions from SU(5), the dashed lines from SU(5) and PS and the (light) blue points from PS (dimension-six operators). The yellow squads are the GUT scale Yukawa ratios without including SUSY threshold corrections for $\tan\beta = 20$, 30, 40, 50 and 60 from top to bottom.

Journal of Physics: Conference Series **259** (2010) 012062 doi:10.1088/1742-6596/259/1/012062

Higgs phenomenology in the minimal $B - L$ extension of the Standard Model at LHC

Lorenzo Basso, Alexander Belyaev, Stefano Moretti and Giovanni Marco Pruna[1]

School of Physics and Astronomy, University of Southampton, Highfield, Southampton SO17 1BJ, UK.

E-mail: g.m.pruna@phys.soton.ac.uk

Abstract. We present some phenomenology of the Higgs sector of the Minimal $B - L$ $U(1)$ Extension of the Standard Model at the Large Hadron Collider. In this model, the existence of an extra gauge boson (Z') and an extra scalar (heavy Higgs) are predicted as naturally related with the breaking of the $B - L$ (baryon minus lepton number) symmetry. For this, we have started by deriving the unitarity bounds in the high energy limit for the Minimal $B - L$ Model parameter space. This was accomplished by analysing the full class of Higgs and would-be Goldstone boson two-to-two scatterings at tree level (exploiting the Equivalence Theorem). Hence, we studied some peculiar signature that could be observed at the CERN machine in the search of both light and heavy Higgs bosons.

1. Introduction

Despite there is no experimental evidence of a Higgs boson, the Higgs mechanism is still considered one of the favourite means for generating the masses of particles.

In the Standard Model (SM) framework this mechanism is realised by one Higgs doublet consisting of four degrees of freedom, three of which, after spontaneous Electro-Weak Symmetry Breaking (EWSB), turn out to be absorbed in the longitudinal polarisation component of each of the three weak gauge bosons, $W^{\pm}$ and Z, while the fourth one gives the physical Higgs state h.

It is clear that the SM represents a minimal choice that is completely arbitrary (as far as we know), and in the past years a big effort has been devoted to explore the implication of more complicated Higgs models, both in the context of the SM and in Beyond the Standard Model (BSM) extended theories.

One of the possible BSM scenarios is the minimal $B - L$ (Barion minus Lepton number) gauge extension of the SM, which has been recently explored (see [1]) as one of the candidates in the description of a rather simple phenomenological framework.

This model has an augmented $B - L$ gauge symmetry, that results in the natural presence of a new vector boson Z', a new Higgs field (related to the $B - L$ symmetry breaking) and, in order to preserve the theory from anomalies, one right-handed neutrino field per fermionic family (related to three heavy neutrinos as particle contents of the model).

[1] Speaker

In the present work we present a brief analysis of the phenomenology of the $B-L$ Higgs sector at the Large Hadron Collider (LHC), with emphasis on one distinctive signature of the model: the heavy neutrino pair production mediated by a light Higgs in proton-proton collision.

Firstly, in order to give a consistent picture of the allowed parameter space of the Higgs sector, we will briefly present the results of Reference [2, 3] where the Higgs parameter space of the minimal $B-L$ model was studied in detail by accounting for both experimental and theoretical constraints.

Thereafter, we will present the production cross-sections and Branching Ratios (BRs), in order to use these results to introduce some peculiar Higgs signatures at the LHC that are not allowed by the SM assumption, and therefore they could be the hallmark of the $B-L$ model.

2. The Model

The model under study is the so-called "pure" or "minimal" $B-L$ model because of the vanishing mixing between the two $U(1)_Y$ and $U(1)_{B-L}$ groups. In this model the classical gauge invariant Lagrangian, obeying the $SU(3)_C \times SU(2)_L \times U(1)_Y \times U(1)_{B-L}$ gauge symmetry, can be decomposed as:

$$\mathscr{L} = \mathscr{L}_{YM} + \mathscr{L}_s + \mathscr{L}_f + \mathscr{L}_Y . \tag{1}$$

The non-Abelian field strengths in $\mathscr{L}_{YM}$ are the same as in the SM whereas the Abelian ones can be intuitively identified. In this field basis, the co-variant derivative is: $D_\mu \equiv \partial_\mu + ig_S T^\alpha G_\mu^{\ \alpha} + igT^a W_\mu^{\ a} + ig_1 Y B_\mu + i(\widetilde{g}Y + g_1' Y_{B-L})B'_\mu$. The "pure" or "minimal" $B-L$ model is defined by the condition $\widetilde{g}(EW) = 0$, that implies no mixing between the Z' and the SM-Z gauge bosons at the tree-level at the EW scale.

The fermionic Lagrangian is the usual SM one, apart from the presence of Right-Handed (RH) neutrinos. The charges are the usual SM and $B-L$ ones (in particular, $B-L=1/3$ for quarks and -1 for leptons). The $B-L$ charge assignments of the fields as well as the introduction of new fermionic RH-neutrinos (ν_R) and scalar Higgs (χ, charged $+2$ under $B-L$) fields are designed to eliminate the triangular $B-L$ gauge anomalies and to ensure the gauge invariance of the theory, respectively. Therefore, the $B-L$ gauge extension of the SM group broken at the Electro-Weak (EW) scale does necessarily require at least one new scalar field and three new fermionic fields which are charged with respect to the $B-L$ group.

The scalar Lagrangian is:

$$\mathscr{L}_s = \left(D^\mu H\right)^\dagger D_\mu H + \left(D^\mu \chi\right)^\dagger D_\mu \chi - V(H,\chi), \tag{2}$$

with the scalar potential given by

$$V(H,\chi) = m^2 H^\dagger H + \mu^2 \mid \chi \mid^2 + \lambda_1 (H^\dagger H)^2 + \lambda_2 \mid \chi \mid^4 + \lambda_3 H^\dagger H \mid \chi \mid^2, \tag{3}$$

where H and χ are the complex scalar Higgs doublet and singlet fields, respectively.

From this potential, with standard algebraic manipulation (see [2]), one finds the explicit expressions for the Higgs bosons masses and mixing angle in terms of λ parameters.

Being h_1 and h_2 the scalar fields with masses m_{h_1} and m_{h_2} respectively (we conventionally choose $m^2_{h_1} < m^2_{h_2}$), we give the explicit expressions for the scalar mass eigenvalues:

$$m^2_{h_1} = \lambda_1 v^2 + \lambda_2 x^2 - \sqrt{(\lambda_1 v^2 - \lambda_2 x^2)^2 + (\lambda_3 x v)^2}\,, \tag{4}$$

$$m^2_{h_2} = \lambda_1 v^2 + \lambda_2 x^2 + \sqrt{(\lambda_1 v^2 - \lambda_2 x^2)^2 + (\lambda_3 x v)^2}\,, \tag{5}$$

and eigenvectors:

$$\left(\begin{array}{c} h_1 \\ h_2 \end{array}\right) = \left(\begin{array}{cc} \cos\alpha & -\sin\alpha \\ \sin\alpha & \cos\alpha \end{array}\right) \left(\begin{array}{c} h \\ h' \end{array}\right), \tag{6}$$

where $-\frac{\pi}{2} \leq \alpha \leq \frac{\pi}{2}$ fulfils:

$$\sin 2\alpha = \frac{\lambda_3 x v}{\sqrt{(\lambda_1 v^2 - \lambda_2 x^2)^2 + (\lambda_3 x v)^2}}. \quad (7)$$

Finally, the Yukawa interactions are: $\mathscr{L}_Y = -y^d_{jk}\overline{q_{jL}}d_{kR}H - y^u_{jk}\overline{q_{jL}}u_{kR}\widetilde{H} - y^e_{jk}\overline{l_{jL}}e_{kR}H - y^\nu_{jk}\overline{l_{jL}}\nu_{kR}\widetilde{H} - y^M_{jk}\overline{(\nu_R)^c_j}\nu_{kR}\chi + \text{h.c.}$, where $\widetilde{H} = i\sigma^2 H^*$ and i, j, k take the values 1 to 3, where the last term is the Majorana contribution and the others the usual Dirac ones.

3. Results

Firstly, we have made an extensive study on the Higgs-sector parameter-space allowed by theoretical constraints exploiting the well-known techniques from consideration on vacuum stability and triviality (renormalisation group equations (RGEs) techniques, see [3]) and perturbative unitarity (PU) (see [2]). The latter, in particular, is not energy-scale dependent, and it results in a simpler description of the $m_{h_1} - m_{h_2} - \alpha$ allowed space, and we will mainly consider it in the following analysis.

By analysing the full class of Higgs and would-be Goldstone boson two-to-two scatterings at tree level (exploiting the Equivalence Theorem) one finds the following result: the theory is PU-stable if $m_{h_1} < 700$ GeV and

$$m_{h_2} < 2\sqrt{\frac{2}{3}} \min\left(\frac{m_W}{\sqrt{\alpha_W}\sin\alpha}, \sqrt{\pi}x\right), \quad (8)$$

where $\alpha_W = \alpha_{em}/\sin^2\theta_W$.

Considering the experimental limits from LEP (which established that the safest choice for the light Higgs boson mass is $m_{h_1} > 115$ GeV) we have a complete definition of the allowed Higgs-sector parameter space.

For completeness, we have also combined the RGEs and PU techniques in order to find a dynamical constraint on the g'_1 domain (see [4]). From this method, assuming that the $B-L$ symmetry breaking occurs at the TeV scale, one finds that the model is PU-stable to the Planck scale only if $g'_1 < 0.23$.

Thereafter, we have analysed the production mechanism channels both for h_1 and h_2 in proton-proton colliders, with emphasis on two LHC energy-luminosity configurations: "early discovery scenario" with $\sqrt{s} = 7$ GeV and $L = 1$ fb^{-1}, "full integrated luminosity" scenario with $\sqrt{s} = 14$ GeV and $L = 300$ fb^{-1}.

As explicit example, in figure 1 we show the full set of production mechanisms for h_1 in proton-proton collision at $\sqrt{s} = 7$ GeV for $\alpha = \pi/5$: as in the SM case, the dominant mode is represented by the gluon-gluon fusion (black line) process, while the inclusive processes as vector boson fusion (VBF) (red line), the H-strahlung processes (blue line for $W^\pm$ and violet line for Z) and the associated production of top and Higgs (green line) represent a significantly smaller contribution. For completeness, we have superimposed the SM-like case ($\alpha = 0$) in dotted lines.

Besides, we have evaluated the BRs both for h_1 and h_2, in the search for configurations in which it could be possible to have any peculiar signature of the model.

Considering the h_1-decay, we have analysed the role that a "light" heavy neutrino mass ($m_Z/2 < m_{\nu_h} < m_W$) could play in the BRs, in order to establish if such decaying-channel could be visible in the early discovery scenario at LHC. In particular, in figure 2 we plot the Branching ratio for $h_1 \to 2\nu_h$ (summing over the three generation of heavy neutrinos) against the light Higgs boson mass m_{h_1} for $\nu_h = 50$ GeV[2] and several values of the mixing angle α in units of $\pi/2$: $\alpha = 0.2$ (blue line), $\alpha = 0.4$ (green line), $\alpha = 0.6$ (red line), $\alpha = 0.8$ (black line).

[2] We assume that the three heavy neutrinos mass eigenstates are degenerate as this does not affect our analysis.

Considering the h_2-decay, instead, we have analysed how the presence of a "light" Z'-boson (the $m_{Z'} = 210$ GeV choice forces us to choose $g_1' < 0.03$ because of the LEP limits, see [5]) could affect the BRs, in order to establish if such decaying-channel could be visible in the full integrated luminosity scenario at LHC. Nevertheless, we have also studied the role of a light Higgs boson ($m_{h_1} = 120$ GeV) in the h_2's BRs, because of the fact that the channel $h_2 \to 2h_1$ represents a peculiar signature of the model, that is distinctive with respect to supersymmetrical Models in which the mixing in the scalar sector is forbidden.

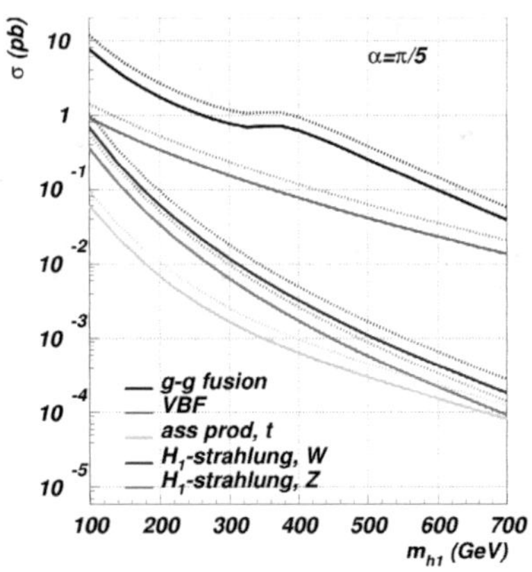

Figure 1. Cross-section for h_1 production at LHC ($\sqrt{s} = 7$ TeV) plotted against the light Higgs boson mass m_{h_1} at $\alpha = \pi/5$ in g-g fusion (black line), VBF (red line), Higgs-strahlung from $W^{\pm}$ (blue line) and Z (violet line) and associated production with top (green line). The SM-like case has been superimposed in dotted lines.

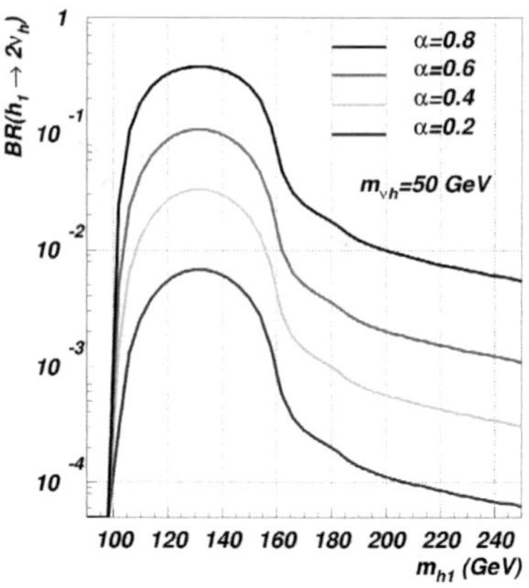

Figure 2. Branching ratio for the light Higgs boson decaying in two heavy neutrinos ($h_1 \to 2\nu_h$) in the minimal $B-L$ model plotted against the light Higgs boson mass m_{h_1} for $\nu_h = 50$ GeV and several values of the mixing angle α in units of $\pi/2$: $\alpha = 0.2$ (blue line), $\alpha = 0.4$ (green line), $\alpha = 0.6$ (red line), $\alpha = 0.8$ (black line).

Finally, by combining the two analysis (Higgses productions and BRs), we have made a detailed study of the cross-section for some peculiar signature of the model: $pp \to h_1 \to \nu_h\nu_h$, $pp \to h_2 \to h_1h_1$ and $pp \to h_2 \to Z'Z'$. The analysis of each of these processes has shown how there is the possibility to observe such signatures both in the early discovery scenario ($pp \to \nu_h\nu_h$) and in the full integrated luminosity scenario ($pp \to h_1h_1$ and $pp \to Z'Z'$).

In particular, in figure 3 we show the explicit result for the $pp \to h_1 \to \nu_h\nu_h$ process at LHC with $\sqrt{s} = 7$ TeV and $m_{\nu_h} = 50$ GeV: a cross-section contour "sliced" in the m_{h_1}-α plane. Several values of the cross-section have been considered: $\sigma = 5$ fb (black line), $\sigma = 10$ fb (red line), $\sigma = 100$ fb (green line), $\sigma = 250$ fb (blue line). The red-shadowed region is excluded by the LEP experiments.

Even if we consider a low-luminosity scenario ($L \simeq 1$ fb^{-1}), it is clear from the plot that there is a noticeable allowed parameter space for which the rate of events is considerably large: when the integrated luminosity reaches $L = 1$ fb^{-1} (that is equivalent to $18 - 24$ months of $\sqrt{s} = 7$ TeV running at LHC according to the official programme) we estimated a collection of ~ 10 heavy neutrino pair productions for 100 GeV$< m_{h_1} <$ 165 GeV and $0.05\pi < \alpha < 0.48\pi$, that scales up to $\sim 10^2$ events for 110 GeV$< m_{h_1} <$ 150 GeV and $0.15\pi < \alpha < 0.46\pi$.

This represents a clear chance to establish a heavy neutrino discovery within the next two years at the CERN machine.

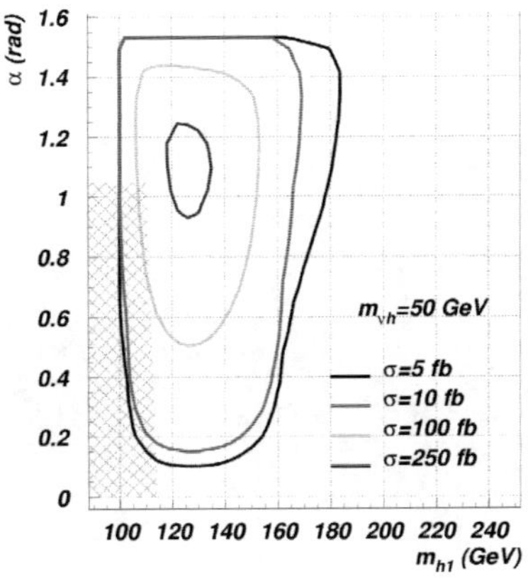

Figure 3. Cross-section contour plot for a heavy neutrino pair production in proton-proton scattering (gluon-gluon fusion) via h_1 ($pp \to h_1 \to \nu_h \nu_h$) at LHC $\sqrt{s} = 7$ TeV, plotted against the light Higgs mass m_{h_1} and the mixing angle α for $m_{\nu_h} = 50$ GeV. Several values of the cross-section have been considered: $\sigma = 5$ fb (black line), $\sigma = 10$ fb (red line), $\sigma = 100$ fb (green line), $\sigma = 250$ fb (blue line). The red-shadowed region is excluded by the LEP experiments.

References

[1] E. E. Jenkins, Phys. Lett. B **192** (1987) 219; W. Buchmuller, C. Greub and P. Minkowski, Phys. Lett. B **267** (1991) 395; S. Khalil, J. Phys. G **35** (2008) 055001.
[2] L. Basso, A. Belyaev, S. Moretti and G. M. Pruna, Phys. Rev. D **81** (2010) 095018 [arXiv:1002.1939 [hep-ph]].
[3] L. Basso, S. Moretti and G. M. Pruna, arXiv:1004.3039 [hep-ph].
[4] L. Basso, S. Moretti and G. M. Pruna, arXiv:1009.4164 [hep-ph].
[5] G. Cacciapaglia, C. Csaki, G. Marandella and A. Strumia, Phys. Rev. D **74** (2006) 033011 [arXiv:hep-ph/0604111].

Journal of Physics: Conference Series **259** (2010) 012063 doi:10.1088/1742-6596/259/1/012063

Neutrino masses and mixing in $\mu\nu$SSM

Pradipta Ghosh

Department of Theoretical Physics and Centre for Theoretical Sciences,
Indian Association for the Cultivation of Science,
2A & 2B Raja S.C. Mullick Road,
Kolkata 700 032, India

E-mail: tppg@iacs.res.in

Abstract. $\mu\nu$SSM is an R-parity violating non-minimal supersymmetric model which uses right chiral neutrino superfields to solve the μ-problem. The R-parity violation together with a TeV scale seesaw mechanism using right handed neutrinos are instrumental for the light neutrino mass generation in $\mu\nu$SSM. We show that it is possible to accommodate three flavour global neutrino data in $\mu\nu$SSM with three massive neutrinos at the tree level. Ingression of the one-loop corrections to neutrino masses and mixing shows certain variations over the tree level analysis depending on the specific hierarchy of neutrino masses involved. In $\mu\nu$SSM some of the R-parity violating decay branching ratios of the lightest neutralino show nice correlation with certain neutrino mixing angle. These correlations along with the presence of displaced vertices in the decay of the lightest neutralino can be further investigated as a test of $\mu\nu$SSM in collider experiments.

1. Introduction

Recent findings of neutrino oscillation experiments has put forward strong evidences for the massive neutrinos. A satisfactory explanation for the non-zero neutrino masses is beyond the Standard Model framework. Weak scale supersymmetry (SUSY) is a well motivated candidate for beyond the Standard Model (SM) physics and is of immense interest with the initiation of the Large Hadron Collider (LHC) era. However, SUSY has its own theoretical problems out of which a well-known name is the μ-problem [1]. Solutions to the μ-problem are addressed in a non-minimal version of SUSY model using a SM gauge singlet superfield ($\hat{S}$). On the other hand, both the minimal supersymmetric standard model (MSSM) and next-to-minimal supersymmetric standard model (NMSSM) predicts massless neutrinos similar to the SM. Massive neutrinos in a SUSY framework can be achieved in two ways, either using the R-parity violation ($\not{R}_P$) or through canonical seesaw mechanism (initially proposed to prevent fast proton decay through sparticle mediated processes). R_p is defined as $R_p = (-1)^{L+3B+2s}$ with $L(B)$ as lepton(baryon) number and s is the spin. MSSM can accommodate neutrino masses through the bilinear or trilinear $\not{R}_P$ without solving the μ-problem. However, these new bilinear $\not{R}_P$ violating terms will again cause *naturalness* problem similar to the μ-problem [2].

The "μ from ν" supersymmetric standard model ($\mu\nu$SSM) [3, 4] invokes right-handed neutrino superfields to solve the μ-problem and at the same time uses the same set of right-handed neutrino superfields to generate light neutrino masses. In a nutshell $\mu\nu$SSM is a *minimalistic extension* of MSSM (includes only right-handed neutrino superfields apart from the MSSM

Journal of Physics: Conference Series **259** (2010) 012063 doi:10.1088/1742-6596/259/1/012063

superfields) for accommodating neutrino masses and simultaneously solving the μ-problem.

The $\mu\nu$SSM can fit the three flavour global neutrino data even with flavour diagonal structure of the neutrino Yukawa couplings for various hierarchies of light neutrino masses [5] at the tree level. All three light neutrinos can acquire non-zero masses at the tree level. $R\!\!\!/_P$ in $\mu\nu$SSM will lead to an unstable Lightest Supersymmetric Particle (LSP). Some ratios of the decay branching ratios of the LSP show nice correlations with certain neutrino mixing angles [5] depending on the LSP nature and the hierarchy in the light neutrino masses. These correlations can be investigated as the experimental signatures of $\mu\nu$SSM with possible discriminating features from other SUSY models. Another important feature of $\mu\nu$SSM is the possibility of having displaced vertices in the decay of the lightest neutralino, which can vary from a few *mm* to ~ 1 *meter* depending on the nature of the lightest neutralino [5]. Tree level results of neutrino masses and mixing show variations with the addition of the one-loop radiative corrections. The amount of variations were observed to be dependent on the choice of the hierarchy in light neutrino masses [6].

We note in passing that various other aspects of $\mu\nu$SSM like LHC phenomenology, spontaneous CP-violation, gravitino dark matter, baryogenesis are discussed in refs.[7, 8, 9, 10] respectively. For a review of $\mu\nu$SSM and seesaw mechanism in $\mu\nu$SSM see refs. [11] and [12].

2. The model

The model superpotential is

$$\begin{aligned} W &= \epsilon_{ab}(Y_u^{ij}\hat{H}_2^b\hat{Q}_i^a\hat{u}_j^c + Y_d^{ij}\hat{H}_1^a\hat{Q}_i^b\hat{d}_j^c + Y_e^{ij}\hat{H}_1^a\hat{L}_i^b\hat{e}_j^c + Y_\nu^{ij}\hat{H}_2^b\hat{L}_i^a\hat{\nu}_j^c) \\ &- \epsilon_{ab}\lambda^i\hat{\nu}_i^c\hat{H}_1^a\hat{H}_2^b + \frac{1}{3}\kappa^{ijk}\hat{\nu}_i^c\hat{\nu}_j^c\hat{\nu}_k^c, \end{aligned} \tag{1}$$

where $\hat{\nu}_i^c$ are the right chiral neutrino superfields ingressed in $\mu\nu$SSM apart from the MSSM $SU(2)$ doublet $(\hat{H}_1, \hat{H}_2, \hat{Q}_i, \hat{L}_i)$ and singlet $(\hat{u}_j^c, \hat{d}_j^c, \hat{e}_j^c)$ superfields. In eq.(1) i, j, k represent generational indices. Appearance of any bilinear term is prohibited by imposing a Z_3 symmetry in the $\mu\nu$SSM superpotential. This superpotential explicitly breaks R_p through lepton number violation by odd units (5th and 6th terms of eq.(1)). There are corresponding entries in the soft SUSY breaking sector too [3].

After the electroweak symmetry breaking (EWSB) when the scalar component of Higgses and left and right sneutrino fields develope the respective Vacuum Expectation Values (VEVs) v_1, v_2, v_i', v_i^c, we have an effective μ-term and bilinear $R\!\!\!/_P$ terms (ϵ_i) as $\mu = \sum \lambda^i v_i^c$ and $\epsilon^i = \sum Y_\nu^{ij} v_j^c$, respectively. The 6th term of eq.(1), with the coefficient κ^{ijk}, is included in order to avoid an unacceptable axion associated to the breaking of a global U(1) symmetry [13]. After EWSB this term generates effective Majorana masses $(m_{\nu_{ij}^c})$ for the right-handed neutrinos at the electroweak (EW) scale and are given by $2\ \kappa^{ijk} v_k^c$. These EW scale right neutrinos are further responsible for light neutrino mass generation through seesaw, thereby lowering the seesaw-scale within the reach of a TeV scale collider.

Here, as a digression, let us mention that the spontaneous breakdown of the Z_3 symmetry through right-sneutrino VEV can in general lead to the formation of domain walls [14, 15, 16]. The associated problems can, however, be ameliorated through well-known methods [17, 18].

3. Tree level neutrino masses and mixing

The lepton number violating interactions in the superpotential and in the soft SUSY breaking part of the scalar potential allows mixing between states having zero lepton number with states having non-zero lepton number. As a consequence mass matrices in the scalar and fermion sectors get enhanced over their MSSM structures [4, 5]. In the neutral fermion sector now the four MSSM neutralinos mix with three generations of left and right-handed neutrinos and hence the neutralino mass matrix enlarges to 10×10. In the basis $\Psi^{0^T} = \left(\widetilde{B}^0, \widetilde{W}_3^0, \widetilde{H}_1^0, \widetilde{H}_2^0, \nu_i^c, \nu_i\right)$, the neutralino mass matrix is given by

$$\mathcal{M}_n = \begin{pmatrix} M_{7\times7} & m_{3\times7}^T \\ m_{3\times7} & 0_{3\times3} \end{pmatrix}. \tag{2}$$

The matrix $M_{7\times7}$ contains a 4×4 block of MSSM neutralinos as well as a 3×3 block of right-neutrinos and mixing terms between them [4, 5]. The null 3×3 block in $\mathcal{M}_n$ signifies the absence of Majorana mass terms for the left handed neutrinos. The elements of $m_{3\times7}$ contain either left handed sneutrino VEVs (v_i') or Higgs VEVs multiplied by neutrino Yukawa couplings (Y_ν^{ij}), and hence, are of much smaller magnitudes compared to the entries of $M_{7\times7}$. This feature ensures a *seesaw*-like structure of $\mathcal{M}_n$[4, 5] and the light neutrino mass matrix using seesaw is given by

$$M^{seesaw} = -m_{3\times7} M_{7\times7}^{-1} m_{3\times7}^T. \tag{3}$$

This seesaw mass matrix can be diagonalized (with m_{ν_i} as mass eigenvalues) using a unitary matrix U_{PMNS} containing neutrino mixing angles (provided that the charged lepton mass matrix is already in the diagonal form) as follows

$$U_{PMNS}^T M^{seesaw} U_{PMNS} = diag(m_{\nu_1}, m_{\nu_2}, m_{\nu_3}). \tag{4}$$

With a few simplifying assumptions [5] one can get an approximate analytical form for the entries of seesaw mass matrix (M^{seesaw}) as

$$M_{ij}^{seesaw} = \frac{2Av^c}{3\Delta} b_i b_j + \frac{1}{6\kappa v^c} a_i a_j (1 - 3\delta_{ij}), \tag{5}$$

where

$$\begin{aligned} \Delta &= \lambda^2 (v_1^2 + v_2^2)^2 + 4\lambda\kappa v^{c2} v_1 v_2 - 4M\lambda A \mu, \ \mu = 3\lambda v^c, \\ A &= (\kappa v^{c2} + \lambda v_1 v_2), \ \frac{1}{M} = \frac{g_1^2}{M_1} + \frac{g_2^2}{M_2}, \\ a_i &= Y_\nu^{ii} v_2, \ b_i = (Y_\nu^{ii} v_1 + 3\lambda v_i'), \end{aligned} \tag{6}$$

with $i, j, k = e, \mu, \tau$. We have used the fact that $v_i' \ll v_1, v_2$.

The seesaw structure in $\mu\nu$SSM can be well understood if we investigate eq.(5) in the following limits[5], (i) $v^c \to \infty$ and $v \to 0$ ($v^2 = (v_1^2 + v_2^2)$) and (ii) $M \to \infty$.

In the limit (i) we end up with

$$M_{ij}^{seesaw} \approx -\frac{v_i' v_j'}{2M}, \tag{7}$$

and in case (ii) we have

$$M_{ij}^{seesaw} \approx \frac{v_2^2}{6\kappa v^c} Y_\nu^{ii} Y_\nu^{jj} (1 - 3\delta_{ij}). \tag{8}$$

The form of eq.(7) is associated with the gaugino seesaw effect where the gauge coupling$\times$left sneutrino VEV (v_i') acts as the *Dirac mass* and the effective gaugino mass (M) plays the role of the *Majorana mass*[6]. In the gaugino seesaw scenario because of the presence of left sneutrino VEV, an effective $\Delta L = 2$ Majorana mass term for light neutrinos is generated from a pair of $\Delta L = 1$ vertex involving left-handed neutrino and neutral wino/bino ($\widetilde{B}^0/\widetilde{W}_3^0$). This is analogous to Type-III seesaw because of the association of the hypercharge zero triplet fermion

[6]. Another feature of gaugino seesaw is that it generates only one massive neutrino at the tree level. On the contrary eq.(8) is associated with the ordinary seesaw effect involving right-handed neutrinos. For this case one can get two massive neutrinos at the tree level. This is the well-known example of Type-I seesaw using singlet fermion (right-handed neutrino for $\mu\nu$SSM), where the $\Delta L = 2$ effect is coming through right chiral neutrino Majorana masses [5].

With a suitable choice of model parameters one can treat the second term of eq.(5) as a perturbation over the first term. We would like to emphasize that for most of the parameter choice $a_i \sim b_i$, hence the relative weight difference is coming from the co-efficient in front ($\frac{2Av^c}{3\Delta}$ and $\frac{1}{6\kappa v^c}$) [5]. Appearance of $b_i b_j$ in the first term of eq.(5) gives only one non-zero neutrino mass $\propto \sum b_i^2$. The other two masses emerge due to the effect of ordinary seesaw [5]. Approximate analytical expressions for the masses and mixing for normal hierarchical schemes of light neutrino masses were obtained in ref. [5] which show good agreement with the complete numerical analysis without any approximation [5]. As an illustrative example, in case of normal hierarchy the atmospheric mixing angle is given by

$$\sin^2\theta_{23} = \frac{b_\mu^2}{b_\mu^2 + b_\tau^2}, \tag{9}$$

with b_i's are given by eq.(6). It is clear from the analytical formula in eq.(9), that the maximal mixing in the atmospheric sector indicates $b_\mu = b_\tau$.

Note that using $b_i = (a_i \cot\beta + 3\lambda c_i)$ with $c_i = v_i'$ and $\tan\beta = \frac{v_2}{v_1}$, one ends up with more elucidate form of eq.(5) as [6]

$$(M_\nu^{seesaw})_{ij} = f_1 a_i a_j + f_2 c_i c_j + f_3(a_i c_j + a_j c_i), \tag{10}$$

where (using eq.(6))

$$f_1 = \frac{1}{6\kappa v^c}(1 - 3\delta_{ij}) + \frac{2Av^c\cot^2\beta}{3\Delta}, \quad f_2 = \frac{2A\lambda\mu}{\Delta}, \quad f_3 = \frac{2A\mu\cot\beta}{3\Delta}. \tag{11}$$

It is very clear from eq.(10), that the 1st and the 2nd term can contribute to only one mass eigenvalue and are $\propto \sum a_i^2$ and $\sum c_i^2$ with suitable co-efficient in front. It is the 3rd term or the mixing term which is responsible for giving masses to all three light neutrinos [6].

4. One-loop corrected neutrino masses and mixing

It is clear from the above discussion that in $\mu\nu$SSM all three light neutrinos get seesaw masses at the tree level even with the flavour diagonal choice of neutrino Yukawa couplings (Y_ν) [5] consistent with the three flavour global neutrino data [19]. With the inclusion of one-loop radiative corrections tree level results of neutrino masses and mixing receive corrections over their tree level values depending on the concerned mass hierarchy [6].

There are seven possible sources [6] of one-loop corrections to the light neutrino masses in $\mu\nu$SSM and they are listed below

(i) neutralino - neutral scalar in the loop,
(ii) neutralino - neutral pseudoscalar in the loop,
(iii) neutralino - Z-boson in the loop,
(iv) chargino - charged scalar in the loop,
(v) chargino - $W^{\pm}$-boson in the loop,
(vi) up-type quark - up-type suqark in the loop,

Journal of Physics: Conference Series **259** (2010) 012063 doi:10.1088/1742-6596/259/1/012063

(vii) down-type quark - down-type suqark in the loop.

The relevant Feynman rules are given in ref.[6]. In the absence of any fine cancellation, the dominant contribution to loop corrected neutrino masses arises from (i) and (ii) above, when right sneutrinos are in the loop. This contribution is proportional to squared mass difference between right sneutrino scalar and pseudoscalar mass eigenstates [7, 20, 21, 22].

Using the dimensional reduction ($\overline{DR}$) scheme in the 't-Hooft-Feynman gauge ($\xi = 1$) one can write down the expression for one-loop corrected neutrino mass matrix in the basis, where the tree level mass matrix is diagonalized. This one-loop corrected neutrino mass matrix can be further diagonalized to get one-loop corrected mass eigenvalues. In order to obtain the neutrino mixing matrix we rotate back to the flavour basis using neutralino mixing matrix [6]. In this way we obtain the one loop corrected neutrino mass matrix in the flavour basis and further apply the known methods of seesaw mechanism to get the one-loop corrected mass eigenvalues (m'_{ν_i}). The corresponding unitary matrix U'_{PMNS}, which diagonalizes this matrix contain the one-loop corrected mixing angles [6].

The one-loop corrected neutrino mass matrix as shown in ref. [6] looks like

$$(M^{seesaw}_{loop-corrected})_{ij} = A_1 a_i a_j + A_2 c_i c_j + A_3(a_i c_j + a_j c_i), \tag{12}$$

where A_is are functions of our model parameters and the Passarino-Veltman functions (B_0, B_1) [23, 24]. The form of the loop corrected mass matrix thus obtained is identical to the tree level one (see, eq.(10)) with different coefficients A_is in front arising because of the one-loop corrections.

The effect of one-loop corrections to neutrino masses and mixing varies with the neutrino mass hierarchy. We found it relatively easier to accommodate the normal hierarchical spectrum of light neutrino masses compared to inverted or quasi-degenerate spectrum when the effect of one-loop correction was taken into account [6]. The allowed region of tree level parameter space were observed to shrink severely in the case of inverted hierarchy of light neutrino mass and thereby only a very little window of allowed parameter space remains open with the inclusion of loop-effect [6]. The effect of one-loop correction produces practically vanishing allowed region in parameter space for the quasi-degenerate scheme. However, we must emphasize here that these conclusion are parameter dependent and the huge parameter space of $\mu\nu$SSM always left us with enough room to observe some different phenomena in a entirely different corner of parameter space.

5. Decays of the LSP

$\not{R}_P$ in $\mu\nu$SSM leaves no room for a stable LSP. The lightest SUSY particle in this model will eventually decay into the SM particles making this model testable in collider experiment. It is an well-known feature of the SUSY models with bilinear $\not{R}_P$ to show nice correlations among neutrino oscillation parameters and LSP decay patterns[25, 26, 27]. In the $\mu\nu$SSM one observes similar correlations among certain neutrino mixing angles with ratio of decay branching ratios [5, 7]. These correlations vary with the concerned neutrino mass hierarchy and with the LSP nature [5]. In $\mu\nu$SSM, apart from a *gaugino* or *higgsino* LSP one can also have a right-handed neutrino or *singlino* like LSP. This third possibility is a special feature of $\mu\nu$SSM and is of great interest because of its direct relation with the seesaw scale. A proper detection may be followed by a faithful mass reconstruction tool to probe the right-handed neutrino mass scale or the seesaw scale which is hitherto unseen for its gauge-singlet nature [5]. Decays of fermionic LSP can produce multiple leptons and jets in the final state decay products, for which a comparative analysis can be performed for obtaining signatures of $\mu\nu$SSM at colliders [7]. In $\mu\nu$SSM right-sneutrino is also an eligible candidate to be the LSP [5].

One more crucial experimental signature of $\mu\nu$SSM can come from the study of displaced vertices. Depending on the LSP composition the decay length can vary several orders of magnitude, which is $\sim$ a few *meters* for a *singlino* like LSP. As a corollary a dedicated investigation of the displaced vertices can provide characteristic signature of $\mu\nu$SSM.

Acknowledgments

I would like to thank the Council of Scientific and Industrial Research, Govt. of India for a Senior Research Fellowship. I gratefully acknowledge P. Dey, B. Mukhopadhyaya and S. Roy for their collaborative supports. I also wish to thank the organizers of the PASCOS 2010, "16th International Symposium on particles, strings and cosmology" for their kind hospitality. I would further like to thank my supervisors U. Chattopadhyay and S. Roy for providing financial assistance through their respective research grants. Finally, I am grateful to Prof. K. Bhattacharya, Director, IACS for his kind approval and for the encouragement I received from him.

References

[1] Kim J E and Nilles H P 1984 *Phys. Lett.* **B138** 150
[2] Nilles H P and Polonsky N 1997 *Nucl. Phys.* **B484** 33–62 (*Preprint* hep-ph/9606388)
[3] Lopez-Fogliani D E and Munoz C 2006 *Phys. Rev. Lett.* **97** 041801 (*Preprint* hep-ph/0508297)
[4] Escudero N, Lopez-Fogliani D E, Munoz C and de Austri R R 2008 *JHEP* **12** 099 (*Preprint* 0810.1507)
[5] Ghosh P and Roy S 2009 *JHEP* **04** 069 (*Preprint* 0812.0084)
[6] Ghosh P, Dey P, Mukhopadhyaya B and Roy S 2010 *JHEP* **05** 087 (*Preprint* 1002.2705)
[7] Bartl A, Hirsch M, Vicente A, Liebler S and Porod W 2009 *JHEP* **05** 120 (*Preprint* 0903.3596)
[8] Fidalgo J, Lopez-Fogliani D E, Munoz C and Ruiz de Austri R 2009 *JHEP* **08** 105 (*Preprint* 0904.3112)
[9] Choi K Y, Lopez-Fogliani D E, Munoz C and de Austri R R 2010 *JCAP* **1003** 028 (*Preprint* 0906.3681)
[10] Chung D J H and Long A J 2010 *Phys. Rev.* **D81** 123531 (*Preprint* 1004.0942)
[11] Munoz C 2010 *AIP Conf. Proc.* **1200** 413–416 (*Preprint* 0909.5140)
[12] Lopez-Fogliani D E 2010 (*Preprint* 1004.0884)
[13] Ellis J R, Gunion J F, Haber H E, Roszkowski L and Zwirner F 1989 *Phys. Rev.* **D39** 844
[14] Ellis J R *et al.* 1986 *Phys. Lett.* **B176** 403
[15] Rai B and Senjanovic G 1994 *Phys. Rev.* **D49** 2729–2733 (*Preprint* hep-ph/9301240)
[16] Abel S A, Sarkar S and White P L 1995 *Nucl. Phys.* **B454** 663–684 (*Preprint* hep-ph/9506359)
[17] Abel S A 1996 *Nucl. Phys.* **B480** 55–72 (*Preprint* hep-ph/9609323)
[18] Panagiotakopoulos C and Tamvakis K 1999 *Phys. Lett.* **B446** 224–227 (*Preprint* hep-ph/9809475)
[19] Schwetz T, Tortola M A and Valle J W F 2008 *New J. Phys.* **10** 113011 (*Preprint* 0808.2016)
[20] Hirsch M, Klapdor-Kleingrothaus H V and Kovalenko S G 1997 *Phys. Lett.* **B398** 311–314 (*Preprint* hep-ph/9701253)
[21] Grossman Y and Haber H E 1997 *Phys. Rev. Lett.* **78** 3438–3441 (*Preprint* hep-ph/9702421)
[22] Hirsch M, Klapdor-Kleingrothaus H V and Kovalenko S G 1998 *Phys. Rev.* **D57** 1947–1961 (*Preprint* hep-ph/9707207)
[23] Passarino G and Veltman M J G 1979 *Nucl. Phys.* **B160** 151
[24] 't Hooft G and Veltman M J G 1979 *Nucl. Phys.* **B153** 365–401
[25] Mukhopadhyaya B, Roy S and Vissani F 1998 *Phys. Lett.* **B443** 191–195 (*Preprint* hep-ph/9808265)
[26] Choi S Y, Chun E J, Kang S K and Lee J S 1999 *Phys. Rev.* **D60** 075002 (*Preprint* hep-ph/9903465)
[27] Romao J C, Diaz M A, Hirsch M, Porod W and Valle J W F 2000 *Phys. Rev.* **D61** 071703 (*Preprint* hep-ph/9907499)

Gravitino dark matter with neutralino NLSP in the constrained NMSSM

Grigoris Panotopoulos

Departamento de Fisica Teorica, University of Valencia and IFIC (CSIC-Univ. of Valencia), Av. Moliner 50, 46100 Valencia, Spain

E-mail: grigoris.panotopoulos@uv.es

Abstract. The gravitino dark matter with neutralino NLSP hypothesis is investigated in the framework of NMSSM. We have considered both the thermal and non-thermal gravitino production mechanisms, and we have taken into account all the collider and cosmological constraints. The maximum allowed reheating temperature after inflation, as well as the maximum allowed gravitino mass are determined.

1. Introduction

There is accumulated evidence both from astrophysics and cosmology that about 1/4 of the energy budget of the universe consists of so called dark matter [1], namely a component which is non-relativistic and neither feels the electromagnetic nor the strong interaction. It is known that in the standard model of particle physics there is not a suitable dark matter candidate. Supersymmetry, a very well-motivated idea, is perhaps the most popular way beyond the standard model, and it provides us with an ideal candidate for playing the role of cold dark matter in the universe.

Out of the several dark matter candidates, certainly the most well-studied case is the one of the neutralino. However, the gravitino is also a very compelling candidate for several reasons. Its interactions are completely fixed by the supergravity Lagrangian, its mass $m_{\tilde{G}}$ is directly related to the scale at which supersymmetry is broken, and finally the reheating temperature after inflation T_R, which is essential for the precise baryon asymmetry generation mechanism, plays a certain role in gravitino cosmology.

The MSSM [2] is the most economical supersymmetric extension of the standarm model. However, it suffers from the so-called μ problem [3]. The natural values for the higgs/higgsino mass parameter μ is either 0 or the Planck mass, while phenomenology requires the μ should be at the electroweak scale. The simplest supersymmetric model that solves the μ problem is the NMSSM [4], where one more singlet chiral superfield is introduced.

It is therefore interesting to try to see whether a particular cosmological scenario, with gravitino LSP and neutralino NLSP in the framework of the NMSSM, is a viable one.

2. The theoretical framework

2.1. Basics of NMSSM

The particle physics model is defined by the superpotential

$$W = \epsilon_{ij}\left(Y_u\, H_2^j\, Q^i\, u + Y_d\, H_1^i\, Q^j\, d + Y_e\, H_1^i\, L^j\, e\right) - \epsilon_{ij}\lambda\, S\, H_1^i H_2^j + \frac{1}{3}\kappa S^3 \qquad (1)$$

as well as the soft breaking masses and couplings

$$\begin{aligned} -\mathcal{L}_{\rm soft} = \quad & m_{\tilde{Q}}^2\, \tilde{Q}^*\, \tilde{Q} + m_{\tilde{U}}^2\, \tilde{u}^*\, \tilde{u} + m_{\tilde{D}}^2\, \tilde{d}^*\, \tilde{d} + m_{\tilde{L}}^2\, \tilde{L}^*\, \tilde{L} + m_{\tilde{E}}^2\, \tilde{e}^*\, \tilde{e} \\ & + m_{H_1}^2\, H_1^*\, H_1 + m_{H_2}^2\, H_2^* H_2 + m_S^2\, S^* S \\ & + \epsilon_{ij}\left(A_u\, Y_u\, H_2^j\, \tilde{Q}^i\, \tilde{u} + A_d\, Y_d\, H_1^i\, \tilde{Q}^j\, \tilde{d} + A_e\, Y_e\, H_1^i\, \tilde{L}^j\, \tilde{e} + \text{H.c.}\right) \\ & + \left(-\epsilon_{ij}\lambda\, A_\lambda S H_1^i H_2^j + \tfrac{1}{3}\kappa\, A_\kappa\, S^3 + \text{H.c.}\right) \\ & - \tfrac{1}{2}\left(M_3\, \lambda_3\, \lambda_3 + M_2\, \lambda_2\, \lambda_2 + M_1\, \lambda_1\, \lambda_1 + \text{H.c.}\right) \end{aligned} \qquad (2)$$

When the singlet acquires a vaccum expectation value, S, we obtain an effective μ parameter, $\mu_{eff} = \lambda S$. Imposing universality at the GUT scale, a small controllable number of free parameters remains, namely

$$tan\beta = v_u/v_d, m_0, A_0, m_{1/2}, \lambda, A_k$$

and the sign of the effective μ parameter can be chosen at will.

Because of the extra singlet superfield, in the NMSSM there is a larger higgs sector and a larger neutralino sector. The neutralino mass matrix is characterized by the appearence of a fifth neutralino state, meaning that the composition of the lightest neutralino has an extra singlino contribution

$$\tilde{\chi}_1^0 = N_{11}\tilde{B}^0 + N_{12}\tilde{W}_3^0 + N_{13}\tilde{H}_1^0 + N_{14}\tilde{H}_2^0 + N_{15}\tilde{S} \qquad (3)$$

In the following, neutralinos with $N_{11}^2 > 0.9$, or $N_{15}^2 > 0.9$, will be referred to as bino- or singlino-like, respectively.

Furthermore, in the Higgs sector we have now two CP-odd neutral, and three CP-even neutral Higgses. We make the assumption that there is no CP-violation in the Higgs sector, and therefore the CP-even and CP-odd states do not mix. We are not interested in the CP-odd states, while the CP-even Higgs interaction and physical eigenstates are related by the transformation

$$h_a^0 = S_{ab} H_b^0 \qquad (4)$$

where S is the unitary matrix that diagonalises the CP-even symmetric mass matrix, $a, b = 1, 2, 3$, and the physical eigenstates are ordered as $m_{h_1^0} < m_{h_2^0} < m_{h_3^0}$.

2.2. Production of gravitinos

After inflation gravitinos can be produced in two ways. One way to produce gravitinos is with scatterings from the thermal bath, and another is from the out-of-equillibrium decays of the NLSP, which decouple from the thermal bath before primordial Big-Bang Nucleosynthesis and decay after the BBN time. Thus, imposing the WMAP bounds [5] we can write for the gravitino abundance

$$0.1097 < \Omega_{3/2}h^2 = \Omega_{3/2}^{TP}h^2 + Y_{3/2}^{NLSP}h^2 < 0.1165 \qquad (5)$$

where

$$\Omega_{3/2}^{NLSP}h^2 = \frac{m_{\tilde{G}}}{m_{NLSP}}\, \Omega_{NLSP}h^2 \qquad (6)$$

Journal of Physics: Conference Series **259** (2010) 012064 doi:10.1088/1742-6596/259/1/012064

with m_{NLSP} the mass of the NLSP, and $\Omega_{NLSP}h^2$ the abundance the NLSP would have, had it not decayed into the gravitino. The thermal contribution is given by (approximately for a light gravitino, $m_{\tilde{G}} \ll m_{\tilde{g}}$) [6]

$$\Omega_{3/2}^{TP} \simeq 0.27 \left(\frac{T_R}{10^{10}\ GeV}\right)\left(\frac{m_{\tilde{g}}}{TeV}\right)^{-2}\left(\frac{m_{\tilde{G}}}{100\ GeV}\right) \tag{7}$$

2.3. Neutralino life-time

The NSLP is unstable with a lifetime that is typically larger than BBN time $t_{BBN} \sim 1$ sec. Energetic particles produced by the NLSP decay may dissociate the background nuclei and significantly affect the primordial abundances of light elements. If such processes occur with sizable rates, the predictions of the standard BBN scenario would be altered and the success of the primordial nucleosynthesis would be spoiled. BBN constraints on cosmological scenarios with exotic long-lived particles predicted by physics beyond the Standard Model have been studied, and here we have used the figure 2 of Ref. [7]. Thus, we need to compute the lifetime of the neutralino. There are three main decay channels, namely neutralino decaying into gravitino and a standard model particle that can be one of the followings: Photon, Z boson and Higgs. The supergravity Lagrangian is known, and the Feynman rules for the gravitino interactions have been derived [8]. Therefore, the neutralino lifetime is given by

$$\tau = \frac{1}{\Gamma} \tag{8}$$

$$\Gamma = \Gamma(\chi \rightarrow \gamma\tilde{G}) + \Gamma(\chi \rightarrow Z\tilde{G}) + \Gamma(\chi \rightarrow h\tilde{G}) \tag{9}$$

and the exact expressions for the diferent contributions can be found in [9].

3. Constraints and results

We have used NMSSMTools [10], a computer software that computes the masses of the Higgses and the superpartners, the couplings, and the relic density of the neutralino, for a given set of the free parameters. We have performed a random scan over the whole parameter space (with fixed $\mu > 0$ motivated by the muon anomalous magnetic moment), and we have selected only those points that satisfy i) theoretical requirements, such as neutralino LSP, correct electroweak symmetry breaking, absence of tachyonic masses etc., and ii) LEP bounds on the Higgs mass, collider bounds on SUSY particle masses, and experimental data from B-physics. Finally, for any given point in the cNMSSM parameter space, the neutralino lifetime is a function of the gravitino mass only. Imposing the BBN constraints we find the maximum allowed gravitino mass, and from the cold dark matter bound we can determine the maximum allowed reheating temperature. For all the acceptable points the lightest neutralino is either a bino or a singlino, and our main results are summarized in the figures below.

For the bino case, figure 1 shows the maximum allowed reheating temperature after inflation versus the maximum allowed gravitino mass, both in GeV. Although it cannot be seen directly from the figures, the maximum possible gravitino mass in the bino case is $m_{\tilde{G}} \simeq 1$ GeV, and the corresponding reheating temperature is $T_R \sim 10^7$ GeV. Therefore, we see that a) the gravitino in this scenario must be much lighter than the rest of superpatners, but nevertheless it is still in the gravity-mediated SUSY-breaking scheme, and b) the reheating temperature after inflation is not large enough for thermal leptogenesis.

For the singlino case, we show in figure 2 the maximum allowed reheating temperature versus gravitino mass, both in GeV. This time the neutralino relic density is even larger than before, and gravitino now must be extremely light. This is due to the smallness of the coefficients N_{11}, N_{12} in the decay rate to gravitino and photon. For the same lifetime as before, the gravitino mass

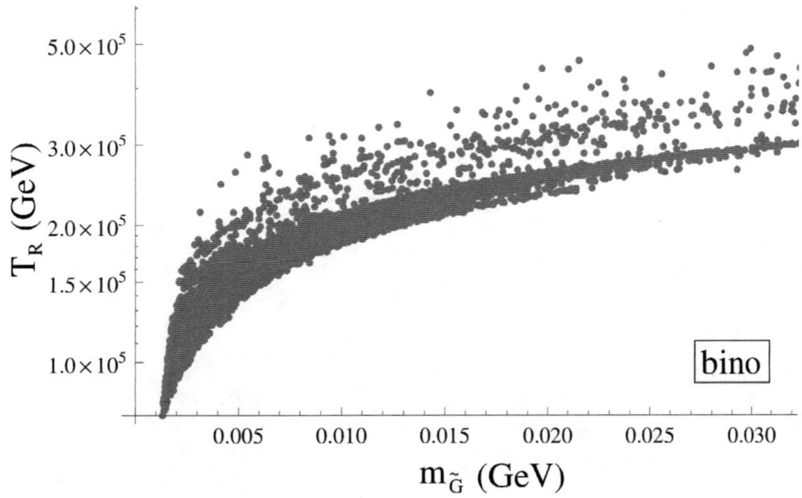

Figure 1. Reheating temperature versus gravitino mass (both in GeV) for the bino case.

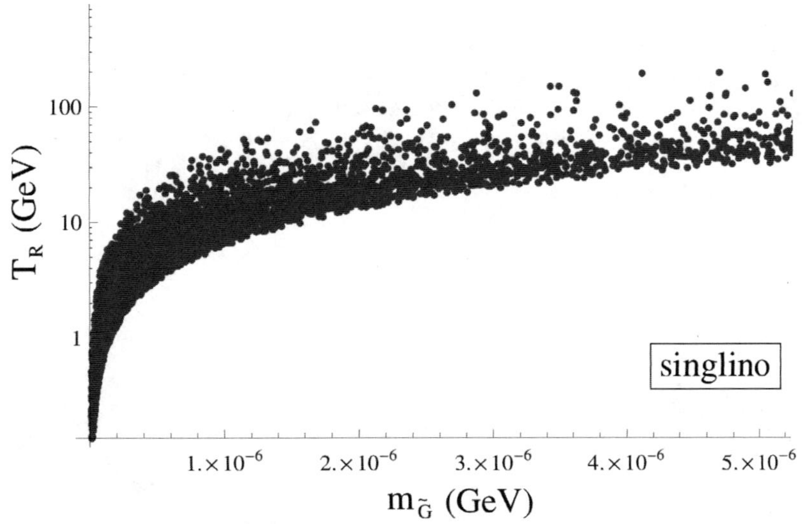

Figure 2. Reheating temperature versus gravitino mass (both in GeV) for the singlino case.

must be several orders of magnitude lower than in the bino case. The last figure shows that in the singlino case the reheating temperature cannot be larger than about 200 GeV. However, this value is much lower than the minimum value required for the computation of the gravitino thermal production, and therefore we conclude that this scenario must be excluded.

4. Conclusion

In the framework of the cNMSSM, we have considered a possible cosmological scenario with the gravitino LSP and the neutralino NLSP. The gravitino is stable and plays the role of cold dark matter in the universe, while the neutralino is unstable and decays to gravitino. We have taken into account the relevant gravitino production mechanisms, which are i) the NLSP decay, and ii) scattering processes from the thermal bath. Our results can be seen in the figures. We have found that i) the gravitino is necessarily very light, and ii) the reheating temperature after inflation is two orders of magnitude lower than the temperature required for thermal leptogenesis. The singlino scenario must be excluded, while in the bino case it is hardly possible to have a gravitino in the gravity-mediated SUSY-breaking scheme.

Acknowledgments

We acknowledge financial support from FPA2008-02878, and Generalitat Valenciana under the grant PROMETEO/2008/004.

References

[1] K. A. Olive, arXiv:astro-ph/0301505;
C. Munoz, Int. J. Mod. Phys. A **19** (2004) 3093 [arXiv:hep-ph/0309346].
[2] P. Fayet, "Spontaneously Broken Supersymmetric Theories Of Weak, Electromagnetic And Strong Interactions," Phys. Lett. B **69**, 489 (1977);
S. Dimopoulos and H. Georgi, "Softly Broken Supersymmetry And SU(5)," Nucl. Phys. B **193** (1981) 150.
[3] J. E. Kim and H. P. Nilles, Phys. Lett. B **138** (1984) 150.
[4] H. P. Nilles, M. Srednicki and D. Wyler, Phys. Lett. B **120** (1983) 346;
J. R. Ellis, J. F. Gunion, H. E. Haber, L. Roszkowski and F. Zwirner, Phys. Rev. D **39** (1989) 844;
M. Drees, Int. J. Mod. Phys. A **4** (1989) 3635.
[5] E. Komatsu *et al.* [WMAP Collaboration], Astrophys. J. Suppl. **180** (2009) 330 [arXiv:0803.0547 [astro-ph]].
[6] M. Bolz, A. Brandenburg and W. Buchmuller, Nucl. Phys. B **606** (2001) 518 [arXiv:hep-ph/0012052].
[7] M. Kawasaki, K. Kohri and T. Moroi, Phys. Lett. B **625** (2005) 7 [arXiv:astro-ph/0402490].
[8] T. Moroi, arXiv:hep-ph/9503210.
[9] G. Barenboim and G. Panotopoulos, JHEP **1009** (2010) 011 [arXiv:1004.4525 [hep-ph]].
[10] U. Ellwanger, J. F. Gunion and C. Hugonie, JHEP **0502** (2005) 066 [arXiv:hep-ph/0406215];
U. Ellwanger and C. Hugonie, Comput. Phys. Commun. **175** (2006) 290 [arXiv:hep-ph/0508022];
G. Belanger, F. Boudjema, C. Hugonie, A. Pukhov and A. Semenov, JCAP **0509** (2005) 001 [arXiv:hep-ph/0505142].

16th International Symposium on Particles Strings and Cosmology (PASCOS 2010) IOP Publishing
Journal of Physics: Conference Series **259** (2010) 012065 doi:10.1088/1742-6596/259/1/012065

Lepton flavor violation in SUSY left-right symmetric theories

A. Vicente
AHEP Group, Institut de Física Corpuscular – C.S.I.C. & Universitat de València
Edificio Institutos de Paterna, Apt 22085, E–46071 Valencia, Spain

E-mail: Avelino.Vicente@ific.uv.es

Abstract. The seesaw mechanism is the most popular explanation for the smallness of neutrino masses. However, its high scale makes direct tests impossible and only indirect signals at low energies are reachable for collider experiments. One of these indirect links with the high scale is lepton flavor violation (LFV). We discuss LFV decays of sleptons in the context of a SUSY left-right symmetric model that naturally incorporates the seesaw mechanism. This non-minimal embedding of the seesaw leads to observable LFV effects in the right-handed sleptons sector, contrary to minimal models where these are found to be totally negligible. Therefore, LFV observables can be used as a powerful tool to study physics right below the GUT scale.

1. Introduction

Neutrino physics has been at the forefront of particle phenomenology in the last years. After the discovery of neutrino oscillations, experiments have entered the precision era [1, 2, 3], measuring the relevant parameters with increasing accuracy [4] and highlighting the need for an explanation of neutrino masses.

The seesaw mechanism is the most popular framework to accommodate neutrino masses [5, 6, 7, 8, 9, 10]. The existence of very heavy fields, right-handed neutrinos in the most common variation, naturally explains the smallness of neutrino masses due to the famous seesaw relation, $m_\nu \sim v_{EW}^2/M_{heavy}$. However, the structure of the seesaw makes direct tests impossible. The new heavy fields cannot be produced at colliders and therefore at best only indirect tests will be possible.

In addition to the question of neutrino masses, the Standard Model has other theoretical problems that need to be addressed. Several extensions have been built with this purpose, filling the literature with a wide variety of ideas that will be put to experimental test in the coming years. Among the different choices, Supersymmetry (SUSY) is the most popular one because it technically solves the hierarchy problem [11] and it has the capability to address many of the other open questions, like the nature of dark matter, gauge coupling unification and radiative symmetry breaking, to mention a few.

One of the key points in supersymmetric model building is R-parity violation/conservation [12, 13]. This discrete symmetry, defined as $R_p = (-1)^{3(B-L)+2s}$ (where B and L stand for baryon and lepton numbers and s for the spin of the particle), is usually imposed to forbid the dangerous baryon and lepton number violating interactions, never seen in nature. If both were simultaneously present, proton decay would be extremely fast, a phenomenological disaster that

Journal of Physics: Conference Series **259** (2010) 012065 doi:10.1088/1742-6596/259/1/012065

is prevented by forbidding the R-parity violating ($\not{R}_p$) couplings. One of the consequences of this new symmetry is the existence of a stable particle, a natural dark matter candidate, and thus its practical importance is beyond any doubt. However, its fundamental origin is totally unknown and in most cases it is introduced by hand in the theory. Compared to the standard model, where baryon and lepton numbers are automatically conserved, this is a step back.

A possible approach to understand the origin of R-parity is to embed the MSSM in an extended model with a gauge group containing a $U(1)_{B-L}$ piece. The original high-energy theory will conserve R-parity due to the gauge symmetry and R-parity will become a remnant subgroup at low energies if the scalar fields responsible for the breaking have even $B-L$ charges.

This picture can be realized in minimal $U(1)_{B-L}$ extensions of the MSSM, see for example [14], or, more ambitiously, in Left-Right (LR) symmetric models [15]. Apart from the conservation of R-parity this type of models have other motivations. The original motivation was the restoration of parity as an exact symmetry at higher energies [15]. In addition, it has been shown that they provide technical solutions to the SUSY CP and strong CP problems [16], they give an understanding of the $U(1)$ charges and they can be embedded in $SO(10)$ Grand Unified Theories (GUTs).

Here we study some phenomenological aspects of a supersymmetric LR model that leads to R-parity conservation at low energies and incorporates a type-I seesaw mechanism to generate neutrino masses. We will concentrate the discussion on lepton flavor violating signals in slepton decays, such as $\tilde{l}_i \to \tilde{\chi}_1^0 l_j$ with $i \neq j$. These signatures have been known to be important in SUSY models for many years [17, 18, 19], and in fact they have been already studied in great detail for minimal seesaw implementations, see for example [20, 21, 22]. This work, however, studies the case of a non-minimal seesaw, what implies new features. In particular, the high-energy restoration of parity enhances the flavor violating effects in the right-handed slepton sector, contrary to the usual expectation. This way, by measuring braching ratios of LFV decays at colliders one can get valuable information on the structure of the high-energy theory.

2. The model

2.1. How to break the LR symmetry

There are many supersymmetric LR models in the literature. From the pioneering works in the 70s, many extensions and variations have been proposed. In all of them the left-right gauge group $SU(3)_c \times SU(2)_L \times SU(2)_R \times U(1)_{B-L}$ breaks down to the SM gauge group $SU(3)_c \times SU(2)_L \times U(1)_Y$. However, one can choose different representations in the scalar sector, responsible for the breaking of the symmetry, and obtain very different low energy effective theories. In addition, there are several ingredients that cannot be forgotten if one wants to have a consistent framework. Therefore, four requirements will be imposed as guidelines for the choice of the model: (a) Automatic conservation of R-parity, (b) Parity conservation at high energies, (c) Seesaw mechanism, and (d) Cancellation of anomalies.

The first LR models used $SU(2)_R$ doublets to break the gauge symmetry. The non-supersymmetric model proposed in references [15] introduced two additional scalar doublets χ_L and χ_R, where $\chi_L \equiv \chi_L(1,2,1,1)$ and $\chi_R \equiv \chi_R(1,1,2,-1)$ under $SU(3)_c \times SU(2)_L \times SU(2)_R \times U(1)_{B-L}$[1]. When the neutral component of χ_R gets a vacuum expectation value (VEV), $\langle \chi_R^0 \rangle \neq 0$, the gauge symmetry is broken down to the SM gauge group and the known low energy phenomenology with broken parity is recovered.

However, such models are not suited for the purpose of this study. The reason comes from the oddness of χ_R under $U(1)_{B-L}$. In a supersymmetric version of the model, the breaking of

[1] Note the duplication in the number of fields. This comes from parity conservation, that implies the same number of $SU(2)_L$ and $SU(2)_R$ charged fields, and anomaly cancellation, that implies that for every group representation with charge $+Q$ under $U(1)_{B-L}$ there must be another one with charge $-Q$. This will be also found in the models discussed below.

the gauge symmetry by $\langle \chi_R^0 \rangle$ also implies the breaking of R-parity. This could be solved by imposing additional discrete symmetries to the model that forbid the dangerous $\not{R}_p$ operators [23], but this cannot be regarded as automatic R-parity conservation. In addition, there is no seesaw mechanism to generate neutrino masses, and additional superfields would be needed to account for them.

The simplest solution is to break the gauge symmetry by $SU(2)_R$ fields with even charge under $U(1)_{B-L}$. This was in fact proposed in reference [24], where four triplets were added to the MSSM spectrum: $\Delta(1,3,1,2)$, $\Delta^c(1,1,3,-2)$, $\bar{\Delta}(1,3,1,-2)$ and $\bar{\Delta}^c(1,1,3,2)$. In these models, sometimes called MSUSYLR (Minimal Supersymmetric Left-Right), $SU(2)_R \times U(1)_{B-L}$ is broken down to $U(1)_Y$ by the VEVs of the scalar components of Δ^c and $\bar{\Delta}^c$. When this occurs, a right-handed neutrino mass is generated from the operator $L^c \Delta^c L^c$, leading to a type-I seesaw mechanism. However, the issue of R-parity conservation is not clear. Although one would naively expect that R-parity is automatically conserved due to the even $B-L$ charges of Δ^c and $\bar{\Delta}^c$, the scalar potential of the theory might not allow to have vanishing sneutrino VEVs [25], favoring the existence of R-parity breaking minima. For many years, MSUSYLR and its minimal extensions were considered to break R-parity. Nevertheless, the authors of the recent reference [26] claimed that 1-loop corrections change the picture, allowing for vanishing sneutrino VEVs in the minimum of the potential. Since this is still a controversial issue that relies on not fully calculated loop corrections, we leave this possibility for future studies.

Finally, Aulakh and collaborators [27, 28] extended MSUSYLR by the addition of two triplets, $\Omega(1,3,1,0)$ and $\Omega^c(1,1,3,0)$. They showed that the scalar potential offers the new possibility, within large regions of parameter space, of having minima that conserve R-parity while breaking the gauge symmetry in the proper way. Moreover, since the Δ triplets are part of the spectrum, the seesaw mechanism is present like in MSUSYLR, generating small masses for the light neutrinos. In conclusion, this model fulfills the requirements that we imposed, and thus we will concentrate on it in the following.

2.2. Model basics

The matter content of the model is [27, 28]

Superfield	generations	$SU(3)_c$	$SU(2)_L$	$SU(2)_R$	$U(1)_{B-L}$
Q	3	3	2	1	$\frac{1}{3}$
Q^c	3	$\bar{3}$	1	2	$-\frac{1}{3}$
L	3	1	2	1	-1
L^c	3	1	1	2	1
Φ	2	1	2	2	0
Δ	1	1	3	1	2
$\bar{\Delta}$	1	1	3	1	-2
Δ^c	1	1	1	3	-2
$\bar{\Delta}^c$	1	1	1	3	2
Ω	1	1	3	1	0
Ω^c	1	1	1	3	0

Here Q, Q^c, L and L^c contain the quark and lepton superfields of the MSSM with the addition of a right-handed neutrino ν^c. The two Φ superfields are $SU(2)_L \times SU(2)_R$ bidoublets and contain the usual H_d and H_u MSSM Higgs doublets. Finally, the rest of the superfields are introduced to break the LR symmetry. With these representations, the most general superpotential compatible with the gauge symmetry and parity is

$$\begin{aligned} \mathcal{W} &= Y_Q Q\Phi Q^c + Y_L L\Phi L^c - \frac{\mu}{2}\Phi\Phi + f L\Delta L + f^* L^c\Delta^c L^c \\ &+ a\Delta\Omega\bar{\Delta} + a^*\Delta^c\Omega^c\bar{\Delta}^c + \alpha\Omega\Phi\Phi + \alpha^*\Omega^c\Phi\Phi \\ &+ M_\Delta\Delta\bar{\Delta} + M_\Delta^*\Delta^c\bar{\Delta}^c + M_\Omega\Omega\Omega + M_\Omega^*\Omega^c\Omega^c \end{aligned} \tag{1}$$

Family and gauge indices have been omitted in equation (1). Y_Q and Y_L are the usual quark and lepton Yukawa couplings. However, note that $Y_Q Q\Phi Q^c \equiv Y_Q^\alpha Q\Phi_\alpha Q^c$ and $Y_L L\Phi L^c \equiv Y_L^\alpha L\Phi_\alpha L^c$, with $\alpha = 1, 2$, and thus there are four 3×3 Yukawa matrices. Conservation of parity implies that they must be symmetric. f is a 3×3 complex symmetric matrix, whereas α is a 2×2 antisymmetric matrix, and thus it only contains one complex parameter, α_{12}.

The breaking of the LR gauge group to the MSSM gauge group happens in two steps.

$$SU(2)_R \times U(1)_{B-L} \longrightarrow U(1)_R \times U(1)_{B-L} \longrightarrow U(1)_Y$$

The first step is due to the VEV of the $\Omega^{c\,0}$ field $\langle\Omega^{c\,0}\rangle = \frac{v_R}{\sqrt{2}}$, which breaks $SU(2)_R$. However, note that, since $T_{3R}(\Omega^{c\,0}) = 0$ there is a $U(1)_R$ symmetry left over. Next, the group $U(1)_R \times U(1)_{B-L}$ is broken by $\langle\Delta^{c\,0}\rangle = \frac{v_{BL}}{\sqrt{2}}$ and $\langle\bar{\Delta}^{c\,0}\rangle = \frac{\bar{v}_{BL}}{\sqrt{2}}$. The remaining symmetry is $U(1)_Y$, with hypercharge defined as $Y = I_{3R} + \frac{B-L}{2}$.

Note that, since the tadpole equations do not link Ω^c, Δ^c and $\bar{\Delta}^c$ with their left-handed counterparts, the left-handed triplets can be taken to have vanishing vevs [27].

Although a hierarchy between the two breaking scales may exist, $v_{BL} \ll v_R$, one cannot neglect the effects of the second breaking stage on the first one. The tadpole equations mix them, and only through the contribution of the $\Delta^c - \bar{\Delta}^c$ fields one can understand a non-vanishing v_R VEV. In fact, there is an inverse hierarchy between the VEVs and the superpotential masses M_Δ, M_Ω, given by

$$v_R = \frac{2M_\Delta}{a} \qquad v_{BL} = \frac{2}{a}(2M_\Delta M_\Omega)^{1/2} \tag{2}$$

And so, for $v_{BL} \ll v_R$ one needs $M_\Delta \gg M_\Omega$ [27].

3. Slepton decays and LFV

Lepton flavor violation is a well known indirect test of the seesaw mechanism [17, 18, 19]. Assuming flavor-blind soft SUSY breaking terms at some high-energy scale, the RGE running down to the SUSY scale generates non-zero off-diagonal entries in the slepton soft squared masses. These flavor violating entries are connected to the effective neutrino mass matrix and thus, by making some assumptions, one can find testable relations. Moreover, they induce LFV decays, such as $l_i \to l_j\gamma$ and $\tilde{l}_i \to \tilde{\chi}_1^0\, l_j$ with $i \neq j$. By studying these decays one can set important constraints and get valuable information on the underlying theory.

In all cases studied in the literature, based on minimal seesaw models, the off-diagonal entries of the soft masses of right-handed sleptons get negligible contributions from the RGE running and thus one expects no visible signal of LFV from the right-handed sector at the LHC. However, in LR models the gauge symmetry makes left- and right-handed sectors behave the same, and then the LFV violating entries of the soft squared masses in the right-handed sector must contain non-negligible contributions. This novel LHC signal would point to a high-energy LR symmetry in a very clean way.

That is the main result of this work. As we will show below, LFV in the right-handed lepton/slepton sector can be observable in this model. In order to prove it we perform a numerical calculation using the code SPheno [29], including 2-loop RGEs and the corresponding 1-loop threshold corrections at the intermediate scales. The Yukawa parameters Y_L are fixed in

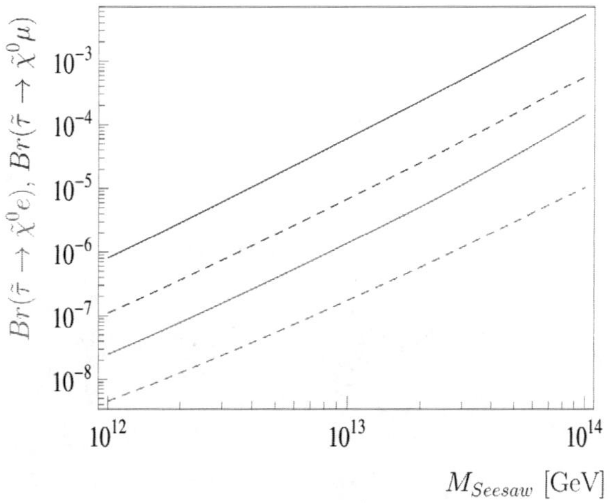

Figure 1. $Br(\tilde{\tau}_i \to \tilde{\chi}_1^0\, e)$ and $Br(\tilde{\tau}_i \to \tilde{\chi}_1^0\, \mu)$ as a function of the seesaw scale, defined as $M_{Seesaw} \equiv f v_{BL}$, for the parameter choice $v_{BL} = 10^{15}$ GeV and $v_R = 5 \cdot 10^{15}$ GeV. The dashed lines correspond to $\tau_1 \simeq \tau_R$, whereas the solid ones correspond to $\tau_2 \simeq \tau_L$. The mSUGRA parameters have been taken as in the SPS3 benchmark point [31] and the right-handed neutrino spectrum has been assumed to be degenerate, $M_{Ri} = M_{Seesaw}$.

order to correctly reproduce neutrino oscillation data. Finally, all analytical computations have been done with the help of the Mathematica package Sarah [30].

Figure 1 shows $Br(\tilde{\tau}_i \to \tilde{\chi}_1^0\, e)$ and $Br(\tilde{\tau}_i \to \tilde{\chi}_1^0\, \mu)$ as a function of the seesaw scale, defined as $M_{Seesaw} \equiv f v_{BL}$. The dependence on the seesaw scale is clearly understood from the seesaw formula. This implies that larger M_{Seesaw} requires larger Yukawa parameters in order to fit neutrino masses which, in turn, leads to larger flavor violating terms due to RGE running.

Furthermore, figure 1 also shows that right-handed staus can also have LFV decays with observable rates. This is the main novelty in this model. One can see that for large M_{Seesaw} values, around $10^{13} - 10^{14}$ GeV, the rates for LFV are measurable for both left- and right-handed staus. See references [32, 33] for the LHC discovery potential in the search for LFV.

The previous result can be understood by using analytical approximations for the slepton soft squared masses. The running from the GUT scale to the SUSY scale generates off-diagonal entries Δm^2 in both left- and right-handed slepton soft masses. In the first step, from the GUT scale to the v_R scale they can be written in leading-log approximation as [34]

$$\Delta m_L^2 = -\frac{1}{4\pi^2}\left(3ff^\dagger + Y_L^{(k)} Y_L^{(k)\dagger}\right)(3m_0^2 + A_0^2)\ln\left(\frac{m_{GUT}}{v_R}\right) \quad (3)$$

$$\Delta m_{L^c}^2 = -\frac{1}{4\pi^2}\left(3f^\dagger f + Y_L^{(k)\dagger} Y_L^{(k)}\right)(3m_0^2 + A_0^2)\ln\left(\frac{m_{GUT}}{v_R}\right) \quad (4)$$

After parity breaking at v_R the Yukawa coupling Y_L splits into Y_e, the charged lepton Yukawa, and Y_ν, the neutrino Yukawa. The later contributes to LFV entries in the running down to the v_{BL} scale

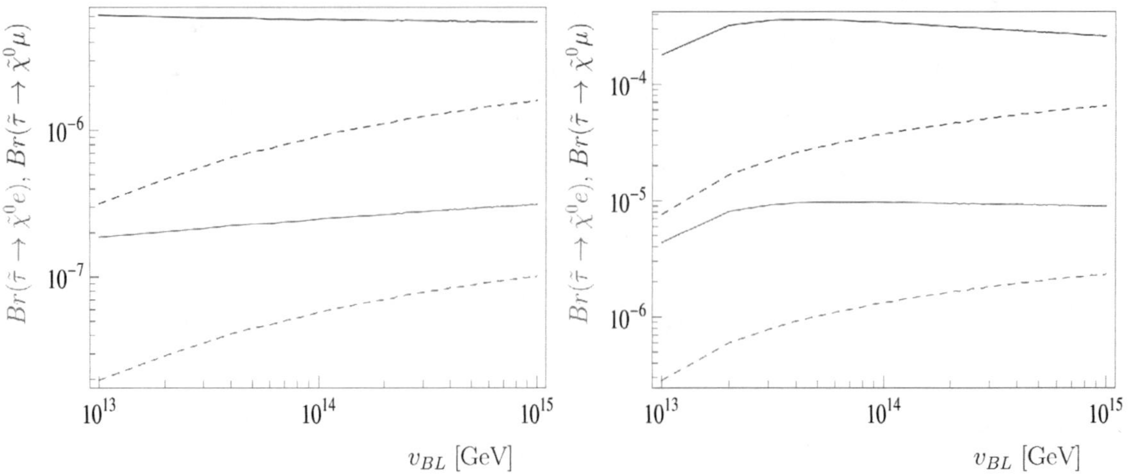

Figure 2. $Br(\tilde{\tau}_i \to \tilde{\chi}_1^0 e)$ and $Br(\tilde{\tau}_i \to \tilde{\chi}_1^0 \mu)$ as a function of v_{BL}, for a fixed $v_R = 3 \cdot 10^{15}$ GeV. To the left, $M_{Seesaw} = 10^{12}$ GeV, whereas to the right $M_{Seesaw} = 10^{13}$ GeV. The dashed lines correspond to $\tau_1 \simeq \tau_R$, whereas the solid ones correspond to $\tau_2 \simeq \tau_L$. The mSUGRA parameters have been taken as in the SPS3 benchmark point [31] and the right-handed neutrino spectrum has been assumed to be degenerate, $M_{Ri} = M_{Seesaw}$.

$$\Delta m_L^2 = -\frac{1}{8\pi^2} Y_\nu Y_\nu^\dagger (3m_0^2 + A_0^2) \ln\left(\frac{v_R}{v_{BL}}\right) \quad (5)$$

$$\Delta m_{\tilde{e}^c}^2 = 0 \quad (6)$$

Finally, from v_{BL} to the SUSY scale one recovers the MSSM RGEs, which do not add any flavor violating effect.

This short discussion shows an important consequence of the symmetry breaking pattern. From the GUT scale to the v_R scale parity is conserved and the magnitude of the LFV entries in the left- and right-handed sectors is the same, see eqs. (3) and (4). However, from v_R to v_{BL} only the left-handed ones keep running, and thus one expects larger flavor violation in this sector. Moreover, if the difference between v_R and v_{BL} is increased, the difference between the LFV entries in the L and R sectors gets increased as well.

This is shown in figure 2, which shows branching ratios for the LFV decays of the staus as a function of v_{BL} for a fixed value of $v_R = 3 \cdot 10^{15}$ GeV. The theoretical expectation is obtained: the difference between $Br(\tilde{\tau}_L)$ and $Br(\tilde{\tau}_R)$ strongly depends on the difference between v_R and v_{BL}.

The question arises whether one can determine the ratio v_{BL}/v_R by measuring both $Br(\tilde{\tau}_L)$ and $Br(\tilde{\tau}_R)$ at the LHC. This is answered in figure 3 where the ratio $Br(\tilde{\tau}_R \to \tilde{\chi}_1^0 \mu)/Br(\tilde{\tau}_L \to \tilde{\chi}_1^0 \mu)$ is plotted as a function of v_{BL}/v_R. A measurement of both branching ratios would allow to constrain the ratio v_{BL}/v_R. However, there is a slight dependence on other important quantities, such as m_{GUT} and M_{Seesaw}. This implies that more experimental information will be needed in order to set reliable constraints.

4. Summary and conclusions

Neutrino masses and R-parity conservation are two issues not addressed in the MSSM. On the one hand, the MSSM does not provide an explanation for the observation of neutrino

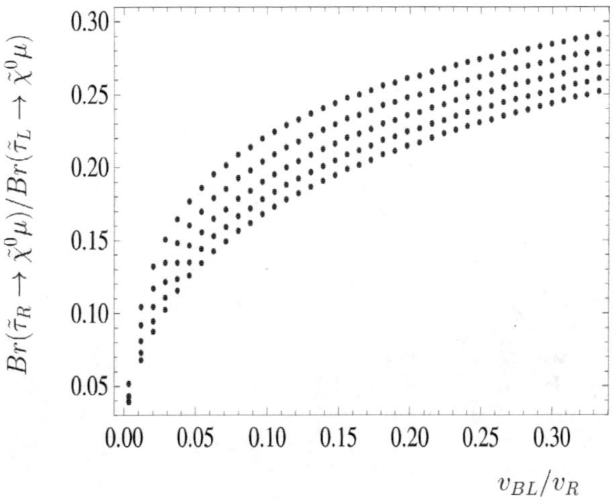

Figure 3. $Br(\tilde{\tau}_R \to \tilde{\chi}_1^0\,\mu)/Br(\tilde{\tau}_L \to \tilde{\chi}_1^0\,\mu)$ as a function of v_{BL}/v_R. The seesaw scale M_{Seesaw} takes values in the range $[10^{12}, 10^{13}]$ GeV. The rest of the parameters have been chosen as in figure 2.

oscillations and the subsequent non-zero neutrino masses. These experimental results require the introduction of a mechanism that can explain the smallness of the neutrino masses, being the seesaw mechanism the most popular choice. On the other hand, R-parity is introduced in the MSSM as an ad-hoc symmetry, without any theoretical motivation. It is therefore interesting to study extended symmetry groups that can lead to R-parity conservation at low energies.

In this work we have studied some phenomenological aspects of a supersymmetric Left-Right model which automatically conserves R-parity and contains the seesaw mechanism to generate neutrino masses. We have found that, contrary to minimal realizations of the seesaw, large lepton flavor violating effects are obtained both in the left- and right-handed slepton sectors. This is a useful signature that allows us to get additional information on the high energy regime and clearly points to an underlying left-right symmetry.

In particular, we have shown that observables like $Br(\tilde{\tau}_R \to \tilde{\chi}_1^0\,l)$ can get strong deviations from the standard seesaw picture, allowing us to constrain the parameters of the high energy theory and get a hint on its structure. Furthermore, there are other observables which are also very sensitive to new right-handed flavor violation. Examples are slepton mass splittings and the polarization of the outgoing electrons in $\mu \to e\gamma$. We plan to address these issues in a future publication [35].

Acknowledgments
This talk was based on work in collaboration with J. Esteves, M. Hirsch, W. Porod, J. C. Romao and F. Staub, and is supported by the Spanish MICINN under grants FPA2008-00319/FPA, FPA2008-04002-E and MULTIDARK Consolider CAD2009-00064, by Prometeo/2009/091 and by the EU grant UNILHC PITN-GA-2009-237920. A.V. thanks the Generalitat Valenciana for financial support.

References

[1] Super-Kamiokande collaboration, Y. Fukuda *et al.*, Phys. Rev. Lett. **81**, 1562 (1998); Q. R. Ahmad *et al.* [SNO Collaboration], Phys. Rev. Lett. **89**, 011301 (2002); K. Eguchi *et al.* [KamLAND Collaboration], Phys. Rev. Lett. **90**, 021802 (2003).

[2] [MINOS Collaboration], arXiv:0708.1495 [hep-ex].
[3] S. Abe *et al.* [KamLAND Collaboration], Phys. Rev. Lett. **100** (2008) 221803 [arXiv:0801.4589 [hep-ex]].
[4] T. Schwetz, M. A. Tortola and J. W. F. Valle, New J. Phys. **10** (2008) 113011 [arXiv:0808.2016 [hep-ph]].
[5] P. Minkowski, Phys. Lett. B **67** (1977) 421.
[6] T. Yanagida, in *KEK lectures*, ed. O. Sawada and A. Sugamoto, KEK, 1979; M Gell-Mann, P Ramond, R. Slansky, in *Supergravity*, ed. P. van Niewenhuizen and D. Freedman (North Holland, 1979);
[7] R.N. Mohapatra and G. Senjanovic, *Phys. Rev. Lett.* **44** 912 (1980).
[8] J. Schechter and J. W. F. Valle, Phys. Rev. D **22**, 2227 (1980).
[9] T. P. Cheng and L. F. Li, Phys. Rev. D **22**, 2860 (1980).
[10] R. Foot, H. Lew, X. G. He and G. C. Joshi, Z. Phys. C **44**, 441 (1989).
[11] S. Dimopoulos and H. Georgi, Nucl. Phys. B **193** (1981) 150.
[12] P. Fayet, Nucl. Phys. B **90** (1975) 104.
[13] G. R. Farrar and P. Fayet, Phys. Lett. B **76** (1978) 575.
[14] P. Fileviez Perez and S. Spinner, arXiv:1005.4930 [hep-ph].
[15] Early references on left-right symmetric models are R. N. Mohapatra and J. C. Pati, Phys. Rev. D **11** (1975) 2558 and G. Senjanovic and R. N. Mohapatra, Phys. Rev. D **12** (1975) 1502.
[16] R. N. Mohapatra and A. Rasin, Phys. Rev. D **54** (1996) 5835 [arXiv:hep-ph/9604445].
[17] F. Borzumati and A. Masiero, Phys. Rev. Lett. **57** (1986) 961.
[18] J. Hisano, T. Moroi, K. Tobe, M. Yamaguchi and T. Yanagida, Phys. Lett. B **357** (1995) 579 [arXiv:hep-ph/9501407].
[19] J. Hisano, T. Moroi, K. Tobe and M. Yamaguchi, Phys. Rev. D **53** (1996) 2442 [arXiv:hep-ph/9510309].
[20] M. Hirsch, J. W. F. Valle, W. Porod, J. C. Romao and A. Villanova del Moral, Phys. Rev. D **78** (2008) 013006 [arXiv:0804.4072 [hep-ph]].
[21] M. Hirsch, S. Kaneko and W. Porod, Phys. Rev. D **78** (2008) 093004 [arXiv:0806.3361 [hep-ph]].
[22] J. N. Esteves, S. Kaneko, J. C. Romao, M. Hirsch and W. Porod, Phys. Rev. D **80** (2009) 095003 [arXiv:0907.5090 [hep-ph]].
[23] M. Malinsky, J. C. Romao and J. W. F. Valle, Phys. Rev. Lett. **95**, 161801 (2005) [arXiv:hep-ph/0506296].
[24] M. Cvetic and J. C. Pati, Phys. Lett. B **135** (1984) 57.
[25] R. Kuchimanchi and R. N. Mohapatra, Phys. Rev. D **48**, 4352 (1993) [arXiv:hep-ph/9306290].
[26] K. S. Babu and R. N. Mohapatra, Phys. Lett. B **668**, 404 (2008) [arXiv:0807.0481 [hep-ph]].
[27] C. S. Aulakh, K. Benakli and G. Senjanovic, Phys. Rev. Lett. **79**, 2188 (1997) [arXiv:hep-ph/9703434].
[28] C. S. Aulakh, A. Melfo, A. Rasin and G. Senjanovic, Phys. Rev. D **58**, 115007 (1998) [arXiv:hep-ph/9712551].
[29] W. Porod, Comput. Phys. Commun. **153**, 275 (2003) [arXiv:hep-ph/0301101].
[30] F. Staub, Comput. Phys. Commun. 181, 1077 (2010).
[31] B. C. Allanach *et al.*, in *Proc. of the APS/DPF/DPB Summer Study on the Future of Particle Physics (Snowmass 2001)* ed. N. Graf, Eur. Phys. J. C **25** (2002) 113 [arXiv:hep-ph/0202233].
[32] Yu. M. Andreev, S. I. Bityukov, N. V. Krasnikov and A. N. Toropin, Phys. Atom. Nucl. **70** (2007) 1717 [arXiv:hep-ph/0608176].
[33] F. del Aguila *et al.*, Eur. Phys. J. C **57** (2008) 183 [arXiv:0801.1800 [hep-ph]].
[34] W. Chao, arXiv:0705.4351 [hep-ph].
[35] J. Esteves, M. Hirsch, W. Porod, J. C. Romao, F. Staub and A. Vicente, in preparation.

Journal of Physics: Conference Series **259** (2010) 012066 doi:10.1088/1742-6596/259/1/012066

Dark matter in minimal supergravity with type-II seesaw

J. N. Esteves

E-mail: joaomest@cftp.ist.utl.pt
Departamento de Física and CFTP, Instituto Superior Técnico
Av. Rovisco Pais 1, 1049-001 Lisboa, Portugal

Abstract. We calculate the relic density of the lightest neutralino in a supersymmetric seesaw type-II ("triplet seesaw") model with minimal supergravity boundary conditions at the GUT scale. The presence of a triplet below the GUT scale, required to explain measured neutrino data in this setup, leads to a characteristic deformation of the sparticle spectrum with respect to the pure mSugra expectations, affecting the calculated relic dark matter (DM) density. We discuss how the DM allowed regions in the $(m_0, M_{1/2})$ plane change as a function of the (type-II) seesaw scale. We also compare the constraints imposed on the models parameter space form upper limits on lepton flavour violating (LFV) decays to those imposed by DM.

1. Introduction

Standard cosmology requires the existence of a non-baryonic dark matter (DM) contribution to the total energy budget of the universe. In the past few years estimates of the DM abundance have become increasingly precise. Indeed, the Particle Data Group now quotes at 1 σ c.l. [1]

$$\Omega_{DM} h^2 = 0.105 \pm 0.008. \tag{1}$$

Neutrino oscillation experiments have shown that neutrinos have non-zero mass and mixing angles and the most recent global fits to all data [2] confirm again that the mixing angles are surprisingly close to the so-called tri-bimaximal mixing (TBM) values [3]. It was shown long ago that if neutrinos are Majorana particles, their mass is described by a unique dimension-5 operator [4]

$$m_\nu = \frac{f}{\Lambda}(HL)(HL). \tag{2}$$

All (Majorana) neutrino mass models reduce to this operator at low energies. Assuming complete flavour blindness in the soft supersymmetry breaking parameters at some large scale, the neutrino Yukawa matrices will, in general, lead to non-zero flavour violating entries in the slepton mass matrices, if the seesaw scale is lower than the scale at which SUSY is broken. This was first pointed out in [5].

In this work [7] we studied neutralino dark matter within a supersymmetric type-II seesaw model with mSugra boundary conditions. For definiteness, the model we consider consists of the MSSM particle spectrum to which we add a single pair of **15**- and $\overline{\mathbf{15}}$-plets. This is the simplest supersymmetric type-II setup, which allows one to maintain gauge coupling unification [6] and explain measured neutrino oscillation data.

In mSugra - assuming a standard thermal history of the early universe - only four very specific regions in parameter space can correctly explain the most recent WMAP data [8]. These are (i) the bulk region; (ii) the co-annihilation line; (iii) the "focus point" line and (iv) the "higgs funnel" region. More details can be found on [7].

2. Setup: mSugra and $SU(5)$ motivated type-II seesaw

We will always refer to minimal Supergravity (mSugra) as the "standard" against which we compare all our results. The model consists in extending the MSSM particle spectrum by a pair of **15** and $\overline{\mathbf{15}}$. It is the minimal supersymmetric seesaw type-II model which maintains gauge coupling unification [6].

One of the triplets in the model described in [6] has the correct quantum numbers to generate neutrino masses. Integrating out the heavy triplets at their mass scale a dimension-5 operator of the form eq. (2) is generated and after electro-weak symmetry breaking the resulting neutrino mass matrix can be written as

$$m_\nu = \frac{v_2^2}{2}\frac{\lambda_2}{M_T}Y_T. \tag{3}$$

Here v_2 is the vacuum expectation value of Higgs doublet H_2 and we use the convention $\langle H_i \rangle = \frac{v_i}{\sqrt{2}}$. m_ν can be diagonalized in the standard way with a unitary matrix U, containing in general 3 angles and 3 phases. Note that $\hat{Y}_T = U^T \cdot Y_T \cdot U$ is diagonalized by *the same matrix as* m_ν. This means that if all neutrino eigenvalues, angles and phases were known, Y_T would be completely fixed up to an overall constant, which can be written as $\frac{M_T}{\lambda_2} \simeq 10^{15}\text{GeV}\left(\frac{0.05\text{ eV}}{m_\nu}\right)$. Thus, current neutrino data requires M_T to be lower than the GUT scale by (at least) an order or magnitude.

The full set of RGEs for the $\mathbf{15} + \overline{\mathbf{15}}$ can be found in [6] and in the numerical calculation, presented in the next section, we solve the exact RGEs.

3. Numerical results

In this section we discuss our numerical results. All the plots shown below are based on the program packages SPheno [12] and micrOMEGAs [14, 13]. We use SPheno V3 [15], including the RGEs for the $\mathbf{15} + \overline{\mathbf{15}}$ case [6, 16] at the 2-loop level for gauge couplings and gaugino masses and at one-loop level for the remaining MSSM parameters and the 15-plet parameters, for a discussion see [16]. For any given set of mSugra and 15-plet parameters SPheno calculates the supersymmetric particle spectrum at the electro-weak scale, which is then interfaced with micrOMEGAs2.2 [17] to calculate the relic density of the lightest neutralino, $\Omega_{\chi_1^0}h^2$. A sketch of the all process can be seen in fig.(1).

For the standard model parameters we use the PDG 2008 values [1], unless specified otherwise. As discussed below, especially important are the values (and errors) of the bottom and top quark masses, $m_b = 4.2 + 0.17 - 0.07$ GeV and $m_t = 171.2 \pm 2.1$ GeV. Note, the m_t is understood to

Journal of Physics: Conference Series **259** (2010) 012066 doi:10.1088/1742-6596/259/1/012066

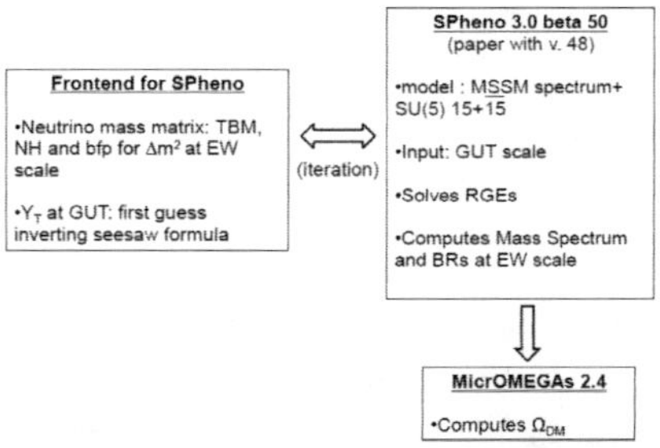

Figure 1. Dark Matter computation.

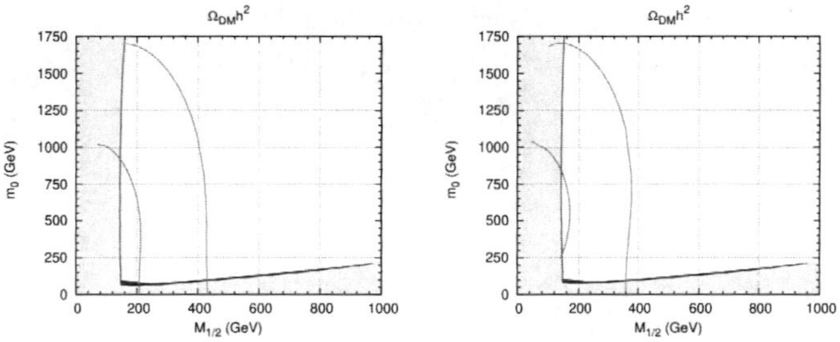

Figure 2. Limits for mSugra with $\tan\beta = 10$, and $\mu > 0$ for $A_0 = -300$ GeV (left panel) and $A_0 = -500$ GeV (right panel). The blue regions are allowed by the DM constraint The yellow bounds refer to the $\tilde{\tau}$ being the LSP (bottom) and to the LEP bound on the chargino mass (left).

be the pole-mass and $m_b(m_b)$ is the $\overline{MS}$ mass. As the allowed range for $\Omega_{DM}h^2$ we always use the 3 σ c.l. boundaries as given in [1], i.e. $\Omega_{DM}h^2 = [0.081, 0.129]$. Note, however that the use of 1 σ contours results in very similar plots, due to the small error bars.

We define our "standard choice" of mSugra parameters as $\tan\beta = 10$, $A_0 = 0$ and $\mu > 0$ and use these values in all plots, unless specified otherwise. We then show our results in the plane of the remaining two free parameters, $(m_0, M_{1/2})$.

In fig. (2) we show two examples for the DM allowed region and the regions disfavored by the Higgs boson mass bound at $m_{h^0} = 114.4$ GeV and $m_{h^0} = 110$ GeV. Larger negative A_0 leads to a less stringent constraint (for $\mu > 0$). Note, that all of the bulk region becomes allowed at $A_0 = -500$ GeV, once the theoretical uncertainty in the Higgs boson mass calculation is taken into account.

We now turn to a discussion of large $\tan\beta$. At large values of $\tan\beta$ the width of the CP-odd Higgs boson A becomes large, $\Gamma_A \sim M_A \tan^2\beta(m_b^2 + m_t^2)$, and a wide s-channel resonance occurs in the region $m_{\chi_1^0} \simeq M_A/2$. The enhanced annihilation cross section reduces $\Omega_{\chi_1^0}h^2$ to acceptable levels, the resulting region is known as the "higgs funnel" region.

The higgs funnel is very sensitive to the exact value of $\tan\beta$ and to the values (and errors) of the top and bottom quark mass. The position of the funnel is especially sensitive to the exact value of m_b. However, future determinations of m_b and m_t could improve the situation considerably.

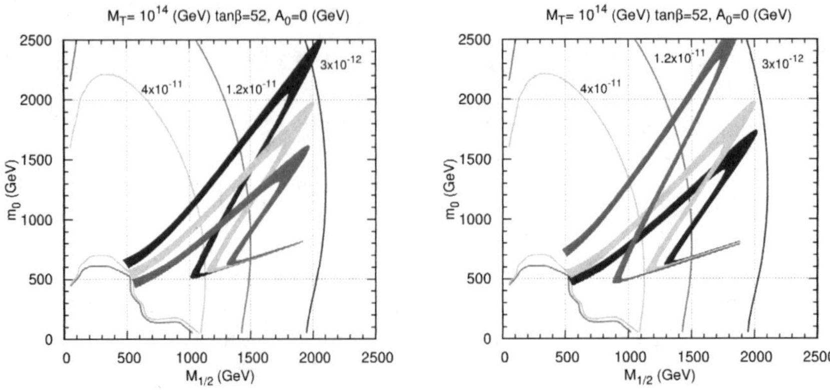

Figure 3. Allowed region for the dark matter density in the $(m_0, M_{1/2})$ plane for $A_0 = 0$, $\mu \geq 0$ and $\tan\beta = 52$, for $M_T = 10^{14}$ GeV and (to the left) for three values of $m_{top} = 169.1$GeV (blue), $m_{top} = 171.2$ GeV (green) and $m_{top} = 173.3$ GeV (red). To the right: The same, but varying m_b. $m_{bot} = 4.13$ GeV (blue), $m_{bot} = 4.2$ GeV (green) and $m_{bot} = 4.37$ GeV (red). Superimposed are contour levels for $\mathrm{Br}(\mu \to e\gamma)$

In the numerical calculation we have chosen neutrino masses to be of the normal hierarchical type and fitted the neutrino angles to exact tri-bimaximal (TBM) values [3], i.e. $\tan^2\theta_{\mathrm{Atm}} = 1$, $\tan^2\theta_{\odot} = 1/2$ and $\sin^2\theta_{\mathrm{R}} = 0$. This has to be done in a simple iterative procedure, since the triplet parameters are defined at the high scale, whereas neutrino masses and angles are measured at low scale. For more details on the fit procedure see [16].

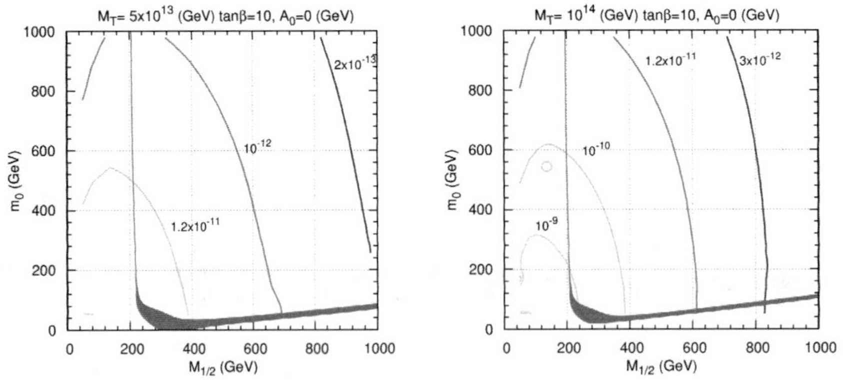

Figure 4. Limits for mSugra with $\tan\beta = 10$, $\mu > 0$ and $A_0 = 0$ GeV for $M_T = 5 \times 10^{13}$ GeV (left panel) and $M_T = 10^{14}$ GeV (right panel). Contour levels for $\mathrm{Br}(\mu \to e\gamma)$ are also shown.

Finally, we will compare the constraints imposed on the parameter space of the model by $\Omega_{DM}h^2$ with the constraints from the current data on non-observation of lepton flavour violating processes. Since LFV within the present model has been studied in some detail in [16], we will not repeat all of the discussion here. Instead, here we concentrate on $\mu \to e\gamma$ exclusively, since the upper bound on $Br(\mu \to e\gamma)$ of $Br(\mu \to e\gamma) \leq 1.2 \cdot 10^{-11}$ [1] has been shown to provide currently the most important constraint.

In fig. (4) we show the DM allowed parameter regions for $\tan\beta = 10$ and two values of M_T,

$M_T = 5 \cdot 10^{13}$ GeV (to the left) and $M_T = 10^{14}$ GeV (to the right), for a fixed choice of all other parameters. Superimposed on this plot are lines of constant branching ratio for $Br(\mu \to e\gamma)$. The latter have been calculated requiring neutrino masses being hierarchical and fitted to solar and atmospheric neutrino mass squared differences and neutrino angles fitted to TBM values. Within the $(m_0, M_{1/2})$ region shown, $Br(\mu \to e\gamma)$ can vary by two to three orders of magnitude, depending on the exact combination of $(m_0, M_{1/2})$, even for all other parameters fixed. The most important parameter determining $Br(\mu \to e\gamma)$, once neutrino data is fixed, however, is M_T, as can be seen comparing the figure to the left with the plot on the right. While for $M_T = 10^{14}$ GeV about "half" of the plane is ruled out by the non-observation of $\mu \to e\gamma$, for $M_T = 5 \cdot 10^{13}$ GeV with the current upper limit nearly all of the plane becomes allowed. The strong dependence of $\mu \to e\gamma$ on M_T can be understood from the analytical formulas presented in [16]. In this paper it was shown that $Br(\mu \to e\gamma)$ scales very roughly as $Br(\mu \to e\gamma) \propto M_T^4 \log(M_T)$, if neutrino masses are to be explained correctly. For $\tan\beta = 10$ one thus concludes that with present data values of M_T larger than (few) 10^{13} GeV - (few) 10^{14} GeV are excluded by $Br(\mu \to e\gamma)$, to be compared with $M_T/\lambda_2 \lesssim 10^{15}$ GeV from the measured neutrino masses. Note, however, that (i) the constraint from neutrino masses is relatively independent of $\tan\beta$, m_0 and $M_{1/2}$, while $\mu \to e\gamma$ shows strong dependence on these parameters; and (ii) allowing the value of the reactor angle $\sin^2\theta_{\rm R}$ to vary up to its experimental upper limit, $\sin^2\theta_{\rm R} = 0.056$ [2], leads to larger values of $Br(\mu \to e\gamma)$ and thus to a tighter upper limit on M_T.

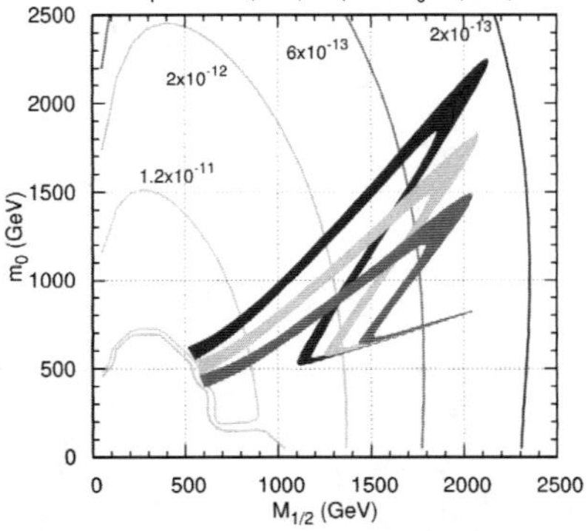

Figure 5. Allowed region for dark matter density $(0.081 < \Omega_{\chi_1^0} h^2 < 0.129)$ in the $(m_0, M_{1/2})$ plane for $A_0 = 0$, $\mu \geq 0$ and $\tan\beta = 52$, for three values of $m_{top} = 169.1$ GeV (blue), $m_{top} = 171.2$ GeV (green) and $m_{top} = 173.3$ GeV (red) for $M_T = 5 \times 10^{13}$. Superimposed are the contour lines for the $Br(\mu \to e\gamma)$.

In figs. (3) and (5) we show the results for a calculation comparing dark matter and LFV in the case of large $\tan\beta$. Here the same constraints as in fig. (4) are shown, however for $\tan\beta = 52$. Again we show the calculation for two values of M_T, since M_T is the most important free parameter. It is known that at large values of $\tan\beta$, LFV decays are enhanced due to an enhanced chargino diagram, which in the limit of large $\tan\beta$ scales approximately as $\tan^2\beta$ [18]. Therefore, constraints on the parameter space from non-observation of LFV decays are more severe in case of large $\tan\beta$, leading to tighter upper limits on M_T. This is clear if we compare fig. (4) on one side and figs. (3) and (5) on the other, noticing the different scales. However, because of the higgs funnel region developing for large $\tan\beta$, the interesting part of the parameter space enlarges compensating for the larger values of the LFV decays. This can be seen in fig. (5), where for $M_T = 5 \times 10^{13}$ GeV (left), most of the $(m_0, M_{1/2})$ plane is allowed by the upper limit on $Br(\mu \to e\gamma)$, while in fig. (3), left, for $M_T = 10^{14}$ GeV about "half" of the plane is ruled out by this limit.

Acknowledgments

J. N. Esteves wishes to thank M. Hirsch, S. Kaneko, W. Porod and J. C. Romao for a fruitful collaboration that leaded to the work [7] that was presented here.

References

[1] C. Amsler et al., Physics Letters B667, 1 (2008); http://pdg.lbl.gov/

[2] For a review of neutrino oscillation data, see: M. Maltoni, T. Schwetz, M. A. Tortola and J. W. F. Valle, New J. Phys. **6**, 122 (2004); an updated analysis has now been published in T. Schwetz, M. Tortola and J. W. F. Valle, New J. Phys. **10**, 113011 (2008) (*Preprint* arXiv:0808.2016 [hep-ph]) where all experimental references can be found.

[3] P. F. Harrison, D. H. Perkins and W. G. Scott, Phys. Lett. **B530**, 167 (2002) (*Preprint* hep-ph/0202074)

[4] S. Weinberg, Phys. Rev. Lett. **43**, 1566 (1979); S. Weinberg, Phys. Rev. D **22**, 1694 (1980)

[5] F. Borzumati and A. Masiero, Phys. Rev. Lett. **57**, 961 (1986)

[6] A. Rossi, Phys. Rev. D **66**, 075003 (2002) (*Preprint* hep-ph/0207006)

[7] J. N. Esteves, S. Kaneko, J. C. Romao, M. Hirsch and W. Porod, Phys. Rev. D **80** (2009) 095003 (*Preprint* arXiv:0907.5090 [hep-ph])

[8] E. Komatsu *et al.* [WMAP Collaboration], Astrophys. J. Suppl. **180**, 330 (2009) (*Preprint* arXiv:0803.0547 [astro-ph])

[9] B. C. Allanach, G. Belanger, F. Boudjema and A. Pukhov, JHEP **0412**, 020 (2004) (*Preprint* hep-ph/0410091)

[10] H. Baer, C. Balazs and A. Belyaev, JHEP **0203**, 042 (2002) (*Preprint* hep-ph/0202076)

[11] J. L. Feng, K. T. Matchev and T. Moroi, Phys. Rev. D **61**, 075005 (2000) (*Preprint* hep-ph/9909334)

[12] W. Porod, Comput. Phys. Commun. **153**, 275 (2003) (*Preprint* hep-ph/0301101)

[13] G. Belanger, F. Boudjema, A. Pukhov and A. Semenov, Comput. Phys. Commun. **176**, 367 (2007) (*Preprint* hep-ph/0607059)

[14] G. Belanger, F. Boudjema, A. Pukhov and A. Semenov, Comput. Phys. Commun. **174**, 577 (2006) (*Preprint* hep-ph/0405253)

[15] The latest version of SPheno can be obtained from: http://theorie.physik.uni-wuerzburg.de/ porod/SPheno.html

[16] M. Hirsch, S. Kaneko and W. Porod, Phys. Rev. D **78**, 093004 (2008) (*Preprint* arXiv:0806.3361 [hep-ph])

[17] http://wwwlapp.in2p3.fr/lapth/micromegas/

[18] J. Hisano, T. Moroi, K. Tobe and M. Yamaguchi, Phys. Rev. D **53**, 2442 (1996) (*Preprint* hep-ph/9510309)

16th International Symposium on Particles Strings and Cosmology (PASCOS 2010) IOP Publishing
Journal of Physics: Conference Series **259** (2010) 012067 doi:10.1088/1742-6596/259/1/012067

Searching Neutrino Physics through Displaced Vertices at Colliders

M. B. Magro

Centro Universitário Fundação Santo André, Santo André – SP, Brazil.

E-mail: magro@if.usp.br

Abstract. We present a brief review on the features of bilinear R-parity violation in mSUGRA models at the LHC. These kind of models can provied a way of probing neutrino oscillation physics at colliders. We show that the LHC will be able to cover such models for a large fraction of the parameter space.

The Large Hadron Collider (LHC) will probe a new land of physics by accessing a new scale of energy, and it is expected that the LHC will shed light upon some of the main open problems in actual particle physics, such as the origin of mass and the nature of the electroweak symmetry breaking. One of the most popular models for breaking electroweak symmetry is supersymmetry (SUSY) which provides an elegant way to solve the hierarchy problem [1] that appears in the symmetry breaking mechanism. On the other hand, extensions for the standard model of particle physics are in evidence nowadays after the discovery of neutrino masses and oscillations [2].

In SUSY models one define a new multiplicative quantum number called R-parity that is even for the standard model particles and odd for their supersymmetrical partners. One natural way of generating masses and mixings for neutrinos compatible with recent neutrino oscillation data is through R-parity breaking by violating leptonic number. We then assume that R-parity is broken by a bilinear R-parity breaking supergravity model (BRpV–mSUGRA) in which the MSSM superpotential is supplemented by the following terms [3]

$$W_{\rm BRpV} = W_{\rm MSSM} + \varepsilon_{ab}\epsilon_i \widehat{L}_i^a \widehat{H}_u^b \,, \tag{1}$$

with three extra parameters (ϵ_i), one for each fermion generation. In addition to these bilinear terms, we must also include new soft supersymmetry breaking terms (B_i).

$$V_{\rm soft} = V_{\rm MSSM} - \varepsilon_{ab}\left(B_i\epsilon_i \widetilde{L}_i^a H_u^b\right) \,. \tag{2}$$

Taken together, the new terms in the BRpV Lagrangian (the three bilinear ϵ_i and the B_i) lead to the explicit violation of lepton number as well as R-parity. These terms also induce vevs $v_i \equiv v_{Li}$, $i = 1, 2, 3$ for the three left-type sneutrinos.

The basic difference between BRpV–mSUGRA and the conventional R-parity conserving mSUGRA scenario is that the lightest supersymmetric particle is no longer stable and it decays, typically inside the detector, consequently modifying the SUSY phenomenology at colliders.

This model contains eleven free parameters: m_0 and $m_{1/2}$ being the scalar soft SUSY breaking masses and the common gaugino mass at the unification scale; A_0, the common trilinear term;

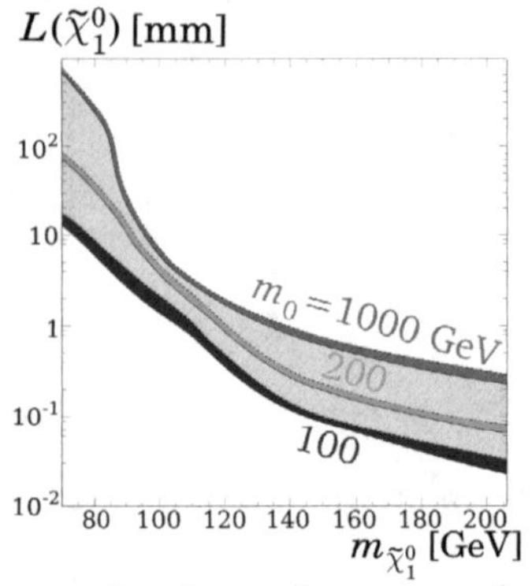

Figure 1. $\tilde{\chi}^0_1$ decay length for several values of m_0 as a function of its mass, for $A_0 = -100$ GeV, $\tan\beta = 10$, $\mu > 0$

$\tan\beta$, the ratio between the Higgs field vev's; sign(μ); ϵ_i and Λ_i. We trade the soft parameters B_i by the "alignment" parameters $\Lambda_i = \epsilon_i v_d + \mu v_i$ which are directly related to the neutrino–neutralino properties [4].

While the BRpV parameters ϵ_i and Λ_i have no effect in the production cross sections of supersymmetric states, they affect the LSP decay length and branching ratios. In Fig. 1 we illustrate the LSP decay length as a function of its mass in its rest frame. As we can see, for a large region of the parameter space the LSP decay length will be large enough to produce displaced vertices at the LHC.

We show in Fig. 2 the main LSP branching ratios as a function of m_0 for several values of $m_{1/2}$. When it is kinematically allowed the LSP decays via two body decays into $W\ell$, $Z\nu$ and $h\nu$ with the gauge bosons further decaying into quarks and leptons. We can see from Fig 2 that $\tilde{\chi}^0_1$ does have a sizeable branching ratio into $\nu b\bar{b}$, specially for heavier LSP and larger m_0. Notice that the main contribution for this decay is via higgs production. From the figures we can also learn that the leptonic decays modes $\nu\tau^+\tau^-$ and $\nu\tau^\pm\ell^\mp$ vary from $\approx 40\%$ at small m_0 to a few percent at large m_0. At moderate and large m_0, these decays originate from the lightest neutralino decaying into the two–body modes $\tau^\pm W^\mp$, $\mu^\pm W^\mp$ and νZ, followed by the leptonic decay of the weak gauge bosons. In general, semi-leptonic decays of the LSP are suppressed at small m_0 and they dominate at large m_0 due to two–body decays.

As can be seen from the previous figures, the LHC will be able to probe the BRpV-mSUGRA model both using standard SUSY searches looking for displaced vertices. We employed PYTHIA version 6.409 [5] to generate the signal and backgrounds, using the SPheno program [6] to generate all BRpV branching ratios in the SLHA format [7].

We searched for SUSY looking for standard signals at the LHC using the same strategy used in [8]. Fig. 3 displays the LHC reach in the three–(multi–)lepton (right–(left–)panel) channels with/without R–parity for integrated luminosity of 100 fb^{-1}, which constitutes the best standard channel for BRpV discovery. For the reach of other topologies, see our full analysis in [9].

The decay of the LSP has an impact of reducing the LHC reach for SUSY when we take into account only the main all inclusive channel. Notwithstanding, the new BRpV interactions lead to a substantial increase in the final states containing isolated charged leptons, specially for pairs of same sign leptons and trileptons, compensating the losses in the inclusive channel [9]. Nevertheless, The existence of long decay lengths is an important feature of the BRpV models since detached vertices are a smoking gun of SUSY with R–parity violation. Furthermore, displaced vertices provide an additional handle that makes possible the SUSY search at the LHC. Since the neutralino decay length is large enough to be detached from the primary vertex of LHC, we expect it to decay producing a displaced vertex inside the ATLAS/CMS detectors.

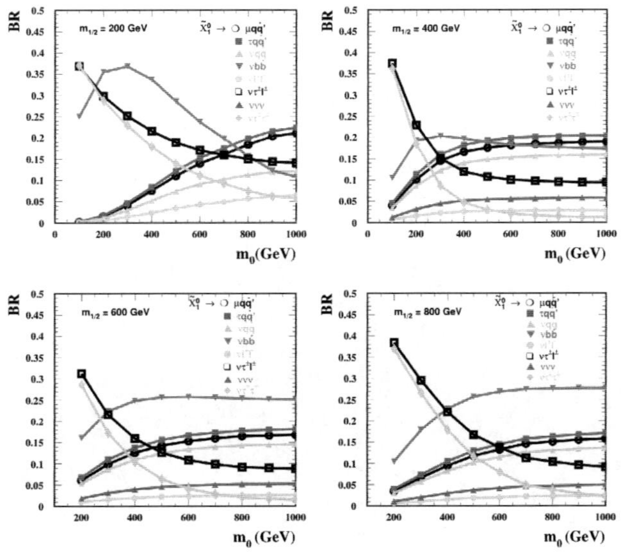

Figure 2. Lightest neutralino branching ratios as a function of m_0 for several values of $m_{1/2}$ and $A_0 = -100$ GeV, $\tan\beta = 10$, and $\mu > 0$

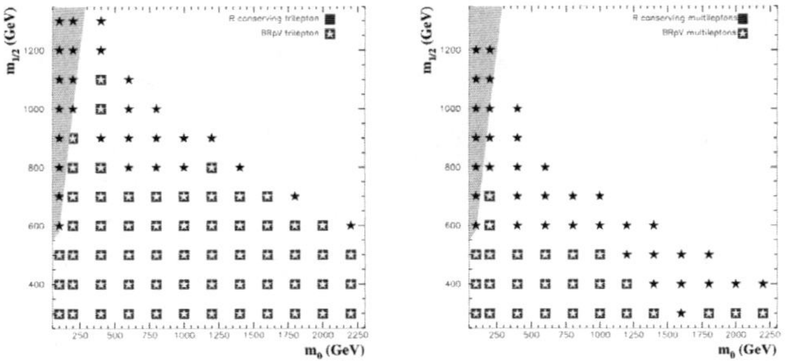

Figure 3. LHC discovery potential in the three lepton channel (left panel) and the multilepton one (right panel) for an integrated luminosity of 100 fb^{-1} [9].

In principle, this topology has little, if none, background expected at LHC.

In Fig. 4 we present the displaced vertex reach. As one can see, the LHC will be able to look for the displaced vertex signal up to $m_{1/2}$ ~800 (1000) GeV for a large range of m_0 values and an integrated luminosity of 10 (100) fb^{-1}. Notice that the reach in this channel is rather independent of m_0 as expected from Fig. 1. Moreover, this signal for BRpV–mSUGRA disappears in the region where the stau is the LSP due to its rather short lifetime.

Since the LHCb experiment has excellent vertex capabilities, it is natural to conclude that it can also play a role in the displaced vertex search for SUSY. Taking into account that the LHCb experiment already has a dimuon trigger, we started our analysis looking for displaced vertices associated to $\tilde{\chi}_1^0 \to \nu\mu^+\mu^-$. Despite the smallness of this branching ratio this is a very clean signal. Since the LHCb has very good vertex system we have also studied its discovery potential using all LSP decays that allow vertex reconstruction, *i.e.* the decays exhibiting two or more charged particles that can be reconstructed as emanating from the same point.

We present in Figure 5 the region of the $m_0 \otimes m_{1/2}$ plane where we expect 5 or more displaced vertices for integrated luminosity of 2 fb^{-1}, $A_0 = -100$ GeV, $\tan\beta = 10$, and $\mu > 0$ for both di–muon and all visible modes. For further details, see [10].

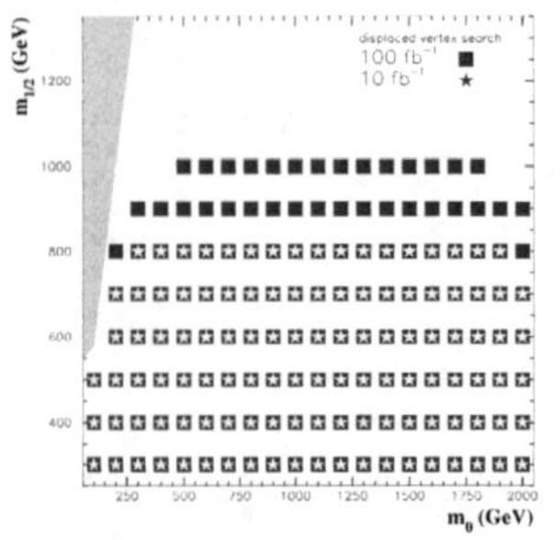

Figure 4. Discovery reach for displaced vertices channel in the $m_0 \otimes m_{1/2}$ plane for $\tan\beta = 10$, $\mu > 0$, $A_0 = -100$ GeV. The stars (squares) stand for points where there are more than 5 displaced vertex signal events for an integrated luminosity of 10 (100) fb^{-1} [9].

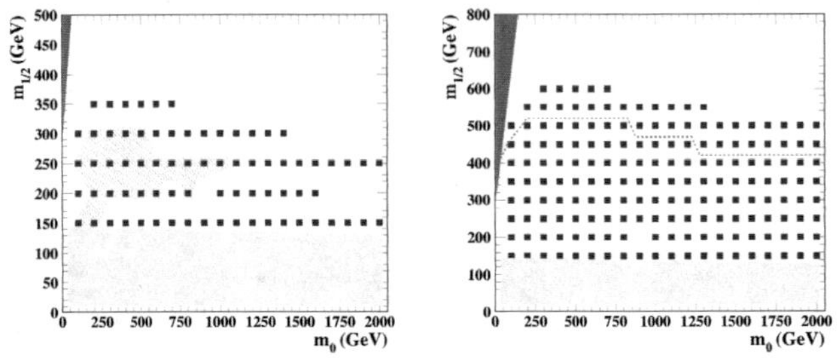

Figure 5. The left(right) panel contains the LHCb discovery potential of BRpV–mSUGRA using dimuons(all visible modes) coming from a detached vertex [10]. We assumed $A_0 = -100$ GeV, $\tan\beta = 10$, and $\mu > 0$. The blue squares stand for the points with 5 or more events for $\mathcal{L} = 2$ fb^{-1}. The magenta areas correspond to a cross section larger than 10 fb^{-1}.

It is interesting to contrast the LHCb discovery potential with the ATLAS/CMS ones in the low luminosity initial run in order to estimate the contribution to BRvP searches that the different experiments can give. There are two main difference between the LHCb and ATLAS/CMS experiments: first of all, the ATLAS/CMS will have a luminosity 5 times the LHCb one in this period. Secondly, ATLAS/CMS vertex detectors have a larger pseudo-rapidity coverage, ranging from $-2.5 < \eta < 2.5$. We can see that ATLAS/CMS reach is a factor of 2 larger than the LHCb reach in the di–muon channel. This larger ATLAS/CMS discovery potential originates mainly from the larger planned integrated luminosity for these detectors.

We also analyze the two–body LSP decay into the lightest Higgs boson h^0 as a Higgs discovery channel $\tilde{\chi}_1^0 \to h\nu$. If the LSP lives long enough it will be detached from the primary interaction point leaving a displaced vertex as signal at the LHC. Since the Higgs boson decays mostly into b–quark pairs we expect a displaced vertex with two b–jets as a characteristic signature for Higgs production. Therefore, we looked for events with at least one displaced vertex containing at least one jet tagged as a b–jet. We considered a b–tagging efficiency up to 50% [11].

For our analysis the Higgs mass lies in the range 110 GeV $\lesssim M_h \lesssim$ 120 GeV when we vary M_0 and $M_{1/2}$. Since we are only interested in detached jets coming from Higgs decays, we have further required that the jet–jet invariant mass is around the Higgs mass value. Physical SM processes do not lead to sizeable backgrounds to the detached Higgs searches due to to large difference in the invariant mass of the visible products. However, BRpV LSP decays into νZ are a potential source of background for the Higgs signal.

From Fig. 6 one can see that the ATLAS and CMS experiments will be able to look for the signal up to $M_{1/2} \sim 700$ (900) GeV for a LHC integrated luminosity of 10 (100) fb^{-1}, for the full analysis see [11]. Notice that the LHC Higgs discovery potential is almost independent of M_0. We have also estimated the reach expected at LHCb for our Higgs search proposal. The hatched region in Fig. 6 indicates the LHCb reach for 10 fb^{-1}. Due to the strong cut on the pseudo–rapidity required by this experiment the reach for 2 fb^{-1} is severely depleted and only a small region of the parameter space is covered.

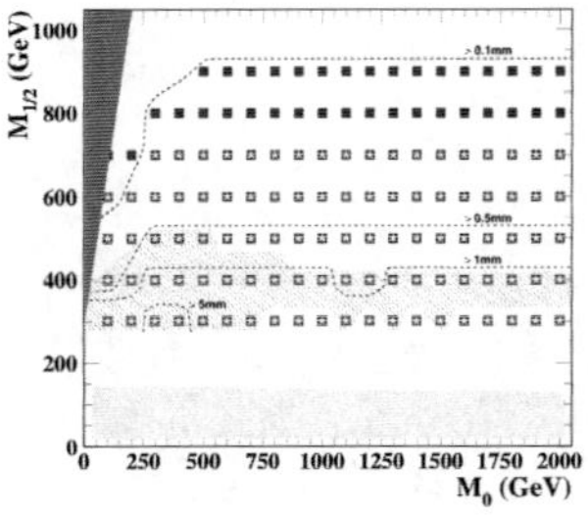

Figure 6. LHC reach for Higgs search in displaced vertices for the BRpV–mSUGRA model in the plane $M_{1/2} \otimes M_0$ assuming $\tan\beta = 10$, $A_0 = -100$ GeV, and $\mu > 0$. The yellow stars (blue squares) represent the reach for an integrated luminosity of 10 (100) fb^{-1} while the hatched region corresponds to the reach of the LHCb experiment for an integrated luminosity of 10 fb^{-1} [11]. Note that the black lines delimit different regimes of LSP decay length.

Finally we analyze the correlation presented in the BRpV-mSUGRA model between neutralino decay properties at LHC and neutrino mixing angles determined in low energy neutrino oscillation experiments. In [12] it is shown that the atmospheric angle is related to hte neutralino branching ratios as

$$\tan^2\theta_{\text{atm}} \simeq \frac{BR(\tilde{\chi}_1^0 \to \mu^\pm W^\mp)}{BR(\tilde{\chi}_1^0 \to \tau^\pm W^\mp)} = R. \tag{3}$$

We present a quantitative analysis of the precision that the LHC can achieve in measuring the ratio R through the reconstruction of the LSP branching ratios, see [13].

We present in Fig. 7 the attainable precision $\sigma(R)/R$ with which the correlation R can be measured as a function of $m_0 \otimes m_{1/2}$ for $A_0 = -100$ GeV, $\tan\beta = 10$ and $\text{sgn}(\mu) = +1$ for an integrated luminosity of 100 fb^{-1} and a center-of-mass energy of 14 TeV. We require in all plots that at least 5 events of reconstructed taus are observed. In the left panel of this figure we present the expected statistical error on the ratio R assuming no systematic errors on the determination of the reconstruction efficiencies, while in the right panel we consider a more conservative scenario, where we anticipate a systematic error of 10% in each of the reconstruction efficiencies. One can see from this panel that the precision drops as $m_{1/2}$ grows since the neutralino production rates from squark/gluino cascade decays also decrease with increasing $m_{1/2}$ values. Therefore, if the systematic errors of the efficiency determination are negligible the LHC collaborations should be able to probe with a very good precision ($\lesssim 10\%$) the ratio R for $m_{1/2} \lesssim 650$ GeV, that correspond to an LSP mass up to $\simeq 270$ GeV. The inclusion of systematic errors at the level assumed in the right panel of Fig. 7 increases the uncertainty in R, however, it is still possible to perform an accurate test of RmSUGRA scenario.

Note that in Fig. 7 we also present results for the 7 TeV run of the LHC. One can see that the LHC has a much more limited capability of probing the ratio R, since the reach of this run covers only up to $m_{1/2} \lesssim 300$ GeV. Still, although large, the statistical errors in this region

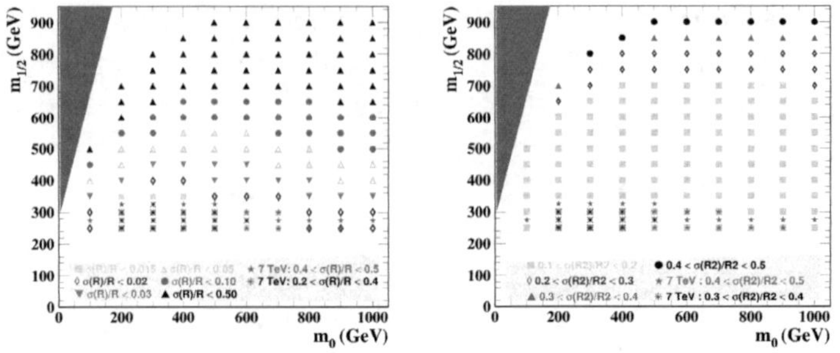

Figure 7. Precision in the determination of the ratio R in the plane $m_{1/2} \times m_0$ for a luminosity of 100 fb^{-1}, center-of-mass energy of 14 TeV, $A_0 = -100$ GeV, $\tan\beta = 10$ and $\text{sgn}(\mu) = +1$. In the right (left) panel we did (not) include a possible systematic uncertainty in the extraction of the efficiencies for the channels μW and τW. The stars in the right panel represent the results for the 7 TeV run with an integrated luminosity of 1 fb^{-1} [13].

$(0.3 \lesssim \sigma(R)/R \lesssim 0.5)$, due mainly to the small anticipated integrated luminosity of 1 fb^{-1}, allow a determination of the atmospheric angle comparable to that obtained at low energies.

In conclusion, we have shown that the LHC will be able to probe the features of the BRpV-mSUGRA model for a sizeable range of the parameter space. SUSY standard searches will be supressed by the high multiplicity of final states due to the LSP decaying into the detectors and the difficulty of isolating the final leptons, while the displaced vertex signal, since it has no background, will be the best channel of probing such models. Looking for displaced vertices at LHC will provide a important tool for probing R-partity in SUSY models. On top of that, measuring accuratelly the neutralino branching ratios can connect the high energy physics at LHC with the neutrino oscillation data from low energy experiments. We have shown here that the error in the determination of the atmospheric angle by the LHC in the BRpV-mSUGRA model is comparable with the errors presented in the regular neutrino oscillation experiments.

Acknowledgments
The author would like to thank the PASCOS 2010 organizers for the invitation and IFIC/C.S.I.C. and IFUSP for their nice hospitality.

References

[1] S. Dimopoulos, S. Raby and F. Wilczek, Phys. Rev. D **24**, 1681 (1981).
[2] For reviews of neutrino oscillations see M. Maltoni, et al, New J. Phys. **6**, 122 (2004); T. Schwetz, M. Tortola and J. W. F. Valle, New J. Phys. **10** (2008) 113011 [arXiv:0808.2016 [hep-ph]].
[3] M. A. Díaz, J. C. Romão, and J. W. F. Valle, Nucl. Phys. **B524**, 23 (1998), hep-ph/9706315.
[4] M. Hirsch *et al.*, Phys. Rev. **D62**, 113008 (2000), hep-ph/0004115, Err-ibid. **D65**:119901,2002.
[5] T. Sjostrand *et al.*, Comput. Phys. Commun. **135**, 238 (2001), hep-ph/0010017; T. Sjostrand, Comput. Phys. Commun. **82**, 74 (1994).
[6] W. Porod, Comput. Phys. Commun. **153**, 275 (2003), hep-ph/0301101.
[7] P. Skands *et al.*, JHEP **07**, 036 (2004), hep-ph/0311123.
[8] H. Baer, J. K. Mizukoshi, and X. Tata, Phys. Lett. **B488**, 367 (2000), hep-ph/0007073.
[9] F. de Campos *et al.*, JHEP **0805**, 048 (2008), arXiv:0712.2156 [hep-ph].
[10] F. de Campos *et al.*, Phys. Rev. **D79**, 055008 (2009), arXiv:0809.0007 [hep-ph].
[11] F. de Campos *et al.*, Phys. Rev. **D80**, 015002 (2009), arXiv:0809.1637 [hep-ph].
[12] W. Porod, M. Hirsch, J. C. Romão and J. W. F. Valle, Phys. Rev. **D63**, 115004 (2001), [hep-ph/0011248].
[13] F. de Campos *et al.*, arXiv:1006.5075 [hep-ph], *to be published in Phys. Rev. D.*

A common source for neutrino and sparticle masses[1]

Andrea Brignole
INFN, Sezione di Padova, I-35131 Padua, Italy

Filipe R. Joaquim
CERN, Theory Division, CH-1211 Geneva 23, Switzerland

Anna Rossi
Dipartimento di Fisica "G. Galilei", Università di Padova, I-35131 Padua, Italy

Abstract. We discuss supersymmetric scenarios in which neutrino masses arise from effective $d = 6$ operators in the Kähler potential (including SUSY-breaking insertions). Simple explicit realizations of those Kähler operators are presented in the context of the type II seesaw. An appealing scenario emerges upon identifying the seesaw mediators with SUSY-breaking messengers.

The smallness of neutrino masses can be explained by the conventional seesaw mechanism in which $m_\nu \sim v^2/M$, where v is the electroweak scale and $M \gg v$ is a heavy mass. It is also conceivable that neutrino masses are suppressed by a higher power of the heavy scale M, like $m_\nu \sim mv^2/M^2$, where $m \ll M$ is another mass parameter. In a SUSY framework this mass behaviour may stem from either superpotential or Kähler $d = 6$, $\Delta L = 2$ effective operators. Two examples of the latter have been proposed in [2], namely $\int d^4\theta\, (H_1^\dagger L)(H_2 L)/M^2$ and $\int d^4\theta\, (H_1^\dagger L)^2/M^2$, which imply $m \sim \mu$ (the superpotential Higgsino mass parameter). We point out the importance of including SUSY-breaking insertions to this approach and find new contributions to neutrino masses. We also present the simplest explicit realization of the Kähler operator $(H_1^\dagger L)^2$, including SUSY-breaking effects, and discuss a predictive scenario where the heavy seesaw mediators are also messengers of SUSY breaking.

1. Neutrino masses from Kähler operators and broken SUSY

Let us focus on the $\Delta L = 2$ effective operator $(H_1^\dagger L)^2/M^2$. In general, this will appear along with similar operators containing insertions of the type X/M_S, $X^\dagger/M_S$, $XX^\dagger/M_S^2$, where $X = \theta^2 F_X$ is a SUSY-breaking spurion superfield and M_S is the SUSY-breaking mediation scale, which can be either larger or smaller than M. Hence, the relevant $\Delta L = 2$ effective lagrangian can be parametrized as

$$\mathcal{L}_{\rm eff} = \int d^4\theta \frac{1}{2M^2} \left(\boldsymbol{\kappa} + \theta^2 \mathbf{B}_\kappa + \bar\theta^2 \tilde{\mathbf{B}}_\kappa + \theta^2\bar\theta^2 \mathbf{C}_\kappa\right)_{ij} (H_1^\dagger L_i)(H_1^\dagger L_j) \; + \text{h.c.}\,, \qquad (1)$$

[1] Based on Ref. [1]

where $\boldsymbol{\kappa}$ is dimensionless, $\mathbf{B}_\kappa$, $\tilde{\mathbf{B}}_\kappa$ and $\mathbf{C}_\kappa$ are dimensionful SUSY-breaking parameters and $i,j = e,\mu,\tau$ are flavour indices. The magnitude and flavour structure of all those parameters depend on the underlying physics which generates the effective operators.

The scale dependence of the above quantities is governed by their renormalization group equations (RGEs), which can be derived using the general expression of the one-loop corrected Kähler potential obtained in [3]. The RGE for $\boldsymbol{\kappa}$ can be found in [2], while the ones for the remaining operator coefficients have been obtained in [1]. For instance, the RGE for $\tilde{\mathbf{B}}_\kappa$ is:

$$\begin{aligned} 8\pi^2 \frac{d\tilde{\mathbf{B}}_\kappa}{dt} &= \left[g^2 + g'^2 + \mathrm{Tr}(\mathbf{Y}_e^\dagger \mathbf{Y}_e + 3\mathbf{Y}_d^\dagger \mathbf{Y}_d)\right] \tilde{\mathbf{B}}_\kappa - \frac{1}{2}\left[\tilde{\mathbf{B}}_\kappa \mathbf{Y}_e^\dagger \mathbf{Y}_e + (\mathbf{Y}_e^\dagger \mathbf{Y}_e)^T \tilde{\mathbf{B}}_\kappa\right] \\ &+ \left[g^2 M_2^* + g'^2 M_1^* - 2\,\mathrm{Tr}(\mathbf{A}_e^\dagger \mathbf{Y}_e + 3\mathbf{A}_d^\dagger \mathbf{Y}_d)\right] \boldsymbol{\kappa} + \boldsymbol{\kappa}\, \mathbf{A}_e^\dagger \mathbf{Y}_e + (\mathbf{A}_e^\dagger \mathbf{Y}_e)^T \boldsymbol{\kappa}\,. \end{aligned} \tag{2}$$

All four operators of eq. (1) contribute to neutrino masses. Those with coefficients $\boldsymbol{\kappa}$ and $\tilde{\mathbf{B}}_\kappa$ contribute to $\mathbf{m}_\nu$ at the tree-level, in such a way that $\mathbf{m}_\nu \simeq \mathbf{m}_\nu^{(\kappa)} + \mathbf{m}_\nu^{(\tilde{B}_\kappa)}$. The $\boldsymbol{\kappa}$-operator leads to a lagrangian term of the form $(F_{H_1}^\dagger L)(H_1^\dagger L)$, which contributes to neutrino masses as [2]

$$\mathbf{m}_\nu^{(\kappa)} = 2\,\boldsymbol{\kappa}\,\mu\,\frac{v^2}{M^2}\sin\beta\cos\beta\,. \tag{3}$$

The $\tilde{\mathbf{B}}_\kappa$-operator gives a lagrangian term of the form $(H_1^\dagger L)^2$, which induces [1]

$$\mathbf{m}_\nu^{(\tilde{B}_\kappa)} = \tilde{\mathbf{B}}_\kappa \frac{v^2}{M^2}\cos^2\beta\,. \tag{4}$$

This novel contribution to $\mathbf{m}_\nu$ can dominate over $\mathbf{m}_\nu^{(\kappa)}$. The remaining operators of eq. (1), with coefficients $\mathbf{B}_\kappa$ and $\mathbf{C}_\kappa$, contribute to neutrino masses through finite one-loop diagrams involving gauginos and sleptons. Considering the soft mass matrix of 'left-handed' sleptons $\tilde{L}$ as $\mathbf{m}_{\tilde{L}}^2 = \tilde{m}_L^2(\mathbb{1} + \boldsymbol{\Delta}_L)$ ($\tilde{m}_L^2$ sets the overall mass scale and the dimensionless matrix $\boldsymbol{\Delta}_L$ accounts for the flavour dependence), we get the following contributions to $\mathbf{m}_\nu$ (at first order in $\boldsymbol{\Delta}_L$)

$$\delta_{B_\kappa}\mathbf{m}_\nu \simeq \frac{1}{32\pi^2}\left[-\left(\frac{g^2}{M_2} f_{L2} + \frac{g'^2}{M_1} f_{L1}\right)\mathbf{B}_\kappa + \left(\frac{g^2}{M_2} h_{L2} + \frac{g'^2}{M_1} h_{L1}\right)(\mathbf{B}_\kappa \boldsymbol{\Delta}_L + \boldsymbol{\Delta}_L^T \mathbf{B}_\kappa)\right] 2\mu \frac{v^2}{M^2}\sin\beta\cos\beta\,, \tag{5}$$

$$\delta_{C_\kappa}\mathbf{m}_\nu \simeq \frac{1}{32\pi^2}\left[-\left(\frac{g^2}{M_2} f_{L2} + \frac{g'^2}{M_1} f_{L1}\right)\mathbf{C}_\kappa + \left(\frac{g^2}{M_2} h_{L2} + \frac{g'^2}{M_1} h_{L1}\right)(\mathbf{C}_\kappa \boldsymbol{\Delta}_L + \boldsymbol{\Delta}_L^T \mathbf{C}_\kappa)\right] \frac{v^2}{M^2}\cos^2\beta\,, \tag{6}$$

where $f_{La} = f(\tilde{m}_L^2/|M_a|^2)$, $h_{La} = h(\tilde{m}_L^2/|M_a|^2)$, $f(x) = (x - 1 - \log x)/(x-1)^2$ and $h(x) = (x^2 - 1 - 2x\log x)/(x-1)^3$. Both the flavour structure and the size of $\delta_{B_\kappa}\mathbf{m}_\nu$, $\delta_{C_\kappa}\mathbf{m}_\nu$ are model dependent.

2. Type II seesaw realizations

Among the three variants of the seesaw mechanism (which generate the familiar $d = 5$ superpotential operator at the tree level), the type II is the natural one in which the above $d = 6$ operators emerge. In fact, the tree-level exchange of type I or type III mediators leads to $\Delta L = 0$ Kähler operators of the form $|H_2 L|^2$, whereas the type II mediators induce both $\Delta L = 0$ and $\Delta L = 2$ operators. The type II seesaw mechanism is realized through the exchange of $SU(2)_W$ triplet states T and $\bar{T}$ in a vector-like $SU(2)_W \times U(1)_Y$ representation, $T \sim (3,1)$, $\bar{T} \sim (3,-1)$. The relevant superpotential is: $W \supset \frac{1}{\sqrt{2}}\mathbf{Y}_T^{ij} L_i T L_j + \frac{1}{\sqrt{2}}\lambda_1 H_1 T H_1 + \frac{1}{\sqrt{2}}\lambda_2 H_2 \bar{T} H_2 + M_T T\bar{T}$, where $\mathbf{Y}_T^{ij}$ is a 3×3 symmetric matrix, $\lambda_{1,2}$ are dimensionless couplings and M_T is the (SUSY) triplet mass. By integrating out the triplets one obtains $\Delta L = 2$ effective operators of dimension

$d = 5$, *i.e.* $W_{\rm eff} \supset \frac{\lambda_2}{2M_T}\mathbf{Y}_T^{ij}(L_iH_2)(L_jH_2)$, and $d = 6$, *i.e.* $K_{\rm eff} \supset \frac{\lambda_1^*}{2|M_T|^2}\mathbf{Y}_T^{ij}(H_1^\dagger L_i)(H_1^\dagger L_j) + h.c.$, both of which can generate neutrino masses. The former operator is usually leading, but we assume it to be strongly suppressed (absent) by a very small (vanishing) value of λ_2, which can be justified by symmetry arguments. In this case the leading $\Delta L = 2$ operator is the Kähler one, which can be matched to the SUSY part of eq. (1) through the identification: $\boldsymbol{\kappa} = \lambda_1^*\mathbf{Y}_T$ and $M^2 = |M_T|^2$. The resulting contribution to neutrino masses is the tree-level term $\mathbf{m}_\nu^{(\kappa)}$ of eq. (3).

One can now ask how the SUSY-breaking operators of eq. (1) arise in the type II seesaw framework. The answer to this question depends on the ordering of the SUSY-breaking mediation scale M_S and the triplet mass M_T: $M_S < M_T$; $M_S > M_T$ or $M_S = M_T$.

• $M_S < M_T$. If SUSY breaking mediation occurs at $M_S < M_T$ via a messenger sector coupled to the MSSM through gauge interactions only (pure gauge mediation), then SUSY-breaking gaugino and (flavour blind) sfermion masses arise at M_S through loop diagrams, while trilinear terms are driven below M_S by gaugino mass terms in the RGEs. As for our $\Delta L = 2$ operators of eq. (1), only the SUSY one with coefficient $\boldsymbol{\kappa}$ is present above M_S (it is generated at M_T), whereas the SUSY-breaking ones receive finite two-loop contributions at M_S, and logarithmic ones below M_S through RGEs. The dominant source of neutrino masses is, generically, the SUSY contribution $\mathbf{m}_\nu^{(\kappa)}$ of eq. (3).

• $M_S > M_T$. Suppose that SUSY-breaking terms are generated at $M_S > M_T$ through, *e.g.*, gravity or gauge mediation. This means that all the MSSM and triplet fields have SUSY-breaking mass parameters at M_T. In this case, the tree-level decoupling of the triplets generates all the $\Delta L = 2$ effective operators of eq. (1), with coefficients $\boldsymbol{\kappa} = \lambda_1^*\mathbf{Y}_T$, $\mathbf{B}_\kappa = \lambda_1^*(\mathbf{Y}_T B_T - \mathbf{A}_T)$, $\tilde{\mathbf{B}}_\kappa = (\lambda_1^* B_T^* - A_1^*)\mathbf{Y}_T$, $\mathbf{C}_\kappa = (\lambda_1^* B_T^* - A_1^*)(\mathbf{Y}_T B_T - \mathbf{A}_T) - \lambda_1^*\mathbf{Y}_T m_{\tilde{T}}^2$.

3. $M_S = M_T$: Seesaw mediators as SUSY-breaking messengers

The $M_S = M_T$ scenario is obtained by identifying T and $\bar{T}$ as being SUSY-breaking mediators. They are embedded in a messenger sector which (in order to generate the gluino mass) should also include coloured fields. We also require that perturbative unification of gauge couplings be preserved and that all messenger masses be of the same order. This implies that the messenger sector should have a common total Dynkin index N for each subgroup of $SU(3)_C \times SU(2)_W \times U(1)_Y$. Since the $T + \bar{T}$ pair has $SU(2)$ index $N_2 = 4$, we are constrained to $N \geq 4$. The minimal case ($N = 4$) can be built by adding a pair of $SU(3)_C$ triplets $(3, 1, -1/3) + (\bar{3}, 1, +1/3)$ and an $SU(3)_C$ adjoint $(8, 1, 0)$ to T and $\bar{T}$.

The MSSM SUSY-breaking parameters are generated at the quantum level by a messenger sector of the type described above, coupled to the MSSM fields through both gauge and Yukawa interactions [4]. At the one-loop level, the gaugino masses M_a, the Higgs B-term B_H and the trilinear terms $\mathbf{A}_x$ are: $M_a = -\frac{NB_T}{16\pi^2}g_a^2$, $B_H = \frac{3B_T}{16\pi^2}|\lambda_1|^2$ $\mathbf{A}_e = \frac{3B_T}{16\pi^2}\mathbf{Y}_e(\mathbf{Y}_T^\dagger\mathbf{Y}_T + |\lambda_1|^2)$, $\mathbf{A}_d = \frac{3B_T}{16\pi^2}\mathbf{Y}_d|\lambda_1|^2$ and $\mathbf{A}_u = 0$, where $g_1^2 = (5/3)g'^2$ and $g_2^2 = g^2$. Non-vanishing $\mathcal{O}(B_T^2)$ contributions for the squared scalar masses arise at the two-loop level. For sleptons these are:

$$\mathbf{m}_{\tilde{L}}^2 = \left(\frac{|B_T|}{16\pi^2}\right)^2 \left[N\left(\frac{3}{10}g_1^4 + \frac{3}{2}g_2^4\right) - \left(\frac{27}{5}g_1^2 + 21g_2^2\right)\mathbf{Y}_T^\dagger\mathbf{Y}_T + 3|\lambda_1|^2(\mathbf{Y}_T^\dagger\mathbf{Y}_T - \mathbf{Y}_e^\dagger\mathbf{Y}_e) \right.$$
$$\left. +3\,\mathbf{Y}_T^\dagger(\mathbf{Y}_e^\dagger\mathbf{Y}_e)^T\mathbf{Y}_T + 18\,(\mathbf{Y}_T^\dagger\mathbf{Y}_T)^2 + 3\,\mathbf{Y}_T^\dagger\mathbf{Y}_T\mathrm{Tr}(\mathbf{Y}_T^\dagger\mathbf{Y}_T)\right], \quad (7)$$

$$\mathbf{m}_{\tilde{e}^c}^2 = \left(\frac{|B_T|}{16\pi^2}\right)^2 \left[N\left(\frac{6}{5}g_1^4\right) - 6\,\mathbf{Y}_e(\mathbf{Y}_T^\dagger\mathbf{Y}_T + |\lambda_1|^2)\mathbf{Y}_e^\dagger\right]. \quad (8)$$

Notice that the flavour structures of $\mathbf{A}_e$, $\mathbf{m}_{\tilde{L}}^2$ and $\mathbf{m}_{\tilde{e}^c}^2$ are controlled by $\mathbf{Y}_T$ and $\mathbf{Y}_e$, which in turn are determined by the low-energy lepton masses and mixing angles. Such minimal LFV

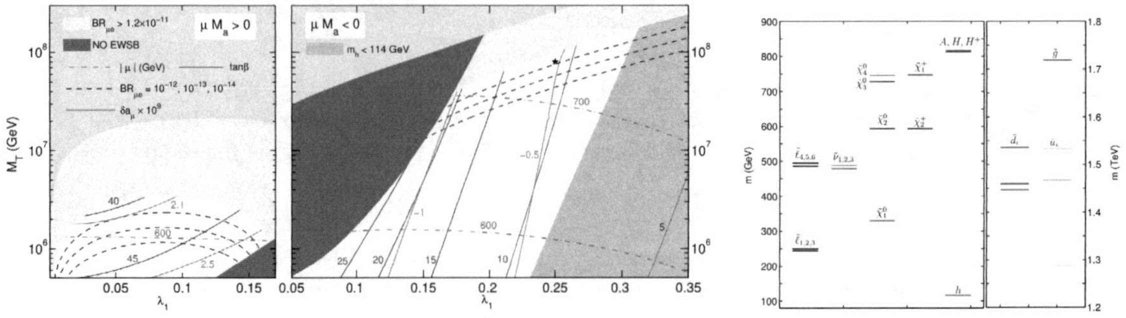

Figure 1. Plots of the $N = 4$ model for $B_T = 60$ TeV and normally ordered neutrino spectrum with $0 = m_1 < m_2 \ll m_3$ and $s_{13} = 0$. *Left panels*: The (λ_1, M_T) parameter space for $\mu M_a > 0$ (left) and $\mu M_a < 0$ (right). The white region is the allowed one. The dashed lines correspond to $\mathrm{BR}(\mu \to e\gamma) = 10^{-12}, 10^{-13}, 10^{-14}$ (from top to bottom.). *Right panel*: Sparticle and Higgs spectrum for $M_T = 8 \times 10^7$ GeV and $\lambda_1 = 0.25$ (point $\star$ of the parameter space).

properties are a characteristic feature of the SUSY type II seesaw [4, 5]. Our scenario is a flavoured variant of gauge mediation and possesses the property of minimal flavour violation in both the quark and lepton sectors.

Upon decoupling the triplets, the MSSM SUSY-breaking masses are generated through finite radiative effects, while the $\Delta L = 2$ SUSY-breaking parameters $\mathbf{B}_\kappa$, $\tilde{\mathbf{B}}_\kappa$ and $\mathbf{C}_\kappa$ arise at the tree level. The latter have a very simple form: $\mathbf{B}_\kappa = B_T\,\boldsymbol{\kappa}$, $\tilde{\mathbf{B}}_\kappa = B_T^*\,\boldsymbol{\kappa}$ and $\mathbf{C}_\kappa = |B_T|^2\,\boldsymbol{\kappa}$, where $\boldsymbol{\kappa} = \lambda_1^* \mathbf{Y}_T$. Eqs. (3) and (4) imply that the tree-level contribution to $\mathbf{m}_\nu$ is:

$$\mathbf{m}_\nu = \boldsymbol{\kappa}\,(B_T^* + 2\,\mu \tan\beta)\cos^2\!\beta\,\frac{v^2}{|M_T|^2}\,. \tag{9}$$

Notice that the SUSY-breaking parameter B_T acts as a common source for both sparticle and neutrino masses.

This scenario has a small number of free parameters, namely M_T, B_T, λ_1 and the messenger index N. Once these are fixed, the remaining parameters $\mathbf{Y}_T$, $\tan\beta$ and μ are determined by the low-energy neutrino data and by requiring proper EWSB. Concerning neutrino data, we recall that the neutrino mass matrix $\mathbf{m}_\nu$ is related to the low-energy observables as $\mathbf{m}_\nu = \mathbf{U}^*\mathbf{m}_\nu^D\mathbf{U}^\dagger$, where $\mathbf{m}_\nu^D = \mathrm{diag}(m_1, m_2, m_3)$, m_a are the neutrino masses and $\mathbf{U}$ is the lepton mixing matrix. Some representative numerical results for the $N = 4$ model are shown in Fig. 1, for $B_T = 60$ TeV and a normally ordered neutrino spectrum, with $0 = m_1 < m_2 \ll m_3$ and $s_{13} = 0$. In the left part of Fig. 1 we show two plots of the (λ_1, M_T) parameter space, including contours of $\tan\beta$ and μ (extracted by imposing EWSB). The left (right) panel corresponds to solutions of the EWSB conditions with $\mu M_a > 0$ ($\mu M_a < 0$). The main phenomenological constraints come from the LFV decay $\mu \to e\gamma$ and the lightest Higgs mass. In the right panel of Fig. 1 we show the sparticle and Higgs spectrum for $M_T = 8 \times 10^7$ GeV and $\lambda_1 = 0.25$ (which corresponds to $\tan\beta \simeq 11$), again for $B_T = 60$ TeV. The Higgs sector is in the decoupling regime, since the states A, H and H^+ are much heavier than h. Gluino and squarks are the heaviest sparticles and the lightest of them is $\tilde{t}_1$ (which is mainly $\tilde{t}_R$). In the electroweak sector, the heaviest chargino and neutralinos ($\tilde{\chi}_2^+$, $\tilde{\chi}_{3,4}^0$) are mainly Higgsino-like, while $\tilde{\chi}_1^+$ and $\tilde{\chi}_2^0$ are mostly Wino-like. The (mainly left-handed) sleptons $\tilde{\ell}_{4,5,6}$ and the sneutrinos $\tilde{\nu}_{1,2,3}$ are somewhat lighter than those states, and the Bino-like neutralino $\tilde{\chi}_1^0$ is even lighter. Finally, the lightest MSSM sparticles are the (mainly right-handed) sleptons $\tilde{\ell}_{1,2,3}$, as generically occurs in gauge mediated models with messenger index $N > 1$ and not too large mediation scale. The slepton $\tilde{\ell}_1$ (which is mainly $\tilde{\tau}_R$) is the next-to-lightest SUSY particle (NLSP), while the gravitino $\tilde{G}$ is the lightest one (LSP).

The scenario described above can be tested at current and future colliders. For instance, if a *pp* collision at the LHC produces two squarks, each of them can decay through well known chains, such as $\tilde{q}_R \to q\tilde{\chi}^0_1 \to q\tau\tilde{\ell}_1$, or $\tilde{q}_L \to q\tilde{\chi}^0_2 \to q\ell\tilde{\ell} \to q\ell^+\ell^-\tilde{\chi}^0_1 \to q\ell^+\ell^-\tau\tilde{\ell}_1$, or similar ones with charginos and/or sneutrinos (and neutrinos). Hence, in general, the final state of such a collision contains SM particles and two NLSPs $\tilde{\ell}_1$, which eventually decay to $\tau\tilde{G}$ with rate $\Gamma = m^5_{\tilde{\ell}_1}/(16\pi F^2)$. The latter decays can occur either promptly, or at a displaced vertex, or even outside the main detector.

Since LFV is intrinsically present in our framework, LFV processes are a crucial tool to discriminate our model from pure gauge mediation ones. For moderate $\tan\beta$, the leading LFV structure $\mathbf{Y}^\dagger_T\mathbf{Y}_T$, which appears in $\mathbf{m}^2_{\tilde{L}}$ (and $\mathbf{A}_e$), can be related to the neutrino parameters as

$$(\mathbf{m}^2_{\tilde{L}})_{ij} \propto B^2_T(\mathbf{Y}^\dagger_T\mathbf{Y}_T)_{ij} \propto \left(\frac{M^2_T\tan^2\beta}{\lambda_1}\right)^2 \left[\mathbf{V}(\mathbf{m}^D_\nu)^2\mathbf{V}^\dagger\right]_{ij} \propto \tan^5\beta\, M^4_T \left[\mathbf{V}(\mathbf{m}^D_\nu)^2\mathbf{V}^\dagger\right]_{ij} . \quad (10)$$

LFV signals can therefore appear either at high-energy colliders or in low-energy processes. Concerning the former possibility, LFV could show up at the LHC in, *e.g.*, neutralino decays, such as $\tilde{\chi}^0_2 \to \ell^\pm_i\tilde{\ell}^\mp_a \to \ell^\pm_i\ell^\mp_j\tilde{\chi}^0_1$ with $i \neq j$, followed by the flavour-conserving decay $\tilde{\chi}^0_1 \to \tau^+\tau^-\tilde{G}$ (or $\tilde{\chi}^0_1 \to \tilde{\ell}^\pm_1\tau^\mp$ if the NLSP is long-lived).

As for low-energy LFV processes, let us focus on the radiative decays $\ell_i \to \ell_j\gamma$. By using eq. (10), we can infer that

$$\mathrm{BR}(\ell_i \to \ell_j\gamma) \propto \frac{|(\mathbf{m}^2_{\tilde{L}})_{ij}|^2}{\tilde{m}^8}\tan^2\beta \propto \left(\frac{M_T}{B_T}\right)^8 (\tan\beta)^{12} \left|\left[\mathbf{V}(\mathbf{m}^D_\nu)^2\mathbf{V}^\dagger\right]_{ij}\right|^2 , \quad (11)$$

where the flavour-dependence is determined by the low-energy neutrino parameters only [5, 4]. If we take ratios of BRs, we have that, for a normal neutrino mass spectrum, $\mathrm{BR}(\tau \to \mu\gamma)/\mathrm{BR}(\mu \to e\gamma) \simeq 400\,(2)\,[3]$ and $\mathrm{BR}(\tau \to e\gamma)/\mathrm{BR}(\mu \to e\gamma) \simeq 0.2\,(0.1)\,[0.3]$ if $s_{13} = 0\,(s_{13} = 0.2, \delta = 0)\,[s_{13} = 0.2, \delta = \pi]$. These approximate results hold for small or moderate $\tan\beta$. Special features may emerge for large $\tan\beta$ [6].

In the left panel of Fig. 2 the three $\mathrm{BR}(\ell_i \to \ell_j\gamma)$ are shown as a function of $\tan\beta$, taking either $B_T = 60\,\mathrm{TeV}$ (solid lines) or $B_T = 100\,\mathrm{TeV}$ (dashed lines), for $M_T = 8 \times 10^7\,\mathrm{GeV}$, normally ordered neutrino spectrum and $s_{13} = 0$. In both examples $\mathrm{BR}(\mu \to e\gamma)$ can be tested at the MEG experiment for suitable ranges of $\tan\beta$. If $\mathrm{BR}(\mu \to e\gamma)$ is close to its present bound, $\mathrm{BR}(\tau \to \mu\gamma)$ is above 10^{-9}, within the reach of future Super Flavour Factories.

The double panel on the right of Fig. 2 illustrates the dependence of the BRs on the least known neutrino parameters, namely s_{13} and δ, for $B_T = 60\,\mathrm{TeV}$, $M_T = 8 \times 10^7\,\mathrm{GeV}$ and two values of $\tan\beta$. Regarding $\mathrm{BR}(\mu \to e\gamma)$ and $\mathrm{BR}(\tau \to e\gamma)$, the solid (dashed) curves correspond to $\delta = 0\,(\pi)$, while the region between such curves is spanned by intermediate values of δ. The dependence of $\mathrm{BR}(\tau \to \mu\gamma)$ on s_{13} and δ is negligible. The first subpanel corresponds to a scenario which could be tested very soon at MEG through the search of $\mu \to e\gamma$, if $s_{13} \ll 0.01$. Notice that $\mathrm{BR}(\tau \to \mu\gamma)$ in this example is around 4×10^{-9}, within the reach of future Super Flavour Factories, while $\tau \to e\gamma$ would be unobservable because $\mathrm{BR}(\tau \to e\gamma) \sim 10^{-12}$. For $s_{13} \sim 0.01$, $\mathrm{BR}(\mu \to e\gamma)$ and $\mathrm{BR}(\tau \to e\gamma)$ can be either enhanced or suppressed since, depending on the value of δ, a cancellation can occur in the LFV quantity $\left[\mathbf{V}(\mathbf{m}^D_\nu)^2\mathbf{V}^\dagger\right]_{ij}$ [4, 6]. The cancellation takes place in $\mathrm{BR}(\mu \to e\gamma)$ $[\mathrm{BR}(\tau \to e\gamma)]$ for $\delta = \pi\,(0)$ in the case of normal ordering, while the opposite occurs for inverted ordering. If $\mathrm{BR}(\mu \to e\gamma)$ is suppressed by that cancellation mechanism, only $\tau \to \mu\gamma$ can be observed. In such a case, we can even obtain

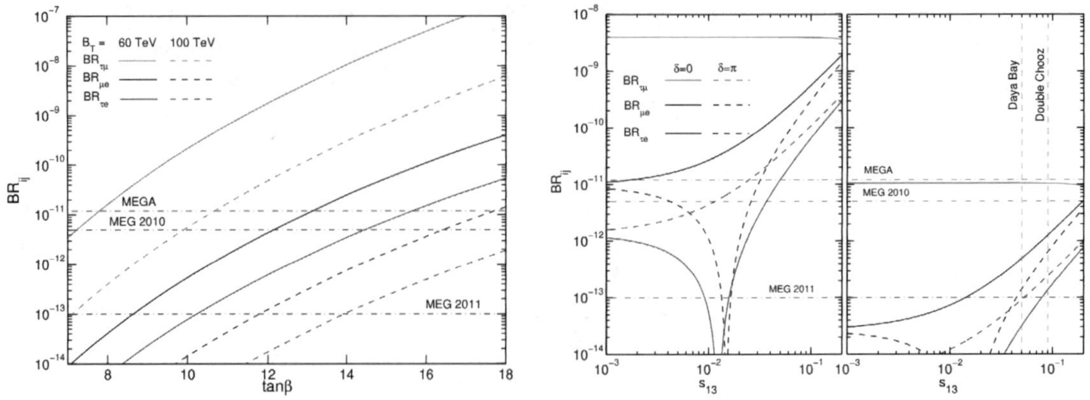

Figure 2. Plots of BR($\ell_i \to \ell_j\gamma$) for $M_T = 8 \times 10^7\,$GeV. *Left panel*: The BRs as a function of $\tan\beta$ for $B_T = 60\,$TeV (solid lines) or $B_T = 100\,$TeV (dashed lines), with $s_{13} = 0$. *Right panel*: The BRs as a function of s_{13} for $B_T = 60\,$TeV, with $\tan\beta = 13\,(8)$ in the first (second) subpanel. For BR($\mu \to e\gamma$) and BR($\tau \to e\gamma$), the solid (dashed) curves correspond to $\delta = 0\,(\pi)$ assuming a normally-ordered neutrino spectrum.

values of BR($\tau \to \mu\gamma$) above 10^{-8} by slightly changing the model parameters. In the case of partial cancellations, $\mu \to e\gamma$ could be still probed by MEG for values of s_{13} up to about 0.03, which are in the potential reach of future Neutrino Factories. The second subpanel shows an alternative possibility, in which LFV τ decays are unobservable, whereas BR($\mu \to e\gamma$) lies in the range $10^{-13} - 5 \times 10^{-12}$ if $0.05 \sim s_{13} < 0.2$. Those values of BR($\mu \to e\gamma$) should be probed by MEG next year, while the indicated range of s_{13} is within the sensitivity of the present and incoming neutrino experiments. This example shows the importance of the interplay between LFV searches and neutrino oscillation experiments.

In conclusion, we have summarized the investigation of Ref. [1] on an interesting alternative neutrino mass mechanism, which relies on the $d = 6$, $\Delta L = 2$ Kähler operator $(H_1^\dagger L)^2/M^2$. We have presented both a general effective-theory description and explicit realizations in the context of the type II seesaw mechanism. If the seesaw mediators are also identified with SUSY-breaking messengers, strong correlations arise between neutrino parameters, sparticle and Higgs masses, as well as LFV processes.

References

[1] Brignole A, Joaquim F R and Rossi A 2010, Beyond the standard seesaw: neutrino masses from Kähler operators and broken supersymmetry, *J. High Energy Phys* JHEP1008(2010)133.

[2] Casas J A, Espinosa J R and Navarro I 2002, New supersymmetric source of neutrino masses and mixings, *Phys. Rev. Lett.* **89** 161801.

[3] Brignole A 2000, One-loop Kähler potential in non-renormalizable theories, Nucl. Phys. B **579** 101.

[4] Joaquim F R and Rossi A 2006, Gauge and Yukawa mediated supersymmetry breaking in the triplet seesaw scenario, *Phys. Rev. Lett.* **97** 181801; Joaquim F R and Rossi A 2007, Phenomenology of the triplet seesaw mechanism with Gauge and Yukawa mediation of SUSY breaking, Nucl. Phys. B **765** 71.

[5] Rossi A 2002, Supersymmetric seesaw without singlet neutrinos: Neutrino masses and lepton-flavour violation, Phys. Rev. D **66** 075003.

[6] Joaquim F R 2009, Predictions for $\ell_i \to \ell_j\gamma$ in the SUSY triplet seesaw mechanism: large $\tan\beta$ effects, Nucl. Phys. Proc. Suppl. **188** 342; Joaquim F R, Running effects on neutrino parameters and $\ell_i \to \ell_j\gamma$ predictions in the triplet-extended MSSM, *J. High Energy Phys* JHEP1006(2010)079.

Journal of Physics: Conference Series **259** (2010) 012069 doi:10.1088/1742-6596/259/1/012069

Tree Level Mediation of Supersymmetry Breaking

Marco Nardecchia
SISSA/ISAS and INFN, I–34013 Trieste, Italy
E-mail: marco.nardecchia@sissa.it

Abstract. We propose a new scheme in which supersymmetry breaking is communicated to the MSSM sfermions by GUT gauge interactions at the tree level. The (positive) contribution of MSSM fields to $\mathrm{Str}(\mathcal{M}^2)$ is automatically compensated by a (negative) contribution from heavy fields. Sfermion masses are flavour universal, thus solving the supersymmetric flavour problem. In the simplest SO(10) embedding, the ratio of different sfermion masses is predicted and differs from mSugra and other schemes, thus making this framework testable at the LHC. Gaugino masses are generated at the loop level but enhanced by model dependent factors.

We finally entered the LHC era. Hopefully, the ongoing experiments at CERN will clarify the nature of the electroweak symmetry breaking. The simplest and most economical solution to obtain the $SU(2)_L \otimes U(1)_Y$ breaking is trough the Higgs mechanism, and a light SM Higgs scalar is also preferred by the EWPT.

The presence of an elementary scalar in the theory poses new questions and an extension of the SM is expected at the TeV scale. In particular, if supersymmetry exists close to the TeV energy scale, it allows for a solution of the naturalness problem of the SM.

In the MSSM, supersymmetry is explicitly broken by the presence of the soft terms. Fully generic flavor-breaking structures in the soft terms are ruled out by experimental constraints. However, these constraints can be used to identify the restricted class of allowed soft terms, providing useful guidelines to understand the mechanism of the supersymmetry breaking and its mediation.

Recently [1] we considered a new option in which spontaneous supersymmetry breaking is communicated to the observable sector at the tree level through GUT gauge interactions.

Tree level supersymmetry breaking is sometimes considered not to be viable because the supertrace formula [2]. This clearly represents a problem if the only fermions in chiral superfields are the SM ones, as the experimental constraints rather require a significantly larger sfermion total squared mass.

In our scheme, the supertrace does vanish (in the full theory at the GUT scale), but the positive contribution from the MSSM matter fields is automatically compensated by a negative contribution from heavier chiral superfields. In order for this to work, a SM-neutral gauge U(1) in addition to the SM hypercharge is needed to avoid a stronger implication of the supertrace formula, which requires the lightest squark in either the up or down sector not to be heavier than the corresponding lightest quark [3]. The main features of our model arise from requiring that such an extra U(1) be part of a unified group.

Before presenting the model, let us motivate its gauge structure and field content. Our aim is to identify the supersymmetry breaking messengers with heavy vector superfields corresponding

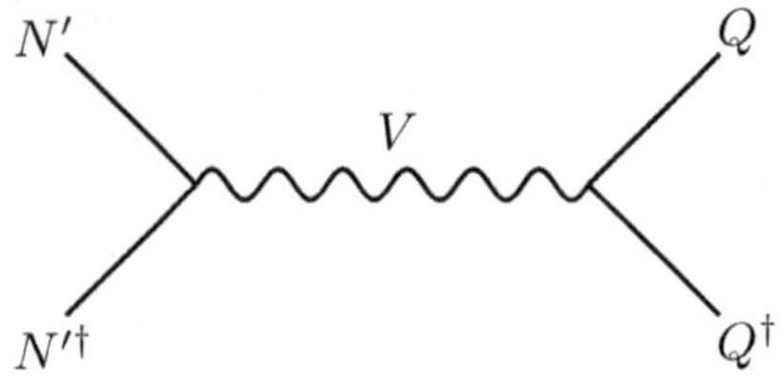

Figure 1: Tree level gauge mediation supergraph inducing a soft mass for the sfermion $\tilde{Q}$.

to broken generators, X, of a simple grand unified group, as illustrated in Fig. 1. There, N' is a SM singlet superfield whose F-term breaks supersymmetry, $\langle N' \rangle = F\,\theta^2$ (the prime is there just for consistency with the notations used below). As N' has to couple to the heavy vector V associated to the broken generator X, N' must belong to a non-trivial multiplet of the unified group. Q represents a generic MSSM superfield. In the effective theory below $M_{\rm GUT}$, the diagram in Fig. 1 induces a non-renormalizable contribution $-2g^2 X_N X_Q (Q^\dagger Q N'^\dagger N')/M_V^2$ to the Kähler potential, analogous to the ones of effective supergravity, but flavour universal ($X_{N,Q}$ are the X-charges of N', Q, M_V is the vector mass). A sfermion mass $\tilde{m}_Q^2 = 2g^2 X_N X_Q (F/M_V)^2$ is then generated. In the full theory at $M_{\rm GUT}$, on the other hand, everything takes place at the renormalizable level. In fact, the sfermion masses arise because N' couples to the broken generator X. As a consequence, its F-term generates a non-vanishing vev for the corresponding D-term

$$\langle D_X \rangle = -2gX_N \left(\frac{F}{M_V}\right)^2, \tag{1}$$

which in turn induces the soft mass

$$\tilde{m}_Q^2 = -gX_Q \langle D_X \rangle = 2g^2 X_N X_Q \left(\frac{F}{M_V}\right)^2 \tag{2}$$

for the sfermion $\tilde{Q}$.

Such a scheme requires specific gauge structures and field contents. First of all, the heavy vector field V in Fig. 1 must be a SM singlet, as N' is. Then, SU(5) does not provide viable candidates for the gauge messenger V and the minimal option is identifying the broken generator with the SU(5) singlet generator X of SO(10). As for the SM singlet N' whose F-term breaks supersymmetry, it must belong to a non-trivial SO(10) multiplet such that N' has a non-vanishing charge under X. Limiting ourselves to representations with dimension $d < 126$, the only possibility is that N' be the singlet component of a spinorial representation, 16 or $\overline{16}$. We also need a $16 + \overline{16}$ participating to SO(10) breaking at the GUT scale. At least two $16 + \overline{16}$ are then required, one getting a vev along the scalar component and the other along the F-term component. Finally, the standard embedding of a whole MSSM family into a 16 of SO(10) would not work, as it would lead to negative sfermion masses for some of the sfermions. That is why we distribute the matter fields in three 16 and three 10 of SO(10).

Having motivated some of its features, we now illustrate a minimal model satisfying the above requirements. The gauge group is SO(10). The matter fields (negative R-parity) are three $16_i = (\bar{5}_i^{16}, 10_i^{16}, 1_i^{16})$ and three $10_i = (5_i^{10}, \bar{5}_i^{10})$, $i = 1,2,3$, where the SU(5) decomposition is also indicated. Supersymmetry and SO(10) breaking to SU(5) are provided by $16 = (\bar{5}^{16}, 10^{16}, N)$,

Journal of Physics: Conference Series **259** (2010) 012069 doi:10.1088/1742-6596/259/1/012069

$\overline{16} = (5^{16}, \overline{10}^{16}, \bar{N})$, $16' = (\bar{5}'^{16}, 10'^{16}, N')$, $\overline{16}' = (5'^{16}, \overline{10}'^{16}, \overline{N}')$ (positive R-parity), with

$$\langle N' \rangle = F\,\theta^2 \qquad \langle \overline{N}' \rangle = 0 \qquad \langle N \rangle = M \qquad \langle \overline{N} \rangle = M, \tag{3}$$

$\sqrt{F} \ll M \sim M_{\rm GUT}$. The D-term condition forces $|\langle N \rangle| = |\langle \overline{N} \rangle|$ and the phases of all the vevs can be taken positive without loss of generality. The MSSM up Higgs h_u is embedded in a $10 = (5^{10}, \bar{5}^{10})$ of SO(10), while the down Higgs h_d is a mixture of the doublets in the 10 and the 16,

$$10 = h_u + c_d h_d + \text{heavy}, \quad 16 = s_d h_d + \text{heavy}, \tag{4}$$

where $c_d = \cos\theta_d$, $s_d = \sin\theta_d$ and $0 < \theta_d < \pi/2$ parametrizes the mixing in the down Higgs sector.

At this point we are in the condition of calculating the sfermion masses induced by integrating out the heavy vector fields:

$$\tilde{m}_Q^2 = \frac{X_Q}{2X_N} m^2, \qquad m \equiv \frac{F}{M}. \tag{5}$$

In the normalization we use for X, $X_N = 5$. In order to determine the X charge of the SM fermions we need to specify their embedding in the matter fields $16_i + 10_i$. We do that by first writing the most general R-parity conserving superpotential, except a possible mass term for the 10_i, as

$$W = \frac{y_{ij}}{2} 16_i 16_j 10 + h_{ij} 16_i 10_j 16 + h'_{ij} 16_i 10_j 16' + W_{\rm vev} + W_{\rm NR}, \tag{6}$$

where $W_{\rm vev} = W_{\rm vev}(16, \overline{16}, 10, \ldots)$ does not involve the matter fields and takes care of the vevs, the doublet triplet splitting, and the Higgs mixing, and $W_{\rm NR}$ contains non-renormalizable contributions to the superpotential needed in order to account for the measured ratios of down quark and charged lepton masses (we will ignore such issue here).

We can now see that the vev of the 16 gives rise to the mass term $h_{ij} M \bar{5}_i^{16} 5_j^{10}$, which makes the $\bar{5}_i^{16}$ and 5_j^{10} heavy. Only the MSSM superfield content survives at the electroweak scale (assuming the three singlets in the 16_i get mass e.g. from non-renormalizable interactions with the $\overline{16}$). Moreover, the three MSSM families turn out to be embedded in the three 10_i^{16}, with $X = 1$ and in the three $\bar{5}_i^{10}$, with $X = 2$. We can then go back to eq. (5) and obtain

$$\tilde{m}_q^2 = \tilde{m}_{u^c}^2 = \tilde{m}_{e^c}^2 = \tilde{m}_{10}^2 = \frac{1}{10} m^2, \quad \tilde{m}_l^2 = \tilde{m}_{d^c}^2 = \tilde{m}_{\bar{5}}^2 = \frac{1}{5} m^2 \tag{7}$$

$$m_{h_u}^2 = -\frac{1}{5} m^2, \quad m_{h_d}^2 = \frac{2c_d^2 - 3s_d^2}{10} m^2 \tag{8}$$

at the GUT scale. The result in eq. (7) is quite general, as it only depends on the choice of the gauge group and on the embedding of the three MSSM families in the $10_i^{16} + \bar{5}_i^{10}$. We note a few interesting features of this result.

- All the sfermion masses turn out to be positive. This is because the negative X charges (which must be there as X is traceless) happen to be associated to the fields that get an heavy supersymmetric mass.
- The sfermions masses are flavour universal, thus solving the supersymmetric flavour problem.
- The sfermions masses belonging to the 10 and $\bar{5}$ of SU(5) are related by

$$\tilde{m}_{q,u^c,e^c}^2 = \frac{1}{2} \tilde{m}_{l,d^c}^2 \tag{9}$$

at the GUT scale, a peculiar prediction that allows to distinguish this model from mSugra, gauge mediation, and other models of supersymmetry breaking.

Let us now consider gaugino masses. While the tree-level prediction for the sfermion masses, eq. (7), only depends on the choice of the unified gauge group and the MSSM embedding, gaugino masses arise at one loop, as in standard gauge mediation, and depend on the superpotential parameters. The chiral multiplets $\bar{5}_i^{16}$ and 5_j^{10} get an heavy supersymmetric mass $h_{ij}M$ and their scalar components get a supersymmetry breaking mass $h'_{ij}F$. They play the role of three pairs of chiral messengers in standard gauge mediation and give rise to one loop gaugino masses. The contribution of each messenger arises at a different scale. In the one loop approximation for the RGE running, the total gaugino masses at lower scales can be calculated by running effective GUT-scale gaugino masses given by

$$M_a = \frac{\alpha}{4\pi}\,\mathrm{tr}(h'h^{-1})\,m \equiv M_{1/2}, \quad a = 1,2,3, \tag{10}$$

where α is the unified coupling.

Let us compare gaugino and sfermion masses. Particularly interesting is the ratio $\tilde{m}_t/M_2$. In fact, the W-ino mass M_2 is at present bounded to be heavier than about 100 GeV, while $\tilde{m}_t$ enters the radiative corrections to the Higgs mass. Therefore, the ratio $\tilde{m}_t/M_2$ should not be too large in order not to increase the fine-tuning and not to push the stops and the other sfermions out of the LHC reach. From

$$\left.\frac{M_2}{\tilde{m}_t}\right|_{M_{\mathrm{GUT}}} = \frac{3\sqrt{10}}{(4\pi)^2}\lambda, \quad \lambda = \frac{g^2\,\mathrm{tr}(h'h^{-1})}{3} \tag{11}$$

we see first of all that the loop factor separating $\tilde{m}_t$ and M_2 is partially compensated by a combination of numerical factors: $(4\pi)^2 \sim 100$ (leading to $\tilde{m}_t \gtrsim 10\,\mathrm{TeV}$ for $\lambda = 1$) becomes $(4\pi)^2/(3\sqrt{10}) \sim 10$ (leading to $\tilde{m}_t \gtrsim 1\,\mathrm{TeV}$ for $\lambda = 1$). Note that the factor $\sqrt{10}$ is related to the ratio of X charges in eq. (5) and the factor 3 corresponds to the number of families ($\mathrm{tr}(h'h^{-1}) = 3$ for $h = h'$).

Next, we comment on the μ problem. Relating the μ-term to supersymmetry breaking is a highly model-dependent issue, but in our scenario we have a simple possibility in which both the F-term, $\langle N'\rangle = F\,\theta^2$ and μ originate from the same parameter $m \sim$ TeV in the superpotential: $W \supseteq mN'\overline{N}$.

Once $\overline{N}$ is forced to get its vev $\langle\overline{N}\rangle = M \sim M_{\mathrm{GUT}}$, N' acquires an F-term $F = mM$ (so that m is indeed the parameter introduced in eq. (5)). In our setup, N' and $\overline{N}$ are part of the SO(10) multiplets $16'$ and $\overline{16}$ respectively. A μ term related to the supersymmetry breaking scale $\mu \sim m$ is then therefore generated if h_u has a component in $\overline{16}$ and h_d has a component in $16'$. Such a situation can be achieved with an appropriate superpotential. Contrary to standard gauge mediation, there is no μ-$B\mu$ problem here, as $B\mu/\mu$ is not enhanced by an inverse loop factor. $B\mu$ can be generated at the tree level, for example as in [4], or it can be generated by the RGE evolution.

We now illustrate an example of low energy spectra that can be obtained in our framework. We neglect the (small, for our purposes) effect of the intermediate scale $\bar{5}_i^{16}$ and 5_j^{10} and use the MSSM RGE equations, with boundary conditions at high energy as in eqs. (7,8), the A-terms set to zero. We assume the messenger mass to coincide with the GUT scale, $M = M_{\mathrm{GUT}}$. The overall normalization of the unified gaugino masses $M_{1/2}$ can be considered as a free parameter due to the presence of the factor $\mathrm{tr}(h'h^{-1})$ in eq. (10), or equivalently of the factor λ in eq. (11). As the size of the parameters μ and B_μ is model dependent, we consider them as free parameters as well and recover them as usual in terms of M_Z and $\tan\beta$.

Table 2 shows the low-energy spectrum corresponding to $\theta_d = \pi/6$, $\tan\beta = 30$ and $\mathrm{sign}(\mu) = +$. The common gaugino mass is $M_{1/2} = 150\,\mathrm{GeV}$, near the minimal value allowed at

Higgs:	m_{h^0}	114
	m_{H^0}	1543
	m_A	1543
	$m_{H^\pm}$	1545
Gluinos:	$M_{\tilde{g}}$	448
Neutralinos:	$m_{\chi_1^0}$	62
	$m_{\chi_2^0}$	124
	$m_{\chi_3^0}$	1414
	$m_{\chi_4^0}$	1415
Charginos:	$m_{\chi_1^\pm}$	124
	$m_{\chi_2^\pm}$	1416
Squarks:	$m_{\tilde{u}_L}$	1092
	$m_{\tilde{u}_R}$	1027
	$m_{\tilde{d}_L}$	1095
	$m_{\tilde{d}_R}$	1494
	$m_{\tilde{t}_1}$	1007
	$m_{\tilde{t}_2}$	1038
	$m_{\tilde{b}_1}$	1069
	$m_{\tilde{b}_2}$	1435
Sleptons:	$m_{\tilde{e}_L}$	1420
	$m_{\tilde{e}_R}$	1091
	$m_{\tilde{\tau}_1}$	992
	$m_{\tilde{\tau}_2}$	1387
	$m_{\tilde{\nu}_e}$	1418
	$m_{\tilde{\nu}_\tau}$	1382

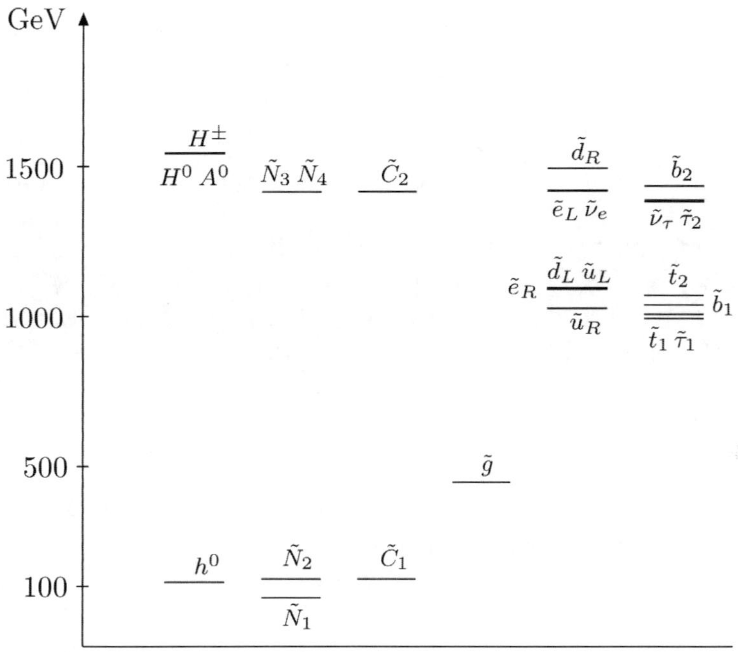

Figure 2: An example of spectrum, corresponding to $m = 3.2\,\text{TeV}$, $M_{1/2} = 150\,\text{GeV}$, $\theta_d = \pi/6$, $\tan\beta = 30$ and $\text{sign}(\mu) = +$, $A = 0$, $\eta = 1$. All the masses are in GeV, the first two families have an approximately equal mass.

present by chargino direct searches. The value of m is near the minimal value allowed by the bound $m_h > 114\,\text{GeV}$. This spectrum corresponds to $\lambda = 2.5$. Given the (moderate) hierarchy between $M_{1/2}$ and the sfermion masses, the sfermion RGEs are not significantly affected by the gaugino masses and the sfermion mass relations characterizing the model, eq. (7), survive, to some extent, at low energy.

Finally, we comment about cosmology. As in loop gauge mediation, the LSP is the gravitino. In fact, the supersymmetry breaking parameter is given by

$$\sqrt{F} \approx 0.8 \cdot 10^{10}\,\text{GeV} \left(\frac{\tilde{m}_{10}}{\text{TeV}} \frac{M}{2 \cdot 10^{16}\,\text{GeV}} \right)^{1/2} \tag{12}$$

and the gravitino mass by

$$m_{3/2} = \frac{F}{\sqrt{3} M_\text{P}} \approx 15\,\text{GeV} \left(\frac{\tilde{m}_{10}}{\text{TeV}} \frac{M}{2 \cdot 10^{16}\,\text{GeV}} \right), \tag{13}$$

Journal of Physics: Conference Series **259** (2010) 012069 doi:10.1088/1742-6596/259/1/012069

where $\tilde{m}_{10}$ is the tree-level mass of the sfermions in the 10 of SU(5) at the GUT scale. Note that F and the gravitino mass are smaller than in loop gauge mediation, for a given messenger scale M, because of the absence of a loop factor in eqs. (12,13). For a stable (on the age of the universe timescale) gravitino with a mass as large as in eq. (13), a dilution mechanism such as inflation is necessary in order for the its energy density not to exceed the dark matter one. The upper bound on the reheating temperature T_R depends on the gravitino and the gaugino masses [5]. The thermal contribution to the gravitino energy density, for a reheating temperature around 10^9 GeV is given by

$$\Omega_{\tilde{G}}^{\rm TP} h^2 \approx 6 \times 10^{-2} \left(\frac{T_{RH}}{10^9\,{\rm GeV}} \right) \left(\frac{15\,{\rm GeV}}{m_{3/2}} \right) \left(\frac{M_{1/2}}{150\,{\rm GeV}} \right)^2 . \tag{14}$$

For the spectrum in Table 2, the bound $\Omega_{\tilde{G}}^{\rm TP} h^2 \leq \Omega_{\rm DM} h^2 = 0.11$ translates in $T_R < 2 \cdot 10^9$ GeV.

We then have to take care of the decays of the NLSP into the gravitino, which might spoil big bang nucleosynthesis (BBN) unless it is fast enough. The fate of BBN depends on what the NLSP is. In the bulk of the parameter space we expect the NLSP to be the lightest neutralino or a stau. In the example in Table 2, the NLSP is essentially a Bino. For $m_{3/2} \sim 15\,{\rm GeV}$, the decay of a Bino NLSP through its coupling to the Goldstino component of the gravitino is way too slow (one would need $m_{3/2} < 100\,{\rm MeV}$ in order not to spoil BBN [6])). A Bino NLSP therefore requires a much faster decay channel. The latter can be provided by a tiny amount of R-parity violation [7]. Such a possibility is also consistent with thermal leptogenesis and gravitino dark matter. The other possibility is that the NLSP is a stau. In this case, all the BBN constraints can be satisfied if the lifetime of the stau is $\tau_{\tilde{\tau}} \approx 48\pi m_{3/2}^2 M_{\rm P}^2 / m_{\tilde{\tau}}^5 \lesssim 6 \cdot 10^3\, s$ [8]. This is a viable possibility, which however requires large $\lambda = \mathcal{O}(100)$ and sizable gaugino masses. For such large values of λ, radiative contributions to sfermion masses (from RGEs and the standard gauge mediation contribution) dominate over the tree level one, the spectrum approaches the usual one loop gauge mediated one, and the peculiar relation between sfermion masses at the messenger scale gets hidden.

In conclusion, we have shown that is possible to communicate the supersymmetry breaking through a tree level renormalizable exchange of a gauge (GUT) messenger, as in Fig. 1. This scheme solves the supersymmetric FCNC problem and, in its simplest implementation, leads to peculiar relations among sfermion masses that can be tested at the LHC.

Acknowledgments
I thank Andrea Romanino, Robert Ziegler and the organizers of PASCOS 2010.

References
[1] M. Nardecchia, A. Romanino and R. Ziegler, JHEP **0911** (2009) 112 [arXiv:0909.3058 [hep-ph]]; M. Nardecchia, A. Romanino and R. Ziegler, JHEP **1003** (2010) 024 [arXiv:0912.5482 [hep-ph]].
[2] S. Ferrara, L. Girardello and F. Palumbo, Phys. Rev. D **20** (1979) 403.
[3] S. Dimopoulos and H. Georgi, Nucl. Phys. B **193**, 150 (1981).
[4] N. Arkani-Hamed, S. Dimopoulos, G. F. Giudice and A. Romanino, Nucl. Phys. B **709** (2005) 3 [arXiv:hep-ph/0409232].
[5] M. Bolz, A. Brandenburg and W. Buchmuller, Nucl. Phys. B **606** (2001) 518 [Erratum-ibid. B **790** (2008) 336] [arXiv:hep-ph/0012052]; V. S. Rychkov and A. Strumia, Phys. Rev. D **75**, 075011 (2007) [arXiv:hep-ph/0701104].
[6] J. L. Feng, S. Su and F. Takayama, Phys. Rev. D **70** (2004) 075019 [arXiv:hep-ph/0404231].
[7] W. Buchmuller, L. Covi, K. Hamaguchi, A. Ibarra and T. Yanagida, JHEP **0703** (2007) 037 [arXiv:hep-ph/0702184].
[8] M. Pospelov, J. Pradler and F. D. Steffen, JCAP **0811** (2008) 020 [arXiv:0807.4287 [hep-ph]].

16th International Symposium on Particles Strings and Cosmology (PASCOS 2010) IOP Publishing
Journal of Physics: Conference Series **259** (2010) 012070 doi:10.1088/1742-6596/259/1/012070

Lepton number, black hole entropy and 10^{32} copies of the Standard Model [1]

Sergey Kovalenko[1], **Heinrich Päs**[2] **and Ivan Schmidt** [3]

[1,3] Departamento de Física, Universidad Técnica Federico Santa María, Casilla 110-V, Valparaíso, Chile and Center of Subatomic Physics, Valparaíso, Chile
[2] Fakultät Physik, Technische Universiät Dortmund, 44221 Dortmund, Germany

E-mail: [1] sergey.kovalenko@usm.cl

E-mail: [2] heinrich.paes@tu-dortmund.de

E-mail: [3] ivan.schmidt@usm.cl

Abstract. Lepton number violating processes are a typical problem in theories with a low quantum gravity scale. In this paper we examine lepton number violation (LNV) in theories with a saturated black hole bound on a large number of species. Such theories have been advocated recently as a possible solution to the hierarchy problem and an explanation of the smallness of neutrino masses. Naively one would expect black holes to introduce TeV scale LNV operators, thus generating unacceptably large rates of LNV processes. We show, however, that this does not happen in this scenario due to a complicated compensation mechanism between contributions of different Majorana neutrino states to these processes. As a result rates of LNV processes are extremely small and far beyond experimental reach, at least for the left-handed neutrino states.

Recently the existence of a large number of copies of Standard Model particles has been proposed as a possibility to lower the Planck scale and solve the electroweak hierarchy problem [1, 2]. In the following we briefly review the main argument of this approach, closely following [2]: Let us assume there exist N copies of the Standard Model, each carrying a separately conserved charge, where N is a large number $\sim 10^{32}$. It is in principle possible to prepare a black hole containing one particle of each species. Now, due to charge conservation, the contained charges have to be revealed in the evaporation process via Hawking radiation. On the other hand, a particle of mass Λ can only be emitted if the Hawking temperature is large enough,

$$T_H \simeq \frac{M_P^2}{M_{BH}} \gtrsim \Lambda. \tag{1}$$

Moreover, energy conservation bounds the maximal number of particles with mass Λ emitted due to the evaporation process of a black hole of mass M_{BH} to be

$$n_{\max} = \frac{M_{BH}}{\Lambda}. \tag{2}$$

[1] Talk presented by Heinrich Päs.

Assuming that the number of states N saturates the black hole bound we obtain with (1)

$$N = n_{\max} = \frac{M_P^2}{\Lambda^2}. \tag{3}$$

Finally, since semiclassically the lifetime of a blck hole is given by

$$\tau_{BH} = \frac{1}{N}\int \frac{1}{T_H^2} dM_{BH} \sim \Lambda^{-1} \tag{4}$$

we conclude that a black hole of size Λ has a life time of order Λ^{-1} implying that Λ is the scale where the semi-classical treatment breaks down and quantum gravity sets in, namely the true Planck scale,

$$\Lambda \simeq \frac{M_P}{\sqrt{N}}, \tag{5}$$

where M_P has to be interpreted as the effective Planck scale, which implies $\Lambda \sim \mathcal{O}(TeV)$ for $N \simeq 10^{32}$ and thus solves the hierarchy problem.

Finally, in [4] this scenario has been advocated also as a mechanism for generating small neutrino masses, providing an attractive alternative for seesaw, extra dimensional and other known mechanisms. It is assumed that there exists one SM singlet right-handed neutrino ν_{Rj} per SM$_j$ copy, so that $j = 1, ..., N$. The mechanism relies on the fact that the right-handed neutrinos, being SM singlets, couple to all the SM copies "democratically". This SM singlet democracy combined with the requirement of unitarity of the theory implies a $1/\sqrt{N}$ suppression of the corresponding Yukawa couplings to the left-handed neutrinos ν_{Lj} and thus a suppression of the Dirac type neutrino masses. The minimalistic approach to the problem of smallness of neutrino mass suggests that in this scenario possible $B-L$ violating Majorana masses of ν_{Rj} are unnecessary and the lepton number is conserved.

However, the assumption of lepton number conservation appears rather ad hoc, as there is no fundamental reason to forbid Majorana masses for the right-handed neutrinos and lepton number conservation appears as accidental symmetry. Quantum gravity breaks global symmetries, and then conserved lepton number requires a gauged $B-L$ symmetry $U_{1(B-L)}$. The latter should be spontaneously broken otherwise there must exist the corresponding massless gauge boson stringently constrained by phenomenology. On the other hand lepton number violation might be helpful for successful baryogenesis.

In the following we analyze some consequences of the N-copies SM without lepton number conservation . Interestingly, even in this case, Majorana masses turn out to be suppressed by an exactly analogous reason as the Dirac masses. We assume a

$$\prod_i (SU_{3c} \times SU_{2W} \times U_y)_i \times U_{1(B-L)} \times Z_N \tag{6}$$

gauge symmetry of the N-copies SM. It includes a common anomaly free gauge factor $U_{1(B-L)}$. We introduce this gauge symmetry in order to prevent the appearance of phenomenologically dangerous LNV operators induced by the TeV black holes. An additional permutation symmetry Z_N acting in the space of the SM$_i$ species is also imposed [4] in the N-copy SM scenario. To be unaffected by black holes, this discrete symmetry should be considered as a gauged symmetry in the sense of being a discrete subgroup of some continuous gauge group. The Lagrangian terms relevant for our discussion are the following:

$$\mathcal{L}_{\nu HS} = \lambda_{ij}\overline{\nu_{Rj}}\,(LH)_i + \beta_{ij}\overline{\nu_{Ri}^c}\nu_{Rj}S + \kappa_i(H^\dagger H)_i S^\dagger S. \tag{7}$$

The model involves N right-handed SM singlet neutrinos ν_{Ri} and one SM singlet complex scalar field S having the $B-L$-charge equal to $+2$. Then the trilinear HHS couplings are forbidden in (7). The $U_{1(B-L)}$ is spontaneously broken by a vacuum expectation value $\langle S \rangle$ resulting in a Majorana mass term from the second term in Eq. (7). We assume that the scale of $B - L$ breaking lies below the gravity cutoff Λ. The Dirac mass term considered in ref. [4] originates from the first term after the electroweak symmetry breaking.

Let us consider the $N \times N$ Yuakawa coupling matrix λ_{ij} of Dirac type following ref. [4]. As the ν_{Ri} fields are not charged under the SM symmetry they cannot be assigned to a single SM-copy, besides respecting the same transformation properties under a permutation symmetry acting on the space of species. This permutation symmetry constrains the Yukawa coupling matrix to the form

$$\lambda_{ij} = \begin{pmatrix} a & b & b & .. \\ b & a & b & .. \\ b & b & a & .. \\ .. & .. & .. & .. \end{pmatrix}. \tag{8}$$

This matrix combined in the first term in (7) with the SM Higgs expectation value $\langle H_i \rangle = \langle H \rangle$ results in the Dirac neutrino mass matrix

$$m^D_{ij} = \lambda_{ij} \langle H \rangle. \tag{9}$$

Following ref. [4] we assume that the electroweak symmetry breaking leaves the permutation symmetry unbroken. This implies that the VEVs of all the Higgs species are equal to the same value $\langle H \rangle$. A key point guaranteeing the smallness of neutrino mass matrix entries is the smallness of the Yukawa coupling matrix (8) which follows from the requirement of unitarity of the theory. In fact, let us consider the right-handed neutrino inclusive production in the scattering of the SM particles as shown in Fig. 1(a). At high energies the rate of this process grows like

$$\Gamma \simeq N b^2 E, \tag{10}$$

as follows from dimensional analysis. Here we assumed $a \sim b$ which is suggested by an observation that these two quantities are of the same nature and have no fundamental reason to be very different in magnitude [4]. Unitarity below the gravity cutoff is preserved only for

$$b \lesssim \frac{1}{\sqrt{N}}. \tag{11}$$

Thus the neutrino mass matrix (9) results in $N - 1$ Dirac neutrinos with tiny masses $m_D \simeq \langle H \rangle / \sqrt{N} \lesssim \mathcal{O}(eV)$ [4] which fulfill the experimental bounds constraining them to the sub-eV scale. One neutrino state in this framework is very heavy with the mass of the order $M^D \simeq \sqrt{N} \langle H \rangle$ which is comparable with the Plank scale.

Now let us turn to the second term of Eq. (7) which after the $B - L$-symmetry breaking leads to the Majorana mass matrix of the right-handed neutrinos

$$m^M_{ij} = \beta_{ij} \langle S \rangle \tag{12}$$

The permutation symmetry constrains the Majorana type Yukawa coupling $N \times N$ matrix, just as the Dirac type Yukawa couplings had been constrained before, to be of the form

$$\beta_{ij} = \begin{pmatrix} c & d & d & .. \\ d & c & d & .. \\ d & d & c & .. \\ .. & .. & .. & .. \end{pmatrix}. \tag{13}$$

Journal of Physics: Conference Series **259** (2010) 012070 doi:10.1088/1742-6596/259/1/012070

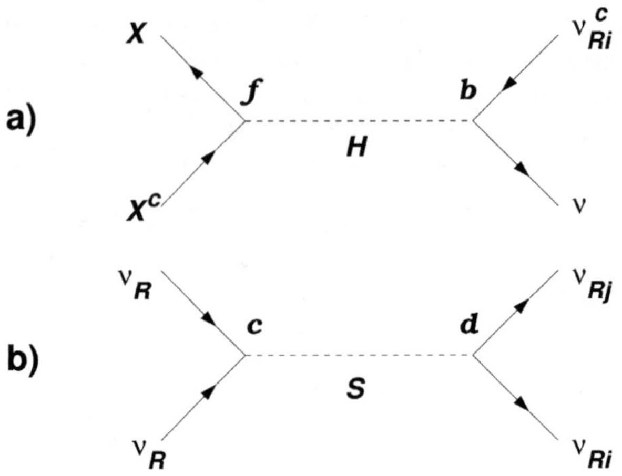

Figure 1. Diagrams relevant for Unitarity constraints on Neutrino Yukawa couplings: (a) to the Higgs doublet H_i (Lepton number conserving); (b) to a Higgs Singlet (Lepton number violating). X_i denotes some of the SM fields coupled to H_i with a strength f.

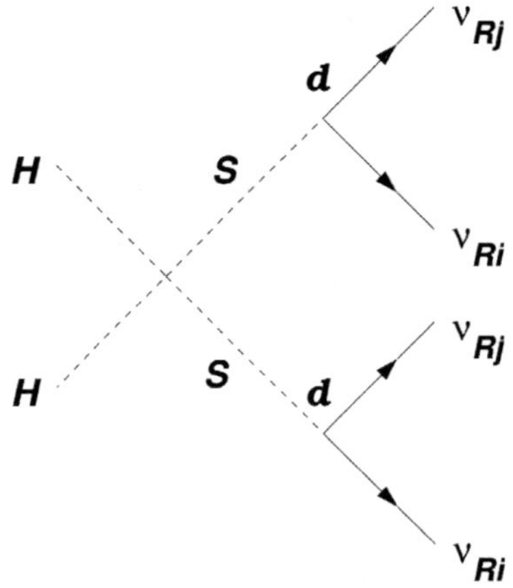

Figure 2. The diagram leading to the strongest Unitarity constraint on the Majorana neutrino mass term.

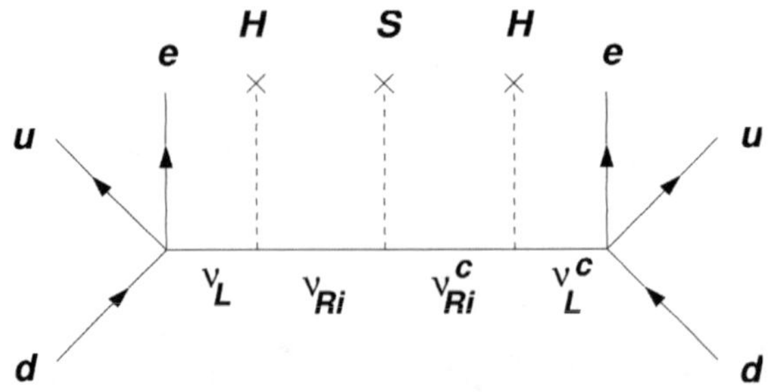

Figure 3. Diagram for neutrinoless double beta decay in the presence of N copies of the SM particle content.

The similarity arguments used above to justify $a \sim b$ in Eqs. (8), (10) can equally be applied to motivate $d \sim c$. Now, when considering the scattering process of right-handed neutrinos, both final states can be any of the N copies as in Fig. 1(b) and, thus, the inclusive rate grows like

$$\Gamma \simeq N^2 c^2 d^2 E, \tag{14}$$

which preserves unitarity below the gravity cutoff only for

$$c \sim d \lesssim \frac{1}{\sqrt{N}}. \tag{15}$$

An even more stringent bound originates from the diagram of Higgs doublet scattering in Fig. 2,

$$\Gamma \simeq N^4 \kappa^2 d^4 E, \tag{16}$$

The scalar quartic couplings κ_i should not be very small since there is no symmetry protecting its smallness. Thus, an assumption $\kappa_i \sim 1$ could be used for rough estimations. Then we have

$$c \sim d \lesssim \frac{1}{N}. \tag{17}$$

This limit remains unaffected by insertion of additional S-branches in the diagram in Fig. 2. As a consequence the neutrino Majorana mass matrix entries (12) are even more strongly suppressed than the Dirac masses discussed above.

In order to discuss the phenomenology of LNV processes we now analyze the mass spectrum of the present scenario. The mass matrix written in the basis of the $2N$ fields $\mathcal{N} = \{\nu_{Li}, \nu_{Ri}\}$ has the following form

$$\mathcal{M}^\nu = \begin{pmatrix} 0 & m^D \\ m^D & m^M \end{pmatrix}. \tag{18}$$

where m^D and m^M are $N \times N$ submatrices given by Eqs. (9)-(12). The set of $2N$ mass eigenstates $\nu_i = U_{ij}\mathcal{N}_j$ of this symmetric matrix split in two groups of $(N-1)$ degenerate states ν^+ and ν^- and another two heavy states $N^\pm$. The details of the diagonalization procedure will be given elsewhere [5].

For consistency with neutrino phenomenology one needs one neutrino at the sub-eV scale, say, $m_\nu \sim 10^{-2}$eV. Then, the states $N^\pm$ are pushed in mass towards the Planck scale and, therefore, their phenomenological impact is negligible. The light Majorana states $\nu^\pm$ having very small mass splitting $\delta_m \sim 1/\sqrt{N}$ form a quasi Dirac state with the mass m_ν. Thus they are expected to induce lepton number violating processes at rates $\sim 1/N$. Moreover due to the structure of the mass matrix (18) with zero submatrix in the upper-left corner LNV processes are even more strongly suppressed. Thus the dominating contributions to neutrinoless double beta decay are

$$\begin{aligned} \langle m_\nu^3 \rangle &= (a-b)^2(c-d)\langle H\rangle^2\langle S\rangle \sim N^{-2}, \\ \left\langle \frac{1}{M_N} \right\rangle &\approx \frac{\langle H\rangle}{\langle S\rangle^2}\frac{d}{b^2N^2} \sim N^{-2}. \end{aligned} \tag{19}$$

Therefore, the amplitude for neutrinoless double beta decay and other, related processes in the studied framework is extremely small.

In this paper we have adressed a problem arising in any scenario with a low quantum gravity scale: do LNV operators induced by TeV scale black holes invalidate the model? For the case of the $N = 10^{32}$-copies SM we have shown that this consequence is avoided due to a non-trivial cancelation mechanism. This property should be considered as an important benefit of the model.

Nevertheless, the presence of a large number of right-handed Majorana states can have interesting phenomenological consequences. For example, a very naive estimate of the right-handed neutrino decay diagrams on tree and one-loop level, which give rise to the baryon asymmetry in leptogenesis, scale as $(\sqrt{N})^2$ from the Yukawa coupling with N ν_{Ri} copies contributing potentially to the decay, and the ν_{Ri} propagator in the loop diagram. So the process may be relevant despite LNV signals being suppressed for the left-handed neutrino states. A similar line of reasoning could apply to single ν_{Ri} production at the LHC.

We thus conclude that the $N = 10^{32}$-copies SM is safe from LNV in the SM sector, and leave the potentially interesting phenomenology of ν_{Ri} production and decay for further study.

Acknowledgments

This work has been partially supported by the DFG grant PA-803/5-1 and by the PBCT project ACT-028 "Center of Subatomic Physics" and CONICYT, Programa de Financiamiento Basal para Centros Científicos y Tecnológicos de Excelencia. HP thanks the Universidad Técnica Federico Santa María for hospitality offered while part of this work was carried out.

References

[1] G. Dvali, arXiv:0706.2050 [hep-th].
[2] G. Dvali and M. Redi, Phys. Rev. D **77**, 045027 (2008) [arXiv:0710.4344 [hep-th]].
[3] G. Dvali and C. Gomez, Phys. Lett. B **674**, 303 (2009) [arXiv:0812.1940 [hep-th]].
[4] G. Dvali and M. Redi, Phys. Rev. D **80**, 055001 (2009), [arXiv:0905.1709 [hep-ph]].
[5] S. Kovalenko, H. Päs, I. Schmidt, to be published

Journal of Physics: Conference Series **259** (2010) 012071 doi:10.1088/1742-6596/259/1/012071

CP-violating MSSM Higgs at Tevatron and LHC

Siba Prasad Das[1] **and Manuel Drees**[2]

[1] AHEP Group, Institut de Física Corpuscular – C.S.I.C./Universitat de València Edificio Institutos de Paterna, Apt 22085, E–46071 Valencia, Spain,

[2] Bethe Center for Theoretical Physics and Physikalisches Institut, Universität Bonn, Nussallee 12, D–53115 Bonn, Germany

E-mail: spdas@ific.uv.es, drees@th.physik.uni-bonn.de

Abstract.

We analyze the prospect for observing the intermediate neutral Higgs boson (h_2) in its decay to two lighter Higgs bosons (h_1) at the presently operating hadron colliders in the framework of the CP violating MSSM using the PYTHIA event generator. We consider the lepton+ 4-jets+ $\not{E}_T$ channel from associate Wh_2 production, with $Wh_2 \rightarrow Wh_1h_1 \rightarrow \ell\nu_\ell b\bar{b}b\bar{b}$. We require two or three tagged b-jets. We explicitly consider all relevant Standard Model backgrounds, treating c-jets separately from light flavor and gluon jets and allowing for mistagging. We find that it is very hard to observe this signature at the Tevatron, even with 20 fb^{-1} of data, in the LEP–allowed region of parameter space due to the small signal efficiency. At the LHC, a priori huge SM backgrounds can be suppressed by applying judiciously chosen kinematical selections. After all cuts, we are left with a signal cross section of around 0.5 fb, and a signal to background ratio between 1.2 and 2.9. According to our analysis this Higgs signal should be viable at the LHC in the vicinity of present LEP exclusion once 20 to 50 fb^{-1} of data have been accumulated at $\sqrt{s} = 14$ TeV.

1. Introduction

The Minimal Supersymmetric Standard Model (MSSM) [1] requires two Higgs doublets, leading to a total of five physical Higgs bosons. At the tree level, these can be classified as two neutral CP–even bosons (ϕ_1 and ϕ_2), one neutral CP–odd boson (a) and two charged bosons. In the presence of CP violation, the three neutral Higgs bosons can mix radiatively [2, 3]. The mass eigenstates h_1, h_2 and h_3 with $m_{h_1} < m_{h_2} < m_{h_3}$ can then be obtained from the interaction eigenstates ϕ_1, ϕ_2 and a with the help of the orthogonal matrix $O_{\alpha i}$, $(\phi_1, \phi_2, a)^T_\alpha = O_{\alpha i}(h_1, h_2, h_3)^T_i$, which diagonalizes the Higgs boson mass matrix. O depends on various parameters of the SUSY Lagrangian.

Due to this mixing, the Higgs mass eigenstates are no longer CP eigenstates. Moreover, the masses of the Higgs bosons, their couplings to SM and MSSM particles, and their decays are significantly modified [3]. For example, the Higgs boson couplings to pairs of gauge bosons are scaled by g_{h_iVV} relative to the SM. These couplings can be expressed as $g_{h_iVV} = \cos\beta\, O_{\phi_1 i} + \sin\beta\, O_{\phi_2 i}$, where $\tan\beta$ is the ratio of Higgs vacuum expectation values (VEVs). The magnitude of g_{h_2WW} is directly related to the production process studied in this paper.

In the absence of mixing between neutral CP–even and CP–odd states the LEP experiments were able to derive absolute lower bounds of about 90 GeV on the masses of both the lighter

CP–even Higgs and the CP–odd boson [4]. However, in the presence of CP violation, the LEP experiment were not able to exclude certain scenarios with very light h_1. In this "LEP hole" h_1 is dominantly a CP–odd state with almost vanishing coupling to the Z boson. One then has to search for Zh_2 or h_1h_2 production. In part of the LEP hole, these cross sections are suppressed by the rather large h_2 mass. Moreover, $h_2 \to h_1h_1$ decays lead to quite complicated final states, which often yield low efficiencies after cuts. One LEP–allowed region has m_{h_1} less than 10 GeV, so that $h_1 \to \tau^+\tau^-$ is dominant; in the other, $m_{h_1} \sim 30 - 50$ GeV so that $h_1 \to b\bar{b}$ is dominant. m_{h_2} lies between slightly below 90 and slightly above 130 GeV. Scenarios with even lighter h_2 are excluded by decay–independent searches for Zh_2 production [4]. If m_{h_2} is much above 130 GeV, the CP–odd component of h_1 becomes subdominant, so that the cross section for Zh_1 production becomes too large. Finally, the LEP hole occurs for $\tan\beta$ in between 3 and 10 [4]. We analyze the prospect for observing a signal for the production of neutral Higgs bosons in the second of these LEP allowed regions.

2. Numerical analysis

In our analysis we took five different benchmark points, denoted by S1 through S5 [m_{h_2}=130-90 GeV with m_{h_1}=30 GeV] and two CPX$_{0.5}$–scenario [CPX–1(2) where m_{h_2}=102 (103) GeV with m_{h_1}=36 (45) GeV, which can be realized for M_{H^+} = 131.8 GeV with $\tan\beta$ = 4.02 (4.39)] of the MSSM [5]. We calculated the spectrum and the couplings for these two benchmark points using `CPsuperH` [3]. In our simulation we used the `PYTHIA v6.408` [6] event generator with the `SLHA` [7] interfacing option. We used `MadGraph/MadEvent v4.2.8` [8] for generating parton level SM backgrounds which were fed to `PYTHIA` for showering. We set the renormalization and factorization scale to $Q = \sqrt{\hat{s}}$ and used CTEQ5L for the parton distribution functions (PDF).

The signal arises from $p\bar{p} \to Wh_2 \to \ell\nu_\ell h_1h_1 \to \ell\nu_\ell b\bar{b}b\bar{b}$, leading to $\ell jjjj\not{E}_T$ events, where $\ell = e$ or μ. The effective cross section for this signal topology can be expressed as,

$$\sigma^{\rm tot}_{\rm signal} = \sigma_{SM}(p\bar{p}/pp \to Wh_2) \times g^2_{h_2WW} \times Br(h_2 \to h_1h_1) \times Br(h_1 \to b\bar{b})^2 \times 2Br(W \to e\nu_e) \; , \quad (1)$$

where g_{h_2WW} is the h_2WW coupling in units of the corresponding SM value, W stands for $W^\pm$ and the factor 2 is for ℓ = e and μ. This process has recently been studied in refs.[9, 10], using parton–level analyses with quite promising results. We instead performed a full hadron–level analysis, including initial and final state showering as well as the underlying event. We take $g^2_{h_2WW} \times Br(h_2 \to h_1h_1) \times Br(h_1 \to b\bar{b})^2 = 0.50$.

We simulate our signal and backgrounds at Tevatron Run-II with $\sqrt{s}$ = 1.96 TeV. We have used the toy calorimeter simulation (`PYCELL`) provided in `PYTHIA` with the following criteria: calorimeter coverage is $|\eta| < 3.64$; the segmentation is given by $\Delta\eta \times \Delta\phi$=0.16 $\times$ 0.098 which resembles the CDF detector [11]; Gaussian smearing of the total energy of jets and leptons; a cone algorithm with $\Delta R(j,j) = \sqrt{\Delta\eta^2 + \Delta\phi^2} = 0.4$ has been used for jet finding; $E^{\rm cell}_{T,\rm min} \geq 1.5$ GeV is considered to be a potential candidate for jet initiator; minimum summed $E^{\rm jet}_{T,\rm min} \geq 10.0$ GeV is accepted as a jet and the jets are ordered in E_T; leptons (ℓ = e, μ) are selected with $E^\ell_T \geq 15.0$ GeV and $|\eta^\ell| \leq 2.0$ and no jet should match with a hard lepton in the event. We reconstructed the missing energy ($\not{E}_T$) from all observed particles.

The tagging of b–jets plays a crucial role in our analysis. A jet with $|\eta_j| \leq 1.2$ and $E^j_T \geq 15$ GeV "matched" with a b–flavored hadron (B–hadron), i.e. with $\Delta R(j, B - \text{hadron}) < 0.2$, is considered to be "taggable". We assume that such jets are actually tagged with with probability $\epsilon_b = 0.50$ [12]. We find that our tagging algorithm agrees well with the $t\bar{t}$ analysis of CDF [13]. We also modeled mistagging of non–b jets as b–jets, treating c–jets differently from those due to gluons or light quarks with $\epsilon_c = 0.10$ and $\epsilon_{u,d,s,g} = 0.01$.

The cross-sections for the signal benchmarks lie in between 0.045 – 0.16 pb while the total background is approximately 455 pb. We have displayed the raw number of events for signal

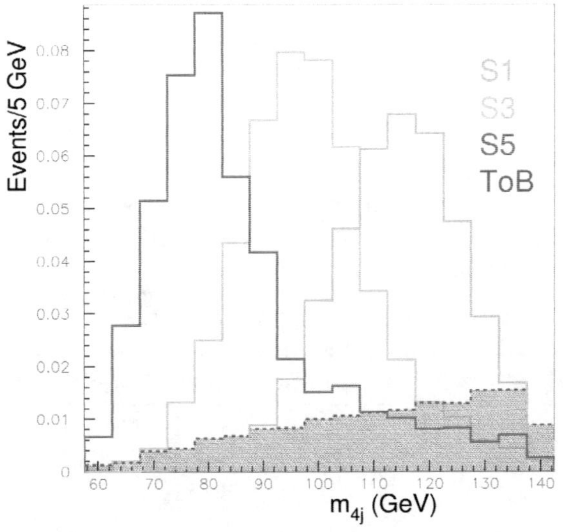

Figure 1. The four–jet invariant mass m_{4j} distribution after all cuts for signal scenarios S1, S3 and S5 and for the total background (ToB), requiring triple $b-$tag following the last column of Table 1 at Tevatron.

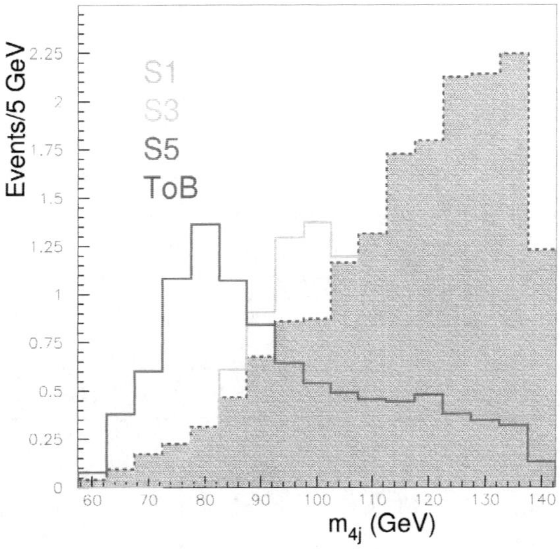

Figure 2. The four–jet invariant mass m_{4j} distribution after all cuts for signal scenarios S1, S3 and S5 and for the total background (ToB), requiring triple $b-$tag following the last column of Table 1 at LHC.

and backgrounds in the first column in Table 1. The second column is the number of events after applying the following basic acceptance cuts: $N_{\rm jet} \geq 4$, $E_T^{j=1-4} > 10$ GeV, $|\eta^{j=1-4}|<3.0$; $N_{\rm lepton} \geq 1$,$E_T^{\ell} > 15$ GeV, $|\eta^{\ell}|<2.0$ and $\not{E}_T \geq 15$ GeV. Not surprisingly, $t\bar{t}$ is the main source of background at this stage. We found that the suppressions by applying $N_{b-tag} \geq 2$ are almost the same for signal and $t\bar{t}$. The signal contains more b quarks, but the $t\bar{t}$ background has much harder b jets, leading to larger tagging probabilities. However, this background can contain a third b jet only due to showering. Hence requiring $N_{b-tag} \geq 3$ greatly reduces the background and increases the signal–to–background ratio. Unfortunately the small triple $b-$tagging probability, which lies between 1.3 and 3.6% depending on m_{h_2}, also reduces the signal rate.

We have also looked for invariant mass peaks to isolate the signal on top of a sizable background. In addition to the basic acceptance cuts (defined above) we demand that the signal contains exactly (rather than at least) four jets. This reduces combinatorial backgrounds for Higgs mass reconstructions. Finally, we pick the jet pairing $(ij)(kl)$ (with $i,j,k,l \in \{1,2,3,4\}$) that minimizes the difference $m_{j_ij_j} - m_{j_kj_l}$ of di–jet invariant masses; in the absence of showering and for perfect energy resolution, the signal would have $m_{j_ij_j} = m_{j_kj_l} = m_{h_1}$. We then demand that both $m_{j_ij_j}$ and $m_{j_kj_l}$ lie between 10 and 60 GeV, where the lower bound comes from the requirement that $h_1 \to b\bar{b}$ decays should be allowed, and the upper bound from the requirement that $h_2 \to h_1h_1$ decays should be open. We next require the four–jet invariant mass to lie between 60 and 140 GeV; this covers the entire “LEP hole” in the MSSM Higgs parameter space. Table 1 shows that after these cuts, the signal actually exceeds the total background.

The results of the column labeled Eff3 in Table 1 have been obtained by including all events that pass the other cuts (not related to tagging) and have at least three *taggable* jets, assigning each event a weight given by its (mis)tagging probability. This greatly increases the statistics. We checked that this gives results that are consistent with the event rejection technique whenever the latter has good statistics; this is the case if at most one $b-$tag results from mistagging.

The distribution of m_{4j} is shown in Fig. 1 for signal scenarios S1, S3 and S5 as well as for the

Table 1. Process column shows the signal benchmarks and the SM backgrounds, where j stands for u, d, s, g and – for background process 7 and 8 – c. RawEvt is stands for the number of events produced in the experiments (for backgrounds we applied basic pre–selections in the generator level: $p_T^{j,b} \geq 5$ GeV, $|\eta^{j,b}|<5.0$ and $\Delta R(jj, bb, bj) \geq 0.3$). N_{acc} is the number after the basic selection cuts, whereas N_{3b} is with at least three jets tagged as $b-$jets, allowing for mistagging. Eff3 is the number of events passing the selection cuts that contain exactly four jets and at least $3b-$ tagged jets; the numbers in parentheses represent the number of events with the inclusion of m_{pair} and m_{4j} cuts. Finally, ToB is the total number of background events.

	Tevatron with 4fb $^{-1}$				LHC with 10fb $^{-1}$			
Process	RawEvt	N_{acc}	N_{3b}	Eff3 (+h1)	RawEvt	N_{acc}	N_{3b}	Eff3(+h1)
S1	38.11	11.09	0.77	0.49 (0.40)	1157	352.5	33.48	13.96(6.85)
S2	51.59	13.85	0.83	0.51 (0.44)	1486	418.3	36.84	15.36(7.76)
S3	68.91	16.58	0.83	0.54 (0.47)	1962	506.5	39.81	17.03(8.91)
S4	94.76	19.88	0.87	0.54 (0.46)	2620	610.6	43.18	17.81(9.61)
S5	133.6	23.92	0.86	0.52 (0.45)	3516	724.7	43.92	18.96(9.63)
CPX-1	89.89	20.27	0.82	0.49 (0.43)	2509	600.2	40.07	17.25(9.07)
CPX-2	87.56	22.46	0.84	0.53 (0.47)	2421	597.2	40.28	16.88(9.64)
$t\bar{t}$	6760	3545	25.62	9.65(0.05)	1,690,000	818,800	7795	1469(5.52)
$b\bar{b}b\bar{b}W^{\pm}$	12.59	1.52	0.06	0.03(0.01)	337.6	31.8	4.10	2.95(0.53)
$b\bar{b}bjW^{\pm}$	0.043	0.01	0	0(0)	23.3	2.3	0.13	0.11(0.01)
$b\bar{b}cjW^{-}$	131.2	17.6	0.05	0.02(0.01)	73,170	7359	77.56	56.50(6.79)
$b\bar{b}c\bar{c}W^{\pm}$	44.51	5.53	0.03	0.01(0.01)	1126	89.9	1.68	1.17(0.25)
$b\bar{b}jjW^{\pm}$	5181	610.2	0.28	0.15(0.06)	535,700	45,830	17.14	17.89(1.93)
$bjjjW^{\pm}$	14.31	2.01	0	0(0)	7194	586.3	0.23	0.05(0.01)
$jjjjW^{\pm}$	384000	47340	0.24	0.02(0)	59,700,000	4,332,000	2.18	4.59(0.68)
$t\bar{t}b\bar{b}$	12.15	6.80	0.44	0.03(0)	10,100	5700	751.5	72.82(1.28)
$t\bar{t}c\bar{c}$	21.69	13.66	0.19	0.02(0)	16,440	9245	259.8	31.54(0.45)
ToB	396200	51550	26.91	9.93(0.14)	62,030,000	5,220,000	8910	1657(17.45)

total background. We observe clear peaks for the signal, which are shifted downwards due to, for example, showering (by 10 to 15 GeV) from the naive expectation $m_{4j} = m_{h_2}$ at the parton level.

By seeing the distributions of m_{pair} (average of the two optimal pairing of di–jet invariant masses) and m_{4j} (see Fig. 1) allow us define the final significance of the signal by counting events that satisfy: $0.6m_{h_1} \leq m_{\text{pair}} \leq m_{h_1} + 5$ GeV and $0.7m_{h_2} \leq m_{4j} \leq m_{h_2} + 10$ GeV. We see that requiring triple $b-$tags leads to very good signal to background ratio, of around 10 for $m_{h_1} = 30$ GeV and slightly less for heavier h_1. However, we expect less than 2 signal events after all cuts for a total integrated luminosity ($\int \mathcal{L}dt$) of 20 fb^{-1}. We tried also to see the effect for two $b-$tag and found that the signal rate increase by a factor between 3.7 and 5 while the background increases by two orders of magnitude and $S/\sqrt{B}$ is well below two.

The significance defined in this way overestimates the true statistical significance of a double peak in the m_{pair} and m_{4j} distributions somewhat, due to the "look elsewhere" effect: since m_{h_1} and m_{h_2} are not known a priori, one would need to try different combinations when looking for peaks. However, given that we use rather broad search windows, there are probably only $\mathcal{O}(10)$ statistically independent combinations within the limits of the LEP hole. Note also that the signal rate is still quite small. Further kinematical cuts, which might slightly increase the signal to background ratio, are therefore not likely to increase the statistical significance of the signal. We are therefore forced to conclude that the search for $Wh_2 \to Wh_1h_1 \to \ell\nu b\bar{b}b\bar{b}$ events at the Tevatron does not seem promising, and turn instead to the LHC.

Our analysis for the LHC follows broadly similar lines as that for the Tevatron. We simulate our signal and backgrounds at the LHC with $\sqrt{s} = 14$ TeV. The PYCELL model is based on the ATLAS detector [14] with calorimeter coverage $|\eta| < 5.0$, segmentation $\Delta\eta \times \Delta\phi = 0.087 \times 0.10$. We use a Gaussian energy resolution for leptons and jets and a cone algorithm for jet finding, with $\Delta\mathrm{R(j,j)} = 0.4$. Calorimeter cells with $\mathrm{E}^{\mathrm{cell}}_{\mathrm{T,min}} \geq 1.0$ GeV are considered to be a potential candidates for jet initiator. All cells with $\mathrm{E}^{\mathrm{cell}}_{\mathrm{T,min}} \geq 0.1$ GeV are treated as part of the would–be jet; minimum summed $\mathrm{E}^{\mathrm{jet}}_{\mathrm{T,min}} \geq 15.0$ GeV is accepted as a jet and the jets are ordered in E_T; leptons (ℓ = e, μ) are selected if they satisfy $\mathrm{E}^{\ell}_{\mathrm{T}} \geq 20$ GeV and $|\eta^{\ell}| \leq 2.5$. The jet–lepton isolation criterion and the missing transverse energy $\not{E}_T$ are adopted similar to our Tevatron analysis.

Only jets with $|\eta_j| < 2.5$ are considered to be taggable as b–jets. If the jet is "matched" to a b–flavored hadron, with $\Delta R(j, \mathrm{hadron}) \leq 0.2$, the tagging efficiency is taken to be 50%. If instead the jet is matched to a c–hadron, the (mis)tagging efficiency is taken to be 10%, whereas jets matched to a τ–lepton have zero tagging probability. All other taggable jets have (mis)tagging probability of 0.25%. These efficiencies follow recent ATLAS analyses [15, 16, 17].

The cross-sections for the signal benchmarks lie between 0.5 – 1.7 pb while the total backgrounds is approximately $2.84 \cdot 10^4$ pb. We have displayed the raw number of events for signal and backgrounds (again with the same generator–level cuts as of Tevatron) in the first column in Table 1. The second column is the number of events after applying the basic acceptance cuts: $N_{\mathrm{jet}} \geq 4$, $E_T^{j=1-4} > 15$ GeV, $|\eta^{j=1-4}| < 5.0$; $N_{\mathrm{lepton}} \geq 1$, $E_T^{\ell} > 20$ GeV $|\eta^{\ell}| < 2.5$ and $\not{E}_T \geq 20$ GeV. They reduce the cross section by about a factor of 5 (3) for $m_{h_2} = 90$ (130) GeV.

The number of events (for $\int \mathcal{L}dt$=10 fb^{-1}) passing the acceptance cuts and containing exactly four jets, at least three of which are tagged (adopting the same strategy like Tevatron), is given by the Eff3 column of Table 1. Similar to Tevatron we require both jet pair invariant masses to lie between 10 and 60 GeV and m_{4j} to lie between 60 and 140 GeV. The m_{4j} distribution is shown in Fig. 2. After these cuts we are left with slightly less than one signal event and slightly less than two background events per fb^{-1} of data. A 5σ signal would then require almost 100 fb^{-1} of data, more than the LHC is likely to collect during "low" luminosity running.

We also checked that requiring a fourth b–tag reduces the signal cross section by another order of magnitude or more. The signal rate then becomes so low that one would have to wait for the high–luminosity phase of the LHC to accumulate enough events to reconstruct invariant mass peaks. We therefore stick to triple b–tag in our LHC analysis.

We found that the background shows a peak in the m_{pair} distribution between 30 and 40 GeV, not far from the peak of the signal in the scenarios we consider. A tighter cut on m_{pair} will nevertheless improve the signal–to–background ratio. Moreover, the four–jet invariant mass distribution (in Fig. 2) of the background peaks at large values, largely due to the contribution from $t\bar{t}$ production. At least for scenarios with h_2 masses in the lower half of the "LEP hole" region a tighter cut on m_{4j} will therefore also improve the significance of the signal. We therefore applied the same "double–peak" cuts as at the Tevatron. The signal then always exceeds the background. Assuming an integrated luminosity of 60 fb^{-1} at the end of "low" luminosity running, we find a final statistical significance of at least 5 standard deviations, and a signal sample of some 30 events.

3. Conclusions

We analyzed the possibility of observing neutral Higgs bosons at currently operating hadron colliders in the framework of the CP violating MSSM. We explored the $\ell jjjj \not{E}_T$ channel with double, triple and quadruple b tag, focusing on the region of parameter space not excluded by LEP searches. We considered a large number of SM backgrounds and employed a full hadron–level Monte Carlo simulation using the PYTHIA event generator. We carefully implemented $b-$tagging, including mistagging of $c-$jets or light flavor or gluon jets. At the Tevatron, requiring $3b-$tag, we can only expect about one signal event per 10 fb^{-1} of $\int \mathcal{L}dt$, on a background of about 0.3 events. If we require only $2b-$tag, the signal increases by a factor of about 4, but the background increases by two orders of magnitude, making the signal unobservable. At the LHC we focussed on events with exactly four jets, cutting simultaneously on the average di–jet invariant mass and the four–jet invariant mass and demanding at least $3b$-tags. We found a signal rate above the background, and a signal significance exceeding 5σ for an $\int \mathcal{L}dt$ of 60 fb^{-1}. One might be able to increase the S:B ratio even more by requiring $4b-$tags with softer tagging criteria (enhancing mistag rate also), possibly simultaneously relaxing the requirement on the number of jets. This could be used to confirm the existence of a signal.

We conclude that searches for Wh_2 production at the LHC with $W \to \ell\nu$ and $h_2 \to h_1 h_1 \to b\bar{b}b\bar{b}$ should be able to close that part of the "LEP hole" in parameter space where $h_1 \to b\bar{b}$ decays dominate. The details of this analysis can be found in our recent paper [18].

3.1. Acknowledgments

This work was partially supported by the Bundesministerium für Bildung und Forschung (BMBF) under Contract No. 05HT6PDA, by the EC contract UNILHC PITN-GA-2009-237920, and by the Spanish grants FPA2008-00319, CSD2009-00064 (MICINN) and PROMETEO/2009/091 (Generalitat Valenciana).

References

[1] For introductions into supersymmetric extensions of the Standard Model in general, and the MSSM in particular, see e.g. M. Drees, R.M. Godbole and P. Roy, *Theory and Phenomenology of Sparticles*, World Scientific, Singapore (2004); H.A. Baer and X.R. Tata, *Supersymmetry: from superfields to scattering events*, Cambridge Press (2006).

[2] A. Pilaftsis, Phys. Rev. **D58** (1998) 096010; Phys. Lett. **B435** (1998) 88; E. Accomando *et al.*, arXiv:hep-ph/0608079.

[3] J.S. Lee *et al.*, Comput. Phys. Commun. **156** (2004) 283; Comput. Phys. Commun. **180**, 312 (2009).

[4] S. Schael *et al.* (ADLO Collab.) Eur. Phys. J. C **47** (2006) 547; OPAL Collab., G. Abbiendi *et al.*, Eur. Phys. J. **C27** (2003) 311; DELPHI Collab., J. Abdallah et al., Eur. Phys. J. **C38** (2004) 1; P. Bechtle *et al.*, Comput. Phys. Commun. **181** (2010) 138.

[5] M. Carena, J.R. Ellis, A. Pilaftsis and C.E.M. Wagner, Phys. Lett. B **495** (2000) 155, hep-ph/0009212.

[6] T. Sjostrand, S. Mrenna and P. Skands, JHEP **0605**, 026 (2006).

[7] P. Skands *et al.*, JHEP **0407**, 036 (2004).

[8] F. Maltoni and T. Stelzer, JHEP **0302**, 027 (2003).

[9] K. Cheung, J. Song and Q.-S. Yan, Phys. Rev. Lett. **99** (2007) 031801, hep-ph/0703149.

[10] M. Carena, T. Han, G.Y. Huang and C.E.M. Wagner, JHEP **0804**, 092 (2008), arXiv:0712.2466 [hep-ph].

[11] **CDF** Collab., T. Aaltonen *et al.*, *Phys. Rev. Lett.* **100** (2008) 091803; **CDF IIb** Collab., P. T. Lukens, *The CDF IIb detector: Technical design report*, FERMILAB-TM-2198.

[12] K. Hanagaki [D0 Collab.], FERMILAB-CONF-05-647-E; C. Neu [CDF Collab.], FERMILAB-CONF-06-162-E; T. Wright [CDF and D0 Collabs.], arXiv:0707.1712 [hep-ex]; M.T. Bowen *et al.*, Phys. Rev. D **72**, 074016 (2005).

[13] D. Acosta *et al.*, Phys. Rev. D **71** (2005) 052003.

[14] The Atlas Collab., CERN-LHCC-99-15, ATLAS-TDR-15 (1999).

[15] G. Aad *et al.* [The ATLAS Collab.], arXiv:0901.0512 [hep-ex].

[16] M. Lehmacher, arXiv:0809.4896 [hep-ex].

[17] E. Alagoez *et al.*, Particle Physics with CMS for the LHC, http://unizh.web.cern.ch/unizh/Activities/cms.htm.

[18] S. P. Das and M. Drees, to appear.

16th International Symposium on Particles Strings and Cosmology (PASCOS 2010) IOP Publishing
Journal of Physics: Conference Series **259** (2010) 012072 doi:10.1088/1742-6596/259/1/012072

Flavour constraints on the Aligned Two-Higgs-Doublet Model and CP violation

Paula Tuzón
Departament de Física Teòrica, IFIC, Universitat de València - CSIC,
Apt. Correus 22085, E-46071 València, Spain.

E-mail: paula.tuzon@ific.uv.es

Abstract. The Aligned Two-Higgs-Doublet Model (A2HDM) [1] describes a particular way of enlarging the scalar sector of the Standard Model (SM), with a second Higgs doublet which is aligned to first the one in flavour space. This implies the absence of flavour-changing neutral currents (FCNCs) at tree level and the presence of three complex parameters. Within this general approach, we analyze the charged Higgs phenomenology, including CP asymmetries in the K and B systems [2].

1. Introduction

The two-Higgs-doublet model (2HDM) is one of the simplest extensions of the SM, only based on the enlargement of the scalar sector by one more doublet. With this small assumption a rich phenomenology is provided, making the model very interesting not only by itself but also as part of some other extensions of the SM. Without any other model-building constraint, the structure of the Yukawa Lagrangian results in

$$-\mathcal{L}_Y = \frac{\sqrt{2}}{v}\left\{\bar{Q}'_L(M'_d\Phi_1 + Y'_d\Phi_2)d'_R + \bar{Q}'_L(M'_u\tilde{\Phi}_1 + Y'_u\tilde{\Phi}_2)u'_R + \bar{L}'_L(M'_l\Phi_1 + Y'_l\Phi_2)l'_R \ + \ \text{h.c.}\right\} . \quad (1)$$

where $\bar{Q}'_L$ and $\bar{L}'_L$ are the left-handed quark and lepton doublets respectively and f'_R $(f = u, d, l)$ the right-handed fermions. The scalar fields are defined in the *Higgs basis*

$$\Phi_1 = \left[\begin{array}{c} G^+ \\ \frac{1}{\sqrt{2}}(v + S_1 + iG^0) \end{array}\right] , \qquad \Phi_2 = \left[\begin{array}{c} H^+ \\ \frac{1}{\sqrt{2}}(S_2 + iS_3) \end{array}\right] ,$$

where only the first doublet acquires a vacuum expectation value (VEV), v, and contains the Goldstone bosons, $G^{\pm}$ and G^0. The physical degrees of freedom of the scalars are five, two charged fields, $H^{\pm}$, and three neutrals that need to be rotated to be mass-eigensates, $\{S_1, S_2, S_3\} \to \{h, H, A\}$. $\tilde{\Phi}_{1,2}(x) \equiv i\tau_2\Phi^*_{1,2}$ are the charge-conjugated scalar fields and M'_f and Y'_f the corresponding Yukawa matrices. Since each right-handed fermion is coupled to two unrelated matrices that in general cannot be diagonalized simultaneously, dangerous FCNC interactions are automatically generated when going to the mass-eigenstate Lagrangian. To avoid (or suppress) these FCNCs, which are strongly constrained by the experiments, many models have been developed from the general 2HDM. A new approach based on the alignment of the Yukawa matrices in flavour space was presented in [1]. It opens an alternative where FCNCs are absent at tree level and, in addition, the presence of three complex parameters preserves the possibility of having new CP violating sources, which is not possible in other models.

2. The Aligned two-Higgs-doublet model

The alignment condition in flavour space means

$$Y'_d = \varsigma_d\, M'_d\,, \qquad Y'_u = \varsigma_u^*\, M'_u\,, \qquad Y'_l = \varsigma_l\, M'_l\,, \tag{2}$$

with ς_f arbitrary complex numbers. These conditions imply that Y'_f are not arbitrary anymore but proportional to M'_f so they can be simultaneously diagonalized: $Y_{d,l} = \varsigma_{d,l} M_{d,l}$ and $Y_u = \varsigma_u^* M_u$. In terms of mass-eigenstate fields, the Yukawa Lagrangian takes then the form

$$-\mathcal{L}_Y = \frac{\sqrt{2}}{v} H^+ \left\{ \bar{u}\left[\varsigma_d V M_d \mathcal{P}_R - \varsigma_u M_u V \mathcal{P}_L\right] d + \varsigma_l\, \bar{\nu} M_l \mathcal{P}_R l \right\} + \frac{1}{v} \sum_{\varphi, f} \varphi_i^0 y_f^{\varphi_i^0}\, \bar{f}\, M_f \mathcal{P}_R f \;+\; \text{h.c.}\,, \tag{3}$$

where V is the CKM matrix, $\mathcal{P}_{R,L} \equiv \frac{1\pm\gamma_5}{2}$ and the neutral couplings $y_f^{\varphi_i^0}$ are given in [1]. This Lagrangian has the following features: all fermionic couplings are proportional to the mass matrices and the neutral Yukawas are diagonal in flavour. The only flavour-changing structure is the matrix V, appearing in the charged current part of the quark sector, like in the SM. There are three new complex parameters, ς_f, encoding all the possible freedom allowed by the alignment conditions. They are universal (flavour blind), do not depend on the scalar basis (contrary to the usual $\tan\beta$), recover in some limits all $\mathcal{Z}_2$-type models, and their phases introduce new sources of CP violation without tree-level FCNCs. This fact represents a counterexample to the very well established idea that the only way of having new CP violation in the electroweak sector of a 2HDM is breaking flavour conservation in neutral current interactions.

2.1. Radiative corrections

The alignment condition is not directly protected by any symmetry, therefore quantum corrections could induce some misalignment generating small FCNCs, that are suppressed by the corresponding loop factors. Nevertheless, the flavour structure of the A2HDM strongly constraints the possible FCNC interactions. The Lagrangian is invariant under flavour dependent phase transformations of the fermion mass eigenstates ($f = u, d, l, \nu$, $X = L, R$, $\alpha_i^{\nu,L} = \alpha_i^{l,L}$), $f_X^i(x) \;\to\; e^{i\alpha_i^{f,X}}\, f_X^i(x)$ while V and M_f transform like $V_{ij} \;\to\; e^{i\alpha_i^{u,L}}\, V_{ij}\, e^{-i\alpha_j^{d,L}}$ and $M_{f,ij} \;\to\; e^{i\alpha_i^{f,L}}\, M_{f,ij}\, e^{-i\alpha_j^{f,R}}$. Due to this symmetry, lepton flavour violation is zero to all orders in perturbation theory, while in the quark sector the V matrix remains the only source of flavour-changing phenomena. The possible FCNC structures, $\bar{u}_L F_u^{nm} u_R$ and $\bar{d}_L F_d^{nm} d_R$, are of the type $F_u^{nm} = V(M_d M_d^\dagger)^n V^\dagger (M_u M_u^\dagger)^m M_u$ and $F_d^{nm} = V^\dagger (M_u M_u^\dagger)^n V (M_d M_d^\dagger)^m M_d$, or similar structures with additional factors of V, $V^\dagger$ and quark mass matrices. Therefore, at the quantum level the A2HDM provides an explicit implementation of the popular Minimal Flavour Violation (MFV) scenarios [3, 4, 5, 6, 7, 8, 9, 10], but allowing at the same time for new CP violating phases.

Using the renormalization group equations [11, 12, 13], one finds that the one-loop gauge corrections preserve the alignment, and the only FCNC operator is [2]

$$\begin{aligned}
\mathcal{L}_{\text{FCNC}}^{1Loop} \;&=\; \frac{C(\mu)}{4\pi^2 v^3}\,(1+\varsigma_u^*\varsigma_d) \sum_i \varphi_i^0(x)\,\times \\
&\times\;\; \Big\{(\mathcal{R}_{i2} + i\,\mathcal{R}_{i3})\,(\varsigma_d - \varsigma_u)\,\left[\bar{d}_L\, F_d^{01}\, d_R\right] - (\mathcal{R}_{i2} - i\,\mathcal{R}_{i3})\,(\varsigma_d^* - \varsigma_u^*)\,\left[\bar{u}_L\, F_u^{10}\, u_R\right]\Big\} \\
&+\;\; \text{h.c.}\,,
\end{aligned} \tag{4}$$

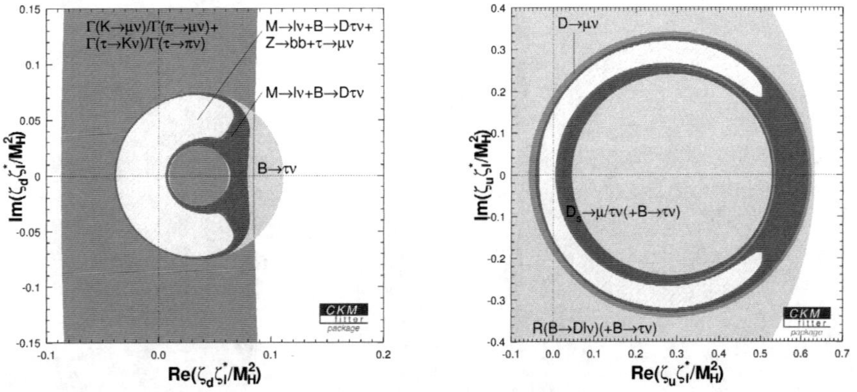

Figure 1. *$\varsigma_d\varsigma_l^*/M_{H^\pm}^2$ (left) and $\varsigma_u\varsigma_l^*/M_{H^\pm}^2$ (right) in the complex plane, in units of GeV^{-2}, constrained by leptonic and semileptonic decays. The inner yellow area shows the allowed region at 95% CL, in the case of $\varsigma_d\varsigma_l^*/M_{H^\pm}^2$ using additional information.*

which of course vanishes in all $\mathcal{Z}_2$-type models. It is suppressed by $m_q m_{q'}/v^3$ and quark-mixing factors, which implies interesting effects in heavy quark systems like $\bar{s}_L b_R$ and $\bar{c}_L t_R$.

3. Phenomenology

One of the most important features of the A2HDM is the presence of a charged Higgs. In [2] we analyzed the most relevant flavor-changing processes that are sensitive to charged-scalar exchange and determined the corresponding constraints on the model parameters. We discussed tree-level processes, $\tau \to \mu/e$, $P^- \to l^-\nu_l$ and $P \to P' l^- \bar{\nu}_l$, where P is a pseudoscalar meson, and loop-induced processes, ΔM_{B_s}, ϵ_K, $Z \to b\bar{b}$ and $\bar{B} \to X_s\gamma$.

Pure leptonic decays give a direct bound on $|\varsigma_l|/M_{H^\pm} \leq 0.40$ GeV^{-1} at 95% CL. Combining this and the information from the other tree-level processes discussed in [2], bounds on $\varsigma_u\varsigma_l^*/M_{H^\pm}^2$ and $\varsigma_d\varsigma_l^*/M_{H^\pm}^2$ parameter space were obtained. Figure 1 shows the resulting limits at 95% CL. These limits are rather weak, allowing the model to fit the experiments in a wide range of its parameter space. Thus, the A2HDM results in a more versatile model than other two-Higgs-doublet models. Loop-induced processes offer direct bounds on ς_u and ς_d. No significative constraints on ς_d can be found from ΔM_{B_s}, ϵ_K and $Z \to b\bar{b}$, because ς_u-terms are enhanced by the top mass in comparison to ς_d-terms. However, quite strong bounds on ς_u are found. Figure 2 shows the allowed $|\varsigma_u| - M_{H^\pm}$ parameter space at 95% CL for values of $\varphi \in [0, 2\pi]$ and $\varsigma_d \in [0, 50]$, given by ϵ_K, which is the most constraining. $\bar{B} \to X_s\gamma$ gives information in both ς_u and ς_d. Figure 3 shows the resulting constraints on $\varsigma_u - \varsigma_d$ plane for complex (left) and real (right) couplings. $M_{H^\pm}$ is scanned over the range $[80, 500]$ GeV while $\varphi \in [0, 2\pi]$ in the complex case. The main conclusions coming from this figure are the following: First, large couplings of opposite sign are forbidden in the real case. Second, the role of a non-vanishing relative phase is to allow some parameter space which is excluded in the real case. And finally, the product $|\varsigma_d\varsigma_u^*| \leq 20$. Again, we find weaker limits than in other models, showing that the A2HDM gives more possibilities to accomplish the experimental constraints.

4. CP violation

The $\bar{B} \to X_s\gamma$ decay is known to be very sensitive to new physics because the SM prediction for the CP rate asymmetry (a_{CP}) is tiny. Requiring that the experimental branching ratio should be correctly reproduced (at 95% CL), we find the results shown in figure 4. We see

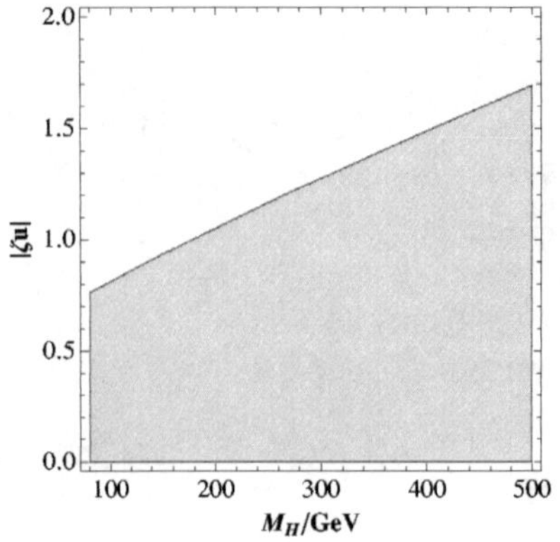

Figure 2. *95% CL constraints from ϵ_K.*

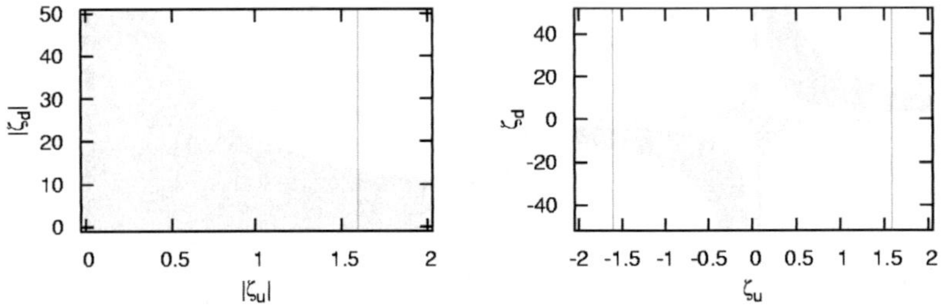

Figure 3. *Constraints on ς_u and ς_d from $\bar{B} \to X_s\gamma$, taking $M_{H^\pm} \in [80, 500]$ GeV. The white areas are excluded at 95% CL. The black line corresponds to the upper limit from $\epsilon_K, Z \to \bar{b}b$ on $|\varsigma_u|$. In the left panel, the relative phase has been varied in the range $\varphi \in [0, 2\pi]$. The right panel assumes real couplings.*

that the maximal asymmetry is compatible with the experimental measurement at 95% CL within the scale dependence of the prediction. Then, we conclude that the A2HDM prediction for this observable does not give new constraints on the parameter space that are not already given by the branching ratio, although it reaches the experimental bound. Therefore, a precise measurement of this asymmetry and a more accurate calculation in both SM and 2HDM parts of the branching ratio would be very interesting.

Some months ago, the D0 experiment measured an enhanced like-sign dimuon charge asymmetry [14] in the B_s system incompatible with a purely SM rate. In [2] we concluded that although the D0 central value is quite unlikely, it is possible to accommodate an enhanced a_{CP} within the A2HDM. Figure 5 shows how large is this enhancement ($a_{CP}^{A2HDM}/a_{CP}^{SM}$) depending on the relative phase φ, where the bound on the product $|\varsigma_u^*\varsigma_d| < 20$ coming from $\bar{B} \to X_s\gamma$ has been taken into account. From the plot we see that the asymmetry can be enhanced even 60 times compared to the SM. The preferred negative sign of a_{sl}^s constrains $\varphi \in [\pi/2, \pi], [3\pi/2, 2\pi]$. A large asymmetry requires large $|\varsigma_d|$ values and small values for the charged Higgs mass.

Due to the different structure of the $b \to s\gamma$ and $B_s^0 - \bar{B}_s^0$ amplitudes, the enhanced asymmetry in each case corresponds to different regions in the A2HDM parameter space, giving complementary information on the relative phase.

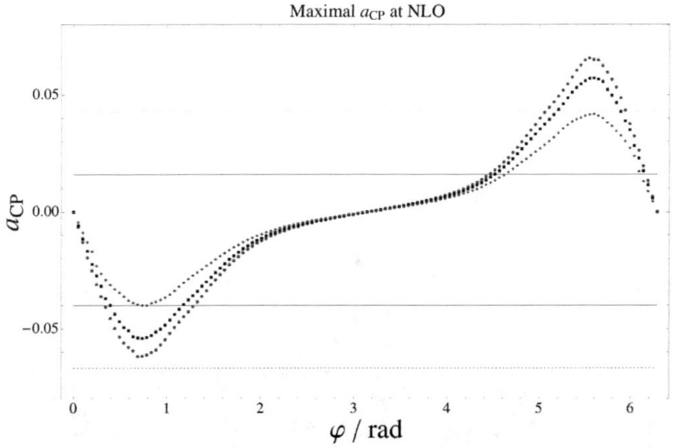

Figure 4. *Maximal a_{CP} over the relative phase φ at NLO for $M_{H^\pm} \in [80, 500]$ GeV, $|\varsigma_u| \in [0, 2]$ and $|\varsigma_d| \in [0, 50]$, taking into account the experimental constraint on $\bar{B} \to X_s\gamma$ branching ratio at 95% CL. The three curves correspond to the maximal a_{CP} at $\mu_b = 2, 2.5, 5$ GeV (outer, center and inner respectively), the minimal a_{CP} (black) is always zero independently on the scale. The dotted (continuous) horizontal lines denote the band of the experimental a_{CP} at 1.96σ (1σ).*

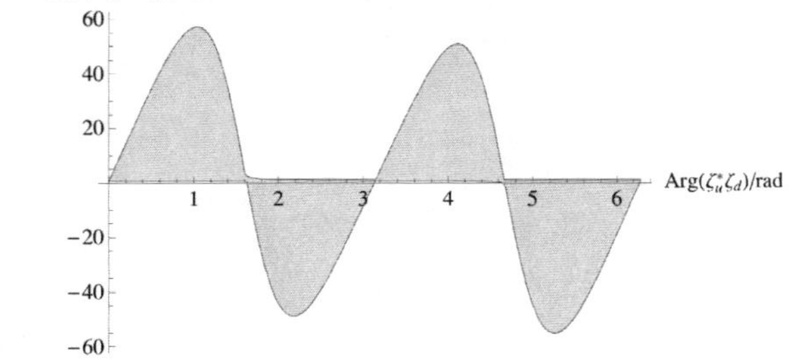

Figure 5. *Dependence of $a_{CP}^{A2HDM}/a_{CP}^{SM}$ on φ, constraining $|\varsigma_u^*\varsigma_d| \leq 20$, for $M_{H^\pm}$, $|\varsigma_u|$ and ς_d scanned in the same ranges as in figure 4.*

5. Conclusions

The A2HDM provides a powerful realization of a 2HDM with no tree-level FCNCs and three complex parameters ς_f. These parameters are new sources of CP violation, flavour blind, scalar basis independent and recover in some limits all the models implemented by discrete $\mathcal{Z}_2$ symmetries. This parametrization allows for more freedom and accomplishes all the experimental constraints. Some misalignment can come from quantum corrections, only as MFV structures which are under control and just occur in the quark sector. Processes involving a charged Higgs give information on the parameters ς_f, although they result in weaker limits compared to the usual scenarios with $\mathcal{Z}_2$, where the freedom introduced by the ς_f phases does not exist. The CP asymmetries generated in $\bar{B} \to X_s\gamma$ and in the B_s systems within the A2HDM enhance the SM prediction in complementary regions. The predicted asymmetry for $\bar{B} \to X_s\gamma$ does not give new bounds on the parameter space compared to the branching ratio, regarding this process, a more precise measurement and a complete calculation reducing the theoretical error are essential. On the other hand, if the experiments confirm a large asymmetry in B_s, it could be perfectly accommodated in this framework.

Journal of Physics: Conference Series **259** (2010) 012072 doi:10.1088/1742-6596/259/1/012072

Acknowledgements

This work has been done in collaboration with Martin Jung and Antonio Pich. It has been supported in part by the EU MRTN network FLAVIAnet [Contract No. MRTN-CT-2006-035482] and by MICINN, Spain [FPU No. AP2006-04522, Grants FPA2007-60323 and Consolider-Ingenio 2010 Program CSD2007-00042 –CPAN–].

References

[1] Pich A and Tuzon P 2009 *Phys. Rev.* **D80** 091702 (*Preprint* 0908.1554)
[2] Jung M, Pich A and Tuzon P 2010 (*Preprint* 1006.0470)
[3] Chivukula R S and Georgi H 1987 *Phys. Lett.* **B188** 99
[4] Hall L J and Randall L 1990 *Phys. Rev. Lett.* **65** 2939–2942
[5] Buras A J, Gambino P, Gorbahn M, Jager S and Silvestrini L 2001 *Phys. Lett.* **B500** 161–167 (*Preprint* hep-ph/0007085)
[6] D'Ambrosio G, Giudice G F, Isidori G and Strumia A 2002 *Nucl. Phys.* **B645** 155–187 (*Preprint* hep-ph/0207036)
[7] Cirigliano V, Grinstein B, Isidori G and Wise M B 2005 *Nucl. Phys.* **B728** 121–134 (*Preprint* hep-ph/0507001)
[8] Kagan A L, Perez G, Volansky T and Zupan J 2009 *Phys. Rev.* **D80** 076002 (*Preprint* 0903.1794)
[9] Buras A J, Carlucci M V, Gori S and Isidori G 2010 (*Preprint* 1005.5310)
[10] Trott M and Wise M B 2010 (*Preprint* 1009.2813)
[11] Cvetic G, Kim C S and Hwang S S 1998 *Phys. Rev.* **D58** 116003 (*Preprint* hep-ph/9806282)
[12] Ferreira P M, Lavoura L and Silva J P 2010 *Phys. Lett.* **B688** 341–344 (*Preprint* 1001.2561)
[13] Braeuninger C B, Ibarra A and Simonetto C 2010 *Phys. Lett.* **B692** 189–195 (*Preprint* 1005.5706)
[14] Abazov V M *et al.* (D0) 2010 *Phys. Rev.* **D82** 032001 (*Preprint* 1005.2757)

Flavour changing neutral currents in two Higgs doublet models with Yukawa alignment

C B Braeuninger

Physik-Department T30d, Technische Universität München,
James-Franck-Straße, 85748 Garching, Germany

Max-Planck-Institut für Physik (Werner-Heisenberg-Institut),
Föhringer Ring 6, 80805 München, Germany

E-mail: `carolin.braeuninger@ph.tum.de`

Abstract. In a general two Higgs doublet model where both Higgses couple to all fermions flavour changing neutral currents (FCNCs) can get arbitrarily large. Such models are thus in conflict with experiment unless there is some mechanism suppressing the FCNCs. A possibility to do so recently brought up is the assumption that the Yukawa couplings are aligned, i.e. proportional to each other. This condition presumably holds at a high-energy scale and is spoiled by radiative corrections. In the work presented in this talk, we computed the size of the radiatively induced flavour violating Higgs couplings at the electroweak scale. We showed that these contributions are well below the experimental bounds in large regions of the parameter space.

Two Higgs doublet models (2HDMs) are amongst the easiest extensions of the Standard Model (SM), consisting in simply adding a second Higgs doublet to the SM particle content. Two complex doublets means eight real degrees of freedom. Three degrees of freedom are the massless Goldstone bosons that are eaten by the W and B bosons, leaving us with five physical mass eigenstates: two CP even neutral scalars h and H, one CP odd neutral scalar A and two charged scalars $H^{\pm}$. In this work we considered a general 2HDM where both Higgses can couple to all fermions so that the Yukawa Lagrangian looks like[1]:

$$\begin{aligned}\mathcal{L}_{\text{Yukawa}} = & (Y_u^{(1)})_{ij}\bar{q}'_{Li}u'_{Rj}\tilde{\phi}_1 + (Y_d^{(1)})_{ij}\bar{q}'_{Li}d'_{Rj}\phi_1 + (Y_e^{(1)})_{ij}\bar{l}'_{Li}e'_{Rj}\phi_1 \\ & +(Y_u^{(2)})_{ij}\bar{q}'_{Li}u'_{Rj}\tilde{\phi}_2 + (Y_d^{(2)})_{ij}\bar{q}'_{Li}d'_{Rj}\phi_2 + (Y_e^{(2)})_{ij}\bar{l}'_{Li}e'_{Rj}\phi_2 + \text{h.c.}\end{aligned} \tag{1}$$

This is problematic since it leads to flavour changing neutral currents (FCNCs) in the Higgs sector. To see this and to be specific, let us consider the down-type quark sector, but the following is equally true for the up-type quarks, of course. The Yukawa Lagrangian expanded in SU(2) components reads:

$$\mathcal{L}_{\text{Yukawa}} \supseteq \bar{u}'_L Y_d^{(1)} d'_R \phi_1^+ + \bar{d}'_L Y_d^{(1)} d'_R \phi_1^0 + \bar{u}'_L Y_d^{(2)} d'_R \phi_2^+ + \bar{d}'_L Y_d^{(2)} d'_R \phi_2^0 + \text{h.c.} \tag{2}$$

In the SM, only the first two terms are present. The coupling to the neutral component of the Higgs gives rise to the mass term for the down-type quarks after electroweak symmetry breaking

[1] This is also sometimes called a type III 2HDM.

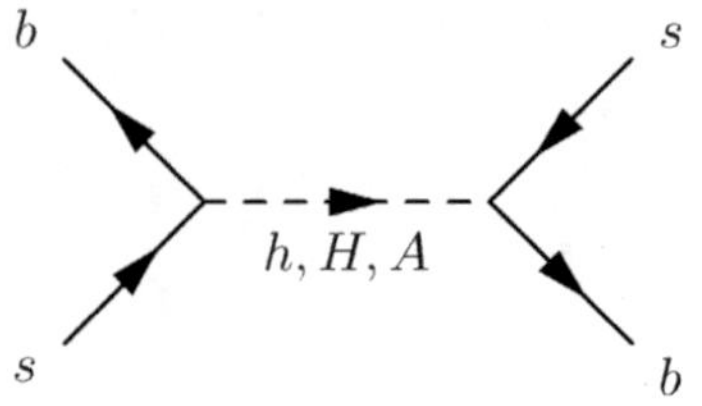

Figure 1: *Tree level mediation of $B_s - \bar{B}_s$-mixing in a 2HDM with flavour violating couplings.*

(EWSB). We can diagonalize its Yukawa matrix by performing bi-unitary transformations on the quark fields:

$$d'_L = V_d^L d_L, \quad d'_R = V_d^R d_R, \quad V_d^{L\dagger} Y_d^{(1)} V_d^R = \text{diag.}(y_d, y_s, y_b) \tag{3}$$

In a type III 2HDM there are also the couplings to the second Higgs and in general these Yukawa matrices are not diagonal in the same basis. The off-diagonal elements will then induce FCNCs. In the SM this cannot happen since there is only one Yukawa coupling per fermion type.
Since FCNCs are known from experiment to be highly suppressed, we must find a way to "tame" these. The standard approach is to impose a discrete symmetry such that all fermions of a given electric charge couple to no more than one Higgs doublet. FCNCs are completely absent at tree level then. This is what is usually called a type I or II 2HDM. Another possibility recently put forward by A. Pich and P. Tuzón [1] is the assumption that the two Yukawa couplings of each fermion type are aligned, i.e. proportional to each other. In the work reported on here [2] we chose to parameterize this condition as:

$$Y_u^{(1)}(\Lambda) = \cos\psi_u Y_u, \qquad Y_u^{(2)}(\Lambda) = \sin\psi_u Y_u, \tag{4}$$
$$Y_d^{(1)}(\Lambda) = \cos\psi_d Y_d, \qquad Y_d^{(2)}(\Lambda) = \sin\psi_d Y_d, \tag{5}$$
$$Y_e^{(1)}(\Lambda) = \cos\psi_e Y_e, \qquad Y_e^{(2)}(\Lambda) = \sin\psi_e Y_e; \tag{6}$$

The assumption of Yukawa alignment can be justified if one assumes that there is an underlying flavour symmetry and that this symmetry is broken only by the Yukawa couplings of the SM (so-called Minimal Flavour Violation) [3].
This ansatz is more general than models with a discrete symmetry in two ways:

- It contains type I and type II models as special cases (type I: $\psi_u = \psi_d = \psi_e = 0$, type II: $\psi_u = 0$, $\psi_d = \psi_e = \pi/2$).
- It allows for new sources of CP violation in the Higgs sector (The Higgs vacuum expectation values (vevs) can be complex and in general only one of these vevs can be made real by an appropriate U(1) phase transition.)

Radiative corrections however introduce a misalignment of the Yukawa couplings at the electroweak scale if the alignment condition is implemented at a high energy scale. In [2] we solved the renormalization group equations (RGEs) [4] describing the running of the Yukawa couplings from the high scale to the electroweak scale numerically and analytically in the so-called leading log approximation. In the analytical approximation it turns out that the off-diagonal parts of the couplings of the neutral Higgs bosons can be parameterized as a parameter E_d which depends on the parameters ψ_u and ψ_d describing the alignment condition times a matrix Q_d which is a product of the (known) quark mass matrices and the CKM matrix:

$$\Delta_u^{\text{off-diag.}} = E_u Q_u, \tag{7}$$
$$\Delta_d^{\text{off-diag.}} = E_d Q_d, \tag{8}$$

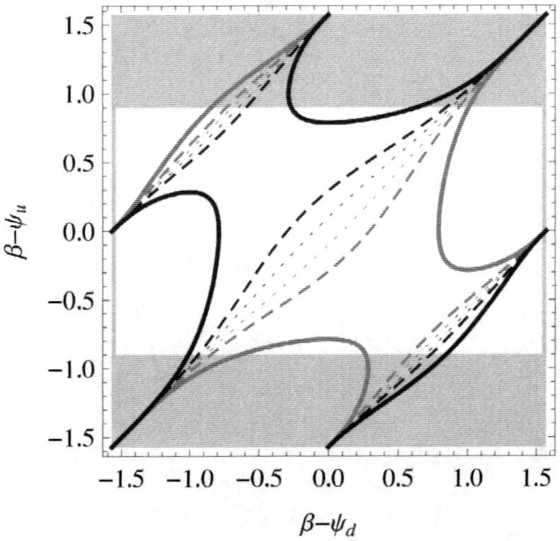

(a) Analytical approximation of E_d

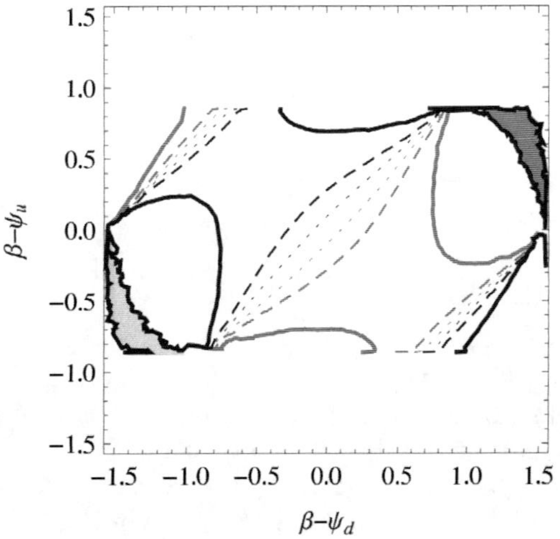

(b) Numerical result for $2.5E_d$

Figure 2: *Contour plots of E_d for $\Lambda = 10^{19}$ GeV. The left figure corresponds to the analytic formula, eq. (12). Solid/dashed/dotted lines correspond to the absolute values of 1/0.3/0.1, blue lines correspond to negative values, green lines to positive ones. The right figure shows $2.5\Delta_{d,23}/Q_{d,23}$ where $\Delta_{d,23}$ has been obtained by numerically solving the RGEs. The rescaling was done in order to make the comparison to the analytical result easier.*

where, assuming real ψ_u, ψ_d:

$$Q_u \equiv \frac{1}{v^3}\left(V_{CKM}\left(M_d^{diag.}\right)^2 V_{CKM}^\dagger M_u^{diag.}\right)^{\text{off-diag.}}, \tag{9}$$

$$E_u \equiv \frac{1}{8\pi^2}\frac{\sin(2(\psi_u-\psi_d))}{\cos^2(\beta-\psi_u)\cos^2(\beta-\psi_d)}\log\left(\frac{m_Z}{\Lambda}\right), \tag{10}$$

$$Q_d \equiv \frac{1}{v^3}\left(V_{CKM}^\dagger\left(M_u^{diag.}\right)^2 V_{CKM} M_d^{diag.}\right)^{\text{off-diag.}}, \tag{11}$$

$$E_d \equiv -E_u. \tag{12}$$

The most stringent experimental bounds on E_d come from $B_s - \bar{B}_s$-mixing[2]. This is a process that in the SM can occur only at loop level while in a general 2HDM there is also a tree level mediation, see fig. 1. The mixing leads to a mass difference of the mesons and antimesons since the flavour eigenstates are no longer mass eigenstates. This mass difference can be calculated using the effective Hamiltonian [5, 6]:

$$H_{\text{eff.}}^{\Delta B=2} = \sum_{i,a} C_i^a(m_Z) Q_i^a(m_Z), \tag{13}$$

where the relevant operators for a 2HDM with flavour changing neutral couplings are:

$$Q_1^{SLL} = (\bar{b}_R s_L)(\bar{b}_R s_L), \quad Q_1^{SRR} = (\bar{b}_L s_R)(\bar{b}_L s_R), \quad Q_2^{LR} = (\bar{b}_R s_L)(\bar{b}_L s_R). \tag{14}$$

and the Wilson coefficients C_i^a can be deduced from the full Hamiltonian by integrating out the Higgs bosons. By demanding that the new contribution to the mass difference be smaller than

[2] Bounds on E_u are in general less stringent than those on E_d.

the theoretical uncertainities on the SM value, we get the following upper limit on the parameters of the 2HDM:

$$\left| \frac{s^2_{\alpha-\beta}}{m_H^2} + \frac{c^2_{\alpha-\beta}}{m_h^2} - \frac{1}{m_A^2} \right| |E_d|^2 \lesssim \frac{1}{(80\ \mathrm{Gev})^2}. \tag{15}$$

The parameter E_d characterizing the size of the flavour changing couplings can thus easily be of the order of one or more, depending on the Higgs masses.
These experimental bounds are satisfied for large portions of the parameter space as can be seen in fig. 2(a) where countour plots of $E_d = 1, 0.3, 0.1$ in the parameter space $\beta - \psi_u$, $\beta - \psi_d$ are shown. For $\psi_u = \psi_d$ and $\psi_u = \psi_d \pm \pi/2$, E_d vanishes completely. This comprises the type I ($\psi_u = \psi_d = 0$) and type II ($\psi_u = 0$, $\psi_d = \pi/2$) 2HDM. In fig. 2(b) a plot of E_d obtained by numerically solving the RGEs is shown. It is rescaled by a factor of 2.5 for better comparison with the analytical result. This factor is due to large, flavour-independent effects in the running of the strong coupling constant and the top Yukawa coupling which are not taken into account by the leading log approximation. The white/gray areas in both plots are not accesible as some Yukawa couplings become non-perturbative below the cut-off scale, which is taken to be the Planck scale in these plots.

Conclusions

With the LHC finally running and searching for the Higgs(es) it is important to investigate different scenarios for the Higgs sector. As said at the beginning, 2HDMs are amongst the easiest extensions of the SM Higgs sector. Without any further protection they however lead to unacceptably large FCNCs. The assumption that the Yukawa couplings are aligned at the high energy scale can provide such a protection and there are then no new sources of flavour violation at tree level. Quantum corrections however introduce a misalignment of the Yukawa couplings at low energies. In the work presented here, we calculated the size of the flavour violating Higgs couplings at the electroweak scale. This also determines the minimal size of the exotic contributions to FCNCs in any 2HDM in the absence of tuning and discrete symmetries. We could show that the exotic contributions are well below experimental bounds for a wide parameter range. The authors of the related recent works [7], [8] came to the same conclusions in the aspects where the analyses overlap.

Acknowledgments

First of all, I would like to thank my collaborators Alejandro Ibarra and Cristoforo Simonetto. Furthermore, I am grateful to Andrzej J. Buras, Gino Isidori and Tillmann Heidsieck for their valuable advice in the calculation of the experimental constraints on the 2HDM from B physics.

References

[1] Pich A and Tuzon P 2009 *Phys. Rev.* **D80** 091702 (*Preprint* 0908.1554)
[2] Braeuninger C B, Ibarra A and Simonetto C 2010 *Phys. Lett.* **B692** 189–195 (*Preprint* 1005.5706)
[3] D'Ambrosio G, Giudice G F, Isidori G and Strumia A 2002 *Nucl. Phys.* **B645** 155–187 (*Preprint* hep-ph/0207036)
[4] Ferreira P M, Lavoura L and Silva J P 2010 *Phys. Lett.* **B688** 341–344 (*Preprint* 1001.2561)
[5] Buras A J, Jager S and Urban J 2001 *Nucl. Phys.* **B605** 600–624 (*Preprint* hep-ph/0102316)
[6] Laiho J, Lunghi E and Van de Water R S 2010 *Phys. Rev.* **D81** 034503 (*Preprint* 0910.2928)
[7] Buras A J, Carlucci M V, Gori S and Isidori G 2010 (*Preprint* 1005.5310)
[8] Jung M, Pich A and Tuzon P 2010 (*Preprint* 1006.0470)

16th International Symposium on Particles Strings and Cosmology (PASCOS 2010) IOP Publishing
Journal of Physics: Conference Series **259** (2010) 012074 doi:10.1088/1742-6596/259/1/012074

Finite density aspects of leptogenesis

Andreas Hohenegger
Max-Planck-Institut für Kernphysik, Saupfercheckweg 1, 69117 Heidelberg, Germany
E-mail: Andreas.Hohenegger@mpi-hd.mpg.de

Abstract. Leptogenesis takes place in the early universe at high temperatures and densities and a deviation from equilibrium in the decay of heavy Majorana neutrinos is a fundamental requirement for the generation of the asymmetry. The equations, commonly used for its description, are largely based on classical Boltzmann equations (BEs) while the source of CP-violation is a quantum interference phenomenon. In view of this clash, it is desirable to study such processes in terms of non-equilibrium quantum field theory. On the other hand, it is simpler to solve BEs rather than the corresponding quantum field theoretical ones. Therefore, we derive modified BEs from first principles in the Kadanoff-Baym (KB) formalism. The results, found for a simple toy model, can be applied to popular phenomenological scenarios by analogy. This approach uncovers structural differences of the corrected equations and leads to different results for the form of the finite density contributions to the CP-violating parameter. In the case of degenerate heavy neutrino masses, corresponding to the popular scenario of resonant leptogenesis, it allows to explicitly distinguish between regimes where BEs are applicable or inapplicable.

1. Introduction

The observation of the matter/anti-matter asymmetry of the universe (baryon to photon ratio $\eta_B \simeq 6 \cdot 10^{-10}$) represents a major puzzle of astro-particle physics. Within standard cosmology (including an inflationary phase) a dynamic generation mechanism is needed for its solution. From an early work by Sakharov it is evident that three conditions need to be fulfilled if this is to be achieved by a perturbative particle-physical process $i \rightarrow f$. In detail these conditions are: violation of *baryon number* ($B(i) \neq B(f)$), violation of *C* and *CP* ($\Gamma(i \rightarrow f) \neq \Gamma(\bar{i} \rightarrow \bar{f})$). Finally, as a consequence of CPT, these processes need to occur in a *non-equilibrium* situation. Assuming the amount of CP-violation present in the standard model, one finds baryon asymmetries much smaller than the observed value η_B. This discrepancy is often interpreted as a major hint for the existence of *physics beyond the standard model*.

1.1. Baryogenesis via Leptogenesis

Among the most popular baryogenesis scenarios, today, are variations in which a lepton asymmetry is produced first and which can therefore be classified as leptogenesis scenarios (see [1] for a recent review). The lepton asymmetry is subsequently converted (partly) into the observed baryon asymmetry by $B - L$ conserving sphaleron processes. The popularity of leptogenesis may be attributed to the comparatively moderate required overhead to the SM (in its simple versions). Moreover, this extension nicely fits into the picture of yet known beyond SM physics as it explains the small observed neutrino masses via the see-saw mechanism. In some scenarios observable predictions for collider and neutrino experiments are possible. We focus here on the lepton number violating decay of the heavy Majorana neutrinos (which act as counterweights on the see-saw), $N_i \rightarrow \ell + \phi$ and $N_i \rightarrow \bar{\ell} + \bar{\phi}$, as implied by their Yukawa interaction with lepton and Higgs fields. If these couplings are generically complex they can induce the required CP-violation. If their magnitude is not too large the required deviation from equilibrium is generated by the expansion of the universe. In the case of *thermal leptogenesis* with hierarchical heavy neutrino masses ($M_1 \ll M_2, M_3$) the baryon asymmetry may be explained if $M_1 \gtrsim 10^9\,\mathrm{GeV}$ which can be derived from its relation to the light neutrino masses. Beyond this simplest phenomenological case there is a multitude of more intricate scenarios such as resonant leptogenesis, flavoured leptogenesis or leptogenesis in the MSSM. In general different processes with different initial and final states can be relevant for the generation of the lepton asymmetry. The results presented here are mainly of a structural kind and may be applied, in slightly modified form, to scenarios in which the CP-violation is due to the interference of the tree-level, one-loop *vertex* and *self-energy contributions* to the (out-of-equilibrium) two particle decay of a heavy species, such as in thermal leptogenesis:

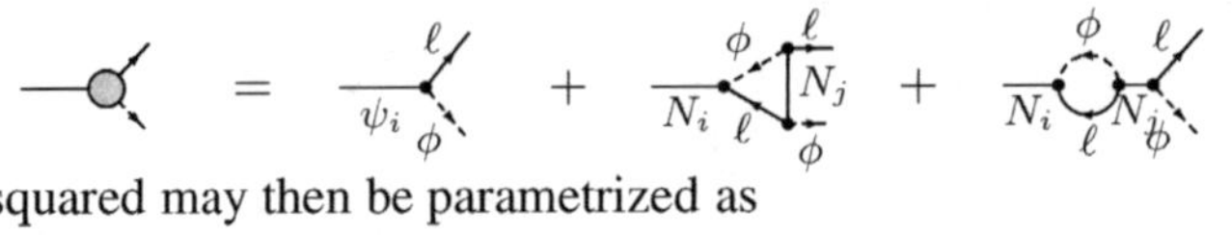

The matrix elements squared may then be parametrized as

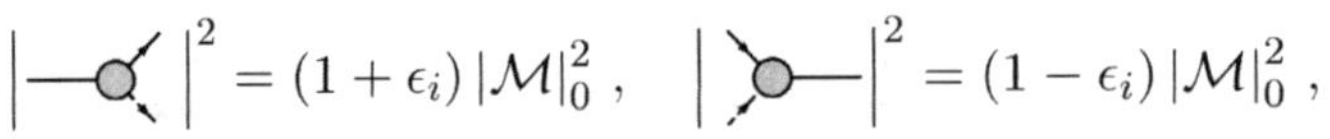

in terms of *CP-violating parameters* $\epsilon_i = (\Gamma_{N_i\to\ell\phi} - \Gamma_{N_i\to\bar{\ell}\bar{\phi}})/(\Gamma_{N_i\to\ell\phi} + \Gamma_{N_i\to\bar{\ell}\bar{\phi}})$:

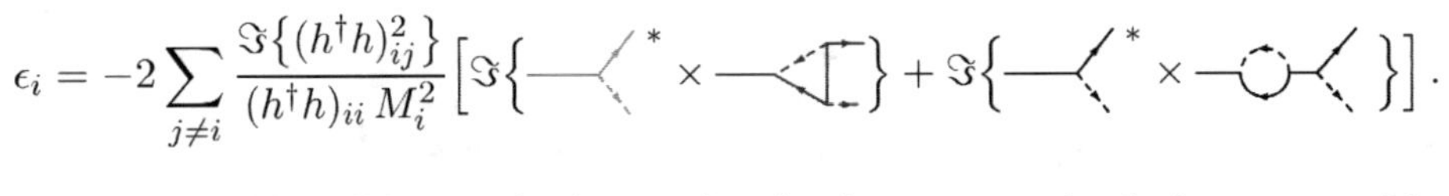

The branch-cut singularities of the graphs (square brackets) can conveniently be computed by means of the *Cutkosky cutting rules* which have an elegant representation as circling rules, e. g.

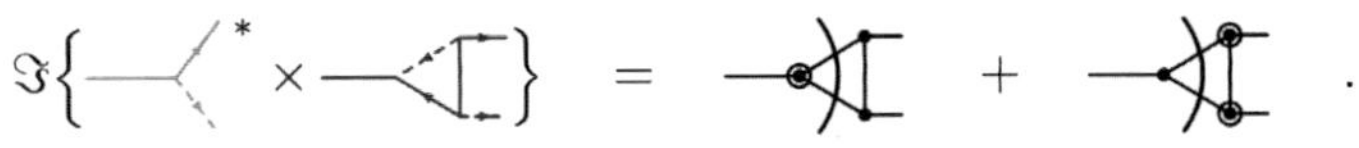

(1)

The "circling rules" implied here are due to 't Hooft and tell to sum all graphs with at least one circled external vertex. Circled vertices change their sign and the propagators connecting them to uncircled ones are replaced by $D_a^{\pm} \propto \Theta(\pm p_0)\delta(p^2 - m_a^2)$ which may be interpreted as a cut.

1.2. Leptogenesis and Kinetic Theory

The first two Sakharov conditions, as stated above, refer to processes (and their S-matrix elements) in vacuum. At the same time, it is evident from the third one that the finite density of the primordial plasma plays a crucial role. Obviously some kind of kinetic equation is needed to study the temporal evolution of baryon/lepton number. The conventional equations are based on kinetic theory which describes the microscopic evolution of the system in terms of averaged classical quantities. The underlying physical picture here is that the plasma is composed of freely propagating (in the context of cosmology there is an additional external gravitational force) particles which occasionally undergo instantaneous interactions. For each species there is a one-particle distribution function which represents the average probability to find an on-shell particle of this kind in a given small space volume and momentum bin. To reflect this picture in our notation we may write ● i → ● f for a general process, or ● N_i → ● ℓ + ● ϕ, ● N_i → ○ $\bar{\ell}$ + ○ $\bar{\phi}$ for the decays – into particles or antiparticles – crucial in thermal leptogenesis. The colored bullets make explicit that a one-particle distribution function, e.g. $f^{\ell}(X,k) = f_k^{\ell}$, for on-shell particles is identified with each of them. Their time dependence is given by *generalized Boltzmann equations* (BEs):

$$k^\mu \mathcal{D}_\mu f^{\ell}(X,k) = \sum_{\text{interactions of } \ell} C_k^{\ell+\cdot\leftrightarrow\cdot+\cdot}[.f^{\ell}.](X,k)\,. \tag{2}$$

There is one equation for each species and a separate collision term on the r.h.s. for each of its interactions such as the important $2 \leftrightarrow 1$ collision term

$$C^{\ell\phi\leftrightarrow N_i}(k) = \frac{1}{2}\int d\Pi_p^{\phi}\, d\Pi_q^{N_i}(2\pi)^4\delta(k+p-q)$$ 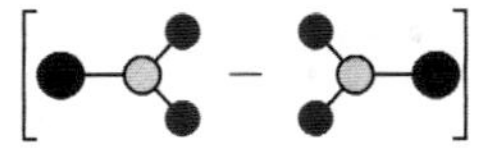 , (3)

where a shortage for the gain- and loss-term has been introduced:

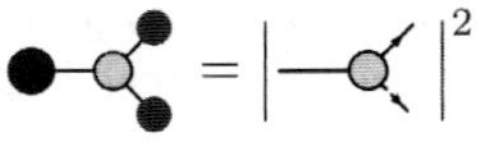 $(1\pm f_k^{\ell})(1\pm f_p^{\phi}) f_q^{N_i}$ and 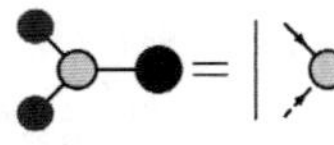 $f_k^{\ell} f_p^{\phi}(1\pm f_q^{N_i})$.

In this notation, the corresponding $2 \leftrightarrow 1$ collision term for antiparticles $C^{\bar\ell\bar\phi\leftrightarrow N_i}$ can be obtained by changing $(\ell, \phi) \to (\bar\ell, \bar\phi)$ in eqn. (3). The quantum statistical terms represent a first finite density effect (Pauli blocking or stimulated emission). The usual procedure continues with the derivation of a rate equation for $n_{B-L} = (n_{\ell} - n_{\bar\ell})$ by subtraction of the BEs for particles and antiparticles and integration over the free momentum k. In thermal equilibrium one finds for the r.h.s.

$$\int \frac{d^3k}{(2\pi)^3 E_k^{\ell}}\left[C^{\ell\phi\leftrightarrow N_i}(k) - C^{\bar\ell\bar\phi\leftrightarrow N_i}(k)\right] \neq 0\,,$$ using 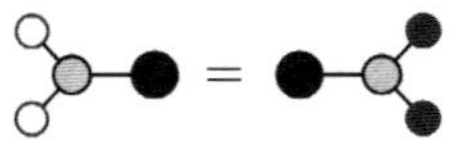 .

This means that, in contradiction to the third Sakharov condition, a non-zero asymmetry is produced even in thermal equilibrium. The physical reason for this discrepancy is a miscounting in eqn. (3) and the corresponding term for antiparticles which fail to take into account the finite probability that a heavy neutrino produced – within a small time interval – in the inverse decay of lepton and Higgs may decay again in the same small time interval. The well-known technical solution to this "double-counting problem" is to subtract the *real intermediate state contribution* to the $2 \leftrightarrow 2$ collision term:

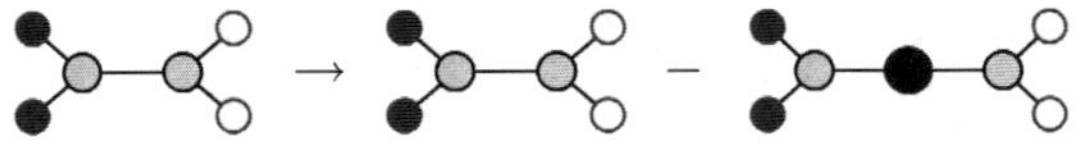 in $C^{\ell\phi\leftrightarrow\bar\ell\bar\phi}$ and $C^{\bar\ell\bar\phi\leftrightarrow\ell\phi}$.

This works because of the *generalized optical theorem*, but only if stimulated emission and blocking factors are dropped.[1] How can consistent equations be obtained which include the quantum statistical factors? One may also be tempted to ask: Why are no distribution functions identified with the internal lines in eqn. (1) which can also be on-shell, as indicated by the cuts. Given the importance of the conclusions often drawn from the requirement of successful baryogenesis (in multiple different models) it seems surprising that comparatively few effort has been made to gain a better understanding of the underlying kinetic description.

2. Systematic Kadanoff–Baym Approach

The weaknesses of the bottom-up approach described above suggest to consider leptogenesis within a more fundamental framework. Our approach was to pursue a consistent *top-down derivation* of BEs from *non-equilibrium QFT* for a *toy-model* in curved space-time [3, 4, 5, 6, 7]. Other recent interesting approaches have been followed in [8, 9, 10, 11]. Our model includes a complex scalar field (toy-baryons) and two mixing real scalars (toy-neutrinos) with Yukawa interaction:

$$S = \int \sqrt{-g}\,d^4x\,\mathscr{L}\,, \quad \text{with} \quad \mathscr{L} \supset -\frac{\lambda}{2!2!}(\bar b b)^2 - \frac{g_i}{2!}\psi_i b b - \frac{g_i^*}{2!}\psi_i \bar b\bar b\,, \quad i=1,2\,.$$

[1] A redefinition of the widths can also lead to consistent RIS-subtracted equations in this case, see e.g. [2].

Despite its simplicity the model captures important features of baryogenesis/leptogenesis which are common to many out-of-equilibrium decay scenarios. In particular, what has been said above remains valid if we make the replacement ($N_i \to \ell\phi$, $N_i \to \bar\ell\bar\phi$) $\Longrightarrow$ ($\psi_i \to bb$, $\psi_i \to \bar b\bar b$).[2]

The first step is the derivation of self-consistent *Schwinger–Dyson equations* for the full propagator of the toy-baryons from the (covariant) generating functional for Greens functions with local and bi-local external sources:

$$D^{-1}(x,y) = \mathscr{D}^{-1}(x,y) - \Sigma(x,y)\,,$$

where the self-energy $\Sigma(x,y) \equiv 2i\delta\Gamma_2[D,\Phi]/\delta D(y,x)$ is obtained from the 2PI functional at 3-loop order

$$\Gamma_2 =$$ 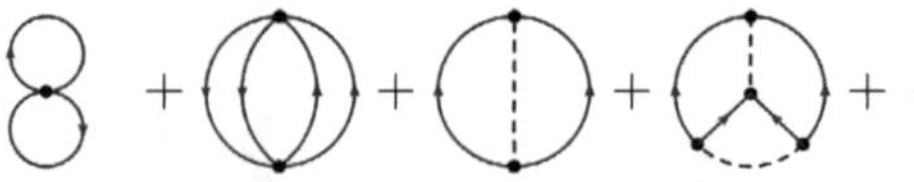 $$+\cdots\,. \qquad (4)$$

There is a similar (matrix-)equation for the full propagators of the toy-neutrinos G^{ij}. Together they lead to a coupled system of *Kadanoff–Baym equations* (KBEs) for quantum correlation functions, i.e. the statistical propagator and the spectral function (here only for the toy-baryons):

$$\left[\Box_x + m^2(x)\right] D_F(x,y) = \int_0^{y^0} \sqrt{-g}d^4z\,\Sigma_F(x,z)D_\rho(z,y) - \int_0^{x^0} \sqrt{-g}d^4z\,\Sigma_\rho(x,z)D_F(z,y)\,,$$

$$\left[\Box_x + m^2(x)\right] D_\rho(x,y) = \int_{x^0}^{y^0} \sqrt{-g}d^4z\,\Sigma_\rho(x,z)D_\rho(z,y)\,.$$

From this set of KBEs, which includes all kinds of quantum effects, we obtain *"quantum-corrected BEs"* following a sequence of systematic approximations. This includes the introduction of a center coordinate X and relative coordinates (covariantly), sending the initial time to $-\infty$, a Wigner transformation wrt. to the relative coordinate, a gradient expansion wrt. to the center coordinate, neglecting the Poisson brackets (i.e. memory effects) as well as the so-called Kadanoff-Baym ansatz and a quasi-particle approximation. Assuming that the full propagator of the toy-neutrinos G^{ij} can be approximated by a diagonal one, the first six steps lead to

$$\left[k^\mu \mathcal{D}_\mu f^{b}(X,k)\right] D_\rho(X,k) = -\frac{1}{2}\left[D_>(X,k)\Sigma_<(X,k) - \Sigma_>(X,k)D_<(X,k)\right],$$

where $D_{\gtrless}(X,p) \equiv D_F(X,p) \pm \frac{1}{2}D_\rho(X,p) = \{(1+f^{b}(X,p))D_\rho(X,p),\ f^{b}(X,p)D_\rho(X,p)\}$, (Kadanoff-Baym ansatz) which defines distribution functions in terms of quantum correlation functions. The relevant contribution to the *self-energy* is

$$\Sigma^{\trianglelefteq}_{\gtrless}(X,k) = -|g_i|^2 \int d\Pi^4_q d\Pi^4_p (2\pi)^4\delta^g(k+p-q)\left[1+\epsilon_i(X,q,p)\right] G^{ii}_{\gtrless}(X,q) D_{\lessgtr}(X,p)$$

If the spectral functions are sharply peaked on-shell they can be approximated by Dirac-deltas (quasi-particle approximation). This yields quantum-corrected BEs of the form (2) where we identify as collision terms

$$C^{bb\leftrightarrow\psi_i}(k) = \frac{1}{2}\int d\Pi^{b}_p\, d\Pi^{\psi_i}_q (2\pi)^4\delta(k+p-q)\left(1+\epsilon^{th}_i\right)$$ 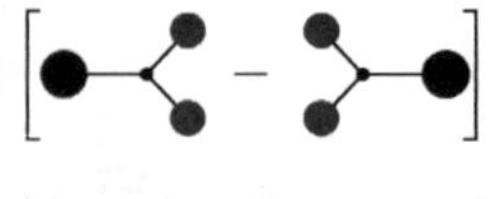 $$, \qquad (5)$$

[2] Surely the toy-model may fail to capture important features of a given phenomenological scenario such as gauge-interactions or processes which involve SM fermions in the case of thermal leptogenesis. However, often minor modifications such as different thermal masses or decay widths will be sufficient. In section (1) different signs in quantum statistical factors and different symmetrization factors may be needed.

and the corresponding equation for antiparticles, obtained by $(\textcircled{b}, \textcircled{b}) \to (\textcircled{\bar b}, \textcircled{\bar b})$, and $(1+\epsilon_i^{th})$ is to be exchanged by $(1-\epsilon_i^{th})$. The CP-violating parameter in eqn. (5) appears as overall prefactor. By detailed balance, this guarantees that no asymmetry is produced in thermal equilibrium, since the square brackets vanish in this case. As before ϵ_i^{th} receives vertex and self-energy contributions, $\epsilon_i^{th} = \epsilon_i^{V,th} + \epsilon_i^{S,th}$. Both of them depend on the distribution functions of the toy-baryons:

$$\epsilon_i^{V,th} = -\frac{1}{8\pi}\frac{|g_j|^2}{M_i^2}\Im\left\{\frac{g_i g_j^*}{g_i^* g_j}\right\}\int\frac{d\Omega_l}{4\pi}\frac{1+f_{E_1}^{\textcircled{b}}+f_{E_2}^{\textcircled{b}}}{M_j^2/M_i^2+\frac{1}{2}(1+\cos\theta_l)}\,,\quad E_{1,2}=\frac{1}{2}\left[E_q^{\psi_1}\mp|\mathbf{q}|\cos\theta_{ql'}\right]. \tag{6}$$

and an analogous result for $\epsilon_i^{S,th}$.

3. Application to Phenomenological Theories

The form of eqn. (6) is reminiscent of older results for the finite density contributions to CP-violating parameters in the phenomenological scenario of thermal leptogenesis, see e.g. [1]. However, an important difference is that the established results exhibit an additional term quadratic in the distribution functions. It turns out that this discrepancy can be resolved, reconsidering the derivation based on thermal QFT in the real-time formalism [3, 12]. Using *causal products* (Kobes 1991) instead of rules for graphs with fixed type external vertices one finds e.g.

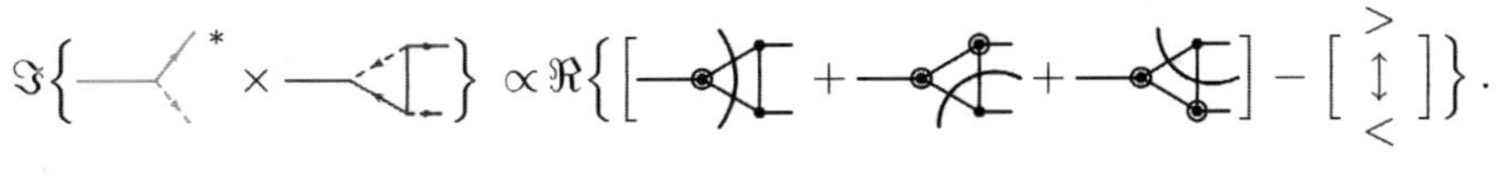

The thermal circling rules employed here differ from their vacuum version used in eqn. (1). In particular the (cut) propagators are to be replaced by their thermal versions. These circling rules lead to the correct results for the toy model, such as eqn. (6).[3] For the phenomenological scenario of thermal leptogenesis it leads to

$$\epsilon_i^{V,th} = \frac{1}{16\pi}\sum_{j\neq i}\frac{\Im\left\{(h^\dagger h)_{ij}^2\right\}}{(h^\dagger h)_{ii}}\frac{M_j}{M_i}\int\frac{d\Omega_l}{4\pi}\frac{1-\cos\theta_l}{M_j^2/M_i^2+\frac{1}{2}(1+\cos\theta_l)}\left\{1-f_{E_1}^{\textcircled{l},eq}+f_{E_2}^{\textcircled{\phi},eq}\right\}.$$

The thermal approach reveals the nature of the finite density effects represented by the distribution functions as absorption and subsequent re-emission of the intermediate on-shell particles by the plasma. Vertex and self-energy contribution exhibit the same $f^{,eq}$ dependence and yield the vacuum result for $f^{,eq}\to 0$. ϵ_i^{th} is momentum dependent and always enhanced by medium corrections. The enhancement can lead to a correction of the final asymmetry by about 20%. These results are in agreement with [9].

4. Resonant Case

The self-energy contribution $\epsilon_i^S \propto M_i M_j/(M_i^2 - M_j^2)$ can be strongly enhanced in the case of degenerate heavy neutrino masses $M_i \simeq M_j$. This enhancement corresponds to the popular scenario of *resonant leptogenesis*. According to the classical computation it allows to lower the temperature scale of leptogenesis significantly, thus (in supersymmetric scenarios) it allows to circumvent the gravitino problem. In this case the finite width needs to be included in the heavy neutrino propagator. For the toy-model the established computation leads to:

$$\epsilon_i^S = -\frac{|g_j|^2}{16\pi}\Im\left\{\frac{g_i g_j^*}{g_i^* g_j}\right\}\frac{M_j^2-M_i^2}{(M_j^2-M_i^2)^2+M_j^2\Gamma_j^2}\,.$$

[3] Moreover, the results for causal products are known to agree with those obtained in the imaginary time formalism for the self-energy loop and the three-point function. The vertex which is always circles has the smallest/largest time component which agrees with the Boltzmann picture.

The derivation of BEs, sketched in section (2), requires sharply peaked spectral functions G^{ij}_{ρ}. In the non-resonant case the on-shell peaks are well separated by the mass splitting, but for $M_i \simeq M_j$ the mass splitting is much smaller than the masses themselves (*"resonant"* case) and may in addition be comparable to the width of the spectral functions (*"maximal resonant"* case). In the latter regime the peaks of the spectral functions may overlap significantly which prevents the unambiguous definition of quasi-particles and their one-particle distribution functions. Also the behaviour of the off-diagonal components of the correlation functions can be important so that the effects cannot be described in terms of on-shell CP-violating parameters. In the resonant regime these concepts can still be employed. However, it is important to take into account that the mass spectrum of the quasi-particle excitations is modified by the finite density of the medium since the enhancement depends sensitively on the mass splitting. An improved analysis [5] leads to modified expressions for the CP-violating parameters:

$$\epsilon_i^S = \frac{|g_j|^2}{16\pi}\Im\left\{\frac{g_i g_j^*}{g_i^* g_j}\right\}\frac{\Delta_{ij}-\delta_{ij}}{c^2[\Delta_{ij}-\delta_{ij}]^2 + s^2\left[\Delta_{ij}^2 + (M_j\Gamma_j)^2\, L_\rho^2\right]}\cdot L_\rho\,,$$

where $c = \cos(\delta_{CP})$, $s = \sin(\delta_{CP})$. $L_\rho(X,p) = \int \frac{d\Omega}{4\pi}[1 + f^{\textcircled{b}}_{E_1} + f^{\textcircled{b}}_{E_2}]$ encodes medium corrections discussed above for the hierarchical case and Δ_{ij}, δ_{ij} describe finite density corrections to the masses. On the other hand, there is also a significant change of the in-medium decay width Γ_i^{med} which can have an opposite effect on final baryon asymmetry. These corrections may already lead to significant quantitative modifications in the resonant case, while there is even more uncertainty in the maximal resonant case where the derivation of BEs fails.

5. Conclusions

In the case of hierarchical heavy neutrino masses quantum-corrected BEs can be derived systematically for leptogenesis from the corresponding set of KBEs. These BEs are free of the double-counting problem and include a different result for the CP-violating parameter. This can be reconciled with thermal QFT if causal products are used in the computation. The corrections can be significant but for a given scenario further thermal effects must be included to obtain quantitative results. The application of the Boltzmann picture can be extended to a regime of intermediate degeneracy but fails in the maximal resonant case where the spectral functions overlap due to their finite width. It is highly important to test the different steps in our systematic derivation in order to find out when BEs are applicable. For example memory effects could be important in the resonant case. Also the next-to-leading order in the gradient expansion leads to a qualitatively different contribution to the rate equations [6]. These can be neglected in the case of the standard scenario of thermal leptogenesis, however. An alternative is to compare directly to solutions of the corresponding set of KBEs. In [13, 8] an approach to solve them explicitly in the special case of leptogenesis has been sketched.

Acknowledgements

I like to thank my collaborators M. Garny and A. Kartavtsev. This would also be the place to thank Alexey Anisimov for very useful discussion at the PASCOS conference. Unfortunately his unexpected death leaves us shocked. I met him first at the conference, but I can say with certainty that his early death is a grievous loss for the field and that we will also miss a particularly humorous colleague.

References

[1] Davidson S, Nardi E and Nir Y 2008 *Physics Reports* **466** 105–177 (*Preprint* 0802.2962)
[2] Kießig C P, Plümacher M and Thoma M H 2010 *Phys. Rev. D* **82** 036007 (*Preprint* 1003.3016)
[3] Hohenegger A, Kartavtsev A and Lindner M 2008 *Phys. Rev. D* **78** 085027 (*Preprint* 0807.4551)
[4] Garny M, Hohenegger A, Kartavtsev A and Lindner M 2009 *Phys. Rev. D* **80** 125027 (*Preprint* 0909.1559)
[5] Garny M, Hohenegger A, Kartavtsev A and Lindner M 2010 *Phys. Rev. D* **81** 085027 (*Preprint* 0911.4122)
[6] Garny M, Hohenegger A and Kartavtsev A 2010 (*Preprint* 1005.5385)

[7] Hohenegger A 2009 *Finite Density Aspects of Leptogenesis* Ph.D. thesis University of Heidelberg

[8] Anisimov A, Buchmüller W, Drewes M and Mendizabal S 2010 *Phys. Rev. Lett.* **104** 121102 (*Preprint* 1001.3856)

[9] Beneke M, Garbrecht B, Herranen M and Schwaller P 2010 *Nuclear Physics B* **838** 1–27 (*Preprint* 1002.1326)

[10] Beneke M, Garbrecht B, Fidler C, Herranen M and Schwaller P 2010 (*Preprint* 1007.4783)

[11] De Simone A and Riotto A 2007 *Journal of Cosmology and Astroparticle Physics* **0708** 002 (*Preprint* hep-ph/0703175)

[12] Garny M, Hohenegger A and Kartavtsev A 2010 *Phys. Rev. D* **81** 085028 (*Preprint* 1002.0331)

[13] Anisimov A, Buchmüller W, Drewes M and Mendizabal S 2008 *Annals of Physics* **324** 1234–1260 (*Preprint* 0812.1934)

16th International Symposium on Particles Strings and Cosmology (PASCOS 2010) IOP Publishing
Journal of Physics: Conference Series **259** (2010) 012075 doi:10.1088/1742-6596/259/1/012075

Supersymmetric leptogenesis and light hidden sectors

Christoph Weniger

Deutsches Elektronen-Synchrotron (DESY), Notkestrasse 85, 22603 Hamburg, Germany

E-mail: christoph.weniger@desy.de

Abstract. Thermal leptogenesis and supergravity are attractive scenarios for physics beyond the standard model. However, it is well known that the super-weak interaction of the gravitino often leads to problems with primordial nucleosynthesis in the standard scenario of matter parity conserving MSSM + three right-handed neutrinos. We will present and compare two related solutions to these problems: 1) The conflict between BBN and leptogenesis can be avoided in presence of a hidden sector with light supersymmetric particles which open new decay channels for the dangerous long-lived particles. 2) If there is a condensate in the hidden sector, such additional decay channels can be alternatively opened by dynamical breaking of matter parity in the hidden sector.

1. Introduction

Thermal leptogenesis [1] via the out-of-equilibrium decay of heavy right-handed neutrinos is one of the most promising models for the generation of the baryon asymmetry in the universe. However, it is known that this mechanism requires a high reheating temperature above 10^9 GeV. In supersymmetric extensions of the standard model, such temperatures can lead to the so-called gravitino problem, which can be circumvented when the gravitino itself is the LSP and hence dark matter. Indeed, it is well known that, given a high reheating temperature $T_R \sim \mathcal{O}(10^9$–$10^{10}\,\mathrm{GeV})$ and gravitino masses around $m_{3/2} \sim \mathcal{O}(10$–$100\,\mathrm{GeV})$, the thermal relic density of gravitinos reproduces the correct dark matter abundance.

However, the above scenario is not free of problems: The next-to-lightest supersymmetric particle (NLSP), which is often the stau, typically decays into the gravitino with lifetimes that are of the order of minutes or days. This is generically in conflict with the successful predictions of standard primordial nucleosynthesis (BBN), since the NLSP decay can destroy some of the primordial elements or lead to catalytic over-production of ^{6}Li and ^{9}Be [2, 3].

Different solutions to the above problem were proposed. In this note we shortly review two possible solutions that are related to extensions of the MSSM with light hidden sectors: 1) We will assume the existence of a hidden sector fermion, X, lighter than the lightest observable supersymmetric particle (LOSP). Then new decay channels are possible for the LOSP, for instance, when the LOSP is the lightest stau or the lightest neutralino, $\widetilde{\tau}_1 \to \tau X$ and $\chi_1^0 \to (Z^0, \gamma, h^0, f\bar{f})X$. If these decays are fast enough, the density of LOSPs at the time of nucleosynthesis can be significantly reduced and thus the successful predictions of the standard BBN scenario will not be jeopardized (for details see Ref. [4]). 2) It is well known in the literature that in cases where matter parity (or R-parity) is weakly violated, the NLSP can decay into standard model particles before the onset of BBN [5], while being compatible with

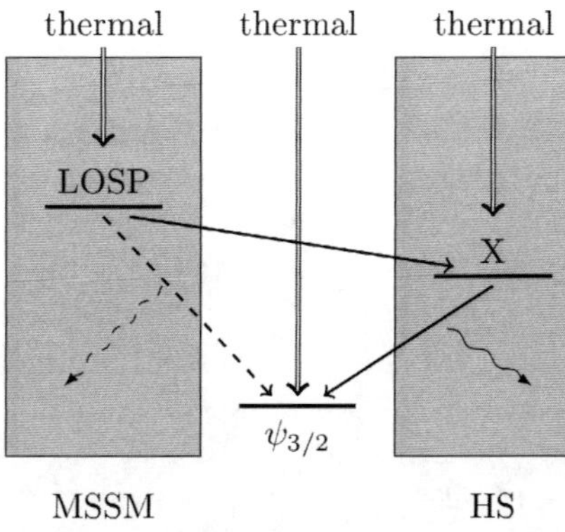

Figure 1. The LOSP, the gravitino and the hidden fermion can be produced thermally or non-thermally through the decays of heavier particles. If the reheating temperature is high, $T_R \gtrsim 10^9\,\mathrm{GeV}$, the LOSP decay into the gravitino occurs during or after the time of primordial nucleosynthesis, altering the predictions of the standard BBN scenario, yielding abundances in conflict with observations. However, if the LOSP coupling to the hidden fermion is large enough, this decay can occur before the onset of the nucleosynthesis reactions, thus avoiding altogether any possible effect of the LOSP on nucleosynthesis. See Ref. [4].

all cosmological constraints. We will present a model which can give rise to the required small matter-parity violation (for details see Ref. [6]) and discuss cosmological constraints.

2. Decay into hidden sector particles

We consider a scenario where the MSSM particle content is extended with a light standard model (SM) singlet superfield (chiral or vector). We will further assume that the fermionic component, the *hidden fermion* X, of this superfield couples to the LOSP and its standard model counterpart via a tiny Yukawa coupling, and we concentrate on the case where the LOSP is a stau. The interaction Lagrangian between the hidden fermion and the lightest stau, $\widetilde{\tau}_1$, is given by the renormalizable term $-\mathcal{L} = \lambda_{\widetilde{\tau}_1} \bar{X}\,\tau\,\widetilde{\tau}_1 + \mathrm{h.c.}$ We will assume below that the gravitino is the LSP and the hidden fermion X the NLSP (see Ref. [4] for other scenarios). The stau can then decay either via $\widetilde{\tau}_1 \to \psi_{3/2}\,\tau$ or via $\widetilde{\tau}_1 \to X\,\tau$ with decay rates given by $\Gamma_{\widetilde{\tau}_1 \to \psi_{3/2}\tau} \sim m^5_{\widetilde{\tau}_1} m^{-2}_{3/2} m_P^{-2}$ and $\Gamma_{\widetilde{\tau}_1 \to X\tau} \sim |\lambda_{\widetilde{\tau}_1}|^2 m_{\widetilde{\tau}_1}$, respectively. If the coupling $\lambda_{\widetilde{\tau}_1}$ is large enough, the stau will decay before BBN, thus preventing the catalytic production of ^{6}Li. The mechanism is sketched in Fig. 1.

2.1. Cosmologically stable hidden fermions

If the hidden fermion is stable, gravitinos as well as hidden fermions contribute to dark matter, each of them having a thermal component and a non-thermal component:

$$\Omega_{\mathrm{dm}} = \Omega^{\mathrm{th}}_{3/2} + \Omega^{\widetilde{\tau}_1}_{3/2} + \Omega^{\mathrm{th}}_X + \Omega^{\widetilde{\tau}_1}_X\,. \tag{1}$$

Here, $\Omega^{\widetilde{\tau}_1}_{3/2} = (m_{3/2}/m_{\widetilde{\tau}_1})\mathrm{BR}(\widetilde{\tau}_1 \to \psi_{3/2}\tau)\,\Omega^{\mathrm{th}}_{\widetilde{\tau}_1}$ and $\Omega^{\widetilde{\tau}_1}_X = (m_X/m_{\widetilde{\tau}_1})\mathrm{BR}(\widetilde{\tau}_1 \to X\tau)\,\Omega^{\mathrm{th}}_{\widetilde{\tau}_1}$, are the non-thermal contributions to the gravitino and hidden fermion relic density, respectively, where $\Omega^{\mathrm{th}}_{\widetilde{\tau}_1}$ and $\Omega^{\mathrm{th}}_{3/2}$ are the stau and the gravitino thermal abundance. If the hidden fermion couples to the observable sector through a renormalizable coupling, the thermal production proceeds dominantly via the decay of thermally produced staus. The corresponding hidden fermion relic abundance is given by $\Omega^{\mathrm{th}}_X h^2 \sim 10^{23}|\lambda_{\widetilde{\tau}_1}|^2(m_X/m_{\widetilde{\tau}_1})(1 - m^2_X/m^2_{\widetilde{\tau}_1})^2$.

In order to sufficiently reduce the number density of staus at the time of BBN it is necessary that $\mathrm{BR}(\widetilde{\tau}_1 \to \tau X) \simeq 1$, and therefore, $\Omega^{\widetilde{\tau}_1}_{3/2} \simeq 0$. Requiring that the total dark matter density does not exceed the measured value by WMAP implies then $\Omega^{\mathrm{th}}_{3/2} + \Omega^{\mathrm{th}}_X + \frac{m_X}{m_{\widetilde{\tau}_1}}\Omega^{\mathrm{th}}_{\widetilde{\tau}_1} \lesssim 0.11h^{-2}$. In the regime where the production of X is sizeable, this gives a strong upper bound of the order of 10^{-12} on the coupling $\lambda_{\widetilde{\tau}_1}$.

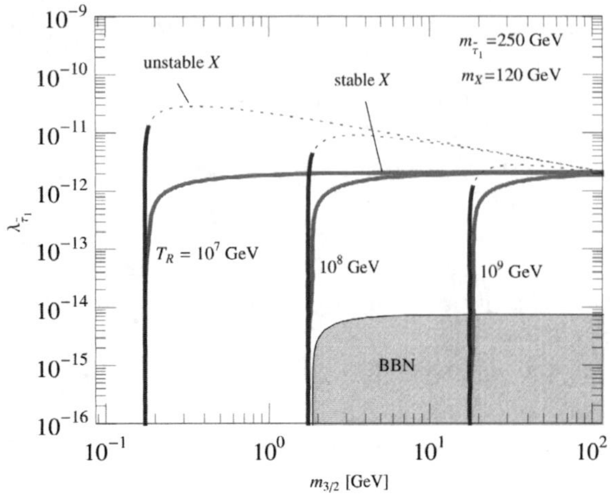

Figure 2. Summary of constraints. The masses of X and $\tilde{\tau}_1$ are fixed as indicated. The *red* and *blue lines* show the values of $m_{3/2}$ and $\lambda_{\tilde{\tau}_1}$ that yield the correct total relic abundance for different reheating temperatures $T_R = 10^7$–10^9 GeV, assuming that X is stable or unstable, respectively (the gluino mass has been set to $m_{\tilde{g}} = 800$ GeV). Furthermore, the *dashed* part of the blue lines is excluded by constraints on mixed warm/cold dark matter as discussed in the text, and the gray region is excluded by ^{6}Li overproduction during BBN. See Ref. [4].

The bounds are illustrated in Fig. 2 (red lines). The lines show, for different reheating temperatures and different masses of the hidden fermion, the value of the coupling $\lambda_{\tilde{\tau}_1}$ as a function of the gravitino mass from the requirement that the *total* dark matter density is equal to the value inferred by the WMAP collaboration. For large enough couplings $\lambda_{\tilde{\tau}_1}$, the BBN bounds (gray region) can be avoided.

2.2. Unstable hidden fermions

If kinematically allowed, the hidden fermion X decays into gravitinos and hidden sector particles (which are not dangerous for BBN) well before matter-radiation equality. In this case, dark matter consists of thermally produced gravitinos, as well as non-thermally produced gravitinos coming from the late decay of hidden fermions X and staus $\tilde{\tau}_1$. The dark matter abundance is then given by

$$\Omega_{\rm dm} = \Omega_{3/2}^{\rm th} + \underbrace{\Omega_{3/2}^{\tilde{\tau}_1} + \frac{m_{3/2}}{m_X}\left(\Omega_X^{\rm th} + \Omega_X^{\tilde{\tau}_1}\right)}_{=\Omega_{\rm WDM}}, \tag{2}$$

where we assumed for simplicity that the hidden-sector particles produced in the decay of X are massless. The component coming from the late decay of X, as well as the small fraction of gravitinos produced directly in $\tilde{\tau}_1$ decays, will typically act as warm dark matter (WDM), with free-streaming lengths $\lambda_{\rm FS} \gtrsim 5$ Mpc. Bounds on the fraction f of the dark matter density that is allowed to be warm with a free-streaming length above 0.5 Mpc were discussed in Ref. [7] in the context of sterile neutrinos. There, using Lyman-α data and WMAP5 results, 2σ-bounds around $f \lesssim 0.1$ were found for a warm component with free-streaming lengths around $\mathcal{O}(10\,\text{Mpc})$, corresponding to $\mathcal{O}(1\,\text{km/s})$ thermal velocities. Allowing a fraction f of dark matter to be warm, and provided that $\Omega_{3/2}^{\tilde{\tau}_1} \simeq 0$, implies then the upper bound $(m_{3/2}/m_X)(\Omega_X^{\rm th} + \Omega_X^{\tilde{\tau}_1}) \lesssim f\; 0.11\, h^{-2}$.

The allowed parameter space is shown in Fig. 2 (blue lines). The dashed part of the lines is excluded by constraints on mixed cold/warm dark matter. Compared to the scenario where the hidden fermion is stable, larger values of the coupling $\lambda_{\tilde{\tau}_1}$ are allowed.

2.3. Experimental signatures

In the above scenario the coupling constant of the LOSP to the hidden fermion X is typically smaller than 10^{-12}, therefore the LOSP decay length is much larger than the size of typical

Journal of Physics: Conference Series **259** (2010) 012075 doi:10.1088/1742-6596/259/1/012075

collider detectors. If the LOSP is the lightest stau, it propagates through the detector leaving a heavily ionizing charged track. Furthermore, if the stau velocity is small enough, it could get trapped in the detector and decay eventually, producing a tau moving in a non-radial direction. This signature, albeit very spectacular, is not specific of this scenario but also arises in scenarios with stau NLSP and gravitino or axino LSP.

However, a hint towards our proposed scenario arises from the measurement of the coupling stau-tau-hidden fermion. More concretely, at colliders it will be possible a determination of the stau mass and the stau lifetime, from which the coupling $\lambda_{\tilde{\tau}}$ could be inferred. If the coupling lies in the range preferred by cosmology, this scenario would gain strength. Further more model dependent predictions can be found in Ref. [4].

3. Dynamical matter-parity violation

As mentioned above, a small violation of matter parity can help to reconcile leptogenesis with gravitino dark matter. If we consider the matter parity to be the $\mathbb{Z}_2$ subgroup of the anomaly-free $\mathrm{U}(1)_{B-L}$ gauge symmetry, a very natural explanation of the required smallness of the breaking is to relate it to the condensation scale Λ of an asymptotically free hidden sector gauge group factor. This scale can be much lower than the $\mathrm{U}(1)_{B-L}$ breaking scale M_S, and if the hidden sector is charged under $B-L$, the condensate induces bilinear matter-parity breaking of order Λ^2/M_S.

3.1. Model

We consider an extension of the supersymmetric standard model with gauge group

$$G = \mathrm{SU}(3)_c \times \mathrm{SU}(2)_L \times \mathrm{U}(1)_Y \times \mathrm{U}(1)_{B-L} \times \mathrm{SU}(2)_{\text{hid}} \ . \tag{3}$$

We assume that light neutrino masses are generated in the standard way by the see-saw mechanism after the breaking of $\mathrm{U}(1)_{B-L}$ at a high scale M_S. Large Majorana masses for the right-handed neutrinos N^c are generated by a singlet field S in our model,

$$W_{\text{see-saw}} = h_{ij}^{(n)} L_i N_j^c H_u + \frac{1}{2}\lambda_i^S S N_i^c N_i^c \ , \tag{4}$$

where we have taken the basis where the Majorana mass matrix for N^c is diagonal. We assume $\langle S \rangle = M_S$. Then, the N_i^c have Majorana masses, $M_i = \lambda_i^S \times M_S$ for $i = 1-3$, respectively, and we define $M_1 < M_2 < M_3$ (we will assume that $M_3 = M_S$ below). For successful thermal leptogenesis, we require $M_1 > 10^9$ GeV. Since $\mathrm{U}(1)_{B-L}$ is gauged initially, the corresponding Nambu–Goldstone (NG) mode will be absorbed by the gauge field, which then decouples from the low-energy effective theory. The field S necessarily has $B-L$ charge -2, hence a global discrete $\mathbb{Z}_2$ subgroup remains unbroken during this process, the 'matter parity'.

Furthermore, the hidden sector is assumed to contain two doublet quarks $\mathcal{Q}_1^\alpha, \mathcal{Q}_2^\alpha$ with $B-L$ charge $+1/2$, and two doublet quarks $\mathcal{Q}_3^\alpha, \mathcal{Q}_4^\alpha$ with charge $-1/2$.[1] Here and in the following $\alpha, \beta = 1,2$ are $\mathrm{SU}(2)_{\text{hid}}$ indices. The low-energy degrees of freedom are the antisymmetric combinations $V_{ij} = -V_{ji} = \Lambda^{-1}\mathcal{Q}_i^\alpha \mathcal{Q}_{j\alpha}$ with convention $\mathcal{Q}_{i\alpha} = \epsilon_{\alpha\beta}\mathcal{Q}_i^\beta$, where $\epsilon_{\alpha\beta}$ is totally antisymmetric and $\epsilon_{12} = 1$. Condensation gives rise to non-vanishing vacuum expectation values for the two charged effective mesons, $\langle V_{12} \rangle = \langle V_{34} \rangle = \Lambda$ (for details see [6]). Since V_{12} and V_{34} have $B-L$ charge $+1$ and -1, respectively, we conclude that matter parity is broken dynamically at the scale Λ in the hidden sector, and we will take $\Lambda \ll M_S$.

The only unsuppressed and renormalizable interaction allowed by the gauge symmetries, connecting hidden and visible sector, is given by the term $W \supset -f_i \mathcal{Q}_3^\alpha \mathcal{Q}_{4\alpha} N_i^c$. Here, f_i with

[1] We also require in our model the existence of five neutral singlets $Z_{13}, Z_{14}, Z_{23}, Z_{24}$ and X, see Ref. [6].

Journal of Physics: Conference Series **259** (2010) 012075 doi:10.1088/1742-6596/259/1/012075

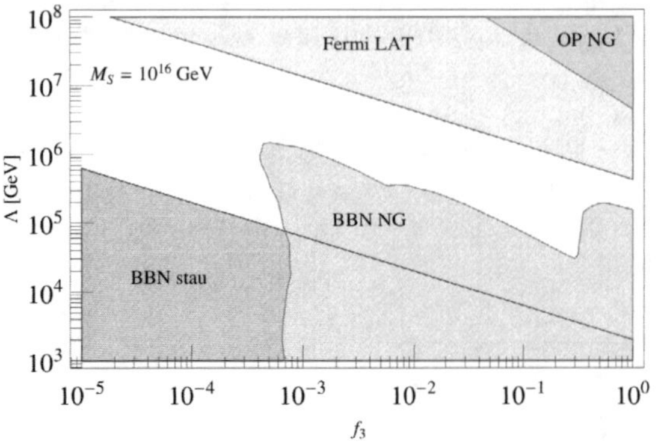

Figure 3. Summary of constraints on condensation scale Λ. We assumed $m_{3/2} = 120\,\text{GeV}$, $h_{ij}^{(n)} \sim 10^{-1}$, $M_S = 10^{16}\,\text{GeV}$, and $T_R = 10^{9.8}\,\text{GeV}$. We show the bounds coming from BBN catalysis by late decaying staus (BBN stau, $m_{\tilde{\tau}} \simeq 150$–$200$ GeV), from the late decay of the fermionic partner of the pseudo NG boson of matter-parity breaking (BBN NG), and from the non-observation of gamma-ray lines from gravitino decay by Fermi LAT. In the red region the pseudo NG boson relic density exceeds the observational limit (OP NG). See Ref. [6].

$i = 1, 2, 3$ are free parameters, and we will assume a simple situation where $f_3 \leq 1$ and $f_1 = f_2 = 0$, to show the presence of a consistent parameter region in the model. After $\text{SU}(2)_{\text{hid}}$ condensation, this becomes a linear term for the right-handed neutrino multiplets, which together with the mass term in Eq. (4) implies a non-vanishing vacuum expectation value for the corresponding sneutrinos,

$$\langle N_i^c \rangle = \frac{f_i}{\lambda_i^S} \frac{\Lambda^2}{M_S} \,. \tag{5}$$

Hence, the matter-parity breaking is mostly bilinear in our model, originating from the Yukawa couplings in (4). Its scale is related to the condensation scale of the hidden sector gauge group.

Since the interactions between hidden and visible sector are suppressed by M_S, an approximate global $\text{U}(1)_{B-L}$ symmetry in the hidden sector remains after S acquires a vev. This $\text{U}(1)_{B-L}$ symmetry is broken by $\mathcal{Q}^\alpha \mathcal{Q}_\alpha$ condensation, producing a nearly massless pseudo NG multiplet in the hidden sector. After supersymmetry breaking, soft mass terms raise the masses of the modes in the pseudo NG multiplet. If we assume pure gravity mediation to the hidden sector, this yields a mass for the pseudo NG boson given by $m_a \sim f_3 \Lambda (m_{3/2}/M_3)^{1/2} \sim 100\,\text{MeV} f_3 (m_{3/2}/100\,\text{GeV})^{1/2} (\Lambda/10^6\,\text{GeV}) (M_3/10^{16}\,\text{GeV})^{-1/2}$, whereas the tree-level masses of the fermion partner ψ and the radial scalar pseudo NG component ρ are given by $m_\psi \approx m_{3/2}$ and $m_\rho \approx 4\, m_{3/2}$. Below we will always assume that $m_\psi > m_{3/2}$, since we are interested in gravitino dark matter.

3.2. Phenomenology

A number of astrophysical observations constrain the lifetime of the stau and the gravitino, which translates into bounds on the matter-parity breaking parameters. Firstly, the $\widetilde{\tau}_1$ has to decay before BBN, with a lifetime shorter than 2×10^3 s, to avoid catalytic overproduction of ^{6}Li. This implies a lower limit on the matter-parity breaking. Furthermore, the gravitino decay into gamma-ray lines is limited by observations of the Fermi LAT satellite to lifetimes $\Gamma^{-1}_{\psi_{3/2} \to \gamma\nu} \gtrsim 10^{29}$ s, yielding an upper limit if $m_{3/2} \sim 100\,\text{GeV}$. The corresponding limits on the condensation scale Λ are summarized in Fig. 3 for a reference scenario.

Even when conservatively assuming that reheating only affects the MSSM sector, hidden sector particles are produced in the early universe by scattering of MSSM particles. The hidden sector contains 10 chiral multiplets with a mass at the condensation scale Λ, as well as one chiral multiplet that remains light and is the pseudo NG mode of the breaking of the accidental global

Journal of Physics: Conference Series **259** (2010) 012075 doi:10.1088/1742-6596/259/1/012075

$B-L$ symmetry in the hidden sector. Depending on the details of the symmetries of the hidden sector, some of the heavy hidden sector states can be stable, which could lead to overclosure of the universe. However, depending on the details of the model, all these potentially stable heavy hidden sector states can either efficiently annihilate into the pseudo NG modes or decay into SM particles. Then, one only has to consider the effects of the NG modes (see Ref. [6] for details).

The most problematic hidden sector particle turns out to be the fermionic pseudo NG boson partner ψ, which mixes with the neutrinos ν_i and inherits their SM interactions. If $m_\psi > M_Z$, the fast two-body decays $\psi \to Z^0\nu_i$ and $\psi \to W^\pm\ell^\mp$ are kinematically allowed, but BBN bounds are still relevant for small mixing angles and masses m_ψ. The bosonic pseudo NG boson partner ρ can perform two-body decays like $\rho \to \tilde{h}_u\nu$, if kinematically allowed, in which case the lifetime of ρ is much smaller than the lifetime of ψ, since its decay is not further suppressed by gauge couplings. The pseudo NG boson a is stable on cosmological time-scales.

In addition to the bounds mentioned above, one can find in Fig. 3 bounds from BBN that come from the late decay of the fermionic and bosonic pseudo NG boson partner. We also show the region where the relic density of the pseudo NG boson would exceed the observational limit. As apparent from Fig. 3, all constraints can be satisfied for condensation scales $\Lambda \sim 10^5$–10^8 GeV.

4. Conclusions

Cosmological scenarios where the observed matter-antimatter asymmetry is generated by the supersymmetric thermal leptogenesis mechanism generically fail to reproduce the observed abundances of primordial elements. We have discussed two possible solutions to this problem.

Firstly, we have shown that the existence of a light hidden sector fermion, which couples very weakly to the lightest observable supersymmetric particle (LOSP, *e.g.* the lightest stau), opens new decay channels for the LOSP. If the coupling is large enough, the LOSP will decay dominantly into hidden sector fermions before the epoch of primordial nucleosynthesis, avoiding all the nucleosynthesis constraints altogether. We have summarized the different constraints on this coupling and commented on the experimental signatures at particle colliders.

Secondly, we have discussed a possible mechanism that generates a small matter-parity violation in the visible sector. Due to the matter-parity breaking, the NLSP can then decay into standard model particles before conflicting with BBN. In our scenario, visible sector and part of the hidden sector are simultaneously charged under a gauged $\mathrm{U}(1)_{B-L}$. This $\mathrm{U}(1)_{B-L}$ is broken at a high scale M_S by right-handed neutrino masses to its matter-parity $\mathbb{Z}_2$ subgroup, and matter parity is then subsequently broken completely in the hidden sector by a $\mathrm{SU}(2)_{\mathrm{hid}}$ quark condensate at a scale Λ. This mechanism gives rise to bilinear matter-parity breaking in the visible sector of the order of Λ^2/M_S. We summarized the constraints coming from pseudo NG modes in the hidden sector and showed that the model can be phenomenologically viable for large enough $B-L$ breaking scales.

References

[1] Fukugita M and Yanagida T 1986 *Phys.Lett.* **B174** 45
[2] Jedamzik K 2006 *Phys.Rev.* **D74** 103509 (*Preprint* `hep-ph/0604251`)
[3] Pospelov M 2007 *Phys.Rev.Lett.* **98** 231301 (*Preprint* `hep-ph/0605215`)
[4] Simone A D, Garny M, Ibarra A and Weniger C 2010 *JCAP* **1007** 017 (*Preprint* `arXiv:1004.4890`)
[5] Buchmuller W, Covi L, Hamaguchi K, Ibarra A and Yanagida T 2007 *JHEP* **0703** 037 (*Preprint* `hep-ph/0702184`)
[6] Schmidt J, Weniger C and Yanagida T T 2010 (*Preprint* `arXiv:1008.0398`)
[7] Boyarsky A, Lesgourgues J, Ruchayskiy O and Viel M 2009 *JCAP* **0905** 012 (*Preprint* `arXiv:0812.0010`)

Journal of Physics: Conference Series **259** (2010) 012076 doi:10.1088/1742-6596/259/1/012076

Dynamical four-form fields

Ufuk Aydemir, Luca Grisa and Lorenzo Sorbo

Department of Physics, 1130 LGRC, UMass at Amherst, Amherst, MA 01003-9337 (USA)

E-mail: uaydemir@physics.umass.edu

E-mail: lgrisa@physics.umass.edu

E-mail: sorbo@physics.umass.edu

Abstract. We present an example of a gauge-invariant Lagrangian that contains four derivatives and describes one massive, non-ghostlike, degree of freedom.

Motivation. Gauge invariance is often assumed to imply the existence of massless degrees of freedom, although there are famous counter-examples, ranging from [1, 2] to [3, 4]. Moreover, it is often assumed that theories with higher derivatives contain ghosts or other pathologies. Recently, [5] has observed that the model, proposed in [6] and therein called "new massive gravity," does provide a counter-example to both these commonly accepted no-goes. Alas, the model [6] is defined only in three-dimensional space-times.

In the present paper we will discuss a model that violates both aforementioned assumptions in any number of dimensions. As we will see, the model shares many of the properties with the new massive gravity of [6], but without the limitation of three space-time dimensions (though we will mostly focus, for the sake of argument, on four dimensions).

The key ingredient of this model is, in d space-time dimensions, a $(d-1)$-form field potential $\mathcal{A}_{\mu_1\mu_2\cdots\mu_{d-1}}$. As for electromagnetism, which is described by a 1-form potential, we require gauge invariance under the transformation $\mathcal{A}_{\mu_1\mu_2\cdots\mu_{d-1}} \rightarrow \mathcal{A}_{\mu_1\mu_2\cdots\mu_{d-1}} + \partial_{[\mu_1}\mathcal{B}_{\mu_2\cdots\mu_{d-1}]}$, where $\mathcal{B}_{\mu_1\cdots\mu_{d-2}}$ is an arbitrary $(d-2)$-form. Thus, the field strength $\mathcal{F}_{\nu\mu_1\mu_2\cdots\mu_{d-1}} \equiv \partial_{[\nu}\mathcal{A}_{\mu_1\mu_2\cdots\mu_{d-1}]}$ is gauge invariant.

Specializing to four space-time dimensions, it is usually assumed, by analogy with electromagnetism, that the Lagrangian for a 4-form field strength should be

$$\mathcal{L} = \sqrt{-g}\Big[-\frac{1}{48}\,\mathcal{F}^{\mu\nu\rho\lambda}\,\mathcal{F}_{\mu\nu\rho\lambda}\Big]\,. \tag{1}$$

Had this Lagrangian – or, more generally, any polynomial in $\mathcal{F}_{\mu\nu\rho\lambda}$ – been chosen, the field $\mathcal{A}_{\mu\nu\rho}$ would not have any local dynamics. We observe that the tensorial structure of the field strength imposes $\mathcal{F}_{\mu\nu\rho\lambda}(x^\alpha) = \sqrt{-g}\,\epsilon_{\mu\nu\rho\lambda}\,\Phi(x^\alpha)$, where $\Phi(x^\alpha)$ is a scalar field and $\epsilon_{\mu\nu\rho\lambda}$ the Levi-Civita symbol with $\epsilon_{0123} = 1$. The equations of motion derived from (1), $\nabla^\mu\,\mathcal{F}_{\mu\nu\rho\lambda} = 0$, imply $\partial_\mu\Phi = 0$, *i.e.*, Φ is constant.

We will show that adding higher derivatives term, $\nabla_\alpha\mathcal{F}_{\mu\nu\rho\lambda}$, in the Lagrangian allows the (dual) scalar field Φ to propagate. Moreover, for an appropriate choice of parameters, the propagating degree of freedom has a well-behaved kinetic term and a positive-definite potential.

The Lagrangian. The most general Lagrangian, quadratic in $\mathcal{F}$ and with at most four derivatives and "dimension six" operators, is (we will assume the "mostly plus" convention for the signature of the metric)

$$\mathcal{L} = \sqrt{-g}\Big[\frac{c_0}{48}\mathcal{F}^{\mu\nu\rho\lambda}\mathcal{F}_{\mu\nu\rho\lambda} + \frac{c_1}{48\,M^2}\nabla^\alpha\mathcal{F}^{\mu\nu\rho\lambda}\nabla_\alpha\mathcal{F}_{\mu\nu\rho\lambda} + \frac{c_2}{12\,M^2}\nabla_\alpha\mathcal{F}^{\alpha\nu\rho\lambda}\nabla^\beta\mathcal{F}_{\beta\nu\rho\lambda} + \\ + \text{terms in } R\mathcal{F}\mathcal{F}\Big], \tag{2}$$

where $R\mathcal{F}\mathcal{F}$ indicates any term built by contracting two field strengths with a Riemann tensor or its contractions. We will neglect these terms to ease the discussion, but their role is discussed in our paper [7]. In (2), the coefficients c_0, c_1, c_2 are dimensionless numbers, and M is a mass scale. A term proportional to $\nabla^\alpha\mathcal{F}^{\beta\nu\rho\lambda}\nabla_\beta\mathcal{F}_{\alpha\nu\rho\lambda}$ can also be added to (2), however it is identical to the term proportional to c_2 after an integration by parts and so it will be neglected.

The properties of our system become more apparent in the first order formalism, where the potential $A_{\mu\nu\rho} \equiv M\mathcal{A}_{\mu\nu\rho}$ and the field strength $F_{\mu\nu\rho\lambda} \equiv M\mathcal{F}_{\mu\nu\rho\lambda}$ are treated as independent variables, and the condition $F_{\mu\nu\rho\lambda} \equiv \partial_{[\mu}A_{\nu\rho\lambda]}$ is imposed by introducing a Lagrange multiplier $q(x^\alpha)$. The Lagrangian then reads

$$\mathcal{L} =\sqrt{-g}\Big[\frac{c_0\,M^2}{48}\,F^{\mu\nu\rho\lambda}\,F_{\mu\nu\rho\lambda} + \frac{c_1}{48}\nabla^\alpha F^{\mu\nu\rho\lambda}\,\nabla_\alpha F_{\mu\nu\rho\lambda} + \frac{c_2}{12}\nabla_\alpha F^{\alpha\nu\rho\lambda}\,\nabla^\beta F_{\beta\nu\rho\lambda}\Big] + \\ + \frac{q}{24}\,\epsilon^{\mu\nu\rho\lambda}\,(F_{\mu\nu\rho\lambda} - 4\,\partial_\mu A_{\nu\rho\lambda})\;. \tag{3}$$

Since $F^{\mu\nu\rho\lambda}$ is now a fundamental field, we can replace it by $\sqrt{-g}\,\epsilon_{\mu\nu\rho\lambda}\,\phi(x^\alpha)$ directly in the Lagrangian:

$$\mathcal{L} = \sqrt{-g}\Big[-\frac{c_1+c_2}{2}\nabla^\alpha\phi\,\nabla_\alpha\phi - \frac{c_0}{2}\,M^2\,\phi^2\, - q\,\phi\Big] - \frac{q}{6}\,\epsilon^{\mu\nu\rho\lambda}\,\partial_\mu A_{\nu\rho\lambda}\,. \tag{4}$$

The equation for $A^{\mu\nu\rho}$, $\partial_\lambda q = 0$, simply implies the Lagrange multiplier q to be a constant.

If the conditions $c_0 > 0$ and $c_1 + c_2 > 0$ are imposed, the Lagrangian (4) describes a healthy, non-tachyonic scalar field of mass:

$$\mu^2 \equiv \frac{c_0}{c_1+c_2}\,M^2\,. \tag{5}$$

We note that the requirement of non-tachyonic mass term, $c_0 > 0$, corresponds to the "wrong" sign for the F^2 term. This is precisely the same phenomenon observed in the new massive gravity model of [6], where the role of F^2 is played by the Ricci scalar. Let us also note, however, that the addition in the Lagrangian of higher powers of the form field, *e.g.*, $F^{\alpha\beta\gamma\delta}F^{\mu\nu\rho\lambda}F_{\alpha\beta\rho\lambda}F_{\mu\nu\gamma\delta}$, can stabilize the vacuum of the theory at non-vanishing values of the form field even if $c_0 < 0$.

Despite this system being very similar to the one described in [8, 9], it is not the same: while in [8, 9] the different vacua have all the same cosmological constant, in our case they do not, since the vacuum energy is proportional to q^2. Further differences emerge when we couple our form field to matter.

Coupled to matter. Given the "wrong" sign of the term F^2, an interesting question to answer is whether this model is stable once coupled to matter. p-form fields couple naturally to $(p-1)$-branes. In four dimensions, a 2-brane sources a 3-form field

$$\mathcal{L}_{\text{brane}} = \int_\partial \mathrm{d}^3\xi\,\Big[-\sigma\sqrt{-\hat{g}} + \frac{e}{6}A_{\mu\nu\rho}\frac{\partial z^\mu}{\partial\xi^a}\frac{\partial z^\nu}{\partial\xi^b}\frac{\partial z^\rho}{\partial\xi^c}\epsilon^{abc}\Big]\,, \tag{6}$$

where e is the charge and σ the tension of the membrane; we denote by ∂ the brane world-volume. Note that e is proportional by a factor M to the original brane charge for the field

Journal of Physics: Conference Series **259** (2010) 012076 doi:10.1088/1742-6596/259/1/012076

$\mathcal{A}_{\mu\nu\rho}$. This coupling could destabilize the vacuum through brane nucleations, described by the following equations of motion (neglecting the role of gravity)

$$z^{\nu}_{,a}\,\partial_{\nu}(z^{\mu}_{,a}) = \frac{e}{6\,\sigma}\,\phi\,\epsilon^{\mu\nu\rho\lambda}\,\frac{\partial z_{\nu}}{\partial\xi^{a}}\frac{\partial z_{\rho}}{\partial\xi^{b}}\frac{\partial z_{\lambda}}{\partial\xi^{c}}\,\epsilon^{abc}\,, \tag{7}$$

$$\epsilon^{\mu\nu\rho\lambda}\,\partial_{\mu}q = -e\int \mathrm{d}^3\xi\,\delta^{(4)}\left(x - z(\xi)\right)\,\frac{\partial z^{\nu}}{\partial\xi^{a}}\frac{\partial z^{\rho}}{\partial\xi^{b}}\frac{\partial z^{\lambda}}{\partial\xi^{c}}\epsilon^{abc}\,, \tag{8}$$

$$\Box\,\phi - \mu^2\phi - \frac{q}{c_1 + c_2} = 0\,, \tag{9}$$

where we have already substituted the 4-form curvature of $A_{\mu\nu\rho}$ with $F_{\mu\nu\rho\lambda} = \epsilon_{\mu\nu\rho\lambda}\,\phi$ and μ^2 is the effective mass found in (5).

The solution for the equation of motion for the 3-brane (7) is that of a hyperboloidal bubble of constant "radius" r_0. From (8), we notice that the Lagrange multiplier q is constant away from the bubble and changes by e, the charge of the 2-brane, when crossing the bubble, *i.e.*, $|\Delta q| = e$.

Since we are interested in instanton solutions for the nucleation of bubbles with different value of q, we will consider the euclidean action $\mathcal{S}_{\mathrm{E}}$ of (4) with coupling (6) on the solution of the equations of motion (7), (8) and (9)

$$\mathcal{S}_{\mathrm{E}} = \int \mathrm{d}^4x \left[\frac{1}{2}(\partial\bar{\phi})^2 + \frac{1}{2}\mu^2\bar{\phi}^2 + \frac{q}{\sqrt{c_1 + c_2}}\bar{\phi}\right] + \sigma\int_{\partial} \mathrm{d}^3\xi\sqrt{\hat{g}}\,, \tag{10}$$

where $\bar{\phi}$ is the canonically normalized scalar field $\bar{\phi} \equiv \sqrt{c_1 + c_2}\,\phi$, and where we have taken into account the appropriate boundary terms in the Lagrangian. The function $q(r)$ is given by $q(r) = q_<\,\theta(r_0 - r) + q_>\,\theta(r - r_0)$, where the step function $\theta(r)$ is equal to 1 when its argument is positive and equal to zero when it is not, and $|q_> - q_<| = e$. By using the spherical symmetry of the problem, we can simplify (10) even further

$$\mathcal{S}_{\mathrm{E}} = \pi^2\int \mathrm{d}r\,r^3\left[(\partial_r\bar{\phi})^2 + \mu^2\bar{\phi}^2 + 2\,\frac{q(r)}{\sqrt{c_1 + c_2}}\bar{\phi}\right] + 2\,\pi^2\sigma\,r_0^3 \tag{11}$$

and the equation of motion for $\bar{\phi}$ becomes

$$\bar{\phi}'' + 3\frac{\bar{\phi}'}{r} - \mu^2\bar{\phi} - \frac{q(r)}{\sqrt{c_1 + c_2}} = 0\,. \tag{12}$$

Regularity for both $r \to 0^+$ and $r \to \infty$ constrains the solution to be

$$\bar{\phi}(r) = \begin{cases} -\frac{q_<}{\mu^2\sqrt{c_1+c_2}} + A\,\frac{\mathrm{I}_1(\mu\,r)}{\mu\,r} & , \quad \text{for } r < r_0 \\ -\frac{q_>}{\mu^2\sqrt{c_1+c_2}} + B\,\frac{\mathrm{K}_1(\mu\,r)}{\mu\,r} & , \quad \text{for } r > r_0 \end{cases} \tag{13}$$

where $\mathrm{I}_n(\mu\,r)$ and $\mathrm{K}_n(\mu\,r)$ are Bessel functions. The constants A and B are found by requiring continuity of the function and its derivative across the bubble at $r = r_0$:

$$A = \frac{q_< - q_>}{\sqrt{c_1 + c_2}}\,r_0^2\,\mathrm{K}_2(\mu\,r_0) \tag{14}$$

$$B = -\frac{q_< - q_>}{\sqrt{c_1 + c_2}}\,r_0^2\,\mathrm{I}_2(\mu\,r_0)\,, \tag{15}$$

The difference between the action for the solution of constant $q_>$ and the action for $q(r)$ is

$$\Delta\mathcal{S}_{\rm E} = 2\,\pi^2\,\sigma\,r_0^3 + \frac{\pi^2\,r_0^4}{4\,\mu^2}\,\frac{q_>^2 - q_<^2}{c_1 + c_2} + \frac{\pi^2\,r_0^4}{\mu^2}\,\frac{(q_< - q_>)^2}{c_1 + c_2}\,\mathrm{I}_2(\mu\,r_0)\,\mathrm{K}_2(\mu\,r_0)\,. \tag{16}$$

This expression is not very illuminating; it is convenient to consider the limits $\mu\,r_0 \ll 1$ and $\mu\,r_0 \gg 1$.

In the first limit, the difference in actions reduces to

$$\Delta\mathcal{S}_{\rm E} \sim 2\,\pi^2\,\sigma\,r_0^3 + \frac{\pi^2\,r_0^4}{2\,\mu^2}\,\frac{q_>\,\varepsilon\,e}{c_1 + c_2}\,, \tag{17}$$

where we have introduced the symbol $\varepsilon = \pm 1$, $q_> - q_< = \varepsilon\,e$, to discriminate between a brane and an anti-brane nucleation. This expression for the change in action can be easily found by direct calculation, observing that, since the bubble is much smaller than the scalar Compton wavelength, $\bar\phi$ stays essentially constant at the value it assumes outside the bubble, *i.e.*, $\bar\phi = -q_>/(\mu^2\sqrt{c_1 + c_2})$.

It is clear that the action (17) can always be minimized, with an appropriate choice of the sign ε, at

$$r_0 \sim 3\,\frac{\sigma\,\mu^2\,(c_1 + c_2)}{|q_>|\;e}\,, \tag{18}$$

$$\Delta\mathcal{S}_{\rm E}(r_0) \sim \frac{27\,\pi^2}{2}\,\frac{\sigma^4\,\mu^6\,(c_1 + c_2)^3}{|q_>|^3\;e^3}\,. \tag{19}$$

In the opposite regime $\mu\,r_0 \gg 1$, the action reads

$$\Delta\mathcal{S}_{\rm E} \sim 2\,\pi^2\,\sigma\,r_0^3 + \frac{\pi^2}{4\,\mu^2}\,\frac{2\,q_>\,\varepsilon\,e - e^2}{c_1 + c_2}\,r_0^4 + \frac{\pi^2}{2\,\mu^3}\,\frac{e^2}{c_1 + c_2}\,r_0^3\,, \tag{20}$$

where the second term corresponds to the change in bulk energy induced by the bubble nucleation, whereas the third term represents a correction to the effective membrane tension induced by the gradient in the field ϕ.

The bubble radius is found by extremizing the action with respect to r_0. Inspection of eq. (16), and of its explicit limits (17) and (20), shows that it is always possible to find a positive definite radius, r_0, that minimizes the effective action. This implies that, unlike the usual Brown-Teitelboim model [10, 11], this system does not have any absolutely stable vacuum. The reason for such an instability lies in the "wrong" sign for the form kinetic term, F^2. Nevertheless, it is easy to find parameters for which the decay rate, proportional to $e^{-\mathcal{S}_E}$, is extremely small and the system can be considered stable for all practical purposes.

In the decoupling limit for the field ϕ, $M \to \infty$ with $q/M \equiv \bar q$ and $e/M \equiv \bar e$ fixed, the action describing membrane nucleation reads

$$\Delta\mathcal{S}_{\rm E} = 2\,\pi^2\,\sigma\,r_0^3 + \frac{\pi^2\,r_0^4}{4\,c_0}\,\left(2\,\bar q_>\,\varepsilon\,\bar e - \bar e^2\right)\,. \tag{21}$$

which gives the Brown-Teitelboim result [10, 11] when $c_0 = -1$.

By solving numerically the equation $\partial_{r_0}\Delta\mathcal{S}_{\rm E} = 0$, we find that the probability to nucleate a bubble of $q_< = q_> - e$ is always smaller than the one corresponding to the Brown-Teitelboim limit, $M \to \infty$. This can be understood by considering that, when the field $\bar\phi$ is dynamical and massive, a part of the energy gained by nucleating a bubble goes into exciting $\bar\phi$-modes and the smaller the mass, the more likely these modes are to be excited.

Equivalently, it can be noted that $\bar{\phi}$ generates an effective attractive force between branes. For two parallel branes with charges e and e' placed at a distance d, the field $\bar{\phi}$ mediates a force per unit surface

$$\frac{F}{s} = \frac{e\,e'}{2\,\mu\,(c_1 + c_2)}\,\exp\left[-\mu\,d\right]\,, \tag{22}$$

that is, two branes of same charge repel each other, two of opposite charge attract each other. To defy this force the branes will have to be nucleated at radius r_0 greater than when such a force is absent, *i.e.*, $\mu \to \infty$, the Brown-Teitelboim case. A larger bubble will be less likely to nucleate than a smaller one.

Schwinger model with a dynamical two-form. The two dimensional version of our scenario is especially interesting. In fact, in two dimensions, it is known, [2], that once a massless fermion – the analog of a light brane in higher dimensions – of charge e is coupled to a vector field, a gauge-invariant mass term is generated

$$\delta\mathcal{L}_{\rm mass} = -\frac{e^2}{2\,\pi}\,\mathcal{A}_\mu\left(\eta^{\mu\nu} - \frac{\partial^\mu\,\partial^\mu}{\partial^2}\right)\mathcal{A}_\nu\,. \tag{23}$$

This mechanism works also when higher derivatives operators are considered – as it can be easily checked – therefore we simply add this gauge-invariant mass to the two dimensional version of the Lagrangian (2). The longitudinal mode of $\mathcal{A}_\mu$ drops out, since the entire Lagrangian is gauge invariant and it depends only on the transverse mode $\mathcal{A}^{\rm T}_\mu$

$$\mathcal{L} = \frac{c_0}{4}\mathcal{F}^2_{\mu\nu} + \frac{c_1}{4\,M^2}(\nabla_\alpha\mathcal{F}_{\mu\nu})^2 + \frac{c_2}{2\,M^2}(\nabla^\alpha\mathcal{F}_{\alpha\mu})^2 - \frac{e^2}{2\pi}\mathcal{A}^{\rm T}_\mu\,\mathcal{A}^{{\rm T}\mu} + \frac{\mathcal{Q}}{2}\,\epsilon^{\mu\nu}(\mathcal{F}_{\mu\nu} - 2\,\partial_\mu\mathcal{A}^{\rm T}_\nu)\,, \tag{24}$$

where we have added a Lagrange multiplier $\mathcal{Q}$ to treat the vector field and the curvature independently; $\mathcal{F}_{\mu\nu}$ is proportional to the two dimensional Levi-Civita symbol ($\mathcal{F}_{\mu\nu} \equiv \epsilon_{\mu\nu}\Phi$) and $\mathcal{A}^{\rm T}_\mu$ can be integrated out in favor of $\mathcal{Q}$, $\mathcal{A}^{\rm T}_\mu = -(\pi/e^2)\,\epsilon_{\mu\alpha}\partial^\alpha\mathcal{Q}$. These substitutions lead to the following effective Lagrangian in terms of the scalar fields Φ and $\mathcal{Q}$,

$$\mathcal{L}^{\rm eff} = -\frac{c_1 + c_2}{2\,M^2}(\partial_\alpha\Phi)^2 - \frac{c_0}{2}\,\Phi^2 - \mathcal{Q}\,\Phi - \frac{\pi}{2\,e^2}(\partial_\alpha\mathcal{Q})^2\,; \tag{25}$$

the mass term for the vector field realized via the Schwinger mechanism effectively generates a kinetic term for the Lagrange multiplier $\mathcal{Q}$, making it dynamical. The limit $e \to 0$ decouples $\mathcal{Q}$ from the other fields; the condition $\partial_\mu\mathcal{Q} = 0$ is recovered and $\mathcal{Q}$ ceases to propagate.

Before canonically normalizing the fields, we notice that, for $c_1 + c_2 > 0$, the sign of the kinetic term for both Φ and $\mathcal{Q}$ is that of a healthy scalar field, that is, not a ghost.

Let us substitute $\bar{\phi} \equiv \sqrt{c_1 + c_2}\,\Phi/M$ and $q \equiv \sqrt{\pi}\mathcal{Q}/e$

$$\mathcal{L}^{\rm eff} = -\frac{1}{2}(\partial_\mu\bar{\phi})^2 - \frac{1}{2}(\partial_\mu q)^2 - \frac{1}{2}\mu^2\bar{\phi}^2 - \frac{e\,\mu}{\sqrt{\pi}\,c_0}\,q\,\bar{\phi}\,, \tag{26}$$

where μ is the effective mass defined in (5). The model does therefore contain two propagating degrees of freedom of masses:

$$\mu^2_\pm = \frac{\mu^2}{2}\left(1 \pm \sqrt{1 + \frac{4\,e^2}{\pi c_0\,\mu^2}}\right)\,. \tag{27}$$

Because the coefficient c_0 and the effective mass μ^2 are positive definite, the square root is always greater than 1, hence one scalar mode is always tachyonic. This tachyonic instability is the perturbative counterpart of the non-perturbative instability for brane nucleation described above.

In the limit $\mu^2 \gg e^2/c_0$, the two masses are approximately μ^2 and $-e^2/(\pi\, c_0)$ and the tachyonic instability has a characteristic time much longer than the timescale μ^{-1}. In the opposite regime $\mu^2 \ll e^2/c_0$, the mass eigenvalues have equal magnitude $\pm 2\, e\, \mu/\sqrt{\pi\, c_0}$, signaling a fast instability.

Discussion and conclusions. We have shown that the Lagrangian (2) is, despite its higher derivatives, a well-behaved Lagrangian: neither ghosts, nor tachyons are propagating degrees of freedom.

When we couple the form field to matter, instabilities do emerge, but, since they are non-perturbative, their characteristic time can be made safely long. Given the similarities between our system and the new massive gravity of [6], it would be interesting to inquire whether analogous instabilities appear in that model when gravity is coupled to matter. As we noted above, the absence of an absolutely stable minimum lies in the "wrong" sign of the would-be kinetic term for the d-form, *i.e.*, F^2; similarly in the model of [6], the would-be kinetic term for the gravitational degrees of freedom, *i.e.*, R, is chosen to have the "wrong" sign in order for the mass of the propagating degree of freedom to be non-tachyonic, hence we can speculate that a non-perturbative instability similar to the one we found for the present model could appear when the new massive gravity is coupled to matter.

As in our case, [12] has also recently obtained a healthy behavior for form fields despite a higher derivative Lagrangian. Our mechanism is however different, as one can see by observing that [12] considers form fields whose rank is different from the one we are considering.

Let us conclude with a speculation. The goal of this paper is to show an example of a higher derivative gauge theory that does not have any unhealthy behavior and that propagates massive degrees of freedom. Our goal is *not* to embed this theory in a UV-complete model; it would be interesting, however, to study whether the Lagrangian (2) can emerge as a limit of some known theories, *e.g.*, as an effective low-energy description arising from strong dynamics at scale M. Within this framework, the model we presented might be a generalization of the models describing a condensation of topological defects, as studied in [13].

Acknowledgments. This work has been supported in part by the NSF grant PHY - 0855119. We thank John Donoghue, Nemanja Kaloper and Albion Lawrence for useful discussions.

References.

[1] Schwinger J S 1962 *Phys. Rev.* **125** 397–398
[2] Schwinger J S 1962 *Phys. Rev.* **128** 2425–2429
[3] Deser S, Jackiw R and Templeton S 1982 *Phys. Rev. Lett.* **48** 975–978
[4] Porrati M 2002 *Phys. Rev.* **D65** 044015 (*Preprint* hep-th/0109017)
[5] Deser S 2009 *Phys. Rev. Lett.* **103** 101302 (*Preprint* 0904.4473)
[6] Bergshoeff E A, Hohm O and Townsend P K 2009 *Phys. Rev. Lett.* **102** 201301 (*Preprint* 0901.1766)
[7] Aydemir U, Grisa L and Sorbo L (*Preprint* 1009.5690)
[8] Kaloper N and Sorbo L 2009 *Phys. Rev.* **D79** 043528 (*Preprint* 0810.5346)
[9] Kaloper N and Sorbo L 2009 *Phys. Rev. Lett.* **102** 121301 (*Preprint* 0811.1989)
[10] Brown J D and Teitelboim C 1987 *Phys. Lett.* **B195** 177–182
[11] Brown J D and Teitelboim C 1988 *Nucl. Phys.* **B297** 787–836
[12] Deffayet C, Deser S and Esposito-Farese G 2010 *Phys. Rev.* **D82** 061501 (*Preprint* 1007.5278)
[13] Quevedo F and Trugenberger C A 1997 *Nucl. Phys.* **B501** 143–172 (*Preprint* hep-th/9604196)

Journal of Physics: Conference Series **259** (2010) 012077 doi:10.1088/1742-6596/259/1/012077

Minimal rates for lepton flavour violation from supersymmetric leptogenesis

A Ibarra and C Simonetto

Physik-Department T30d, Technische Universität München, James-Franck-Straße, 85748 Garching, Germany

E-mail: alejandro.ibarra@ph.tum.de, cristoforo.simonetto@ph.tum.de

Abstract. The see-saw is a very attractive model for neutrino mass generation in particular in association with supersymmetry as a solution to the hierarchy problem. Under the plausible assumptions of hierarchical neutrino Yukawa eigenvalues and the absence of cancellations, we derive an upper bound on the lightest right-handed neutrino mass from the non-observation of $\mu \to e\gamma$ and μ-e conversion in nuclei. The ongoing experiment MEG as well as the planned experiments Mu2e, COMET and PRISM/PRIME will improve this bound if no evidence of lepton flavour violation is found. We lastly comment on the possibility of ruling out minimal leptogenesis if these experiments find no signal.

1. Introduction

In the see-saw the particle content of the Standard Model (SM) is augmented by three right-handed neutrinos. Being uncharged they have a Majorana mass term – unlike all other particles of the SM. Electroweak symmetry breaking generates also Dirac masses that lead to mixing of left- and right-handed neutrinos and finally to the observed small neutrino masses [1, 2, 3, 4].

Although very appealing theoretically, the see-saw mechanism faces the disadvantage of lacking predictability. To be specific the see-saw can accommodate any neutrino mass and the neutrino mass operator is the only mass-dimension 5 operator of the SM. Thus there is no low energy prediction of the see-saw.

In addition, the existence of such heavy neutrinos leads to quadratically divergent quantum corrections to the Higgs mass, claiming for a solution to the hierarchy problem. Hence the see-saw is naturally built into Supersymmetry (SUSY). In broken SUSY the slepton mass matrices can be affected by the see-saw mechanism [5]. Those flavour structures lead to lepton flavour violating processes, such as $\mu \to e\gamma$ or μ-e conversion in nuclei which can be searched for by experiment.

Unfortunately, the see-saw mechanism does not predict any model independent connection between lepton flavour violating processes and the neutrino mass matrix [6]. On the other hand, imposing any additional assumption will lead to constraints among the low energy parameters. Following this guideline the challenge is to find as plausible assumptions as possible having an impact on low energy observables.

In section 2 we will make use of the two plausible assumptions of hierarchical neutrino Yukawa eigenvalues and the absence of cancellations. This will allow to derive upper bounds on lightest right-handed neutrino mass from the non-observation of lepton flavour violation. Afterwards we will show in section 3 how this might open the possibility of probing the minimal leptogenesis mechanism [7] where the observed baryon asymmetry is generated by the decay of the lightest right-handed neutrino.

2. Constraints on the see-saw parameters

2.1. The framework

Adding right-handed neutrinos ν_R to the Minimal Supersymmetric SM (MSSM) the super-potential includes the following additional terms:

$$W \supset \bar{\nu}_{Ri}\mathbf{Y}_{\boldsymbol{\nu} ij}L_j H_u - \frac{1}{2}\bar{\nu}_{Ri}\mathbf{D}_{\mathbf{M}ij}\bar{\nu}_{Rj} \ . \tag{1}$$

We will work in a basis of diagonal heavy neutrino masses $\mathbf{D_M} = \mathrm{diag}(\mathrm{M}_1, \mathrm{M}_2, \mathrm{M}_3)$ and write the neutrino Yukawa according to the singular value decomposition as:

$$\mathbf{Y}_{\boldsymbol{\nu}} = \mathbf{V_R}\mathbf{D_Y}\mathbf{V_L^\dagger} \ , \qquad\qquad \mathbf{D_Y} = \mathrm{diag}(\mathrm{y}_1, \mathrm{y}_2, \mathrm{y}_3) \ , \tag{2}$$

where $\mathbf{V_R}$, $\mathbf{V_L}$ are unitary matrices. In this parametrisation the see-saw formula for the light neutrino mass matrix $\mathcal{M}$ reads:

$$\mathcal{M} = \langle H_u^0 \rangle^2 \mathbf{Y}_{\boldsymbol{\nu}}^{\mathrm{T}} \mathbf{D_M^{-1}} \mathbf{Y}_{\boldsymbol{\nu}} = \langle H_u^0 \rangle^2 \mathbf{V_L^*} \mathbf{D_Y} \mathbf{V_R^{\mathrm{T}}} \mathbf{D_M^{-1}} \mathbf{V_R} \mathbf{D_Y} \mathbf{V_L^\dagger} \ , \tag{3}$$

being $\langle H_u^0 \rangle$ the vacuum expectation value of the Higgs H_u. We aim to solve for $\mathbf{D_M}$, $\mathbf{V_R}$. Therefore we rewrite the last equation

$$\mathbf{V_R^\dagger}\mathbf{D_M^{-2}}\mathbf{V_R} = \frac{1}{\langle H_u^0 \rangle^4}\mathbf{D_Y^{-1}}\widehat{\mathcal{M}}^\dagger\mathbf{D_Y^{-2}}\widehat{\mathcal{M}}\mathbf{D_Y^{-1}} \ , \tag{4}$$

where we have defined for convenience $\widehat{\mathcal{M}} = \mathbf{V_L^{\mathrm{T}}}\mathcal{M}\mathbf{V_L}$.

Throughout this work we will assume hierarchical neutrino Yukawa eigenvalues $y_1 \ll y_2 \ll y_3$. This assumption is motivated by the fact that the three observed Yukawa matrices of quarks and leptons do follow this pattern. For the moment let us also consider the case of a less hierarchical matrix $\widehat{\mathcal{M}}$. In particular if $\widehat{\mathcal{M}}_{11}$ and $\widehat{\mathcal{M}}_{12}$ are sizeable, one can identify the largest entry on the r.h.s. of eq. (4) and read off its largest mixing (see also [8, 9, 10, 11]):

$$M_1 \simeq y_1^2 \langle H_u^0 \rangle^2 \frac{1}{|\widehat{\mathcal{M}}_{11}|} \ , \qquad\qquad \mathbf{V}_{\mathbf{R}12} \simeq \frac{y_1}{y_2}\frac{\widehat{\mathcal{M}}_{12}}{\widehat{\mathcal{M}}_{11}} \ . \tag{5}$$

As from experiment we know that $\mathcal{M}_{11}$, $\mathcal{M}_{12}$ are sizeable, eq.(5) can only be circumvented for large mixing in $\mathbf{V_L}$. In particular the only possibly natural choice seems to be $(\mathbf{V_L})_{k1} \sim \mathbf{U}_{k1}^{\mathrm{MNS}}$. This case has to be treated separately but will not modify the conclusions [12]. Thus we can assume eq. (5) to hold true independently of $\mathbf{V_L}$ unless there are unnatural cancellations.

2.2. Rare lepton decays

In the case of the MSSM the strongest bound on lepton flavour violation is presently set by the MEGA experiment $\mathrm{BR}(\mu \to e\gamma) \leq 1.2 \times 10^{-11}$ [13]. Very soon an improved bound can be expected from MEG at PSI where the final goal is to go down to $\mathrm{BR}(\mu \to e\gamma) \sim 10^{-13}$ [14]. This value seems to be the best one achievable with present techniques. On the other hand there are experiments in preparation for the search for μ-e conversion in nuclei. Within the next few years Mu2e at Fermilab [15] and COMET at J-PARC [16] could be build which aim to a sensitivity of $\mathrm{R}(\mu\,\mathrm{Al} \to e\,\mathrm{Al}) \sim 10^{-16}$. Extending COMET with a FFAG muon storage ring could finally provide a beam pure enough to lower this sensitivity again by two orders of magnitude in the PRISM/PRIME project [17].

These lepton flavour violating processes are in the MSSM mainly induced by slepton loops and can be approximated as [18]:

$$\mathrm{R}(\mu\,\mathrm{Ti}\to e\,\mathrm{Ti}) \simeq \alpha \mathrm{BR}(\mu\to e\gamma) \simeq \frac{\alpha^4}{G_F^2}\frac{|\mathbf{m^2_{L}}_{12}|^2}{m_S^8}\tan^2\beta\,. \tag{6}$$

In a target medium different from Ti the conversion rate can differ by a factor of 2 [19]. We have denoted $\mathbf{m^2_L}$ the slepton soft mass matrix, m_S some average soft mass and $\tan\beta$ the ratio of the Higgs vacuum expectation values.

Under the standard assumption of flavour blind SUSY breaking mediation – which consists in the most conservative assumption of lepton flavour violation – at some energy scale Λ where SUSY is broken it holds:

$$\mathbf{m^2_L} \propto \mathbb{1} \qquad \text{at } \Lambda\,. \tag{7}$$

Typically Λ is a very high scale like e.g. the Planck scale in Supergravity theories. As long as $\mathbf{D_M} < \Lambda$, RGE effects will transmit lepton flavour violation from the see-saw couplings to the slepton masses. This results into flavour off-diagonal mass entries at low energies that read roughly:

$$\mathbf{m^2_{L}}_{12} \simeq -\frac{3}{8\pi^2}m_S^2\mathbf{P}_{12}\,, \qquad \text{with} \qquad \mathbf{P} = \mathbf{Y}_{\boldsymbol{\nu}}^\dagger \log\frac{\Lambda}{\mathbf{D_M}}\mathbf{Y}_{\boldsymbol{\nu}}\,. \tag{8}$$

2.3. The minimal amount of flavour violation

The aim of this work is to set an upper bound on M_1 from searches for the μ–e transition as in $\mu\to e\gamma$ or μ-e conversion in nuclei. This corresponds to finding the minimal rate for rare lepton decays given fixed M_1. According to eqs. (6), (8) the rate of the μ-e flavour transition is minimised for minimal $\mathbf{P}_{12}$. Therefore let us express $\mathbf{P}_{12}$ in the parametrisation eq. (2):

$$\mathbf{P}_{12} = \left(\mathbf{V_L}\mathbf{D_Y}\mathbf{V_R}^\dagger \log\frac{\Lambda}{\mathbf{D_M}}\mathbf{V_R}\mathbf{D_Y}\mathbf{V_L}^\dagger\right)_{12}\,. \tag{9}$$

If $\mathbf{V_L}$ has sizeable off-diagonal entries one can expect $\mathbf{P}_{12}$ to be dominated by $\mathbf{V_L}_{13}y_3^2\mathbf{V}^*_{\mathbf{L}23}\log\frac{\Lambda}{M_3}$ as we assume $\mathbf{D_Y}$ to be hierarchical. But $\mathbf{P}_{12}$ could naturally be much smaller for $\mathbf{V_L}\sim\mathbb{1}$. In this case

$$\mathbf{P}_{12}^{\min} = y_1\mathbf{V}^\dagger_{\mathbf{R}1i}\log\frac{\Lambda}{M_i}\mathbf{V}_{\mathbf{R}i2}y_2 \simeq y_1y_2\mathbf{V_R}_{12}\log\frac{M_2}{M_1}\,, \tag{10}$$

where insertions of two off-diagonal entries of $\mathbf{V_R}$ have been neglected – a good assumption as $\mathbf{V_R}_{12}$ is the largest off-diagonal entry. Applying once again eq. (5) this reads:

$$|\mathbf{P}_{12}^{\min}| = \frac{M_1|\mathcal{M}_{12}|}{\langle H_u^0\rangle^2}\log\frac{M_2}{M_1}\,. \tag{11}$$

However, an even smaller value for $\mathbf{P}_{12}$ could possibly be achieved if small off-diagonal entries in $\mathbf{V_L}$ cancel this contribution. In order this to happen naturally, a model is necessary that relates $\mathbf{V_L}$ to $\mathbf{V_R}$, $\mathbf{D_Y}$ and the logarithm of the masses. Let us for example assume that the contribution from $\mathbf{V_L}_{12}$ cancels the above $\mathbf{P}_{12}^{\min}$, i.e. $\mathbf{V_L}_{12} = -\frac{y_1}{y_2}\mathbf{V_R}_{12}\log\frac{M_2}{M_1}\log^{-1}\frac{\Lambda}{M_2}$. We find the existence of such a scenario very unplausible in particular because of the appearance of the logarithms and the SUSY breaking scale Λ.

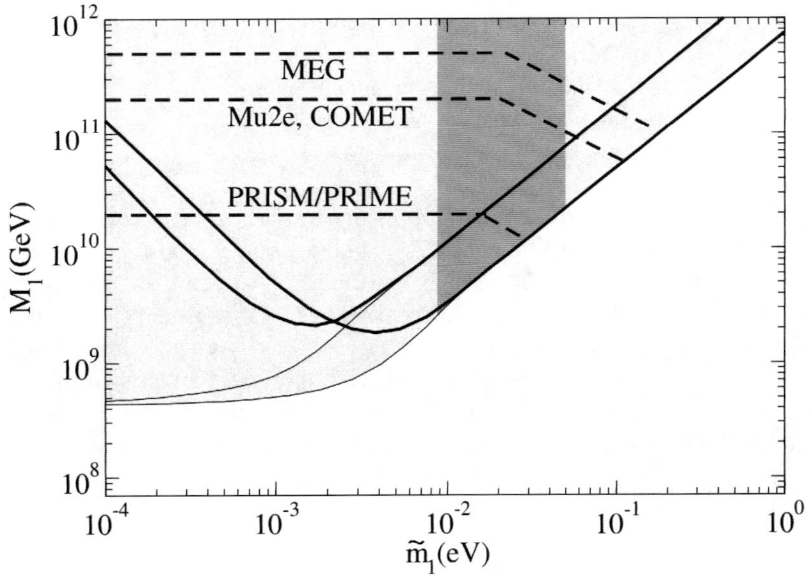

Figure 1. Allowed parameter space of thermal leptogenesis (in yellow, adopted from [20]), including the constraints on M_1 which stem from the assumed non-observation of the μ-e flavour transition in future experiments, under the assumption of hierarchical neutrino Yukawa eigenvalues and barring cancellations. The orange region corresponds to the range of $\widetilde{m}_1$ for generic neutrino parameters. In this plot it is assumed $m_S \simeq 200$ GeV and $\tan\beta \simeq 10$.

2.4. The constraints

Now we have argued that eq. (11) indeed gives the minimal mixing of μ-e, we can calculate the minimal conversion rate using eqs. (6), (8):

$$\mathrm{R}(\mu\,\mathrm{Ti}\to e\,\mathrm{Ti}) \gtrsim \frac{\alpha^4}{G_F^2}\left(\frac{3}{8\pi^2}\frac{\tan\beta}{m_S^2}\right)^2\left(\frac{M_1|\mathcal{M}_{12}|}{\langle H_u^0\rangle^2}\log\frac{M_2}{M_1}\right)^2 . \tag{12}$$

This can of course be inverted to bound M_1 from above. For simplicity assume conservatively $\log\frac{M_2}{M_1} \sim \log 10$ and take $\sin\beta \sim 1$, giving:

$$M_1 \lesssim 5\times 10^{12}\mathrm{GeV}\left(\frac{\mathrm{R}(\mu\,\mathrm{Ti}\to e\,\mathrm{Ti})}{10^{-13}}\right)^{1/2}\left(\frac{m_S}{200\,\mathrm{GeV}}\right)^2\left(\frac{\tan\beta}{10}\right)^{-1} . \tag{13}$$

Here we have normalised the conversion rate by the maximal possible rate in agreement with data from $\mu\to e\gamma$. We believe that it is interesting on its own to shrink the possible range of a fundamental parameter. But if future searches for SUSY and μ-e transitions bring down the constraint on M_1 to $\sim 10^9$ it could also serve as a probe of leptogenesis.

3. Probing leptogenesis

The baryon asymmetry generated through the leptogenesis mechanism depends, under the assumption of hierarchical neutrinos, essentially on two parameters: the lightest right-handed neutrino mass, M_1, and an effective neutrino mass $\widetilde{m}_1$ [21, 22], defined as

$$\widetilde{m}_1 = \frac{(\mathbf{Y}_\nu\mathbf{Y}_\nu^\dagger)_{11}}{M_1}\langle H_u^0\rangle^2 , \tag{14}$$

which measures the strength of the coupling of the lightest right-handed neutrino to the thermal bath. Naively one expects $\widetilde{m}_1$ to be of the order of the light neutrino mass splittings but strictly it is a free parameter. In the situation of hierarchical neutrino Yukawa couplings and minimal lepton flavour violation $\widetilde{m}_1$ equals the solar mass splitting [12].

We show in Fig. 1 the impact of the bounds on the lightest right-handed neutrino mass stemming from the non-observation of the μ-e flavour transition on the parameter space of thermal leptogenesis. The yellow/orange region corresponds to the allowed region found by Blanchet and di Bari, and shown in Fig. 1 of [20]. The orange region marks the range of $\widetilde{m}_1$ for generic neutrino parameters, i.e. the interval from the solar to the atmospheric mass splitting.

As thick dashed lines we show the most stringent upper bound on M_1, eq. (13), for the projected sensitivity of different experiments (cf. sec. 2.2). The solid lines bound the allowed region for certain leptogenesis scenarios: thermal (thin) vs. zero initial abundance (thick) and maximal (right) vs. no flavour effects (left). Further details can be inferred from [12].

We find that for generic $\widetilde{m}_1$ the window for M_1 could be closed with PRISM/PRIME for $\frac{m_S^2}{\tan\beta} \lesssim \frac{(200\text{ GeV})^2}{50}$. Otherwise a better sensitivity is necessary.

4. Conclusions

The see-saw mechanism is a very elegant explanation for the smallness of neutrino masses, which in addition might account for the baryon asymmetry of the Universe, through the mechanism of leptogenesis. The supersymmetric version, which is probably the most natural arena to implement it, endows us with another clue to the see-saw parameters. Namely, the existence of the neutrino Yukawa couplings affects the flavour structure of the slepton mass matrices through quantum effects, leading to lepton flavour violation maybe visible in ongoing or future searches.

Working under very general and well motivated assumptions, namely the absence of cancellations and a hierarchical pattern in the neutrino Yukawa eigenvalues, we have identified the scenario yielding the minimal rate for the rare decay $\mu \to e\gamma$ and μ-e conversion in nuclei. In this scenario, the rate depends essentially on the lightest right-handed neutrino mass, M_1, and on supersymmetric parameters. Using the experimental constraint on $\text{BR}(\mu \to e\gamma)$ we have derived an upper bound $M_1 \lesssim 5 \times 10^{12}$ for typical soft SUSY breaking terms of 200 GeV and $\tan\beta = 10$. This should be compared with the lower bound required by the thermal leptogenesis scenario, $M_1 \gtrsim 10^9$ GeV. The upper bound derived in this paper scales as $\text{BR}(\mu \to e\gamma)^{1/2}$, therefore, future improvements in sensitivity to the process $\mu \to e\gamma$ and to μ-e conversion in nuclei can have important implications for the thermal leptogenesis scenario if no positive signal is found. But only in the case of large $\tan\beta$ a sensitivity of $R(\mu\,\text{Ti} \to e\,\text{Ti}) \sim 10^{-18}$, as expected from the PRISM/PRIME project, can rule out thermal leptogenesis models without fine-tuning and based on the decay of the lightest right-handed neutrino. On the other hand, our bound is very conservative and in general one can expect much larger lepton flavor violating rates in the case of supersymmetric leptogenesis.

Acknowledgments

CS wishes to thank the organizers for this very pleasant conference. This work was partially supported by the DFG cluster of excellence Origin and Structure of the Universe and by the Graduiertenkolleg "Particle Physics at the Energy Frontier of New Phenomena".

References

[1] Minkowski P 1977 *Phys. Lett.* **B67** 421
[2] Gell-Mann M, Ramond P and Slansky R Print-80-0576 (CERN)
[3] Mohapatra R N and Senjanovic G 1980 *Phys. Rev. Lett.* **44** 912
[4] Yanagida T In Proceedings of the Workshop on the Baryon Number of the Universe and Unified Theories, Tsukuba, Japan, 13-14 Feb 1979

[5] Borzumati F and Masiero A 1986 *Phys. Rev. Lett.* **57** 961
[6] Davidson S and Ibarra A 2001 *JHEP* **09** 013 (*Preprint* hep-ph/0104076)
[7] Fukugita M and Yanagida T 1986 *Phys. Lett.* **B174** 45
[8] Davidson S and Ibarra A 2003 *Nucl. Phys.* **B648** 345–375 (*Preprint* hep-ph/0206304)
[9] Davidson S 2003 *JHEP* **03** 037 (*Preprint* hep-ph/0302075)
[10] Branco G C, Gonzalez Felipe R, Joaquim F R and Rebelo M N 2002 *Nucl. Phys.* **B640** 202–232 (*Preprint* hep-ph/0202030)
[11] Akhmedov E K, Frigerio M and Smirnov A Y 2003 *JHEP* **09** 021 (*Preprint* hep-ph/0305322)
[12] Ibarra A and Simonetto C 2009 *JHEP* **08** 113 (*Preprint* 0903.1776)
[13] Brooks M L *et al.* (MEGA) 1999 *Phys. Rev. Lett.* **83** 1521–1524 (*Preprint* hep-ex/9905013)
[14] Mori, T *et al* 1999 *Search for* $\mu \rightarrow e\gamma$ *down to* 10^{14} *branching ratio*, Research Proposal to Paul Scherrer Institut, see http://meg.web.psi.ch
[15] Bernstein, R H and Miller, J P *et al* 2008 *Proposal to Search for* $\mu^- N \rightarrow e^- N$ *with a Single Event Sensitivity below* 10^{-16}, Mu2e Proposal, see http://mu2e.fnal.gov
[16] Kuno, Y *et al* 2007 *An Experimental Search for Lepton Flavor Violating* μ*-e Conversion at Sensitivity of* 10^{16} *with a Slow-Extracted Bunched Proton Beam*, COMET Proposal, see http://comet.phys.sci.osaka-u.ac.jp
[17] Kuno Y 2005 *Nucl. Phys. Proc. Suppl.* **149** 376–378
[18] Paradisi P 2005 *JHEP* **10** 006 (*Preprint* hep-ph/0505046)
[19] Kitano R, Koike M and Okada Y 2002 *Phys. Rev.* **D66** 096002 (*Preprint* hep-ph/0203110)
[20] Blanchet S and Di Bari P 2007 *JCAP* **0703** 018 (*Preprint* hep-ph/0607330)
[21] Buchmuller W, Peccei R D and Yanagida T 2005 *Ann. Rev. Nucl. Part. Sci.* **55** 311–355 (*Preprint* hep-ph/0502169)
[22] Davidson S, Nardi E and Nir Y 2008 *Phys. Rept.* **466** 105–177 (*Preprint* 0802.2962)

Journal of Physics: Conference Series **259** (2010) 012078 doi:10.1088/1742-6596/259/1/012078

Right-Handed Sneutrinos as Thermal Dark Matter

Frank Deppisch

E-mail: frank.deppisch@manchester.ac.uk
School of Physics and Astronomy, University of Manchester, Manchester M13 9PL, UK

Apostolos Pilaftsis

E-mail: apostolos.pilaftsis@manchester.ac.uk
School of Physics and Astronomy, University of Manchester, Manchester M13 9PL, UK

Abstract. We analyse the abundance of right-handed sneutrinos in the supersymmetric F_D-term model of hybrid inflation. As well as providing a natural solution to the μ- and gravitino overabundance problems, the model offers the lightest right-handed sneutrino as a candidate for thermal dark matter. The F_D-term model predicts a new quartic coupling of purely right-handed sneutrinos to the Higgs doublets that thermalises the sneutrinos and makes them annihilate sufficiently fast to a level compatible with the current cosmic microwave background data. We discuss this scenario and identify favourable regions of the parameter space within mSUGRA.

1. F_D-Term Model of Hybrid Inflation

Hybrid inflation [1], along with its supersymmetric realisations [2], remains one of the most predictive models of inflation. In such a scenario, inflation terminates at some critical point when a so called waterfall field acquires a vacuum expectation value (VEV), thereby ending inflation. Recently, a new supersymmetric hybrid inflationary model was proposed [3], which realises F-term hybrid inflation and includes a subdominant non-anomalous Fayet-Iliopoulos D-term that arises from a U(1) gauge symmetry of the inflation waterfall sector. It has therefore been called the F_D-term model of hybrid inflation. The F_D-term model can naturally accommodate the currently favoured red-tilted spectrum, along with the actual value of the power spectrum of curvature perturbations, and the required number of e-folds.

The F_D-term model is defined through the superpotential

$$\begin{aligned} \mathcal{W} &= \kappa\,\widehat{S}\left(\widehat{X}_1\widehat{X}_2 - M^2\right) + \lambda\,\widehat{S}\widehat{H}_u\widehat{H}_d \\ &+ \frac{\rho_{ij}}{2}\,\widehat{S}\,\widehat{N}_i\widehat{N}_j + h^{\nu}_{ij}\widehat{L}_i\widehat{H}_u\widehat{N}_j + W^{(\mu=0)}_{\rm MSSM}\,, \end{aligned} \qquad (1)$$

where $\widehat{S}$ is the gauge-singlet inflaton superfield and $\widehat{X}_{1,2}$ is a chiral multiplet pair of the so-called waterfall fields which have opposite charges under the $\mathrm{U}(1)_X$ gauge group, i.e. $Q(\widehat{X}_1) = -Q(\widehat{X}_2) = 1$. In addition, $W^{(\mu=0)}_{\rm MSSM}$ indicates the MSSM superpotential without the μ-term. The model Lagrangian is supplemented by soft SUSY breaking terms, which contribute to the scalar potential, leading to an inflaton VEV naturally of $\mathcal{O}$ (TeV).

Constraints on the model can be derived from cosmological inflation through the number of e-folds and the power spectrum and spectral index of curvature perturbations [3]. This implies the upper limit $\lambda(M_{\rm SUSY}) \lesssim 1.14(1.82) \times 10^{-2}$ for an inflaton sector with a minimal (next-to-minimal) Kähler potential [5].

The presence of the Fayet-Iliopoulos term in the F_D-term model is necessary to approximately break a discrete symmetry in the waterfall sector. The late decays of the GUT-scale waterfall particles produce an enormous entropy that can reduce the gravitino abundance below the limits imposed by big bang nucleosynthesis (BBN), providing a viable solution to the gravitino overabundance problem [3].

In the F_D-term model the μ-parameter of the MSSM can be generated effectively when the scalar inflaton receives a non-zero VEV, $\mu = \lambda\langle S\rangle$. Moreover, the inflaton VEV will also produce an effective Majorana mass matrix as well, $M_N = \rho\langle S\rangle$. As a consequence, the resulting heavy Majorana neutrinos are expected to have masses of order μ. A possible explanation of the observed baryon asymmetry in the Universe may be obtained by thermal electroweak-scale resonant leptogenesis [4].

In this report, which summarizes the results of Ref. [5], we focus on the properties of the lightest right-handed sneutrino (LRHS) and identify favourable regions of the parameter space within mSUGRA, for which it becomes a viable candidate for thermal dark matter.

2. Sneutrino Mass Spectrum

Ignoring the terms proportional to the small neutrino-Yukawa couplings, the 6×6 right-handed sneutrino mass matrix $\mathcal{M}^2_{\widetilde{N}}$ is given in the weak basis $(\widetilde{N}_{1,2,3}, \widetilde{N}^*_{1,2,3})$ by

$$\mathcal{M}^2_{\widetilde{N}} = \frac{1}{2}\begin{pmatrix} \rho^2 v_S^2 + M^2_{\widetilde{N}} & \rho A_\rho v_S + \rho\lambda v_u v_d \\ \rho A^*_\rho v_S + \rho\lambda v_u v_d & \rho^2 v_S^2 + M^2_{\widetilde{N}} \end{pmatrix}, \tag{2}$$

where $v_S = \langle S\rangle$, $v_{u,d} = \langle H_{u,d}\rangle$. Neglecting the possible flavor structure contained in the soft SUSY-breaking sneutrino mass matrix $M^2_{\widetilde{N}}$ and trilinear coupling matrix A_ρ, the sneutrino spectrum will consist of 3 light (3 heavy) right-handed sneutrinos with masses

$$m^2_{\widetilde{N}_{L(H)}} = \rho^2 v_S^2 + M^2_{\widetilde{N}} - (+)\left|\rho A_\rho v_S + \rho\lambda v_u v_d\right|. \tag{3}$$

All mass terms in (3) are $\mathcal{O}(100\text{–}1000)$ GeV, so a proper choice of model parameters can accommodate a lightest right-handed sneutrino to act as LSP. Unless the trilinear coupling A_ρ is small compared to μ, the off-diagonal elements in (2) will induce a sizeable mixing between the heavy and light right-handed sneutrino states, suppressing the light masses to values smaller than $(\mu^2 + M^2_{\widetilde{N}})^{1/2}$.

In Figure 1 we plot the LHRS mass $m_{\widetilde{N}_1}$ as contours in the mSUGRA parameter plane $(m_0, m_{1/2})$, $A_0 = 300$ GeV, $\mu > 0$ and two different values of $\tan\beta$. For the inflaton couplings λ, ρ we simply choose $\lambda = \rho = 10^{-2}$, in accordance with the bounds derived from inflation. The connection between the LRHS mass $\widetilde{N}_1$ and μ generally points towards a low-energy SUSY spectrum. This coincidentally includes the H_1-boson funnel region, where $m_{H_1} \approx 2m_{\widetilde{N}_1}$. Very large and small values for A_0 and $\tan\beta$ are disfavoured as they exclude a sneutrino LSP. These correlations may be somewhat relaxed if non-universal inflaton couplings λ and ρ are considered.

3. Sneutrino Relic Density

As was first observed in [3], there exists a new quartic coupling between right-handed sneutrinos and Higgs fields in the F_D-term model described by the Lagrangian

$$\mathcal{L}^{\rm LSP}_{\rm int} = \frac{1}{2}\lambda\rho\tilde{N}^*_i\tilde{N}^*_i H_u H_d + \text{H.c.}\,, \tag{4}$$

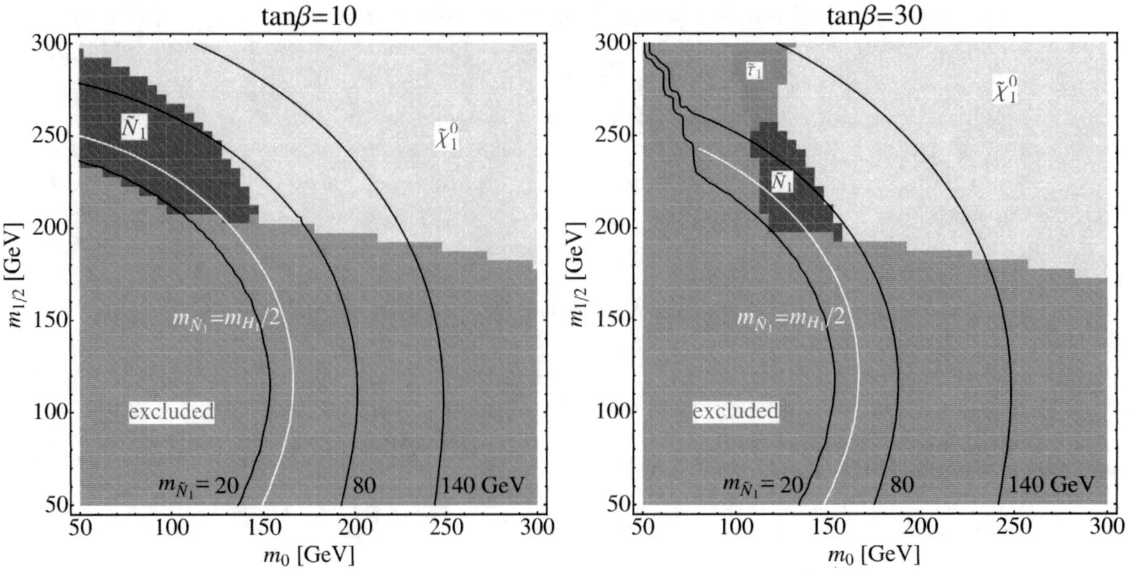

Figure 1. Allowed $(m_0, m_{1/2})$ parameter space for $A_0 = 300$ GeV, $\mu > 0$, $\lambda = \rho = 10^{-2}$ and $\tan\beta = 10$ (left), 30 (right). The black contours show the predicted LRHS mass, while a sneutrino $\tilde{N}_1$/neutralino $\tilde{\chi}_1^0$/stau $\tilde{\tau}_1$ LSP is present in the blue/green/orange area. The red area is excluded by direct SUSY mass searches. The white contour is defined by the condition $m_{\tilde{N}_1} = m_{H_1}/2$.

resulting from the F-term of the inflaton field. This leads to several channels of sneutrino annihilation into Higgses, gauge bosons and fermions via the direct quartic coupling as well as s-channel Higgs and t/u-channel sneutrino exchange once one of the Higgs fields in Eq. 4 acquires a VEV [5]. The most effective annihilation process is $b\bar{b}$ production in the lightest Higgs boson resonance, $m_{\tilde{N}_1} = m_{H_1}/2$. In order to compute the sneutrino relic density remaining after the sneutrino freezes out of thermal equilibrium and to analyze the constraints on the effective annihilation coupling $\lambda\rho$, we adopt the mSUGRA scenario

$$\begin{aligned} &m_0 = 125\text{ GeV},\ m_{1/2} = 212\text{ GeV}, \\ &A_0 = 300\text{ GeV},\ \tan\beta = 30,\ \mu = 263\text{ GeV}\ , \end{aligned} \tag{5}$$

while keeping the LRHS mass as a free parameter. The effective annihilation coupling $\lambda\rho$, Eq. 4, is then calculated so as to obtain a sneutrino relic density $\Omega_{\rm DM} h^2 = 0.11$, consistent with observations. Numerical estimates of the allowed parameters in the $(m_{\tilde{N}_1}, \lambda\rho)$-plane are shown in Figure 2. In order to account for the observed DM relic abundance in the H_1-boson funnel region, where $m_{\tilde{N}_1} \approx m_{H_1}/2$, the effective coupling $\lambda\rho$ should be

$$\lambda\rho \gtrsim 2 \times 10^{-4}\ . \tag{6}$$

This has to be compared with the upper limit derived from successful inflation [5],

$$\lambda\rho \lesssim 2.3(5.8) \times 10^{-4}\,, \tag{7}$$

for a minimal (next-to-minimal) Kähler potential. In general, we find that LRHS masses larger than about 100 GeV are not possible within a mSUGRA realisation of the F_D-term model. This

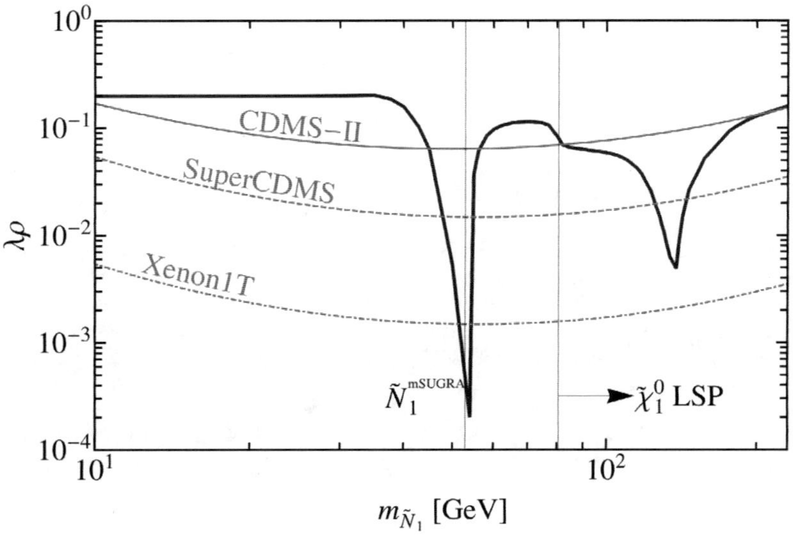

Figure 2. Effective annihilation coupling $\lambda\rho$ as a function of the LRHS mass $m_{\tilde{N}_1}$ for a relic density $\Omega_{\mathrm{DM}}h^2 = 0.11$ (blue curve) in the mSUGRA Scenario, Eq. 5. The actual sneutrino and neutralino masses in the scenario are indicated by vertical lines. The red curves denote the upper bound on $\lambda\rho$ as obtained by the CDMS-II experiment and as expected by the projected sensitivities of SuperCDMS and Xenon1T.

is indicated by the value of the neutralino mass in the given mSUGRA scenario as displayed by vertical line in Figure 2.

Further constraints on the $(m_{\tilde{N}_1}, \lambda\rho)$-plane may be obtained by taking into account the limits from direct searches of experiments which look for scattering between Weakly Interacting Massive Particles and nuclei. Upper limits on $\lambda\rho$ are derived by comparing the calculated elastic cross section between the sneutrino LSP and the nucleon with the current bound on the spin-independent nucleon cross section from the CDMS-II experiment and the expected sensitivities of the SuperCDMS extension [8] and the Xenon1T experiment [9]. These limits are included in Figure 2. The current bound already excludes large parts of the $(m_{\tilde{N}_1}, \lambda\rho)$-parameter plane. The upgraded experiment SuperCDMS and the proposed Xenon1T experiment will cover most of the parameter space, but will leave open the lightest Higgs-boson pole region which is theoretically favoured by inflation within the mSUGRA framework.

4. Conclusion

In this report we discussed the possibility of a right-handed sneutrino LSP as candidate for dark matter in the supersymmetric F_D-term model of hybrid inflation. By virtue of a new quartic coupling with the Higgs fields the sneutrino $\widetilde{N}_{\mathrm{LSP}}$ can efficiently annihilate via the lightest Higgs-boson resonance H_1 into pairs of b-quarks, in the kinematic region $m_{H_1} \approx 2m_{\widetilde{N}_{\mathrm{LSP}}}$, and so drastically reduce its relic density to the observed value $\Omega_{\mathrm{DM}}\, h^2 \approx 0.11$. It might seem that to obtain this particular relation between the masses of the H_1 boson and $\widetilde{N}_{\mathrm{LSP}}$, a severe tuning of the model parameters is required. However, it is worth stressing here that such a mass relation may easily be achieved within a mSUGRA framework of the F_D-term model that successfully realizes hybrid inflation.

Relatively large couplings are required for the sneutrino DM scenario, that could make Higgs bosons decay invisibly, e.g. $H \to 2\widetilde{N}_{\mathrm{LSP}}$. Also, right-handed sneutrinos could be present in the cascade decays of the heavier supersymmetric particles. The F_D-term hybrid inflationary model therefore gives rise to rich phenomenology which can be probed at high-energy colliders [10], as well as in low-energy experiments of lepton flavour and number violation [11].

References

[1] A. D. Linde, Phys. Lett. B **259** (1991) 38.
[2] E. J. Copeland, A. R. Liddle, D. H. Lyth, E. D. Stewart and D. Wands, Phys. Rev. D **49** (1994) 6410; G. R. Dvali, Q. Shafi and R. K. Schaefer, Phys. Rev. Lett. **73** (1994) 1886.
[3] B. Garbrecht and A. Pilaftsis, Phys. Lett. B **636** (2006) 154; B. Garbrecht, C. Pallis and A. Pilaftsis, JHEP **0612** (2006) 038.
[4] A. Pilaftsis, Phys. Rev. D **56** (1997) 5431; A. Pilaftsis and T.E.J. Underwood, Nucl. Phys. B **692** (2004) 303; T. Hambye, J. March-Russell and S. M. West, JHEP **0407** (2004) 070; E. J. Chun, Phys. Rev. D **72** (2005) 095010; A. Pilaftsis, Phys. Rev. Lett. **95** (2005) 081602; A. Pilaftsis and T.E.J. Underwood, Phys. Rev. D **72** (2005) 113001.
[5] F. Deppisch and A. Pilaftsis, JHEP **0810** (2008) 080.
[6] J. S. Lee, A. Pilaftsis, M. S. Carena, S. Y. Choi, M. Drees, J. R. Ellis and C. E. M. Wagner, Comput. Phys. Commun. **156** (2004) 283; J. S. Lee, M. Carena, J. Ellis, A. Pilaftsis and C. E. M. Wagner, arXiv:0712.2360 [hep-ph].
[7] W. Porod, Comput. Phys. Commun. **153** (2003) 275.
[8] R. W. I. Ogburn, *In the Proceedings of SNIC 2006, Menlo Park, California, 3-6 Apr 2006, pp 0150.*
[9] E. Aprile *et al.*, J. Phys. Conf. Ser. **39**, 107 (2006).
[10] A. Pilaftsis, Z. Phys. C **55** (1992) 275; A. Datta, M. Guchait and A. Pilaftsis, Phys. Rev. D **50** (1994) 3195; T. Han and B. Zhang, Phys. Rev. Lett. **97** (2006) 171804; F. del Aguila, J. A. Aguilar-Saavedra and R. Pittau, JHEP **0710** (2007) 047; S. Bray, J. S. Lee and A. Pilaftsis, Nucl. Phys. B **786** (2007) 95; S. Bar-Shalom, G. Eilam, T. Han and A. Soni, Phys. Rev. D **77** (2008) 115019; W. Buchmuller and C. Greub, Nucl. Phys. B **363** (1991) 345; G. Cvetic, C. S. Kim and C. W. Kim, Phys. Rev. Lett. **82** (1999) 4761; F. del Aguila, J. A. Aguilar-Saavedra, A. Martinez de la Ossa and D. Meloni, Phys. Lett. B **613** (2005) 170; F. del Aguila and J. A. Aguilar-Saavedra, JHEP **0505** (2005) 026.
[11] T.P. Cheng and L.F. Li, Phys. Rev. Lett. **45** (1980) 1908; A. Ilakovac and A. Pilaftsis, Nucl. Phys. B **437** (1995) 491; F. Deppisch, T. S. Kosmas and J. W. F. Valle, Nucl. Phys. B **752** (2006) 80.

Journal of Physics: Conference Series **259** (2010) 012079 doi:10.1088/1742-6596/259/1/012079

Fermionic Quasiparticles in Higgs Boson and Heavy Neutrino Decay in Leptogenesis

Clemens P. Kießig, Michael Plümacher

Max-Planck-Institut für Physik (Werner-Heisenberg-Institut), Föhringer Ring 6
D-80805 München, Germany

E-mail: ckiessig@mpp.mpg.de, pluemi@mpp.mpg.de

Markus H. Thoma

Max-Planck-Institut für extraterrestrische Physik, Giessenbachstraße,
D-85748 Garching, Germany

E-mail: mthoma@mpe.mpg.de

Abstract. We calculate the decay rate of the lightest heavy Majorana neutrino in a thermal bath using finite temperature cutting rules and effective Green's functions according to the hard thermal loop resummation technique. Analogously, we calculate the Higgs decay rate at higher temperature. Compared to the usual approach where thermal masses are inserted into the kinematics of final states, we find that deviations arise through two different leptonic dispersion relations. The decay rates differ from the usual approach by more than one order of magnitude in the temperature range which is interesting for the weak washout regime. Some aspects of this work have been published in Ref. [1], to which we refer the interested reader.

1. Introduction

Leptogenesis [2] is an extremely successful theory in explaining the baryon asymmetry of the universe by adding three heavy right-handed neutrinos N_i to the standard model,

$$\delta\mathcal{L} = i\bar{N}_i\partial_\mu\gamma^\mu N_i - \lambda_{\nu,i\alpha}\bar{N}_i\phi^\dagger\ell_\alpha - \frac{1}{2}M_i\bar{N}_iN_i^c + h.c.\,, \tag{1}$$

with masses M_i at the scale of grand unified theories (GUTs) and Yukawa couplings $\lambda_{\nu,i\alpha}$ similar to the other fermions. This also solves the problem of the light neutrino masses via the see-saw mechanism without fine-tuning [3].

The heavy neutrinos decay into lepton and Higgs boson after inflation, the decay is out of equilibrium since there are no gauge couplings to the standard model. If the CP asymmetry in the Yukawa couplings is large enough, a lepton asymmetry is created by the decays which is then partially converted into a baryon asymmetry by sphaleron processes. As temperatures are high, interaction rates and the CP asymmetry need to be calculated using thermal field theory [4] rather than vacuum quantum field theory.

2. Hard Thermal Loops and Thermal Masses

When using bare thermal propagators in TFT [5], one can encounter IR singularities and gauge dependent results. In order to cure this problem, the hard thermal loop (HTL) resummation technique has been invented [6, 7]. If g is the coupling to the thermal bath, then for soft momenta $K \lesssim gT$, resummed propagators are used. For a scalar field e.g. this reads

$$i\Delta^* = i\Delta + i\Delta(-i\Pi)i\Delta + \cdots = \frac{i}{\Delta^{-1} - \Pi} = \frac{i}{Q^2 - m_0^2 - \Pi}. \tag{2}$$

The self energy $\Pi \sim gT$ then acts as a thermal mass $m_{\rm th}^2 = \Pi$ and gives a correction $m_{\rm tot}^2 := m_0^2 + m_{\rm th}^2$.

3. Decay and Inverse Decay Rate

Since we are interested in regimes where both the Higgs boson and the lepton momentum can be soft, we resum both propagators. The HTL resummation technique has been considered in Ref. [8] for the case of a Dirac fermion with Yukawa coupling.

In order to calculate the interaction rate Γ of $N \leftrightarrow \ell\phi$, we cut the N self energy and use the HTL resummation for the lepton and Higgs boson propagators. Since $\lambda_{\nu,i\alpha} \ll 1$, it is justified to neglect the coupling of the neutrino to the thermal bath. According to finite-temperature cutting rules [9, 10], the interaction rate reads

$$\Gamma(P) = -\frac{1}{2p_0}\,{\rm tr}[(\not{P} + M)\;{\rm Im}\,\Sigma(P)]. \tag{3}$$

At finite temperature, the self-energy reads

$$\Sigma(P) = -g^2 T \sum_{k_0 = i(2n+1)\pi T} \int \frac{{\rm d}^3 k}{(2\pi)^3}\, P_L\, S^*(K)\, P_R\, D^*(Q), \tag{4}$$

where P_L and P_R are the projection operators on left- and right-handed states, $Q = P - K$ and we have summed over neutrino and lepton spins. We also sum over the two components of the doublets, particles and antiparticles and the three lepton flavors, such that $g^2 = 4(\lambda_\nu^\dagger \lambda_\nu)_{11}$.

The HTL-resummed Higgs boson propagator is $D^*(Q) = 1/(Q^2 - m_\phi^2)$, where $m_\phi^2/T^2 = (3/16\, g_2^2 + 1/16\, g_Y^2 + 1/4\, y_t^2 + 1/2\,\lambda)$ is the thermal mass of the Higgs boson. The couplings denote the SU(2) coupling g_2, the U(1) coupling g_Y, the top Yukawa coupling y_t and the Higgs boson self coupling λ, where we assume a Higgs boson mass of 115 GeV. The other Yukawa couplings can be neglected since they are much smaller than unity and the remaining couplings are renormalized at the first Matsubara mode $2\pi T$ as explained in Ref. [4].

The effective lepton propagator in the helicity-eigenstate representation is given by [11]

$$S^*(K) = \frac{1}{2}\Delta_+(K)(\gamma_0 - \hat{\mathbf{k}}\cdot\boldsymbol{\gamma}) + \frac{1}{2}\Delta_-(K)(\gamma_0 + \hat{\mathbf{k}}\cdot\boldsymbol{\gamma}), \tag{5}$$

where

$$\Delta_\pm(K) = \left[-k_0 \pm k + \frac{m_\ell^2}{k}\left(\pm 1 - \frac{\pm k_0 - k}{2k}\ln\frac{k_0 + k}{k_0 - k}\right)\right]^{-1} \tag{6}$$

and $m_\ell^2/T^2 = (3/32\, g_2^2 + 1/32\, g_Y^2)$.

The trace can be evaluated as

$${\rm tr}[(\not{P} + M)P_L S^*(K) P_R] = \Delta_+(p_0 - p\eta) + \Delta_-(p_0 + p\eta), \tag{7}$$

where $\eta = \mathbf{p}\cdot\mathbf{k}/pk$ is the angle between neutrino and lepton. We evaluate the sum over Matsubara frequencies by using the Saclay method [12]. For the Higgs boson propagator, the Saclay representation reads

$$D^*(Q) = -\int_0^\beta \mathrm{d}\tau e^{q_0\tau}\frac{1}{2\omega_q}\{[1+n_B(\omega_q)]e^{-\omega_q\tau} + n_B(\omega_q)e^{\omega_q\tau}\}, \tag{8}$$

where $\beta = 1/T$, $n_B(\omega_q) = 1/(e^{\omega_q\beta}-1)$ is the Bose-Einstein distribution and $\omega_q^2 = q^2 + m_\phi^2$. For the lepton propagator it is convenient to use the spectral representation [13]

$$\Delta_\pm(K) = -\int_0^\beta \mathrm{d}\tau' e^{k_0\tau'}\int_{-\infty}^{\infty}\mathrm{d}\omega\,\rho_\pm(\omega,k)[1-n_F(\omega)]e^{-\omega\tau'}, \tag{9}$$

where $n_F(\omega) = 1/(e^{\omega\beta}+1)$ is the Fermi-Dirac distribution and $\rho_\pm$ the spectral density [11].

The lepton propagator in Eq. (5) has two different poles for $1/\Delta_\pm = 0$, which correspond to two leptonic quasiparticles with a positive (Δ_+) or negative (Δ_-) ratio of helicity over chirality [14, 15, 16]. The spectral density $\rho_\pm$ has a contribution from the poles and a discontinuous part. We are interested in the pole contribution

$$\rho_\pm^{\text{pole}}(\omega,k) = \frac{\omega^2-k^2}{2m_\ell^2}(\delta(\omega-\omega_\pm)+\delta(\omega+\omega_\mp)), \tag{10}$$

where $\omega_\pm$ are the dispersion relations for the two quasiparticles, i.e. the solutions for $1/\Delta\pm(\omega_\pm,\mathbf{k}) = 0$, shown in Fig. 1 (a). An analytical solution for $\omega_\pm$ can be found in the appendix of Ref. [1]. One can assign a momentum-dependent thermal mass $m_\pm(k)^2 = \omega_\pm(k)^2 - k^2$ to the two modes as shown in Fig. 1 (b) and for very large momenta the heavy mode m_+ approaches $\sqrt{2}\,m_\ell$, while the light mode becomes massless.

After evaluating the sum over k_0, carrying out the integrations over τ and τ' and integrating over the pole part of $\rho_\pm$ in Eq. (10), we get

$$\begin{aligned} T\sum_{k_0} D^*\Delta_\pm = -\frac{1}{2\omega_q}\Bigg\{&\frac{\omega_\pm^2-k^2}{2m_\ell^2}\left[\frac{1+n_B-n_F}{p_0-\omega_\pm-\omega_q}+\frac{n_B+n_F}{p_0-\omega_\pm+\omega_q}\right] \\ &+\frac{\omega_\mp^2-k^2}{2m_\ell^2}\left[\frac{n_B+n_F}{p_0+\omega_\mp-\omega_q}+\frac{1+n_B-n_F}{p_0+\omega_\mp+\omega_q}\right]\Bigg\}, \end{aligned} \tag{11}$$

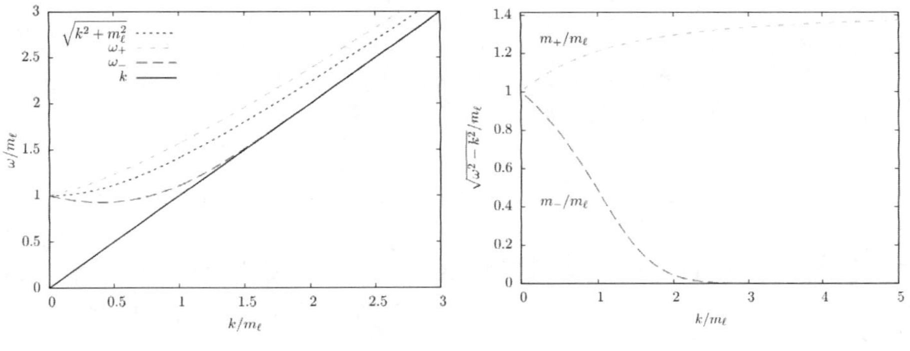

Figure 1. Left: The two leptonic dispersion relations compared with the standard dispersion relation. Right: The momentum-dependent quasiparticle masses $m_\pm^2 = \omega_\pm^2 - k^2$.

where $n_B = n_B(\omega_q)$ and $n_F = n_F(\omega_\pm)$ or $n_F(\omega_\mp)$, respectively.

The four terms in Eq. (11) correspond to the processes with the energy relations indicated in the denominator, i.e. the decay $N \to \phi\ell$, the production $N\phi \to \ell$, the production $N\ell \to \phi$ and the production of $N\ell\phi$ from the vacuum, as well as the four inverse reactions [9]. We are only interested in the process $N \leftrightarrow \phi\ell$, where the decay and inverse decay are illustrated by the statistical factors $1 + n_B - n_F = (1 + n_B)(1 - n_F) + n_B n_F$, given by the first term of Eq. (11).

For carrying out the integration over the angle η, we use

$$\mathrm{Im}\frac{1}{p_0 - \omega_\pm - \omega_q} = -\pi\delta(p_0 - \omega_\pm - \omega_q) = -\pi\frac{\omega_q}{kp}\delta(\eta - \eta_\pm). \tag{12}$$

After integrating over η we get

$$\begin{aligned}\Gamma(P) =& \frac{1}{16\pi p_0 p}\sum_\pm \int_{-1\le\eta_\pm\le 1} \mathrm{d}k\,\frac{k}{\omega_\pm}\,|\mathcal{M}_\pm(P,K)|^2[1 + n_B(\omega_{q\pm}) - n_F(\omega_\pm)]\\ =& \frac{1}{2p_0}\int \mathrm{d}\tilde{k}\,\mathrm{d}\tilde{q}\,(2\pi)^4\delta^4(P - K - Q)\,|\mathcal{M}_\pm(P,K)|^2[1 + n_B - n_F],\end{aligned} \tag{13}$$

where $\omega_{q\pm} = p_0 - \omega_\pm$, we only integrate over regions with $-1 \le \eta \le 1$, $\mathrm{d}\tilde{k} = \mathrm{d}^3k/((2\pi)^3 2\,k_0)$ and $\mathrm{d}\tilde{q}$ analogously and the matrix elements are

$$|\mathcal{M}_\pm(P,K)|^2 = g^2\frac{\omega_\pm^2 - k^2}{2m_\ell^2}\omega_\pm\,(p_0 \mp p\eta_\pm)\,. \tag{14}$$

In order to compare our result to the conventional approximation [4], we do the same calculation for an approximated lepton propagator $S^*_{\text{approx}}(K) = 1/(\not{K} - m_\ell)$. This amounts to setting $\omega^2 = k^2 + m_\ell^2$, $\omega_q = p_0 - \omega$ and we get $|\mathcal{M}|^2 = \frac{g^2}{2}(M^2 + m_\ell^2 - m_\phi^2)$ as matrix element.

This result resembles the zero temperature result with zero temperature masses m_ℓ, m_ϕ. The missing factor $1 + n_B - n_F = (1 + n_B)(1 - n_F) + n_B n_F$ accounts for the statistical distribution of the initial or final particles. As pointed out in more detail in Ref. [17], we have shown that the approach to treat thermal masses like zero temperature masses in the final state [4] is justified since it equals the HTL treatment with an approximate lepton propagator. However this approach does not equal the full HTL result.

The decay rate of the Higgs boson into neutrino and lepton, $\phi \to N\ell$, is calculated analogously. The matrix element is the same as in Eq. (14), the momentum conservation and the statistical factors have to be modified accordingly.

4. Decay Density

The quantity which enters the Boltzmann equations is the decay density integrated over all neutrino momenta. In equilibrium it reads

$$\gamma_D^{\mathrm{eq}} = \int \frac{\mathrm{d}^3p}{(2\pi)^3} f_N^{\mathrm{eq}}(E)\,\Gamma_D = \frac{1}{2\pi^2}\int_M^\infty \mathrm{d}E\,E\,p\,f_N^{\mathrm{eq}}\,\Gamma_D, \tag{15}$$

where $E = p_0$, $f_N^{\mathrm{eq}}(E) = [\exp(E\beta) - 1]^{-1}$ is the equilibrium distribution of the neutrinos and $\Gamma_D = [1 - f_N^{\mathrm{eq}}(E)]\,\Gamma$. The integrated Higgs decay rate, or decay density, is calculated analogously, with the appropriate statistical factors.

In Fig. 2, we compare our result to the conventional approximation [4]. We evaluate the decay rate for $M_1 = 10^{10}$ GeV and $\tilde{m}_1 = (\lambda_\nu\lambda_\nu^\dagger)_{11}\,v^2/M_1 = 0.06$ eV, where $v = 174$ GeV is the vacuum expectation value of the Higgs field.

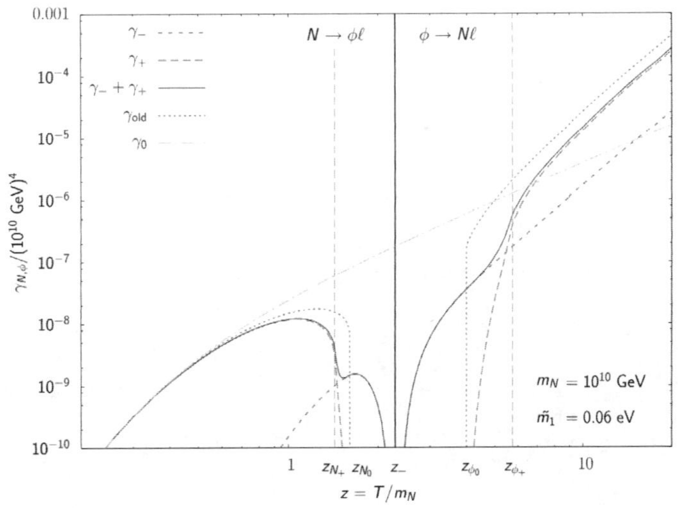

Figure 2. The neutrino and Higgs decay density with the one lepton mode approach γ_0 and the two-mode treatment $\gamma_\pm$ for $M_1 = 10^{10}$ GeV and $\tilde{m}_1 = 0.06$ eV. The thresholds for the two modes (z_+), (z_-) and one mode (z_0) are indicated.

In the one-mode approach, the decay is forbidden when $M < m_\ell + m_\phi$. Considering two modes, the phase space is reduced for the positive mode due to the larger quasi-mass and at $M = m_+(\infty) + m_\phi$, the decay is only possible into leptons with small momenta, thus the rate drops dramatically. The decay into the negative, quasi-massless mode is suppressed due to its much smaller residue. However, the decay is possible up to $M = m_\phi$. The situation is analogous for the Higgs decay which opens up when the thermal Higgs mass is larger than the neutrino mass. These rates differ from the one mode approach by more than one order of magnitude in the interesting temperature regime of $z = T/M \gtrsim 1$.

5. Conclusions

As discussed in detail in Ref. [17], we have, by employing HTL resummation and finite temperature cutting rules, confirmed that treating thermal masses as kinematic masses as in Ref. [4] is a reasonable approximation. We have calculated the decay density of the lightest heavy Majorana neutrino and its behavior can be explained by considering the dispersion relations $\omega_\pm$ of the lepton modes and assigning momentum-dependent quasi-masses to them. The thresholds for neutrino decay reported in Ref. [4] are shifted and the decay density shows deviations of more than an order of magnitude in the interesting temperature regime $T/M \sim 1$. In order to arrive at a minimal consistent treatment, also the decay $\phi \to N\ell$ at high temperatures needs to be included as well as Higgs boson and neutrino CP asymmetries which are corrected for lepton modes.

This contribution summarizes the results of an earlier work [1] and we refer the interested reader to the more elaborate treatment there.

Acknowledgements We thank Georg Raffelt, Florian Hahn-Wörnle, Steve Blanchet, Matthias Garny, Marco Drewes, Wilfried Buchmüller, Martin Spinrath and Philipp Kostka for fruitful and inspiring discussions.

References

[1] C. P. Kießig, M. Plümacher and M. H. Thoma, arXiv:1003.3016 .
[2] M. Fukugita and T. Yanagida, *Phys. Lett. B* **174**, 45 (1986).
[3] P. Minkowski, *Phys. Lett. B* **67**, 421 (1977).
[4] G. F. Giudice, A. Notari, M. Raidal, A. Riotto and A. Strumia, *Nucl. Phys. B* **685**, 89 (2004) [arXiv:hep-ph/0310123].
[5] M. Le Bellac, *Thermal field theory* (Cambridge University Press, Cambridge, UK, 1996).
[6] E. Braaten and R. D. Pisarski, *Nucl. Phys. B* **337**, 569 (1990).
[7] E. Braaten and R. D. Pisarski, *Nucl. Phys. B* **339**, 310 (1990).
[8] M. H. Thoma, *Z. Phys. C* **66**, 491 (1995). [arXiv:hep-ph/9406242].
[9] H. A. Weldon, *Phys. Rev. D* **28**, 2007 (1983).
[10] R. L. Kobes and G. W. Semenoff, *Nucl. Phys. B* **272**, 329 (1986).
[11] E. Braaten, R. D. Pisarski and T. C. Yuan, *Phys. Rev. Lett.* **64**, 2242 (1990).
[12] R. D. Pisarski, *Nucl. Phys. B* **309**, 476 (1988).
[13] R. D. Pisarski, *Physica A* **158**, 146 (1989).
[14] V. V. Klimov, *Sov. J. Nucl. Phys.* **33**, 934 (1981) [*Yad. Fiz.* **33**, 1734 (1981)].
[15] H. A. Weldon, *Phys. Rev. D* **26**, 2789 (1982).
[16] H. A. Weldon, *Phys. Rev. D* **40**, 2410 (1989).
[17] C. P. Kießig and M. Plümacher, arXiv:0910.4872 [hep-ph].

Journal of Physics: Conference Series **259** (2010) 012080 doi:10.1088/1742-6596/259/1/012080

The Dark Energy Survey

E. Sánchez for the DES collaboration
CIEMAT, Ed. 2; Avda. Complutense 22; E-28040 Madrid, Spain
E-mail: eusebio.sanchez@ciemat.es

Abstract. The Dark Energy Survey (DES) is a next generation optical and near infrared survey that will image 5000 deg^2 of the South Galactic Cap in five broad bandpass filters. In order to perform such a survey, a CCD camera of 3 deg^2 field of view is being assembled at Fermilab and will be mounted on the Blanco 4m telescope at Cerro Tololo (Chile). The survey will start in the fall of 2011 and will study the dark energy properties using four independent methods: galaxy clusters counts and distributions, weak gravitational lensing tomography, baryon acoustic oscillations and supernovae Ia distances. Obtaining the four measurements from the same data set will allow a strict control of the systematic uncertainties to obtain a robust and precise determination of the cosmological parameters.

1. Introduction

The current set of cosmological observations solidly establishes an amazing result, the cosmic acceleration. At cosmological distances, the receding velocities of galaxies are increasing. The implications of this observed fact are profound. Either gravity behaves far diferently than what is expected from the general theory of relativity, or some mysterious fluid with negative pressure, the dark energy, fills the universe and produce repulsive gravity. Anyway, new physics is needed. Therefore, the cosmic acceleration puzzle motivates an important fraction of current research which will grow in the near future. Usually, the cosmic acceleration problem is called dark energy, but we should not forget that the observed effect can be due to an unexpected behaviour of gravity at very large scales.

The current standard model of cosmology, ΛCDM, describes the cosmic acceleration using a non-zero cosmological constant, the Λ of its name. CDM stands for Cold Dark Matter, the other dark component of the universe. This model is compatible with all the present cosmological observations [1]. However, in this model about 96% of the universe (dark energy but also dark matter) remains unexplained, despite the succesful theories of fundamental physics. It is generally expected that the next advance in knowledge about dark energy come from the observations, and a rich program of different cosmological surveys has already started and is planned for the future. One of the most important projects for the near future is the Dark Energy Survey (DES) [2].

2. The Dark Energy Survey

DES[1] is a next generation survey aimed to unveil the nature of the dark energy. The DES collaboration consists of approximately 120 scientists from USA, UK, Spain, Brazil and Germany.

[1] http://www.darkenergysurvey.org

The survey will image 5000 square degrees of the south galactic cap, using five wide bandpass filters, $grizY$, from visible to near infrared, up to magnitude 24th. These filters have been chosen to provide accurate photometric redshifts for all the detected objects. The area and location of DES have been chosen to overlap with the South Pole Telescope[2] (SPT) [3], which will provide Sunyaev-Zeldovich effect measurements for a large number of galaxy clusters. To perform such a survey, three elements are needed: A telescope, a camera and a data management system. These elements will be described in the next sections.

3. The Telescope

The DES project has an agreement with the National Optical Astronomical Observatory (NOAO) to have the use of a significant amount of time on the Blanco 4m telescope of the Cerro Tololo Interamerican Observatory (CTIO), in Chile, which is shown in Figure 1. Correspondingly, DES provides a new advanced prime focus instrument. The DES survey will have 525 nights in five years to complete its observations. This means a 30% of the telescope time. The rest of the time the instrument will be available for the community use.

Figure 1. View of the Cerro Tololo observatory (left). The Blanco 4m telescope (right) is located inside the largest building. The black cylinder at the prime focus will be replaced by a new one containing the DES camera, DECam.

4. DECam: The Dark Energy Camera

The camera for the DES survey is called the Dark Energy Camera (DECam). Its design was optimized to maximize the scientific potential and to minimize the contruction time and cost. DECam will replace the entire prime focus cage of the Blanco telescope with a new one, containing the whole DECam system, see Figure 1.

The major components of DECam are a 520 Megapixel optical CCD focal plane, which is housed in a dewar that provides vacuum and cryogenics, a low noise CCD readout electronic system placed in crates around the main dewar that are actively cooled, a shutter, a filter system to house and exchange the DES filters, and a wide-field optical corrector (which gives a field of view of 2.2 degrees of diameter). The CCD vessel and corrector are supported by a hexapod that provides adjustability in all degrees of freedom. A scheme of the DECam structure is shown in Figure 2.

[2] http://pole.chicago.edu

Figure 2. Scheme of DECam (left), where all its components can be seen. A prototype of the dewar is shown at the right, including several CCDs in the focal plane and the corresponding readout electronics.

The major DECam systems are complete, and are in process of assembly at Fermilab.

To have a good detection efficiency for high redshift objects, CCDs that are extremely sensitive at the red part of the spectrum are used. They are fully depleted, 250 μm thick CCDs, that have been developed at the Lawrence Berkeley National Laboratory. The quantum efficiency of these devices in the z band is larger than 50%, almost an order of magnitude higher than traditional thinned devices.

The DECam focal plane will contain 62 CCDs of 2048x4096 pixels for scientific imaging, and 12 CCDs of 2048x2048 pixels for guiding, alignment and focus. A prototype of this focal plane, within the dewar and partially instrumented, can be seen in Figure 2.

5. The DES Data Management System

The DES data management system is being developed to process the raw DECam images into science ready data products, and make those data available to the DES collaboration and to the public community. The design of the data management system is driven by the scientific goal of the DES project and also by the need to automatically and reliably process, calibrate and archive the survey dataset.

The data management system must be able to process and archive around 400 Gb of data every night, for a total of $\sim$ 1Pb at the end of the survey. Including the processed and reduced data, the total amount will be on the several Pb size. The DES data management system consists of an integrated archive, a processing framework, an ensemble of astronomy codes and a data access framework. It uses community high performance computing resources. Image and catalog data are tracked within a distributed archive, and data movement among archive nodes employs high efficiency grid tools. Operators control the system, submit processing requests and monitor data quality using portals. Collaboration scientists and the public retrieve data using the archive portal or web service, as shown in Figure 3.

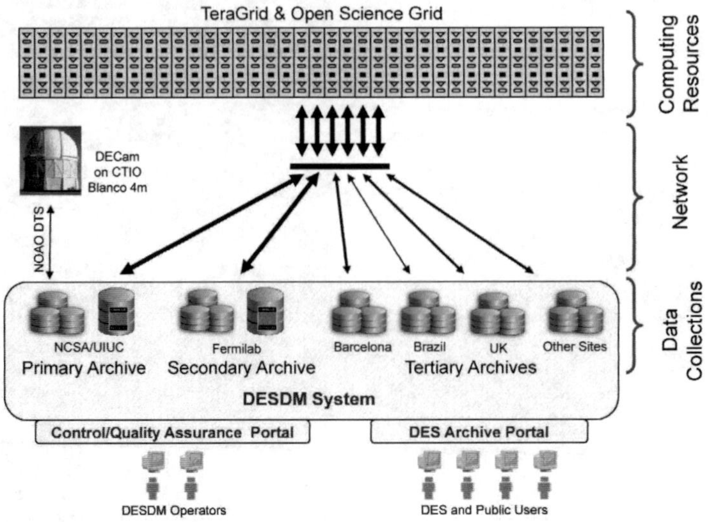

Figure 3. Structure of the DES data management system. It will use grid technology and the data will be distributed in several secondary and tertiary archives.

6. Scientific Objectives and Forecasts

The main goal of DES is, as has been stated before, to study the nature of the dark energy. More precisely, the measurement of the parameter ω of the dark energy equation of state to a precision of $< 5\%$ and its variation with the redshift to a precision of $\sim 30\%$ in each of four separate and independent techniques. These results will be then combined to a final constraint of much higher precision. Obtaining all the measurements with the same system makes possible to cross-compare results and control the systematic errors.

The four methods DES will use to study the dark energy are:

- Galaxy cluster counting and spatial distribution of clusters at $0.1 < z < 1$.
- Weak lensing measurements on several redshift shells up to $z \sim 1$.
- Baryon acoustic oscillations and general study of the power spectrum of galaxies.
- 2000 supernovae Ia at $0.3 < z < 0.8$.

These are the main scientific goals. However, DES will be able to make other dark energy studies. For example, the cross-correlation of the cosmic microwave background measurements with the galaxies identified by DES, which will also probe the dark energy using the integrated Sachs-Wolfe effect, and many others. These additional studies will give more constraints and better precision to the final measurement.

6.1. Galaxy Cluster Distribution

Clusters of galaxies are considered a very promising cosmological tool, because the formation of its gravitational potential wells is only due to the dynamics of dark matter to a good approximation. The redshift distribution, the spatial distribution, and the evolution of the mass function of clusters of galaxies are sensitive to ω.

DECam will be able to identify and measure the redshifts of clusters of galaxies out to $z \sim 1$. This range includes all the time where the dark energy is important to describe the expansion history of the universe and even goes beyond that limit. The survey will identify $\sim$ 20000 clusters, and their redshifts will be determined photometrically with enough precision to do cosmology. Moreover, the overlap of the DES area with the SPT area allows the use of the Sunyaev-Zeldovich (SZ) effect to detect clusters out to large distances, increasing the sensitivity to the dark energy parameters.

The dominant systematics errors of this measurements are the selection of the sample and the determination of the cluster mass. The selection of the clusters using methods as SZ or X-ray identifications, the detailed simulations, and the determination of masses using weak lensing allow the control of systematic errors.

6.2. *Weak Gravitational lensing*

DES will study weak lensing both using the shear-shear correlations and the galaxy-shear correlations, that are sensitive to ω through geometry and through the growth of cosmic structure. Around 300 millions of galaxies will be measured.

The main systematic errors that affect this measurement are the separation of the different components of the shear (intrinsic or coming from the instrument) and possible biases in the photometric redshift. A extremely careful control of the focus and alignment, and detailed simulations of the whole system are necessary.

6.3. *Galaxy Angular Clustering*

Baryon acoustic oscillations are expected to have very small systematics in the determination of dark energy parameters, since they are not affected by astrophysical considerations on the measured objects. The characteristic length scale of these oscillations can be used as a standard ruler to probe the ω parameter through a geometrical test. The angular power spectrum or the angular correlation function of galaxies are used as the observables for this test.

The main systematic uncertaintiess of this measurement arise from the bias for galaxies, the calibration of photometry and the possible biases in the photometric redshifts. To control them, the angular bispectrum, the measurement of clustering for different colors and type of galaxy, a careful calibration strategy and spectroscopic calibration sets will be used.

6.4. *Supernovae*

Supernovae of the type Ia are standarizable candles. They are sensitive to the dark energy properties through the luminosity distance. DES will identify and measure around 2000 supernovae Ia up to $z \sim 0.8$. The observations must be repeated over time, to obtain the luminosity curves of the supernovae. The selection of the required type of supernovae will be performed using color information.

The dominant systematic errors in the measurement are the possible evolution of supernovae with the redshift, the extinction in the host galaxy of the supernova and the calibration of the photometric redshift. There are several methods to control them, namely, the comparison of supernovae at high and low redshift, the information on color and host galaxy of the supernova and a spectroscopic subsample of supernovae to obtain precise calibrations.

6.5. *Forecasts*

The Dark Energy Task Force (DETF) defined a figure of merit to evaluate the sensitivity of a given measurement to dark energy [4]. The expected DES sensitivity for each measurement technique and for their combination is shown in Figure 4. To obtain this result, the equation of state of the dark energy is written as $p/\rho = w_0 + w_a(1 - a)$. DES is a stage III project in the DETF classification. Therefore, an improvement of a factor 4 or 5 with respect to the current result is expected.

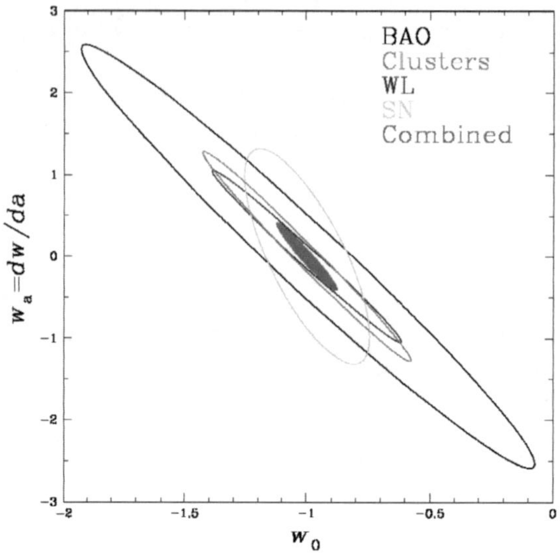

Figure 4. Expected DES sensitivity to dark energy parameters for an equation of state $p/\rho = w_0 + w_a(1-a)$, in each of the measurement techniques: BAO (black), galaxy clusters (magenta), weak lensing (blue) and supernovae Ia (green). The combination of the four methods is also shown (red).

7. Schedule

All of the major components of DECam are complete. They will be assembled at Fermilab during the next few months, except the optics that will be send directly to CTIO. Then the full system will be tested till the end of 2010 with engineering grade CCDs and during the beginning of 2011 with the science grade CCDs. After that, it will be shipped to CTIO. The installation in the Blanco telescope will take place during the summer of 2011. The first light and the beginning of commissioning are expected for the autumn 2011.

8. Conclusions

DES will measure the dark energy using four independent and complementary methods: Galaxy clusters, baryon acoustic oscillations, weak gravitational lensing and supernovae Ia. Having all these measurements with the same system allows to set robust constraints and have a high control of systematic errors. The goal is to improve current measurements by a factor of at least 4 or 5.

In order to reach these results, a new instrument (DECam) is being assembled and will be installed in the prime focus of the 4m Blanco telescope in CTIO. It is a 520 Megapixels CCD camera for optical and near IR observations.

Using this instrument and telescope, a survey of 5000 square degrees in the south galactic cap will be performed, imaging this area with the *grizY* filters. It will last for 525 nights distributed in 5 years. The first light is expected for autumn 2011.

References

[1] Nakamura N *et al* (The Particle Data Group) 2010 *J. Phys. G* **37** 075021
[2] The Dark Energy Survey Collaboration 2005 (*Preprint* `arXiv:astro-ph/0510346`)
[3] Ruhl J E *et al.* (The SPT) 2004 *Proc. SPIE Int. Soc. Opt. Eng.* **5498** 11 (*Preprint* `astro-ph/0411122`)
[4] Albrecht A *et al.* 2006 (*Preprint* `astro-ph/0609591`)

16th International Symposium on Particles Strings and Cosmology (PASCOS 2010) IOP Publishing
Journal of Physics: Conference Series **259** (2010) 012081 doi:10.1088/1742-6596/259/1/012081

Dark Energy and Structure Formation

Anupam Singh
Physics Department, L.N. Mittal I.I.T, Jaipur, Rajasthan, India.
E-mail: morisi@ific.uv.es

Abstract.
We study the gravitational dynamics of dark energy configurations. We report on the time evolution of the dark energy field configurations as well as the time evolution of the energy density to demonstrate the gravitational collapse of dark energy field configurations. We live in a Universe which is dominated by Dark Energy. According to current estimates about 75% of the Energy Density is in the form of Dark Energy. Thus when we consider gravitational dynamics and Structure Formation we expect Dark Energy to play an important role. The most promising candidate for dark energy is the energy density of fields in curved space-time. It therefore become a pressing need to understand the gravitational dynamics of dark energy field configurations. We develop and describe the formalism to study the gravitational collapse of fields given any general potential for the fields. We apply this formalism to models of dark energy motivated by particle physics considerations. We solve the resulting evolution equations which determine the time evolution of field configurations as well as the dynamics of space-time. Our results show that gravitational collapse of dark energy field configurations occurs and must be considered in any complete picture of our universe.

1. Introduction

Increasing accuracy in astronomical observations has led us to an increased precision in the determination of cosmological parameters. This in turn led us to critically re-examine our cosmological models. In particular, the accurate determination of the Hubble constant and the independent determination of the age of the universe forced us to critically re-examine the simplest cosmological model - a flat universe with a zero cosmological constant[1, 2]. These observations forced the current author[3] to consider the idea of a small non-vanishing vacuum energy due to fields as playing an important role in the Universe. Subsequently there has been a large body of work both on the observational and theoretical side that has firmed up our belief in what we now call dark energy.

Given that we now believe that dark energy is the dominant component of the Universe, it is now a pressing need to understand the dynamics of dark energy and in particular the gravitational dynamics of dark energy. In the following discussion, we describe the gravitational dynamics of dark energy and show the gravitational collapse of dark energy field configurations.

Before we start our detailed look at the gravitational dynamics of dark energy field configurations we would like to quickly introduce field theory models of Dark Energy so that all readers can readily relate to the discussion that follows.

Discussions of field theory models for dark energy and connections to particle physics were discussed in detail earlier by the current author[3]. It was noted in that article that these fields must have very light mass scales in order to be cosmologically relevant today. Discussions of realistic particle physics models with particles of light masses capable of generating interesting cosmological consequences have been carried out by several authors[6]. It has been pointed out that the most natural class of models in which to realise these ideas are models of neutrino masses with Pseudo Nambu Goldstone Bosons (PNGB's). The reason for this is that the mass scales associated with such models can be related to the neutrino masses, while any tuning that needs to be done is protected from radiative corrections by the symmetry that gave rise to the Nambu-Goldstone modes[7].

Holman and Singh[8] studied the finite temperature behaviour of the see-saw model of neutrino masses and found phase transitions in this model which result in the formation of topological defects. In fact, the critical temperature in this model is naturally linked to the neutrino masses.

In our next section describing the formalism and evolution equations for studying the gravitational dynamics of dark energy field configurations we will first write down the equations for fields with any general potential to keep the initial discussion general. However, in light of the above discussion on realistic models with light masses we will soon specialize to PNGB models which have a potential with simple and explicit form. Thus, for this purpose we will take the simplest PNGB potential[6].

2. The Evolution Equations

We wish to study the gravitational dynamics of Dark Energy field configurations. The dynamics of fields in cosmological space-times has been extensively discussed elsewhere (see e.g. Kolb and Turner[4]). Likewise, gravitational collapse in the context of general relativity has also been extensively discussed elsewhere (see e.g. Weinberg[5]). These ideas can be pulled together to write down the evolultion equations describing the coupled dynamics of the field configurations and space-time interacting with each other. This is what we will do in this section. In the next section we will discuss the solutions to these evolution equations.

To study gravitational collapse in a cosmological setting we can start with a metric of the form:

$$ds^2 = dt^2 - U(r,t)dr^2 - V(r,t)\left[d\theta^2 + \sin^2\theta d\phi^2\right] \tag{1}$$

where, (t, r, θ, ϕ) are the comoving coordinates describing a space time point. This metric, is of course a generalization of the usual FRW metric used to study cosmological space-times. Note that the functions $U(r,t)$ and $V(r,t)$ are functions of both space and time and can capture both homogeneous cosmological expansion as well as inhomogeneous gravitational collapse under appropriate circumstances.

We are interested in studying the dynamics of fields in space-time. For this purpose and to keep the initial discussion as general as possible we will consider a field Φ with a Lagrangian L given by

$$L = \frac{1}{2}\partial^\mu\Phi\partial_\mu\Phi - \mathcal{V}(\Phi) \tag{2}$$

where $\mathcal{V}$ is the potential for the field Φ and is for now a general function. Later, when we consider the physically motivated PNGB models this potential will take on a specific functional form.

By using standard techniques one can then write down the evolution equations for the metric and the field. These can be written down in the form:

$$\ddot{V} = 2\left[-1 + \frac{V''}{2U} - \frac{V'U'}{4U^2} - \frac{\dot{V}\dot{U}}{4U} + 8\pi GV\left(\frac{\rho}{2} - \frac{P}{2} - \frac{(\Phi')^2}{3U}\right)\right] \tag{3}$$

$$\ddot{U} = 2U\left[-\frac{\ddot{V}}{V} + \frac{\dot{U}^2}{4U^2} + \frac{\dot{V}^2}{2V^2} - 4\pi G\left(\rho + 3P\right)\right] \tag{4}$$

Journal of Physics: Conference Series **259** (2010) 012081 doi:10.1088/1742-6596/259/1/012081

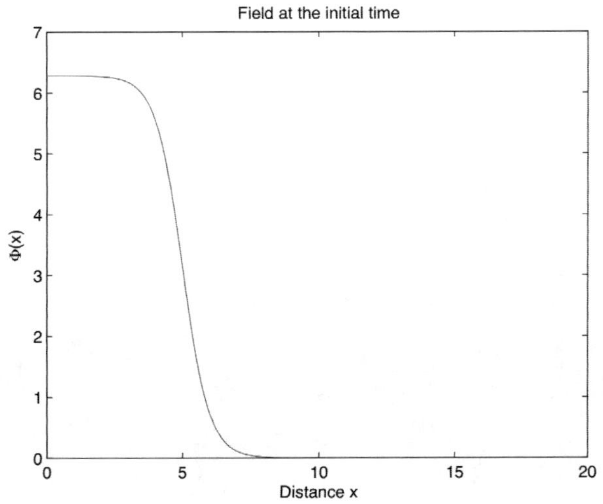

Figure 1. Initial Field configuration

$$\ddot{\Phi} = \frac{\Phi''}{U} - \dot{\Phi}\left[\frac{\dot{V}}{V} + \frac{\dot{U}}{2U}\right] + \frac{\Phi'}{U}\left[\frac{V'}{V} - \frac{U'}{2U}\right] - \frac{\partial \mathcal{V}(\Phi)}{\partial \Phi} \tag{5}$$

where a dot represents a partial derivative w.r.t. t and a prime represents a partial derivative w.r.t. r. Further,

$$\rho = \frac{1}{2}\dot{\Phi}^2 + \mathcal{V}(\Phi) + \frac{(\Phi')^2}{2U} \tag{6}$$

and

$$P = \frac{1}{2}\dot{\Phi}^2 - \mathcal{V}(\Phi) + \frac{(\Phi')^2}{6U}. \tag{7}$$

The above equations are true for any general potential $\mathcal{V}(\Phi)$. One can of course write down the corresponding equations for PNGB fields. The simplest potential one can write down for the physically motivated PNGB fields [6] can be written in the form:

$$\mathcal{V}(\Phi) = m^4\left[K - \cos(\frac{\Phi}{f})\right] \tag{8}$$

As discussed in [6] m is of order the neutrino mass and K is of order 1. We will consider such a potential for studying the dynamics in the next section.

3. Solution to the Evolution Equations

The evolution equations described above can be solved numerically to study the gravitational collapse of the field configurations. The key issue we want to understand and address is the timescale on which the gravitational collapse happens. Clearly if this timescale for collapse is equal to or larger than the age of the Universe then the gravitational collapse has no practical significance or implications. On the other hand, if the gravitational collapse happens on a timescale shorter than the age of the Universe then we must consider the gravitational collapse of the dark energy field configurations.

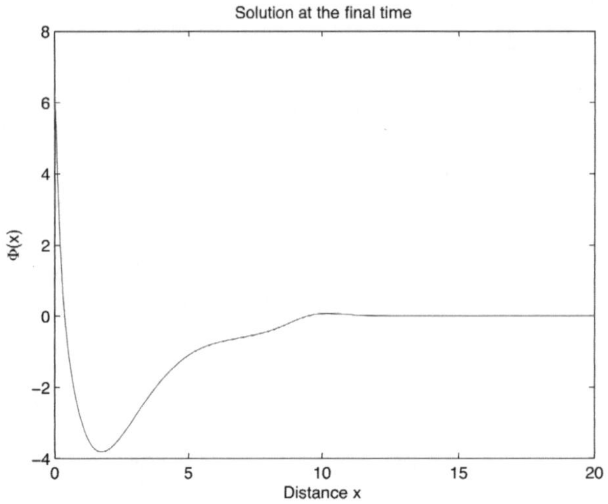

Figure 2. Final Field configuration

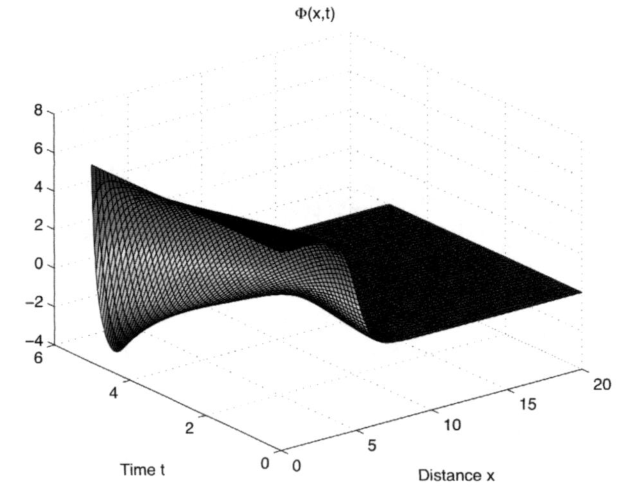

Figure 3. Field configuration in space-time

Guided by the evolution equations given in the previous section we define dimensionless quantitites such that the field is measured in units of f and time and space are measured in units of $\frac{f}{m^2}$.

The initial field configuration in terms of these dimensionless valiables is given in the figure 1 that follows.

The final field configuration in terms of these dimensionless valiables is given in figure 2.

From this it can be clearly seen that field configuration has collapsed and the timescale for collapse can be seen by studying the figure 3. Since the units of time are given by $\frac{f}{m^2}$, we note that gravitational collapse happens on timescales of $\sim \frac{f}{m^2}$. This timescale is much shorter than the age of the Universe.

4. Summary and Conclusions

It is now a well established fact that the Universe is dominated by Dark Energy. The most promising candidate for Dark Energy is the energy density due to fields. Understanding the dynamics of these Dark Energy fields is of paramount importance. We developed and described the formalism to study the gravitational collapse of fields given any general potential for the fields. We then applied this formalism to the simplest and most natural model of dark energy fields motivated by Particle Physics considerations. We reported on the results obtained by numerically solving the resulting evolution equations. Our results show that gravitational collapse of dark energy fields happens on a timescale which is short compared to the age of the Universe.

ACKNOWLEDGEMENTS

This work was supported in part by a research grant from L.N.Mittal I.I.T.

[1] M. J. Pierce, D. L. Welch, R. D. McClure, S. van den Bergh, R. Racine and P. B. Stetson, Nature, 371, (September 1994), pp. 385-389.
[2] Freedman, W.L., et al , Nature, **371**, 757-762 (1994).
[3] Singh, Anupam, Phys. Rev. D 52, 6700-6707, (1995).
[4] E. W. Kolb and M. S. Turner : The Early Universe, 73. (Addison-Wesley Pub. Co. , 1990).
[5] Weinberg, S. : Gravitation and Cosmology. (John Wiley & Sons, Inc. , 1972).
[6] A. K. Gupta, C. T. Hill, R. Holman and E. W. Kolb, Phys. Rev. D45, 441 (1992).
[7] see, e. g. , G. 'tHooft "Naturalness, Chiral Symmetry, and Spontaneous Chiral Symmetry Breaking, " in *Recent Developments in Gauge Theories*, (Plenum Press, New York, 1980), G. 't Hooft, et al. , eds.
[8] Richard Holman and Anupam Singh, Phys. Rev. D 47, 421 (1993).

Journal of Physics: Conference Series **259** (2010) 012082 doi:10.1088/1742-6596/259/1/012082

Relation between inflation and dark energy[1]

Christophe Ringeval

Institute of Mathematics and Physics, Centre for Cosmology, Particle Physics and Phenomenology, Louvain University, 2 Chemin du Cyclotron, 1348 Louvain-la-Neuve, Belgium

Teruaki Suyama

Institute of Mathematics and Physics, Centre for Cosmology, Particle Physics and Phenomenology, Louvain University, 2 Chemin du Cyclotron, 1348 Louvain-la-Neuve, Belgium, and

Research Center for the Early Universe, Graduate School of Science, The University of Tokyo, Tokyo 113-0033, Japan

Tomo Takahashi

Department of Physics, Saga University, Saga 840-8502, Japan

Masahide Yamaguchi[2]

Department of Physics, Tokyo Institute of Technology, Tokyo 152-8551, Japan

Shuichiro Yokoyama

Department of Physics and Astrophysics, Nagoya University, Aichi 464-8602, Japan

Abstract. In this talk, we have shown that current cosmic acceleration can be explained by an almost massless scalar field experiencing quantum fluctuations during primordial inflation. Provided its mass does not exceed the Hubble parameter today, this field has been frozen during the cosmological ages to start dominating the universe only recently. By using supernovae data, completed with baryonic acoustic oscillations from galaxy surveys and cosmic microwave background anisotropies, we infer the energy scale of primordial inflation to be around a few TeV, which implies a negligible tensor-to-scalar ratio of the primordial fluctuations. Moreover, our model suggests that inflation lasted for an extremely long period. Dark energy could therefore be a natural consequence of cosmic inflation close to the electroweak energy scale.

1. Introduction

In the past decade, various cosmological observations have accumulated evidence that the universe is currently undergoing accelerated expansion. Although cosmic acceleration can be triggered by a non-vanishing cosmological constant, one still has to explain why it is so small and why it has started to dominate the energy content of the universe only recently. These questions have motivated the exploration of alternative explanations all referred to as dark

[1] This talk is based on Ref. [1]. Please also see the references therein.
[2] This talk was given by Masahide Yamaguchi.

Journal of Physics: Conference Series **259** (2010) 012082 doi:10.1088/1742-6596/259/1/012082

energy. Among them, the quintessence models consider a scalar field rolling down a potential in a way similar to the mechanism at work during primordial inflation. A quintessence field yields a time-dependent equation of state $w(t) = P/\rho > -1$, P and ρ being its pressure and energy density. A cosmological constant being equivalent to $w = -1$, quintessence models generically predict a different expansion history of the universe, and this can be tested by observations. The quintessence models can be divided into "freezing" and "thawing" types. The former yields a decreasing $w(t)$ which approaches -1 at low redshifts. This type of model has been intensively used in the literature to address the coincidence problem. Indeed, assuming the potential to be of the Ratra–Peebles kind, the quintessence field tracks a cosmological attractor which erases memory of the initial conditions. Once the field energy scale is adjusted to the current dark energy value, its recent domination is automatic. On the other hand, the thawing type gives $w \simeq -1$ at high redshifts but can evolve and start deviating from this value at low redshifts, exactly as during the inflationary graceful exit. These models have been less explored than the freezing type due to their dependence on the initial field values. Initial conditions are not washed out by an attractor mechanism and therefore constitute additional and unwanted model parameters. In this letter, we show that inflationary cosmology solves this problem by giving natural initial values for the quintessence field which can explain the current acceleration. They are set by the quantum generation of field fluctuations during inflation. Assuming the accelerated expansion today to be sourced by the quintessence field, we use the supernovae data to infer the energy scale of inflation which ends up being at a few TeV, i.e. close to the electroweak symmetry breaking energy scale. Such a scenario implies a negligible tensor-to-scalar ratio of the primordial cosmological perturbations and a reheating temperature also around a few TeV. Combined with cosmic microwave background (CMB) bounds, the class of allowed inflationary models is thus severely constrained. Moreover, the total number of e-folds of inflation has to be extremely large, as it could be in a self-reproducing inflationary scenario.

Linking primordial inflation and dark energy has been originally discussed in the context of anthropic selection effects. In these approaches, inflationary quantum fluctuations are used to randomize the possible dark energy density values while anthropic selection effects favour the ones compatible with our own existence. Inflationary stochastic effects in freezing-type quintessence models have been explored to determine how inflation influences the likely initial conditions of the field. They indeed have a tendency to push the freezing type quintessence field away from the region where the tracker behavior can efficiently wash out the initial conditions. As a result, freezing quintessence on the tracker today suggests that inflation lasted a low number of e-folds.

In our model, the initial value of the thawing quintessence field is directly determined by the energy scale of inflation and keeps its initial value until low redshifts. Therefore today's energy density of the quintessence field is almost equal to its initial energy density, and hence is directly related to the inflationary energy scale. Checking the consistency of the model provides us a way to probe primordial inflation from dark energy observations. This is our main point and an essential difference from the freezing type quintessence scenario.

2. Initial field values from inflation

Let us consider a (real) free massive scalar field φ of mass m such that m is less than the present Hubble scale H_0. Clearly, since the inflationary Hubble parameter $H_{\text{inf}} \gg H_0$, we are in the presence of an almost massless scalar field which will acquire quantum fluctuations during inflation. By decomposing the scalar field as a homogeneous mode plus linear perturbations (at the onset of inflation),

$$\varphi(\boldsymbol{x}, t) = \phi(t) + \delta\phi(\boldsymbol{x}, t), \tag{1}$$

Journal of Physics: Conference Series **259** (2010) 012082 doi:10.1088/1742-6596/259/1/012082

and assuming an almost constant value for $H_{\rm inf}$, its Fourier modes after Hubble exit read

$$|\delta\phi_{\boldsymbol{k}}|^2 \simeq \frac{H_{\rm inf}^2}{2k^3}\left(\frac{k}{aH_{\rm inf}}\right)^{2m^2/(3H_{\rm inf}^2)} . \tag{2}$$

The comoving wavevector is $\boldsymbol{k}$ and a stands for the scale factor. If inflation starts at $a = a_{\rm s}$, after $N = \ln(a/a_{\rm s})$ e-folds, super-Hubble fluctuations induce a real space field variance given by

$$\left\langle \delta\phi^2 \right\rangle \simeq \int_{a_{\rm s}H_{\rm inf}}^{aH_{\rm inf}} \frac{{\rm d}^3\boldsymbol{k}}{(2\pi)^3}\left|\delta\phi_{\boldsymbol{k}}\right|^2 = \frac{3H_{\rm inf}^4}{8\pi^2 m^2}\left[1-\exp\left(-\frac{2m^2}{3H_{\rm inf}^2}N\right)\right] \to \frac{3H_{\rm inf}^4}{8\pi^2 m^2} , \tag{3}$$

where the last limit is reached if inflation lasts long enough for the exponential term to cancel. This result can be reproduced using the stochastic inflation formalism. The homogeneous mode evolves as $\phi = \phi_{\rm s}\exp\left[-Nm^2/(3H_{\rm inf}^2)\right]$ and becomes completely suppressed compared to the field fluctuations. The same holds for the mean squared field derivative (in e-folds) which is suppressed by a factor $m^4/H_{\rm inf}^4$ compared to Eq. (3). For $m \ll H_{\rm inf}$ the required number of e-folds could actually be extremely large. Here, one should notice that the long-wave fluctuation $\delta\phi$ is almost homogeneous and determines the typical value of the classical field φ. Notice that backreaction effects induced by the field fluctuations over the expansion rate can produce $\mathcal{O}(1)$ variations in the total number of e-folds. However, one can show that they induce a correction factor to Eq. (3) given by $1+\mathcal{O}\left(H_{\rm inf}^2/M_{\rm P}^2\right)$ which ends up being completely negligible as soon as $H_{\rm inf} \ll M_{\rm P}$. Since after inflation $H \gg m$, Hubble damping prevents the field from rolling down the potential until the time at which $m \simeq H \lesssim H_0$. Before going into a detailed comparison with observations, let us derive some order of magnitude results. The present energy density of this quintessence field is roughly

$$V(\phi) \simeq \frac{1}{2}m^2\langle\delta\phi^2\rangle \simeq \frac{3H_{\rm inf}^4}{16\pi^2}, \tag{4}$$

and does not depend on m. In order to explain dark energy, one needs

$$H_{\rm inf} \simeq (\Omega_{_\Lambda})^{1/4}\sqrt{4\pi H_0 M_{\rm P}} , \tag{5}$$

where $M_{\rm P}$ stands for the reduced Planck mass and $\Omega_{_\Lambda}$ is the current density parameter associated with a cosmological constant. By using fiducial values for H_0 and $\Omega_{_\Lambda}$, one gets $H_{\rm inf} \simeq 6\times10^{-3}\,{\rm eV}$. The energy scale of inflation is thus[3]

$$E_{\rm inf} \equiv \rho_{\rm inf}^{1/4} = \left(3M_{\rm P}^2 H_{\rm inf}^2\right)^{1/4} \simeq 5\,{\rm TeV}. \tag{6}$$

As a result, in the context of our scenario, we rephrase the question on the smallness of the dark energy density into the relatively small energy scale (TeV) of inflation, which may be more tractable to address and surprisingly close to the electroweak energy scale. Notice that although the mass m does not appear in Eq. (4) and (6), it still determines when the quintessence field starts rolling down the potential and how the equation of state of dark energy will deviate from $w = -1$. For TeV scale inflation, our model predicts a negligible tensor-to-scalar ratio. Using the above estimates, with $m \lesssim H_0$, the number of e-folds required to reach the Bunch–Davies fluctuations is $N \simeq H_{\rm inf}^2/m^2 \gtrsim 10^{60}$, which may therefore suggest a self-reproducing inflationary model. In the following, we assume that $N > H_{\rm inf}^2/m^2 > 10^{60}$. For such a large N, the homogeneous mode of the scalar field is also suppressed significantly so that the typical value of the classical field φ is completely determined by the long-wave fluctuation $\delta\phi$.

[3] However, we should keep in mind that this constraint is based on the assumption that the cosmological constant is adjusted to zero by some unknown mechanism.

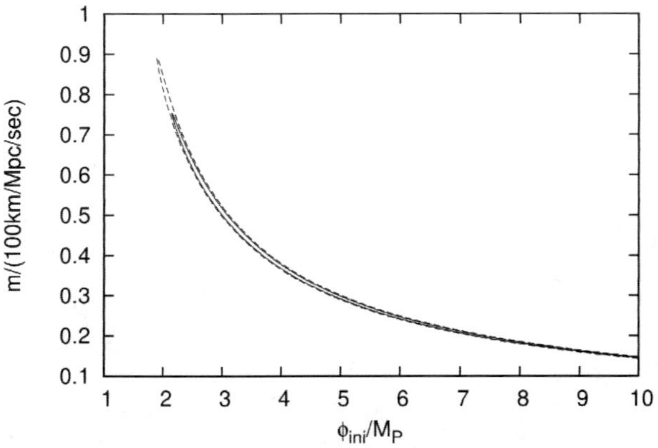

Figure 1. One and two-sigma contours of the two-dimensional marginalised probability distribution in the plane $(\phi_{\rm ini}, m)$ from SN+BAO+CMB data.

3. Observational constraints

In order to constrain our model from current observations, we have numerically solved the Einstein and Klein–Gordon equations

$$H^2 = \frac{1}{3M_{\rm P}^2}\left(\rho_{\rm r} + \rho_{\rm m} + \rho_\phi\right), \quad \ddot{\phi} + 3H\dot{\phi} + m^2\phi = 0, \tag{7}$$

for various values of the parameters $(m, \phi_{\rm ini}, H_0)$. The value $\phi_{\rm ini}$ denotes one realisation of the quintessence field fluctuations in our Hubble patch and $\rho_{\rm r}$ and $\rho_{\rm m}$ are the energy density associated with radiation and matter, respectively. The integration is started at high redshift, when the dark energy is sub-dominant, to the time at which $H = H_0$. We have fixed the current radiation energy density to $\Omega_{\rm r}h^2 = 4.15 \times 10^{-5}$ and dropped the decaying mode for the field. By solving the evolution equations, we obtain H as a function of the redshift z which can be used to compare the model with observational data. We have used type Ia supernovae (SN) redshift-distance modulus relations complemented with the redshift-distance relations at $z = 0.2$ and $z = 0.35$ coming from the baryon acoustic oscillations (BAO) data. Finally, we have used the angular scale and height of the first peak of the CMB power spectrum given by the WMAP team. Our likelihood has therefore been defined as the product of those three, namely by summing the chi-squared of SN, BAO and CMB data.

In order to test how the model can fit cosmic acceleration, let us first consider a Jeffreys' prior for the initial field values $\phi_{\rm ini} \geq M_{\rm P}$. Here we are not yet assuming that the initial conditions are set during inflation, but simply requiring super-Planckian field values (necessary for triggering acceleration). Together with flat priors on $m \leq 100\,{\rm km/s/Mpc}$ and H_0 around the current measured value, we have explored the three-dimensional parameter space with gridding methods. In Fig. 1, we have represented the one and two-sigma contours of the two-dimensional marginalised probability distribution (over H_0) in the plane $(\phi_{\rm ini}, m)$. By marginalising over $\phi_{\rm ini}$, we find the mass of the quintessence field to be constrained by $m < 75\,{\rm km/s/Mpc}$ (at 95% of confidence), which is consistent with our earlier estimation that it cannot exceed H_0 too much. We also find, as expected, that the data favour arbitrarily high super-Planckian initial field values. In order to check our results, we have also derived the field energy density parameter Ω_ϕ and its equation of state w_ϕ, both evaluated at the present time. The contours plotted in Fig. 1 end up being centered around the cosmological constant case $\Omega_\phi = 0.73$, $w_\phi = -1$.

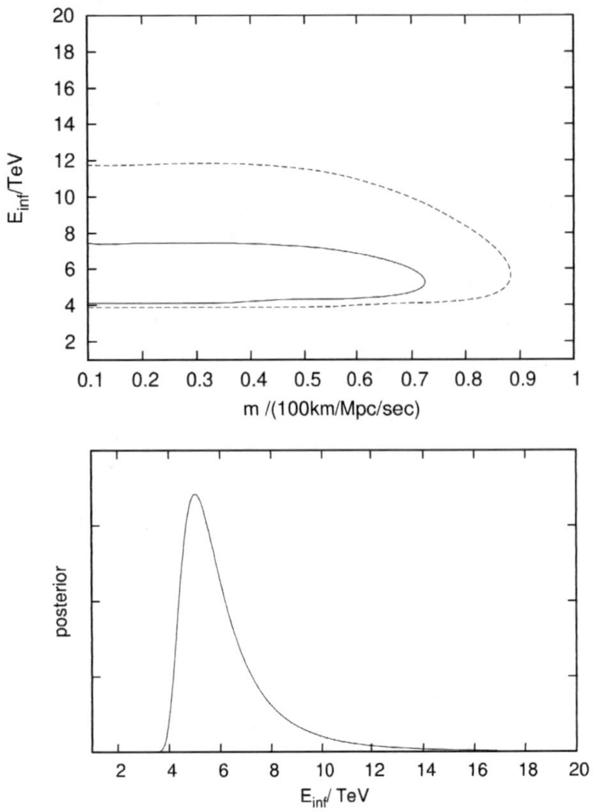

Figure 2. One and two-sigma contours in the plane $(m, E_{\rm inf})$ from SN+BAO+CMB data. The lower panel shows the marginalised posterior for $E_{\rm inf}$. At 95% of confidence, we find $3.8\,\mathrm{TeV} < E_{\rm inf} < 12.1\,\mathrm{TeV}$.

We are now in a position to infer the energy scale of inflation when the initial field values are set by inflationary quantum fluctuations. Compared to the above data analysis, our mechanism gives a prior probability distribution of $\phi_{\rm ini}$ which is Gaussian[4] with a standard deviation given by Eq. (3)

$$\sigma(H_{\rm inf}, m) = \sqrt{\frac{3}{8}\frac{H_{\rm inf}^2}{\pi m}}\,. \tag{8}$$

Denoting by D our data sets, and I the prior space, using Bayes' theorem and marginalising over $\phi_{\rm ini}$ and H_0 yields

$$P(H_{\rm inf}, m|D, I) \propto P(H_{\rm inf}, m|I) \int\int \mathrm{d}H_0 \frac{\mathrm{d}\phi_{\rm ini}}{\sqrt{2\pi}\,\sigma} \exp\left(-\frac{\phi_{\rm ini}^2}{2\sigma^2}\right) P(H_0|I)\mathcal{L}(D|m, \phi_{\rm ini}, H_0, I), \tag{9}$$

where $\mathcal{L}$ is the overall SN+BAO+CMB likelihood and $P(H_{\rm inf}, m|I)$ the prior probability distribution on $H_{\rm inf}$ and m. The scale of inflation being a priori unknown, we have chosen a flat prior on the logarithm of $H_{\rm inf}$ while the other parameters assume the same prior as before.

[4] Notice that embedding our mechanism into an explicit eternal inflationary model may change this prior, either due to volume effects or by a choice of peculiar probability measure. The scale factor cut-off measure might however preserve such a prior.

In Fig. 2, we have represented the one and two-sigma confidence intervals of this two-dimensional posterior, up to the change of variable $H_{\mathrm{inf}} \to E_{\mathrm{inf}}$. The lower panel is the fully marginalised probability distribution for E_{inf} and the energy scale of inflation verifies

$$3.8\,\mathrm{TeV} < E_{\mathrm{inf}} < 12.1\,\mathrm{TeV}, \tag{10}$$

at 95% of confidence. Though this constraint depends on the priors on H_{inf} and m, the dependence is small. As discussed below, our model works for a more general potential, which can slightly change the constraint. Thus, it is safe to say that our model suggests the energy scale of inflation E_{inf} to be around a few TeV for a wide class of thawing quintessential models.

4. Discussion

Although we have focused on a free scalar field so far, the mechanism discussed here could also be applied to any potential having an absolute minimum by using the inflationary stochastic formalism. However, the prior probability distribution for the field ϕ_{ini} will no longer be Gaussian and this could therefore shift the favoured values of E_{inf}. Since our scenario needs an extremely large e-folding number $N \simeq H_{\mathrm{inf}}^2/m^2 \gtrsim 10^{60}$, the running of H_{inf} along the inflaton potential must be extremely small. Thus, as a candidate of such a TeV scale inflation, new inflation might be preferred, which can indeed realize a self-reproducing era. We also would like to mention the case that the total number of e-folds is much smaller than H_{inf}^2/m^2, in which case Eq. (3) yields $H_{\mathrm{inf}}/M_{\mathrm{P}} \propto \sqrt{\Omega_{\Lambda}/N}$. In this limit, we have only a lower bound $E_{\mathrm{inf}} > \mathrm{TeV}$ and would have to explain why the homogeneous value $\phi_{\mathrm{s}} \simeq 0$ after inflation. Notice that CMB anisotropies imposing $H_{\mathrm{inf}}/M_{\mathrm{P}} < 10^{-5}$, we get an absolute lower limit for the total number of e-folds $N > 10^9$, under which the scale of inflation would actually be too high. Finally, it is worth stressing that we are in presence of a transient dark energy model. By rolling down its potential, the thawing quintessence field will acquire kinetic energy and the current cosmic acceleration will come to an end.

5. Acknowledgments

We are grateful to Takeshi Chiba, André Füzfa, Kiyotomo Ichiki, Masahiro Kawasaki, Hideo Kodama, Jérôme Martin, Shinji Mukohyama and Jun'ichi Yokoyama for enlightening discussions. This work is partially supported by the Belgian Federal Office for Scientific, Technical, and Cultural Affairs through the Inter-University Attraction Pole Grant No. P6/11 (C. R. and T. S.); by the Grant-in-Aid for Scientific research from the Ministry of Education, Science, Sports, and Culture, Japan, No. 19740145 (T. T.) and 21740187 (M. Y.). T. S. is supported by the JSPS. The work of S. Y. is supported in part by Grant-in-Aid for Scientific Research on Priority Areas No. 467 "Probing the Dark Energy through an Extremely Wide and Deep Survey with Subaru Telescope". He also acknowledges support from the Grant-in-Aid for the Global COE Program "Quest for Fundamental Principles in the Universe: from Particles to the Solar System and the Cosmos" from MEXT, Japan.

References

[1] C. Ringeval, T. Suyama, T. Takahashi, M. Yamaguchi and S. Yokoyama, Phys. Rev. Lett. **105**, 121301 (2010) [arXiv:1006.0368 [astro-ph.CO]].

Journal of Physics: Conference Series **259** (2010) 012083 doi:10.1088/1742-6596/259/1/012083

The cosmological constant and the relaxed universe

Florian Bauer

High Energy Physics Group, Dept. ECM, and Institut de Ciències del Cosmos, Universitat de Barcelona, Av. Diagonal 647, E-08028 Barcelona, Catalonia, Spain

E-mail: `fbauerphysik@eml.cc`

Abstract. We study the role of the cosmological constant (CC) as a component of dark energy (DE). It is argued that the cosmological term is in general unavoidable and it should not be ignored even when dynamical DE sources are considered. From the theoretical point of view quantum zero-point energy and phase transitions suggest a CC of large magnitude in contrast to its tiny observed value. Simply relieving this disaccord with a counterterm requires extreme fine-tuning which is referred to as the old CC problem. To avoid it, we discuss some recent approaches for neutralising a large CC dynamically without adding a fine-tuned counterterm. This can be realised by an effective DE component which relaxes the cosmic expansion by counteracting the effect of the large CC. Alternatively, a CC filter is constructed by modifying gravity to make it insensitive to vacuum energy.

1. Introduction

The old CC problem is of theoretical nature [1]. It is not forbidden to explain the current accelerated expansion of the universe by a small CC or vacuum energy density $\Lambda_{\text{obs}} \sim 10^{-47}\,\text{GeV}^4$. The cosmological term can be considered as the zero-order term in almost every action functional and it is not absent unless further assumptions are made. If it were a free parameter its observationally favoured smallness would not pose any problem. And due to its minimalist nature it would be a perfect candidate to explain the late-time expansion history.

However, modern theories suggest the existence of large contributions to the CC Λ which have to sum up to the tiny value Λ_{obs}. This is a highly non-trivial requirement because all those parts are usually not related to each other and they can be of different magnitude and sign. Perhaps the worst part comes from quantum field theory in the form of vacuum or zero-point energy. Since every field mode provides a contribution related to its energy ω, the integral over all modes in the energy spectrum results into an infinite CC contribution. For a massless field ω is just the mode momentum p and one finds the quartically divergent result

$$\Lambda_0 \sim \int_0^\infty d^3p\, p. \tag{1}$$

Common estimations impose an ultra-violet (UV) cutoff at a high-energy scale p_{max}, which replaces the upper bound of the integral. This naive procedure yields the often presented number $|\Lambda_0/\Lambda_{\text{obs}}| \sim 10^{123}$ when the maximal momentum is assumed to be around the Planck mass $M_P \sim 10^{19}\,\text{GeV}$. So far no evidence has been found for the existence of a cutoff indicating that p_{max} should be at least above $M_{\text{ew}} \sim 10^2\,\text{GeV}$, which is roughly the energy scale accessible to current accelerator experiments. Obviously, this value is still too high. The energy cutoff

Journal of Physics: Conference Series **259** (2010) 012083 doi:10.1088/1742-6596/259/1/012083

compatible with $\Lambda_{\rm obs}$ would be of the order of neutrino masses $p_{\rm max} \sim 10^{-3}\,$GeV, which is in clear contrast to quantum field theory tested up to the electro-weak scale $M_{\rm ew}$. However, it should be noted that vacuum energy[1] cannot be measured in the lab because it couples only to gravity. Therefore, a low UV cutoff just for the zero-point energy part could be allowed by observations, but it would be awkward from the theoretical point of view. Finally, using renormalisation the divergence in Eq. (1) can be formally removed, but finite terms remain typically of the order m^4 for fields with mass m. Despite the fact that we do not know how to handle the term in (1), there is no reason to believe that it is small or zero. Instead, it might be a large quantity according to the above considerations.

Next, a more quantitative CC contribution comes from the symmetry breaking of the electro-weak sector of the standard model. As the Higgs field acquires a non-zero vacuum expectation value in the broken phase the value of the potential in the new minimum is different from its minimal value in the unbroken phase. Thus, the phase transition induces a finite vacuum energy shift $\mathcal{O}(M_{\rm ew}^4)$ again much larger than $\Lambda_{\rm obs}$. Apart from this classical term, quantum corrections render the situation worse as new terms occur at every order in perturbation theory [2].

Another source of vacuum energy could be primordial inflation, where a large positive value of the inflaton potential drives the rapid accelerated expansion in the early universe. At the end of the inflationary period the inflaton scalar approaches the minimum of its potential, where the effective vacuum energy is much lower than before. However, this does not imply that it is zero or of the observed CC value, it could be much larger in magnitude and even negative. In that case the well known Big Bang evolution had stopped already far in the past.

It is unlikely that the sum Λ of all contributions mentioned above is of the order $\Lambda_{\rm obs}$, instead it is expected to be of the order of the largest contribution. To obtain consistency between the theoretical value Λ and the observed one, the easiest way is adding a counterterm $\Lambda_{\rm ct}$ which matches $(-\Lambda)$ to very high precision such that $\Lambda_{\rm ct} + \Lambda \approx \Lambda_{\rm obs}$. This procedure is considered to be highly unnatural since it requires an enormous amount of fine-tuning in $\Lambda_{\rm ct}$, e.g. fixing up to 122 decimal digits.

Motivated by the problems of understanding the zero-point energy in Eq. (1), it has become popular to "replace" the CC by dynamical DE. Note that in general this does not address the old CC problem. Instead it introduces more steps. First, any CC originating e.g. from the sources mentioned above has to be removed somehow. Clearly, without further explanation it means that a fine-tuned counterterm is added which cancels Λ. This step is often suppressed in the description of DE models. Second, a new dynamical source is introduced, which drives the late-time accelerated expansion [3]. Obviously, the second step requires the first one but not vice-verse. The reason is that making the CC vanish or much smaller than $\Lambda_{\rm obs}$ requires even more fine-tuning than adjusting it to match exactly the observed value. On the other hand, one could argue that a yet unknown theory of quantum gravity (or an undiscovered feature of the known candidates) makes the CC vanish and clears the path for dynamical DE models. However, this unknown but probably powerful theory could also arrange the CC to match its observed value and no additional dynamical DE component would be needed. Moreover, it seems that at the moment observations do not require a dynamical nature of DE [4]. Of course, the situation might change in the future when observational data becomes better and more precise. But even in that case the problem of getting rid of a large Λ persists and the CC as a static DE component is by no way ruled out. Accordingly, the relevant question about DE and the accelerated expansion is not whether there is either a static CC or a dynamical component, but if we need only a CC or a CC plus something dynamical. More complicated but completely analogous to dynamical DE components are modified gravity theories with built-in late-time acceleration [5] or unconventional cosmological models, where the observed acceleration is just

[1] We mean its absolute value, not finite differences related to the Casimir effect.

an apparent effect. It will be interesting to know whether nature exhibits more complexity than the simplest explanation. In the "worst" case we have to deal with a mixture of many sources for acceleration in addition to the old CC problem.

In the following we concentrate on the latter problem, i.e. we accept the existence of a presumed large value $|\Lambda| \gg \Lambda_{\rm obs}$ as a part of the energy content of the universe. Conventional energy sources like dust and radiation dilute with the expansion of the universe, and without counter-measures the large CC term eventually starts dominating the cosmic expansion very early in the cosmic history. As a result, $\Lambda < 0$ initiates a collapse and the universe dies in a Big Crunch. For $\Lambda > 0$ an inflationary epoch with a large expansion rate H starts and continues forever because the constant Λ does not offer a graceful exit. In order to avoid these catastrophic scenarios and to permit a reasonable Big Bang-style universe, we have to neutralise the effect of Λ. Of course, we want to avoid the introduction of a fine-tuned counterterm, instead we look for a dynamical mechanism or a suitable modification of gravity, which tames the large CC term. Here, we discuss three recent approaches in this context. A relaxation model with a variable CC (including the large term Λ), a similar model implemented as modified gravity in the metric formalism, and finally a CC filter in a setup using the Palatini formalism. All these models have in common that they do not need a fine-tuned counterterm despite the existence of a large CC.

2. Relaxing a large CC

Let us begin with a simple example of the relaxation mechanism. To neutralise the effect of the large CC we introduce an additional component of opposite sign, which adjusts to the value of Λ dynamically. In our first model the total DE density is given by $\rho_\Lambda = \Lambda + \beta/f$ with a constant parameter β and a function f. According to the Friedmann equation the Hubble expansion rate H is sensitive to the sum of all energy densities

$$\rho_c = \frac{3H^2}{8\pi G_N} = \rho_m + \rho_\Lambda = \rho_m + \Lambda + \frac{\beta}{f}, \tag{2}$$

where ρ_c is the critical energy density, G_N is Newton's constant, and ρ_m denotes the energy density of matter and radiation. Without the β/f term the constant Λ would control the expansion when ρ_m becomes sufficiently small. However, we want to have a well-behaved late-time evolution with H of the order of the currently observed Hubble rate H_0. For this purpose consider the dynamical term defined by $f = H^2$ and β having the opposite sign of Λ [6]. As a result of the small Hubble rate at late times only the term $(\Lambda + \beta/H^2)$ is relevant, which is obvious after dividing Eq. (2) by $|\Lambda| \gg \rho_c, \rho_m$,

$$1 + \frac{\beta}{\Lambda H^2} = \frac{\rho_c - \rho_m}{\Lambda}. \tag{3}$$

The right-hand side is much smaller than unity and can be safely neglected. Therefore, we find the consistent solution $H^2 \to H_e^2 = -\beta/\Lambda$ describing a de Sitter cosmos with a low Hubble rate H_e, which complies with current observations. Without the term β/f the expected value of H^2 would be proportional to Λ and thus very large. Here, H_e^2 is inversely proportional to the large CC indicating a low Hubble rate. Note also the absence of fine-tuning in the parameter β. Small changes in β only lead to small changes in H_e^2, hence, it is not necessary to fix a lot of decimal digits.

The example $f = H^2$ works well at late times when H is small such that the dynamical term β/H^2 becomes large and dynamically relaxes the large CC term. At earlier times in the cosmic history, the Hubble rate was much larger and we have to enhance the function f for the relaxation mechanism to work. First, consider the radiation era, where the scale factor $a \propto t^{1/2}$ scales like the square root of the cosmic time t and the deceleration variable $q = -\ddot{a}/(aH^2) \lesssim 1$ is very close to unity. Using this property the function $f \propto H^2(q-1)$ would be sufficiently small to allow $\beta/f \propto (q-1)^{-1}$ to counteract the large constant Λ in a similar way as at late times. Also here $\rho_c, \rho_m \ll |\Lambda|$ can be neglected in the Friedmann equation yielding $\beta/(H^2(q-1))+\Lambda = 0$. As a result, the large value of H in the radiation era implies $q \approx 1$ as the solution of this equation, which means that the cosmological evolution will be that of a radiation universe despite the large CC term. Moreover, since H decreases with time the deceleration q will slightly change its value but it stays very close to unity. It is clear that replacing $(q-1)$ in f by $(q-\frac{1}{2})$ yields a cosmological expansion behaviour with $q \approx \frac{1}{2}$ for large Hubble rates. Consequently, the effect of the CC is neutralised also in the matter era, where $q = \frac{1}{2}$ follows from the scale factor $a \propto t^{2/3}$.

In summary, a large CC can be relaxed in all major epochs of a Big Bang universe by constructing an adequate function f in Eq. (2). In Ref. [7] the following model was proposed,

$$f = \frac{4(2-q)}{(1-q)}\left(\frac{1}{2}-q\right)H^2 + y\,72(1+q^2)(1-q)H^6, \tag{4}$$

which allows the relaxation of the large CC in all cosmological epochs. The different powers of H in f ensures the correct temporal sequence. For very large H the last term $\propto H^6(1-q)$ is responsible for the radiation era, while the first term $\propto H^2(\frac{1}{2}-q)$ relaxes the CC in the matter and the final de Sitter eras. The parameter y is related to the radiation-matter transition. Note that Eq. (4) can be written in terms of the Ricci scalar $R = 6H^2(1-q)$, the squared Ricci tensor $Q = R_{ab}R^{ab} = 12H^4(q^2-q+1)$ and the squared Riemann tensor $T = R_{abcd}R^{abcd} = 12H^4(q^2+1)$, i.e. $f = (R^2 - Q + yR^2T)/R$. Moreover, the model features an implicit interaction with dark matter [8] as well as interesting tracking properties [7] and a reasonable evolution of perturbations [9].

3. The relaxation mechanism in modified gravity

In the previous section the large term Λ was enhanced directly by a dynamical term β/f, which provides a reasonable cosmological evolution. Now, we show that this relaxation mechanism can be implemented in an action functional for modified gravity. Consider the action

$$\mathcal{S} = \int d^4x\sqrt{|g|}\left[\frac{R}{16\pi G_N} + \mathcal{L}_{\text{mat}} - \Lambda - \beta\, F(R,G)\right], \tag{5}$$

which contains the Einstein-Hilbert term, the matter Lagrangian $\mathcal{L}_{\text{mat}}$, the large CC term Λ and the modified gravity functional $F(R,G)$ in terms of the Ricci scalar and the Gauß-Bonnet invariant $G = R^2 - 4Q + T$. As before β is a constant parameter. In the metric formalism the variation of $\mathcal{S}$ with respect to the metric g_{ab} yields the Einstein equations

$$G_{ab} = -8\pi G_N\left[T_{ab} + g_{ab}\Lambda + 2\beta E_{ab}\right], \tag{6}$$

where T_{ab} is the energy-stress tensor of matter, and E_{ab} emerges from the new term $F(R,G)$. On a flat Friedmann-Lemaître-Robertson-Walker (FLRW) background E_{ab} is completely defined by the effective energy density ρ_F and pressure p_F. One finds

$$\rho_F = 2\beta\left[\frac{1}{2}F - 3(\dot{H}+H^2)F^R + 3H\dot{F}^R - 12H^2(\dot{H}+H^2)F^G + 12H^3\dot{F}^G\right], \tag{7}$$

where $F^{R,G}$ correspond to partial derivatives of F with respect to R and G, respectively. Next, we consider the model from Ref. [10], where $F(R,G) = 1/B$ with $B := \frac{2}{3}R^2 + \frac{1}{2}G + (y\,R)^3$ is very similar to f in Eq. (4) if written in terms of H and q,

$$B = 24H^4\left(q-\frac{1}{2}\right)(q-2) + H^6\left[6y(1-q)\right]^3. \tag{8}$$

Since every derivative of F in Eq. (7) yields another factor B in the denominator of ρ_F we obtain $\rho_F = N/B^d$ with $d \geq 1$. N denotes terms involving H and its time derivatives. As a result, the structure of the energy density ρ_F is very similar to β/f from the previous section and it behaves in a similar way. ρ_F becomes sufficiently large in the radiation, matter and respectively the late-time DE eras because the terms $(q-1)$, $(q-\frac{1}{2})$ and H^4 in the denominator B are small in the corresponding epochs. Therefore, the Friedmann equation reduces to $\rho_F + \Lambda = 0$ with corrections much smaller than $|\Lambda|$. As before, this yields in a dynamical way a relaxed cosmological expansion behaviour despite the large term Λ. Note that the numerator N in ρ_F does not harm the working principle of the CC relaxation mechanism, but it provides new solutions for the late-time behaviour, e.g. accelerating power-law expansion and future singularities. We refer the reader to Ref. [2] for a detailed study of the above model and its generalisations.

4. Filtering out the CC in the Palatini formalism

In the previous section we discussed the CC relaxation mechanism via modified gravity in the metric formalism, where the connection Γ^a_{bc} is the Levi-Civita connection of the metric g_{ab}. Now, we modify gravity in the Palatini formalism in which g_{ab} and the connection Γ are treated independently. Recently, we have shown that this property allows the construction of a filter for the CC [11], which is based on the action

$$\mathcal{S} = \int d^4x\, \sqrt{|g|}\left[\frac{1}{2}f(R,Q) + \mathcal{L}_{\text{mat}}[g_{ab}] - \Lambda\right], \tag{9}$$

where $\mathcal{L}_{\text{mat}}$ does not depend on Γ. The Ricci scalar $R = g^{ab}R_{ab}$ and $Q = R^{ab}R_{ab}$ are quantities which depend on g_{ab} and Γ, while the Ricci tensor $R_{ab} = \Gamma^e_{ab,e} - \Gamma^e_{eb,a} + \Gamma^e_{ab}\Gamma^f_{fe} - \Gamma^e_{af}\Gamma^f_{eb}$ is defined only by Γ. Here, we assume that Γ and R_{ab} are symmetric, which is not the most general case [12]. The variation of $\mathcal{S}$ with respect to g_{ab} yields the Einstein equations

$$f^R R_{mn} + 2f^Q R_m^{\ a} R_{an} - \frac{1}{2}g_{mn}f = T_{mn}, \tag{10}$$

where we include the large CC Λ in the stress tensor T_{mn} in addition to ordinary matter with the energy density ρ and the pressure p. In the Palatini formalism the variation $\delta\mathcal{S}/\delta\Gamma^a_{bc} = 0$ provides the equation for the connection

$$\nabla_a[\Gamma]\left[\sqrt{|g|}(f^R g^{mn} + 2f^Q R^{mn})\right] = 0, \tag{11}$$

where ∇_a is the covariant derivative in terms of the yet unknown Γ. For solving this set of equations we used the method described in Ref. [13]. In the following we summarise the results for the CC filter model defined by the action functional

$$f(R,Q) = \kappa R + z \quad \text{with} \quad z := \beta\left(\frac{R^{\frac{2}{3}}}{B}\right)^m, \tag{12}$$

where κ, β are parameters and $m > 0$ is dimensionless. Here, the function $B := R^2 - Q$ is similar to f in Sec. 2, however, note that R, Q and B are in general different from their metric versions. From Eq. (10) we obtain two algebraic equations for the two unknowns R and B. First, its trace $(-2-\frac{4}{3}m)z = \kappa R - 4\Lambda + 3p - \rho \approx -4\Lambda$ tells us that z is approximately constant and of the order of the large CC. The second equation reads $\kappa R + r = \frac{2}{9}mz\frac{B}{R^2} + \epsilon$, where we introduced $r := \rho + p$. ϵ denotes suppressed corrections which are neglected in the following. Then, we combine both equations with z from Eq. (12) and find

$$(\kappa R + r)^3 (\kappa R)^4 = \rho_e^7 := \kappa^4\left(\frac{2}{9}mz\left(\frac{\beta}{z}\right)^{\frac{1}{m}}\right)^3, \tag{13}$$

indicating that the Ricci scalar R can be expressed only by the energy density ρ and the pressure p of matter in r and by the constant ρ_e. Moreover, we observe that vacuum energy terms with equation of state $p = -\rho$ do not contribute to R. Obviously, the large CC Λ is filtered out, it appears only in the constant ρ_e, which can be adjusted by the parameter β.

In the following we restrict the cosmological discussion to negative values of ρ_e. In the early universe, when $-\rho_e \ll r \ll |\Lambda|$ we find $\kappa R = -r + \mathcal{O}(\epsilon)$ from Eq. (13). This implies that the parameter κ is negative since $r > 0$ for ordinary matter. At late times, when $r \ll -\rho_e$, the Ricci scalar becomes constant, $\kappa R = \rho_e - \frac{3}{7}r + \mathcal{O}(\epsilon)$. Next, let us express the matter energy density by a power-law in the cosmic scale factor, $\rho \propto a^{-3(\omega+1)}$, with the matter equation of state ω. Thus, R and B can be expressed in terms of $a(t)$, which allows calculating the Palatini connection Γ via Eq. (11) as a function of $a(t)$ and its derivatives. From Γ the Ricci tensor $R_{ab}[\Gamma]$ can be determined and the corresponding Ricci scalar $R = g^{ab}R_{ab}[\Gamma]$ must be equal to $R(r)$, which we found from Eq. (13). With the ansatz $a(t) \propto t^s$ for the scale factor we obtain the following equations for the Hubble rate H. In the era where r is dominated by dust matter with $\omega = 0$ we find $(-\kappa)H^2 = \frac{4}{27}\rho_{\text{dust}}$, whereas in the radiation dominated epoch with $\omega = \frac{1}{3}$ we obtain $(-\kappa)H^2 = \frac{4}{25}\rho_{\text{radiation}}$. Obviously, these results are very close to general relativity when the parameter $\kappa \sim 1/G_N$ is suitably chosen. Moreover, at late times we find the de Sitter solution $\kappa R = \kappa(3H^2) = \rho_e$, where the late-time Hubble rate can be adjusted by the parameter β in (13). As in the previous sections it is not necessary to fine-tune its value. Note in the solutions above the cosmic expansion is controlled by matter only, but not by the large CC term, which has been filtered out. It can be shown that the CC filter is active in black hole-like environments, too [11]. As a closing remark, our setup in the Palatini formalism leads to second order equations of motion just as general relativity.

5. Conclusions

We have discussed several arguments for the existence of a presumed large CC, which has to be neutralised to permit a reasonable evolution of the universe. We have also argued that naively replacing the cosmological term by other sources for late-time acceleration corresponds to the introduction of a fine-tuned counterterm. As an alternative to this method we have reviewed three recent approaches to relax the cosmic expansion in the presence of a large CC.

Acknowledgments

This work has been supported by DIUE/CUR Generalitat de Catalunya under project 2009SGR502, by MEC and FEDER under project FPA2007-66665 and by the Consolider-Ingenio 2010 program CPAN CSD2007-00042.

References

[1] Weinberg S 1989 *Rev. Mod. Phys.* **61** 1–23
[2] Bauer F, Sola J and Stefancic H 2010 (*Preprint* `1006.3944`)
[3] Copeland E J, Sami M and Tsujikawa S 2006 *Int. J. Mod. Phys.* **D15** 1753–1936 (*Preprint* `hep-th/0603057`)
[4] Amanullah R *et al.* 2010 *Astrophys. J.* **716** 712–738 (*Preprint* `1004.1711`)
[5] Nojiri S and Odintsov S D 2006 *ECONF* **C0602061** 06 (*Preprint* `hep-th/0601213`)
[6] Stefancic H 2009 *Phys. Lett.* **B670** 246–253 (*Preprint* `0807.3692`)
[7] Bauer F, Sola J and Stefancic H 2009 *Phys. Lett.* **B678** 427–433 (*Preprint* `0902.2215`)
[8] Grande J, Sola J and Stefancic H 2006 *JCAP* **0608** 011 (*Preprint* `gr-qc/0604057`)
[9] Bauer F 2010 *Class. Quant. Grav.* **27** 055001 (*Preprint* `0909.2237`)
[10] Bauer F, Sola J and Stefancic H 2010 *Phys. Lett.* **B688** 269–272 (*Preprint* `0912.0677`)
[11] Bauer F 2010 (*Preprint* `1007.2546`)
[12] Vitagliano V, Sotiriou T P and Liberati S 2010 (*Preprint* `1007.3937, 1008.0171`)
[13] Olmo G J, Sanchis-Alepuz H and Tripathi S 2009 *Phys. Rev.* **D80** 024013 (*Preprint* `0907.2787`)

16th International Symposium on Particles Strings and Cosmology (PASCOS 2010) IOP Publishing
Journal of Physics: Conference Series **259** (2010) 012084 doi:10.1088/1742-6596/259/1/012084

Low redshift probes and coupled dark matter-dark energy models

Olga Mena
Instituto de Fisica Corpuscular, IFIC, CSIC and Universidad de Valencia, Spain
E-mail: omena@ific.uv.es

Abstract. Coupled cosmologies, in order to satisfy CMB constraints, predict values for the cosmological parameters today which may differ substantially from the parameters values within non-interacting cosmologies. In order to fit high-precision CMB data available today, coupled cosmologies can *hide* their effects at very low redshifts. Therefore, low redshift probes are highly complementary and thus powerful to constrain interacting dark sector models. In this *talk* we focus on near-universe, low-redshift constraints in a variety of coupled dark matter-dark energy models. We find that current data constrain the dimensionless coupling to be $|\xi| < 0.2$. We also explore the forecasts for future low redshift probes.

1. Introduction

Cosmological probes indicate that the universe we observe today possesses a flat geometry and a mass energy density made of $\sim 30\%$ baryonic plus cold dark matter and 70% dark energy, responsible for the late-time accelerated expansion. While the ΛCDM model (a flat universe with a cosmological constant) provides a well fit to current observational data, there exist also dynamical options, as the quintessence fluid, in which a cosmic scalar field is slowly approaching its ground state. This quintessence field, in principle, may couple to other fields in nature. Observations strongly constrain the couplings to ordinary matter [1]. However, interactions within the dark sectors, i.e. between dark matter and dark energy, are still allowed by observations [2, 3, 4, 5, 6, 7, 8, 9, 10, 11, 12, 13, 14]. In this *talk*, based on a recent work [15], we explore the phenomenology of coupled models and consider what type of low-redshift observations are most suitable to improve present constraints.

1.1. Classification of dark coupled models

At the level of the stress-energy tensor it is always possible to introduce an interaction between the fluids of the dark sector in the following way [16]:

$$\nabla_\mu T^\mu_{(\mathrm{dm})\nu} = Q_\nu \quad \text{and} \quad \nabla_\mu T^\mu_{(\mathrm{de})\nu} = -Q_\nu. \tag{1}$$

The 4-vector Q_ν governs the energy-momentum transfer between the dark components and $T^\mu_{(\mathrm{dm})\nu}$ and $T^\mu_{(\mathrm{de})\nu}$ are the energy-momentum tensors for the dark matter and dark energy fluids, respectively. In the first family of models we consider [15] (DEvel), the momentum exchange Q_ν is parallel to the dark energy four velocity $u_\nu^{(\mathrm{de})}$. In the second family of models (DMvel), Q_ν is parallel to the dark matter four velocity $u_\nu^{(\mathrm{dm})}$ [15]. DEvel models include all quintessence coupled models and are effectively "modified gravity" models, implying the presence of a "fifth force" effect (only for the dark matter), that is, a violation of the equivalence principle. For both DEvel and DMvel models the evolution equations for the dark matter and dark energy background energy densities reduce to:

$$\dot{\rho}_{\rm dm} + 3\mathcal{H}\rho_{\rm dm} = Q\,, \tag{2}$$

$$\dot{\rho}_{\rm de} + 3\mathcal{H}\rho_{\rm de}(1+w) = -Q\,. \tag{3}$$

where the dot indicates derivative with respect to conformal time $d\tau = dt/a$, $\mathcal{H} = \dot{a}/a$ and w is the dark-energy equation of state. For $Q > 0$, the energy flows from the dark energy system to dark matter one. For $Q < 0$, the energy flow is reversed. For each of the two families DEvel and DMvel, we propose the definition of two sub-classes of models. In class I models, $Q \propto \rho_{\rm dm}$, while in class II models, $Q \propto \rho_{\rm de}$.

Notice that the expansion history does not depend on the choice DEvel or DMvel. However, even if DMvel and DEvel models provide the same background history, the perturbation evolution is dramatically different. Therefore, while geometrical probes alone are unable to distinguish among the two of them, growth of structure will make these two models fundamentally different. In the Newtonian gauge and neglecting dark energy perturbations, we obtain, for the dark matter perturbations, at linear order

$$\dot{\delta}_{\rm dm} = -(\theta_{\rm dm} - 3\dot{\Phi}) + \frac{Q}{\rho_{\rm dm}}\left(\frac{\delta Q}{Q} - \delta_{\rm dm} + \Psi\right)\,, \tag{4}$$

$$\dot{\theta}_{\rm dm} = -\mathcal{H}\theta_{\rm dm} + (b-1)\frac{Q}{\rho_{\rm dm}}\theta_{\rm dm} + k^2\Psi\,, \tag{5}$$

see [15], where $b = 0$ refers to DEvel models and $b = 1$ refers to DMvel models. Notice that the Euler equation Eq. (5) for the dark matter fluid is *only* modified in the first family of models (DEvel) considered here, where $Q_\nu \propto u_\nu^{(\rm de)}$, violating therefore the weak equivalence principle. Another aspect of coupled models is that they can show non adiabatic, early time instabilities [7, 8, 17, 10, 9, 18, 13] due to the dark coupling term which appears in the dark energy pressure perturbations. In the following, we shall restrict our analysis to coupled models which satisfy the stability criterion of Ref. [10] and therefore are free of early-time, non adiabatic instabilities.

2. Dark coupled models versus $w(z)$ cosmologies

We study uncoupled models with a dynamical dark energy component $\tilde{w}(z)$ versus interacting DMvel and DEvel class II cosmologies with $Q = \xi\mathcal{H}\rho_{\rm de}$ (with $\xi < 0$, in order to avoid early-time instabilities). We investigate how to distinguish among them, even if the background evolution in these cosmologies is identical.

Figure 1 from Ref. [15] illustrates the values of $\Omega^{(0)}_{\rm dm}h^2$ necessary to fit WMAP 5 year angular diameter distance data, i.e. the first acoustic peak position [19], as a function of the dimensionless coupling constant ξ. We analyze now the linear growth in the three possible cosmologies that lie along the one-dimensional degeneracy defined by Fig. 1. The three possible cosmologies are: *a)* an uncoupled, albeit dynamical dark energy cosmology, with a varying equation of state $\tilde{w}(z)$, *b)* the coupled class II ($Q \propto \rho_{\rm de}$) DMvel model with constant $w = -0.9$ and *c)* the coupled class II DEvel model. While in the DMvel model the baryon densities and velocities nearly trace those of the dark matter, this is not the case for the DEvel model. We consider observables probed by lensing ($\propto \Omega_{\rm dm}\delta_{\rm dm}$, if one ignores geometrical factors) and linear peculiar velocities ($\propto \theta_{\rm dm}$). Figure 2, from Ref. [15], shows the ratio of the values of $\Omega_{\rm dm}\delta_{\rm dm}$ and $\theta_{\rm dm}$ in cases *b)* and *c)* to their values in an uncoupled model with the same background expansion history. In both the

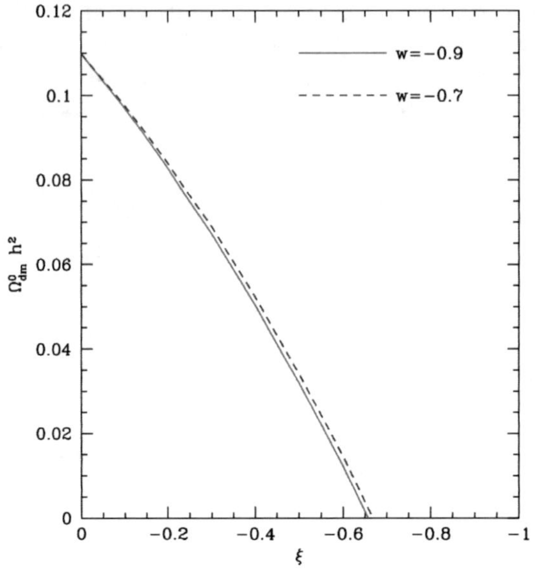

Figure 1. Extracted from Ref. [15], see text for details.

DMvel and the DEvel models the dark coupling can, in principle, be distinguished from a generic, uncoupled, dark energy described by $\tilde{w}(z)$ with perturbations growing as in standard GR. Note however that in the DMvel model GR is not modified. Although extremely small in this example, this class of models displays a mis-match between the reconstructed expansion history and the growth of structures, representing an exception to the commonly accepted interpretation that such a mis-match would be a tell-tale sign of modification of GR.

2.1. Matter abundance in voids

The matter content in voids has received recent attention (see e.g. Refs. [20, 21] and references therein) because it seems that the voids are more empty than expected from ΛCDM model predictions. Reference [21] argue that in the local volume, at a distance from us of about 10 Mpc, there are too few small galaxies with circular velocities V_c below $\sim$ 35 km/s. They estimate a factor 10 discrepancy with the ΛCDM model predictions. Coupled models in which a given amount of dark matter per unit volume turns into dark energy would provide a mechanism to clear dark matter (and galaxies) from the voids. On the other hand, the void phenomenon disfavors coupled models in which the dark matter content per unit volume is enhanced. We use [15] recent void results to derive constraints on DMvel and DEvel class II models where $Q = \xi H \rho_{\rm de}$. Figure 1 shows the current cold dark matter energy density in the universe as a function of the coupling for the former dark sector interaction necessary to fit WMAP 5 year data [19]. Notice that from the figure, $\xi > -0.6$ in order to not to have voids completely empty of matter. It is possible to improve this first limit if we assume that dark matter halos associated to galaxies with circular velocities below 35 km/s are depleted by a factor of 10. After transforming the velocity function into a mass function using spherical collapse, we find that the total dark matter mass in voids can at most be depleted by 20% [15]. This translates into a lower bound for the coupling of $\xi = -0.2$, see Fig. 1.

Journal of Physics: Conference Series **259** (2010) 012084 doi:10.1088/1742-6596/259/1/012084

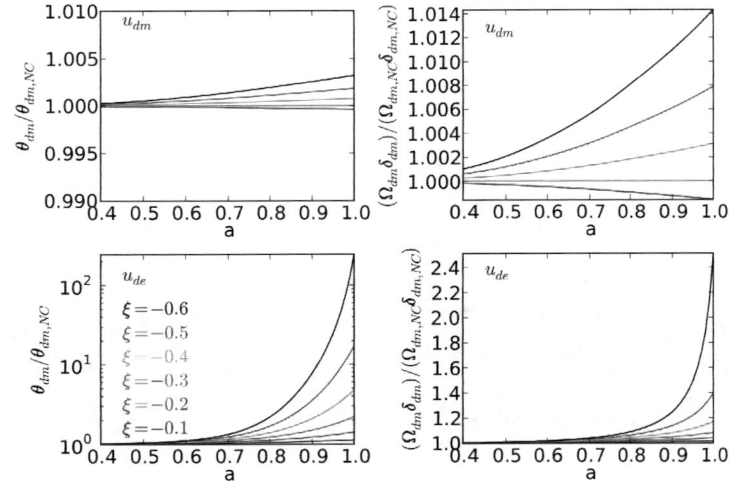

Figure 2. (From Ref. [15]) The ratio of the amplitude of dark matter peculiar velocities, $\theta_{\rm dm}$, and the lensing signal, $\Omega_{\rm dm}\delta_{\rm dm}$ in coupled models compared to models with standard GR growth with identical expansion histories (labelled with NC for "Not Coupled"). The curves show $\xi = 0, -0.1, \ldots - 0.6$.

2.2. Constraints from the peculiar velocity field

The velocity field offers a test of the growth of structure, which is complementary to galaxy clustering, as, for example, is less sensitive to non-linearities and bias. Below we consider constraints that can be obtained from present bulk flow data in the local universe and forecasted constraints achievable from future galaxy redshift surveys. Recently Watkins *et al.* [22] have reported the observation of anomalously large bulk flows on 100 h^{-1} Mpc scales. In a Gaussian window of radius $50h^{-1}$ Mpc, they find a coherent bulk motion of 407 ± 81 km/s in conflict with the ΛCDM expectation of ~ 200 km/s at the 2σ level. Reference [23] pointed out that these results, if confirmed, would favour models with a growth of perturbations larger than ΛCDM model predictions. In Ref. [15] we study the bulk flow results for the DMvel and DEvel class II models ($Q \propto \rho_{\rm de}$) as a function of the coupling ξ. While the DMvel model has dark matter peculiar velocities which are similar to those of non interacting models (the Euler equation is unmodified in the DMvel case), for the DEvel model values of the coupling $\xi < -0.35$ lead to an effective 3 σ level deviation from observations, assuming that the galaxy peculiar velocities trace the dark matter peculiar velocity field, see Ref. [15] for a detailed discussion of the topic.

A different velocity related probe arises from redshift space distortions [24, 25], which provide a measurement of the growth of structure through the quantity $\tilde{f}\sigma_8$, being $\tilde{f}$ the logarithmic derivative of the dark matter growth factor and σ_8 the rms matter fluctuations in spheres of 8 h^{-1} Mpc. Figure 3 from Ref. [15] illustrates the current and future constraining power of this method, where the red crosses show the $\tilde{f}\sigma_8$ predictions for the DMvel class I coupled model of Ref. [12]. The blue triangles (squares) refer to the predictions of the DMvel (DEvel) class II model explored here $Q = \xi\mathcal{H}\rho_{\rm de}$, with $\xi = -0.5$ and $w = -0.9$. The black circles in the first panel of Fig. 3 represent current $\tilde{f}\sigma_8$ measurements [25], in the other two panels they show the $\tilde{f}\sigma_8$ forecasts for BOSS and EUCLID-like galaxy surveys (which also include future Planck CMB experiment information, see Ref. [15] for more details), assuming a ΛCDM fiducial cosmology. Notice that the $\tilde{f}\sigma_8$ values for the DMvel class I model are smaller than current observations. However, current errors on $f\sigma_8$ measurements are large and do not rule out these models at a

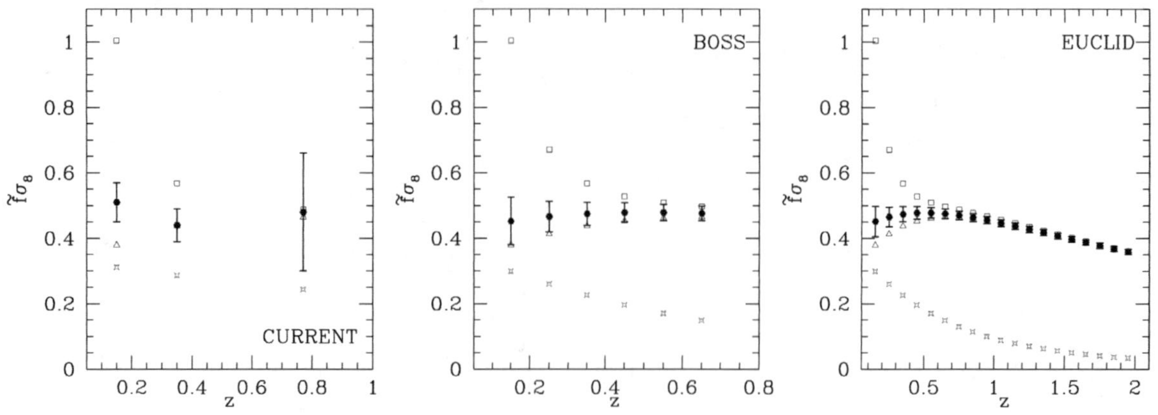

Figure 3. Extracted From Ref. [15], see text for details.

high significance level. While the predictions for the DMvel class II model are very close to those of a non interacting model, DEvel class II models will lead to huge values of $\tilde{f}\sigma_8$ especially at low redshifts, when dark energy starts to be dominant and the effect of the energy-momentum transfer starts to be important in the dark matter velocity differential equation. Such an effect can not be obtained in the two DMvel class I and class II models explored here because the Euler equation for the dark matter velocity is not modified in these interacting cosmologies. In the DEvel class II model, couplings $\xi < -0.4$ will be ruled out at more than 3σ with existing $\tilde{f}\sigma_8$ data. Accurate measurements of $\tilde{f}\sigma_8$ from future BAO surveys as BOSS or EUCLID may rule out significantly the form of couplings explored above, see the central and right panels of Fig. 3.

2.3. Weak equivalence principle constraints

Kesden and Kamionkowski (K&K) [26, 27], analyzed the consequences of WEP violation for dark-matter on galactic scales, focusing on dark-matter dominated satellite galaxies orbiting much larger host galaxies. They inferred that a difference among dark matter and baryonic accelerations larger than 10% is severely disfavoured. In Ref. [15] we have shown that for the DEvel model $Q_\nu = \xi H \rho_{\rm de} u_\nu^{\rm de}$ couplings ξ smaller than ~ -0.01 are strongly disfavoured by K&K numerical analysis.

3. Summary

We summarize the classification and current constraints on dark coupling models in Fig. 4, where d_L denotes the luminosity distance as obtained from e.g. Supernovae Ia observations and D_a denotes the angular diameter distance as measured e.g., by angular BAO.

Acknowledgements

It is a pleasure to thank R. Jimenez, L. Lopez-Honorez, B. Reid and L. Verde, authors of the original work on which the present talk is based. O. M. work is supported by a MICINN Ramón y Cajal contract, AYA2008-03531 and CSD2007-00060.

References

[1] Carroll S M 1998 *Phys. Rev. Lett.* **81** 3067–3070 (*Preprint* astro-ph/9806099)
[2] Amendola L 2000 *Phys. Rev.* **D62** 043511 (*Preprint* astro-ph/9908023)
[3] Amendola L 2000 *Mon. Not. Roy. Astron. Soc.* **312** 521 (*Preprint* astro-ph/9906073)

DEvel		DMvel		
~~GR~~ WEPV		GR		
$\propto \rho_{\rm dm}$	$\propto \rho_{\rm de}$	$\propto \rho_{\rm dm}$	$\propto \rho_{\rm de}$	
				Identity in background
Euler eq.	Euler and continuity eqs	background only	continuity eq.	Modifications compared to LCDM
Coupled quintessence	$\xi < 0$ $w_{\rm const.} > -1$	$w_{\rm const.} < -1$ or $w(z)$	$\xi < 0$ $w_{\rm const.} > -1$	Ensure stability
yes	yes	tiny	v. small	Mis-match Expansion/growth
WEP tests	Pec. Velocities WEP tests voids	H(z), dL, Da ≡w(z) fluid model	Pec. velocities voids	Key observables
$\lvert\beta\rvert < 0.2$	$-0.2 < \xi < 0$	$-0.3 < \frac{\Gamma}{H_0} < 0.15$	$-0.2 < \xi < 0$	Current constraints

Figure 4. Extracted from Ref. [15].

[4] Amendola L 1999 *Phys. Rev.* **D60** 043501 (*Preprint* astro-ph/9904120)
[5] Amendola L and Tocchini-Valentini D 2001 *Phys. Rev.* **D64** 043509 (*Preprint* astro-ph/0011243)
[6] Amendola L 2004 *Phys. Rev.* **D69** 103524 (*Preprint* astro-ph/0311175)
[7] Valiviita J, Majerotto E and Maartens R 2008 *JCAP* **0807** 020 (*Preprint* 0804.0232)
[8] He J H, Wang B and Abdalla E 2009 *Phys. Lett.* **B671** 139–145 (*Preprint* 0807.3471)
[9] Jackson B M, Taylor A and Berera A 2009 *Phys. Rev.* **D79** 043526 (*Preprint* 0901.3272)
[10] Gavela M B, Hernandez D, Honorez L L, Mena O and Rigolin S 2009 *JCAP* **0907** 034 (*Preprint* 0901.1611)
[11] Caldera-Cabral G, Maartens R and Schaefer B M 2009 *JCAP* **0907** 027 (*Preprint* 0905.0492)
[12] Valiviita J, Maartens R and Majerotto E 2009 (*Preprint* 0907.4987)
[13] Majerotto E, Valiviita J and Maartens R 2009 (*Preprint* 0907.4981)
[14] Gavela M B, Lopez-Honorez L, Mena O and Rigolin S 2010 (*Preprint* 1005.0295)
[15] Honorez L L, Reid B A, Mena O, Verde L and Jimenez R 2010 (*Preprint* 1006.0877)
[16] Kodama H and Sasaki M 1984 *Prog. Theor. Phys. Suppl.* **78** 1–166
[17] Chongchitnan S 2009 *Phys. Rev.* **D79** 043522 (*Preprint* 0810.5411)
[18] Corasaniti P S 2008 *Phys. Rev.* **D78** 083538 (*Preprint* 0808.1646)
[19] Komatsu E *et al.* (WMAP) 2009 *Astrophys. J. Suppl.* **180** 330–376 (*Preprint* 0803.0547)
[20] Peebles P J E 2001 *Astrophys. J.* **557** 495–504 (*Preprint* astro-ph/0101127)
[21] Tikhonov A and Klypin A 2008 (*Preprint* 0807.0924)
[22] Watkins R, Feldman H A and Hudson M J 2008 (*Preprint* 0809.4041)
[23] Ayaita Y, Weber M and Wetterich C 2009 (*Preprint* 0908.2903)
[24] Kaiser N 1987 *Mon. Not. Roy. Astron. Soc.* **227** 1–27
[25] Song Y S and Percival W J 2008 (*Preprint* 0807.0810)
[26] Kesden M and Kamionkowski M 2006 *Phys. Rev. Lett.* **97** 131303 (*Preprint* astro-ph/0606566)
[27] Kesden M and Kamionkowski M 2006 *Phys. Rev.* **D74** 083007 (*Preprint* astro-ph/0608095)

Journal of Physics: Conference Series **259** (2010) 012085 doi:10.1088/1742-6596/259/1/012085

A strong first order phase transition in the UMSSM

Amine Ahriche
Laboratory of Theoretical Physics, Department of Physics, University of Jijel, PB 98 Ouled Aissa, DZ-18000 Jijel, Algeria.
E-mail: ahriche@univ-jijel.dz

Abstract. In this work, the electroweak phase transition (EWPT) strength has been investigated within the $U(1)$ extended Minimal Supersymmetric Standard Model (UMSSM) without introducing any exotic fields. We found that the EWPT could be strongly first order for reasonable values of the lightest Higgs and neutralino masses.

1. Introduction

The origin of matter-antimatter asymmetry is one of the main problems of both particle physics and cosmology. The explanation of this asymmetry $n_b/n_\gamma \sim 10^{-10}$ [1], requires the Sakharov criteria, which can be summarized in the existence of such interactions in the early universe that violate the baryon number B, the symmetries C and CP and occur out of equilibrium. It appears that the Standard Model (SM) fulfills all these criteria; the baryon number is not conserved at quantum level due to the $B + L$ anomaly [2], a CP violation source does exist in CKM matrix, and a departure from thermal equilibrium could be reached through a strong first order phase transition [3], but its realization was not possible numerically due to smallness of CP violation effect and the weakness of the electroweak phase transition (EWPT) [4]. In the SM, the EWPT is so weak unless the Higgs mass is less than 45 GeV [5], which is in conflict with present data [6], but it could be strong within some SM extensions (for e.g SM with a singlet [7]).

In spite of its success and popularity, the MSSM with R-parity still has two major problems: the μ-problem [8] and the potential proton decay problem due to dimension 5 operators [9]. A natural solution to these problems would probably require that the MSSM be extended by a new mechanism or a new symmetry. The $U(1)'$-extended MSSM (UMSSM) [10] is a straightforward extension of the MSSM with a non-anomalous TeV scale Abelian gauge symmetry. In the minimal supersymmetric standard model (MSSM), the EWPT could be strongly first order if the light stop is lighter than the top quark [11], and in the UMSSM, it is also strongly first order but with the price of introducing 3 new extra singlet scalars [12], or by adding new extra heavy singlet fermions [13]. In this work, we will investigate the possibility of getting a strong first order phase transition within the minimal gauge extension of the MSSM, UMSSM without adding any new field beside the usual singlet.

This paper is organized as follow: in the second section, we give a brief review of the UMSSM model, define the scalar potential and discuss different constraints on the parameters. After that, we discuss the EWPT dynamics and show how to get a first order phase transition, and discuss our numerical results. Finally, we summarize our results.

2. The UMSSM model

In the $U(1)'$MSSM (or UMSSM), the gauge group is extended to $G = SU(3)_c \times SU(2)_L \times U(1)_Y \times U(1)'$ with the couplings g_3, g_2, g_1 and g', respectively [10]. The superpotential is given by

$$W = \lambda S \epsilon_{ij} H_1^i H_2^j + Y_U \epsilon_{ij} Q^i U^c H_2^j + Y_D \epsilon_{ij} Q^i D^c H_1^j + Y_L \epsilon_{ij} L^i E^c H_1^j, \tag{1}$$

where ϵ_{ij} is the anti-symmetry 2×2 tensor, Y_U, Y_D and Y_L are Yukawa couplings, and λ is a coupling constant in which $\lambda \langle S \rangle$ replaces the μ-term in the MSSM. The particle content, here, is the same as the MSSM in addition to a new singlet and gauge boson; and the superpartners.

2.1. The scalar potential

In case where both sneutrinos or/and squarks do not develop vevs, the scalar potential is the combination of the so-called D, F and soft terms, which are given by

$$\begin{aligned} V_D &= \frac{g_2^2 + g_1^2}{8}\left(H_2^+ H_2 - H_1^+ H_1\right)^2 + \frac{g_2^2}{2}\left|H_1^+ H_2\right|^2 + \frac{g'^2}{2}\left|Q_1 H_1^+ H_1 + Q_2 H_2^+ H_2 + Q_S |S|^2\right|^2, \\ V_F &= |\lambda|^2 \left\{\left|\epsilon_{ij} H_1^i H_2^j\right|^2 + |S|^2 \left[H_1^+ H_1 + H_2^+ H_2\right]\right\} \\ V_{soft} &= m_{H_1}^2 H_1^+ H_1 + m_{H_2}^2 H_2^+ H_2 + m_S^2 |S|^2 + \left\{A_\lambda S \epsilon_{ij} H_1^i H_2^j + h.c\right\}. \end{aligned} \tag{2}$$

Here $m_{H_1}^2$, $m_{H_2}^2$, m_S^2 and A_λ are usually called soft parameters. The charges Q's should be chosen in a way that ensures the anomaly cancelations. The structure of the tree-level potential seems to allow two explicit CP violating relative phases between the scalar vevs, but these phases could be canceled by such gauge transformation, and the ground state is independent of any CP violations phases. However, these relative phases could appear at one loop through the superpartner masses (see Appendix A in [14]).

In this setup, the scalar potential (2) may admit another minimum $(0, 0, x)$, This wrong vacuum could play a very important role in the EWPT dynamics.

2.2. The parameters

In this model, we have many parameters, some of them are free like: g', λ, v_x, $\tan\beta = v_1/v_2$, and the soft terms: m_Q, m_U, A_t, A_λ, M_2, M_1 and M_1'; and others are fixed by such measured physical quantities: g_1, g_2, v and y_t; or such conditions like the elimination m_1^2, m_2^2 and m_S^2 by imposing (v_1, v_2, v_x) to be the absolute minimum of the effective potential at zero temperature. In addition to the constrains that are coming from the extra $U(1)'$ gauge interactions, like mixing between the neutral gauge boson Z and the new one Z' [15]

$$2M_{ZZ'}^2/(M_{Z'Z'}^2 - M_{ZZ}^2) < 10^{-3}, \tag{3}$$

and the bound on the heavy Z' mass [16]

$$M_{Z'} > (500 - 800)\ GeV. \tag{4}$$

Keeping the new charges as free parameters beside the condition $Q_1 + Q_2 + Q_s = 0$, we take random values for the free parameters, taking into account such conditions like the elimination of m^2's parameters in (2), the perturbativity of the quartic couplings in (2), and the vacuum stability. Another constrain could be derived from the upper bound on the mixing between the gauge boson Z and the new one Z' (3), and the lower bound on the new gauge boson Z' mass (4). The condition (4), could be achieved by considering relatively large v_x, or large $g'Q'$s. The condition (3) could be fulfilled by vanishing the mixing term $M_{ZZ'}^2$, i.e., if

$$Q_1 = Q_2 \tan^2\beta, \tag{5}$$

or if $M^2_{Z'Z'} >> M^2_{ZZ}, M^2_{ZZ'}$, which roughly means

$$g' \left|Q_S\right| v_x \gtrsim (500 - 800)\ GeV, \tag{6}$$

or a serious tuning in the values of $Q_{1,2}$ and $tan\beta$.

In our search for the parameter's space that fulfills the strong first-order phase transition criterion: $v(T_c)/T_c > 1$, we will focus our search on two regions:

a) Moderate values for the parameters $Q_{1,2}$ and $tan\beta$, where (3) is nearly satisfied. In this case, the singlet vev v_x, could be the order v or even relatively smaller.

b) The two terms M^2_{ZZ} and $M^2_{ZZ'}$ in the mass-squared matrix of (Z, Z'), should be suppressed with respect to the mass term $M^2_{Z'Z'}$. In this case, the values of $U'(1)$ charge and the vev of the singlet, Q_s and v_x, should be large enough (6).

The mass parameters: m_Q, m_U, A_t, M_2, M_1 and M'_1 appear at one-loop level in the effective potential, therefore we expect that their role is less important in the EWPT dynamics. But each of these parameters: g', λ, v_x and $\tan\beta$, as well the charges Q's that appear multiplied by g', seems to be very important, therefore we focus on these parameters while fixing the mass parameters: $g' = g_1$, $m_Q = m_U = 1\ TeV$, and giving different values for the rest A_t, M_2, M_1 and M'_1.

3. The electroweak phase transition

3.1. The phase dynamics

Due the condition (3) and (4), there could exist a hierarchy between the vev of the singlet and those of the doublets, i.e., $v_x >> v_{1,2}$. In this case, the gauge symmetry could be broken in two step: $(0,0,0) \overset{T'_c}{\to} (0,0,x) \overset{T_c}{\to} (v_1, v_2, v_x)$, or just in once $(0,0,0) \overset{T_c}{\to} (v_1, v_2, v_x)$. Since the singlet dynamics does not affect the $SU(2)$ sphaleron processes, we will not be interested in distinguishing between the one- and two-steps symmetry breaking. We will treat our field dynamics using the effective potential where the singlet is replaced by it thermal vev. At the critical temperature T_c, the two minima get degenerate

$$V_{eff}\left(v_1^c, v_2^c, v_x^c, T_c\right) = V_{eff}\left(0, 0, x_c, T_c\right). \tag{7}$$

Below this temperature, the new minimum becomes the absolute one, and the system has to move from the old (false) vacuum to the new (true) one. In the case where a barrier does exist between the two minima, this transition has to occur via tunneling trough bubbles nucleation at certain points, which expand and fill the whole space by the new vacuum $v_{1,2}(T) \neq 0$, i.e, the symmetry is broken.

The $B+L$ anomalous interactions [2], that violate the baryon number have not the same rate in the symmetric and broken phases (i.e., at both sides of the bubble wall). In the symmetric phase, this rate behaves like $\sim T^4$ [17], and suppressed as $exp(-E_{Sp}/T)$ [18], in the broken phase, where E_{Sp} is the system static energy within such field configuration called the sphaleron [20]. Therefore any generated baryon number at the symmetric phase will erase at the broken phase, unless these interactions are switched off at the broken phase, this is translated to the famous criterion [19].

Here, we are interested in the in the value of the EW vev at the transition temperature, and since the singlet field does not play an important role in the sphaleron processes [7], the criterion of a strong first order phase transition in our case is given by [19]

$$v\left(T_c\right)/T_c \equiv \sqrt{v_1^2\left(T_c\right) + v_2^2\left(T_c\right)}/T_c > 1. \tag{8}$$

In the general case where the relative phases $\theta_{1,2} \neq 0$, the field ground state should be written as $\{v_i\}_{i=1,5} = (v_1,\ v_2\cos\theta_1,\ v_2\sin\theta_1,\ v_x\cos\theta_2,\ v_x\sin\theta_2)$ instead of (v_1, v_2, v_x) and the two relative phases. These 5 variables should be treated independently when looking for $v_{1,2,x}$ and $\theta_{1,2}$ at any temperature T. Then the phase transition could be defined through the equations

$$\frac{\partial}{\partial v_i} V_{eff}(v_i, T_c) = 0, \; V_{eff}(v_1, v_2, v_3, v_4, v_5, T_c) = V_{eff}(0, 0, 0, x_1, x_2, T_c), \tag{9}$$

where $x_{1,2}$ are the real and imaginary parts of the singlet vev in the wrong vacuum.

3.2. Numerical results

In the following figures, we show the dependence of the quantity $v(T_c)/T_c$ in (8) on the lightest Higgs mass (in Fig. 1 left), and on the lightest neutralino mass (in Fig. 1 right), for a random choice of about 2000 cases in both regions **(a)** and **(b)**, where different conditions are fulfilled.

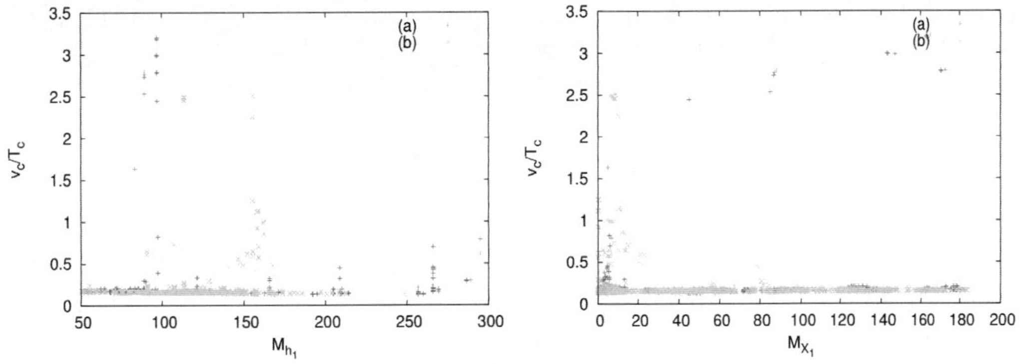

Figure 1. *The dependence of the quantity v_c/T_c on the lightest Higgs (right) and the lightest neutralino (left) masses. The green points refer to cases in region* **(a)**, *and the red ones to cases in region* **(b)**.

As it is clear from Fig. 1, the EWPT could be strongly first order in the two regions (a) and (b), for different values of the lightest Higgs mass. The fact that the lightest Higgs mass is around 90∼100 GeV might be consistent with experiment because the doublet couplings will be modified due to the mixing with singlet. The EWPT could be also, strongly first order for different the lightest neutralino masses. This leads to the conclusion that the EWPT strength is a factor beyond the fields contributions to the effective potential at zero temperature.

The critical temperature is, in general, higher when comparing with minimal SM (∼100 *GeV*), the generic value is larger than 500 *GeV*, this is a consequence of the the interaction of the doublets with the singlet that has, in general, a very large vev. However for the benchmarks, giving a strong first order EWPT, it is relatively smaller than the generic values. As it is small, the EWPT is stronger. In order to understand this point, we take a benchmark from Fig. 1, and study the dependance of the scalar vevs on the temperature T. We will see also how could this behavior be changed with respect to the charges Q's, and some parameters like A_t and $\theta_{1,2}^{(v)}$, that appear in the effective potential at one-loop level. Therefore we consider the following modifications:

From this table, it is clear that the EWPT strength is sensitive to the relative phases $\theta_{1,2}^{(v)}$, that appear in the effective potential at one-loop through the superpartners masses. For this parameters set, these relative phase shift the critical temperature by some units. The effect of the parameter A_t, that appears in the effective potential through the one-loop corrections of the stops, is more important. The critical temperature, in this case (4) is shifted by 200 GeV, which makes the EWPT so weak. From case (3), the effect of the charges Q's is extremely important.

	Q_1	Q_2	A_t	$\theta_1^{(v)}$	$\theta_2^{(v)}$	T_c	v_c/T_c	m_{h_1}	m_{χ_1}
(1)	20/7	−4/7	300	0	0	73.25	3.454	116.34	7.70
(2)	20/7	−4/7	300	$\pi/9$	$-\pi/9$	75.72	3.338	116.34	8.32
(3)	2	1	300	0	0	629.17	0.149	112.31	7.70
(4)	20/7	−4/7	1500	0	0	274.13	0.198	116.34	7.70

Table 1. *Values of the parameters used to study the scalar vevs dependance with respect to the temperature. We used common values for the parameters:* $\lambda = 0.01$, $tan\beta = 1$, $A_\lambda = 1210\ GeV$, $v_x = 1046.4\ GeV$, $M_1 = M_2 = 100\ GeV$, $M_1' = 300\ GeV$. *The mass-dimension parameters are given in GeV.*

In Fig. 2, we show the dependance of the ground state on the temperature below the critical temperature for the benchmark (1) in Table-1, and its modifications (2)-(4). From Fig. 2, one remarks that the common feature between all these cases is that the dependance of the singlet vev on the temperature is very weak around and below the critical temperature.

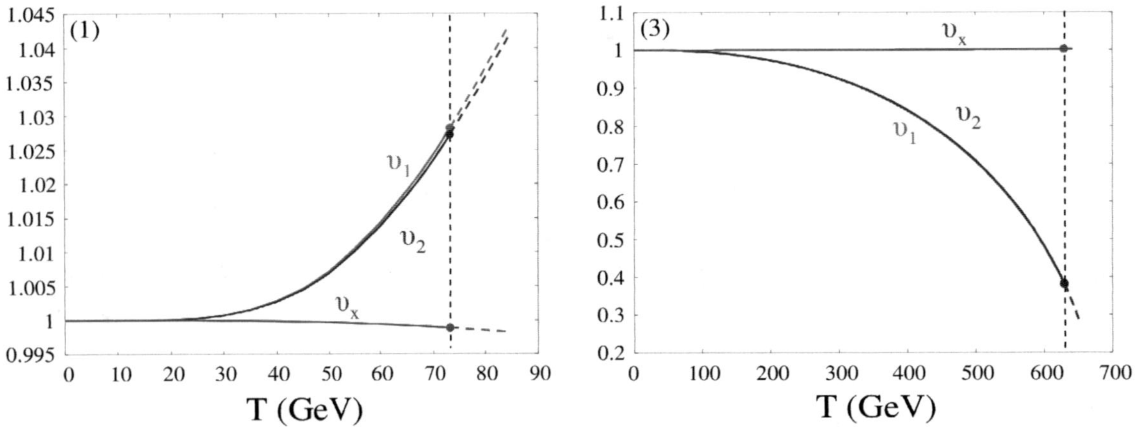

Figure 2. *The dependence of the scalar vevs on the temperature T. These quantities are scaled by their zero temperature values. The solid curves refer to the broken phase while the dashed ones refer to the symmetric phase, where (0,0,x(T)) is still the global minimum.*

Another important remark, is that the Higgs vevs for this benchmark (Fig. 2-1), are increasing when the Universe gets cooled unlike the SM, or MSSM. In case (3) (Fig. 2-3), the Higgs vevs decrease, but slower than the SM or MSSM. This can be seen clearly in Fig. 1-3, for the benchmarks whose critical temperature is between $250 \sim 400\ GeV$, where the corresponding value of $v(T_c)/T_c$ is larger when comparing with the SM. The fact that the doublets vev values at high temperatures are larger than their zero temperature values has been mentioned in a similar work [13]. This behavior: increasing Higgs vevs w.r.t the temperature or decreasing slower than the SM case, is a consequence of the interaction of the singlet with the doublet. These interactions lead to relax the shape of the potential in the direction of the doublets, and therefore enhances the ratio in (8), and strengthen the EWPT. This is a common feature for models with singlets [7].

Unlike the SM and MSSM, the wrong vacuum $(0, 0, x(T))$, is evolving w.r.t the temperature, therefore its evolution could be a very important factor that strengthens the EWPT. It could delay the transition, and then enhances the ratio (8). The EWPT dynamics seems to be less sensitive to some input parameters like m_Q, m_U, A_t, M_2, M_1, M_1' and especially the relative phases $\theta_{1,2}^{(v)}$, since they appear in the effective potential at one loop, however, we have shown in Table-1 that the EWPT dynamics is sensitive to these parameters more than expected. This sensitivity is a consequence of the dependance of the two vacua (instead one) on these input parameters.

The possible one-loop spontaneous CP violation due to the relative phases $\theta_{1,2}^{(v)}$, could have an important phenomenological impact. An important feature in this model is that the dark matter candidate (lightest neutralino) could be dominated by the new ingredient $\tilde{B}'$. This possibility will be investigated in further work [21].

4. Conclusion

In this work, the electroweak phase transition nature within the minimal $U(1)$ extension of the MSSM (UMSSM) has been investigated. We found that the EWPT could be strongly first order for reasonable masses of scalar Higgs bosons and the lightest neutralino, without adding extra singlet scalars or fermions. We evaluated the effective potential at one-loop taking into account the whole particle spectrum, and its temperature dependant corrections were estimated exactly using the known techniques. We found that EWPT strength could be enhanced due to two factors: first, the interactions of the singlet scalars with the doublets that relax the shape of the effective potential in the doublets directions, which lead to a large value for the ratio $v(T_c)/T_c$, at the critical temperature. The second factor is that the temperature-dependant local minimum $(0, 0, x(T))$, could play an important role during the EWPT dynamics. It can delay the phase transition until relatively low temperatures (even below 100 GeV), which favor the ratio $v(T_c)/T_c$ to be large enough.

We mention also that the reliability of the parameter choice, as well as the EWPT strength are more sensitive to the input parameters that appear in the effective potential at one-loop.

Acknowledgments

This work was supported by the Algerian Ministry of Higher Education and Scientific Research under the cnepru-project D01720090023.

[1] D.N. Spergel et al. Astrophys. J. Suppl. 377 (2007) 170.
[2] S. Adler, Phys. Rev. 177 (1969) 2426; J.S. Bell and R. Jackiw, Nuovo Cimento A60 (1969) 47.
[3] V.A. Kuzmin, V.A. Rubakov and M.E. Shaposhnikov, Phys. Lett. B155 (1985) 36.
[4] A. Riotto and M. Trodden, Ann. Rev. Nucl. Part. Sci. 49 (1999) 35.
[5] A. I. Bochkarev and M.E. Shaposhnikov, Mod. Phys. Lett A2 (1987) 417.
[6] Particle Data Group (C. Amsler et al.), Phys. Lett. B667 (2008) 1.
[7] A. Ahriche, Phys. Rev. D75 (2007) 083522.
[8] J. E. Kim and H. P. Nilles, Phys. Lett. B138 (1984) 150.
[9] R. Harnik, D.T. Larson, H. Murayama and M. Thormeier, Nucl. Phys. B706 (2005) 372.
[10] M. Cvetic, D. A. Demir, J. R. Espinosa, L. L. Everett and P. Langacker, Phys. Rev. D56 (1997) 2861 [Erratum-ibid. D58 (1998) 119905]; P. Langacker and J. Wang, Phys. Rev. D58 (1998) 115010.
[11] G.F. Giudice, Phys. Rev. D45 (1992) 3177; A. Brignole, J.R. Espinosa, M. Quiros and F. Zwirner, Phys. Lett. B324 (1994) 181; D. Delepine, J.-M. Gerard, R.G. Felipe and J. Weyers, Phys. Lett. B386 (1996) 183; J.R. Espinosa, Nucl. Phys. B475 (1996) 273.
[12] J. Kang, P. Langacker, T. Li and T. Liu, Phys. Rev. Lett. 94 (2005) 061801; C.-W. Chiang and E. Senaha, JHEP 1006 (2010) 030.
[13] S.W. Ham, E.J. Yoo, S.K. OH, Phys. Rev. D76 (2007) 075011.
[14] A. Ahriche and S. Nasri, arXiv:1008.3106.
[15] M. Cvetic and P. Langacker, Mod. Phys. Lett. A11 (1996) 1247.
[16] F. Abe et al., Phys. Rev. Lett. 79 (1997) 2192.
[17] D. Bodeker, Phys. Lett. B426 (1998) 351; Nucl. Phys. B559 (1999) 502; P. Arnold, D. Son and L. Yaffe, Phys. Rev. D55 (1997) 6264; Phys. Rev. D59 (1999) 105020.
[18] A.I. Bochkarev, S.V. Kuzmin and M.E. Shaposhnikov, Phys. Rev. D43 (1991) 369.
[19] M.E. Shaposhnikov, Nucl. Phys. B287 (1987) 757; B299 (1988) 797.
[20] R.F. Klinkhamer and N.S. Manton, Phys. Rev. D30 (1984) 2212.
[21] A. Ahriche and S. Nasri, in progress.

Journal of Physics: Conference Series **259** (2010) 012086 doi:10.1088/1742-6596/259/1/012086

Imprint of cosmic neutrino decoupling in the spectrum of inflationary gravitational waves

Massimiliano Lattanzi[1], **Riccardo Benini**[1] **and Giovanni Montani**[1,2]

[1] ICRA and Physics Department, University of Rome, "La Sapienza", P.le Aldo Moro 2, 00185 Rome, Italy

[2] ENEA – C.R. Frascati (Department F.P.N.), Via Enrico Fermi, 45 (00044), Frascati (Rome), Italy

E-mail: `lattanzi@icra.it`

Abstract. In this paper we study the effect of cosmic neutrino decoupling on the spectrum of cosmological gravitational waves (GWs). Before decoupling, namely at temperatures $T \gg 1\,\mathrm{MeV}$, neutrinos constitute a perfect fluid and do not hinder GW propagation, while well after decoupling, for $T \ll 1\,\mathrm{MeV}$, they free-stream generating an effective viscosity that damps cosmological GWs by a constant amount. In the intermediate regime, corresponding to the time of neutrino decoupling, the damping is frequency-dependent. GWs entering the horizon during neutrino decoupling have a frequency $f \sim 1\,\mathrm{nHz}$; in particular, we show how neutrino decoupling induces a step in the spectrum of cosmological GWs just below 1 nHz, a frequency region that will be probed by Pulsar Timing Arrays (PTAs). We briefly discuss the conditions for a detection of this feature and conclude that it is unlikely to be observed by PTAs.

Introduction The presence in the Universe today of a stochastic background of gravitational waves (GWs) is a quite general prediction of several early cosmology scenarios. In fact, the production of GWs is the outcome of many processes that could have occurred in the early phases of the cosmological evolution, like the amplification of vacuum fluctuations in inflationary [1] and pre-big-bang cosmology scenarios [2], phase transitions [3], and finally the oscillation of cosmic strings loops [4]. In most of these cases, the predicted spectrum of GWs extends over a very large range of frequencies.

The detection of GWs produced in the early Universe would be a major breakthrough in cosmology and high energy physics. This is because GWs decouple from the cosmological plasma at very early times, when the temperature of the Universe is of the order of the Planck energy. In this way, relic GWs provide us a "snapshot" of the Universe at the time of their production, in a similar way as the cosmic microwave background (CMB) radiation images the Universe at the time of recombination. The extremely low frequency region ($f \lesssim 10^{-15}\,\mathrm{Hz}$) in the spectrum of primordial GWs can be probed through the anisotropies of the CMB. In particular, GWs leave a distinct imprint in the so-called magnetic or B-modes of its polarization field [5, 6].

The frequency region between ~ 10 Hz and few kHz is probed by operating large scale interferometric GW detectors like LIGO [7] and VIRGO [8], that, although designed with the aim to detect astrophysical signals, can possibly also detect signals of cosmological origin [9]. The LISA space interferometer [10], that will hopefully operate in the 2020s, will probe the frequency region between 10^{-4} and 1 Hz. Finally, pulsar observations can be used to obtain information on the stochastic GW background, through the technique known as pulsar timing. The so-called "Pulsar Timing Arrays" will probe the region from ~1 to ~ 100 nHz, with a maximum sensitivity between ~3 and ~10 nHz [11].

In order to compare the theoretical predictions with the expected instrument sensitivities, one needs to evolve the GWs from the time of their production to the present. This is often done by assuming that GWs propagate freely across the Universe. In this case, the only effect is that the amplitude of the wave decreases due to the expansion of the Universe. However, GWs are sourced by the anisotropic stress part of the energy-momentum tensor of matter, so that the above assumption is well-motivated only when anisotropic stress can be neglected. It is already known that the anisotropic stress of free streaming neutrinos acts as an effective viscosity, absorbing GWs in the low frequency region, thus resulting in a damping of the B-modes of CMB [12, 13, 14, 15, 16]. In Ref. [18], the evolution of cosmological GWs in the presence of neutrino free-streaming was studied up to second order in perturbation theory. Moreover, features in the spectrum are also induced by entropy generating events (like the e^+e^- annihilation) occurring during the cosmological evolution [17]. The authors of Ref. [17] also consider the effect of neutrino free streaming but assume an instantaneous neutrino decoupling.

Here we briefly summarize our results concerning a possible signature of neutrino decoupling on the spectrum of cosmological GWs, as reported in Refs. [25, 26]. During decoupling, the effective viscosity of neutrinos is increasing from zero (its value at high temperatures, when the neutrinos are tightly coupled to the cosmological plasma) to a finite value. Neutrino decoupling happens when the temperature of the Universe is $T \simeq 1$ MeV, or redshift $z \simeq 10^{10}$; GWs entering the horizon at that time have a frequency $f \simeq 1$ nHz. Since the viscosity results in a damping of the wave, we expect to have a frequency-dependent absorption of GWs in this frequency range. A spectral feature in the nHz range would be potentially interesting for PTAs. PTAs combine the fact that pulsars are very stable clocks with the fact that the time of arrival of the pulse also depends on the GW background between the pulsar and the Earth [11]. The use of a pulsar array allows to correlate the electromagnetic signals from different pulsars, thus eliminating variations in the arrival times that do not depend on the GW background. The dominant contribution to the GW background in the PTAs sensitivity range is that of supermassive black hole binaries following galaxy mergers [19, 20, 21]. This has an astrophyisical origin and would obviously not be affected by the decoupling of cosmological neutrinos. The dominant *cosmological* constribution to the GW signal in the nHz range is expected to be that of cosmic strings, that are topological defects left after symmetry-breaking phase transitions occurring in the early Universe. Other, possibly important, contributions in this frequency range are those of cosmic superstrings (i.e., topological defects arising in string-theory inspired inflationary models) and of inflation-generated GWs (through the usual amplification of quantum zero-point fluctuations). For our purposes, one important difference between the cosmic (super)strings-generated GWs on one side, and the inflationary GWs on the other, is that the former are created by causal processes at horizon and subhorizon scales, while the latter are generated by the stretching of quantum fluctuations to super-horizon scales. In the following we will concentrate on GWs that were outside the horizon at some time before neutrino decoupling, so that our results rigorously applies to inflationary GWs but not necessarily to string-generated GWs. It should be noted that the inflationary signal is, in the nHz range, the weakest of the cosmological signals mentioned above, and is probably beyond the reach of the ongoing PTA projects. It could be however within the reach of future instruments [11, 22] like the Square Kilometer Array (SKA) [23].

Basic statements. We consider a GW, described by a metric perturbation h_{ij}, propagating on the background of a flat Friedmann Universe. In synchronous gauge, the spatial components of the perturbed metric are written as[1]

$$g_{ij} = a^2(\eta)\left[\delta_{ij} + h_{ij}\right] \tag{1}$$

while the other components are left unperturbed: $g_{00} = -1$ and $g_{0i} = 0$. We will consider only the transverse traceless part of h_{ij}. With this restriction, h_{ij} only has two degrees of freedom, corresponding to the two polarizations of the GW. Here a is the cosmological scale factor and η is the conformal time, related to the synchronous time t by $dt = a(\eta)d\eta$. With this choice, a GW h_{ij} with wave number k evolves according to[2] [24]

$$\ddot{h}_{ij} + 2\mathcal{H}\dot{h}_{ij} + k^2 h_{ij} = 16\pi G \Pi_{ij} \ , \tag{2}$$

where dots denote derivatives with respect to η, $\mathcal{H}$ is the conformal Hubble constant $\mathcal{H} = \dot{a}/a$, and $\Pi^i_j = T^i_j - \delta^i_j T^k_k/3$ is the anisotropic stress, i.e. the traceless part of the three dimensional energy-momentum tensor T^i_j of the cosmological fluid, representing dissipative effects. We assume that the anisotropic stress is relevant only for one component of the cosmological fluid (the neutrinos), providing a fraction f_ν of the total background density $\bar{\rho}$, while for the other components (e.g., the photons) it can be safely neglected. The anisotropic stress can be computed by solving the Boltzmann equation for the neutrino phase space density $f(x^i, P_i, \eta)$, from which the energy-momentum tensor is obtained by integrating over momenta. Since the Boltzmann equation itself contains the metric h_{ij}, it forms a coupled system together with the Einstein equation (2). In the following, we will restrict our attention to waves entering the horizon well before the time of matter-radiation equality, corresponding to a redshift $z \simeq 10^4$. This corresponds to waves with a present frequency $\nu \gg 10^{-16}$ Hz.

Coupled evolution of GWs and neutrinos. The thermal evolution of neutrinos can be divided into two distinct regimes. In the early Universe, when the temperature was sufficiently high, neutrinos were kept in thermal equilibrium with the cosmological plasma by frequent reactions like $\nu\bar{\nu} \leftrightarrow e^+e^-$, $\nu e \leftrightarrow \nu e$, etc. As the Universe expanded and cooled down, the interaction rate Γ_ν eventually became smaller than the expansion rate H and the neutrinos decoupled from the cosmological plasma. From this point on, neutrinos can be considered as freely streaming through the Universe. The temperature $T_{\rm dec}$ that separates the collisional from the non-collisional regime, defined as the temperature when the interaction rate is equal to the expansion rate, i.e. $\Gamma_\nu(T_{\rm dec}) \simeq H(T_{\rm dec})$, is found to be $T_{\rm dec} \simeq 1\,\rm MeV$, corresponding to $z \simeq 10^{10}$. A wave that enters the horizon at that time has a frequency $f \sim 1\,\rm nHz$. Higher frequency waves enter the horizon *before* neutrino decoupling, and viceversa. Then, the following three regimes for the evolution of a GW can be correspondingly identified:

- When $T \gg 1\,\rm MeV$, the neutrinos are tightly coupled to the cosmological plasma, that behaves like a single, perfect fluid. The anisotropic stress of neutrinos is negligible and the GW evolves according to the homogeneous version of Eq. (2). This regime is relevant for waves with $f \gg 1\,\rm nHz$.
- When $T \simeq 1\,\rm MeV$, the neutrinos are decoupling from the plasma. They cannot be considered as a perfect fluid because the mean free path of particles is getting large, so that the anisotropic stress cannot be neglected. However, the collisions have still to be taken into account. The GW evolution has to be calculated from the full Einstein-Boltzmann system, taking into account neutrino interactions. This regime is relevant for waves with $f \simeq 1\,\rm nHz$.

[1] We use the $(-+++)$ signature for the metric. We shall also use, all throughout the paper, natural units in which $c = \hbar = k_B = 1$.

[2] The spatial dependence of all quantities has been Fourier transformed.

- When $T \ll 1\,\mathrm{MeV}$ the neutrinos are collisionless and behave as free particles. The anisotropic stress cannot be neglected: the GW evolution has to be calculated from the full Einstein-Boltzmann system, but neutrino interactions can be neglected. This regime is relevant for waves with $f \ll 1\,\mathrm{nHz}$.

In the first, high-temperature regime, the metric perturbation evolves as $h_{ij} = \bar{h}_{ij} \equiv \lambda_{ij} \sin(k\eta)/k\eta$ (assuming the initial velocity vanishes), where λ_{ij} is the initial value of the perturbation. In other words, the GW oscillates with a time-dependent amplitude $\propto 1/k\eta$. This can be seen as the energy loss of the wave due to the expansion of the Universe. The third, low-temperature regime was first studied in Ref. [14], where it was realized that the anisotropic stress of neutrinos would lead to a partial absorption of the GW. In this limit (waves entering the horizon well after neutrino decoupling) the absorption is frequency independent, and only depends on the neutrino fraction f_ν; in particular, for the standard value $f_\nu = 0.4052$ (corresponding to three neutrino families in thermal equilibrium with a present temperature $T_\nu = 1.9\mathrm{K}$) the intensity of the wave is reduced to $\sim 75\%$ of its value in the absence of stress [25]. In the second, intermediate regime the effect of neutrino interactions should be taken into account. We have modeled neutrino collisions by means of a suitable collision term in the Boltzmann equation, and integrated the Boltzmann-Einstein system for different values of the GW frequency between $10^{-12}\,\mathrm{Hz}$ and $10^{-8.5}\,\mathrm{Hz} \simeq 3\,\mathrm{nHz}$, as described in Refs. [25, 26]. We have assumed the standard value for f_ν. We have defined a frequency-dependent damping factor $D_f \equiv \sqrt{2\langle (k\eta h_{ij})^2 \rangle}$ where the brackets $\langle \ldots \rangle$ denote the average over many periods, taken when the wave is well inside the horizon ($k\eta \gg 1$). The definition is chosen so that when $h_{ij} = \sin(k\eta)/k\eta$ then $D_f = 1$. The damping factor basically quantifies how much the amplitude of a GW produced in the early Universe is reduced due to the anisotropic stress of neutrinos, with respect to propagation in a perfect fluid. We express our results in terms of the GW density parameter $\Omega_{\mathrm{GW}}(f) = (d\rho_{\mathrm{GW}}/d\ln f)/\rho_c$, where ρ_{GW} is the energy density of GWs and ρ_c is the critical density of the Universe. The quantity Ω_{GW} is used by theorists to quantify the intensity of the cosmological GW background predicted in a given scenario, and experimental sensitivities are also often quoted in terms of it. The present intensity is usually computed simply by rescaling the GW amplitude at the source by a factor $1/k\eta$ to take into account the redshift due to the expansion. This should be corrected to take into account also the anisotropic stress of neutrinos; since $\rho_{\mathrm{GW}} \propto h_{ij}^2$, the correction factor is D_f^2. Calling $\bar{\Omega}_{\mathrm{GW}}(f)$ the present intensity in absence of stress, we have that the observed spectrum $\Omega_{\mathrm{GW}}(f) = D_f^2 \bar{\Omega}_{\mathrm{GW}}(f)$. The damping factor introduces an additional dependence on frequency that was not present at the source. In the left panel of Fig. 1 we show D_f^2 as a function of frequency.

The fact that the collisions introduce a frequency dependence on the amplitude of the GW background (in addition to any dependence that could be already present at the source) is a more interesting case with respect to a frequency independent suppression, since the latter could not be disentagled from our ignorance on the amplitude of the original spectrum (that usually depends on the free parameters of the theory and thus cannot be determined *a priori* without any additional experimental input). On the other hand, a frequency-dependent effect can be disentangled by using this same dependence. To better quantify this, let us suppose that the GW spectrum at the source, has a featureless power law behaviour $\propto f^\alpha$. After correcting for the expansion, the shape is still the same. Then the spectrum, once the effect of anisotropic stress has been taken into account, will be $\Omega_{GW}(f) \propto D_f^2 f^\alpha$. The logarithmic slope of the spectrum in a given point is given by $\alpha' = d\log \Omega_{GW}/d\log f$, so that, defining the deviation $\Delta\alpha = \alpha' - \alpha$, we have:

$$\Delta\alpha = \frac{d\log \Omega_{\mathrm{GW}}}{d\log f} - \alpha = \frac{d\log D_f^2}{d\log f}. \tag{3}$$

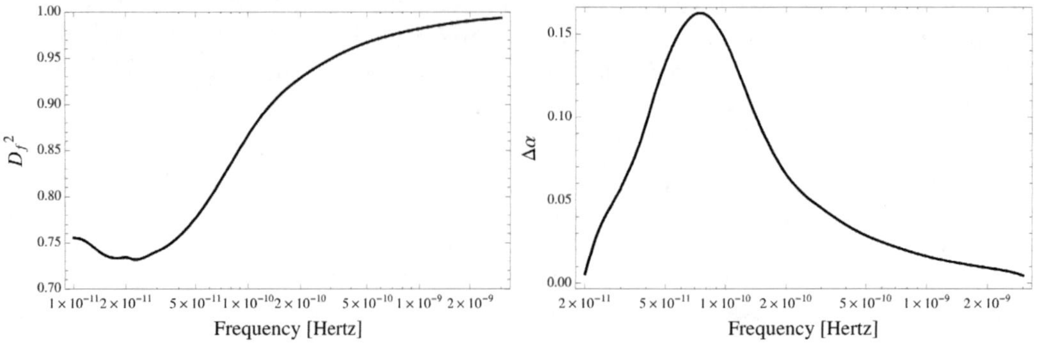

Figure 1. Left panel: correction factor $D_f^2 = \Omega_{\rm GW}/\bar{\Omega}_{\rm GW}$. The highest frequencies in the plot correspond to waves entering the horizon when the neutrinos are tightly coupled to the plasma, so that there is no dissipation. The lowest frequencies correspond to waves entering the horizon when the neutrinos are freely streaming, so that D_f^2 is constant and roughly equal to 0.75. Right panel: Change $\Delta\alpha$ in the logarithmic slope of a featureless power-law spectrum f^α at the source, induced by the anisotropic stress.

In the right panel of Fig. 1, we show $\Delta\alpha$ as a function of frequency. We find that it has a maximum value of $\Delta\alpha = 0.15$ at $f \simeq 0.1\,$nHz.

Observational prospects The nHz region of the spectrum of GWs can be probed by PTAs, combining the fact that pulsars are very stable clocks with the fact that the time of arrival of the pulse also depends on the GW background between the pulsar and the Earth [11]. Correlating the electromagnetic signals from different pulsars could, in principle, enable a positive detection of the GW background in the galaxy. PTAs are sensitive to GWs with frequency between roughly 1 and 100 nHz, with a maximum sensitivity in the region between 3 and 10 nHz. The lower bound comes from the fact that standard pulsar timing techniques absorb any low-frequency signal, and so the time span of the data (currently $\sim$ 30 years) gives a lower bound on the observable frequencies. The principal source of GWs in the region of maximum sensitivity are believed to be coalescing supermassive binary black-hole systems in the centre of merging galaxies. However, a signal of cosmological origin could also be present in that same frequency range, for example GWs produced during the inflationary era or from the oscillation of cosmic string loops.

Can the effect of anisotropic stress leave an imprint that could be, in the future, detected using PTAs? This could be possible, at some conditions. First of all, obviously, a signal of cosmological origin has to be present in the nHz range and it has to be strong enough to be detectable by PTAs. As noted in the introduction, the oscillation of cosmic (super)strings loops could give a detectable signal in the nHz range. However, GWs generated in this way are produced at scales smaller or equal than the horizon, so that the analysis made here does not rigorously apply. Although it is sensible to expect that GWs with wavelengths comparable to the horizon would behave in a similar way to the super-horizon GWs considered here, we cannot assess the effect on the overall spectrum and we defer this generalization to a future work. This leaves the GWs produced during inflation by the amplification of quantum vacuum fluctuations, that are too weak to be detected by ongoing PTA projects but could be within the reach of future experiments like SKA. In any case, an independent observation of the signal at larger frequencies, like those probed by interferometers, would also be useful to normalize the spectrum at the source. Secondly, the cosmological signal should be clearly separated from the astrophysical signal, like that produced by black hole binaries. The third point is the more problematic. We have shown that the effects of the damping are more evident at frequencies between 0.1 and 1 nHz. As explained above, the lower limit to the detectable frequency is given by the time span of the observations, currenly 30 years. In 70 years from now, with a total of 100 years of observations, the lower limit will be $f = 1/(100\text{years}) \simeq 0.3\,\text{nHz}$, where the intensity of the wave is reduced by only 5% and the deviation from a featureless power law is $\Delta\alpha = 0.05$. The maximum change of the slope is at $f \simeq 0.1\,\text{nHz}$, corresponding to 300 years of observations. It should also be noted that, even if we were willing to wait such a long time, it is not certain that it would actually increase the sensitivity, because of intrinsic low-frequency instabilities in the timing data. In conclusion, we think that the possibility to detect this effect are quite small, and that they will depend crucially on the existence of a sizeable inflationary-like GW background at the frequencies of interest and on the capability of estimating the logarithimic slope of the spectrum in the region just below 1 nHz.

Acknowledgments. This work has been developed in the framework of the CGW collaboration (www.cgwcollaboration.it).

[1] L.P. Grishchuk, *Sov. Phys. JETP* **40**, 409 (1975); A. Starobinski, *JETP Lett.* **30**, 682 (1979); R. Barkana, *Phys. Rev. D* **50**, 1157 (1994).
[2] M. Gasperini and M. Giovannini, *Phys. Lett. B* **282**, 36 (1992); M. Gasperini and M. Giovannini, *Phys. Rev. D* **47**, (1993); R. Brustein, et al., *Phys. Lett. B* **361**, 45 (1995); A. Buonanno, et al., *Phys. Rev. D* **55**, 3330 (1997); V. Mandic & A. Buonanno, *Phys. Rev. D* **73**, 063008 (2006).
[3] C. Hogan, *Mon. Not. R. Astr. Soc.* **218**, 629 (1986); R. Apreda, et al. Nucl. Phys. B 631, 342 (2002).
[4] T. W. B. Kibble, J. Phys. A 9, 1387 (1976); A. Vilenkin, *Phys. Lett. B* **107**, 47 (1981); S. Sarangi & S. H. H. Tye, Phys. Lett. B 536, 185 (2002). T. Damour, & A. Vilenkin, Phys. Rev. D 71, 063510 (2005). X. Siemens, V. Mandic, & J. Creighton, J., Phys. Rev. Lett. 98, 111101 (2007).
[5] W. Hu and M. White, *New Astron.* **2**, 323 (1997).
[6] J.R. Pritchard and M. Kamionkowski, *Annals Phys.* **318**, 2 (2005).
[7] A. Abramovici et al., *Science* **256**, 325 (1992); B, Abbott, et al., *Nucl. Instrum. Meth. A* **517**, 154 (2004). See also `http://www.ligo.org/`.
[8] B Caron et al., *Class. Quantum Grav.* **14** 1461 (1997). F. Acernese et al. *Class. Quant. Grav.* **25** 114045 (2008). See also `http://www.virgo.infn.it/`.
[9] M. Maggiore, *Phys. Rep.* **331**, 283 (2000).
[10] K. Danzmann for the LISA Study Team, *Class. Quantum Grav.* **14** 1399 (1997); P. L. Bender, K. Danzmann, & the LISA study team, *Laser Interferometer Space Antenna for the Detection and Observation of Gravitational Waves: Pre-Phase A Report 2nd edn* (MPQ233, Max-Plank Institut fur Quantenoptik, 1998). See also `http://sci.esa.int/science-e/www/area/index.cfm?fareaid=27`
[11] J. P. W. Verbiest *et al.*, arXiv:0908.0244 [astro-ph.GA]; G. Hobbs, arXiv:0911.0943 [astro-ph.SR]; G. Hobbs *et al.*, arXiv:0911.5206 [astro-ph.SR]; J. P. W. Verbiest *et al.*, arXiv:0912.2692 [astro-ph.GA]; R. N. Manchester, arXiv:1004.3602 [astro-ph.HE].
[12] J.R. Bond, in *Cosmology and Large Scale Structure, Les Houches Session LX*, edited by R. Schaeffer, J. Silk, and J. Zinn-Justin (Elsevier, Amsterdam, 1996)
[13] R. Durrer and T. Kahniashvili, *Helv.Phys.Acta*, **71**, 445, (1998).
[14] S. Weinberg, *Phys. Rev. D* **69**, 023503, (2004).
[15] T.Y. Xia and Y. Zhang, *Phys. Rev. D* **78**, 123005 (2008)
[16] W. Zhao, Y. Zhang and T.Y. Xia, *Phys. Lett. B* **677**, 235 (2009).
[17] Y. Watanabe and E. Komatsu, Phys. Rev. D **73**, 123515 (2006)
[18] A. Mangilli, N. Bartolo, S. Matarrese and A. Riotto, Phys. Rev. D **78**, 083517 (2008)
[19] A.H. Jaffe and D.C. Backer, *Astrophys. J.* **583**, 616 (2003)
[20] F. A. Jenet, G. B. Hobbs, K. J. Lee and R. N. Manchester, Astrophys. J. **625**, L123 (2005)
[21] A. Sesana, A. Vecchio and C. N. Colacino, Mon. Not. Roy. Astron. Soc. **390**, 192 (2008)
[22] F. Jenet *et al.*, arXiv:0909.1058 [astro-ph.IM].
[23] M. Kramer et al., New Astron. Rev. **48**, 993 (2004) [PoS **MRU**, 020 (2007)]
[24] S. Weinberg, *Cosmology* (Oxford University Press, Oxford, 2008).
[25] R. Benini, M. Lattanzi, G. Montani, *Gen. Rel. Grav.*, in press. Available on-line at `http://dx.doi.org/10.1007/s10714-010-0994-4`
[26] M. Lattanzi, R. Benini, G. Montani, *Class. Quant. Grav.* **27**, 194008 (2010).

Aspects of noncanonical inflation

Paul Franche[1], **Rhiannon Gwyn**[2], **Bret Underwood**[1] **and Alisha Wissanji**[1]

[1] Department of Physics, McGill University
3600 University Street, Montréal, Québec, Canada H3A 2T8,
[2] Department of Physics, King's College London
Strand, London, U.K. WC2R 2LS

E-mail: franchep@hep.physics.mcgill.ca, rhiannon.gwyn@kcl.ac.uk, bjwood@hep.physics.mcgill.ca, wissanji@hep.physics.mcgill.ca

Abstract. This paper reviews recent work by the authors on inflation in single-field models with noncanonical (higher-order) kinetic terms. Such terms arise naturally in an EFT approach to modelling inflation, and also from certain string theory embeddings, such as in DBI inflation. Theories containing such terms can support a period of noncanonical inflation which is smoothly connected to the usual slow-roll regime. We give a general formalism for finding solutions of the equations of motion corresponding to noncanonical inflation. However, not every Lagrangian with higher-power kinetic operators supports noncanonical inflation, and not every set of initial conditions in a suitable theory will give rise to an inflationary trajectory. We give some sufficient conditions required for a Lagrangian to support inflation in the non-canonical regime. Further, we investigate the problem of fine-tuning in these models. We find that noncanonical kinetic terms tend to reduce the severity of the fine-tuning problem in general, as long as a regime of noncanonical inflation is present.

1. Noncanonical inflation

1.1. Inflation as an effective field theory

Inflation is an appealing solution to the homogeneity, isotropy, and flatness problems, and is well supported by observation as providing the dominant mechanism for structure formation. For these reasons inflation is widely believed to have taken place , and has been the subject of intense research. However, because the UV-complete theory in which it should take place is as yet unknown, all meaningful discussion is necessarily at the effective-field-theory (EFT) level.

The EFT description is valid only up to some energy cutoff Λ, with the physics above this scale being integrated out. The resulting higher-dimensional operators are suppressed by powers of Λ, and can usually be safely neglected. However, the EFT can be sensitive to the UV physics, since in inflationary theories the self-coupling of the inflaton is small.

One well-known example is the η problem: dimension 6 corrections of the form $\frac{\mathcal{O}_6}{M_p^2} \sim \frac{\mathcal{O}_4}{M_p^2}\phi^2$, where ϕ is the inflaton, can spoil the flatness of the potential, leading to $\mathcal{O}(1)$ corrections to the slow-roll parameter $\eta = M_p^2 \frac{V''}{V}$ when $\mathcal{O}_4$ has a VEV comparable to the inflationary energy density. Inflation, which depends on η being small, can be spoilt by these corrections. This is a generic problem for supergravity models of inflation in which the F-term potential is nonzero (see e.g. [1]).

1.2. Noncanonical kinetic terms

We saw above that the underlying UV theory can affect the potential; it can also lead to important corrections to the kinetic terms. Consider the following two-field toy model:

$$\mathcal{L} = \frac{1}{2}(\partial\phi)^2 + \frac{1}{2}(\partial\rho)^2 + \frac{\rho}{M}(\partial\phi)^2 - \frac{1}{2}M^2\rho^2.$$

At energy scales $H << M$, the heavy field ρ can be integrated out. Because of its coupling to ϕ, a higher-order kinetic term for ϕ results: $\mathcal{L}_{eff} = \frac{1}{2}(\partial\phi)^2 + \frac{(\partial\phi)^4}{M^4}$. The DBI action, which contains exactly such higher-order kinetic terms in a closed-form expression, can be analogously understood as arising when massive degrees of freedom (W-bosons) are integrated out [2].

Such couplings and massive degrees of freedom will be generically present in any UV-complete description of inflation. Thus noncanonical kinetic terms are generically expected to be present. While remaining completely ignorant of the correct UV-complete description, we can ask what the effect of these noncanonical kinetic terms will be. We consider general single-field noncanonical Lagrangians, by which we mean Lagrangians of the form

$$S = \int d^4x\sqrt{-g_4}\left[\frac{M_p^2}{2}\mathcal{R}_4 + p(X,\phi)\right], \tag{1.1}$$

where $p(X,\phi)$ is a general function of ϕ and the canonical kinetic term $X = \frac{1}{2}(\dot{\phi})^2$. It is consistent to ignore higher derivative corrections of the form $\partial^n\phi$, as shown in [3]. Lagrangians of the form $\mathcal{L}_{eff} = p(X,\phi)$, with $p(X,\phi) \to 0$ as $X \to 0$ and zero potential energy have been shown to support non-slow roll inflation, and are known as k-inflation models [4]. These have been studied before [5, 6, 7, 8], but no general formalism for finding inflationary solutions to the equations of motion was given. In the following section we outline the formalism for finding general inflationary solutions to Lagrangians of the form (1.1), in which both kinetic and potential energy terms enter and play a role, and which encompass "noncanonical" inflation (NCI) as well as the canonical slow-roll solutions. In Section 1.4 we give conditions which a given Lagrangian must satisfy in order to guarantee that it supports a period of NCI. The work in Sections 1.3 and 1.4 was published in [3].[1]

Models of NCI can give rise to an observable amount of non-gaussianity in the CMB, making them especially interesting. The leading order non-gaussianity from noncanonical single-field models is of the equilateral type [6], with $f_{NL}^{equil} \sim c_s^{-2} = (1 + 2Xp_{XX}/p_X)$ a function of how far we are from the canonical limit. Current constraints are fairly weak ($|f_{NL}| \lesssim \mathcal{O}(100)$; $c_s \gtrsim 0.1$)[9, 10, 11], but are expected to improve with more sensitive data in the near future.

1.3. Noncanonical inflationary solutions

To find solutions of noncanonical inflation, we first generalise the canonical slow-roll parameters and then invert the equations of motion in the limit that our new generalised inflationary parameters are small. This procedure reproduces the inflationary solutions of canonical slow-roll inflation in the canonical limit (see [3] for details). By definition, H must be approximately constant on inflationary trajectories. This condition defines ϵ, as in the canonical case:

$$\epsilon \equiv -\frac{\dot{H}}{H^2} = -\frac{d}{dN}\log H << 1 \tag{1.2}$$

[1] We consider only Lagrangians which obey the null-energy condition, i.e. $\frac{\partial p}{\partial X} \geq 0$, but our formalism can also be extended to Lagrangians with negative coefficients for the higher-dimensional operators in some region of the phase space. In this case we reproduce the results of [4] (along with subleading corrections) and furthermore find an additional inflationary parameter which must be small and is not seen in the analysis of [4].

Furthermore, in order for inflation to last long enough, the time variation of ϵ must be small over several e-foldings:

$$\eta = \frac{\dot{\epsilon}}{H\epsilon} = 4\epsilon - \eta_X - \eta_\Pi << 1, \tag{1.3}$$

where

$$\eta_X \equiv \epsilon - \frac{\dot{X}}{2HX;} \quad \eta_\Pi \equiv \epsilon - \frac{\dot{\Pi}}{H\Pi}.$$

Rewriting the EOM for $\epsilon, \eta_X, \eta_\Pi$ small, we find the **general non-canonical inflationary solutions**

$$p_{inf} = -\rho_{inf}\left(1+\frac{2}{3}\epsilon\right) \approx -\rho_{inf} \tag{1.4}$$

$$\Pi_{inf}(\phi) = \frac{\partial p}{\partial\phi}\frac{1}{3H}\left(1+\frac{\epsilon-\eta_\Pi}{3}\right)^{-1} \approx \frac{\partial p}{\partial\phi}\frac{1}{3H}. \tag{1.5}$$

Here p_X indicates a derivative with respect to X, $\rho \equiv 2Xp_X - p$, and $\Pi \equiv \frac{\partial p}{\partial\phi} = \dot{\phi}\frac{\partial p}{\partial X}$. The inflationary trajectories described by these solutions can be shown to be attractors in both the small and large field case for all models of noncanonical inflation, including the canonical limit of small and large field inflation [3]. Note that (1.4) and (1.5) reduce to the canonical solution in that limit, while $\epsilon \to M_p^2/2(V'/V)^2$ and $\eta_X, \eta_\Pi \to M_p^2 V''/V$. Whether we are in the canonical or non-canonical regime depends on the parameter

$$A = \frac{V'}{3H}\frac{1}{\Lambda^2} = \left(\frac{2}{3}\epsilon_{SR}\frac{V}{\Lambda^4}\right)^{\frac{1}{2}}; \tag{1.6}$$

the limit $A \ll 1 \Rightarrow \epsilon_{SR} \ll 1$ corresponds to canonical inflation, while in the case $A \gg 1 \Rightarrow \epsilon_{SR} \geq 1$ inflation will be highly non-canonical. The inflationary parameters are suppressed by powers of A compared to the usual slow-roll results: $\epsilon, \eta_X \approx \frac{\epsilon_{SR}}{A^m}$ and $\eta_\Pi \approx \frac{\eta_{SR}}{A^m}$.

1.4. Conditions for noncanonical inflation

Not all Lagrangians of the form (1.1) support NCI. A detailed discussion of the conditions sufficient for a Lagrangian to have non-canonical inflationary solutions is given in [3]; here we give a brief summary. First, we impose two physicality constraints, requiring that the action satisfy the null-energy condition $\frac{\partial p}{\partial X} \geq 0$, and that superluminal propagation of perturbations be disallowed, i.e. $\frac{\partial^2 p}{\partial X^2} > 0$. We begin by assuming that the Lagrangian can be written in the separable form

$$p(X,\phi) \approx \Lambda^4 q\left(\tfrac{X}{\Lambda^4}\right) - V(\phi). \tag{1.7}$$

For power series Lagrangians with a nonzero radius of convergence $X = R$, NCI will be supported within the radius of convergence if A and V/Λ^4 are large and $\partial p/\partial X$ diverges at R. For a closed-form Lagrangian we must have $A, V/\Lambda^4 \gg 1$ as well as a positively curved kinetic term $\partial^2 q/\partial X^2 > 0$. This condition is just the condition hat the speed of propagation of perturbations be real and subluminal, which we have already imposed. If a Lagrangian is not separable, noncanonical inflationary solutions exist only in the limit where the coefficients of X^n do not depend strongly on ϕ. These conditions are illuminating, since they make it clear that the existence of a speed limit on ϕ, seen in DBI inflation for instance, is neither necessary nor sufficient for the existence of NCI solutions [3].

2. Initial Conditions

2.1. The Initial Conditions Fine Tuning Problem

We now know how to tell if a given Lagrangian supports NCI, how to find the general inflationary solutions, and that inflationary trajectories are attractors in phase space [3]. However, not all initial conditions give rise to trajectories that reach the inflationary attractors. Trajectories with an arbitrarily large initial momentum can miss the attractor altogether, a situation known as the initial conditions fine-tuning (ICFT) or overshoot problem. Typically, if $\Delta\phi$ is the distance the inflaton travels during inflation, small-field ($\Delta\phi \ll M_p$) models have an ICFT problem while large-field ($\Delta\phi \gg M_p$) models do not [12]. In this section we summarise the effect of a period of NCI on the ICFTP, finding that trajectories generically decay towards the inflationary solution at a steeper angle for noncanonical kinetic terms, compared to canonical kinetic terms, so that a larger fraction of the initial-coniditons space leads to inflation. Thus noncanonical kinetic terms can reduce the ICFT problem in small-field inflation models, when NCI is relevant in the allowed phase space. This work was published in [13].

2.2. Regional Analysis

To begin with, we divide phase space into regions of interest, depending on the dominant terms in the EOM: for

$$\rho = K.E + P.E; \qquad \dot{\Pi} = -3H\Pi + V'(\phi), \tag{2.8}$$

we define

- **Region A**: $H\Pi \gg V'$ and K.E. $\gg V(\phi)$.
- **Region B**: $H\Pi \ll V'$ and K.E. $\ll V(\phi)$
- **Region C**: $H\Pi \gg V'$ and K.E. $\ll V(\phi)$
- **Region D**: $H\Pi \ll V'$ and K.E. $\gg V(\phi)$

as shown in Figure 1. [2] The inflationary trajectory, which exists for only part of the phase space range, is indicated by the dashed line Γ. Notice how Regions A and B shrink when going to the non-canonical case. The dot-dashed line $\Pi = \Lambda^2$ corresponds to $A = 1$; inflation below this line is non-canonical while inflation above this line is noncanonical. Arrows denote the general direction of flow of trajectories, and are generally steeper for non-canonical kinetic terms. The trend is clearly visible for the specific example, shown in Figure 2, of an inflection-point potential (with parameters chosen to be consistent with current observations of the power spectrum and spectral index). Previously overshooting trajectories for a canonical kinetic term are attracted strongly to the inflationary solution. In Section 2.3 we will study this behaviour analytically.

2.3. Parametrising the dynamical behaviour

We can analytically calculate the angle the trajectory makes in phase space:

$$\tan\theta = \frac{1}{M_p}\frac{d\Pi}{d\phi} = \frac{1}{M_p}\frac{\dot{\Pi}}{\dot{\phi}}; \tag{2.9}$$

$$\tan\theta_{canon} = -\frac{\dot{\Pi}}{\Pi M_p} = \frac{3H\Pi - V'}{\Pi M_p}; \tag{2.10}$$

$$\tan\theta_{non-canon} = -\frac{\dot{\Pi}}{M_p\sqrt{2X}} = \frac{3H\Pi - V'}{\sqrt{2R}M_p\Lambda^2}. \tag{2.11}$$

[2] Note that in this section, Π has an additional minus sign compared to the usual definition because we work in only one quadrant of phase space where $\phi \geq 0$ and $\dot{\phi} \leq 0$ and it is convenient to choose this to be the purely positive (first) quadrant where $\phi, \Pi \geq 0$.

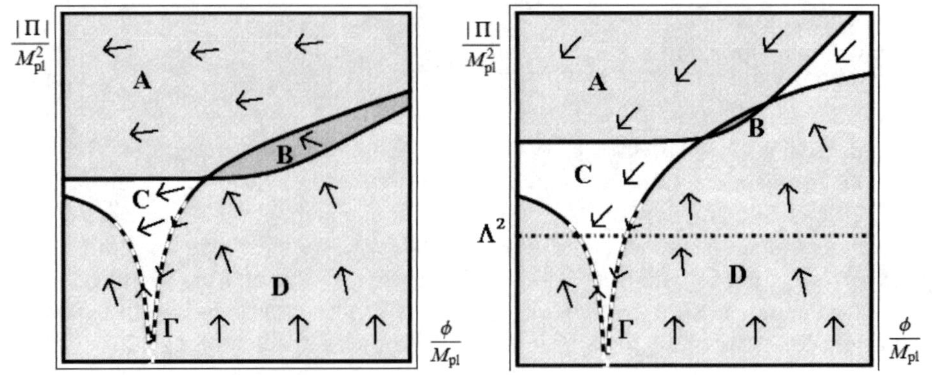

Figure 1. Regions A-D of (ϕ, Π) phase space for an inflection-point type potential with canonical (left) and non-canonical (right) kinetic terms.

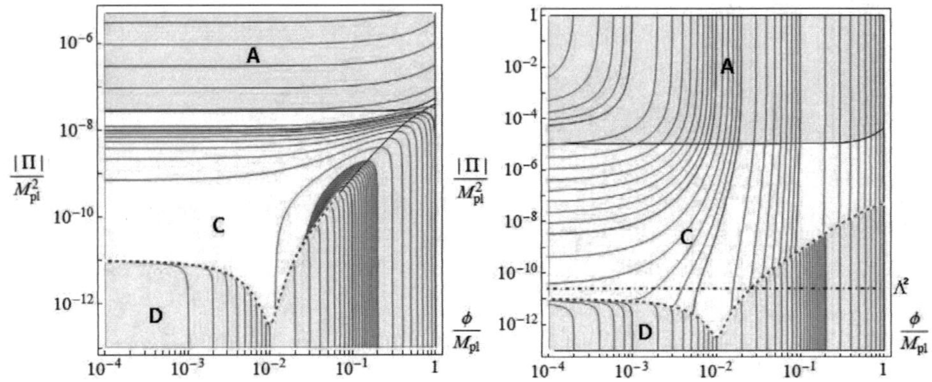

Figure 2. Left: Phase-space plot for a canonical Lagrangian with an inflection-point potential. Right: Phase-space plot for the non-canonical Lagrangian with the same inflection-point potential.

By this definition we find, as expected, that trajectories in Region A are approximately horizontal in the canonical case. These are the overshoot trajectories. In all regions it follows from (2.9)-(2.11) that **non-canonical trajectories are steeper**:

$$\frac{\tan\theta_{non-can}}{\tan\theta_{canon}} \gg 1. \qquad (2.12)$$

Furthermore, we can parametrise the amount of overshoot using the overshoot parameter α, defined as the ratio of the typical distance a trajectory must travel in order to exit Region A, $(\Delta\phi)_A$, to the size of field space in which inflation occurs, $(\Delta\phi)_{inf}$:

$$\alpha \equiv \frac{(\Delta\phi)_A}{(\Delta\phi)_{inf}}. \qquad (2.13)$$

When $\alpha \ll 1$, there is no fine-tuning problem, while for $\alpha \gg 1$, fine tuning will be necessary to avoid overshoot. Advantages of using α to parametrise the fine-tuning are that it can be calculated analytically and is independent of potential. Calculating $(\Delta\phi)_A^{canon}$ (by integrating $\left.\frac{d\Pi}{d\phi}\right|_A = \sqrt{\frac{3}{2}\frac{\Pi}{M_p}}$) gives $(\Delta\phi)_A^{canon} \sim M_p$ for canonical models, independent of the potential. This makes explicit why small-field models, for which $(\Delta\phi)_{inf} \ll M_p \sim (\Delta\phi)_A$, have a fine-tuning problem while large-field models, for which $(\Delta\phi)_{inf} \sim \mathcal{O}(10-100) \times M_p$, do not. Similarly, $(\Delta\phi)_A^{noncanon}$ can be computed for noncanonical inflationary models. Comparing to canonical kinetic terms, the ratio $\alpha_{non-canon}/\alpha_{canon}$ is found to be

$$\frac{\alpha_{non-canon}}{\alpha_{canon}} \sim \frac{(2R)^{1/2}}{2\sqrt{3}}\sqrt{\frac{\Lambda^4}{V}}, \tag{2.14}$$

confirming that the overshoot problem is dramatically reduced by the presence of noncanonical kinetic terms exactly when $A \sim \sqrt{V/\Lambda^4}$ is large, i.e. the theory is strongly noncanonical. This confirms our intuition from the plots above.

3. Conclusion

In this article we have reviewed the results of [3] and [13] on the nature of noncanonical inflationary solutions. The results can be summarised as follows: we have given a general formalism for constructing NCI soliutions in terms of the generalised inflationary parameters ϵ, η_X and η_Π, which smoothly reduce to the usual SR parameters in the canonical limit. We have given (sufficient) conditions for constructing Lagrangians which support NCI. We have also analysed the effect of noncanonical kinetic terms on the dynamical behaviour in (homogeneous) phase space, finding that trajectories decay towards the inflationary attractor at a steeper angle in the noncanonical case. Thus noncanonical kinetic terms tend to reduce the severity of the overshoot problem, parametrised by the overshoot parameter α, as long as some period of NCI occurs in the allowed phase space of the theory.

Acknowledgments

We would like to thank the organisers and participants of the 2010 PASCOS conference. B.U. is supported in part through an IPP (Institute of Particle Physics, Canada) Postdoctoral Fellowship, and by a Lorne Trottier Fellowship at McGill University. The work of P.F. is supported by the Natural Sciences and Engineering Research Council (NSERC) of Canada. The work of A.W. is supported by the Fonds Québécois de la Recherche sur la Nature et les Technologies (FQRNT). R.G. is supported by an NSERC Postdoctoral Fellowship and a Canada-UK Millennium Research Award.

[1] Baumann D and McAllister L 2009 *Ann. Rev. Nucl. Part. Sci.* **59** 67–94 (*Preprint* 0901.0265)
[2] Silverstein E and Tong D 2004 *Phys. Rev.* **D70** 103505 (*Preprint* hep-th/0310221)
[3] Franche P, Gwyn R, Underwood B and Wissanji A 2010 *Phys. Rev.* **D81** 123526 (*Preprint* 0912.1857)
[4] Armendariz-Picon C, Damour T and Mukhanov V F 1999 *Phys. Lett.* **B458** 209–218 (*Preprint* hep-th/9904075)
[5] Garriga J and Mukhanov V F 1999 *Phys. Lett.* **B458** 219–225 (*Preprint* hep-th/9904176)
[6] Chen X, Huang M x, Kachru S and Shiu G 2007 *JCAP* **0701** 002 (*Preprint* hep-th/0605045)
[7] Salopek D S and Bond J R 1990 *Phys. Rev.* **D42** 3936–3962
[8] Bean R, Chung D J H and Geshnizjani G 2008 *Phys. Rev.* **D78** 023517 (*Preprint* 0801.0742)
[9] Komatsu E *et al.* (WMAP) 2009 *Astrophys. J. Suppl.* **180** 330–376 (*Preprint* 0803.0547)
[10] Smith K M, Senatore L and Zaldarriaga M 2009 (*Preprint* 0901.2572)
[11] Yadav A P S and Wandelt B D 2008 *Phys. Rev. Lett.* **100** 181301 (*Preprint* 0712.1148)
[12] Dodelson S, Kinney W H and Kolb E W 1997 *Phys. Rev.* **D56** 3207–3215 (*Preprint* astro-ph/9702166)
[13] Franche P, Gwyn R, Underwood B and Wissanji A 2010 (*Preprint* 1002.2639)

Journal of Physics: Conference Series **259** (2010) 012088 doi:10.1088/1742-6596/259/1/012088

Suppression of Lepton Flavor Violation from Corrections above M_{GUT}

Mario E. Gómez[1], **S. Lola**[2], **P. Naranjo**[1] **and J. Rodríguez Quintero**[1]

[1]Departamento de Física Aplicada, Facultad de Ciencias Experimentales, Universidad de Huelva, 21071 Huelva, Spain

[2]Department of Physics, University of Patras, 26500 Patras, Greece

E-mail: mario.gomez@dfa.uhu.es

Abstract. We show how quantum corrections above the GUT scale can result in supersymmetric models where lepton flavour violation is naturally suppressed, in agreement with all experimental bounds. Within this framework, a soft term structure as predicted by an Abelian flavour symmetry combined with $SU(5)$ RGEs for scales above M_{GUT}, results to an efficient suppression of the off-diagonal terms in the scalar soft matrices, particularly for $m_0 < 100$ GeV.

1. Introduction

The purpose of this presentation is to analyse the predictions for lepton flavor violation (LFV) in models where SU(5) unification is combined with flavour symmetries, taking into account the RGE evolution above the GUT scale from Ref. [1] and focusing on the WMAP preferred area presented in [2]. We show that, by postulating the mass matrices at a high scale M_X and evolving them down to M_{GUT}, the pathological situation encountered in the conventional models [3, 4] may be remedied. For regions of the parameter space with a low m_0, the pattern we end up with exhibits a sizable suppression in the off-diagonal terms as compared to the textures at M_X, yielding acceptable LFV predictions. We show that this is true, even in the case of maximal mixing in the charged lepton sector, which is the most dangerous one as far as LFV is concerned.

2. LFV in SUSY-SU(5) with see-saw neutrinos

In supersymmetric theories, charged lepton flavour violation may be generated at the loop-level, even in models with universal soft terms at a high scale. This is also true for the MSSM, when extended with a see-saw mechanism to generate small neutrino masses [5]. In this case, LFV terms are generated radiatively, since it is not possible to simultaneously diagonalise neutrino, charged lepton and slepton mass matrices [6].

In this section we consider a SUSY-SU(5) theory with see-saw neutrinos. Our starting point is the following SUSY $SU(5)$ superpotential:

$$\mathcal{W}_X = T_1^T\,\mathcal{Y}_u^\delta\,T_1\,H + T_1^T\,\not{\mathcal{Y}}_d\,\bar{F}_1\,\bar{H} + \bar{F}_1{}^T\,\mathcal{Y}_\nu^\delta\,S_1\,H + S_1^T\,\bar{M}_R\,S_1, \tag{1}$$

where $\mathcal{Y}_\alpha$ ($\alpha = u, d, \nu$) are the Yukawa matrices for the up-type quarks, down-quarks/charged-leptons and Dirac neutrinos, respectively. M_R is the heavy Majorana mass matrix. The symbol

Journal of Physics: Conference Series **259** (2010) 012088 doi:10.1088/1742-6596/259/1/012088

δ stands for *diagonal*, the original fields rotated as

$$T = U_{10}T_1, \qquad \bar{F} = U_{\nu L}\bar{F}_1, \qquad S = U_{\nu R}S_1, \tag{2}$$

with the rotating matrices defined as

$$\mathcal{Y}_u = U_{10}\mathcal{Y}_u^\delta U_{10}^T, \qquad \mathcal{Y}_d = U_{5L}^*\mathcal{Y}_d^\delta U_{5R}^\dagger, \qquad \mathcal{Y}_\nu = U_{\nu L}^*\mathcal{Y}_\nu^\delta U_{\nu R}^\dagger, \tag{3}$$

and

$$\mathcal{Y}_d = V_{CKM}^*\mathcal{Y}_d^\delta V_E^\dagger, \tag{4}$$

Here, $V_{CKM} = U_{10}^\dagger U_{5L}$ and $V_E = U_{\nu L}^\dagger U_{5R}$ denote the mixings in the quark and lepton sectors, while $\bar{M}_R = U_{\nu R}^T M_R U_{\nu R}$.

The off-diagonal contributions to slepton mass matrices, when the superfields are rotated so that charged leptons become diagonal, can be understood through three rotations at different energy scales:

- M_X. The rotations in the superpotential fields lead to the following transformation of the soft terms:
$$\bar{m}_{10}^2 = U_{10}^\dagger m_{10}^2 U_{10} \qquad \bar{m}_5^2 = U_{\nu L}^\dagger m_5^2 U_{\nu L} \tag{5}$$
- M_{GUT}. Assuming that $SU(5)$ is broken down to the MSSM gauge group, the superpotential becomes
$$\mathcal{W}_{MSSM} = Q^T\mathcal{Y}_u^\delta U H_2 + Q^T(V_{CKM}^*\mathcal{Y}_d^\delta)DH_2 + L^T(V_E^*\mathcal{Y}_d^\delta)EH_2 + L^T\mathcal{Y}_\nu^\delta SH_2 + S^T\bar{M}_R S \tag{6}$$
where we have absorbed the matrices V_E^* and V_{CKM}^* in the definitions of the superfields E and D respectively. The scalar soft masses then become:
$$m_E^2 = V_{CKM}^\dagger \bar{m}_{10}^2 V_{CKM}, \qquad m_L^2 = \bar{m}_5^2, \tag{7}$$
$$m_Q^2 = m_U^2 = \bar{m}_{10}^2, \qquad m_D^2 = V_E^\dagger \bar{m}_5^2 V_E, \tag{8}$$
- M_N. M_N is the scale at which the heavy right handed neutrinos decouple. Below this scale, the particle content is the one of the MSSM complemented with the neutrino mass operator resulting from the see-saw mechanism. Consequently, the superpotential can be written in a basis where the charged-lepton mass matrix becomes diagonal and the left slepton mass matrix becomes
$$\bar{m}_L^2 = V_E^\dagger m_L^2 V_E \simeq U_{5R}^\dagger m_5^2 U_{5R} \tag{9}$$

3. Runs above M_{GUT}: *Umbrella Effect.*

The introduction of a non-trivial flavour structure for the slepton soft terms at M_{GUT}, as predicted by the family symmetry that also generates Yukawa couplings, typically results to a large violation of the bounds on $l_j \to l_i\gamma$ [3, 4, 7]. This picture may be remedied by taking into account RGE effects from a scale $M_X > M_{GUT}$. In this case, the cosmological requirement of having a neutral particle as the LSP imposes low values on m_0, such that $m_{\tilde{\tau}} > m_\chi$ [2, 8, 9] (diagonal terms in the soft mass matrices have a large RGE growth, while non-diagonal elements remain almost unaffected by the runs). Thus, even assuming non-diagonal soft terms with matrix elements of the same order of magnitude at M_X, the corresponding matrix at M_{GUT} exhibits dominant diagonal elements.

To some extent, the RGE effect is similar to the action of closing an umbrella: the general non-universal soft terms at M_X resemble an open umbrella that approaches a diagonal matrix at the GUT scale. The sleptons tend then to be aligned by the flavour blind gauge interaction,

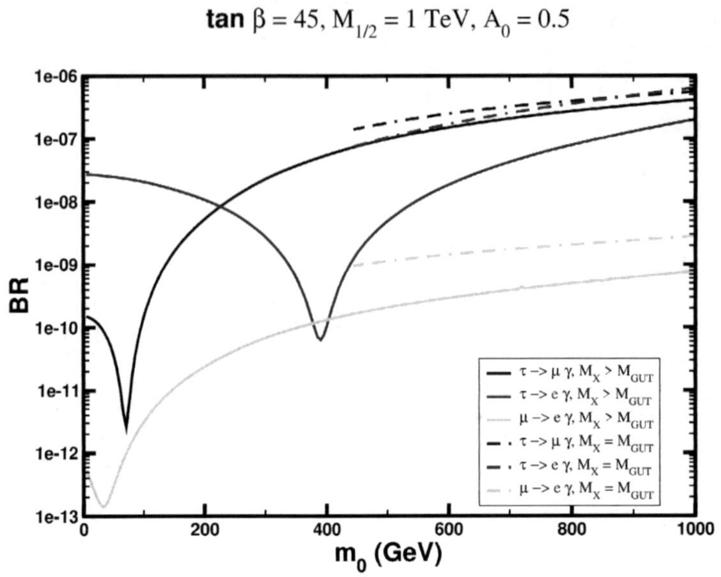

Figure 1. *Prediction for the charged-lepton flavour violating branching ratios showing the difference of taking either M_X or M_{GUT} as the starting point of the runs.*

driven by $\alpha_{\rm GUT}$ above the GUT scale. An analogous effect is well known to take place in the squark sector [10] where the strong radiative corrections to the squark masses, proportional to the gluino mass, are family independent and counteract the splitting effect on the top Yukawa coupling (the effect is commonly referred to as the "gluino focusing effect").

In Fig. 1, we show the differences between the following: i) SU(5) RGE evolution of the soft terms from a high scale M_X down to M_{GUT} and then to the MSSM with see-saw neutrinos (solid lines), and ii) Soft SUSY breaking terms given at M_{GUT} and then the MSSM with see-saw neutrinos (dash-lines). In case ii) we stop the lines at the value of m_0 below which $m_{\tilde{\tau}}$ becomes the LSP. In contrast, m_0 can even vanish at M_X in case i). The textures and soft terms we use are similar to Ref. [4]; in our work, however, we decouple the right-handed neutrinos below M_{GUT}. As a result, the predicted BR's do not vanish in the limit $m_0 = 0$.

4. $SU(5)$ textures

Let us consider an specific example of $SU(5)$ Yukawa textures that match the fermion data and may also predict the pattern of soft terms to be expected.

Following the $U(1)_F$ charge assignment in [11], the Yukawa matrices have the form

$$\mathcal{Y}_u \propto \begin{pmatrix} \varepsilon^6 & \varepsilon^5 & \varepsilon^3 \\ \varepsilon^5 & \varepsilon^4 & \varepsilon^2 \\ \varepsilon^3 & \varepsilon^2 & 1 \end{pmatrix}, \quad \mathcal{Y}_\ell^T \propto \mathcal{Y}_d \propto \begin{pmatrix} \varepsilon^4 & \varepsilon^3 & \varepsilon^3 \\ \varepsilon^3 & \varepsilon^2 & \varepsilon^2 \\ \varepsilon & 1 & 1 \end{pmatrix}, \quad \mathcal{Y}_\nu \propto \begin{pmatrix} \varepsilon^{|1\pm n_1|} & \varepsilon^{|1\pm n_2|} & \varepsilon^{|1\pm n_3|} \\ \varepsilon^{|n_1|} & \varepsilon^{|n_2|} & \varepsilon^{|n_3|} \\ \varepsilon^{|n_1|} & \varepsilon^{|n_2|} & \varepsilon^{|n_3|} \end{pmatrix} \tag{10}$$

where n_i stand for the heavy Majorana neutrino charges. We will assume that the entire lepton mixing is arising from the charged-lepton sector, which is potentially the most dangerous case as far as LFV is concerned.

In addition, flavour symmetries generally imply non-universal soft terms [12], since the structure of the soft terms is linked to the family charges. For the Yukawa textures in Eq.(10)

the soft mass matrices m_{10}^2 and $m_{\bar{5}}^2$ become

$$m_{10}^2 \propto \begin{pmatrix} 1 & \varepsilon & \varepsilon^3 \\ \varepsilon & 1 & \varepsilon^2 \\ \varepsilon^3 & \varepsilon^2 & 1 \end{pmatrix} m_0^2, \quad m_{\bar{5}}^2 \propto \begin{pmatrix} 1 & \varepsilon & \varepsilon \\ \varepsilon & 1 & 1 \\ \varepsilon & 1 & 1 \end{pmatrix} m_0^2, \tag{11}$$

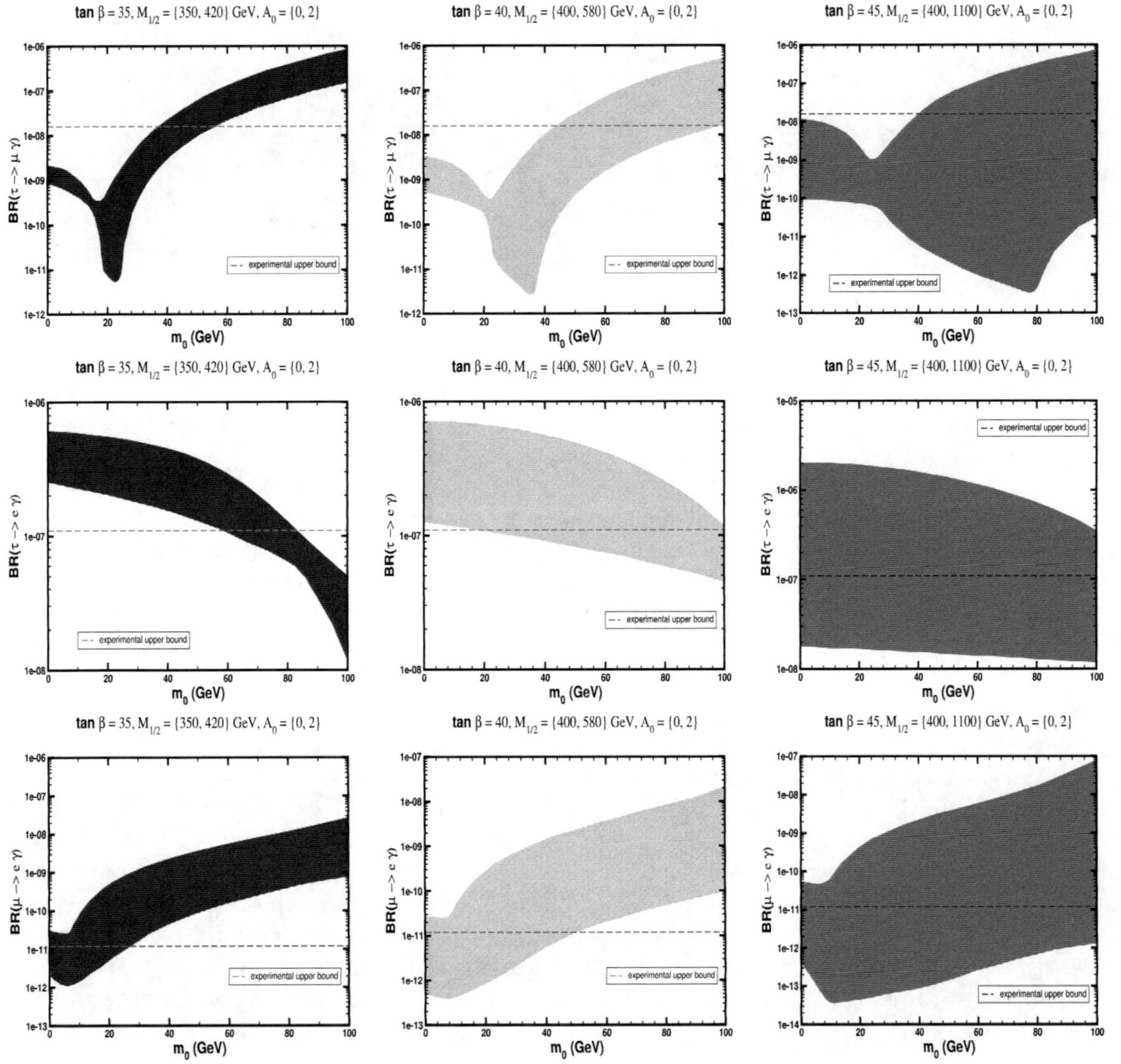

Figure 2. *Prediction for the $\tau \to \mu\gamma$ (top), $\tau \to e\gamma$ (middle) and $\mu \to e\gamma$ branching ratios for the cosmologically preferred area of values of $M_{1/2}$, A_0, for three different values of* $\tan\beta$.

Some LFV predictions of the above model are displayed in Fig. 3. We should stress that the depicted ranges for both m_0 and $M_{1/2}$ are the cosmologically preferred parameter space, as found in [2]. We can see that the case $\tan\beta = 35$ is ruled out, as there is no overlapping region for the three decays. This is because of the RR sector-induced cancellations mentioned above, which arises for much larger values of m_0 for $\tau \to e\gamma$ than for the other two processes. The case $\tan\beta$

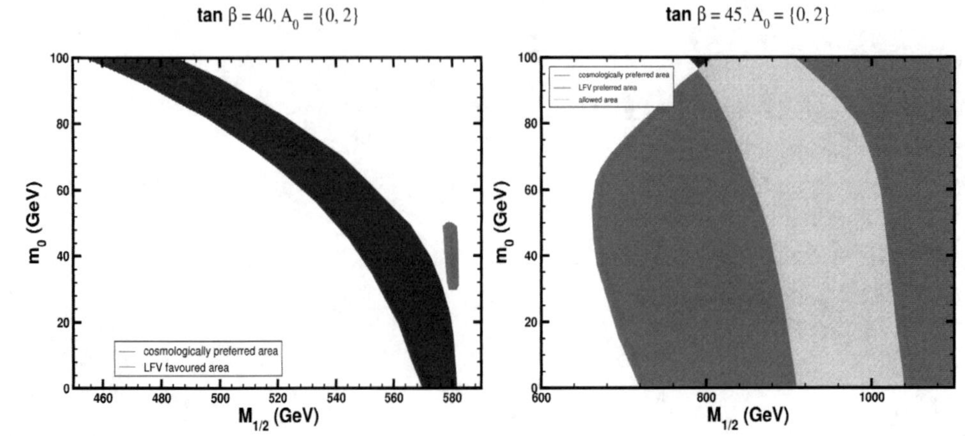

Figure 3. *Allowed parameter space when both LFV and cosmological constraints are taken into account for* $\tan\beta = 40$ *and* 45.

= 40 does possess a common area for $20 < m_0 < 50$ GeV. However, such an area lies outside the cosmologically preferred region, as shown on the left panel of Fig. 3. Thus, we conclude that the case $\tan\beta = 40$ is only marginally allowed. Finally, for $\tan\beta = 45$ the whole range for $m_0 < 100$ GeV is allowed, when suitable values for $M_{1/2}$, A_0 are chosen. Moreover, as shown on the right panel of Fig. 3, there exists an overlapping region when considering the cosmologically favoured parameter space (values of any parameters involved in these plots (m_0, $M_{1/2}$, A_0) beyond the ranges shown lead to tachyonic soft masses. Thus, GUT runs efficiently suppress the off-diagonal entries, yielding charged-LFV rates that render the $SU(5)$ model *compatible* with current experimental bounds.

5. Conclusions

We have shown that $SU(5)$ runs above the GUT scale, naturally suppress the off-diagonal entries of the soft matrices, through what has been called the *umbrella effect*, leading to a nearly flavour-independent contribution for $m_0 < 100$ GeV. Such low values of m_0 are discarded if $SU(5)$ runs are not taken into account, since in this case the lightest stau becomes the LSP. Within this framework, $SU(5)$ runs lead to acceptable LFV rates for the cosmologically preferred parameter space found in [2], favoring values of $\tan\beta$ around 45. Significant deviations from this value, in either way, are harder to reconcile with cosmological data. Our results indicate that even in this case, the umbrella effect leads to suppression to LFV rates, leading to a very significant enhancement of the available parameter space, as compared to the conventional schemes, with runs below M_{GUT}.

Acknowledgments

The research of S. Lola and P. Naranjo has been funded by the FP6 Marie Curie Excellence Grant MEXT-CT-2004-014297. M.E. Gómez and J. Rodríguez-Quintero acknowledge support from the project P07FQM02962 funded by "Junta de Andalucia", the Spanish MICINN projects FPA2009-10773 and MULTIDARK Consolider-Ingenio: CSD2009-00064.

References

[1] M. E. Gomez, S. Lola, P. Naranjo and J. Rodriguez-Quintero, JHEP **1006** (2010) 053 [arXiv:1003.4937 [hep-ph]].

[2] M. E. Gomez, S. Lola, P. Naranjo and J. Rodriguez-Quintero, JHEP **0904** (2009) 043 and AIP Conf. Proc. **1115** (2009) 273; E. Carquin et al., PoS IDM2008 (2008) 090.
[3] G. K. Leontaris and N. D. Tracas, Phys. Lett. B **431** (1998) 90; M. E. Gomez, G. K. Leontaris, S. Lola and J. D. Vergados, Phys. Rev. D **59** (1999) 116009.
[4] P. H. Chankowski, K. Kowalska, S. Lavignac and S. Pokorski, arXiv:hep-ph/0507133, proceedings of 40th Rencontres de Moriond on Electroweak Interactions and Unified Theories, La Thuile, Aosta Valley, March 2005.
[5] J. Hisano, T. Moroi, K. Tobe and M. Yamaguchi, Phys. Rev. D **53** (1996) 2442; J. Hisano and D. Nomura, Phys. Rev. D **59** (1999) 116005.
[6] F. Borzumati and A. Masiero, Phys. Rev. Lett. **57** (1986) 961; L. J. Hall, V. A. Kostelecky and S. Raby, Nucl. Phys. B **267** (1986) 415.
[7] R. Barbieri, L. J. Hall and A. Strumia, Nucl. Phys. B **445** (1995) 219; M. E. Gomez and H. Goldberg, Phys. Rev. D **53** (1996) 5244; G. Barenboim, K. Huitu and M. Raidal, Phys. Rev. D **63**(2001) 055006; L. Calibbi, A. Faccia, A. Masiero and S. K. Vempati, Phys. Rev. D **74** (2006) 116002.
[8] L. Calibbi, Y. Mambrini and S. K. Vempati, JHEP **0709** (2007) 081.
[9] J. Ellis, A. Mustafayev, K. Olive, arXiv:hep-ph/1003.3677.
[10] M. Dine, A. Kagan and S. Samuel, Phys. Lett. B **243** (1990) 250.
[11] J. R. Ellis, M. E. Gomez and S. Lola, JHEP **0707** (2007) 052.
[12] M. Raidal et al., Report of Working Group 3 of the CERN Workshop Flavor in the Era of the LHC, Eur.Phys.J. **C57** (2008) 13 and references therein.

Journal of Physics: Conference Series **259** (2010) 012089 doi:10.1088/1742-6596/259/1/012089

Right-handed neutrino magnetic moments

Alberto Aparici[1], Kyungwook Kim[2], Arcadi Santamaria[1] and José Wudka[2]

[1] Departament de Física Teòrica, Universitat de València and IFIC, Universitat de València-CSIC, Doctor Moliner 50, E-46100 Burjassot, València, Spain
[2] Department of Physics and Astronomy, University of California, Riverside, California 92521-0413, USA

E-mail: alberto.aparici@uv.es

Abstract. We consider the most general dimension 5 effective Lagrangian that can be built using only Standard Model fields plus right-handed neutrinos, and find that there exists a term that provides electroweak moments (i.e., couplings to the Z and photon) for the right-handed neutrinos. Such term has not been described previously in the literature. We discuss its phenomenology and the bounds that can be derived from LEP results and from the observation of the cooling process of red giants and supernovae.

1. Motivation

Neutrino physics has been a hot topic of research and discussion in the last thirty years, and especially since we have compelling evidence that the structure of their masses is highly nontrivial (for a review on the subject, see [1, 2]). The remarkable smallness of these masses, at least a factor 10^5 lighter than the electron mass, is usually regarded as an indication that new physics should be involved in their generation.

The definite nature of this new physics, of course, depends on the specific model one wishes to consider, and there are plenty of them: the seesaw mechanism [2], which is one of the most popular proposals, the several models for radiative generation of masses [3] and many more. Precisely this abundance of proposals makes appealing the possibility of studying the new physics associated to neutrinos in a model-independent way. This can be done most easily if the new particles are very heavy, by eliminating them from our description of low-energy physics; the result is called an *effective theory*[1], and has been for long a powerful tool for examining neutrino physics (see, for example, two early but interesting applications in [5, 6]).

However, in our current situation, and given the present knowledge and unknowns about neutrino masses, a piece of the puzzle has been mainly ignored: as we don't know if the neutrino masses are Dirac or Majorana, right-handed neutrinos might be among the light degrees of freedom of the theory. If they are, they must be included in the effective theory, or otherwise it will be incomplete. In any case, we don't know yet the nature of neutrino masses, so it seems sensible to include them for the sake of generality.

This constitutes our starting point: we want to inspect an effective theory which can describe all possible neutrino mass structures, and see what insight it can cast upon new physics effects;

[1] For an example of the procedure, see, for instance, [4].

for that aim we include right-handed neutrinos among the low-energy fields. This we do in section 2, and promptly find a dimension five operator that describes Majorana electroweak moments for the right-handed neutrinos. This operator had not been previously discussed in the literature. We spend the remaining of this work, namely sections 3 and 4, analysing the bounds and prospects of observation for such interactions. All along this work we follow closely that of [7], where the reader may find wider and more detailed discussions on all the topics covered here.

2. The effective Lagrangian

This section is devoted to the discussion of an effective Lagrangian capable of describing the most general neutrino masses. For that purpose, we consider as low-energy fields those of the Standard Model (SM) plus a number of right-handed neutrinos. Following the rules of effective field theory, we construct with our low-energy fields all the terms compatible with our low-energy symmetries (namely, the SM symmetries). We don't have to stop at the renormalisable operators, but can continue to build operators of dimension five, six, etc.; such operators have couplings suppressed by inverse powers of a certain energy scale, the 'new physics scale', that we denote by $\Lambda_{\rm NP}$. This scale is identified as the mass scale of the heavy particles that we have abstracted from the theory; it is the trace that these particles leave in the effective theory.

Among all the operators found, we want to focus in this work on a certain dimension five effective operator:

$$\mathcal{L}_\zeta = \overline{\nu_{\rm R}^{\rm c}}\,\zeta\,\sigma^{\mu\nu}\nu_{\rm R}\,B_{\mu\nu} + {\rm H.c.} \qquad (1)$$

which describes Majorana electroweak moments for the right-handed neutrinos. The field B in (1) is the gauge boson of the SM hypercharge, $B_\mu = \cos\theta_W\,A_\mu - \sin\theta_W\,Z_\mu$, while ζ is a matrix of dimensionful couplings in the right-handed neutrino flavour space. Fermi statistics impose that ζ is antisymmetric, and so the electroweak moments are all *transition* moments. As the operator is five-dimensional, the couplings will be suppressed by one power of the new physics scale; we will generically write $\zeta \sim {}^1\!/_{\Lambda_{\rm NP}}$.

Before we go on with the analysis of the effective interaction, we want to present (1) also in terms of the mass eigenfields. In the course of this work we will try to avoid assuming particular mass structures for the neutrinos; sometimes, however, the calculations will require us to do so. In those cases, we will assume a seesaw-like scenario for neutrino masses, and will use the following form for the electroweak moments:

$$\mathcal{L}_\zeta = \left(\overline{N} - \overline{\nu}\,\epsilon^*\right)\left(\zeta\,{\rm P_R} - \zeta^*\,{\rm P_L}\right)\sigma^{\mu\nu}\left(N - \epsilon^{\rm T}\,\nu\right)B_{\mu\nu} \qquad (2)$$

where ν represents the light, usual neutrinos, whereas N represents the heavy neutrinos, which under seesaw are composed mainly of $\nu_{\rm R}$; the extent to which light and heavy neutrinos are mixed is measured by the matrix ϵ, which is composed of small entries, and we can write generically $\epsilon \sim \sqrt{m_\nu/m_N}$, where m_ν represents the order of the light neutrino masses, $m_\nu \sim \mathcal{O}(1\ {\rm eV})$, and m_N represents the order of the heavy neutrino masses. Notice that through this change of basis, the right-handed neutrino moments generate a whole set of heavy-heavy, light-light and heavy-light electroweak moments, each suppressed by the appropriate number of powers of ϵ. Note also that the change of basis does not change the character of the moments: all the couplings in (2) are transition moments.

3. Bounds on the electroweak moment

In this section we review the most stringent bounds that can be placed upon the new electroweak moments, both arising from collider experiments and astrophysical observations. Several astrophysical scenarios allow us to constrain tightly the new interactions if the masses of the particles associated with the right-handed neutrinos are light. Collider bounds extend to heavier

Journal of Physics: Conference Series **259** (2010) 012089 doi:10.1088/1742-6596/259/1/012089

masses, but are poorer. Further comments on other bounds, either from collider or astrophysics, can be found in [7].

3.1. Collider bounds from invisible Z decays

Z boson decays were thoroughly studied at CERN's electron-positron collider LEP. In particular, comparing the width of the Z line-shape with the width deduced from observed Z-decay events, one obtains the *invisible width* of the Z boson [8]. Within the SM only neutrinos can escape unobserved, so the invisible width is interpreted as the partial decay width to neutrinos; but were the electroweak moments to exist, the Z would also decay to the massive particles composed of right-handed neutrinos (if such particles are not heavier than the Z itself). Then the question is whether LEP data leave room for the new moments, especially bearing in mind that the observed invisible Z width is in very good agreement with the value expected from theoretical SM calculations with three neutrino species.

To answer this question we calculate the remaining width that can not be accounted for by SM decays: from the invisible width, $\Gamma_{\text{inv}} = 499 \pm 1.4$ MeV, we subtract the SM contribution, which we deduce from the experimental partial width to charged leptons, $\Gamma_{\ell^+\ell^-} = 83.984 \pm 0.086$ MeV, a quantity that is not affected by the new moments. Using the known theoretical neutrino-to-charged-lepton width ratio, we can transform $\Gamma_{\ell^+\ell^-}$ into a mixed theoretical-experimental quantity that we take as the SM contribution to the invisible decay width. Then:

$$\Delta\Gamma_{\text{inv}} = \Gamma_{\text{inv}} - 3\left(\frac{\Gamma_{\bar{\nu}\nu}}{\Gamma_{\ell^+\ell^-}}\right)^{\text{SM}} \Gamma_{\ell^+\ell^-} = -2.6 \pm 1.5 \text{ MeV} \tag{3}$$

which tells us that the contribution of new physics to the invisible width is compatible with zero. But of course a negative value for a quantity that is positive-definite is meaningless; we use the Feldman & Cousins prescription [9] to transform (3) into an upper bound on $\Delta\Gamma_{\text{inv}}$, resulting in

$$\Delta\Gamma_{\text{inv}} < 0.72 \text{ MeV} \qquad 90\% \text{ C.L.} \tag{4}$$

We can use this result to constrain our new interactions. The width of the Z going to right-handed neutrinos through the electroweak moments, neglecting the masses of the products, is

$$\Gamma\left(Z \to \nu_{\text{R}i}\nu_{\text{R}j}\right) = \frac{2}{3\pi}\,|\zeta_{ij}|^2 \sin^2\theta_W\, m_Z^3$$

and from that, assuming that the new moments account for all of (4), we have

$$\Lambda_{\text{NP}} = \frac{1}{|\zeta_{ij}|} > 7 \text{ TeV}$$

This bound applies whenever $m_i + m_j < m_Z$ and gets somehow degraded if we consider massive final states with masses comparable to m_Z. For a picture of the excluded region, see figure 5.

3.2. Astrophysical bounds from the helium flash of red giants

In 1990, Georg Raffelt pointed out [10] that the observational limits on the luminosity of red giant stars just before and after the ignition of helium could be converted into limits on the cooling mechanisms of the star core. In particular, magnetic moments for weakly interacting particles pose one of such cooling mechanisms; the reason is that photons immersed in a plasma acquire an effective mass term due to the interaction with the surrounding charged particles. Such massive photons, known as *plasmons*, can decay as any other massive particle can, and if they have couplings to, say, right-handed neutrinos, those right-handed neutrinos could escape the star core, thus providing a mechanism for energy loss. The magnetic moment of any such weakly interacting particle light enough to be produced in plasmon decays is, then, bounded; specifically, it is bounded to very small values, $\mu_{\text{eff}} < 3 \times 10^{-12}\,\mu_B$.

We have calculated the plasmon decay width to right-handed neutrinos through the electroweak moments, and obtain, in the plasma rest frame,

$$\Gamma\left(\gamma^* \to \nu_{\text{R}i}\nu_{\text{R}j}\right) = \frac{2}{3\pi}\,|\zeta_{ij}|^2 \cos^2\theta_W \,\frac{m_P^4}{\omega} = \frac{\mu_{\text{eff}}^2}{24\pi}\,\frac{m_P^4}{\omega}$$

where m_P is the plasmon effective mass, ω is the plasmon energy in the plasma rest frame, and we have neglected the mass of the final state particles. The maximum plasmon mass is given by the maximum plasma energy, which in the case of the core of red giants is around 5 keV. μ_{eff} is the quantity bounded by [10]; from this expression we obtain that

$$\Lambda_{\text{NP}} > 4 \times 10^6 \text{ TeV}$$

applicable whenever the mass of the final state neutrinos is lighter than the plasmon mass, that is to say, $m_i + m_j < 5$ keV. The bound degrades slightly, due to kinematical suppression of the decay, if the masses are very close to m_P.

3.3. Astrophysical bounds from the anomalous cooling of supernova cores

The bound described in the previous section is impressively stringent, but can only apply to rather light right-handed neutrinos. As the mass of the plasmons is essentially determined by the energy of the surrounding plasma, seems wise to look for denser, hotter objects to which similar cooling limits can be applied. Among the several possibilities [11], the one that imposes the tightest bounds is the anomalous cooling of supernova cores.

The core of a supernova is so dense that even the usual, light neutrinos become trapped; thus, a cooling mechanism for such objects requires something interacting even more weakly. Right-handed neutrinos can be a good candidate, and (2) allows us to produce them through a transition process like $\nu + \gamma \longrightarrow N$; such process has attracted much attention in the literature, and the magnetic moment involved is bounded to be $\mu_{\text{eff}} < 3 \times 10^{-12}\,\mu_B$ [11]. The mass of the final N here will be limited by the maximum energy of the plasma in the supernova core, which is around 30 MeV.

Notice that in the two previous sections we could neglect the mass of the final ν_{R}, and so forget about the mass and mixing structure of the neutrinos. The process that we are considering now, however, requires of a $\nu - N$ transition moment, which only arises from (1) through mass mixing; therefore, to make this calculation we will have to assume something about the mass structure. As we stated in section 2, we will use the popular seesaw mechanism, which yields that the $\nu - N$ magnetic moment is to be suppressed by a factor of ϵ, the heavy-light mixing, with $\epsilon \sim \sqrt{m_\nu/m_N}$.

With all these elements, we can make the calculation and obtain the following bound:

$$\Lambda_{\text{NP}} > 4 \times 10^6 \times \sqrt{\frac{m_\nu}{m_N}} \text{ TeV}$$

which is also depicted in figure 5. Remember, this bound only applies for $m_N < 30$ MeV.

4. Prospects for the observation of the electroweak moments in colliders

The electroweak moments open a new window for the production of right-handed neutrinos in colliders. These particles, being singlets under all the symmetries of the SM, are frequently regarded as 'sterile'; if (1) is present, however, the situation is different: the right-handed neutrinos can be produced through a Drell-Yan process (i.e., through an s-channel Z or photon), and also can decay with very clear signatures. Of course, many searches have been conducted in the different collider experiments for such decaying neutral fermions [8], all of which have yielded negative results. The bound from invisible decays of the Z boson (see figure 5) somewhat covers the excluded region provided by such searches, though for particular right-handed neutrino spectra the bounds can be extended to higher $\Lambda_{\rm NP}$ [7].

In any case, newer machines such as Tevatron and LHC, with enhanced capabilities both in terms of energy and luminosity, must allow us to probe further regions of the parameter space. To estimate the size of these regions, we have calculated the production cross section of heavy neutrinos in proton-proton and proton-antiproton collisions, first calculating the quark-antiquark process and then convoluting over the parton distribution functions of the proton (for which we used the CTEQ6.6M pdf set [12]). We have translated those cross sections into explorable parameter space regions in figure 5; in these plots, we have shaded the areas where we expect 100 or more pairs of heavy neutrinos to be produced, assuming an integrated luminosity of 10 fb^{-1} for Tevatron and 1000 fb^{-1} for LHC, and a center-of-mass energy of 1,96 TeV for Tevatron and 14 TeV for LHC.

Once produced, the heavy neutrinos must be detected; fortunately, our interactions provide a mechanism which can generate very clear signals. As we said before, all the moments in (2) are transition moments; this means that if the heavy neutrinos are to be generated in pairs, all these pairs will be composed of dissimilar particles; in particular, they will have different masses. But (2) also provides a way for massive particles to decay to lighter ones with emission of a photon; this decays are expected to be very fast, at least for $m_N \gtrsim 1$ GeV and $\Lambda_{\rm NP} < 1000$ TeV. So, what we expect to see when a production event occurs is a cascade of hard photons with fixed, definite energies which can be easy to spot and identify.

After that, all which is left is a population of N_1, the lightest heavy neutrino, whose fate depends on the mass mixing with the light states; if the mixings are large, they will decay fast to SM particles and gauge bosons; if they are small, the N_1's might escape the collider undetected. In between there's the possibility of the N_1 decaying away from the collision point, thus providing a very nice detection signal, but the analysis is not simple because the mean free path depends also on the boost of the N_1, which in turn depends on the kinematical conditions of the production and the energy of the parent partons. We believe that the photon cascade is a safer signal for the detection of the electroweak moments, even leaving this last possibility open.

5. Conclusions

We have presented here some of the results discussed in [7]; we have argued that, given the current knowledge of neutrino masses, an effective field theory analysis of the Standard Model should include right-handed neutrinos, to account for possible Dirac terms in the neutrino mass matrix. By doing so, we have found a dimension five operator which provides electroweak moments for the $\nu_{\rm R}$; such interactions might open up a series of completely new scenarios in which the $\nu_{\rm R}$ are no longer 'sterile', and so an analysis of their phenomenology is in order. We have discussed what we can say about these electroweak moments from past collider experiments and astrophysical observations, and presented what insight can be gained from current collider experiments.

Figure 5 summarises all these issues. We see that for right-handed neutrino masses below roughly 10 keV a very stringent bound is cast upon the electroweak moments from the observational limits on energy loss in red giant cores. A milder but still tight bound can be extended up to 30 MeV masses from limits on energy losses in supernova cores, but this one depends on the mixing between the right-handed and left-handed neutrinos, so it must be considered less safe and more model-dependent. Finally, the fact that the decay width of the Z boson to undetected particles agrees so exactly with the theoretical expectations for the SM with three neutrino species also imposes a limit on the possible decays of Z into $\nu_{\rm R}$; this bound is definitely weaker, but the experimental setup is solid and well understood, and it extends to masses up to m_Z.

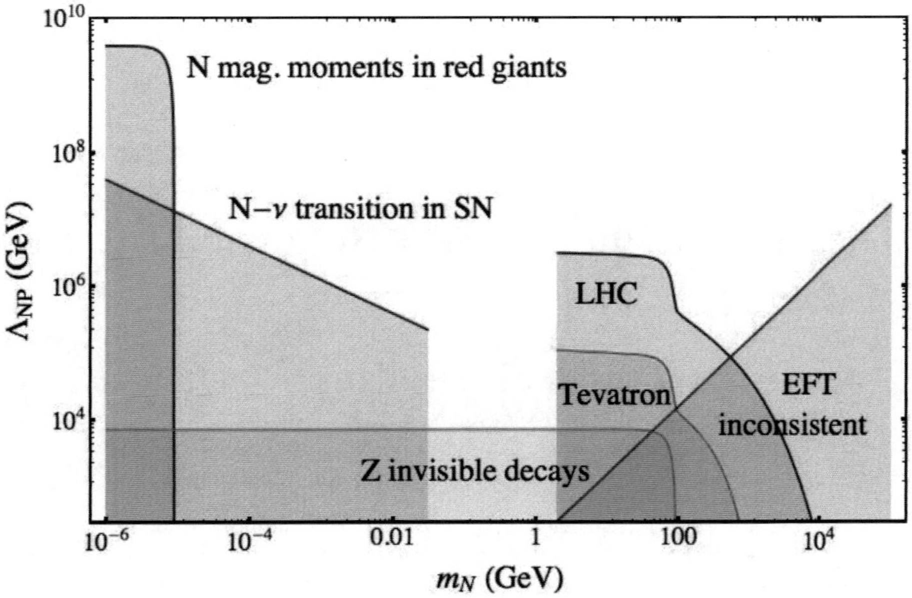

Figure 1. A summary of all the results presented. The shaded regions marked as 'N magnetic moments in red giants', '$N - \nu$ transition in SN' and 'Z invisible decays' represent exclusion regions, while the 'Tevatron' and 'LHC' ones represent regions to be probed. The rightmost red area is the region where the effective field theory loses its validity because the low-energy particles (in this case, heavy neutrinos) have masses of the order of the new physics scale.

Finally, the prospects for current collider experiments show that important regions of the parameter space are to be probed both in Tevatron and LHC, especially for masses below the Z mass. We have explained that the collider production, via a Drell-Yan process, should be followed by a cascade of hard photons of definite energies, which constitutes a neat detection signal. Thus, we feel optimistic about the prospects for this interaction in ongoing collider experiments.

This work has been supported by the Ministry of Science and Innovation (MICINN) Spain, under Grant No. FPA2008-03373, by the European Union within the Marie Curie Research & Training Networks, MRTN-CT-2006-035482 (FLAVIAnet), and by the U.S. Department of Energy Grant No. DE-FG03-94ER40837. A.A. is supported by the MICINN under the FPU program.

References

[1] Nakamura K and Petcov S T (Particle Data Group) 2010 *J. Phys. G: Nucl. Part. Phys.* **37** 075021, Chapter 13 URL `http://pdg.lbl.gov/2010/reviews/rpp2010-rev-neutrino-mixing.pdf`
[2] King S F 2004 *Rept. Prog. Phys.* **67** 107–158 (*Preprint* `hep-ph/0310204`)
[3] Babu K S and Ma E 1989 *Mod. Phys. Lett.* A **4** 1975
[4] Bilenky M S and Santamaria A 1994 *Nucl. Phys.* B **420** 47–93 (*Preprint* `hep-ph/9310302`)
[5] Weinberg S 1979 *Phys. Rev. Lett.* **43** 1566–1570
[6] Weldon H A and Zee A 1980 *Nucl. Phys.* B **173** 269
[7] Aparici A, Kim K, Santamaria A and Wudka J 2009 *Phys. Rev.* D **80** 013010 (*Preprint* `0904.3244`)
[8] Nakamura K *et al.* (Particle Data Group) 2010 *J. Phys. G: Nucl. Part. Phys.* **37** 075021
[9] Feldman G J and Cousins R D 1998 *Phys. Rev.* D **57** 3873–3889 (*Preprint* `physics/9711021`)
[10] Raffelt G G 1990 *Phys. Rev. Lett.* **64** 2856–2858
[11] Raffelt G G 1999 *Ann. Rev. Nucl. Part. Sci.* **49** 163–216 (*Preprint* `hep-ph/9903472`)
[12] Nadolsky P M *et al.* 2008 *Phys. Rev.* D **78** 013004 (*Preprint* `0802.0007`)

Journal of Physics: Conference Series **259** (2010) 012090 doi:10.1088/1742-6596/259/1/012090

Large 1-3 leptonic mixing and Renormalization Group Effects

Srubabati Goswami,
Physical Research Laboratory, Ahmedabad 380 009, India,
E-mail: sruba@prl.res.in

Abstract. Starting from tri-bimaximal mixing at high scale, we investigate if it is possible to generate a sizable value of $|U_{e3}|$, while at the same time keeping solar neutrino mixing near its measured value, which is close to $\sin^2\theta_{12} = \frac{1}{3}$ through renormalization group corrections. Generation of indicated values of $|U_{e3}| \approx 0.1$ via this effect requires the neutrinos to be quasi-degenerate in mass. If we consider Standard Model (SM) as the low energy effective theory then the required value of the mass scale is beyond 2.3 eV which is the current constraint from tritium β-decay. On the other hand considering the Minimal Supersymmetric Standard Model (MSSM) as the effecitve theory at low energy it is possible to generate the non-zero U_{e3} values hinted by the current global analysis for lower m_0 values. The consistency with the allowed range of $\sin^2\theta_{12}$ together with large running of $|U_{e3}|$ forces one of the Majorana phases to be close to π. This implies large cancellations in the effective Majorana mass governing neutrino-less double beta $((\beta\beta)_{0\nu}$-)decay, constraining it to lie near its minimum allowed value of $m_0 \cos 2\theta_{12}$, where $m_0 \gtrsim 0.1$ eV.

1. Introduction

Assuming three neutrino flavours there are 9 parameters that define the low energy neutrino mass matrix - three masses, three mixing angles and three phases. Experimental data from neutrino oscillation experiments have determined the two mass squared differences as $\Delta m^2_{\odot} = 7.67^{+0.16,\,0.52}_{-0.19,\,0.53} \times 10^{-5}\text{eV}^2, |\Delta m^2_{\rm A}| = 2.39^{+0.11,\,0.42}_{-0.08,\,0.33} \times 10^{-3}\text{eV}^2$. Note that the sign of $|\Delta m^2_{\rm A}| \simeq |\Delta m^2_{31}| \simeq |\Delta m^2_{32}|$, i.e., the ordering of neutrino masses is still not known and there can be three options: normal hierarchy (NH) with $m_1 \ll m_2 \ll m_3$, inverted hierarchy (IH) with $m_3 \ll m_1 \simeq m_2$, or quasi-degenerate neutrinos (QD) with $m_0^2 = m_1^2 \simeq m_2^2 \simeq m_3^2 \gg \Delta m^2_{\odot}, |\Delta m^2_{\rm A}|$. The latter requires that $m_{1,2,3} \gtrsim 0.10$ eV. For the QD case also one can still ask the question whether m_1 or m_3 is the lowest mass, i.e. whether $\Delta m^2_{\rm A} > 0$ or $\Delta m^2_{\rm A} < 0$.

The current values of the mixing angles are close to tri-bimaximal mixing (TBM) [1] with $\theta_{12} = \sin^{-1}(1/\sqrt{3})$, $\theta_{23} = \pi/4$ and $\theta_{13} = 0$.

Recently the global fit of the solar and KamLAND experiment data showed a preference for a non-zero θ_{13}. The origin of this can be understood by observing that for $\theta_{13} = 0$ solar and KamLAND prefer different values of θ_{12} with no overlap at 1σ level. For $\theta_{13} > 0$ solar prefer higher θ_{12}, KamLAND prefer lower θ_{12} and the disagreement is reduced. From a combined analysis the following best-fit value and 1σ range was obtained in [2]:

$$\sin^2\theta_{13} = 0.016 \pm 0.010\,, \tag{1}$$

or $|U_{e3}| = \sin\theta_{13} = 0.126^{+0.035}_{-0.049}$, or $\theta_{13} = \left(7.3^{+2.0}_{-2.8}\right)^\circ$. In [3] the global fit gives. However, the hint in the atmospheric data has been questioned [4], but the recent MINOS data show a 0.7σ excess of electron events [5]. Thus the value of θ_{13} still very much remains an open question to be decided by future experimental data. We explore the effect of renormalization group (RG) corrections to TBM [7, 8, 9] assuming this to have been generated at some high energy scale. Our aim is to check if the RG effects can accommodate the non-zero θ_{13}indicated by the current data at the same time keeping all other parameters within their allowed ranges.

2. Renormalization Group Effects on Tri-bimaximal mixing

In general, the corrections to the mixing angles can be expressed as [6, 7, 8]:

$$\theta_{ij}^{\lambda} \simeq \theta_{ij}^{\Lambda} + C\, k_{ij}\, \Delta_\tau + \mathcal{O}(\Delta_\tau^2)\,, \tag{2}$$

where Λ is the high scale at which TBM is implemented and λ is the low energy scale at which measurements take place. We will indicate high scale values by a superscript Λ in the following, and omit, for simplicity, the superscript λ, which would indicate low scale values. Hence we have $\theta_{12}^{\Lambda} = \sin^{-1}\sqrt{1/3}$, $\theta_{23}^{\Lambda} = \pi/4$ and $\theta_{13}^{\Lambda} = 0$. We consider the RG evolution of the neutrino masses and the mixing parameters in the effective theory and take the high scale to be $\Lambda = 10^{12}$ GeV. The low scale is taken to be $\lambda = 10^2$ GeV when the effective theory is the Standard Model (SM), while we take $\lambda = 10^3$ GeV when the effective theory at low energy is the MSSM. The constant C in Eq. (2) is given by $C = -3/2$ for the SM and $C = +1$ for the MSSM. The result in Eq. (2) is obtained in first order in the parameter

$$\Delta_\tau \equiv \begin{cases} \frac{m_\tau^2}{8\pi^2 v^2}\,(1+\tan^2\beta)\,\ln\frac{\Lambda}{\lambda} \simeq 1.4\cdot 10^{-5}\,(1+\tan^2\beta) & (\text{MSSM})\,, \\ \frac{m_\tau^2}{8\pi^2 v^2}\,\ln\frac{\Lambda}{\lambda} \simeq 1.5\cdot 10^{-5} & (\text{SM})\,, \end{cases} \tag{3}$$

with $\Delta_{e,\mu}$ having been neglected since $m_{e,\mu} \ll m_\tau$ and the vev of the Higgs is taken to be $v/\sqrt{2} = 174$ GeV. The k_{ij}'s are given as [10, 7, 8],

$$\begin{aligned} k_{12} &= \frac{\sqrt{2}}{6}\frac{\left|m_1 + m_2\, e^{i\alpha_2}\right|^2}{\Delta m_{21}^2}\,, \\ k_{23} &= -\left(\frac{1}{3}\frac{\left|m_2 + m_3\, e^{i(\alpha_3-\alpha_2)}\right|^2}{\Delta m_{32}^2} + \frac{1}{6}\frac{\left|m_1 + m_3\, e^{i\alpha_3}\right|^2}{\Delta m_{31}^2}\right)\,, \\ k_{13} &= -\frac{\sqrt{2}}{6}\left(\frac{\left|m_2 + m_3\, e^{i(\delta+\alpha_3-\alpha_2)}\right|^2}{\Delta m_{32}^2} - \frac{\left|m_1 + m_3\, e^{i(\delta+\alpha_3)}\right|^2}{\Delta m_{31}^2} - \frac{4\, m_3^2\, \Delta m_{21}^2}{\Delta m_{31}^2\, \Delta m_{32}^2}\sin^2\frac{\delta}{2}\right)\,, \end{aligned} \tag{4}$$

where we have used $\theta_{ij} = \theta_{ij}^{\Lambda}$, which is correct up to $\mathcal{O}(\Delta_\tau)$, and $\Delta m_{ij}^2 \equiv m_i^2 - m_j^2$. The running of the masses has been neglected in the above expressions for the k_{ij}. The masses are decreasing from high to low scale and run as

$$|m_i| = I_K\left(|m_i^{\Lambda}| + \mu_i\,\Delta_\tau\right)\,, \tag{5}$$

where I_K is a scalar factor that depends on the SU(2) and U(1) gauge coupling constants and the Yukawa matrix in the up quark sector [11, 12, 10] and μ_i are $\mathcal{O}(1)$ numbers. Thus, neglecting the running of masses introduces an error $\mathcal{O}(\Delta_\tau)$ in k_{ij} and hence $\mathcal{O}(\Delta_\tau^2)$ in θ_{ij}. One also observes that for $|\Delta_\tau| \gtrsim ((m_2^{\Lambda})^2 - (m_1^{\Lambda})^2)/(m_0^{\Lambda})^2$, the $\mathcal{O}(\Delta_\tau^2)$ terms dominate over the $\mathcal{O}(\Delta_\tau)$ terms in the evolution of $m_2^2 - m_1^2$ [10]. For such cases Eqs. (4) will no longer be valid. Thus, for the validity of these equations, we require $(m_0^{\Lambda})^2\,\Delta_\tau \lesssim (m_2^{\Lambda})^2 - (m_1^{\Lambda})^2$, which may not be satisfied if $(m_2^\Lambda)^2 - (m_1^\Lambda)^2$ is indeed very small. We will therefore use the full running equations for the mass matrix itself for the plots and numerical values to be presented. Analytical estimates are made with the expressions of the k_{ij} and, as we show, these estimates can explain the numerical results with a sufficient degree of correctness. There is a subtle issue involved when we consider k_{13} in Eq. (4). As is seen, k_{13} depends on the Dirac CP phase δ which is unphysical for the case of $\theta_{13} = 0$ at the high scale Λ. However as discussed in [10, 13], the value of δ at this point depends on the values of the masses and the Majorana phases and RG evolution takes care of that automatically. For analytical estimates, it is convenient to consider the shift of the mixing angles θ_{ij} from their initial values. From the above expressions for the k_{ij}, and in the limit of $|k_{ij}\,\Delta_\tau| \ll 1$, one obtains the following expressions for the observables:

$$|\sin\theta_{13}| \simeq |C\,k_{13}\,\Delta_\tau| \ , \ \ \sin^2\theta_{23} \simeq \frac{1}{2} - C\,k_{23}\,\Delta_\tau \ , \ \ \sin^2\theta_{12} - \frac{1}{3} \simeq \frac{2\sqrt{2}}{3}\,C\,k_{12}\,\Delta_\tau \ . \qquad (6)$$

In the spirit of our analysis we require 1)) that $|C\,k_{13}\,\Delta_\tau| = 0.077 - 0.161$, while $-C\,k_{12}\,\Delta_\tau = 2.8 \cdot 10^{-3} - 4.2 \cdot 10^{-2}$. Note that for the 1σ range we are taking, $C\,k_{12}\,\Delta_\tau$ (and therefore C) is supposed to be negative. Hence, within the MSSM the required deviation from TBM cannot be realized. Therefore we use the 3σ ranges for $\sin^2\theta_{12}$ and $\sin^2\theta_{23}$ for the purposes of illustration [14]. If indeed the trend of $\sin^2\theta_{12} < \frac{1}{3}$ continues then it will not be possible to account for a high scale value of $\sin^2\theta_{12} = \frac{1}{3}$ solely by RG effects within the MSSM From Eqs. (2) and (4) it is evident that whether the angles θ_{ij} will decrease or increase during evolution will depend on the effective theory (SM or MSSM, through the factor C) and also on the sign of $\Delta m^2_{31} (\simeq \Delta m^2_{32})$ for θ_{23}. Since k_{12} is always positive, θ_{12} at the low scale is always larger (smaller) than that at high scale for the MSSM (SM). The size of the RG corrections will depend on the values of the Majorana phases, the neutrino masses and, in case of the MSSM, $\tan\beta$. One can at the outset make some interesting observations from Eqs. (4). It is easy to see that only quasi-degenerate neutrinos will be able to lead to values of $|U_{e3}|$ around 0.1. Note also that in this case the running of solar neutrino mixing is in general enhanced by a factor $|\Delta m^2_{\rm A}|/\Delta m^2_\odot$ with respect to the running of the other mixing angles. We further can expect that the deviation from maximal θ_{23} is of the same order than the deviation from zero $|U_{e3}|$. In the following we will quantify these statements.

3. Resuls

We have performed a detailed analysis, by numerically solving the RG running equations for the effective neutrino mass matrix and then diagonalizing it to extract the masses, mixing angles and phases at low scale. At the high scale Λ, the angles θ_{ij}^Λ are fixed by the requirement of the TBM scenario, while the masses and the CP phases are chosen randomly so that after the RG evolution at low scale the parameters are consistent with the chosen ranges of the current experimental data. We have used the following ranges for the high scale values of the mass-squared differences: $(\Delta m^2_\odot)^\Lambda = 10^{-6} - 10^{-3}$ eV^2 and $|\Delta m^2_{\rm A}|^\Lambda = 1.5 \times 10^{-3} - 10^{-2}$ eV^2, while the phases are varied over the full range of $0 - 2\pi$.

Starting with the SM, Fig. 1 shows the allowed region in the $m_0 - \sin^2\theta_{13}$ plane at the low scale λ, after performing the RG evolution, for both NH (left panel) and IH (right panel). Recall that $m_0^2 \gg |\Delta m^2_{\rm A}|$ is the common neutrino mass scale for quasi-degenerate neutrinos. As can be seen, to generate values of $|U_{e3}|$ within the range of interest, neutrino masses should exceed the direct limit of 2.3 eV from tritium decay [15], and hence also the more stringent but model-dependent limits from cosmology. We conclude that a high scale value of $\theta_{13} = 0$ is incompatible with the indicated range of $|U_{e3}|$. The dependence of this statement on the initial values of θ_{12}^Λ and θ_{23}^Λ is moderate and hence this statement is valid in general.

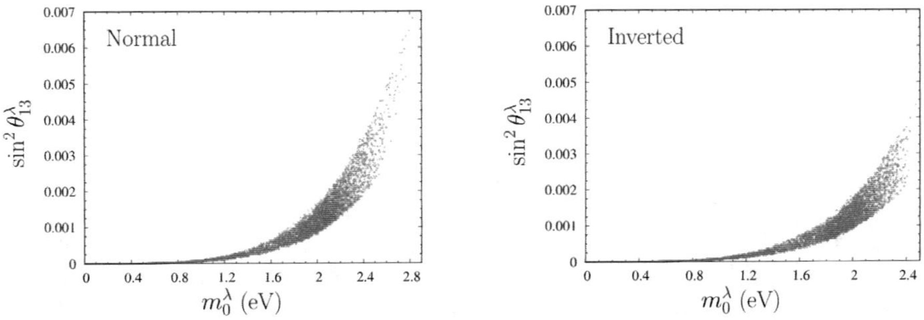

Figure 1. The running of $|U_{e3}|^2 = \sin^2\theta_{13}$ in SM for both NH (left panel) and IH (right panel). The high scale values of mixing angles are kept fixed at TBM values while the masses and phases are varied randomly such that after RG evolution the parameter values are within current experimental ranges.

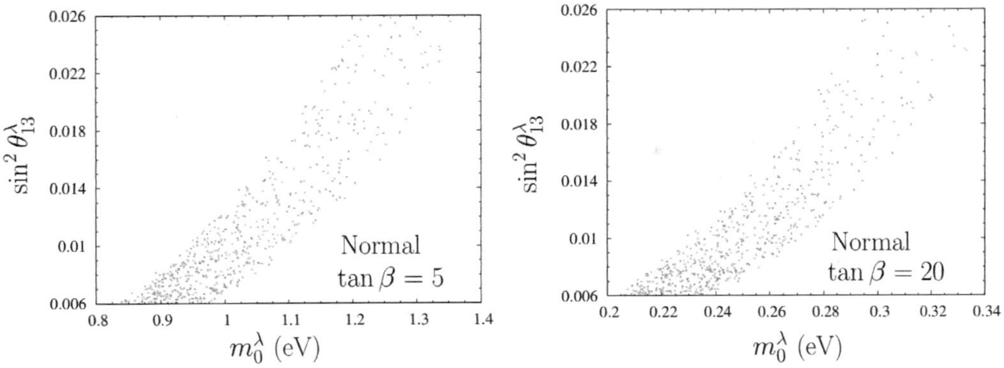

Figure 2. Scatter plots showing the running of $\sin^2\theta_{13}$ with m_0, for MSSM with NH and $\tan\beta = 5, 20$. For a given $\tan\beta$, the allowed regions are the same as above for IH.

Fig. 2 shows the allowed region in the $m_0 - \sin^2\theta_{13}$ plane, when the effective theory is the MSSM, for $\tan\beta = 5, 20$ and NH. The left panel shows that $\sin^2\theta_{13}$ lies in the required range when 0.8 eV $\lesssim m_0 \lesssim$ 1.4 eV for $\tan\beta = 5$, while the allowed mass range becomes 0.2 eV $\lesssim m_0 \lesssim$ 0.34 eV for $\tan\beta = 20$, as can be seen from the right panel. Thus, the relevant range of $m_0 \tan\beta$ is given by (see below for analytical estimates) $4.1 \lesssim (m_0/\text{eV})\tan\beta \lesssim 6.9$. Hence the allowed mass ranges depend strongly on $\tan\beta$ and for higher values of $\tan\beta$, lower values of m_0 are sufficient to produce the required running of θ_{13}. It has been checked that for a fixed $\tan\beta$ value, there is no significant dependence on the mass ordering, other than the direction of the correction to θ_{23}. From the allowed mass ranges obtained in Fig. 2 it is seen that to have $\sin^2\theta_{13}$ in the 1σ range under consideration, we need the neutrinos to be quasi-degenerate even for the MSSM with $\tan\beta = 20$. Fig. 3 shows scatter plots of the allowed region of the neutrino mass scale m_0 and the Majorana phase α_2, which is particularly important for the running of θ_{12} [10] (see also [16]). We compare the allowed regions at high and low scale for NH and $\tan\beta = 5, 20$. The scattered plots obtained for IH show the same characteristics. We see that $|\alpha_2|$ is restricted in a narrow region around $|\alpha_2| = \pi$ for all cases.

Fig. 4 shows the correlation between the low scale values of $\sin^2\theta_{13}$ and $\sin^2\theta_{23}$. For NH $\sin^2\theta_{23} > \frac{1}{2}$, whereas for IH $\sin^2\theta_{23} < \frac{1}{2}$. For NH θ_{13} and θ_{23} are correlated, i.e., a higher value of θ_{13} requires a higher value of θ_{23}. For IHthe predicted values of the two angles are

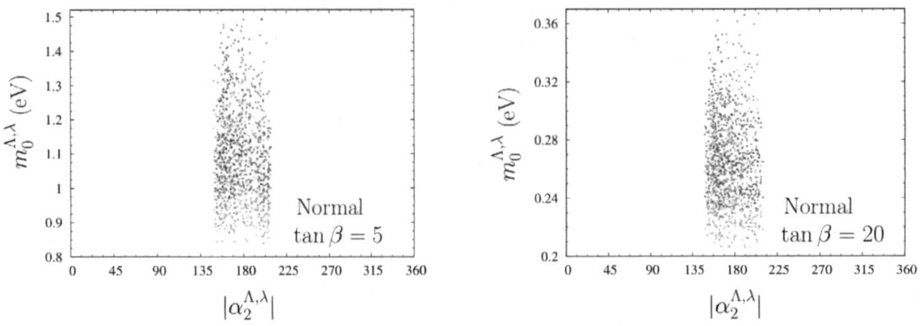

Figure 3. Scatter plots in the $m_0 - |\alpha_2|$ plane for MSSM ($\tan\beta = 5, 20$) and NH, both for high (black circles) and low (red squares) energy scales. Same variation is obtained for IH.

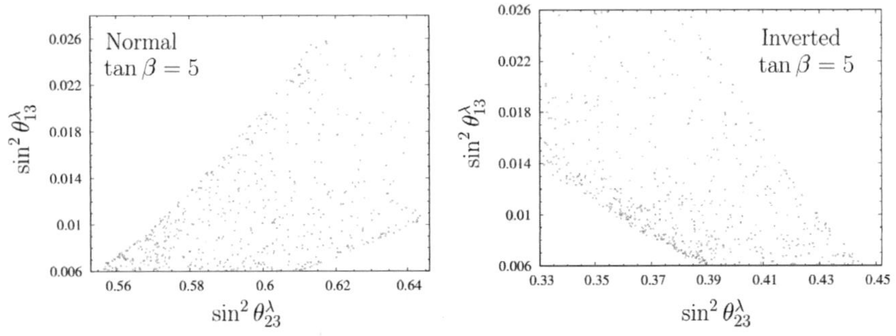

Figure 4. Scatter plots showing the correlation between $\sin^2\theta_{13}$ and $\sin^2\theta_{23}$ for MSSM with $\tan\beta = 5, 20$. The left panel shows the case with NH while the right panel is for IH.

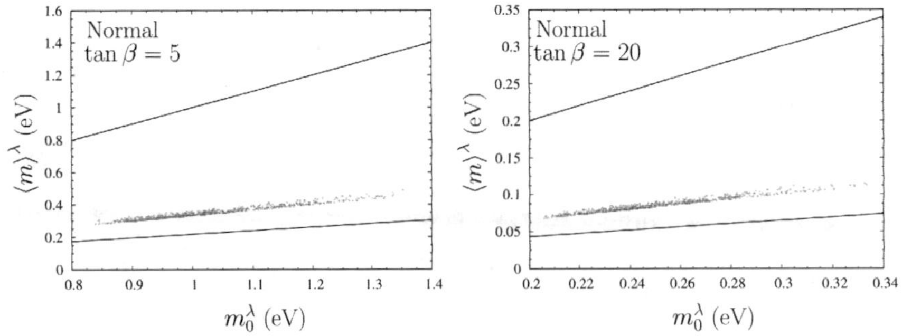

Figure 5. Scatter plot for the effective neutrino mass $\langle m\rangle$ that contributes to neutrino-less double beta decay as a function of m_0, in MSSM with $\tan\beta = 5, 20$ and NH. The solid (black) lines indicate the maximum and minimum possible values of $\langle m\rangle$ for given m_0, obtained by varying the oscillation parameters in their current 3σ range and the phases between 0 to 2π. The cases with IH show same characteristics.

anti-correlated. The plots obtained with $\tan\beta = 20$ are identical to those shown in Fig. 4 for $\tan\beta = 5$, when the mass ordering is the same. For a different $\tan\beta$ the value of m_0 adjusts itself to comply with the low energy cuts on the parameters and the allowed points in the $\sin^2\theta_{23} - \sin^2\theta_{13}$ plane remain same. We note here that maximal atmospheric neutrino mixing is not possible.

In Fig. 5 we plot the effective Majorana mass which governs the rate of $(\beta\beta)_{0\nu}$-decay at low energy. The scatter points show the values of $\langle m \rangle$ allowed by the low energy neutrino oscillation data after RG analysis. The solid (black) lines indicate the maximum and minimum possible values of $\langle m \rangle$ at low scale for a given m_0, obtained by varying the oscillation parameters in their current 3σ range and the phases between 0 to 2π. The plots show that the effective mass obtained after RG analysis lies close to its minimum allowed range. As can also be seen from Fig. 5, for $\tan\beta = 5$, $\langle m \rangle$ takes values between 0.26 and 0.50 eV, to be compared with the general upper and lower limits of 0.2 eV and 1.4 eV. If $\tan\beta = 20$, then 0.07 eV $\lesssim \langle m \rangle \lesssim$ 0.11 eV, while in general the effective Majorana mass could be in between 0.05 eV and 0.34 eV.

4. Conclusions

Global oscillation data imply two large (θ_{12} and θ_{23}) one small (θ_{13}) mixing angle Seesaw mechanism which postulates a new particle with a heavy mass scale can explain the smallness of neutrino mass. However since the measurements are at low energy one needs to include RG effects which even if small are important in view of the onset of precision era in neutrino physics. In this work, we study the RG effects starting with TBM at a high scale and specifically examine whether the current non-zero hint of θ_{13} can be accommodated. RG effects are found to be small in SM In MSSM there is an enhancement by $\tan^2\beta$. Non-zero θ_{13} from current global fit cannot be accommodated in TBM + RG effects if the effective theory is SM but it is possible in MSSM for large m_0 and high $\tan\beta$ Compatibility with current 3σ range of oscillation parameters implies the Majorana phase $\alpha_2 \approx \pi$. Precision measurements of m_0, θ_{12},θ_{13} are expected to play an important role according to our analysis.

References

[1] P. F. Harrison, D. H. Perkins and W. G. Scott, Phys. Lett. B **530**, 167 (2002).
[2] G. L. Fogli, E. Lisi, A. Marrone, A. Palazzo and A. M. Rotunno, Phys. Rev. Lett. **101**, 141801 (2008); arXiv:0809.2936 [hep-ph].
[3] T. Schwetz, M. A. Tortola and J. W. F. Valle, New J. Phys. **10**, 113011 (2008) [arXiv:0808.2016 [hep-ph]].
[4] M. Maltoni and T. Schwetz, arXiv:0812.3161 [hep-ph].
[5] MINOS collaboration, Nu 2010; Talks by J. DeJong and M. Sanchez, PASCOS 2010.
[6] A. Dighe, S. Goswami and P. Roy, Phys. Rev. D **73**, 071301 (2006) [arXiv:hep-ph/0602062].
[7] A. Dighe, S. Goswami and W. Rodejohann, Phys. Rev. D **75**, 073023 (2007), [arXiv:hep-ph/0612328].
[8] A. Dighe, S. Goswami and P. Roy, Phys. Rev. D **76**, 096005 (2007), [arXiv:0704.3735 [hep-ph]].
[9] S. Luo and Z. Z. Xing, Phys. Lett. B **632**, 341 (2006) [arXiv:hep-ph/0509065]; R. N. Mohapatra, S. Nasri and H. B. Yu, Phys. Lett. B **639**, 318 (2006) [arXiv:hep-ph/0605020]; M. Hirsch, E. Ma, J. C. Romao, J. W. F. Valle and A. Villanova del Moral, Phys. Rev. D **75**, 053006 (2007) [arXiv:hep-ph/0606082]; I. d. M. Varzielas, G. G. Ross and M. Serna, arXiv:0811.2226 [hep-ph].
[10] S. Antusch, J. Kersten, M. Lindner and M. Ratz, Nucl. Phys. B **674**, 401 (2003) [arXiv:hep-ph/0305273].
[11] J. R. Ellis and S. Lola, Phys. Lett. B **458**, 310 (1999) [arXiv:hep-ph/9904279].
[12] P. H. Chankowski and S. Pokorski, Int. J. Mod. Phys. A **17**, 575 (2002) [arXiv:hep-ph/0110249].
[13] A. Dighe, S. Goswami and S. Ray, Phys. Rev. D **79**, 076006 (2009), [arXiv:0810.5680 [hep-ph]].
[14] S. Goswami, S. T. Petcov, S. Ray and W. Rodejohann, Phys. Rev. D **80**, 053013 (2009) [arXiv:0907.2869 [hep-ph]].
[15] C. Kraus *et al.*, Eur. Phys. J. C **40**, 447 (2005) [arXiv:hep-ex/0412056]; V. M. Lobashev, Nucl. Phys. A **719**, 153 (2003).
[16] S. T. Petcov, T. Shindou and Y. Takanishi, Nucl. Phys. B **738**, 219 (2006) [arXiv:hep-ph/0508243].
[17] S. M. Bilenky, S. Pascoli and S. T. Petcov, Phys. Rev. D **64**, 053010 (2001) [arXiv:hep-ph/0102265].

Journal of Physics: Conference Series **259** (2010) 012091 doi:10.1088/1742-6596/259/1/012091

Constraining nonstandard neutrino-quark interactions with solar, reactor and accelerator data

F. J. Escrihuela[1,*], O. G. Miranda[2,+], M. A. Tórtola[1,#] and J. W. F. Valle[1,$]

[1] Instituto de Física corpuscular-C.S.I.C/Universitat de València, Campus de Paterna, Apt 22085, E-46017 València, Spain

[2] Departamento de Física, Centro de Investigación y de Estudios Avanzados del IPN, Apt 14-740, 07000 México DF, México

[*] franesfe@alumni.uv.es **(speaker)**
[+] Omar.Miranda@fis.cinvestav.mx
[#] mariam@ific.uv.es
[$] valle@ific.uv.es

Abstract. We present an analysis of nonstandard neutrino interactions (NSI) of electron and tau neutrinos with down-quark using solar, reactor and accelerator data. We include new solar data from SNO phase III and Borexino, as well as new KamLAND data and solar fluxes. We combine these results with the CHARM data allowing us to better constrain the axial and axial-vector electron and tau-neutrino NSI parameters.

1. Introduction

Current solar neutrino data, in conjunction with reactor data from the KamLAND experiment show that the neutrino oscillation mechanism is the correct picture to explain the solar neutrino physics. Solar neutrino experiments are also sensitive to matter effects, and the combination of both solar and KamLAND data determines the so-called Large Mixing Angle (LMA) solution as the correct explanation to the data. The different cases involving nonstandard effects can only play a subleading role.

However, while constrained by the solar and KamLAND data in an important way, neutrino nonstandard interactions (NSI) still provide an important exception to the robustness of the neutrino oscillation interpretation. Indeed, it has been found that they might even shift the solution to the so--called dark side region of the neutrino parameter space [1].

In what follows we are going to analyze the robustness of the oscillation interpretation of the solar neutrino data in the presence of nonstandard interactions (NSI) using the next effective low-energy neutral currents four-fermion operator:

$$L_{NSI} = -\varepsilon_{\alpha\beta}^{fP} 2\sqrt{2}\, G_F\, (\bar{\nu}_\alpha \gamma_\mu L \nu_\beta)\, (\bar{f} \gamma^\mu P f) \qquad (1)$$

where P = L, R and f is a first generation fermion: *e, u, d*. The coefficients $\varepsilon_{\alpha\beta}^{fP}$ denote the strength of the NSI between the neutrinos of flavours α and β and the P--handed component of the fermion f. For definiteness, we take for f the down-type quark. However, one can also consider the presence of

NSI with electrons and up and down quarks simultaneously. Here we confine ourselves to NSI couplings involving only electron and tau neutrinos.

Nonstandard interactions may in principle affect neutrino propagation properties in matter as well as detection cross sections. NSI effects in neutrino propagation affect the analysis of data from solar neutrino experiments and to some extent also KamLAND, through the vectorial NSI couplings $\varepsilon_{\alpha\beta}^{dV} = \varepsilon_{\alpha\beta}^{dL} + \varepsilon_{\alpha\beta}^{dR}$. On the other hand, detection shows sensitivity also to the axial NSI couplings $\varepsilon_{\alpha\beta}^{dA} = \varepsilon_{\alpha\beta}^{dL} - \varepsilon_{\alpha\beta}^{dR}$ in the SNO experiment.

2. Effects of NSI in neutrino propagation

We first analyze the determination of the oscillation parameters in the presence of nonstandard interactions. The Hamiltonian describing solar neutrino evolution in the presence of NSI contains, in addition to the standard oscillations term, a term involving NSI, accounting for an effective potential induced by the NSI with matter:

$$H_{vac} = \begin{pmatrix} -\frac{\Delta m^2}{4E}\cos 2\theta & \frac{\Delta m^2}{4E}\sin 2\theta \\ \frac{\Delta m^2}{4E}\sin 2\theta & \frac{\Delta m^2}{4E}\cos 2\theta \end{pmatrix} + H_{mat} = \sqrt{2}\, G_F\, N_e \begin{pmatrix} 1 & 0 \\ 0 & 0 \end{pmatrix} + H_{NSI} = \sqrt{2}\, G_F\, N_d \begin{pmatrix} 0 & \varepsilon \\ \varepsilon & \varepsilon' \end{pmatrix} \quad (2)$$

Here ε and ε' are two effective parameters related with the vectorial components of NSI which affect the neutrino propagation by:

$$\varepsilon = -\sin\theta_{23}\, \varepsilon_{e\tau}^{dV} \qquad \varepsilon' = \sin\theta_{23}\, \varepsilon_{\tau\tau}^{dV} - \varepsilon_{ee}^{dV} \quad (3)$$

The quantity N_d is the density of the down-type quark along the neutrino path, and θ_{23} is the atmospheric neutrino mixing angle.

In the adiabatic regime the survival probability can be approximated by Parke's formula [1]:

$$P(\nu_e \to \nu_e) = \frac{1}{2}[1 + \cos 2\theta \cos 2\theta_m) \quad (4) \quad \Rightarrow \quad \cos 2\theta_m = \frac{\Delta m^2 \cos 2\theta - 2\sqrt{2} E G_F N_e}{[\Delta m^2]_{matter}} \quad (5)$$

$$[\Delta m^2]_{matter} = \sqrt{[\Delta m^2 \cos 2\theta - 2\sqrt{2} E G_F (N_e - \varepsilon' N_d)]^2 + [\Delta m^2 \sin 2\theta + 4\sqrt{2}\, \varepsilon\, E G_F N_d]^2} \quad (6)$$

where θ_m is the effective mixing angle at the neutrino production point inside the sun. In this case we have already introduced NSI in addition to oscillations.

Using data from solar experiments as Chlorine, Super-K spectrum, SAGE, SNO day/night, SNO salt data and combining them with KamLAND reactor experiment we obtain the figure 1 [1] where NSI are included. As we can see the presence of NSI will lead to the appearance of the LMA-D solution and thus to the ambiguous determination of the solar mixing angle.

2.1. Updating data

In a new analysis we have included for the solar data the most updated results from the radiochemical experiments Homestake, SAGE, and GALLEX/GNO, the zenith-spectra data set from Super-Kamiokande I, and the two first phases of SNO.

As main updates have been included:

- The third phase of SNO.
- The latest measurements of the ^{7}Be solar neutrino rate performed by Borexino .

Journal of Physics: Conference Series **259** (2010) 012091 doi:10.1088/1742-6596/259/1/012091

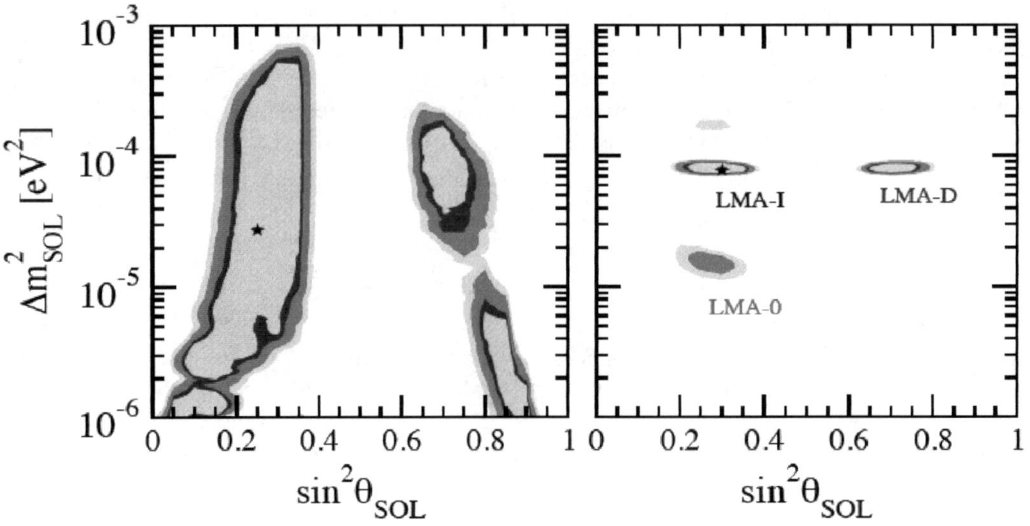

Figure 1: Allowed regions for the generalized OSC + NSI case, determined from previous data (2006): left panel correspond to a solar only analysis, while the right panel corresponds to solar+KamLAND.

- The fluxes of the latest version of SSM: BPS08 high metallicity.
- The latest data from KamLAND reactor experiment.

Combining again solar + KamLAND we obtain the result shown in the figure 2 [2]. One can see that the region in the so-called dark side of the neutrino parameters remains even after the inclusion of the new data. Note however that its status is worse than previously. In contrast, as seen in the figure 2, the other solutions LMA-0 and LMA-II [1], which were present before, have disappeared as a result of the new KamLAND data, that now provide a very precise measurement of Δm^2.

By analyzing the goodness of the neutrino oscillation solutions in the presence of NSI we can also constrain the NSI parameters ε and ε':

$$-0.41 < \varepsilon < 0.06 \tag{7}$$

$$-0.50 < \varepsilon' < 0.19 \quad \& \quad 0.89 < \varepsilon' < 0.99 \tag{8}$$

We can see that the new data allow us to constrain ε, the flavor changing parameter, while for the flavor conserving case there is still room for relatively large values ε' that correspond to the solution in the dark side of the neutrino oscillation parameters.

For the flavor-changing parameter ε we can use the Eq. (3) and the Eq.(7) in order to obtain a limit over the vectorial coupling $\varepsilon_{e\tau}^{dV}$:

$$-0.08 < \varepsilon_{e\tau}^{dV} < 0.58 \qquad (90\%\ C.L.) \tag{9}$$

where we have used the best fit value for the atmospheric mixing angle [1].

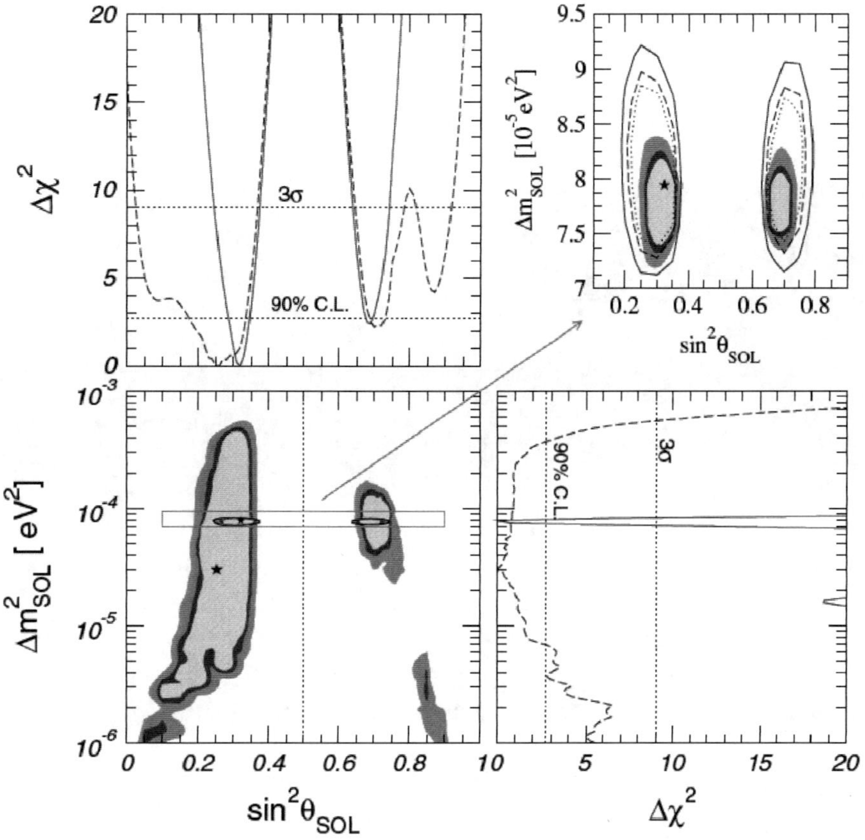

Figure 2: The left bottom panel indicates 90%, 95% and 99% C.L. allowed regions from the solar and solar + KamLAND combined analysis, which is presented as a zoom in the top right panel, where the shaded regions show the result of the current analysis while the lines indicate the earlier region [1]. The other two panels indicate the corresponding χ^2.

3. Effects of NSI in neutrino detection

The presence of NSI can also affect the detection processes at some experiments. In particular, the cross section for the neutral current detection reaction at SNO is proportional to g_A^2, which is the coupling of the neutrino current to the axial isovector hadronic current, giving an extra contribution to the NC signal at SNO experiment. This nonstandard contribution can be parameterized as:

$$\phi_{NC} \approx f_B(1+2\varepsilon_A) \quad (10) \quad \Rightarrow \quad \varepsilon_A = -\sum_{\alpha=e,\mu,\tau} \langle P_{e\alpha}\rangle_{NC}\, \varepsilon^{dA}_{\alpha\alpha} \quad (11)$$

where f_B is the boron neutrino flux and the nonstandard axial couplings with up-type quark are set to zero.

The results obtained in a generalized 5-parameters analysis including the axial component of NSI are summarized in the figure 3 [2]. We can observe that the previous analysis (Sec. 2) without axial component and the present analysis are consistent, thought, as expected, the inclusion of the axial parameter in the analysis extends the allowed region.

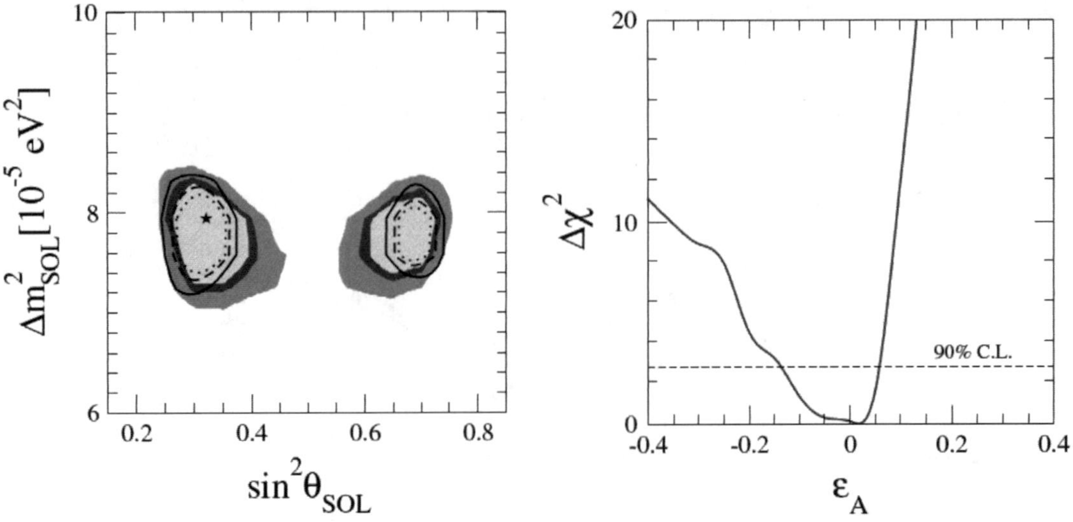

Figure 3: The shaded regions of the left panel show the updated analysis including the axial parameter combining solar+KamLAND while the lines show the updated regions obtained only with vectorial NSI couplings. In the right panel we show $\chi^2 - \varepsilon_A$ where we can see how the data prefer $\varepsilon_A \approx 0$.

From the χ^2 analysis we obtain the following allowed range at 90% C.L.

$$-0.14 < \varepsilon_A < 0.06 \tag{12}$$

4. Constraints on non-universal NSI: CHARM+solar+KamLAND

Laboratory experiments measuring neutrino-nucleon scattering show sensitivity to neutrino NSI on d-type quarks. In particular, here we will combine the results of the accelerator experiment CHARM together with the ones in Sec. 2 in order to obtain stronger constraints on the NSI parameters focusing our analysis on the flavor-conserving NSI couplings.

The regions for the vector (left) and axial-vector (right) NSI couplings allowed by the global analysis are given in the figure 4 [2], where they are also compared with the constraints coming only from the CHARM analysis (vertical bands in Fig. 4 [2]) and that from solar+KamLAND analysis (diagonal band in the Fig. 4 [2]).

In the vectorial case we can appreciate the existence of two allowed islands corresponding to the usual LMA solution (lower one) and to the dark solution (upper one). On the other hand, using the average probabilities reported by SNO [2] we have reanalyzed the results obtained for the effective axial coupling ε_A in Eq. (12) obtaining a constraint in the axial couplings ε_{ee}^{dA} and $\varepsilon_{\tau\tau}^{dA}$. The results are summarized in the table I.

5. Conclusions

The neutrino oscillation interpretation of solar neutrino data is still fragile in the presence of NSI.

We have updated the solar NSI analysis using new data from SNO-III, Borexino, KamLAND and solar fluxes but the dark solution (LMA-D) still survives, while the previous LMA-0 and LMA-II have disappeared as a result of new data.

Further information relevant to lift the degeneracy may come from atmospheric and laboratory data. Indeed the LMA-D solution may become inconsistent with atmospheric and laboratory data [1] but new improved data will be needed.

In summary, we have studied the limits on the NSI couplings combining solar, KamLAND (reactor experiment) and CHARM (accelerator experiment) and we have obtained improved bounds on the vector and axial parameters involving electron and tau neutrino on down-type quarks.

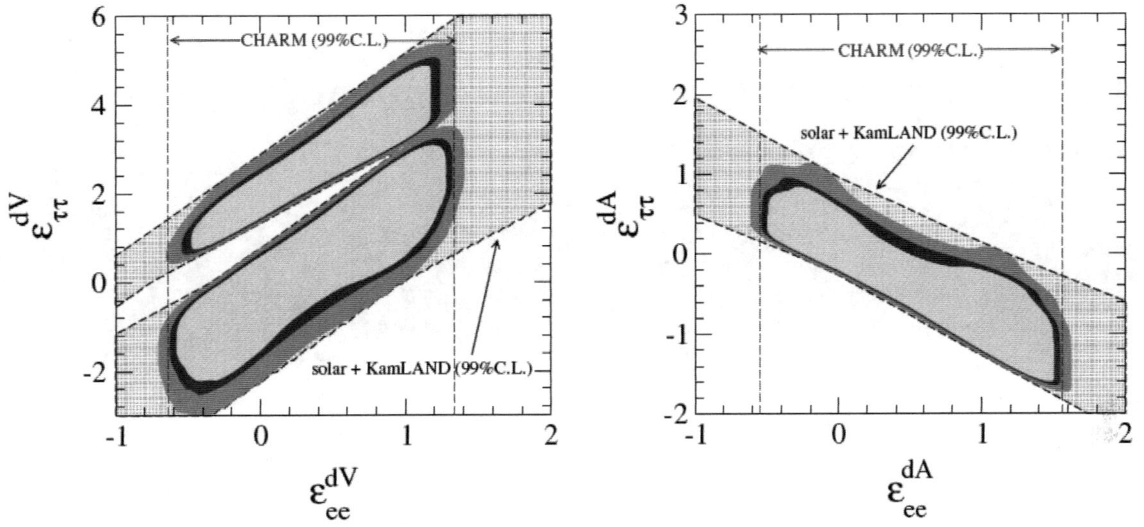

Figure 4: Constraints on the vector (left panel) and axial-vector (right panel) NSI coupling from our global analysis at 90%, 95% and 99% C.L.

	Present results		Previous results	
	Vectorial couplings			
Global	$-0.5 < \varepsilon_{ee}^{dV} < 1.2$	$-1.8 < \varepsilon_{\tau\tau}^{dV} < 4.4$		
	One parameter at a time			
CHARM	$-0.5 < \varepsilon_{ee}^{dV} < 1.2$			
Global	$-0.2 < \varepsilon_{ee}^{dV} < 0.5$	$-1.1 < \varepsilon_{\tau\tau}^{dV} < 0.4$ & $1.6 < \varepsilon_{\tau\tau}^{dV} < 2.2$	$-0.9 < \varepsilon_{ee}^{dV} < 0.8$	$-2.7 < \varepsilon_{\tau\tau}^{dV} < 2.7$
	Axial couplings			
Global	$-0.4 < \varepsilon_{ee}^{dA} < 1.4$	$-1.5 < \varepsilon_{\tau\tau}^{dA} < 0.7$		
	One parameter at a time			
CHARM	$-0.4 < \varepsilon_{ee}^{dA} < 1.4$			
Global	$-0.2 < \varepsilon_{ee}^{dA} < 0.3$	$-0.2 < \varepsilon_{\tau\tau}^{dA} < 0.4$	$-0.8 < \varepsilon_{ee}^{dA} < 0.9$	$-1.8 < \varepsilon_{\tau\tau}^{dA} < 1.8$

Table I: New constraints on the vectorial and axial NSI couplings at 90% C.L. obtained from the CHARM data alone and combining it with solar+KamLAND data.

Acknowledgments
I would like to thank O.G. Miranda, M. A. Tortola and J. W. F. Valle as well as the hospitality of AHEP group (Valencia, Spain) and CINVESTAV (Mexico).

References

[1] O.G. Miranda, M.A. Tortola and J.W.F. Valle, J. High Energy Phys. 10 (2006) 008.
[2] F.J. Escrihuela, O.G. Miranda, M.A. Tortola and J.W.F. Valle, Phys.Rev.D 80, 105009 (2009).

16th International Symposium on Particles Strings and Cosmology (PASCOS 2010) IOP Publishing
Journal of Physics: Conference Series **259** (2010) 012092 doi:10.1088/1742-6596/259/1/012092

Antimatter spectra from a time-dependent modeling of supernova remnants

M. Kachelrieß[1], S. Ostapchenko[1,2], R. Tomàs[3]
[1]Institutt for fysikk, NTNU, Trondheim, Norway
[2]D. V. Skobeltsyn Institute of Nuclear Physics, Moscow State University, Russia
[3]II. Institut für Theoretische Physik, Universität Hamburg, Germany

E-mail: `ricard.tomas.bayo@desy.de`

Abstract. We calculate the energy spectra of cosmic rays (CR) and their secondaries produced in a supernova remnant (SNR), taking into account the time-dependence of the SNR shock. We model the trajectories of charged particles as a random walk with a prescribed diffusion coefficient, accelerating the particles at each shock crossing. Secondary production by CRs colliding with gas is included as a Monte Carlo process. We find that SNRs produce less antimatter than suggested previously: The positron/electron ratio $\mathcal{F}_{e^+}/\mathcal{F}_{e^+ + e^-}$ and the antiproton/proton ratio $\mathcal{F}_{\bar{p}}/\mathcal{F}_{\bar{p}+p}$ are a few percent and few $\times 10^{-5}$, respectively. Both ratios do not rise with energy.

1. Introduction

Measurements of the antimatter fraction of cosmic rays (CR) provide not only insight into CR physics itself [1], as e.g. their propagation in the galaxy, but are also valuable probes for cosmology and particle physics. In particular, the annihilation of dark matter (DM) leads to an equal injection rate of matter and antimatter particles into the Galaxy, while the CR flux from astrophysical sources is matter-dominated. A possible way to detect DM is therefore to estimate carefully the expected antimatter fluxes from astrophysical sources and to search then for any excess [2].

The PAMELA collaboration presented recently results of their measurement of the positron fraction in CRs, which is rising rapidly from 10 to 100 GeV [3]. At the same time, the antiproton ratio measured by PAMELA declines above 10 GeV [4], consistent with expectations. The conventional estimate for antimatter fluxes from astrophysical sources uses as only production mechanism of antimatter the scattering of CRs on interstellar gas [1]. As discussed e.g. in Ref. [5], the energy dependence of the Galactic diffusion coefficient, $D \propto E^\delta$ with $\delta = 0.5 - 0.6$, is inconsistent with an increase of the antimatter fraction with energy. By contrast, the spectral shape of fragmentation functions leads quite naturally to such a rise in the case of DM annihilations.

The DM interpretation of the PAMELA excess faces however several difficulties [2]: First, the required rate of positron production is larger than expected for a stable thermal relic. As a consequence, either the annihilation rate has to be enhanced, or the DM particle should be unstable with the appropriate life-time. Second, in gauge boson or quark fragmentation, positron, antiproton and photon production are tied together and thus one has to postulate a DM particle annihilating only into electrons and muons. More importantly, assuming antimatter production by diffusing CRs as the only astrophysical source for antimatter falls short: Since electrons lose fast energy, the high-energy part of the $e^- + e^+$ spectrum should be dominated by local sources as nearby pulsars, as pointed out already 20 years ago [6]. Moreover, electromagnetic pair cascades in pulsars result naturally in a large positron fraction together with a "standard" antiproton flux.

More recently, supernova remnants (SNR) were put forward as an alternative astrophysical explanation for a rising positron fraction [7]: Positrons created as secondaries of hadronic interactions in the shock vicinity participate in the acceleration process and, according to Ref. [7], should thus have a flatter energy spectrum than primary electrons. It was estimated that the resulting positron fraction can explain the PAMELA excess and rise up to 50% at higher energies [7], while subsequently a similar mechanism for antiprotons was suggested in Ref. [8]. Since shock acceleration in SNR is expected to be the main source for Galactic CRs [9], such a scenario has also important consequences for the interpretations of CR data as, e.g., the boron-to-carbon ratio [10].

The present work examines the production of secondary $\bar{p}$ and e^+ in SNRs, improving on previous studies [7, 8, 11] in two respects: First, we use a Monte Carlo (MC) approach calculating the trajectory of each particle individually in a random walk picture. This makes it easy to include interactions and the production of secondaries. Second, our approach allows us to include the time (and spatial) dependence of relevant parameters describing the evolution of a SNR as, e.g., the shock radius and its velocity, the magnetic field or the CR injection rate and to test their influence on the CR spectra. We should also stress what are *not* the aims of the present work: We do address neither the problem of acceleration from a microscopic point of view nor consider any feedback of CRs on the shock or the magnetic field. Although the latter processes are important to obtain accurate CR escape fluxes, we shall show that our simplified treatment leads to an upper limit on the secondary fluxes.

2. Simulation procedure

Shocks around SNRs are supposed to be collisionless, with charged particle scattering mainly on inhomogeneities of the turbulent magnetic field. We model such trajectories by a random walk in three dimensions with step size $l_0(E)$ determined by an energy-dependent diffusion coefficient D. Diffusion close to the shock is usually assumed to proceed in the Bohm regime with the mean free path l_0 proportional to the Larmor radius R_L. Thus $D(E) = cl_0/3 = c^2p/(3ef_BB)$, where f_B denotes the ratio of the energy density in the turbulent and in the total magnetic field. We neglect the coupling between CRs and the turbulent magnetic field, assuming that a layer with Bohm diffusion extends far enough into the up-stream region. For a constant magnetic field, CRs do not escape but are confined in the SNR, corresponding to an "age-limited" scenario for the CR flux from SNRs.

We describe the evolution of the shock in the rest frame of the SNR. Then the (yet unshocked) up-stream region is at rest, $v_1 = 0$, and has the density of the surrounding interstellar medium (ISM), $\rho_1 = \rho_{\rm ISM}$. Assuming a strong shock with Mach number $\mathcal{M} \gg 1$, the shocked down-stream region flows with the velocity $v_2 = 3v_{\rm sh}/R$ and has the density $\rho_2 = R\rho_1$. Here, R denotes the compression ratio $R = (\gamma+1)/(\gamma-1) = 4$ for a mono-atomic gas with $\gamma = 5/3$. We account for the flow, adding in the down-stream region on top of the random walk an ordered movement of the particle with velocity v_2 that is directed radially outwards. Thus a particle trajectory evolves during the time step $\Delta t = l_0/c$ as

$$\mathbf{x}(t + \Delta t) = \mathbf{x}(t) + v_2\,\Delta t\,\vartheta(r_{\rm sh} - r)\,\mathbf{e}_r + \mathbf{l}_0\,, \qquad (1)$$

where $\mathbf{l}_0$ denotes a random step, $r_{\rm sh}$ the position of the shock, and $\vartheta(x)$ the step function.

Crossing the shock front, particles are accelerated. We neglect that the relative energy gain $\xi = (E_{k+1} - E_k)/E_k$ per cycle k depends on the angle of the trajectory to the shock front, and use for simplicity that on average for a non-relativistic shock $\xi = \frac{4}{3c}(v_2 - v_1) = v_{\rm sh}/c$. For the position $r_{\rm sh}$ and the velocity $v_{\rm sh}$ of the SNR shock we use the $n = 0$ case of the analytical solutions derived in Ref. [12]. These solutions connect smoothly the ejecta-dominated phase with free expansion $r_{\rm sh} \propto t$ and the Sedov-Taylor stage $r_{\rm sh} \propto t^{2/5}$. The acceleration of CRs is assumed to cease after the transition to the radiative phase at the time $t_{\rm max}$.

As the injected particles diffuse, electrons lose energy via synchrotron radiation and inverse Compton scattering, while protons can scatter on gas of the ISM producing secondaries that include antiprotons and positrons. Cross sections and the final state of pp-interactions are simulated using QGSJET-II [13], while we use SIBYLL 2.1 [14] for decays of unstable particles.

The last ingredient for our simulation procedure is an injection model. To ease the comparison with the results of [7], we fix the electron/proton ratio K_{ep} at injection to $K_{ep} = 7 \times 10^{-3}$. As injection energy we use $E_0 = 10\,\text{GeV}$. In the first model used, the injection rate

$$\dot{N} \propto r_{\rm sh}^2\, v_{\rm sh}^\alpha\, \delta(E - E_0)\, \delta(r - r_{\rm sh}) \tag{2}$$

is proportional to the volume swept out per time by the shock, i.e. $\alpha = 1$ (thermal leakage model [15]). In the second model, the injection rate $\dot{N}$ is proportional to the CR pressure [16] and $\alpha = 3$. In the case of model 2, the fraction of particles injected very early is significantly larger than in model 1.

We use the following parameters to describe the SNR: We choose the injected mass as $M_{\rm ej} = 4M_\odot$, the mechanical explosion energy as $E_{\rm snr} = 5 \times 10^{51}\,\text{erg}$, and the density of the ISM as $n_{\rm ISM} = 2\,\text{cm}^{-3}$. The end of the Sedov-Taylor phase follows then as $t_{\rm max} = 13.000\,\text{yr}$ [12]. We use two scenarios for the magnetic field evolution: In the first one we consider a static field with $B = 1\mu$G and $f_B = 1$ to ease comparison with the stationary approach of Refs. [7, 8, 11]. In the second one, we assume that non-linear effects amplify magnetic fields in the early phase, while in the late stage magnetic fields are damped [17]. We model this case by a time-dependent diffusion coefficient with $f_B B = 100\mu$G before the transition to the Sedov-Taylor phase at $t_* = 240\,\text{yr}$ and $f_B B = 1/20\mu$G afterwards.

3. Numerical results

In Fig. 1, we show the energy dependence of the proton spectra in model 1 and 2 for a constant magnetic field, $f_B B = 1\mu$G. Additionally to the total spectra, the contribution of protons staying at $t_{\rm max}$ in the up-stream region is shown in red, while the spectra of protons advected down-stream are shown in blue. The spectra of electrons are not shown, since they have the same shape as the proton spectra apart from a somewhat lower cutoff energy. While the total energy spectra at low energies agree well with a $1/E^2$ power-law, changing the injection model leads to large differences at high energies. The strong dependence of the spectra close to $E_{\rm max}$ on the injection model is expected, since this part is sensitive to how many particles are injected early, when shock acceleration is most effective.

We switch now to the produced secondaries and show in Fig. 2 for injection model 2 and $f_B B = 1\mu$G the antiproton flux split into a part produced in the acceleration zone (A) and a part produced in the inner part of the SNR (B). More exactly, we define the contribution A as all secondaries that crossed at least once the shock. This contribution increases fast, since the time $t_{\rm acc}$ primary protons stay in the acceleration zone and can interact increases as $t_{\rm acc} \propto D(E) \propto E$. Hence for the *relative* rise of A not the acceleration of secondaries but of primary protons is important. Indeed, the component A (defined by being produced up-stream) of neutral primaries as e.g. photons rises in the same way as the one of antiprotons. Since the inelasticity, i.e. the energy fraction $\langle z_{\bar{p}} \rangle \approx 0.02$ transferred to *all* antiprotons is practically constant in the relevant

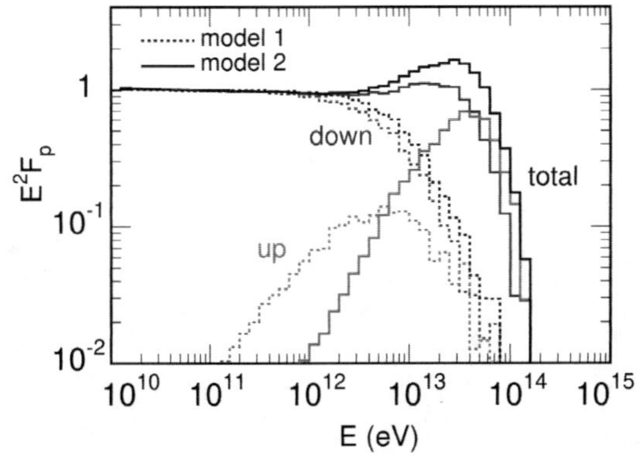

Figure 1. Proton spectra (black) as function of energy for two different injection models and $f_B B = 1\mu$G. Additionally the contribution of protons staying until $t_{\max}$ in the up-stream (red) and of protons in down-stream region (blue) are shown.

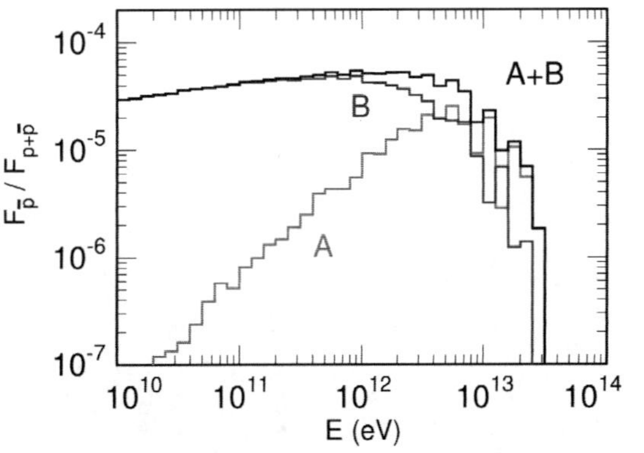

Figure 2. The total flux of antiprotons together with the contribution A and B in model 2 as function of energy.

energy range, $E_0/\langle z_{\bar{p}}\rangle \gtrsim 10^{11}$ eV, its does not influence the shape of the antiproton flux [1]. At $E_{\rm b} \approx 2 \times 10^{12}$ eV, the increase of contribution A stops, the total flux retains its approximate E^{-2} slope and stays small in contrast to the result of Ref. [8].

How can we understand this behavior and the maximal value of $\mathcal{F}_{\bar{p}}/\mathcal{F}_{\bar{p}+p}$? We may assume in a gedankenexperiment that in each $pp \to \bar{p} + X$ interaction the most energetic antiproton carries away all the energy, $E_{\bar{p}} \approx \langle z\rangle E_p$ with $\langle z\rangle \approx 1$. Then interactions just convert part of the p into a $\bar{p}$ flux. But since p and $\bar{p}$ diffuse and are accelerated in the same way, the total $\bar{p}$ flux is not affected if the $\bar{p}$ or the parent proton is accelerated. Hence the total flux of antiprotons produced in the acceleration zone and inside the SNR, i.e. the sum of A and B, should be simply the proton flux scaled down by a constant factor. In particular, the secondary flux of the species i is bounded by the proton interaction depth τ and the (spectrally averaged) energy fraction $\langle z_i\rangle$ transferred to i. The maximal conversion rate during the life-time of a SNR is with $\sigma_{\rm inel} = 30$ mb as inelastic pp cross section at 100 GeV given by $\tau = c\, t_{\max}\, R\, n_{\rm ISM}\, \sigma_{\rm inel} \approx 3 \times 10^{-3}$. The mean energy fraction of antiprotons (plus antineutrons) is $\langle z_{\bar{p}}\rangle \approx 0.02$, so we may expect a maximal ratio of $\mathcal{F}_{\bar{p}}/\mathcal{F}_{\bar{p}+p} \sim \langle z_{\bar{p}}\rangle\, \tau \sim 6 \times 10^{-5}$. The obtained $\mathcal{F}_{\bar{p}}/\mathcal{F}_{\bar{p}+p}$ ratio shown in Fig. 2 is indeed close to this estimate.

The relative size of the partial contributions A and B can be understood considering the relation between the time $t_{\rm acc}$ spent by protons in A, their final energy $E \propto t_{\rm acc}$ and thus the interaction depth τ_A in A as function of energy, $\tau_A \propto t_{\rm acc} \propto E$. In particular, it takes all the life-time $t_{\max}$ of the SNR to accelerate protons to the highest energies, cf. the up-stream component in Fig. 1. For the component B, the optical depth τ_B of the parent proton is $\tau_B \propto (t_{\max} - t_{\rm acc})$, which explains why the two components A and B sum up to a flat spectrum. Note that the relative normalization of component A and B in the stationary approach of Refs. [7, 8, 11] has to be imposed by hand, since B is formally infinite.

The same discussion applies to the case of positrons, with the sole exception that the primary electron flux is scaled down by the factor K_{ep} and that the energy fraction transferred to positrons

[1] Note also that, in contrast to the assumptions of Refs. [7, 8, 11], the average energy fraction per *single* antiproton (or positron) $\xi_i = \langle z_i\rangle/\langle n_i\rangle$ decreases strongly with energy, since the multiplicity n_i in pp interaction increases fast.

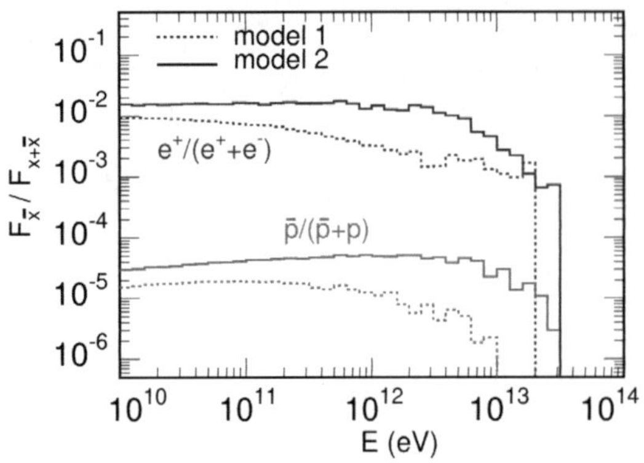

Figure 3. The positron ratio $\mathcal{F}_{e^+}/\mathcal{F}_{e^++e^-}$ (blue) and antiproton ratio $\mathcal{F}_{\bar{p}}/\mathcal{F}_{\bar{p}+p}$ (red) in model 1 (dotted) and 2 (solid).

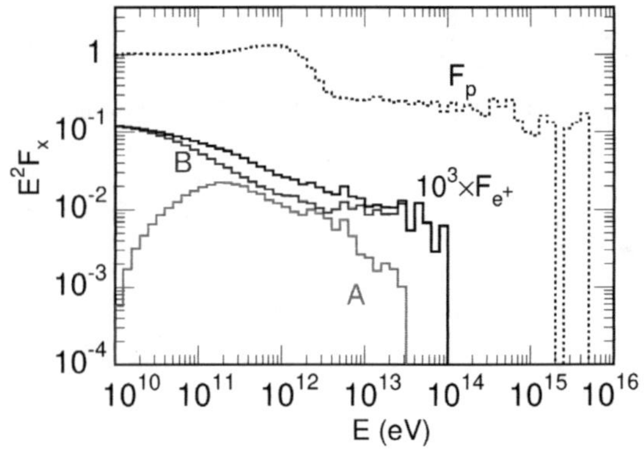

Figure 4. Spectra of cosmic rays protons and positron spectra (scaled up by 1000) together with the partial contributions A and B in model 2 for a time-dependent diffusion coefficient.

is $\langle z_{e^+}\rangle \approx 0.05$. The results of our simulation are shown in Fig. 3, confirming with the maximal value of $\mathcal{F}_{e^+}/\mathcal{F}_{e^++e^-} \lesssim$ few % this simple picture. Note that while above $\sim 100\,\mathrm{GeV}$ the ratio $\mathcal{F}_{e^+}/\mathcal{F}_{e^++e^-}$ from SNR starts to be larger than the conventional prediction using only secondary production on the ISM, it cannot explain the rise to $\mathcal{F}_{e^+}/\mathcal{F}_{e^++e^-} \approx 10\%$ at $10\,\mathrm{GeV}$ in the PAMELA data [3].

Up to now, we have discussed only our numerical results for constant $f_B B = 1\mu\mathrm{G}$ and one may wonder if a "better" choice of parameters can increase the antimatter fluxes. In particular, the analytical formula of Ref. [7, 8, 11] seem to imply that the contribution A increases for weaker diffusion, i.e. larger D. However, the term D/v_1^2 regulating the importance of A limits also via $t_{\mathrm{acc}} \propto D/v_1^2$ the maximal proton energy. Using a constant value $f_B = 1/20$ as in Ref. [7, 8, 11] thus reduces $E_{\max}$ by the same factor. In the stationary approach, however, $E_{\max}$ is an external parameter and by increasing $E_{\max}$ relative to its natural value given by $t_{\mathrm{acc}} = t_{\max}$ one enlarges the relative contribution of A. This approach has been justified as a method to account in an effective way for amplification and damping of the magnetic field.

In our time-dependent approach, we test this suggestion in our second scenario for the diffusion coefficient with $f_B B = 100\mu\mathrm{G}$ for $t < t_*$ and $f_B B = 1/20\mu\mathrm{G}$ afterwards. In Fig. 4, we show for this case the proton and positron spectra using the injection model 2. Protons that were injected early are accelerated up to few $\times 10^{15}\,\mathrm{eV}$, while the bulk of CRs injected when the tubulent magnetic field is damped has a cutoff at few $\times 10^{12}\,\mathrm{eV}$. The contribution A to the positron flux saturates at $E \sim$ few $\times\, 10^{11}\,\mathrm{eV}$, i.e. at the energy expected fo $f_B B = 1/20\mu\mathrm{G}$. In contrast to Fig. 2, the component B dominates now the high-energy end of the positron flux: Most CRs escape from the acceleration zone during the transition $f_B B/\mu\mathrm{G} = 100 \to 1/20$ and those advected downward contribute the new high-energy extension of component B, while those escaping up-stream contribute to A but are not longer accelerated.

Finally, we stress that the splitting between contribution A and B is artificial and depends as well as the total flux on the definition of the escape flux: If the diffusion coefficient drops above a certain energy and/or outside a sufficiently small radius $r_{\mathrm{sh}} + \delta r$ to a value close to the one typical for the Galaxy, then CRs can escape up-stream instead of being confined down-stream. Clearly, this effect reduces the contribution B. On the other hand, the bounds $\mathcal{F}_{e^+}/\mathcal{F}_{e^++e^-} \lesssim$ few % and $\mathcal{F}_{\bar{p}}/\mathcal{F}_{\bar{p}+p} \lesssim 6\times 10^{-5}$ will become stronger, since also the time for interactions in the acceleration zone will be shortened. Since our maximal values of $\mathcal{F}_{\bar{p}}/\mathcal{F}_{\bar{p}+p}$ and $\mathcal{F}_{e^+}/\mathcal{F}_{e^++e^-}$ depend only

Journal of Physics: Conference Series **259** (2010) 012092 doi:10.1088/1742-6596/259/1/012092

on $t_{\max}$, which is lower in escape-limited than in age-limited models, we conclude that the contribution of SNR to the observed antimatter in CRs does not lead to rising antimatter fractions and is smaller than estimated in earlier works.

4. Summary

We calculated the energy spectra of CRs and their secondaries produced in a supernova remnant using a simple random walk picture. In contrast to a previous prediction that the positron fraction $\mathcal{F}_{e^+}/\mathcal{F}_{e^++e^-}$ can rise up to 40%–50% for $K_{ep} = 7 \times 10^{-3}$, we found that the ratio levels off at a few percent. This value corresponds to the expectation combining the interaction depth $\tau \approx 3 \times 10^{-3}$ of a proton during the life-time of a SNR with the energy fraction $z \sim 0.05$ transferred to positrons. Similarly, the antiproton ratio $\mathcal{F}_{\bar{p}}/\mathcal{F}_{\bar{p}+p}$ does not rise beyond few$\times 10^{-5}$. Our results suggest that antimatter production in SNRs cannot explain the rise of the positron fraction observed by PAMELA. Since a rising antiproton fraction is neither expected from CR interactions with the ISM nor from pulsars, such a measurement could be used as a signature for dark matter.

References

[1] A. W. Strong and I. V. Moskalenko, Adv. Space Res. **27**, 717 (2001); N. Prantzos, arXiv:1003.2317 [astro-ph.GA].
[2] L. Bergström, New J. Phys. **11**, 105006 (2009).
[3] O. Adriani *et al.*,Nature **458**, 607 (2009).
[4] O. Adriani *et al.*, Phys. Rev. Lett. **102**, 051101 (2009).
[5] P. D. Serpico, Phys. Rev. D **79**, 021302 (2009).
[6] A. K. Harding and R. Ramaty, Proc. 20th ICRC, Moscow, **2**, 92 (1987); A. Boulares, Astrophys. J. **342**, 807 (1989); F. A. Aharonian, A. M. Atoyan and H. J. Völk, Astron. Astrophys. **294**, L41 (1995).
[7] P. Blasi, Phys. Rev. Lett. **103**, 051104 (2009).
[8] P. Blasi and P. D. Serpico, Phys. Rev. Lett. **103**, 081103 (2009).
[9] S. P. Reynolds, Ann. Rev. Astr. Astrophys. **46**, 89 (2008).
[10] P. Mertsch and S. Sarkar, Phys. Rev. Lett. **103**, 081104 (2009).
[11] M. Ahlers, P. Mertsch and S. Sarkar, Phys. Rev. D **80**, 123017 (2009).
[12] J. K. Truelove and Ch. F. McKee, Astrophys. J. Suppl. **120**, 299 (1994).
[13] S. Ostapchenko, Nucl. Phys. Proc. Suppl. **151**, 143 (2006); Phys. Rev. D **74**, 014026 (2006).
[14] R. Engel *et al.*, Proc. 26th ICRC, Salt Lake City, **1**, 415 (1999); E.-J. Ahn *et al.*, Phys. Rev. D **80**, 094003 (2009).
[15] M. A. Malkov and H. J. Völk, A & A **300**, 605 (1995).
[16] V. S. Ptuskin and V. N. Zirakashvili, Astron. Astrophys. **429**, 755 (2005).
[17] A. R. Bell and S. G. Lucek, Mon. Not. R. Astron. Soc. **321**, 433 (2001).

16th International Symposium on Particles Strings and Cosmology (PASCOS 2010) IOP Publishing
Journal of Physics: Conference Series **259** (2010) 012093 doi:10.1088/1742-6596/259/1/012093

An A4 model for neutrinos

Eduardo Peinado
AHEP Group, Institut de Física Corpuscular – C.S.I.C./Universitat de València
Edificio Institutos de Paterna, Apt 22085, E–46071 Valencia, Spain
E-mail: epeinado@ific.uv.es

Abstract. The study of an extension of the standard model based on the flavor symmetry A_4 is presented. Neutrino Majorana mass terms arise from dimension five operator and charged lepton masses from renormalizable Yukawa couplings. We introduce three Higgs doublets that belong to one triplet irreducible representation of A_4. We study the most general A_4-invariant scalar potential and the phenomenological consequences of the model. We find that the reactor angle could be as large as, $\sin^2\theta_{13\text{max}} \sim 0.03$, while the atmospheric mixing angle θ_{23} is close to maximal, $\sin^2\theta_{23} = 1/2$.

1. Introduction

Experiments on neutrino oscillations confirmed that neutrinos are massive and mix among themselves like the quarks do. In contrast to the quark sector, neutrino oscillations have two large mixing angles [1]. In the literature continuous as well as discrete (Abelian and non-Abelian) flavor symmetries have been extensively studied. However so far we do not have a unique top-down hint for the choice of the flavor symmetry. In a bottom-up approach simplicity and predictivity are possible criteria that we can use. In the quark sector the heaviness of the top quark suggests that first and second families could belong to a doublet irreducible representation and the third family belongs to a singlet representation of the flavor symmetry group. Differently the large solar and atmospheric neutrino mixing angles suggest that the three neutrino families belong to one triplet irreducible representation of a flavor group. The group of even permutation of four objects, A_4[2, 3], is the smallest non-Abelian discrete group with triplet irreducible representation. Most of the models based on A_4 need to introduce extra auxiliar Abelian symmetries and supersymmetry (or extra dimensions) in order to reproduce tri-bimaximal mixing [4].

In this paper we use the A_4 symmetry group as the flavor group but we renounce to predict the tri-bimaximal mixing. The Dirac mass matrices arise from renormalizable operator coupled with the Higgs, that is the Yukawa interactions. Neutrino Majorana mass terms are generated from a dimension five Weinberg operator. It is a known fact that A_4 can be broken spontaneously into its Z_3 or Z_2 subgroups assuming the vacuum expectation values (vevs) to be real [1]. Recently in [5], a model for quarks mixing has been suggested where the most generic A_4-invariant potential has complex solutions. In this case A_4 is completely broken. This solution open new possibilities for the description of the leptonic sector that deserve further investigation.

[1] When A_4 is broken into Z_3 in the charged lepton sector and into Z_2 in the neutrino sector the lepton mixing is tri-bimaximal.

Journal of Physics: Conference Series **259** (2010) 012093 doi:10.1088/1742-6596/259/1/012093

In the next section we present the model, in section 3 and 4 we discuss the charged lepton and neutrino mass matrices respectively. In section 5 we discuss the implication of the model and finally, in section 6 we summarize the results of the model.

2. The Model

In our model [6] the Higgs sector is extended from one $SU(2)_L$-doublet to three $SU(2)_L$-doublets belonging to a triplet irreducible representation of A_4. The left-handed as well as the right-handed charged leptons, belong to the triplet irreducible representation of A_4. The irreducible representation assignment for the particles is given in Table 1.

fields	L_i	l_i^c	ϕ_i
$SU(2)_L$	2	1	2
A_4	3	3	3

Table 1. Lepton multiplet structure of the model

A_4 is the group of even permutations of 4 objects and is isomorphic to the symmetries of the tetrahedron. A_4 can be generated by two generators S and T with the properties

$$S^2 = T^2 = (ST)^3 = 1. \tag{1}$$

A_4 contain one 3-dimensional representation, 3, and three one-dimensional, 1, $1'$ and $1''$. The product of two 3 gives $3 \otimes 3 = 1 \oplus 1' \oplus 1'' \oplus 3 \oplus 3$ and $1' \otimes 1' = 1''$, $1' \otimes 1'' = 1$, $1'' \otimes 1'' = 1'$ etc. For two triplets $3_a \sim (a_1, a_2, a_3)$, $3_b \sim (b_1, b_2, b_3)$ the irreducible representations obtained from their product are:

$$1 = a_1b_1 + a_2b_2 + a_3b_3, \tag{2}$$

$$1' = a_1b_1 + \omega^2 a_2b_2 + \omega a_3b_3, \tag{3}$$

$$1" = a_1b_1 + \omega a_2b_2 + \omega^2 a_3b_3, \tag{4}$$

$$3 \sim (a_2b_3, a_3b_1, a_1b_2), \tag{5}$$

$$3 \sim (a_3b_2, a_1b_3, a_2b_1), \tag{6}$$

this in the basis of S diagonal and where $\omega = e^{i2\pi/3}$.

The most general renormalizable Yukawa Lagrangian for the charged leptons in the model is

$$\begin{aligned} L_{\text{Yukawa}} &= y_1 \left(\bar{L_1}\phi_3 l_2^c + \bar{L_2}\phi_1 l_3^c + \bar{L_3}\phi_2 l_1^c\right) + \\ &\quad + y_2 \left(\bar{L_1}\phi_2 l_3^c + \bar{L_2}\phi_3 l_1^c + \bar{L_3}\phi_1 l_2^c\right). \end{aligned} \tag{7}$$

Once the electroweak symmetry (EW) is broken, the charged lepton mass matrix are obtained from this Yukawa Lagrangian:

$$M_l = \begin{pmatrix} 0 & y_1\langle\phi_3\rangle & y_2\langle\phi_2\rangle \\ y_2\langle\phi_3\rangle & 0 & y_1\langle\phi_1\rangle \\ y_1\langle\phi_2\rangle & y_2\langle\phi_1\rangle & 0 \end{pmatrix}. \tag{8}$$

The most general neutrino dimension five operator invariant under A_4 (see for instance [?]), is given by

$$\begin{aligned} \mathcal{L}_{5d} &= \beta\,(LL)_3\,(HH)_3 + k\,(LL)_1\,(HH)_1 + \alpha'\,(LL)_{1'}\,(HH)_{1''} + \alpha''\,(LL)_{1''}\,(HH)_{1'} + \\ &+ \left[a\,(LH)_{3^a}\,(LH)_{3^a} + b\,(LH)_{3^a}\,(LH)_{3^b} + c\,(LH)_{3^b}\,(LH)_{3^a} + d\,(LH)_{3^b}\,(LH)_{3^b}\right] + \\ &+ l\,(LH)_1\,(LH)_1 + l'\left[(LH)_{1'}\,(LH)_{1''} + (LH)_{1''}\,(LH)_{1'}\right], \end{aligned} \tag{9}$$

Journal of Physics: Conference Series **259** (2010) 012093 doi:10.1088/1742-6596/259/1/012093

where $\beta, k, \alpha, \alpha', \alpha'', a, b, c, d$ are arbitrary complex couplings. Once the EW symmetry is broken, the Majorana neutrino mass matrix is given by

$$M_\nu = \begin{pmatrix} x\langle\phi_1\rangle^2 + y\langle\phi_2\rangle^2 + z\langle\phi_3\rangle^2 & \kappa\langle\phi_1\rangle\langle\phi_2\rangle & \kappa\langle\phi_1\rangle\langle\phi_3\rangle \\ \kappa\langle\phi_1\rangle\langle\phi_2\rangle & z\langle\phi_1\rangle^2 + x\langle\phi_2\rangle^2 + y\langle\phi_3\rangle^2 & \kappa\langle\phi_2\rangle\langle\phi_3\rangle \\ \kappa\langle\phi_1\rangle\langle\phi_3\rangle & \kappa\langle\phi_2\rangle\langle\phi_3\rangle & y\langle\phi_1\rangle^2 + z\langle\phi_2\rangle^2 + x\langle\phi_3\rangle^2 \end{pmatrix}, \tag{10}$$

where x, y, z and κ are functions of the couplings in eq. (9).

The most general renormalizable scalar potential invariant under the symmetry A_4 [5, 6] has the solution

$$\langle\phi_1\rangle = v_1, \quad \langle\phi_2\rangle = ve^{i\alpha/2}, \quad \langle\phi_3\rangle = ve^{-i\alpha/2}. \tag{11}$$

3. Charged leptons

With the vevs in eq. (11), the charged lepton mass matrix in eq. (8) takes the form

$$M_l = \begin{pmatrix} 0 & ae^{i\alpha} & be^{-i\alpha} \\ be^{i\alpha} & 0 & ar \\ ae^{-i\alpha} & br & 0 \end{pmatrix}, \tag{12}$$

with the parameters in the matrix (12) defined by $a = y_1 v$, $b = y_2 v$ and $r = v_1/v$. We write the symmetric matrix, $M_l M_l^T$

$$M_l M_l^T = \begin{pmatrix} a^2 + b^2 & abr & abr \\ abr & b^2 + a^2 r^2 & ab \\ abr & ab & a^2 + b^2 r^2 \end{pmatrix}, \tag{13}$$

which is diagonalized by an orthogonal matrix O_l. It is straightforward to obtain the analytical expressions for a, b and r as function of the charged lepton masses [6], it can be written as

$$\begin{aligned} r &\approx \tfrac{m_\tau}{\sqrt{m_e m_\mu}} \sqrt{1 - \tfrac{m_e^2 m_\mu^2}{m_\tau^4}}, \\ a &\approx \tfrac{m_\mu}{m_\tau} \sqrt{m_e m_\mu} \left[1 + \tfrac{1}{2} \tfrac{m_\mu^2}{m_\tau^2}\right], \\ b &\approx \sqrt{m_e m_\mu} \left[1 - \tfrac{1}{2} \tfrac{m_\mu^2}{m_\tau^2}\right]. \end{aligned} \tag{14}$$

With

$$m_e = 0.511006 \text{ MeV}, \quad m_\mu = 105.656 \text{ MeV}, \quad m_\tau = 1776.96 \text{ MeV},$$

we have $a = 0.43474$, $b = 7.3471$ and $r = 241.8582$. Note that $a < b \ll r$ thus the orthogonal matrix O_l diagonalizing $M_l M_l^T$ is approximatively

$$O_{l_{12}} \approx \frac{b}{a} r^{-1}, \quad O_{l_{13}} \approx \frac{a}{b} r^{-1}, \quad O_{l_{23}} \approx \frac{a}{b} r^{-2}. \tag{15}$$

The element, $O_{l_{12}}$, give a contribution to the reactor mixing angle, θ_{13}, see section 5.1. The analytical expression for this element is given as

$$O_{l_{12}} \approx \sqrt{\frac{m_e}{m_\mu}} \left[1 - \left(\frac{m_\mu}{m_\tau}\right)^2\right]. \tag{16}$$

The numerical expression for the matrix O_l is

$$O_l = \begin{pmatrix} 0.997 & 0.069 & 2.44 \times 10^{-4} \\ -0.069 & 0.997 & 1.075 \times 10^{-6} \\ -2.439 \times 10^{-4} & -1.800 \times 10^{-5} & 0.999 \end{pmatrix}. \tag{17}$$

4. Neutrinos

The mass matrix for the neutrinos in eq. (10) with the vevs in eq. (11) takes the form

$$M_\nu = \begin{pmatrix} xr^2 + ye^{-2i\alpha} + ze^{2i\alpha} & \kappa re^{-i\alpha} & \kappa re^{i\alpha} \\ \kappa re^{-i\alpha} & zr^2 + xe^{-2i\alpha} + ye^{2i\alpha} & \kappa \\ \kappa re^{i\alpha} & \kappa & yr^2 + ze^{-2i\alpha} + xe^{2i\alpha} \end{pmatrix}. \tag{18}$$

From the charged lepton sector we know that r is fixed as shown in eq. (14), and $r \gg 1$, then we can neglect in the diagonal the terms not proportional to r^2. With this the mass matrix in eq. (18) can be written as

$$M_\nu = \begin{pmatrix} xr^2 & \kappa re^{-i\alpha} & \kappa re^{i\alpha} \\ \kappa re^{-i\alpha} & zr^2 & \kappa \\ \kappa re^{i\alpha} & \kappa & yr^2 \end{pmatrix}. \tag{19}$$

Note that there are 4 complex free parameters, x, y, z, κ and one extra phase α coming from the Higgs sector. We can absorb two phases in the fields, then it remains 7 free parameters in the neutrino mass matrix. Neutrino oscillation experiments determine two mass square difference $\Delta m^2_{12} \equiv m^2_2 - m^2_1$ and $\Delta m^2_{13} \equiv |m^2_3 - m^2_1|$ with the corresponding three mixing angle [1]. If θ_{13} is different from zero, the Dirac phases could be probed in future experiments [7]. The absolute neutrino mass scale can be probed in future tritium beta decay [8] and neutrinoless double beta decay [9] experiments. While it will be hard to measure the two Majorana phase. There are seven measurable physical observable plus two Majorana physical phases.

5. Phenomenology of the model

In our Model we have six free parameters, x, y, θ, ϕ_{xy}, δ and α in the neutrino sector for nine physical parameters, Δm^2_{12}, Δm^2_{13}, m_{ee}, three mixing angles and two Majorana phases and the Jarlskog invariant J, for Dirac CP violation in the neutrino sector. We can construct the expressions for Δm^2_{12} and Δm^2_{13} and find x and y as functions of the observables, Δm^2_{12}, Δm^2_{13}, the mixing parameter θ and its relative phase ϕ_{xy}. This model is only compatible with inverted hierarchy neutrino mass spectrum [6].
In the next subsections we present the predictions for the allowed region for mixing angles, the Jarlskog invariant as well as the neutrinoless double beta decay.

5.1. Neutrino mixing angles

Recently has been given an indication that the reactor neutrino angle θ_{13} could be different from zero [1]

$$\sin^2\theta_{13} = 0.016 \pm 0.010\,(1\sigma). \tag{20}$$

The θ^ν_{13} angle coming from the diagonalization of the neutrino mass matrix, is exactly zero in the limit $\delta, \alpha = 0$. However the reactor angle resulting from the product of the unitary matrix that diagonalize the charged lepton matrix, eqs. (16) and (17) and the neutrino mass matrix, is different from zero and we have

$$\sin^2\theta_{13} \approx \frac{m_e}{2m_\mu} \approx 0.0024. \tag{21}$$

The solar mixing angle is given by the θ parameter up to corrections coming from the charged lepton sector of the order $\mathcal{O}(\sqrt{m_e/m_\mu})$. The parameters α and δ are related with the deviations of θ_{13} and θ_{23} from the zero and maximal values respectively.

In the left side of figure 1, we show the allowed region for the atmospheric mixing angle. The deviation from its maximal value is small. The reactor mixing angle, θ_{13}, can be large. In the right figure 1 side of we show the magnitude of the Jarlskog invariant [10], J, of CP violation in neutrino oscillation defined as

$$J = Im(V_{11}V_{22}V_{12}^*V_{21}^*). \tag{22}$$

J is correlated to the $\sin^2\theta_{13}$ that can be measured in next experiments like Double Chooz [7].

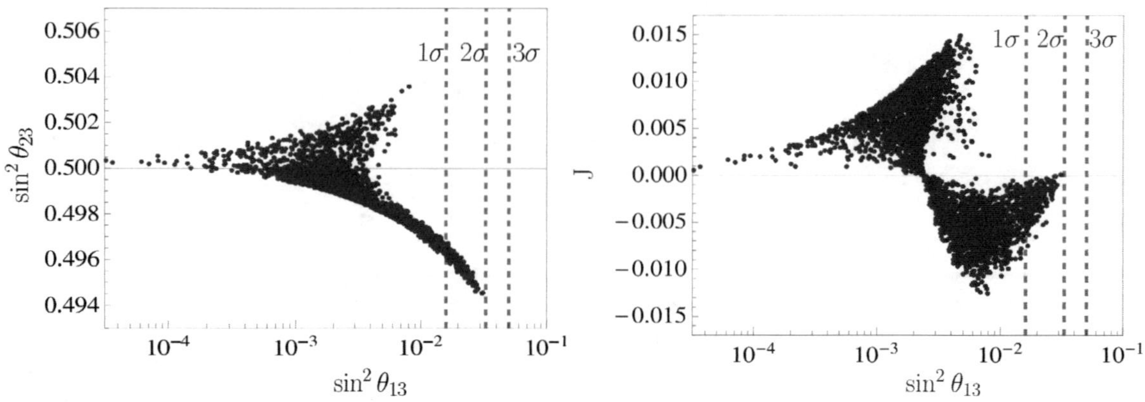

Figure 1. The figure in the left side shows the allowed region for θ_{23} vs. θ_{13}. The figure in the right side shows allowed region for J vs. θ_{13}.

5.2. Neutrinoless double beta decay

Neutrinos are guaranteed to have a non-zero Majorana mass if the neutrinoless double beta decay $0\nu\beta\beta$ is observed [11]. The $0\nu\beta\beta$ decay rate m_{ee} is proportional to the $\nu_e - \nu_e$ entry of the Majorana neutrino mass matrix M_ν. For an introduction to the phenomenology of $0\nu\beta\beta$ see for instance [12].

We observed that the parameter ϕ_{xy} is related with the Majorana phase β and the m_{ee} is predicted in our model. The allowed region for the neutrino double beta decay is shown in figure 2. We show the dependence of the m_{ee} as function of the Majorana physical phase β. As can be seeing from the figure, the values for $\beta = 0$ and $\beta = \pi$ of m_{ee} are forbidden by the upper limits obtained by the HM collaboration. Other models in literature excluding $\beta = 0$ but not $\beta = \pi$ has been studied [13]. We note that $m_{\text{light}} > 0.008$.

6. Conclusion

We have studied a model for lepton mixing based on a A_4 flavor symmetry. This constraints the model, reducing the number of free parameters with respect to the case of the Standard Model. In the scalar sector we introduce three Higgs doublets that belong to a triplet representation of A_4. If the vevs of the Higgs fields are assumed to be real, there are only two possible solutions: i) the three vevs are all equal, or, ii) two vevs are equal to zero. In this paper we consider the most general case with complex vevs. This solution is different from that of real vevs, it is found that one of the vevs is real and the other two are the complex conjugate one of each other, that is, $\langle H\rangle = (v_1, v, v^*)$, where v_1 is different from v as noted also in [5]. This fact opens an interesting scenarios in the model building due to the extra CP phase in the Higgs sector. We studied the phenomenological implications of the neutrino masses and mixings. The charged lepton mass matrix arises only from renormalizable Yukawa interactions, while the Majorana neutrino mass matrix arises from a dimension five operator. We do not enter into details how this dimension five operator is generated. In order to fit the data we assumed a moderate fine tuning between the free parameters in the neutrino mass matrix. We found that the model is

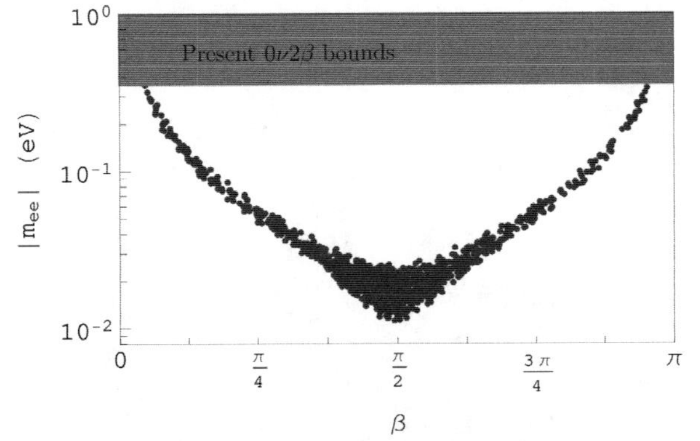

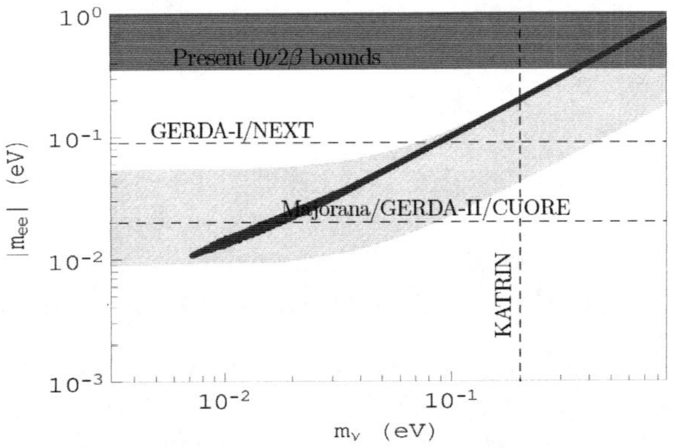

Figure 2. The figure in the left side shows the allowed range for the $0\nu\beta\beta$ as function of the physical Majorana phase β. The figure in the right side shows $|m_{ee}|$ as function of the lightest neutrino mass $m_{\nu 3}$. We also present here the future experimental sensitivity [9] .

compatible with inverse hierarchy only. The atmospheric angle is very close to the maximal value, $\sin^2\theta_{23} \sim 0.5$ and the maximum allowed value for the reactor angle is close to the current 2σ upper bound, that is $\sin^2\theta_{13} \sim 0.03$. The solar mixing angle can be fitted in the allowed experimental range at 3σ. The maximal value for the CP Jarlskog invariant is $|J| \approx 0.015$. We also found that the current $0\nu\beta\beta$ upper bound restricts the physical Majorana phase β, to be slightly different from zero and π. The power predictivity of this model in the leptonic sector make it interesting enough to make an attempt to extend this model to the quark sector. An important point is that the alignment we obtain here is incompatible with the hierarchy of the quark masses and the charged lepton masses simultaneously. A supersymetric version of this model is being invastigated [14] in order to solve this problem.

7. Acknowledgments

This work was supported by the Spanish grant FPA2008-00319/FPA and PROMETEO/2009/091.

References

[1] T. Schwetz, M. A. Tortola and J. W. F. Valle, New J. Phys. **10** (2008) 113011 [arXiv:0808.2016 [hep-ph]].
[2] K. S. Babu, E. Ma and J. W. F. Valle, Phys. Lett. B **552** (2003) 207 [arXiv:hep-ph/0206292].
[3] G. Altarelli and F. Feruglio, Nucl. Phys. B **720** (2005) 64 [arXiv:hep-ph/0504165].
[4] P. F. Harrison, D. H. Perkins and W. G. Scott, Phys. Lett. B **530** (2002) 167 [arXiv:hep-ph/0202074].
[5] L. Lavoura and H. Kuhbock, Eur. Phys. J. **C55**, 303 (2008), 0711.0670.
[6] S. Morisi and E. Peinado, Phys. Rev. D **80** (2009) 113011 [arXiv:0910.4389 [hep-ph]].
[7] F. Ardellier *et al.* [Double Chooz Collaboration], arXiv:hep-ex/0606025.
[8] G. Drexlin [KATRIN Collaboration], Nucl. Phys. Proc. Suppl. **145** (2005) 263.
[9] A. Osipowicz *et al.* [KATRIN Collaboration], arXiv:hep-ex/0109033; V. E. Guiseppe *et al.* [Majorana Collaboration], arXiv:0811.2446 [nucl-ex]; A. A. Smolnikov [GERDA Collaboration], arXiv:0812.4194 [nucl-ex]; A. Giuliani [CUORE Collaboration], J. Phys. Conf. Ser. **120**, 052051 (2008).
[10] C. Jarlskog, Z. Phys. C **29**, 491 (1985).
[11] J. Schechter and J. W. F. Valle, Phys. Rev. D **25**, 2951 (1982); J. Schechter and J. W. F. Valle,Phys. Rev. D **25**, 774 (1982).
[12] M. Hirsch, arXiv:hep-ph/0609146.
[13] M. Hirsch, S. Morisi and J. W. F. Valle, Phys. Rev. D **78**, 093007 (2008) [arXiv:0804.1521 [hep-ph]].
[14] S. Morisi, E. Peinado, Y. Shimizu and J. W. F. Valle, Work in progress.

Journal of Physics: Conference Series **259** (2010) 012094 doi:10.1088/1742-6596/259/1/012094

Relating neutrino mass and dark matter with discrete flavor symmetry

S. Morisi

AHEP Group, Institut de Física Corpuscular – C.S.I.C./Universitat de València
Edificio Institutos de Paterna, Apt 22085, E–46071 Valencia, Spain

E-mail: morisi@ific.uv.es

Abstract.
We propose a new motivation for the stability of dark matter (DM). We suggest that the same non-abelian discrete flavor symmetry which accounts for the observed pattern of neutrino oscillations, spontaneously breaks to a Z_2 subgroup which renders DM stable. The simplest scheme leads to a scalar doublet DM potentially detectable in nuclear recoil experiments, inverse neutrino mass hierarchy, hence a neutrinoless double beta decay rate accessible to upcoming searches, while $\theta_{13} = 0$ gives no CP violation in neutrino oscillations.

1. Introduction

The existence of dark matter (DM) plays a central role in the modeling of structure formation and galaxy evolution, affecting also the cosmic microwave background. Despite the strong evidence in favor of DM, its detailed nature remains rather elusive. Viable particle physics candidates for dark matter must be electrically neutral, and provide the correct relic abundance. Therefore they must be stable over cosmological time scales. A simple way to justify the stability of the DM is by *assuming* some parity symmetry Z_2, which might arise from the spontaneous breaking of an abelian U(1) gauge symmetry [1, 2, 3] [1], or from a non-abelian discrete symmetry, as might be the case in some string models.

Non abelian discrete symmetries are motivated by neutrino oscillation data. Here we propose that the same symmetry explaining neutrino mixing angles is also responsible for the dark matter stability. In our simplest type-I seesaw [4] realization the flavor symmetry A_4 spontaneously breaks to Z_2 providing a stable DM candidate. We extend the scalar sector of the standard model by adding three Higgs doublets transforming as a triplet of A_4 we show that there is a consistent pattern of vacuum expectation values (vevs) for which only one of the three extra Higgs doublets takes a vev, while the other two give rise to the dark matter candidate. The model accounts for the observed pattern of mixing angles [5] indicated by current neutrino oscillation data, predicting $\theta_{13} = 0$ and inverted spectrum of neutrino masses. It will therefore be tested in upcoming double beta and neutrino oscillation searches [6], while the dark matter has potentially detectable rates within reach of nuclear recoil experiments.

[1] In supersymmetry a viable DM particle is the neutralino, whose stability stems from the imposition of the so-called R-parity.

2. Model

We assign matter fields to irreducible representations of A_4, see [7, 8], as in table in table 1. The

	L_e	L_μ	L_τ	l_e^c	l_μ^c	l_τ^c	N_T	N_4	H	η
$SU(2)$	2	2	2	1	1	1	1	1	2	2
A_4	1	$1'$	$1''$	1	$1''$	$1'$	3	1	1	3

Table 1. Summary of relevant model quantum numbers

resulting Yukawa Lagrangian is

$$\begin{aligned}\mathcal{L} &= y_e L_e l_e^c H + y_\mu L_\mu l_\mu^c H + y_\tau L_\tau l_\tau^c H + y_1^\nu L_e (N_T\eta)_1 + y_2^\nu L_\mu (N_T\eta)_{1''} + \\ &+ y_3^\nu L_\tau (N_T\eta)_{1'} + y_4^\nu L_e N_4 H + M_1 N_T N_T + M_2 N_4 N_4 + \text{h.c.}\end{aligned}$$

This way H is responsible for quark and charged lepton masses, the latter automatically diagonal. Note that we do not discuss the quark sector, assumed to be blind to A_4, namely all left and right-handed up and down-type quarks transform trivially under A_4, their mass and mixing hierarchies might arise from an extra family symmetry, for example, Frogatt-Nielsen-like. Neutrino masses arise from H and η, see below. The scalar potential is of the form

$$\begin{aligned}V &= \mu_\eta^2 \eta^\dagger\eta + \mu_H^2 H^\dagger H + \lambda_1 [H^\dagger H]^2 + \lambda_2 [\eta^\dagger\eta]_1^2 + \\ &+ \lambda_3 [\eta^\dagger\eta]_{1'} [\eta^\dagger\eta]_{1''} + \lambda_4 \left([\eta^\dagger\eta^\dagger]_{1'}[\eta\eta]_{1''} + h.c\right) + \\ &+ \lambda_5 \textstyle\sum_i [\eta^\dagger\eta]_{3_i}[\eta^\dagger\eta]_{3_i} + \lambda_5'([\eta^\dagger\eta]_{3_1}[\eta^\dagger\eta]_{3_2} + h.c.) + \\ &+ \lambda_6 \left(\textstyle\sum_{i,j}[\eta^\dagger\eta^\dagger]_{3_i}[\eta\eta]_{3_j} + h.c\right) + \lambda_7 [\eta^\dagger\eta]_1 H^\dagger H + \\ &+ \lambda_8 \left([\eta^\dagger\eta^\dagger]_1 HH + h.c\right) + \lambda_9 \left([\eta^\dagger\eta]_{3_{1,2}}\eta^\dagger H + h.c\right) + \\ &+ \lambda_{10}\left([\eta^\dagger\eta^\dagger]_{3_{1,2}}\eta H + h.c\right)\end{aligned} \tag{1}$$

where $i, j = 1, 2$, and $[...]_{3_i}$ means the product of two triplets contracted into a triplet of A_4, $[...]_1$ means the product of two triplets contracted into a singlet of A_4 and so on. Note that $[\eta^\dagger\eta]_{1,1',1''} \equiv [\eta\eta^\dagger]_{1,1',1''}$, $[\eta\eta]_{3_1} \equiv [\eta\eta]_{3_2}$ and so on.

We have studied the minimization of the potential for the scalar fields H, η_1, η_2 and η_3 and we have checked that for suitable parameter choices an allowed minimum is, see [9]

$$\left\langle H^0 \right\rangle = v_h \neq 0, \qquad \left\langle \eta_1^0 \right\rangle = v_\eta \neq 0 \qquad \left\langle \eta_{2,3}^0 \right\rangle = 0\,. \tag{2}$$

Note that the alignment $\langle\eta\rangle \sim (1, 0, 0)$ breaks spontaneously A_4 to Z_2 since $(1, 0, 0)$ is invariant under the S generator. The Z_2 is defined as

$$\begin{array}{lllllllll} N_2 & \to & -N_2\,, & h_2 & \to & -h_2\,, & A_2 & \to & -A_2\,, \\ N_3 & \to & -N_3\,, & h_3 & \to & -h_3\,, & A_3 & \to & -A_3\,. \end{array} \tag{3}$$

This residual symmetry is responsible for the stability of our DM candidate. This atractive feature reminds of the inert doublet model [?], with the difference that here it follows naturally from the underlying flavor symmetry which accounts for neutrino oscillations. We have four Higgs doublets[2] giving three physical charged scalar bosons, four neutral scalars, and three

[2] Lepton flavor violating processes are suppressed by the large right-handed neutrino scale.

pseudoscalars. After electroweak symmetry breaking we can write

$$H = \begin{pmatrix} 0 \\ v_h + h \end{pmatrix}, \quad \eta_1 = \begin{pmatrix} \eta_1^+ \\ v_\eta + h_1 + iA_1 \end{pmatrix},$$
$$\eta_2 = \begin{pmatrix} \eta_2^+ \\ h_2 + iA_2 \end{pmatrix}, \quad \eta_3 = \begin{pmatrix} \eta_3^+ \\ h_3 + iA_3 \end{pmatrix}. \tag{4}$$

The mass of the neutral scalar fields is block diagonal with the standard model Higgs h mixed with h_1, but not with the scalar fields with zero vev's $h_{2,3}$.

3. Dark matter

The lightest combination of the stable scalar fields h_2, h_3 plays the role of our dark matter particle, which we will denote generically by η_{DM}. We list below all interactions of η_{DM}:

(i) Yukawa interactions

$$\eta_{DM}\, \overline{\nu}_i N_{2,3}\,, \tag{5}$$

where $i = e,\ \mu,\ \tau$.

(ii) Higgs-Vector boson couplings

$$\begin{array}{ll} \eta_{DM}\eta_{DM} ZZ\,, & \eta_{DM}\eta_{DM} WW\,, \\ \eta_{DM}\eta^{\pm}_{2,3} W^{\pm} Z\,, & \eta_{DM}\eta^{\pm}_{2,3} W^{\pm}\,, \\ \eta_{DM} A_{2,3} Z\,. & \end{array} \tag{6}$$

(iii) Scalar interactions from the Higgs potential:

$$\begin{array}{ll} \eta_{DM}\, A_1 A_2 h\,, & \eta_{DM}\, A_1 A_3 h_1\,, \\ \eta_{DM}\, A_1 A_2 h_1\,, & \eta_{DM}\, A_1 A_3 h\,, \\ \eta_{DM}\, A_2 A_3 h_3\,, & \eta_{DM}\, h_1 h_3 h \\ \eta_{DM}\eta_{DM} hh\,, & \eta_{DM}\eta_{DM} h_1 h_1\,. \end{array} \tag{7}$$

After electroweak symmetry breaking, the vevs v_h and v_η are generated, so that additional terms are obtained from those in Eq. (7) by replacing $h \to v_h$ and $h_1 \to v_\eta$. The flavor symmetry A_4 is broken down to the residual Z_2 symmetry in Eq. (3), implying the stability of our dark matter candidate.

Relic Density Assuming that our DM candidate arises as thermal relic in the early universe, one of the most important requirements one must check is its relic abundance. For definiteness we require that η_{DM} makes up all the observed DM. For $M_\eta \gg M_z$ the most important annihilation and coanihilation processes are those with vector bosons, though for large $\lambda > g_2$, where $\lambda = \lambda_7 + 2\lambda_8$, annihilation into Higgs bosons plays an important role, see Eq. (9). The DM abundance can be approximated as [10]

$$\begin{aligned} \frac{n_{\rm DM}(T)}{s(T)} &\approx \sqrt{\frac{180}{\pi\ g_*}} \frac{1}{M_{\rm Pl}\ T_f \langle\sigma_A v\rangle}, \\ \frac{M_\eta}{T_f} &\approx 26, \end{aligned} \tag{8}$$

where $g_* = 106.75 + n$ is the number of SM degrees of freedom plus $1 \leq n \leq 12$ degrees of freedom arising from the extra scalars, and $M_{Pl} = 1.22 \times 10^{19} GeV$ is the Planck scale. The cross section for $\eta_{DM}\eta_{DM} \to VV$ where V are vector bosons is given by [10]

$$\langle\sigma_A v\rangle \simeq \frac{3g_2^4 + g_Y^4 + 6g_2^2 g_Y^2 Y^2 + 4\lambda^2}{256\pi\ M_\eta^2}, \tag{9}$$

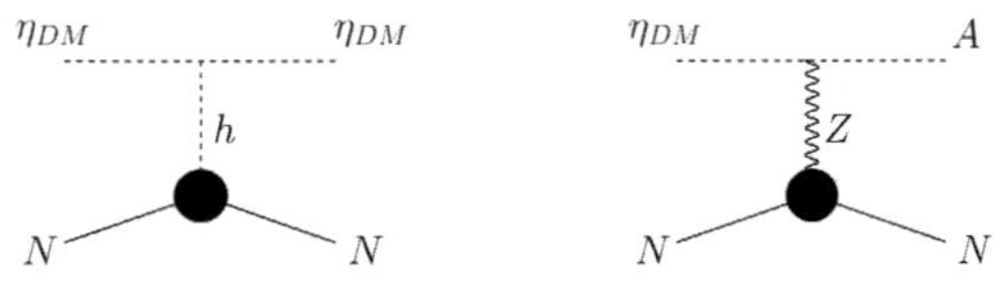

Figure 1. Feynman diagrams relevant for direct DM detection. Elastic scattering (left) is generically more important than inelastic (right).

where $Y = 1/2$ is the weak hypercharge, $g_2 = \sqrt{4\pi/\alpha}/(1-M_W^2/M_Z^2)$ and $g_Y = \sqrt{4\pi/\alpha}M_Z/M_W$. From these equations it follows that, in order to provide the correct relic abundance $\Omega_{\rm DM}h^2 = 0.110 \pm 0.006$ i.e. $n_{\rm DM}/s = (0.40 \pm 0.02)eV/M_\eta$, a correlation between the mass of the dark matter M_η and the quartic coupling constant λ is required. For simplicity if we take the limit of small λ we obtain a mass for the DM candidate of $M_\eta \approx 0.51$ TeV. For large λ values we have that the DM mass M_η scales as λ.

Direct detection The quartic couplings $\eta^\dagger\eta H^\dagger H$ and $\eta^\dagger\eta^\dagger HH$ give an interaction of the DM candidate with the nucleon through the interchange of the SM Higgs boson. Hence our DM candidate can be detected through the elastic scattering with a nucleus $\eta_{DM}N \to \eta_{DM}N$ via the exchange of a Higgs, or through inelastic scattering with a nucleus $\eta_{DM}N \to AN$ with the exchange of a Z boson, see Fig. 1, where A is the lightest pseudoscalar, in general a mixture of A_2 and A_3.

Barring fine-tuned choices of parameters for which the threshold for inelastic scattering opens up, the detection will be dominated by the elastic process, whose cross section is given by [12]

$$\sigma_{\rm el}(\text{nucleon}) \approx \lambda^2 \frac{1}{1+(\tan\beta)^2}\left(\frac{100\ \text{GeV}}{M_h}\right)^4 \times \\ \times \left(\frac{50\ \text{GeV}}{M_\eta}\right)^2 \left(5\times 10^{-42}\ \text{cm}^2\right), \tag{10}$$

where $\tan\beta = v_h/v_\eta$. Note that all uncertainties associated with the nuclear form factor in Eq. (10), have been neglected. From the requirement of correctly reproducing the relic density, Eqs. (8) and (9), one can find an expression for λ as function of the DM mass, M_η. Using this relation and eq. (10) one can plot the estimated cross section for the direct detection for each value of $\tan\beta$ and mass of the Higgs, M_h, as illustrated in Fig. 2 [3]. The figure compares the experimental sensitivities with our model expectations, fixing $m_H = 120$ GeV and three $\tan\beta$ values. This choice of Higgs mass is motivated by the LEP bounds $m_H > 114$ GeV, which however is not strictly valid in our model due to the additional Higgs doublets.

4. Neutrino phenomenology

Our model has four heavy right-handed neutrinos, and is a special case, called (3,4), of the general type-I seesaw mechanism [13, 14]. After electroweak symmetry breaking, it is characterized by

[3] Here we focus on the region $M_\eta \gg M_z$. The interesting case of light DM will be treated elsewhere.

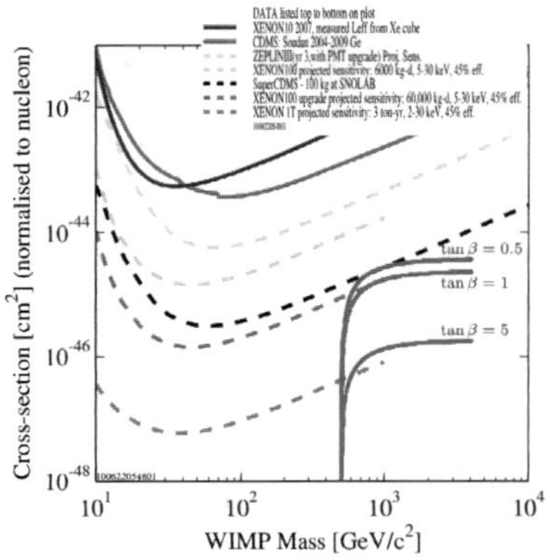

Figure 2. Elastic DM scattering cross section with a nucleon versus DM mass. We compare present and future sensitivities with our model expectations, for $m_H = 120\,\mathrm{GeV}$ and $\tan\beta = 0.5,\ 1,\ 5$ (grey solid lines)

Dirac and Majorana mass terms given as

$$m_D = \begin{pmatrix} x_1 & 0 & 0 & y_1 \\ x_2 & 0 & 0 & 0 \\ x_3 & 0 & 0 & 0 \end{pmatrix}, M_R = \mathrm{diag}(M_1, M_1, M_1, M_2)\ , \tag{11}$$

so that the light neutrinos get Majorana mass by means of the type-I seesaw relation $m_\nu = -m_{D_{3\times4}} M^{-1}_{R_{4\times4}} m^T_{D_{3\times4}}$ the light-neutrinos mass matrix M_ν being given as

$$\begin{pmatrix} \frac{x_1^2}{M_1} + \frac{y_1^2}{M_2} & \frac{x_1 x_2}{M_1} & \frac{x_1 x_3}{M_1} \\ \frac{x_1 x_2}{M_1} & \frac{x_2^2}{M_1} & \frac{x_2 x_3}{M_1} \\ \frac{x_1 x_3}{M_1} & \frac{x_2 x_3}{M_1} & \frac{x_3^2}{M_1} \end{pmatrix} = \begin{pmatrix} y^2 & ab & ac \\ ab & b^2 & bc \\ ac & bc & c^2 \end{pmatrix}. \tag{12}$$

It falls in the class of scaling matrix introduced in Ref. [15]. This form of the light neutrino mass matrix has an inverse hierarchical neutrino mass spectrum and a zero eigenvalue with $m_3 = 0$ and corresponding eigenvector $(0, -b/c, 1)^T$ implying a vanishing reactor mixing angle $\theta_{13} = 0$. One can see explicitly that the solar and atmospheric square mass differences and mixing angles indicated by neutrino oscillation data [5] can indeed be fitted by taking, as an example, the tri-bimaximal (TBM) *ansatz.* When $b = c$ and $y^2 = 2c^2 - ac$ the neutrino mass matrix Eq. (12) is $\mu - \tau$ invariant yielding maximal atmospheric mixing, $\sin^2\theta_{23} = 1/2$ and $M_{\nu,11} + M_{\nu,13} = M_{\nu,22} + M_{\nu,23}$, which gives the TBM value of the solar angle, $\sin\theta_{12}^2 = 1/3$, in good agreement with experimental data within one σ. The eigenvalues are $\{m_1, m_2, m_3\} = \{2ac + 2c^2, 2c^2 - ac, 0\}$, which can fit the two mass-squared differences required to account for the observed pattern of neutrino oscillations. By relaxing the condition $b = c$ and $y^2 = 2c^2 - ac$ one generates deviations from the TBM limit, while keeping $\theta_{13} = 0$. Note the above imples a neutrinoless double beta decay effective mass parameter in the range 0.03 to 0.05 eV at 3 σ, within reach of upcoming experiments.

5. Conclusions

In summary we have suggested that DM stability follows from the same non-abelian discrete flavor symmetry which accounts for the observed pattern of neutrino oscillations. In our example we have an A_4 symmetry which spontaneously breaks to a Z_2 parity that stabilizes a scalar doublet dark matter, potentially detectable in nuclear recoil experiments, as well as accelerators. Our simple example gives rates accessible to upcoming experiments and no CP violation in neutrino oscillations. One may construct alternative models, based on different A_4 realizations or different discrete flavor symmetry groups.

Acknowledgments We thank Marco Taoso for useful discussions. This work was supported by the Spanish MICINN under grants FPA2008-00319/FPA and MULTIDARK Consolider CAD2009-00064, by Prometeo/2009/091, by the EU grant UNILHC PITN-GA-2009-237920. S. M. is supported by a Juan de la Cierva contract.

References

[1] M. Frigerio and T. Hambye, Phys. Rev. D **81**, 075002 (2010) [arXiv:0912.1545 [hep-ph]].
[2] M. Kadastik, K. Kannike and M. Raidal, Phys. Rev. D **81** (2010) 015002 [arXiv:0903.2475 [hep-ph]].
[3] B. Batell, arXiv:1007.0045 [hep-ph].
[4] J. W. F. Valle, J. Phys. Conf. Ser. **53**, 473 (2006) [arXiv:hep-ph/0608101].
[5] T. Schwetz, M. A. Tortola and J. W. F. Valle, New J. Phys. **10**, 113011 (2008) [arXiv:0808.2016 [hep-ph]].
[6] H. Nunokawa, S. J. Parke and J. W. F. Valle, Prog. Part. Nucl. Phys. **60**, 338 (2008) [arXiv:0710.0554 [hep-ph]].
[7] K. S. Babu, E. Ma and J. W. F. Valle, Phys. Lett. B **552**, 207 (2003) [arXiv:hep-ph/0206292].
[8] G. Altarelli and F. Feruglio, Nucl. Phys. B **720**, 64 (2005) [arXiv:hep-ph/0504165].
[9] M. Hirsch, S. Morisi, E. Peinado and J. W. F. Valle, arXiv:1007.0871 [hep-ph].
[10] M. Cirelli, N. Fornengo and A. Strumia, Nucl. Phys. B **753**, 178 (2006) [arXiv:hep-ph/0512090].
[11] E. Komatsu *et al.* [WMAP Collaboration], Astrophys. J. Suppl. **180**, 330 (2009) [arXiv:0803.0547 [astro-ph]].
[12] C. P. Burgess, M. Pospelov and T. ter Veldhuis, Nucl. Phys. B **619**, 709 (2001) [arXiv:hep-ph/0011335].
[13] J. Schechter and J. W. F. Valle, Phys. Rev. D **25**, 774 (1982).
[14] J. Schechter and J. W. F. Valle, Phys. Rev. D **22**, 2227 (1980).
[15] R. N. Mohapatra and W. Rodejohann, Phys. Lett. B **644**, 59 (2007) [arXiv:hep-ph/0608111].

Journal of Physics: Conference Series **259** (2010) 012095 doi:10.1088/1742-6596/259/1/012095

The uses of singlets

Bohdan Grzadkowski[1], **José Wudka**[2]

[1]Institute of Theoretical Physics, University of Warsaw, Hoża 69, PL-00-681 Warsaw, Poland.
[2]Department of Physics, University of California, Riverside CA 92521-0413, USA.

E-mail: bohdan.grzadkowski@fuw.edu.pl

E-mail: jose.wudka@ucr.edu

Abstract. A simple extension of the standard model is proposed that can mitigate some of the difficulties associated with the little hierarchy problem and simultaneously provide a viable dark matter candidate. Possible interactions with the neutrino sector and their consequences are also briefly described.

1. The little hierarchy problem

Despite its enormous experimental success the Standard Model (SM) is not believed to provide a complete description of Nature at all energy scales; the SM then represents then an effective theory valid below a UV cutoff Λ. Being a renormalizable theory, no observable prediction of the SM will depend on Λ. However, one can also examine the naturality of the theory and within this context determine to what extent radiative corrections are sensitive to the UV cutoff. The corrections to the Higgs mass play an important role in this approach since they are the only ones depending quadratically on Λ:

$$\delta^{(SM)} m_h^2 = \frac{3\Lambda^2}{8\pi^2 v^2}\left[4m_t^2 - 2m_W^2 - m_Z^2 - m_h^2\right] \tag{1}$$

as first calculated by Veltman [1] Consistency and naturality of the model require $\delta^{(SM)} m_h^2 < m_h^2 < \Lambda^2$ which limits the allowed values of m_h and Λ to the shaded region in the graph below

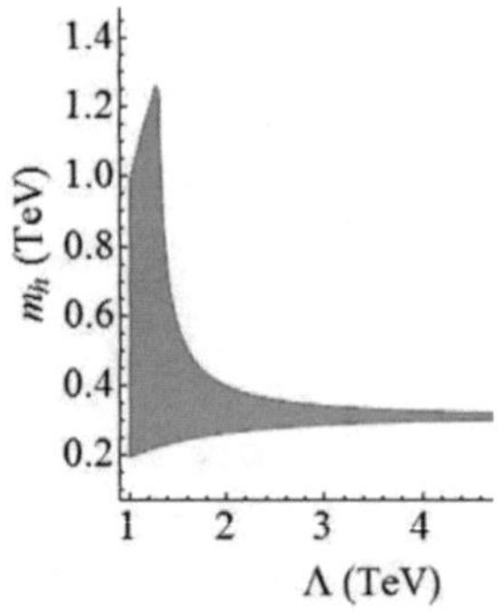

In particular a light Higgs ($m_h < 200$ GeV) is disallowed. For example if we require $\delta^{(\mathrm{SM})} m_h^2 \simeq m_h^2 = (130\ \mathrm{GeV})^2$ then $\Lambda \simeq 580$ GeV, while the current electroweak precision

limits (mostly from the oblique T-parameter[2]) require $\Lambda >$ few TeV. To remedy this situation two approaches have been followed in the literature: *(i)* cancel the SM $\ln m_h$ contributions to $T_{\rm oblique}$ so that heavy Higgs boson is allowed by the experimental constraints; or *(ii)* cancel SM contributions to δm_h (e.g. SUSY). Here we follow the second approach.

The goal of the model to be proposed is to relax the tension between naturality and the EW precision measurements for a UV cutoff $\Lambda \lesssim 3-10$ TeV. Our approach is to cancel the top-quark contributions to $\delta^{(SM)} m_h^2$ through the introduction of additional scalars. Since we also want to preserve the wealth of electroweak predictions derived from the SM, we assume such scalars are gauge singlets. We will then consider the SM with an extended scalar sector consisting of one isodoublet and N_φ scalar singlets φ_i, and arrange the model parameters so that all naturality problems occur at scales above the $O(10$ TeV) range.

The scalar potential we use is

$$V(H,\varphi_i) = -\mu_H^2|H|^2 + \lambda_H|H|^4 + \mu_\varphi^2\vec{\varphi}^{\,2} + \frac{1}{24}\lambda_\varphi\left(\vec{\varphi}^{\,2}\right)^2 + \lambda_x|H|^2\vec{\varphi}^{\,2} \qquad (2)$$

where $\vec{\varphi} = \left(\varphi_1, \dots \varphi_{N_\varphi}\right)$ and we assume μ_φ^2, λ_x, $\lambda_H > 0$ so that the $\vec{\varphi}$ vacuum expectation values (VEV) vanishes $\langle\vec{\varphi}\rangle = 0$. Upon spontaneous symmetry breaking the physical Higgs mass becomes $m_h^2 = 2\mu_H^2$ while the singlet mass equals $m^2 = 2\mu_\varphi^2 + (\lambda_x/\lambda_H)\mu_H^2$

We demand that the potential V be symmetric under $\varphi \leftrightarrow -\varphi$. For simplicity we also assume V to have a $O(N_\varphi)$ symmetry, but this is not essential for the considerations below.

It is now straightforward to calculate the new one-loop corrections to m_h generated by $\vec{\varphi}$:

$$\delta^{(\varphi)} m_h^2 = -[N_\varphi\lambda_x/(8\pi^2)]\left[\Lambda^2 - m^2\ln\left(1+\Lambda^2/m^2\right)\right] \qquad (3)$$

and, with this additional contribution it now becomes possible to tame the little hierarchy problem; see Fig. 1. For example, for 130 GeV $< m_h <$ 230 GeV we find $|\delta_{\rm tot} m_h^2| \equiv \left|\delta^{(SM)} m_h^2 + \delta^{(\varphi)} m_h^2\right| \lesssim m_h^2$ when $\Lambda \lesssim 5$ TeV and $\lambda_x \lesssim 1.5$. Similar considerations can be made at the two loop level [4]; see Fig. 2.

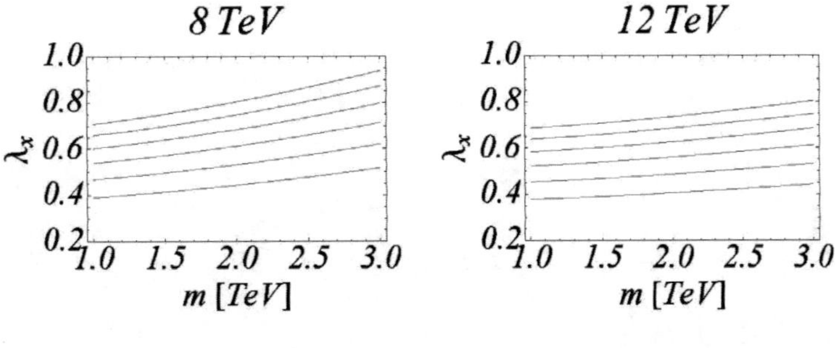

Figure 1. Values of m, λ_x that insure $\delta_{\rm tot} m_h^2 = 0$ for $N_\varphi = 6$, $\Lambda = 8$ TeV (left) and $\Lambda = 12$ TeV (right) and various choices of m_h, ranging from 130 GeV (top line) to 230 GeV (bottom line).

2. Dark matter

The existence of a parity symmetry allows us to use the singlets also as dark matter (DM) candidates (for $N_\varphi = 1$ see [3]) . In this case the relevant averaged cross section $\varphi\varphi \to$ SM SM is given by

$$\langle\sigma v\rangle \simeq \frac{\lambda_x^2}{8\pi m^2} + \frac{\lambda_x^2 v^2 \Gamma_h(2m)}{8m^5} \simeq \frac{1.73}{8\pi}\frac{\lambda_x^2}{m^2} \qquad (4)$$

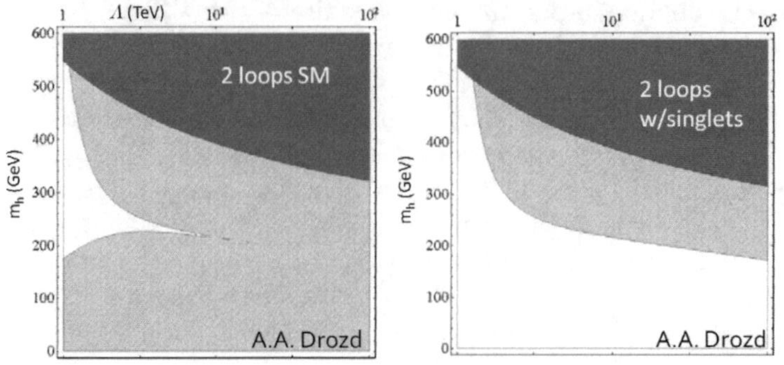

Figure 2. Regions in the $m_h - \Lambda$ plane for which $m_h^2 < \delta m_h^2$ (white) or $m_h^2 > \delta m_h^2$ (gray), for $0.1 < \lambda_x < 5$ at two loops for the pure SM (left), and the SM with one additional singlet (right). Dark gray regions correspond to parameters that also violate the triviality constraint. These graphs are courtesy of A.A. Drozd.

where the first term represents the s-channel h exchange contribution, and the second term the sum of the other channels. For this process the Higgs width is well approximated by $\Gamma(2m) \simeq 1.92m^2$ TeV, with m in TeV units. The freeze-out temperature and corresponding abundance can now be obtained using standard procedures [5]:

$$\frac{m}{T_{\rm freeze}} \simeq \ln\left[0.038\, M_{\rm Pl} \frac{m\,\langle\sigma v\rangle}{(g_\star x_f)^{1/2}}\right] \quad \Rightarrow \quad \Omega_\varphi h^2 = N_\varphi \frac{1.06\cdot 10^9 x_f}{(g_\star^{1/2} M_{\rm Pl}\langle\sigma v\rangle\ {\rm GeV})} \tag{5}$$

As illustrated in fig. 3 it is quite remarkable that there is a large region in parameter space for which the scalars have the required DM abundance and the little hierarchy problem is postponed to scales $\Lambda \sim 10$ TeV. Note that for a fixed abundance Ω_φ the scalar mass scales as $N_\varphi^{-3/2}$ so one can decrease the φ mass by augmenting the number of species.

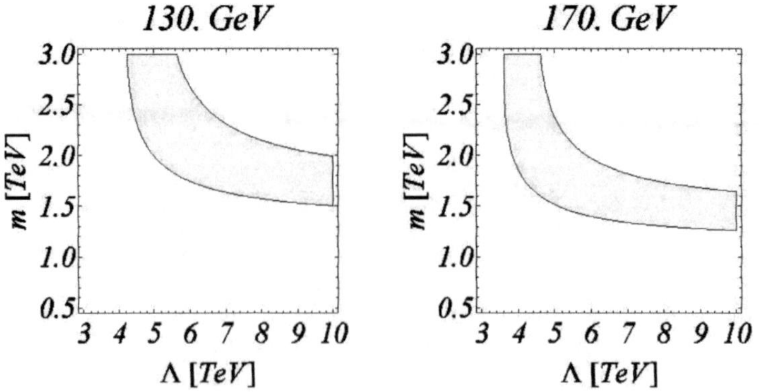

Figure 3. Allowed m and Λ values for which $\delta_{\rm tot} m_h^2 = 0$ and $\Omega_\varphi h^2 = 0.106 \pm 0.008$ (3σ) [2], for $N_\varphi = 6$ and $m_h = 130$ GeV (left), 170 GeV (right).

3. Neutrino interactions

In the presence of right-handed neutrinos ν_R the scalars can have additional couplings:

$$\mathcal{L}_Y = -\bar{L}Y_e H e_R - \bar{L}Y_\nu \tilde{H}\nu_R - \frac{1}{2}\overline{(\nu_R)^c}M\nu_R - \varphi\overline{(\nu_R)^c}Y_\varphi \nu_R + \text{H.c.} \tag{6}$$

(L denotes the left-handed lepton isodoublet and e_R the right-handed charged-lepton isosinglet). For $M \gg \langle H \rangle$ there will be a set of light Majorana neutrinos that we denote by n, and a corresponding set of heavy Majorana neutrinos we call N. In terms of the n and N fields the mass terms and the neutrino field ν become

$$\begin{aligned} \mathcal{L}_m &= -\bar{n}M_n n - \frac{1}{2}\bar{N}MN\,; \qquad M_n = \mu^* P_R + \mu P_L \\ \nu &= \left(n_L + M_D M^{-1} N_L + \cdots\right) + \left(N_R - M^{-1}M_D^T n_R + \cdots\right) \end{aligned} \tag{7}$$

where $M_D = Y_\nu \langle H \rangle$ and $\mu = -4M_D M^{-1} M_D^T$.

The discrete symmetry $\varphi \to -\varphi$ can be maintained even after the fermions couplings are introduced provided we require $L \to S_\ell L$, $e_R \to S_\ell e_R$, $\nu_R \to S_\nu \nu_R$ with (we assume 3 neutrino flavors)

$$S_\nu = \epsilon \begin{pmatrix} 1 & \cdot & \cdot \\ \cdot & 1 & \cdot \\ \cdot & \cdot & -1 \end{pmatrix} \quad S_\ell = \begin{pmatrix} s_1 & \cdot & \cdot \\ \cdot & s_2 & \cdot \\ \cdot & \cdot & s_3 \end{pmatrix} \quad Y_\varphi = \begin{pmatrix} \cdot & \cdot & b_1 \\ \cdot & \cdot & b_2 \\ b_1 & b_2 & \cdot \end{pmatrix} \tag{8}$$

where $\epsilon = \pm 1$, $|s_i| = 1$, and the b_i are complex numbers. In addition the Yukawa matrix Y_ν is restricted to a set of 10 possible choices, but requiring *(i)* consistency with tri-bimaximal mixing [6], *(ii)* the presence of no more than one massless n, and *(iii)* absence of $\varphi \to n_i n_j$, there remains a single possibility (up to base permutations): $s_{1,2,3} = \epsilon$, corresponding to $S_\ell = \epsilon \mathbb{1}$, and

$$Y_\nu = \begin{pmatrix} a & b & \cdot \\ -a/2 & b & \cdot \\ -a/2 & b & \cdot \end{pmatrix} \qquad \begin{aligned} m_{n_1} &= -6\langle H \rangle^2 a^2/M_1 \\ m_{n_2} &= -12\langle H \rangle^2 b^2/M_2 \\ m_{n_3} &= 0 \end{aligned} \tag{9}$$

where, for simplicity, a and b are real parameters. The resulting neutrino masses and mixing parameters are consistent with current data (see, e.g. [7]).

For the above choices only φ and ν_{R3} are odd under the discrete symmetry, while all other fields are even. For example, if we choose $\epsilon = 1$, the only $\varphi - \nu$ coupling is

$$\mathcal{L}_Y \simeq -2\varphi \sum_{i=1,2} b_i\, N_3^T C P_R N_i + \text{H.c.} + \cdots \tag{10}$$

so that the φ remain stable if $m < M_3$, so that the singlets can still function as DM candidates even in the presence of this type of neutrino couplings.

4. Comments

The above arguments show that the introducing of a number o gauge singlets φ can both ameliorate the little hierarchy problem and introduce a viable DM candidate. In addition, if right-handed neutrinos exist they can have non-trivial couplings to the scalars, though the LHC signatures of such interactions would be difficult to observe (at least within the context of the present model).

Acknowledgments

The work of B.G. is supported in part by the Ministry of Science and Higher Education (Poland) as research project N N202 006334 (2008-11); B.G. acknowledges support of the European Community within the Marie Curie Research & Training Networks: "HEPTOOLS" (MRTN-CT-2006-035505) and "UniverseNet" (MRTN-CT-2006-035863). The work of J.W. is supported in part by the U. S. Dept. of Energy Grant No. DE-FG03-94ER40837.

References

[1] Veltman M J G 1981, *Acta Phys. Polon. B* **12**, 437.
[2] Amsler C *et al.* (Particle Data Group) 2008, *Phys. Lett. B* **667**, 1.
[3] Silveira V and Zee A 1985, *Phys. Lett. B* **157**, 191; McDonald J 1994, *Phys. Rev. D* **50**, 3637 (*Preprint* hep-ph/0702143); Burgess C P, Pospelov M and ter Veldhuis T 2001, *Nucl. Phys. B* **619**, 709 (*Preprint* hep-ph/0011335); Davoudiasl H, Kitano R, Li T and Murayama H 2005, *Phys. Lett. B* **609**, 117 (*Preprint* hep-ph/0405097); van der Bij J J 2006, *Phys. Lett. B* **636**, 56 (*Preprint* hep-ph/0603082); Bertolami O and Rosenfeld R 2007, (*Preprint* arXiv:0708.1784); Barger V, Langacker P, McCaskey M, Ramsey-Musolf M J and Shaughnessy G 2008, *Phys. Rev. D* **77**, 035005 (*Preprint* arXiv:0706.4311); Andreas S, Hambye T and Tytgat M H G 2008, *JCAP* **0810**, 034 (*Preprint* arXiv:0808.0255).
[4] Drozd A A 2010, University of Warsaw M.Sc. thesis (unpublished).
[5] Kolb E W and Turner M S 1990, *The Early universe*, (Front. Phys. vol. **69**) (Reading, Massachusetts, Addison Wesley, 1990)
[6] Harrison P F, Perkins D H and Scott W G 2002, *Phys. Lett. B* **530**, 167 (*Preprint* hep-ph/0202074).
[7] Altarelli G 2007, *Preprint* arXiv:0711.0161.

Journal of Physics: Conference Series **259** (2010) 012096 doi:10.1088/1742-6596/259/1/012096

Constraints on fourth generation Majorana neutrinos[1]

Alexander Lenz, Heinrich Päs and Dario Schalla

Fakultät Physik, Technische Universiät Dortmund, 44221 Dortmund, Germany

E-mail: dario.schalla@tu-dortmund.de

Abstract. We investigate the possibility of a fourth sequential generation in the lepton sector. Assuming neutrinos to be Majorana particles and starting from a recent - albeit weak - evidence for a non-zero admixture of a fourth generation neutrino from fits to weak lepton and meson decays we discuss constraints from neutrinoless double beta decay, radiative lepton decay and like-sign di-lepton production at hadron colliders

1. Introduction

The addition of a fourth family of fermions to the known three generations has recently become a popular extension of the standard model (see [1, 2] for a review). Interesting features and benefits of such scenarios include for example:

- weakening of the tension between direct and indirect bounds on the Higgs mass, see e.g. [3, 4, 5],
- a sizeable enhancement of the measure of CP violation, see e.g. [6],
- gauge coupling unification [7],
- new strong dynamic effects due to large Yukawa couplings allowing for dynamical symmetry breaking, see e.g. [8, 9],
- a solution of flavor problems like the observed 3.8 σ deviation of the measured value of B_d-B_s mixing from the standard model ($SM3$) prediction [10] which also enhances the dimuon asymmetry of the $SM3$ [11] towards the value measured by the D0 collaboration [12].

In the following we consider the addition of a fourth family

$$\begin{pmatrix} t' \\ b' \end{pmatrix}_L \quad t'_R \quad b'_R \quad \begin{pmatrix} \nu_4 \\ \ell_4 \end{pmatrix}_L \quad \ell_{4R} \quad \nu_{4R}, \tag{1}$$

while the gauge and Higgs sector remains unchanged compared to the $SM3$. Naturally the addition of right-handed neutrinos allows for both Dirac and Majorana mass terms for all four generations of neutrinos. Moreover, tiny neutrino masses are most naturally explained for Majorana neutrinos in a seesaw scheme, so we do not adopt lepton number conservation at this point.

[1] Talk presented by Dario Schalla.

Journal of Physics: Conference Series **259** (2010) 012096 doi:10.1088/1742-6596/259/1/012096

A lower mass bound on SU(2) doublet fourth generation neutrinos can be obtained by the invisible Z-decay width [13]:

$$N_\nu = 2.984 \pm 0.008 \tag{2}$$

constraining the number of light neutrinos with masses $< M_Z/2$ to be three. Consequently

$$m_4 \geq \frac{m_Z}{2} \approx 45.6\,\text{GeV} \tag{3}$$

provides a lower mass bound on fourth generation neutrinos. As the light mass eigenvalues are bounded from above by the Dirac masses – at least in a typical seesaw model, this provides a bound on the Dirac-type mass m_4^D as well. Assuming perturbativity of the fourth generation neutrino Yukawa couplings then constrains Dirac type neutrino masses approximately to the interval

$$45\,\text{GeV} < m_4^D < 1000\,\text{GeV}. \tag{4}$$

Moreover, recent fits to electroweak precision data in a four generation framework ($SM4$) lead to the following constraints on the fourth generation particle spectrum [14]:

$$|m_{t'} - m_{b'}| < 80\,\text{GeV} \qquad |m_{l_4} - m_4| < 140\,\text{GeV}. \tag{5}$$

Finally a recent fit to a set of experimental data in the $SM4$ framework has provided some evidence for a non-zero admixture of a fourth generation neutrino, resulting in a PMNS matrix [15]

$$U_{PMNS} = \begin{pmatrix} * & * & * & {}^{<0.089}_{>0.021} \\ * & * & * & <0.029 \\ * & * & * & <0.085 \\ <0.115 & <0.115 & <0.115 & {}^{<0.9998}_{>0.9934} \end{pmatrix}. \tag{6}$$

In the following we use this evidence – as weak as it may be – as a starting point to reconsider bounds on fourth generation neutrino masses.

2. Neutrinoless double beta decay

The most sensitive probe for neutrino Majorana masses is generally neutrinoless double beta decay ($0\nu\beta\beta$). $0\nu\beta\beta$ decay can be realized by the exchange of a Majorana neutrino (see Fig. 1). In the presence of additional heavy neutrino states the usual effective Majorana mass $\langle m_\nu \rangle$ has to be complemented by an effective heavy neutrino mass $\langle m_N \rangle^{-1}$:

$$\langle m_\nu \rangle = \sum_{i=1}^{3} U_{ei}^2 m_i \qquad \langle m_N \rangle^{-1} = \sum_N U_{eN}^2 m_N^{-1}. \tag{7}$$

The half-life of the decay is then given by [16]

$$\left[T_{1/2}^{0\nu\beta\beta}\right]^{-1} = \left(\frac{\langle m_\nu \rangle}{m_e}\right)^2 C_{mm}^{LL} + \left(\frac{m_p}{\langle m_N \rangle}\right)^2 C_{mm}^{NN} + \left(\frac{\langle m_\nu \rangle}{m_e}\right)\left(\frac{\langle m_p \rangle}{\langle m_N \rangle}\right) C_{mm}^{NL}, \tag{8}$$

where the C_{mm} factors include phase-space factors and nuclear matrix elements [17, 18] and m_p (m_e) the proton (electron) mass. Considering only the heavy neutrino contribution and using the PMNS matrix obtained in Eq. (6) one obtains stringent bounds on the allowed mass range from the current experimental lower half-life bound $T_{1/2}^{\text{Ge}} > 1.57 \cdot 10^{25}$ years [19]. The allowed region is shown in Fig.2.

This leads to the following mass bounds for a single fourth generation Majorana neutrino

$$U_{e4}^{max} = 0.089 \quad \Rightarrow \quad m_4^{max} = 6.8 \cdot 10^5\,\text{GeV} \tag{9}$$

$$U_{e4}^{min} = 0.021 \quad \Rightarrow \quad m_4^{min} = 3.8 \cdot 10^4\,\text{GeV}, \tag{10}$$

Journal of Physics: Conference Series **259** (2010) 012096 doi:10.1088/1742-6596/259/1/012096

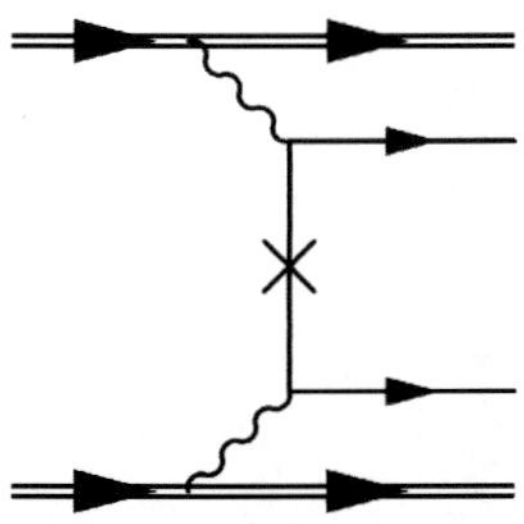

Figure 1. Feynman diagram of $0\nu\beta\beta$ induced by the exchange of a heavy fourth generation Majorana neutrino.

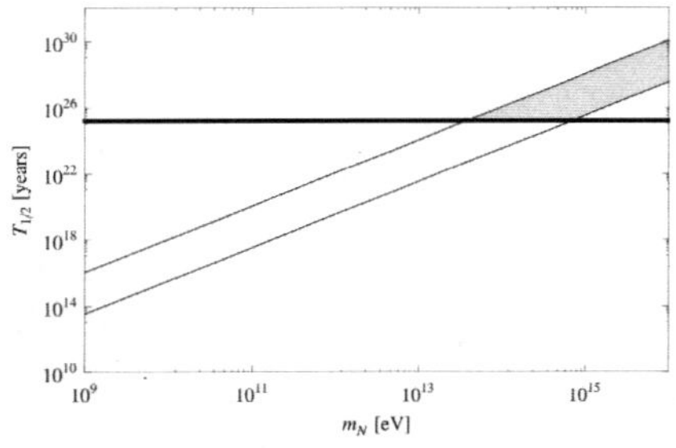

Figure 2. Contribution of heavy neutrino on $0\nu\beta\beta$ half-life (thin lines) within the mixing region given by Eq. (6) and IGEX lower bound (tick line). The gray area indicates the allowed region.

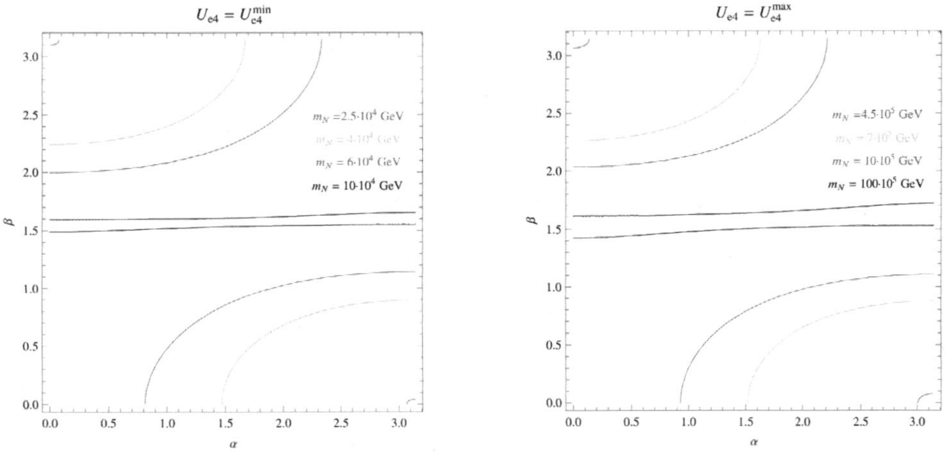

Figure 3. Masses that reproduce the experimental $0\nu\beta\beta$ bound. α is the light and β the heavy neutrino contribution phase.

which are far above the perturbativity constraint of Eq. (4).
Relative phases between light (α) and heavy (β) contributions may cancel each other and thus loosen this bound (see Fig. for several phases α and β). Maximizing the light neutrino contribution by using the largest possible allowed light neutrino masses consistent with the large scale structure of the universe ($\sum m_\nu < 0.66\,\text{eV}$) the mass region of the heavy neutrino can be lowered to

$$2.50 \cdot 10^4\,\text{GeV} < m_4 < 4.49 \cdot 10^5\,\text{GeV}, \tag{11}$$

which remains several orders of magnitude above the desired range (5). In principle there are three different ways to save the possibility of a heavy fourth generation neutrino:

(i) neutrinos are Dirac particles and therefore $0\nu\beta\beta$ is forbidden, which would come at the cost of seesaw neutrino mass suppression and leptogenesis as a successful way to generate the baryon asymmetry of the universe

(ii) some other physics beyond the standard model is involved and cancels the heavy neutrino contribution, which would require fine tuning

(iii) neutrinos are pseudo-Dirac particles.

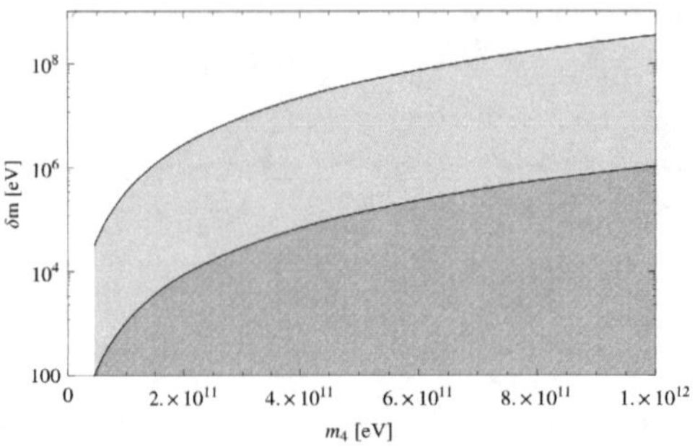

Figure 4. Maximal mass splitting for heavy pseudo-Dirac neutrinos. The upper (lower) curve corresponds to the lower (upper) bound on U_{e4} according to Eq. (6). The marked area represents the allowed parameter space.

In the following we will focus on the latter alternative which may provide useful guidance for future model building.
Pseudo-Dirac neutrinos arise when the Majorana mass is small compared to the Dirac mass. The two resulting mass eigenstates (m_+, m_-) are nearly degenerate with tiny mass splitting δm and the active and sterile component exhibit practically maximal mixing.
The $0\nu\beta\beta$ half-life of such a neutrino then reads

$$\left[T_{1/2}^{0\nu\beta\beta}\right]^{-1} = \left(\frac{m_p}{\langle m_-\rangle}\right)^2 C_{mm}^{NN} - \left(\frac{m_p}{\langle m_+\rangle}\right)^2 C_{mm}^{NN}. \tag{12}$$

The allowed mass splittings are shown in Fig. 4 and vary from 32 keV to 350 MeV.

3. Radiative lepton decays

The analysis on neutrino mixing used [15] is dominated by the radiative lepton flavor violating decays of charged leptons. Here we shortly reconsider this bound for the case of pseudo-Dirac neutrinos with masses in the 100 GeV range. The decay width of these processes is given by [20]:

$$\Gamma_{\ell\to\ell'\gamma} = \frac{1}{2}\frac{G_F^2 m_\ell^5}{(32\pi^2)^2}\alpha \left|U_{\ell\alpha}U_{\ell'\alpha}\right|^2 F^2(x), \tag{13}$$

where $F(x)$ is a function of the neutrino masses. It is easy to see that the analysis holds for a pseudo-Dirac neutrino as well. As the fourth generation active and sterile states mix maximally and the masses are close to degenerate $F(x)$ does not change considerably compared to the pure Dirac case. As can be seen from Fig. 5, the decay rate is suppressed by the tiny masses for the first three generations, while the contribution of a fourth heavy generation has to be suppressed due to small mixing.
In the analysis [15] the neutrino mass was fixed to 45 GeV. While a mass dependent study is encouraged, the conclusions of this work will remain unchanged, as the size of the allowed region is anticipated to vary only slightly for different neutrino masses. The decays of the τ lepton do not provide further information as the experimental constraints in this channel are much weaker [21, 22].

4. Like-sign dilepton production

Finally, a process very similar to $0\nu\beta\beta$ is the production of two charged leptons of the same charge at hadron colliders:

$$pp \to \ell_1^+\ell_2^+ X. \tag{14}$$

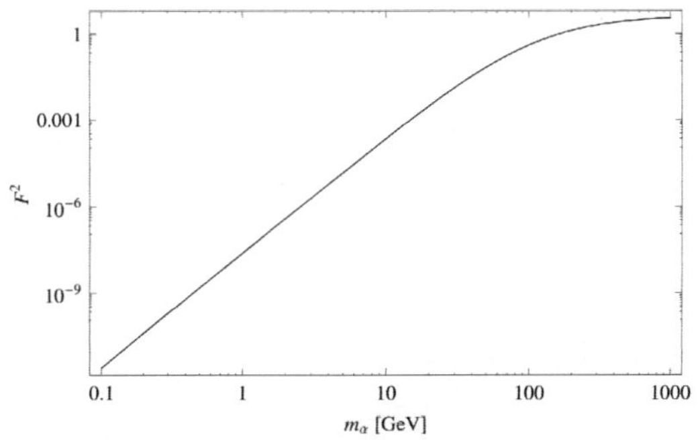

Figure 5. F^2 as a function of the mass of the exchanged neutrino.

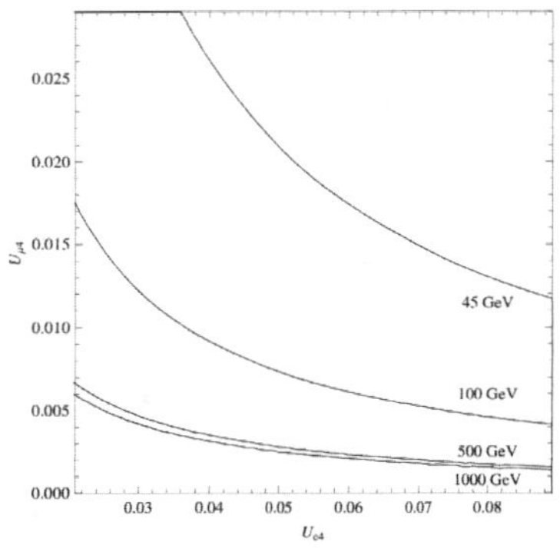

Figure 6. Constraint on the $U_{\mu 4} - U_{e4}$ parameter space obtained from the bound on the branching ratio for $\mu \to e\gamma$. Allowed is the region to the lower left. The boundaries of the intervalls plotted are given by the allowed values for $m_4 = 45\,\text{GeV}$ according to Eq. (6).

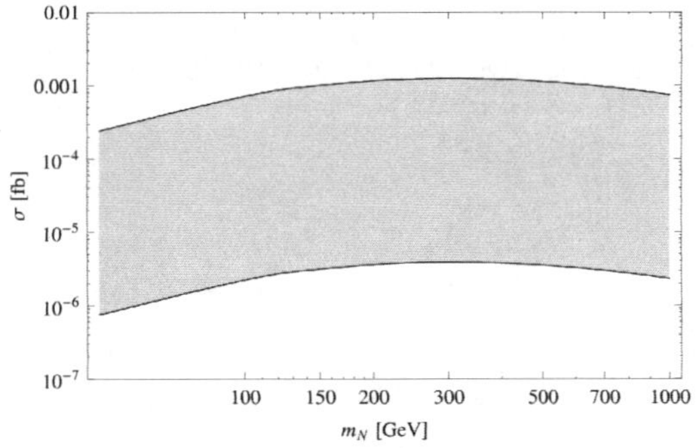

Figure 7. Cross section for like-sign dilepton production by an electroweak scale Majorana neutrino without $0\nu\beta\beta$ constraints.

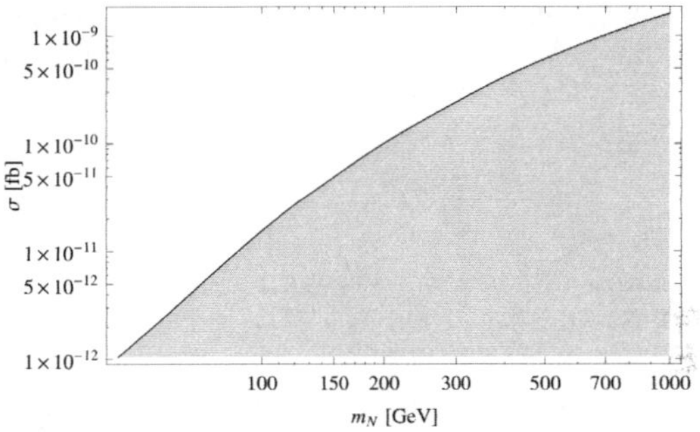

Figure 8. Cross section for like-sign dilepton production by an electroweak scale pseudo-Dirac neutrino with $0\nu\beta\beta$ constraints.

As shown in Fig. 9 a heavy Majorana neutrino exchange drives the process whose cross section is [23]:

$$\sigma\left(pp \to \ell_1^+ \ell_2^+ X\right) = \frac{G_F^4 m_W^6}{8\pi^5}\left(1 - \frac{1}{2}\delta_{\ell_1 \ell_2}\right)\left|U_{\ell_1 4} U_{\ell_2 4}\right|^2 F\left(E, m_4\right), \tag{15}$$

where $F\left(E, m_4\right)$ is a function of beam energy and neutrino mass. To describe the exchange of a pseudo-Dirac neutrino the cross section has to be modified by introducing a suppression factor (12):

$$\Delta_{pD} \approx 2\frac{\delta m}{m_4}. \tag{16}$$

Here the mass splitting δm follows from the $0\nu\beta\beta$ constraint and results in a suppression of the mass dependent cross section as shown in Fig. 7 and 8. The resulting cross sections are far to small to be observed at expected LHC luminosities.

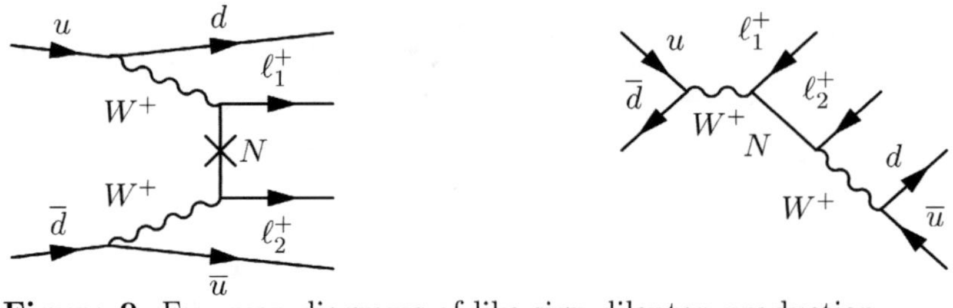

Figure 9. Feynman diagrams of like-sign dilepton production

5. Summary

In this note we have revisited bounds on additional Majorana neutrinos, in order to provide a useful guide for fourth generation neutrino model building. We have shown that a fourth generation Majorana neutrino is not yet excluded if it has a mass of several hundred GeV and the Majorana states pair up to form a pseudo-Dirac state. The mixing of such a neutrino is dominantly constrained by the radiative decay of the muon. Due to the pseudo-Dirac nature lepton number violating processes like like-sign dilepton production turn out to be strongly suppressed. Besides being potentially observable in next generation $0\nu\beta\beta$ experiments, the pseudo-Dirac neutrinos could be directly produced at the LHC, as discussed in [24]. In this paper a 5 σ discovery reach for heavy neutrino masses up to 100 GeV was advocated with 30 fb^{-1}. While for larger masses the production cross section would decrease, new decay channels open up once the heavy neutrino mass exceeding the Higgs mass, which would require a detailed simulation.

References

[1] Frampton P H, Hung P Q and Sher M 2000 *Phys. Rept.* **330** 263 (*Preprint* hep-ph/9903387)
[2] Holdom B *et al.* 2009 *PMC Phys.* **A3** 4 (*Preprint* 0904.4698)
[3] Novikov V A, Rozanov A N and Vysotsky M I 2010 *Phys. Atom. Nucl.* **73** 636–642 (*Preprint* 0904.4570)
[4] Kribs G D, Plehn T, Spannowsky M and Tait T M P 2007 *Phys. Rev.* **D76** 075016 (*Preprint* 0706.3718)
[5] Chanowitz M S 2010 (*Preprint* 1007.0043)
[6] Hou W S, Mao Y Y and Shen C H 2010 (*Preprint* 1003.4361)
[7] Hung P Q 1998 *Phys. Rev. Lett.* **80** 3000–3003 (*Preprint* hep-ph/9712338)
[8] Hung P Q and Xiong C 2009 (*Preprint* 0911.3890)
[9] Holdom B 2006 *JHEP* **08** 076 (*Preprint* hep-ph/0606146)
[10] Lenz A *et al.* 2010 (*Preprint* 1008.1593)
[11] Lenz A and Nierste U 2007 *JHEP* **06** 072 (*Preprint* hep-ph/0612167)
[12] Abazov V M *et al.* (D0) 2010 (*Preprint* 1005.2757)
[13] 2006 *Phys. Rept.* **427** 257 (*Preprint* hep-ex/0509008)
[14] Eberhardt O, Lenz A and Rohrwild J 2010 (*Preprint* 1005.3505)
[15] Lacker H and Menzel A 2010 (*Preprint* 1003.4532)
[16] Hirsch M, Klapdor-Kleingrothaus H V and Panella O 1996 *Phys. Lett.* **B374** 7–12 (*Preprint* hep-ph/9602306)
[17] Hirsch M and Klapdor-Kleingrothaus H V Prepared for International Workshop on Neutrinoless Double Beta Decay and Related Topics, Trento, Italy, 24 Apr - 5 May 1995
[18] Muto K and Klapdor H V IN *KLAPDOR, H.V. (ED.): NEUTRINOS* 183-237
[19] Aalseth C E *et al.* (IGEX) 2002 *Phys. Rev.* **D65** 092007 (*Preprint* hep-ex/0202026)
[20] Cheng T P and Li L F 1980 *Phys. Rev. Lett.* **45** 1908
[21] Brooks M L *et al.* (MEGA) 1999 *Phys. Rev. Lett.* **83** 1521–1524 (*Preprint* hep-ex/9905013)
[22] Aubert B *et al.* (BABAR) 2010 *Phys. Rev. Lett.* **104** 021802 (*Preprint* 0908.2381)
[23] Ali A, Borisov A V and Zamorin N B 2001 *Eur. Phys. J.* **C21** 123–132 (*Preprint* hep-ph/0104123)
[24] del Aguila F and Aguilar-Saavedra J A 2009 *Phys. Lett.* **B672** 158–165 (*Preprint* 0809.2096)

Journal of Physics: Conference Series **259** (2010) 012097 doi:10.1088/1742-6596/259/1/012097

Finite Unification: phenomenology

S. Heinemeyer[1], E. Ma[2], M Mondragón[3] and G. Zoupanos[4]

[1] Instituto de Física de Cantabria (CSIC-UC), Edificio Juan Jorda, Avda. de Los Castros s/n 39005 Santander, Spain

[2] Physics Department, University of California, Riverside, California 92521, U.S.A.

[3] Inst. de Física, Universidad Nacional Autónoma de México, Apdo. Postal 20-364, México 01000 D.F., México

[4] Physics Department, National Technical University of Athens, Zografou Campus: Heroon Polytechniou 9, 15780 Zografou, Athens, Greece

E-mail: sven.heinemeyer@cern.ch, ma@phyun8.ucr.edu, myriam@fisica.unam.mx, george.zoupanos@cern.ch

Abstract. We study the phenomenological implications of Finite Unified Theories (FUTs). In particular we look at the predictions for the lightest Higgs mass and the s-spectra of two all-loop finte models with $SU(5)$ as gauge group. We also consider a two-loop finite model with gauge group $SU(3)^3$, which is finite if and only if there are exactly three generations. In this latter model we concetrate here only on the predictions for the third generation of quark masses.

1. Introduction

The traditional way to reduce the independent parameters of a theory is the introduction of a symmetry. Supersymmetric Grand Unified Theories (GUTs) are representative examples of such attempts. Finite Unified Theories (FUTs) are $N = 1$ supersymmetric Grand Unified Theories (GUTs) which can be made finite to all-loop orders, including the soft supersymmetry breaking sector. The constructed *finite unified* $N = 1$ supersymmetric SU(5) GUTs predicted correctly from the dimensionless sector (Gauge-Yukawa unification), among others, the top quark mass [1, 2]. Eventually, the full theories can be made all-loop finite and their predictive power is extended to the Higgs sector and the s-spectrum [3]. For a detailed discussion see [4–6].

Consider a chiral, anomaly free, $N = 1$ globally supersymmetric gauge theory based on a group G with gauge coupling constant g. The superpotential of the theory is given by

$$W = \frac{1}{2} m^{ij} \, \Phi_i \, \Phi_j + \frac{1}{6} C^{ijk} \, \Phi_i \, \Phi_j \, \Phi_k \; , \tag{1}$$

where m^{ij} (the mass terms) and C^{ijk} (the Yukawa couplings) are gauge invariant tensors, and the matter field Φ_i transforms according to the irreducible representation R_i of the gauge group G. All the one-loop β-functions of the theory vanish if the β-function of the gauge coupling $\beta_g^{(1)}$, and the anomalous dimensions of the Yukawa couplings $\gamma_i^{j(1)}$, vanish, i.e.

$$\sum_i \ell(R_i) = 3C_2(G) \, , \; \frac{1}{2} C_{ipq} C^{jpq} = 2\delta_i^j g^2 C_2(R_i) \; , \tag{2}$$

where $\ell(R_i)$ is the Dynkin index of R_i, and $C_2(G)$ is the quadratic Casimir invariant of the adjoint representation of G. A theorem given in [7–10] then guarantees the vanishing of the β-functions to all-orders in perturbation theory. This requires that, in addition to the one-loop finiteness conditions (2), the Yukawa couplings are reduced in favour of the gauge coupling.

In the soft breaking sector, the one- and two-loop finiteness for the trilinear terms h^{ijk} can be achieved by [11]

$$h^{ijk} = -MC^{ijk} + \ldots = -M\rho^{ijk}_{(0)}\, g + O(g^5)\ . \tag{3}$$

It was also found that the soft supersymmetry breaking (SSB) scalar masses in Gauge-Yukawa and finite unified models satisfy a sum rule [12, 13]

$$\frac{(\ m_i^2 + m_j^2 + m_k^2\ )}{MM^\dagger} = 1 + \frac{g^2}{16\pi^2}\,\Delta^{(2)} + O(g^4) \tag{4}$$

for i, j, k, where $\Delta^{(2)}$ is the two-loop correction, which vanishes when all the soft scalar masses are the same at the unification point.

2. $SU(5)$ Finite Unified Theories

We will examine here all-loop Finite Unified theories with $SU(5)$ gauge group, where the reduction of couplings has been applied to the third generation of quarks and leptons. An extension to three families, and the generation of quark mixing angles and masses in Finite Unified Theories has been addressed in [14], where several examples are given. These extensions are not considered here. The particle content of the models we will study consists of the following supermultiplets: three $(\overline{\mathbf{5}} + \mathbf{10})$, needed for each of the three generations of quarks and leptons, four $(\overline{\mathbf{5}} + \mathbf{5})$ and one $\mathbf{24}$ considered as Higgs supermultiplets. When the gauge group of the finite GUT is broken the theory is no longer finite, and we will assume that we are left with the MSSM.

A predictive Gauge-Yukawa unified $SU(5)$ model which is finite to all orders, should have the following properties:

(i) One-loop anomalous dimensions are diagonal, i.e., $\gamma_i^{(1)\,j} \propto \delta_i^j$.

(ii) Three fermion generations, in the irreducible representations $\overline{\mathbf{5}}_i, \mathbf{10}_i$ $(i = 1, 2, 3)$, which obviously should not couple to the adjoint $\mathbf{24}$.

(iii) The two Higgs doublets of the MSSM should mostly be made out of a pair of Higgs quintet and anti-quintet, which couple to the third generation.

In the following we discuss two versions of the all-order finite model. The model of ref. [1, 2], which will be labeled **A**, and a slight variation of this model (labeled **B**), which can also be obtained from the class of the models suggested in ref. [15] with a modification to suppress non-diagonal anomalous dimensions.

The superpotential which describes the two models takes the form [1, 2, 16]

$$\begin{aligned} W &= \sum_{i=1}^{3} [\ \frac{1}{2} g_i^u\, \mathbf{10}_i \mathbf{10}_i H_i + g_i^d\, \mathbf{10}_i \overline{\mathbf{5}}_i\, \overline{H}_i\ ] \\ &+ g_{23}^u\, \mathbf{10}_2 \mathbf{10}_3 H_4 + g_{23}^d\, \mathbf{10}_2 \overline{\mathbf{5}}_3\, \overline{H}_4 + g_{32}^d\, \mathbf{10}_3 \overline{\mathbf{5}}_2\, \overline{H}_4 \\ &+ \sum_{a=1}^{4} g_a^f\, H_a\, \mathbf{24}\, \overline{H}_a + \frac{g^\lambda}{3}\, (\mathbf{24})^3\ , \end{aligned} \tag{5}$$

where H_a and $\overline{H}_a$ $(a = 1, \ldots, 4)$ stand for the Higgs quintets and anti-quintets.

The main difference between model **A** and model **B** is that two pairs of Higgs quintets and anti-quintets couple to the **24** in **B**, so that it is not necessary to mix them with H_4 and $\overline{H}_4$ in order to achieve the triplet-doublet splitting after the symmetry breaking of $SU(5)$ [16]. Thus, although the particle content is the same, the solutions to the finiteness equations and the sum rules are different, which will reflect in the phenomenology.

The non-degenerate and isolated solutions to $\gamma_i^{(1)} = 0$ for model **FUTA**, which are the boundary conditions for the Yukawa couplings at the GUT scale, are:

$$\begin{aligned}
&(g_1^u)^2 = \frac{8}{5}\, g^2 \ , \ (g_1^d)^2 = \frac{6}{5}\, g^2 \ , \ (g_2^u)^2 = (g_3^u)^2 = \frac{8}{5}\, g^2 \ , \\
&(g_2^d)^2 = (g_3^d)^2 = \frac{6}{5}\, g^2 \ , \ (g_{23}^u)^2 = 0 \ , \ (g_{23}^d)^2 = (g_{32}^d)^2 = 0 \ , \\
&(g^\lambda)^2 = \frac{15}{7} g^2 \ , \ (g_2^f)^2 = (g_3^f)^2 = 0 \ , \ (g_1^f)^2 = 0 \ , \ (g_4^f)^2 = g^2 \ .
\end{aligned} \tag{6}$$

In the dimensionful sector, the sum rule gives us the following boundary conditions at the GUT scale for this model [16–18]:

$$m_{H_u}^2 + 2m_{\mathbf{10}}^2 \ = \ m_{H_d}^2 + m_{\overline{\mathbf{5}}}^2 + m_{\mathbf{10}}^2 = M^2 \ , \tag{7}$$

and thus we are left with only three free parameters, namely $m_{\overline{\mathbf{5}}} \equiv m_{\overline{\mathbf{5}}_3}$, $m_{\mathbf{10}} \equiv m_{\mathbf{10}_3}$ and M.

For the model **FUTB** the non-degenerate and isolated solutions to $\gamma_i^{(1)} = 0$ give us:

$$\begin{aligned}
&(g_1^u)^2 = \frac{8}{5}\, g^2 \ , \ (g_1^d)^2 = \frac{6}{5}\, g^2 \ , \ (g_2^u)^2 = (g_3^u)^2 = \frac{4}{5}\, g^2 \ , \\
&(g_2^d)^2 = (g_3^d)^2 = \frac{3}{5}\, g^2 \ , \ (g_{23}^u)^2 = \frac{4}{5}\, g^2 \ , \ (g_{23}^d)^2 = (g_{32}^d)^2 = \frac{3}{5}\, g^2 \ , \\
&(g^\lambda)^2 = \frac{15}{7} g^2 \ , \ (g_2^f)^2 = (g_3^f)^2 = \frac{1}{2}\, g^2 \ , \ (g_1^f)^2 = 0 \ , \ (g_4^f)^2 = 0 \ ,
\end{aligned} \tag{8}$$

and from the sum rule we obtain:

$$m_{H_u}^2 + 2m_{\mathbf{10}}^2 = M^2 \ , \ m_{H_d}^2 - 2m_{\mathbf{10}}^2 = -\frac{M^2}{3} \ , m_{\overline{\mathbf{5}}}^2 + 3m_{\mathbf{10}}^2 = \frac{4M^2}{3} \ , \tag{9}$$

i.e., in this case we have only two free parameters $m_{\mathbf{10}} \equiv m_{\mathbf{10}_3}$ and M.

3. Predictions of the $SU(5)$ models

We confront now the predictions of the four models with the experimental data, starting with the heavy quark masses (see refs. [3,6] for more details). Since the gauge symmetry is spontaneously broken below $M_{\rm GUT}$, the finiteness conditions do not restrict the renormalization properties at low energies, and all it remains are boundary conditions on the gauge and Yukawa couplings (6), the $h = -MC$ relation eq.(3), and the soft scalar-mass sum rule (4) at $M_{\rm GUT}$, as applied in each of the models. Thus we examine the evolution of these parameters according to their RGEs up to two-loops for dimensionless parameters and at one-loop for dimensionful ones with the relevant boundary conditions. Below $M_{\rm GUT}$ their evolution is assumed to be governed by the MSSM. We further assume a unique supersymmetry breaking scale M_s and therefore below that scale the effective theory is just the SM.

As a first step, we compare the predictions of the two models, **FUTA** and **FUTB** (both with $\mu > 0$ and $\mu < 0$), with the experimental values of the top and bottom quark masses. We use for the top quark the value for the pole mass [19]

$$M_t^{exp} = (173.1 \pm 1.3) \tag{10}$$

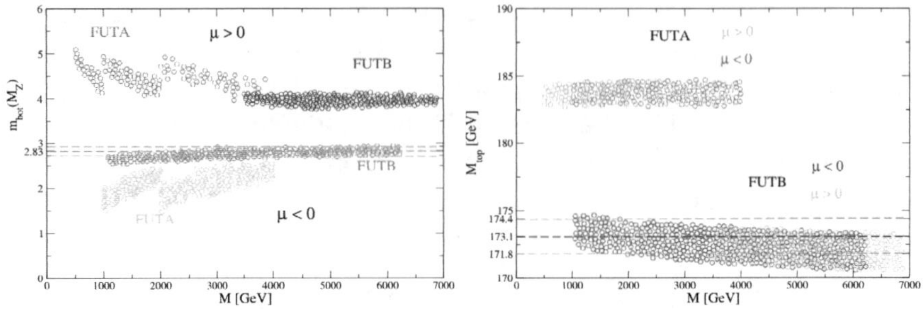

Figure 1. $m_b(M_Z)$ as function of M (left) and M_{top} as function of M (right) for models **FUTA** and **FUTB**, for $\mu < 0$ and $\mu > 0$.

where we notice that the theoretical values for M_t may suffer from a correction of $\sim 4\%$ [4]. For the bottom quark mass we use the value at M_Z [20]

$$m_b(M_Z) = 2.83 \pm 0.10 GeV \tag{11}$$

to avoid errors that come from the furhter running from the M_Z to the m_b mass, and where we have taken the Δb effects into account.

From fig. 1 it is clear that the model **FUTB** with $\mu < 0$ is the only one where both top and bottom quark masses lie within experimental limits. In this case the value of $\tan\beta$ is found to be ~ 48. Thus, we will concentrate now on the results for **FUTB**, $\mu < 0$. In the case where all the soft scalar masses are universal at the unfication scale, there is no region of M below $\mathcal{O}$(few TeV) in which $m_{\tilde{\tau}} > m_{\chi^0}$ is satisfied (where $m_{\tilde{\tau}}$ is the lightest $\tilde{\tau}$ mass, and m_{χ^0} the lightest neutralino mass). This problem can be solved naturally, thanks to the sum rule (9).

Furthermore, we impose the conditions of successful radiative electroweak symmetry breaking, $m^2_{\tilde{\tau}} > 0$ and $m_{\tilde{\tau}} > m_{\chi^0}$, plus the constraints coming from B physics, namely the experimental bounds on BR(b $\to$ sγ) and BR($B_s \to \mu^+\mu^-$) (which we have evaluated with Micromegas [21]). This way, we find a prediction for the lightest Higgs mass and the s-spectra. From the analysis we find that the lightest observable particle (LOSP) is either the stau or the second lightest neutralino, with mass starting around ~ 500 GeV. The prediction for the lightest Higgs mass is

$$M_h \sim 121 - 126\text{GeV} , \tag{12}$$

as shown in fig.2, where the red points satisfy the loose CDM constraint $\Omega h^2 < 0.3$. To this value one has to add ± 3 GeV coming from unknown higher order corrections [22]. See [3, 6] for a more detailed analysis.

4. Finite $SU(3)^3$ model

We now examine the possibility of constructing realistic FUTs based on product gauge groups. Consider an $N = 1$ supersymmetric theory, with gauge group $SU(N)_1 \times SU(N)_2 \times \cdots \times SU(N)_k$, with n_f copies of the supersymmetric multiplet $(N, N^*, 1, \ldots, 1) + (1, N, N^*, \ldots, 1) + \cdots + (N^*, 1, 1, \ldots, N)$. The one-loop β-function coefficient in the renormalization-group equation of each $SU(N)$ gauge coupling is simply given by

$$b = \left(-\frac{11}{3} + \frac{2}{3}\right) N + n_f \left(\frac{2}{3} + \frac{1}{3}\right)\left(\frac{1}{2}\right) 2N = -3N + n_f N \, . \tag{13}$$

This means that $n_f = 3$ is the only solution of eq. (13) that yields $b = 0$. Since $b = 0$ is a necessary condition for a finite field theory, the existence of three families of quarks and leptons is natural in such models, provided the matter content is exactly as given above.

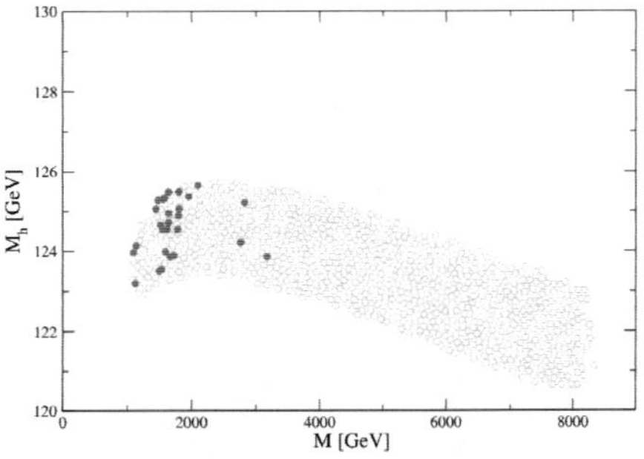

Figure 2. The lighest Higgs mass M_h as function of the unified gaugino mass M. The red (dark) points satisfy the loose CDM constraint.

The model of this type with best phenomenology is the $SU(3)^3$ model discussed in ref. [23], where the details of the model are given. It corresponds to the well-known example of $SU(3)_C \times SU(3)_L \times SU(3)_R$ [24–27]. Thus, this $SU(3)^3$ model is finite between the Planck M_P and the unification M_{GUT} scales, then breaks spontaneously down to the MSSM at M_{GUT} [26]. Notice that this model has extra exotic particles above M_{GUT}, which are Higgs-like and down-quark like.

With three families, the most general superpotential contains 11 f couplings, and 10 f' couplings, subject to 9 conditions, due to the vanishing of the anomalous dimensions of each superfield. The conditions are the following

$$\sum_{j,k} f_{ijk}(f_{ljk})^* + \frac{2}{3}\sum_{j,k} f'_{ijk}(f'_{ljk})^* = \frac{16}{9} g^2 \delta_{il}\,, \tag{14}$$

The assumptions below the GUT scale are the same as for the $SU(5)$ models, where now we have the boundary conditions given by eq. (14), the $h = -MC$ relation, and the soft scalar-mass sum rule (4) at $M_{\rm GUT}$.

We will concentrate here on a two-loop finite solution, in which we keep f' non-vanishing and we use it to introduce the lepton masses. This model is finite only up to two-loops since the corresponding solution of eq. (14) is not isolated. In this case we have the following boundary conditions for the Yukawa couplings

$$f^2 = r\left(\frac{16}{9}\right) g^2\,, \quad f'^2 = (1-r)\left(\frac{8}{3}\right) g^2\,. \tag{15}$$

As for the boundary conditions for the soft scalars, we have the universal case.

As with the case of the $SU(5)$ models, we first look at the predictions for the top and quark masses, noticing that we have here an extra parameter, namely r. We found that there is a small region where both top and bottom quark masses are in the experimental range for the same value of r. But if we take some of the exotic particles into account, decoupling below the unification scale, the situation improves [28]. This can be seen in Fig. 1, where we took only one down-like exotic particle decoupling at 10^{14}, below that the usual MSSM. In this case, for $r \sim 0.5 \sim 0.62$ we have reasonable agreement with experimental data for both top and bottom quark masses, where the red points in the figure are the ones that satisfy the B physics constraints [20].

Acknowledgements
This work is partially supported by the NTUA programmes for basic research PEVE 2008 and 2009, the European Union's RTN programme under contract MRTN-CT-2006-035505 and the

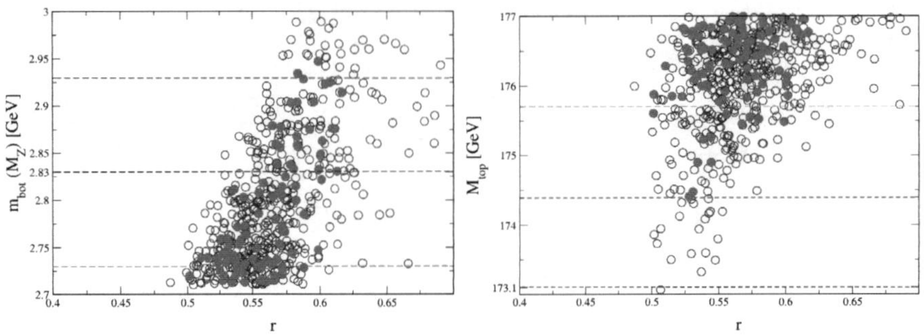

Figure 3. The figure shows the values for the top and bottom quark masses for the FUT model $SU(3)^3$, with $\mu < 0$, where the darker horizontal line is the experimental central value, and the lighter ones are the one and two sigma limits. The red points are the ones that satisfy the B physics constraints.

European Union's ITN programme "UNILHC" PITN-GA-2009-237920. Supported also by a mexican PAPIIT grant IN112709 and by Conacyt grants 82291 and 51554-F.

References

[1] Kapetanakis D, Mondragon M and Zoupanos G 1993 *Z. Phys.* **C60** 181–186 (*Preprint* hep-ph/9210218)
[2] Mondragon M and Zoupanos G 1995 *Nucl. Phys. Proc. Suppl.* **37C** 98–105
[3] Heinemeyer S, Mondragon M and Zoupanos G 2008 *JHEP* **07** 135 (*Preprint* 0712.3630)
[4] Kubo J, Mondragon M and Zoupanos G 1997 *Acta Phys. Polon.* **B27** 3911–3944 (*Preprint* hep-ph/9703289)
[5] Kobayashi T, Kubo J, Mondragon M and Zoupanos G 2001 *Surveys High Energ. Phys.* **16** 87–129
[6] Heinemeyer S, Mondragon M and Zoupanos G 2010 *SIGMA* **6** 049 (*Preprint* 1001.0428)
[7] Lucchesi C, Piguet O and Sibold K 1988 *Helv. Phys. Acta* **61** 321
[8] Piguet O and Sibold K 1986 *Int. J. Mod. Phys.* **A1** 913
[9] Piguet O and Sibold K 1986 *Phys. Lett.* **B177** 373
[10] Lucchesi C and Zoupanos G 1997 *Fortschr. Phys.* **45** 129–143 (*Preprint* hep-ph/9604216)
[11] Jack I and Jones D R T 1994 *Phys. Lett.* **B333** 372–379 (*Preprint* hep-ph/9405233)
[12] Kobayashi T, Kubo J and Zoupanos G 1998 *Phys. Lett.* **B427** 291–299 (*Preprint* hep-ph/9802267)
[13] Kubo J, Mondragon M and Zoupanos G 1994 *Nucl. Phys.* **B424** 291–307
[14] Babu K S, Enkhbat T and Gogoladze I 2003 *Phys. Lett.* **B555** 238–247 (*Preprint* hep-ph/0204246)
[15] Kazakov D I, Kalmykov M Y, Kondrashuk I N and Gladyshev A V 1996 *Nucl. Phys.* **B471** 389–408 (*Preprint* hep-ph/9511419)
[16] Kobayashi T, Kubo J, Mondragon M and Zoupanos G 1998 *Nucl. Phys.* **B511** 45–68 (*Preprint* hep-ph/9707425)
[17] Jones D R T and Raby S 1984 *Phys. Lett.* **B143** 137
[18] Leon J, Perez-Mercader J, Quiros M and Ramirez-Mittelbrunn J 1985 *Phys. Lett.* **B156** 66
[19] 2009 Tevatron Electroweak Working Group, arXiv:0903.2503 [hep-ex] (*Preprint* 0903.2503)
[20] Amsler C *et al.* (Particle Data Group) 2008 *Phys. Lett.* **B667** 1
[21] Belanger G, Boudjema F, Pukhov A and Semenov A 2002 *Comput. Phys. Commun.* **149** 103–120 (*Preprint* hep-ph/0112278)
[22] Degrassi G, Heinemeyer S, Hollik W, Slavich P and Weiglein G 2003 *Eur. Phys. J.* **C28** 133–143 (*Preprint* hep-ph/0212020)
[23] Ma E, Mondragon M and Zoupanos G 2004 *JHEP* **12** 026 (*Preprint* hep-ph/0407236)
[24] De Rújula A, Georgi H and L G S 1984 88 fifth Workshop on Grand Unification, K. Kang, H. Fried, and P. Frampton eds., World Scientific, Singapore
[25] Lazarides G, Panagiotakopoulos C and Shafi Q 1993 *Phys. Lett.* **B315** 325–330 (*Preprint* hep-ph/9306332)
[26] Lazarides G and Panagiotakopoulos C 1994 *Phys. Lett.* **B336** 190–193 (*Preprint* hep-ph/9403317)
[27] Ma E 1987 *Phys. Rev.* **D36** 274
[28] Heinemeyer S, Ma E, Mondragón M and Zoupanos G to appear.

16th International Symposium on Particles Strings and Cosmology (PASCOS 2010) IOP Publishing
Journal of Physics: Conference Series **259** (2010) 012098 doi:10.1088/1742-6596/259/1/012098

The quantum vacuum of the minimal SO(10) GUT

Stefano Bertolini[1], **Luca Di Luzio**[1], **Michal Malinský**[2]

[1] INFN, Sezione di Trieste, and SISSA, via Bonomea 265, 34136 Trieste, Italy

[2] Theoretical Particle Physics Group, Department of Theoretical Physics, Royal Institute of Technology (KTH), Roslagstullsbacken 21, SE-106 91 Stockholm, Sweden

E-mail: malinsky@kth.se

Abstract. We reexamine the longstanding no-go excluding all potentially viable $SO(10) \to SU(3)_c \otimes SU(2)_L \otimes U(1)_Y$ symmetry breaking patterns within the minimal renormalizable non-supersymmetric SO(10) GUT framework featuring the 45-dimensional adjoint representation in the Higgs sector. A simple symmetry argument indicates that quantum effects do change the vacuum structure of the model dramatically. A thorough analysis of the one-loop effective potential reveals that the phenomenologically favoured symmetry breaking chains passing through the $SU(4)_C \otimes SU(2)_L \otimes U(1)_R$ or $SU(3)_c \otimes SU(2)_L \otimes SU(2)_R \otimes U(1)_{B-L}$ intermediate stages are, indeed, supported at the quantum level. This brings the class of minimal non-supersymmetric $SO(10)$ GUTs back from oblivion, providing a new ground for a potentially realistic model building.

1. Introduction

In spite of the enormous progress the field of grand unified theories (GUTs) experienced since the seminal works of Georgi and Glashow [1] it is still not clear whether the basic scheme is by any means reflected by nature. Nevertheless, the inherent testability of the simplest GUT scenarios makes the idea of a unified description of the strong and electroweak interactions so popular that there has been a lot of efforts spent recently in attempts to pin down the simplest potentially viable scenarios, in particular those based on the $SU(5)$ [2] and/or $SO(10)$ [3, 4] gauge symmetry. Apart from the canonical test imposed by the required high-energy convergence of all three Standard Model (SM) gauge couplings, a successful candidate should obey also the constraints inflicted by the proton longevity (stretching over 10^{33} years) and the absolute neutrino mass scale, which are indeed closely related to the generic need for a compatibility with the observed SM flavour structure.

However, the situation is not very clear at the moment as even the very concept of minimality admits a certain number of interpretations. For instance, it can be understood as a mere "simplicity" related to e.g. a minimum rank of the gauge group, the Higgs sector dimensionality, the naturalness of the doublet-triplet splitting, or even the complexity of the gauge unification pattern. In this respect, the models based on $SU(5)$ gauge group are frequently regarded to as more minimal than the $SO(10)$ ones. Alternatively, predictivity (i.e., the number of independent couplings) is often invoked as a measure of minimality; what matters in such a case is namely the renormalizability of the theory, the number of Higgs multiplets, the complexity of the GUT-scale Yukawa sector, the presence or absence of supersymmetry (SUSY) etc. In this respect, the $SO(10)$ GUTs (especially their SUSY variants) usually score better than the $SU(5)$ models, see

Journal of Physics: Conference Series **259** (2010) 012098 doi:10.1088/1742-6596/259/1/012098

e.g. [4, 5] and references therein. To us, even though potentially troublesome on the naturalness side, this is a very physical option - besides giving generally better prospects for an ultimate testability the preference of $SO(10)$ also offers a natural relief from the persistent trouble with the simplest $SU(5)$ models, failing either on the unification side (like e.g. the original Georgi-Glashow model) or suffering from an overly fast proton decay when SUSY is invoked[1].

Sticking to the $SO(10)$ case, minimality is essentially equivalent to the complexity of the Higgs sector. In the class of $SO(10)$ GUTs, the choice of the Higgs representation is subject to several basic requirements. First, there is the need to break consistently the rank=5 $SO(10)$ gauge symmetry down to the rank=4 $SU(3)_c \otimes SU(2)_L \otimes U(1)_Y$ of the SM. Remarkably, none of the basic candidate Higgs representations (namely, **16** and/or **126**) capable of breaking the $U(1)_{B-L}$ subgroup of $SO(10)$, can account for the full $SO(10) \to$ SM breakdown on its own, since their SM singlets leave a full $SU(5)$ subgroup of $SO(10)$ intact. From the group theory point of view, the simplest solution one can devise is an extra adjoint **45** in the Higgs sector whose additional pair of VEVs brings in enough freedom to provide the full $SO(10) \to SU(3)_c \otimes SU(2)_L \otimes U(1)_Y$ breakdown. Unfortunately, this approach fails in the SUSY case where the extra F-flatness conditions force the VEVs in **45** to align along the direction of $\langle \mathbf{16} \rangle$ and thus $SU(5)$ still remains intact [6]. Thus, **210** is usually invoked in the SUSY $SO(10)$ context, giving rise to the minimal SUSY $SO(10)$ GUT with a Higgs sector spanning over the reducible representation $\mathbf{10} \oplus \overline{\mathbf{126}} \oplus \mathbf{126} \oplus \mathbf{210}$, see e.g. [7, 8] and also [4].

However, though successful in accounting for some of the basic features of the lepton flavour mixing [9], the minimal renormalizable SUSY SO(10) model fails due to a generic tension between the absolute neutrino mass scale and the gauge coupling unification constraints, [10, 11] and there is a general need to go beyond the simplest setting, see e.g. [12, 13, 14, 15] and references therein.

On the other hand, if predictivity rather than naturalness is at stakes, supersymmetry is not a mandatory ingredient of the $SO(10)$ GUTs because the unification constraints refuting the simplest non-SUSY $SU(5)$ settings can be satisfied with an intermediate energy scale M_I below the scale of grand unification M_G. In particular, the recent analysis of the relevant symmetry breaking chains [16] reveals an interesting picture associated to the two basic single-step chains

$$SO(10) \xrightarrow{M_G} SU(4)_C \otimes SU(2)_L \otimes U(1)_R \xrightarrow{M_I} \text{SM}, \tag{1}$$

$$SO(10) \xrightarrow{M_G} SU(3)_c \otimes SU(2)_L \otimes SU(2)_R \otimes U(1)_{B-L} \xrightarrow{M_I} \text{SM}, \tag{2}$$

driven by the simplest conceivable Higgs sector spanned over just $\mathbf{45} \oplus \mathbf{16}$ of $SO(10)$. Unfortunately, this scenario, requiring in practice a few orders of magnitude hierarchy between M_I and M_G, has been long ago [17, 18, 19] claimed incompatible with the dynamics of the scalar sector. The reason is an emergence of tachyonic masses in the Higgs spectrum unless M_I essentially coincides with M_G, at odds with the running constraints.

All these arguments are, however, based on the tree-level analysis of the Higgs potential. Remarkably enough, the recent work [20] reveals a dramatic qualitative change of the picture once quantum effects are properly taken into account, with a potential of bringing the class of minimal non-supersymmetric $SO(10)$ GUTs back to life.

2. The classical failure of the minimal non-SUSY $SO(10)$ GUT - a closer look

In order to fully appreciate this fact one has to inspect in detail the available vacua of the renormalizable non-SUSY scalar potential spanned over $\mathbf{45} \oplus \mathbf{16}$ in the Higgs sector. However, before entering the realm of quantum effects one has to understand the origin of the tree-level no-go implied by [17, 18, 19].

[1] For a nice account for the proton decay issue in the minimal SUSY $SU(5)$ see e.g. [2].

The most generic classical renormalizable scalar potential one can write with only $\mathbf{45} \oplus \mathbf{16}$ at hand reads $V_0 = V_{45} + V_{16} + V_{\rm int}$, where [21]

$$V_{45} = -\frac{\mu^2}{2}\,\mathrm{Tr}\,\mathbf{45}^2 + \frac{a_1}{4}(\mathrm{Tr}\,\mathbf{45}^2)^2 + \frac{a_2}{4}\,\mathrm{Tr}\,\mathbf{45}^4\,, \tag{3}$$

$$V_{16} = -\frac{\nu^2}{2}\mathbf{16}^\dagger\mathbf{16} + \frac{\lambda_1}{4}(\mathbf{16}^\dagger\mathbf{16})^2 + \frac{\lambda_2}{4}(\mathbf{16}_+^\dagger\Gamma_j\mathbf{16}_-)(\mathbf{16}_-^\dagger\Gamma_j\mathbf{16}_+) \tag{4}$$

correspond to the mass and self-interaction terms for **45** and **16**, respectively, while

$$V_{\rm int} = \alpha(\mathbf{16}^\dagger\mathbf{16})\,\mathrm{Tr}\,\mathbf{45}^2 + \beta\mathbf{16}^\dagger\mathbf{45}^2\mathbf{16} + \tau\mathbf{16}^\dagger\mathbf{45}\,\mathbf{16} \tag{5}$$

provides the desired interaction between the two sectors. Employing an $SU(3)_c \otimes SU(2)_L \otimes SU(2)_R \otimes U(1)_{B-L}$ notation the vacuum manifold is parametrized by the VEVs of the three different SM singlets

$$\langle(1,1,1,0)_{45}\rangle \equiv \omega_Y\,, \quad \langle(1,1,3,0)_{45}\rangle \equiv \omega_R\,, \quad \langle(1,1,2,+1/2)_{16}\rangle \equiv \chi_{B-L}\,. \tag{6}$$

Since the gauge unification in this minimal scenario typically requires the $B-L$ breaking scale χ_{B-L} to be lower than the higher from the pair ω_R, ω_Y (to be identified with M_G), the heavy Higgs spectrum is governed namely by the VEVs of **45**. Apart from the generic setting with non-zero $\omega_R \neq \omega_Y$ corresponding to an intermediate $SU(3)_c \otimes SU(2)_L \otimes U(1)_R \otimes U(1)_{B-L}$ stage, there are several important limiting cases exhibiting an enhanced intermediate symmetry, in particular

$$\begin{aligned} \omega_R \neq 0,\ \omega_Y = 0 &\ :\ SU(4)_C \otimes SU(2)_L \otimes U(1)_R\,, \\ \omega_R = 0,\ \omega_Y \neq 0 &\ :\ SU(3)_c \otimes SU(2)_L \otimes SU(2)_R \otimes U(1)_{B-L}\,, \\ \omega_R = \pm\omega_Y \neq 0 &\ :\ \text{standard } (+) \text{ or flipped } (-)\ SU(5) \otimes U(1)\,. \end{aligned} \tag{7}$$

The point is that while the first two cases (corresponding to the simplest conceivable chains (1), (2)) are the only ones from list (7) which are compatible with the non-SUSY unification constraints [20], they seem to be disfavoured by the detailed Higgs sector dynamics [17, 18, 19] due to the emergence of tachyons within the corresponding vacua.

2.1. The emergence of tachyons in the non-$SU(5) \otimes U(1)$ vacua

This is due to the fact that a full-fledged analysis of the classical Higgs potential (3)-(5) reveals a pair of scalar multiplets with the classical SM-level masses given by the formulae

$$\begin{aligned} M^2(1,3,0)_{45} &= 2a_2(\omega_Y - \omega_R)(\omega_Y + 2\omega_R)\,, \\ M^2(8,1,0)_{45} &= 2a_2(\omega_R - \omega_Y)(\omega_R + 2\omega_Y)\,. \end{aligned} \tag{8}$$

It is clear that in the majority of the parametric space, in particular, everywhere outside the domain

$$-2 \leq \omega_Y/\omega_R \leq -1/2\,, \tag{9}$$

at least one of them develops a negative mass-squared, indicating an instability of the corresponding vacuum configuration. Consequently, all stable settings exhibit an intermediate flipped $SU(5) \otimes U(1)$ stage, at odds with unification constraints. This, indeed, is the core of the classical argument against the potential viability of the minimal non-SUSY $SO(10)$ GUT framework.

Since the simplicity of the formulae (8) is crucial for the above reasoning it is important to understand their structure in as much detail as possible. In this respect, it is particularly illuminating to discuss the global symmetries of the scalar potential V_0.

Journal of Physics: Conference Series **259** (2010) 012098 doi:10.1088/1742-6596/259/1/012098

2.2. Understanding the structure of the key mass formulae in terms of symmetries

When only trivial invariants (i.e. moduli) of both **45** and **16** are considered, the global symmetry of V_0 is $O(45) \otimes O(32)$. It is spontaneously broken down to $O(44) \otimes O(31)$ by the **45** and **16** VEVs yielding 44+31=75 Goldstone bosons in the scalar spectrum. Since in this case also the gauge $SO(10)$ symmetry is broken to the SM gauge group, $45 - 12 = 33$ would-be Goldstone bosons, with the quantum numbers of the coset $SO(10)/\mathrm{SM}$ algebra, decouple from the physical spectrum, while $75 - 33 = 42$ pseudo-Goldstone bosons remain. Their masses are generally expected to receive contributions from the explicitly breaking terms a_2, λ_2, β and τ.

However, the tree-level masses of the crucial $(1,3,0)_{45}$ and $(8,1,0)_{45}$ multiplets, which belong among these pseudo-Goldstone bosons, turn out to depend only on the parameter a_2 but *not* on the other parameters expected, namely λ_2, β and τ. While the λ_2 and τ terms cannot obviously contribute at the tree level to **45** mass terms, one would generally expect a contribution from the β term proportional to χ^2_{B-L} to pop up in relations (8). This, however, is not the case. One can understand that on general grounds by observing that the relevant scalar interaction in (5) resembles the the covariant-derivative interaction responsible for the contribution of $\langle \mathbf{16} \rangle$ to the gauge boson masses. As a consequence, no tree-level mass contribution from the β coupling can be generated for the scalars in **45** which carry the SM quantum numbers of the $W^\pm$ and Z^0 bosons, i.e., $(1,3,0)_{45}$, and gluons $(8,1,0)_{45}$ that are protected by the residual $SU(5)$ symmetry left unbroken even with $\langle \mathbf{16} \rangle \neq 0$.

3. The quantum structure of the minimal non-SUSY $SO(10)$ GUT vacuum

On the other hand, we should expect the τ and β interactions to contribute to the masses of $(1,3,0)_{45}$ and $(8,1,0)_{45}$ at the quantum level. While for the former the loops simply admit to contract the two **16**'s sticking out of the τ-vertex in (5), a similar loop due to a pair of β-vertices brings into play the $SU(5)$-breaking VEVs of **45**. Similar contributions should also arise from the gauge interactions which break explicitly the independent global transformations on the **45** and **16** discussed in the previous section.

In the diagrammatic language, the relevant one-loop self-energies are depicted in Figure 1 where the individual components of the scalars transforming as **45** and **16** of $SO(10)$ have been denoted by generic symbols ϕ and χ, respectively. While the exchange of the components of **16** is crucial, the χ_{B-L} does not need to enter; in the phenomenologically allowed unification patterns it gives actually negligible contributions.

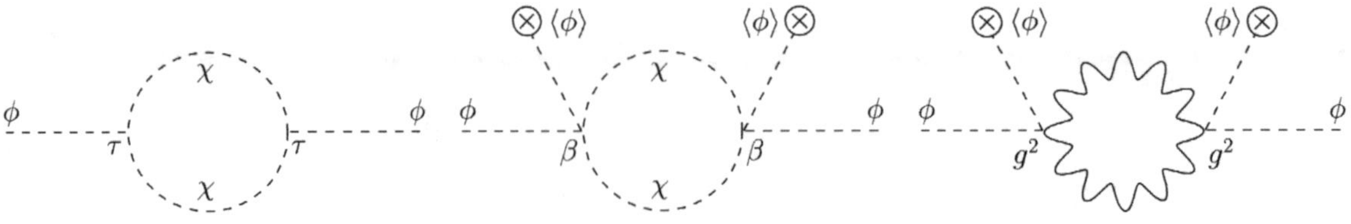

Figure 1. The diagrammatic structure of the three important types of quantum corrections which stabilise the vacua supporting the desired $SU(4)_C \otimes SU(2)_L \otimes U(1)_R$ (case A) or $SU(3)_c \otimes SU(2)_L \otimes SU(2)_R \otimes U(1)_{B-L}$ (case B) symmetry breaking chains at the one-loop level. The dangerous tachyonic mass parameters of the $(1,3,0)_{45}$ and/or $(8,1,0)_{45}$ Higgs multiplets in **45** are lifted provided $8\pi^2|a_2| < \tau^2/\omega_R^2 + \beta^2 + 13g^4$ in case A, or $8\pi^2|a_2| < \tau^2/\omega_Y^2 + 2\beta^2 + 19g^4$ in case B. The interested reader can find further details in the original work [20].

One is thus lead to the conclusion that any result based on the particular shape of the tree-level **45** vacuum is dramatically altered at the quantum level. Let us emphasise that although one may in principle avoid the τ-term by means of e.g. an extra Z_2 symmetry, no symmetry can forbid the β-term and the gauge loop contributions.

3.1. The one-loop effective potential analysis

The quantum structure of the model vacuum is best assessed by the minimisation of the effective potential [22]. It can be conveniently written as $V_{\rm eff} = V_0 + \Delta V$ where the leading one-loop contribution to ΔV can be cast in the form

$$\Delta V(\phi,\chi,\mu) = \frac{1}{64\pi^2}\,{\rm Tr}\left[\mathcal{M}_s^4(\phi,\chi)\left(\log\frac{\mathcal{M}_s^2(\phi,\chi)}{\mu^2}-\frac{3}{2}\right)+3\mathcal{M}_g^4(\phi,\chi)\left(\log\frac{\mathcal{M}_g^2(\phi,\chi)}{\mu^2}-\frac{5}{6}\right)\right]. \tag{10}$$

Here the terms in the bracket correspond to the contribution from the (real) scalars and gauge bosons, respectively, and $\mathcal{M}_s$ and $\mathcal{M}_g$ denote the relevant field-dependent tree-level mass matrices. Note also that the contribution due to the SM fermions has been neglected because of the hierarchical suppression of their masses with respect to the GUT-scale Higgs and gauge bosons. For completeness, let us also mention that the specific form of ΔV corresponds to the Landau gauge and the $\overline{\rm MS}$ renormalization scheme.

Using a generic symbol ψ for any of the components of $\mathbf{45}\oplus\mathbf{16}$ the first derivative of the one-loop part of the effective potential (10) is given by

$$\frac{\partial\Delta V}{\partial\psi} = \frac{1}{64\pi^2}\,{\rm Tr}\left[\left\{\frac{\partial\mathcal{M}_s^2}{\partial\psi},\mathcal{M}_s^2\right\}\left(\log\frac{\mathcal{M}_s^2}{\mu^2}-\frac{3}{2}\right)+\mathcal{M}_s^2\frac{\partial\mathcal{M}_s^2}{\partial\psi}\right]+\text{gauge part}\,, \tag{11}$$

where, for the sake of simplicity, only the scalar contribution due to the first term in eq. (10) has been written explicitly; the gauge part derived from the $\mathcal{M}_g$-piece of (10) is completely analogous. Let us also note that the formula above holds regardless of the commutation properties of $\mathcal{M}$ and its first derivatives.

Similarly, the second derivatives of ΔV can be written in the form

$$\begin{aligned}
\frac{\partial^2\Delta V}{\partial\psi_a\partial\psi_b} &= \frac{1}{64\pi^2}\,{\rm Tr}\Bigg\{\\
&\times\left[\left\{\frac{\partial^2\mathcal{M}_s^2}{\partial\psi_a\partial\psi_b},\mathcal{M}_s^2\right\}+\left\{\frac{\partial\mathcal{M}_s^2}{\partial\psi_a},\frac{\partial\mathcal{M}_s^2}{\partial\psi_b}\right\}\right]\left(\log\frac{\mathcal{M}_s^2}{\mu^2}-\frac{3}{2}\right)+\frac{\partial\mathcal{M}_s^2}{\partial\psi_a}\frac{\partial\mathcal{M}_s^2}{\partial\psi_b}+\mathcal{M}_s^2\frac{\partial^2\mathcal{M}_s^2}{\partial\psi_a\partial\psi_b}\\
&+\sum_{m=1}^{\infty}(-1)^{m+1}\frac{1}{m}\sum_{k=1}^{m}\binom{m}{k}\left\{\mathcal{M}_s^2,\frac{\partial\mathcal{M}_s^2}{\partial\psi_a}\right\}\left[\mathcal{M}_s^2,..\left[\mathcal{M}_s^2,\frac{\partial\mathcal{M}_s^2}{\partial\psi_b}\right]..\right]\left(\mathcal{M}_s^2-1\right)^{m-k}\Bigg\}\\
&+\text{gauge part}\,,
\end{aligned} \tag{12}$$

where the infinite series of nested commutators (taken $k-1$ times for each k) looks after the general non-commutativity of $\mathcal{M}_s$ and $\partial\mathcal{M}_s^2/\partial\psi$ which becomes relevant here.

Note also that such terms are also crucial for a consistent cancellation of the infrared divergencies emerging in the one-loop mass matrix $\overline{m}^2_{ab}\equiv\partial^2V_{\rm eff}/\partial\psi_a\partial\psi_b$ when evaluated for the solutions of the one-loop stationary equations $\partial V_{\rm eff}/\partial\psi_a=0$. These are, of course, due to the presence of Goldstone bosons in the Landau gauge inherent to formula (10). Let us recall that the effective potential, however, is defined at zero external momenta, and the physical Higgs spectrum is derived from the "shifted" mass matrix $M^2_{ab}(p^2)=\overline{m}^2_{ab}+\Delta\Sigma_{ab}(p^2)$ for $p^2=M_a^2$ where $\Delta\Sigma_{ab}(p^2)\equiv\Sigma_{ab}(p^2)-\Sigma_{ab}(0)$ is the difference of the relevant one-loop self-energies evaluated at two different four-momenta.

3.2. The one-loop corrections to the masses of $(1,3,0)_{45}$ and $(8,1,0)_{45}$

Putting all this together, the quantum shifts in the masses of the two critical multiplets read

$$\begin{aligned}
\Delta M^2(1,3,0)_{45} &= \frac{1}{4\pi^2}\left[\tau^2+\beta^2(2\omega_R^2-\omega_R\omega_Y+2\omega_Y^2)+g^4\left(16\omega_R^2+\omega_Y\omega_R+19\omega_Y^2\right)\right]+\text{logs}\,,\\
\Delta M^2(8,1,0)_{45} &= \frac{1}{4\pi^2}\left[\tau^2+\beta^2(\omega_R^2-\omega_R\omega_Y+3\omega_Y^2)+g^4\left(13\omega_R^2+\omega_Y\omega_R+22\omega_Y^2\right)\right]+\text{logs}\,,
\end{aligned}$$

where the symbols "logs" denote the sub-leading logarithmic contributions. It is easy to see that with these terms added to the tree-level structure (8) both $(1,3,0)_{45}$ and $(8,1,0)_{45}$ mass-squares can be made simultaneously positive for wide ranges of the ω_R and ω_Y parameters, well outside the classical domain (9). This holds, especially, for the configurations with either $\omega_Y \ll \omega_R$ or $\omega_R \ll \omega_Y$ corresponding to the classically forbidden $SU(4)_C \otimes SU(2)_L \otimes U(1)_R$ (case A) or $SU(3)_c \otimes SU(2)_L \otimes SU(2)_R \otimes U(1)_{B-L}$ (case B) intermediate symmetries. In particular, the dangerous tachyonic mass parameters of the $(1,3,0)_{45}$ and/or $(8,1,0)_{45}$ Higgs multiplets in **45** are lifted for $8\pi^2|a_2| < \tau^2/\omega_R^2 + \beta^2 + 13g^4$ in case A, or $8\pi^2|a_2| < \tau^2/\omega_Y^2 + 2\beta^2 + 19g^4$ in case B.

Thus, the quantum effects restore the full symmetry breaking potential of the minimal non-SUSY $SO(10)$ Higgs sector spanned over **45**$\oplus$**16**, bringing back into play the symmetry breaking chains favoured by the gauge coupling unification constraints! Let us also stress that the fact that $U(1)_{B-L}$ has been broken by the spinorial **16** does not play a major role in the argument so an alternative Higgs model based on **45** $\oplus$ **126** representation exhibits the same qualitative behaviour. Similarly, an extra **10** added to the Higgs sector in order to get potentially realistic Yukawa couplings makes no harm because it can not develop any GUT-scale VEVs.

4. Conclusions and outlook

Typically, only the minimal GUTs can be subject to a thorough scrutiny by complementary methods such as the proton lifetime constraints, the absolute neutrino mass scale or the match between the effective SM flavour structure and the underlying GUT-scale Yukawa sector. In this respect, the recent failure of the minimal renormalizable SUSY $SO(10)$ GUT re-opened the question of the potential viability of the simplest non-SUSY $SO(10)$ scenarios. Remarkably, the generic no-go obtained about 30 years ago, which plagued the simplest model with **45** $\oplus$ **16** in the Higgs sector, has been recently identified as a mere artefact of the classical approach that is lifted at the quantum level. This reiterates the intriguing question whether a potentially viable and intrinsically consistent non-SUSY $SO(10)$ theory can be spanned over a renormalizable Higgs sector with only **45**$\oplus$**16** (or, alternatively, **45**$\oplus$**126**) participating at the GUT symmetry breakdown. The final answer, however, is left to a future analysis.

References

[1] Georgi H and Glashow S L 1974 *Phys. Rev. Lett.* **32** 438–441
[2] Bajc B, Fileviez Perez P and Senjanovic G 2002 (*Preprint* hep-ph/0210374)
[3] Albright C H and Barr S M 2000 *Phys. Rev.* **D62** 093008 (*Preprint* hep-ph/0003251)
[4] Aulakh C S, Bajc B, Melfo A, Senjanovic G and Vissani F 2004 *Phys. Lett.* **B588** 196–202 (*Preprint* hep-ph/0306242)
[5] Fukuyama T, Ilakovac A, Kikuchi T, Meljanac S and Okada N 2004 *JHEP* **09** 052 (*Preprint* hep-ph/0406068)
[6] Aulakh C S, Bajc B, Melfo A, Rasin A and Senjanovic G 2001 *Nucl. Phys.* **B597** 89–109 (*Preprint* hep-ph/0004031)
[7] Clark T E, Kuo T K and Nakagawa N 1982 *Phys. Lett.* **B115** 26
[8] Aulakh C S and Mohapatra R N 1983 *Phys. Rev.* **D28** 217
[9] Bajc B, Senjanovic G and Vissani F 2003 *Phys. Rev. Lett.* **90** 051802 (*Preprint* hep-ph/0210207)
[10] Aulakh C S and Garg S K 2006 *Nucl. Phys.* **B757** 47 (*Preprint* hep-ph/0512224)
[11] Bertolini S, Schwetz T and Malinsky M 2006 *Phys. Rev.* **D73** 115012 (*Preprint* hep-ph/0605006)
[12] Bertolini S, Frigerio M and Malinsky M 2004 *Phys. Rev.* **D70** 095002 (*Preprint* hep-ph/0406117)
[13] Aulakh C S and Garg S K 2006 (*Preprint* hep-ph/0612021)
[14] Bajc B, Dorsner I and Nemevsek M 2008 *JHEP* **11** 007 (*Preprint* 0809.1069)
[15] Melfo A, Ramirez A and Senjanovic G 2010 (*Preprint* arXiv:1005.0834)
[16] Bertolini S, Di Luzio L and Malinsky M 2009 *Phys. Rev.* **D80** 015013 (*Preprint* arXiv:0903.4049 [hep-ph])
[17] Yasue M 1981 *Phys. Rev.* **D24** 1005 ; Yasue M 1981 *Phys. Lett.* **B103** 33
[18] Anastaze G, Derendinger J P and Buccella F 1983 *Z. Phys.* **C20** 269–273
[19] Babu K S and Ma E 1985 *Phys. Rev.* **D31** 2316
[20] Bertolini S, Di Luzio L and Malinsky M 2010 *Phys. Rev.* **D81** 035015 (*Preprint* arXiv:0912.1796 [hep-ph])
[21] Buccella F, Ruegg H and Savoy C A 1980 *Phys. Lett.* **B94** 491
[22] Coleman S R and Weinberg E J 1973 *Phys. Rev.* **D7** 1888–1910

Journal of Physics: Conference Series **259** (2010) 012099 doi:10.1088/1742-6596/259/1/012099

The interplay between grand unified and flavour symmetries in a Pati-Salam $\times$ S_4 model

Reinier de Adelhart Toorop
Nikhef Theory Group, Sciencepark 105, 1098 XG Amsterdam, The Netherlands
E-mail: reintoorop@nikhef.nl

Abstract. Both discrete flavour symmetries and Grand Unified symmetries explain apparent structures in the mass sector of the Standard Model. A model that combines both symmetries is therefore very appealing. We construct a model with the S_4 flavour symmetry and the Pati-Salam unification. We show that this model can indeed explain many observable relations between the masses of the quarks and leptons and that it is predictive in the neutrino sector. However, the combination of the two symmetries leads to new complications in the Higgs sector and in the running of the renormalisation group equations.

1. Introduction

The Standard Model (SM) of particle physics has 19 free parameters. 13 of these are related to the mass sector as the masses and mixing angles and phases of the quarks and leptons. If righthanded neutrinos are added, this number even grows to 22 out of 28.

When the masses and mixing angles are studied in detail, patterns seem to emerge. Some of these can be explained by assuming a discrete flavour symmetry at an high energy scale, others by assuming the symmetry of a grand unified theory (GUT). In section 2, we introduce these patterns and in section 3 we show how they are related to new symmetries. In section 4, we present a concrete implementation of this idea in a model [1] with an S_4 family symmetry and the Pati-Salam (PS) unificartion symmetry. This model has a positive and a negative message ("a beauty and a beast"). The positive point (section 4.1) is that it is indeed possible to build a model that combines discrete and GUT symmetries and explains many observables in the mass sector. The negative point (section 4.2) is that there are unexpected effects in the combination of the two symmetries. This leads to a very enlarged Higgs sector and to a strong effect on the renormalisation group (RG) behaviour. Due to RG effects, a large part of the naively accessible parameter space is excluded. Section 5 will conclude.

2. Structure in the fermion masses

The mass sector of the Standard Model seems to contain a lot of structure. Firstly the mass gaps between down and strange and between strange and bottom quarks are more or less of the same logarithmic size and of order λ^2, with $\lambda \sim 0.2$ the Cabibbo angle. The same holds for the charged leptons, while for the up-type quarks, the gaps are somewhat larger, being more or less of order λ^4. See also figure 1.

A second place where patterns show up, is in the mixing between flavour and mass eigenstates. This mixing is relatively small in the quark sector, where there is one moderate angle (θ^q_{12}, the

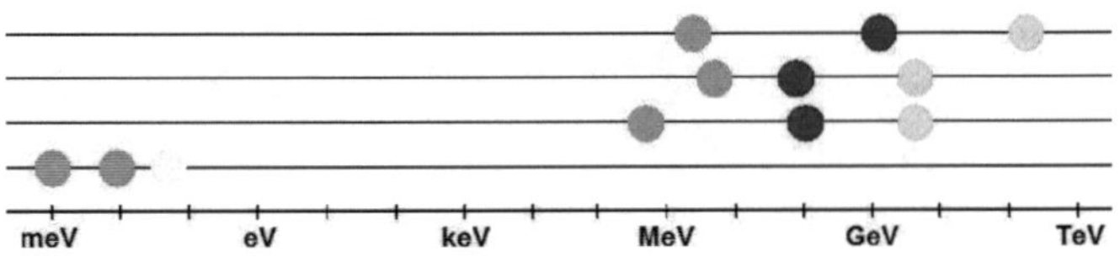

Figure 1. The masses of the SM fermions on a logarithmic scale. From first to last line: uptype quarks, downtype quarks, charged leptons and neutrinos (here pictured in the normal hierarchy)

Cabibbo angle), as well as two small angles ($\theta_{23}^q \sim \lambda^2$ and $\theta_{13}^q \sim \lambda^3$). In the lepton sector, the mixing is much larger and seems to show a particular structure. One neutrino mass eigenstate is an almost equal mixture of all three flavour eigenstates, while another mass eigenstate is a mixture one of two eigenstates (the so called tribimaximal mixing pattern [2]). See also figures 2 and 3. Furthermore, we notice that both in the 12 and in the 23 sector the neutrino and the quark mixing angles add to a maximal angle (45 degrees).

Figure 2. Pie charts showing the flavor content of the three quark mass eigenstates.

Figure 3. Pie charts showing the flavor content of the three neutrino mass eigenstates.

3. Flavour symmetries and GUTs

Flavour symmetries (for an introduction, see e.g. [4]) are symmetries that work between the three family copies of a certain particle type ("a horizontal symmetry"), which is to be contrasted with the usual gauge symmetries that work between the different particle types of one family ("a vertical symmetry").

Family symmetries can explain many of the structures mentioned in the previous section. A simple example is the Froggatt-Nielsen mechanism [3]. In this case, some of the particles are charged under an Abelian $U(1)$ charge, that differs between the families. The $U(1)$ charges are such that the original Yukawa Lagrangian is not invariant. This can be solved by introducing a new scalar field, the so-called flavon θ, that has the opposite charge as some of the fermions. Now terms that include some powers $\theta/\Lambda_{\text{cut-off}}$, are again invariant. Here $\Lambda_{\text{cut-off}}$ is the cut-off scale of the theory. If the scalar gets a vacuum expectation value (VEV) that is somewhat smaller than the cut-off, the mass gaps of section 2 can be generated with only parameters that are "naturally" of order 1. Using non-Abelian groups, mixing patterns such as the tribimaximal one can easily be generated. In particular the discrete groups A_4 and S_4 are often used.

A second reason to consider flavour symmetries is that these can unify the description of the fermions. If the symmetry group has three-dimensional representations (as for instance A_4 and S_4 do), the fermions can be represented as one representation of a large group instead of as multiple disjunct copies. The same can be accomplished by some grand unified groups. Their name states that they unify the gauge couplings, but they also do so for the fermions. For instance in the Pati-Salam GUT, instead of six representations to describe one family of quarks and leptons (including a righthanded neutrino), one can do with two. We will see that in the Pati-Salam $\times S_4$ model, instead of $3 \times 6 = 18$ representations, one can use only four. See figure 4.

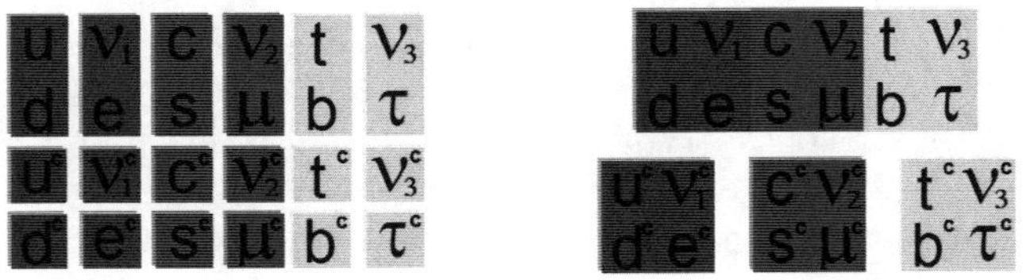

Figure 4. Fermion representations in the Standard Model and in the PS $\times S_4$ model.

On the other hand, relations between different particle types of the same generation can often be produced by GUTs. Examples are bottom-tau unification ($m_b = m_\tau$) and the Georgi-Jarlskog relation ($m_\mu = 3m_s$). We conclude that a model that combines flavour and GUT symmetries is appealing, both on grounds of predictivity and of unification.

4. The model

We build an extension of the Standard Model, where the family symmetry group is S_4 and the GUT group is Pati-Salam. For more details, see [1]. The model uses a supersymmetrical context, although this is not strictly necessary: a non-supersymmetric version of the model with the same main ingredients could be constructed as well.

Technically, the Pati-Salam gauge group $SU(4)_c \times SU(2)_L \times SU(2)_R$ is not a unified group, as even at high energies, there are still three gauge groups: the $SU(4)$ of extended colour and $SU(2)$s for left- and righthanded particles. However, we see that for the fermion representations, there is an unifying aspect, as the lefthanded quark and lepton ($SU(2)_L$)-doublets are combined in one PS representation $(\mathbf{4}, \mathbf{2}, \mathbf{1})$, as are the four righthanded ones (into $(\bar{\mathbf{4}}, \mathbf{1}, \mathbf{2})$). For each family, the fermions are now grouped in two representations (that we will denote as F and f^c), instead of in six (Q, L, u^c, d^c, l^c and ν^c).

The discrete group S_4 is the permutation group of four objects and has 24 elements, grouped in five conjugate classes. It has two one-dimensional representations (denoted 1_1 and 1_2), one two-dimensional representation (2) and two three-dimensional representations (3_1 and 3_2). In particular the fact that there are three-dimensional representations is important, as this allows us to put the lefthanded multiplets of the three families in one representation, thereby providing more unification. We do not put the righthanded multiplets together in a three dimensional multiplet as well. Instead put them in three separate one-dimensional representations (of type 1_2, 1_2 and 1_1 respectively). This has the advantage that we can charge the families under $U(1)_{FN}$, such that the Froggatt-Nielsen mechanism, alluded to in section 3 can give suppression to the masses of the first and second generation. The fields are also charged under an extra Abelian Z_4 that is used to prevent unwanted couplings in the superpotential. The assignment for the matter fields is summarized in figure 4 and table 1.

Table 1. Transformation properties of the matter fields.

	F_L	F_1^c	F_2^c	F_3^c
PS	$(4,2,1)$	$(\bar{4},1,2)$	$(\bar{4},1,2)$	$(\bar{4},1,2)$
S_4	3_1	1_2	1_2	1_1
$U(1)_{FN}$	0	2	2	0
Z_4	1	1	i	$-i$

Next to the Froggatt-Nielsen messenger θ, we need four more flavons, three S_4 triplets and one singlet. Their properties are given in table 2. The vacuum expectation values mentioned follow from the minimization and the typical magnitude is approximately $\lambda \sim 0.2$ times the cut-off scale of the theory. The VEVs are valid to first order in the number of flavons present in that potential. Subleading terms in this potential have one extra flavon inserted; due to the relative small VEVs, these give relative corrections of order λ. In fact, these corrections are very well calculable and are crucial in reproducing the observed patterns.

Table 2. The flavon field content and their transformation properties under the flavour symmetries. All flavon fields are singlet of the gauge group.

	θ	φ	φ'	χ	σ
S_4	1_1	3_1	3_2	3_1	1_1
$U(1)_{FN}$	-1	0	0	0	0
Z_4	1	i	i	1	1
VEV $\propto$	1	$\begin{pmatrix} 0 \\ 1 \\ 1 \end{pmatrix}$	$\begin{pmatrix} 0 \\ 1 \\ -1 \end{pmatrix}$	$\begin{pmatrix} 0 \\ 0 \\ 1 \end{pmatrix}$	1

Lastly, we will need a number of Higgs fields that break the gauge symmetry and are crucial for obtaining fermion masses. The Higgs spectrum will include at least a field ϕ that acts as the electroweak Higgs boson and some Higgs fields that break the PS gauge group to the (supersymmetric) standard model at high energies. The Higgs fields that are relevant for the mass matrices are given in table 3. More details of the Higgs sector are given in section 4.2.

Table 3. The Higgs fields responsible of generating fermion mass matrices and their transformation under the gauge and the Abelian flavour symmetries. All Higgs fields are singlets under $S_4 \times U(1)_{FN} \times U(1)_R$, while they can transform under the Z_4 factor.

	ϕ, ϕ'	ρ	Δ_L	Δ_R
PS	$(1,2,2)$	$(15,2,2)$	$(\overline{10},3,1)$	$(10,1,3)$
Z_4	1	-1	1	-1

4.1. The beauty

As mentioned in the introduction, the model has both a good and a bad side. The good news is that it can explain the patterns mentioned in sections 2 and 3, as we will show in this section.

At leading order (LO) in the number of flavons, we find mass matrices that can well produce the masses of the second and the third generation, while the first generation is massless. Given that these particles are very light, this is a satisfying first approximation.

The fermion mixing matrices (V_{CKM} and U_{PMNS}) however, do not fit the data well: the CKM matrix is the unity matrix, while the PMNS matrix is the so-called bimaximal mixing matrix. The bimaximal mixing matrix has both a maximal atmospheric angle and a maximal solar angle. This result is relatively fine in for the atmospheric angle, where the deviation seems to be small (at most of order λ^2), while a correction of order λ is needed for the solar angle. By definition, we also need a correction of size λ to introduce the Cabibbo angle in the CKM matrix. Fortunately, next-to-leading order (NLO) corrections are tailored to be exactly of that order and these can indeed be such that the 12 angles of in both the quark and lepton sector are corrected, while the 23 angles only receive next-to-next-to-leading order corrections. The quark-lepton complementarity of section 2 is reproduced: the quark and lepton angles add in both cases to a maximal angle. It turns out that the last mixing angle θ_{13} is relatively large in the neutrino sector (giving an reactor mixing angle of order λ), while it is indeed tiny ($\mathcal{O}(\lambda^3)$) in the quark sector.

In a Pati-Salam context furthermore, the matrices of downquarks and charged leptons are connected, as they both particles are in the same Pati-Salam representations. This allows the bottom-tau unification and the Georgi-Jarlskog relations and it helps in connecting the NLO corrections of quark and lepton mixing. The idea of bimaximal mixing with quark complementarity is shown in figures 5 and 6.

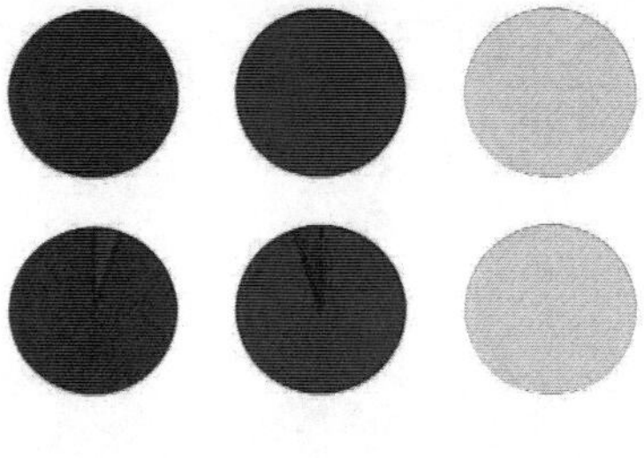

Figure 5. Pie charts showing the flavor content of the three quark mass eigenstates at LO (upper line) and at NLO, when the Cabibbo angle is introduced (lower line). The other two quark mixing angles are generated at even higher order

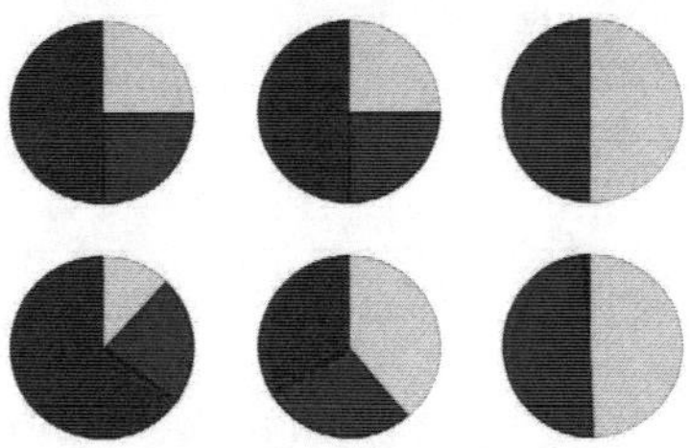

Figure 6. Pie charts showing the bimaximal mixing prediction in the lepton sector at LO (upper line) and a typical NLO result (lower line)

Apart from explaining quark and lepton mass relations, the model is also predictive in the neutrino sector. Above, it was already mentioned that we predict the third neutrino mixing angle to be relatively large (of the order of the Cabibbo angle). For almost all points in parameter space, the reactor mixing angle should be measureable by next generation experiments such as Daya Bay or Double CHOOZ. A second prediction is that the quasi-degenerate hierarchy is favoured in most of parameter space. Also the normal and inverted hierarchy are possible, albeit only for slightly unnatural values of the parameters. However, we will see that this prediction is severely affected by "interference" effects between the two types of symmetries that are analysed in the next section.

4.2. The beast

The positive results of the previous section are somewhat overshadowed by the fact that the combination of the flavour and the GUT symmetry leads to extra effects, which are less appealing. These appear at three points: the Higgs sector and the gauge and Yukawa renormalization group behaviour.

The first point is the Higgs sector, which is very non-minimal in our model. A minimal Pati-Salam model can have only four Higgs fields. These are labeled "minimal" in table 4. Other Higgs fields only occur in more extended versions of Pati-Salam models. Examples are the Δ_L that features in the type II seesaw and ρ that couples to some particles of the second generation and as such generates the Georgi-Jarlskog relation and the stronger suppression of the charm quark. Our construction features even more fields, which is due to the extra Z_4 symmetry that forces to double some fields and to the fact that some Higgs fields need to have VEV structures that require the presence of extra fields.

Table 4. The Higgs fields of our model. Fields marked "minimal" are present in minimal Pati-Salam models; fields marked "extended" only in more extended models and fields marked "new" are only used in our construction.

	A	B	Δ_R	$\bar{\Delta}_R$	ϕ	ϕ'
PS	(15,1,1)	(15,1,1)	(10,1,3)	($\bar{1}0$,1,3)	(1,2,2)	(1,2,2)
Z_4	1	-1	-1	-1	1	1
	minimal	new	minimal	minimal	minimal	new
	ρ	Δ_L	$\bar{\Delta}_L$	Σ	Σ'	ξ
PS	(15,2,2)	($\bar{1}0$,3,1)	(10,3,1)	(1,3,3)	(1,3,3)	(1,1,1)
Z_4	-1	1	1	1	-1	-1
	extended	extended	extended	extended	new	new

This large Higgs sector has a strong effect on the running of the gauge and Yukawa couplings. In particular in the neutrino sector the effects are large. We see that the atmospheric angle might get corrections that drive it too far away from the maximal angle that is predicted at LO and NLO. The bad news is that this happens in particular when then the neutrinos are in the quasi-degenerate hierarchy and in section 4.1 we concluded that this is the case in most of parameter space. We conclude that the quasi-degenerate hierarchy is excluded, but that the slightly less natural inverse and normal hierarchies are still possible, with now the inverse hierarchy favoured in most of the remaining phase space.

5. Conclusions

We conclude by stating once again that a model that combines flavour symmetries and grand unified symmetries is very appealing. A possible choice is the Pati-Salam $\times$ S_4 model described here. We have seen that on first sight, the model can explain many observed patterns in the fermion mass sector. However, on second sight, this is not without a price. The simplicity in the fermion sector (4 instead of 18 representations needed) is partly lost to a much larger Higgs sector. Also a detailed analysis of the running of the gauge and Yukawa couplings showed that much of the naiely preferred phase space is in fact no longer viable.

References

[1] De Adelhart Toorop R, Bazzocchi F and Merlo L 2010 "The Interplay Between GUT and Flavour Symmetries in a Pati-Salam x S4 Model" *JHEP* **1008** (2010) 001
[2] Harrison P F, Perkins D H and Scott W G 2002, "Tri-bimaximal mixing and the neutrino oscillation data" *Phys. Lett.* B **530** (2002) 167
[3] Froggatt C D and Nielsen H B 1979 "Hierarchy Of Quark Masses, Cabibbo Angles And CP Violation," *Nucl. Phys.* B **147** (1979) 277.
[4] Altarelli G 2006, "Models of neutrino masses and mixings," arXiv:hep-ph/0611117.

16th International Symposium on Particles Strings and Cosmology (PASCOS 2010) IOP Publishing
Journal of Physics: Conference Series **259** (2010) 012100 doi:10.1088/1742-6596/259/1/012100

Neutrino oscillometry

J.D.Vergados[1], Y. Giomataris[2] and Yu. N. Novikov[3]

1 Physics Department, University of Ioannina, Ioannina, Greece

E-mail: vergados@uoi.gr

2 CEA, Saclay, DAPNIA, Gif-sur-Yvette, Cedex,France

3 Petersburg Nuclear Physics Institute, 188300, Gatchina,Russia

Abstract. Neutrino oscillations are studied employing sources of low energy monoenergetic neutrinos following electron capture by the nucleus and measuring electron recoils. Since the neutrino energy is very low the oscillation length L_{23} appearing in this electronic neutrino disappearance experiment can be so small that the full oscillation can take place inside the detector so that one may determine very accurately the neutrino oscillation parameters. In particular, since the oscillation probability is proportional to $\sin^2 2\theta_{13}$, one can measure or set a better limit on the unknown parameter θ_{13}. One, however, has to pay the price that the expected counting rates are very small. Thus one needs a very intensive neutrino source and a large detector with as low as possible energy threshold and high energy and position resolution. Both spherical gaseous and cylindrical liquid detectors are studied. Different source candidates are considered

1. Introduction.

The discovery of neutrino oscillations can be considered as one of the greatest triumphs of modern physics. It began with atmospheric neutrino oscillations [1]interpreted as $\nu_\mu \rightarrow \nu_\tau$ oscillations, as well as ν_e disappearance in solar neutrinos [2]. These results have been recently confirmed by the KamLAND experiment [3], which exhibits evidence for reactor antineutrino disappearance. As a result of these experiments we have a pretty good idea of the neutrino mixing matrix and the two independent quantities Δm^2, e.g $|m_2^2 - m_1^2|$ and $|m_3^2 - m_2^2|$. Fortunately these two Δm^2 values are vastly different,

$$\Delta m_{21}^2 = |m_2^2 - m_1^2| = (7.65^{+0.23}_{-0.20}) \times 10^{-5}(eV)^2, \quad \Delta m_{32}^2 = |m_3^2 - m_2^2| = (2.4^{+0.12}_{-0.11}) \times 10^{-3}(eV)^2.$$

This means that the relevant L/E parameters are very different. Thus for a given energy the experimental results can approximately be described as two generation oscillations. For an accurate description of the data, however, a three generation analysis [4],[5] is necessary.

In all of these analyses the oscillation length is much larger than the size of the detector. So one is able to see the effect, if the detector is placed in the right distance from the source.

The most precise and unambiguous way to measure neutrino oscillations would be to determine changes in the flux of the given flavor of neutrinos over the entire oscillation length. Since the oscillation length is proportional to neutrino energy, the proper neutrino oscillometry would require a detector hundreds or even thousands of kilometers long if used with the present or proposed neutrino beams! As this is unrealistic, all beam experiments aiming at neutrino oscillations consider just a single or at most two point measurements instead of the full oscillometric approach. Also when using reactor neutrinos, the distance from the source to the first minimum is about 2 km - still beyond the current technological and financial boundaries for a detector. To be able to perform neutrino oscillometry using a realistic-size detector like LENA (100 m long) one needs a strong source of monoenergetic neutrinos with the energy of a few hundred of keV. Such a source could be produced in a nuclear reactor making neutrino oscillometry with LENA possible [6]. Neutrino oscillometry provides a competitive and considerably less expensive alternative to long baseline neutrino beams.

The best way to detect low energy electron neutrinos is by measuring electron recoils from neutrino-electron scattering. The total neutrino electron scattering cross section be cast in the form:

$$\sigma(L,x,y_{th}) = \sigma(0,x,y_{th})\,(1-\chi(x,y_{th})p(L,x)) \tag{1}$$

with $x = \frac{E_\nu}{m_e}$ and $y_{th} = \frac{(T_e)_{th}}{m_e}$, with $(T_e)_{th}$ the threshold electron energy imposed by the detector and

$$p(L,x) = \sin^2\left(\frac{0.595922L}{33x}\right)\sin^2(2\theta_{solar}) + \sin^2\left(\frac{0.595922L}{x}\right)\sin^2(2\theta_{13}) \tag{2}$$

with L the source detector distance in meters. The functions $\sigma(0,x,y_{th})$, the cross section in the absence of oscillation, and $\chi(x,y_{th})$, which takes care of the other neutrino flavors, have been previously described [6]. The oscillation length of interest to us take the form:

$$L_{32} = \frac{2.48[\mathrm{m}]E_\nu}{\Delta m^2_{32}([\mathrm{eV}])^2} \Rightarrow L_{32}[\mathrm{m}] \approx E_\nu[\mathrm{keV}] \tag{3}$$

The values in the square brackets in Eqs (3) indicate the dimensions used.

The neutrino sources of interest are divided into two categories: Those which have $L_{32} \leq 50$ m and those with $L_{32} > 110$ m. For the former nuclides the TPC counting method can be used in the gas-filled NOSTOS sphere approach [7], [8], whereas for both and mainly for the latter category with the larger L_{32}, the long liquid scintillator (LS) detector [9] is preferable. One of the main advantages of the spherical TPC detectors is the very low energy threshold [10] they can achieve (0.1 keV), which allows them to take advantage of the very low energy neutrinos. From this point of view a comparison between the two types of detectors is given in Fig. 1.

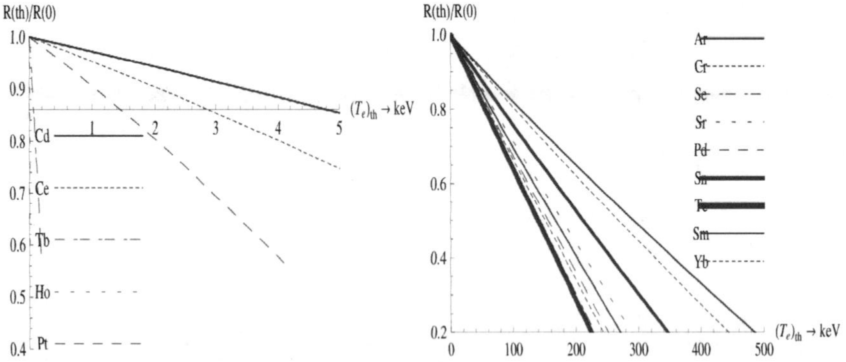

Figure 1. The dependence of the rate on the energy threshold, T_{th}, in the case of a gaseous spherical TPC detector on the left and the LENA detector on the right.

For a spherical detector two typical examples, obtained with a threshold of 0.1keV, are shown in Fig. 2. Clearly a compromise has to be made to achieve as large as possible portion of the oscillation inside the detector with a reasonable detection rate. Since the beautiful results of

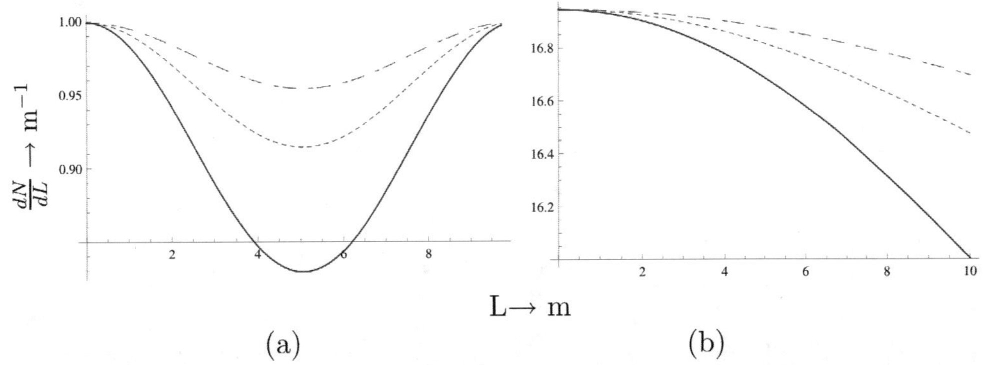

Figure 2. The rate $\frac{dN}{dL}$ (per meter) for Ar at 10 Atm with 1 Kg of ^{157}Tb (a) and ^{193}Pt (b) as a function of the source-detector distance (in m). The results shown correspond to $\sin^2 2\theta_{13} = 0.170, 0.085$ and 0.045 (decreasing from bottom to top). This rate was obtained for a running period equal to the half life of the source. The analysis is much simpler than that for the cylindrical geometry since the geometric factor g_{av} for the spherical detector is unity.

the first category have been previously discussed [7], [6], in this paper we will concentrate on the second category.

2. Short baseline neutrino oscillations-The detection principle

Any change of scattering rate as a function of position, in excess of the geometric factor would give, for the first time, a continuous (oscillometric) measure of flavor disappearance.

The detector for $\nu - e$ scattering events should be as long as possible and should have a large fiducial volume. It's energy registration threshold should be as law as possible since to have short oscillation length one needs low energy neutrinos. Presently only liquid scintillator (LS) technology can provide the required low detection threshold of 200 keV [11]. The proposed LENA detector would match these requirements. Moreover, with the suggested length of about 100 m LENA would be the longest LS detector ever build. Due to the improved signal processing and timing characteristics as compared to the operating LS detectors [11] the expected position sensitivity of LENA will be better than 50 cm even at the level of a few hundreds of keV of the recoil energy of the electron. The energy resolution would be ~10 % in this energy region [9].

As the cross-sections for $\nu - e$ scattering are tiny, a very strong neutrino source should be used to provide adequate statistics. Fortunately, there are many nuclei decaying via electron capture (EC). Since EC is a two body process the emitted electron neutrino is monoenergetic and carries most of the transition energy. Table 1 lists some of the isotopes decaying via EC with suitable Q values to produce monoenergetic neutrinos of a few hundreds of keV and with half-lifes of a few months allowing for convenient handling. They are relatively easy to produce via neutron capture reaction, see, e.g., the GALLEX experiment [12] for the 62 PBq ^{51}Cr source.

The number of events in between L and $L + dL$, where L is the distance between the center of the source and the detection point, can be written in the following form [6]:

$$R_0 \frac{dN}{dL} = f_\Phi \Lambda g_{av}(u, L/R_0) \tilde{\sigma}(L, x, y_{th}), \tag{4}$$

where

$$\Lambda = \frac{G_F^2 m_e^2}{2\pi} R_0 N_\nu n_e \tag{5}$$

Table 1. Neutrino sources which could be produced by irradiation in the neutron reactors. The intensities of neutrino sources per second have been estimated per 1 kg of the target element with the natural isotope abundances and assuming a 10 day irradiation with the neutron flux of 5×10^{14} n/cm^2/s. Neutron capture cross sections were taken from [http://ie.lbl.gov/].

Nuclide	$T_{1/2}$, d	Q_ε(keV) (keV)	E_ν (keV)	$E_{e,max}$ (keV)	Ir. target (10 d)	ν-intensity (s^{-1}) (per kg)
^{37}Ar	35	814	811 (100%)	617	Ar	8.3x10^{15}
^{51}Cr	28	753	747 (90%)	560	^{50}Cr	2.3x10^{16}
^{75}Se	120	863	450 (96%)	287	Se	1.1x10^{14}
^{113}Sn	116	1037	617 (98%)	436	Sn	8x10^{11}
^{145}Sm	340	616	510 (91%)	340	Sm	2x10^{12}
^{169}Yb	32	910	470 (83%)	304	Yb	1.1x10^{15}

with N_ν the number of neutrinos emitted by the source, n_e the density of electrons in the target ($n_e = 3 \times 10^{29}$ m^{-3} for LENA), R_0 the radius of the target and $\tilde{\sigma}(L, x, y_{th})$ is the neutrino - electron cross section in units of $G_F^2 m_e^2/2\pi$. The quantity f_Φ reflects the fraction of the total flux relevant for the detector (f_Φ=1, and 1/2 for a spherical detector (with the source at the center) and a cylindrical detector (with the source at the center of one of its bases respectively). The geometric factor $g_{av}(u, L/R_0)$ in the case of a spherical detector is unity, while for a cylindrical geometry has been previously given [6].

If one is content in extracting the value of the mixing angle only, this can be achieved by integrating the event rates over all L in the detector. This essentially involves integrating the cross section over L, folded with the function $g_{av}(u, L/R_0)$, i.e. integrating Eq. (4) over the L-values allowed by the detector. For sufficiently small mixing angle one can show that the event rate (in units of Λ) takes the form:

$$\frac{N}{\Lambda} = -A \sin^2 2\theta_{13} + B \qquad (6)$$

for ^{51}Cr $A = 0.048304$ and B=0.982456. Then, depending on the specifics of the experiment total number of events, N_0 can be presented in the form:

$$N_0 = -a \sin^2 2\theta_{13} + b \qquad (7)$$

3. The physics case for neutrino oscillometry

Neutrino oscillometry offers an elegant way to solve a number of questions related to neutrino oscillations: a precise determination of the mixing angle $\theta_{13}$and the oscillation length L_{23}, confirmation of the results of the "global" analysis of the oscillation data, and determination of the neutrino mass hierarchy. The latter would require a simultaneous long baseline measurement with the same detector.

3.1. Determination of the mixing angle θ_{13}

The big advantage of the short baseline oscillometry is that there is no matter influence in the observed events. As it is well known [5],[13] this matter effect gives degeneracy in the determination of the oscillation parameters in the long baseline experiments and should also be taken into consideration in some oscillation experiments with the reactor antineutrinos.

The angle θ_{13} can be determined from the analysis of both differential number of $\nu - e$ scattering events related to the length dL and the total number of events N_0 collected during the time of data acquisition in the full volume of LENA-detector. Differential curves for the neutrino scattering from the source ^{51}Cr are shown in Fig. 3. As can be seen from this figure, the curves for different mixing angles $\sin^2 2\theta_{13}$ are well separated within the length of LENA detector.

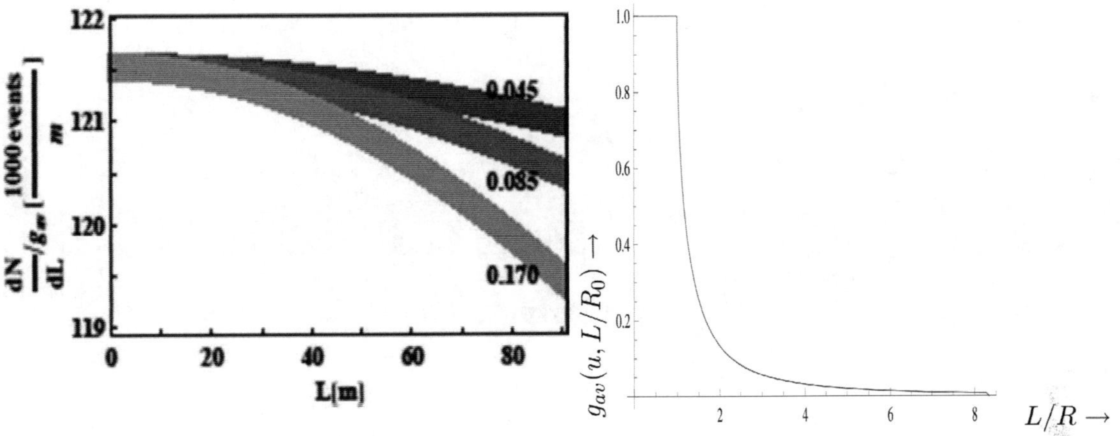

Figure 3. The number of the $\nu - e$ scattering events over the length dL divided by the geometric factor g_{av}. The calculated values correspond to 5 data acquisition cycles, 55 days each with the ^{51}Cr source installed at the top of LENA tank. The bottom (green), the middle (red) and the top (blue) correspond to $\sin^2 2\theta_{13}$= 0.170, 0.085 and 0.045, respectively. The geometric factor for $u = 11/90$ is shown on the right.

3.2. The short oscillation length L_{23}.

The value L_{23} can also be deduced from the oscillometry curves. This value can be compared with the neutrino energy which is usually well known, or can be measured independently very precisely [14]. For ^{51}Cr the neutrino energy is presently known with the precision of 0.05 %. Since Eq. (3) is valid, if the value of global analysis for $\Delta m^2_{23} = 2.5 \times 10^{-3}$ (eV)^2 is used, this comparison will be helpful for assessment of the global analysis itself.

3.3. Neutrino mass hierarchy.

As the oscillometry method provides precise determination of θ_{13} free of 8-fold degeneracy, the long baseline measurements in the same detector (LENA) with -neutrino beam would yield the information on the neutrino mass hierarchy without the need to change neither the energy of the neutrino beam nor the detector position. In this case the CERN- Pyhasälmi combination [15] looks quite promising for such type of measurements. Since short and long baseline experiments are disentangled by the energy region, both θ_{13} and the sign of $\{\Delta m^2_{13}\}$ experiments can be implemented using the same detector – LENA in Pyhasälmi.

4. Conclusions

We have discussed the importance of neutrino oscillometry involving low energy monocromatic neutrinos. Ideally one would like to employ gaseous TPC detectors with an extremely low energy threshold of 0.1 keV and neutrino sources with energy less than 50 keV. At present, however, one may have to content with a compromise, i.e. employ liquid detectors and use neutrino sources with energy of a few hundreds of keV. To this end the LENA detector is exceptionally well suited to perform precise determination of neutrino oscillation parameters thanks to the relatively low detection threshold (~200 keV) and considerable length (~100m). The needed electron-capture source emitting high-intensity monoenergetic and low-energy neutrinos can be manufactured by neutron irradiation in the core of a reactor. The disappearance of electron neutrinos can be followed over the full length of the detector by registering neutrino-electron scattering events. The resulting oscillometric curve and the total number of the events will provide accurate determination of the mixing angle θ_{13}. The main advantages of the gaseous TPC detectors are:

- The energy threshold can be very low.
- One can explore real low energy neutrinos.
- The geometry is simple. The only L-dependence of the event rate comes from the oscillation.

The disadvantage is that, for at present realistic neutrino sources, the event rate is small. Furthermore the solar neutrino background may be serious. It does not, however, depend on L and, if necessary, it can be measured.

The main advantages of neutrino oscillometry with LENA are summarized as follows:

- The short oscillation length L_{23} can be determined directly and the value of θ_{13} very precisely, without being affected by the 8-folded degeneracy.
- The mass hierarchy can be measured simultaneously with the same detector by performing a long baseline experiment (preferably CERN-Pyhas*ä*lmi) and using the determined θ_{13}.
- The background from the solar neutrino events (whose total number is by a factor of two less than the expected effect) can be directly measured (by removing the source) and systematic uncertainty can be determined from the measurements with a different source of Table 1.

The disadvantage is that some spurious L-dependence of the event rate comes from the geometry. This, however, can be taken care of by the geometric factor g_{av}.

Acknowledgments: JDV is indebted to Prof. Jose Valle, the PASCOS10 organizing committee and Dr M. Gomez for their hospitality during PASCOS10, while Yu. N. N. to T. Enqvist, A. Erykalov, F.v. Feilitzsch, J. Hissa, K. Loo, J. Maalampi, D. Nesterenko, L. Oberauer, F. Thurne, W. Trzaska and M. Wurm for useful discussions and private communications.

5. References

[1] Y. Fukuda *et al*, The Super-Kamiokande Collaboration, *Phys. Rev. Lett.* **86**, (2001) 5651; *ibid* **81** (1998) 1562 & 1158; *ibid* **82** (1999) 1810 ;*ibid* **85** (2000) 3999.

[2] Q.R. Ahmad *et al*, The SNO Collaboration, *Phys. Rev. Lett.* **89** (2002) 011302; *ibid* **89** (2002) 011301; *ibid* **87** (2001) 071301.
K. Lande *et al*, Homestake Collaboration, *Astrophys, J* **496**, (1998) 505
W. Hampel *et al*, The Gallex Collaboration, *Phys. Lett. B* **447**, (1999) 127;
J.N. Abdurashitov *al*, Sage Collaboration, *Phys. Rev. C* **80** (1999) 056801;
G.L Fogli *et al*, *Phys. Rev. D* **66** (2002) 053010.

[3] K. Eguchi *et al*, The KamLAND Collaboration, Phys. Rev. Lett. 90 (2003) 021802, hep-exp/0212021.

[4] Bahcall J N, Gonzalez-Garcia M and Peña-Garay C 2003 *JHEP* **0302** 009 (hep-ph/0212147)

[5] Barger V and Marfatia D 2002 *Phys. Lett. B* **555** 144 (arXiv:hep-ph/0212126)

[6] Vergados J and Novikov Y 2010 *Nucl. Phys.* **B 839** 1 ; [arXiv:1006.3862, hep-ph]

[7] Giomataris Y and Vergados J 2004 *Nucl. Instr. Meth. A* **530** 330

[8] Giomataris Y *et al.* 1996 *Nucl. Instr. and Meth. A* **376** 29

[9] Oberauer L, von Feilitzsch F and Potzel W 2005 *Nucl. Phys. B* **138** 108

[10] E. Bougamont1, P. Colas1, J. Derrel, I. Giomataris, G. Gerbier, M. Gros, P. Magnier,X.F. Navick, P. Salin, I. Savvidis, G. Tsiledakis and J. D. Vergados, Ultra low energy results and their impact to dark matter and low energy neutrino physics, (to be published)

[11] Borexino C 2009 *Nucl. Instr. and Methods* **A 600** 568

[12] Hampel W and Others 1998 *Phys. Lett.* **B 420**

[13] V Barger D Marfatia K W 2002 *Phys. Rev.* **D 66** 053007 ; arXiv:hep-ph/0206038

[14] Blaum K, Novikov Y and Werth G 2010 *Contemp. Phys.* **51** 149 [arXiv:0909.1095](physics.atom-ph)

[15] Peltoniemi J and Sarkamo J 2006 *Nucl. Phys. B (Proc. Suppl.)* **155** 201

Journal of Physics: Conference Series **259** (2010) 012101 doi:10.1088/1742-6596/259/1/012101

Dark Matter detection via lepton cosmic rays

Roberto A. Lineros

INFN, sezione di Torino, I-10122 Torino, Italy and Dipartimento di fisica teorica, Università di Torino, I-10122 Torino, Italy.

E-mail: `lineros@to.infn.it`

Abstract. Recent observations of lepton cosmic rays, coming from the PAMELA and FERMI experiments, have pushed our understanding of the interstellar medium and cosmic rays sources to unprecedented levels. The imprint of dark matter on lepton cosmic rays is the most exciting explanation of both PAMELA's positron excess and FERMI's total flux of electrons. Alternatively, supernovae are astrophysical objects with the same potential to explain these observations. In this work, we present an updated study of the astrophysical sources of lepton cosmic rays and the possible trace of a dark matter signal on the positron excess and total flux of electrons.

Proceeding for the 16th International Symposium on Particles, Strings and Cosmology. PASCOS 2010, Valencia, Spain.
Preprint DFTT 16/2010

1. Introduction

Last decades, different cosmic ray experiments like HEAT [1] and AMS01 [2] have presented very interesting cosmic rays results, specially for cosmic rays of *electrons*[1], in which the theoretical prediction of the positron fraction (number of positron divided by the number of *electrons* per unit of energy) based on standard sources mismatched the experimental results, and observing an increment in the fraction for energies larger than 10 GeV. Recently, newer and refined experiments like PAMELA [3] and FERMI [4, 5] present more accurate measurements of the positron fraction and the *electrons* flux, confirming the positron excess and other kinds of anomalies present in the observations which are not been completely explained by theoretical predictions.

In this work, we analyze the potential of dark matter and astrophysical sources to explain these anomalies. Moreover, we consider the effect of uncertainties in the propagation model and in the sources to verify the robustness of a possible discovery.

2. Propagation of cosmic rays

Cosmic rays propagation is rather different of propagation of photons and neutrinos, which is a straight line among the source and the observer. Cosmic rays have to deal with inhomogeneous magnetic fields, radiation fields from stellar activity and from the microwave background, and with the interstellar gas. The continuous interaction with the environment makes them to loose

[1] in this work, we refer to electrons and positrons as *electrons*

(or gain) energy and to propagate following a diffusion pattern.

Precise description of possibles cosmic ray interaction with the environment is far to be reached due to the uncertainties related to environmental quantities in the whole Galaxy. For instance, the galactic center is a very dynamic region that makes hard to model it and to provide under-control estimations. In opposition, regions outside the Galactic disk – which are very important for cosmic ray propagation – have few astrophysical objects, and then big uncertainties in the estimation of magnetic fields, radiation field, etc.

The model of propagation of cosmic rays is based on a continuity equation for the number density of cosmic rays, known as transport equation, which contains most of the processes between cosmic rays and the environment.
In our case, the transport equation for cosmic rays of electrons and positrons is:

$$\frac{\partial \psi}{\partial t} - \nabla\big(D(\varepsilon)\nabla\psi\big) - \frac{\partial}{\partial \varepsilon}\big(b(\varepsilon)\psi\big) = q(\vec{x},\varepsilon)\,, \tag{1}$$

where ψ is the number density of electrons or positrons per unit of energy, $D(\varepsilon)$ is the diffusion term, $b(\varepsilon)$ the energy loss term, and $q(\vec{x},\varepsilon)$ is the source term.

The diffusion term is considered homogeneous in space with a power laws dependence in energy:

$$D(\varepsilon) = K_0 \left(\frac{\varepsilon}{\varepsilon_0}\right)^{\delta}\,, \tag{2}$$

where K_0 and δ are phenomenological parameters inspired by magnetohydrodynamics models of the interstellar medium. The normalization energy scale ε_0 here is fixed at the value of 1 GeV.

The energy loss term depends on the interaction between the galactic environment and the type of cosmic rays. The process that dominates the energy evolution of electrons and positrons at GeV–TeV scale is the inverse Compton scattering with the Interstellar Radiation Field and the synchrotron emission related to the Galactic magnetic field.
For simplicity, the inverse Compton scattering can be used in the Thomson regime which is valid for *electrons* with energy lower than the equivalent photon energy (details in [6, 7]). In this case, the energy loss term corresponds to:

$$b(\varepsilon) = \frac{\varepsilon_0}{\tau_E}\left(\frac{\varepsilon}{\varepsilon_0}\right)^{2}\,, \tag{3}$$

where τ_E is the energy loss scale time which is calculated from the total photon energy density. However, this formula is not accurate for high energy *electrons* due to the scattering with the photon bath becomes less frequent. This regime is known as Klein-Nishina regime [8] and becomes non-negligible for the interaction of GeV-TeV *electrons* with the ultraviolet, infrared and microwave component of the interstellar radiation field [7, 6].

For generic functions of the diffusion and energy loss terms, solutions of the transport equation (Equation 1) are fully analytical. Analytical solutions for continuous and burst injection cases (details in [9, 6]) are fully of physical meaning. We remark that the analytical approach allows us to study the importance of different sources and to scan the propagation space of parameters in rather short time compared to more sophisticated methods.

2.1. Uncertainties

Most of the physical processes are contained in the propagation model. Nevertheless, these processes depend on environmental variables that are constrained by observations but not completely determined.

The principal propagation parameters are the diffusion parameters (K_0, δ), and the geometry of the zone of propagation. It is usually a cylinder of radius equal to the galactic one (20 kpc) and half-thickness L_z which is a parameter of the model [10]. The model also includes the escape of cosmic rays from the propagation zone by imposing that cosmic rays density equal to zero at the boundaries. As well, the escape of cosmic rays can be archived when a non-homogeneous diffusion term is considered. Some models [11] use a diffusion term that grows exponentially,

$$D_{\text{exp}}(z, \varepsilon) = D(\varepsilon) \exp\left(\frac{|z|}{z_t}\right), \tag{4}$$

where z_t is a vertical scale. This family of models provides a smooth transition among the diffusive and the straight–line propagation zones but loosing analyticity and needs to be solved numerically.

One method to constrain the propagation space of parameter is based on observation of the Boron/Carbon ratio (B/C) [12]. For the scope of our analysis, we assume that propagation space of parameters is common for all species of cosmic rays.

3. Sources of cosmic rays electrons

As important as the propagation, sources of *electrons* present a challenge for models of production and injection. In principle, each source is different from the others in the following aspects: spatial and temporal distribution, energy spectrum, and intensity. Also, if we consider sources in the Galaxy, physical processes regarding to cosmic rays acceleration, nuclear physics, and physics beyond standard model would have an impact into the observed flux of *electrons*.

At this point, we describe some of the principal sources of *electrons* at galactic scale. Our intention is to show how the observations by PAMELA and FERMI would be described as different combination of sources.

3.1. Dark matter

This is one of the most interesting and active topic. The possibility that cosmic rays anomalies are due to the annihilation or decay of dark matter particles it is far to be the most excited solution to the puzzle. The evidences obtained from cosmology and galaxy dynamics, among other, point towards the presence of dark matter. However, its imprints in cosmic rays, photons, neutrinos, etc. are not well defined, yet.

In the case of annihilation of dark matter, the source term of *electrons* is:

$$q_{\text{DM}}(\vec{x}, \varepsilon) = \eta \langle \sigma v \rangle \left(\frac{\rho_{\text{DM}}(\vec{x})}{m_\chi}\right)^2 \frac{dn}{d\varepsilon}(\varepsilon), \tag{5}$$

where η is a statistical coefficient which depends on dark matter nature, $\langle \sigma v \rangle$ is thermally averaged cross section which related with the dark matter relic abundance, ρ_{DM} is the dark matter distribution, m_χ is the dark matter mass, and $\frac{dn}{d\varepsilon}$ is the multiplicity distribution of electrons or

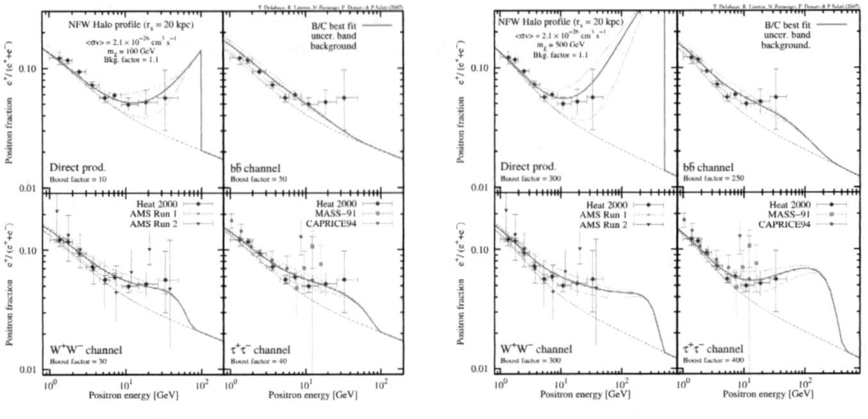

Figure 1. Positron fraction versus energy for 4 different annihilation channels: direct production of positrons, $b\bar{b}$, W^+W^-, and $\tau^+\tau^-$. (Left-panel) case with dark matter mass of 100 GeV. (Right-panel) case with dark matter mass of 500 GeV. Yellow bands in each plot corresponds to different propagation models compatible with B/C analysis. Details in Ref. [13].

positrons produced in an annihilation event.

In principle, the signal from annihilation of dark matter depends on the annihilation channel and the dark matter mass. In figure 1, we presents the positron fraction calculated for different annihilation channels and masses [13]. We note that channels which produce mainly leptons are favorite to explain the signal. In addition, the yellow bands represent the propagation models compatible with the B/C analysis. It is remarkable that a possible dark matter signal is not vanished by the uncertainties in the propagation [13]. Nevertheless, uncertainty in the propagation should also affect the background of *electrons*, which should be from astrophysical origin.

3.2. Secondaries

This component is the result of the spallation of nuclear cosmic rays (protons and alpha particles) on the interstellar gas, which is mainly composed by hydrogen and helium.
In figure 2, we present the calculation for both type of secondaries. Electrons from secondary processes are not dominant in the electron flux instead of positrons that encompass the available low energy flux [14]. As in the dark matter case, yellow bands correspond to the compatible B/C space of parameter. We denote that the compatibility among observation and propagation space of parameter reaffirm the idea that propagation models are common for every specie of cosmic rays.

3.3. Supernovae remnants and pulsars

Supernovae are the astrophysical objects responsible of most of the galactic cosmic ray. Moreover, those may give birth to two astrophysical objects with potential to produce electrons and positrons: Supernova remnants and pulsars.
Supernova remnants can expel a big fraction of the electron cosmic rays which is contained inside the former star, but very few positrons. Different mechanisms have been proposed to enhance the positron production [15], however, it seems to be not enough to explain the both anomalies.

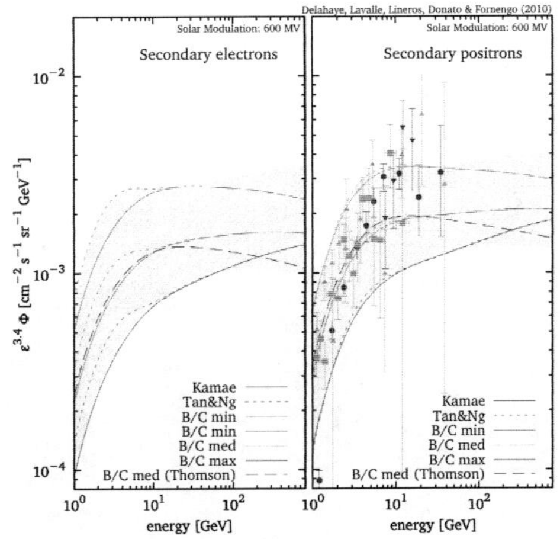

Figure 2. Secondary electrons (left) and positrons (right) versus energy. Yellow bands correspond to propagation models compatible with B/C analysis. Secondary positrons encompass the available data which is expected at low energies. Details in [6, 14].

On the other hand, pulsars are able to produce in same amount positrons and electrons due to the interaction of pulsar's magnetic field with ambient photons.

In both cases, the injection spectra is expected to follow a power-law like function:

$$q_{\rm SN} \propto Q_0 \varepsilon^{-\gamma} \exp\left(-\frac{\varepsilon}{\varepsilon_c}\right), \tag{6}$$

where Q_0 is the injection intensity of electrons/positrons, which is calculated from the averaged energy released by the supernova (or pulsar) and the efficiency to convert this energy into electron kinetic energy (details in [6]), γ is the power index, and ε_c is a cut-off energy (1–10 TeV) suggested by FERMI and HESS [16].

Let us remark that supernovae are distributed non-smoothly in space and time. Nevertheless, the diffusive behavior of propagated *electrons* smooths any peak at low energy, because this range is dominated by the older and farther supernovae. On the other hand, it is expected to be observed some features at high energies due to the contribution of younger and closer objects.

In figure 3, we present some of the results of a detailed study of astrophysical sources of *electrons* at the GeV-TeV scale [6]. We remark that under conservative assumption of injection profiles, the observations by FERMI and PAMELA can be encompassed with just the astrophysical component i.e. supernova remnants, pulsars and secondaries. Also, the presence of inhomogeneities in the local source distribution naturally explains the features observed by FERMI at the TeV range.

4. Conclusions

The cosmic rays anomalies observed by PAMELA and FERMI, have triggered a revolution in our understanding about the galactic environment. Dark matter is far the most exiting solution to the puzzle, but this supposition considers that the astrophysical background is absolutely known. Uncertainties in the propagation model and in the sources make harder the task to

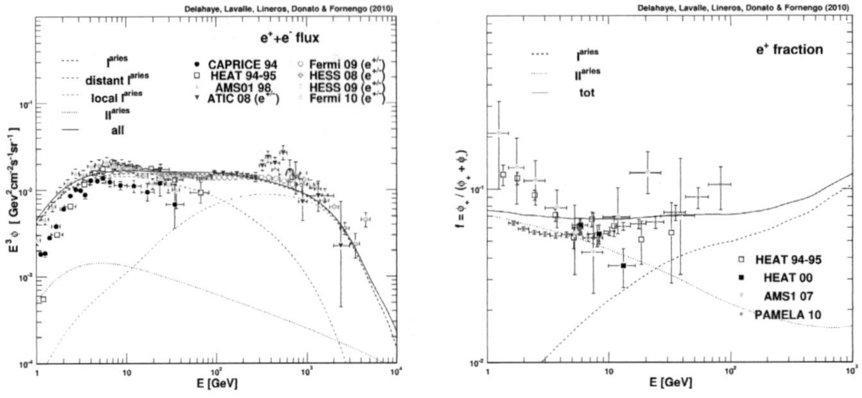

Figure 3. *Electrons* flux (left) and positron fraction (right) versus energy. *Electrons* are calculated using information about supernova remnants, pulsars, and also contribution from secondaries. Both observables are encompassed by the conservative assumption about astrophysical sources. Further details in [6].

discriminate a possible non-astrophysical source. After some detailed study regarding galactic supernova population, it seems natural that supernova remnants and pulsars may be the solution to these anomalies. It is indispensable to look for dark matter in other ways, but also to refine the propagation model using other types of observables like diffuse gamma emission and radio observations.

Acknowledgements

R.L. is grateful to T. Delahaye, J. Lavalle, N. Fornengo, P. Salati, and F. Donato for the rich collaboration and discussion at different stages of the publications referred in this proceeding. Also, R.L. acknowledges financial support given by Ministero dell'Istruzione, dell'Università e della Ricerca (MIUR), by the University of Torino (UniTO), by the Istituto Nazionale di Fisica Nucleare (INFN) within the Astroparticle Physics Project, and by the Italian Space Agency (ASI) under contract Nro: I/088/06/0.

References

[1] Barwick S W and The HEAT Collaboration 1997 *ApJL* **482** L191+ (*Preprint* arXiv:astro-ph/9703192)
[2] Alcaraz J *et al.* 2000 *Physics Letters B* **484** 10–22
[3] Adriani O *et al.* 2009 *Nature* **458** 607–609 (*Preprint* 0810.4995)
[4] Fermi LAT collaboration 2010 *ArXiv e-prints* (*Preprint* 1008.3999)
[5] The Fermi LAT Collaboration 2010 *ArXiv e-prints* (*Preprint* 1008.5119)
[6] Delahaye T, , Fornengo N *et al.* 2010 *ArXiv e-prints* (*Preprint* 1002.1910)
[7] Blumenthal G R and Gould R J 1970 *Reviews of Modern Physics* **42** 237–271
[8] Klein O and Nishina T 1929 *Zeitschrift für Physik A Hadrons and Nuclei* **52**(11) 853–868 ISSN 0939-7922 10.1007/BF01366453 URL http://dx.doi.org/10.1007/BF01366453
[9] Lineros R A 2008 *ArXiv e-prints* (*Preprint* 0812.4272)
[10] Ginzburg V L, *et al.* 1980 *Astrophysics and Space Science* **68** 295–314
[11] Di Bernardo G, , Maccione L *et al.* 2009 *ArXiv e-prints* (*Preprint* 0909.4548)
[12] Maurin D *et al.* 2001 *ApJ* **555** 585–596 (*Preprint* arXiv:astro-ph/0101231)
[13] Delahaye T, *et al.* 2008 *PRD* **77** 063527–+ (*Preprint* 0712.2312)
[14] Delahaye T, , , Taillet R *et al.* 2009 *A&A* **501** 821–833 (*Preprint* 0809.5268)
[15] Blasi P 2009 *Physical Review Letters* **103** 051104–+ (*Preprint* 0903.2794)
[16] Aharonian F *et al.* 2008 *Physical Review Letters* **101** 261104–+ (*Preprint* 0811.3894)

Constraints on light asymmetric dark matter from solar neutrinos

Marco Taoso

IFIC (CSIC-Universitat de Valencia), Ed.Instituts, Apt.22085, 46071 Valencia, Spain

E-mail: taoso@ific.uv.es

Abstract. We study the effect of dark matter (DM) particles in the Sun, focusing in particular on the possible reduction of the solar neutrinos flux due to the energy carried away by DM particles from the innermost regions of the Sun, and to the consequent reduction of the temperature of the solar core. In the very low-mass range between 4 and 10 GeV, recently advocated to explain the findings of the DAMA and CoGent experiments, the effects on neutrino fluxes are detectable only for DM models with very small, or vanishing, self-annihilation cross section, such as the so-called asymmetric DM models, and we study the combination of DM masses and Spin Dependent cross sections which can be excluded with current solar neutrino data.

1. Introduction

Dark Matter (DM) particles crossing a star can be scattered to velocities lower than the star escape velocity and be therefore gravitationally trapped by the celestial body, the efficiency of the process depending on the DM scattering cross section and DM ambient density. Once captured, DM particles can still scatter with the star nuclei transferring energy inside the object and eventually annihilate, providing therefore an exotic source of energy. These effects have been shown to significantly change the evolution and the properties of stars placed in environments with high DM density like main sequence stars at the Galactic center [1, 2], first stars [3, 4, 5] and compact objects [6, 7]. In these scenarios it is however impossible to set robust constraints on the DM cross sections and mass because of the large uncertainties on the DM densities and/or the lack of precise enough observations of the star targets.

Here instead we focus on the Sun, which properties have been measured with good precision. In addition to that, recent studies [8, 9, 10] have shown that current observations constrain the DM density in the solar system within a factor two, assuming spherical DM profiles (significant deviations may occur for other DM distributions, see [11] and references therein). For these reasons, the Sun can be used as a diagnostic tool to test small modifications of its structure induced by DM. The most recent works in this direction have focused on modifications of the solar neutrino fluxes and helioseismology data [12, 13]. Here we perform a complete and self-consistent calculation of the Sun evolution inside the galactic DM halo. In particular we extend previous works considering light DM candidates, with masses in the range suggested by DAMA [14] CoGent [15] and CDMSII [16] experiments.

In this paper, we focus on scenarios with negligible annihilations which are the most promising in order to detect the small modifications of the Sun structure induced by DM particles. In fact,

Journal of Physics: Conference Series **259** (2010) 012102 doi:10.1088/1742-6596/259/1/012102

for Standard Weakly Interacting Massive Particles (WIMPs) models the number of WIMPs inside the Sun, N_χ, is limited by annihilations and after a transient τ_χ, typically much shorter than the age of the Sun, an equilibrium between capture and annihilation is reached. The number of WIMPs in the Sun stays then constant $N_\chi = C\tau_\chi$, where C is the capture rate. Instead, for annihilation cross section much smaller than those found in WIMPs model, i.e. for $\sigma v \leq 10^{-33}$ cm^{3}s^{-1}, the equilibrium between capture and annihilations is not yet reach in the Sun and the number of trapped DM particles is simply $N_\chi = Ct_\odot$, with $t_\odot$ the age of the Sun. The number of DM particles inside the Sun is therefore greatly enhanced with respect to the that obtained obtained in WIMPs models and detectable modifications on the Sun properties can be obtained.

For the rest of the paper we neglect any DM annihilation. This possibility can be realized for instance in the so-called asymmetric DM models. In these scenarios, despite the DM candidates can in principle have weak-scale interactions and therefore sizable scattering cross sections off baryons, annihilations do not occour because of the presence of an asymmetry in the DM sector between particles and anti-particles. Concrete realizations of this idea are for example models where the dark sector contains a conserved $U(1)_X$ symmetry, analogue of the baryon number, responsible for the stability of the lightest particle in the DM sector. If this quantum charge is shared between baryons and DM it can link the asymmetries in the two sectors and this may naturally explain why the baryons and DM abundances are of the same order of magnitude (see e.g.[17, 18, 19]).

The results here presented are based on [27], where also other dark matter models have been considered.

The paper is organised as follows: in Sec 2 we discuss how the solar neutrino fluxes can be used to constrain the DM parameter space. In Sec. 3 we briefly describe the methodogy used and we present our results.

2. Diagnostic tools

The modification of the stellar structure produced by DM induces changes in the frequencies of stellar oscillations modes and in the neutrino fluxes. The first signature can in principle be observed with helioseismic measurements [12, 13], however, the neutrino flux is much more sensitive to the variation of temperature and density profile of the innermost regions of the Sun, and hence it is a much more powerful diagnostic tool.

For the standard solar model, Ref. [20] indicates that the ^{8}B neutrino flux varies as T^{25}, therefore a 1% temperature change will produce a 25% change in the ^{8}B neutrino flux. However, as it has been in shown in Ref. [13] , this simple scaling law is not valid to describe the peculiar modifications of the temperature profile induced by the DM particles. Therefore, the use of a stellar code is mandatory to study the effects on DM on the solar neutrino fluxes.

The distribution of DM inside the Sun is crucial to determine the modifications of the neutrino fluxes. Once captured, the DM particles get redistributed inside a small spatial scale, r_χ, of the order of 10^9 cm $\sim 0.01\ R_\odot$, for a DM mass of 100 GeV (r_χ scales as $m_\chi^{-1/2}$). The DM energy transport will be therefore more efficient in the innermost regions of the Sun core, as it can be appreciated in the left panel of Fig. 2. The ^{8}B and ^{7}Be neutrinos, which are mostly produced at $\sim 0.04R_\odot$ and $\sim 0.06R_\odot$, will be more affected by the presence of DM than the *pp* neutrinos. In fact, although the *pp* neutrinos are the most abundant solar neutrinos, they are mainly produced at $\sim 0.1R_\odot$, thus is a region well outside the one affected by the DM energy transfer. For the same reason and considering the experimental uncertainties in the determination of ^{8}B and ^{7}Be neutrino fluxes, we conclude that the ^{8}B neutrino flux is the best diagnostic tool in order to test the effects of DM on the Sun.

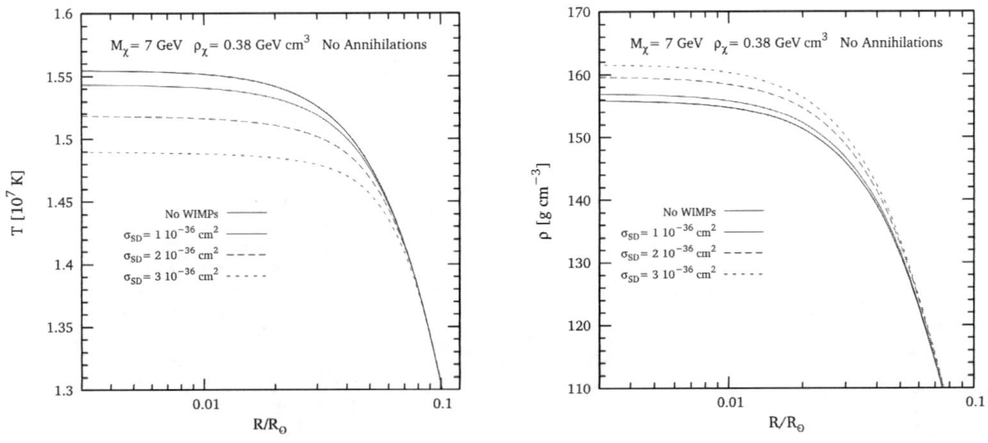

Figure 1. Impact of asymmetric DM on the Sun. *Left Panel:* Temperature profile of the Sun for different SD scattering cross-sections at $t_\odot = 4.57$ Gyrs. *Right Panel:* Density profile (baryons only) inside the Sun.

The ^{8}B flux has been determined with good accuracy by SNO [21]:

$$\phi_B^\nu = 5.046^{+0.159}_{-0.152}(\text{stat})^{+0.107}_{-0.123}(\text{syst})\ 10^6\ \text{cm}^{-2}\text{s}^{-1}.$$

For completeness, we report also the ^{7}Be neutrino flux inferred by Borexino[22]:

$$\phi_{Be}^\nu = (5.18 \pm 0.51) \times 10^9\ \text{cm}^{-2}\text{s}^{-1}.$$

The solar model we use, as described in [27], predicts at the solar age $t_\odot$:

$$\phi_B^\nu = 4.56 \times 10^6\ \text{cm}^2\text{s}^{-1}\phi_{Be}^\nu 7 = 4.47 \times 10^9\ \text{cm}^{-2}\text{s}^{-1},$$

values that are well in agreement with the experimental results to within the theoretical uncertainties of solar model calculations [23, 24].

Despite its great success in explaining a large variety of observations, the standard solar model suffers nowadays from the so-called *solar composition problems.* Recent analysis points to a lower surface heavy element content than previously thought (see [25] for recent results) and solar models incorporating these revised metallicities conflict dramatically with helioseismological measurements, in particular right below the solar convective envelope. With the solar abundances of [25], the radius of the convective zone, R_{CZ}, is more than 10 σ higher than the measured value: $R_{CZ} = 0.713 \pm 0.001 R_\odot$ (see Ref.[23] for an analysis of comparison of different solar models with helioseismology measurements.) High and low heavy elements models also produce differences on ϕ_B^ν and ϕ_{Be}^ν respectively of $\sim$ 20% and 10%. However, the theoretical uncertainties obtained within a particular solar model (with high or low metallicity estimations) are significantly smaller, $\sim$ 10% and 6% for ϕ_B^ν and ϕ_{Be}^ν [24]. Considering the theoretical and experimental uncertainties on ϕ_B^ν, we study the region of the DM parameter space where significant deviations on ϕ_B^ν are found (the maximum allowed deviation of the ^{8}B flux depends on the value of the theoretical uncertainties considered. We define this thereshold in Sec.3). As reference value, we consider the neutrino flux obtained from our solar model without DM particles and in the Sec. 3 we define the maximum deviations compatible with present data.

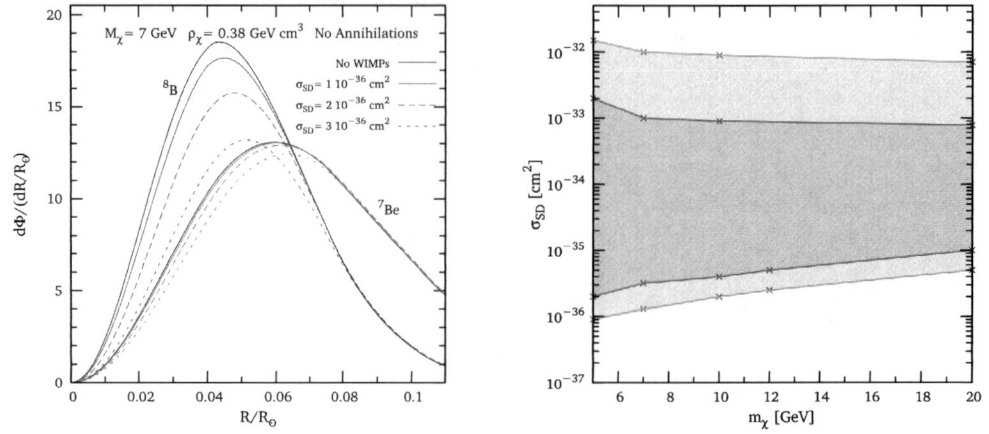

Figure 2. Solar neutrino fluxes in presence of asymmetric DM in the Sun. Left Panel: Differential ^{8}B and ^{7}Be neutrino fluxes as a function of radius. The curves referring to our solar model are normalized to unity. Right panel: isocontours of 25% (red) and 5% (yellow) ϕ^ν_B deviations with respect our solar model prediction in the m_χ-σ_{SD} plane. The red area show the region of the parameter space with ϕ^ν_B modifications larger than 25% and therefore in tension with present ϕ^ν_B data. The weakening of the DM transport effects at high cross sections is due to the transition to local transport regime. See text for details.

3. Results

In order to model the presence of DM in the Sun we have implemented the DM capture and evaporation in the GENEVA code. The total number of DM particles N_χ inside the Sun is thus obtained solving

$$\dot{N}_\chi = C - EN_\chi, \tag{1}$$

where C is the particle capture rate over the Sun and E the evaporation one. The energy transported by DM particles inside the Sun is then computed following ref.[26]. We have evolved the Sun from the ZAMS up to its current age $t_\odot = 4.57 \times 10^9$ years with the GENEVA stellar code. More details on the equations and the code used can be found in [27].

The first two panels of Figure 1 shows the temperature and density profiles of the Sun in presence of asymmetric DM. Note the change in temperature with increasing SD scattering cross section: the temperature decreases at the center and (although this is difficult to appreciate in the plot) slightly increases close to the external edge of the stellar core. The reason of the decrease in temperature can be understood in terms of energy transported away by DM from the solar core. For these values of scattering cross sections, the DM mean free path is much higher than the typical radius of the DM cloud, this indicating that DM transport effects are non local. In this regime, the DM particles scatters can efficiently transport the heat from the inside of the stellar out to colder regions, thus operating toward a flattening of the temperature profile, and a consequent readjustement of the entire stellar structure.

In general, any amount of energy removed from the core leads to its contraction. Naively, one would expect a warming of the internal regions of the Sun since part of the energy extracted from the gravitational energy reservoir goes into internal energy. However, the energy carried away by the DM particles is well above the energy released by the core contraction, resulting therefore in a cooling of the central regions.

Left panel of Figure 2 shows the differential ^{8}B and ^{7}Be neutrino fluxes as a function of the stellar radius, in presence of different DM models. As expected, the reduction of the neutrino

Journal of Physics: Conference Series **259** (2010) 012102 doi:10.1088/1742-6596/259/1/012102

production, due to the cooling of the baryons inside the Sun, is more efficient at small radii, where the DM particles are concentrated. As noticed in Sec. 2, this leads to a larger modification on the total ^{8}B neutrino flux, ϕ^ν_B, which is the integral over the whole Sun of the corresponding differential quantity plotted in the left panel of Figure 2, than those on the ^{7}Be neutrino flux, ϕ^ν_{Be}.

To study the impact of these structural variations on the solar neutrino flux, we have performed a systematic study of the DM parameter space, varying the DM scattering cross section and mass. For Spin Independent interactions we find sensible variations of the neutrino fluxes only for very large values of the scattering cross section σ_{SI}, already severely excluded by direct detection experiments, therefore we focus for the rest of the section on Spin Dependent interactions.

In the right panel of Fig. 2, we show in the $m_\chi - \sigma_{SD}$ plane the isocontours corresponding to ϕ^ν_B variations of 25% and 5% with respect to our solar model without DM particles.

The DM mass plays a relevant role in determining the effects on DM in stars since lowering the DM mass goes in the direction of maximizing the transport effects, but also the evaporation rate. For masses below 5 GeV the evaporation becomes relevant and the number of DM particles inside the star is strongly suppressed so that the changes induced on the solar neutrinos fluxes are negligible. On the contrary, above this mass threshold evaporation can be safetely neglected. Above $m_\chi = 20$ GeV the DM transport starts to become inefficient and even for high scattering cross section the DM energy transport is relevant only at the very center of the star, providing a local dip of energy. However, for increasing DM masses the existing constraints from direct detection experiments become more severe, and the region of the parameter space able to produce sizeable modifications of ϕ^ν_B is already excluded. Because of that, we do not explore that region any further.

Increasing the DM scattering cross section the DM mean free path reduced and at large annihilation cross section (i.e. $\geq 10^{-33}$cm^2), DM particles remain progressively "trapped" in the interior of the star. This implies that the heat transport from DM particles becomes local and the the modifications of ϕ^ν_B tend to decrease. This is also the reason of the non specularity of the exclusion curves in Figure 3: the low and high cross-section regions are characterized by different physics (non-local vs local transport effects, respectively).

Combining experimental and conservative (20%) theoretical uncertainties we derive that modifications of the ϕ^ν_B above $\sim$ 30% are excluded at 95% CL. The maximum modifications that we obtained in our computations are slightly below this level: further increasing σ_{SD}, problems are encountered in solving the stellar structure at ages well below the solar one. We have not further investigated if these difficulties are merely a numerical artifact or are instead related to the non existence of a solution of the stellar structure. However, we notice that once we obtain variations of ϕ^ν_B of the order of 20%, further small changes of σ_{SD} induce rapid modifications of ϕ^ν_B. Because of that, the isocontours corresponding to 30% ϕ^ν_B variations should be closed to the ones corresponding to $\delta\phi^\nu_B = 25\%$ shown in Fig. 2.

Considering a more optimistic value for the theoretical uncertainties on the ϕ^ν_B predictions, i.e. 10 %, the threshold for exclusion at 95% CL is lowered to 18% variations from our solar model prediction. Most of the SD cross-sections inside the 25% region in Fig.2 are excluded by the direct detection experiments constraints, apart from a small region at low masses which is somewhat in tension with those bounds. A reduction of the theoretical and experimental uncertainties on ϕ^ν_B in the next years may in principle improve the sensitivities on σ_{SD}. We show however in Fig.2 that considering a 5 % modification of ϕ^ν_B the region of the $m_\chi - \sigma_{SD}$ parameter space which can be probed enlarges very little.

4. Conclusions

By the use of a stellar evolution code, we have studied the modifications on the Sun structure induced by the DM particles captured by the Sun. We have focused on scenarios with null or very small annihilation cross sections, like asymmetric dark matter models, for which the number of DM particles trapped in the Sun can be sufficiently high to induce detectable modifications of the Sun observables. In fact, in presence of sizeable DM scattering cross sections off nuclei, the transport of energy by the DM from the interior of the Sun to the outer shells can dramatically reduce the core temperature, the most important consequence being a reduction of the ^8B neutrino flux. Considering the present theoretical and experimental uncertainties on ϕ^ν_B, we have studied the combination of DM masses and SD scattering cross sections which can be ruled out with this argument, finding only a small region of the parameter space which is not already excluded by direct detection bounds. Even with a significant decrease of the uncertainties on ϕ^ν_B, the region of the parameter space which can be probed remains approximately the same, therefore future experimental advances will not significantly change the situation.

4.1. Acknowledgments

We acknowledge support of the Spanish MICINNs ConsoliderIngenio 2010 Programme under grant MULTIDARK CSD2009-00064. Our work is partly supported by the Spanish grants FPA2008-00319 (MICINN) and PROM- ETEO/2009/091 (Generalitat Valenciana) and the European Council (Contract Number UNILHC PITN-GA-2009-237920).

References

[1] M. Fairbairn, P. Scott and J. Edsjo, Phys. Rev. D 77 (2008) 047301
[2] P. Scott, M. Fairbairn and J. Edsjo, Mon. Not. Roy. Astron. Soc. **394**, 82 (2008) [arXiv:0809.1871 [astro-ph]].
[3] F. Iocco, Astrophys. J. **677** (2008) L1 [arXiv:0802.0941 [astro-ph]].
[4] K. Freese, D. Spolyar and A. Aguirre, 2008 [arXiv:0802.1724].
[5] M. Taoso, G. Bertone, G. Meynet and S. Ekstrom, Phys. Rev. D **78** (2008) 123510 [arXiv:0806.2681 [astro-ph]].
[6] I.V. Moskalenko and L.L. Wai, Astrophys. J. 659:L29-L32, 2007 [arXiv:astro-ph/0702654].
[7] G. Bertone and M. Fairbairn, Phys. Rev. D **77** (2008) 043515 [arXiv:0709.1485 [astro-ph]].
[8] R. Catena and P. Ullio, arXiv:0907.0018 [astro-ph.CO].
[9] P. Salucci, F. Nesti, G. Gentile and C. F. Martins, arXiv:1003.3101 [astro-ph.GA].
[10] M. Weber and W. de Boer, arXiv:0910.4272 [astro-ph.CO].
[11] M. Pato, O. Agertz, G. Bertone, B. Moore and R. Teyssier, arXiv:1006.1322 [astro-ph.HE].
[12] I. Lopes and J. Silk, Phys. Rev. Lett. **88**, 151303 (2002) [arXiv:astro-ph/0112390].
[13] A. Bottino, G. Fiorentini, N. Fornengo, B. Ricci, S. Scopel and F. L. Villante, Phys. Rev. D **66**, 053005 (2002) [arXiv:hep-ph/0206211].
[14] R.Bernabei *et al.*, Riv.Nuovo Cim.**26N1** (2003) 1.
[15] C. E. Aalseth *et al.* [CoGeNT collaboration], arXiv:1002.4703 [astro-ph.CO].
[16] Z. Ahmed *et al.* [The CDMS-II Collaboration], arXiv:0912.3592 [astro-ph.CO].
[17] D. Hooper, J. March-Russell and S. M. West, Phys. Lett. B **605** (2005) 228 [arXiv:hep-ph/0410114].
[18] D. E. Kaplan, M. A. Luty and K. M. Zurek, Phys. Rev. D **79** (2009) 115016 [arXiv:0901.4117 [hep-ph]].
[19] F. Sannino and R. Zwicky, Phys. Rev. D **79**, 015016 (2009) [arXiv:0810.2686 [hep-ph]].
[20] J. N. Bahcall Phys. Rev. C**65**, 015802 (2002)
[21] B. Aharmim *et al.* [SNO Collaboration], arXiv:0910.2984 [nucl-ex].
[22] C. Arpesella *et al.* [The Borexino Collaboration], Phys. Rev. Lett. **101** (2008) 091302 [arXiv:0805.3843 [astro-ph]].
[23] A. Serenelli, S. Basu, J. W. Ferguson and M. Asplund, arXiv:0909.2668 [astro-ph.SR].
[24] C. Pena-Garay and A. Serenelli, arXiv:0811.2424 [astro-ph].
[25] M. Asplund, N. Grevesse, A. J. Sauval and P. Scott, Ann. Rev. Astron. Astrophys. **47** (2009) 481 [arXiv:0909.0948 [astro-ph.SR]].
[26] A. Gould and G. Raffelt, Astrophys. J. **352** (1990) 654.
[27] M. Taoso, F. Iocco, G. Meynet, G. Bertone and P. Eggenberger, arXiv:1005.5711 [astro-ph.CO].

16th International Symposium on Particles Strings and Cosmology (PASCOS 2010) IOP Publishing
Journal of Physics: Conference Series **259** (2010) 012103 doi:10.1088/1742-6596/259/1/012103

PAMELA's cosmic positron from decaying LSP in SO(10) SUSY GUT[1]

Bumseok Kyae
Department of Physics, Pusan National University, Busan 609-735, Republic of Korea
E-mail: bkyae@pusan.ac.kr

Abstract. We propose two viable scenarios explaining the recent observations on cosmic positron excess. In both scenarios, the present relic density in the Universe is assumed to be still supported by thermally produced WIMP or LSP (χ). One of the scenarios is based on two dark matter (DM) components scenario, and the other is on SO(10) SUSY GUT. In the two DM components scenario, extremely small amount of non-thermally produced meta-stable DM component [$\mathcal{O}(10^{-10}) < n_X/n_\chi$] explains the cosmic positron excess. In the SO(10) model, extremely small R-parity violation for LSP decay to $e^\pm$ is naturally achieved with a non-zero VEV of $\tilde{\nu}^c$ and a global symmetry.

1. Introduction

For a long time thermally produced weakly interacting massive particles (WIMPs) have been believed to be the most promising dark matter (DM) candidates. It is because the correct order of the magnitude of the cross section for explaining the present relic density of the Universe is naturally possible with WIMPs. Particularly, the lightest supersymmetric particle (LSP), which is a well-motivated particle coming from the promising particle physics model, i.e. the minimal supersymmetric standard model (MSSM), has attracted much attentions as an excellent example of WIMP. Actually, the Universe relic density by WIMP or LSP DM is the traditional scenario, which has been believed so far.

However, recently some astrophysical experimental groups including PAMELA [3], ATIC [4], and the Fermi-LAT collaborations [5] reported the very challenging observations in cosmic ray: PAMELA observed positron fractions [$e^+/(e^+ + e^-)$] exceeding the theoretical expectation [6] above 10 GeV upto 100 GeV. On the other hand, the PAMELA's observations on anti-proton/proton flux ratio were quite consistent with the theoretical calculation. The ATIC and Fermi-LAT's observations exhibit excesses of ($e^+ + e^-$) flux in cosmic ray from 100 GeV to 1 TeV. They would result from the positron flux that keeps rising upto 1 TeV.

Apparently the above observational results are very hard to be interpreted in view of the conventional MSSM cold DM scenario: explaining the excess positrons with annihilations of Majorana fermions such as the LSP needs a too huge boost factor. Moreover, ATIC and Fermi-LAT's observations seem to require a TeV scale DM, if they are caused indeed by DM annihilation or decay. However, TeV scale DM seems to be disfavored by the gamma ray data, if the excess positron flux should be originated from the same physics explaining DM creation in the early Universe, i.e. from DM annihilation [7].

[1] This article is based on Refs. [1, 2].

Such astrophysical observations seem to destroy our traditional scenario on DM. In this article, however, it will be pointed out that such observations do *not* necessarily imply the presence of a new DM theory replacing the traditional one. We will propose two viable scenarios explaining them, in which the present relic density in the Universe is assumed to be still supported by thermally produced WIMP.

2. Two dark matter components scenario

Let us consider DM decay scenario to explain the cosmic positron excess. "Helicity suppression" is not valid in DM decay any longer. Unlike the DM annihilation scenario, DM decay scenario is relatively free from the gamma ray constraint, since the positron flux is just linearly proportional to the number density of DM [8]. In the DM decay scenario, however, there are some serious hurdles to overcome: (1) one is to naturally obtain the extremely small decay rate of the DM, $\Gamma_{\rm DM} \sim 10^{-26}$ sec.$^{-1}$, and (2) the other is to *naturally* explain the relic density of the DM.

The first hurdle could be somehow resolved by introducing an extra symmetry, an extra DM component with a TeV scale mass, and grand unified theory (GUT) scale superheavy particles, which mediate DM decay into the standard model (SM) charged leptons (and the LSP) [2]. It is because the required decay rate of 10^{-26} sec.$^{-1}$ can be achieved, if the dominant operator for decay to e^+e^- is dimension 6 suppressed by $M_{\rm GUT}^2$, i.e. a four fermion interaction:

$$\Gamma_{\rm DM} \sim \frac{m_{\rm DM}^5}{196\pi^3 M_{\rm GUT}^4} \sim 10^{-26} \text{ sec.}^{-1} , \tag{1}$$

where $m_{\rm DM} \sim$ a few TeV. The fact that the GUT scale particles are involved in the DM decay might be an important hint supporting GUT [9]. However, since the interaction between the new DM and the SM charged lepton are made extremely weak by introducing superheavy particles mediating the DM decay, it is hard to thermally produce the new DM, and so non-thermal production of DM with a carefully tuned reheating temperature should be necessarily assumed for the required relic density. One way to avoid it is to consider a SUSY model with two DM components (χ, X), where χ is just the ordinary WIMP such as the LSP and X indicates a new DM component [2, 9]:

χ : **main component of DM explaining the relic density, thermally produced, absolutely stable.**
X : **(extremely) small number density, non-thermally produced, meta-stable, decay to e^+e^- explaining PAMELA/Fermi-LAT.**

In this class of models, a (global) symmetry should be introduced to forbid all unwanted dimension 4 and 5 operators contributing to DM decay. The required number density of X $(\equiv n_X)$ turns out to be just $\mathcal{O}(10^{-10}) < n_X/n_\chi$. It is possible only if the suppression of the dimension 6 operator is smaller than $M_{\rm GUT}$, 10^{12} GeV $< M_* <$ 10^{16} GeV [2], since the positron flux in the case of DM decay is proportional to $n_X \cdot \Gamma_X$. The low energy field spectrum in this class of models [2] is the same as that of the MSSM except for the neutral singlet extra DM component. Moreover, the models of [2] can be embedded in the flipped SU(5) GUT.

3. LSP DM scenario: SO(10) SUSY GUT model

In the second scenario, we suppose again that just the conventional bino-like LSP is the main component of the DM. Since the "bino" is a WIMP, thermally produced binos could explain well the relic density of the Universe. Without introducing a new DM component and interaction, we will attempt to explain the PAMELA's observation within the framework of the already existing particle physics model, SO(10) SUSY GUT.

3.1. SU(5) vs. SU(2)$_R$ scales

In terms of the SM's quantum numbers, the SO(10) generator ($= \mathbf{45}_G$) is split into the generators of the SM gauge group plus $\{(\mathbf{1},\mathbf{1})_{-1},(\mathbf{1},\mathbf{1})_1\}$, $(\mathbf{1},\mathbf{1})_0$, $\{(\mathbf{3},\mathbf{2})_{-5/6},(\overline{\mathbf{3}},\mathbf{2})_{5/6}\}$, and $\{(\mathbf{3},\mathbf{2})_{1/6},(\overline{\mathbf{3}},\mathbf{2})_{-1/6};\ (\mathbf{3},\mathbf{1})_{2/3},(\overline{\mathbf{3}},\mathbf{1})_{-2/3}\}$. We will simply write them as

$$\{E,E^c\}\ ,\quad N\ ,\quad \{Q',Q'^c\}\ ,\quad \{Q,Q^c\ ;\ U,U^c\}\ , \tag{2}$$

respectively. The SM gauge group's generators and $\{E,E^c\}$, N compose the generators of $SU(3)_c\times SU(2)_L\times SU(2)_R\times U(1)_{B-L}$ ($\equiv$ LR), where $\{E,E^c\}$ and a linear combination of the SM hypercharge generator and N ($\equiv N_R$) correspond to the $SU(2)_R$ generators. The other combination orthogonal to it is the $U(1)_{B-L}$ generator ($\equiv N_{BL}$). *Note that $\{E,E^c\}$ and N don't carry color charges.* By the VEV of the adjoint Higgs $\langle\mathbf{45}_H\rangle$, the SO(10) gauge symmetry may break to LR. On the other hand, the SM gauge group's generators and $\{Q',Q'^c\}$ compose the SU(5) generators. The VEVs of the Higgs in the spinorial representations $\langle\mathbf{16}_H\rangle$, $\langle\overline{\mathbf{16}}_H\rangle$ breaks SO(10) down to SU(5). Hence, $\langle\mathbf{45}_H\rangle$ and $\langle\mathbf{16}_H\rangle$ determine the SU(5) and LR breaking scales, respectively.

Non-zero VEVs of both $\langle\mathbf{45}_H\rangle$ and $\langle\mathbf{16}_H\rangle$ eventually give the SM gauge symmetry at low energies.[2] Thus, if $\langle\mathbf{45}_H\rangle > \langle\mathbf{16}_H\rangle = \langle\overline{\mathbf{16}}_H\rangle \neq 0$ ($\langle\mathbf{16}_H\rangle = \langle\overline{\mathbf{16}}_H\rangle > \langle\mathbf{45}_H\rangle \neq 0$), the gauge bosons and gauginos of $\{Q',Q'^c\}$ achieve heavier (lighter) masses than those of $\{E,E^c\}$ and N. The masses of the gauge sectors for $\{Q,Q^c\ ;\ U,U^c\}$ would be given dominantly by the heavier masses in any cases, since both $\langle\mathbf{45}_H\rangle$ and $\{\langle\mathbf{16}_H\rangle,\langle\overline{\mathbf{16}}_H\rangle\}$ contribute to their masses.

3.2. LSP decay in SO(10)

To achieve the needed extremely small decay rate of the bino-like LSP χ, we need extremely small R-parity violation *naturally.* χ can *never* decay, if (1) R-parity is absolutely preserved *and* (2) χ is really the LSP. We mildly relax these two conditions such that χ *can* decay by assuming

(1) a non-zero VEV of the superpartner of one family of RH neutrino, $\tilde{\nu}_1^c$ (i.e. R-parity violation), *or*

(2) a mass of $\tilde{\nu}_1^c$ lighter than the χ's mass, m_χ (i.e. $\tilde{\nu}_1^c$ LSP).

By introducing a global symmetry, one can forbid the (renormalizable) Yukawa couplings between $\tilde{\nu}_1^c$ and the MSSM fields. Then, *$\tilde{\nu}_1^c$ can interact with the MSSM fields only through the superheavy gauge fields and gauginos of SO(10)*, since the (s)RH neutrino ν_1^c ($\tilde{\nu}_1^c$) is a neutral singlet under the SM gauge symmetry but it is charged under SO(10). It is embedded e.g. in **16** of SO(10). Consequently, decay of χ is possible but extremely suppressed. For instance, refer to the diagram of Figure 1-(a). We will discuss how this diagram can be dominant for the χ decay.

3.3. The conditions for leptonic decay of χ

Let us consider the interactions of the superheavy gauginos.[3] In Table 1, we list all the gauge interactions between the superheavy gauginos of SO(10) and two MSSM fields. They are, of course, the renormalizable operators. As seen in Table 1, $\tilde{\nu}_i^c$ or ν_i^c couples to the superheavy SO(10) gauginos, $\{\tilde{E},\tilde{E}^c\}$, $\tilde{N}$, $\{\tilde{Q},\tilde{Q}^c\}$, and $\{\tilde{U},\tilde{U}^c\}$.

According to PAMELA data [3], the branching ratio of the hadronic DM decay modes should not exceed 10 %. To make the leptonic interactions, i.e. $\tilde{e}_i^{c*}\nu_i^c\tilde{E}^c$, $\tilde{\nu}_i^{c*}e_i^c\tilde{E}$, and $\tilde{\nu}_i^{c*}\nu_i^c\tilde{N}$, $\tilde{e}_i^{c*}e_i^c\tilde{N}$ dominant over the other interactions in Table 1, we assume that

[2] Alternatively, one can employ the large representations, $\mathbf{126}_H$, $\overline{\mathbf{126}}_H$, and $\mathbf{210}_H$, instead of $\mathbf{16}_H$, $\overline{\mathbf{16}}_H$, and $\mathbf{45}_H$. $\mathbf{126}_H$ and $\overline{\mathbf{126}}_H$ break SO(10) to SU(5), while $\mathbf{210}_H$ breaks SO(10) to $SU(4)_c\times SU(2)_L\times SU(2)_R$. In our discussion throughout this article, $\mathbf{16}_H$ ($\overline{\mathbf{16}}_H$) and $\mathbf{45}_H$ can be replaced by $\mathbf{126}_H$ ($\overline{\mathbf{126}}_H$) and $\mathbf{210}_H$, respectively.

[3] In this article we don't discuss the cases in which χ decays through the mediation of the superheavy *gauge bosons*. However, it turns out that the decay channels of χ through the mediation of the superheavy gauge fields are relatively suppressed, compared to the mediation of the superheavy gauginos discussed here [1].

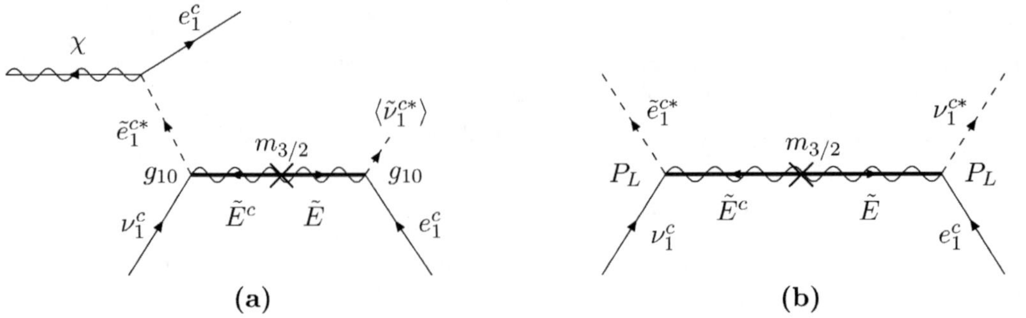

Figure 1. Dominant diagram of the bino decay (a) and the gauge interaction between electrically charged superheavy LR gauginos and the MSSM lepton singlets (b).

Interactions of the MSSM fields and heavy gauginos
$\tilde{e}_i^{c*}\nu_i^c\tilde{E}^c$, $\tilde{d}_i^{c*}u_i^c\tilde{E}^c$, $h_u^{+*}\tilde{h}_d^0\tilde{E}^c$, $h_u^{0*}\tilde{h}_d^-\tilde{E}^c$
$\tilde{\nu}_i^{c*}e_i^c\tilde{E}$, $\tilde{u}_i^{c*}d_i^c\tilde{E}$, $h_d^{0*}\tilde{h}_u^+\tilde{E}$, $h_d^{-*}\tilde{h}_u^0\tilde{E}$
$\tilde{\nu}_i^{c*}\nu_i^c\tilde{N}$, $\tilde{u}_i^{c*}u_i^c\tilde{N}$, $h_u^{+*}\tilde{h}_u^+\tilde{N}$, $h_u^{0*}\tilde{h}_u^0\tilde{N}$
$\tilde{e}_i^{c*}e_i^c\tilde{N}$, $\tilde{d}_i^{c*}d_i^c\tilde{N}$, $h_d^{-*}\tilde{h}_d^-\tilde{N}$, $h_d^{0*}\tilde{h}_d^0\tilde{N}$
$\tilde{e}_i^{c*}q_i\tilde{Q}'^c$, $\tilde{d}_i^{c*}l_i\tilde{Q}'^c$, $\tilde{q}_i^*u_i^c\tilde{Q}'^c$
$\tilde{q}_i^*e_i^c\tilde{Q}'$, $\tilde{l}_i^*d_i^c\tilde{Q}'$, $\tilde{u}_i^{c*}q_i\tilde{Q}'$
$\tilde{\nu}_i^{c*}q_i\tilde{Q}^c$, $\tilde{u}_i^{c*}l_i\tilde{Q}^c$, $\tilde{q}_i^*d_i^c\tilde{Q}^c$
$\tilde{q}_i^*\nu_i^c\tilde{Q}$, $\tilde{l}_i^*u_i^c\tilde{Q}$, $\tilde{d}_i^{c*}q_i\tilde{Q}$
$\tilde{u}_i^{c*}\nu_i^c\tilde{U}^c$, $\tilde{l}_i^*q_i\tilde{U}^c$, $\tilde{d}_i^{c*}e_i^c\tilde{U}^c$
$\tilde{\nu}_i^{c*}u_i^c\tilde{U}$, $\tilde{q}_i^*l_i\tilde{U}$, $\tilde{e}_i^{c*}d_i^c\tilde{U}$

Table 1. Gauge interactions between two MSSM fields and a heavy gaugino in the SO(10) GUT

• The LR (or B − L) breaking scale should be lower than the SU(5) breaking scale, i.e. $\langle\mathbf{16}_H\rangle \ll \langle\mathbf{45}_H\rangle$. Then $M_{Q'}$, M_Q, M_U[4] become much heavier than M_E and M_N, and so most of hadronic decay modes of χ can be easily suppressed except those by $\tilde{E}^c$, $\tilde{E}$, and $\tilde{N}$ in Table 1.

• The slepton $\tilde{e}_1^c$, which composes an $SU(2)_R$ doublet together with ν_1^c, needs to be lighter than the squarks. Then the decay channels of χ by $\tilde{d}_i^{c*}u_i^c\tilde{E}^c$, $\tilde{u}_i^{c*}d_i^c\tilde{E}$, and $\tilde{u}_i^{c*}u_i^c\tilde{N}$, $\tilde{d}_i^{c*}d_i^c\tilde{N}$ become suppressed. We also require that χ and $\tilde{e}_1^c$ are much lighter than the charged MSSM Higgs. So the leptonic interactions, $\tilde{e}_1^{c*}\nu_1^c\tilde{E}^c$, $\tilde{\nu}_1^{c*}e_1^c\tilde{E}$, and $\tilde{\nu}_1^{c*}\nu_1^c\tilde{N}$, $\tilde{e}_1^{c*}e_1^c\tilde{N}$ can dominate over the others.

• At least one RH neutrino, i.e. the $SU(2)_L$ singlet neutrino ν_1^c (and its superpartner $\tilde{\nu}_1^c$) must be lighter than χ so that χ decays to charged leptons. It is because ν_i^c is always accompanied by $\tilde{\nu}_i^c$ in the effective operators leading to the leptonic decay of χ, composed of $\tilde{e}_1^{c*}\nu_1^c\tilde{E}^c$, $\tilde{\nu}_1^{c*}e_1^c\tilde{E}$, and $\tilde{\nu}_1^{c*}\nu_1^c\tilde{N}$, $\tilde{e}_1^{c*}e_1^c\tilde{N}$. If all the sneutrino masses are heavier than χ, $\tilde{\nu}_1^c$ must develop a VEV for decay of χ. Once ν_1^c is light enough, $\tilde{\nu}_1^c$ can achieve a VEV much easily.

To be consistent with PAMELA's observations on high energy galactic positron excess [3], the DM mass should be around 300 – 400 GeV [11].[5] Thus, we simply take the following values;

[4] The triplets Higgs contained in $\mathbf{10}_h$ can achieve the mass proportional to $\langle\mathbf{45}_H\rangle$ via $\mathbf{10}_h\mathbf{45}_H\mathbf{10}_h$ [10].

[5] The $(e^+ + e^-)$ excess observed by Fermi-LAT could be explained by astrophysical sources such as nearby pulsars (and/or with the sub-dominant extra TeV scale DM component [2]). In fact, pulsars can explain both the PAMELA and Fermi-LAT's data in a suitable parameter range [12]. However, this does not imply that DM in

1. $\langle \mathbf{16}_H \rangle \ll \langle \mathbf{45}_H \rangle$. (If $m_\chi < m_{\tilde{\nu}_i^c}$, then $\langle \tilde{\nu}_1^c \rangle \neq 0$.) We will assume $\langle \mathbf{16}_H \rangle \sim 10^{14}$ GeV.
2. squarks, charged Higgs, higgsinos and other typical soft masses are of $\mathcal{O}(1)$ TeV.
3. $m_{\nu_1^c} < m_\chi \sim 300 - 400$ GeV $< m_{\tilde{e}_1^c} \ll \mathcal{O}(1)$ TeV.

Consequently, SO(10) is broken first to LR, which would be the effective gauge symmetry valid below the GUT scale. As seen from Table 1, the gauge interactions by the LR gauginos (and also gauge fields) preserve the baryon numbers. Even if the masses of the LR gauginos and gauge fields are relatively light, their gauge interactions don't give rise to proton decay.

3.4. Seesaw mechanism
Although one RH neutrino is light enough, the seesaw mechanism for obtaining the three extremely light physical neutrinos still may work. Let us consider the following superpotential;

$$W_\nu = y_{ij}^{(\nu)}\ l_i h_u \nu_j^c (j \neq 1) + \frac{1}{2} M_{ij}\ \nu_i^c \nu_j^c (i, j \neq 1), \tag{3}$$

where the Majorana mass term of ν_i^c could be generated from the non-renormalizable superpotential $\langle \overline{\mathbf{16}}_H \rangle \langle \overline{\mathbf{16}}_H \rangle \mathbf{16}_i \mathbf{16}_j / M_P$ $(i, j \neq 1)$. Thus, M_{ij} $(\gg \langle h_u \rangle)$ could be determined, if the LR breaking scale by $\langle \overline{\mathbf{16}}_H \rangle$ is known. In this superpotential, we note that one RH neutrino ν_1^c does not couple to the MSSM lepton doublets and Higgs. For instance, by assigning an exotic U(1) R-charge to ν_1^c, one can forbid its Yukawa couplings to the MSSM superfields. Thus, ν_1^c would be decoupled from the other MSSM fields, were it not for the heavy gauge fields and gauginos of the SO(10) SUSY GUT.

Taking into account only Eq. (3), one neutrino remains massless. The two heavy Majorana mass terms of ν_2^c and ν_3^c are sufficient for the other two neutrinos to achieve extremely small physical masses through the constrained seesaw mechanism [13]:

$$m_\nu = m_\nu^T = -\begin{pmatrix} 0 & v_{12} & v_{13} \\ 0 & v_{22} & v_{23} \\ 0 & v_{32} & v_{33} \end{pmatrix} \begin{pmatrix} 0 & 0 & 0 \\ 0 & M_{22}^{-1} & M_{23}^{-1} \\ 0 & M_{23}^{-1} & M_{33}^{-1} \end{pmatrix} \begin{pmatrix} 0 & 0 & 0 \\ v_{12} & v_{22} & v_{32} \\ v_{13} & v_{23} & v_{33} \end{pmatrix}, \tag{4}$$

where $v_{ij} \equiv y_{ij}^{(\nu)} \langle h_u \rangle$, and M_{ij}^{-1} denotes the inverse matrix of M_{ij}. One of the eigenvalues of m_ν is zero and the other two are of order v^2/M. Through the diagonalization of the mass matrix in Eq. (4), the three left-handed neutrinos from the lepton doublet l_1, l_2, and l_3 can be maximally mixed, whereas the mixing of the RH neutrinos is only between ν_2^c and ν_3^c. A complex phase in $y_{ij}^{(\nu)}$ could make the leptogenesis possible [13].

3.5. LSP decay rate and the seesaw scale
Let us consider the following terms in the superpotential;

$$W \supset \frac{1}{M_P} \langle \overline{\mathbf{16}}_H \rangle \mathbf{16}_1 \Sigma^2 + \kappa \Sigma^3, \tag{5}$$

where $M_P = 2.4 \times 10^{18}$ GeV and κ is a dimensionless coupling constant. Σ is an SO(10) singlet. We assign e.g. the U(1) R-charge of 2/3 to $\mathbf{16}_1$ and Σ, and 0 to $\overline{\mathbf{16}}_H$. This charge assignment

addition to pulsars can not be the source of the galactic positrons. In fact, we don't know yet a complete pulsar model, in which all the free parameters would be fixed by the fundamental physical constants.

Alternatively, one could assume $m_\chi \approx 3.5$ TeV and the model is slightly modified such that χ decays dominantly to $\mu^\pm$, ν_2^c rather than to $e^\pm$, ν_1^c, which is straightforward, the Fermi-LAT's data as well as the PAMELA's can be also explained [11]. In this case, however, the soft SUSY breaking scale should be higher than 3.5 TeV.

forbids the renormalizable Yukawa couplings between ν_1^c and other MSSM fields carrying integer R-charges.

The scale of $\langle\overline{\mathbf{16}}_H\rangle$ $(= \langle\mathbf{16}_H\rangle = M_E/\sqrt{2}g_{10})$ can be determined such that it is consistent with PAMELA data. The soft mass term of Σ and the A-term of $\kappa\Sigma^3$ in the scalar potential permit a VEV $\langle\tilde{\Sigma}\rangle \sim m_{3/2}/\kappa$. Then, the scalar potential generates a linear term of $\tilde{\nu}_1^c$ coming from the A-term corresponding to the first term of Eq. (5), $V \supset m_{3/2}^3(\langle\overline{\mathbf{16}}_H\rangle/\kappa^2 M_P)\tilde{\nu}_1^c$. The linear term and the soft mass term of $\tilde{\nu}_1^c$ in the scalar potential can induce a non-zero VEV of $\tilde{\nu}_1^c$:

$$\langle\tilde{\nu}_1^c\rangle \sim \frac{m_{3/2}}{\kappa^2} \times \frac{M_E}{M_P}. \tag{6}$$

Thus, the decay rate of χ in Figure 1-(a) can be estimated:

$$\Gamma_\chi = \frac{\alpha_{10}^2\alpha_Y m_\chi^5}{96M_E^4}\left(\frac{m_{3/2}\langle\tilde{\nu}_1^c\rangle}{m_{\tilde{e}_1^c}^2}\right)^2 \sim \frac{\alpha_{10}^2\alpha_Y m_\chi^5}{96M_E^2 M_P^2}\left(\frac{m_{3/2}}{\kappa m_{\tilde{e}_1^c}}\right)^4 \sim 10^{-26}\ \text{sec.}^{-1}, \tag{7}$$

where α_{10} $(\equiv g_{10}^2/4\pi)$ and α_Y $[\equiv g_Y^2/4\pi = (3/5)\times g_1^2/4\pi$, where g_1 is the SO(10) normalized gauge coupling of $g_Y]$ are approximately 1/24 and 1/100, respectively. Here, we ignore the RG correction to α_{10}. 300 – 400 GeV fermionic DM decaying to $e^\pm$ and a light neutral particle can fit the PAMELA data [11]. For $m_\chi \approx 300 - 400$ GeV, $(m_{3/2}/\kappa m_{\tilde{e}_1^c}) \sim 10$, M_E or $\langle\mathbf{16}_H\rangle$ is estimated to be of order 10^{14} GeV. This is consistent with the assumption $\langle\mathbf{16}_H\rangle \ll \langle\mathbf{45}_H\rangle \sim 10^{16}$ GeV. Therefore, the masses of the other two heavy RH neutrinos, which do not contribute to the process of Figure 1-(a), are around 10^{10} GeV or smaller in this case: $W \supset y_{ij}(\langle\overline{\mathbf{16}}_H\rangle\langle\overline{\mathbf{16}}_H\rangle/M_P)\mathbf{16}_i\mathbf{16}_j(i,j\neq 1) \supset y_{ij}(10^{10}\ \text{GeV})\times\nu_i^c\nu_j^c(i,j\neq 1)$.

4. Conclusion

In this article, we pointed out that the traditional DM scenario based on thermally produced WIMP do not necessarily conflict with the cosmic positron excess observed by PAMELA and Fermi-LAT. We have proposed two viable scenarios based on two DM components and SO(10) SUSY GUT. Particularly in the SO(10) model, extremely small R-parity violation for LSP decay could be naturally achieved with a non-zero VEV of $\tilde{\nu}_1^c$ and a global symmetry.

[1] B. Kyae, Phys. Lett. B **685** (2010) 19 [arXiv:0909.3139 [hep-ph]].
[2] K. J. Bae and B. Kyae, JHEP **0905**, 102 (2009); B. Kyae, JCAP **0907**, 028 (2009) [arXiv:0902.0071 [hep-ph]].
[3] O. Adriani *et al.* [PAMELA Collaboration], Nature **458**, 607 (2009); Phys. Rev. Lett. **102**, 051101 (2009).
[4] J. Chang *et al.* [ATIC Collaboration], Nature **456**, 362 (2008).
[5] A. A. Abdo *et al.* [The Fermi LAT Collaboration], Phys. Rev. Lett. **102**, 181101 (2009).
[6] I. V. Moskalenko and A. W. Strong, Astrophys. J. **493** (1998) 694.
[7] G. Bertone, M. Cirelli, A. Strumia and M. Taoso, JCAP **0903**, 009 (2009) [arXiv:0811.3744 [astro-ph]].
[8] E. Nardi, F. Sannino and A. Strumia, JCAP **0901** (2009) 043 [arXiv:0811.4153 [hep-ph]].
[9] J. H. Huh, J. E. Kim and B. Kyae, Phys. Rev. D **79**, 063529 (2009) [arXiv:0809.2601 [hep-ph]]; K. J. Bae, J. H. Huh, J. E. Kim, B. Kyae and R. D. Viollier, Nucl. Phys. B **817**, 58 (2009) [arXiv:0812.3511 [hep-ph]].
[10] S. M. Barr and S. Raby, Phys. Rev. Lett. **79**, 4748 (1997) [arXiv:hep-ph/9705366]; B. Kyae and Q. Shafi, Phys. Rev. D **72**, 063515 (2005) [arXiv:hep-ph/0504044].
[11] A. Ibarra, D. Tran and C. Weniger, JCAP **1001** (2010) 009; A. Ibarra and D. Tran, JCAP **0902**, 021 (2009).
[12] D. Grasso *et al.* [FERMI-LAT Collaboration], Astropart. Phys. **32** (2009) 140.
[13] P. H. Frampton, S. L. Glashow and T. Yanagida, Phys. Lett. B **548**, 119 (2002) [arXiv:hep-ph/0208157].

AUTHOR INDEX

AUTHOR INDEX

Institute of Physics
Dirac House, Temple Back
Bristol BS1 6BE UK